Digital Modulation Techniques

Second Edition

For a complete listing of recent titles
in the *Artech House Telecommunications Library*,
turn to the back of this book.

Digital Modulation Techniques

Second Edition

Fuqin Xiong

ARTECH
HOUSE
BOSTON | LONDON
artechhouse.com

Library of Congress Cataloging-in-Publication Data
A catalog record of this book is available from the U.S. Library of Congress.

British Library Cataloguing in Publication Data
A catalogue record of this book is available from the British Library.

Cover design by Igor Valdman

685 Canton Street
Norwood, MA 02062

International Standard Book Number: 1-58053-863-0

10 9 8 7 6 5 4 3 2 1

Contents

Preface

Digital modulation techniques are essential to many digital communication systems, whether it is a telephone system, a mobile cellular communication system, or a satellite communication system. In the past 20 years or so, research and development in digital modulation techniques have been very active and have yielded many promising results. However, these results are scattered all over the literature. As a result, engineers and students in this field usually have difficulty locating particular techniques for applications or for research topics.

This book provides readers with complete, up-to-date information on all modulation techniques in digital communication systems. There exist numerous textbooks of digital communications, each of them containing one or more chapters of digital modulation techniques covering either certain types of modulation, or only principles of the techniques. There are also a few books specializing in certain modulations. This book presents principles and applications information on all currently used digital modulation techniques, as well as new techniques now being developed. For each modulation scheme, the following topics are covered: historical background, operation principles, symbol and bit error performance (power efficiency), spectral characteristic (bandwidth efficiency), block diagrams of modulator, demodulator, carrier recovery (if any), clock recovery, comparison with other schemes, and applications. After we fully understand the modulations and their performance in the AWGN channel, we will discuss their performance in multipath-fading channels.

This second edition of *Digital Modulation Techniques* is an expanded version of the first edition. In recent years, multicarrier modulation (MCM), in particular, orthogonal frequency division multiplexing (OFDM) has been getting a lot of attention, due to its advantages in multipath fading channels in comparison with traditional single-carrier modulations. Many research results about OFDM have been published. Some systems have adopted OFDM. In order to help readers understand this emerging technique, a comprehensive coverage of OFDM is the major addition to this second edition of the book. About 280 pages in five chapters are devoted to OFDM. All aspects of OFDM, from basics to advanced topics such as peak-to-average power ratio reduction, synchronization, channel estimation and equalization, are described in detail. Another major change is the addition of a chapter about amplitude shift keying (ASK). There was a section of ASK in Chapter 8 (quadrature amplitude modulation (QAM)) of the first edition. To enhance the coverage of ASK, it was separated from QAM and expanded to include noncoherent M-ary ASK and other contents. The old Chapter 10 was about performance of modulations in fading channels. This chapter is now Chapter 11, which is greatly expanded to address remedial measures against channel impairments, including channel estimation, equalization, and diversity reception techniques. Other minor revisions include an added section in Chapter 2 for pulse time modulation (PTM), which is most suitable for optical communications; and newly published accurate bit error rate (BER) formulas for Gray-coded M-ary PSK in Chapter 4, and for QAM in Chapter 9. Chapter 1 is also enhanced by adding a modulation comparison figure for classical modulations. Some typographic errors in the first edition are also corrected. Finally, an appendix of mathematical tables is added.

Organization of the Book

This book is organized into 16 chapters. Chapter 1 is an introduction for those requiring basic knowledge about digital communication systems, and modulation methods. A brief overview of various modulation techniques is provided. This chapter also includes a table of abbreviations of modulations, a chart showing relations between various modulation schemes, and a figure of bandwidth efficiency versus power efficiency for classical modulations. All these provide the reader with a concise understanding of digital modulations and prepare the reader for studying the rest of the book.

Chapter 2 is about baseband signal modulation that does not involve a carrier. It is usually called baseband signal formatting or line coding. Traditionally the term *modulation* refers to "impression of message on a carrier"; however, if we widen the definition to "impression of message on a transmission medium," this formatting is also a kind of modulation. Baseband modulation is important not only because it is used in short distance data communications, magnetic recording, optical recording, and so forth, but also because it is the front end of bandpass modulations. This chapter also includes a section on pulse time modulation, which is specifically suitable for optical transmissions.

Chapters 3 and 4 cover classical frequency shift keying (FSK) and phase shift keying (PSK) techniques, including coherent and noncoherent. These techniques are currently used in many digital communication systems, such as cellular digital telephone systems, and satellite communication systems.

Chapters 5–7 are advanced phase modulation techniques, which include minimum shift keying (MSK), continuous phase modulation (CPM), and multi-*h* phase modulation (MHPM). These techniques are the research results of recent years, and some of them are being used in the most advanced systems; for example, MSK has been used in NASA's Advanced Communications Technology Satellite (ACTS) launched in 1993, and the others are being perfected for future applications.

Chapter 8 discusses amplitude shift keying (ASK). ASK is one of the simplest modulations and is used in systems where simplicity is the priority. Noncoherent ASK schemes, including OOK, are particularly suitable for fading channels where carrier phase synchronization is difficult, if not impossible.

Chapter 9 is about quadrature amplitude modulation (QAM). QAM schemes are widely used in telephone modems. For instance, CCITT (Consultative Committee for International Telephone and Telegraph)-recommended V.29 and V.33 modems use 16- and 128-QAM, reaching speeds of 9,600 bps and 14,400 bps respectively, over four-wire leased telephone lines.

Chapter 10 covers nonconstant-envelope bandwidth-efficient modulation schemes. We will study eight schemes, namely, QBL, QORC, SQORC, QOSRC, IJF-OQPSK, TSI-OQPSK, SQAM, and Q^2PSK. These schemes improve the power spectral density with little loss in error probability. They are primarily designed for satellite communications.

Chapter 11 first briefly introduces characteristics of channels with fading and multipath propagation. Then all modulations discussed in Chapters 2–9 are examined under the fading-multipath environment. The latter half of the chapter is devoted to remedial measures against channel impairments, including channel estimation, equalization, and diversity reception techniques. Even the newest measures like space-time coding and MIMO techniques are briefly introduced.

Starting from Chapter 12 we turn our attention to OFDM.

Chapter 12 covers the basics of OFDM, including modulation, demodulation, FFT-based implementation, various FFT algorithms that are pertinent to OFDM, cyclic extension, and spectrum shaping.

Chapter 13 discusses the peak-to-average power ratio issue and the techniques that can reduce it.

Chapter 14 describes the synchronization techniques for OFDM, including symbol timing synchronization and frequency offset estimation.

Chapter 15 examines the OFDM performance in various types of fading channels, and presents the methods of combating its effect. Channel estimation, equalization, diversity reception, and so forth, are discussed. Like in Chapter 11, space-time coding and MIMO techniques are briefly introduced in the context of OFDM.

Chapter 16 introduces the new developments in OFDM technology, which are the wavelet OFDM and M-ary ASK OFDM. These techniques do not use FFT pair for modulation and demodulation. Instead, the discrete wavelet transform and discrete cosine transform are used, respectively.

Appendixes A and B provide a basic knowledge of signal spectra and classical signal detection and estimation theory. Readers are strongly urged to read these two appendixes if they are not familiar with these theories or have already forgotten what they learned a long time ago.

Appendix C is a collection of tables of trigonometry identities, Fourier transform pairs and theorems, and Q-function and error functions.

This book can be used as a reference book for engineers and researchers. It also can be used as a textbook for graduate students. The material in the book can be covered in a half-year course. For short course use, the instructor may select relevant chapters to cover.

Acknowledgments

First I would like to thank the reviewers and editors at Artech House, whose many critiques and suggestions based on careful reviews contributed to the improvement of the manuscript.

I would like to thank Cleveland State University and Fenn College of Engineering for granting me the sabbatical leave in 1997 during which I wrote a substantial part of the book and also for granting me the sabbatical leave in 2003 during which I started writing the second edition. I am grateful for the support and encouragement from many colleagues at the Department of Electrical and Computer Engineering.

I am grateful to NASA Glenn Research Center for providing me with several research grants, particularly, the grant for investigating various modulation schemes that resulted in a report that was well received by NASA engineers and researchers. Encouraged by their enthusiastic response to the report, I published the tutorial article "Modem Techniques in Satellite Communications" in the *IEEE Communication Magazine*, August 1994. Further encouraged by the positive response to the tutorial paper, I developed the idea of writing a book detailing all major modulation schemes.

I would like to thank Professor Djamal Zeghlache of the Institut National des Télécommunications of France for his support and encouragement for the book.

I would like to thank City University of Hong Kong (CUHK), Tsinghua University of China, and Nanyang Technological University of Singapore, for supporting my sabbatical leaves. Particularly, I would like to thank Professor Li Ping of CUHK for his suggestions for the book and Professor Cao Zhigang of Tsinghua for his support of the book writing, and Professor Koh Soo Ngee of Nanyang Technological University for his support of the writing of the second edition.

I am very grateful for the excellent education that I received from Tsinghua University and the University of Manitoba, Canada. Particularly, I would like to thank my doctoral program advisor, Professor Edward Shwedyk of the Department of Electrical and Computer Engineering, University of Manitoba, and Professor John B. Anderson of the Electrical, Computer, and Systems Engineering Department, Rensselaer Polytechnic Institute, who served on my doctoral dissertation committee, for their guidance and encouragement.

I also appreciate the support and suggestions from my graduate students during the past few years.

Finally, the support and help for the book from my family are also deeply appreciated.

Fuqin Xiong
Cleveland, Ohio
April 2006

Chapter 1

Introduction

In this chapter we briefly discuss the role of modulation in a typical digital communication system, basic modulation methods, and criteria for choosing modulation schemes. Also included is a brief description of various communication channels, which will serve as a background for the later discussion of the modulation schemes.

1.1 DIGITAL COMMUNICATION SYSTEMS

Figure 1.1 is the block diagram of a typical digital communication system. The message to be sent may be from an analog source (e.g., voice) or from a digital source (e.g., computer data). The analog-to-digital (A/D) converter samples and quantizes the analog signal and represents the samples in digital form (bit 1 or 0). The source encoder accepts the digital signal and encodes it into a shorter digital signal. This is called source encoding, which reduces the redundancy hence the transmission speed. This in turn reduces the bandwidth requirement of the system. The channel encoder accepts the output digital signal of the source encoder and encodes it into a longer digital signal. Redundancy is deliberately added into the coded digital signal so that some of the errors caused by the noise or interference during transmission through the channel can be corrected at the receiver. Most often the transmission is in a high-frequency passband; the modulator thus impresses the encoded digital symbols onto a carrier. Sometimes the transmission is in baseband; the modulator is a baseband modulator, also called formator, which formats the encoded digital symbols into a waveform suitable for transmission. Usually there is a power amplifier following the modulator. For high-frequency transmission, modulation and demodulation are usually performed in the intermediate frequency (IF). If this is the case, a frequency up-convertor is inserted between the modulator and the power amplifier. If the IF is too low compared with the carrier frequency, several stages of carrier frequency conversions are needed. For wireless systems an antenna is the final stage of the trans-

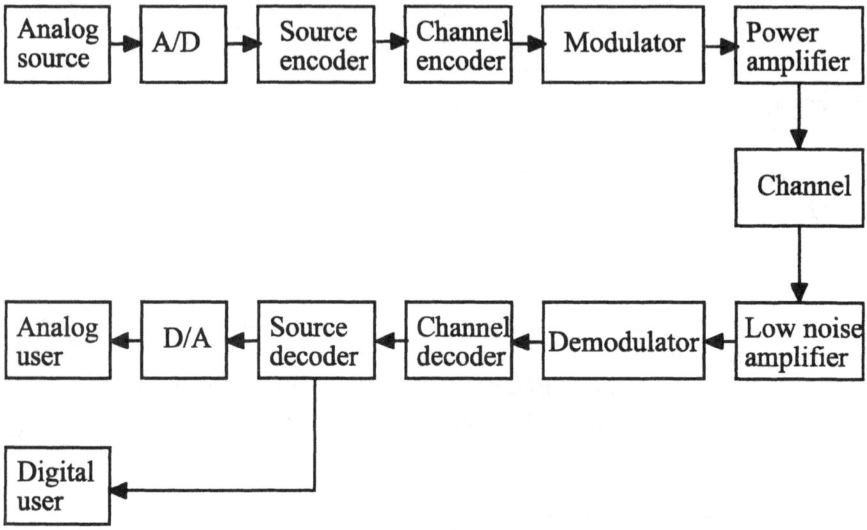

Figure 1.1 Block diagram of a typical digital communication system.

mitter. The transmission medium is usually called the channel, where noise adds to the signal and fading and attenuation effects appear as a complex multiplicative factor on the signal. The term noise here is a wide-sense term, which includes all kinds of random electrical disturbance from outside or from within the system. The channel also usually has a limited frequency bandwidth so that it can be viewed as a filter. In the receiver, virtually the reverse signal processing happens. First the received weak signal is amplified (and down-converted if needed) and demodulated. Then the added redundancy is taken away by the channel decoder and the source decoder recovers the signal to its original form before being sent to the user. A digital-to-analog (D/A) converter is needed for analog signals.

The block diagram in Figure 1.1 is just a typical system configuration. A real system configuration could be more complicated. For a multiuser system, a multiplexing stage is inserted before the modulator. For a multistation system, a multiple access control stage is inserted before the transmitter. Other features like frequency spread and encryption can also be added into the system. A real system could be simpler too. Source coding and channel coding may not be needed in a simple system. In fact, only the modulator, channel, demodulator, and amplifiers are essential in all communication systems (with antennas for wireless systems).

For the purpose of describing modulation and demodulation techniques and an-

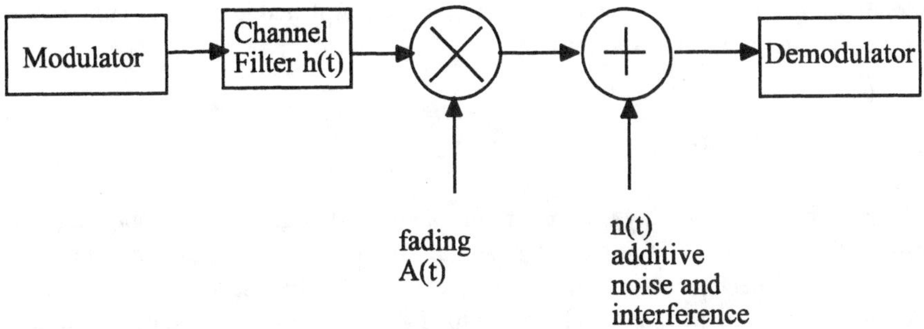

Figure 1.2 Digital communication system model for modulation and demodulation.

alyzing their performance, the simplified system model shown in Figure 1.2 will often be used. This model excludes irrelevant blocks with regard to modulation so that relevant blocks stand out. However, recently developed modem techniques combine modulation and channel coding together. In these cases the channel encoder is part of the modulator and the channel decoder is part of the demodulator. From Figure 1.2, the received signal at the input of the demodulator can be expressed as

$$r(t) = A(t)\left[s(t) * h(t)\right] + n(t) \tag{1.1}$$

where $*$ denotes convolution. In Figure 1.2 the channel is described by three elements. The first is the channel filter. Because of the fact that the signal $s(t)$ from the modulator must pass the transmitter, the channel (transmission medium), and the receiver before it can reach the demodulator, the channel filter therefore is a composite filter whose transfer function is

$$H(f) = H_T(f)H_C(f)H_R(f) \tag{1.2}$$

where $H_T(f)$, $H_C(f)$, and $H_R(f)$ are the transfer function of the transmitter, the channel, and the receiver, respectively. Equivalently, the impulse response of the channel filter is

$$h(t) = h_T(t) * h_C(t) * h_R(t) \tag{1.3}$$

where $h_T(t)$, $h_C(t)$, and $h_R(t)$ are the impulse responses of the transmitter, the channel, and the receiver, respectively. The second element is the factor $A(t)$, which is generally complex. This factor represents fading in some types of channels, such as mobile radio channel. The third element is the additive noise and interference term $n(t)$. We will discuss fading and noise in more detail in the next section. The channel model in Figure 1.2 is a general model. It may be simplified in some circumstances, as we will see in the next section.

1.2 COMMUNICATION CHANNELS

Channel characteristic plays an important role in studying, choosing, and designing modulation schemes. Modulation schemes are studied for different channels in order to know their performance in these channels. Modulation schemes are chosen or designed according to channel characteristic in order to optimize their performance. In this section we discuss several important channel models in communications.

1.2.1 Additive White Gaussian Noise Channel

Additive white Gaussian noise (AWGN) channel is a universal channel model for analyzing modulation schemes. In this model, the channel does nothing but add a white Gaussian noise to the signal passing through it. This implies that the channel's amplitude frequency response is flat (thus with unlimited or infinite bandwidth) and phase frequency response is linear for all frequencies so that modulated signals pass through it without any amplitude loss and phase distortion of frequency components. Fading does not exist. The only distortion is introduced by the AWGN. The received signal in (1.1) is simplified to

$$r(t) = s(t) + n(t) \tag{1.4}$$

where $n(t)$ is the additive white Gaussian noise.

The whiteness of $n(t)$ implies that it is a stationary random process with a flat power spectral density (PSD) for all frequencies. It is a convention to assume its PSD as

$$N(f) = N_o/2, \quad -\infty < f < \infty \tag{1.5}$$

This implies that a white process has infinite power. This of course is a mathematical idealization. According to the Wiener-Khinchine theorem, the autocorrelation function of the AWGN is

$$\begin{aligned} R(\tau) &\triangleq E\{n(t)n(t-\tau)\} = \int_{-\infty}^{\infty} N(f)e^{j2\pi f\tau}df \\ &= \int_{-\infty}^{\infty} \frac{N_o}{2}e^{j2\pi f\tau}df = \frac{N_o}{2}\delta(\tau) \end{aligned} \tag{1.6}$$

where $\delta(\tau)$ is the Dirac delta function. This shows the noise samples are uncorrelated no matter how close they are in time. The samples are also independent since the process is Gaussian.

At any time instance, the amplitude of $n(t)$ obeys a Gaussian probability density function given by

$$p(\eta) = \frac{1}{\sqrt{2\pi\sigma^2}} \exp\{-\frac{\eta^2}{2\sigma^2}\} \tag{1.7}$$

where η is used to represent the values of the random process $n(t)$ and σ^2 is the variance of the random process. It is interesting to note that $\sigma^2 = \infty$ for the AWGN process since σ^2 is the power of the noise, which is infinite due to its "whiteness."

However, when $r(t)$ is correlated with an orthonormal function $\phi(t)$, the noise in the output has a finite variance. In fact

$$r = \int_{-\infty}^{\infty} r(t)\phi(t)dt = s + n$$

where

$$s = \int_{-\infty}^{\infty} s(t)\phi(t)dt$$

and

$$n = \int_{-\infty}^{\infty} n(t)\phi(t)dt$$

The variance of n is

$$\begin{aligned} E\{n^2\} &= E\left\{\left[\int_{-\infty}^{\infty} n(t)\phi(t)dt\right]^2\right\} \\ &= E\left\{\int_{-\infty}^{\infty}\int_{-\infty}^{\infty} n(t)\phi(t)n(\tau)\phi(\tau)dtd\tau\right\} \\ &= \int_{-\infty}^{\infty}\int_{-\infty}^{\infty} E\{n(t)n(\tau)\}\phi(t)\phi(\tau)dtd\tau \\ &= \int_{-\infty}^{\infty}\int_{-\infty}^{\infty} \frac{N_o}{2}\delta(t-\tau)\phi(t)\phi(\tau)dtd\tau \\ &= \frac{N_o}{2}\int_{-\infty}^{\infty} \phi^2(t)dt = \frac{N_o}{2} \end{aligned} \tag{1.8}$$

Then the probability density function (PDF) of n can be written as

$$p(n) = \frac{1}{\sqrt{\pi N_o}} \exp\{-\frac{n^2}{N_o}\} \tag{1.9}$$

This result will be frequently used in this book.

Strictly speaking, the AWGN channel does not exist since no channel can have an infinite bandwidth. However, when the signal bandwidth is smaller than the channel bandwidth, many practical channels are approximately an AWGN channel. For example, the line-of-sight (LOS) radio channels, including fixed terrestrial microwave links and fixed satellite links, are approximately AWGN channels when the weather is good. Wideband coaxial cables are also approximately AWGN channels since there is no other interference except the Gaussian noise.

In this book, all modulation schemes are studied for the AWGN channel. The reason for doing this is twofold. First, some channels are approximately an AWGN channel; the results can be used directly. Second, additive Gaussian noise is ever present regardless of whether other channel impairments such as limited bandwidth, fading, multipath, and other interferences exist or not. Thus the AWGN channel is the best channel that one can get. The performance of a modulation scheme evaluated in this channel is an upper bound on the performance. When other channel impairments exist, the system performance will degrade. The extent of degradation may vary for different modulation schemes. The performance in AWGN can serve as a standard in evaluating the degradation and also in evaluating effectiveness of impairment-combatting techniques.

1.2.2 Bandlimited Channel

When the channel bandwidth is smaller than the signal bandwidth, the channel is bandlimited. Severe bandwidth limitation causes intersymbol interference (ISI) (i.e., digital pulses will extend beyond their transmission duration (symbol period T_s)) and interfere with the next symbol or even more symbols. The ISI causes an increase in the bit error probability (P_b) or bit error rate (BER), as it is commonly called. When increasing the channel bandwidth is impossible or not cost-efficient, channel equalization techniques are used for combatting ISI. Throughout the years, numerous equalization techniques have been invented and used. New equalization techniques are appearing continuously. We will not cover them in this book. For introductory treatment of equalization techniques, the reader is referred to [1, Chapter 6] or any other communication systems books.

1.2.3 Fading Channel

Fading is a phenomenon that occurs when the amplitude and phase of a radio signal change rapidly over a short period of time or travel distance. Fading is caused by interference between two or more versions of the transmitted signal which arrive at the receiver at slightly different times. These waves, called multipath waves, combine at the receiver antenna to give a resultant signal which can vary widely in amplitude and phase. If the delays of the multipath signals are longer than a symbol period, these multipath signals must be considered as different signals. In this case, we have individual multipath signals.

In mobile communication channels, such as terrestrial mobile channel and satellite mobile channel, fading and multipath interferences are caused by reflections from surrounding buildings and terrains. In addition, the relative motion between the transmitter and receiver results in random frequency modulation in the signal due to different Doppler shifts on each of the multipath components. The motion of surrounding objects, such as vehicles, also induces a time-varying Doppler shift on multipath components. However, if the surrounding objects move at a speed less than the mobile unit, their effect can be ignored [2].

Fading and multipath interference also exist in fixed LOS microwave links [3]. On clear, calm summer evenings, normal atmospheric turbulence is minimal. The troposphere stratifies with inhomogeneous temperature and moisture distributions. Layering of the lower atmosphere creates sharp refractive index gradients which in turn create multiple signal paths with different relative amplitudes and delays.

Fading causes amplitude fluctuations and phase variations in received signals. Multipath causes intersymbol interference. Doppler shift causes carrier frequency drift and signal bandwidth spread. All these lead to performance degradation of modulations. Analysis of modulation performances in fading channels is given in Chapter 11 where characteristics of fading channels will be discussed in more detail.

1.3 BASIC MODULATION METHODS

Digital modulation is a process that impresses a digital symbol onto a signal suitable for transmission. For short distance transmissions, baseband modulation is usually used. Baseband modulation is often called line coding. A sequence of digital symbols is used to create a square pulse waveform with certain features which represent each type of symbol without ambiguity so that they can be recovered upon reception. These features are variations of pulse amplitude, pulse width, and pulse position. Figure 1.3 shows several baseband modulation waveforms. The first one is the nonreturn to zero-level (NRZ-L) modulation which represents a symbol 1 by a posi-

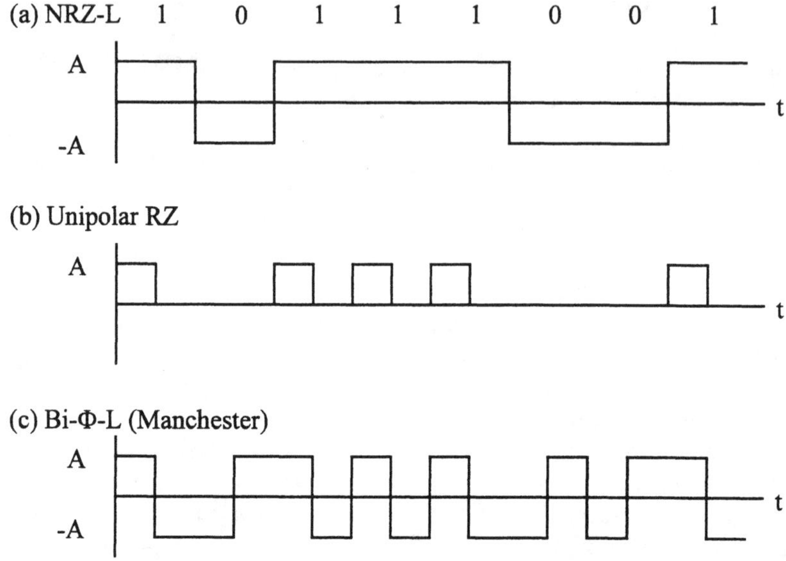

Figure 1.3 Baseband digital modulation examples (a–c).

tive square pulse with length T and a symbol 0 by a negative square pulse with length T. The second one is the unipolar return to zero modulation with a positive pulse of T/2 for symbol 1 and nothing for 0. The third is the biphase level or Manchester, after its inventor, modulation which uses a waveform consisting of a positive first-half T pulse and a negative second-half T pulse for 1 and a reversed waveform for 0. These and other baseband schemes will be discussed in detail in Chapter 2.

For long distance and wireless transmissions, bandpass modulation is usually used. Bandpass modulation is also called carrier modulation. A sequence of digital symbols are used to alter the parameters of a high-frequency sinusoidal signal called carrier. It is well known that a sinusoidal signal has three parameters: amplitude, frequency, and phase. Thus amplitude modulation, frequency modulation, and phase modulation are the three basic modulation methods in passband modulation. Figure 1.4 shows three basic binary carrier modulations. They are amplitude shift keying (ASK), frequency shift keying (FSK), and phase shift keying (PSK). In ASK, the modulator puts out a burst of carrier for every symbol 1, and no signal for every symbol 0. This scheme is also called on-off keying (OOK). In a general ASK scheme, the amplitude for symbol 0 is not necessarily 0. In FSK, for symbol 1 a higher frequency burst is transmitted and for symbol 0 a lower frequency burst

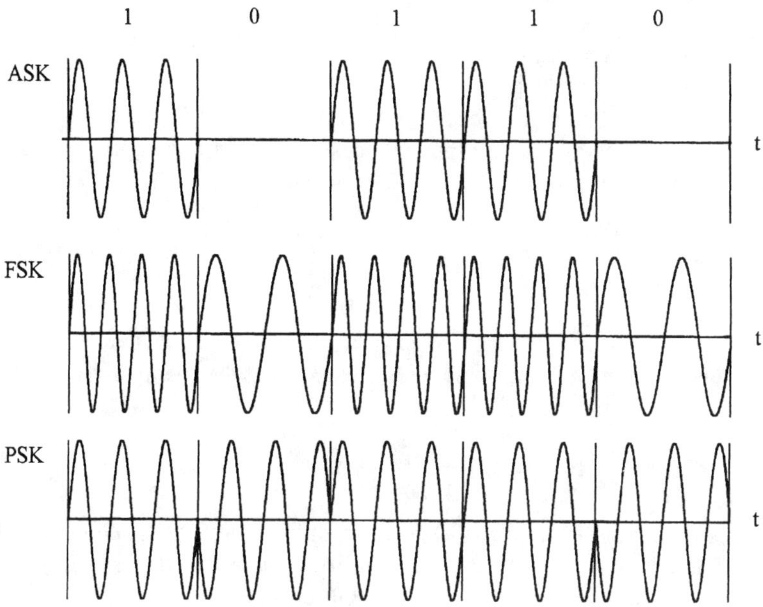

Figure 1.4 Three basic bandpass modulation schemes.

is transmitted, or vice versa. In PSK, a symbol 1 is transmitted as a burst of carrier with 0 initial phase while a symbol 0 is transmitted as a burst of carrier with 180° initial phase.

Based on these three basic schemes, a variety of modulation schemes can be derived from their combinations. For example, by combining two binary PSK (BPSK) signals with orthogonal carriers a new scheme called quadrature phase shift keying (QPSK) can be generated. By modulating both amplitude and phase of the carrier, we can obtain a scheme called quadrature amplitude modulation (QAM), and so forth.

1.4 CRITERIA OF CHOOSING MODULATION SCHEMES

The essence of digital modem design is to efficiently transmit digital bits and recover them from corruptions from the noise and other channel impairments. There are three primary criteria of choosing modulation schemes: power efficiency, bandwidth efficiency, and system complexity.

1.4.1 Power Efficiency

The bit error rate or bit error probability of a modulation scheme is inversely related to E_b/N_o, the bit energy to noise spectral density ratio. For example, P_b of ASK in the AWGN channel is given by

$$P_b = Q\left(\sqrt{\frac{E_b}{N_o}}\right) \tag{1.10}$$

where E_b is the average bit energy, N_o is the noise power spectral density (PSD), and $Q(x)$ is the Gaussian integral, sometimes referred to as the Q-function. It is defined as

$$Q(x) = \int_x^\infty \frac{1}{\sqrt{2\pi}} e^{-\frac{u^2}{2}} du \tag{1.11}$$

which is a monotonically decreasing function of x. Therefore the power efficiency of a modulation scheme is defined straightforwardly as the required E_b/N_o for a certain bit error probability (P_b) over an AWGN channel. $P_b = 10^{-5}$ is usually used as the reference bit error probability.

1.4.2 Bandwidth Efficiency

The determination of bandwidth efficiency is a bit more complex. The bandwidth efficiency is defined as the number of bits per second that can be transmitted in one Hertz of system bandwidth. Obviously it depends on the requirement of system bandwidth for a certain modulated signal. For example, the one-sided power spectral density of an ASK signal modulated by an equiprobable independent random binary sequence is given by

$$\Psi_s(f) = \frac{A^2 T}{4} \text{sinc}^2 \left[T(f - f_c)\right] + \frac{A^2}{4} \delta(f - f_c)$$

and is shown in Figure 1.5, where T is the bit duration, A is the carrier amplitude, and f_c is the carrier frequency. From the figure we can see that the signal spectrum stretches from $-\infty$ to ∞. Thus to perfectly transmit the signal an infinite system bandwidth is required, which is impractical. The practical system bandwidth requirement is finite, which varies depending on different criteria. For example, in Figure 1.5, most of the signal energy concentrates in the band between two nulls, thus a null-to-null bandwidth requirement seems adequate. Three bandwidth efficiencies

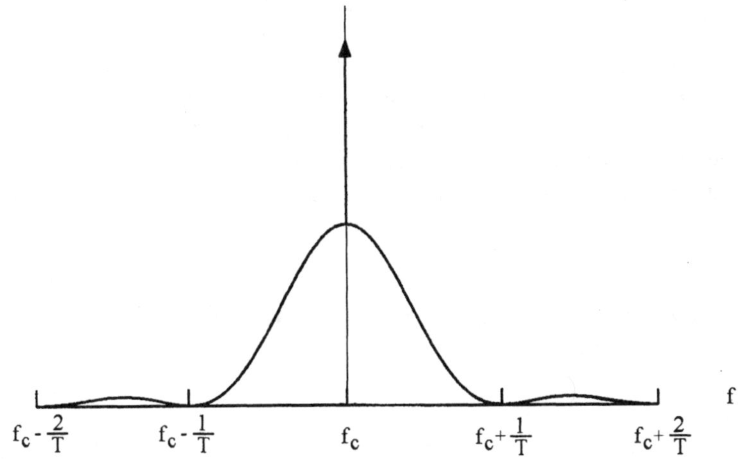

Figure 1.5 Power spectral density of ASK.

used in the literature are as follows:

Nyquist Bandwidth Efficiency—Assuming the system uses Nyquist (ideal rectangular) filtering at baseband, which has the minimum bandwidth required for intersymbol interference-free transmission of digital signals, then the bandwidth at baseband is $0.5R_s$, R_s is the symbol rate, and the bandwidth at carrier frequency is $W = R_s$. Since $R_s = R_b/\log_2 M$, R_b = bit rate, for M-ary modulation, the bandwidth efficiency is

$$R_b/W = \log_2 M \tag{1.12}$$

Null-to-Null Bandwidth Efficiency—For modulation schemes that have power density spectral nulls such as the one of ASK in Figure 1.5, defining the bandwidth as the width of the main spectral lobe is convenient.

Percentage Bandwidth Efficiency—If the spectrum of the modulated signal does not have nulls, as in general continuous phase modulation (CPM), null-to-null bandwidth no longer exists. In this case, energy percentage bandwidth may be used. Usually 99% is used, even though other percentages (e.g., 90%, 95%) are also used.

1.4.3 System Complexity

System complexity refers to the amount of circuits involved and the technical difficulty of the system. Associated with the system complexity is the cost of manufacturing, which is of course a major concern in choosing a modulation technique. Usually the demodulator is more complex than the modulator. Coherent demodulator is much more complex than noncoherent demodulator since carrier recovery is required. For some demodulation methods, sophisticated algorithms like the Viterbi algorithm are required. All of these are the basis for complexity comparison.

Since power efficiency, bandwidth efficiency, and system complexity are the main criteria of choosing a modulation technique, we will always pay attention to them in the analysis of modulation techniques in the rest of the book.

1.5 OVERVIEW OF DIGITAL MODULATION SCHEMES AND COMPARISON

To provide the reader with an overview, we list the abbreviations and descriptive names of various digital modulations that we will cover in Table 1.1 and arrange them in a relationship tree diagram in Figure 1.6. Some of the schemes can be derived from more than one "parent" scheme. The schemes where differential encoding can be used are labeled by letter D and those that can be noncoherently demodulated are labeled with a letter N. All schemes can be coherently demodulated.

The modulation schemes listed in the table and the tree are classified into two large categories: constant envelope and nonconstant envelope. Under constant envelope class, there are three subclasses: FSK, PSK, and CPM. Under nonconstant envelope class, there are three subclasses: ASK, QAM, and other nonconstant envelope modulations.

Among the listed schemes, ASK, PSK, and FSK are basic modulations, and MSK, GMSK, CPM, MHPM, and QAM, and so forth are advanced schemes. The advanced schemes are variations and combinations of the basic schemes.

The constant envelope class is generally suitable for communication systems whose power amplifiers must operate in the nonlinear region of the input-output characteristic in order to achieve maximum amplifier efficiency. An example is the TWTA (traveling wave tube amplifier) in satellite communications. However, the generic FSK schemes in this class are inappropriate for satellite application since they have very low bandwidth efficiency in comparison with PSK schemes. Binary FSK is used in the low-rate control channels of first generation cellular systems, AMPS (advance mobile phone service of the United States) and ETACS (European total access communication system). The data rates are 10 kbps for AMPS and 8 kbps for ETACS. The PSK schemes, including BPSK, QPSK, OQPSK, and MSK have been used in satellite communication systems.

The $\pi/4$-QPSK is worth special attention due to its ability to avoid 180° abrupt phase shift and to enable differential demodulation. It has been used in digital mobile

Abbreviation	Alternate Abbr.	Descriptive Name
Frequency Shift Keying (FSK)		
BFSK	FSK	Binary Frequency Shift Keying
MFSK		M-ary Frequency Shift Keying
Phase Shift Keying (PSK)		
BPSK	PSK	Binary Phase Shift Keying
QPSK	4PSK	Quadrature Phase Shift Keying
OQPSK	SQPSK	Offset QPSK, Staggered QPSK
$\pi/4$-QPSK		$\pi/4$ Quadrature Phase Shift Keying
MPSK		M-ary Phase Shift Keying
Continuous Phase Modulations (CPM)		
SHPM		Single-h (modulation index) Phase Modulation
MHPM		Multi-h Phase Modulation
LREC		Rectangular Pulse of Length L
CPFSK		Continuous Phase Frequency Shift Keying
MSK	FFSK	Minimum Shift Keying, Fast FSK
SMSK		Serial Minimum Shift Keying
LRC		Raised Cosine Pulse of Length L
LSRC		Spectrally Raised Cosine Pulse of Length L
GMSK		Gaussian Minimum Shift Keying
TFM		Tamed Frequency Modulation
Amplitude and Amplitude/Phase Modulations		
ASK		Amplitude Shift Keying (generic name)
OOK	ASK	Binary On-Off Keying
MASK	MAM	M-ary ASK, M-ary Amplitude Modulation
QAM		Quadrature Amplitude Modulation
Nonconstant Envelope Modulations		
QORC		Quadrature Overlapped Raised Cosine Modulation
SQORC		Staggered QORC
QOSRC		Quadrature Overlapped Squared Raised Cosine Modulation
Q^2PSK		Quadrature Quadrature Phase Shift Keying
IJF-OQPSK		Intersymbol-Interference/Jitter-Free OQPSK
TSI-OQPSK		Two-Symbol-Interval OQPSK
SQAM		Superposed-QAM
XPSK		Crosscorrelated QPSK

Table 1.1 Digital modulation schemes (Abbr.=Abbreviation).

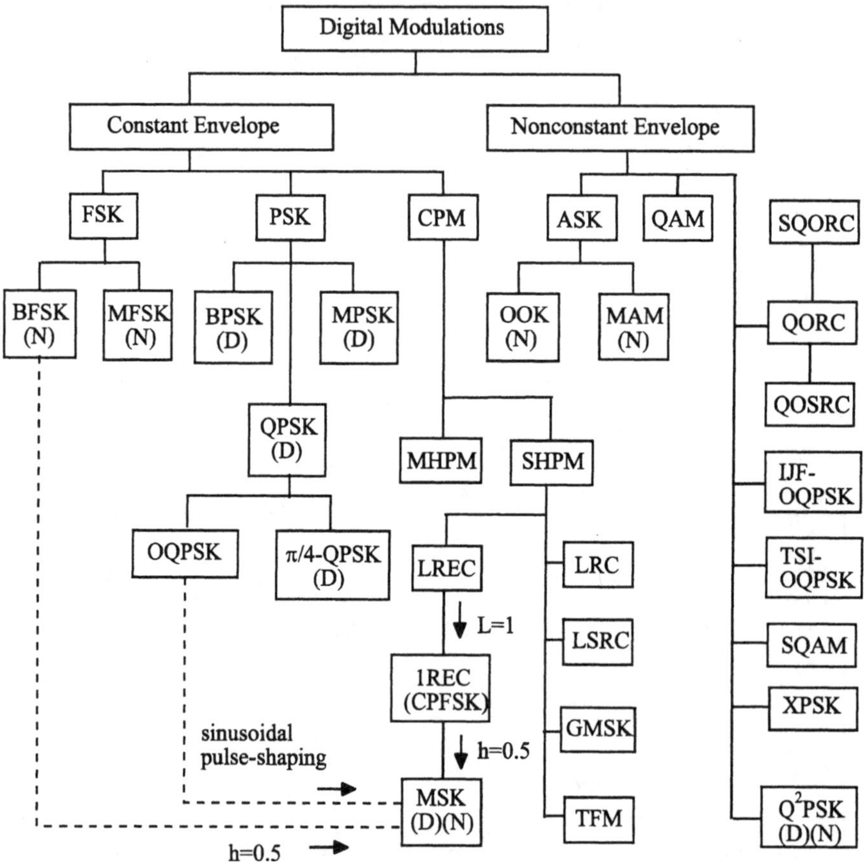

Solid lines indicate "can be derived from"

Dashed lines indicate "alternatively can be derived from"

(D) Can be differentially encoded and decoded

(N) Can be noncoherently detected

Figure 1.6 Digital modulation tree. After [4].

cellular systems, such as the United States digital cellular (USDC) system.

The PSK schemes have constant envelope but discontinuous phase transitions from symbol to symbol. The CPM schemes have not only constant envelope, but also continuous phase transitions. Thus they have less side lobe energy in their spectra in comparison with the PSK schemes. The CPM class includes LREC, LRC, LSRC, GMSK, and TFM. Their differences lie in their different *frequency pulses* which are reflected in their names. For example, LREC means the frequency pulse is a rectangular pulse with a length of L symbol periods. MSK and GMSK are two important schemes in the CPM class. MSK is a special case of CPFSK, but it also can be derived from OQPSK with extra sinusoidal pulse-shaping. MSK has excellent power and bandwidth efficiency. Its modulator and demodulator are also not too complex. MSK has been used in NASA's Advanced Communication Technology Satellite (ACTS). GMSK has a Gaussian frequency pulse. Thus it can achieve even better bandwidth efficiency than MSK. GMSK is used in the U.S. cellular digital packet data (CDPD) system and the European GSM (global system for mobile communication) system.

MHPM is worth special attention since it has better error performance than single-h CPM by cyclically varying the modulation index h.

The generic nonconstant envelope schemes, such as ASK and QAM, are generally not suitable for systems with nonlinear power amplifiers. However QAM, with a large signal constellation, can achieve extremely high bandwidth efficiency. QAM has been widely used in modems used in telephone networks, such as computer modems. QAM can even be considered for satellite systems. In this case, however, back-off in TWTA's input and output power must be provided to ensure the linearity of the power amplifier.

The third class under nonconstant envelope modulation includes quite a few schemes. These are primarily designed for satellite applications since they have very good bandwidth efficiency and the amplitude variation is minimal. All of them except Q^2PSK are based on $2T_s$ amplitude pulse shaping and their modulator structures are similar to that of OQPSK. The scheme Q^2PSK is based on four orthogonal carriers.

Figure 1.7 is a comparison of classical modulation schemes, including coherent MPSK (M-ary PSK), noncoherent MFSK (M-ary FSK), and coherent QAM. Their Nyquist bandwidth efficiency and power efficiency are compared at a BER level of 10^{-5}. Each modulation is represented by a dot of a different shape. The horizontal coordinate of a dot is the SNR (E_b/N_o) required for BER $= 10^{-5}$. The vertical coordinate of a dot is the bandwidth efficiency (bits/second/Hz) that can be achieved by that modulation with that SNR and 10^{-5} BER. The label R/W of the vertical axis is the ratio of bit rate (R) over required Nyquist bandwidth (W).

From the figure, it is clear that BPSK and QPSK/OQPSK have the same power

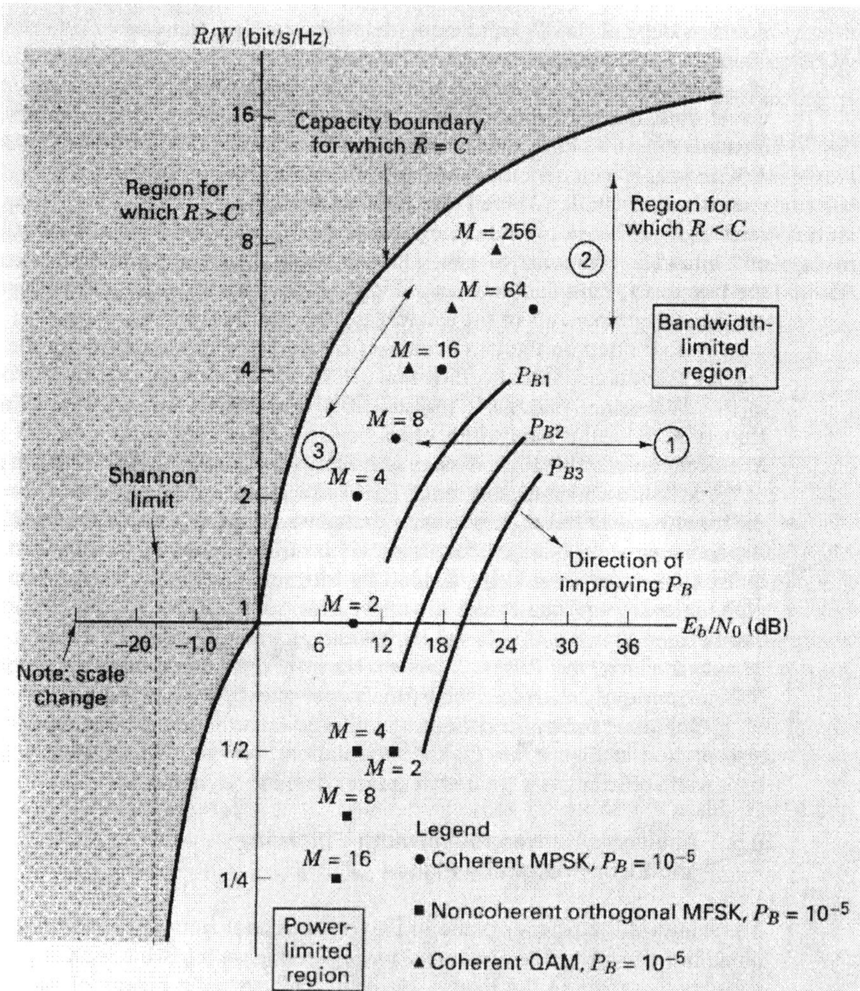

Figure 1.7 Bandwidth and power efficiency plane. From [5]. Copyright © 1988. Adapted by permission of Pearson Education.

efficiency, but QPSK/OQPSK have twice the bandwidth efficiency. In general, compared with the other two, MFSKs have better power efficiency but poorer bandwidth efficiencies. Compared with MPSKs of the same order, QAMs have the same bandwidth efficiencies but better power efficiencies.

Similar power-bandwidth trade-off planes for CPM and multi-h CPM (MHPM) will be presented in Chapters 6 and 7, respectively.

While most communication systems have been using a single carrier to transmit signals, multicarrier modulations (MCM), in particular, the orthogonal frequency division multiplexing (OFDM), have been getting attention recently and are already used in some systems, such as the IEEE 802.11a standard. OFDM was first proposed in the 1960s. In the 1970s and the 1980s, it was discovered that its modulation and demodulation can be implemented by inverse discrete Fourier transform (IDFT) and discrete Fourier transform (DFT), respectively. Its implementation becomes economically feasible recently due to the advances in digital signal processing devices. The concept of OFDM is to convert serial data into parallel data symbols and transmit them on a set of orthogonal subcarriers simultaneously. In this way the symbol length is multiplied by the number of subcarriers. The much longer symbol duration reduces the effect of intersymbol interference introduced by the channel, particularly a multipath fading channel.

Because of the need to maintain orthogonality between subcarriers, not all modulations for single carrier systems can be used for OFDM. Popular modulations for OFDM are QAM and PSK. Lately, M-ary ASK and wavelets are considered for OFDM too. It is possible that other modulations may be used for OFDM in the future.

OFDM will be treated in great detail in Chapters 12 to 16.

References

[1] Proakis, J., *Digital Communication,* New York: McGraw-Hill, 1983.

[2] Rappaport, T., *Wireless Communications: Principles and Practice,* Upper Saddle River, New Jersey: Prentice Hall, 1996.

[3] Siller, C., "Multipath propagation," *IEEE Communications Magazine,* vol. 22, no. 2, February 1984, pp. 6–15.

[4] Xiong, F., "Modem techniques in satellite communications," *IEEE Communications Magazine,* vol. 32, no. 8, August 1994, pp. 84–98.

[5] Sklar, B., *Digital Communications: Fundamentals and Applications,* Upper Saddle River, New Jersey: Prentice Hall, 1988.

Selected Bibliography

- Feher, K., ed., *Advanced Digital Communications*, Englewood Cliffs, New Jersey: Prentice Hall, 1987.
- Proakis, J., and M. Salehi, *Communication Systems Engineering*, Englewood Cliffs, New Jersey: Prentice Hall, 1994.

Chapter 2

Baseband Modulation (Line Codes)

Baseband modulation is defined as a direct transmission without frequency transform. It is the technology of representing digital sequences by pulse waveforms suitable for baseband transmission. A variety of waveforms have been proposed in an effort to find ones with some desirable properties, such as good bandwidth and power efficiency, and adequate timing information. These baseband modulation waveforms are variably called *line codes*, *baseband formats* (or *waveforms*), and *PCM waveforms* (or *formats*, or *codes*). PCM (pulse code modulation) refers to the process that a binary sequence representing a digitized analog signal is coded into a pulse waveform. For a data signal PCM is not needed. Therefore the terms line code and baseband format (or waveform) are more pertinent and the former one is more often used. Line codes were mainly developed in the 1960s by engineers at AT&T, IBM, or RCA for digital transmission over telephone cables or digital recording on magnetic media [1–5]. Recent developments in line codes mainly concentrate on fiber optic transmission systems [6–11].

In this chapter we first introduce the differential coding technique which is used in the later part of the chapter in constructing line codes. Then we describe various basic line codes in Section 2.2. Their power spectral densities are discussed in Section 2.3. The demodulation of these waveforms is in effect a detection problem of signals in noise. In Section 2.4 we first describe optimum detection of binary signals in additive white Gaussian noise (AWGN) and then apply the resultant general formulas to obtain expressions for bit error probabilities or bit error rates (BER) of various line codes. The general results can be used for any binary signal, including bandpass signals which will be described in later chapters. It also should be pointed out that practical detectors for line codes are often not optimum in order to simplify circuitry. However, the performance of an optimum detector can always serve as a reference for comparison. Substitution codes and block line codes are more complicated codes with improved performance over basic line codes. They are discussed in Sections 2.5 and 2.6. Pulse time modulation is a family of "pulse analog" baseband modulation, primarily designed for optical transmission. It is introduced in Section 2.7. Section 2.8 summarizes this chapter.

2.1 DIFFERENTIAL CODING

Since some of the baseband binary waveforms use a technique called *differential encoding*, we need to become familiar with this simple yet important baseband technique. This technique is not only used in baseband modulation but also in bandpass modulation where it is used to encode the baseband data before modulating it onto a carrier. The benefit of using differential coding will become clear when we discuss the schemes that use it. We now study this technique and it will be used throughout the rest of the book.

Let $\{a_k\}$ be the original binary data sequence, then a differentially encoded binary data sequence $\{d_k\}$ is produced according to the rule

$$d_k = a_k \oplus d_{k-1} \tag{2.1}$$

where $\oplus$ indicates *modulo-2 addition*. Modulo-2 addition is also called *exclusive-OR (XOR)*. The modulo-2 addition rules are $0+0=0$, $0+1=1$, $1+0=1$, and $1+1=0$. From (2.1) and the modulo-2 rules we can see that the current output bit of the encoder is determined by the current input bit and the previous output bit. If they are different the output bit is 1, otherwise the output bit is 0. This gives the name differential encoding.

To perform differential encoding an initial bit is needed and it is called a reference bit. For example, if $\{a_k\}$ and $\{d_k\}$ both start with $k=1$, then we need a d_0 as the reference bit. Since d_0 could be chosen as 0 or 1, then $\{a_k\}$ can be encoded into two different data sequences. They are complementary to each other.

The decoding rule is

$$\widehat{a}_k = \widehat{d}_k \oplus \widehat{d}_{k-1} \tag{2.2}$$

where the hat indicates the received data at the receiver. The received $\{\widehat{d}_k\}$ could be the same as or different from $\{d_k\}$. For example channel noise might have altered some of the bits in $\{d_k\}$ when it is received. Even if noise is light so that no bits have been altered by noise, the polarity reversals in various stages of the transmitter and receiver might have reversed the polarity of the entire sequence. One of the important uses of differential coding is to eliminate the effect of polarity reversal. This is clear from (2.2) since the decoder output depends on the difference of the two consecutive received bits, not their polarities. When the polarity of the entire sequence is altered, the difference between two consecutive bits remains the same.

Encoding													
a_k		1	0	1	1	0	0	0	1	1	1	0	0
$d_k = a_k \oplus d_{k-1}$	0	1	1	0	1	1	1	1	0	1	0	0	0
Decoding													
$\widehat{d}_k$ (correct polarity)	0	1	1	0	1	1	1	1	0	1	0	0	0
$\widehat{a}_k = \widehat{d}_k \oplus \widehat{d}_{k-1}$		1	0	1	1	0	0	0	1	1	1	0	0
Decoding													
$\widehat{d}_k$ (reversed polarity)	1	0	0	1	0	0	0	0	1	0	1	1	1
$\widehat{a}_k = \widehat{d}_k \oplus \widehat{d}_{k-1}$		1	0	1	1	0	0	0	1	1	1	0	0

Table 2.1 Examples of differential coding.

Table 2.1 is an example which illustrates the encoding and decoding processes with or without polarity reversal. The results are the same. Note that no errors caused by noise are assumed in the example. The first bit of $\{d_k\}$ is the reference bit which is 0 in the example.

Figure 2.1 shows the block diagrams of the differential encoder and decoder defined by (2.1) and (2.2).

The probability distribution of the differentially encoded sequence is of interest. It is useful when the autocorrelation function of the coded sequence is calculated later in the chapter. Assume the data sequence $\{a_k\}$ is stationary, its bits are independent, and it has a distribution of (p_0, p_1), where $p_0 = \Pr(0)$ and $p_1 = \Pr(1), p_0 + p_1 = 1$. Assume the distribution of the kth bit of the coded sequence $\{d_k\}$ is $(q_0^{(k)}, q_1^{(k)})$, where $q_0^{(k)} = \Pr(0)$ and $q_1^{(k)} = \Pr(1)$, $q_0^{(k)} + q_1^{(k)} = 1$. According to (2.1) we have

$$q_0^{(k)} = p_0 q_0^{(k-1)} + p_1 q_1^{(k-1)} \tag{2.3}$$

and

$$q_1^{(k)} = p_0 q_1^{(k-1)} + p_1 q_0^{(k-1)} \tag{2.4}$$

Since an initial bit is specified when encoding, $q_0^{(0)}$ and $q_1^{(0)}$ are known. They are either 0 or 1, depending on what is chosen. For instance, if the reference is 0, then $q_0^{(0)} = 1$, $q_1^{(0)} = 0$. It is easy to verify that if $p_0 = p_1 = 1/2$, then $q_0^{(k)} = q_1^{(k)} = 1/2$, for all k. That is, differential encoding does not change data distribution for equally likely data. However, when the distribution of the original data is not equal, differential encoding does change the data distribution. Further, we can show that $q_0^{(k)} = q_1^{(k)} = 1/2$ asymptotically regardless of values of p_0 and p_1.

We start with any one of the above two equations, say, (2.3), from which we have

$$q_0^{(k)} = (1 - p_0) + (2p_0 - 1) q_0^{(k-1)}$$

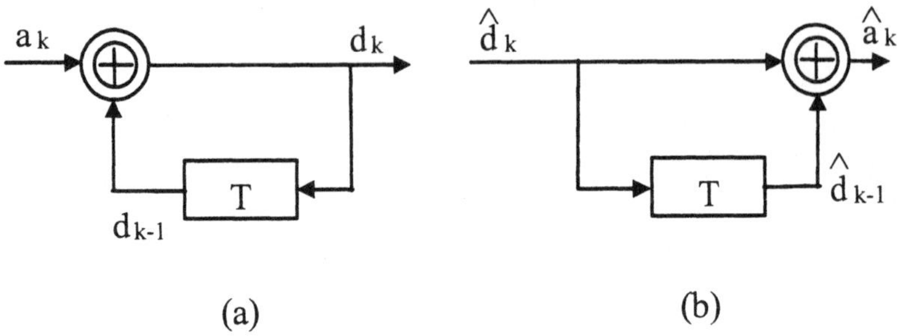

Figure 2.1 Differential encoder (a), and decoder (b).

Taking a z-transform of both sides of the above equation, we obtain

$$Q_0(z) = \frac{1 - p_0}{(1 - z^{-1})} + (2p_0 - 1)z^{-1}Q_0(z)$$

where $Q_0(z)$ is the z-transform of sequence $\{q_0^{(k)}\}$. Rearranging terms we have

$$Q_0(z) = \frac{1 - p_0}{(1 - z^{-1})[1 - (2p_0 - 1)z^{-1}]}$$

Using the final-value theorem we obtain the limit of $q_0^{(k)}$ as

$$\begin{aligned} \lim_{k \to \infty} q_0^{(k)} &= \lim_{z \to 1} [(1 - z^{-1})Q_0(z)] \\ &= \lim_{z \to 1} \frac{1 - p_0}{1 - (2p_0 - 1)z^{-1}} = \frac{1}{2} \end{aligned}$$

Thus we can conclude that regardless of the distribution of the original data, the distribution of the differentially encoded data is always asymptotically equal.

To see how fast $q_0^{(k)}$ converges to 1/2, we define two ratios as

$$r_k = q_1^{(k)}/q_0^{(k)}$$

$$l = p_1/p_0$$

and substituting (2.3) and (2.4) into the r_k expression we have

$$r_k = \frac{l + r_{k-1}}{1 + l r_{k-1}} \tag{2.5}$$

Next we define the ratio difference as

$$\Delta r_k = r_k - r_{k-1}$$

If $\Delta r_k \to 0$, then $r_k \to r_{k-1}$, solving (2.5) will give $r_k = 1$ (i.e., $q_1^{(k)} = q_0^{(k)}$ for $k \to \infty$).

Calculations show that for $p_0 = 0.3$ and 0.1, $q_1^{(k)}$ virtually equals $q_0^{(k)}$ ($\Delta r_k < 0.001$) at $k = 10$ and 38, respectively. For a very skewed distribution (e.g., $p_0 = 0.01$), to reach $\Delta r_k < 0.001$, 411 iterations are needed. All these k values are small when compared with numbers of data in practical systems. Thus we can see that the distribution of the differentially encoded data becomes virtually equal very quickly, regardless of the distribution of the original data.

Differential encoding can also be done by taking the binary complement of the modulo-2 adder as the output, that is

$$d_k = \overline{a_k \oplus d_{k-1}} \tag{2.6}$$

where $\overline{x}$ denotes a binary complement of x. Again this second rule can generate two complementary sequences with the two different choices of the reference bit. The corresponding decoding rule is

$$\widehat{a}_k = \overline{\widehat{d}_k \oplus \widehat{d}_{k-1}} \tag{2.7}$$

which is also capable of correcting polarity reversals. The block diagrams of the encoder and the decoder defined by this set of rules are similar to that in Figure 2.1 except that an inverter is needed at the output of both the encoder and decoder.

The above argument about distribution also applies to data encoded this way since this coded sequence is just a complement of the previous one.

Another type of differential encoding is

$$d_k = a_{k-1} - a_k \tag{2.8}$$

which produces a three-level sequence $(-1, 0, +1)$. An arbitrary initial reference bit a_o must be specified. It is obvious that the distribution of d_k is

$$d_k = \begin{cases} 1, & q_1 = p_1 p_0, & \text{for } a_{k-1} = 1 \text{ and } a_k = 0 \\ -1, & q_{-1} = p_0 p_1, & \text{for } a_{k-1} = 0 \text{ and } a_k = 1 \\ 0, & q_0 = p_0^2 + p_1^2, & \text{for } a_{k-1} = a_k = 0 \text{ or } 1 \end{cases} \tag{2.9}$$

Decoding can be done as follows. First $\{\widehat{d}_k\}$ is converted to unipolar $\{\widehat{u}_k\}$ by full-wave rectification, then $\{\widehat{a}_k\}$ is recovered from $\{\widehat{u}_k\}$ by XOR operation:

$$\widehat{a}_k = \widehat{a}_{k-1} \oplus \widehat{u}_k \tag{2.10}$$

where the initial $\widehat{a}_o = a_o$ is known. This coding scheme is also immune from a polarity inversion-ambiguity problem since after full-wave rectification, the waveform would be the same.

2.2 DESCRIPTION OF LINE CODES

Many binary line codes have been proposed in the literature and some of them have been used in practical systems [1–11]. Basic codes can be classified into four classes: *nonreturn-to-zero (NRZ)*, *return-to-zero (RZ)*, *pseudoternary (PT)*, and *biphase*. NRZ and RZ classes can be further divided into *unipolar* and *polar* subclasses. Advanced codes include *substitution codes* and *block codes*. There are some other codes which do not belong to any of the classes. Some codes may belong to more than one class. Figure 2.2 is a quite complete collection of waveforms of various basic line codes. Each of them will be studied in detail shortly. Figure 2.2 does not include substitution codes and block codes. They will be studied separately in the latter part of this chapter.

The reason for the large selection of line codes is because of their differences in performance which will lead to different applications. The features to look for in choosing a line code are as follows. For a particular application, some of the features may be important while others may not be.

(1) *Adequate timing information.* Bit or symbol timing is usually recovered from the received data sequence. This requires that the line code format provides adequate transition density in the coded sequence. Formats with higher transition density are preferable since the timing recovery will have fewer problems with these kinds of signals. A long string of binary 1s and 0s in the data should not cause a problem in timing recovery.

(2) *A spectrum that is suitable for the channel.* For example, line codes with no dc component and small near-dc components in their power spectral density (PSD) are desirable for magnetic recording systems, ac coupled channels, or systems using transformer coupling which have very poor low frequency responses. In addition the PSD of the line code should have sufficiently small bandwidth compared with the channel bandwidth so ISI will not be a problem.

(3) *Narrow bandwidth.* The bandwidth of the line code should be as narrow as possible. The transmission bandwidth may be reduced by filtering and

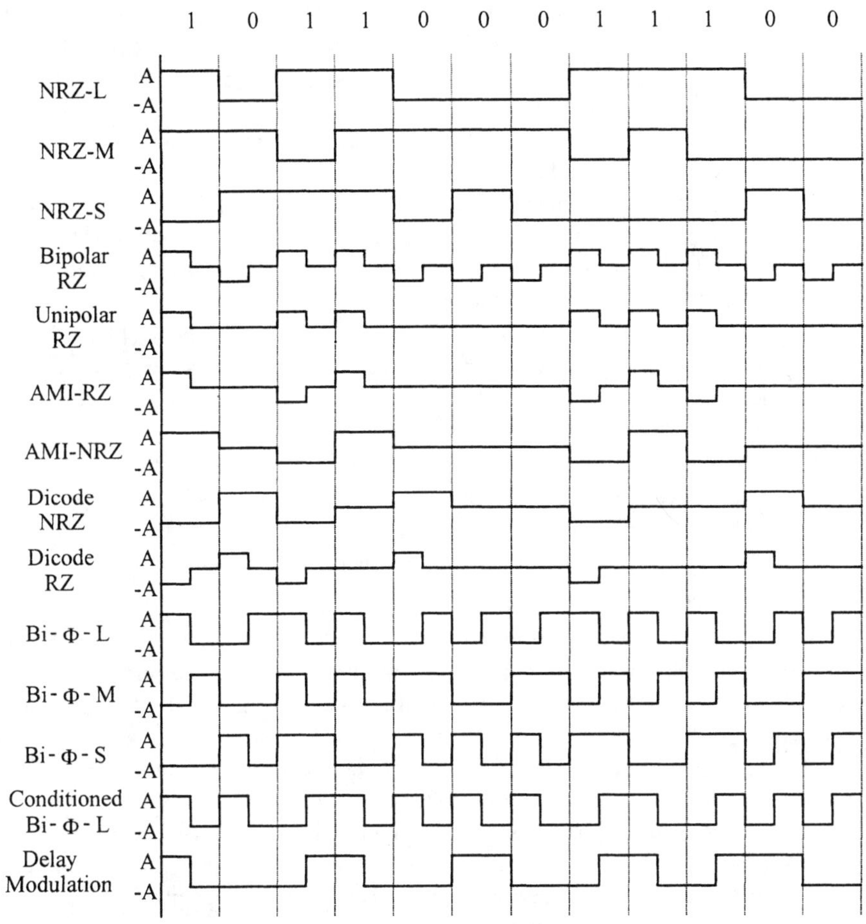

Figure 2.2 Line codes. After [12].

multilevel transmission schemes. The penalty is an increase in P_b due to an increase in ISI and a decrease in signal-to-noise ratio. Some line codes may suffer less degradation than others.

(4) *Low error probability*. The line code can be recovered with low bit error probability (P_b) from noise and/or ISI corrupted received signal. The ones with lower P_b for the same average bit energy are usually preferable, but consideration should be given to other characteristics, such as bandwidth and self-timing capability.

(5) *Error detection capability.* Some schemes have the capability of detecting errors in the received sequence without introducing extra bits like in channel coding schemes. This error detection capability can be used as a means of performance monitoring. However, error correction is not possible; this can only be achieved through channel coding techniques or automatic retransmission schemes.

(6) *Bit sequence independence (transparency)*. The line code must be able to encode any data sequence from any source and the decoder must be able to decode it back to original data. In other words, attributes of the code are independent of the source statistics.

(7) *Differential coding*. This feature is useful since differentially coded sequences are immune from polarity inversion as we studied in the previous section. However, if differential coding is not inherent in the line code itself, a separate differential coding scheme can be incorporated in the system.

In the following we describe the various line codes basically in groups, but some of the line codes are singled out due to their importance or unique features. When we study these codes the above criteria should be kept in mind, and we will refer to them from time to time.

We put emphases on coding rules and characteristics. We generally omit coder and decoder implementations. Simple codes can be implemented by simple combinational and sequential digital circuits, while complex codes can be implemented by digital signal processing techniques. A comprehensive coverage on coder and decoder implementation is not necessary and also is beyond the page limit of this book. Interested readers may refer to listed references for circuits. However many reported circuits are obsolete already, new circuits should be designed based on new products of IC chips.

2.2.1 Nonreturn-to-Zero Codes

The nonreturn-to-zero group includes the first three codes in Figure 2.2. Two levels

$(\pm A)$ of the pulse amplitude are used to distinguish binary 1 and 0 in the *NRZ-L* format. This waveform has no dc component for an equiprobable binary data sequence. In the *NRZ-M* format a level change (A to $-$A or $-$A to A) is used for a mark (1) in the binary sequence and no change for a space (0). The *NRZ-S* waveform is similar except that the level change is used to indicate a space (0). Both NRZ-M and NRZ-S are differentially encoded waveforms. They can be produced by modulating the differentially encoded binary sequence using the NRZ-L format. The NRZ-M waveform is generated with the encoding rule of (2.1) and the NRZ-S waveform with the encoding rule of (2.6). The reference bit is 0. The corresponding coded sequence for NRZ-M is already given in Table 2.1.

Recovery of NRZ-L[1] from NRZ-M or NRZ-S is accomplished by differential decoding.

The main advantage of NRZ-M and NRZ-S over NRZ-L is their immunity to polarity reversals owing to the differential coding.

All the above three formats can be made *unipolar* by changing the lower level $-$A to level 0. For a binary sequence with equally likely 1s and 0s, which is the usual assumption, the unipolar waveforms have a dc component at a level of A/2, whereas the polar ones do not.

Since a string of 1s in NRZ-S, a string of 0s in NRZ-M, and a string of 1s or 0s in NRZ-L does not contain any transitions, this class of waveforms may not provide adequate timing information for data with long strings of 1s and 0s. Solutions to this shortcoming include precoding the data sequence to eliminate long strings of 1s and 0s or transmission of a separate synchronizing sequence.

NRZ-L is used extensively in digital logic as we all know. NRZ-M is used primarily in magnetic tape recording. In telecommunication, NRZ format applications are limited to short-haul links due to their timing characteristic.

2.2.2 Return-to-Zero Codes

The lack of timing information of the NRZ formats can be overcome by introducing more transitions into the waveform. This leads to the RZ formats shown in Figure 2.2. However, the bandwidth of an RZ format is wider than that of an NRZ format, as we will see shortly.

In the *unipolar RZ* format, a binary 1 is represented by a positive pulse for a half-bit period then returns to zero level for the next half period, resulting in a transition in the middle of the bit. A binary 0 is represented by the zero level for the entire bit period. Since there are no transitions in a string of 0s, precoding or scrambling is needed to eliminate long strings of 0s. This format also has a dc component since it

---

1 Since NRZ-L waveform is the most common waveform in the digital circuit, it is chosen as the ultimate decoding result of line codes in our discussion.

is unipolar.

In the *polar RZ* format, 1 and 0 are represented by positive and negative half-period pulses, respectively. This waveform ensures two transitions per bit. It has no dc component.

2.2.3 Pseudoternary Codes (Including AMI)

This group of line codes use three levels $\pm A$ and 0. *AMI* (alternative mark inversion) *codes* are in this group. They are often called *bipolar codes* in the telecommunication industry. In *AMI-RZ* (return-to-zero AMI) format, a 1 is represented by an RZ pulse with alternative polarities if 1s are consecutive. A 0 is represented by the zero level. In *AMI-NRZ* (nonreturn-to-zero AMI), the coding rule is the same as AMI-RZ except that the symbol pulse has a full length of T. They have no dc component but like unipolar RZ their lack of transitions in a string of 0s may cause a synchronization problem. Therefore AMI codes with zero extraction (substitution codes) are proposed as will be discussed later in this chapter.

Recovery of NRZ-L from the AMI-NRZ code is accomplished by simple full-wave rectification [1]. Similarly AMI-RZ code can be full-wave rectified to form RZ-L waveform which can be easily converted to NRZ-L waveform.

These formats are used in baseband data transmission and magnetic recording. The AMI-RZ format is most often used in telemetry systems. It is used by AT&T for T1 carrier systems.

Other members of this group include the *dicode NRZ* and *dicode RZ*. Dicode formats are also called *twinned binary* in the literature [1,2]. In dicode NRZ the 1 to 0 or 0 to 1 transition changes the pulse polarity, a zero level represents no data transition. In dicode RZ, the same coding rule is used except that the pulse is only half-bit wide (i.e., it returns to zero for the second half bit).

Dicodes and AMI codes are related by differential coding [2]. If data sequence $\{a_k\}$ is the sequence used directly to construct a dicode, then sequence $\{d_k\}$, where $d_k = a_{k-1} - a_k$, can be used to construct an AMI code which is a dicode to the original sequence $\{a_k\}$. This can be seen from the example in Figure 2.2. Assuming $a_0 = 0$, we convert $\{a_k\} = (1,0,1,1,0,0,0,1,1,1,0,0)$ to $\{d_k\} = (-1,1,-1,0,1,0,0,-1,0,0,1,0)$. Using $\{d_k\}$ and AMI rules we can construct exactly the dicodes in the figure (with replacements A and $-$A for 1 and -1, respectively).

Dicodes can be decoded as follows. First $\{\widehat{d}_k\}$ is converted to unipolar $\{\widehat{u}_k\}$ by full-wave rectification, then $\{\widehat{a}_k\}$ is recovered from $\{\widehat{u}_k\}$ by modulo-2 summation: $\widehat{a}_k = \widehat{a}_{k-1} \oplus \widehat{u}_k$ $(\widehat{a}_0 = a_0)$. Readers can easily verify this using the above example.

2.2.4 Biphase Codes (Including Manchester)

This group of line codes uses half-period pulses with different phases according to certain encoding rules in the waveform. Four waveforms of this group are shown in Figure 2.2.

The *Bi-Φ-L* (biphase-level) format is better known as *Manchester*, and is also called *diphase*, or *split-phase*. In this format, a 1 is represented as a pulse with the first half bit at a higher level and the second half bit at a lower level. A 0 is represented as a pulse with the opposite phase (i.e., a lower level for the first half bit and a higher level for the second half bit). The pulse shapes for 1 and 0 can be exchanged.

The *Bi-Φ-M* (biphase-mark) format requires that a transition is always present at the beginning of each bit. A 1 is coded as a second transition in the middle of the bit and a 0 is coded as no second transition in the bit. This results in representing a 1 by one of the two phases of the pulse. In the *Bi-Φ-S* (biphase-space) format the opposite coding rules are applied to 1 and 0. The above three biphase formats ensure that there is at least one transition in a bit duration, thus providing adequate timing information to the demodulator.

The fourth format in this group is the *conditioned Bi-Φ-L*. In fact it is a differentially encoded Bi-Φ-L (i.e., the data sequence used for modulation is generated from the original binary sequence with differential encoding). Like NRZ-M and NRZ-S, this format is immune from polarity inversions in the circuit.

Biphase formats are used in magnetic recording, optical communications, and in some satellite telemetry links. Manchester code has been specified for the IEEE 802.3 standard for baseband coaxial cable using carrier sense multiple access and collision detection (CSMA/CD) (i.e., Ethernet [13, 14]). It has also been used in MIL-STD-1553B, which is a shielded twisted-pair bus system designed for high-noise environments [14]. Differential Manchester has been specified for the IEEE 802.5 standard for token ring, using either baseband coaxial cable or twisted-pair. Because it uses differential coding, differential Manchester is preferred for a twisted-pair channel.

2.2.5 Delay Modulation (Miller Code)

Delay modulation (DM) [3] or *Miller code* also can be classified into the biphase group since there are two phases in the waveform. However, it has some unique features. A 1 is represented by a transition in the midpoint of the bit. A 0 is represented by no transition unless it is followed by another 0. Then a transition is placed at the end of the first 0 bit. This format has a very small bandwidth and, most importantly, a very small dc component as we will see shortly. This makes it suitable for magnetic recording since magnetic recorders have no dc response [3].

2.3 POWER SPECTRAL DENSITY OF LINE CODES

In this section we first present a general formula for power spectral density (PSD) calculation for digitally modulated baseband waveforms. It can be used for most of the binary line codes. Therefore for most of the time we will simply apply this formula to various codes in the rest of this section. However, this formula is not applicable to some codes so other methods to calculate the PSD will be discussed.

We know that most signals like voice signal and image signal are essentially random. Therefore digital signals derived from these signals are also random. Data signals are also essentially random.

Assume the digital signal can be represented by

$$s(t) = \sum_{k=-\infty}^{\infty} a_k g(t - kT) \tag{2.11}$$

where a_k are discrete random data bits and $g(t)$ is a signal of duration T (i.e., nonzero only in $[0, T]$). Let us name $g(t)$ as *symbol function*. It could be any signal with a Fourier transform. For example it could be a baseband symbol shaping pulse or a burst of modulated carrier at passband. The random sequence $\{a_k\}$ could be binary or nonbinary.

In Appendix A (A.16) we show that the power spectral density of $s(t)$ is

$$\Psi_s(f) = \frac{|G(f)|^2}{T} \sum_{n=-\infty}^{\infty} R(n) e^{-jn\omega T} \tag{2.12}$$

where $\omega = 2\pi f$. $G(f)$ is the Fourier transform of $g(t)$ and $R(n)$ is the autocorrelation function of random sequence $\{a_k\}$, defined as $R(n) = E\{a_k a_{k+n}\}$, where $E\{x\}$ is the probabilistic average of x. Equation (2.12) shows that the PSD of a digitally modulated signal is not only determined by its symbol function but also is affected by the autocorrelation function of the data sequence.

In the following we always assume that the original binary data sequence has 1s and 0s equally likely. That is, $p_0 = Pr(0) = p_1 = Pr(1) = 1/2$. However, in order to write the modulated waveform in the form of (2.11), sequence $\{a_k\}$ of (2.11) is usually not the original sequence, rather it is derived from the original. Therefore its probability distribution needs to be calculated.

For uncorrelated sequence $\{a_k\}$,

$$R(n) = \begin{cases} \sigma_a^2 + m_a^2, & n = 0 \\ m_a^2, & n \neq 0 \end{cases} \tag{2.13}$$

where σ_a^2 is the variance and m_a is the mean of the sequence $\{a_k\}$. Using the Poisson sum formula, the PSD expression can be written as (see (A.17))

$$\Psi_s(f) = \frac{|G(f)|^2}{T}\left[\sigma_a^2 + m_a^2 R_b \sum_{n=-\infty}^{\infty} \delta(f - nR_b)\right] \tag{2.14}$$

where $R_b = 1/T$ is the data bit rate.

For line codes with $R(n) = 0$ for $n \neq 0$, (2.12) is more convenient. For line codes with $R(n) \neq 0$ for $n \neq 0$, (2.14) is more convenient.

Among baseband modulated signals, NRZ-L, NRZ-M, NRZ-S, RZ (polar or unipolar), AMI-RZ, AMI-NRZ, Bi-Φ-L, and dicode (RZ or NRZ) can be written in the form of (2.11). Therefore their PSD can be found quite easily using the above series of equations. However, there are some digital signals which cannot simply be represented by (2.11). Among the line codes, Bi-Φ-M, Bi-Φ-S, DM, and the substitution codes and block codes (which will be described later) belong to this group. If the signal is wide sense stationary (WSS), to find its PSD, the approach is to find the autocorrelation $R(\tau)$ first, then take the Fourier transform to find the PSD (Wiener-Khinchine theorem). If the signal is cyclostationary, then $R(\tau)$ is the time average of the time-dependent $R(t, \tau)$ in a period. The Wiener-Khinchine theorem is still applicable when time average of $R(t, \tau)$ is used for a nonstationary (including cyclostationary) process.

Some coded sequences, like Bi-Φ-M and delay modulation, can be described as a first order Markov random process. Their $R(\tau)$ can be found by using the method provided in [3]. We will use this method when we encounter the calculation of PSDs of Bi-Φ-M and delay modulation.

For more complex coded sequences one can use the general formula given by [15, 16]:

$$\Psi_s(f) = \frac{1}{T}\sum_i \sum_j G_i(-f)G_j(f)\left[p_i U_{ij}(f) + p_j U_{ji}(f) - p_i \delta_{ij}\right] \tag{2.15}$$

where $G_i(f)$ is the Fourier transform of the state i pulse waveform,[2] T is the pulse width, and δ_{ij} is the Kronecker delta function. p_i is the state i steady-state occurrence probability, and $U_{ji}(f)$ is the transform probability of state j occurring after state i. Probability p_i is found by taking residues of $U_{ji}(f)$ at its poles when $f = 0$. The $U_{ji}(f)$ is calculated by a signal flow graph and Mason's formula [17]. We do not attempt to use this method in this chapter in order not to make our discussion too

[2] A state of a coded sequence is determined by the symbols representing an information bit. For example, Manchester code has two states: 10 or 01, corresponding to the two half-positive and half-negative pulses.

mathematical.[3] Instead, results obtained using this method may be quoted when necessary. Or we may use the computer Monte-Carlo simulation to find the $R(\tau)$ and PSD.

Now we are ready to discuss the PSDs of the binary line codes described in the previous section.

2.3.1 PSD of Nonreturn-to-Zero Codes

Recall that NRZ-M and NRZ-S are generated by NRZ-L modulation with differentially encoded data sequences. Assume that original binary data are equally likely. Then according to Section 2.1, the differentially encoded sequences with (2.1) or (2.6) are also equally likely. In other words, the statistic properties of the sequences used directly for modulation are the same for NRZ-L, NRZ-M, and NRZ-S. And their symbol functions are also the same. *As a result their PSDs are the same.*

NRZ formats' symbol function is a square pulse with amplitude A in the interval $[0, T]$, which can be expressed as

$$g(t) = \begin{cases} A, & 0 \le t \le T \\ 0, & \text{elsewhere} \end{cases} \tag{2.16}$$

Its Fourier transform can be easily found as

$$\begin{aligned} G(f) &= AT\left(\frac{\sin \pi f T}{\pi f T}\right) e^{-j\pi f T} \\ &= AT \text{sinc}(\pi f T) e^{-j\pi f T} \end{aligned} \tag{2.17}$$

where $sinc(x) = sin(x)/x$ is the *sinc function.* Next we need to find the autocorrelation function $R(n)$ of the binary data sequence $\{a_k\}$. For this waveform

$$a_k = \begin{cases} 1, & \text{for binary 1}, \quad p_1 = 1/2 \\ -1, & \text{for binary 0}, \quad p_0 = 1/2 \end{cases} \tag{2.18}$$

Thus

$$R(n) = \begin{cases} p_1(1)^2 + p_0(-1)^2 = 1, & n = 0 \\ (p_0)^2(1)^2 + (p_1)^2(1)^2 + 2p_0p_1(1)(-1) = 0, & n \neq 0 \end{cases} \tag{2.19}$$

---

[3] We also do not expect the reader to be able to use this method by just reading this formula. This is to provide a reference in case the reader is interested in finding out the PSD of a complex coded sequence.

Substituting expressions of $G(f)$ and $R(n)$ into (2.12) we have,

$$\Psi_s(f) = A^2T\left(\frac{\sin \pi fT}{\pi fT}\right)^2, \quad \text{(NRZ)} \tag{2.20}$$

Figure 2.3(a) shows the plot of $\Psi_s(f)$ and the plot of the fractional out-of-band power, $P_{ob}(B)$, defined as[4]

$$P_{ob}(B) = 1 - \frac{\int_{-B}^{B} \Psi_s(f)df}{\int_{-\infty}^{\infty} \Psi_s(f)df} \tag{2.21}$$

In the figure $P_{ob}(B)$ is in decibels and the horizontal axis is normalized frequency $fT = f/R_b$. In the figure we set $A = 1$ and $T = 1$ for unity symbol pulse energy. This PSD is a squared sinc function with its first null at $fT = 1$. The signal energy concentrates near 0 frequency and the null bandwidth is $B_{null} = 1.0R_b$, 90% energy bandwidth is $B_{90\%} \approx 0.85R_b$, and the 99% bandwidth is $B_{99\%} \approx 10R_b$.

We mentioned that all three NRZ waveforms can be made unipolar. Assume the pulse amplitude is A, then there is a dc component of A/2 in the signals and it appears as an impulse function with strength of $A^2/4$ at 0 frequency in the power spectral density as we will show next. In this case the pulse function is still the one in (2.16), and the data are

$$a_k = \begin{cases} 1, & \text{for binary } 1, p_1 = 1/2 \\ 0, & \text{for binary } 0, p_0 = 1/2 \end{cases} \tag{2.22}$$

From this we have

$$m_a = E\{a_k\} = p_1(1) + p_0(0) = 1/2 \tag{2.23}$$

and

$$\sigma_a^2 = E\{(a_k - m_a)^2\} = p_1(1 - 1/2)^2 + p_0(0 - 1/2)^2 = 1/4 \tag{2.24}$$

Substitute expressions of $G(f)$, m_a, and σ_a^2 into (2.14), we have

$$\Psi_s(f) = \frac{A^2T}{4}\left(\frac{\sin \pi fT}{\pi fT}\right)^2 + \frac{A^2}{4}\delta(f), \quad \text{(unipolar NRZ)} \tag{2.25}$$

For unity average symbol energy we must set $A = \sqrt{2}$. This PSD has the same shape as that of the polar NRZs. The only difference is the impulse at the 0 frequency. The

---

[4] $\Psi_s(f)$ may be singular. For instance, $\Psi_s(f)$ may contain $\delta(f)$. In these cases we must integrate the singular part separately. $\int_{-\infty}^{\infty} \Psi_s(f)df$ will equal unity when $\Psi_s(f)$ is normalized. In this book $P_{ob}(B)$ is evaluated by numerical integration.

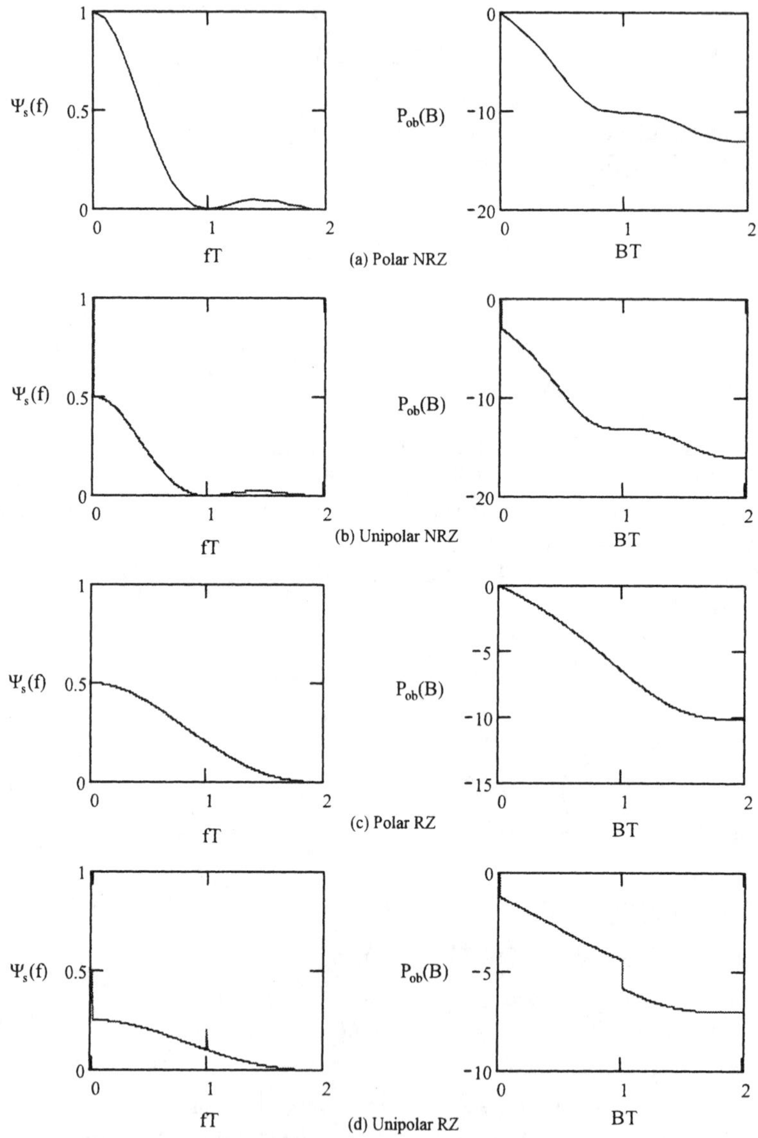

Figure 2.3 PSD of NRZ and RZ line codes (a–d) ($P_{ob}(B)$ is in dB).

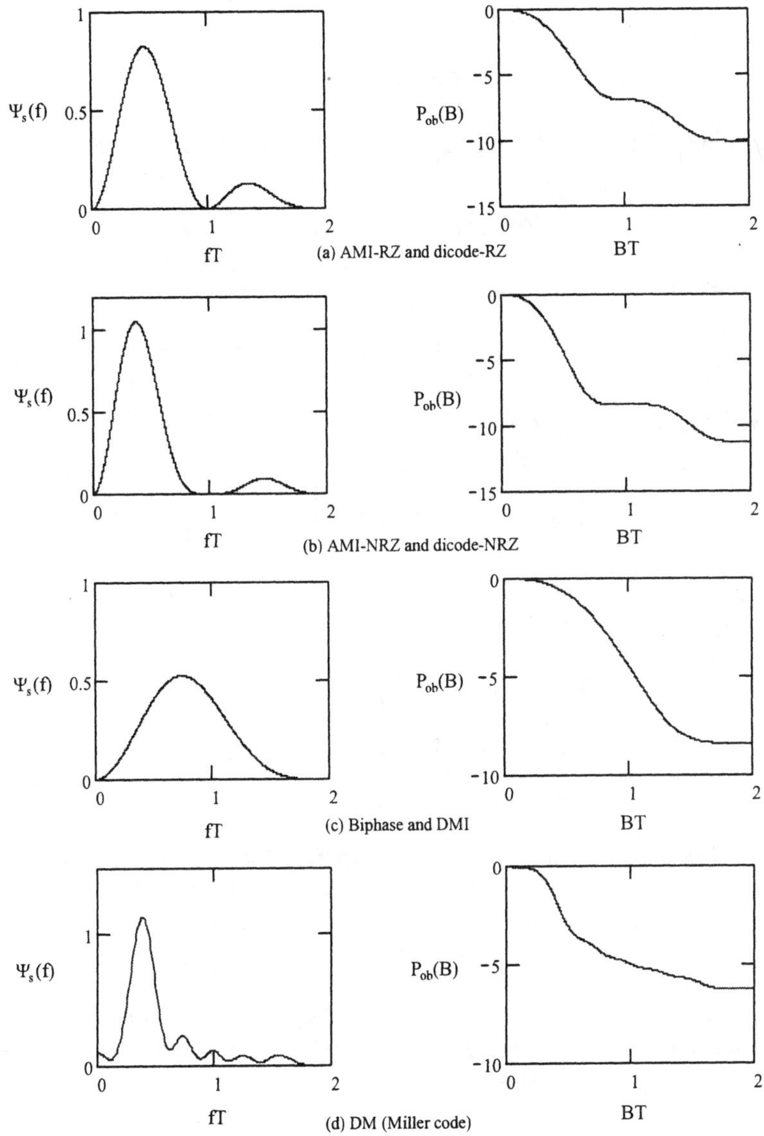

Figure 2.4 PSD of more line codes (a–d) ($P_{ob}(B)$ is in dB).

PSD and the out-of-band power are shown in Figure 2.3(b). The bandwidths are $B_{null} = 1.0R_b$, $B_{90\%} \approx 0.54R_b$, and $B_{99\%} \approx 5R_b$.

2.3.2 PSD of Return-to-Zero Codes

For RZ formats, the pulse function is a square pulse with half-bit duration

$$g(t) = \begin{cases} A, & 0 \leq t \leq \frac{T}{2} \\ 0, & \text{elsewhere} \end{cases} \tag{2.26}$$

The corresponding Fourier transform is

$$G(f) = \frac{AT}{2}\left(\frac{\sin \pi fT/2}{\pi fT/2}\right) e^{-j\pi fT/2} \tag{2.27}$$

For *polar RZ* waveform

$$a_k = \begin{cases} 1, & \text{for binary } 1, \quad p_1 = 1/2 \\ -1, & \text{for binary } 0, \quad p_0 = 1/2 \end{cases} \tag{2.28}$$

which is the same as polar NRZ. Thus

$$R(n) = \begin{cases} 1, & n = 0 \\ 0, & n \neq 0 \end{cases} \tag{2.29}$$

Substituting (2.27) and (2.29) into (2.12), the result is

$$\Psi_s(f) = \frac{A^2T}{4}\left(\frac{\sin \pi fT/2}{\pi fT/2}\right)^2, \quad \text{(polar RZ)} \tag{2.30}$$

For unity average symbol energy we must set $A = \sqrt{2}$. The PSD and the out-of-band power are shown in Figure 2.3(c). Compared with the PSD of NRZ format, this PSD is a stretched version with frequency axis scaled up twice. Therefore all its bandwidths are double that of NRZ. The bandwidths are $B_{null} = 2.0R_b$, $B_{90\%} \approx 1.7R_b$, and $B_{99\%} \approx 22R_b$.

For *unipolar RZ* format the symbol pulse is the same as in (2.26). The data sequence, its mean, and variance are the same as those of unipolar NRZ as given in (2.22), (2.23), and (2.24). Substituting $G(f)$ in (2.27), m_a and σ_a^2 into (2.14), we have

$$\Psi_s(f) = \frac{A^2T}{16}\left(\frac{\sin \pi fT/2}{\pi fT/2}\right)^2 \left[1 + R_b \sum_{k=-\infty}^{\infty} \delta(f - nR_b)\right], \quad \text{(unipolar RZ)} \tag{2.31}$$

For unity average symbol energy we must set $A = 2$. The PSD and the out-of-band power are shown in Figure 2.3(d). From the figure we can see that besides the continuous spectrum resembling the PSD of polar RZ, there are spikes at all odd integer frequencies. Their strengths are determined by the second term of (2.31) and are given by

$$\frac{A^2}{16}\left(\frac{\sin \pi n/2}{\pi f n/2}\right)^2$$

For $A = 2$, the strengths of the components form $f = 0, R_b, 3R_b, 5R_b, 7R_b$, and $9R_b$ are: 0.25, 0.101, 0.011, 0.004, 0.002, and 0.001 in the two-side spectrum. All discrete harmonics add up to 0.5 and the rest of energy (0.5) is in the continuous part of the spectrum.

The bandwidths are $B_{null} = 2.0R_b$, $B_{90\%} \approx 1.6R_b$, and $B_{99\%} \approx 22R_b$, which are almost the same as those of the polar RZ format.

2.3.3 PSD of Pseudoternary Codes

For *AMI* codes the data sequence $\{a_k\}$ takes on three values:

$$a_k = \begin{cases} 1, & \text{for binary 1,} \quad p_1 = 1/4 \\ -1, & \text{for binary 1,} \quad p_{-1} = 1/4 \\ 0, & \text{for binary 0,} \quad p_0 = 1/2 \end{cases} \tag{2.32}$$

We can find $R(0)$ as follows

$$R(0) = E\{a_k^2\} = \frac{1}{4}(1)^2 + \frac{1}{4}(-1)^2 + \frac{1}{2}(0)^2 = \frac{1}{2}$$

Adjacent bits in $\{a_k\}$ are correlated due to the alternate mark inversion. The adjacent bit pattern in the original binary sequence must be one of these: (1,1), (1,0), (0,1), and (0,0). The possible $a_k a_{k+1}$ products are –1, 0, 0, 0. Each of them has a probability of 1/4. Thus

$$R(1) = E\{a_k a_{k+1}\} = \frac{1}{4}(-1) + \frac{1}{4}\cdot 0 + \frac{1}{4}\cdot 0 + \frac{1}{4}\cdot 0 = -\frac{1}{4}$$

For $n > 1$, a_k and a_{k+n} are not correlated. The possible $a_k a_{k+n}$ products are ± 1, 0, 0, 0. Each case occurs with a probability of $1/4$. Thus

$$R(n > 1) = E\{a_k a_{k+n}\} = \frac{1}{8}(1) + \frac{1}{8}(-1) = 0$$

Summarizing the above results, we have

$$R(n) = \begin{cases} \frac{1}{2}, & n = 0 \\ -\frac{1}{4}, & |n| = 1 \\ 0, & |n| > 1 \end{cases}$$

Substitute this $R(n)$ and the symbol pulse spectrum of (2.27) into (2.12), we have the PSD of *AMI-RZ* code:

$$\begin{aligned} \Psi_s(f) &= \frac{1}{T}|G(f)|^2(\frac{1}{2} - \frac{1}{4}e^{j\omega T} - \frac{1}{4}e^{-j\omega T}) \\ &= \frac{1}{T}|G(f)|^2(\frac{1}{2} - \frac{1}{2}\cos(\omega T)) \\ &= \frac{A^2T}{4}\left(\frac{\sin \pi fT/2}{\pi fT/2}\right)^2 \sin^2 \pi fT, \quad \text{(AMI-RZ)} \end{aligned} \tag{2.33}$$

The PSD is shown in Figure 2.4(a) where we set $A = 2$ to normalize the PSD. The bandwidths are $B_{null} = 1.0R_b$, $B_{90\%} \approx 1.71R_b$, and $B_{99\%} \approx 20R_b$, which are narrower than those of other RZ formats; especially the null bandwidth is only half of the others.

The PSD of *AMI-NRZ* can be obtained by replacing $T/2$ with T in $G(f)$ of AMI-RZ (2.27) since both of them have the same coding rules and the only difference is the pulse width. Thus the PSD of AMI-NRZ is given as

$$\Psi_s(f) = A^2T\left(\frac{\sin \pi fT}{\pi fT}\right)^2 \sin^2 \pi fT, \quad \text{(AMI-NRZ)} \tag{2.34}$$

The PSD is shown in Figure 2.4(b) where we set $A = \sqrt{2}$ to normalize the PSD. The bandwidths are $B_{null} = 1.0R_b$, $B_{90\%} \approx 1.53R_b$, and $B_{99\%} \approx 15R_b$ which are narrower than those of AMI-RZ.

The PSDs of other members of this group (i.e., *dicode NRZ* and *dicode RZ* or *twinned binary codes*) are derived in the following. As we described before, the dicodes can be constructed using AMI rules and a differentially coded sequence, that is, the dicodes can be written in the form of (2.11) as

$$s(t) = \sum_{k=-\infty}^{\infty} d_k g(t - kT) \tag{2.35}$$

where,

$$d_k = a_{k-1} - a_k = \begin{cases} 1, & p_1 = 1/4 \\ -1, & p_{-1} = 1/4 \\ 0, & p_0 = 1/2 \end{cases}$$

Sequence $\{d_k\}$ is a pseudoternary sequence derived from the original binary data sequence $\{a_k\}$. Its probability distribution is exactly the same as that of AMI (2.32). Therefore the PSDs of dicodes are the same as those of AMI codes and bandwidths are also the same as those of corresponding AMI codes.

2.3.4 PSD of Biphase Codes

For *Bi-Φ-L* (*Manchester*), the symbol function is half-positive and half-negative pulse defined by

$$g(t) = \begin{cases} A, & 0 \le t \le \frac{T}{2} \\ -A, & \frac{T}{2} \le t \le T \\ 0, & \text{elsewhere} \end{cases} \tag{2.36}$$

The Fourier transform of $g(t)$ is

$$G(f) = AT\left(\frac{\sin \pi f T/2}{\pi f T/2}\right)(\sin \pi f T/2)\, j e^{-j\omega T/2} \tag{2.37}$$

The data probability distribution is

$$a_k = \begin{cases} 1, & \text{for binary 1}, p_1 = 1/2 \\ -1, & \text{for binary 0}, p_0 = 1/2 \end{cases}$$

which is the same as that of NRZ. We have shown that $R(n) = 1$ for $n = 0$ and $R(n) = 0$ for $n \neq 0$ (2.19). Using (2.12) we obtain

$$\Psi_s(f) = A^2 T\left(\frac{\sin \pi f T/2}{\pi f T/2}\right)^2 \sin^2 \pi f T/2, \qquad \text{(Bi-}\Phi\text{-L)} \tag{2.38}$$

This PSD is shown in Figure 2.4(c) where we set $A = 1$ for unity symbol energy.

The PSD of the conditioned Bi-Φ-L is the same as that of the Bi-Φ-L since it is merely a differentially encoded Bi-Φ-L, and differential encoding does not change the probability distribution of the equally likely data.

It is obvious that Bi-Φ-M and Bi-Φ-S have the same PSD since the marks and spaces are equally likely in an equiprobable data sequence. We also observe that their waveforms are very similar to that of Bi-Φ-L in terms of pulse shapes and number of transitions. We therefore may intuitively guess that their PSD is the same as that of Bi-Φ-L. In fact this guess is right. We will prove it in the following.

We will use the method used in deriving the PSD of the delay modulation or Miller code whose PSD will be discussed in the next section [3]. We will base our derivation on Bi-Φ-M and the result is applicable to Bi-Φ-S as we mentioned already.

The coding rule of Bi-Φ-M can be described as a first-order Markov random process. Each bit interval can be divided into two half-bit intervals, then each bit interval can be described by the levels occurring in the two half-bit intervals. Temporarily assume the amplitude A to be 1, then the two levels are $+1$ and -1. There are four types of bit intervals or states that can occur in a bit interval: $(-1, +1)$, $(+1, -1)$, $(-1, -1)$, and $(+1, +1)$. They are equally likely (i.e., $p_i = 1/4$, $i = 1, 2, 3, 4$). A state of a bit interval depends only on the state of the previous bit interval. This is a first-order Markov process. The process is then completely described by $P = [p_{ij}]$, the probability-of-transition matrix, in which an element $p_{ij} = p(j/i)$ equals the conditional probability of the state j occurring after a given state i has occurred in the previous bit interval. For Bi-Φ-M, from the coding rules we can find the transition matrix as

$$P = \begin{bmatrix} 0.5 & 0 & 0.5 & 0 \\ 0 & 0.5 & 0 & 0.5 \\ 0 & 0.5 & 0 & 0.5 \\ 0.5 & 0 & 0.5 & 0 \end{bmatrix}$$

which is defined in Figure 2.5(a).

The autocorrelation function $R(\tau)$ at $\tau = nT$ $(n = 0, 1, 2...)$ is [3][5]

$$\begin{aligned} R(nT) &= \sum_{i,j} p(j/i, n) p_i \int_0^T g_i(t) g_j(t) dt/T \\ &= \frac{1}{4} \text{trace}[P^n W^T] \end{aligned} \tag{2.39}$$

where

$g_i(t) =$ waveform of state i.

$p(j/i, n) =$ probability of occurrence of state j at $t = nT$, given state i at $t = 0$, which is equal to ijth element of the matrix P^n.

$W =$ a matrix with elements $w_{ij} = \int_0^T g_i(t) g_j(t) dt/T$, W^T is the transpose of W.

The integral $w_{ij} = \int_0^T g_i(t) g_j(t) dt/T = \int_{nT}^{(n+1)T} g_i(t) g_j(t) dt/T$ accounts for "time averaging" the time-varying $R(nT, t)$ over a bit interval in order to obtain $R(nT)$. Based on the waveforms of the states, we can find W as

$$W = \begin{bmatrix} 1 & -1 & 0 & 0 \\ -1 & 1 & 0 & 0 \\ 0 & 0 & 1 & -1 \\ 0 & 0 & -1 & 1 \end{bmatrix}$$

5 The transpose of W is missing in [3].

which is defined in Figure 2.5(b), where shaded areas are areas of integration. Substituting P and W into (2.39), the result is

$$R(nT) = \begin{cases} 1, & n = 1 \\ 0, & n \neq 0 \end{cases}$$

Similarly we can find the autocorrelation for $\tau = (n + 1/2)T$ as follows:

$$\begin{aligned} R((n+1/2)T) &= \sum_{i,j} p(j/i, n)p_i \int_0^{T/2} g_i(t)g_j(t + \frac{T}{2})dt/T \\ &+ \sum_{i,j} p(j/i, n+1)p_i \int_{T/2}^{T} g_i(t)g_j(t + \frac{T}{2})dt/T \\ &= \frac{1}{4}\text{trace}[P^n W_1^T] + \frac{1}{4}\text{trace}[P^{n+1} W_2^T] \end{aligned} \quad (2.40)$$

where

$$W_1 = \begin{bmatrix} -0.5 & 0.5 & 0.5 & -0.5 \\ 0.5 & -0.5 & -0.5 & 0.5 \\ -0.5 & 0.5 & 0.5 & -0.5 \\ 0.5 & -0.5 & -0.5 & 0.5 \end{bmatrix}$$

and

$$W_2 = \begin{bmatrix} -0.5 & 0.5 & -0.5 & 0.5 \\ 0.5 & -0.5 & 0.5 & -0.5 \\ 0.5 & -0.5 & 0.5 & -0.5 \\ -0.5 & 0.5 & -0.5 & 0.5 \end{bmatrix}$$

which are defined in Figure 2.5(c, d). The results for $n \geq 0$ are

$$R((n+1/2)T) = \begin{cases} -0.5, & n = 0 \\ 0, & n > 0 \end{cases}$$

Due to symmetry of the $R(\tau)$, it is clear that $R(-T/2) = -0.5$ and $R(nT/2) = 0$ for all $n < -1$. The autocorrelation at intermediate values of τ is obtained exactly by joining these points. $R(\tau)$ of Bi-Φ-M is shown in Figure 2.5(e). By taking the Fourier transform of $R(\tau)$, the PSD is easily found as

$$\Psi_s(f) = A^2T\left(\frac{\sin \pi fT/2}{\pi fT/2}\right)^2 \sin^2 \pi fT/2, \qquad \text{(Bi-}\Phi\text{-M and S)} \quad (2.41)$$

which is exactly the same as that of Bi-Φ-L or Manchester.

The bandwidths of all biphase codes are $B_{null} = 2.0R_b$, $B_{90\%} \approx 3.05R_b$, and $B_{99\%} \approx 29R_b$.

2.3.5 PSD of Delay Modulation

The spectral analysis of delay modulation is also based on the first-order Markov process [3]. The autocorrelation $R(nT)$ and $R((n+1/2)T)$ are also given by (2.39) and (2.40).

The probability-of-transition matrix is

$$P = \begin{bmatrix} 0 & 0.5 & 0 & 0.5 \\ 0 & 0 & 0.5 & 0.5 \\ 0.5 & 0.5 & 0 & 0 \\ 0.5 & 0 & 0.5 & 0 \end{bmatrix}$$

which is defined in Figure 2.6(a). The W matrices are

$$W = \begin{bmatrix} 1 & 0 & 0 & -1 \\ 0 & 1 & -1 & 0 \\ 0 & -1 & 1 & 0 \\ -1 & 0 & 0 & 1 \end{bmatrix}$$

$$W_1 = \begin{bmatrix} 0.5 & -0.5 & 0.5 & -0.5 \\ 0.5 & -0.5 & 0.5 & -0.5 \\ -0.5 & 0.5 & -0.5 & 0.5 \\ -0.5 & 0.5 & -0.5 & 0.5 \end{bmatrix}$$

$$W_2 = \begin{bmatrix} 0.5 & 0.5 & -0.5 & -0.5 \\ -0.5 & -0.5 & 0.5 & 0.5 \\ 0.5 & 0.5 & -0.5 & -0.5 \\ -0.5 & -0.5 & 0.5 & 0.5 \end{bmatrix}$$

All of them are defined in Figure 2.6.

By using these matrices we can determine $R(\tau)$ for $\tau = nT$ and $(n+1/2)T$. The intermediate $R(\tau)$ is obtained exactly by joining these points. In addition it can be easily verified that for DM, $P^4W = -\frac{1}{4}W$ and $P^4W_i = -\frac{1}{4}W_i, i = 1, 2$. Therefore

$$R(\tau + 4T) = -\frac{1}{4}R(\tau), \qquad \tau > 0 \tag{2.42}$$

Thus, the first eight values of $R(\tau)$ from $R(0)$ to $R(3.5T)$ given in Figure 2.6(e) completely specify $R(\tau)$. Taking the Fourier transform of $R(\tau)$, using (2.42), and

P

i \ j	-1 +1	+1 -1	-1 -1	+1 +1
-1 +1	0.5	0	0.5	0
+1 -1	0	0.5	0	0.5
-1 -1	0	0.5	0	0.5
+1 +1	0.5	0	0.5	0

(a)

W

i \ j	-1 +1	+1 -1	-1 -1	+1 +1
-1 +1	1	-1	0	0
+1 -1	-1	1	0	0
-1 -1	0	0	1	-1
+1 +1	0	0	-1	1

(b)

W_1

i \ j	-1 +1	+1 -1	-1 -1	+1 +1
-1 +1	-0.5	0.5	0.5	-0.5
+1 -1	0.5	-0.5	-0.5	0.5
-1 -1	-0.5	0.5	0.5	-0.5
+1 +1	0.5	-0.5	-0.5	0.5

(c)

W_2

i \ j	-1 +1	+1 -1	-1 -1	+1 +1
-1 +1	-0.5	0.5	-0.5	0.5
+1 -1	0.5	-0.5	0.5	-0.5
-1 -1	0.5	-0.5	0.5	-0.5
+1 +1	-0.5	0.5	-0.5	0.5

(d)

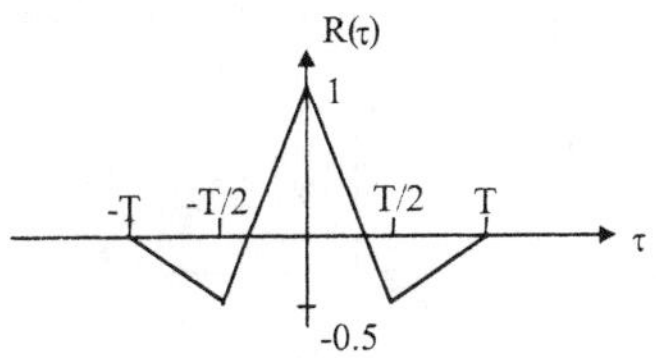

(e)

Figure 2.5 Bi-Φ-M matrices: (a) the probability-of-transition matrix, (b) the matrix W, (c) the matrix W_1, (d) the matrix W_2, and (e) autocorrelation $R(\tau)$.

P

i \ j	+1 +1	+1 -1	-1 +1	-1 -1
+1 +1	0	0.5	0	0.5
+1 -1	0	0	0.5	0.5
-1 +1	0.5	0.5	0	0
-1 -1	0.5	0	0.5	0

(a)

W

i \ j	+1 +1	+1 -1	-1 +1	-1 -1
+1 +1	1	0	0	-1
+1 -1	0	1	-1	0
-1 +1	0	-1	1	0
-1 -1	-1	0	0	1

(b)

W_1

i \ j	+1 +1	+1 -1	-1 +1	-1 -1
+1 +1	0.5	-0.5	0.5	-0.5
+1 -1	0.5	-0.5	0.5	-0.5
-1 +1	-0.5	0.5	-0.5	0.5
-1 -1	-0.5	0.5	-0.5	0.5

(c)

W_2

i \ j	+1 +1	+1 -1	-1 +1	-1 -1
+1 +1	0.5	0.5	-0.5	-0.5
+1 -1	-0.5	-0.5	0.5	0.5
-1 +1	0.5	0.5	-0.5	-0.5
-1 -1	-0.5	-0.5	0.5	0.5

(d)

τ	$R(\tau)$
0	1
0.5T	0.25
T	-0.5
1.5T	-0.5
2T	0
2.5T	0.375
3T	0.25
3.5T	-0.125

(e)

Figure 2.6 *DM* matrices: (a) the probability-of-transition matrix, (b) the matrix W, (c) the matrix W_1, (d) the matrix W_2, and (e) autocorrelation $R(\tau)$.

the relation that $\Psi_s(-f) = \Psi_s^*(f)$, where * indicates conjugate, the PSD of DM is

$$\begin{aligned}
\Psi_s(f) &= \int_{-\infty}^{\infty} R(\tau)e^{-j2\pi f\tau}d\tau \\
&= 2\,\mathrm{Re}\left[\int_{0}^{\infty} R(\tau)e^{-j2\pi f\tau}d\tau\right] \\
&= 2\,\mathrm{Re}\left[\frac{1}{1+\frac{1}{4}e^{-j2\pi f(4T)}}\int_{0}^{4T} R(\tau)e^{-j2\pi f\tau}d\tau\right]
\end{aligned}$$

where Re indicates real part. The integral is easily, though tediously, evaluated [3]:

$$\begin{aligned}
\Psi_s(f) &= \frac{2A^2}{(2\pi f)^2\,T(17+8\cos\theta)}(23-2\cos\theta-22\cos 2\theta \\
&\quad -12\cos 3\theta+5\cos 4\theta+12\cos 5\theta+2\cos 6\theta \\
&\quad -8\cos 7\theta+2\cos 8\theta), \qquad \text{(DM or Miller)} \qquad (2.43)
\end{aligned}$$

where $\theta = \pi fT$. The PSD is shown in Figure 2.4(d) where we set $A = 1$ for unity symbol energy. The PSD has a peak at $f = 0.4R_b$, and it has a very narrow main lobe bandwidth of about $0.5R_b$. However, it converges to zero very slowly. As a result its energy within a bandwidth of $2R_b$ is only 76.4% and within $250R_b$ is only 83.7%.

2.4 BIT ERROR RATE OF LINE CODES

In this section we discuss optimum detection of line codes transmitted through an AWGN channel and their error probabilities. We should keep in mind that the AWGN channel model implies that the channel frequency response is flat and has infinite bandwidth. The only distortion is introduced by the additive white Gaussian noise. This of course is not accurate for many practical channels. However, it is a reasonably accurate model as long as the signal bandwidth is much narrower than that of the channel. It is also important to note that the optimum receiver might not be the practical solution. Other nonoptimum receivers might be just as good as the optimum one in certain circumstances (e.g., high signal-to-noise ratio) and their structures are simpler. Nonetheless the error probability in the AWGN channel serves as a reference for performance comparison.

We start with the binary codes and will progress to pseudoternary codes and other complex codes. The fundamental theory of detection and estimation of signals in noise is provided in Appendix B. Here we will first present the signal model and use the corresponding results in Appendix B to obtain the error probabilities of the line codes.

2.4.1 BER of Binary Codes

A binary line code consists of two kinds of signals, or from the point of view of detection theory, we have two hypotheses:

$$\begin{aligned} H_1 &: s_1(t), \quad 0 \le t \le T, \quad \text{is sent with probability of } p_1 \\ H_2 &: s_2(t), \quad 0 \le t \le T, \quad \text{is sent with probability of } p_2 \end{aligned}$$

where p_1 and p_2 are called *a priori probability*. The energies of the two signals are

$$E_1 = \int_0^T s_1^2(t)dt$$

and

$$E_2 = \int_0^T s_2^2(t)dt$$

In general these two signals may be correlated. We define

$$\rho_{12} = \frac{1}{\sqrt{E_1 E_2}} \int_0^T s_1(t)s_2(t)dt \tag{2.44}$$

as the *correlation coefficient* of $s_1(t)$ and $s_2(t)$. $|\rho_{12}| \le 1$. The received signal is

$$r(t) = s_i(t) + n(t),\ i = 1, 2$$

where the noise $n(t)$ is the AWGN with zero mean and a two-sided spectral density of $N_o/2$.

The optimum receiver consists of a correlator or a matched filter matched to the difference signal (see Appendix B)

$$s_d(t) = s_1(t) - s_2(t)$$

These two forms of receiver are shown in Figure 2.7(a, b) and they are equivalent in terms of error probability. The decision regions of binary signal detection are shown in Figure 2.7(c) where $\mu_i = \mu_i(T), i = 1, 2$.

The threshold detector compares the integrator or sampler output $z = z(T)$ to

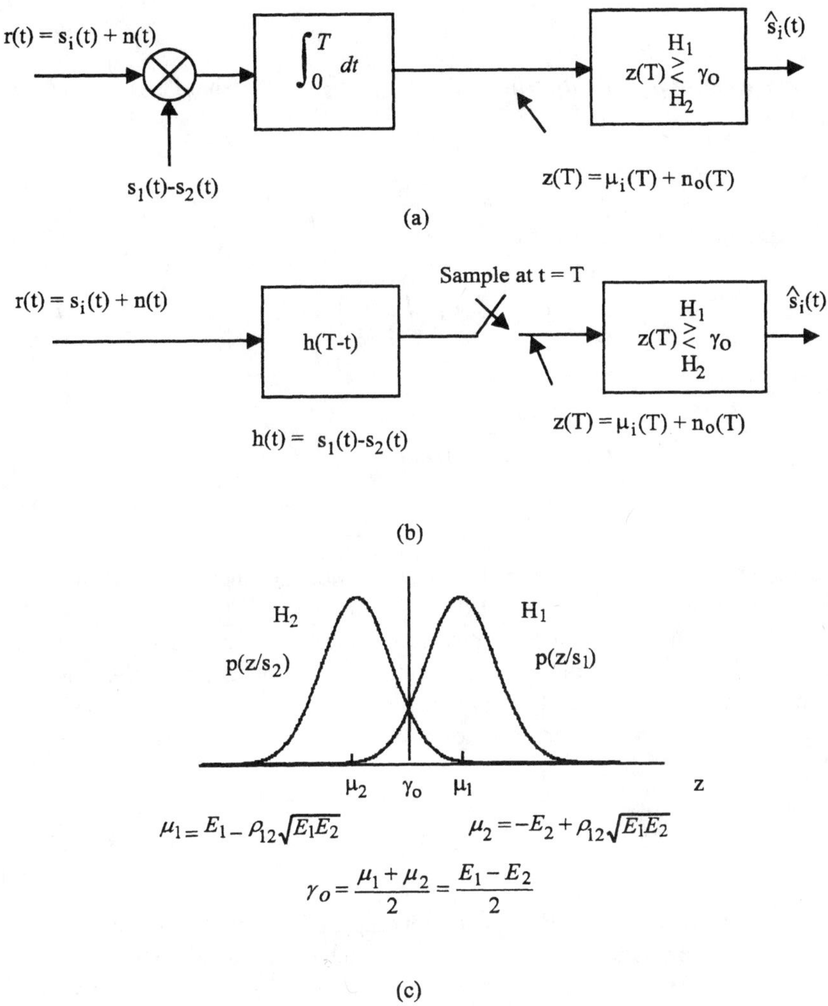

Figure 2.7 Optimum receiver for binary signals: (a) correlator, (b) matched filter, and (c) decision regions.

threshold γ_o and decides which hypothesis is true, that is, the decision rule is

$$z \underset{H_2}{\overset{H_1}{\gtrless}} \gamma_o$$

For *minimum error probability criterion* and $p_1 = p_2$, the threshold is chosen as (see (B.22))

$$\gamma_o = \frac{E_1 - E_2}{2} \tag{2.45}$$

The bit error probability is given by

$$P_b = Q\left(\sqrt{\frac{E_d}{2N_o}}\right) \tag{2.46}$$

where

$$\begin{aligned} E_d &= \int_0^T s_d^2(t)dt \\ &= E_1 + E_2 - 2\rho_{12}\sqrt{E_1 E_2} \end{aligned}$$

is the energy of the difference signal, and $Q(x)$ is the Q-function, which we have defined in Chapter 1 already (see (1.11)).

Equation (2.46) shows that the larger the distance (E_d) between the two signals $s_1(t)$ and $s_2(t)$, the smaller the P_b. This is intuitively convincing since the larger the distance, the easier it is for the detector to distinguish them. In terms of each signal's energy, the above P_b expression becomes

$$P_b = Q\left(\sqrt{\frac{E_1 + E_2 - 2\rho_{12}\sqrt{E_1 E_2}}{2N_o}}\right) \tag{2.47}$$

This expression indicates that P_b depends not only on the individual signal energies, but also on the correlation between them. It is interesting to discover that when $\rho_{12} = -1$, P_b is the minimum. Binary signals with $\rho_{12} = -1$ are called *antipodal*. When $\rho_{12} = 0$, the signals are *orthogonal*.

2.4.1.1 BER of Nonreturn-to-Zero Codes

NRZ-L. The NRZ-L is antipodal with

$$s_1(t) = A, \quad 0 \le t \le T$$

$$s_2(t) = -A, \quad 0 \le t \le T$$

and

$$s_d(t) = 2A, \quad 0 \le t \le T$$

Then $\rho_{12} = -1$, $E_1 = E_2 = A^2T$, and $E_b = (E_1 + E_2)/2 = A^2T$. From (2.47) its P_b is

$$P_b = Q\left(\sqrt{\frac{2A^2T}{N_o}}\right) = Q\left(\sqrt{\frac{2E_b}{N_o}}\right), \quad \text{(NRZ-L)} \tag{2.48}$$

which is plotted in Figure 2.8. The optimum threshold is

$$\gamma_o = \frac{E_1 - E_2}{2} = 0, \quad \text{(NRZ-L)}$$

Unipolar NRZ. For unipolar NRZ,

$$\begin{aligned} s_1(t) &= A, \quad 0 \le t \le T \\ s_2(t) &= 0, \quad 0 \le t \le T \end{aligned}$$

and

$$s_d(t) = A, \quad 0 \le t \le T$$

Thus $E_1 = A^2T$, $E_2 = 0$, $E_b = A^2T/2$, and $\rho_{12} = 0$. Its error probability is

$$P_b = Q\left(\sqrt{\frac{A^2T}{2N_o}}\right) = Q\left(\sqrt{\frac{E_b}{N_o}}\right), \quad \text{(unipolar-NRZ)} \tag{2.49}$$

which is plotted in Figure 2.8. The optimum threshold is

$$\gamma_o = \frac{A^2T}{2}, \quad \text{(unipolar-NRZ)}$$

NRZ-M or S. It is modulated by differentially coded data sequence. To the coded sequence the optimum receiver produces an error probability of (2.48) with $\gamma_o = 0$. After detection the coded sequence is differentially decoded back to the original data sequence. The present bit and the previous bit of the coded sequence are used to produce the present bit of the original sequence (see (2.2)). Therefore error probability is

$$\begin{aligned} P_b' &= \Pr\text{(present bit correct and previous bit incorrect)} \\ &\quad + \Pr\text{(present bit incorrect and previous bit correct)} \end{aligned}$$

$$
\begin{aligned}
&= (1-P_b)P_b + P_b(1-P_b) \\
&= 2(1-P_b)P_b \\
&\approx 2P_b, \qquad \text{for small } P_b
\end{aligned}
$$

That is

$$P_b \approx 2Q\left(\sqrt{\frac{2A^2T}{N_o}}\right) = 2Q\left(\sqrt{\frac{2E_b}{N_o}}\right), \qquad \text{(NRZ-M or S)} \tag{2.50}$$

which is plotted in Figure 2.8.

2.4.1.2 BER of Return-to-Zero Codes

Polar RZ. The signals are antipodal with

$$
\begin{aligned}
s_1(t) &= \begin{cases} A, & 0 \le t \le \frac{T}{2} \\ 0, & \text{elsewhere} \end{cases} \\
s_2(t) &= -s_1(t)
\end{aligned}
$$

and

$$s_d(t) = \begin{cases} 2A, & 0 \le t \le \frac{T}{2} \\ 0, & \text{elsewhere} \end{cases}$$

Then $\rho_{12} = -1$, $E_1 = E_2 = A^2T/2$, and $E_b = (E_1 + E_2)/2 = A^2T/2$. From (2.47) its P_b is

$$P_b = Q\left(\sqrt{\frac{A^2T}{N_o}}\right) = Q\left(\sqrt{\frac{2E_b}{N_o}}\right), \qquad \text{(polar-RZ)} \tag{2.51}$$

which is the same as that of NRZ-L in terms of E_b/N_o and the optimum threshold γ_o is also 0. P_b is plotted in Figure 2.8.

Unipolar RZ. The signals are

$$
\begin{aligned}
s_1(t) &= \begin{cases} A, & 0 \le t \le \frac{T}{2} \\ 0, & \text{elsewhere} \end{cases} \\
s_2(t) &= 0, \quad 0 \le t \le T
\end{aligned}
$$

and

$$s_d(t) = \begin{cases} A, & 0 \le t \le \frac{T}{2} \\ 0, & \text{elsewhere} \end{cases}$$

Then $\rho_{12} = 0$, $E_1 = A^2T/2, E_2 = 0$, and $E_b = (E_1 + E_2)/2 = A^2T/4$. From (2.47) its P_b is

$$P_b = Q\left(\sqrt{\frac{A^2T}{4N_o}}\right) = Q\left(\sqrt{\frac{E_b}{N_o}}\right), \quad \text{(unipolar-RZ)} \qquad (2.52)$$

which is plotted in Figure 2.8. This is the same as that of unipolar NRZ in terms of E_b/N_o and optimum threshold γ_o is $A^2T/4$. Note that for the E_b of unipolar RZ to be the same as that of unipolar NRZ, its amplitude must be $\sqrt{2}$ times that of unipolar NRZ. This will make their thresholds the same when their E_bs are the same. However, if the amplitude is fixed, the unipolar NRZ pulse will have twice the energy of the unipolar RZ pulse, thus the error probability is lower. From this discussion we can see that different conclusions will be drawn for different comparison bases. In the following, our comparison is always based on the same E_b/N_o. However, the reader should be aware of different conclusions if comparison is based on the same amplitude.

2.4.2 BER of Pseudoternary Codes

According to the results of the detection theory in Appendix B ((B.37), see also [18, Chapter 4]), for M-ary signals, the *minimum error probability* receiver computes

$$l_j = \ln P_j - \frac{1}{N_o}\sum_{i=1}^{N}(r_i - s_{ij})^2, \quad j = 1, 2, ..., M \qquad (2.53)$$

and chooses the largest, where

$$r_i = \int_0^T r(t)\phi_i(t)dt, \quad i = 1, 2, ..., N$$

r_i are statistically independent Gaussian random variables with variance of $N_o/2$. Their mean values depend on hypotheses, that is,

$$\begin{aligned} s_{ij} &= E\{r_i/H_j\} = \int_0^T s_j(t)\phi_i(t)dt \\ i &= 1, 2, ..., N\,;\, j = 1, 2, ..., M \end{aligned}$$

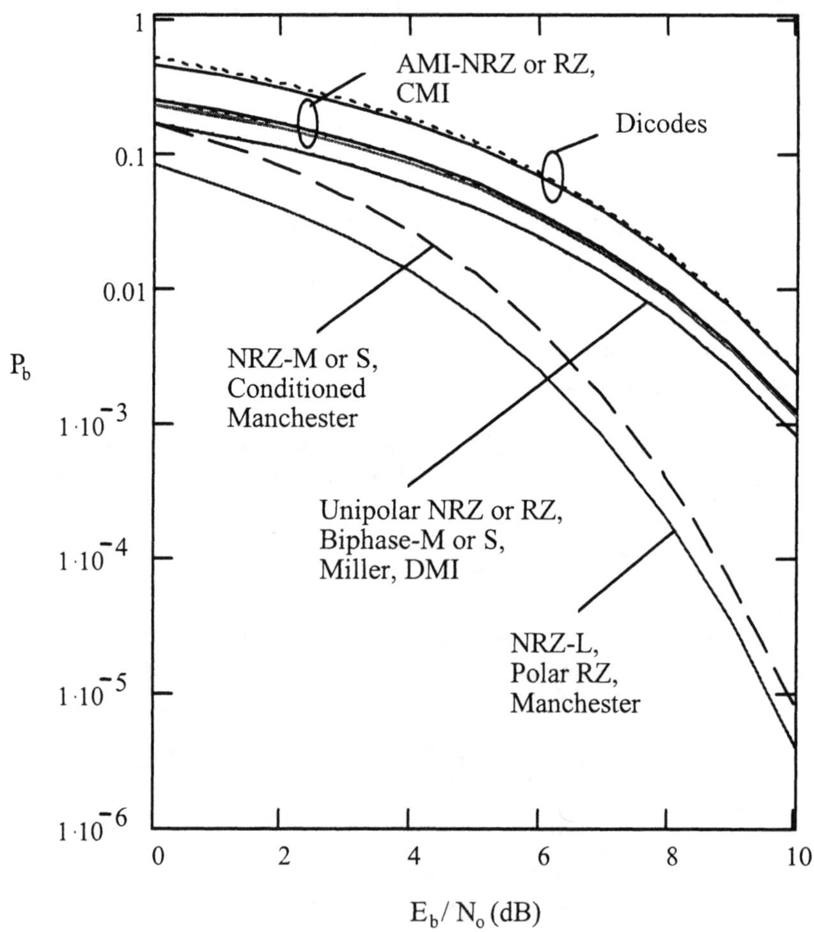

Figure 2.8 Bit error probabilities of some line codes.

$\phi_i(t)$ are the orthonormal coordinates for projecting $r(t)$ onto them. N is the dimension of the vector space spanned by $\phi_i(t)$.

AMI-NRZ codes consist of three types of signals, or we have three hypotheses:

$$\begin{aligned} H_1 &: s_1(t) = A, \qquad 0 \le t \le T, \quad p_1 = \frac{1}{4} \\ H_2 &: s_2(t) = -A, \quad 0 \le t \le T, \quad p_2 = \frac{1}{4} \\ H_3 &: s_3(t) = 0, \qquad 0 \le t \le T, \quad p_3 = \frac{1}{2} \end{aligned}$$

We choose $\phi_1(t) = s_1(t)/\sqrt{E_1} = s_1(t)/A\sqrt{T}$, and the rest of the signals are linearly related to it. Therefore $N = 1$ in this case and the optimum receiver consists of one correlator and a threshold detector. The detection rule (2.53) reduces to

$$\begin{aligned} l_1 &= \ln\frac{1}{4} - \frac{1}{N_o}(r - \sqrt{E_1})^2 \\ l_2 &= \ln\frac{1}{4} - \frac{1}{N_o}(r + \sqrt{E_1})^2 \\ l_3 &= \ln\frac{1}{2} - \frac{1}{N_o}r^2 \end{aligned}$$

The decision rule is to choose the largest. For l_1 to be the largest we must have $l_1 > l_2$ and $l_1 > l_3$, from these relations we can deduce that

$$H_1 \text{ is true if } r > \frac{N_o \ln 2 + E_1}{2\sqrt{E_1}} \triangleq \gamma_{13}$$

Similarly we can deduce that

$$H_2 \text{ is true if } r < -\gamma_{13} \triangleq \gamma_{23}$$

and

$$H_3 \text{ is true if } \gamma_{13} > r > \gamma_{23}$$

Therefore there are two thresholds and three decision regions as shown in Figure 2.9. The optimum receiver consists of just one correlator and a threshold detector with two thresholds as shown in Figure 2.9.

Using $\Pr(e/s_i)$ to denote the probability of error when signal $s_i(t)$ is transmitted, the average bit error probability can be calculated as follows:

$$P_b = p_1 \Pr(e/s_1) + p_2 \Pr(e/s_2) + p_3 \Pr(e/s_3)$$

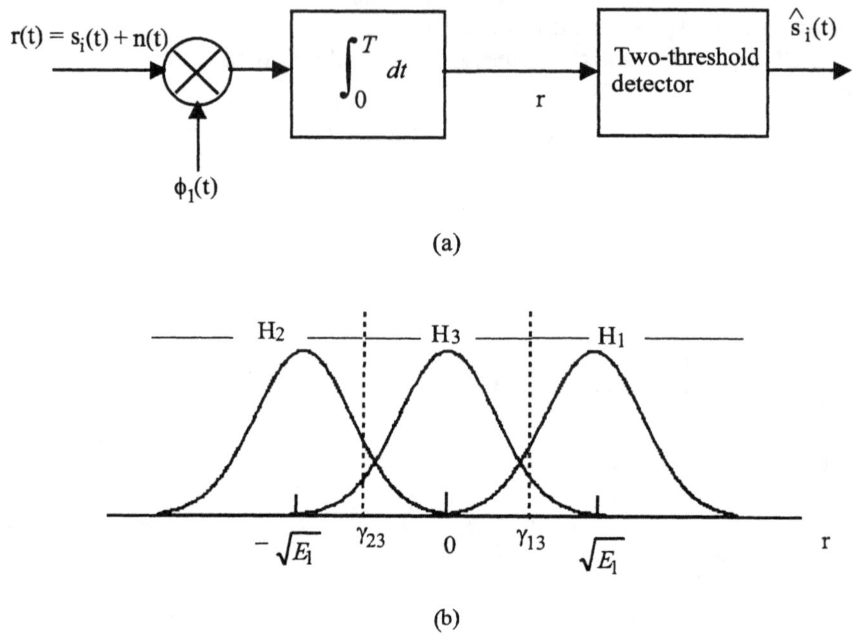

Figure 2.9 Optimum receiver (a), and decision regions (b), of AMI signals.

$$
\begin{aligned}
&= p_1 \int_{-\infty}^{\gamma_{13}} p(r/s_1)dr + p_2 \int_{\gamma_{23}}^{\infty} p(r/s_2)dr + 2p_3 \int_{\gamma_{13}}^{\infty} p(r/s_3)dr \\
&= 2p_1 \int_{-\infty}^{\gamma_{13}} p(r/s_1)dr + 2p_3 \int_{\gamma_{13}}^{\infty} p(r/s_3)dr \\
&= \frac{1}{2}Q\left(\frac{E_1 - N_o \ln 2}{\sqrt{2N_oE_1}}\right) + Q\left(\frac{E_1 + N_o \ln 2}{\sqrt{2N_oE_1}}\right) \qquad (2.54)
\end{aligned}
$$

Given $E_1 = A^2T$ and $E_b = 0.5E_1$, we have

$$
\begin{aligned}
P_b &= \frac{1}{2}Q\left(\frac{A^2T - N_o \ln 2}{A\sqrt{2N_oT}}\right) + Q\left(\frac{A^2T + N_o \ln 2}{A\sqrt{2N_oT}}\right) \\
&= \frac{1}{2}Q\left(\frac{2E_b/N_o - \ln 2}{2\sqrt{E_b/N_o}}\right) + Q\left(\frac{2E_b/N_o + \ln 2}{2\sqrt{E_b/N_o}}\right), \quad \text{(AMI-NRZ)} \qquad (2.55)
\end{aligned}
$$

This is the bit error probability of the optimum reception.

However, when signal-to-noise ratio is high, then $N_o \ln 2 \ll E_1$, the thresholds can be set in the midway between signals with very little loss of error performance. That is, $\gamma_{13} \approx \sqrt{E_1}/2$ and $\gamma_{23} \approx -\sqrt{E_1}/2$. Thus the BER expression (2.55) reduces to

$$P_b \approx \frac{3}{2} Q\left(\sqrt{\frac{E_b}{N_o}}\right), \quad \text{(AMI-NRZ)} \tag{2.56}$$

which is 3/2 times that of unipolar NRZ. Both accurate and approximate values of P_b for AMI-NRZ are plotted in Figure 2.8, where the upper curve is the approximation. It can be seen that they are very close, even at low signal-to-noise ratios.

For *AMI-RZ* whose $E_1 = A^2T/2$ and average $E_b = A^2T/4$, the BER can be found from (2.54)

$$\begin{aligned} P_b &= \frac{1}{2} Q\left(\frac{A^2T/2 - N_o \ln 2}{A\sqrt{N_oT}}\right) + Q\left(\frac{A^2T/2 + N_o \ln 2}{A\sqrt{N_oT}}\right) \\ &= \frac{1}{2} Q\left(\frac{2E_b/N_o - \ln 2}{2\sqrt{E_b/N_o}}\right) + Q\left(\frac{2E_b/N_o + \ln 2}{2\sqrt{E_b/N_o}}\right), \quad \text{(AMI-RZ)} \end{aligned} \tag{2.57}$$

and

$$P_b \approx \frac{3}{2} Q\left(\sqrt{\frac{E_b}{N_o}}\right), \quad \text{(AMI-RZ)} \tag{2.58}$$

which is the same as that of AMI-NRZ in terms of E_b/N_o.

Since recovery of NRZ-L from AMI codes is accomplished by simple full-wave rectification, the bit error probability of the final recovered data sequence remains the same.

Dicodes and AMI codes are related by differential coding. As described in Section 2.2.3, data sequence $\{\widehat{a}_k\}$ is recovered from $\{\widehat{u}_k\}$ using modulo-2 addition: $\widehat{a}_k = \widehat{a}_{k-1} \oplus \widehat{u}_k$, where $\{\widehat{u}_k\}$ is the unipolar sequence recovered from the dicode $\{\widehat{d}_k\}$ by full-wave rectification. The P_b of $\{\widehat{d}_k\}$ is the same as AMI and so is the P_b of $\{\widehat{u}_k\}$. $\widehat{a}_k$ is incorrect when either $\widehat{a}_{k-1}$ or $\widehat{u}_k$ is incorrect. This is the same situation as we discussed for NRZ-M or NRZ-S codes in Section 2.4.1. Therefore the bit error probability of $\{\widehat{a}_k\}$ is two times the bit error probability of their AMI counterparts. Using (2.58) we have

$$P_b \approx 3Q\left(\sqrt{\frac{E_b}{N_o}}\right), \quad \text{(dicodes)} \tag{2.59}$$

which is shown in Figure 2.8.

2.4.3 BER of Biphase Codes

Bi-Φ-L (*Manchester*) signals are antipodal binary with

$$\begin{aligned} s_1(t) &= \begin{cases} A, & 0 \le t \le \frac{T}{2} \\ -A, & \frac{T}{2} \le t \le T \end{cases} \\ s_2(t) &= -s_1(t) \end{aligned} \tag{2.60}$$

Bi-Φ-L signals can be detected using the optimum receiver in Figure 2.7. The correlation coefficient between them is $\rho_{12} = -1$. The signal energies are

$$\begin{aligned} E_1 &= E_2 = A^2T = E_b \\ E_d &= E_1 + E_2 - 2\rho_{12}\sqrt{E_1E_2} \\ &= 4A^2T = 4E_b \end{aligned}$$

Thus the bit error probability is

$$P_b = Q\left(\sqrt{\frac{2A^2T}{N_o}}\right) = Q\left(\sqrt{\frac{2E_b}{N_o}}\right), \quad \text{(Bi-Φ-L (Manchester))} \tag{2.61}$$

which is the same as that of NRZ-L. This is not surprising since they have the same bit energy and both are antipodal.

Conditioned Bi-Φ-L has a bit error probability of about two times that of Bi-Φ-L since it is just differentially coded Bi-Φ-L. This is to say that conditioned Bi-Φ-L has the same BER as that of NRZ-M or S.

Bi-Φ-M and *Bi-Φ-S* have the same error probability. So it suffices to consider Bi-Φ-M. There are four signals in Bi-Φ-M:

$$\begin{aligned} s_1(t) &= A, & 0 \le t \le T, & \quad p_1 = \frac{1}{4} \\ s_2(t) &= \begin{cases} A, & 0 \le t \le \frac{T}{2} \\ -A, & \frac{T}{2} \le t \le T \end{cases}, & & \quad p_2 = \frac{1}{4} \\ s_3(t) &= -s_1(t), & 0 \le t \le T, & \quad p_3 = \frac{1}{4} \\ s_4(t) &= -s_2(t), & 0 \le t \le T, & \quad p_4 = \frac{1}{4} \end{aligned}$$

Each of them has energy of $E = A^2T$. So average bit energy is also the same. We can choose $\phi_1(t) = s_1(t)/\sqrt{E}$ and $\phi_2(t) = s_2(t)/\sqrt{E}$ as basis functions. Thus from (2.53) we have

$$l_j = \ln P_j - \frac{1}{N_o}\sum_{i=1}^{2}(r_i - s_{ij})^2$$

where

$$\begin{array}{ll} s_{11} = \sqrt{E}, & s_{21} = 0 \\ s_{12} = 0, & s_{22} = \sqrt{E} \\ s_{13} = -\sqrt{E}, & s_{23} = 0 \\ s_{14} = 0, & s_{24} = -\sqrt{E} \end{array}$$

The decision rule is to choose the largest l_j. Since P_j are equal for all j, the above rule becomes computing

$$d_j = \sum_{i=1}^{2}(r_i - s_{ij})^2$$

and choosing the minimum. Thus the optimum receiver is as shown in Figure 2.10(a) and the decision space is two-dimensional as shown in Figure 2.10(b). Since the problem is symmetrical, it is sufficient to assume that $s_1(t)$ is transmitted and compute the resulting $\Pr(e/s_1)$ which is equal to average P_b. We also can see that the answer would be invariant to a 45 degree rotation of the signal set because the noise is circularly symmetric. Thus we can use the simple diagram in Figure 2.10(c) to calculate the BER.

$$\begin{aligned} P_b &= \Pr(e/s_1) = 1 - \int\int_{Z_1} p(r_1, r_2/s_1)dr_1 dr_2 \\ &= 1 - \left[\int_0^{\infty} \frac{1}{\sqrt{\pi N_o}} \exp\left(-\frac{(r_1 - \sqrt{E/2})^2}{N_o}\right)dr_1\right]^2 \end{aligned}$$

where the two-dimensional integral is converted to the squared one-dimensional integral because r_1 and r_2 are independent identical Gaussian variables. Changing variables we have

$$\begin{aligned} P_b &= 1 - \left[\int_{-\sqrt{E/N_0}}^{\infty} \frac{1}{\sqrt{2\pi}} \exp\left(-\frac{x^2}{2}\right)dx\right]^2 \\ &= 1 - Q\left(-\sqrt{\frac{E}{N_o}}\right) \end{aligned}$$

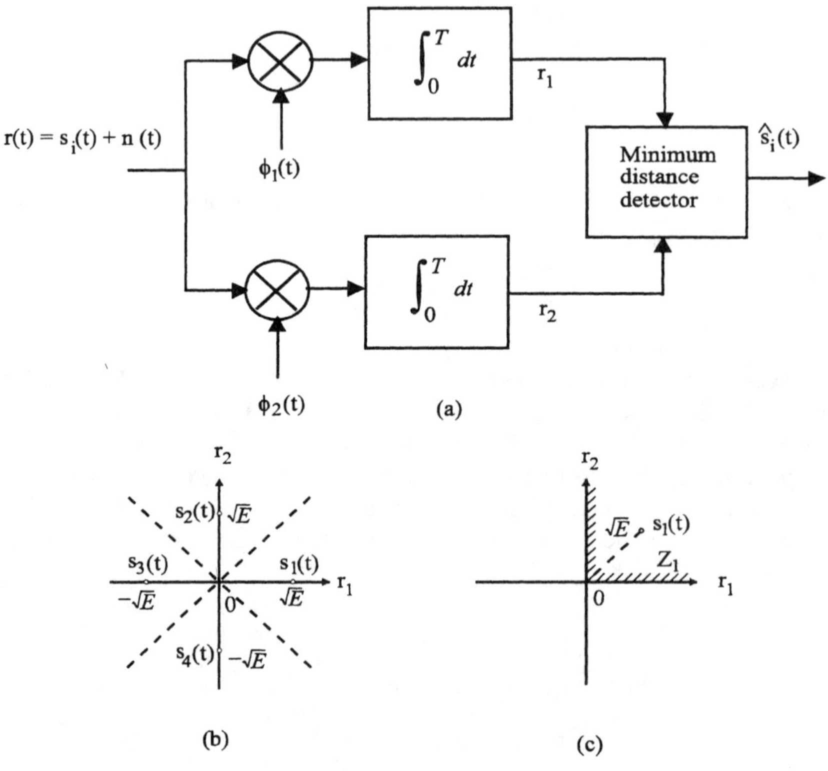

Figure 2.10 Optimum receiver and decision regions of Bi-Φ-M signals (a–c).

$$= Q\left(\sqrt{\frac{E}{N_o}}\right) = Q\left(\sqrt{\frac{E_b}{N_o}}\right), \quad \text{(Bi-}\Phi\text{-M,S)} \qquad (2.62)$$

The BER of Bi-Φ-M is identical to that of unipolar codes and is 3 dB worse than that of NRZ-L or polar-RZ. This BER is plotted in Figure 2.8.

2.4.4 BER of Delay Modulation

The four symbol signals in delay modulation are the same as those of the Bi-Φ-M, and their probability distribution is also the same. Therefore the BER of delay modulation is the same of that of Bi-Φ-M

$$P_b = Q\left(\sqrt{\frac{E_b}{N_o}}\right), \quad \text{(DM or Miller)} \qquad (2.63)$$

Like Bi-Φ-M, the BER of delay modulation is identical to that of unipolar codes and is 3 dB worse than that of NRZ-L or polar-RZ.

2.5 SUBSTITUTION LINE CODES

The AMI code is a preferable choice due to its many advantages. It has a narrow bandwidth and no dc component. It has error detection capability due to its alternate mark inversion. The occurrence of consecutive positive or negative amplitudes indicates transmission errors and is called a *bipolar violation*. Synchronization is made easier due to transitions in each binary 1 bit.

Even though synchronization of AMI code is better than that of NRZ code, it is still not satisfactory. A string of 0s will result in a long period of zero level which will cause loss of synchronization. In the T1 system, by eliminating the all-zero code word from the 8-bit source encoder, the maximum number of consecutive zeros is limited to 14. However, for data signals even this is impractical. A solution to this is to substitute the block of N consecutive zeros with a special sequence with intentional bipolar violations. These violations enable the zero-substitution sequence to be identified and replaced by spaces (zeros) at the receiving end of the line. The pulse density is at least $1/N$. This will improve bit timing recovery. Two popular zero substitution codes are the *binary N-zero substitution* (BNZS) and the *high density bipolar n* (HDBn) codes.

2.5.1 Binary N-Zero Substitution Codes

BNZS code was proposed by Johannes et al. in 1969 [4]. It is the most popular substitution code which replaces a string of N 0s in the AMI waveform with a special N-bit waveform with at least one bipolar violation. All BNZS formats are dc free and retain the balanced feature of AMI, which is achieved by choosing the substitution sequences properly so that the conditioned sequences have an equal number of positive and negative pulses.

There are two kinds of BNZS codes. One is called *nonmodal code* in which two substitution sequences are allowed and the choice between them is based solely on the polarity of the pulse immediately preceding the zeros to be replaced.

For balance purposes, substitution sequences for nonmodal codes must contain an equal number of positive and negative pulses. They also may contain zeros, and the total number of zeros may be odd. The last pulse in the substitution sequences must have the same polarity as the pulse immediately preceding the sequence. If this property is not fulfilled, a unipolar pattern consisting of one, N zeros, a one, N zeros, and so on, would be converted to a sequence consisting of a + (say) for the one, the substitution sequence, another + (to follow alternate polarity rule),[6] the same substitution sequence, and so on, and since the substitution sequence is balanced the signal would have a dc component. A sequence of two opposite polarity pulses would satisfy the above requirements. However, it would include no bipolar violations by which it may be recognized at the receiving end, so it is necessary to add another two pulses. Thus for nonmodal codes, N must be at least 4. Table 2.2 shows some practical nonmodal codes: B6ZS and B8ZS. Both of them are balanced and the last pulse is the same as the preceding pulse.

Another substitution code is called *modal code* in which more than two substitution sequences are provided, and the choice of sequence is based on the polarity of the pulse immediately preceding the zeros to be replaced as well as the previous substitution sequence used. For modal codes, N can be two or three. Modal code substitution sequences need not be balanced, and balance is achieved by properly alternating the sequences. To illustrate this, we refer to Table 2.2 where B3ZS is a modal code (also see Figure 2.11). Let B represent a normal bipolar that conforms to the AMI rule, V represent a bipolar violation, and 0 represent no pulse. Then in the B3ZS code, a block of 000 is replaced by B0V or 00V. The choice of B0V or 00V is made so that the number of B pulses between two consecutive V pulses is odd. Thus if we consider the contribution to dc component by the signal segment since last substitution, there will be two "extra" pulses that contribute to the dc component due to

6 This bit would be a − if the coding rule is to simply replace the N-zero sequence. However, to make more transitions in the signal, it is better to follow the alternate polarity rule right after the substitution sequence.

	(a) B3ZS	
	Number of B pulses since last substitution	
Preceding pulse	odd	even
−	00−	+0+
+	00+	−0−

	(b) B6ZS
Preceding pulse	Substitution sequence
−	0−+0+−
+	0+−0−+

	(c) B8ZS
Preceding pulse	Substitution sequence
−	000−+0+−
+	000+−0−+

	(d) HDB3	
	Number of B pulses since last substitution	
Preceding pulse	odd	even
−	000−	+00+
+	000+	−00−

Table 2.2 Substitution codes. From [19]. Copyright © 1993 Kluwer. Reprinted with permission.

the bipolar violation. However, the polarity of these two pulses will be alternating since it depends on the polarity of the preceding pulse and whether the number of B pulses since last substitution is odd or even. All these conditions happen equally likely in a random sequence. Therefore in a long time average these "extra" pulses will cancel each other so that there will be no dc component.

B3ZS and B6ZS are specified for DS-3 and DS-2 North American standard rates, respectively, and B8ZS is specified as an alternative to AMI for the DS-1 and DS-1C rates. Figure 2.11 shows examples of some substitution codes.

Spectrum calculation of BNZS codes is based on a flow graph of the pulse states and is quite involved. The spectrum obviously depends on the substitution sequence used and the statistical property of the data sequence. Refer to [4] for details. Figure 2.12 shows spectra of some substitution codes [4], where F/F_{bit} is the frequency normalized to the bit rate F_{bit}.

No results of bit error probability of BNZS codes are available in the literature. Since they are conditioned AMI codes we may conjecture that their bit error rates at the detector must be very close to those of AMI codes. However, there are more errors due to failures to recognize the substitution sequences in the decoder.

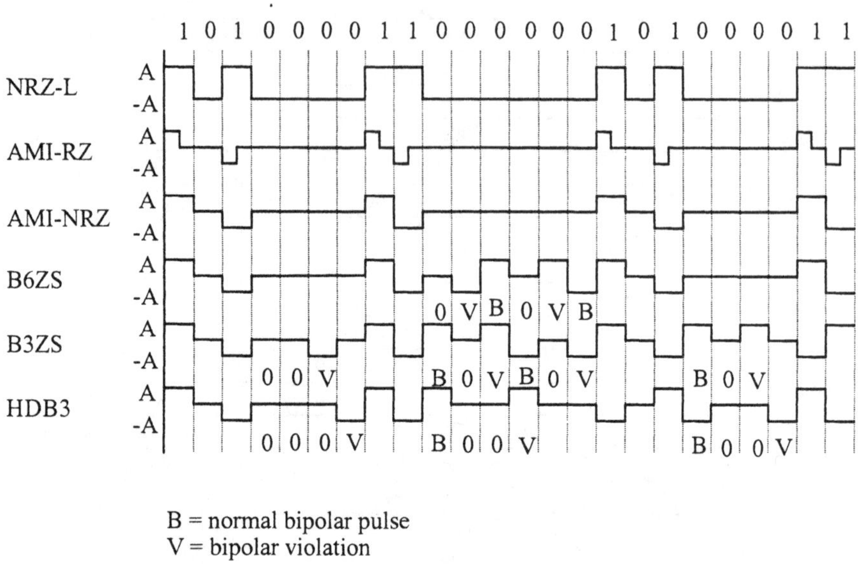

Figure 2.11 Substitution codes. From [19]. Copyright © 1993 Kluwer. Reprinted with permission.

2.5.2 High Density Bipolar *n* Codes

Croisier proposed HDBn code and *compatible high density bipolar n (CHDBn)* code in 1970 [2]. CHDBn may be considered as an improved version of HDBn since CHDBn coding and decoding hardware is somewhat simpler [2].

The common feature of these two codes is that they limit the number of consecutive 0s to n by replacing the $(n+1)$th 0 by a bipolar violation. In addition, to avoid the dc component, they are made modal. That is, they each have more than one possible substitution sequence. The substitution sequences are:

$$\begin{array}{llll} \text{HDB} & \text{B}\,0\,0,\cdots,0\,\text{V} & \text{or} & 0\,0\,0,\cdots,0\,\text{V} \\ \text{CHDB} & 0\,0,\cdots,0\,\text{B}\,0\,\text{V} & \text{or} & 0\,0,\cdots,0\,0\,0\,\text{V} \end{array}$$

where bits omitted are all zeros. There are total n zeros in each of the sequences.

The choices of the sequences must be such that the number of B pulses between two consecutive V pulses is always odd. It can be easily verified that the polarity of V pulses will always alternate so that a long sequence will produce virtually no dc

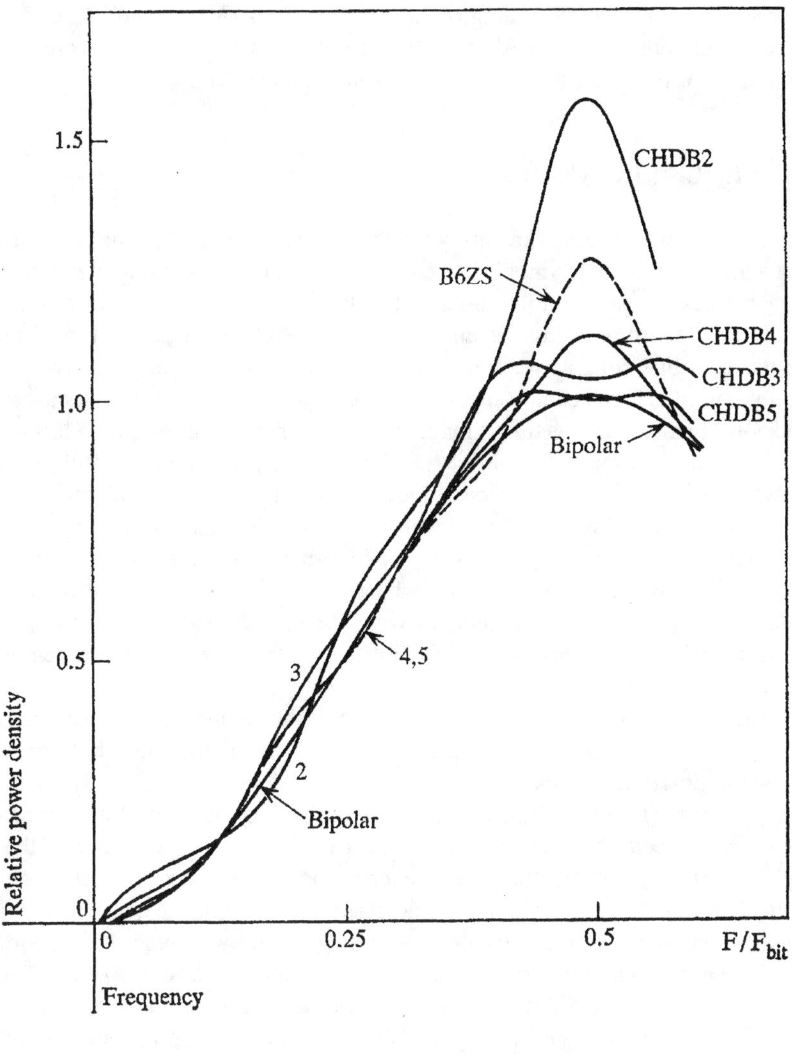

Figure 2.12 PSDs of some substitution codes. From [2]. Reprinted by permission from *IBM Journal of Research and Development*, copyright 1970 by International Business Machines Corporation.

component.

Two commonly used HDBn codes are HDB2 and HDB3. HDB2 is identical to B3ZS. HDB3 is used for coding of 2.048 Mbps, 8.448 Mbps, and 34.368 Mbps multiplex within the European digital hierarchy [20]. Its substitution rules are shown in Table 2.2(d). An example of HDB3 is shown in Figure 2.11.

2.6 BLOCK LINE CODES

So far the codes we discussed are bit-by-bit codes, in which each input bit is translated one at a time to an output symbol. In a *block code* input bits are grouped into blocks and each block is translated into another block of symbols. The purpose of using block codes is to introduce redundancy in order to meet one or more of the desired attributes of a line code as we stated at the beginning of this chapter. Two basic techniques used in block coding are (1) insertion of additional binary pulses to create a block of n binary symbols that is longer than the number of information bits m, or (2) translation of a block of input bits to a block of output symbols that uses more than two levels per symbol. The first technique is mainly used in optical transmission where modulation is limited to two-level (on-off) but is relatively insensitive to a small increase in transmission rate since optical fiber has a very wide bandwidth. The second technique applies to cases where bandwidth is limited but multilevel transmission is possible, such as metallic wires used for digital subscriber loops.

All basic line codes described in Section 2.2 can be viewed as special cases of block codes.

There are some technical terms which should be defined before we describe various block codes. Some of them are also used for nonblock codes, but some are only used in describing block codes.

Digital sum (DS) of a digital sequence is defined as the numerical sum of the symbols in the sequence. Digital sum is also called *disparity*. To suppress the dc component of a coded sequence, the sequence should have a zero digital sum or disparity. Because in many cases the DS decreases as the sequence gets longer, long-term DS tends to be zero even though short-term DS varies with time. It is important to know the maximum *digital sum variation* (DSV) which leads to *dc wander*. When a sequence has a finite DSV, it is said to be *balanced*, otherwise it is *unbalanced*.

A block code which chooses symbols from more than one alphabet is called *alphabetic*, otherwise it is *nonalphabetic*.

The *efficiency* of a code is defined as [21]

$$\eta = \frac{\text{actual information rate}}{\text{theoretical maximum information rate}} \times 100\%$$

For example, NRZ codes encode 1 bit into 1 binary symbol, the efficiency is

$$\eta = \frac{\log_2 2}{\log_2 2} = 100\%$$

AMI codes encode 1 bit into 1 ternary symbol, the efficiency is

$$\eta = \frac{\log_2 2}{\log_2 3} = 63.09\%$$

The Manchester code encodes 1 bit into 2 binary symbols, the efficiency is

$$\eta = \frac{\log_2 2}{2\log_2 2} = 50\%$$

2.6.1 Coded Mark Inversion Codes

Fiber optical communication systems use baseband modulation, not bandpass modulation, since the transmission of a symbol is represented by the intensity of the light of the optical source, namely the laser diode. Even though AMI has been widely used in digital coaxial or pair cable systems due to its merit described before, it can not be used in fiber optical systems since it uses three levels[7] and thus suffers from nonlinearity of the laser diode. A promising solution for avoiding this difficulty is to replace the zero level of AMI with two-level waveforms. This leads to two-level AMI codes including coded mark inversion (CMI) scheme and differential mode inversion (DMI) scheme.

CMI was first proposed by Takasaki et al. in 1976 [6] for optical fiber systems. CMI uses A or –A for a binary 1 for a full-bit period. The levels A and –A are alternated for each occurrence of 1. The 0s are represented by a pulse with level A for the first half bit and –A for the second half bit, or vice versa. CMI can be also viewed as 1 bit to 2 bits coding (1B2B) with 1→00 or 1→11, alternatively and 0→01 only (or 0→10 only). An example of CMI is shown in Figure 2.13.

CMI improves the transition density significantly. It also has error detection feature through the monitoring of coding rule violation. Decoding is done by comparing the second half bit with the first half bit so that it is insensitive to polarity reversals. Compared with AMI codes, the disadvantage of this code is that the transmission rate is twice that of the binary input signal.

CMI is chosen for the coding of the 139.246-Mbps multiplex within the European digital hierarchy [20].

---

[7] A negative 1 corresponds to laser diode off, zero to 1/2 intensity, and positive 1 to full intensity.

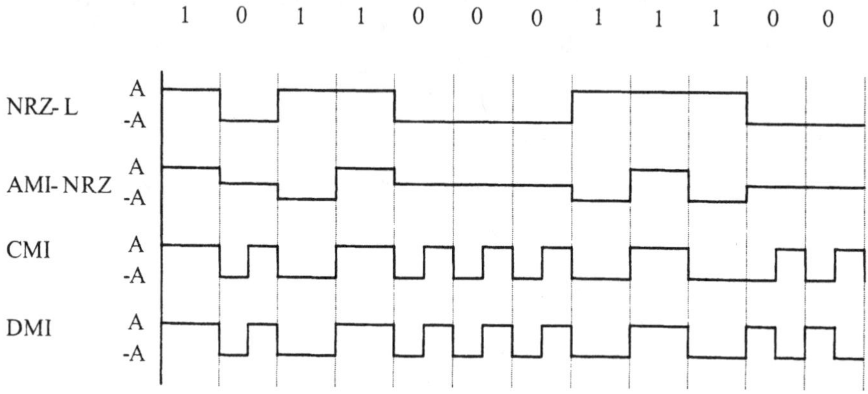

Figure 2.13 CMI and DMI waveforms.

The signal format of CMI cannot be described by (2.11) since it uses two waveforms (1,1) or (−1,1) for each bit. Nor can it be described by first-order Markov process since the mark is alternately inverted which implies that the current bit state may depend on a bit state long before the current one. Therefore its $R(\tau)$ is found by a computer Monte Carlo simulation as shown in Figure 2.14(a). Also shown is the $R(\tau)$ (Figure 2.14(b)) found by approximating CMI by first-order Markov process, which is quite close to the simulated one. The simulated $R(\tau)$ can be decomposed into two parts as shown in Figure 2.14(c, d), where (d) is periodical. By taking the Fourier transforms of (c) and (d) analytically the PSD is given by

$$\begin{aligned}\Psi_s(f) &= A^2\frac{6-4\cos(\pi fT)-3\cos(2\pi fT)+\cos(4\pi ft)}{8\pi^2 f^2 T}\\ &\quad +A^2\sum_{k=-\infty}^{\infty}\frac{1}{\pi^2(2k+1)^2}\delta(f-\frac{2k+1}{T}), \qquad \text{(CMI)} \qquad (2.64)\end{aligned}$$

which is plotted in Figure 2.14(e), where $A=1$ for unity average bit energy. The periodical part of $R(\tau)$ manifests itself as a series of impulses at odd multiples of data rate in the spectrum. The sudden drop in the $P_{ob}(B)$ curve is due to the impulse at $f=R_b$ in the PSD. The bandwidths are $B_{null}=2.0R_b$, $B_{90\%}\approx 2.92R_b$, and $B_{99\%}\approx 28R_b$.

Note that the above spectrum of CMI is calculated with two levels of $\pm A$, that is, the code waveform is bipolar. For optical transmission these two levels would be A and 0, that is, the code waveform is unipolar. This level shift will not change the

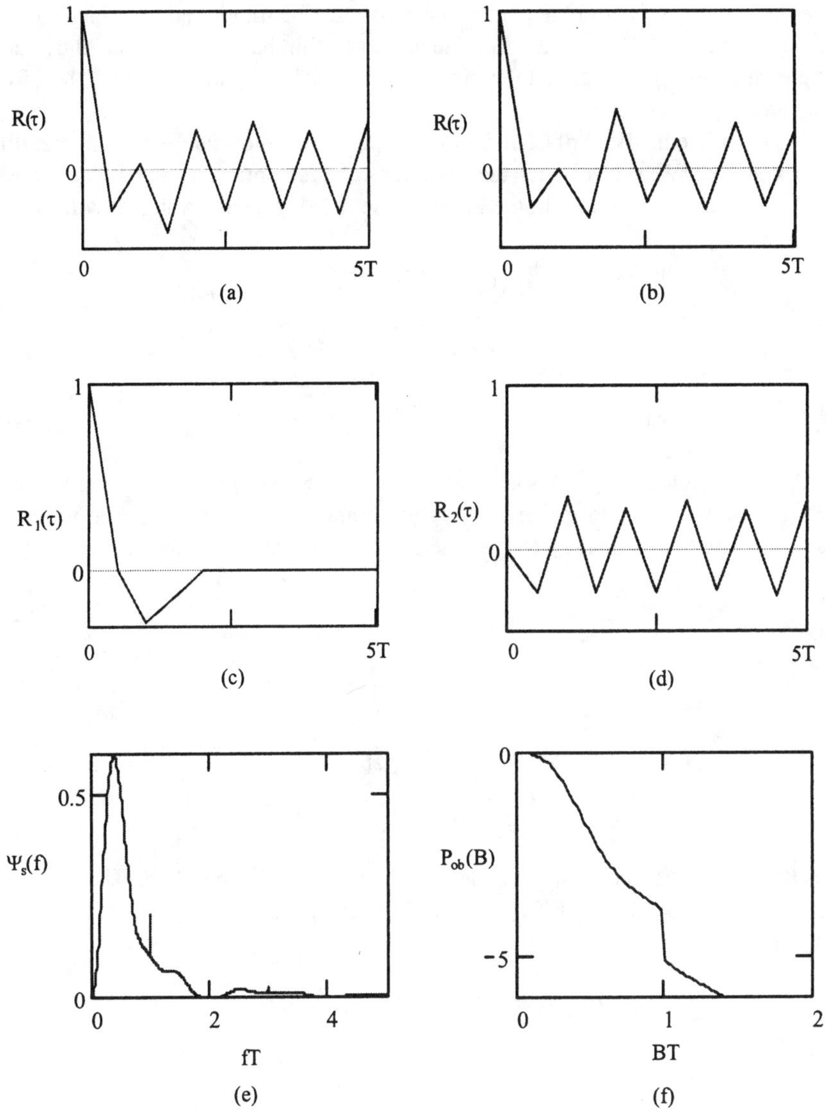

Figure 2.14 CMI correlations and spectrum: (a) simulated $R(\tau)$, (b) approximate $R(\tau)$, (c) nonperiodical part of the simulated $R(\tau)$, (d) periodical part of the simulated $R(\tau)$, (e) PSD, and (f) out-of-band power.

spectral shape, but only causes the spectrum to have a dc component which can be represented by $\frac{1}{2}\delta(f)$, and the continuous part must be multiplied by a factor of $\frac{1}{2}$. The null bandwidth $B_{null} = 2.0R_b$. The energy within bandwidth R_b is 79.4% and just above R_b is 99.7% due to the jump caused by $0.101\delta(f)$ at $f = R_b$. Therefore $B_{90\%}$ and $B_{99\%}$ cannot be found.

We calculate the BER of CMI for the bipolar case which can be used in metallic wired systems. We will discuss the unipolar case for optical systems where CMI finds its primary use shortly. In the bipolar case, CMI is a ternary signal with

$$\begin{aligned} s_1(t) &= A, \qquad 0 \le t \le T, \qquad p_1 = \frac{1}{4} \\ s_2(t) &= \begin{cases} -A, & 0 \le t \le \frac{T}{2} \\ A, & \frac{T}{2} \le t \le T \end{cases} \qquad p_2 = \frac{1}{2} \\ s_3(t) &= -A, \qquad 0 \le t \le T, \qquad p_3 = \frac{1}{4} \end{aligned} \tag{2.65}$$

All three signals have the same energy: $E = A^2T$. $s_2(t)$ is orthogonal to $s_1(t)$ and $s_3(t)$. So we can choose $\phi_1(t) = s_1(t)/\sqrt{E}$ and $\phi_2(t) = s_2(t)/\sqrt{E}$ as the basis functions. The third signal $s_3(t) = -\sqrt{E}\phi_1(t)$. Thus from (2.53) we have

$$l_j = \ln P_j - \frac{1}{N_o}\sum_{i=1}^{2}(r_i - s_{ij})^2$$

where

$$\begin{aligned} &s_{11} = \sqrt{E}, \quad && s_{21} = 0 \\ &s_{12} = 0, && s_{22} = \sqrt{E} \\ &s_{13} = -\sqrt{E}, && s_{23} = 0 \end{aligned}$$

Thus the decision space is two-dimensional as shown in Figure 2.15. The decision rule is to choose the largest. For l_1 to be the largest we must have $l_1 > l_2$ and $l_1 > l_3$, from these relations we can deduce that

$$H_1 \text{ is true if } r_1 > 0 \text{ and } r_1 > r_2 + \beta$$

where

$$\beta = \frac{N_o \ln 2}{2\sqrt{E}}$$

Similarly we can deduce that

$$H_2 \text{ is true if } -(r_2 + \beta) < r_1 < (r_2 + \beta)$$

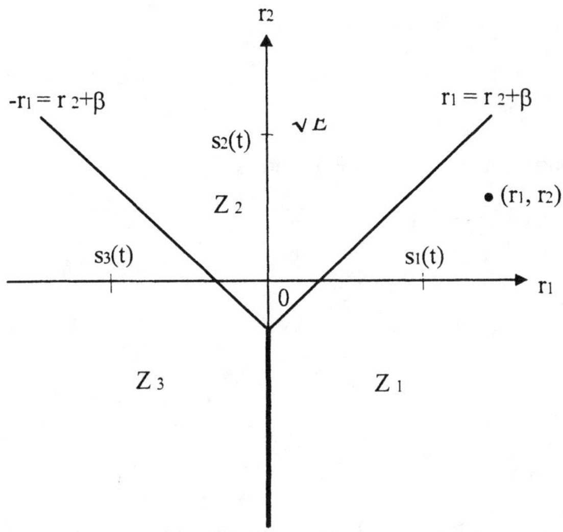

Figure 2.15 CMI decision space.

and

$$H_3 \text{ is true if } r_1 < 0 \text{ and } r_1 < -(r_2 + \beta)$$

Therefore there are three decision boundaries as shown in Figure 2.15. The optimum receiver consists of two correlators for computing r_1 and r_2 (see Figure 2.10(a)).

The average bit error probability is

$$\begin{aligned} P_b &= p_1 \Pr(e/s_1) + p_2 \Pr(e/s_2) + p_3 \Pr(e/s_3) \\ &= p_1[1 - \iint_{Z_1} p(r_1, r_2/s_1) dr_1 dr_2] + p_2[1 - \iint_{Z_2} p(r_1, r_2/s_2) dr_1 dr_2] \\ &\quad + p_3[1 - \iint_{Z_3} p(r_1, r_2/s_3) dr_1 dr_2] \end{aligned}$$

Due to the geometrical symmetry of decision regions Z_1 and Z_3 and the fact that

$p_3 = p_1$, we have

$$P_b = 2p_1[1 - \int\int_{Z_1} p(r_1, r_2/s_1)dr_1dr_2] + p_2[1 - \int\int_{Z_2} p(r_1, r_2/s_2)dr_1dr_2]$$

Integrating and omitting detail, the P_b is found as

$$\begin{aligned} P_b &= 1 - \frac{1}{2}\int_{-\alpha}^{\infty} \frac{1}{\sqrt{2\pi}} \exp(-\frac{x^2}{2})Q(\frac{\ln 2}{\alpha} - \alpha - x)dx \\ &- \int_{0}^{\infty} \frac{1}{\sqrt{2\pi}} \exp(-\frac{x^2}{2})Q(-\frac{\ln 2}{\alpha} - \alpha + x)dx, \quad \text{(CMI)} \quad (2.66) \end{aligned}$$

where

$$\alpha = \sqrt{\frac{2E_b}{N_o}}$$

If we ignore the boundary bias for high SNR and set boundaries as bisectors between signals, then we can get the approximate P_b by replacing $ln2$ with 0 in the above expression:

$$\begin{aligned} P_b &\approx 1 - \frac{1}{2}\int_{-\alpha}^{\infty} \frac{1}{\sqrt{2\pi}} \exp(-\frac{x^2}{2})Q(-\alpha - x)dx \\ &- \int_{0}^{\infty} \frac{1}{\sqrt{2\pi}} \exp(-\frac{x^2}{2})Q(-\alpha + x)dx, \\ &\text{(CMI, for high SNR)} \quad (2.67) \end{aligned}$$

Moreover, if we ignore the error probability between s_1 and s_3, then only error probability between s_1 and s_2 or s_3 and s_2 is considered. Thus P_b can be approximated as

$$P_b \approx Q\left(\sqrt{\frac{E_b}{N_o}}\right), \quad \text{(CMI, approximation)} \quad (2.68)$$

which is the same as that of unipolar codes. Equations (2.66) and (2.67) are numerically evaluated. They are plotted in Figure 2.8 under the same label of AMI codes since they are very close to each other and are indistinguishable.

For optical transmission, the $-A$ level in the CMI signal set would be 0. Not only are the pulses unipolar, but also many other factors that are particular to optical systems must be considered in calculating the BER. The required power p_o to achieve a desired error rate has been studied in [22] for straight binary formats (unipolar NRZ-

L). It can be expressed as

$$p_o = EZ^{\frac{x}{2+2x}} \Sigma^{\frac{1}{1+x}} \Gamma(x, r) \tag{2.69}$$

where

$$\Gamma(x, r) = \Gamma_1(x, r)^{\frac{x}{2+2x}} \Gamma_2(x, r)^{\frac{2+x}{1+x}}$$

$$\Gamma_1(x, r) = \frac{-(2-r) + \sqrt{(2-r)^2 + 16(1+x)x^{-2}(1-r)}}{2(1-r)}$$

$$\Gamma_2(x, r) = \sqrt{\Gamma_1^{-1}(x, r) + 1 - r} + \sqrt{\Gamma_1^{-1}(x, r) + 1}$$

$$r = I_o/\Sigma$$

and where

E = constant that depends on the error rate and bit rate.

Z = thermal noise contribution.

I_i = shot noise contribution that would be caused by the optical pulse in the ith time slot ($i = 0$ corresponds to the time slot under decision).

Σ = total shot noise contribution.

x = excess noise factor for APD (avalanche photodiode).

Takasaki et al. [6] showed that a slight modification of the above expressions provides the required optical power for other pulse formats, including CMI and DMI. For CMI and DMI the following substitutions in (2.69) will provide the required optical power for direct transmission

$$\left.\begin{array}{c} \Sigma_1 \rightarrow \Sigma + I_{-1/2} \\ I_1 \rightarrow I_o - I_{1/2} \\ Z \rightarrow Z \end{array}\right\} \tag{2.70}$$

The BER of edge-detected CMI data transmission over the AWGN channel was given in [23] as

$$P_b \approx \frac{BT}{4\sqrt{3}} \exp\left(\frac{-E_b/N_o}{2BT}\right)$$

where B is the equivalent noise bandwidth of the receiver low-pass filter.

2.6.2 Differential Mode Inversion Codes

DMI is another two-level AMI scheme proposed by Takasaki et al. in 1976 [6] for optical fiber systems. Its coding rule for binary 1s is the same as that of CMI. Its coding rule for binary 0s is different: 0→01 or 0→10 so that no pulses in a sequence have pulse widths wider than T, the bit duration. An example of DMI is shown in Figure 2.13. For optical applications, the DMI code is unipolar. If DMI is used for metallic applications, it could be made bipolar.

The PSD of the unipolar DMI was found using a code flow graph method by Yoshikai in 1986 [16] as

$$\Psi_s(f) = \frac{|G(f)|^2}{T_l} \cdot \frac{2p(1-p)(1-\cos 4\pi f T_l)(1-\cos 2\pi f T_l)}{4p^2 - 6p + 3 - 4(p-1)^2 \cos 4\pi f T_l + (1-2p)\cos 8\pi f T_l}$$

where $G(f)$ is the Fourier transform of the line pulse waveform which has the width of $T_l = T/2$ and p is the occurrence probability of mark. For equiprobable data sequence, $p = 0.5$, the above expression reduces to

$$\Psi_s(f) = \frac{|G(f)|^2}{T/2} \sin^2 \pi f T/2$$

For the rectangular pulse (NRZ pulse) with amplitude A, it becomes

$$\begin{aligned} \Psi_s(f) &= \frac{1}{T/2}\left(\frac{AT}{2}\frac{\sin \pi f T/2}{\pi f T/2}\right)^2 \sin^2 \pi f T/2 \\ &= \frac{A^2 T}{2}\left(\frac{\sin \pi f T/2}{\pi f T/2}\right)^2 \sin^2 \pi f T/2, \end{aligned}$$

(continous part of the PSD of the unipolar DMI) (2.71)

The energy under the PSD curve is only half of the bit energy. This implies that the above PSD expression is the continuous part of the PSD of the unipolar DMI. Another half of the energy is in the dc component which can be represented by a delta function $\frac{1}{2}\delta(f)$. It has the first null at $fT = 2$. The bandwidths are $B_{null} = 2.0R_b$, $B_{90\%} \approx 1.27R_b$, and $B_{99\%} \approx 15R_b$.

When DMI is bipolar, then (2.71) should be multiplied by a factor 2 to represent the entire PSD. In this case its PSD has the same form of the PSD of biphase codes (2.41). It has the first null at $fT = 2$. The bandwidths are $B_{null} = 2.0R_b$, $B_{90\%} \approx 3.05R_b$, and $B_{99\%} \approx 29R_b$.

It is not surprising that the PSD of the bipolar DMI is the same as that of biphase codes. In fact bipolar DMI is the same code as Bi-Φ-S code even though their coding rules are presented differently. One can easily verify this claim by simply coding a sample sequence using their rules. DMI is also called frequency shift code (FSC) in a paper by Morris [24], where other simple codes for optical transmission are also presented.

Since bipolar DMI code is equal to Bi-Φ-S code, we can therefore detect it using

Figure 2.10(a) and use (2.62) for its BER calculation. For unipolar DMI which is used primarily in optical systems we can use (2.69) and (2.70) to calculate required optical power for a given BER and bit rate.

2.6.3 mBnB Codes

CMI and DMI can be interpreted as two special cases of a larger family of block codes called *mBnB* codes. An mBnB code converts m binary digits into n binary digits with $m < n$. CMI and DMI are 1B2B codes. Biphase codes are also 1B2B codes.

When $n = m + 1$, the codes are *mB (m+1)B* codes which are popular in high-speed-fiber optical transmission. Besides CMI, DMI and biphase codes which belong to this class, there are other mB(m+1)B codes. CMI and DMI have a poor efficiency. CMI and DMI encode a logic 1 into a binary symbol and a logic 0 into two binary symbols; the efficiency is

$$\eta = \frac{\log_2 2}{\frac{1}{2}\log_2 2 + 2 \times \frac{1}{2}\log_2 2} \times 100\% = 66.7\%$$

To increase the efficiency, a larger m must be chosen.

Following is a rather complete collection of mBnB codes proposed in literature.

2.6.3.1 Carter Code

Carter proposed a code in 1965 for PCM systems [5]. It is an mB(m+1)B code. The coding rules are as follows. The eight-digit character of a PCM channel is transmitted either unchanged or with the digits inverted (i.e., marks for spaces and spaces for marks), depending on which condition will reduce the total disparity since transmission commenced. Thus there will never be any continually increasing disparity, and the dc component will be zero over a long period. In order to indicate to the receiver whether the character is to be passed to the PCM terminal unchanged or reinverted, the encoder at the transmitter precedes the character by a parity-control digit, the polarity of which gives this information. Thus this Carter code is an 8B9B code. Its efficiency is 94.64%. It is clear that this code can also apply to characters of any length. To apply it to an input having no character structure, the only additional thing to do is to define "characters" or blocks.

With Carter code, completely unrestricted binary input digits can be transmitted. This is not strictly true for the AMI codes since synchronization will be lost if long strings of zeros or marks are present. However, the substitution codes proposed later (in 1969) [4], described in Section 2.5, can also overcome the problem of AMI.

m+1	Words having disparity of					
	0	+2	+4	+6	+8	+10
2	2	1				
4	6	4	1			
6	20	15	6	1		
8	70	56	28	8	1	
10	252	210	120	45	10	1

Table 2.3 Number of code words with various disparities. From [25]. Copyright © 1969 IEE.

2.6.3.2 Griffiths Code

This code is a type of mB(m+1)B codes. It was proposed by Griffiths (1969) [25] as an improvement on the Carter code in which strings of similar digits of length $2m+2$ may still occur for a block length of m. Griffiths checked numbers of possible code words with various lengths and disparities. If m is odd, the number of zero-disparity words of length $m+1$ is $\binom{m+1}{\frac{(m+1)}{2}}$ and the number of words having positive disparity is $2^m - \binom{m+1}{\frac{(m+1)}{2}}/2$. There are the same number of words having negative disparity. If words of opposite disparity are paired for transmission of alternate disparity words, there will be a total of $2^m + \binom{m+1}{\frac{(m+1)}{2}}/2$ words and pairs available. This is sufficient to translate a block of m binary digits, leaving $\binom{m+1}{\frac{(m+1)}{2}}/2$ code words spare. The number of code words for the first several values of m is listed in Table 2.3.

For $m+1$ equals two, there are two zero-disparity words (i.e., 0 and 1 may be translated to 01 and 10). This is in fact the Manchester code. The bit rate is doubled after coding.

Particular attention is drawn to the cases in which $m+1$ equals four and six. In both cases, translation may be achieved using only zero-disparity words and double-disparity words. For $m+1 = 6$ (i.e., 5B6B code), we can use all zero-disparity words and only those double-disparity words not having four same successive digits. The coding rules are shown in Table 2.4. For a block of five digits containing two ones, transmit it followed by a one. For a block containing three ones, transmit it followed by a zero. Twenty of the 32 possible blocks may be transmitted this way with zero disparity. The remaining 12 blocks are translated into double-disparity words. On alternate occasions, these double-disparity words are transmitted inverted. Words of opposite sign are sent alternatively. Thus the dc component of the line signal will be

11000		00111	
10100		01011	
10010		01101	
10001		01110	
01100	transmitted as	10011	transmitted as
01010	× × × × ×1	10101	× × × × ×0
01001		10110	
00110		11001	
00101		11010	
00011		11100	

11111		111010		000101	
11110		110110		001001	
11101		101110		010001	
11011		111001		000110	
10111		110101		001010	
01111	transmitted as	101101	or	010010	alternately
10000		011101		100010	
01000		110011		001100	
00100		101011		010100	
00010		011011		100100	
00001		100111		011000	
00000		010111		101000	

Table 2.4 Translation table for 5B6B Griffiths code. From [25]. Copyright © 1969 IEE.

zero. This code never has more than six consecutive digits of the same type. The rate increase is only 20%.

When this method is used to encode blocks of three bits (i.e., 3B4B), a set of suitable translation rules are shown in Table 2.5. The increase of transmission rate is 33%, higher than that of 5B6B Griffiths code. Similarly seven bits may be encoded into eight. However, use of quadruple-disparity words would have to be made to provide sufficient translations. In general, the complexity of higher-order encoders would appear to be excessive.

2.6.3.3 PAM-PPM Code

This code was proposed by Bosotti and Pirani in 1978 [26] for optical communications. Coded bits 00, 10, and 01 are mapped to signals a, b, and c with one corresponding to a positive pulse for a duration of θT $(0 < \theta < 1)$ and zero corresponding to zero level for $(1-\theta)T$. Each of them is a pulse-amplitude and pulse-position mod-

001 010 100	transmitted as × × × 1
110 101 011	transmitted as × × × 0
000 111	transmitted as {0100 or 1011; 0010 or 1101} alternately

Table 2.5 Translation table for 3B4B Griffiths code. From [25]. Copyright © 1969 IEE.

ulated signal, hence the name. The coding rules are shown in Table 2.6. This is in fact a 3B4B code with 0 and 1 in code words occupying different lengths.

The PSD of this code with $\theta = 1/2$ has the first null at $f = 2R_b$, which is the same as a straight PAM signal with pulse width of $T/2$. However, the PSD shows strongly reduced low-frequency components. The rectified PAM-PPM pulse train contains an abundant timing frequency component. The longest time interval without signal level change is $3T$.

2.6.3.4 2B3B dc-Constrained Code

This code was proposed by Takasaki et al. in 1976 [6]. Conventional codes try to suppress the dc component [5]. It is shown in [27] that such codes cannot provide the capability of error monitoring, since they use up most of the redundancy in suppressing the dc component. The 2B3B dc-constrained code uses a different approach. It constrains the dc component instead of suppressing it and thus makes it possible to use the redundancy for error detection also. Table 2.7 shows the translation rules. The data bits 1 and 0 are converted to + and –, respectively. Then a third symbol + or – is added to make combinations of one + and two –s (mode 1). The data pair 11 cannot meet this requirement. Therefore modes 1 and 2 are used alternately for this pair to produce one + and two –s on the average. This code produces a dc component of –1/3. However, it does not suffer from dc wander since the dc component is constant regardless of data pattern. In terms of timing information, the average number of changes of levels per block ranges from 1 to 2. The longest succession of same levels is 7.

2.6.4 mB1C Codes

This code belongs to a class of codes called *bit insertion codes* which are popular in high-speed optical transmission.

Binary signal	PAM-PPM signal
000	a c = 0001
001	a b = 0010
010	c a = 0100
011	b a = 1000
100	c c = 0101
101	c b = 0110
110	b c = 1001
111	b b = 1010

Table 2.6 PAM-PPM code. From [26]. Copyright © 1978 IEE.

Binary	2B3B code	
	Mode 1	Mode 2
00	− − +	
01	− + −	
10	+ − −	
11	+ + −	− − −

Table 2.7 2B3B dc-constrained code. From [6]. Copyright © 1983 IEEE.

The *mB1C* code was proposed by Yoshikai et al. for optical fiber transmission in 1984 [8] in order to raise the speed limit achieved by CMI, DMI, and mBnB codes. In the coding process, the speed of input signal is increased by $(m+1)/m$, then a complementary bit is inserted at the end of every block of m information bits. The inserted bit is complementary to the last bit of the m information bits. The coding rule is very simple therefore sophisticated electronic circuits at high speed required by mBnB codes can be avoided. The increase in code rate is smaller, only $1/m$, when compared with CMI and DMI which require a code rate twice as fast as the information rate.

With the mB1C code the maximum number of consecutive identical symbols is $m+1$ which occurs when the inserted bit and the m succeeding bits are the same. In-service error can be monitored using an XOR applied to the last information bit and the complementary bit. This code has been adopted in a Japanese 400 Mbps optical system in the form of 10B1C.

The spectrum of this code contains continuous part and discrete components. If the mark rate in the information sequence is $1/2$, only a dc component exists. The dc component is due to the fact that the code is unipolar. If the mark rate deviates from $1/2$, the PSD contains harmonic spectra with a fundamental component of $1/(m+1)T$. The harmonic spectra are caused by the fact that when the mark rate is not $1/2$, the occurrence probability for a mark in the C bit is also not $1/2$; instead it is inversely proportional to that in the information bit. The harmonic spectra cause jitter. A scrambler can be used to scramble the mB1C coded sequence hence suppress these harmonics.

2.6.5 DmB1M Codes

Differential m binary with 1 mark insertion (*DmB1M*) code is proposed by Kawanishi and Yoshikai et al. (1988) for very high-speed optical transmission [9]. The coding rule is rather simple. In the coding process, the speed of input signal (P) is increased by $(m+1)/m$, then a mark bit is inserted at the end of every block of m information bits. The mark-inserted signal (Q) is converted to the DmB1M code (S) by the following differential encoding equation:

$$S_k = S_{k-1} \oplus Q_k$$

where S_k and Q_k denote the kth signal bit of S and Q, respectively. The symbol $\oplus$ denotes XOR. A C bit (complementary bit) is generated automatically in the coding process. This can be seen as follows (the inserted bit Q_{m+1} is always equal to 1):

$$S_{m+1} = S_m \oplus Q_{m+1} = S_m \oplus 1 = \overline{S_m}$$

That is, the $(m+1)$th bit is always complementary to the mth bit. So the S sequence is an mB1C code. Decoding is accomplished by equation

$$Q_k = S_k \oplus S_{k-1}$$

and the original signal P is then recovered through mark deletion and $m/(m+1)$ speed conversion.

Like mB1C code, DmB1M limits the maximum length of consecutive identical digits to $(m+1)$ and provides error detection through monitoring of the inserted bit every $(m+1)$ bits. Unlike mB1C code, however, it can be shown that the mark probability of the coded sequence is always 1/2 regardless of the mark probability of the information sequence [9]. Thus there are no jitter-causing discrete harmonics in the spectrum except the dc component.

When m is small, the continuous part of the spectra of both mB1C code and DmB1M code is similar to that of AMI, which ensures that high-frequency and low-frequency components are suppressed. As the block length m increases, however, the spectrum flattens and assumes the shape of the spectrum of a random signal, having nonzero components near the zero frequency. The problem of spectrum control in mB(m+1)B codes can be solved by adding further coding [10] or using scrambling techniques [11]. And the PFmB(m+1)B code described next can overcome the spectrum problem to some extent.

2.6.6 PFmB(m+1)B Codes

This partially flipped code or *PFmB(m+1)B* code was proposed by Krzymien in 1989 [10]. It is balanced, with minimum rate increase $(1/m)$, and easy to encode and decode. Thus it is suitable for high-speed optical systems.

The coding process of a PFmB(m+1)B code consists of two stages: precoding and balanced coding. In the precoding stage the input binary sequence $\{B_i\}$ is grouped into blocks of m bits (m is an odd number) to which an additional bit B_{m+1} is added according to the rule

$$B_{m+1} = \begin{cases} 1 & \text{if } \sum_{i=1}^{m} B_i < \frac{m}{2} \\ 0 & \text{if } \sum_{i=1}^{m} B_i > \frac{m}{2} \end{cases} \tag{2.72}$$

The added bit B_{m+1} will serve a check bit in decoding as will be seen shortly.

In the second stage, the precoded input words $\mathbf{B} = [B_1, B_2, \cdots, B_{m+1}]$ are mapped into the binary codewords $\mathbf{C} = [C_1, C_2, \cdots, C_{m+1}]$ according to the rule

$$C_i = \begin{cases} B_i & \text{if } D(n) \times \text{WRDS}(n-1) \le 0 \\ \overline{B}_i & \text{if } D(n) \times \text{WRDS}(n-1) > 0 \\ & \quad \text{for } i = 1, 2, \cdots, m \end{cases} \tag{2.73}$$

and

$$C_{m+1} = B_{m+1}$$

where

$\overline{B}_i$ is the complement of B_i.

$D(n)$ is the disparity of the nth precoded word $\mathbf{B}_n$, which is determined by assigning -0.5 to a binary zero and 0.5 to a binary one and summing them up.

WRDS$(n-1)$ is the value of the running digital sum (total disparity) at the end of the $(n-1)$th precoded word $\mathbf{B}_{n-1}$

$$\text{WRDS}(n-1) = \sum_{k=-\infty}^{n-1} D(n)$$

From (2.73) we have the following observations. If the total disparity and the disparity of the current precoded word have the same sign, the precoded word **B** is partially flipped, only leaving the last bit unchanged because it will serve as a check bit in decoding. Otherwise the precoded word **B** is copied to **C** without any change. This will balance the total disparity or running digital sum.

Block synchronization in the receiver is accomplished by checking whether the received coded sequence $\{\widehat{C}_i\}$ satisfies the coding rules (2.72) and (2.73). High rate of code rule violations indicates that the system is out of synchronism. After block synchronism is established, decoding is performed by checking a received block $\widehat{\mathbf{C}}$ against the code rule (2.72) (with B_i replaced by $\widehat{C}_i$). If (2.72) is satisfied, no inversion of the bits of $\widehat{\mathbf{C}}$ is necessary. Otherwise, bits of $\widehat{\mathbf{C}}$ have to be inverted to yield **B**.

There are error extension cases in decoding this PFmB(m+1)B code. However, this extension is very small. An example given in [10] is 0.3 dB deterioration of SNR for $m = 7$, a crossover probability of 10^{-9} in an AWGN channel. The digital sum variation (DSV) is bounded and is $(3m-1)/2$. The maximum number of consecutive like bits is also bounded and is $2(m+1)$. The spectrum of the code has a roll-off to zero at zero frequency and $f = R_b$. However, when input data probability is unbalanced sharp peaks appear at both edges of the band. This disadvantage can be eliminated by scrambling.

2.6.7 kBnT Codes

A class of codes using three levels is called *kBnT* codes where k binary bits are coded into n $(n < k)$ ternary symbols. AMI can be considered as a 1B1T code. The most important codes of this class are 4B3T type and 6B4T type. Their efficiencies are 84.12% and 94.64%, respectively.

Binary	Ternary transmitted when total disparity is		
	Negative (-3, -2, -1)	Positive (0, 1, 2)	Word disparity
0000	+0−	+0−	0
0001	−+0	−+0	0
0010	0−+	0−+	0
0011	+−0	+−0	0
0100	++0	−−0	±2
0101	0++	0−−	±2
0110	+0+	−0−	±2
0111	+++	−−−	±3
1000	++−	−−+	±1
1001	−++	+−−	±1
1010	+−+	−+−	±1
1011	+00	−00	±1
1100	0+0	0−0	±1
1101	00+	00−	±1
1110	0+−	0+−	0
1111	−0+	−0+	0

Table 2.8 Translation table for 4B3T. From [21]. Copyright © 1983 International Journal of Electronics.

2.6.7.1 4B3T Code

The code proposed by Waters [21] is the simplest of the 4B3T class. The translation table is shown in Table 2.8, where disparities are calculated by assigning a weight of 1 to a positive mark and a weight of −1 to a negative mark. This is an alphabetic code since two alphabets are used.

There are 27 (3^3) possible combinations of three ternary digits. In order to maintain bit sequence independence, 000 is not used. All other combinations are used. We can allocate six binary four-bit blocks to the six zero disparity words. The remaining ten are allocated both a positive disparity word and its negative disparity inverse. During translation, a count is kept of the total disparity. When this is negative, positive disparity words are selected for transmission and vice versa. This ensures that the transmitted code has zero dc content. The disparity is bounded and at the end of a word only six values of total disparity are possible (−3, −2, −2, 0, 1, and 2, where 0 is counted as positive). Total disparity +3 is not possible despite that word disparity could be +3, since the word of +3 disparity is sent only when total disparity is negative.

Binary	Ternary transmitted when total disparity is		
	-2	-1 or 0	+1
0 0 0 0	+ + +	− + −	− + −
0 0 0 1	+ + 0	0 0 −	0 0 −
0 0 1 0	+ 0 +	0 − 0	0 − 0
0 0 1 1	0 − +	0 − +	0 − +
0 1 0 0	0 + +	− 0 0	− 0 0
0 1 0 1	− 0 +	− 0 +	− 0 +
0 1 1 0	− + 0	− + 0	− + 0
0 1 1 1	− + +	− + +	− − +
1 0 0 0	+ − +	+ − +	− − −
1 0 0 1	0 0 +	0 0 +	− − 0
1 0 1 0	0 + 0	0 + 0	− 0 −
1 0 1 1	0 + −	0 + −	0 + −
1 1 0 0	+ 0 0	+ 0 0	0 − −
1 1 0 1	+ 0 −	+ 0 −	+ 0 −
1 1 1 0	+ − 0	+ − 0	+ − 0
1 1 1 1	+ + −	+ − −	+ − −

Table 2.9 Translation table for MS43. From [21].

2.6.7.2 MS43 Code

MS43 code proposed by Franaszek in 1968 [28] is also a 4B3T code. It has a more sophisticated three-alphabet translation table as shown in Table 2.9.

The word disparities range from –3 to +3. There are only four possible total disparity states at the end of a word: –2, –1, 0, or 1. In the –2 state, only zero or positive disparity words are sent; in the +1 state, only zero or negative disparity words; and in the 0 or –1 states, only zero and unit disparity words. This allocation reduces the low-frequency content compared with 4B3T, making it more tolerant to ac coupling [29].

2.6.7.3 6B4T Code

This code was proposed by Catchpole in 1975 [30]. The translation table is shown in Table 2.10. It is similar in concept to 4B3T, using two alphabets with both the zero and unit disparity words uniquely allocated to 50 of 64 possible six-bit binary blocks. Disparity control is achieved by pairing the ± 2 and ± 3 disparity words

Ternary words			
+ − + −			18 zero disparity and 32 unit disparity ternary words always selected
− + − +			
+ 0 − 0	And 3 cyclic shifts of each		
+ + − −			
+ 0 0 −			
− 0 0 +			
+ 0 + −			
+ + − 0			
+ + 0 −			
+ 0 0 0			
− 0 − +			
− − + 0			
− − 0 +			
− 0 0 0			
+ 0 + 0	10 +2 disparity words		
0 + 0 +			
+ + 0 0			
0 + + 0		0 + + +	4 +3 disparity words
0 0 + +		+ 0 + +	
+ 0 0 +		+ + 0 +	
− + + +		+ + + 0	
+ − + +			
+ + − +			
+ + + −			
This set is inverted when the total disparity is positive			

Table 2.10 Translation table for 6B4T. From [21]. Copyright © 1975 IEE.

allocated to the other 14 binary combinations. This code is more efficient than the 4B3T codes, but has unbounded total disparity. However, it can be shown that with practical values of low frequency cut at the repeaters, this does not significantly affect the transmission performance provided a scrambler is used [21].

2.7 PULSE TIME MODULATION

Besides the binary line codes and binary block line codes described earlier, a family called *pulse time modulation* (PTM), has seen wide applications in optical transmissions, both in fiber optical [31] and infrared wireless systems [32], and magneto-optical recording [33]. In this PTM family, one of the time-dependent features, rather than amplitude, of a pulsed carrier is used to convey information. PTM family includes *pulse width modulation* (PWM), *pulse position modulation* (PPM), *pulse interval modulation* (PIM), *pulse interval and width modulation* (PIWM), *pulse frequency modulation* (PFM), and *square wave frequency modulation* (SWFM). PPM is the most popular one and some of its variations include *multipulse PPM* (MPPM) [34] and *differential PPM* (DPPM) [35].

PTM can be used for analog signals as well as digital signals (PCM codes or data). With PTM, one of the time-dependent features of the pulse carrier is used to represent the signal amplitude of the analog signal or the value of the digital signal.

PTM produces a sequence of pulses with a spectrum around the sampling frequency and its harmonics. Therefore it is a type of baseband modulation. This pulse train is used to modulate an optical carrier using intensity modulation or coherent modulation in optical transmission.

PTM is a family of schemes between strictly analog and digital modulation (except for the case of PTM for digital signals). It has a constant amplitude. Therefore it avoids the problem of nonlinearity of the analog modulation and improves the signal-to-noise ratio over analog modulation. On the other hand, it is not purely digital so that no PCM coding is required and bandwidth requirement is reduced.

PWM and PPM are long-established techniques and have been used widely in optical communications [36–42]. Due to their fixed symbol interval, they are much easier to multiplex and demultiplex in the time domain. PFM has been used extensively for the optical fiber transmission of video and broadcast quality TV signals [43–48]. SWFM are employed for the fiber transmission of HDTV and other wideband instrumentation signals [49–54]. However, relatively less work has been published on PIM and PIWM. PIM and PIWM applied to wideband fiber transmission can be found in [44, 55–58]. Some work on digital PPM has been aimed primarily at long-haul fiber applications [59, 60].

In this section, we describe PTM formats and their spectra and summarize their performance using the results presented in the review paper by Wilson [31].

2.7.1 Formats of Pulse Time Modulation

Figure 2.16 shows PWM and PPM. This group is called *isochronous*, meaning all of them have the same duration for a symbol at all times. Figures 2.16(a, b) are for analog signals. In PWM, the pulse width of the pulse carrier in a fixed symbol duration is proportional to the amplitude of the signal sample. As shown in Figure 2.16(a), in this example, the half of the symbol width (T_s) corresponds to 0 sampled value (first pulse), and more than half of T_s corresponds to a positive sample and less than half of T_s corresponds to a negative sample. In PPM, the delay of a narrow pulse with respect to the symbol's start is proportional to the amplitude of the signal

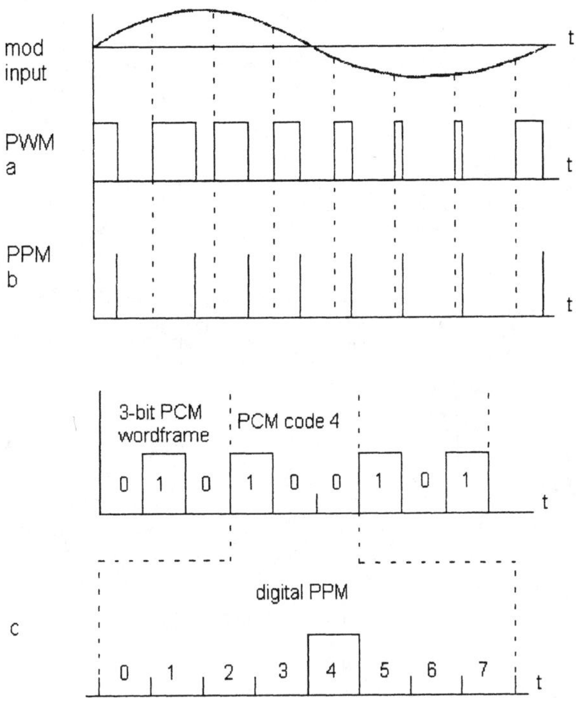

a PWM
b PPM
c Digital PPM

Figure 2.16 Isochronous PTM technique. From [31]. Copyright © 1993 IEE.

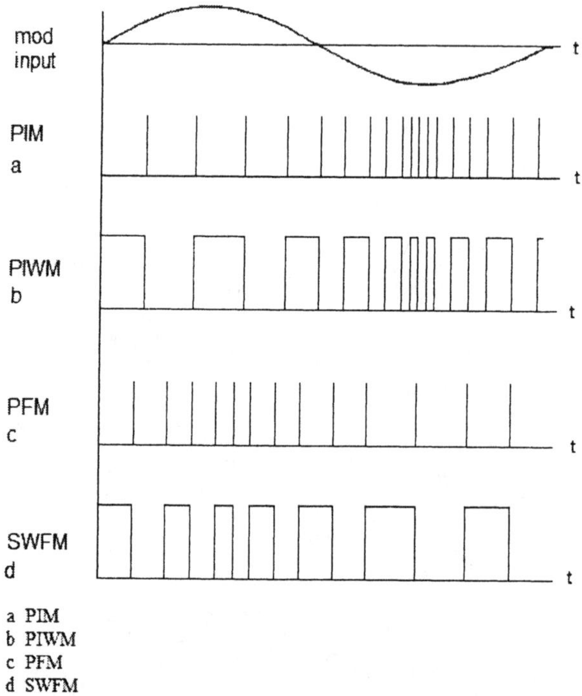

Figure 2.17 Anisochronous PTM technique. From [31]. Copyright © 1993 IEE.

sample. As shown in Figure 2.16(b), a delay of a half of T_s corresponds to a 0 sample, and a delay of more than half of T_s corresponds to a positive sample and less than half of T_s corresponds to a negative sample. In fact the delayed narrow pulse in PPM is exactly at the trailing edge of the corresponding PWM pulse.

Figure 2.16(c) is for a digital PCM signal where the expanded figure shows that a PCM code with a value of 4 is represented by a pulse at the 4th chip of a PPM symbol.

Figure 2.17 shows another group of different PTM techniques. This group includes PIM, PIWM, PFM and SWFM. This group is *anisochronous* for they have variable symbol durations. In PIM, the intervals of pulses are proportional to the signal amplitudes (Figure 2.17(a)). Again, in this example, 0 amplitude is represented by an interval of a half symbol width. The longest interval corresponds to the highest positive amplitude and the shortest interval corresponds to the lowest (the most negative) amplitude. PIWM is constructed by converting a PIM signal into a square wave sequence using mark and space alternatively, with the edges corresponding to the PIM pulses (Figure 2.17(b)). In PFM, the frequency of the narrow pulse train is proportional to the signal amplitude (Figure 2.17(c)). SWFM is derived from PFM with the same manner as converting PIM to PIWM (Figure 2.17(d)). SWFM is closely related to FM (frequency modulation) with edges occurring at zero-crossing points of the FM signal.

From above we can see that PTM for analog signal is neither a conventional analog modulation since modulated signal is discrete in time, nor is it a conventional digital modulation since each pulse does not represent a bit or a group of bits of a digital sequence. Therefore Wilson called it a "pulse analog" modulation [31].

For the digital PPM in Figure 2.16(c), the modulation is digital in the sense that a pulse at a certain position is used to represent a group of digital bits. A PCM code of k bits is represented by a narrow pulse at one of the $L = 2^k$ positions. The bandwidth expansion ratio is $2^k/k$. Compared to PCM, digital PPM can achieve 5–11 dB improvement in sensitivity at the cost of transmission bandwidth [59–61]. However, the bandwidth expansion could be substantial. For example, for $k = 7$, the expansion ratio is about 18.28. To overcome this problem, newer digital PPMs are being proposed, such as the dicode PPM in [61]. Digital PPM is aimed primarily for long-haul fiber backbone and deep space communications where it is desirable to trade transmission bandwidth with repeater spacing, and where the cost of coding complexity can be readily absorbed.

A PWM (and PPM) signal can be generated by uniform sampling or natural sampling. A naturally-sampled PWM signal can be generated by comparison of the modulating signal directly with a linear ramp waveform, or triangular in the case of double-edge modulation, in a comparator. In uniformly-sampled PWM, the input signal is first routed through a sample-and-hold circuit and then compared with the reference waveform. The sampling points of the naturally sampled PWM are at variable intervals, whereas they are evenly spaced in a uniformly sampled PWM.

In terms of pulse position, PWM and PPM signal can be leading-edge, trailing-edge, or double-edge modulated. Figure 2.16(a) shows a trailing-edge-modulated PWM, where the leading edge is fixed at the symbol's starting point, and the trailing edge position varies with modulating signal's amplitude at the time of sampling. A leading edge PWM will fix the trailing edge at symbol's ending point and vary the leading edge with modulating signal's amplitude at the time of sampling. A double edge PWM will fix the center of the pulse at the center of the symbol and vary the pulse width according to the modulating signal. A reference ramp with a positive slope will produce a trailing-edge PWM whereas a negative-slope ramp will produce a leading-edge PWM.

For PPM, the same terminology can be used as long as we consider the pulse's reference point is one edge and the pulse's position is another edge.

For the anisochronous group, there is also the distinction of being naturally or uniformly sampled. However, there is no such "edge" distinction since symbol durations are not fixed and there are no "empty" gaps in a symbol.

A modulation index M is defined for each PTM scheme. For all styles of PWM (and PPM), a modulation index M $(0 < M < 1)$ can be defined as ratio of the maximum pulse width over the symbol (or sampling) period. The maximum modulation $(M = 1)$ occurs when the positive peak amplitude produces a full width pulse and the negative peak amplitude produces a zero-width pulse. If $M < 1$, the positive peak amplitude will produce a pulse shorter than the symbol period, say the width is $T_s - \Delta T$, the negative peak amplitude will produce a pulse with a width of ΔT, where from definition of M, we can easily determine $\Delta T = (1 - M)T_s$. Similar argument applies to PPM with "pulse width" replaced by "pulse position." For PIM and PIWM, M $(0 < M < 1)$ is defined as the peak-to-peak modulating signal swing divided by twice the positive dc level which is used to shift the input signal before entering a comparator to compare with a linear ramp in the modulator. For PFM and SWFM, $M = \Delta f / f_m$ where Δf is the frequency deviation.

Spectra and distortion of PTM are influenced by M.

2.7.2 Spectra of Pulse Time Modulation

The spectra of PTM waveforms are given in the literature as a Fourier series decomposition of a pulse train modulated by a single sine wave, rather than assuming a random input signal and employing the autocorrelation of the process to obtain the power spectral density. In a broad range of applications, such as TV signal, this approach is justified since the analog TV signals display a dominant subcarrier.

Figures 2.18 and 2.19 show spectra of various PTM formats [31]. All figures

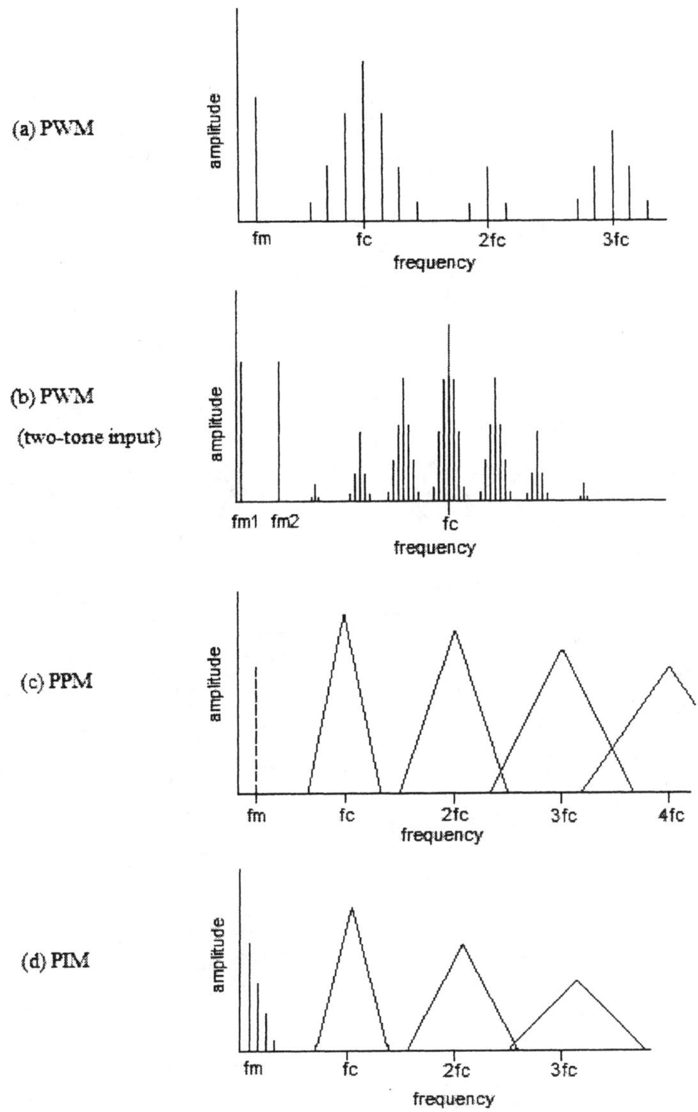

Figure 2.18 Spectra of (a, b) PWM, (c) PPM, and (d) PIM. Edited from [31]. Copyright © 1993 IEE.

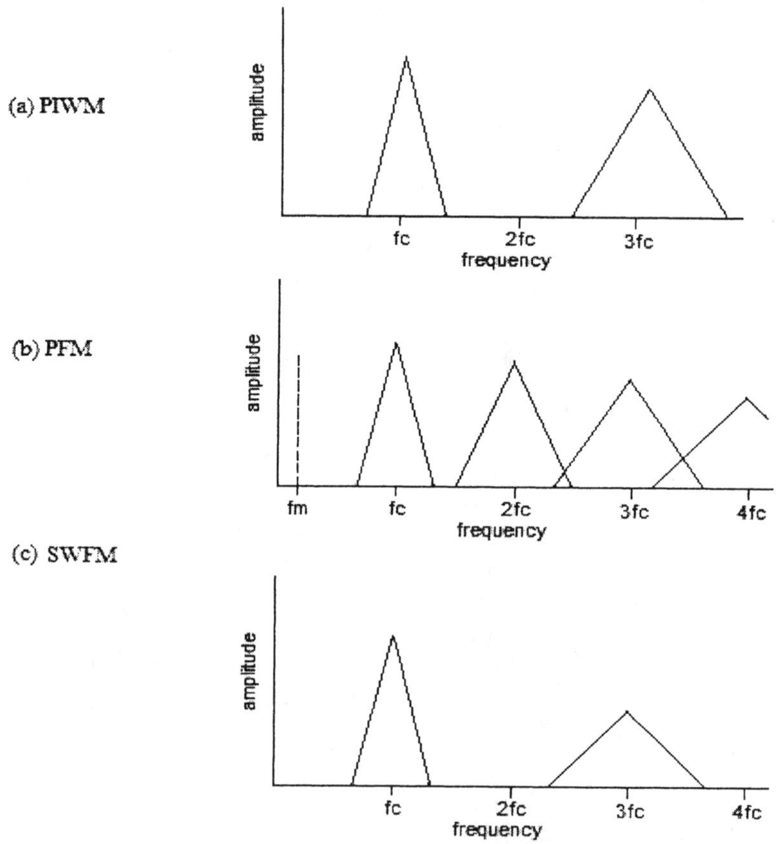

Figure 2.19 Spectra of (a) PIWM, (b) PFM, and (c) SWFM. Edited from [31]. Copyright © 1993 IEE.

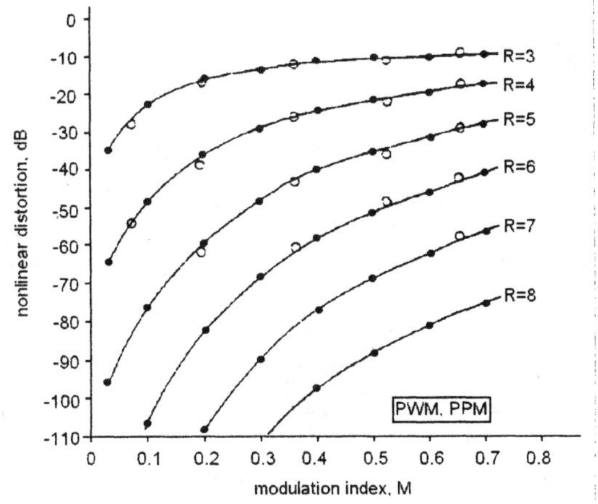

Figure 2.20 Distortion behavior of PWM and PPM. —•— theoretical, o experimental. From [31]. Copyright © 1993 IEE.

except for Figure 2.18(b) are for single-tone input modulating signal. Details of harmonics are shown for PWM. For other formats, only profiles of the harmonics are shown. A common feature of these spectra is that there are diminishing harmonics separated by f_m around diminishing harmonics of the carrier (sampling) frequency f_c. The baseband components exist for some formats whereas they are absent for others.

2.7.3 Performance of Pulse Time Modulation

Performance of PTM is measured in terms of a nonlinear distortion level which is defined as the total sidetone power over the modulating signal power. Distortion level reduces as sampling ratio $R = f_c/f_m$ increases, which is intuitively understandable, and increases as modulation index M increases. Figure 2.20 illustrates the distortion behavior of PWM and PPM. This figure can be recast to produce design curves in Figure 2.21, where the minimum required sampling ratio is plotted for a range of distortion levels as a function of modulation index. For a given distortion level and modulation index (or sampling ratio), one can choose a sampling ratio (or modulation index). Similarly, design curves can be obtained for PFM and SWFM as shown in Figure 2.22. SWFM is clearly superior to PFM in this respect. In turn, PFM is superior to PWM under similar modulation conditions, making SWFM the best choice in terms of minimum sampling ratio and bandwidth overhead.

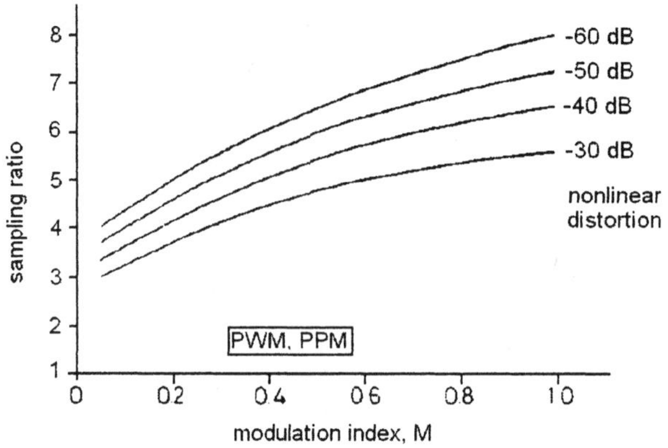

Figure 2.21 Sampling ratios for isochronous PTM. From [31]. Copyright © 1993 IEE.

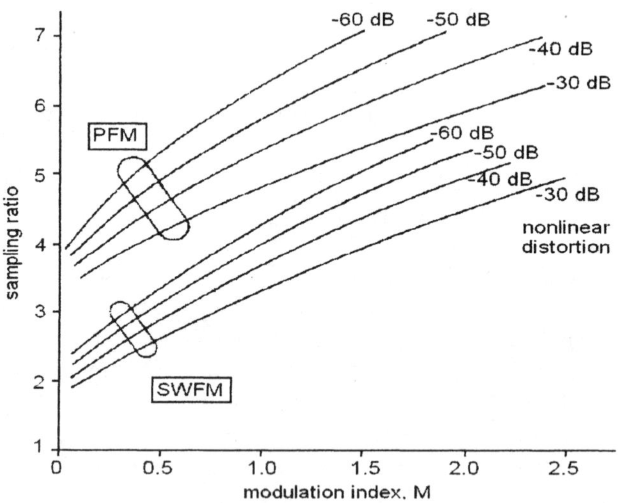

Figure 2.22 Sampling ratios for anisochronous PTM. From [31]. Copyright © 1993 IEE.

2.8 SUMMARY

In this chapter we first described the differential coding technique often used in line codes. Then before discussing various line codes we presented a list of line coding criteria. They include timing information, spectral characteristic, bandwidth, error probability, error detection capability, bit sequence independence, and use of differential coding. Armed with these criteria we took a close look at the classical line codes, including NRZ, RZ, PT, biphase, and delay modulation.

NRZ codes include NRZ-L, NRZ-M, and NRZ-S. NRZ-M and NRZ-S are differentially encoded forms of NRZ-L. They are the simplest in terms of coding rules and have a narrow bandwidth $(B_{null} = R_b)$, but lack most of the other desired characteristics. All NRZ codes have a similar BER with NRZ-L's the lowest and NRZ-M's and NRZ-S's slightly higher.

RZ codes increase the density of transitions which is good for timing recovery, but their bandwidth is doubled $(B_{null} = 2R_b)$. Polar-RZ has the same BER as that of NRZ-L. Both unipolar-NRZ and unipolar-RZ have the same BER which is 3 dB inferior than that of NRZ-L. Spectra of NRZ and RZ codes have major energy near dc, which is not suitable for ac-coupled circuits.

PT codes include AMI (NRZ and RZ), and dicodes (NRZ and RZ). AMI codes have a narrow bandwidth $(B_{null} = R_b)$, and most importantly, have no dc frequency component and near-dc components are also small. This makes AMI codes suitable for ac-coupled circuits. However, the three-level signaling of AMI codes makes BER performance about 3 dB worse than polar NRZ codes. Dicodes are related to AMI by differential encoding. So they have the same spectra as those of AMI codes and their BER performance is also close to but slightly higher than that of AMI codes. The lack of transitions in a string of 0s in AMI codes and dicodes may cause synchronization problems. Thus AMI codes with zero extraction were proposed to overcome this disadvantage.

Biphase codes include Bi-Φ-L, or Manchester, Bi-Φ-M, Bi-Φ-S, and conditioned (differential) Manchester. All of them have the same spectrum which has a $B_{null} = 2R_b$. However, the spectral shape is better than RZ's in that the dc and near-dc components are eliminated. In terms of BER, Manchester and conditioned Manchester are similar. The BER of Manchester is identical to that of NRZ-L. The BER of conditioned Manchester is slightly higher. The BER performance of Bi-Φ-M and Bi-Φ-S are the same, which is 3 dB worse than that of Manchester. Biphase codes always have at least one transition in a symbol, thus timing information is adequate. Due to its superb merit, Manchester code, especially the differential form, is widely used.

From the above discussion we see that AMI and Manchester have stood out as the two most favorite codes. Both have small near-dc components. Manchester is 3 dB better than AMI in terms of BER, but AMI requires only half of the bandwidth required by Manchester. Manchester also provides better timing information than AMI.

Delay modulation has a very narrow bandwidth with a main lobe of about $0.5R_b$. However, its BER performance is 3 dB worse than that of NRZ-L. Its timing information is not as much as that of biphase codes, but is more than what other codes can provide. It is a potential competitor to AMI and Manchester, even though its practical use has not been reported in literature.

In the latter part of this chapter we discussed more complex codes, including substitution codes and block codes. Substitution codes are designed for suppression of long all-zero strings. Two important substitution codes, BNZS and HDBn (including CHDBn) are discussed. All of them are based on AMI codes. By suppressing consecutive zeros, the lack of timing information of original AMI codes are overcome. Thus these codes are very competitive against Manchester, especially in bandlimited systems.

Block codes are designed to introduce redundancy to meet one or more of the desirable characteristics of a code. Their drawback is the increase in the transmission rate. They are widely used in optical fiber transmission systems where wide bandwidth is available. There are a great number of block codes. Important codes which are used in practical systems include the following codes: CMI and DMI which are two-level AMI codes, are designed to replace AMI in optical fiber system. A general class of binary block codes which can replace AMI are the mBnB codes. For high-speed optical transmission, simple coder and decoder are desirable and rate increase should be kept at minimum, thus simple codes in the class, like mB(m+1)B, mB1C, and DmB1M are proposed. All of them only add one redundant bit to the original bit block, thus keep the rate increase at the minimum. All of them have simple coding and decoding rules, thus high speed can be achieved.

Block codes are not limited to binary. Multiple-level block codes can increase the efficiency of the line codes. Popular multiple-level codes are ternary codes such as 4B3T, MS43, and 6B4T. They are designed to have a zero or constant dc component and bounded disparity. They may not be used for optical systems due to multiple-level signaling.

At the end of this chapter, we introduce pulse time modulation. It is a family of baseband modulations between pure analog and pure digital. It is mainly for optical transmission due to its binary nature. They all have similar spectral structure with diminishing harmonics separated by modulating frequency around diminishing harmonics of sampling frequency. Of course the speed of diminishing is slower when the pulse is narrower such as for PPM, PIM, and PFM, thus they require a larger transmission bandwidth. However these narrow pulse schemes save transmission power substantially. PWM and PPM have a fixed symbol duration so that they are easier to multiplex and demultiplex in time. PFM, and particularly SWFM, suffer from less distortion for the same modulation conditions (sampling ratio and modulation index).

References

[1] Aaron, M. R., "PCM transmission in the exchange plant," *Bell System Technical Journal,* vol. 41, no. 99, January 1962.

[2] Croisier, A., "Introduction to pseudoternary transmission codes," *IBM J. Res. Develop.*, July 1970, pp. 354–367.

[3] Hecht, M., and A. Guida, "Delay modulation," *Proc. IEEE*, vol. 57, no. 7, July 1969, pp. 1314–1316.

[4] Johannes, V. I., A. G. Kaim, and T. Walzman, "Bipolar pulse transmission with zero extraction," *IEEE Trans. Comm. Tech.* vol. 17, 1969, p. 303.

[5] Carter, R. O., "Low disparity binary coding system," *Electron. Lett.*, vol. 1, no. 3, 1965, p. 67.

[6] Takasaki, Y., et al., "Optical pulse formats for fiber optic digital communications," *IEEE Trans. Comm.* vol. 24, April 1976, pp. 404–413.

[7] Takasaki, Y., et al., "Two-level AMI line coding family for optic fiber systems," *Int. J. Electron.*, vol. 55, July 1983, pp. 121–131.

[8] Yoshikai, N., K. Katagiri, and T. Ito, "mB1C code and its performance in an optical communication system," *IEEE Trans. Comm.* vol. 32, no. 2, 1984, pp. 163–168.

[9] Kawanish, S., et al., "DmB1M code and its performance in a very high-speed optical transmission system," *IEEE Trans. Comm.* vol. 36, no. 8, August 1988, pp. 951–956.

[10] Krzymien, W., "Transmission performance analysis of a new class of line codes for optical fiber systems," *IEEE Trans. Comm.* vol. 37, no. 4, April 1989, pp. 402–404.

[11] Fair, I. J. et al., "Guided scrambler: a new line coding technique for high bit rate fiber optic transmission systems," *IEEE Trans. Comm.* vol. 39, no. 2, February 1991, pp. 289–297.

[12] Sklar, B., *Digital Communications: Fundamentals and Applications,* Englewood Cliffs, New Jersey: Prentice Hall, 1988.

[13] Killen, H., *Fiber Optic Communications*, Englewood Cliffs, New Jersey: Prentice Hall, 1991.

[14] Stallings, W., "Digital signaling techniques," *IEEE Communications Magazine*, vol. 22, no.12, December 1984, pp. 21–25.

[15] Tsujii, S., and H. Kasai, "An algebraic method of computing the power spectrum of coded pulse trains," *Electron. Commun. Japan*, vol. 51-A, no. 7, 1968.

[16] Yoshikai, N., "The effects of digital random errors on the DMI code power spectrum," *IEEE Trans. Comm.* vol. 34, July 1986, pp. 713–715.

[17] Huggins, W. H., "Signal-flow graphs and random signals," *Proc. IRE*, vol. 45, January 1957, pp. 74–86.

[18] Van Trees, H. L., *Detection, Estimation, and Modulation Theory, Part I*, New York: John Wiley & Sons, Inc., 1968.

[19] Smith, D. R., *Digital Transmission Systems*, 2nd ed., New York: Van Nostrand Reinhold, 1993.

[20] CCITT Yellow Book, vol. III.3 *Digital Networks—Transmission Systems and Multiplexing Equipment*, Geneva, ITU, 1981.

[21] Waters, D. B.,"Line codes for metallic systems," *Int. J. Electron.*, vol. 55, July 1983, pp. 159–169.

[22] Personick, S. D., "Receiver design for digital fiber optic communication systems, I-II," *Bell Syst. Tech. J.*, vol. 52, July–August 1973, pp. 843–886.

[23] Chew, Y. H., and T. T. Tjhung, "Bit error rate performance for edge-detected CMI data transmission," *IEEE Trans. Comm.* vol. 41, June 1993, pp. 813–816.

[24] Morris, D. J.,"Code your fiber-optic data for speed, without losing circuit simplicity," *Electronic Design*, vol. 22, October 1978, pp. 84–91.

[25] Griffiths, J. M., "Binary code suitable for line transmission," *Electron. Lett.*, vol. 5, February 1969, p. 79.

[26] Bosotti, L., and G. Pirani, "A new signaling format for optical communications," *Electronics Letters*, vol. 14, no. 3, 1978, pp. 71–73.

[27] Takasaki, Y., et al., "Line coding plans for fiber optic communication systems, " *Proc. Int. Conf. Commun.*, 1975, paper 32E.

[28] Franaszek, P. A., "Sequence state coding for digital data transmission," *Bell Syst. Tech. J.*, vol. 47, 1968, pp. 143–157.

[29] Buchner, J. B., "Ternary line codes," *Philips Telecommun. Rev.*, vol. 34, 1976, pp. 72–86.

[30] Catchpole, R. J., "Efficient ternary transmission codes," *Electronics Letters*, vol. 11, 1975, pp. 482–484.

[31] Wilson, B., and Z. Ghassemlooy, "Pulse time modulation techniques for optical communications: a review," *IEE Proceedings-J*, vol. 140, no. 6, December 1993, pp. 346–357.

[32] Kahn, J. and J. Barry, "Wireless infrared communications," *Proceedings of IEEE*, vol. 85, no. 2, February 1997, pp. 265–298.

[33] Lynch, R., "Channels and codes for magneto-optical recording," *IEEE Journal on Selected Areas in Communications*, vol. 10, no. 1, January 1992, pp. 57–72.

[34] Hisayoshi S., and K. Nosu, "MPPM: a method for improving the band-utilization efficiency in optical PPM," *Journal of Lightwave Technology*, vol. 7, no. 3, March 1989, pp. 465–471.

[35] Shiu, D., and J. M. Kahn, "Differential pulse-position modulation for power-efficient optical communication," *IEEE Transactions on Communications*, vol. 47, no. 8, August 1999, pp. 1201–1210.

[36] Schrocks, C. B., "Proposal for a hub controlled cable television system using optical fiber," *IEEE Trans.*, CATV-4, 1979, pp. 70–77.

[37] Berry, M. C., "Pulse width modulation for optical fibre transmission," Ph.D. thesis, Nottingham University, England, 1983.

[38] Berry, M. C., and J. M. Arnold, "Pulse width modulation for optical fibre transmission of video." *IEE Int. Conf. on the Impact of VLSI Technology on Communication Systems*, London, 1983.

[39] Suh, S.Y., "Pulse width modulation for analog fiber-optic communications," *IEEE J. of Lightwave Technology*, vol. 5, no. 1, 1987, pp. 102–112.

[40] Wilson, B., and Z. Ghassemlooy, "Optical pulse width modulation for electrically isolated analogue transmission," *J. Phys. (E)*, vol. 18, 1985, pp. 954–958.

[41] Wilson, B., and Z. Ghassemlooy, "Optical PWM data link for high quality analogue and video signals," *J. Phys. (E)*, 1987, vol. 20, no. 7, pp. 841–845.

[42] Wilson, B., and Z. Ghassemlooy, "Optical fibre transmission of multiplexed video signals using PWM," *Int. J. Optoelectron.*, vol. 4, 1989, pp. 3–17.

[43] Heatley, D. J. T., "Video transmission in optical fibre local networks using pulse time modulation." *ECOC 83 – 9th European Conference on Optical Communication*, Geneva, September 1983, pp. 343–346.

[44] Okazaki, A., "Still picture transmission by pulse interval modulation," *IEEE Transactions on Cable Television*, 1979, CATV-4, pp. 17–22.

[45] Heatley, D. J. T., "Unrepeatered video transmission using pulse frequency modulation over 100 km of monomode optical fibre," *Electron. Lett.*, vol. 18, 1982, pp. 369–371.

[46] Heatley, D. J. T. and T. G. Hodgkinson,"Video transmission over cabled monomode fibre at 1.5 pm using PFM with 2-PSK heterodyne detection," *Electron. Lett.*, vol. 20, 1984, pp, 110–112.

[47] Heker, S. F., G. J. Herskowitz, H. Grebel, and H. Wichansky, "Video transmission in optical fiber communication systems using pulse frequency modulation," *IEEE Trans. Commun.*, vol. 36, no. 2, 1988, pp. 191–194.

[48] Kanada, T., K. Hakoda, and E. Yoneda, "SNR fluctuation and non-linear distortion in PFM optical NTSC video transmission systems," *IEEE Trans. Commun.,* COM-30, no. 8, 1982, pp. 1868–1875.

[49] Lu, C., "Optical transmission of wideband video signals using SWFM," Ph.D. thesis, University of Manchester Institute of Science and Technology, Manchester, England, 1990.

[50] Pophillat, L., "Video transmission using a 1.3 pm LED and monomode fiter," *10th European Conf. on Optical Communications*, Stuttgart, W. Germany, 1984, pp. 238–239.

[51] Sato, K., S. Aoygai, and T. Kitami, "Fiber optic video transmission employing square wave frequency modulation," *IEEE Trans. Commun.*, COM-33, no. 5, 1985, pp. 417–423.

[52] Wilson, B., et al., "Optical squarewave frequency modulation for wideband instrumentation and video signals," *IEE Colloquium on Analogue Optical Communications,* London, 1989, Digest 1989/165, Paper 9.

[53] Wilson, B., Z. Ghassemlooy, and C. Lu, "Optical fibre transmission of high-definition television signals using squarewave frequency modulation," *Third Bangor Symposium on Communications,* University of Wales, Bangor, May 1991, pp. 258–262.

[54] Wilson, B., Z. Ghassemlooy, and C. Lu, "Squarewave FM optical fibre transmission for high definition television signals," *(Fibre Optics 90, 1990, London), Proc. Int. Soc. Optical Eng.*, vol. 1314, 1990, pp. 90–97.

[55] Okazaki, A., "Pulse interval modulation applicable to narrowband transmission," *IEEE Trans.,* 1978, CATV-3, pp. 155–164.

[56] Sato, M., M. Murata, and T. Namekawa, "Pulse interval and width modulation for video transmission," *IEEE Transactions on Cable Television,* CATV-3, no. 4, 1978, pp. 166–173.

[57] Sato, M., M. Murata, and T. Namekawa, "A new optical communication system using the pulse interval and width modulated code," *IEEE Trans.*, CATV-4, no. 1, 1979, pp. 1–9.

[58] Wilson, B., Z. Ghassemlooy, and J. C. S. Cheung, "Spectral predictions for pulse interval and width modulation," *Electron. Lett.,* vol. 27, no. 7, 1991, pp. 580–581.

[59] Calvert, N. M., M. J. N. Sibley, and R. T. Unwin, "Experimental optical fibre digital pulse position modulation system," *Electron. Lett.*, vol. 24, no. 2, 1988, pp. 129–131.

[60] Cryan, R. A., et al., "Optical fibre digital pulse-position-modulation assuming a Gaussian received pulse shape," *IEE Proc. J.*, vol. 137, 1990, pp. 89–96.

[61] Sibley, M. J. N., "Dicode pulse-position modulation: a novel coding scheme for optical-fibre communications," *IEE Proc.-Optoelectron.*, vol. 150, no. 2, April 2003, pp. 125–131.

Selected Bibliography

- "Binary line codes for digital transmission on optical fibers," *CCITT contrib. Study Group XVIII*, no. 291, 1979.
- Brooks, R. M., and Jessop A., "Coding for optical systems," *Int. J. Electron.*, vol. 55, July 1983, pp. 81–120.
- Couch II, L. W., *Digital and Analog Communication Systems*, 3rd ed., New York: Macmillan, 1990.
- Haykin, S., *Digital Communications*, New York: John Wiley & Sons, Inc., 1988.

Chapter 3

Frequency Shift Keying

We have seen in chapter one that there are three basic forms of digital bandpass modulations: amplitude shift keying (ASK), frequency shift keying (FSK), and phase shift keying (PSK). FSK is probably the earliest type of digital modulation used in the communication industry. In this chapter we first describe in Section 3.1 binary FSK signal, modulator, and its power spectrum density. Then we present the coherent demodulator and error probability in Section 3.2. Next we discuss noncoherent demodulation and error probability in Section 3.3. M-ary FSK (MFSK) is given in Section 3.4. Section 3.5 discusses FSK demodulators using a conventional discriminator and other simple techniques. Section 3.6 is a brief discussion of synchronization. Finally we summarize the chapter with Section 3.7.

3.1 BINARY FSK

3.1.1 Binary FSK Signal and Modulator

In its most general form, the binary FSK scheme uses two signals with different frequencies to represent binary 1 and 0.

$$\begin{aligned} s_1(t) &= A\cos(2\pi f_1 t + \Phi_1), \quad kT \le t \le (k+1)T, \text{ for } 1 \\ s_2(t) &= A\cos(2\pi f_2 t + \Phi_2), \quad kT \le t \le (k+1)T, \text{ for } 0 \end{aligned} \tag{3.1}$$

where Φ_1 and Φ_2 are initial phases at $t = 0$, and T is the bit period of the binary data. These two signals are not coherent since Φ_1 and Φ_2 are not the same in general. The waveform is not continuous at bit transitions. This form of FSK is therefore called *noncoherent* or *discontinuous* FSK. It can be generated by switching the modulator output line between two different oscillators as shown in Figure 3.1. It can be noncoherently demodulated.

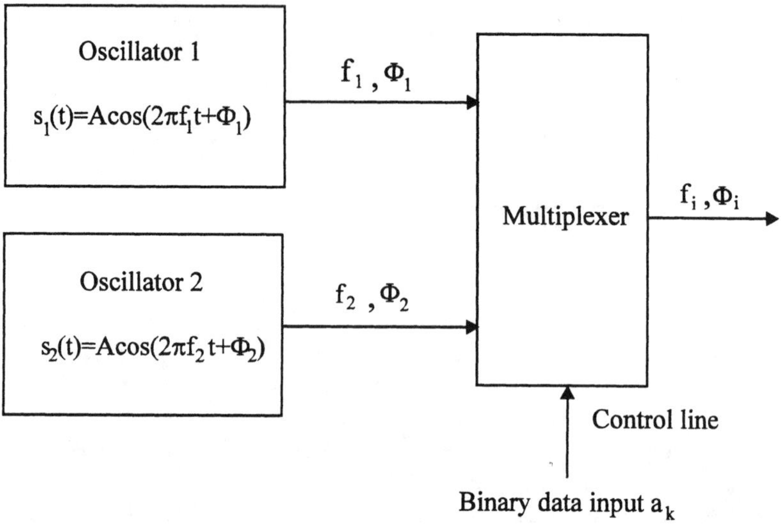

Figure 3.1 Noncoherent FSK modulator.

The second type of FSK is the *coherent* one where two signals have the same initial phase Φ at $t = 0$:

$$\begin{aligned} s_1(t) &= A\cos(2\pi f_1 t + \Phi), \quad kT \leq t \leq (k+1)T, \text{ for } 1 \\ s_2(t) &= A\cos(2\pi f_2 t + \Phi), \quad kT \leq t \leq (k+1)T, \text{ for } 0 \end{aligned} \tag{3.2}$$

This type of FSK can be generated by the modulator as shown in Figure 3.2. The frequency synthesizer generates two frequencies, f_1 and f_2, which are synchronized. The binary input data controls the multiplexer. The bit timing must be synchronized with the carrier frequencies. The details will be discussed shortly. If a 1 is present, $s_1(t)$ will pass and if a 0 is present, $s_2(t)$ will pass. Note that $s_1(t)$ and $s_2(t)$ are always there regardless of the input data. So when considering their phase in any bit interval $kT \leq t \leq (k+1)T$, the starting point of time is 0, not kT.

For coherent demodulation of the coherent FSK signal, the two frequencies are chosen so that the two signals are orthogonal:

$$\int_{kT}^{(k+1)T} s_1(t)s_2(t)dt = 0$$

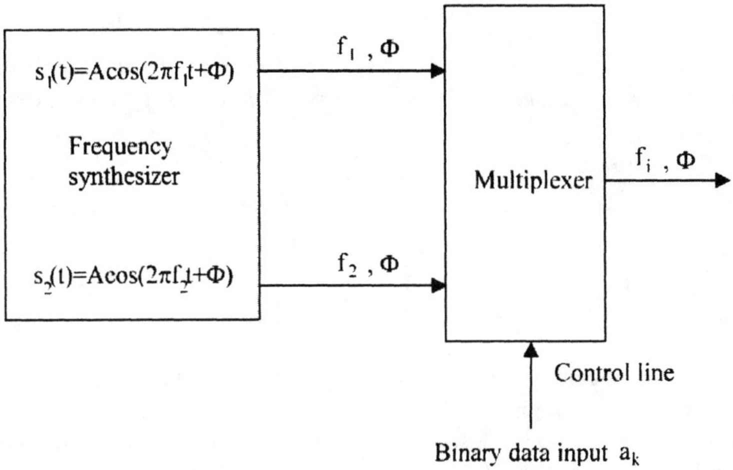

Figure 3.2 Coherent FSK modulator.

That is

$$
\begin{aligned}
& \int_{kT}^{(k+1)T} \cos(2\pi f_1 t + \Phi)\cos(2\pi f_2 t + \Phi)dt \\
= \; & \frac{1}{2}\int_{kT}^{(k+1)T} [\cos[2\pi(f_1+f_2)t + 2\Phi] + \cos 2\pi(f_1 - f_2)t]dt \\
= \; & \frac{1}{4\pi(f_1+f_2)}[\cos 2\Phi \sin 2\pi(f_1+f_2)t + \sin 2\Phi \cos 2\pi(f_1+f_2)t]\Big|_{kT}^{(k+1)T} \\
& + \frac{1}{4\pi(f_1-f_2)} \sin 2\pi(f_1-f_2)t]\Big|_{kT}^{(k+1)T} \\
= \; & 0
\end{aligned}
$$

This requires that $2\pi(f_1+f_2)T = n\pi$ and $2\pi(f_1-f_2)T = m\pi$, where n and m are integers. This leads to

$$f_1 = \frac{n+m}{4T}$$

$$f_2 = \frac{n-m}{4T}$$

$$2\Delta f = f_1 - f_2 = \frac{m}{2T}$$

Thus we conclude that for orthogonality f_1 and f_2 must be integer multiple of $1/4T$ and their difference must be integer multiples of $1/2T$. Using Δf we can rewrite the two frequencies as

$$f_1 = f_c + \Delta f$$

$$f_2 = f_c - \Delta f$$

$$f_c = \frac{f_1 + f_2}{2} = \frac{n}{2T}$$

where f_c is the nominal (or apparent) carrier frequency which must be an integer multiple of $1/2T$ for orthogonality.

When the separation is chosen as $1/T$, then the phase continuity will be maintained at bit transitions and the FSK is called *Sunde's FSK*. It is an important form of FSK and will be discussed in detail in this chapter. As a matter of fact, if the separation is k/T, where k is an integer, the phase of the coherent FSK signal of (3.2) is always continuous.

Proof: at $t = nT$, the phase of $s_1(t)$ is

$$\begin{aligned} 2\pi f_1 nT + \Phi &= 2\pi(f_2 + k/T)nT + \Phi \\ &= 2\pi f_2 nT + 2\pi kn + \Phi \\ &= 2\pi f_2 nT + \Phi \text{ (Modulo-}2\pi) \end{aligned}$$

which is exactly the phase of $s_2(t)$. Thus at $t = nT$, if the input bit switches from 1 to 0, the new signal $s_2(t)$ will start at exactly the same amplitude where $s_1(t)$ has ended.

The minimum separation for orthogonality between f_1 and f_2 is $1/2T$. As we have just seen above, this separation cannot guarantee continuous phase. A particular form of FSK called minimum shift keying (MSK) *not only* has the minimum separation *but also* has continuous phase. However, MSK is much more than an ordinary FSK, it has properties that an ordinary FSK does not have. It must be generated by methods other than the one described in Figure 3.2. MSK is an important modulation scheme which will be covered in Chapter 5.

Figure 3.3(a) is an example of Sunde's FSK waveform where bit 1 corresponds to a higher frequency f_1 and bit 0 a lower f_2. Since f_1 and f_2 are multiples of $1/T$, the ending phase of the carrier is the same as the starting phase, therefore the waveform has continuous phase at the bit boundaries. Sunde's FSK is a continuous phase FSK.

A coherent FSK waveform might have discontinuous phase at bit boundaries. Figure 3.3(b) is an example of such a waveform, where $f_1 = 9/4T$, $f_2 = 6/4T$, and

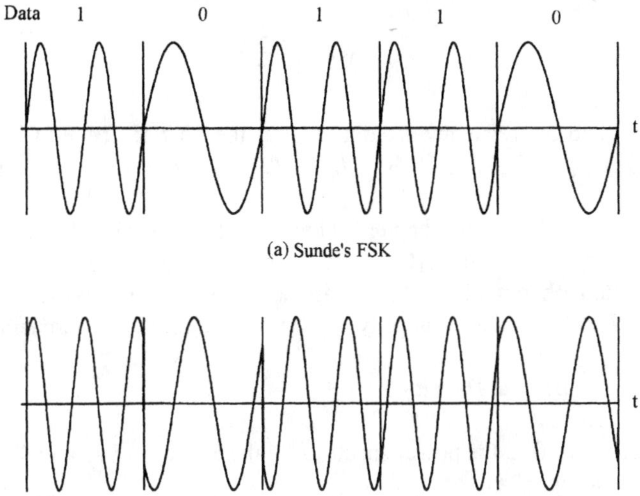

(a) Sunde's FSK

(b) FSK signal with discontinuous phase

Figure 3.3 FSK waveforms: (a) Sunde's FSK, $f_1 = 2/T$, $f_2 = 1/T$, $2\Delta f = 1/T$, and (b) FSK with discontinuous phase, $f_1 = 9/4T$, $f_2 = 6/4T$, $2\Delta f = 3/4T$.

the separation is $2\Delta f = 3/4T$.

Discontinuity in waveform will broaden the signal bandwidth. Therefore the third type of FSK, continuous phase FSK (CPFSK), is desired. To ensure continuous phase in the FSK, the signal must be in the form

$$s(t) = A\cos(2\pi f_c t + \frac{\pi h a_k (t - kT)}{T} + \pi h \sum_{i=0}^{k-1} a_i), \qquad kT \leq t \leq (k+1)T \quad (3.3)$$

where h is called the modulation index and $a_k = \pm 1$ is the kth data bit. Logic 0 and 1 correspond to binary data $+1$ and -1, respectively. In the parentheses, the second term represents the linearly changing phase or a constant frequency deviation required by an FSK signal. The third term represents the accumulated phase. Since the phase increment is proportional to $\Delta t = t - kT$, the phase is continuous in $(kT, (k+1)T)$. When $t = (k+1)T$, the phase accumulation is increased by $\pi h a_k$ and as t increases further, the phase changes continuously again. The frequency deviation is the derivative of the second term divided by 2π: $\Delta f = |ha_k/2T| = h/2T$. Thus

$$h = 2\Delta f T = \frac{2\Delta f}{R_b}$$

which is the ratio of frequency separation over the bit rate $R_b = 1/T$. The two frequencies are $f_1 = f_c - h/2T$, $f_2 = f_c + h/2T$. When $h = 1$, it becomes Sunde's FSK.

We will revisit CPFSK in Chapter 6 where it will be described from a perspective of continuous phase modulation (CPM). In this chapter we will concentrate on simple coherent and noncoherent FSK schemes. In the following we always assume that the initial phase $\Phi = 0$. This will simplify derivations without loss of generality.

3.1.2 Power Spectral Density

Now we proceed to find the power spectrum of the Sunde's FSK signal. We expand the Sunde's FSK signal as

$$\begin{aligned} s(t) &= A\cos 2\pi(f_c + a_k \frac{1}{2T})t \\ &= A\cos(a_k \frac{\pi t}{T})\cos(2\pi f_c t) - A\sin(a_k \frac{\pi t}{T})\sin(2\pi f_c t) \\ &= A\cos(\frac{\pi t}{T})\cos(2\pi f_c t) - Aa_k\sin(\frac{\pi t}{T})\sin(2\pi f_c t) \\ kT &\leq t \leq (k+1)T \end{aligned} \tag{3.4}$$

where the last expression is derived using the fact that $a_k = \pm 1$. The inphase component $A\cos(\frac{\pi t}{T})$ is independent of the data. The quadrature component $Aa_k\sin(\frac{\pi t}{T})$ is directly related to data. The inphase and quadrature components are independent of each other.

In Appendix A we have shown that the PSD of a bandpass signal

$$s(t) = \text{Re}[\widetilde{s}(t)e^{j2\pi f_c t}]$$

is the shifted version of the equivalent baseband signal or complex envelope $\widetilde{s}(t)$'s PSD $\Psi_{\widetilde{s}}(f)$

$$\Psi_s(f) = \frac{1}{2}[\Psi_{\widetilde{s}}(f - f_c) + \Psi_{\widetilde{s}}^*(-f - f_c)]$$

where $*$ denotes a complex conjugate. Therefore it suffices to determine the PSD of the equivalent baseband signal $\widetilde{s}(t)$. Since the inphase component and the quadrature component of the FSK signal of (3.4) are independent of each other, the PSD for the complex envelope is the sum of the PSDs of these two components.

$$\Psi_{\tilde{s}}(f) = \Psi_I(f) + \Psi_Q(f)$$

$\Psi_I(f)$ can be found easily since the inphase component is independent of data. It is defined on the entire time axis. Thus

$$\Psi_I(f) = |\mathcal{F}\{A\cos(\frac{\pi t}{T})\}|^2 = \frac{A^2}{4}[\delta(f - \frac{1}{2T}) + \delta(f + \frac{1}{2T})]$$

where $\mathcal{F}$ stands for Fourier transform. It is seen that the spectrum of the inphase part of the FSK signal are two delta functions.

To find $\Psi_Q(f)$, refer to (A.19) of Appendix A. It shows that the PSD of a binary, bipolar (± 1), equiprobable, stationary, and uncorrelated digital waveform is just equal to the energy spectral density of the symbol shaping pulse divided by the symbol duration. The shaping pulse here is $A\sin(\frac{\pi t}{T})$, therefore

$$\begin{aligned} \Psi_Q(f) &= \frac{1}{T}|\mathcal{F}\{A\sin(\frac{\pi t}{T})\}|^2, \quad 0 \le t \le T \\ &= \frac{1}{T}\left(\frac{2AT[\cos \pi T f]}{\pi[1-(2Tf)^2]}\right)^2 \end{aligned}$$

The complete baseband PSD of the binary FSK signal is the sum of $\Psi_I(f)$ and $\Psi_Q(f)$:

$$\Psi_{\tilde{s}}(f) = \frac{A^2}{4}[\delta(f - \frac{1}{2T}) + \delta(f + \frac{1}{2T})] + T\left(\frac{2A[\cos \pi T f]}{\pi[1-(2Tf)^2]}\right)^2 \quad (3.5)$$

Figure 3.4(a, b) shows the baseband PSD of this Sunde's FSK for positive frequency only, where we set $A = \sqrt{2}$ in (3.5) for a unity signal energy. So the total energy under the PSD curve is 1 in one side and 2 in two sides. A spectral line arises at $fT = 0.5$. This means the passband spectrum should have spectral lines at $f = f_c \pm \frac{1}{2T}$ which are the two frequencies of binary FSK. The null bandwidth is $B_{null} = 1.5R_b$ in baseband, thus the null-to-null bandwidth at f_c is $3R_b$. Figure 3.4(c) is the out-of-band power, P_{ob} (dB), which is defined in Chapter 2. The abscissa of Figure 3.4(c) is the two-sided bandwidth normalized to data rate. The abrupt drop in P_{ob} is due to the delta function in the spectrum. The two-sided bandwidths at carrier frequency are $B_{90\%} \approx 1.23R_b$ and $B_{99\%} \approx 2.12R_b$. The transmission bandwidth thus is usually set as $B_T = 2R_b$.

The general spectral expression of CPFSK for index values other than 1 is more difficult to determine. It will be given in Section 3.4 for arbitrary index values and even M values for M-ary FSK. The derivation will be given in Chapter 6 in the context of continuous phase modulation.

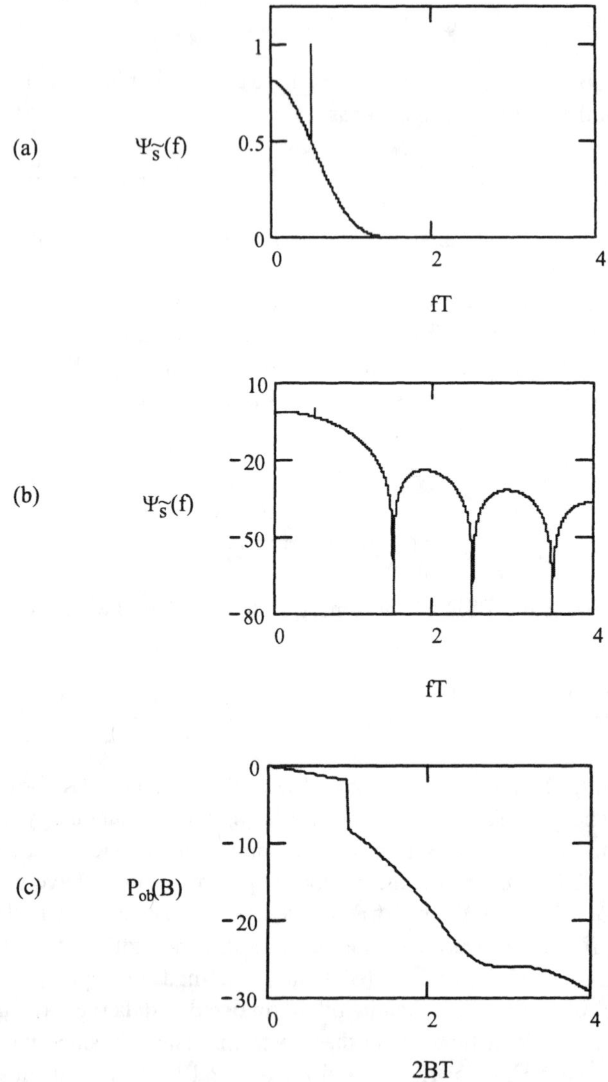

Figure 3.4 PSD of Sunde's FSK: (a) linear scale, (b) logarithmic scale (dB), and (c) out-of-band power (dB).

3.2 COHERENT DEMODULATION AND ERROR PERFORMANCE

The coherent demodulator for the coherent FSK signal falls in the general form of coherent demodulators described in Appendix B. The demodulator can be implemented with two correlators as shown in Figure 3.5, where the two reference signals are $\cos(2\pi f_1 t)$ and $\cos(2\pi f_2 t)$. They must be synchronized with the received signal. The receiver is optimum in the sense that it minimizes the error probability for equally likely binary signals. Even though the receiver is rigorously derived in Appendix B, some heuristic explanation here may help understand its operation. When $s_1(t)$ is transmitted, the upper correlator yields a signal l_1 with a positive signal component and a noise component. However, the lower correlator output l_2, due to the signals' orthogonality, has only a noise component. Thus the output of the summer is most likely above zero, and the threshold detector will most likely produce a 1. When $s_2(t)$ is transmitted, opposite things happen to the two correlators and the threshold detector will most likely produce a 0. However, due to the noise nature that its values range from $-\infty$ to ∞, occasionally the noise amplitude might overpower the signal amplitude, then detection errors will happen.

An alternative to Figure 3.5 is to use just one correlator with the reference signal $\cos(2\pi f_1 t) - \cos(2\pi f_2 t)$ (Figure 3.6). The correlator in Figure 3.6 can be replaced by a matched filter that matches $\cos(2\pi f_1 t) - \cos(2\pi f_2 t)$ (Figure 3.7). All implementations are equivalent in terms of error performance (see Appendix B).

Assuming an AWGN channel, the received signal is

$$r(t) = s_i(t) + n(t), \qquad i = 1, 2$$

where $n(t)$ is the additive white Gaussian noise with zero mean and a two-sided power spectral density $N_o/2$. From (B.33) the bit error probability for any equally likely binary signals is

$$P_b = Q\left(\sqrt{\frac{E_1 + E_2 - 2\rho_{12}\sqrt{E_1 E_2}}{2N_o}}\right)$$

where $N_o/2$ is the two-sided power spectral density of the additive white Gaussian noise. For Sunde's FSK signals $E_1 = E_2 = E_b$, $\rho_{12} = 0$ (orthogonal), thus the error probability is

$$P_b = Q\left(\sqrt{\frac{E_b}{N_o}}\right) \tag{3.6}$$

where $E_b = A^2T/2$ is the average bit energy of the FSK signal. The above P_b is plotted in Figure 3.8 where P_b of noncoherently demodulated FSK, whose expression

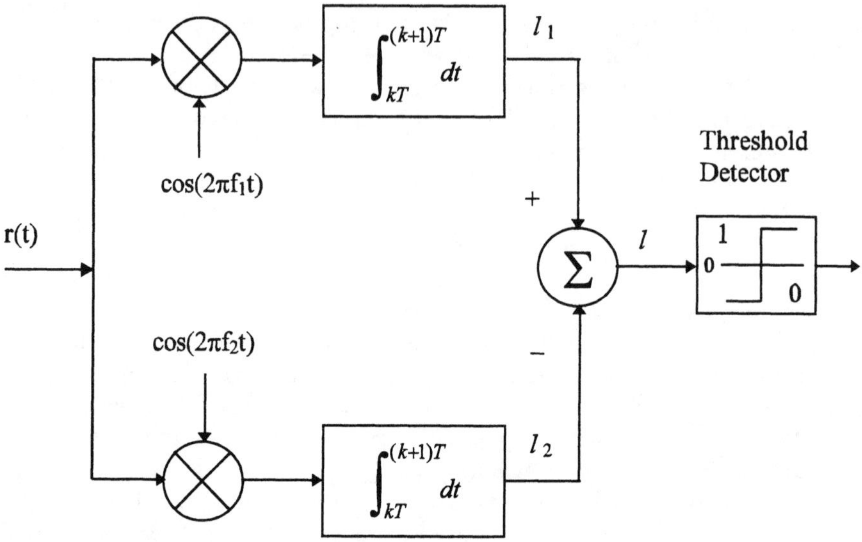

Figure 3.5 Coherent FSK demodulator: correlator implementation.

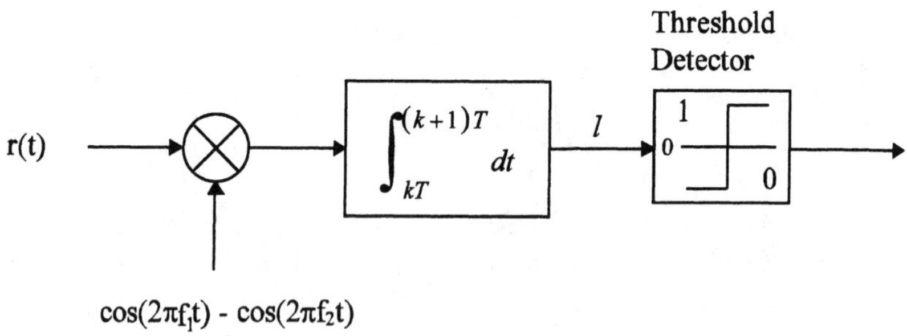

Figure 3.6 Coherent FSK demodulator: one correlator implementation.

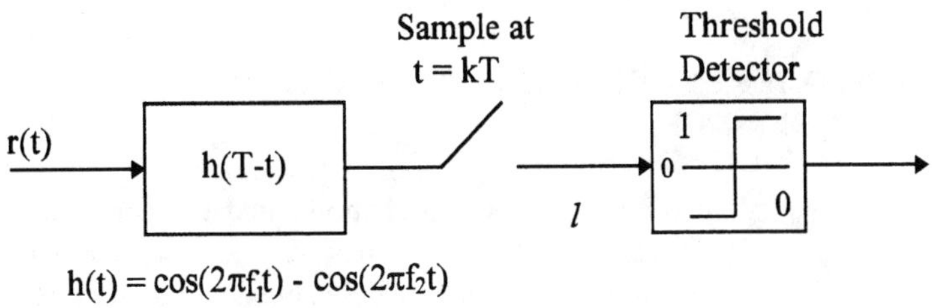

Figure 3.7 Coherent FSK demodulator: matched filter implementation.

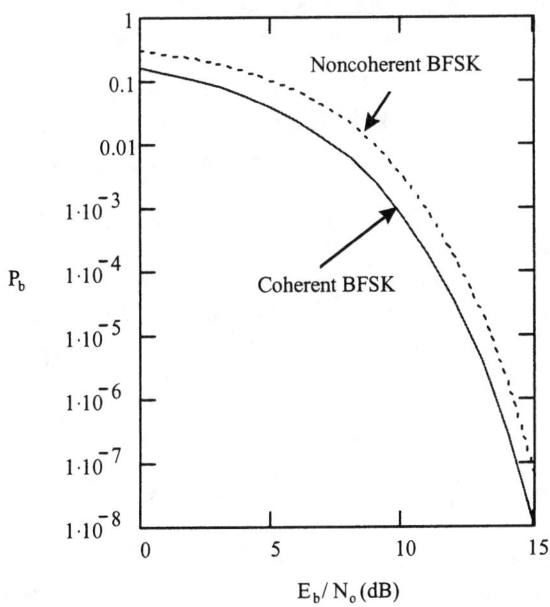

Figure 3.8 P_b of coherently and noncoherently demodulated FSK signal.

will be given shortly, is also plotted for comparison.

3.3 NONCOHERENT DEMODULATION AND ERROR PERFORMANCE

Coherent FSK signals can be noncoherently demodulated to avoid the carrier recovery. Noncoherently generated FSK can only be noncoherently demodulated. We refer to both cases as noncoherent FSK. In both cases the demodulation problem becomes a problem of detecting signals with unknown phases. In Appendix B we have shown that the optimum receiver is a quadrature receiver. It can be implemented using correlators or equivalently, matched filters.

Here we assume that the binary noncoherent FSK signals are equally likely and with equal energies. Under these assumptions, the demodulator using correlators is shown in Figure 3.9. Again, like in the coherent case, the optimality of the receiver has been rigorously proved (Appendix B). However, we can easily understand its operation by some heuristic argument as follows.

The received signal (ignoring noise for the moment) with an unknown phase can be written as

$$\begin{aligned} s_i(t,\theta) &= A\cos(2\pi f_i t+\theta), \qquad i=1,2 \\ &= A\cos\theta\cos 2\pi f_i t - A\sin\theta\sin 2\pi f_i t \end{aligned}$$

The signal consists of an inphase component $A\cos\theta\cos 2\pi f_i t$ and a quadrature component $A\sin\theta\sin 2\pi f_i t\sin\theta$. Thus the signal is *partially* correlated with $\cos 2\pi f_i t$ and *partially* correlated with $\sin 2\pi f_i t$. Therefore we use two correlators to collect the signal energy in these two parts. The outputs of the inphase and quadrature correlators will be $\frac{AT}{2}\cos\theta$ and $\frac{AT}{2}\sin\theta$, respectively. Depending on the value of the unknown phase θ, these two outputs could be anything in $(-\frac{AT}{2},\frac{AT}{2})$. Fortunately the squared sum of these two signals is not dependent on the unknown phase. That is

$$(\frac{AT}{2}\cos\theta)^2+(\frac{AT}{2}\sin\theta)^2=\frac{A^2T^2}{4}$$

This quantity is actually the signal part of the statistics l_i^2 when signal $s_i(t)$ is transmitted and noise is taken into consideration. When $s_i(t)$ is not transmitted the signal part of l_i^2 is 0. The comparator decides which signal is sent by checking these l_i^2.

The matched filter equivalence to Figure 3.9 is shown in Figure 3.10 which has the same error performance. For implementation simplicity we can replace the matched filters by bandpass filters centered at f_1 and f_2, respectively (Figure 3.11). However, if the bandpass filters are not matched to the FSK signals, degradation to

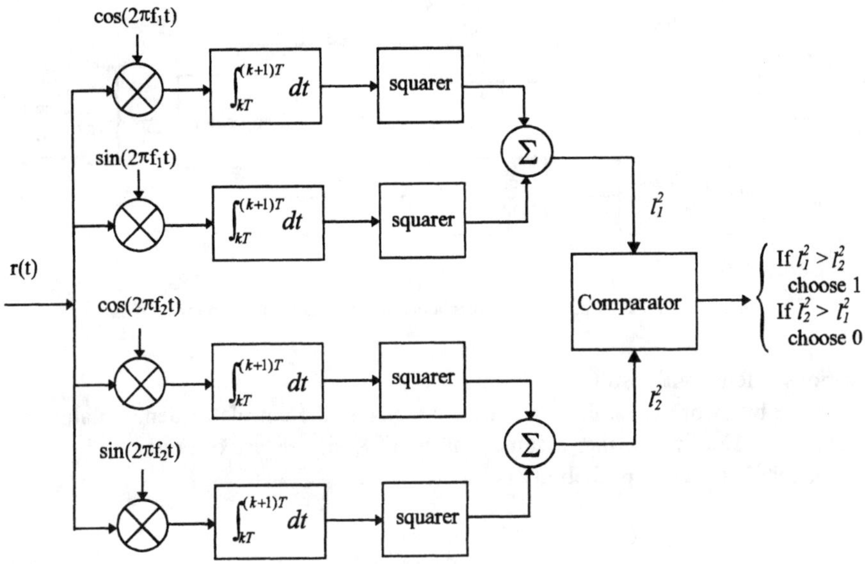

Figure 3.9 FSK noncoherent demodulator: correlator implementation.

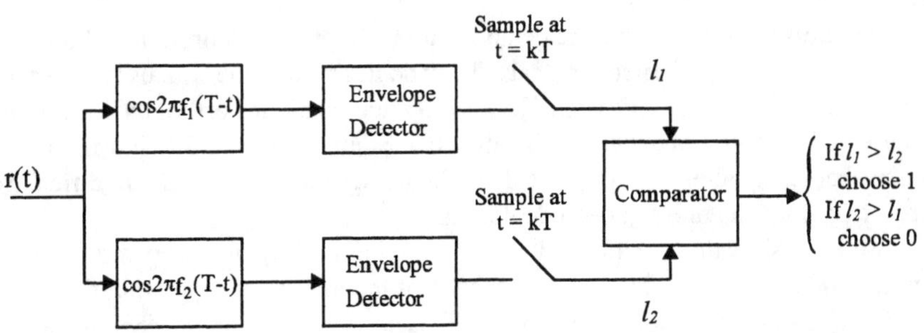

Figure 3.10 FSK noncoherent demodulator: matched filter implementation.

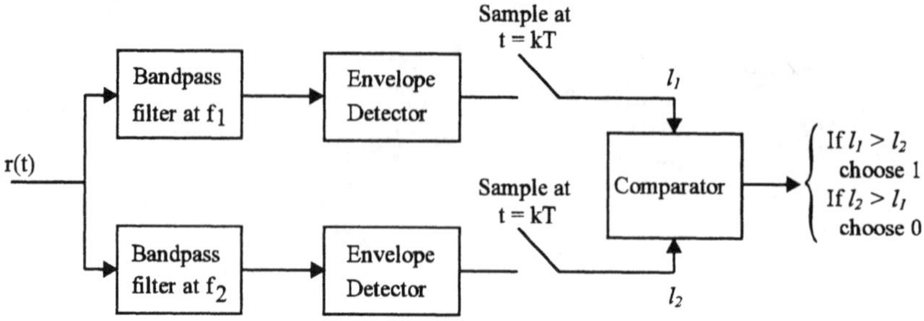

Figure 3.11 FSK noncoherent demodulator: bandpass filter implementation.

various extents will result.

The bit error probability can be derived using the correlator demodulator (Appendix B). Here we further assume that the FSK signals are orthogonal, then from Appendix B the error probability is

$$P_b = \frac{1}{2} e^{-E_b/2N_o} \tag{3.7}$$

The plot is shown in Figure 3.8. It is seen that the noncoherent FSK requires, at most, only 1 dB more E_b/N_o than that for coherent FSK for $P_b \leq 10^{-4}$. The noncoherent FSK demodulator is considerably easier to build since coherent reference signals need not be generated. Therefore in practical systems almost all of the FSK receivers use noncoherent demodulation.

It is worth reiterating that the demodulators in Figures 3.9 through 3.11 are good for equiprobable, equal-energy signals. They do not require the signals to be orthogonal. However, the P_b expression (3.7) is only applicable for orthogonal, equiprobable, equal-energy, noncoherent signals. If the noncoherent FSK signals are not equiprobable, equal-energy, then the demodulators need to be slightly modified. See Appendix B for the more general demodulators.

We have shown in Section 3.1 that the minimum frequency separation for coherent FSK signals is $1/2T$. We now show that the minimum separation for noncoherent FSK signals is $1/T$ instead of $1/2T$. The two noncoherent FSK signals are $s_1(t) = \cos 2\pi f_1 t$ and $s_2(t) = \cos(2\pi f_2 t + \Phi)$. For them to be orthogonal, we need

$$\int_{kT}^{(k+1)T} \cos 2\pi f_1 t \cos(2\pi f_2 t + \Phi) dt = 0$$

That is

$$\begin{aligned} &\cos\Phi \int_{kT}^{(k+1)T} \cos 2\pi f_1 t \cos 2\pi f_2 t dt \\ &-\sin\Phi \int_{kT}^{(k+1)T} \sin 2\pi f_2 t \cos 2\pi f_1 t dt \quad = 0 \end{aligned}$$

$$\begin{aligned} &\cos\Phi \int_{kT}^{(k+1)T} [\cos 2\pi(f_1+f_2)t + \cos 2\pi(f_1-f_2)t]dt \\ &-\sin\Phi \int_{kT}^{(k+1)T} [\sin 2\pi(f_1+f_2)t - \sin 2\pi(f_1-f_2)t]dt \quad = 0 \end{aligned}$$

$$\begin{aligned} &\cos\Phi \left[\frac{\sin 2\pi(f_1+f_2)t}{2\pi(f_1+f_2)t} + \frac{\sin 2\pi(f_1-f_2)t}{2\pi(f_1-f_2)t}\right]_{kT}^{(k+1)T} \\ &+\sin\Phi \left[\frac{\cos 2\pi(f_1+f_2)t}{2\pi(f_1+f_2)t} - \frac{\cos 2\pi(f_1-f_2)t}{2\pi(f_1-f_2)t}\right]_{kT}^{(k+1)T} \quad = 0 \end{aligned}$$

For arbitrary Φ, this requires that the sums inside the brackets be zero. This, in turn, requires that $2\pi(f_1+f_2)T = k\pi$ for the first term and $2\pi(f_1-f_2)T = l\pi$ for the second term in the first bracket, that $2\pi(f_1+f_2)T = 2m\pi$ for the first term and $2\pi(f_1-f_2)T = 2n\pi$ for the second term in the second bracket, where k, l, m, and n are integers and $k > l$, $m > n$. The $k\pi$ and $l\pi$ cases are included in the $2m\pi$ and $2n\pi$ case, respectively. Therefore, all these requirements can be satisfied if and only if

$$2\pi(f_1+f_2)T = 2m\pi$$

$$2\pi(f_1-f_2)T = 2n\pi$$

This leads to

$$f_1 = \frac{m+n}{2T}$$

$$f_2 = \frac{m-n}{2T}$$

$$f_1 - f_2 = \frac{n}{T}$$

This is to say that for two noncoherent FSK signals to be orthogonal, the two frequencies must be integer multiples of $1/2T$ and their separation must be a multiple of $1/T$. When $n = 1$, the separation is the $1/T$, which is the minimum. Comparing with a coherent FSK case, the separation of noncoherent FSK is double that of FSK. Thus more system bandwidth is required for noncoherent FSK for the same symbol rate.

3.4 *M*-ARY FSK

3.4.1 MFSK Signal and Power Spectral Density

In an M-ary FSK modulation, the binary data stream is divided into n-tuples of $n = \log_2 M$ bits. We denote all M possible n-tuples as M messages: m_i, $i = 1, 2, \cdots, M$.[1] There are M signals with different frequencies to represent these M messages. The expression of the ith signal is

$$s_i(t) = A\cos(2\pi f_i t + \Phi_i), \qquad kT \leq t \leq (k+1)T, \text{ for } m_i \tag{3.8}$$

where T is the symbol period which is n times the bit period.

If the initial phases are the same for all i, the signal set is coherent. As in the binary case we can always assume $\Phi_i = 0$ for coherent MFSK. The demodulation could be coherent or noncoherent. Otherwise the signal set is noncoherent and the demodulation must be noncoherent.

From the discussion of binary FSK, we know that in order for the signals to be orthogonal, the frequency separations between any two of them must be $m/2T$ for coherent case and m/T for noncoherent case. Thus the minimum separation between two adjacent frequencies is $1/2T$ for an orthogonal case and $1/T$ for a noncoherent case. These are the same as those of the binary case. Usually a uniform frequency separation between two adjacent frequencies is chosen for MFSK.

The derivation of the power spectral density of MFSK schemes is much more complicated than that for the binary case [1]. Now we quote the PSD expression of the complex envelope of the MFSK with the following parameters: (1) The frequency separations are uniform but arbitrary, which is denoted as $2\Delta f$. We express the separation in terms of modulation index $h = 2\Delta fT$. (2) The M-ary messages are

$$\begin{aligned} m_i &= 2i - (M+1), \qquad i = 1, 2, ..., M \\ &= \pm 1, \pm 3, ..., \pm(M-1) \end{aligned}$$

and the M-ary signals are

$$s_i(t) = A\cos(2\pi f_c t + m_i h \frac{\pi}{T}(t - kT) + \Phi_i), \qquad kT \leq t \leq (k+1)T, \text{ for } m_i$$

where A is the signal amplitude and all signals have equal energies. For *equiprobable*

---

[1] M may be odd in some applications. Then n can be set as the nearest integer greater than $\log_2 M$. Some of the m_i must not be used.

messages, the PSD expression is given in [1] as

$$\Psi_{\tilde{s}}(f) = \frac{A^2T}{M}\sum_{i=1}^{M}\left[\frac{1}{2}\frac{\sin^2\gamma_i}{\gamma_i^2} + \frac{1}{M}\sum_{j=1}^{M}A_{ij}\frac{\sin\gamma_i}{\gamma_i}\frac{\sin\gamma_j}{\gamma_j}\right] \tag{3.9}$$

Other parameters are defined as

$$A_{ij} = \frac{\cos(\gamma_i+\gamma_j) - C_a\cos(\gamma_i+\gamma_j-2\pi fT)}{1+C_a^2-2C_a\cos 2\pi fT}$$

$$\gamma_i = (fT - m_ih/2)\pi, \qquad i = 1, 2, ..., M$$

$$C_a = \frac{2}{M}\sum_{i=1}^{M/2}\cos[h\pi(2i-1)]$$

We have plotted curves for $h = 0.5$ to 1.5 for $M = 2, 4$, and 8 in Figures 3.12, 3.13, and 3.14, respectively, where each abscissa is normalized frequency fT [1]. Curves are presented for various h values to show how the spectral shape changes with h. Figure 3.12 is for binary FSK where $h = 1$ is a special case that an impulse arises at $fT = 0.5$, that is, at the two FSK frequencies. This is the Sunde's FSK. We have calculated and plotted its PSD in Section 3.1.2. Figures 3.13 and 3.14 are for 4-ary and 8-ary MSK, respectively. The multilevel cases show considerable similarity to the binary ones. For small values of h, the spectra are narrow and decrease smoothly towards zero. As h increases towards unity the spectrum widens and spectral power is increasingly concentrated around $fT = 0.5$ and its odd multiples. These are the frequencies of the M signals in the scheme. At $h = 1$, spectral lines arise at these frequencies. As h increases further the concentration is again broadened and reduced in intensity. For coherent orthogonal case, $h = 0.5$, most spectral components are in a bandwidth of $\frac{M}{2T}$. Thus the transmission bandwidth can be set as $B_T = \frac{M}{2T}$. Similarly for noncoherent orthogonal case, $h = 1.0$, then $B_T = \frac{M}{T}$. This can be easily seen from the curves.

In some applications, M is *odd* and/or messages are *not equiprobable*. For this case the PSD expression in (3.9) is still applicable except that the parameter C_a must be in its original form as [1]

$$C_a = \sum_{i=1}^{M} P_i \exp\{jh\pi m_i)$$

where P_i is the a priori probability of message m_i.

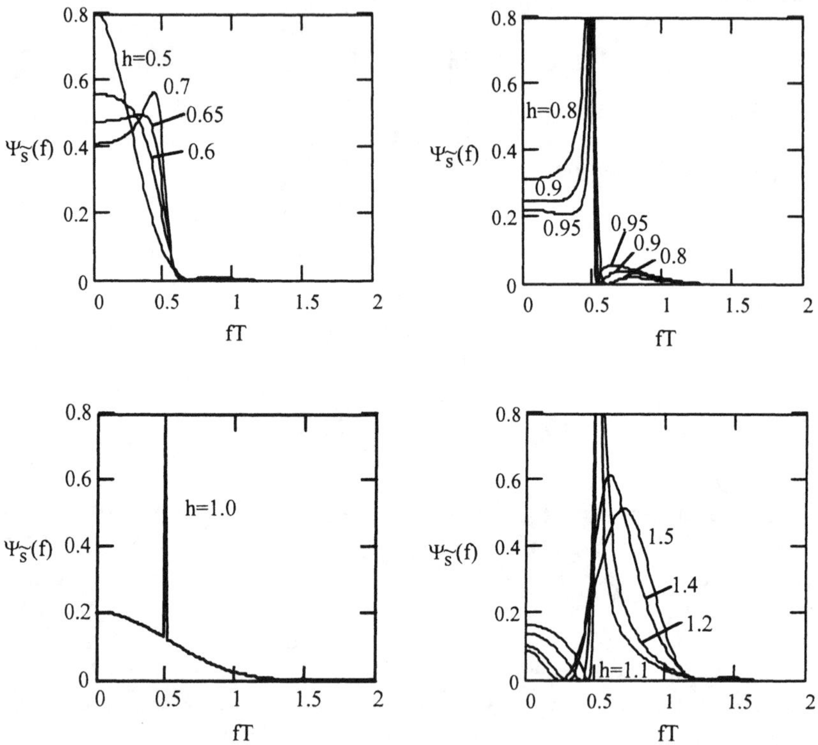

Figure 3.12 PSD of MFSK: M = 2. After [1].

3.4.2 Modulator, Demodulator, and Error Performance

The noncoherent modulator for binary FSK in Figure 3.1 can be easily extended to noncoherent MFSK by simply increasing the number of independent oscillators to M (Figure 3.15). The coherent modulator for binary FSK in Figure 3.2 can also be extended to MFSK (Figure 3.16). Then the frequency synthesizer generates M signals with the designed frequencies and coherent phase, and the multiplexer chooses one of the frequencies, according to the n data bits.

The coherent MFSK demodulator falls in the general form of detector for M-ary equiprobable, equal-energy signals with known phases as described in Appendix B. The demodulator consists of a bank of M correlators or matched filters (Figures 3.17 and 3.18). At sampling times $t = kT$, the receiver makes decisions based on

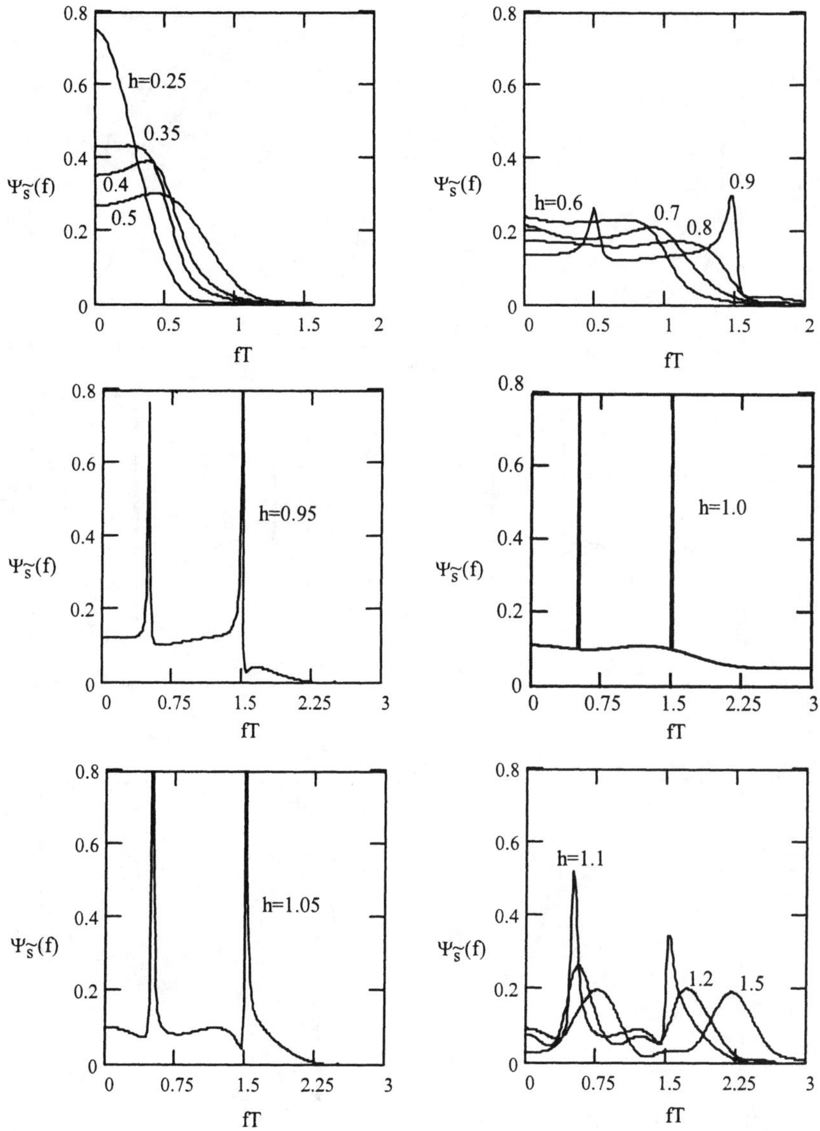

Figure 3.13 PSD of MFSK: M = 4. After [1].

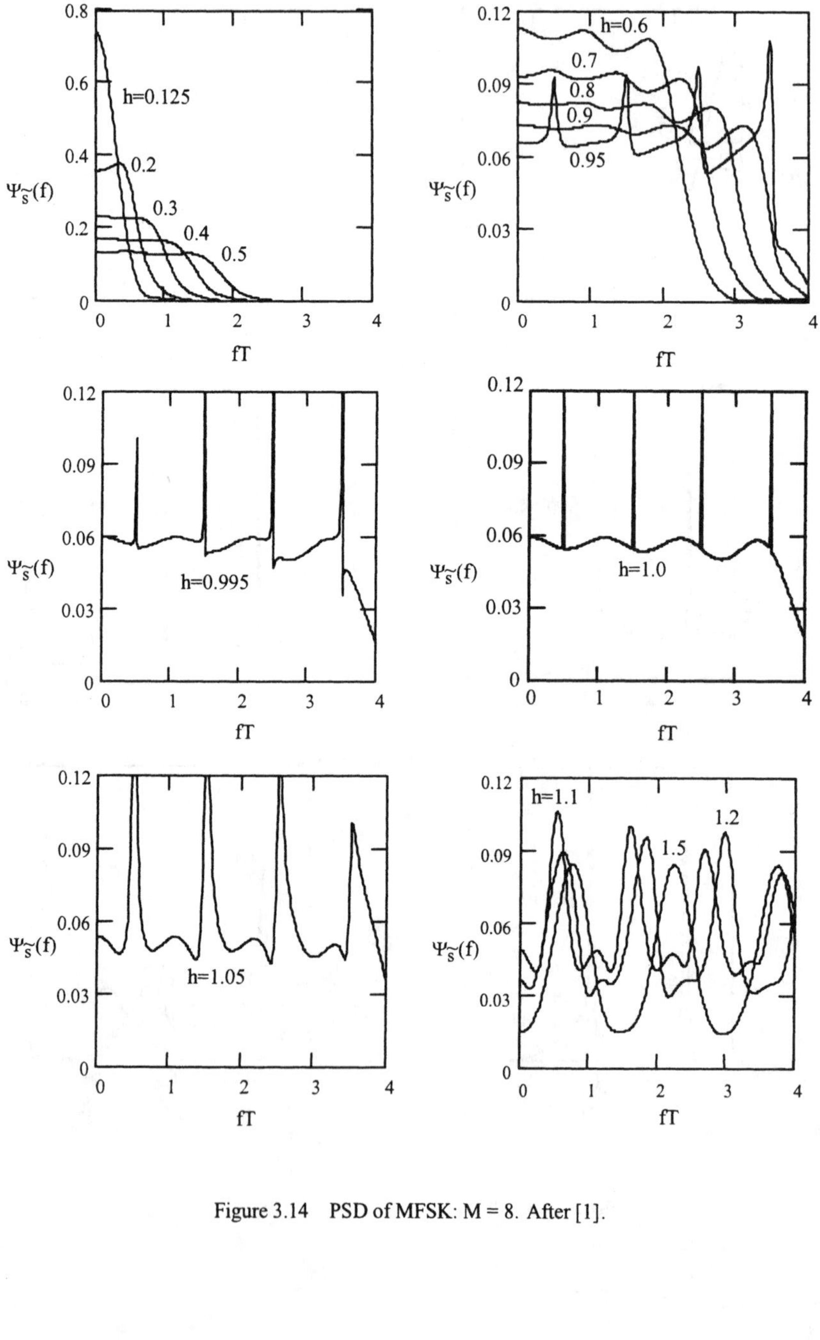

Figure 3.14 PSD of MFSK: M = 8. After [1].

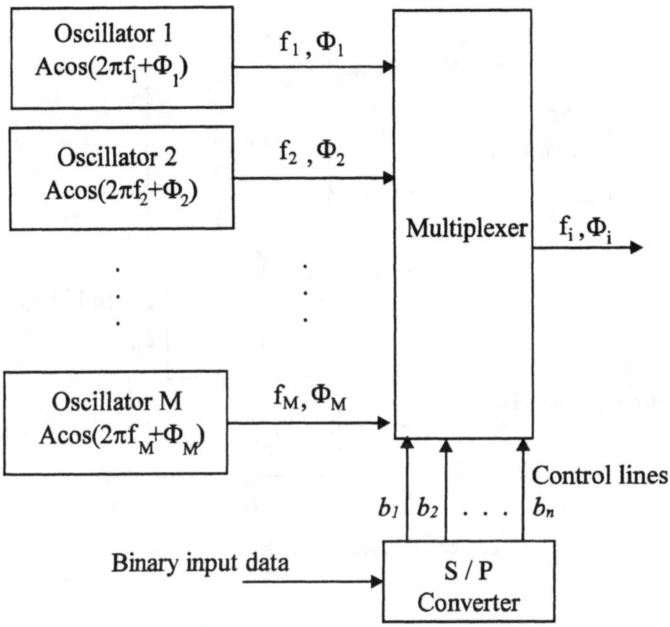

Figure 3.15 Noncoherent MFSK modulator.

the largest output of the correlators or matched filters. We have to point out that the coherent MFSK receivers in Figures 3.17 and 3.18 only require that the MFSK signals be equiprobable, equal-energy, and do not require them to be orthogonal. For more general cases where the M-ary signal set is not equiprobable and/or not equal in energy, a bias term

$$B_i = \frac{N_o}{2} \ln P_i - \frac{1}{2} E_i$$

must be included in the sufficient statistics l_i in the receiver (see Figures B.6 and B.7).

The exact expression for the symbol error probability for a symmetrical signal set (equal-energy, equiprobable) has been given in Appendix B (B.42) as

$$P_s = 1 - \int_{-\infty}^{\infty} dx \frac{1}{\sqrt{2\pi}} \exp\{-\frac{(x - \sqrt{2E_s/N_o})^2}{2}\}[1 - Q(x)]^{M-1} \tag{3.10}$$

This expression does not require the signal set to be orthogonal. This expression

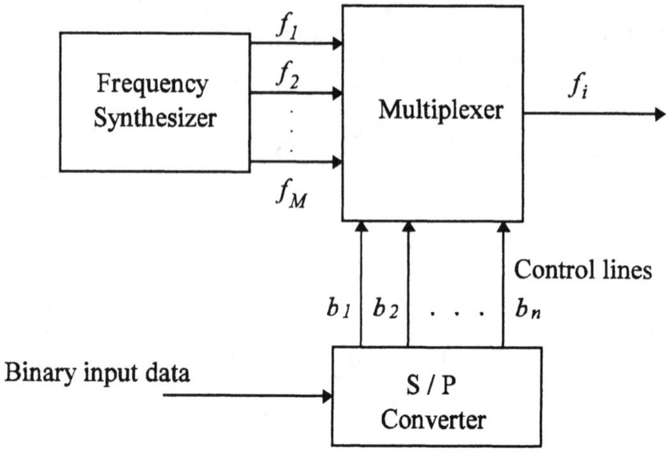

Figure 3.16 Coherent M-ary FSK modulator.

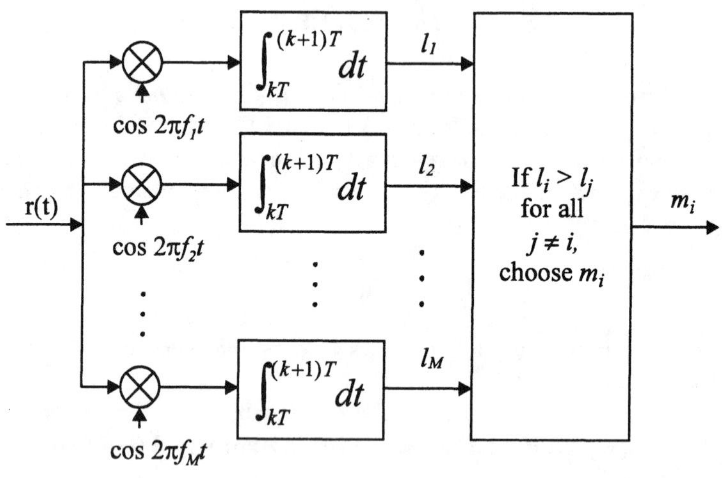

Figure 3.17 Coherent M-ary FSK demodulator: correlator implementation.

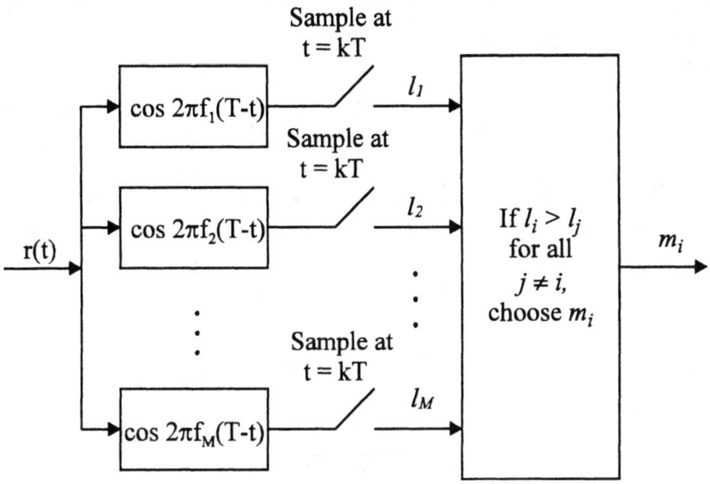

Figure 3.18 Coherent M-ary FSK demodulator: matched filter implementation.

cannot be analytically evaluated.

If the signal set is not symmetrical, an upper bound has been obtained in Appendix B (B.43) as

$$P_s \leq \sum_{j \neq i} Q\left(\frac{d_{ij}}{\sqrt{2N_o}}\right)$$

If the signal set is equal-energy *and* orthogonal (not necessarily equiprobable), all distances between any two signals are equal. The distance $d = \sqrt{2E_s}$, and the upper bound becomes

$$P_s \leq (M-1)Q\left(\sqrt{\frac{E_s}{N_o}}\right) \tag{3.11}$$

For fixed M this bound becomes increasingly tight as E_s/N_o is increased. In fact, it becomes a good approximation for $P_s \leq 10^{-3}$. For $M = 2$, it becomes the exact expression.

If the signaling scheme is such that each signal represents $n = \log_2 M$ bits, as we stated in the very beginning of this section, then the P_s can be expressed as a function of $E_b = E_s/\log_2 M$.

At this point we need to derive a relation between bit error probability P_b and symbol error probability P_s. For binary signals P_b is simply equal to P_s. For equally likely orthogonal M-ary signals, all symbol errors are equiprobable. That is, the demodulator may choose any one of the $(M-1)$ erroneous orthogonal signals with equal probability:[2]

$$\frac{P_s}{M-1} = \frac{P_s}{2^n - 1}$$

There are $\binom{n}{k}$ ways in which k bits out of n bits may be in error. Hence the average number of bit errors per n-bit symbol is

$$\sum_{k=1}^{n} k \binom{n}{k} \frac{P_s}{2^n - 1} = n \frac{2^{n-1}}{2^n - 1} P_s$$

and the average bit error probability is just the above result divided by n. Thus

$$P_b = \frac{2^{n-1}}{2^n - 1} P_s \tag{3.12}$$

For large values of M, $P_b \approx \frac{1}{2} P_s$.

If M is odd then $M \neq 2^n$. If the M messages are still equiprobable, the above expression is modified as

$$P_b = \frac{2^{n-1}}{M-1} P_s$$

P_s and P_b for coherently demodulated, *equal-energy, equiprobable, and orthogonal* MFSK are shown in Figures 3.19 and 3.20, respectively. The solid lines are accurate error curves and the dotted lines are upper bounds. Note that the curves in these two figures are very close at high SNRs, but the differences are clear at low SNRs. It can be seen that the upper bound is very tight for E_b/N_o over about 5 dB. We also observe that for the same E_b/N_o error probability reduces when M increases, or for the same error probability the required E_b/N_o decreases as M increases. However, the speed of decrease in E_b/N_o slows down when M gets larger.

The noncoherent demodulator for M-ary FSK falls in the general form of detector for M-ary equiprobable, equal-energy signals with unknown phases as described in Appendix B. The demodulator can be implemented in correlator-squarer form or matched filter-squarer or matched filter-envelope detector form (Figures 3.21, 3.22, and 3.23). They are the extensions of their binary counterparts. Their operations are the same except for the increase of number of signals. Therefore we do not repeat the description here.

---

[2] This is, however, not true for M-ary PSK signals since they are not orthogonal. Therefore the P_b versus P_s relation is different, as we will see in Chapter 4.

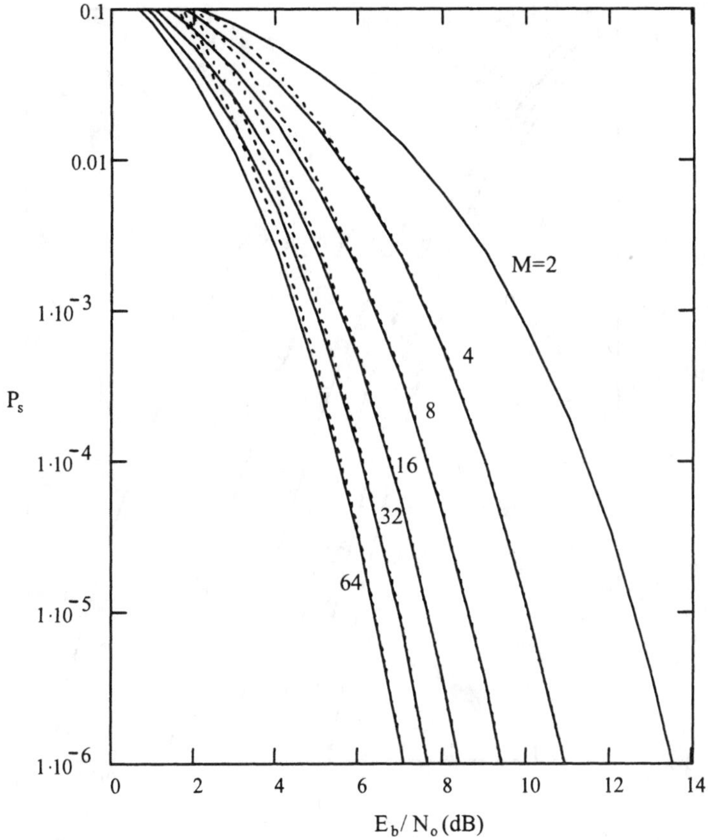

Figure 3.19 P_s of coherently demodulated, equiprobable, equal-energy, and orthogonal MFSK (dotted line: upper bound).

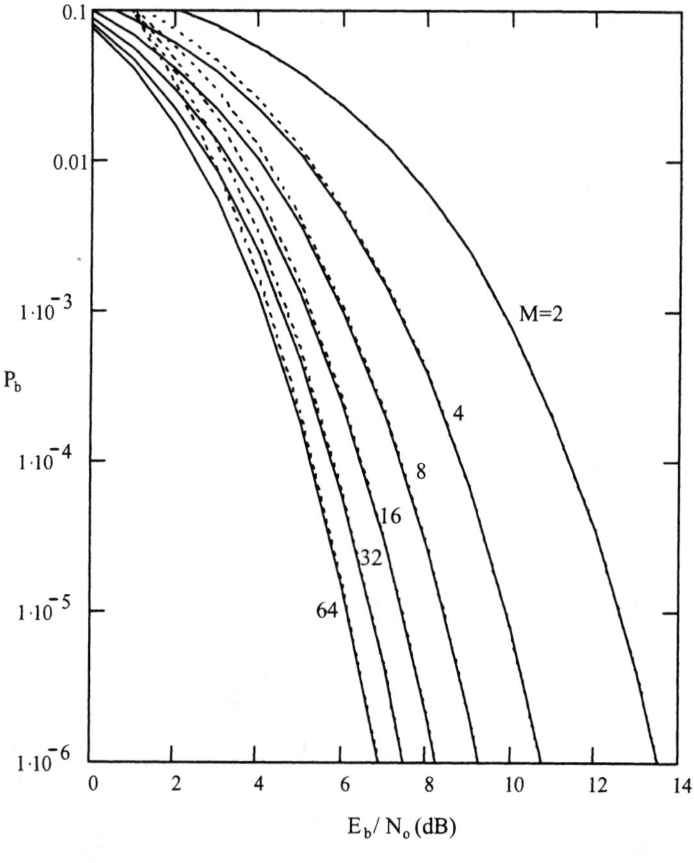

Figure 3.20 P_b of coherently demodulated, equiprobable, equal-energy, and orthogonal MFSK (dotted line: upper bound).

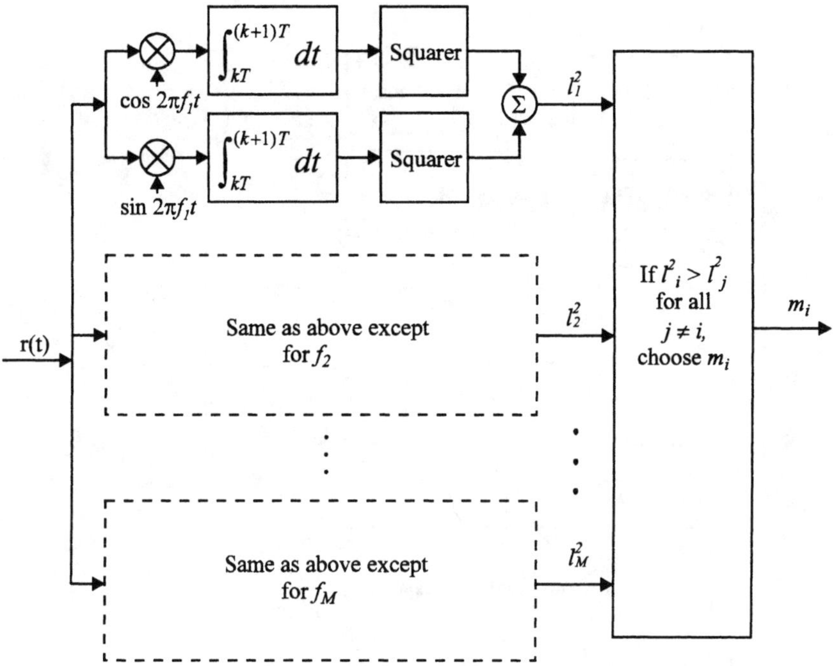

Figure 3.21 Noncoherent MFSK demodulator: correlator-squarer implementation.

Symbol error probability expression for noncoherently demodulated, *equiprobable, equal-energy, and orthogonal* MFSK has been given in Appendix B (B.57) as

$$P_s = 1 - P(c/H_1) = \sum_{k=1}^{M-1} \frac{(-1)^{k+1}}{k+1} \binom{M-1}{k} \exp[-\frac{kE_s}{(k+1)N_o}] \tag{3.13}$$

where $\binom{M-1}{k}$ is the binomial coefficient, defined by

$$\binom{M-1}{k} = \frac{(M-1)!}{(M-1-k)!k!}$$

The first term of the summation in (3.13) provides an upper bound (B.58) as

$$P_s \leq \frac{M-1}{2} \exp[-\frac{E_s}{2N_o}] \tag{3.14}$$

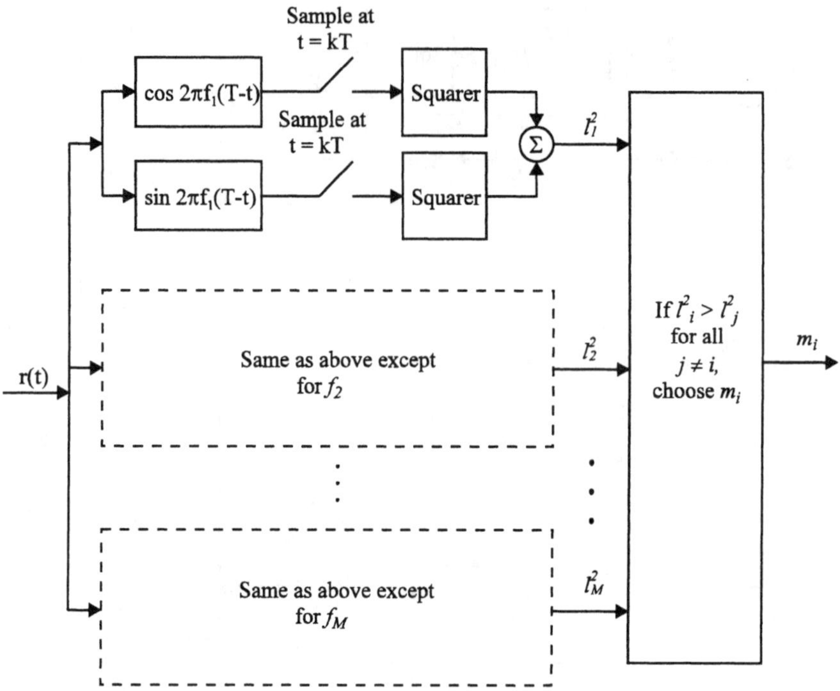

Figure 3.22 Noncoherent MFSK demodulator: matched filter-squarer detector implementation.

For fixed M this bound becomes increasingly close to the actual value of P_s as E_s/N_o is increased. In fact when $M = 2$, the upper bound becomes the exact expression.

Figures 3.24 and 3.25 show P_s and P_b for noncoherently demodulated, equiprobable, equal-energy, and orthogonal MFSK for $M = 2, ..., 32$. Note that the curves in these two figures are very close at high SNRs, but the differences are clear at low SNRs. Like the coherent case, again the P_b versus P_s relation is given by (3.12). The behavior of the curves with values of M is very similar to that of the coherent case. The only difference is that the noncoherent one requires slightly more E_b/N_o for the same error probability. The increase of E_b/N_o is only a fraction of a decibel for $P_b \leq 10^{-4}$ for all values of M and becomes smaller as M becomes larger.

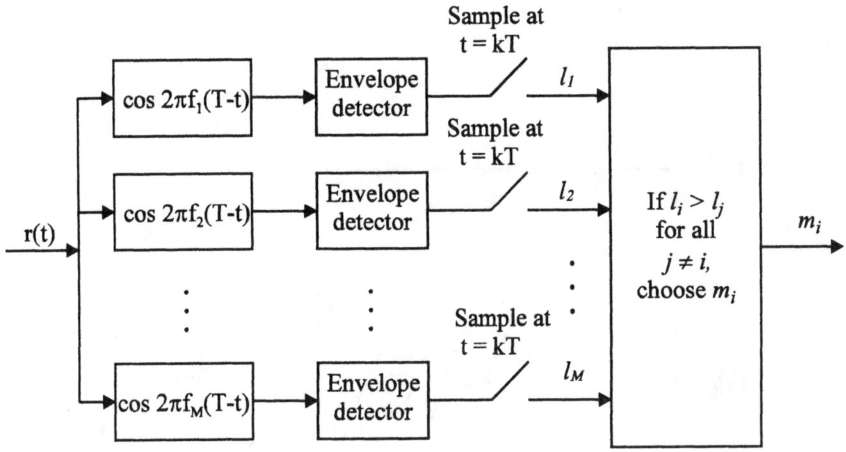

Figure 3.23 Noncoherent MFSK demodulator: matched filter-envelope detector implementation.

3.5 DEMODULATION USING DISCRIMINATOR

In comparison with other modulation schemes, such as PSK and QAM, the most important advantage of FSK is that demodulation can be very simple. We already have seen that FSK can be noncoherently demodulated and the error performance degradation is less than 1 dB in a meaningful range of bit error rate. Demodulation can be even simpler by using the conventional frequency discriminator which is widely used in analog FM demodulation.

Figure 3.26 is the block diagram of a typical binary FSK demodulator using limiter-discriminator detection and integrate-and-dump post-detection filtering. The IF filter is a narrow-band filter, with a frequency response $H(f)$ centered at f_c. This filter is to reject the out-of-band noise and restrict the frequency band of the signal. Although this filter is shown in Figure 3.26 as a part of the receiver, it should be considered as the overall filtering characteristic of the transmitter filter, channel, and receiver for analysis purposes. The demodulation is done by the limiter-discriminator combination which for the purpose of analysis is assumed to have an ideal characteristic, that is, the output is proportional to the angle of the input signal. The post-detection filter is an integrate-and-dump filter with an integration time of T. At the end of each bit interval the output of the integrate-and-dump filter is sampled and the polarity of the sample determines whether a 1 or 0 was sent.

For a narrow-band IF filter, a set of expressions for bit error probability was first published in [2]. The results in [2] show that for both the Gaussian and rectangular

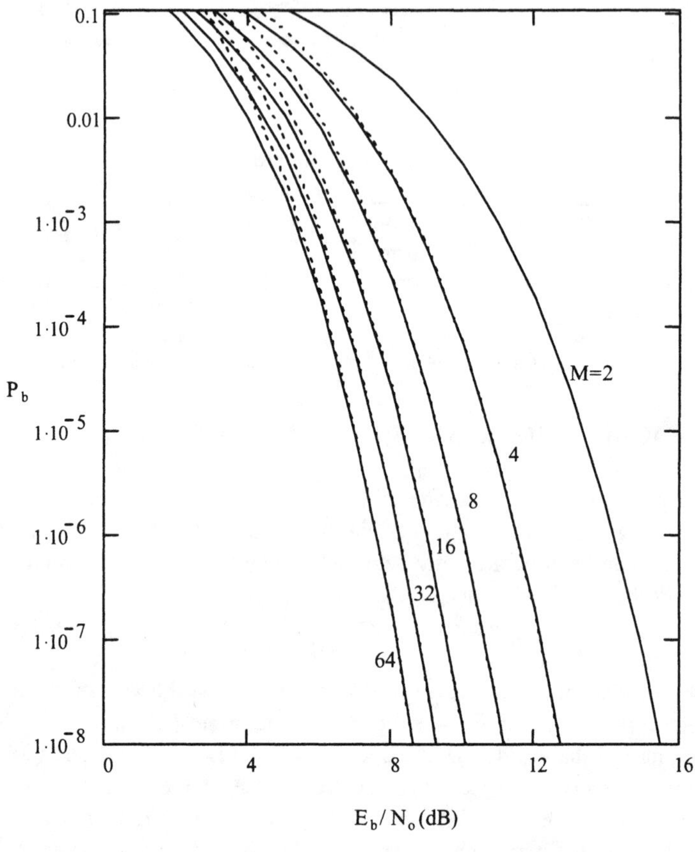

Figure 3.24 P_s of noncoherently demodulated, equiprobable, equal-energy, and orthogonal MFSK (dotted line: upper bound).

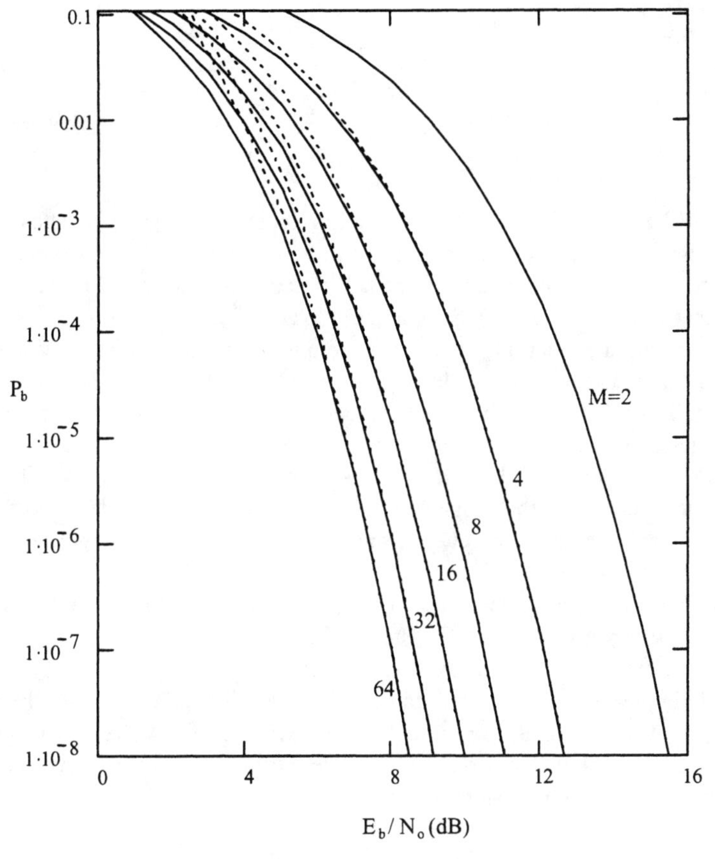

Figure 3.25 P_b of noncoherently demodulated, equiprobable, equal-energy, and orthogonal MFSK (dotted line: upper bound).

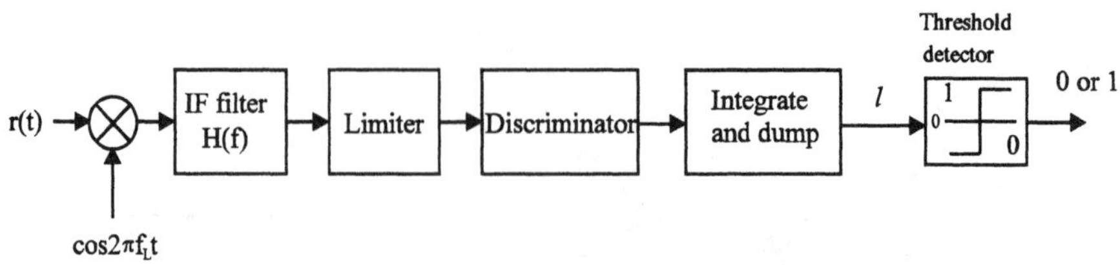

Figure 3.26 Binary FSK demodulator using limiter-discriminator detection and integrate-and-dump postdetection filtering.

bandpass filters, the system with a modulation index of $h = 0.7$ is the best in that it requires the least signal energy in a given noise environment to achieve a 10^{-4} bit error rate for bandwidths ranging from $BT = 0.8$ to 3.0, where B is the *equivalent noise bandwidth.*[3] For a given SNR and modulation index there is a bandwidth that gives a minimum probability of error. It might be argued heuristically that as the bandwidth is increased more noise power is accepted and errors increase. On the other hand, as the bandwidth is reduced below a certain point, error arising out of the distortion of the signal increases. In general it appears that a bandpass filter bandwidth of about $B = 1/T$ gives a minimum probability of error. The precise value would depend on the shape of the filter, the modulation index, and the SNR. The theoretical results were generally in good agreement with experiment results reported in [2].

The work in [3] provided a set of simpler expressions by restricting the noise equivalent bandwidth B of the IF filter in the range of $1 \leq BT \leq 3$ and the frequency deviation ratio $h < 1.5$. Also the SNR is assumed large (≥ 6 dB). However, even the simpler expressions are too complicated to be included here. Instead we present the curves from [3] in Figure 3.27 for several IF filter characteristics. They are the Gaussian filter, the six-element Butterworth filter, and two-stage synchronously tuned filters. Their characteristics are, respectively,

$$H(f) = e^{-\pi f^2/2B^2}, \qquad \text{(Gaussian)}$$

$$H(f) = \frac{2}{1 + (1 + j\sqrt{8}f/B)^2}, \qquad \text{(Butterworth)}$$

[3] The equivalent noise bandwidth of a filter $H(f)$ is defined as

$$B = \frac{1}{|H(0)|}\int_{-\infty}^{\infty} |H(f)|^2 df$$

$$H(f) = \frac{2}{(1+jf/\alpha)^2}, \quad \alpha = \frac{B/2}{\sqrt{\sqrt{2}-1}}$$

(two-stage synchronously tuned)

Figure 3.27 shows the curves for the optimum index $h = 0.707$, and the optimum bandwidth $BT = 1$ for all three filters. The BER curves with the Gaussian and Butterworth filters are shown as a single curve since there is only a few percent difference in the performance with the Gaussian filter being sightly better. The performance of the synchronously tuned IF filters is about 0.75 dB inferior to the Gaussian IF filter. Also shown in the figure for comparison is the performance curve for noncoherent orthogonal FSK which is about 0.75 dB inferior to the synchronously tuned IF filters. These results are consistent with those reported in [2]. We would expect that the error performance of the optimum noncoherent FSK be better than the narrow-band demodulator. However, the above results show the opposite. The reason why this happens was not given either in [2] or in [3]. Recall that for noncoherent demodulation, the optimum demodulator consists of matched filters followed by envelope detectors (or equivalently, correlators). It is optimum in the sense that the matched filter will give a maximum SNR. But this optimization is based on the assumption that the filter is linear. In the narrow-band demodulator, the discriminator is a nonlinear device. What is optimum for linear filters may not be optimum for nonlinear filters. This might explain why the error performance of the "optimum" noncoherent FSK is inferior to the narrow-band discriminator demodulator with the best set of parameters $(BT = 1$ and $h = 0.707)$.

If the IF filter's bandwidth is sufficiently broad so that distortion of the signal can be ignored, and the post-detection low-pass filter is approximated by an ideal integrator with an integration time of T, then a simple expression of the symbol error probability for M-ary orthogonal FSK signals demodulated by the discriminator is given in [4] as

$$P_s = [\frac{1}{2} + \frac{1}{4}(\frac{M}{2}+1)]\exp(-\frac{2}{M}\frac{E_s}{N_o}) \tag{3.15}$$

When compared with the optimum noncoherent receiver whose error rate behaves as $\exp[-E_s/2N_o]$ (see (3.14)) we have lost a factor of $M/4$ in the error exponent by substituting discriminator detection for matched filter detection. When $M \leq 4$, there is actually no loss, rather there is a gain in symbol error performance. When $M = 2$, (3.15) becomes $exp(-E_s/N_o)$ which is better by 3 dB than the matched filter case. And it is 1.5 dB better than the Gaussian filter case shown in Figure 3.27. The previous reasoning that explains why narrow-band discriminator detection performs better than optimum noncoherent detection can also be applied here to explain why the wide-band discriminator performs better.

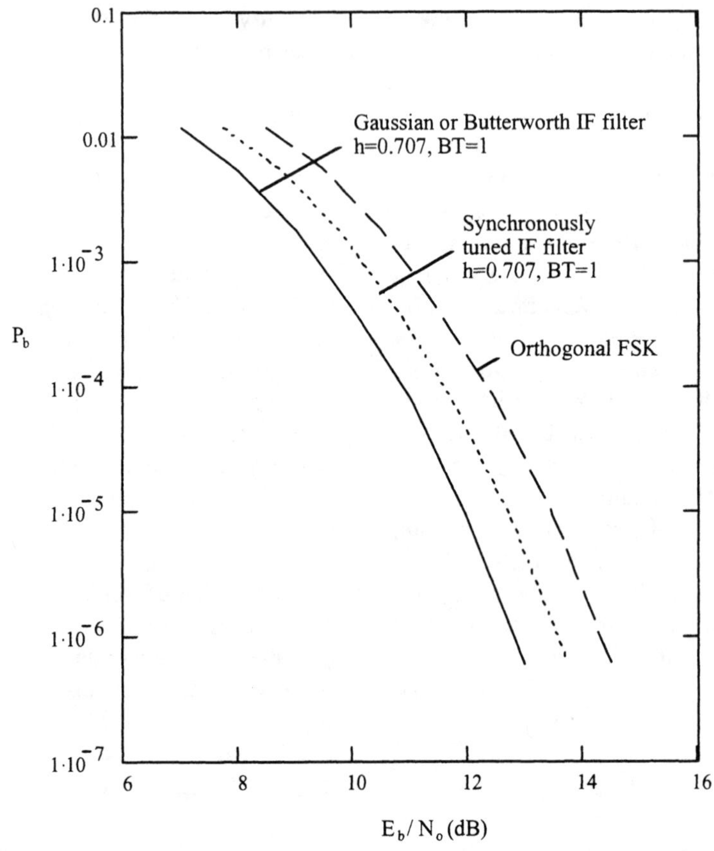

Figure 3.27 P_b of FSK demodulated by limiter-discriminator detection and integrate-and-dump postdetection filtering.

3.6 SYNCHRONIZATION

For coherent detection of the FSK signal the reference signals $\cos 2\pi f_i t, i = 1, 2$ and the bit timing signal must be recovered from the received signal. However, since practical systems never use coherent demodulation, there are no carrier recovery schemes found in the literature. For the usual Sunde's FSK, since the signal has two strong spectral components at f_1 and f_2, two phase lock loops may be used to extract f_1 and f_2 from the FSK signal.

For noncoherent demodulation, only symbol synchronization is needed. Symbol timing can be achieved by generating the symbol clock signal using an early/late-gate circuit. Since a symbol timing recovery problem is common to FSK and PSK modulation schemes, we leave it to be discussed at the end of the next chapter on PSK.

3.7 SUMMARY

In this chapter we have discussed coherent and noncoherent binary FSK and M-ary FSK schemes. For both cases we looked into the conditions for phase continuity and signal orthogonality. For phase continuity of coherent FSK (including MFSK), we found that the sufficient condition is that the frequency separation is an integer multiple of $1/T$. However, this is not necessary since there is a special form of binary FSK, namely, MSK, which has a separation of $1/2T$. MSK is an important modulation scheme which is covered in great detail in Chapter 5. For orthogonality of the coherent FSK (including MFSK) the signal frequencies must be integer multiples of $1/4T$ and the separation must be an integer multiple of $1/2T$. The phase of noncoherent FSK signal is not continuous and its orthogonality requires that the frequency separation be an integer multiple of $1/T$. So the minimum separation for noncoherent FSK is double that of the coherent one.

The power spectral densities of FSK, both binary and M-ary, in the form of expressions or curves, were presented in this chapter.

We also presented modulators and demodulators, both coherent and noncoherent, binary and M-ary, in this chapter. The coherent demodulators consist of correlators or equivalently, matched filters and samplers. They require reference signals that are synchronized with the transmitted signals. The noncoherent demodulators consist of correlators and squarers, or equivalently, matched filters and envelope detectors. They do not require that reference signals be synchronous to the transmitted signals. Therefore additional squarers or envelope detectors are used to eliminate the adverse effect of random phase differences between the reference signals and the received signals. Since circuits of generating synchronous reference signals are very costly, most of FSK receivers use noncoherent demodulation.

It is expected that the error performance of the noncoherent receivers is inferior to that of the coherent ones. However, the degradation is only a fraction of a dB. The expressions and curves for the error probabilities are also presented in great detail.

Finally we explored other possible demodulations. The discriminator demodulator is simple and efficient. It is even better than the noncoherent optimum demodulator for BFSK.

References

[1] Anderson, R. R., and J. Salz, "Spectra of digital FM," *Bell System Technical Journal*, vol. 44, July–August 1965, pp.1165–1189.

[2] Tjhung, T. T., and P. H. Wittke, "Carrier transmission of binary data in a restricted band," *IEEE Trans. Comm. Tech.*, vol. 18, no. 4, August 1970, pp. 295–304.

[3] Pawula, R. F., "On the theory of error rates for narrow-band digital FM," *IEEE Trans. Comm.*, vol. 29, no. 11, November 1981, pp. 1634–1643.

[4] Mazo, J. E., "Theory of error rates for digital FM," *Bell System Technical Journal,* vol. 45, November 1966, pp. 1511–1535.

Selected Bibliography

- Couch II, L. W., *Digital and Analog Communication Systems*, 3rd ed., New York: Macmillan, 1990.
- Haykin, S., *Digital Communications*, New York: John Wiley & Sons, Inc., 1988.
- Salz, J., "Performance of multilevel narrow-band FM digital communication systems," *IEEE Trans. Comm. Tech.*, vol. 13, no. 4, December 1975, pp. 420–424.
- Sklar, B., *Digital Communications: Fundamentals and Applications*, Englewood Cliffs, New Jersey: Prentice Hall, 1988.
- Smith, D. R., *Digital Transmission Systems*, 2nd ed., New York: Van Nostrand Reinhold, 1993.
- Sunde, E. D., "Ideal binary pulse transmission by AM and FM," *Bell System Technical Journal*, vol. 38, November 1959, pp. 1357–1426.
- Van Trees, H. L., *Detection, Estimation, and Modulation Theory, Part I*, New York: John Wiley & Sons, Inc., 1968.

Chapter 4

Phase Shift Keying

Phase shift keying (PSK) is a large class of digital modulation schemes. PSK is widely used in the communication industry. In this chapter we study each PSK modulation scheme in a single section where signal description, power spectral density, modulator/demodulator block diagrams, and receiver error performance are all included. First we present coherent binary PSK (BPSK) and its noncoherent counterpart, differential BPSK (DBPSK), in Sections 4.1 and 4.2. Then we discuss in Section 4.3 M-ary PSK (MPSK) and its PSD in Section 4.4. The noncoherent version, differential MPSK (DMPSK) is treated in Section 4.5. We discuss in great detail quadrature PSK (QPSK) and differential QPSK (DQPSK) in Sections 4.6 and 4.7, respectively. Section 4.8 is a brief discussion of offset QPSK (OQPSK). An important variation of QPSK, the $\pi/4$-DQPSK which has been designated as the American standard of the second-generation cellular mobile communications, is given in Section 4.9. Section 4.10 is devoted to carrier and clock recovery. Finally, we summarize the chapter with Section 4.11.

4.1 BINARY PSK

Binary data are represented by two signals with different phases in BPSK. Typically these two phases are 0 and π, the signals are

$$
\begin{aligned}
s_1(t) &= A\cos 2\pi f_c t, \quad 0 \le t \le T, \quad \text{for } 1 \\
s_2(t) &= -A\cos 2\pi f_c t, \quad 0 \le t \le T, \quad \text{for } 0
\end{aligned} \tag{4.1}
$$

These signals are called *antipodal.* The reason that they are chosen is that they have a correlation coefficient of -1, which leads to the minimum error probability for the same E_b/N_o, as we will see shortly. These two signals have the same frequency and energy.

As we will see in later sections, all PSK signals can be graphically represented

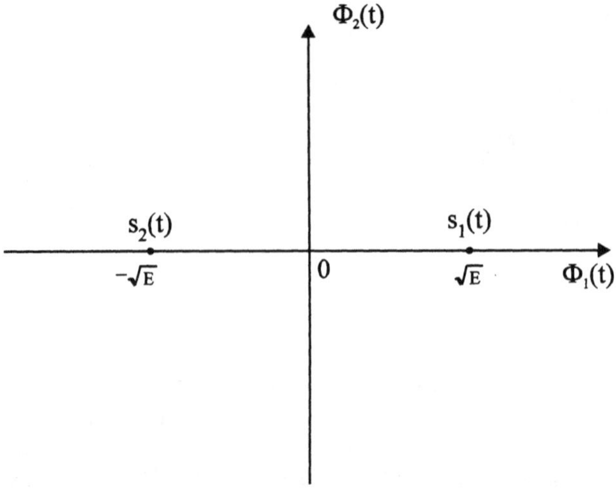

Figure 4.1 BPSK signal constellation.

by a *signal constellation* in a two-dimensional coordinate system with

$$\phi_1(t) = \sqrt{\frac{2}{T}} \cos 2\pi f_c t, \qquad 0 \leq t \leq T \tag{4.2}$$

and

$$\phi_2(t) = -\sqrt{\frac{2}{T}} \sin 2\pi f_c t, \qquad 0 \leq t \leq T \tag{4.3}$$

as its horizontal and vertical axis, respectively. Note that we deliberately add a minus sign in $\phi_2(t)$ so that PSK signal expressions will be a sum instead of a difference (see (4.14)). Many other signals, especially QAM signals, can also be represented in the same way. Therefore we introduce the signal constellation of BPSK here as shown in Figure 4.1 where $s_1(t)$ and $s_2(t)$ are represented by two points on the horizontal axis, respectively, where

$$E = \frac{A^2 T}{2}$$

The waveform of a BPSK signal generated by the modulator in Figure 4.3 for a data stream $\{10110\}$ is shown in Figure 4.2. The waveform has a constant envelope like FSK. Its frequency is constant too. In general the phase is not continuous at

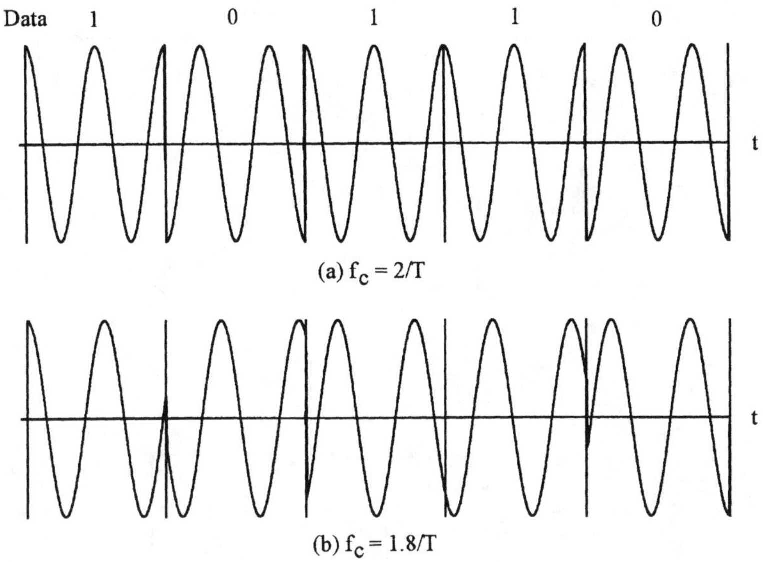

Figure 4.2 BPSK waveforms (a, b).

bit boundaries. If the $f_c = m\ R_b = m/T$, where m is an integer and R_b is the data bit rate, and the bit timing is synchronous with the carrier, then the initial phase at a bit boundary is either 0 or π (Figure 4.2(a)), corresponding to data bit 1 or 0. However, if the f_c is not an integer multiple of R_b, the initial phase at a bit boundary is neither 0 nor π (Figure 4.2(b)). In other words, the modulated signals are not the ones given in (4.1). We will show next in the discussion of demodulation that condition $f_c = m\ R_b$ is necessary to ensure minimum bit error probability. However, if $f_c >> R_b$, this condition can be relaxed and the resultant BER performance degradation is negligible.[1]

The modulator which generates the BPSK signal is quite simple (Figure 4.3(a)). First a bipolar data stream $a(t)$ is formed from the binary data stream

$$a(t) = \sum_{k=-\infty}^{\infty} a_k p(t - kT) \tag{4.4}$$

---

1 This is true for all PSK schemes and PSK-derived schemes, including QPSK, MSK, and MPSK. We will not mention this again when we discuss other PSK schemes.

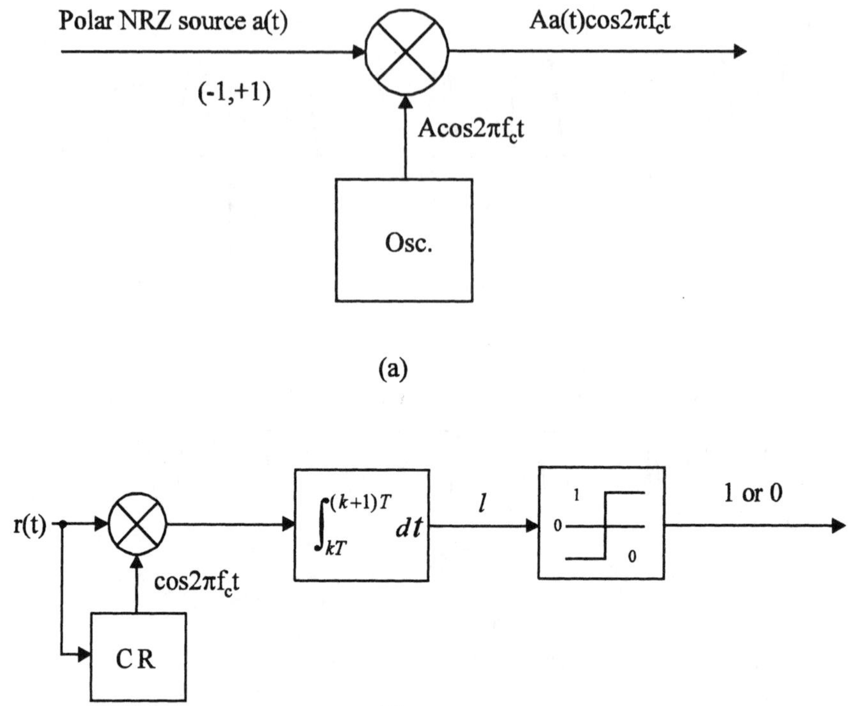

Figure 4.3 BPSK modulator (a), and coherent BPSK demodulator (b).

where $a_k \in \{+1, -1\}$, $p(t)$ is the rectangular pulse with unit amplitude defined on $[0, T]$. Then $a(t)$ is multiplied with a sinusoidal carrier $A\cos 2\pi f_c t$. The result is the BPSK signal

$$s(t) = Aa(t)\cos 2\pi f_c t, \quad -\infty < t < \infty \tag{4.5}$$

Note that the bit timing is not necessarily synchronous with the carrier.

The coherent demodulator of BPSK falls in the class of coherent detectors for binary signals as described in Appendix B. The coherent detector could be in the form of a correlator or matched filter. The correlator's reference signal is the difference signal ($s_d(t) = 2A\cos 2\pi f_c t$). Figure 4.3(b) is the coherent receiver using a correlator where the reference signal is the scaled-down version of the difference signal. The reference signal must be synchronous to the received signal in frequency and phase.

It is generated by the carrier recovery (CR) circuit. Using a matched filter instead of a correlator is not recommended at passband since a filter with $h(t) = \cos 2\pi f_c(T-t)$ is difficult to implement.

In the absence of noise, setting $A = 1$, the output of the correlator at $t = (k+1)T$ is

$$\begin{aligned}
&\int_{kT}^{(k+1)T} r(t)\cos 2\pi f_c t dt \\
= &\int_{kT}^{(k+1)T} a_k \cos^2 2\pi f_c t dt \\
= &\frac{1}{2}\int_{kT}^{(k+1)T} a_k(1+\cos 4\pi f_c t)dt \\
= &\frac{T}{2}a_k + \frac{a_k}{8\pi f_c}[\sin 4\pi f_c(k+1)T - \sin 4\pi f_c kT]
\end{aligned}$$

If $f_c = m\ R_b$, the second term is zero, thus the original signal $a(t)$ is perfectly recovered (in the absence of noise). If $f_c \neq m\ R_b$, the second term will not be zero. However, as long as $f_c >> R_b$, the second term is much smaller than the first term so that its effect is negligible.

The bit error probability can be derived from the formula for general binary signals (Appendix B):

$$P_b = Q\left(\sqrt{\frac{E_1 + E_2 - 2\rho_{12}\sqrt{E_2E_1}}{2N_o}}\right)$$

For BPSK $\rho_{12} = -1$ and $E_1 = E_2 = E_b$, thus

$$P_b = Q\left(\sqrt{\frac{2E_b}{N_o}}\right), \text{ (coherent BPSK)} \tag{4.6}$$

A typical example is that, at $E_b/N_o = 9.6$ dB, $P_b = 10^{-5}$. Figure 4.4 shows the P_b curve of BPSK. The curves of coherent and noncoherent BFSK are also shown in the figure. Recall the P_b expression for coherent BFSK is $P_b = Q\left(\sqrt{\frac{E_b}{N_o}}\right)$ which is 3 dB inferior to coherent BPSK. However, coherent BPSK requires that the reference signal at the receiver to be synchronized in phase and frequency with the received signal. This will be discussed in Section 4.10. Noncoherent detection of BPSK is also possible. It is realized in the form of differential BPSK which will be studied in the next section.

Next we proceed to find the power spectral density of the BPSK signal. It suf-

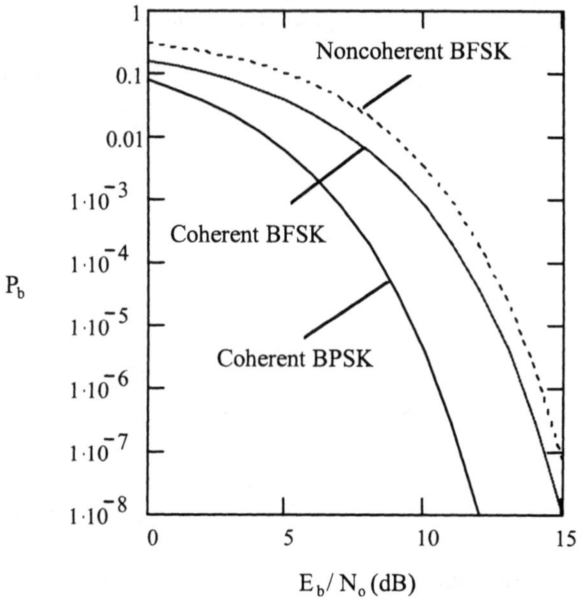

Figure 4.4 P_b of BPSK in comparison with BFSK.

fices to find the PSD of the baseband shaping pulse. As shown in Appendix A, the PSD of a binary, bipolar, equiprobable, stationary, and uncorrelated digital waveform is just equal to the energy spectral density of the pulse divided by the symbol duration (see (A.19)). The basic pulse of BPSK is just a rectangular pulse[2]

$$p(t) = \begin{cases} A, & 0 < t < T \\ 0, & \text{otherwise} \end{cases} \tag{4.7}$$

Its Fourier transform is

$$G(f) = AT\frac{\sin \pi fT}{\pi fT}e^{-j2\pi fT/2}$$

Thus the PSD of the baseband BPSK signal is

$$\Psi_{\tilde{s}}(f) = \frac{|G(f)|^2}{T} = A^2T\left(\frac{\sin \pi fT}{\pi fT}\right)^2, \text{ (BPSK)} \tag{4.8}$$

[2] The bipolarity of the baseband waveform of BPSK is controlled by the bipolar data $a_k = \pm 1$.

which is plotted in Figure 4.5. From the figure we can see that the null-to-null bandwidth $B_{null} = 2/T = 2R_b$. (Keep in mind that the PSD at the carrier frequency is two-sided about f_c.) Figure 4.5(c) is the out-of-band power curve which is defined by (2.21). From this curve we can estimate that $B_{90\%} \approx 1.7R_b$ (corresponding to –10 dB point on the curve). We also calculated that $B_{99\%} \approx 20R_b$.

4.2 DIFFERENTIAL BPSK

In Chapter 2 we first introduced differential encoding and decoding of binary data. This technique can be used in PSK modulation. We denote differentially encoded BPSK as DEBPSK. Figure 4.6(a) is the DEBPSK modulator. DEBPSK signal can be coherently demodulated or differentially demodulated. We denote the modulation scheme that uses differential encoding and differential demodulation as DBPSK, which is sometimes simply called DPSK.

DBPSK does not require a coherent reference signal. Figure 4.6(b) is a simple, but suboptimum, differential demodulator which uses the previous symbol as the reference for demodulating the current symbol.[3] The front-end bandpass filter reduces noise power but preserves the phase of the signal. The integrator can be replaced by an LPF. The output of the integrator is

$$l = \int_{kT}^{(k+1)T} r(t)r(t-T)dt$$

In the absence of noise and other channel impairment,

$$l = \int_{kT}^{(k+1)T} s_k(t)s_{k-1}(t)dt = \begin{cases} E_b, \text{ if } s_k(t) = s_{k-1}(t) \\ -E_b, \text{ if } s_k(t) = -s_{k-1}(t) \end{cases}$$

where $s_k(t)$ and $s_{k-1}(t)$ are the current and the previous symbols. The integrator output is positive if the current signal is the same as the previous one, otherwise the output is negative. This is to say that the demodulator makes decisions based on the difference between the two signals. Thus information data must be encoded as the difference between adjacent signals, which is exactly what the differential encoding can accomplish. Table 4.1 shows an example of differential encoding, where an arbitrary reference bit 1 is chosen. The encoding rule is

$$d_k = \overline{a_k \oplus d_{k-1}}$$

[3] This is the commonly referred DPSK demodulator. Another DPSK demodulator is the optimum differentially coherent demodulator. Differentially encoded PSKs can also be coherently detected. These will be discussed shortly.

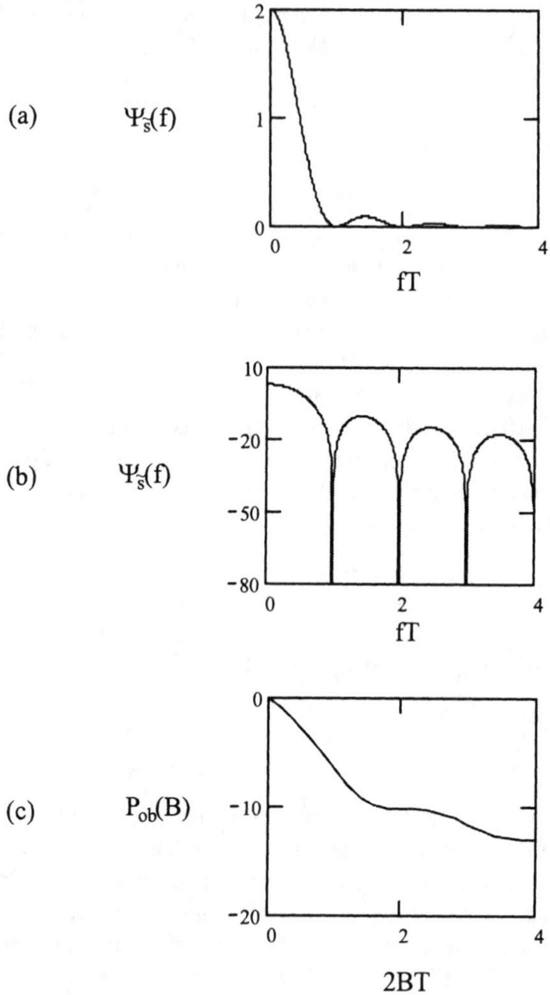

Figure 4.5 Power spectral density of BPSK: (a) linear, (b) logarithmic, and (c) out-of-band power.

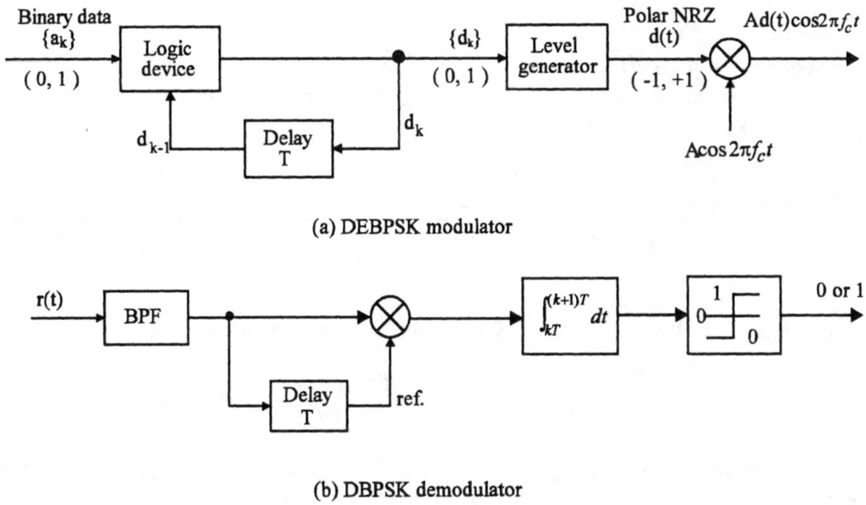

Figure 4.6 DBPSK modulator (a), and demodulator (b).

Inversely we can recover a_k from d_k using

$$a_k = \overline{d_k \oplus d_{k-1}}$$

If d_k and d_{k-1} are the same, then they represent a 1 of a_k. If d_k and d_{k-1} are different, they represent a 0 of a_k. This can be verified by comparing sequences $\{d_k\}$ and $\{a_k\}$ in the table. The sequence $\{d_k\}$ is modulated onto a carrier with phase 0 or π. In the absence of noise and other channel impairment, the demodulator output $\widehat{a}_k$ is identical to the message sequence.

The preceding receiver is suboptimum, since the reference signal is the previous symbol which is noisy. The optimum noncoherent, or differentially coherent, demodulation of DEBPSK is presented now. As discussed above, a message bit is represented by two modulated symbols. If the transmitted bit is 1, the two symbols are the same. Thus we can define a signal with a duration of $2T$ as follows to represent binary 1

$$\xi_1(t) = \begin{cases} A\cos 2\pi f_c t, & 0 \le t \le T \\ A\cos 2\pi f_c t, & T \le t \le 2T \end{cases}, \quad \text{for binary 1}$$

Modulation	ref.									
Message a_k		1	0	1	1	0	0	0	1	1
Encoding $d_k = \overline{a_k \oplus d_{k-1}}$	1	1	0	0	0	1	0	1	1	1
Signal phase θ	0	0	π	π	π	0	π	0	0	0
Demodulation										
$\frac{l}{E_b} = \frac{1}{E_b}\int_{kT}^{(k+1)T} s_k(t)s_{k-1}(t)dt$		1	-1	1	1	-1	-1	-1	1	1
Demodulator output $\widehat{a}_k$		1	0	1	1	0	0	0	1	1

Table 4.1 Examples of differential coding.

If the transmitted bit is 0, the two symbols are different. Thus we can define

$$\xi_2(t) = \begin{cases} A\cos 2\pi f_c t, & 0 \le t \le T \\ -A\cos 2\pi f_c t, & T \le t \le 2T \end{cases}, \quad \text{for binary 0}$$

Note that in the modulated signal stream, the $2T$-symbols are overlapped by T seconds.

Since we desire an optimum noncoherent demodulation, the DBPSK receiver may be implemented in the general forms for signals with unknown phases as depicted in Appendix B. However, a simpler form is possible due to the special property of the signals. The simpler form avoids the squarers or matched filters. We derive this receiver starting from (B.55). Assuming the received signal is $r(t)$, the sufficient statistic for $\xi_1(t)$ is

$$\begin{aligned} l_1^2 &= \left(\int_0^{2T} r(t)\xi_1(t)dt\right)^2 + \left(\int_0^{2T} r(t)\xi_1(t,\frac{\pi}{2})dt\right)^2 \\ &= \left(\int_0^{2T} r(t)A\cos 2\pi f_c t dt\right)^2 + \left(\int_0^{2T} r(t)A\sin 2\pi f_c t dt\right)^2 \\ &= (w_0 + w_1)^2 + (z_0 + z_1)^2 \end{aligned}$$

where

$$w_0 \triangleq \int_0^T r(t)A\cos 2\pi f_c t dt$$

$$w_1 \triangleq \int_T^{2T} r(t)A\cos 2\pi f_c t dt$$

$$z_0 \triangleq \int_0^T r(t)A\sin 2\pi f_c t dt$$

$$z_1 \triangleq \int_T^{2T} r(t) A \sin 2\pi f_c t dt$$

Similarly, the sufficient statistic for $\xi_2(t)$ is

$$l_2^2 = (w_0 - w_1)^2 + (z_0 - z_1)^2$$

The decision rule is

$$l_1^2 \underset{0}{\overset{1}{\gtrless}} l_2^2$$

Substituting expressions for l_1^2 and l_2^2 into the above expression and cancelling like terms, we obtain

$$x \triangleq w_1 w_0 + z_1 z_0 \underset{0}{\overset{1}{\gtrless}} 0$$

For the kth symbol period, this rule is

$$x_k \triangleq w_k w_{k-1} + z_k z_{k-1} \underset{0}{\overset{1}{\gtrless}} 0 \tag{4.9}$$

This rule can be implemented by the receiver shown in Figure 4.7. The reference signals are locally generated since phase synchronization between $r(t)$ and the reference signals is not required. However, the frequency of the reference signals must be the same as the received signal's. This can be maintained by using stable oscillators, such as crystal oscillators, in both transmitter and receiver. However, in the case where Doppler shift exists in the carrier frequency, such as in mobile communications, frequency tracking is needed to maintain the same frequency. In this case the local oscillator must be synchronized in frequency to the received signal. The reference signals' amplitude A is set as 1 in Figure 4.7. In fact A could be any value since its value will not affect the decision rule in (4.9). The correlators produce w_k and z_k. The x_k is calculated by the delay-and-multiply circuits or differential decoders.

To derive the error probability of the optimum demodulator, we observe that two DBPSK symbols are orthogonal over $[0, 2T]$ since

$$\int_0^{2T} \xi_1(t)\xi_2(t)dt = \int_0^T (A\cos 2\pi f_c t)^2 dt - \int_T^{2T} (A\cos 2\pi f_c t)^2 = 0$$

In other words DBPSK is a special case of noncoherent orthogonal modulation with $T_s = 2T$ and $E_s = 2E_b$. Hence using the result of Appendix B (B.56) we have the

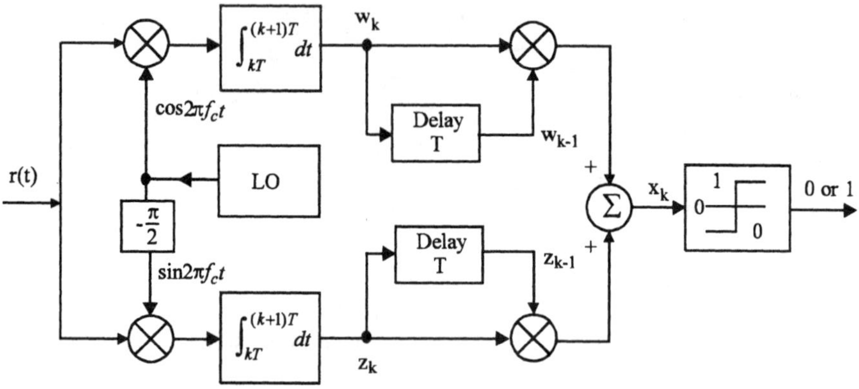

Figure 4.7 Optimum demodulator for DBPSK.

bit error probability

$$P_b = \frac{1}{2}e^{-E_b/N_o}, \text{ (optimum DBPSK)} \tag{4.10}$$

Note that the demodulator of Figure 4.7 does not require phase synchronization between the reference signals and the received signal. But it does require the reference frequency be the same as the received signal. Therefore the suboptimum receiver in Figure 4.6(b) is more practical, and indeed it is the usual-sense DBPSK receiver. Its error performance is slightly inferior to that of the optimum given in (4.10).

The performance of the suboptimum receiver is given by Park in [1]. It is shown that if an ideal narrow-band IF filter with bandwidth W is placed before the correlator in Figure 4.6(b), the bit error probability is

$$P_b = \frac{1}{2}e^{-0.76E_b/N_o}, \quad \text{for } W = 0.5/T$$

or

$$P_b = \frac{1}{2}e^{-0.8E_b/N_o}, \quad \text{for } W = 0.57/T$$

which amounts to a loss of 1.2 dB and 1 dB, respectively, with respect to the optimum.

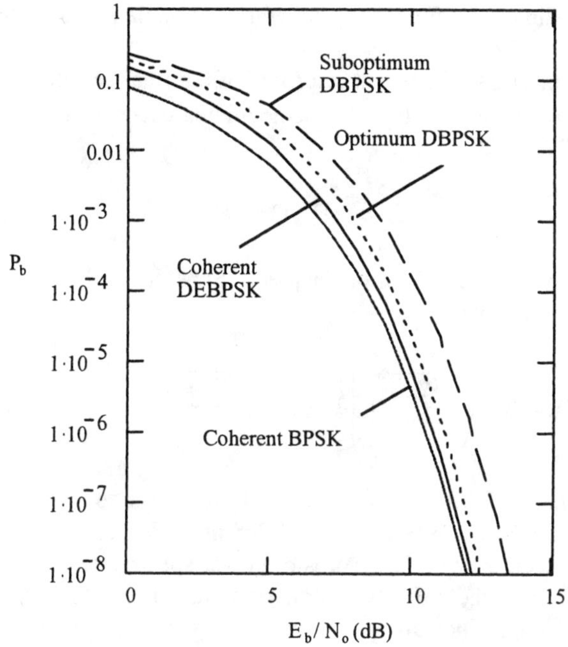

Figure 4.8 P_b of the differential BPSK in comparison with the coherent BPSK scheme.

If an ideal wide-band IF filter is used, then

$$\begin{aligned} P_b &\approx Q\left(\sqrt{\frac{E_b}{N_o}}\right), \quad \text{for } W > 1/T \\ &\approx \frac{1}{2\sqrt{\pi}\sqrt{E_b/2N_o}} e^{-E_b/2N_o}, \quad \text{for } W > 1/T \text{, (suboptimum DBPSK)} \end{aligned}$$

Typical value of W is 1.5/T. If W is too large or too small the above expression does not hold [1]. The P_b for the wide-band suboptimum receiver is about 2 dB worse than the optimum at high SNR. The bandwidth should be chosen as 0.57/T for the best performance. P_b curves of DBPSK are shown in Figure 4.8.

A differentially encoded BPSK signal can also be demodulated coherently (denoted as DEBPSK). It is used when the purpose of differential encoding is to eliminate phase ambiguity in the carrier recovery circuit for coherent PSK (see Section

4.10). This is not usually meant by the name DBPSK. DBPSK refers to the scheme of differential encoding and differentially coherent demodulation as we have discussed above.

In the case of DEBPSK, the bit error rate of the final decoded sequence $\{\widehat{a}_k\}$, P_b is related to the bit error rate of the demodulated encoded sequence $\{\widehat{d}_k\}$, $P_{b,d}$, by

$$P_b = 2P_{b,d}(1 - P_{b,d}) \tag{4.11}$$

as we have shown in Section 2.4.1 of Chapter 2. Substituting $P_{b,d}$ as in (4.6) into the above expression we have

$$P_b = 2Q\left(\sqrt{\frac{2E_b}{N_o}}\right)\left[1 - Q\left(\sqrt{\frac{2E_b}{N_o}}\right)\right], \text{ (DEBPSK)} \tag{4.12}$$

for coherently detected differentially encoded PSK. For large SNR, this is just about two times that of coherent BPSK without differential encoding.

Finally we need to say a few words of power spectral density of differentially encoded BPSK. Since the difference of differentially encoded BPSK from BPSK is differential encoding, which always produces an asymptotically equally likely data sequence (see Section 2.1), the PSD of the differentially encoded BPSK is the same as BPSK which we assume is equally likely. The PSD is shown in Figure 4.5. However, it is worthwhile to point out that if the data sequence is not equally likely the PSD of the BPSK is not the one in Figure 4.5, but the PSD of the differentially encoded PSK is still the one in Figure 4.5.

4.3 *M*-ARY PSK

The motivation behind MPSK is to increase the bandwidth efficiency of the PSK modulation schemes. In BPSK, a data bit is represented by a symbol. In MPSK, $n = \log_2 M$ data bits are represented by a symbol, thus the bandwidth efficiency is increased to n times. Among all MPSK schemes, QPSK is the most-often-used scheme since it does not suffer from BER degradation while the bandwidth efficiency is increased. We will see this in Section 4.6. Other MPSK schemes increase bandwidth efficiency at the expense of BER performance.

M-ary PSK signal set is defined as

$$s_i(t) = A\cos(2\pi f_c t + \theta_i), \quad 0 \le t \le T, \quad i = 1, 2, \ldots, M \tag{4.13}$$

where

$$\theta_i = \frac{(2i-1)\pi}{M}$$

The carrier frequency is chosen as an integer multiple of the symbol rate, therefore in any symbol interval, the signal initial phase is also one of the M phases. Usually M is chosen as a power of 2 (i.e., $M = 2^n$, $n = \log_2 M$). Therefore, the binary data stream is divided into n-tuples. Each of them is represented by a symbol with a particular initial phase.

The above expression can be written as

$$\begin{aligned} s_i(t) &= A\cos\theta_i \cos 2\pi f_c t - A\sin\theta_i \sin 2\pi f_c t \\ &= s_{i1}\phi_1(t) + s_{i2}\phi_2(t) \end{aligned} \tag{4.14}$$

where $\phi_1(t)$ and $\phi_2(t)$ are orthonormal basis functions (see (4.2) and (4.3)), and

$$s_{i1} = \int_0^T s_i(t)\phi_1(t)dt = \sqrt{E}\cos\theta_i$$

$$s_{i2} = \int_0^T s_i(t)\phi_2(t)dt = \sqrt{E}\sin\theta_i$$

where

$$E = \frac{1}{2}A^2T$$

is the symbol energy of the signal. The phase is related with s_{i1} and s_{i2} as

$$\theta_i = \tan^{-1}\frac{s_{i2}}{s_{i1}}$$

The MPSK signal constellation is therefore two-dimensional. Each signal $s_i(t)$ is represented by a point (s_{i1}, s_{i2}) in the coordinates spanned by $\phi_1(t)$ and $\phi_2(t)$. The polar coordinates of the signal are $(\sqrt{E}, \theta_i)$. That is, its magnitude is $\sqrt{E}$ and its angle with respect to the horizontal axis is θ_i. The signal points are equally spaced on a circle of radius $\sqrt{E}$ and centered at the origin. The bits-signal mapping could be arbitrary provided that the mapping is one-to-one. However, a method called Gray coding is usually used in signal assignment in MPSK. Gray coding assigns n-tuples with only one-bit difference to two adjacent signals in the constellation. When an M-ary symbol error occurs, it is more likely that the signal is detected as the adjacent signal on the constellation, thus only one of the n input bits is in error. Figure 4.9 is the constellation of 8-PSK, where Gray coding is used for bit assignment. Note that

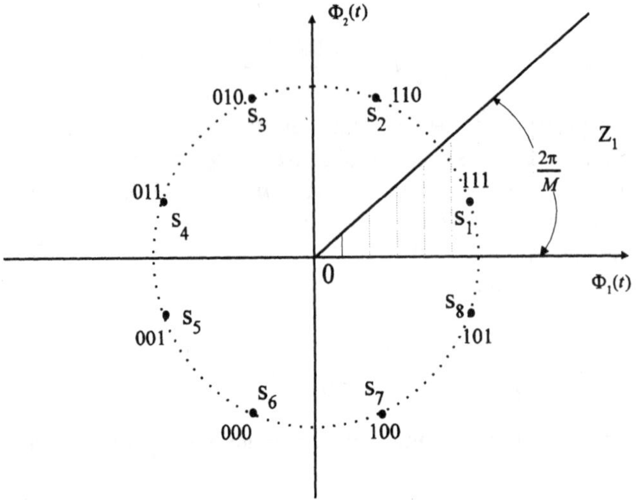

Figure 4.9 8-PSK constellation with Gray coded bit assignment.

BPSK and QPSK are special cases of MPSK with $M = 2$ and 4, respectively. On the entire time axis, we can write MPSK signal as

$$s(t) = s_1(t)\cos 2\pi f_c t - s_2(t)\sin 2\pi f_c t, \quad -\infty < t < \infty \tag{4.15}$$

where

$$s_1(t) = A\sum_{k=-\infty}^{\infty}\cos(\theta_k)p(t-kT) \tag{4.16}$$

$$s_2(t) = A\sum_{k=-\infty}^{\infty}\sin(\theta_k)p(t-kT) \tag{4.17}$$

where θ_k is one of the M phases determined by the input binary n-tuple and $p(t)$ is the rectangular pulse with unit amplitude defined on $[0, T]$. Expression (4.15) implies that the carrier frequency is an integer multiple of the symbol timing so that the initial phase of the signal in any symbol period is θ_k.

Since MPSK signals are two-dimensional, for $M \geq 4$, the modulator can be implemented by a quadrature modulator. The MPSK modulator is shown in Figure 4.10. The only difference for different values of M is the level generator. Each n-tuple of the input bits is used to control the level generator. It provides the I- and

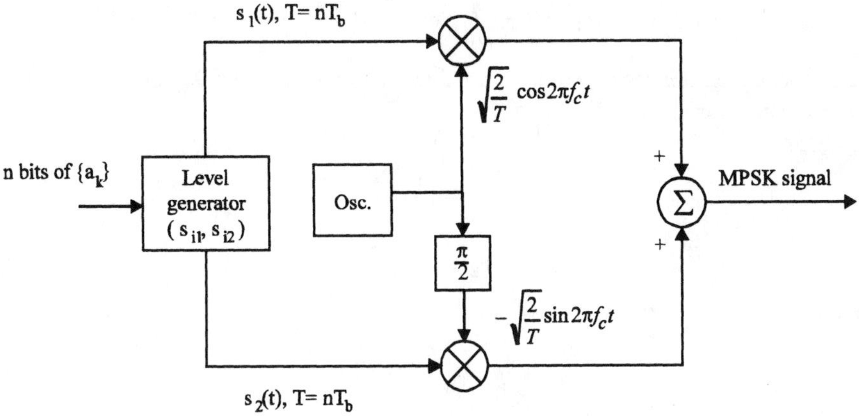

Figure 4.10 MPSK modulator.

Q-channels with the particular sign and level for a signal's horizontal and vertical coordinates, respectively. For QPSK, the level generator is particularly simple, it is simply a serial-to-parallel converter (see Section 4.6).

Modern technology intends to use completely digital devices. In such an environment, MPSK signals are digitally synthesized and fed to a D/A converter whose output is the desired phase modulated signal.

The coherent demodulation of MPSK could be implemented by one of the coherent detectors for M-ary signals as described in Appendix B. Since the MPSK signal set has only two basis functions, the simplest receiver is the one that uses two correlators (Figure B.8 with $N = 2$). Due to the special characteristic of the MPSK signal, the general demodulator of Figure B.8 can be further simplified. For MPSK the sufficient statistic is

$$\begin{aligned} l_i &= \int_0^T r(t)s_i(t)dt = \int_0^T r(t)[s_{i1}\phi_1(t) + s_{i2}\phi_2(t)]dt \\ &= \int_0^T r(t)[\sqrt{E}\cos\theta_i\phi_1(t) + \sqrt{E}\sin\theta_i\phi_2(t)]dt \\ &= \sqrt{E}\left[r_1\cos\theta_i + r_2\sin\theta_i\right] \end{aligned} \tag{4.18}$$

where

$$r_1 \triangleq \int_0^T r(t)\phi_1(t)dt = \int_0^T [s(t)+n(t)]\phi_1(t)dt = s_{i1} + n_1$$

$$r_2 \triangleq \int_0^T r(t)\phi_2(t)dt = \int_0^T [s(t)+n(t)]\phi_2(t)dt = s_{i2}+n_2$$

are independent Gaussian random variables with mean values s_{i1} and s_{i2}, respectively. Their variance is $N_o/2$.

Let

$$r_1 = \rho\cos\widehat{\theta}$$

$$r_2 = \rho\sin\widehat{\theta}$$

then

$$\rho = \sqrt{r_1^2 + r_2^2} \tag{4.19}$$

$$\widehat{\theta} \triangleq \tan^{-1}\frac{r_2}{r_1} \tag{4.20}$$

$$\begin{aligned} l_i &= \sqrt{E}[\rho\cos\widehat{\theta}\cos\theta_i + \rho\sin\widehat{\theta}\sin\theta_i] \\ &= \sqrt{E}\rho\cos(\theta_i - \widehat{\theta}) \end{aligned}$$

In the absence of noise, $\widehat{\theta} = \tan^{-1} r_2/r_1 = \tan^{-1} s_{i2}/s_{i1} = \theta_i$. With noise, $\widehat{\theta}$ will deviate from θ_i. Since ρ is independent of any signal, then choosing the largest l_i is equivalent to choosing the smallest $|\theta_i - \widehat{\theta}|$. This rule is in fact to choose signal $s_i(t)$ when $\mathbf{r} = \begin{bmatrix} r_1 \\ r_2 \end{bmatrix}$ falls inside the pie-shaped decision region of the signal (see Figure 4.9). Figure 4.11 is the demodulator based on the above decision rule where subscript k indicates the kth symbol period and CR stands for carrier recovery. Note that the amplitude of the reference signals can be any value, which is $\sqrt{2/T}$ in the figure, since the effect of the amplitude is cancelled when computing $\widehat{\theta}_k$.

The symbol error probability can be derived as follows. Given $s_i(t)$ is transmitted (or hypothesis H_i is true), the received vector $\mathbf{r} = \begin{bmatrix} r_1 \\ r_2 \end{bmatrix}$ is a point in the $\phi_1(t) - \phi_2(t)$ plane. Its joint probability density function is two-dimensional.

$$p(\mathbf{r}/H_i) = \frac{1}{\pi N_o}\exp\{-\frac{1}{N_o}[(r_1 - \sqrt{E}\cos\theta_i)^2 + (r_2 - \sqrt{E}\sin\theta_i)^2]\}$$

Geometrically, the PDF is a bell-shape surface centered at $\mathbf{s}_i = \begin{bmatrix} s_{i1} \\ s_{i2} \end{bmatrix}$ (Figure 4.12).

An error occurs when $\mathbf{r}$ falls outside the decision region Z_i (see Figure 4.9). Thus

$$P_s = 1 - \int_{Z_i} p(\mathbf{r}/H_i)d\mathbf{r}$$

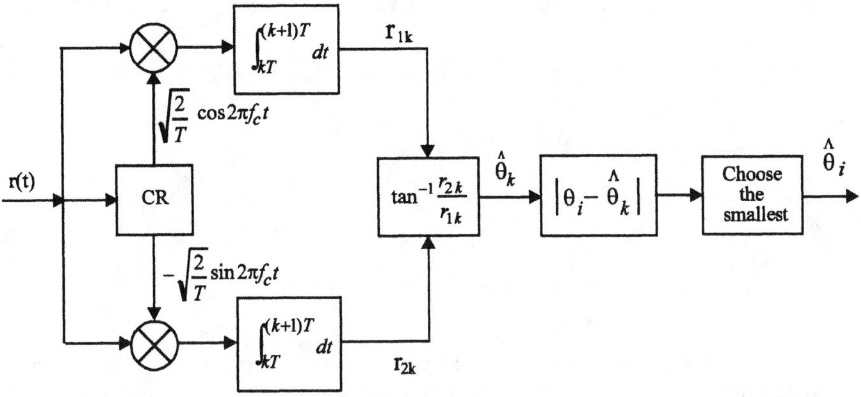

Figure 4.11 Coherent MPSK demodulator using two correlators.

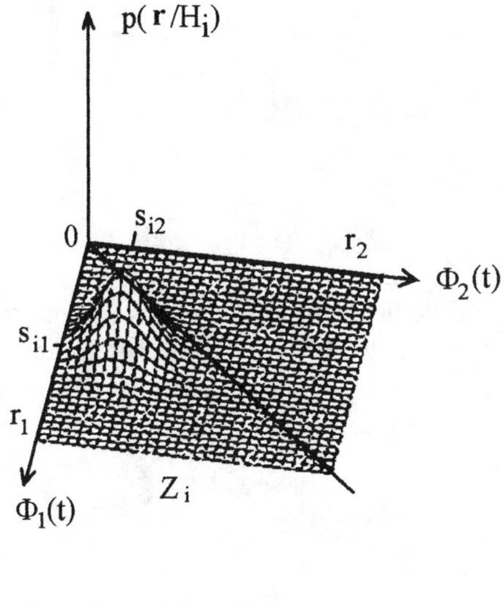

Figure 4.12 Joint PDF of **r** given $s_i(t)$ is transmitted.

Using (4.19) and (4.20) we can transform the above into polar coordinates.[4]

$$\begin{aligned} P_s &= 1 - \int_{Z_i} \frac{1}{\pi N_o} \exp\{-\frac{1}{N_o}[\rho^2 + E - 2\rho\sqrt{E}\cos(\theta_i - \widehat{\theta})]\}\rho d\rho d\theta \\ &= 1 - \int_{Z_i} p(\rho, \widehat{\theta}/H_i) d\rho d\theta \end{aligned}$$

where

$$p(\rho, \widehat{\theta}/H_i) = \frac{\rho}{\pi N_o} \exp\{-\frac{1}{N_o}[\rho^2 + E - 2\rho\sqrt{E}\cos(\theta_i - \widehat{\theta})]\}$$

is the joint probability density of ρ and θ. We define $\varphi = \widehat{\theta} - \theta_i$, which represents the phase deviation of the received signal from the transmitted one. Integrating both sides of the above with respect to ρ yields the PDF of $\varphi \in [-\pi, \pi]$ (see Appendix 4A for derivation).

$$\begin{aligned} p(\varphi/H_i) &= \frac{e^{-E/N_o}}{2\pi}\left\{1 + \sqrt{\frac{\pi E}{N_o}}(\cos\varphi)e^{(E/N_o)\cos^2\varphi}\right. \\ &\quad \left.\cdot\left[1 + \text{erf}\left(\sqrt{\frac{E}{N_o}}\cos\varphi\right)\right]\right\} \\ &= p(\varphi) \end{aligned}$$

where

$$\text{erf}(x) \triangleq \frac{2}{\sqrt{\pi}}\int_0^x e^{-t^2} dt$$

is the *error function*. Note the distribution of φ is independent of index i. This is intuitively correct since φ is the phase deviation, not the absolute phase.

The symbol error probability is the probability that $\widehat{\theta}$ is outside the decision region, or the deviation φ is greater than π/M in absolute value.

$$P_s = 1 - \int_{-\pi/M}^{\pi/M} p(\varphi)d\varphi \tag{4.21}$$

When $M = 2$ (BPSK) and $M = 4$ (QPSK) this integration results in the formulas given by (4.6) and (4.37). For $M > 4$, this expression cannot be evaluated in a closed form and the symbol error probability can be obtained by numerically integrating

4 Note that $dr_1 dr_2 = \rho d\rho d\widehat{\theta}$.

(4.21). Another form of P_s is given in [2, p. 209]. The derivation is very complicated and is omitted here. The result is

$$\begin{aligned} P_s &= \frac{M-1}{M} - \frac{1}{2}\,\text{erf}\left[\sqrt{\frac{E}{N_o}}\sin\frac{\pi}{M}\right] \\ &\quad -\frac{1}{\sqrt{\pi}}\int_0^{\sqrt{E/N_o}\sin\pi/M} e^{-y^2}\,\text{erf}(y\cot\frac{\pi}{M})dy \end{aligned} \tag{4.22}$$

This again can only be numerically evaluated for $M > 4$.

Figure 4.13 shows P_s curves for $M = 2, 4, 8, 16$, and 32 given by the exact expression (4.22). Beyond $M = 4$, doubling the number of phases, or increasing one bit in the n-tuples represented by the phases, requires a substantial increase in SNR. For example, at $P_s = 10^{-5}$, the SNR difference between $M = 4$ and $M = 8$ is approximately 4 dB and the difference between $M = 8$ and $M = 16$ is approximately 5 dB. For large values of M, doubling the number of phases requires an SNR increase of 6 dB to maintain the same performance.

For $E/N_o \gg 1$, we can derive an approximation of the P_s expression. First we can use the approximation[5]

$$\text{erf}(x) \approx 1 - \frac{e^{-x^2}}{\sqrt{\pi}x}, \quad x \gg 1$$

to obtain the approximation of the PDF of the phase deviation

$$p(\varphi/H_i) \approx \sqrt{\frac{E}{\pi N_o}}(\cos\varphi)e^{-(E/N_o)\sin^2\varphi} \tag{4.23}$$

Finally substituting (4.23) into (4.21) we arrive at the result

$$\begin{aligned} P_s &\approx \text{erfc}\left(\sqrt{\frac{E}{N_o}}\sin\frac{\pi}{M}\right) \\ &= 2Q\left(\sqrt{\frac{2E}{N_o}}\sin\frac{\pi}{M}\right), \text{ (coherent MPSK)} \end{aligned} \tag{4.24}$$

where

$$\text{erfc}(x) = 1 - \text{erf}(x) = 2Q(\sqrt{2}x)$$

---

5 In fact $1 - \frac{e^{-x^2}}{\sqrt{\pi}x}$ is a lower bound of $\text{erf}(x)$, however they are extremely close for $x >> 1$.

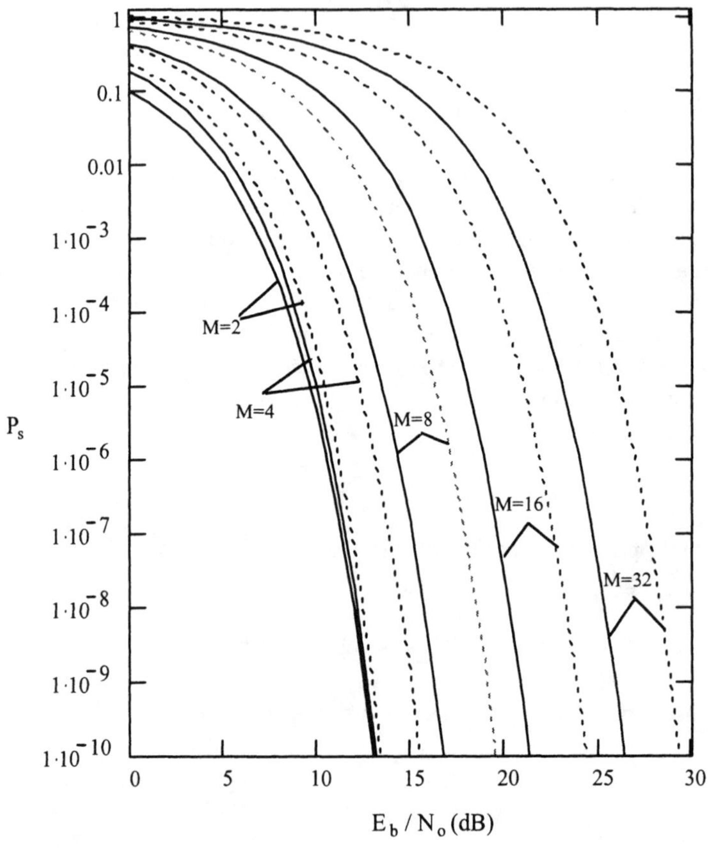

Figure 4.13 P_s of MPSK (solid lines) and DMPSK (dotted lines).

is the *complementary error function*. Note that only high signal-to-noise ratio assumption is needed for the approximation. Therefore (4.24) is good for any values of M, even though it is not needed for $M = 2$ and 4 since precise formulas are available.

Expression (4.24) can be derived geometrically. Consider Figure 4.14. Due to symmetry of the signal constellation, P_s is equal to the error probability of detecting $\mathbf{s}_1$, which is the probability that the received signal vector $\mathbf{r}$ does not fall in the decision region Z_1. This is bounded below and above as follows

$$\Pr(\mathbf{r} \in A_1) \leq P_s < \Pr(\mathbf{r} \in A_1) + \Pr(\mathbf{r} \in A_2) = 2\Pr(\mathbf{r} \in A_2)$$

where the equal sign on the left part of the inequality accounts for the case of $M = 2$. The distance from $\mathbf{s}_1$ to the nearest signal is

$$d_{12} = d_{18} = 2\sqrt{E}\sin\frac{\pi}{M}$$

Since white Gaussian noise is identically distributed along any set of orthogonal axes [3, Chapter 3], we may temporarily choose the first axis in such a set as one that passes through the points $\mathbf{s}_1$ and $\mathbf{s}_2$, then for high SNR

$$\begin{aligned} \Pr(\mathbf{r} \quad &\in \quad A_1) = \Pr(\mathbf{r} \in A_2) \approx \Pr(\mathbf{s}_2/H_1) \\ &= \int_{-\infty}^{-d_{12}/2} \frac{1}{\sqrt{\pi N_o}} \exp\{-\frac{x^2}{N_o}\}dx = Q\left(\sqrt{\frac{2E}{N_o}}\sin\frac{\pi}{M}\right) \end{aligned}$$

Thus

$$Q\left(\sqrt{\frac{2E}{N_o}}\sin\frac{\pi}{M}\right) \leq P_s \lesssim 2Q\left(\sqrt{\frac{2E}{N_o}}\sin\frac{\pi}{M}\right)$$

Since the lower and upper bounds differ only by a factor of two, which translates into a very small difference in terms of SNR, these bounds are very tight.

The bit error rate can be related to the symbol error rate by

$$P_b \approx \frac{P_s}{\log_2 M} \tag{4.25}$$

for Gray coded MPSK signals since most likely the erroneous symbols are the adjacent signals which only differ by one bit.

The approximation given by (4.24) and (4.25) is not accurate at low SNRs. A more accurate BER expression for Gray coded MPSK signals is derived in [4], which

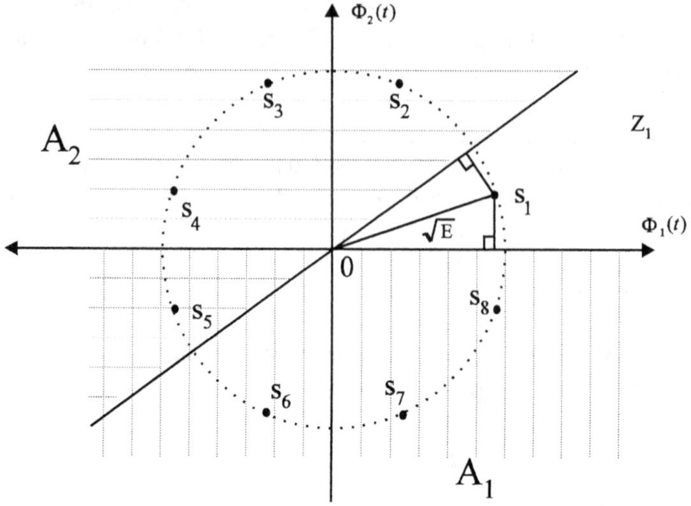

Figure 4.14 Decision regions for bounding P_s of MPSK signals.

is

$$P_b \approx \frac{2}{\max(\log_2 M, 2)} \sum_{i=1}^{\max(M/4,1)} Q\left(\sqrt{\frac{2E_b \log_2 M}{N_o}} \sin \frac{(2i-1)\pi}{M}\right)$$

For $M = 2$ and 4, the modulations are BPSK and QPSK and exact BER expressions exist. Therefore the above expression is really meaningful for $M > 4$. For example, $M = 16, 32$, the number of the terms in the sum will be $4, 8$, respectively. However, simulations in [4] show that when two terms are included, the calculated results are almost identical with the simulation results. If all terms are included, the calculated results are upper bounds. The approximation given by (4.24) and (4.25) is in fact a lower bound. These upper and lower bounds are very tight at $E_b/N_o \geq 5$ dB for 16-PSK and $E_b/N_o \geq 9$ dB for 32-PSK. However they are loose below those SNRs. Thus for $M \geq 8$, based on the fact that two-term approximation is the best, we can obtain the most accurate BER expression for MPSK as

$$P_b \approx \frac{2}{\log_2 M} \sum_{i=1}^{2} Q\left(\sqrt{\frac{2E_b \log_2 M}{N_o}} \sin \frac{(2i-1)\pi}{M}\right)$$

4.4 PSD OF MPSK

The PSD of MPSK is similar to that of BPSK except that the spectral is narrower on a frequency scale normalized to the bit rate. As for all carrier modulated signals, it suffices to find the PSD of the complex envelope (Appendix A).

Substituting (4.16) and (4.17) into (4.15), we can write the MPSK signal as

$$s(t) = \mathrm{Re}\left\{\left[\sum_{k=-\infty}^{\infty} A\exp(j\theta_k)p(t-kT)\right]\exp(j2\pi f_c t)\right\}$$

Thus the complex envelope of MPSK is

$$\begin{aligned}\widetilde{s}(t) &= \sum_{k=-\infty}^{\infty} A\exp(j\theta_k)p(t-kT) \\ &= \sum_{k=-\infty}^{\infty} A\cos\theta_k p(t-kT) + j\sum_{k=-\infty}^{\infty} A\sin\theta_k p(t-kT)\end{aligned}$$

where

$$\cos\theta_k \in \left(\cos\frac{(2i-1)\pi}{M}, i = 1, 2...M\right)$$

is a random variable which has $M/2$ different values with equal probabilities $(\frac{2}{M})$. Refer to the example of 8-PSK in Figure 4.9. We can see that $\cos\theta_k = \cos\frac{\pi}{8}$ or $\cos\frac{3\pi}{8}$ or $\cos\frac{5\pi}{8}$ or $\cos\frac{7\pi}{8}$, which is $\cos\theta_k = 0.924$ or 0.383 or -0.383 or -0.924. These values are symmetrical about zero. Thus the mean value is zero. The mean square value is

$$\sigma^2 = \sum_{i=1}^{M/2}\frac{2}{M}\cos^2\frac{(2i-1)\pi}{M} = \frac{1}{2}$$

Note that the mean square value is always $\frac{1}{2}$ for $M = 2^n, n > 1$. The distribution of $\sin\theta_k$ is the same.

Thus the complex envelope can be written as

$$\widetilde{s}(t) = \sum_{k=-\infty}^{\infty} x_k p(t-kT) + j\sum_{k=-\infty}^{\infty} y_k p(t-kT)$$

where $\{x_k = \cos\theta_k\}$ and $\{y_k = \sin\theta_k\}$ are independent, identically distributed random sequences with zero means and a mean square value of $1/2$. The PSD of this type of complex envelope has been derived in Appendix A. The result (A.21) can be directly used here.

Since $\sigma_x^2 = \sigma_y^2 = \sigma^2 = 1/2$ and

$$|P(f)| = \left| AT \frac{\sin \pi f T}{\pi f T} \right|$$

then from (A.21) we have

$$\begin{aligned} \Psi_{\tilde{s}}(f) &= 2\sigma^2 A^2 T \left(\frac{\sin \pi f T}{\pi f T} \right)^2 \\ &= A^2 T \left(\frac{\sin \pi f T}{\pi f T} \right)^2 \\ &= A^2 n T_b \left(\frac{\sin \pi f n T_b}{\pi f n T_b} \right)^2, \text{ (MPSK)} \end{aligned} \tag{4.26}$$

where $n = \log_2 M$. This is exactly the same as that of BPSK in terms of symbol rate. However, in terms of bit rate the PSD of MPSK is n-times narrower than the BPSK. Figure 4.15 is the PSDs ($A = \sqrt{2}$ and $T_b = 1$ for unit bit energy: $E_b = 1$) for different values of M where the frequency axis is normalized to the bit rate (fT_b).

Since the passband minimum (Nyquist) bandwidth required to transmit the symbols is $1/T$, the maximum bandwidth efficiency is

$$\frac{R_b}{B_{\min}} = \frac{(\log_2 M)/T}{1/T} = \log_2 M$$

4.5 DIFFERENTIAL MPSK

In Section 4.2 we discussed DBPSK, which is in fact a special case of differential MPSK (DMPSK). The term DMPSK refers to "differentially encoded and differentially coherently demodulated MPSK." The differentially coherent demodulation is in fact noncoherent in the sense that phase coherent reference signals are not required. It is used to overcome the adversary effect of the random phase in the received signal.

Differentially encoded MPSK can also be coherently demodulated (denoted as DEMPSK). In this case, the purpose of differential encoding is to eliminate phase ambiguity in the carrier recovery process. This is not usually meant by the term DMPSK.

In both cases, the modulation processes are the same. In other words, the trans-

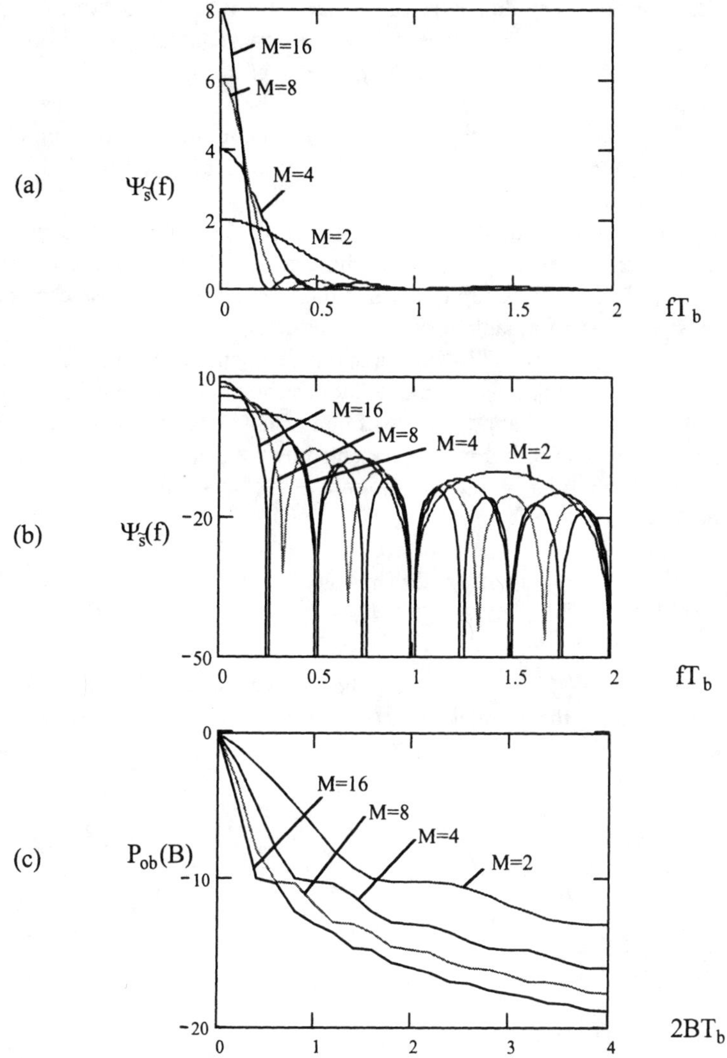

Figure 4.15 PSDs of MPSK: (a) linear, (b) logarithmic, and (c) out-of-band power.

mitted MPSK signals are the same. Only demodulations are different.

In the modulator the information bits are first differentially encoded. Then the encoded bits are used to modulate the carrier. In a DEMPSK signal stream, information is carried by the phase difference $\Delta\theta_i$ between two consecutive symbols. There are M different values of $\Delta\theta_i$ and each represents an n-tuple $(n = \log_2 M)$ of information bits.

For $M = 2$ and 4, encoding, modulation, and demodulation are simple, as we have seen in Section 4.2 for DBPSK and will see shortly for differentially encoded QPSK.

In light of the modern digital technology, DEMPSK signals can be generated by a digital frequency synthesis technique. A phase change from one symbol to the next is simply controlled by the n-tuple which is represented by the phase change. This technique is particularly suitable for large values of M.

In DMPSK scheme, the DEMPSK signal is demodulated by a differentially coherent (or optimum noncoherent) demodulator as shown in Figure 4.16.

The derivation of the demodulator is similar to that of binary DPSK. In DEMPSK a message m_i of $n = \log_2 M$ bits is represented by the phase difference of two consecutive symbols. In other words, m_i is represented by a symbol with two symbol periods defined as

$$\xi_i(t) = \begin{cases} A\cos\left[2\pi f_c t + \Phi_0\right], & 0 \le t \le T \\ A\cos\left[2\pi f_c t + \Phi_0 + \Delta\theta_i\right], & T \le t \le 2T \end{cases}$$

where $\Delta\,\theta_i = \frac{2(i-1)\pi}{M}, i = 1, 2, \ldots M$. The received signal $r(t)$ has an unknown phase θ introduced by the channel and is corrupted by AWGN. Consider the first two symbol durations $[0 \le t \le 2T]$, from (B.55), the sufficient statistic for detecting $\xi_i(t)$ is

$$l_i^2 = \left(\int_0^{2T} r(t)\xi_i(t)dt\right)^2 + \left(\int_0^{2T} r(t)\xi_i(t, \frac{\pi}{2})dt\right)^2 \tag{4.27}$$

the first integral is

$$\begin{aligned} &\int_0^{2T} r(t)\xi_i(t)dt \\ = &\int_0^{T} r(t)A\cos(2\pi f_c t + \Phi_0)dt \\ &+ \int_T^{2T} r(t)A\cos(2\pi f_c t + \Phi_0 + \Delta\theta_i)dt \end{aligned}$$

$$
\begin{aligned}
&= \int_0^T r(t)A\left[\cos 2\pi f_c t \cos\Phi_0 - \sin 2\pi f_c t \sin\Phi_0\right] dt \\
&\quad + \int_T^{2T} r(t)A\left[\cos 2\pi f_c t \cos(\Phi_0 + \Delta\theta_i) - \sin 2\pi f_c t \sin(\Phi_0 + \Delta\theta_i)\right] dt \\
&= w_0 \cos\Phi_0 + z_0 \sin\Phi_0 + w_1 \cos(\Phi_0 + \Delta\theta_i) + z_1 \sin(\Phi_0 + \Delta\theta_i)
\end{aligned}
$$

where

$$
\begin{aligned}
w_0 &\triangleq \int_0^T r(t)A\cos 2\pi f_c t dt \\
z_0 &\triangleq -\int_0^T r(t)A\sin 2\pi f_c t dt \\
w_1 &\triangleq \int_T^{2T} r(t)A\cos 2\pi f_c t dt \\
z_1 &\triangleq -\int_T^{2T} r(t)A\sin 2\pi f_c t dt
\end{aligned}
$$

and the second integral is

$$
\begin{aligned}
&\int_0^{2T} r(t)\xi_i(t, \frac{\pi}{2})dt \\
&= \int_0^T r(t)A\sin(2\pi f_c t + \Phi_0)dt + \int_T^{2T} r(t)A\sin(2\pi f_c t + \Phi_0 + \Delta\theta_i)dt \\
&= -z_0\cos\Phi_0 + w_0\sin\Phi_0 - z_1\cos(\Phi_0 + \Delta\theta_i) + w_1\sin(\Phi_0 + \Delta\theta_i)
\end{aligned}
$$

Then by substituting these two integrals into (4.27), expanding the squares, discarding squared terms since they are independent of transmitted signals and dropping a factor of two, we have the following new sufficient statistic

$$
L_i = (w_1 w_0 + z_1 z_0)\cos\Delta\theta_i + (z_1 w_0 - w_1 z_0)\sin\Delta\theta_i
$$

For the kth symbol duration this is

$$
\begin{aligned}
L_i &= \underbrace{(w_k w_{k-1} + z_k z_{k-1})}_{x_k}\cos\theta_i + \underbrace{(z_k w_{k-1} - w_k z_{k-1})}_{y_k}\sin\theta_i \\
&= x_k\cos\Delta\theta_i + y_k\sin\Delta\theta_i
\end{aligned}
\tag{4.28}
$$

The decision rule is to choose the largest. Or we can write (4.28) as

$$
L_i = A\cos\Delta\widehat{\theta}_k\cos\Delta\theta_i + A\sin\Delta\widehat{\theta}_k\sin\Delta\theta_i
$$

$$= \quad A\cos(\Delta\theta_i - \Delta\widehat{\theta}_k)$$

where

$$\Delta\widehat{\theta}_k = \tan^{-1}\frac{y_k}{x_k}$$

Thus the decision rule is to choose the smallest $|\Delta\theta_i - \Delta\widehat{\theta}_k|$. Figure 4.16 implements this rule. As we stated in the binary DPSK case, the local oscillator output must have the same frequency, but not necessarily the same phase, as the received signal. The amplitude of the reference signals can be any value, which is unit in the figure, since the effect of the amplitude is cancelled when computing $\Delta\widehat{\theta}_k$.

The symbol error probability is given by [2]

$$P_s = \frac{\sin\frac{\pi}{M}}{2\pi}\int_{-\pi/2}^{\pi/2}\frac{\exp\{-\frac{E}{N_o}[1-\cos\frac{\pi}{M}\cos x]\}}{1-\cos\frac{\pi}{M}\cos x}dx \qquad (4.29)$$

which can be evaluated in a closed form for $M = 2$ (see (4.10)). For other values of M, it can only be numerically evaluated. Many approximate expressions have been found [2,5], a simple one is

$$P_s \approx 2Q\left(\sqrt{\frac{2E}{N_o}}\sin\frac{\pi}{\sqrt{2}M}\right), \text{ (optimum DMPSK)} \qquad (4.30)$$

for large SNR. The exact curves as obtained from (4.29) are given in Figure 4.13 together with those of coherent MPSK. Compared with coherent MPSK, asymptotically the DMPSK requires 3 dB more SNR to achieve the same error performance. This also can be quite easily seen by comparing the arguments of the Q-function in (4.30) and (4.24), using $sin(x) \approx x$ for small x.

For the purpose of phase ambiguity elimination, the DEMPSK signal is coherently demodulated. The optimum demodulator is shown in Figure 4.17 which is similar to Figure 4.11, the demodulator for coherent MPSK, except that a differential decoder is attached as a final stage. This is intuitively convincing since at carrier frequency the DEMPSK signal is the same as the MPSK signal, thus the correlator part is the same as that of coherent MPSK. The additional differential decoder recovers the differential phase $\Delta\theta_i$ from phases of two consecutive symbols. The $\Delta\theta_i$ then is mapped back to the corresponding n-tuple of bits. Rigorous derivation of this optimum demodulator and its equivalent forms can be found in [2,5].

The symbol error probability of coherently demodulated DEMPSK is given by

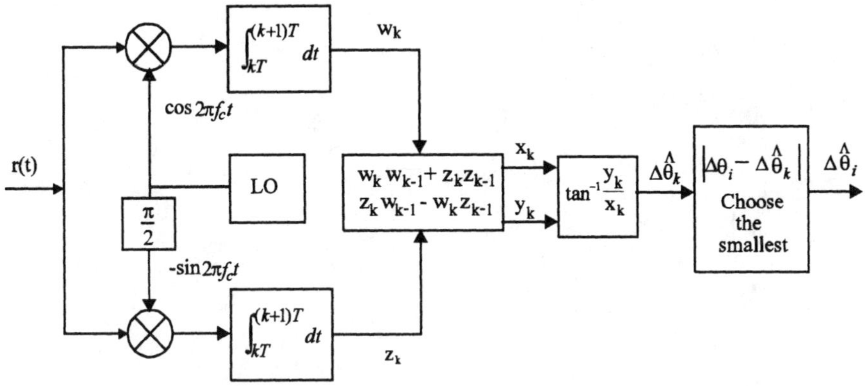

Figure 4.16 Differentially coherent demodulator for differentially encoded MPSK signals.

[2,5]

$$P_s = 2P_{s-MPSK}\left[1 - \frac{1}{2}P_{s-MPSK} - \frac{1}{2}\frac{\sum_{i=1}^{M-1} P_i(C)}{P_{s-MPSK}}\right], \text{ (DEMPSK)} \quad (4.31)$$

where P_{s-MPSK} is the symbol error probability for MPSK without differential encoding and is given by (4.21), and

$$P_i(C) = \left[\int_{(2i-1)\pi/M}^{(2i+1)\pi/M}\int_0^{\infty}\frac{r}{\pi}\exp\left\{-\left[r^2 - 2r\sqrt{\frac{E}{N_o}}\cos\Theta + \frac{E}{N_o}\right]\right\} dr d\Theta\right]^2$$

For $M = 2$, we have coherent detection of differentially encoded BPSK for which (4.31) reduces to (4.12). For $M = 4$, it reduces to

$$\begin{aligned} P_s &= 4Q\left(\sqrt{\frac{E}{N_o}}\right) - 8\left[Q\left(\sqrt{\frac{E}{N_o}}\right)\right]^2 \\ &\quad + 8\left[Q\left(\sqrt{\frac{E}{N_o}}\right)\right]^3 - 4\left[Q\left(\sqrt{\frac{E}{N_o}}\right)\right]^4 \end{aligned} \quad (4.32)$$

For large SNR, the second, third, and fourth terms can be ignored. Thus the above is just about two times that of coherent QPSK without differential encoding. In fact for any value of M when the SNR is large, the terms in the bracket of (4.31) are close to one, thus the P_s of the coherently demodulated DEMPSK is about two times that

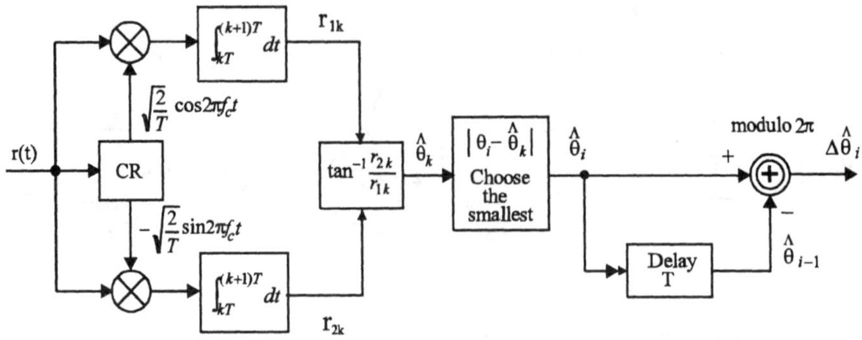

Figure 4.17 Coherent demodulation of differentially encoded MPSK signals.

of coherent MPSK without differential encoding. This translates to 0.5 dB or less degradation in SNR. This is the price paid for removing the phase ambiguity.

A DEMPSK signal's PSD would be the same as its nonencoded counterpart if the encoding process does not change the statistic characteristic of the baseband data, since the final signal from the modulator is just an MPSK signal. We always assume that the original data have an equally likely distribution. This results in that the distribution of $\Delta\theta_i$ is equally likely too. In turn the absolute phases of the DEMPSK signals are also equally likely. This satisfies the condition for deriving (4.26). Thus the PSD of DEMPSK is the same as that of MPSK given in (4.26) for an equally likely original data sequence.

As we have proved in Chapter 2 and mentioned in Section 4.2, that differential encoding in DEBPSK always produces an equally likely data sequence asymptotically regardless of the distribution of the original data. This leads to a PSD given by (4.8) for DEBPSK even if the original data is not evenly distributed.

4.6 QUADRATURE PSK

Among all MPSK schemes, QPSK is the most often used scheme since it does not suffer from BER degradation while the bandwidth efficiency is increased. Other MPSK schemes increase bandwidth efficiency at the expenses of BER performance. In this section we will study QPSK in great detail.

Since QPSK is a special case of MPSK, its signals are defined as

$$s_i(t) = A\cos(2\pi f_c t + \theta_i), \qquad 0 \le t \le T, \quad i = 1, 2, 3, 4 \tag{4.33}$$

Dibit	Phase θ_i	$s_{i1} = \sqrt{E}\cos\theta_i$	$s_{i2} = \sqrt{E}\sin\theta_i$
11	$\pi/4$	$+\sqrt{E/2}$	$+\sqrt{E/2}$
01	$3\pi/4$	$-\sqrt{E/2}$	$+\sqrt{E/2}$
00	$-3\pi/4$	$-\sqrt{E/2}$	$-\sqrt{E/2}$
10	$-\pi/4$	$+\sqrt{E/2}$	$-\sqrt{E/2}$

Table 4.2 QPSK signal coordinates.

where

$$\theta_i = \frac{(2i-1)\pi}{4}$$

The initial signal phases are $\frac{\pi}{4}, \frac{3\pi}{4}, \frac{5\pi}{4}, \frac{7\pi}{4}$. The carrier frequency is chosen as an integer multiple of the symbol rate, therefore in any symbol interval $[kT, (k+1)T]$, the signal initial phase is also one of the four phases.

The above expression can be written as

$$\begin{aligned} s_i(t) &= A\cos\theta_i \cos 2\pi f_c t - A\sin\theta_i \sin 2\pi f_c t \\ &= s_{i1}\phi_1(t) + s_{i2}\phi_2(t) \end{aligned} \tag{4.34}$$

where $\phi_1(t)$ and $\phi_2(t)$ are defined in (4.2) and (4.3),

$$s_{i1} = \sqrt{E}\cos\theta_i$$

$$s_{i2} = \sqrt{E}\sin\theta_i$$

and

$$\theta_i = \tan^{-1}\frac{s_{i2}}{s_{i1}}$$

where $E = A^2T/2$ is the symbol energy. We observe that this signal is a linear combination of two orthonormal basis functions: $\phi_1(t)$ and $\phi_2(t)$. On a coordinate system of $\phi_1(t)$ and $\phi_2(t)$ we can represent these four signals by four points or vectors: $\mathbf{s}_i = \begin{bmatrix} s_{i1} \\ s_{i2} \end{bmatrix}, i = 1, 2, 3, 4$. The angle of vector $\mathbf{s}_i$ with respect to the horizontal axis is the signal initial phase θ_i. The length of the vectors is $\sqrt{E}$.

The signal constellation is shown in Figure 4.18. In a QPSK system, data bits are divided into groups of two bits, called dibits. There are four possible dibits, 00, 01, 10, and 11. Each of the four QPSK signals is used to represent one of them. The mapping of the dibits to the signals could be arbitrary as long as the mapping is one to one. The signal constellation in Figure 4.18 uses the Gray coding. The coordinates of signal points are tabulated in Table 4.2.

In the table, for convenience of modulator structure, we map logic 1 to $\sqrt{E/2}$ and logic 0 to $-\sqrt{E/2}$. We also map odd-numbered bits to s_{i1} and even-numbered bits to s_{i2}. Thus from (4.34) the QPSK signal on the entire time axis can be written as

$$s(t) = \frac{A}{\sqrt{2}} I(t) \cos 2\pi f_c t - \frac{A}{\sqrt{2}} Q(t) \sin 2\pi f_c t, \quad -\infty < t < \infty \tag{4.35}$$

where $I(t)$ and $Q(t)$ are pulse trains determined by the odd-numbered bits and even-numbered bits, respectively.

$$I(t) = \sum_{k=-\infty}^{\infty} I_k p(t - kT)$$

$$Q(t) = \sum_{k=-\infty}^{\infty} Q_k p(t - kT)$$

where $I_k = \pm 1$ and $Q_k = \pm 1$, the mapping between logic data and I_k or Q_k is $1 \rightarrow 1$ and $0 \rightarrow -1$. $p(t)$ is a rectangular pulse shaping function defined on $[0, T]$.

The QPSK waveform using the signal assignment in Figure 4.18 is shown in Figure 4.19. Like BPSK, the waveform has a constant envelope and discontinuous phases at symbol boundaries. But unlike BPSK, the symbol interval is $2T_b$ instead of T_b. If the transmission rate of the symbols is the same in QPSK and BPSK, it is intuitively clear that QPSK transmits data twice as fast as BPSK does. Also we observe that the distance of adjacent points of the QPSK constellation is shorter than that of the BPSK. Does this cause the demodulator more difficulty, in comparison with BPSK, to distinguish those symbols, therefore symbol error performance is degraded and consequently bit error rate is also degraded? Surprisingly, it turns out that even though symbol error probability is increased, the bit error probability remains unchanged, as we will see shortly.

The modulator of QPSK is based on (4.35). This leads to the modulator in Figure 4.20(a). The channel with cosine reference is called inphase (I) channel and the channel with sine reference is called quadrature (Q) channel. The data sequence is separated by the serial-to-parallel converter (S/P) to form the odd-numbered-bit sequence for I-channel and the even-numbered-bit sequence for Q-channel. Then logic 1 is converted to a positive pulse and logic 0 is converted to a negative pulse, both have the same amplitude and a duration of T. Next the odd-numbered-bit pulse train is multiplied to $\cos 2\pi f_c t$ and the even-numbered-bit pulse train is multiplied to $\sin 2\pi f_c t$. It is clear that the I-channel and Q-channel signals are BPSK signals with a symbol duration of $2T_b$. Finally a summer adds these two waveforms together to

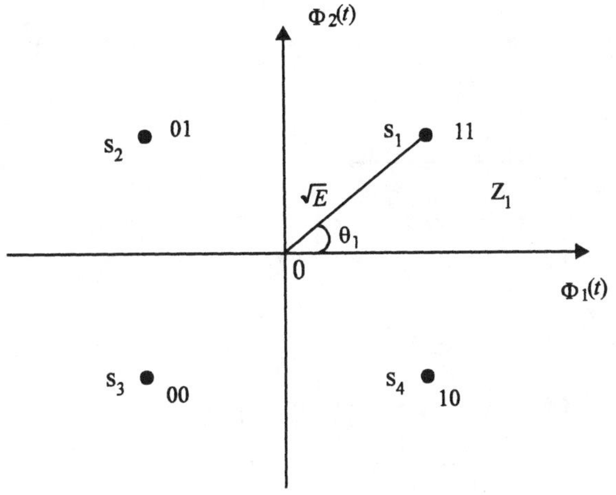

Figure 4.18 QPSK signal constellation.

produce the final QPSK signal. (See Figure 4.19 for waveforms at various stages.)

Since QPSK is a special case of MPSK, the demodulator for MPSK (Figure 4.11) is applicable to QPSK. However, due to the special property of the QPSK constellation, a simpler demodulator is possible. It is shown in Figure 4.20(b) which is equivalent to Figure 4.11. I- and Q-channel signals are demodulated separately as two individual BPSK signals. A parallel-to-serial converter (P/S) is used to combine two sequences into a single sequence. This is possible because of the one-to-one correspondence between data bits and I- and Q-channel signals and their orthogonality. For $M > 4$, the optimum receiver can only be the form in Figure 4.11, since the signal in the I-channel or Q-channel does not correspond to a single bit, as we have seen in Section 4.3.

The bit error probability of the optimum demodulators can be derived using the demodulator of Figure 4.20. Since $E\{r_j\}, j = 1, 2$, is either $\sqrt{E/2}$ or $-\sqrt{E/2}$, corresponding to a bit of 1 or 0 (Table 4.2), the detection is a typical binary detection with a threshold of 0. The average bit error probability for each channel is

$$
\begin{aligned}
P_b &= \Pr(e/1 \text{ is sent}) = \Pr(e/0 \text{ is sent}) \\
&= \int_0^\infty \frac{1}{\sqrt{\pi N_o}} \exp\left[-\frac{(R_j + \sqrt{E/2})^2}{N_o}\right] dR_j
\end{aligned}
$$

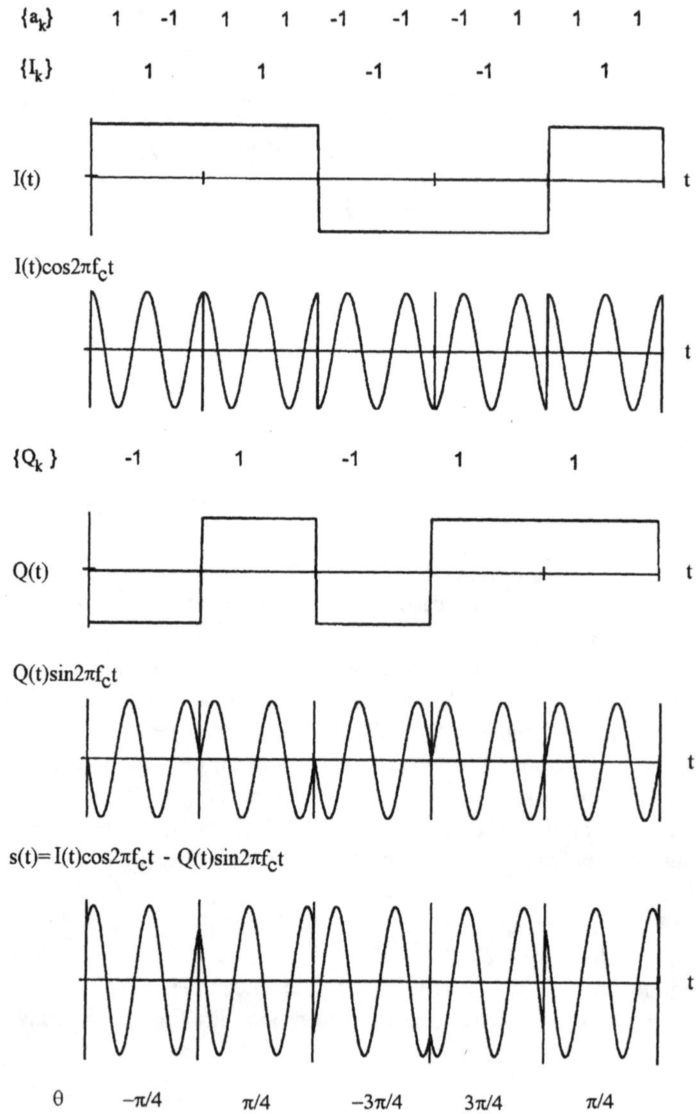

Figure 4.19 QPSK waveforms.

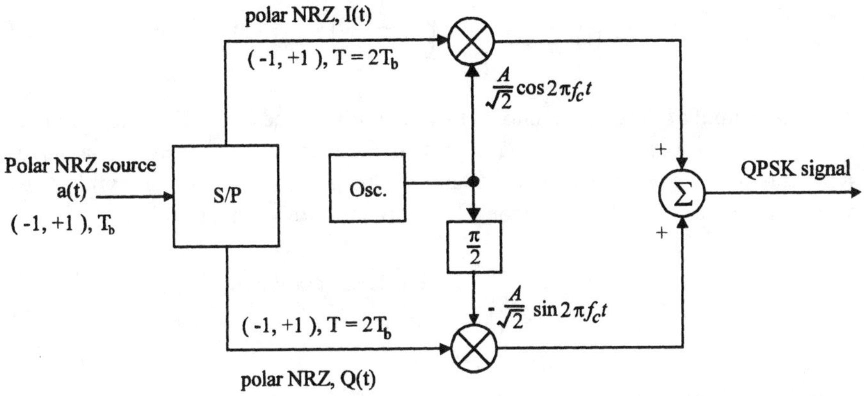

(a)

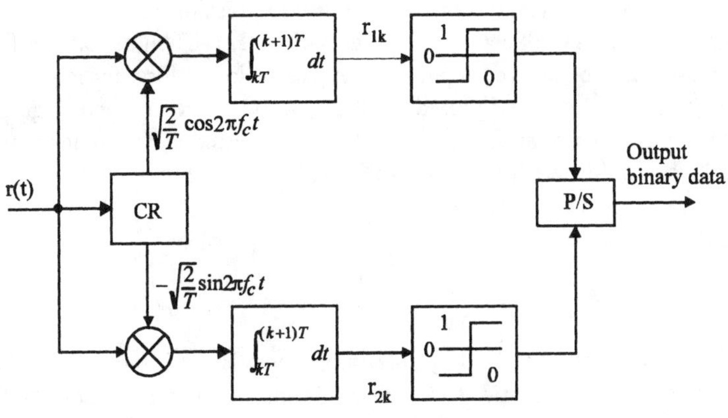

(b)

Figure 4.20 (a) QPSK modulator, and (b) QPSK demodulator.

$$= \int_{\sqrt{\frac{E}{N_o}}}^{\infty} \frac{1}{\sqrt{2\pi}} \exp\left[-\frac{x^2}{2}\right] dx$$

$$= Q\left(\sqrt{\frac{E}{N_o}}\right) = Q\left(\sqrt{\frac{2E_b}{N_o}}\right), \text{ (coherent QPSK)} \tag{4.36}$$

The final output of the demodulator is just the multiplexed I- and Q-channel outputs. Thus the bit error rate for the final output is the same as that of each channel. A symbol represents two bits from the I- and Q-channels, respectively. A symbol error occurs if any one of them is in error. Therefore the symbol error probability is

$$\begin{aligned} P_s &= 1 - \Pr(\text{ both bits are correct}) \\ &= 1 - (1 - P_b)^2 \\ &= 2P_b - P_b^2 \\ &= 2Q\left(\sqrt{\frac{E}{N_o}}\right) - \left[Q\left(\sqrt{\frac{E}{N_o}}\right)\right]^2 \end{aligned} \tag{4.37}$$

The above symbol error probability expression can also be derived from the general formula in Section 4.3 for MPSK (4.21). Then the bit error probability expression can be derived in another way as follows. First for large SNR, the second term in (4.37) can be ignored. Second, for Gray coding and large SNR, a symbol error most likely causes the symbol being detected as the adjacent symbol which is only one bit different out of two bits. Thus

$$P_b \approx \frac{1}{2} P_s \approx Q\left(\sqrt{\frac{2E_b}{N_o}}\right)$$

This expression is derived by approximations. But it is the same as the one obtained by the accurate derivation. We have made approximations twice. The first is to ignore the second term in (4.37). This increases the estimate of P_b slightly. The second is to ignore the symbol errors caused by choosing the nonadjacent symbols which may cause two bit errors for a symbol error. This decreases the estimate of P_b slightly. The fact that the final estimate is exactly equal to the accurate one shows that these two approximations happen to cancel each other. It is purely a coincidence.

The P_b curve of QPSK is shown in Figure 4.21, which is the same as that of BPSK. The P_s curve of QPSK is shown in Figure 4.13 together with other MPSK schemes.

The PSD of QPSK is similar to that of BPSK except that the spectral is narrower

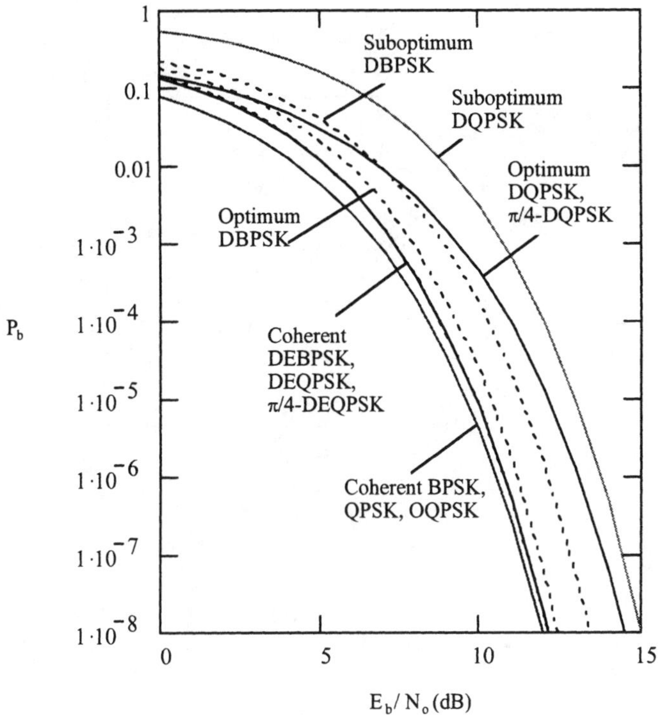

Figure 4.21 P_b of QPSK and DQPSK.

on a frequency scale normalized to the bit rate. From (4.26) we have

$$\Psi_{\tilde{s}}(f) = 2A^2T_b\left(\frac{\sin 2\pi fT_b}{2\pi fT_b}\right)^2, \text{ (QPSK)} \tag{4.38}$$

Figure 4.22(a, b) is the PSD curves of the QPSK. The null-to-null bandwidth $B_{null} = 1/T_b = R_b$. Figure 4.22(c) is the out-of-band power curve from which we can estimate that $B_{90\%} \approx 0.85R_b$. We also calculated that $B_{99\%} \approx 10R_b$.

4.7 DIFFERENTIAL QPSK

Now we study an important special case of DEMPSK, the DEQPSK. In DEQPSK

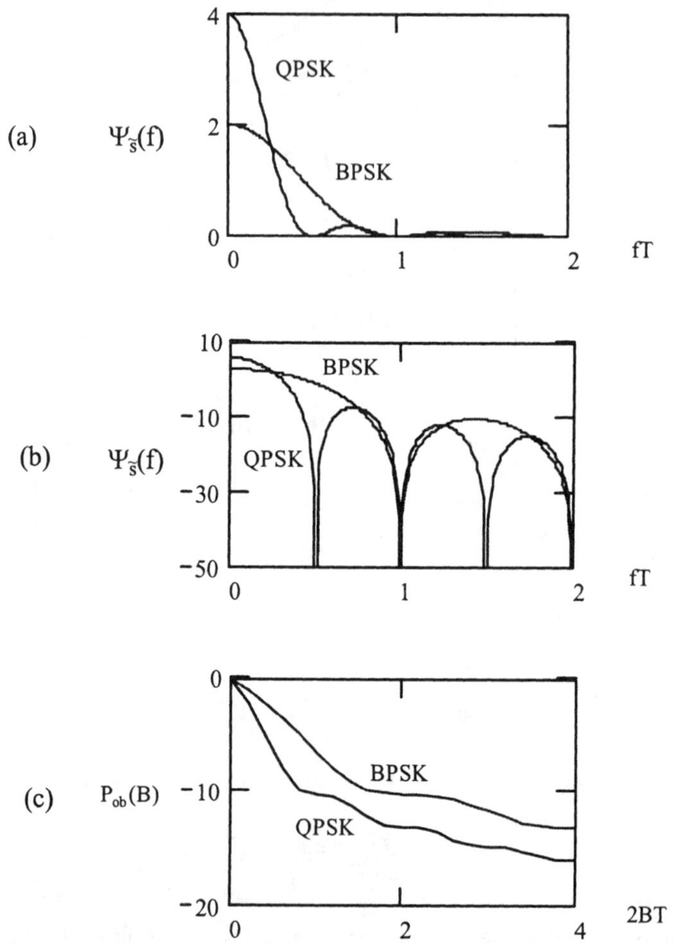

Figure 4.22 PSD of QPSK in comparison with BPSK: (a) linear, (b) logarithmic, and (c) out-of-band power.

Dibit	$\Delta\theta_i$	$\cos\Delta\theta_i$	$\sin\Delta\theta_i$
00	0	1	0
01	$\pi/2$	0	1
10	$-\pi/2$	0	-1
11	π	-1	0

Table 4.3 DEQPSK signal phase assignment.

information dibits are represented by the phase differences $\Delta\theta_i$ from symbol to symbol. There are different phase assignments between $\Delta\theta_i$ and logic dibits. A possible phase assignment is listed in Table 4.3. Our discussion in this section is based on this phase assignment choice (later when we study $\pi/4$–QPSK, the phase assignment is different). An example for this choice is shown in Table 4.4.

The coding rules are as follows [5].

$$\begin{aligned} u_k &= \overline{(I_k \oplus Q_k)}(I_k \oplus u_{k-1}) + (I_k \oplus Q_k)(Q_k \oplus v_{k-1}) \\ v_k &= \overline{(I_k \oplus Q_k)}(Q_k \oplus v_{k-1}) + (I_k \oplus Q_k)(I_k \oplus u_{k-1}) \end{aligned} \tag{4.39}$$

where $\oplus$ denotes exclusive OR operation. $I_k \in (0,1)$ and $Q_k \in (0,1)$ are odd-numbered and even-numbered original information bits, respectively; $u_k \in (0,1)$ and $v_k \in (0,1)$ are coded I-channel and Q-channel bits, respectively. Pairs (I_k, Q_k) and (u_{k-1}, v_{k-1}) are used to produce pair (u_k, v_k) which is used to control the absolute phase of the carrier. The resultant signal is a QPSK signal as shown in Figure 4.18 for (u_k, v_k), but it is a DEQPSK signal for (I_k, Q_k). Therefore the modulator is basically the same as the QPSK modulator (Figure 4.20) except that two differential encoders must be included in each channel before the carrier multiplier. The modulator is shown in Figure 4.23.

When DEQPSK is differentially coherently demodulated, the scheme is DQPSK. The optimum DQPSK demodulator can be derived from Figure 4.16 as a special case of $M = 4$. The symbol error probability is given by (4.29) or (4.30). The bit error probability can be approximately calculated using (4.25). Thus

$$P_b \approx Q\left(\sqrt{\frac{4E_b}{N_o}}\sin\frac{\pi}{4\sqrt{2}}\right), \text{ (optimum DQPSK)} \tag{4.40}$$

which is plotted in Figure 4.21. It is about 2 to 3 dB inferior to coherent QPSK. Alternately, the bit error probability of DQPSK can be evaluated using an expression given by [6] as

$$P_b = e^{-2\gamma_b}\sum_{k=0}^{\infty}(\sqrt{2}-1)^k I_k(\sqrt{2}\gamma_b) - \frac{1}{2}I_0(\sqrt{2}\gamma_b)e^{-2\gamma_b} \tag{4.41}$$

Modulation		ref.								
Information sequence	I_k		1	0	1	0	1	1	0	1
	Q_k		0	1	0	1	1	0	0	1
Encoded sequence	u_k	1	1	1	1	1	0	0	0	1
	v_k	1	0	1	0	1	0	1	1	0
Transmitted absolute phases		$\frac{\pi}{4}$	$\frac{7\pi}{4}$	$\frac{\pi}{4}$	$\frac{7\pi}{4}$	$\frac{\pi}{4}$	$\frac{5\pi}{4}$	$\frac{3\pi}{4}$	$\frac{3\pi}{4}$	$\frac{7\pi}{4}$

Table 4.4 Differential coding for DEQPSK.

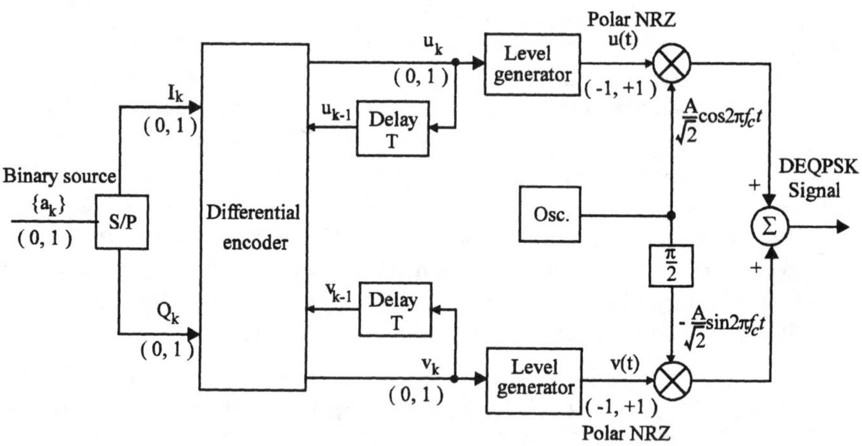

Figure 4.23 DEQPSK modulator.

where $\gamma_b = E_b/N_o$ and $I_\alpha(x)$ is the αth order modified Bessel function of the first kind which may be represented by the infinite series

$$I_\alpha(x) \triangleq \sum_{k=0}^{\infty} \frac{(x/2)^{\alpha+2k}}{k!\Gamma(\alpha+k+1)}, \quad x \geq 0$$

and the gamma function is defined as

$$\Gamma(p) \triangleq \int_0^{\infty} t^{p-1}e^{-t}dt, \quad p > 0$$

Like in the binary DPSK case, a suboptimum demodulator using previous symbols as references is shown in Figure 4.24 where the integrator can be replaced by a lowpass filter [7]. The front-end bandpass filter reduces noise power but preserves the phase of the signal. In the absence of noise, the I-channel integrator output is

$$\begin{aligned}
&\int_{kT}^{(k+1)T} A^2\cos(2\pi f_c t+\theta_k)\cos(2\pi f_c t+\theta_{k-1})dt \\
&= \frac{1}{2}\int_{kT}^{(k+1)T} A^2[\cos(4\pi f_c t+\theta_k+\theta_{k-1})+\cos(\theta_k-\theta_{k-1})]dt \\
&= \frac{1}{2}A^2 T\cos\Delta\theta_k
\end{aligned}$$

Similarly the Q-channel integrator output is $\frac{1}{2}A^2T\sin\Delta\theta_k$. The arctangent operation extracts the $\Delta\widehat{\theta}_k$ (estimate of $\Delta\theta_k$ with the presence of noise) and a comparator compares it to the four $\Delta\theta_i$ and chooses the closest. The dibit is then recovered from the detected $\Delta\theta_i$. For special dibits-$\Delta\theta_i$ assignment, such as the one for $\pi/4$-QPSK, as will be seen in the next section, the angle detector can be replaced by two threshold detectors. The bit error probability of the suboptimum demodulator in Figure 4.24 is given by [8, p. 260]

$$P_b \approx e^{-(A^2/2\sigma^2)(1-1/\sqrt{2})} \tag{4.42}$$

where $A^2/2\sigma^2$ is the carrier-to-noise power ratio. In order to compare this to other error probabilities, we need to relate $A^2/2\sigma^2$ to E_b/N_o. In the derivation of this expression [8], the narrow-band noise has a variance of σ^2 for the inphase and quadrature component at the output of the front-end bandpass filter. The total noise variance is also equal to σ^2 [9, p. 76]. The baseband signal is bandlimited to B. The bandwidth of the bandpass filter is just the same. Thus there is no intersymbol interference and the signal amplitude at sampling instances is A. So far in this chapter, the baseband pulse shape is always assumed as rectangular. For this pulse shape, the intersym-

Transmitted absolute phases		$\frac{\pi}{4}$	$\frac{7\pi}{4}$	$\frac{\pi}{4}$	$\frac{7\pi}{4}$	$\frac{\pi}{4}$	$\frac{5\pi}{4}$	$\frac{3\pi}{4}$	$\frac{3\pi}{4}$	$\frac{7\pi}{4}$
Demodulation										
Estimated absolute phases		$\frac{3\pi}{4}$	$\frac{\pi}{4}$	$\frac{3\pi}{4}$	$\frac{\pi}{4}$	$\frac{3\pi}{4}$	$\frac{7\pi}{4}$	$\frac{5\pi}{4}$	$\frac{5\pi}{4}$	$\frac{\pi}{4}$
Detected digits	$\widehat{u}_k$	0	1	0	1	0	1	0	0	1
	$\widehat{v}_k$	1	1	1	1	1	0	0	0	1
Detected information digits	$\widehat{I}_k$		1	0	1	0	1	1	0	1
	$\widehat{Q}_k$		0	1	0	1	1	0	0	1

Table 4.5 Differential decoding for DEQPSK.

bol interference free filter is the Nyquist filter which has a bandwidth of $B = 1/T$ at carrier frequency. Thus the noise power $\sigma^2 = N_o/T$. The signal symbol energy $E_s = \frac{1}{2}A^2T$. Thus

$$A^2/2\sigma^2 = \frac{2E_s/T}{2N_o/T} = \frac{E_s}{N_o} = \frac{2E_b}{N_o}$$

This is also stated in [2, p. 444, eqn. (7.6)]. Thus (4.42) can be written as

$$P_b \approx e^{-\frac{2E_b}{N_o}(1-1/\sqrt{2})} = e^{-0.59\frac{E_b}{N_o}}, \text{(suboptimum DQPSK)} \tag{4.43}$$

This is plotted in Figure 4.21. Seen from the figure, the degradation to the optimum DQPSK is less than 1 dB for high SNRs (>12 dB). But at the lower SNRs, the degradation is much larger. This is intuitively convincing since the reference signal is the previous signal which has noise. Compared to the suboptimum DBPSK, the degradation is about 1.75 dB for all SNRs.

For the purpose of phase ambiguity elimination, the DEQPSK signals are coherently demodulated. The demodulator is thus basically the same as the QPSK demodulator except that a differential decoder must be inserted after demodulation. The demodulator can be in the form of Figure 4.17 where differential decoding is carried out on signal phases. However, since there are only two levels in the I- and Q-channels, the demodulator can be in a simpler form as shown in Figure 4.25, where the differential decoding is carried out on digital signal levels. The decoding rules are

$$\begin{aligned}
\widehat{I}_k &= \overline{(\widehat{u}_k \oplus \widehat{v}_k)}(\widehat{u}_k \oplus \widehat{u}_{k-1}) + (\widehat{u}_k \oplus \widehat{v}_k)(\widehat{v}_k \oplus \widehat{v}_{k-1}) \\
\widehat{Q}_k &= \overline{(\widehat{u}_k \oplus \widehat{v}_k)}(\widehat{v}_k \oplus \widehat{v}_{k-1}) + (\widehat{u}_k \oplus \widehat{v}_k)(\widehat{u}_k \oplus \widehat{u}_{k-1})
\end{aligned} \tag{4.44}$$

Table 4.5 shows the decoding process assuming a phase ambiguity of $\pi/2$.

Note that the coherent demodulator for DEQPSK in Figure 4.25 is not suitable for $\pi/4$–QPSK since its dibits-$\Delta\theta_i$ assignment is different (see the $\pi/4$–QPSK section).

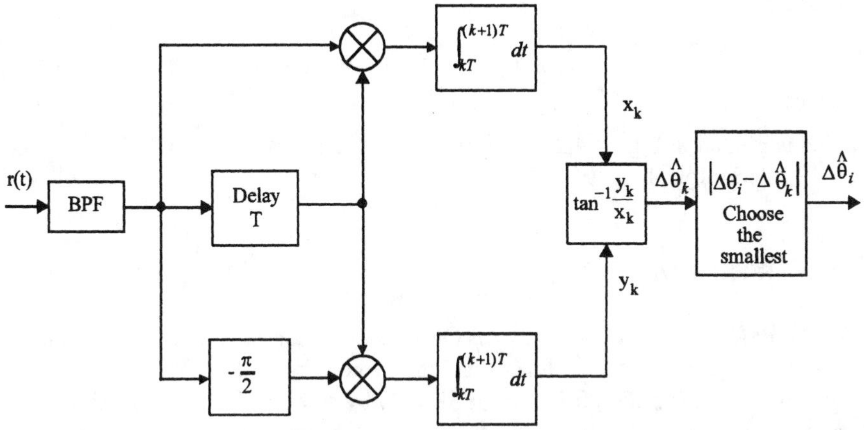

Figure 4.24 Suboptimum DQPSK demodulator (see Figure 4.16 for optimum demodulator).

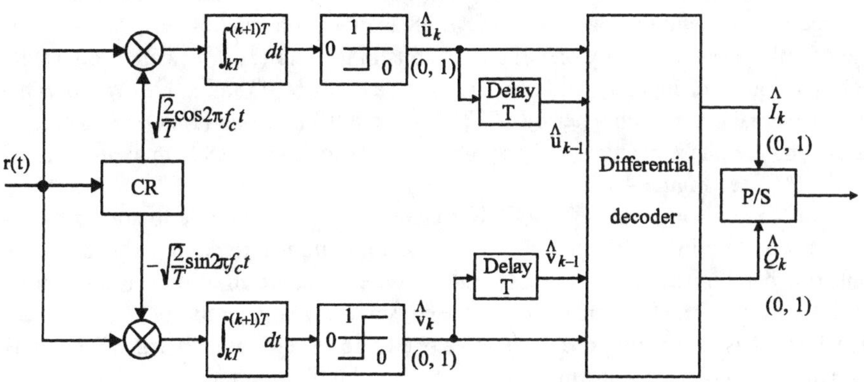

Figure 4.25 Coherent demodulation of DEQPSK.

The symbol error probability has been given in (4.32). For Gray coded constellation and at high SNR, this translates to a bit error probability of

$$P_b \approx 2Q\left(\sqrt{\frac{2E_b}{N_o}}\right), \text{ (DEQPSK)} \tag{4.45}$$

which is plotted in Figure 4.21. It is seen from the figure that DQPSK is less than 0.5 dB inferior to coherent QPSK.

4.8 OFFSET QPSK

Offset QPSK is essentially the same as QPSK except that the I- and Q-channel pulse trains are staggered. The modulator and the demodulator of OQPSK are shown in Figure 4.26, which differs from the QPSK only by an extra delay of $T/2$ seconds in the Q-channel. Based on the modulator, the OQPSK signal can be written as

$$s(t) = \frac{A}{\sqrt{2}}I(t)\cos 2\pi f_c t - \frac{A}{\sqrt{2}}Q(t-\frac{T}{2})\sin 2\pi f_c t, \quad -\infty < t < \infty$$

Since OQPSK differs from QPSK only by a delay in the Q-channel signal, its power spectral density is the same as that of QPSK, and its error performance is also the same as that of QPSK.

The OQPSK waveforms are shown in Figure 4.27. We observe that due to the staggering of I- and Q-channels, the OQPSK signal has a symbol period of $T/2$. At any symbol boundary, only one of the two bits in the pair (I_k, Q_k) can change sign. Thus the phase changes at symbol boundaries can only be $0°$ and $\pm 90°$. Whereas the QPSK signal has a symbol period of T, both two bits in the pair (I_k, Q_k) can change sign, and the phase changes at the symbol boundaries can be $180°$ in addition to $0°$ and $\pm 90°$ (see Figure 4.19).

In comparison to QPSK, OQPSK signals are less susceptible to spectral sidelobe restoration in satellite transmitters. In satellite transmitters, modulated signals must be bandlimited by a bandpass filter in order to conform to out-of-band emission standards. The filtering degrades the constant-envelope property of QPSK, and the $180°$ phase shifts will cause the envelope to go to zero momentarily. When this signal is amplified by the final stage, usually a highly nonlinear power amplifier, the constant envelope will be restored. But at the same time the sidelobes will be restored. Note that arranging the bandpass filter after the power amplifier is not feasible since the bandwidth is very narrow compared with the carrier frequency, the Q-value of the filter must be extremely high such that it cannot be implemented by the current technology. In OQPSK, since the $180°$ phase shifts no longer exist, the

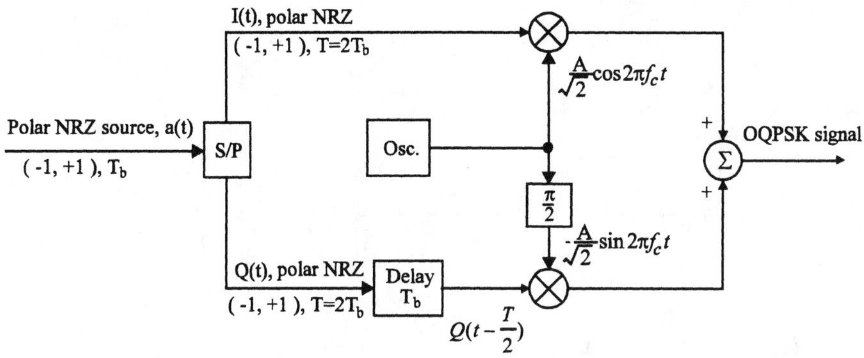

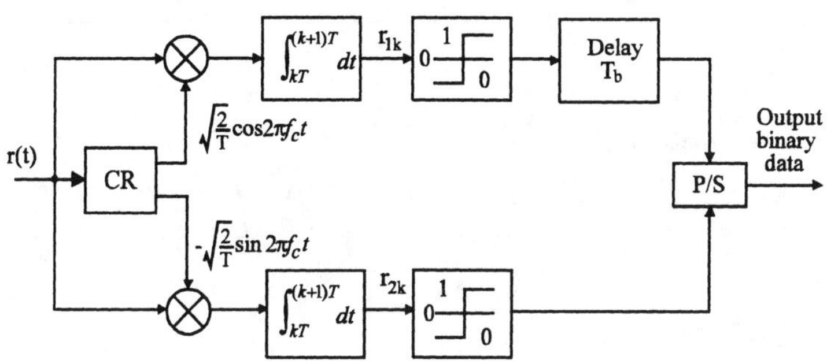

Figure 4.26 OQPSK modulator (a) and demodulator (b).

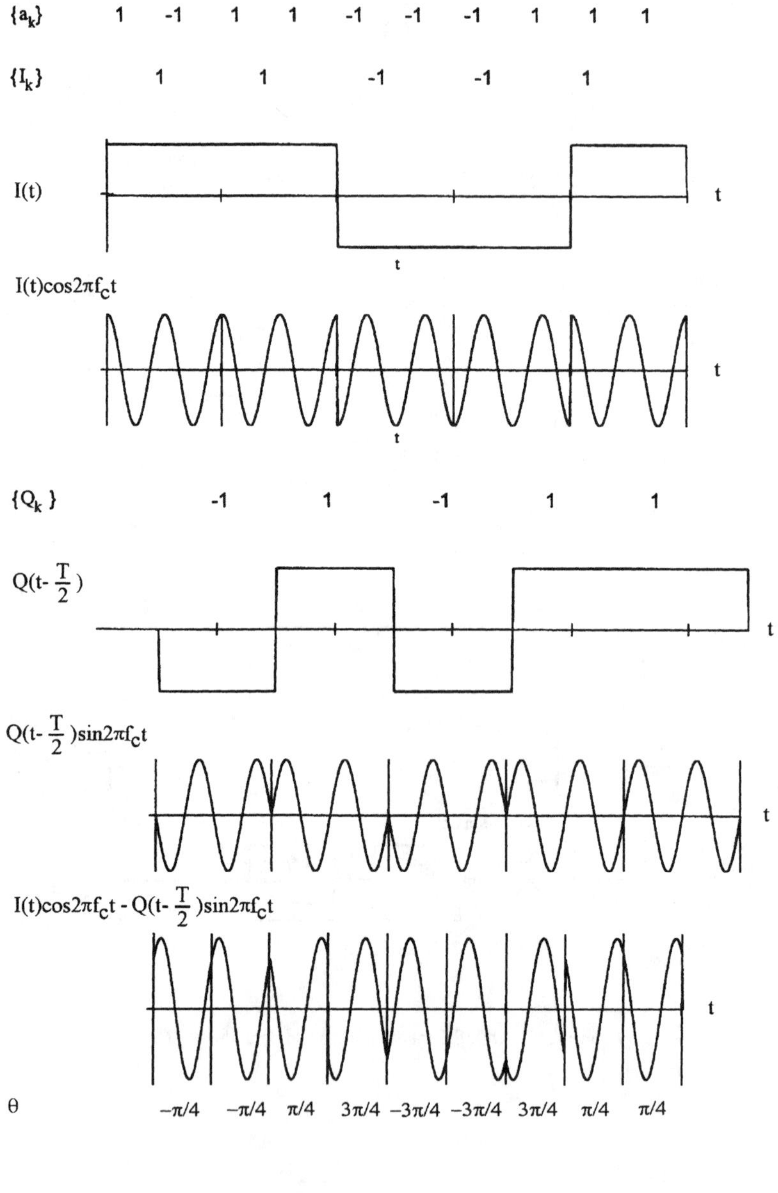

Figure 4.27 OQPSK waveforms.

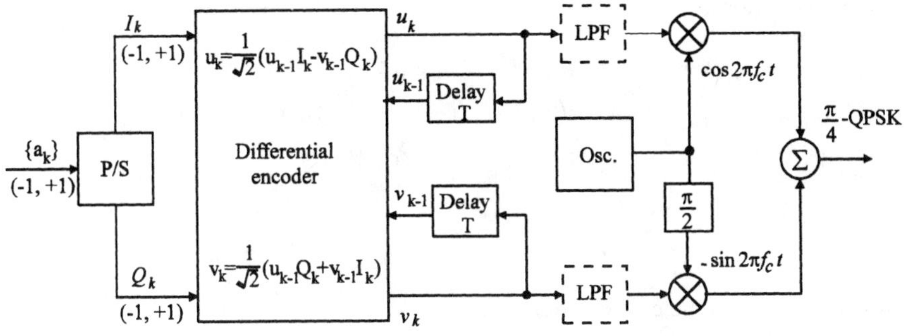

Figure 4.28 $\pi/4$-QPSK modulator.

sidelobe restoration is less severe [10].

4.9 $\pi/4$-QPSK

Although OQPSK can reduce spectral restoration caused by nonlinearity in the power amplifier, it cannot be differentially encoded and decoded. $\pi/4$-QPSK is a scheme which not only has no 180° phase shifts like OQPSK, but also can be differentially demodulated. These properties make it particularly suitable to mobile communications where differential demodulation can reduce the adversary effect of the fading channel. $\pi/4$- QPSK has been adopted as the standard for the digital cellular telephone system in the United States and Japan.

$\pi/4$-QPSK was first introduced by Baker in 1962 [11] and studied in [12, 13] and other articles.

The $\pi/4$-QPSK is a form of differentially encoded QPSK. But it differs from the DEQPSK described in the previous section by the differential coding rules. Figure 4.28 is the $\pi/4$-QPSK modulator. $(I(t), Q(t))$ and $(u(t), v(t))$ are the uncoded and coded I-channel and Q-channel bits. The differential encoder of $\pi/4$-QPSK modulator encodes $I(t)$ and $Q(t)$ into signals $u(t)$ and $v(t)$ according to the following rules

$$\begin{aligned} u_k &= \frac{1}{\sqrt{2}}(u_{k-1}I_k - v_{k-1}Q_k) \\ v_k &= \frac{1}{\sqrt{2}}(u_{k-1}Q_k + v_{k-1}I_k) \end{aligned} \tag{4.46}$$

where u_k is the amplitude of $u(t)$ in the kth symbol duration and so on. We assume that I_k, Q_k takes values of $(-1, 1)$. If we initially specify that $u_0 = 1$ and $v_0 = 0$, then u_k and v_k can take the amplitudes of $\pm1, 0$, and $\pm1/\sqrt{2}$. The output signal of the modulator is

$$\begin{aligned} s(t) &= u_k \cos 2\pi f_c t - v_k \sin 2\pi f_c t \\ &= A\cos(2\pi f_c t + \Phi_k), \quad kT \le t \le (k+1)T \end{aligned}$$

where

$$\Phi_k = \tan^{-1}\frac{v_k}{u_k}$$

which depends on the encoded data, and

$$A = \sqrt{u_k^2 + v_k^2}$$

is independent of time index k, that is, the signal has a constant envelope. This can be easily verified by substituting (4.46) into the expression of A and it turns out $A_k = A_{k-1}$. In fact $A = 1$ for initial values $u_0 = 1$ and $v_0 = 0$. It can be proved that the phase relationship between two consecutive symbols is

$$\begin{aligned} \Phi_k &= \Phi_{k-1} + \Delta\theta_k \\ \Delta\theta_k &= \tan^{-1}\tfrac{Q_k}{I_k} \end{aligned} \tag{4.47}$$

where $\Delta\theta_k$ is the phase difference determined by input data.

Proof: By definition

$$\begin{aligned} \tan\Phi_k &= \frac{v_k}{u_k} \\ &= \frac{u_{k-1}Q_k + v_{k-1}I_k}{u_{k-1}I_k - v_{k-1}Q_k} = \frac{Q_k + \frac{v_{k-1}}{u_{k-1}}I_k}{I_k - \frac{v_{k-1}}{u_{k-1}}Q_k} \\ &= \frac{Q_k + \tan\Phi_{k-1}I_k}{I_k - \tan\Phi_{k-1}Q_k} = \frac{Q_k\cos\Phi_{k-1} + I_k\sin\Phi_{k-1}}{I_k\cos\Phi_{k-1} - Q_k\sin\Phi_{k-1}} \end{aligned}$$

Now let

$$\begin{aligned} I_k &= \sqrt{2}\cos\Delta\theta_k \\ Q_k &= \sqrt{2}\sin\Delta\theta_k \end{aligned} \tag{4.48}$$

then

$$\Delta\theta_k = \tan^{-1}\frac{Q_k}{I_k}$$

I_kQ_k	$\Delta\theta_k$	$\cos\Delta\theta_k$	$\sin\Delta\theta_k$
1 1	$\pi/4$	$1/\sqrt{2}$	$1/\sqrt{2}$
-1 1	$3\pi/4$	$-1/\sqrt{2}$	$1/\sqrt{2}$
-1 -1	$-3\pi/4$	$-1/\sqrt{2}$	$-1/\sqrt{2}$
1 -1	$-\pi/4$	$1/\sqrt{2}$	$-1/\sqrt{2}$

Table 4.6 $\pi/4$-QPSK signal phase assignment.

and we have

$$\begin{aligned}\tan\Phi_k &= \frac{\sin\Delta\theta_k\cos\Phi_{k-1}+\cos\Delta\theta_k\sin\Phi_{k-1}}{\cos\Delta\theta_k\cos\Phi_{k-1}-\sin\Delta\theta_k\sin\Phi_{k-1}}\\ &= \frac{\sin(\Phi_{k-1}+\Delta\theta_k)}{\cos(\Phi_{k-1}+\Delta\theta_k)}=\tan(\Phi_{k-1}+\Delta\theta_k)\end{aligned}$$

thus we have proved (4.47). Using (4.48) we can write (4.46) as

$$\begin{aligned}u_k &= u_{k-1}\cos\Delta\theta_k - v_{k-1}\sin\Delta\theta_k\\ v_k &= u_{k-1}\sin\Delta\theta_k + v_{k-1}\cos\Delta\theta_k\end{aligned} \tag{4.49}$$

Table 4.6 shows how $\Delta\theta_k$ is determined by the input data.

Referring to the values of $\Delta\theta_k$ in Table 4.6, we can see clearly from (4.47) that the phase changes are confined to odd-number multiples of $\pi/4$ (45°). There are no phase changes of 90° or 180°. In addition, information is carried by the phase changes $\Delta\theta_k$, not the absolute phase Φ_k. The signal constellation is shown in Figure 4.29. The angle of a vector (or symbol) with respect to the positive direction of axis u is the symbol phase Φ_k. The symbols represented by • can only become symbols represented by $\times$, and vice versa. Transitions among themselves are not possible. The phase change from one symbol to the other is $\Delta\theta_k$.

Since information is carried by the phase changes $\Delta\theta_k$, differentially coherent demodulation can be used. However, coherent demodulation is desirable when higher power efficiency is required. There are four ways to demodulate a $\pi/4$–QPSK signal:[6]

1. Baseband differential detection;
2. IF band differential detection;
3. FM-discriminator detection;
4. Coherent detection.

The first three demodulators are reported to be equivalent in error performance

[6] As a matter of fact, these methods are also applicable to other differential MPSK schemes.

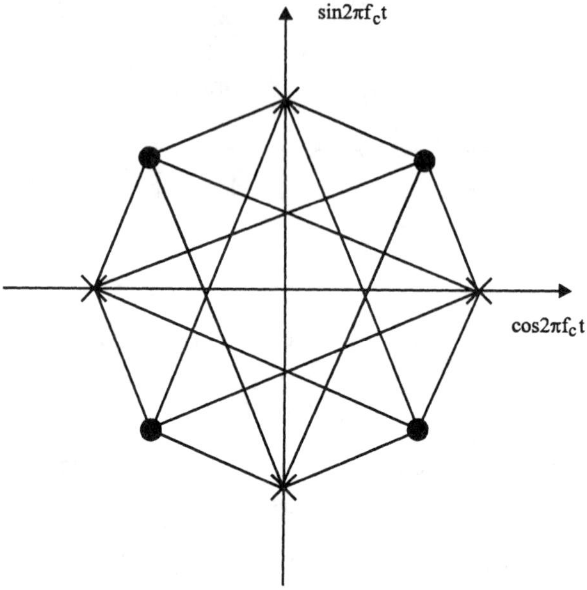

Figure 4.29 $\pi/4$-QPSK signal constellation.

[12]. The coherent demodulator is 2 to 3 dB better.

Figure 4.30 is the baseband differential demodulator which is just a special case of the DMPSK demodulator in Figure 4.16. The LPF in Figure 4.30 is equivalent to the integrator in Figure 4.16. The angle calculation and comparison stages in Figure 4.16 are equivalently replaced by two threshold detectors. The bandpass filter (BPF) at the front end is used to minimize the noise power. However, the carrier phase must be preserved for the proper differential detection. A square-root raised-cosine roll-off BPF can achieve this goal [13]. The local oscillation has the same frequency as the unmodulated carrier but its phase is not synchronous with the received signal. It is assumed that the difference phase θ in the received signal remains essentially the same from the previous symbol duration to the current symbol duration. This phase difference will be cancelled in the baseband differential decoder.

In the absence of noise, the output of the BPF in the kth symbol duration is

$$r(t) = A_k \cos(2\pi f_c t + \Phi_k + \theta), \qquad kT \le t \le (k+1)T \tag{4.50}$$

where θ is the random phase introduced by the channel. We assume that θ changes very slowly in comparison to the symbol rate so that it is considered constant in two

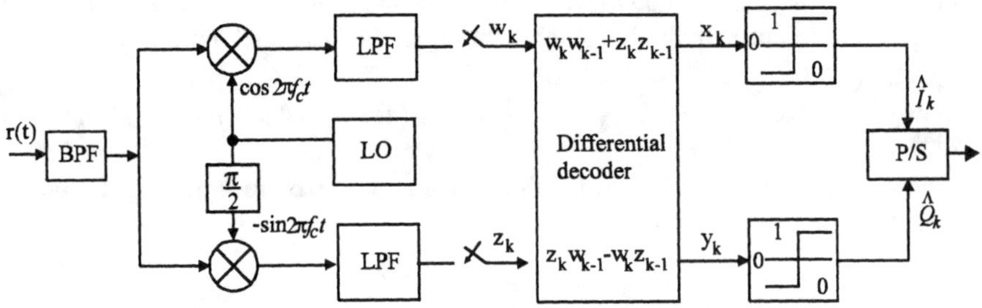

Figure 4.30 Baseband differential demodulator for $\pi/4$-QPSK.

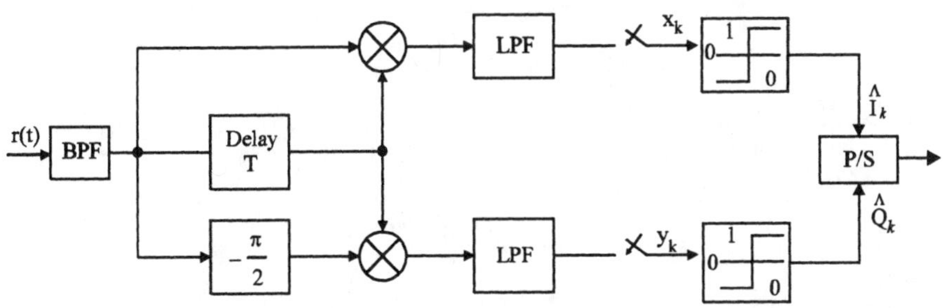

Figure 4.31 IF band differential demodulator for $\pi/4$-QPSK.

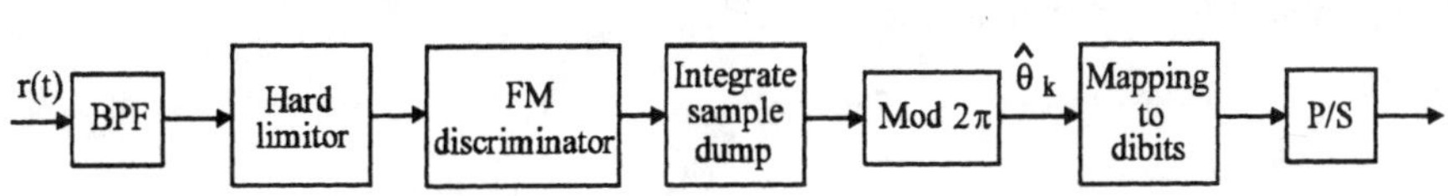

Figure 4.32 FM-discriminator demodulator for $\pi/4$-QPSK.

consecutive symbols. The time-varying amplitude A_k has replaced the constant amplitude in the transmitted signal. The variation in amplitude might be due to channel fading or interference. In the kth symbol duration, the I-channel multiplier output is

$$A_k \cos(2\pi f_c t)\cos(2\pi f_c t + \Phi_k + \theta) = \frac{1}{2} A_k \left[\cos(4\pi f_c t + \Phi_k + \theta) + \cos(\Phi_k + \theta)\right]$$

The low-pass filter (LPF) output for the I-channel is therefore (ignoring the factor $1/2$ and the LPF loss)

$$w_k = A_k \cos(\Phi_k + \theta)$$

Similarly the Q-channel LPF output is

$$z_k = A_k \sin(\Phi_k + \theta)$$

Since θ has not been changed from the previous symbol duration, then

$$w_{k-1} = A_{k-1} \cos(\Phi_{k-1} + \theta)$$

$$z_{k-1} = A_{k-1} \sin(\Phi_{k-1} + \theta)$$

The decoding rule is

$$\begin{aligned} x_k &= w_k w_{k-1} + z_k z_{k-1} \\ y_k &= z_k w_{k-1} - w_k z_{k-1} \end{aligned}$$

which is

$$\begin{aligned} x_k &= A_k A_{k-1}\left[\cos(\Phi_k + \theta)\cos(\Phi_{k-1} + \theta) + \sin(\Phi_k + \theta)\sin(\Phi_{k-1} + \theta)\right] \\ &= A_k A_{k-1}\cos(\Phi_k - \Phi_{k-1}) = A_k A_{k-1}\cos\Delta\theta_k \\ y_k &= A_k A_{k-1}\left[\sin(\Phi_k + \theta)\cos(\Phi_{k-1} + \theta) - \cos(\Phi_k + \theta)\sin(\Phi_{k-1} + \theta)\right] \\ &= A_k A_{k-1}\sin(\Phi_k - \Phi_{k-1}) = A_k A_{k-1}\sin\Delta\theta_k \end{aligned}$$

From Table 4.6, the decision devices decide[7]

$$\begin{aligned} \widehat{I}_k &= 1, \text{if } x_k > 0 \text{ or } \widehat{I}_k = -1, \text{if } x_k < 0 \\ \widehat{Q}_k &= 1, \text{if } y_k > 0 \text{ or } \widehat{Q}_k = -1, \text{if } y_k < 0 \end{aligned}$$

The symbol error probability is given in (4.29) or (4.30), and P_b is given in (4.40).

---

[7] In Figure 4.16, the decision rule is to choose the smallest $|\Delta\theta_i - \Delta\widehat{\theta}_k|$. This rule is simplified for $\pi/4$-QPSK as described here.

The IF band differential demodulator (Figure 4.31) cancels the phase difference θ in the IF band. The I-channel multiplier output is

$$\begin{aligned} & A^2 \cos(2\pi f_c t + \Phi_k + \theta) \cos(2\pi f_c t + \Phi_{k-1} + \theta) \\ = & \frac{1}{2} A^2 \left[\cos(4\pi f_c t + \Phi_k + \Phi_{k-1} + 2\theta) + \cos(\Phi_k - \Phi_{k-1})\right] \end{aligned}$$

Again ignoring the factor $A^2/2$ and the LPF loss, LPF output is $\cos(\Phi_k - \Phi_{k-1}) = x_k$. Similarly the Q-channel LPF output is found to be $\sin(\Phi_k - \Phi_{k-1}) = y_k$. The rest is the same as the baseband differential detection. The advantage of this demodulator is that no local oscillator is needed.

The discriminator demodulator is shown in Figure 4.32. The ideal bandpass hard limiter keeps the envelope of the received signal constant without changing its phase. The ideal frequency discriminator output is proportional to the instantaneous frequency deviation of the input signal. That is

$$v(t) = \frac{d}{dt}\left(\Phi(t) + \theta\right)$$

The integrate-sample-dump (ISD) circuit output is

$$\begin{aligned} \int_{(k-1)T}^{kT} v(t)dt &= \Phi_k - \Phi_{k-1} \\ &= \Delta\theta_k + 2n\pi \\ &= \Delta\theta_k, \text{(mod } 2\pi) \end{aligned}$$

where $2n\pi$ is caused by click noise [14]. The modulo-2π operation removes the $2n\pi$ term and the output is $\Delta\theta_k$ which in turn will be mapped to a corresponding dibit.

The error probabilities of the above three demodulators are reported to be equal [12].[8] Like the DQPSK in the last section, the symbol error probability for the above three equivalent demodulators is given by (4.29) or (4.30). The bit error probability can be approximately calculated using (4.25). Alternately, bit error probability can be evaluated using (4.40).

The coherent $\pi/4$-QPSK demodulator can be in the form of Figure 4.17 where differential decoding is performed on the signal phases. The coherent demodulator of DEQPSK in Figure 4.25 is not suitable for $\pi/4$-QPSK since its dibits-$\Delta\theta_i$ assignment is different. A novel coherent $\pi/4$-QPSK demodulator (Figure 4.33) has been proposed in [12] where differential decoding is performed on the baseband signal levels. In $\pi/4$-QPSK, assuming $A_k = 1$ in (4.50), the demodulated signals are

[8] But it is not clear why the IF band differential demodulator could be equivalent to the baseband differential demodulator, because the latter is optimum and the former is not, according to our discussion in the section discussing DQPSK. The same doubt should arise regarding the discriminator demodulator.

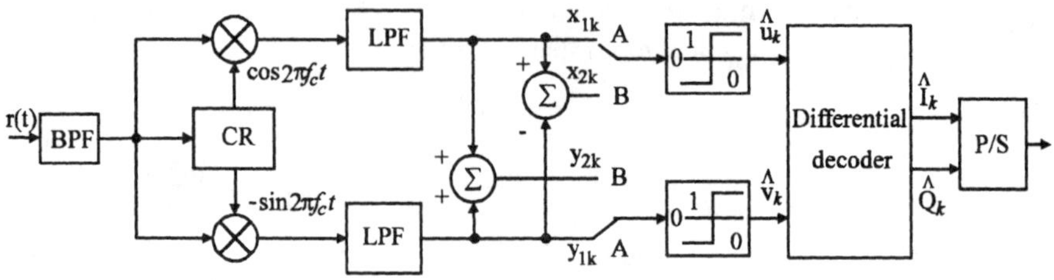

Figure 4.33 $\pi/4$-QPSK coherent demodulator.

two-level ($\pm 1/\sqrt{2}$) at every other sampling instant. In between, the signals are three-level $(0, \pm 1)$. This can be seen from the signal constellation (Figure 4.29), where • signals are two-level and × signals are three-level, and a • signal must be followed by a × signal or vice versa. If three-level detection is employed, the performance degrades compared with two-level detection. The three-level signals are converted to two-level signals in Figure 4.33. The converted two-level signals are detected by a two-level threshold detector. When the signals are two-level, the switches are in position A and the detection is the same as in QPSK. When the signals are three-level, the switches are in position B, the signals are converted to two-level by the following simple operations

$$\begin{aligned} x_{2k} &= x_{1k} - y_{1k} \\ y_{2k} &= x_{1k} + y_{1k} \end{aligned} \tag{4.51}$$

It is easy to verify that the conversions are (ignoring noise)

$$\begin{array}{ccc} (x_{1k}, y_{1k}) & \longrightarrow & (x_{2k}, y_{2k}) \\ (+1, 0) & \longrightarrow & (+1, +1) \\ (-1, 0) & \longrightarrow & (-1, -1) \\ (0, +1) & \longrightarrow & (-1, +1) \\ (0, -1) & \longrightarrow & (+1, -1) \end{array}$$

This is equivalent to rotate vector (x_{1k}, y_{1k}) by $+\pi/4$ and amplify its amplitude by $\sqrt{2}$. In other words, it is to rotate a × vector to the next • vector position with an amplitude gain of $\sqrt{2}$. This makes signal power doubled. However, the noise power is also doubled since the in-phase and quadrature channel noise are uncorrelated (see (4.51)). Thus the BER performance of the coherent $\pi/4$-QPSK is the same as that of the coherent QPSK.

The detected signals $\widehat{u}_k$ and $\widehat{v}_k$ are decoded by a DEQPSK differential decoder

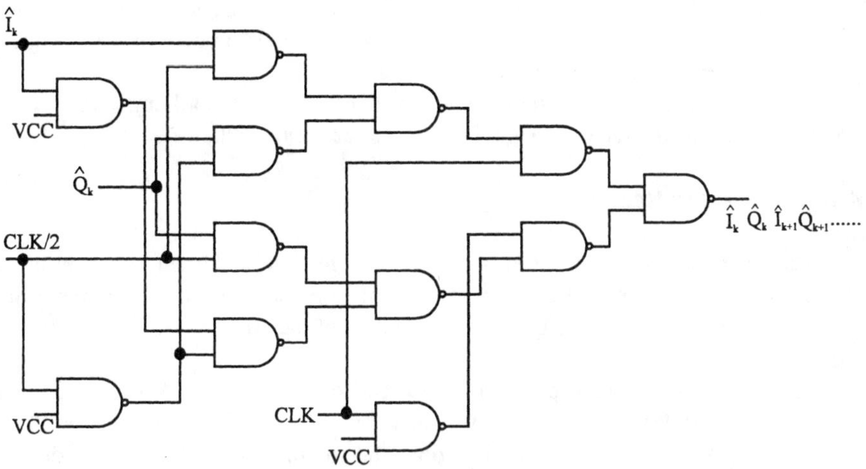

Figure 4.34 The parallel-to-serial converter of the $\pi/4$-QPSK coherent demodulator.

as in (4.44). Then the signals must be passed through a special P/S converter. The circuit is shown in Figure 4.34. The clock is derived from the symbol clock by dividing the frequency by two. The phase of this clock is synchronous to the switch for selecting sampled signals. The S/P converter delivers inphase and quadrature data alternately to the output. Refer to [12] for details.

As mentioned above in terms of error probability in the AWGN channel, the $\pi/4$-QPSK is the same as ordinary DEQPSK or DQPSK. That is, the coherently demodulated $\pi/4$-QPSK has the same BER as that of DEQPSK and the noncoherently demodulated $\pi/4$-QPSK has the same BER as that of DQPSK (see Figure 4.21).

4.10 SYNCHRONIZATION

Coherent demodulation requires that the reference signal at the receiver be synchronized in phase and frequency with the received signal. Both coherent and noncoherent demodulations require symbol timing at the receiver to be synchronized in phase and frequency with the received signal.

Carrier synchronization can be achieved by sending a pilot tone before message signals. Because the pilot tone has a strong spectral line at the carrier frequency, the receiver can easily lock on it and generates a local coherent carrier. However, this requires extra transmission bandwidth.

Carrier synchronization also can be achieved with a carrier recovery circuit which extracts the phase and frequency information from the noisy received signal and use it to generate a clean sinusoidal reference signal.

Symbol synchronization usually is achieved by a clock (symbol timing) recovery circuit which uses the received signal to control the local oscillator.

4.10.1 Carrier Recovery

The PSK signals have no spectral line at carrier frequency. Therefore a nonlinear device is needed in the carrier recovery circuit to generate such a line spectrum. There are two main types of carrier synchronizers, the Mth power loop, and the Costas loop.

Figure 4.35 is the Mth power loop for carrier recovery for M-ary PSK. For BPSK (or DEBPSK), $M = 2$, thus it is a squaring loop. For QPSK (or OQPSK, DEQPSK), $M = 4$, it is a quadrupling loop, and so on. It is the Mth power device that produces the spectral line at Mf_c. The phase lock loop consisting of the phase detector, the LPF, and the VCO, tracks and locks onto the frequency and phase of the Mf_c component. The divide-by-M device divides the frequency of this component to produce the desired carrier at frequency f_c and with almost the same phase of the received signal. Before locking, there is a phase difference in the received signal relative to the VCO output signal. We denote the phase of the received signal as θ and the phase of the VCO output as $M\widehat{\theta}$.

For BPSK, using (4.5), setting $A = 1$, and noticing that $a^2(t) = 1$, we have

$$\begin{aligned} z(t) &= [s(t) + n(t)]^2 = a^2(t)\cos^2(2\pi f_c t + \theta) + \text{ noise terms} \\ &= \frac{1}{2}[1 + \cos(4\pi f_c t + 2\theta)] + \text{noise terms} \end{aligned}$$

which contains a spectral line at $2f_c$ that can be tracked by the phase lock loop (PLL). The VCO output is divided by two in frequency to provide the desired carrier. It is obvious that the loop will produce a carrier with the same phase when the phase θ is either 0 or π. Then the demodulator output could be $+a(t)$ or $-a(t)$. We say that the loop has a phase ambiguity of π. Differential coding can eliminate phase ambiguity, as we described in previous sections in this chapter.

For QPSK, using (4.35), setting $A = \sqrt{2}$, and noticing that $I^2(t) = Q^2(t) = 1$, we have

$$\begin{aligned} z(t) &= [s(t) + n(t)]^4 \\ &= \{[I(t)\cos(2\pi f_c t + \theta) - Q(t)\sin(2\pi f_c t + \theta) + n(t)]^2\}^2 \\ &= [1 - I(t)Q(t)\sin(4\pi f_c t + 2\theta)]^2 + \text{noise terms} \end{aligned}$$

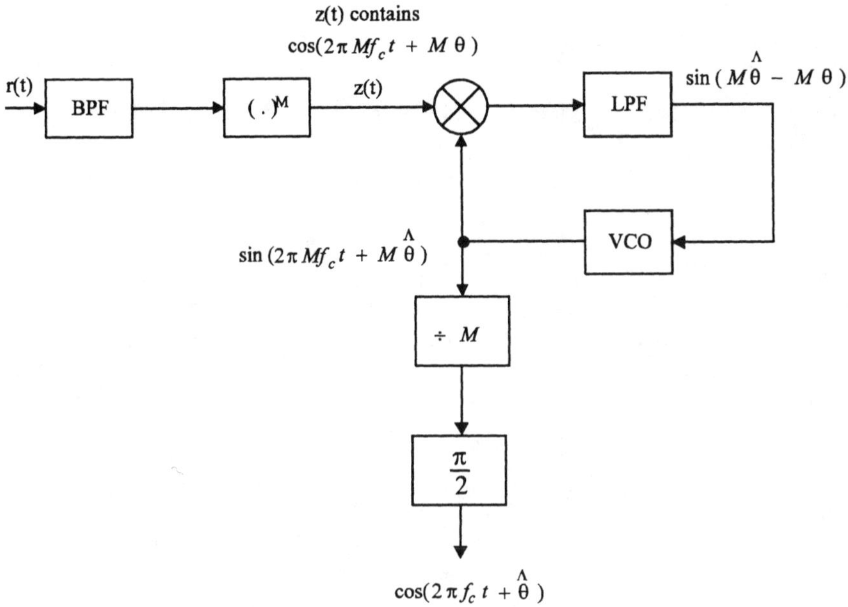

Figure 4.35 *M*th power synchronizer for carrier recovery.

$$= \quad 1 - 2I(t)Q(t)\sin(4\pi f_c t + 2\theta) + \frac{1}{2} - \frac{1}{2}\cos(8\pi f_c t + 4\theta) + \text{noise terms}$$

The last signal term contains a spectral line at $4f_c$ which is locked onto by the PLL. A divide-by-four device is used to derive the carrier frequency. Note that the $I(t)Q(t)\sin(4\pi f_c t + 2\theta)$ term resulting from squaring operation cannot produce a line spectrum since $I(t)Q(t)$ has a zero mean value. Therefore fourth power operation is needed for QPSK (and OQPSK). The last term will have a 0 initial phase for $\theta = 0$, or $\pm\pi/2$. The demodulator output could be $\pm I(t)$ or $\pm Q(t)$. This is to say that there is a $\pi/2$ phase ambiguity in the carrier recovery. It can be eliminated by differential coding as we discussed before.

For general MPSK, where $M = 2^n$, the Mth power operation will produce a spectral line at Mf_c and the phase ambiguity is $2\pi/M$.

The performance of the Mth power loop is generally measured by the phase tracking error. Under the usual small angle approximation (i.e., the phase error is small so that $M\widehat{\theta} - M\theta \approx \sin(M\widehat{\theta} - M\theta)$), the variance of such error is given by

[15]

$$\sigma_\theta^2 = 2N_o B_L S_L \tag{4.52}$$

where B_L is the loop bandwidth defined in terms of the loop transfer function $H(f)$ as follows

$$B_L = \int_0^\infty |H(f)|^2 df, \quad \text{(Hz)}$$

The parameter S_L is the upper bound of squaring loss in the BPSK case and quadrupling loss in the QPSK case. S_L is a number without unit that reflects the increase in the variance of phase error due to squaring or quadrupling operations in the phase tracking loop. They are given by [15]

$$S_L = 1 + \frac{1}{2\rho_i}, \quad \text{for } M = 2$$

and

$$S_L = 1 + \frac{9}{\rho_i} + \frac{6}{\rho_i^2} + \frac{3}{2\rho_i^3}, \quad \text{for } M = 4$$

where ρ_i is the input signal-to-noise ratio of the carrier recovery circuit. Observing (4.52), the unit of σ_θ^2 is seemingly watt instead of radian^2. This is due to the small-angle approximation in deriving the phase error variance where volt is replaced by radian. Therefore the unit of σ_θ^2 should be radian^2.

A difficulty in circuit implementation of the Mth power loop is the Mth power device, especially at high frequencies. Costas loop design avoids this device.

Figure 4.36 is the Costas loop for carrier recovery for BPSK. Initially the VCO generates a sinusoid with a frequency close to the carrier frequency f_c and some initial phase. The frequency difference and the initial phase are accounted for by the phase $\widehat{\theta}$. The multipliers in the I- and Q-channels produce $2f_c$ terms and zero frequency terms. The LPFs attenuate the $2f_c$ terms and their outputs are proportional to $a(t)\cos(\widehat{\theta}-\theta)$ or $a(t)\sin(\widehat{\theta}-\theta)$. Then these two terms multiply again to give the term $\frac{1}{2}a^2(t)\sin 2(\widehat{\theta}-\theta)$ which is low-pass filtered one more time to get rid of any amplitude fluctuation in $a^2(t)$, thus the control signal to the VCO is proportional to $\sin 2(\widehat{\theta}-\theta)$, which drives the VCO such that the difference $\widehat{\theta}-\theta$ becomes smaller and smaller. For sufficiently small $\widehat{\theta}-\theta$, the I-channel output is the demodulated signal.

The Costas loop for QPSK is shown in Figure 4.37. The figure is self-explanatory and its working principle is similar to that of BPSK. The limiters are bipolar, which are used to control the amplitude of the two channels' signal to maintain balance.

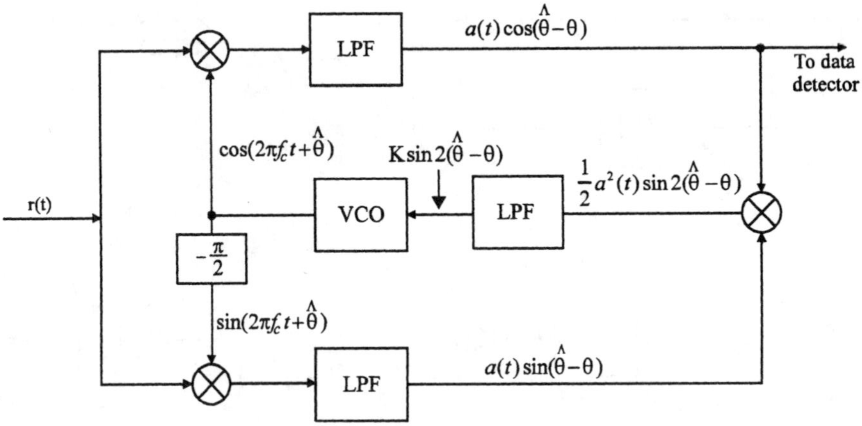

Figure 4.36 Costas loop for carrier recover for BPSK.

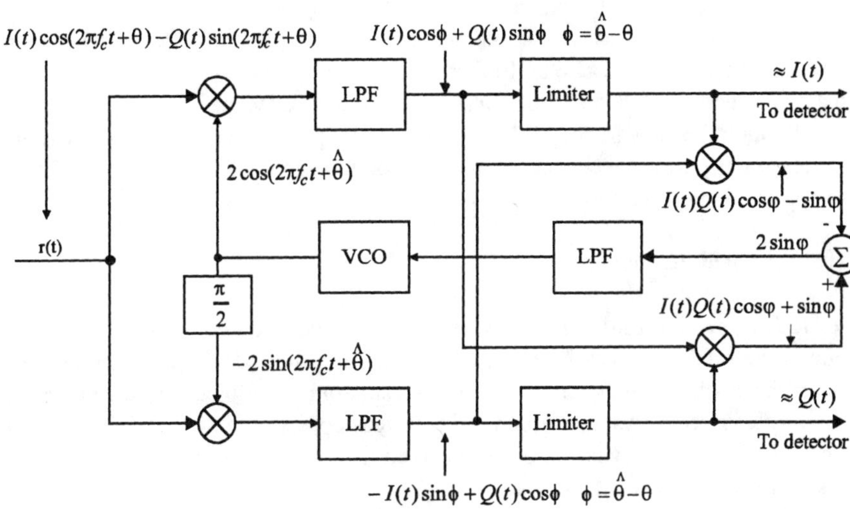

Figure 4.37 Costas loop for carrier recovery for QPSK.

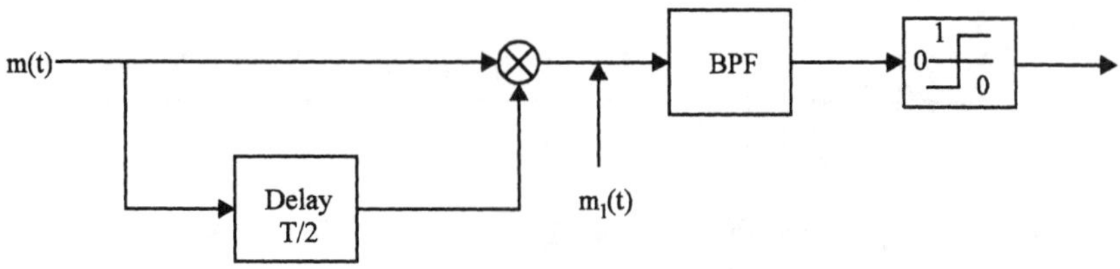

(a)

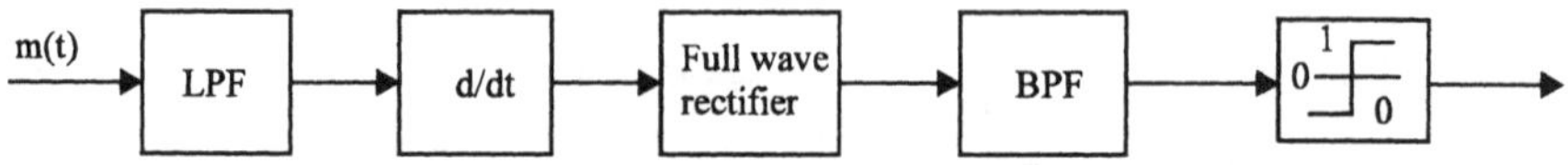

(b)

Figure 4.38 Two types of open-loop symbol synchronizers.

When the phase difference $\phi = \widehat{\theta} - \theta$ is sufficiently small, the I- and Q-channel outputs are the demodulated signals.

A difficulty in Costas loop implementation is to maintain the balance between the I- and Q-channel. The two multipliers and low-pass filters in these two channels must be perfectly matched in order to achieve the theoretical performance.

Although the appearance of the Mth power loop and the Costas loop are quite different, their performance can be shown to be the same [15]. Therefore (4.52) is also applicable to Costas loops.

4.10.2 Clock Recovery

The clock or symbol timing recovery can be classified into two basic groups. One group is the open loop synchronizer which uses nonlinear devices. These circuits recover the clock signal directly from the data stream by nonlinear operations on the received data stream. Another group is the closed-loop synchronizers which attempt to lock a local clock signal onto the received data stream by use of comparative measurements on the local and received signals.

Two examples of the open-loop synchronizer are shown in Figure 4.38. The data stream that we use in the phase shift keying modulation is NRZ waveform. Recall in Chapter 2 we have shown that this waveform has no spectral energy at the clock frequency (see Figure 2.3(a)). Thus in the open-loop synchronizers in Figure

4.38, the first thing that one needs to do is to create spectral energy at the clock frequency. In the first example, a Fourier component at the data clock frequency is generated by the delay-and-multiply operation on the demodulated signal $m(t)$. This frequency component is then extracted by the BPF that follows and shaped into square wave by the final stage. The second example generates the clock frequency component by using the differentiator-rectifier combination. The differentiator is very sensitive to wideband noise, therefore a low-pass filter is placed in the front end of the synchronizer.

An early/late-gate circuit shown in Figure 4.39 is an example of the class of closed-loop synchronizers. The working principle is easily understood by referencing Figure 4.40. The time zero point is set by the square wave clock locally generated by the VCO. If the VCO square wave clock is in perfect synchronism with the demodulated signal $m(t)$, the early-gate integrator and the late-gate integrator will accumulate the same amount of signal energy so that the error signal $e = 0$. If the VCO frequency is higher than that of $m(t)$, then $m(t)$ is late by $\Delta < d$, relative to the VCO clock. Thus the integration time in the early-gate integrator will be $T - d - \Delta$, while the integration time in the late-gate integrator is still the entire $T - d$. The error signal will be proportional to $-\Delta$. This error signal will reduce the VCO frequency and retard the VCO timing to bring it back toward the timing of $m(t)$. If the VCO frequency had been lower and the timing had been late, the error signal would be proportional to $+\Delta$, and the reverse process would happen, that is, the VCO frequency would be increased and its timing would be advanced toward that of the incoming signal.

4.10.3 Effects of Phase and Timing Error

As an example, we want to check the effects of phase and timing error on the bit error probability of coherent BPSK in the AWGN channel.

If the local carrier is in error by ϕ radians, the correlator output amplitude will be reduced by a factor $\cos\phi$. Thus the conditional bit error probability of BPSK will be

$$P_b(\phi) = Q\left(\sqrt{\frac{2E_b}{N_o}}\cos\phi\right) \tag{4.53}$$

If ϕ is Gaussianly distributed with variance σ_ϕ^2, the bit error probability is [16, pp. 270–271]

$$P_b = \int_{-\infty}^{\infty} p(\phi)P_b(\phi)d\phi$$

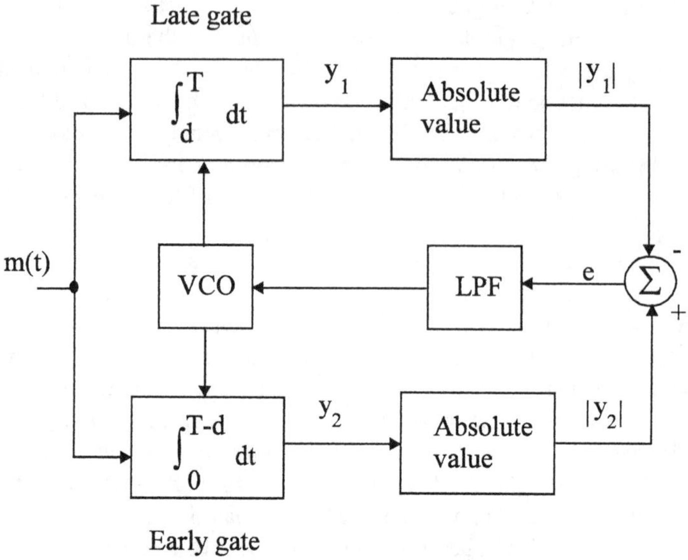

Figure 4.39 Early/late-gate clock synchronizer.

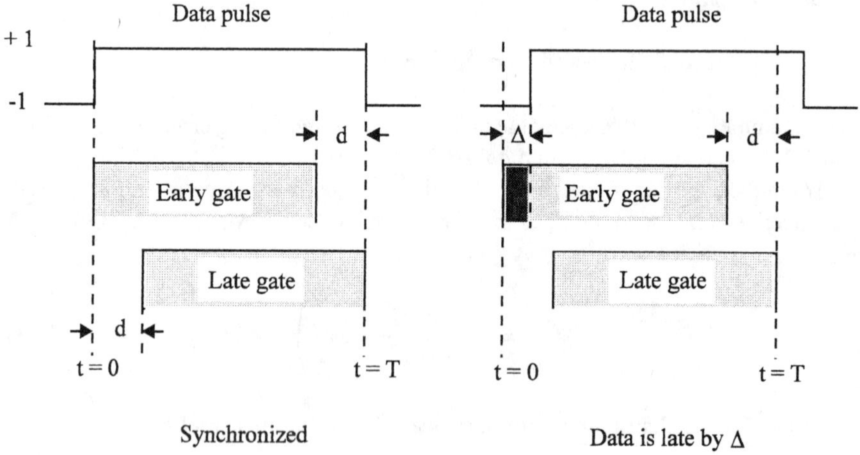

Figure 4.40 Early/late-gate timing illustration.

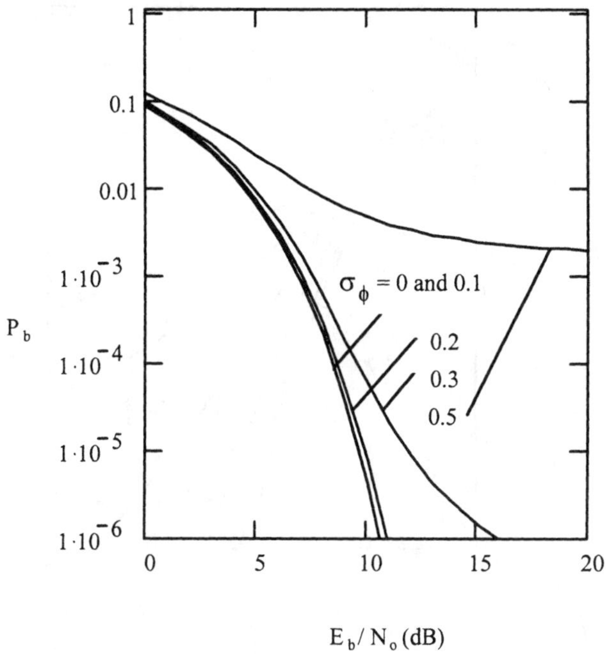

Figure 4.41 Effect of imperfect carrier synchronization on P_b of BPSK.

$$= \frac{2}{\sqrt{2\pi}\sigma_\phi}\int_0^\infty \exp\left(-\frac{\phi}{2\sigma_\phi^2}\right) Q\left(\sqrt{\frac{2E_b}{N_o}}\cos\phi\right) d\phi \qquad (4.54)$$

This is plotted in Figure 4.41 where different curves are labeled according to values of the standard deviation of the phase error σ_ϕ in radians. It is seen that when $\sigma_\phi < 0.2$, the degradation is not significant.

The effect of a symbol synchronization error on the bit error probability of BPSK depends on the presence or absence of a symbol transition. If two successive symbols are identical, an incorrect symbol reference will have no effect on the error probability. If two successive symbols differ, the magnitude of the correlator output is reduced by a factor of $1 - (2|\Delta|/T)$ where Δ is the timing error. Thus given a timing error, the conditional P_b is

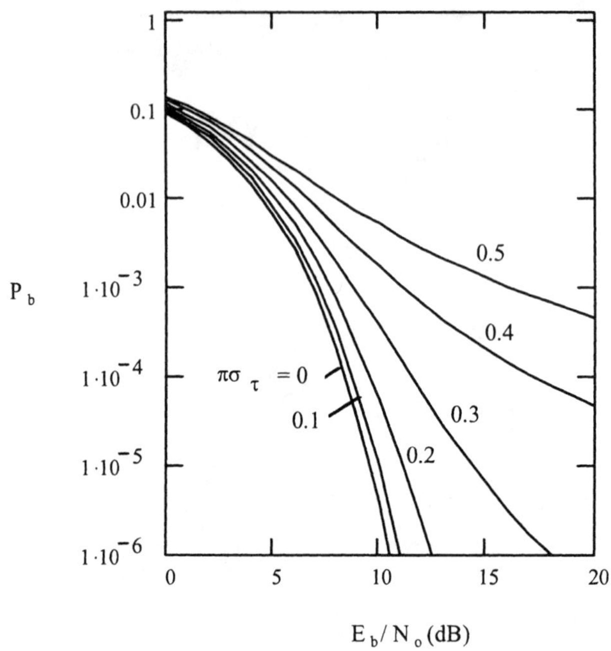

Figure 4.42 Effect of imperfect symbol synchronization on P_b of BPSK.

$$\begin{aligned} P_b(\Delta) &= \Pr(\text{error} \mid \text{transition}, \Delta)\Pr(\text{transition} \mid \Delta) \\ &\quad + \Pr(\text{error} \mid \text{no transition}, \Delta)\Pr(\text{no transition} \mid \Delta) \end{aligned}$$

If the successive symbols are independent and equally likely to be either of the two binary symbols, the probability of transition is one-half, and, if the normalized timing error $\tau = \Delta/T$ is Gaussianly distributed, then

$$\begin{aligned} P_b &= \frac{1}{2\sqrt{2\pi}\sigma_\tau}\int_{-0.5}^{0.5}\exp\left(-\frac{\tau}{2\sigma_\tau^2}\right)Q\left(\sqrt{\frac{2E_b}{N_o}}(1-2|\tau|)\right)d\tau \\ &\quad + \frac{1}{2}Q\left(\sqrt{\frac{2E_b}{N_o}}\right) \end{aligned} \tag{4.55}$$

where the integration limits are 0.5 to –0.5 which is the range of the timing errors [17, sections 6-4 and 9-4]. This is plotted in Figure 4.42, where different curves are labeled according to a value of the $\pi\sigma_\tau$ where σ_τ is the standard deviation of the timing error τ. When $\pi\sigma_\tau$ is less than 0.2, the degradation is not significant.

It should be pointed out that (4.54) and (4.55), and consequently Figures 4.41 and 4.42, are based on the assumption that the errors are Gaussianly distributed [16, pp. 270–271], which may not be accurate, depending on the carrier and clock recovery systems. More accurate, but also more complicated, error distribution models and error probability results are given in [17, sections 6-4 and 9-4].

4.11 SUMMARY

In this chapter we have covered all important PSK modulation schemes. We described all aspects, including signal expressions, waveforms, power spectral density, modulator, demodulator, and symbol or bit error probability. We started from the basic BPSK and its noncoherent version, DBPSK. Then we proceed to M-ary PSK and its differential version. For them we established general optimum demodulator block diagrams, error probability formulas, and PSD expressions. These results were later used for QPSK and DQPSK. However, QPSK and DQPSK are not merely special cases of MPSK and MDPSK. Their signal constellations allow for further simplification of the demodulators. Noticeably, phase calculation and comparison stages of the demodulators are replaced by level detectors. Offset QPSK, as a solution to suppress sidelobe spread after bandlimiting and nonlinear amplifications, was briefly described due to its historical value. $\pi/4$-QPSK, as standard modulation in several digital cellular systems, was covered in great detail. Its modulator, baseband differential demodulator, IF-band differential demodulator, FM-discriminator demodulator, and coherent demodulator were described. Error performance was described. Finally we covered the synchronization. Carrier synchronization is needed for coherent PSK schemes. Symbol synchronization is needed for any digital modulation schemes. Table 4.7 summarizes and compares the various aspects of PSK schemes described in this chapter. The error performance degradation is measured in increase of E_b/N_o for achieving the same error probability, reference to the coherent demodulation. BPSK and QPSK are the most widely used PSK schemes. This is due to their system simplicity and excellent power and bandwidth efficiency. Higher order MPSK can be used if higher bandwidth efficiency is desired and higher signal-to-noise ratio is available.

Up to this point, we have studied classical frequency and phase shift keying schemes. In the rest of this book, we will study more bandwidth efficient and/or power efficient modulation schemes. In the next chapter, a scheme with important practical applications, minimum shift keying (MSK), which can be considered as a

Modulation	P_b	Degradation	PSD and B_{null}
BPSK	$Q\left(\sqrt{\frac{2E_b}{N_o}}\right)$	0 dB (ref.)	$A^2T_b\left(\frac{\sin\pi fT_b}{\pi fT_b}\right)^2$ $B_{null}=2/T_b$
DEBPSK	$\approx 2Q\left(\sqrt{\frac{2E_b}{N_o}}\right)$	< 0.5 dB	
DBPSK (Optimum)	$\frac{1}{2}e^{-E_b/N_o}$	0.5-1 dB	
DBPSK (Suboptimum)	$\frac{1}{2}e^{-0.8E_b/N_o}$	1.5-2 dB	
QPSK	$Q\left(\sqrt{\frac{2E_b}{N_o}}\right)$	0 dB (ref.)	$2A^2T_b\left(\frac{\sin 2\pi fT_b}{2\pi fT_b}\right)$ $B_{null}=1/T_b$
DEQPSK	$\approx 2Q\left(\sqrt{\frac{2E_b}{N_o}}\right)$	< 0.5 dB	
DQPSK (Optimum)	$\approx Q\left(\sqrt{\frac{4E_b}{N_o}}\sin\frac{\pi}{4\sqrt{2}}\right)$	2-3 dB	
DQPSK (Suboptimum)	$\approx e^{-0.59\frac{E_b}{N_o}}$	3-5 dB	
$\pi/4$-QPSK	$\approx 2Q\left(\sqrt{\frac{2E_b}{N_o}}\right)$ (= DEQPSK)	< 0.5 dB	$2A^2T_b\left(\frac{\sin 2\pi fT_b}{2\pi fT_b}\right)$ $B_{null}=1/T_b$
$\pi/4$-DQPSK	$\approx Q\left(\sqrt{\frac{4E_b}{N_o}}\sin\frac{\pi}{4\sqrt{2}}\right)$ (= optimum DQPSK)	2-3 dB	
MPSK (M>4) (n = $\log_2$M) or	$\approx \frac{2}{n}Q\left(\sqrt{\frac{2nE_b}{N_o}}\sin\frac{\pi}{M}\right)$ $\approx \frac{2}{n}\sum_{i=1}^{2}Q\left(\sqrt{\frac{2E_bn}{N_o}}\sin\frac{(2i-1)\pi}{M}\right)$	0 dB (ref.)	$nA^2T_b\left(\frac{\sin n\pi fT_b}{n\pi fT_b}\right)$ $B_{null}=2/(nT_b)$
DEMPSK	Eqn.(4.31)	< 0.5 dB	
DMPSK (Optimum)	$\approx \frac{2}{n}Q\left(\sqrt{\frac{2nE_b}{N_o}}\sin\frac{\pi}{\sqrt{2}M}\right)$	3 dB (asympt.)	

Table 4.7 PSK schemes comparison.

development from OQPSK or a special continuous phase FSK, will be studied.

APPENDIX 4A DERIVATION OF $p(\varphi/H_i)$

To derive the expression for $p(\varphi/H_i)$, we first complete the square in exponent of $p(\rho,\widehat{\theta}/H_i)$ by writing

$$\begin{aligned}\rho^2+E-2\rho\sqrt{E}\cos\varphi &= (\rho-\sqrt{E}\cos\varphi)^2-E\cos^2\varphi+E\\ &= (\rho-\sqrt{E}\cos\varphi)^2+E\sin^2\varphi\end{aligned}$$

Then we integrate $p(\rho,\widehat{\theta}/H_i)$ with respect to ρ to obtain the PDF of φ

$$\begin{aligned}p(\varphi/H_i) &= \int_0^\infty \frac{\rho}{\pi N_o}\exp\{-\frac{1}{N_o}[(\rho-\sqrt{E}\cos\varphi)^2+E\sin^2\varphi]\}d\rho\\ &= \exp\{-\frac{E}{N_o}\sin^2\varphi\}\int_0^\infty \frac{\rho}{\pi N_o}\exp\{-\frac{1}{N_o}(\rho-\sqrt{E}\cos\varphi)^2\}d\rho\end{aligned}$$

Now change the variable to make the exponent be $-t^2$. This requires

$$t=\frac{1}{\sqrt{N_o}}(\rho-\sqrt{E}\cos\varphi)$$

or inversely

$$\rho=\sqrt{N_o}t+\sqrt{E}\cos\varphi$$

Noting that $d\rho=\sqrt{N_o}dt$, and when $\rho=0$, $t=-\sqrt{E/N_o}\cos\varphi$, the integral becomes

$$\begin{aligned}&p(\varphi/H_i)\\ &= \exp\{-\frac{E}{N_o}\sin^2\varphi\}\int_{-\sqrt{E/N_o}\cos\varphi}^\infty \frac{1}{\pi\sqrt{N_o}}\left[\sqrt{N_o}t+\sqrt{E}\cos\varphi\right]\exp\{-t^2\}dt\end{aligned} \tag{4.56}$$

The first term of the integral is

$$\begin{aligned}\int_{-\sqrt{E/N_o}\cos\varphi}^\infty \frac{t}{\pi}\exp\{-t^2\}dt &= -\frac{1}{2\pi}\exp\{-t^2\}\Big|_{-\sqrt{E/N_o}\cos\varphi}^\infty\\ &= \frac{1}{2\pi}\exp\{-\frac{E}{N_o}\cos^2\varphi\}\end{aligned} \tag{4.57}$$

The second term of the integral is

$$\begin{aligned}
&\int_{-\sqrt{E/N_o}\cos\varphi}^{\infty} \frac{\sqrt{E}\cos\varphi}{\pi\sqrt{N_o}} \exp\{-t^2\}dt \\
&= \frac{\cos\varphi}{2\sqrt{\pi}}\sqrt{\frac{E}{N_o}} \int_{-\sqrt{E/N_o}\cos\varphi}^{\infty} \frac{2}{\sqrt{\pi}} \exp\{-t^2\}dt \\
&= \frac{\cos\varphi}{2\sqrt{\pi}}\sqrt{\frac{E}{N_o}} \left[1+\operatorname{erf}\left(\sqrt{\frac{E}{N_o}}\cos\varphi\right)\right] \qquad (4.58)
\end{aligned}$$

where

$$\operatorname{erf}(x) \triangleq \frac{2}{\sqrt{\pi}}\int_0^x e^{-t^2}dt$$

is the *error function* which has the following properties

$$\operatorname{erf}(x) = 1 - \int_x^{\infty} \frac{2}{\sqrt{\pi}} \exp\{-t^2\}dt$$

and

$$\operatorname{erf}(-x) = -\operatorname{erf}(x)$$

We have used these properties in deriving (4.58). Substituting (4.57) and (4.58) into (4.56) we have

$$\begin{aligned}
p(\varphi/H_i) &= \exp\{-\frac{E}{N_o}\sin^2\varphi\}\left\{\frac{1}{2\pi}\exp\{-\frac{E}{N_o}\cos^2\varphi\} + \frac{\cos\varphi}{2\sqrt{\pi}}\sqrt{\frac{E}{N_o}}\right. \\
&\quad \left.\cdot\left[1+\operatorname{erf}\left(\sqrt{\frac{E}{N_o}}\cos\varphi\right)\right]\right\}
\end{aligned}$$

Factorizing $\frac{1}{2\pi}\exp\{-\frac{E}{N_o}\cos^2\varphi\}$ out of the bracket we obtain

$$\begin{aligned}
p(\varphi/H_i) &= \frac{e^{-E/N_o}}{2\pi}\left\{1+\sqrt{\frac{\pi E}{N_o}}(\cos\varphi)e^{(E/N_o)\cos^2\varphi}\right. \\
&\quad \left.\cdot\left[1+\operatorname{erf}\left(\sqrt{\frac{E}{N_o}}\cos\varphi\right)\right]\right\} \\
&= p(\varphi)
\end{aligned}$$

References

[1] Park, J. H., Jr., "On binary DPSK detection," *IEEE Trans. Commun.*, vol. 26, no. 4, April 1978, pp. 484–486.

[2] Simon, K. M., S. M. Hinedi, and W. C. Lindsey, *Digital Communication Techniques: Signal Design and Detection*, Englewood Cliffs, New Jersey: Prentice Hall, 1995.

[3] Van Trees, H. L., *Detection, Estimation, and Modulation Theory, Part I,* New York: John Wiley & Sons, Inc., 1968.

[4] Lu, J., et al., "M-PSK and M-QAM BER computation using signal-space concepts," *IEEE Trans. Commun.*, vol. 47, no. 2, February 1999, pp. 181–184.

[5] Benedetto, S., E. Biglieri, and V. Castellani, *Digital Transmission Theory*, Englewood Cliffs, New Jersey: Prentice Hall, 1987.

[6] Proakis, J., *Digital Communications*, 2nd ed., New York: McGraw-Hill, 1989.

[7] Feher, K., *Digital Communications: Satellite/Earth Station Engineering*, Englewood Cliffs, New Jersey: Prentice Hall, 1983.

[8] Lucky, R., J. Salz, and J. Weldon, *Principles of Data Communications*, New York: McGraw-Hill, 1968.

[9] Whalen, A. D., *Detection of Signals in Noise*, New York and London: Academic Press, 1971.

[10] Pasupathy, S., "Minimum shift keying: a spectrally efficient modulation," *IEEE Communications Magazine*, July 1979.

[11] Baker, P. A., "Phase Modulation Data Sets for Serial Transmission at 2000 and 2400 Bits per Second, Part 1," *AIEE Trans. Comm. Electron.,* July 1962.

[12] Liu, C. L., and K. Feher, "$\pi/4$-QPSKModems for Satellite Sound/Data Broadcast Systems," *IEEE Trans. Broadcasting,* March 1991.

[13] Feher, K., "MODEMS for Emerging Digital Cellular-Mobile Radio System," *IEEE Trans. on Vehicular Technology,* vol. 40, no. 2, May 1991, pp. 355–365.

[14] Roden, M., *Analog and Digital Communications*, 3rd ed., Englewood Cliffs, New Jersey: Prentice Hall, 1991.

[15] Gardner, F. M., *Phaselock Techniques*, 2nd ed., New York: John Wiley, 1979.

[16] Stiffler, J. J., *Theory of Synchronous Communications*, Englewood Cliffs, New Jersey: Prentice Hall, 1971.

[17] Lindsay, W. C. and M. K. Simon, *Telecommunication Systems Engineering,* Englewood Cliffs, New Jersey: Prentice Hall, 1973.

Selected Bibliography

- Couch II, L. W., *Digital and Analog Communication Systems*, 3rd ed., New York: Macmillan, 1990.
- Divsalar, D., and M. K. Simon, "On the implementation and performance of single and double

differential detection schemes," *IEEE Trans. Commun.*, vol. 40, no. 2, February 1992, pp. 278–291.

- Haykin, S., *Digital Communications*, New York: John Wiley, 1988.
- Haykin, S., *Communication Systems*, 3rd ed., New York: John Wiley, 1994.
- Liu, C. L., and K. Feher, "Bit error performance of $\pi/4$-DQPSK in a frequency-selective fast Rayleigh fading channel," *IEEE Trans. Vehicular Technology*, vol. 40, no. 3, August 1991.
- Sklar, B., *Digital Communications: Fundamentals and Applications*, Englewood Cliffs, New Jersey: Prentice Hall, 1988.
- Smith, D. R., *Digital Transmission Systems*, 2nd ed., New York: Van Nostrand Reinhold, 1993.
- Ziemer, R. E., and R. L. Peterson, *Introduction to Digital Communication*, New York: Macmillan, 1992.

Chapter 5

Minimum Shift Keying and MSK-Type Modulations

In the previous chapter we saw that the major advantage of OQPSK over QPSK is that it exhibits less phase changes at symbol transitions, thus out-of-band interference due to band limiting and amplifier nonlinearity is reduced. This suggests that further improvement is possible if phase transitions are further smoothed or even become completely continuous. Minimum shift keying (MSK) is such a continuous phase modulation scheme. It can be derived from OQPSK by shaping the pulses with half sinusoidal waveforms, or can be derived as a special case of continuous phase frequency shift keying (CPFSK).

MSK was first proposed by Doelz and Heald in their patent in 1961 [1]. DeBuda discussed it as a special case of CPFSK in 1972 [2]. Gronemeyer and McBride described it as sinusoidally weighted OQPSK in 1976 [3]. Amoroso and Kivett simplified it by an equivalent serial implementation (SMSK) in 1977 [4]. Now MSK has been used in actual communication systems. For instance, SMSK has been implemented in NASA's Advanced Communications Technology Satellite (ACTS) [5] and Gaussian MSK (GMSK) has been used as the modulation scheme of the European GSM (global system for mobile) communication system [6].

This chapter is organized as follows: Section 5.1 describes the basic MSK (i.e., parallel MSK) in great detail in order for the readers to grasp the fundamental concept and important properties of MSK thoroughly. Section 5.2 discusses its power spectral density and bandwidth. MSK modulator, demodulator, and synchronization are presented in Sections 5.3, 5.4, and 5.5, respectively. MSK error probability is discussed in Section 5.6. Section 5.7 is devoted to SMSK in a great detail because of its importance in practical applications. MSK-type schemes which are modified MSK schemes for better bandwidth efficiency or power efficiency, are discussed in detail in Sections 5.8 through 5.13. However, GMSK is not covered in this chapter, instead, it is discussed in Chapter 6 in the context of continuous phase modulation. Finally, the chapter is concluded with a summary in Section 5.14.

5.1 DESCRIPTION OF MSK

5.1.1 MSK Viewed as a Sinusoidal Weighted OQPSK

In OQPSK modulation, $I(t)$ and $Q(t)$, the staggered data streams of the I-channel and Q-channel are directly modulated onto two orthogonal carriers. Now we weight each bit of $I(t)$ or $Q(t)$ with a half period of cosine function or sine function with a period of $4T$, $A\cos(\pi/2T)$ or $A\sin(\pi/2T)$, respectively, then modulate them onto one of two orthogonal carriers, $\cos 2\pi f_c t$ or $\sin 2\pi f_c t$, by doing these we create an MSK signal

$$s(t) = AI(t)\cos(\frac{\pi t}{2T}) \cos 2\pi f_c t + AQ(t)\sin(\frac{\pi t}{2T})\sin 2\pi f_c t \qquad (5.1)$$

where T is the bit period of the data.

Figure 5.1 shows the waveforms of MSK at each stage of modulation. Figure 5.1(a) is the $I(t)$ waveform for the sample symbol stream of $\{1, -1, 1, 1, -1\}$. Note that each $I(t)$ symbol occupies an interval of $2T$ from $(2n-1)T$ to $(2n+1)T$, $n = 0, 1, 2....$ Figure 5.1(b) is the weighting cosine waveform with a period of $4T$, whose half period coincides with one symbol of $I(t)$. Figure 5.1(c) is the cosine weighted symbol stream. Figure 5.1(d) is the modulated I-channel carrier that is obtained by multiplying the waveform in Figure 5.1(c) by the carrier $\cos 2\pi f_c t$. This signal is the first term in (5.1).

Figure 5.1(e–h) shows the similar modulation process in Q-channel for the sample $Q(t)$ stream of $\{1, 1, -1, 1, -1\}$. Note that $Q(t)$ is delayed by T with respect to $I(t)$. Each symbol starts from $2nT$ and ends at $(2n+2)T$, $n = 0, 1, 2, ...$ The weighting signal is sine instead of cosine, thus each half period coincides with one symbol of $Q(t)$. Figure 5.1(h) is the second term in (5.1).

Figure 5.1(i) shows the composite MSK signal $s(t)$, which is the sum of waveforms of Figure 5.1(d) and Figure 5.1(h).[1]

From Figure 5.1(i) we observe the following properties of MSK. First, its envelope is constant. Second, the phase is continuous at bit transitions in the carrier. There are no abrupt phase changes at bit transitions like in QPSK or OQPSK. Third, the signal is an FSK signal with two different frequencies and with a symbol duration

[1] The MSK defined in (5.1) and illustrated in Figure 5.1 is called Type I MSK where the weighting is alternating positive and negative half-sinusoid. Another type is called Type II MSK where the weighting is always a positive half-sinusoid [7]. These two types are the same in terms of power spectral density, which is determined by the shape of the half-sinusoid, and error probability, which is determined by the energy of the half-sinusoid. The only difference between them is the weighting signal in the modulator and the demodulator. Therefore it suffices to analyze Type I only in the rest of this chapter.

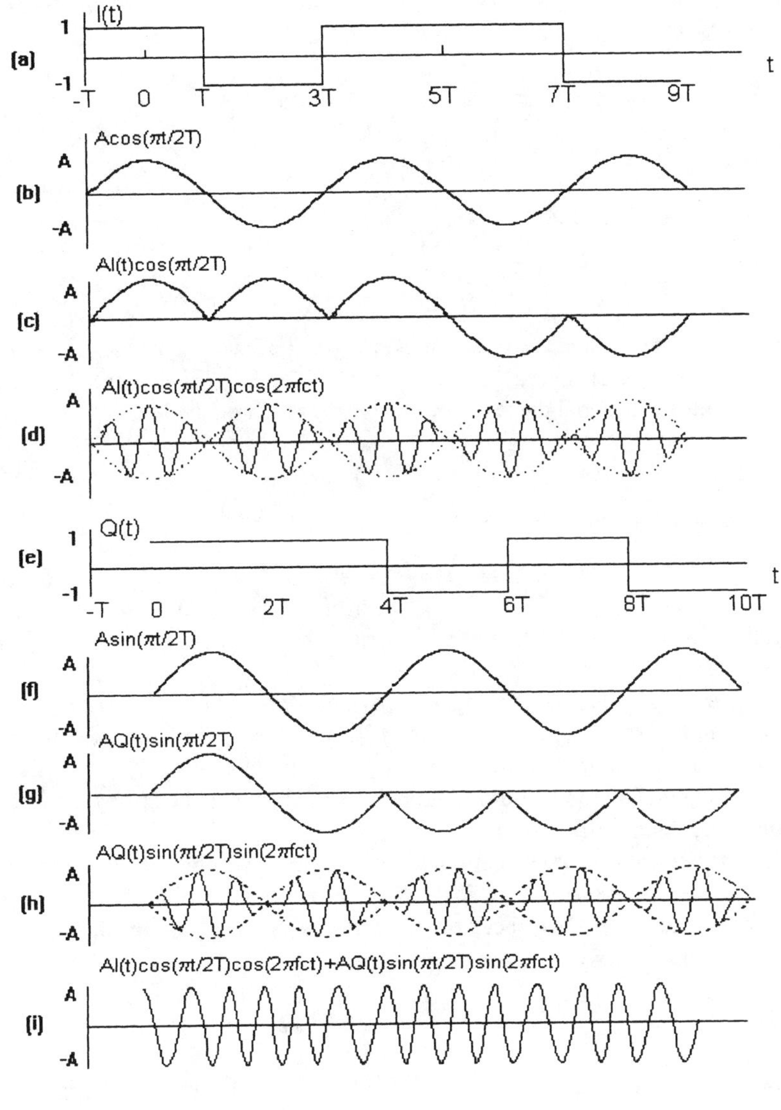

Figure 5.1 MSK waveforms (a–i).

of T.[2]

To better understand the above properties we rewrite (5.1) in a different form. In the kth bit period of T seconds, $I(t)$ and $Q(t)$ is either 1 or –1, we denote them as I_k and Q_k, thus

$$\begin{aligned} s(t) &= \pm A\cos(\frac{\pi t}{2T})\cos 2\pi f_c t \pm A\sin(\frac{\pi t}{2T})\sin 2\pi f_c t \\ &= \pm A\cos[2\pi f_c t + d_k\frac{\pi t}{2T}] \\ &= A\cos[2\pi f_c t + d_k\frac{\pi t}{2T} + \Phi_k] \\ &= A\cos[2\pi(f_c + d_k\frac{1}{4T})t + \Phi_k], \qquad kT \le t \le (k+1)T \end{aligned} \tag{5.2}$$

where $d_k = 1$ when I_k and Q_k have opposite signs (i.e., successive bits in the serial data stream are different), and $d_k = -1$ when I_k and Q_k have same signs (i.e., successive bits in the serial data stream are the same). Or equivalently

$$d_k = -I_k Q_k \tag{5.3}$$

$\Phi_k = 0$ or π corresponding to $I_k = 1$ or -1. Or equivalently

$$\Phi_k = \frac{\pi}{2}(1 - I_k) \tag{5.4}$$

Both d_k and Φ_k are constant in a bit period of T seconds since I_k and Q_k are constant in T.

It is clear from (5.2) that MSK signal is a special FSK signal with two frequencies $f_+ = f_c + 1/4T$ or $f_- = f_c - 1/4T$, where f_+ is called space frequency, f_- mark frequency, and f_c apparent carrier frequency. The frequency separation is $\Delta f = 1/2T$. This is the minimum separation for two FSK signals to be orthogonal, hence the name "minimum shift keying."

Ordinary coherent FSK signal could have continuous phase or discontinuous phase at bit transitions (see Figure 3.3). MSK carrier phase is always continuous at bit transitions. To see this, we check the excess phase of the MSK signal, referenced to the carrier phase, which is given by

$$\Theta(t) = d_k\frac{\pi t}{2T} + \Phi_k = \pm\frac{\pi t}{2T} + \Phi_k, \qquad kT \le t \le (k+1)T \tag{5.5}$$

Because Φ_k is constant in the interval $[kT, (k+1)T]$, $\Theta(t)$ is linear and continuous

[2] Note that the MSK signal has a symbol duration of T instead of $2T$ despite that the symbol durations are $2T$ for $I(t)$ and $Q(t)$. This property is the same as that of OQPSK since both of them have a staggered Q-channel symbol stream. QPSK has a symbol duration of $2T$.

in the interval $[kT, (k+1)T]$. However, to ensure phase continuity at bit transitions, at the end of the kth bit period, we must require

$$d_k \frac{\pi(k+1)T}{2T} + \Phi_k = d_{k+1} \frac{\pi(k+1)T}{2T} + \Phi_{k+1} \pmod{2\pi} \tag{5.6}$$

In the following we will show that this requirement is always satisfied for the MSK signal in (5.1).

Note that since $d_k = -I_k Q_k$, and $\Phi_k = (1 - I_k)\pi/2$, the left-hand side (LHS) and the right-hand side (RHS) of (5.6) become

$$LHS = -I_k Q_k (k+1)\frac{\pi}{2} + \frac{\pi}{2}(1 - I_k) \tag{5.7}$$

$$RHS = -I_{k+1} Q_{k+1} (k+1)\frac{\pi}{2} + \frac{\pi}{2}(1 - I_{k+1}) \tag{5.8}$$

Because I_k and Q_k each occupies $2T$ and are staggered, we can assume $I_k = I_{k+1}$ for odd k and $Q_k = Q_{k+1}$ for even k (or vice versa). Thus, if k is odd, $I_k = I_{k+1}$,

$$RHS = -I_k Q_{k+1} (k+1)\frac{\pi}{2} + \frac{\pi}{2}(1 - I_k) \tag{5.9}$$

Compare (5.9) to (5.7), we can see that to make them equal is to satisfy the requirement

$$-I_k Q_k (k+1)\frac{\pi}{2} = -I_k Q_{k+1} (k+1)\frac{\pi}{2}$$

This obviously is true when $Q_k = Q_{k+1}$. When $Q_k \neq Q_{k+1}$, then $Q_k = -Q_{k+1}$, the above requirement becomes

$$-I_k Q_k (k+1)\frac{\pi}{2} = I_k Q_k (k+1)\frac{\pi}{2} \pmod{2\pi}$$

Since k is odd, $(k+1)$ is even, and note that $I_k = \pm 1$ and $Q_k = \pm 1$, the above requirement becomes

$$-m\pi = m\pi \pmod{2\pi}$$

If m is odd, $\pm m\pi = \pi (\bmod\, 2\pi)$. If m is even, $\pm m\pi = 0 (\bmod\, 2\pi)$. Thus, in any case, the requirement is satisfied.

If k is even, $Q_k = Q_{k+1}$, again we have two cases. In the first case, $I_k = I_{k+1}$, it is easy to see that (5.7) is equal to (5.8). In the second case, $I_k \neq I_{k+1}$ (i.e., $I_k = -I_{k+1}$), then

$$LHS = -I_k Q_k (k+1)\frac{\pi}{2} + \frac{\pi}{2}(1 - I_k)$$

I_k	Q_k	LHS	RHS
1	1	$-(k+1)\frac{\pi}{2}$	$(k+3)\frac{\pi}{2} = (k+2)\pi + LHS = LHS \, (\text{mod}\, 2\pi)$
1	-1	$(k+1)\frac{\pi}{2}$	$(-k+1)\frac{\pi}{2} = -k\pi + LHS = LHS \, (\text{mod}\, 2\pi)$
-1	1	$(k+1)\frac{\pi}{2}$	$(-k+1)\frac{\pi}{2}$ same as the second case
-1	-1	$(-k+1)\frac{\pi}{2}$	$(k+1)\frac{\pi}{2}$ same as the second case

Table 5.1 Possible cases (note that k is even).

$$RHS = I_k Q_k (k+1)\frac{\pi}{2} + \frac{\pi}{2}(1 + I_k)$$

Table 5.1 shows all possible cases of the above two expressions. As seen from Table 5.1, in all cases $LHS = RHS \, (\text{mod}\, 2\pi)$.

The above proof shows that the excess phase $\Theta(t)$ is always continuous. The phase of the apparent carrier is $2\pi f_c t$ which is also continuous at any time. Therefore the total phase, $2\pi f_c t + \Theta(t)$, is always continuous at any time. Note that in the above discussion we did not specify any relationship between f_c and the symbol rate $1/T$. In other words, for the MSK signal phase to be continuous, no specific relation between f_c and $1/T$ is required. However, as we will show in Section 5.4, preferably f_c should be chosen as a multiple of $1/4T$, but it is for orthogonality of its I-channel and Q-channel signal components, not for continuous phase purpose.

From the above discussion we also can see that $\Theta(kT)$ is a multiple of $\pi/2$. However, the total phase at bit transitions (or initial phase of the bit), $2\pi f_c kT +$ $\Theta(kT)$, is not necessarily a multiple of $\pi/2$. It could be any value depending on the value of f_c in relation to the bit period T. If f_c is a multiple of $1/4T$ (i.e., $f_c = m/4T$) for a positive integer m, then $2\pi f_c kT = mk\pi/2$, which is a multiple of $\pi/2$. Thus the total phase at bit transitions is also a multiple of $\pi/2$. If f_c is not a multiple of $1/4T$, then the total phase at bit transitions is usually not a multiple of $\pi/2$. As we have pointed out above, f_c is indeed usually chosen as a multiple of $1/4T$ for orthogonality of its I-channel and Q-channel signals. Consequently, the total phase at bit transitions is a multiple of $\pi/2$.

The excess phase $\Theta(t)$ increases or decreases linearly with time during each bit period of T seconds (see (5.5)). If $d_k = 1$ in the bit period, the carrier phase is increased by $\pi/2$ by the end of the bit period. This corresponds to an FSK signal at the higher frequency f_+. If $d_k = -1$ in the bit period, the carrier phase is decreased by $\pi/2$ by the end of the bit period. This corresponds to an FSK signal at the lower frequency f_-. Figure 5.2 is the phase tree of MSK signal's excess phase $\Theta(t)$. The bold-faced path represents the data sequence $d_k = -I_k Q_k$ for I_k and Q_k in Figure 5.1. The excess phase values at bit transitions are always a multiple of $\pi/2$. If f_c happens to be a multiple of $1/T$, then the excess phase values at bit transitions in the phase tree are also the total phase values of the carrier at bit transitions. From (5.5)

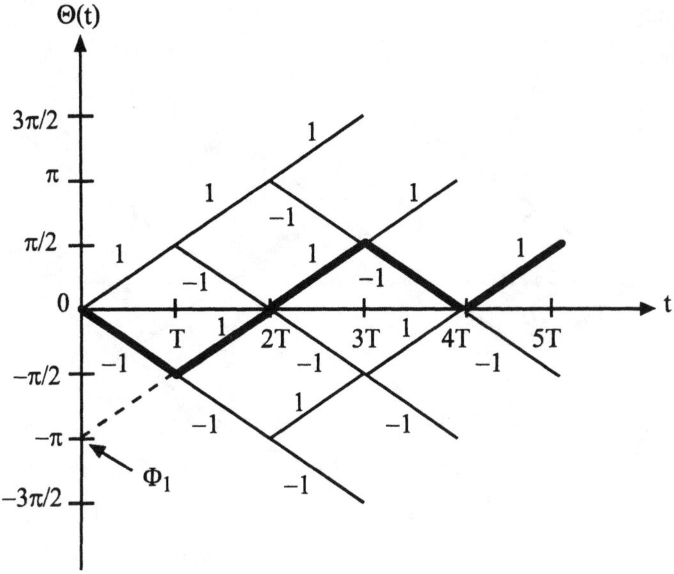

Figure 5.2 MSK excess phase tree.

we can infer that Φ_k is not the initial phase of the kth bit period, since $t \neq 0$ at the bit starting point. Rather it represents the ordinate intercept of the excess phase since $\Phi_k = \Theta(0)$. In Figure 5.2 we show $\Phi_1 = -\pi$ as an example, which is the ordinate intercept of the excess phase at time $t = T$ for the bold face path.

Figure 5.3 is the phase trellis of $\Theta(t)$. A trellis is a tree-like structure with merged branches. In Figure 5.3 nodes with the same phases in a modulo-2π sense are merged. The only possible phases at bit transitions are $\pm\pi/2$ and $\pm\pi$. The data sequence of Figure 5.1 is again shown as the bold-faced path.

5.1.2 MSK Viewed as a Special Case of CPFSK

MSK can also be viewed as a special case of CPFSK with modulation index $h = 0.5$. In Chapter 3 we express CPFSK signal as (see (3.3))

$$s(t) = A\cos(2\pi f_c t + \frac{\pi h d_k(t - kT)}{T} + \pi h \sum_{i=0}^{k-1} d_i), \qquad kT \leq t \leq (k+1)T$$

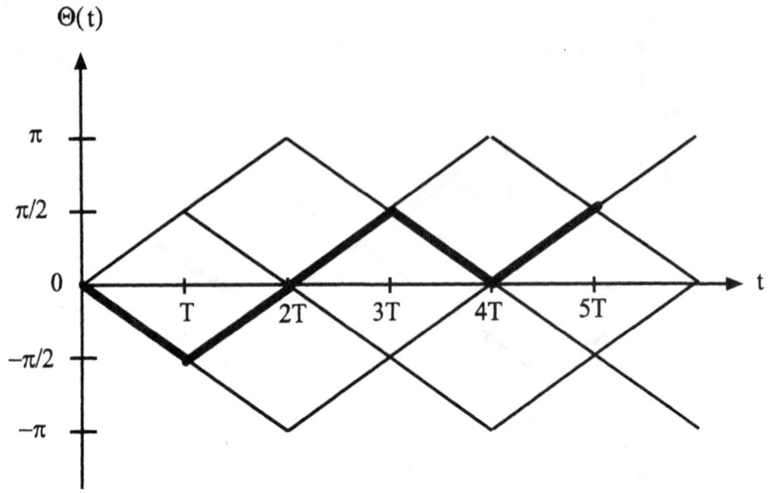

Figure 5.3 MSK excess phase trellis.

which can be written as

$$s(t) = A\cos[2\pi f_c t + hd_k\frac{\pi t}{T} + \Phi_k], \qquad kT \le t \le (k+1)T \tag{5.10}$$

where

$$\Phi_k = \pi h\left(\sum_{i=0}^{k-1} d_i - kd_k\right) \tag{5.11}$$

where d_k is input data (±1) transmitted at rate $R_b = 1/T$. h is modulation index which determines the frequency shift in the bit interval. In fact the frequency shift is $hd_k/2T$. Φ_k is constant in the bit interval; but it is not the initial phase of the bit. It represents the ordinate intercept of the excess phase $\Theta(t) = hd_k\pi t/T + \Phi_k$, as we mentioned before.

With $h = 0.5$ the signal becomes

$$s(t) = A\cos[2\pi f_c t + d_k\frac{\pi t}{2T} + \Phi_k], \qquad kT \le t \le (k+1)T \tag{5.12}$$

which is exactly the signal in (5.2).

To maintain continuous phase at bit transition $t = kT$, the following conditions must be met

$$d_{k-1}\frac{\pi}{2}k + \Phi_{k-1} = d_k\frac{\pi}{2}k + \Phi_k \pmod{2\pi}$$

This is

$$\Phi_k = \Phi_{k-1} + \frac{\pi k}{2}(d_{k-1} - d_k) \pmod{2\pi}$$

or

$$\Phi_k = \begin{cases} \Phi_{k-1} \pmod{2\pi}, & d_k = d_{k-1} \\ \Phi_{k-1} \pm \pi k \pmod{2\pi}, & d_k \neq d_{k-1} \end{cases}$$

Assume that $\Phi_0 = 0$, then $\Phi_k = 0$ or π depending on the value of Φ_{k-1} and the relation between d_k and d_{k-1}.

In the previous section we have shown that when d_k is derived from the staggered I-channel and Q-channel bit streams by $d_k = -I_k Q_k$, then $\Phi_k = 0$ or π, and phase is continuous at any time, including bit transitions. Equivalently, d_k can be generated by differential encoding of the data stream [7]

$$d_k = d_{k-1} \oplus a_k$$

where $\{a_k\}$ is the original data stream, $\oplus$ represents exclusive-OR (XOR) operation (refer to Section 4.2 discussing DPSK for differential encoding). This equivalence can be verified by examining some arbitrary examples. This fact implies that MSK signal can be generated as CPFSK signal with $h = 0.5$, and the sign of frequency shift of each bit is controlled by the differentially encoded input bit stream. The MSK signal realized in this manner is called fast frequency shift keying, or FFSK [2].

5.2 POWER SPECTRUM AND BANDWIDTH

5.2.1 Power Spectral Density of MSK

In Appendix A we have shown that the PSD of a bandpass signal is the shifted version of the equivalent baseband signal or complex envelope's PSD. Therefore it suffices to determine the PSD of the equivalent baseband signal $\tilde{s}(t)$. The MSK signal of (5.1) consists of the in-phase component and the quadrature component which are independent from each other. The PSD of the complex envelope is the sum of the PSDs of these two components (A.21).

$$\Psi_{\tilde{s}}(f) = \Psi_I(f) + \Psi_Q(f)$$

To find $\Psi_I(f)$ and $\Psi_Q(f)$, refer to (A.19). It shows that the PSD of a binary, bipolar (± 1), equiprobable, stationary, and uncorrelated digital waveform is just equal to the energy spectral density of the symbol shaping pulse divided by the symbol duration. In MSK, the symbol shaping pulses are

$$p(t) = \begin{cases} A\cos\frac{\pi t}{2T}, & -T \le t \le T \\ 0, & \text{elsewhere} \end{cases} \tag{5.13}$$

for I-channel and

$$q(t) = p(t-T) = \begin{cases} A\sin\frac{\pi t}{2T}, & 0 \le t \le 2T \\ 0, & \text{elsewhere} \end{cases}$$

for Q-channel. Note their durations are $2T$, not T. Since there is only a phase factor between their Fourier transforms, their energy spectral densities are the same. By taking a Fourier transform of either function, say, $p(t)$, and square the magnitude, divide by $2T$, we have

$$\Psi_I(f) = \Psi_Q(f) = \frac{1}{2T}\left(\frac{4AT[\cos 2\pi Tf]}{\pi[1-(4Tf)^2]}\right)^2$$

Therefore

$$\Psi_{\tilde{s}}(f) = 2\Psi_I(f) = \frac{16A^2T}{\pi^2}\left[\frac{\cos 2\pi Tf}{1-(4Tf)^2}\right]^2 \tag{5.14}$$

Figure 5.4 shows the $\Psi_{\tilde{s}}(f)$ of MSK along with those of BPSK, QPSK, and OQPSK. They are plotted as a function of f normalized to the data rate $R_b = 1/T$. The MSK spectrum falls off at a rate proportional to $(f/R_b)^{-4}$for large values of f/R_b. In contrast, the QPSK or OQPSK spectrum falls off at a rate proportional to only $(f/R_b)^{-2}$. The BPSK spectrum also falls off at a rate proportional to $(f/R_b)^{-2}$ even though its spectral lobe widths are double that of QPSK or OQPSK. The main lobe of the MSK spectrum is narrower than that of BPSK spectrum and wider than that of the QPSK or OQPSK spectrum. The first nulls of BPSK, MSK, and QPSK or OQPSK spectrum fall at $f/R_b = 1.0, 0.75$, and 0.5, respectively. Therefore the null-to-null bandwidth is $2.0R_b$ for BPSK, $1.5R_b$ for MSK, and $1.0R_b$ for QPSK or OQPSK.

5.2.2 Bandwidth of MSK and Comparison with PSK

Another useful measure of the compactness of a modulated signal's spectrum is the fractional out-of-band power, P_{ob}, defined by (2.21).

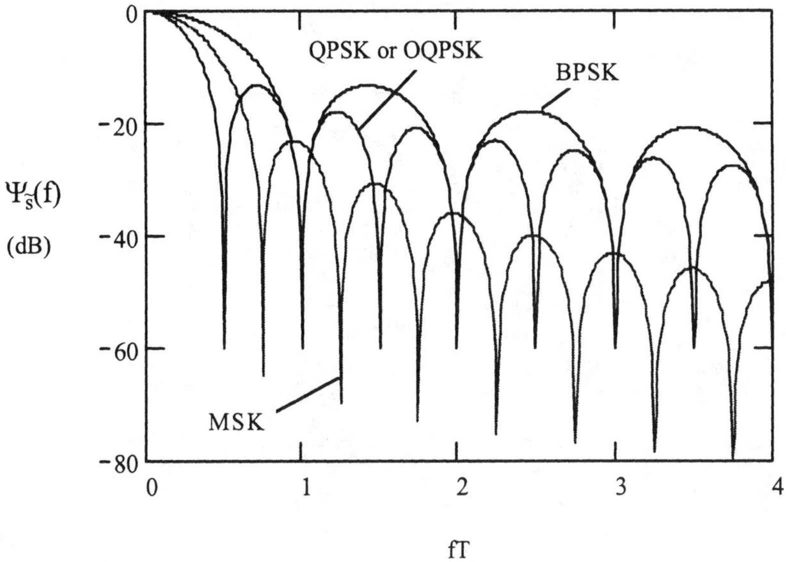

Figure 5.4 MSK power spectral density.

Figure 5.5 shows the $P_{ob}(B)$ for BPSK, QPSK or OQPSK, and MSK as a function of two-sided bandwidth $2B$ normalized to the binary data rate. From this figure we can see that MSK has a bit more out-of-band power than QPSK or OQPSK for $2B < 0.75R_b$, and less out-of-band power for $2B > 0.75R_b$. The bandwidths containing 90% of the power for these modulation schemes can be obtained by numerical calculations. The results are as follows

$$B_{90\%} \approx 0.76R_b \qquad \text{(MSK)}$$

$$B_{90\%} \approx 0.8R_b \qquad \text{(QPSK,OQPSK)}$$

$$B_{90\%} \approx 1.7R_b \qquad \text{(BPSK)}$$

These can also be approximately obtained by noting the bandwidths on the curves corresponding to $P_{ob} = -10$ dB. Because the MSK spectrum falls off much faster, a more stringent in-band power specification, such as 99%, results in a much smaller

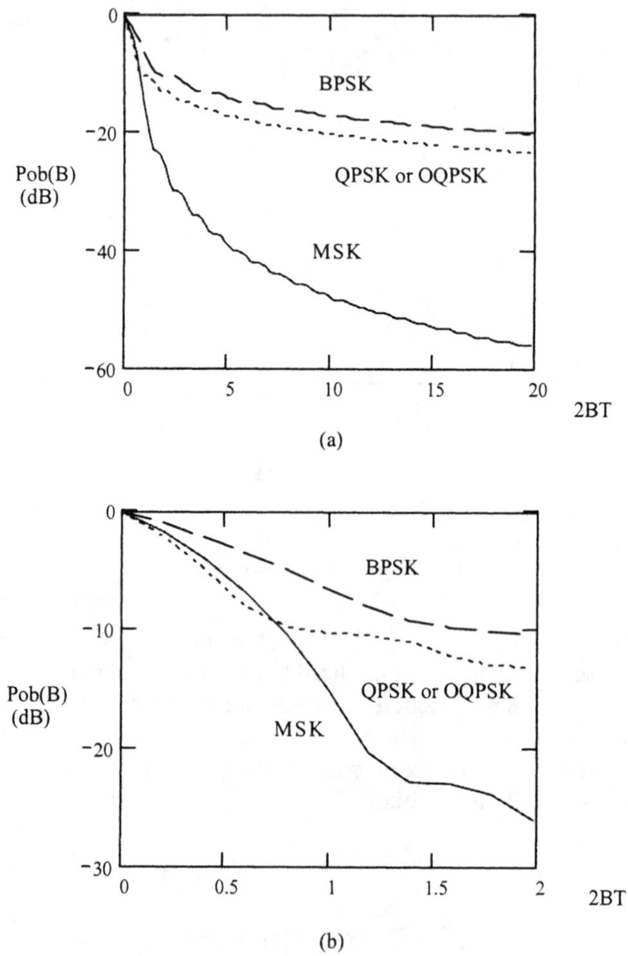

Figure 5.5 Fractional out-of-band power of MSK, BPSK, and QPSK or OQPSK.

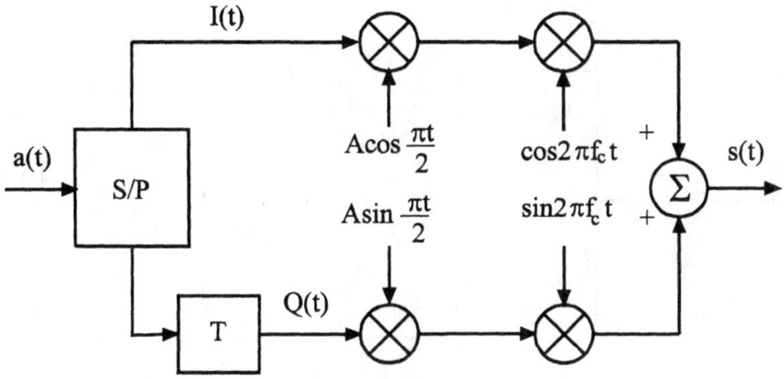

Figure 5.6 MSK modulator (I).

bandwidth for MSK than for BPSK, QPSK or OQPSK. The numerical results are

$$B_{99\%} \approx 1.2R_b \qquad \text{(MSK)}$$

$$B_{99\%} \approx 10R_b \qquad \text{(QPSK or OQPSK)}$$

$$B_{99\%} \approx 20R_b \qquad \text{(BPSK)}$$

These can also be obtained by noting the bandwidths on the curves corresponding to $P_{ob} = -20$ dB.

These comparisons suggest that for system bandwidths exceeding about $1.2R_b$, MSK should provide lower BER performance than QPSK or OQPSK. However, as system bandwidths decrease to $0.75R_b$, their BER performance should be very close since all of them have 90% in-band power. As system bandwidths decrease below $0.75R_b$, the BER performance of QPSK or OQPSK should be better. As system bandwidth is increased, their BER performance converges to infinite bandwidth case, that is, they have the same BER performance. The precise boundaries of regions of superior performance for each modulation scheme are difficult to determine in practical situations, since the detailed channel characteristics must be considered.

5.3 MODULATOR

Figure 5.6 is the MSK modulator implemented as a sinusoidal weighted OQPSK. It

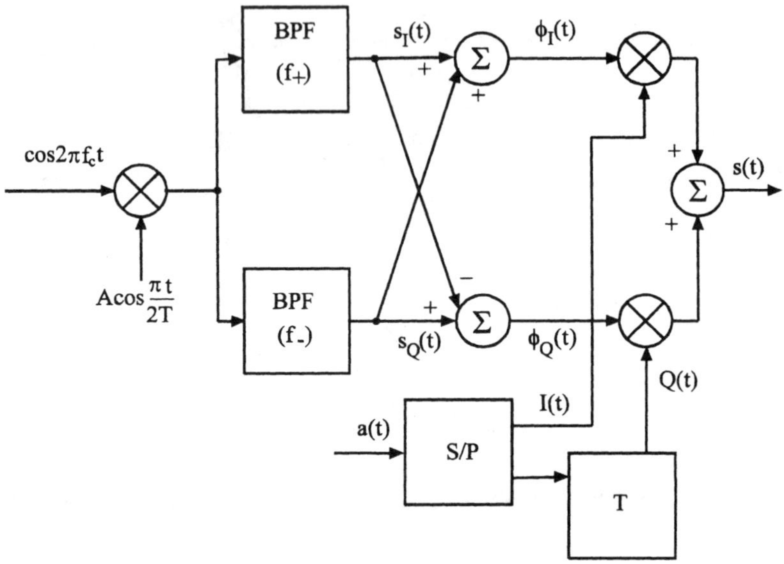

Figure 5.7 MSK modulator (II).

is directly based on (5.1). The data stream signal $a(t)$ is demultiplexed into $I(t)$ and $Q(t)$ by the serial-to-parallel converter (S/P). The in-phase channel signal $I(t)$ consists of even-numbered bits, and the quadrature channel signal $Q(t)$ consists of odd-numbered bits. Each bit in $I(t)$ and $Q(t)$ has a duration of $2T$. $Q(t)$ is delayed by T with respect to $I(t)$. $I(t)$ is multiplied by $A\cos\pi t/2T$ and $\cos 2\pi f_c t$ in the two subsequent multipliers in the I-channel. $Q(t)$ is multiplied by $A\sin\pi t/2T$ and $\sin 2\pi f_c t$ in the two subsequent multipliers in the Q-channel. $A\sin\pi t/2T$ and $\sin 2\pi f_c t$ are obtained through $\pi/2$ phase shifters from $A\cos\pi t/2T$ and $\cos 2\pi f_c t$, respectively. In the summer, the I-channel and Q-channel modulated signals are added to obtain the MSK signal. Previous discussion has shown that $A\cos\pi t/2T$ and $\cos 2\pi f_c t$ need not be synchronized. Therefore $A\cos\pi t/2T$ and $\cos 2\pi f_c t$ can be generated by two independent oscillators.

Figure 5.7 is an alternate implementation. The advantage of it is that carrier coherence and the frequency deviation ratio are largely unaffected by variations in the data rate [8]. The first stage is a high-frequency multiplier which produces two phase coherent frequency components

$$s_I(t) = \frac{1}{2}A\cos 2\pi f_+ t \tag{5.15}$$

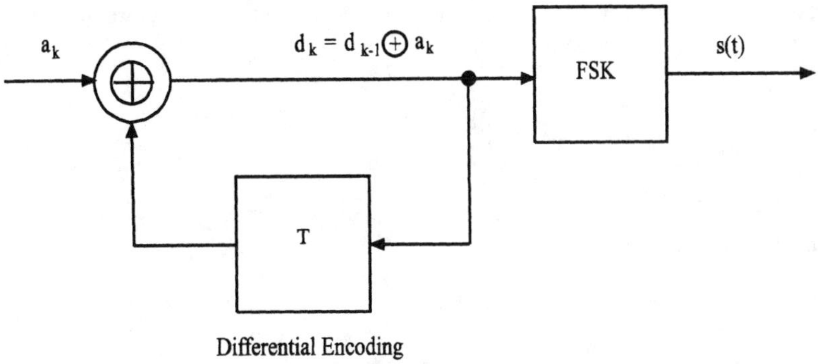

Figure 5.8 MSK modulator (III).

and

$$s_Q(t) = \frac{1}{2} A \cos 2\pi f_- t \tag{5.16}$$

where $f_+ = f_c + 1/4T$ and $f_- = f_c - 1/4T$. These two sinusoidal waves are separated from each other by two narrow-band filters centered at f_+ and f_-, respectively. At the two summers' outputs, the signals are

$$\begin{aligned} \phi_I(t) &= s_I(t) + s_Q(t) \\ &= \frac{1}{2} A \cos 2\pi f_+ t + \frac{1}{2} A \cos 2\pi f_- t \\ &= A \cos \pi t/2T \cos 2\pi f_c t \end{aligned} \tag{5.17}$$

for I-channel and

$$\begin{aligned} \phi_Q(t) &= s_I(t) - s_Q(t) \\ &= -\frac{1}{2} A \cos 2\pi f_+ t + \frac{1}{2} A \cos 2\pi f_- t \\ &= A \sin \pi t/2T \sin 2\pi f_c t \end{aligned} \tag{5.18}$$

for Q-channel. These two signals are the sinusoidally weighted carriers. They are further modulated by $I(t)$ and $Q(t)$, respectively, and then summed to form the final MSK signal.

Figure 5.8 is the MSK modulator implemented as a differentially encoded FSK with $h = 0.5$ (FFSK). The FSK modulator can be any type as described in Chapter 3. The only difference here is the differential encoder which consists of an exclusive

OR gate (XOR) plus a delay device of T seconds which returns the previous encoder output d_{k-1} to the XOR gate.

5.4 DEMODULATOR

Using the two basis functions defined in the previous section, the MSK signal in the kth bit interval can be written as

$$s(t) = I_k\phi_I(t) + Q_k\phi_Q(t), \quad kT \le t \le (k+1)T$$

It can be shown that $\phi_I(t)$ and $\phi_Q(t)$ are orthogonal for $f_c = n/4T$, n integer ($n \neq 1$), over a period of T.

Proof:

$$\begin{aligned}
&\int_{kT}^{(k+1)T} \phi_I(t)\phi_Q(t)dt \\
&= \int_{kT}^{(k+1)T} A^2\cos(\frac{\pi t}{2T})\cos 2\pi f_c t \sin(\frac{\pi t}{2T})\sin 2\pi f_c t dt \\
&= \frac{1}{4}A^2\int_{kT}^{(k+1)T} \sin(\frac{\pi t}{T})\sin 4\pi f_c t dt \\
&= \frac{1}{8}A^2\int_{kT}^{(k+1)T} [\cos(4\pi f_c t - \frac{\pi t}{T}) - \cos(4\pi f_c t + \frac{\pi t}{T})]dt
\end{aligned}$$

The first term integrates to

$$\frac{A^2}{8}\frac{1}{4\pi f_c - \frac{\pi}{T}}\sin\left((4\pi f_c - \frac{\pi}{T})t\right)\Big|_{kT}^{(k+1)T} \tag{5.19}$$

This will be zero when[3]

$$4\pi f_c - \frac{\pi}{T} = \frac{m\pi}{T}, \quad m \text{ integer } (m \neq 0)$$

This is

$$f_c = \frac{(m+1)}{4T} = \frac{n}{4T}, \quad n \text{ integer } (n \neq 1)$$

This obviously also holds for the second term. This concludes the proof.

---

3 When $m = 0$, we have a limit $\lim_{x\longrightarrow 0}\sin x/x = 1$ and the integral in (5.19) evaluates to $A^2/8$.

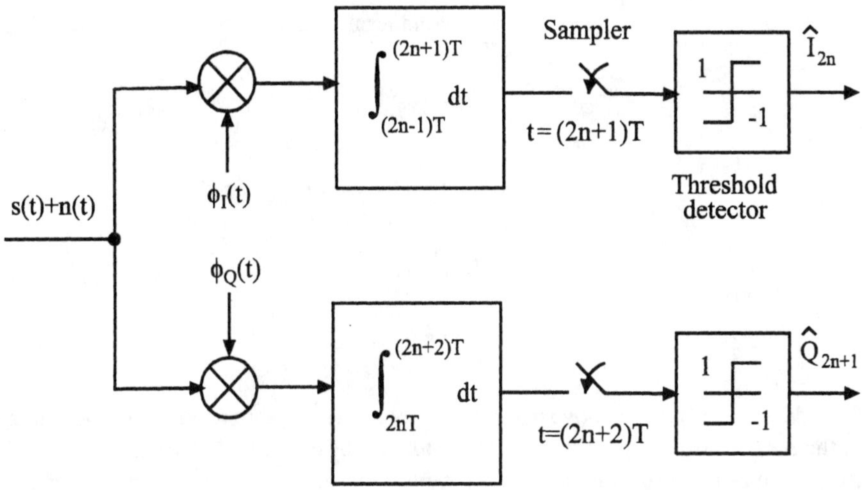

Figure 5.9 MSK demodulator (I).

When n is not an integer, $\phi_I(t)$ and $\phi_Q(t)$ are essentially orthogonal for $f_c >> 1/T$, which is the usual case. This is because the coefficient in front of the sine function in (5.19) is very small when $f_c >> 1/T$. From now on for all practical purposes, we consider that $\phi_I(t)$ and $\phi_Q(t)$ are orthogonal in a period of T. It follows that they are also orthogonal in a period of $2T$.

Since $\phi_I(t)$ and $\phi_Q(t)$ are orthogonal, the optimum coherent demodulation of MSK is very much similar to that of QPSK. Figure 5.9 is the optimum coherent MSK demodulator (the method of obtaining the reference signals and bit timing will be discussed in the next section). Since each data symbol in $I(t)$ or $Q(t)$ occupies a period of $2T$, the demodulator operates on a $2T$ basis. We now denote symbols as $\{I_k, k = 0, 2, 4, ...\}$ and $\{Q_k, k = 1, 3, 5, ...\}$. For kth symbol interval, the integration interval in the I-channel is from $(2n-1)T$ to $(2n+1)T$ and in the Q-channel is from $2nT$ to $(2n+2)T$, where $n = 0, 1, 2, ...$ These intervals correspond to the respective data symbol periods (see Figure 5.1). In the I-channel the integrator output is

$$\int_{(2n-1)T}^{(2n+1)T} s(t)\phi_I(t)dt$$

$$= \int_{(2n-1)T}^{(2n+1)T} \left[I_k\phi_I(t) + Q_k\phi_Q(t)\right]\phi_I(t)dt$$

$$
\begin{aligned}
&= \int_{(2n-1)T}^{(2n+1)T} I_k \phi_I^2(t)dt, \quad \text{(the second term vanishes due to orthogonality)} \\
&= \int_{(2n-1)T}^{(2n+1)T} A^2 I_k \cos^2(\frac{\pi t}{2T}) \cos^2 2\pi f_c t dt \\
&= \int_{(2n-1)T}^{(2n+1)T} A^2 I_k \frac{1}{2}(1+\cos(\frac{\pi t}{T}))\frac{1}{2}(1+\cos 4\pi f_c t)dt \\
&= \int_{(2n-1)T}^{(2n+1)T} \frac{1}{4} A^2 I_k [1+\cos(\frac{\pi t}{T}) + \cos 4\pi f_c t + \cos(\frac{\pi t}{T})\cos 4\pi f_c t]dt \\
&= \frac{1}{2} A^2 T I_k
\end{aligned}
$$

Only the first term in the above integration produces a nonzero result. The integration of the second term is exactly zero. The integrations of the third term and the fourth term are exactly zero only when f_c is a multiple of $1/4T$ (i.e., when the carriers of two channels are orthogonal). Therefore we usually choose f_c as a multiple of $1/4T$. However, even if f_c is not a multiple of $1/4T$, the integrations of the third term and the fourth term are not exactly zero; but they are very small in comparison with the first term for $f_c \gg 1/T$, which is usually the case. Therefore we can conclude that the sampler output of I-channel is essentially $A^2 T I_k/2$ regardless of the carrier orthogonality. Similarly we can show that the sampler output of Q-channel is $A^2 T Q_k/2$. These two signals are detected by the threshold detectors to finally put out I_k and Q_k. The thresholds of detectors are set to zero.

Figure 5.10 is an alternate MSK demodulator where demodulation is accomplished in two steps (the method of obtaining the reference signals and bit timing will be discussed in the next section). It is equivalent to the one in Figure 5.9. In the absence of noise, in I-channel, the output of the first multiplier is

$$
\begin{aligned}
& s(t)\cos 2\pi f_c t \\
=\ & A[I(t)\cos(\frac{\pi t}{2T})\cos 2\pi f_c t + Q(t)\sin(\frac{\pi t}{2T})\sin 2\pi f_c t]\cos 2\pi f_c t \\
=\ & \frac{1}{2} A I(t)\cos(\frac{\pi t}{2T}) + \frac{1}{2} A I(t)\cos(\frac{\pi t}{2T})\cos 4\pi f_c t \\
& + \frac{1}{2} A Q(t)\sin(\frac{\pi t}{2T})\sin 4\pi f_c t
\end{aligned}
$$

After the low-pass filter the two high-frequency terms are rejected and its output is only the first term. It is then multiplied by the weighting signal and integrated for $2T$ which is the symbol length of $I(t)$ and $Q(t)$. Since the symbols of $I(t)$ and $Q(t)$ are staggered, the integrations limits are also staggered. For I-channel, the output of

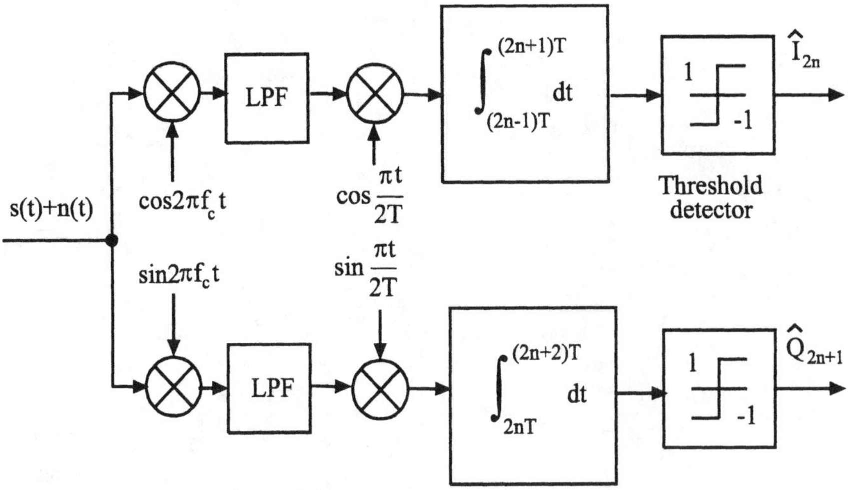

Figure 5.10 MSK demodulator (II).

the integrator at $t = (2n+1)T$ is

$$
\begin{aligned}
& \int_{(2n-1)T}^{(2n+1)T} \frac{1}{2} A I_k \cos(\frac{\pi t}{2T}) \cos(\frac{\pi t}{2T}) dt \\
= & \int_{(2n-1)T}^{(2n+1)T} \frac{1}{4} A [I_k + I_k \cos(\frac{2\pi t}{2T})] dt \\
= & \frac{1}{2} A T I_k
\end{aligned}
$$

which is proportional to the data bit I_k. It is then sent to the detector with a threshold of zero. Without noise or other channel impairment, the detector output is definitely the data bit I_k. Similarly the output of the Q-channel integrator is $\frac{1}{2}ATQ_k$. Thus Q_k can be recovered by the Q-channel detector. When noise or other channel impairment, such as bandlimiting and fading, are present, detection errors will occur. The bit error probability for an AWGN channel will be discussed in a later section.

Since MSK is a type of CPM, it can also be demodulated as a CPM scheme with trellis demodulation using the Viterbi algorithm [7]. This will be discussed in the next chapter. Since MSK is a type of FSK, it can be demodulated noncoherently with about 1 dB asymptotic loss in power efficiency. The demodulated sequence is $\{d_k\}$, which can be converted back to the original data $\{a_k\}$ by the decoding rule

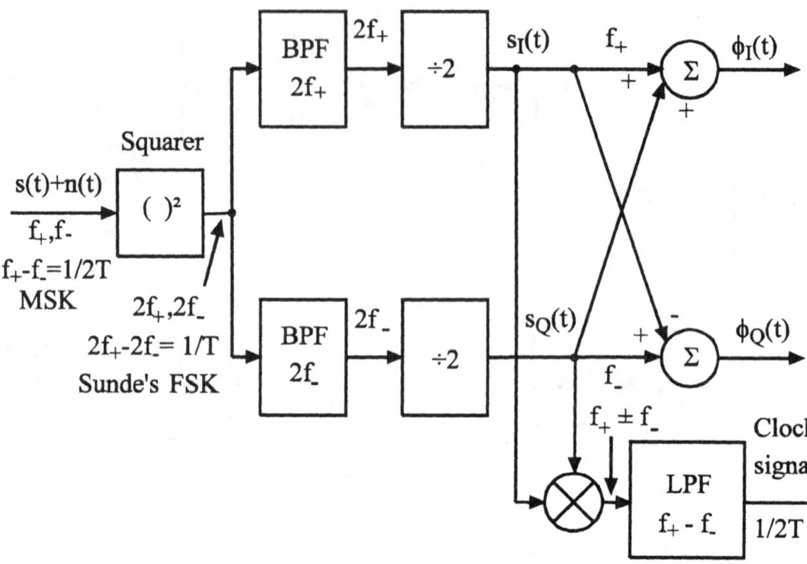

Figure 5.11 MSK carrier and symbol-timing recovery. From [9]. Copyright © 1979 IEEE.

$a_k = d_k \oplus d_{k-1}$. This possibility of noncoherent demodulation permits inexpensive demodulation of MSK when the received signal-to-noise ratio is adequate and provides a low-cost alternative in some systems.

5.5 SYNCHRONIZATION

For the demodulator in Figure 5.9, the reference carriers $\phi_I(t)$, $\phi_Q(t)$, and the clock signal at $1/2$ the bit rate needed at the samplers, are recovered from the received signal by the synchronization circuits in Figure 5.11. (With a little extra circuit it can be used for carrier recovery for the demodulator in Figure 5.10 too. This will be shown shortly.) The MSK signal $s(t)$ has no discrete components which can be used for synchronization (Figure 5.4). However, it produces strong discrete spectral components at $2f_+$ and $2f_-$ when passed through a squarer.

$$\begin{aligned} s^2(t) &= A^2 \cos^2[2\pi f_c t \pm \frac{\pi t}{2T} + \Phi_k] \\ &= \frac{1}{2}A^2[1 + \cos(4\pi f_c t \pm \frac{\pi t}{T} + 2\Phi_k)] \end{aligned}$$

$$= \frac{1}{2}A^2[1+\cos(4\pi f_c t \pm \frac{\pi t}{T})]$$

where $2\Phi_k = 0(\text{mod}\,2\pi)$. The second term is the so-called Sunde's FSK signal with $h = 1$ and the two frequencies are $2f_+$ and $2f_-$. This signal has two strong discrete spectral components at $2f_+$ and $2f_-$, which contain one-half of the total power of the FSK signal (refer to Chapter 3). These components are extracted by bandpass filters (in practice, by phase-lock loops) and then divided by two in frequency to give $s_I(t) = \cos 2\pi f_+ t$ and $s_Q(t) = \cos 2\pi f_- t$. (Suppose their amplitudes are scaled to 1.) The sum $s_I(t)+s_Q(t)$ and difference $s_I(t)-s_Q(t)$ produce the reference carriers $\phi_I(t)$ and $\phi_Q(t)$ (except a factor A), respectively (see (5.17) and (5.18)).

By multiplying $s_I(t)$ and $s_Q(t)$ we have

$$\begin{aligned} s_I(t)s_Q(t) &= \cos 2\pi f_+ t \cos 2\pi f_- t \\ &= \frac{1}{2}[\cos\frac{\pi t}{T} + \cos 4\pi f_c t] \end{aligned} \tag{5.20}$$

By passing this signal through a low-pass filter, the output is $\frac{1}{2}\cos \pi t/T$, a sinusoidal signal at $1/2$ the bit rate, which can be easily converted into a square-wave timing clock for the integrators and the samplers in the demodulator.

By passing the product signal through a high-pass filter the output is $\frac{1}{2}\cos 4\pi f_c t$. Dividing its frequency by 2 and scaling up its amplitude, we can get $\cos 2\pi f_c t$ which is the carrier needed in the demodulator in Figure 5.10. The baseband sinusoidal weighting signal $\cos \pi t/2T$ needed in Figure 5.10 can also be extracted from the signal in (5.20) by a low-pass filter and a divide-by-two frequency divider. Thus the carrier and bit-timing recovery circuit in Figure 5.11 can be used for both demodulators in Figures 5.9 and 5.10, with little extra circuit for the one in Figure 5.10.

There is a 180° phase ambiguity in carrier recovery because of the squaring operation. Since $[\pm s(t)]^2 = s^2(t)$, both $s(t)$ and $-s(t)$ generate the same references $\phi_I(t)$ and $\phi_Q(t)$. This is the 180° phase ambiguity. Therefore the demodulator outputs in the I- and Q-channels will be $-I(t)$ and $-Q(t)$, respectively, if the received signal is $-s(t)$. One method to solve this problem is to differentially encode the data stream before modulation, as described in Chapter 4 for the DPSK.

Recall that if MSK is implemented as FFSK, a differential encoder is needed at modulation and a differential decoder is needed at demodulation. If MSK is implemented as weighted OQPSK, the differential encoder and decoder are not needed. However, due to the 180° phase ambiguity in the carrier recovery operation, differential encoding and decoding are needed in both cases. In this sense FFSK and MSK are essentially the same.

5.6 ERROR PROBABILITY

The derivation of MSK bit error rate is very similar to that of QPSK in the previous chapter.

Assume the channel is AWGN, the received signal is

$$r(t) = s(t) + n(t)$$

where $n(t)$ is the additive white Gaussian noise. Refer to the demodulator in Figure 5.9 or Figure 5.10. The MSK signal is demodulated in I-channel and Q-channel. Due to the orthogonality of the I and Q components of the MSK signal, they do not interfere with each other in the demodulation process. However, the noise will cause the detectors to put out erroneous bits. The probability of bit error (P_b) or bit error rate (BER) is of interest. Because of symmetry, the I- and Q-channels have the same probability of bit error (i.e., $P_{bI} = P_{bQ}$). In addition, the errors in the I-channel and Q-channel are statistically independent (it will be shown shortly) and the detected bits from both channels are *directly* multiplexed to form the final data sequence. Therefore it suffices to consider only P_{bI} and this P_{bI} is the P_b for the entire demodulator.[4]

Refer to Figure 5.10 (same results will be obtained by using Figure 5.9), at the threshold detector input the I-channel signal is

$$y_{Ik} = \frac{1}{2}ATI_k + n_{Ik}$$

where $k = 2n$ and the noise

$$n_{Ik} = \int_{(2n-1)T}^{(2n+1)T} n(t)\cos(\frac{\pi t}{2T})\cos 2\pi f_c t dt$$

which is Gaussian with zero mean (refer to Appendix A). Its variance is

$$\begin{aligned}
\sigma^2 &= E\{n_{Ik}^2\} = E\left\{\left[\int_{(2n-1)T}^{(2n+1)T} n(t)\cos(\frac{\pi t}{2T})\cos 2\pi f_c t dt\right]^2\right\} \\
&= \int_{(2n-1)T}^{(2n+1)T}\int_{(2n-1)T}^{(2n+1)T} E\{n(t)n(\tau)\}
\end{aligned}$$

---

[4] Some authors calculate the symbol error probability P_s first, then derive the bit error probability P_b from P_s. This is not necessary and strictly speaking, it is not right, since for MSK, we never detect symbols, instead, we detect bits in the demodulation process.

$$
\begin{aligned}
&\cdot\cos(\frac{\pi t}{2T})\cos 2\pi f_c t\cos(\frac{\pi\tau}{2T})\cos 2\pi f_c\tau dtd\tau \\
&= \int_{(2n-1)T}^{(2n+1)T}\int_{(2n-1)T}^{(2n+1)T}\frac{N_o}{2}\delta(t-\tau) \\
&\cdot\cos(\frac{\pi t}{2T})\cos 2\pi f_c t\cos(\frac{\pi\tau}{2T})\cos 2\pi f_c\tau dtd\tau \\
&= \frac{N_o}{2}\int_{(2n-1)T}^{(2n+1)T}\cos^2(\frac{\pi t}{2T})\cos^2 2\pi f_c tdt \\
&= \frac{N_o T}{4}
\end{aligned}
$$

The detector has a threshold of zero. The probability of bit error in the I-channel is

$$
\begin{aligned}
P_{bI} &= \Pr[\frac{1}{2}AT + n_{Ik} < 0|I_k = +1] \\
&= \Pr[-\frac{1}{2}AT + n_{Ik} > 0|I_k = -1] \\
&= \Pr[n_{Ik} > \frac{1}{2}AT] \\
&= \int_{\frac{1}{2}AT}^{\infty}\frac{1}{\sqrt{2\pi}\sigma}\exp(-\frac{u^2}{2\sigma^2})du = \int_{\frac{AT}{2\sigma}}^{\infty}\frac{1}{\sqrt{2\pi}}\exp(-\frac{x^2}{2})dx \\
&= Q\left(\frac{AT}{2\sigma}\right) = Q\left(\frac{AT}{2\sqrt{N_oT/4}}\right) = Q\left(\sqrt{\frac{A^2T}{N_o}}\right) \qquad (5.21)
\end{aligned}
$$

The bit energy E_b of the transmitted MSK signal is

$$
\begin{aligned}
E_b &= \int_{(k-1)T}^{kT} A^2\cos^2[2\pi(f_c + d_k\frac{1}{4T})t + \Phi_k]dt \\
&= \frac{1}{2}A^2\int_{(k-1)T}^{kT}\left\{1 + \cos[4\pi(f_c + d_k\frac{1}{4T})t + 2\Phi_k]\right\}dt \\
&= \frac{1}{2}A^2T \qquad (5.22)
\end{aligned}
$$

since the integration of the second term is zero.[5] Thus the P_{bI} expression can be

[5] Note that in one-bit duration, the MSK signal energy is constant even though the signal may have different frequencies from bit to bit.

written as

$$P_{bI} = Q\left(\sqrt{\frac{2E_b}{N_o}}\right) \tag{5.23}$$

Similarly we can derive the expression of P_{bQ}. The result is identical to P_{bI}. Now it remains to show that errors on the I- and Q-channels are statistically independent so that the overall bit error probability is $P_b = (P_{bI} + P_{bQ})/2 = P_{bI} = P_{bQ}$. This requires that noise n_{Ik} and n_{Qk} are uncorrelated, for they are Gaussian and uncorrelated Gaussian random variables are statistically independent. The noise component at the input of the Q-channel threshold detector is defined similarly to n_{Ik}

$$n_{Qk} = \int_{2nT}^{(2n+2)T} n(t)\sin(\frac{\pi t}{2T})\sin 2\pi f_c t dt$$

The correlation of n_{Ik} with n_{Qk} is given by

$$\begin{aligned}
&E\{n_{Ik}n_{Qk}\}\\
= \ & E\left\{\int_{(2n-1)T}^{(2n+1)T} n(t)\cos(\frac{\pi t}{2T})\cos 2\pi f_c t dt\right.\\
&\left.\cdot \int_{2nT}^{(2n+2)T} n(t)\sin(\frac{\pi t}{2T})\sin 2\pi f_c t dt\right\}\\
= \ & \int_{(2n-1)T}^{(2n+1)T}\int_{2nT}^{(2n+2)T} E\{n(t)n(\tau)\}\\
&\cdot\cos(\frac{\pi t}{2T})\sin(\frac{\pi \tau}{2T})\cos 2\pi f_c t\sin 2\pi f_c \tau dt d\tau\\
= \ & \frac{N_o}{2}\int_{(2n-1)T}^{(2n+1)T}\int_{2nT}^{(2n+2)T} \delta(t-\tau)\\
&\cdot\cos(\frac{\pi t}{2T})\sin(\frac{\pi \tau}{2T})\cos 2\pi f_c t\sin 2\pi f_c \tau dt d\tau\\
= \ & \frac{N_o}{2}\int_{2nT}^{(2n+1)T} \cos(\frac{\pi t}{2T})\sin(\frac{\pi t}{2T})\cos 2\pi f_c t\sin 2\pi f_c t dt\\
= \ & 0
\end{aligned}$$

where the limits of the last integral follow due to the fact that $\delta(t-\tau) = 0$ for $t \neq \tau$. Thus n_{Ik} and n_{Qk} are uncorrelated and hence independent since they are Gaussian.

As a result, the bit error probability of the entire demodulator $P_b = P_{bI}$ is given by

$$P_b = Q\left(\sqrt{\frac{2E_b}{N_o}}\right) \tag{5.24}$$

which is exactly the same as that of BPSK, QPSK, and OQPSK whose P_b curve can be found in Figure 4.21.

5.7 SERIAL MSK

The implementation of MSK modulation and demodulation discussed in Sections 5.3 and 5.4 is in a parallel fashion. That is, the serial data stream is demultiplexed into even- and odd-indexed bits which are modulated and demodulated in two parallel channels. The MSK modulation and demodulation can also be implemented in a serial fashion [4, 10]. These two techniques are equivalent in performance theoretically. However, the serial implementation has advantages over the parallel one at high data rates. Serial MSK modulation and demodulation have the advantage that all operations are performed serially. The precise synchronization and balancing required for the quadrature signals of the parallel structures are no longer needed. This is especially beneficial at high data rates. The serial technique is described in this section.

5.7.1 SMSK Description

Serial MSK modulator and demodulator are illustrated in Figure 5.12. The modulator consists of a BPSK modulator with carrier frequency of $f_- = f_c - 1/4T$ and a bandpass conversion filter with impulse response

$$h(t) = \begin{cases} \sin 2\pi f_+ t = \sin 2\pi (f_c + \frac{1}{4T})t, & 0 \le t \le T \\ 0, \quad \text{elsewhere} \end{cases} \tag{5.25}$$

which corresponds to a $sin(x)/x$-shaped transfer function. We will show shortly why these operations produce an MSK modulated signal.

The serial demodulator structure is essentially the reverse of that of the modulator. It consists of a bandpass matched filter followed by a coherent demodulator and a low-pass filter which eliminates the double-frequency component generated by the mixer. The matched filter is not necessary for demodulation, but it can improve the signal-to-noise ratio, hence the error performance [4].

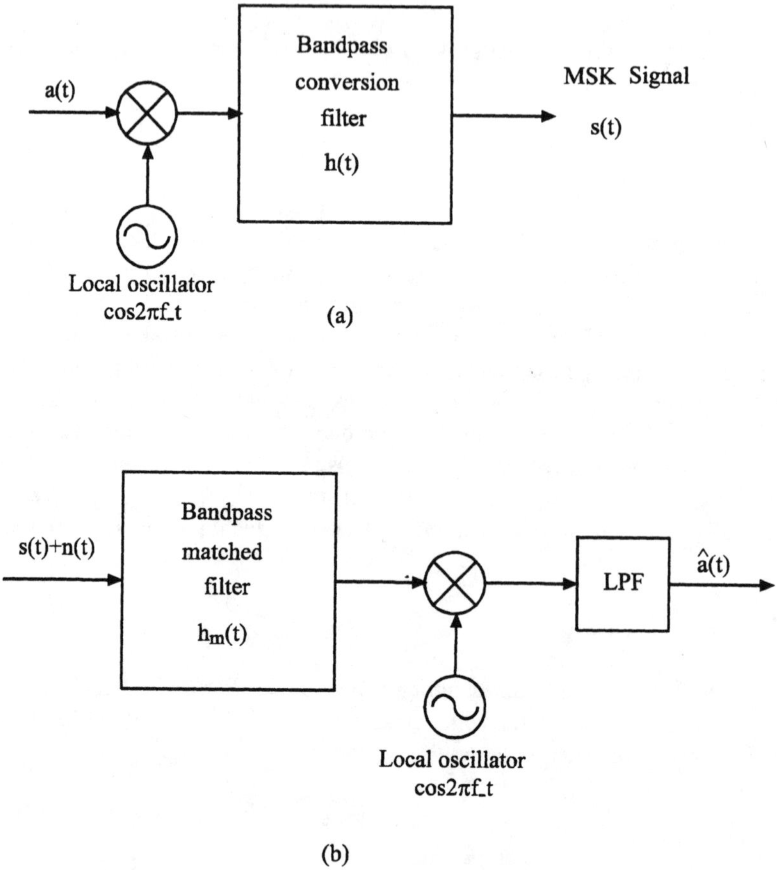

Figure 5.12 Serial modulator and demodulator for MSK (a) modulator, and (b) demodulator.

The matched filter impulse response is

$$h_m(t) = \cos(\frac{\pi t}{2T})\cos(2\pi f_c t) \qquad 0 \le t \le 2T$$

which has a transfer function proportional to the square root of the power spectrum of the MSK signal. We will show shortly how $h_m(t)$ is determined.

5.7.2 SMSK Modulator

The following derivation basically follows that of Amoroso et al. [4]. We will prove that the output of the conversion filter in Figure 5.12(a) is the MSK signal. In general we can assume the local oscillation is

$$f(t) = \sin(2\pi f_- t + \theta)$$

where θ represents the relative phase of f_- and the data transition ($\theta = \pi/2$ in Figure 5.12(a) since the oscillator output is $\cos(2\pi f_- t) = \sin(2\pi f_- t + \pi/2)$). The typical input burst for a bit to the conversion filter is thus

$$a_k f(t) = a_k \sin(2\pi f_- t + \theta), \qquad kT \le t \le (k+1)T$$

where the $a_k \in (-1, +1)$ represents the data.

The output of the conversion filter is the convolution of $h(t)$ with $a_k f(t)$, which we denote as $a_k p_\theta(t)$ (it is denoted as $s(t)$ in Figure 5.12(a) for $\theta = \pi/2$ case).

$$\begin{aligned} a_k p_\theta(t) &= a_k \int_{-\infty}^{\infty} f(\tau) h(t-\tau) d\tau \\ &= \begin{cases} a_k \int_{kT}^{t} \sin(2\pi f_- \tau + \theta)\sin 2\pi f_+(t-\tau) d\tau, \\ \qquad\qquad kT \le t \le (k+1)T \\ a_k \int_{t-T}^{(k+1)T} \sin(2\pi f_- \tau + \theta)\sin 2\pi f_+(t-\tau) d\tau, \\ \qquad\qquad (k+1)T < t < (k+2)T \\ 0, \qquad\qquad \text{elsewhere} \end{cases} \end{aligned}$$

which spans a duration of $2T$. The integration limits result from the fact that $f(\tau)$ and $h(t-\tau)$ are overlapped in $[kT, t]$ for $kT \le t \le (k+1)T$ and in $[t-T, (k+1)T]$ for $(k+1)T \le t \le (k+2)T$. We will see shortly that when $\theta = \pi/2$, $a_k p_\theta(t)$ becomes the MSK signal.

Working out the integrals, $p_\theta(t)$ can be reduced to the form

$$p_\theta(t) = \frac{1}{4\pi(f_+ - f_-)}[\sin(2\pi f_- t + \theta) - (-1)^k \sin(2\pi f_+ t + \theta)]$$

$$+\frac{1}{4\pi(f_+ + f_-)}[\sin(2\pi f_- t + \theta) + (-1)^k \sin(2\pi f_+ t - \theta)],$$
$$kT \leq t \leq (k+2)T$$

For odd values of k ($k = 2n+1$), this reduces to

$$p_{\theta_o}(t) = \frac{T}{\pi}\left[\cos\frac{\pi t}{2T}\sin(2\pi f_c t + \theta) - \frac{1}{4Tf_c}\sin(\frac{\pi t}{2T} - \theta)\cos 2\pi f_c t\right] \quad (5.26)$$

where $f_c = (f_+ + f_-)/2$ is the apparent frequency. For even values of k ($k = 2n$), $p_\theta(t)$ reduces to

$$p_{\theta e}(t) = \frac{T}{\pi}\left[-\sin\frac{\pi t}{2T}\cos(2\pi f_c t + \theta) + \frac{1}{4Tf_c}\cos(\frac{\pi t}{2T} - \theta)\sin 2\pi f_c t\right] \quad (5.27)$$

Only when $\theta = \pi/2$, the above expressions become

$$p_o(t) = \frac{T}{\pi}\frac{4Tf_c + 1}{4Tf_c}\cos\frac{\pi t}{2T}\cos 2\pi f_c t, \quad (2n+1)T \leq t \leq (2n+3)T \quad (5.28)$$

and

$$p_e(t) = \frac{T}{\pi}\frac{4Tf_c + 1}{4Tf_c}\sin\frac{\pi t}{2T}\sin 2\pi f_c t, \quad 2nT \leq t \leq (2n+2)T \quad (5.29)$$

Note that both $p_o(t)$ and $p_e(t)$ spans $2T$. There is an overlap of T between them. In any one-bit interval, the final output of the conversion filter is one of the possible sums of these two components:

$$s(t) = \pm[p_o(t) + p_e(t)] = \pm\frac{T}{\pi}\frac{4Tf_c + 1}{4Tf_c}\cos 2\pi f_- t$$

and

$$s(t) = \pm[p_o(t) - p_e(t)] = \pm\frac{T}{\pi}\frac{4Tf_c + 1}{4Tf_c}\cos 2\pi f_+ t$$

These are exactly the MSK signals. It is clear from comparison that $p_o(t)$ and $p_e(t)$ are equivalent to the I- and Q-channel components of the parallel MSK. For $p_o(t)$ and $p_e(t)$ to be perfectly orthogonal, f_c must be a multiple of $1/4T$.

Now one can realize the essence of serial MSK. The conversion filter spreads a burst of one bit of the BPSK signal over two-bit periods, weights the envelope by half cycle of cosine or sine, and modifies the carrier frequency from f_- to f_c. All these effects are accomplished through convolution. Due to the fact that the filter responds to odd bits and even bits differently, the final output can be viewed as the superposition of these two outputs which are equivalent to I- or Q-channel signals in parallel MSK, respectively.

Note that the BPSK signal phase θ must be $\pi/2$ for SMSK to be accurate. If $\theta \neq \pi/2$, the terms

$$-\frac{1}{4Tf_c}\sin(\frac{\pi t}{2T}-\theta)\cos 2\pi f_c t$$

and

$$+\frac{1}{4Tf_c}\cos(\frac{\pi t}{2T}-\theta)\sin 2\pi f_c t$$

in (5.26) and (5.27), respectively, become undesirable. They cause the final SMSK signal envelope to fluctuate. However, for large ratio of f_c to data rate, the factor $1/4Tf_c$ causes the undesired terms to vanish, leaving the resulting SMSK signal independent of θ.

We can also verify the validity of SMSK in frequency-domain. The BPSK's single-sided spectrum is

$$\Psi_{BPSK}(f) = 2T\text{sinc}^2[(f-f_c)T+0.25]$$

The conversion filter transfer function is

$$H(f) = T\text{sinc}[(f-f_c)T-0.25]\exp(-j\pi fT)$$

The power spectrum of the output of the filter is the product of $\Psi_{BPSK}(f)$ and $|H(f)|^2$ which can be simplified to

$$\Psi_{MSK}(f) = \frac{8T^3}{\pi^4}\left[\frac{\cos 2\pi T(f-f_c)}{1-(4T(f-f_c))^2}\right]^2$$

which is equivalent to the baseband power spectrum for MSK obtained earlier (5.14), except for a scaling factor.

5.7.3 SMSK Demodulator

The demodulation can be simply done by a coherent demodulator, that is, simply multiply the received signal with a local oscillation of frequency f_- and low-pass filter the mixer's output to remove the double-frequency component. This simple coherent demodulator is the one shown in Figure 5.12(b) without the matched filter. The result is the recovered data bits. This can be shown as follows.

Since the SMSK signal is the same as parallel MSK signal, the mixer input signal expression is the one in (5.2). The mixer output is

$$\begin{aligned} & A\cos[2\pi f_c t + d_k \frac{\pi t}{2T} + \Phi_k] \cos 2\pi f_- t \\ = \; & A\cos[2\pi f_c t + d_k \frac{\pi t}{2T} + \Phi_k] \cos(2\pi f_c t - \frac{\pi t}{2T}) \\ = \; & \frac{1}{2} A[\cos(4\pi f_c t + d_k \frac{\pi t}{2T} + \Phi_k - \frac{\pi t}{2T}) + \cos(d_k \frac{\pi t}{2T} + \Phi_k + \frac{\pi t}{2T})] \end{aligned}$$

The first term is the double-carrier-frequency term which will be eliminated by the low-pass filter. The second term (ignoring the constant $\frac{1}{2}A$) is

$$m(t) = \cos(d_k \frac{\pi t}{2T} + \Phi_k + \frac{\pi t}{2T}) = \begin{cases} \cos \frac{\pi t}{T}, & \text{for } d_k = 1, \Phi_k = 0 \\ -\cos \frac{\pi t}{T}, & \text{for } d_k = 1, \Phi_k = \pi \\ 1, & \text{for } d_k = -1, \Phi_k = 0 \\ -1, & \text{for } d_k = -1, \Phi_k = \pi \end{cases}$$

Recall that $d_k = -I_k Q_k$ and $\Phi_k = \frac{\pi}{2}(1 - I_k)$. In serial MSK, $I_k = a_k$ and $Q_k = a_{k-1}$. We can establish the following relations

$I_k = a_k$	$Q_k = a_{k-1}$	d_k	Φ_k	$m(t)$
1	1	−1	0	1
1	−1	1	0	$\cos \frac{\pi t}{T}$
−1	1	1	π	$-\cos \frac{\pi t}{T}$
−1	−1	−1	π	−1

It is clear that four different forms of the low-pass filter output $m(t)$, uniquely determines four different possible data pairs (a_k, a_{k-1}), respectively, thus the demodulation is accomplished.

Figure 5.13 shows how a BPSK coherent demodulator recovers the data stream from an MSK signal [4]. Figure 5.13(a) is the data stream to be sent. Figure 5.13(b) is the frequency transmitted. Note that the frequency of each bit is determined by $d_k = -a_{k-1}a_k$. Figure 5.13(c) shows the MSK signal phase and the local oscillation phase referenced to the phase of f_c. Figure 5.13(d) shows the phase at mixer output which is the difference of the MSK signal phase and the local oscillation phase. Figure 5.13(e) is the final demodulated signal which is the cosine function of the phase difference at the mixer output and resembles the data stream transmitted.

Even though we discuss this demodulation method in the context of SMSK, it is apparent that it can be used for parallel MSK too since the MSK signals are the same regardless of how they are generated.

As mentioned earlier, the matched filter can improve the SNR even though it is not essential for SMSK demodulation. Now we determine its impulse response.

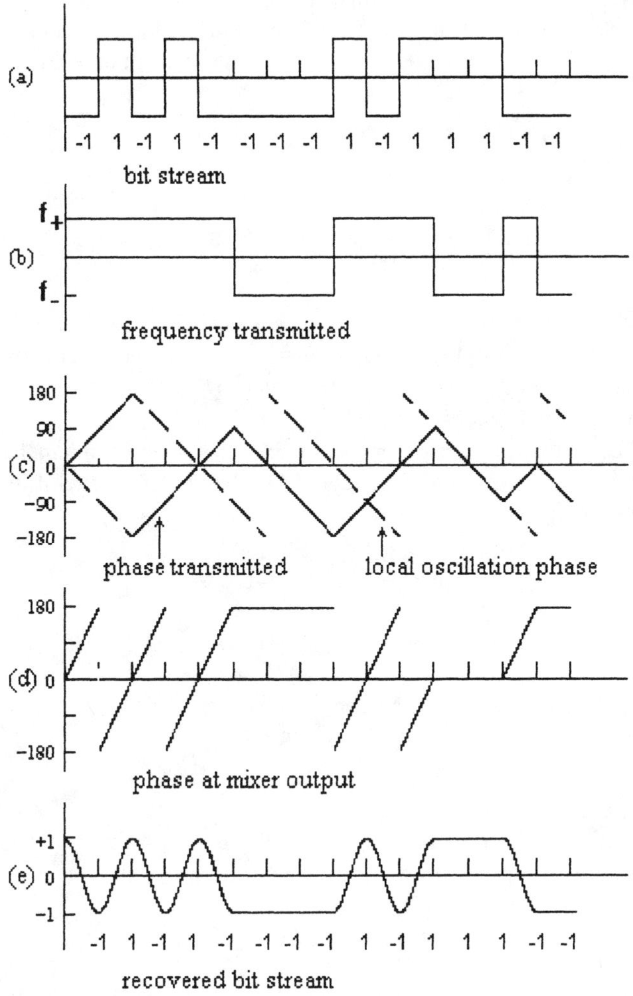

Figure 5.13 Demodulation of MSK signal by coherent BPSK demodulator (a–e). From [4]. Copyright © 1977 IEEE.

From the point of view of serial MSK, the MSK signal can be viewed as a sum of odd- and even-bit components, which are equivalent to I- and Q-channel components of the parallel MSK signal. The matched filter should match to the I- and Q-channel components at the same time. Consider a symbol of the MSK signal. As shown in (5.1) it is the result of superposition of the I- and the Q-channel component where

$$\varphi_I(t) = \cos(\frac{\pi t}{2T})\cos(2\pi f_c t), \quad -T \leq t \leq T$$

and

$$\varphi_Q(t) = \sin(\frac{\pi t}{2T})\sin(2\pi f_c t), \quad 0 \leq t \leq 2T$$

are the basic symbol functions for the I- and Q-channel, respectively. It seems that the matched filter can only be chosen to match to one of them. However, we now show that the matched filter chosen to match to $\varphi_I(t)$ also matches to $\varphi_Q(t)$ as far as the baseband output is concerned.

According to the matched filter theory, to match to the I-channel symbol, the matched filter impulse response should be the scaled mirror image of it, delayed by the signal duration $2T$:

$$\begin{aligned} h_m(t) &= \alpha\cos(\frac{\pi}{2T}(2T-t))\cos(2\pi f_c(2T-t)), \\ &= \alpha\cos(\pi - \frac{\pi t}{2T})\cos(m\pi - 2\pi f_c t), \\ &= \alpha(-1)^{m+1}\cos\frac{\pi t}{2T}\cos 2\pi f_c t, \\ &= \cos\frac{\pi t}{2T}\cos 2\pi f_c t, \quad T \leq t \leq 3T \end{aligned}$$

where $f_c = m/4T$ is assumed and $\alpha(-1)^{m+1}$ is merely a scaling constant which can be set to unity. The response of the matched filter to $\varphi_I(t)$ is

$$y_I(t) = \varphi_I(t) * h_m(t)$$

where $*$ denotes convolution. Then this signal is multiplied by the local carrier and low-pass filtered to give the final demodulated baseband symbol (detail omitted)

$$m_I(t) = \begin{cases} \frac{1}{4}\cos\frac{\pi t}{2T}\int_T^{T+t}\cos\frac{\pi}{2T}\tau\cos\frac{\pi}{2T}(t-\tau)d\tau, & 0 \leq t \leq 2T \\ \frac{1}{4}\cos\frac{\pi t}{2T}\int_{t-T}^{3T}\cos\frac{\pi}{2T}\tau\cos\frac{\pi}{2T}(t-\tau)d\tau, & 2T \leq t \leq 4T \end{cases} \tag{5.30}$$

Working out the integrals (factor $1/4$ omitted) the final demodulated baseband symbol is

$$m_I(t) = \begin{cases} \cos\frac{\pi t}{2T}[\frac{1}{2}t\cos\frac{\pi t}{2T} - \frac{T}{\pi}\sin\frac{\pi t}{2T}], & 0 \le t \le 2T \\ \cos\frac{\pi t}{2T}[\frac{1}{2}(4T-t)\cos\frac{\pi t}{2T} + \frac{T}{\pi}\sin\frac{\pi t}{2T}], & 2T \le t \le 4T \end{cases}$$

which is shown in Figure 5.14(a). It is clear that matched filtering spreads the symbol to occupy $4T$ periods. Fortunately the output has zero values at all sampling instants except for at the center of the pulse. In other words, it introduces no intersymbol interference (ISI) if timing is accurately maintained.

By similar procedure, the Q-channel demodulated baseband symbol is found as (a constant factor $1/4$ is omitted)

$$m_Q(t) = \begin{cases} \sin\frac{\pi t}{2T}[\frac{1}{2}(t-T)\sin\frac{\pi t}{2T} + \frac{T}{\pi}\cos\frac{\pi t}{2T}], & T \le t \le 3T \\ \sin\frac{\pi t}{2T}[\frac{1}{2}(5T-t)\sin\frac{\pi t}{2T} - \frac{T}{\pi}\cos\frac{\pi t}{2T}], & 3T \le t \le 5T \end{cases}$$

which is also shown in Figure 5.14(a) (dotted line). It can be seen that it is merely a time shifted version of the I-channel demodulated symbol. It is easy to verify that $m_I(t-T) = m_Q(t)$.

Intuitively, why the matched filter treats I- and Q-channel symbols equally can be explained as follows. If we shift $\varphi_Q(t)$ to the left by T to coincide with $\varphi_I(t)$ on the time axis, it becomes

$$\varphi_Q(t+T) = \cos(\frac{\pi t}{2T})\sin(2\pi f_c t + \theta),$$

where $\theta = 2\pi f_c T$. This means $\varphi_Q(t)$ and $\varphi_I(t)$ have the same envelope despite that their carriers may have phase difference. Therefore (5.30) is the same for the Q-channel since it only involves the convolution of envelopes. Consequently the results are the same except for a time delay of T.

Figure 5.14(b) shows the resultant demodulated signal for the same sequence in Figure 5.13.

The transfer function of the low-pass equivalent matched filter is

$$\widetilde{H}_m(f) = \mathcal{F}\{\widetilde{h}_m(t)\} = \mathcal{F}\{\cos(\frac{\pi t}{2T})\} = \frac{4T[\cos 2\pi T f]}{\pi[1-(4Tf)^2]}$$

which is proportional to the square root of the PSD of the MSK signal.

5.7.4 Conversion and Matched Filter Implementation

The critical system components of the serial MSK are the biphase modulator, the bandpass conversion and matched filters, and the coherent demodulator (including carrier recovery). The modulator and the demodulator are no more special than the

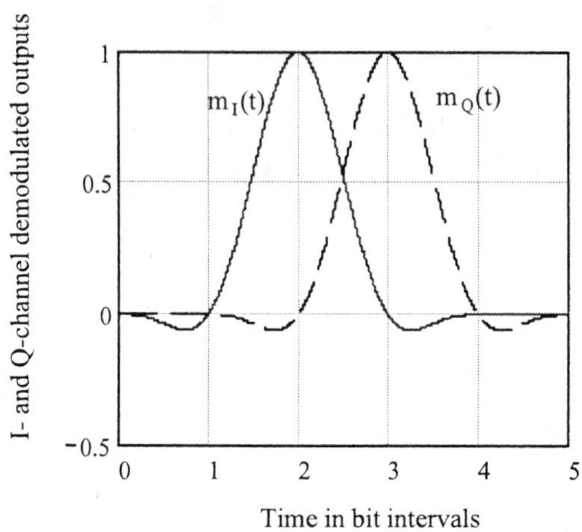

(a)

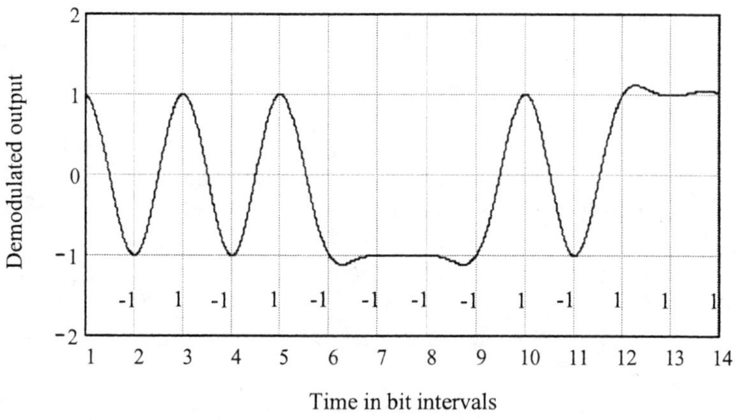

(b)

Figure 5.14 Demodulation of MSK signal with matched filter and coherent BPSK demodulator: (a) I-channel symbol, (b) Q-channel symbol, and (c) demodulator output for the data sequence of Figure 5.13.

ones for BPSK. What are special here are the bandpass conversion and matched filters. They have been implemented with surface acoustic wave (SAW) devices. The maximum bandwidth of SAW devices is about 10% to 30% of the center frequency. For production SAWs, center frequencies of a few hundred MHz represent the upper limit, assuming normal fabrication techniques, with 1 GHz representing the upper limit if laser trimming and other special techniques are used. Thus the use of SAW filters implies an upper limit on data rate of about 100 Mbps, assuming the use of normal fabrication techniques [10].

An alternative to SAW implementation is the one utilizing baseband I/Q equivalents for the conversion and matched filters as shown in Figure 5.15. In Figure 5.15(a), data are demultiplexed into I/Q channels and filtered in baseband before being modulated onto two orthogonal carriers. Basically the reverse is done in the demodulator in Figure 5.15(b). Note that all reference signal frequencies in Figure 5.15 are the lower frequency of the two MSK frequencies: f_-.

The frequency responses of the baseband conversion and matched filters were derived in [10]. The frequency response of the conversion filter is

$$H_I(f) = \frac{T}{2}\left[\frac{\sin\pi(fT-0.5)}{\pi(fT-0.5)} + \frac{\sin\pi(fT+0.5)}{\pi(fT+0.5)}\right]\exp(-j\pi fT)$$

for I-channel and

$$H_Q(f) = j\frac{T}{2}\left[\frac{\sin\pi(fT-0.5)}{\pi(fT-0.5)} - \frac{\sin\pi(fT+0.5)}{\pi(fT+0.5)}\right]\exp(-j\pi fT)$$

for Q-channel. The overall equivalent low-pass transfer function of the conversion filter is

$$\begin{aligned}\widetilde{H}(f) &= H_I(f) - jH_Q(f)\\ &= T\frac{\sin\pi(fT-0.5)}{\pi(fT-0.5)}\exp(-j\pi fT)\end{aligned}$$

These filter transfer functions are shown in Figure 5.16 (factor $\exp(-j\pi fT)$ or $j\exp(-j\pi fT)$ not included). From the figure we can see that the I-channel transfer function is low-pass and an even function of frequency and that the Q-channel transfer function is high-pass and an odd function of frequency. The total transfer function is a bandpass response with even symmetry about the frequency $f_+ = f_c + 1/4T$.

The frequency response of the matched filter is

$$H_{mI}(f) = \frac{T}{2}\left[\frac{2\sin(2\pi fT)}{2\pi fT} + \frac{\sin(\pi(2fT-1))}{\pi(2fT-1)} + \frac{\sin(\pi(2fT+1))}{\pi(2fT+1)}\right]$$

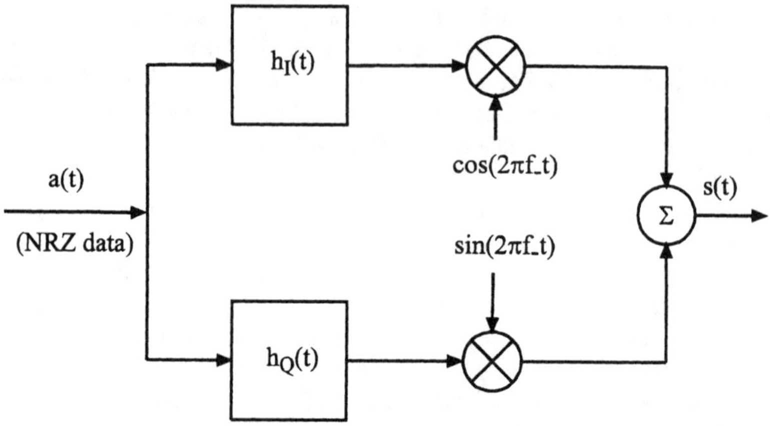

(a)

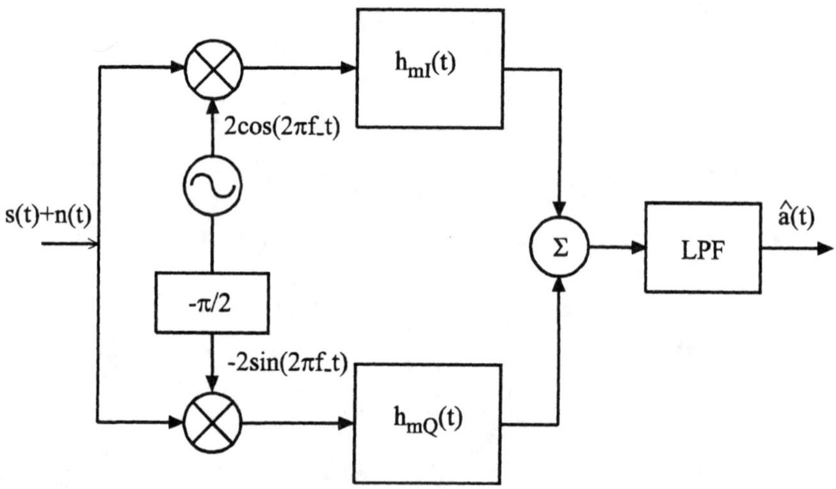

(b)

Figure 5.15 Baseband implementation of SMSK: (a) modulator, and (b) demodulator.

for I-channel and

$$H_{mQ}(f) = -j\frac{T}{2}\left[\frac{\sin(\pi(2fT-1))}{\pi(2fT-1)} - \frac{\sin(\pi(2fT+1))}{\pi(2fT+1)}\right]$$

for Q-channel. The total response of the low-pass equivalent matched filter is

$$\begin{aligned}\widetilde{H}_m(f) &= H_{mI}(f) + jH_{mQ}(f) \\ &= T\left[\frac{\sin(2\pi fT)}{2\pi fT} + \frac{\sin(\pi(2fT-1))}{\pi(2fT-1)}\right]\end{aligned}$$

The matched filters' responses are shown in Figure 5.17 (factor j not included). As in the case of the conversion filters, $H_{mI}(f)$ is even, low-pass, and $H_{mQ}(f)$ is odd, high-pass. The total transfer function $\widetilde{H}_m(f)$ is a bandpass response with even symmetry about the frequency f_c.

These baseband filters can be implemented by transversal filter configuration as described in [10]. At very high data rates (e.g., 550 Mbps), these filters can be implemented by microwave stripline as reported in [10] for the advanced communications technology satellite of NASA.

5.7.5 Synchronization of SMSK

Two approaches can be used to recover the carrier for the SMSK signal. One is the synchronization circuit for the parallel MSK described in Section 5.1, which involves squaring the received signal to produce spectral components at $2f_-$ and $2f_+$. Since SMSK is mainly devised for high data rates, this approach is not satisfactory as it involves doubling the already high-frequency components.

Another approach is to use a Costas loop as shown in Figure 5.18. This structure is especially suitable for the I/Q demodulator structure in Figure 5.15(b), since the I- and Q-channel demodulated signals are already available for the matched filter implementation. The circuitry on the right of the dotted line in Figure 5.18 is the extra needed to form the Costas loop [10].

There exists a 180° phase ambiguity in the carrier acquisition loop due to the baseband multiplier [10]. One method to remove this ambiguity is to deferentially encode the data before modulation, as described in Chapter 4.

5.8 MSK-TYPE MODULATION SCHEMES

After MSK was introduced, a lot of research effort was devoted to finding even more bandwidth-efficient modulation schemes. To improve bandwidth efficiency while

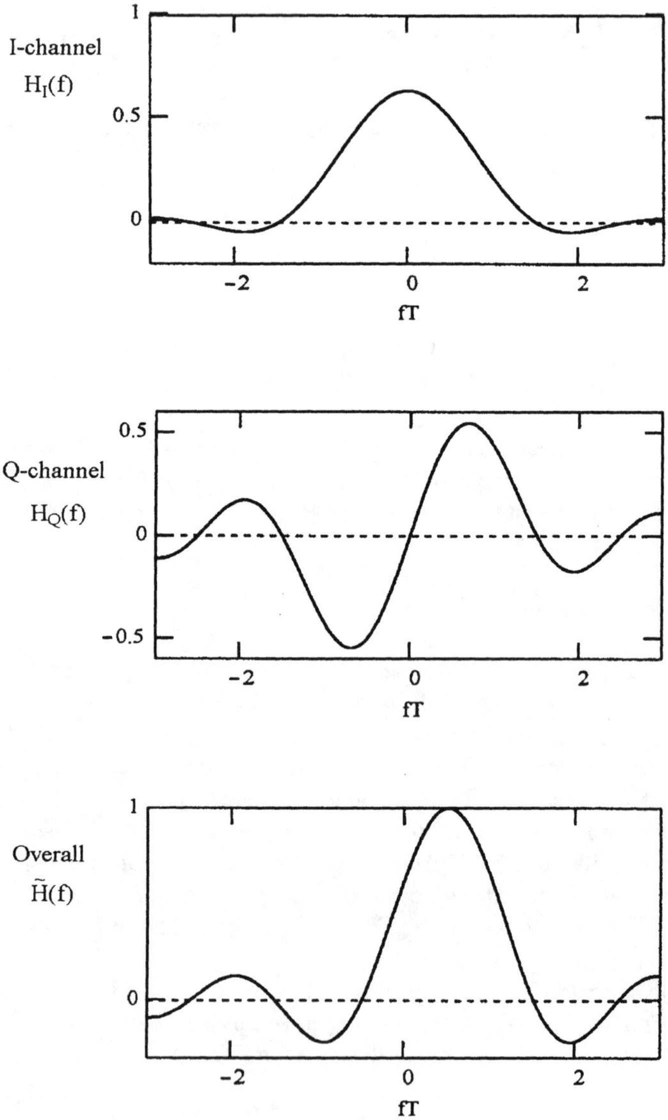

Figure 5.16 Baseband conversion filter frequency response.

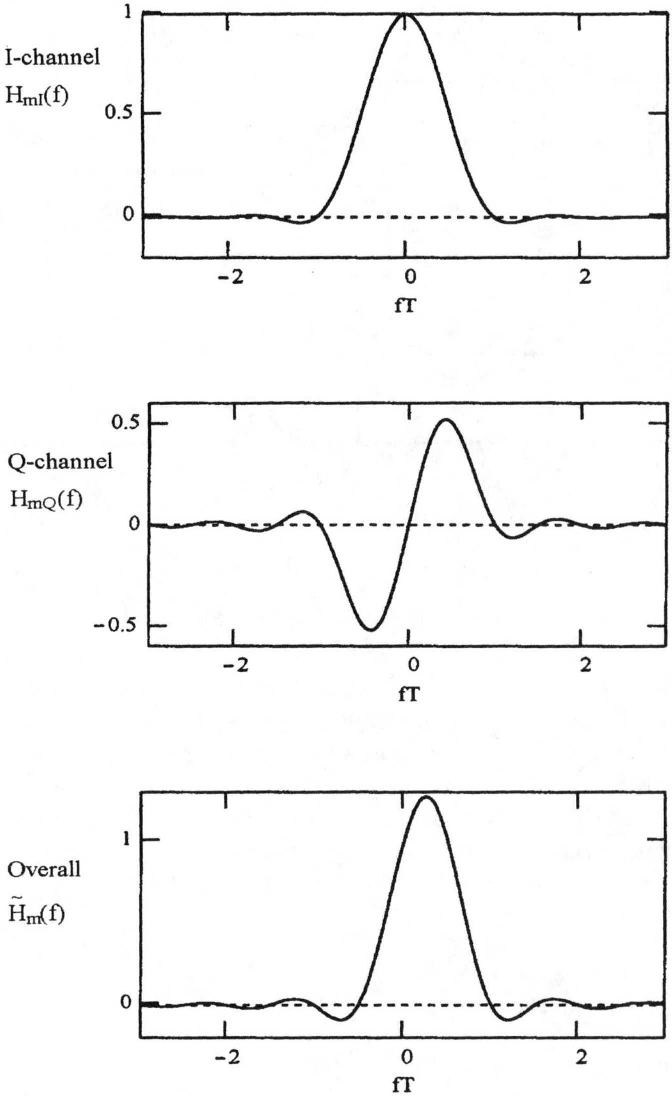

Figure 5.17 Baseband matched filter frequency response.

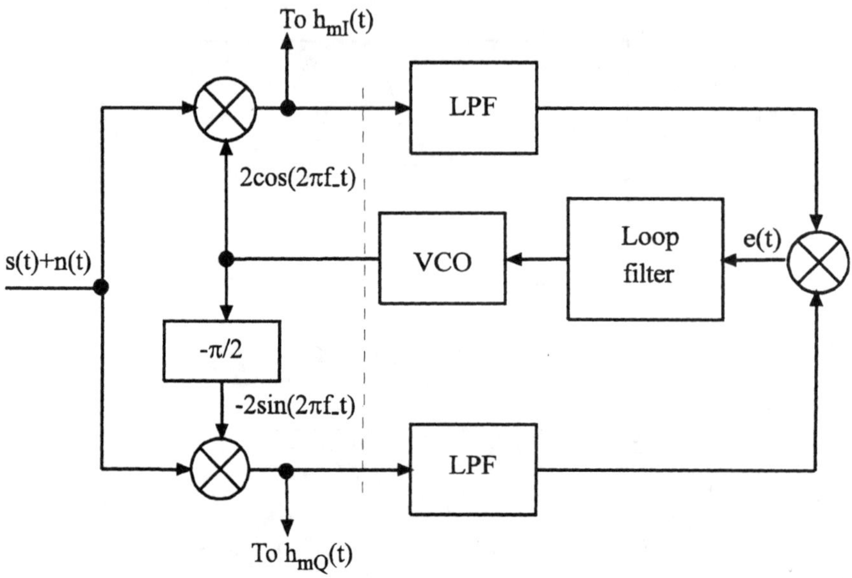

Figure 5.18 Costas loop for SMSK.

maintaining a constant envelope, one development direction is the continuous phase modulation with constant envelope, which has evolved into a large class of modulation schemes as we will study in Chapters 6 and 7. Another direction is to improve the spectra by using pulse shaping in quadrature modulation. In other words, new schemes are still based on two quadrature carriers like in MSK. However, the symbol shaping pulses are no longer half cosine. Instead, other pulse shapes are used. These schemes are sometimes called MSK-type schemes. All MSK-type signals can be expressed as

$$s(t) = s_I(t)\cos 2\pi f_c t + s_Q(t)\sin 2\pi f_c t, \quad -\infty \le t \le \infty \tag{5.31}$$

where $s_I(t)$ and $s_Q(t)$ are

$$s_I(t) = \sum_{k=-\infty}^{\infty} I_k p(t - 2kT) \tag{5.32}$$

$$s_Q(t) = \sum_{k=-\infty}^{\infty} Q_k p(t - 2kT - T) \tag{5.33}$$

where T is the bit time interval corresponding to the input data sequence $\{a_k \in (-1, +1)\}$ that has been demultiplexed into $\{I_k\}$ and $\{Q_k\}$. It is clear that each data symbol lasts for a duration of $2T$ in I- and Q-channels. Each data is weighted by a pulse-shaping function $p(t)$ which has a finite duration of $2T$. Just like MSK, a delay of T is introduced in the Q-channel, that is, the I- and Q-channel modulating signals are staggered. Also just like MSK, due to the staggering of the I- and Q-channels, the symbol duration of the MSK-type signal is T instead of $2T$ despite that the symbol durations are $2T$ for $s_I(t)$ and $s_Q(t)$. However, demodulation must be performed in a $2T$ duration.

For MSK-type schemes, the basic MSK modulator and demodulator in Figures 5.6 and 5.9 are applicable except that the pulse-shaping function must be replaced accordingly. Therefore for these schemes we will not repeat the description of the modulator and demodulator. Further, the serial MSK modulator and demodulator (Figure 5.12) can also be used for these MSK-type schemes, provided that the conversion filter in the modulator must be redesigned so that

$$\Psi_s(f) = |H(f)|^2 \Psi_{BPSK}(f) \tag{5.34}$$

where $\Psi_s(f)$ is the MSK-type signal spectrum, $H(f)$ is the conversion filter transfer function, and $\Psi_{BPSK}(f)$ is the spectrum of the BPSK signal which enters the filter. In the serial demodulator the matched filter must match to the pulse shape.

If the suboptimum receiver is acceptable, then all MSK-type schemes can be demodulated by an OQPSK-type demodulator where baseband correlation with $p(t)$ or matched filtering matched to $p(t)$ is omitted.

By choosing different $p(t)$, a variety of modulation schemes can be obtained. Sometimes $p(t)$ is indirectly determined by choosing the frequency pulse of the signal. The spectrum of the signal is determined by the pulse $p(t)$. In the following sections we will study a variety of pulse-shaping techniques that are primarily designed for satellite communications. What we are looking for from these modulation schemes are compact spectrum, low spectral spreading caused by nonlinear amplification, good error performance, and simple hardware implementation. We will describe a particular scheme or a class of schemes in each section. The emphasis is on the pulse shape and spectral properties. The error performance is also evaluated and often compared with that of MSK.

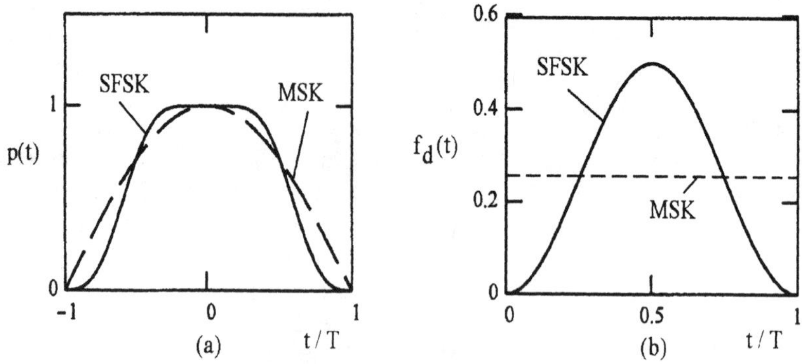

Figure 5.19 SFSK pulses: (a) amplitude pulse, and (b) frequency deviation pulse.

5.9 SINUSOIDAL FREQUENCY SHIFT KEYING

Amoroso first proposed an alternate symbol shaping pulse as [11]

$$p(t) = \begin{cases} \cos\left[\frac{\pi t}{2T} - \frac{1}{4}\sin\frac{2\pi t}{T}\right], & -T \le t \le T \\ 0, & \text{elsewhere} \end{cases} \tag{5.35}$$

This scheme was named as sinusoidal frequency shift keying (SFSK) since the signal can be synthesized by applying a keyed sine wave to a linear integrator followed by a linear frequency modulator. Sensitivity of the spectrum to pulse shaping was examined by varying the factor value in front of the sine function. By comparing spectra for different factor values, it was found that $1/4$ is the value for the lowest sidelobes [11].

Figure 5.19(a) shows the pulse shape in comparison with that of MSK.

This scheme has a constant envelope, because in any symbol period, say, $[0, T]$, the envelope

$$\begin{aligned} A(t) &= \sqrt{[I_0 p(t)]^2 + [Q_0 p(t-T)]^2} \\ &= \sqrt{p^2(t) + p^2(t-T)} \\ &= \sqrt{\cos^2\left[\frac{\pi t}{2T} - \frac{1}{4}\sin\frac{2\pi t}{T}\right] + \sin^2\left[\frac{\pi t}{2T} - \frac{1}{4}\sin\frac{2\pi t}{T}\right]} \\ &= 1 \end{aligned}$$

This scheme also has a continuous phase. Similar to the result for MSK (see (5.2)), in the duration $[kT, (k+1)T]$, the SFSK signal in (5.31) can be written as

$$s(t) = \cos[2\pi f_c t + d_k \phi(t) + \Phi_k] \tag{5.36}$$

where

$$d_k = -I_k Q_k$$

is determined by the staggered I- and Q-channel data, and

$$\phi(t) = \frac{\pi t}{2T} - \frac{1}{4}\sin\frac{2\pi t}{T} \tag{5.37}$$

which is the argument of the pulse-shaping function. The phase $\Phi_k = 0$ or π, corresponding to $I_k = 1$ or -1. From (5.36) it is clear that the total phase in the duration is continuous. At the bit transition $\phi(kT) = \frac{k\pi}{2}$, the total excess phase is

$$\Theta(kT) = d_k \frac{k\pi}{2} + \Phi_k$$

This satisfies the same condition for phase continuity of MSK (see (5.6)). Thus like MSK, the phase of the SFSK signal is continuous at any time.

The frequency deviation is

$$f_d(t) = \frac{1}{2\pi}\frac{d}{dt}[d_k \phi(t)] = d_k \frac{1}{4T}(1 - \cos\frac{2\pi t}{T}) \tag{5.38}$$

which is shown in Figure 5.19(b).

Expression (5.36) shows that the SFSK can be generated by a frequency modulator with $\phi(t)$ as its phase deviation.[6] Figure 5.20 is such a generator (modulator). Assuming the input is $d_k A \sin 2\pi t/T$, the output of the integrator is

$$\int_{kT}^{kT+t} d_k A \sin\frac{2\pi t}{T} dt = \frac{d_k AT}{2\pi}[1 - \cos\frac{2\pi t}{T}]$$

---

6 As we pointed out in the end of Section 5.1.2, having $d_k = -I_k Q_k$ is equivalent to having

$$d_k = d_{k-1} \oplus a_k$$

In other words, if we want the SFSK signal generated by the frequency modulator to be equivalent to the signal generated by a quadrature modulator, d_k must be differentially encoded using the original data sequence $\{a_k\}$. An SFSK signal so generated thus can be demodulated by the MSK-type quadrature demodulator. The outputs are directly I_k and Q_k. If d_k is the original data, the SFSK signal is a variation of the FFSK, which can still be quadraturely demodulated (first $\Theta(kT)$ is found and from which the data can be recovered) [2].

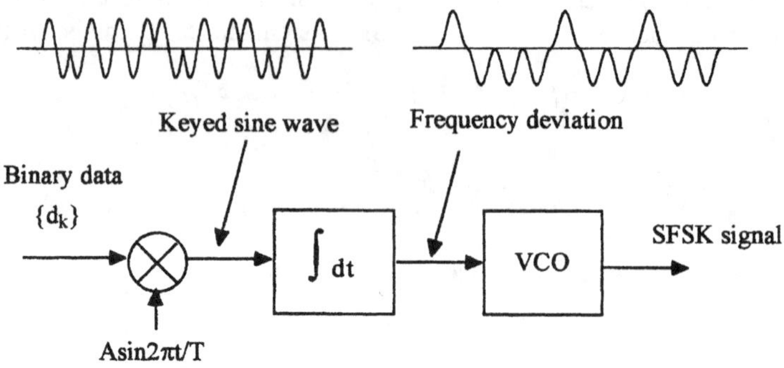

Figure 5.20 SFSK modulator.

This pulse is used as the control input to the VCO whose (radian) frequency deviation is proportional to the input voltage

$$\frac{d\Phi(t)}{dt} = \frac{d_k A K_V T}{2\pi}[1 - \cos\frac{2\pi t}{T}]$$

where K_V is the VCO sensitivity. The VCO output excess phase is

$$\begin{aligned}\Phi(t) &= \int_{kT}^{kT+t} \frac{d_k A K_V T}{2\pi}[1 - \cos\frac{2\pi t}{T}]dt \\ &= \frac{d_k A K_V T}{2\pi}[t - \frac{T}{2\pi}\sin\frac{2\pi t}{T}]\end{aligned}$$

Making $AK_V = \pi^2/T^2$, the above becomes

$$\begin{aligned}\Phi(t) &= d_k\left[\frac{\pi t}{2T} - \frac{1}{4}\sin\frac{2\pi t}{T}\right] \\ &= d_k\phi(t)\end{aligned}$$

Setting the VCO center frequency as f_c, the VCO output is then

$$s(t) = \cos\left[2\pi f_c t + d_k\phi(t) + \Phi_k\right] \tag{5.39}$$

The PSD of the modulated signal is determined by the pulse shape. The spectrum of the pulse shape can be analytically expressed as a sum of Bessel functions [11]. We only present the results here as shown in Figure 5.21. It can be seen that the sidelobes are considerably lower than those of MSK.

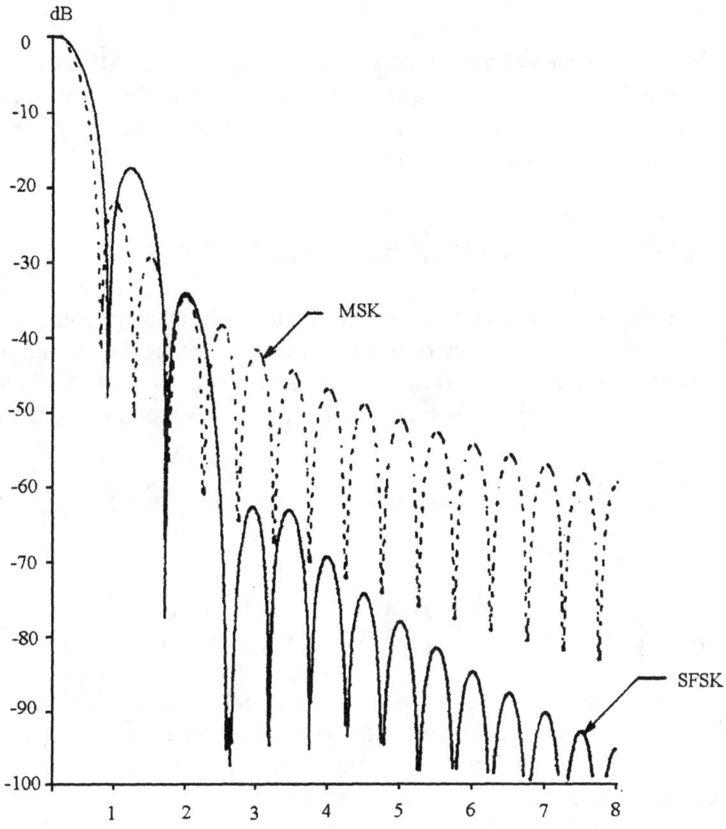

Figure 5.21 SFSK power spectral density. From [11]. Copyright © 1976 IEEE.

The error performance is the same as MSK since the bit energy of the SFSK signal is the same as that of MSK. This can be easily verified as follows (using MathCad, for example)

$$E_b = \int_0^T [p(t)\cos(2\pi f_c t) + p(t-T)\sin(2\pi f_c t)]^2\, dt = \frac{1}{2}T$$

which is the same as (5.22) when $A = 1$.

In general, modulator and demodulator are the same as those of MSK except that baseband weighting pulse (which can be replaced by a matched filter in the receiver) is the one in (5.13). As we have shown above, the modulator can be implemented as the simple frequency modulator in Figure 5.20.

5.10 SIMON'S CLASS OF SYMBOL-SHAPING PULSES

When MSK is viewed as a special case of FSK, (5.10) implies that the data sequence $\{d_k\}$ is first translated into a binary data waveform with rectangular shaped pulses and then frequency modulated onto the carrier. A generalization of (5.10) which allows for other than rectangular frequency shaping pulses is proposed by Simon in [12] as

$$s(t) = A\cos[2\pi f_c t + \frac{\pi}{2T}(t-kT)f_k(t) + \Phi_k], \quad (k-1)T \le t \le (k+1)T \quad (5.40)$$

where

$$f_k(t) = \begin{cases} d_{k-1}g_1(t-(k-1)T), & (k-1)T \le t \le kT \\ d_k g_2(t-kT), & kT \le t \le (k+1)T \end{cases} \quad (5.41)$$

is the frequency shaping pulse which is nonzero only on the interval $[0, T]$. Note that since $s(t)$ is specified over a $2T$ interval, the phase Φ_k must remain constant over that interval and k is restricted to take on only-even or only-odd values. It is clear from (5.40) that the signal is of constant envelope.

It was found that for even symmetrical carrier envelopes in the I- and Q-channels, as is true in the original MSK, the two pulses $g_1(t)$ and $g_2(t)$ must be mirror images of each other around the point $t = T/2$ (i.e., $g_2(t) = g_1(T-t), 0 \le t \le T$). Thus only one of them need be specified. Suppose $g_2(t)$ is specified, then

$$\begin{aligned} f_0(t) &= \begin{cases} d_{-1}g_1(t+T), & -T \le t \le 0 \\ d_0 g_2(t), & 0 \le t \le T \end{cases} \\ &= \begin{cases} d_{-1}g_2(-t), & -T \le t \le 0 \\ d_0 g_2(t), & 0 \le t \le T \end{cases} \end{aligned} \quad (5.42)$$

and

$$f_k(t) = \begin{cases} d_{k-1} g_2(-t+kT), & (k-1)T \le t \le kT \\ d_k g_2(t-kT), & kT \le t \le (k+1)T \end{cases} \tag{5.43}$$

From this expression we can see that when $d_{k-1} = d_k$, the frequency pulses in the kth and $(k-1)$th intervals are symmetrical about $t = kT$. Consider interval $[-T, T]$, the frequency deviation is

$$\begin{aligned} f_d(t) &= \frac{1}{2\pi}\left[\frac{d}{dt}(\frac{\pi}{2T} t f_k(t))\right] \\ &= \frac{1}{4T}\left[t\frac{d}{dt} f_k(t) + f_k(t)\right] \\ &= \begin{cases} \frac{1}{4T} d_{k-1}\left[t\frac{d}{dt} g_2(-t) + g_2(-t)\right], & -T \le t \le 0 \\ \frac{1}{4T} d_k \left[t\frac{d}{dt} g_2(t) + g_2(t)\right], & 0 \le t \le T \end{cases} \end{aligned} \tag{5.44}$$

Excluding the data bits, the expressions of $f_d(t)$ in $[-T, 0]$ and $[0, T]$ are the same if $g_2(t)$ is even, or differ only by a negative sign if $g_2(t)$ is odd. In these cases we only need to consider interval $[0, T]$ or $[kT, (k+1)]$ in general.

Further, since continuous rate of change of in-phase and quadrature envelopes leads to sharper spectral sidelobe roll-off, a constraint

$$g_1(T) = g_2(0) = 0 \tag{5.45}$$

may be imposed on $g_1(t)$ and $g_2(t)$ to ensure this desired property.

For identical I- and Q-channel envelopes further constraints are needed:

$$(1 - \frac{t}{T}) g_2(T - t) = 1 - \frac{t}{T} g_2(t), \quad 0 \le t \le T \tag{5.46}$$

or in terms of $g_1(t)$:

$$(1 - \frac{t}{T}) g_1(t) = 1 - \frac{t}{T} g_1(T - t), \quad 0 \le t \le T \tag{5.47}$$

Substituting $t = T/2$ into (5.46) and (5.47) we get

$$g_1(\frac{T}{2}) = g_2(\frac{T}{2}) = 1 \tag{5.48}$$

Based on these constraints we need only specify $g_2(t)$ in the interval $0 \le t \le T/2$ to determine $g_2(t)$ for $T/2 \le t \le T$. Similarly for $g_1(t)$. Assume

$$g_2(t) = \begin{cases} g_{21}(t), & 0 \le t \le T/2 \\ g_{22}(t), & T/2 \le t \le T \end{cases} \tag{5.49}$$

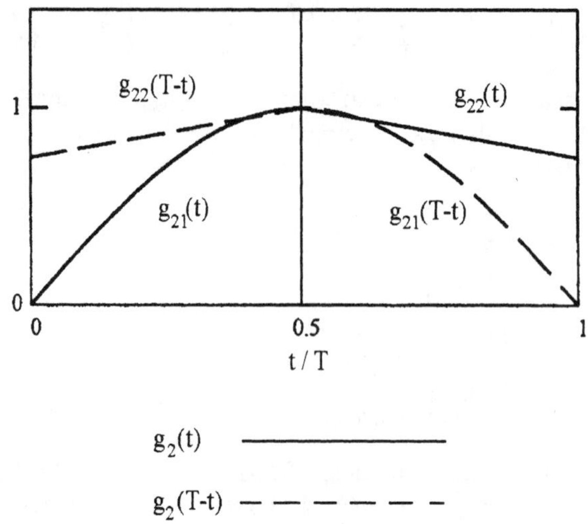

Figure 5.22 Relation between $g_2(T-t)$ and $g_2(t)$.

From Figure 5.22 we can see that

$$g_2(T-t) = \begin{cases} g_{22}(T-t), & 0 \le t \le T/2 \\ g_{21}(T-t), & T/2 \le t \le T \end{cases} \tag{5.50}$$

From (5.46) and (5.49) we can see that

$$\begin{aligned} g_2(T-t) &= \frac{T - tg_2(t)}{T-t} \\ &= \begin{cases} \frac{T-tg_{21}(t)}{T-t}, & 0 \le t \le T/2 \\ \frac{T-tg_{22}(t)}{T-t}, & T/2 \le t \le T \end{cases} \end{aligned}$$

Comparing to (5.50) we obtain

$$g_{21}(T-t) = \frac{T - tg_{22}(t)}{T-t}, \quad T/2 \le t \le T$$

or

$$g_{22}(t) = \frac{T - (T-t)g_{21}(T-t)}{t}, \quad T/2 \le t \le T \tag{5.51}$$

This changes (5.49) into

$$g_2(t) = \begin{cases} g_{21}(t), & 0 \le t \le T/2 \\ \frac{T-(T-t)g_{21}(T-t)}{t}, & T/2 \le t \le T \end{cases} \tag{5.52}$$

Thus the second part is found from the first part $g_{21}(t)$ using the above expression.

Under all the above constraints, we can express (5.41) in terms of $g_2(t)$, and expand (5.40) into I- and Q-channel summation, then we can easily identify the symbol shaping pulses as

$$p(t) = \begin{cases} \cos\left(\frac{\pi t}{2T} g_2(-t)\right), & -T \le t \le 0 \\ \cos\left(\frac{\pi t}{2T} g_2(t)\right), & 0 \le t \le T \end{cases} \tag{5.53}$$

for I-channel and

$$p(t-T) = \begin{cases} \sin\left(\frac{\pi t}{2T} g_2(t)\right), & 0 \le t \le T \\ \sin\left(\frac{\pi(2T-t)}{2T} g_2(2T-t)\right), & T \le t \le 2T \end{cases} \tag{5.54}$$

for Q-channel [12, 13], where $p(t-T)$ can be verified by use of (5.46). Note that when $g_2(t) = 1$, $p(t)$ falls back to that of the original MSK.

Several pulse shapes of $g_2(t)$ were examined in [12]. The first one is

$$g_2(t) = \begin{cases} 1 - \frac{\sin(2\pi t/T)}{(2\pi t/T)}, & 0 \le t \le T \\ 0, & \text{elsewhere} \end{cases} \tag{5.55}$$

which turns out to generate the SFSK. It can be easily verified that (5.55) satisfies (5.45) and (5.46). The corresponding frequency deviation pulse $f_d(t)$ and amplitude pulse are given earlier in (5.38) and (5.35) already.

The second is of polynomial-type

$$g_2(t) = \left(\frac{2t}{T}\right)^n, \quad 0 \le t \le T/2, \quad n = 1, 2, \cdots \tag{5.56}$$

and it satisfies (5.45) and (5.46). The pulse in the interval $[T/2, T]$ can be determined by the relation given in (5.52). This pulse function is odd when n is odd, and even when n is even. The corresponding frequency deviation pulse $f_d(t)$ can be obtained from (5.44) and is given by

$$f_d(t) = \begin{cases} \frac{d_k(n+1)}{4T}\left[\frac{2(t-kT)}{T}\right]^n, & kT \leq t \leq (k+\frac{1}{2})T \\ \frac{d_k(n+1)}{4T}\left[\frac{2[T-(t-kT)]}{T}\right]^n, & (k+\frac{1}{2})T \leq t \leq (k+1)T \end{cases}$$

The amplitude pulse can be found from (5.53) and (5.54).

The third is

$$g_2(t) = \frac{1}{2}\frac{1-\cos \pi t/T}{t/T}, \quad 0 \leq t \leq T/2 \tag{5.57}$$

and it satisfies (5.45) and (5.46). The pulse shape in the interval $[T/2, T]$ can be determined by the relation given in (5.52). This pulse function is odd. The corresponding frequency deviation pulse $f_d(t)$ can be obtained from (5.44) and is given by

$$f_d(t) = \frac{\pi d_k}{8T}\sin\frac{\pi(t-kT)}{T}, \quad kT \leq t \leq (k+1)T$$

The amplitude pulse can be found from (5.53) and (5.54).

Figure 5.23 shows these frequency pulses, corresponding frequency deviation pulses, and amplitude pulse shapes (assuming $d_0 = 1$), where $n = 1$ for the second type. The frequency deviation pulse for the second type is a triangle and that for the third type is a half-sine function. The amplitude pulses are quite similar despite that the frequency deviation pulses are quite different.

The spectra (in the form of fractional out-of-band power P_{ob}) of these amplitude pulses are numerically computed in [12] and the results are shown in Figure 5.24. The results show that the $n = 1$ case polynomial-type pulse and the third type pulse have similar sidelobe roll-offs with the latter being slightly better. They roll off faster than SFSK within a bandwidth of approximately $BT = 2.0$, but slower thereafter. The roll-offs for higher values of n are significantly worse than SFSK. The $n = 1$ case polynomial-type pulse has spectral properties which are in a sense a compromise between those of MSK and SFSK. However, it has the practical advantage of offering easy transmitter implementation, that is, the transmitter oscillator is linearly swept in frequency.

One final comment is that when the receiver matched filters are matched over the $2T$ decision interval to the envelopes $p(t)$ and $p(t-T)$ of (5.53) and (5.54), the error performance is the same as that of MSK since the bit energy of the signals using the pulse shapes given in (5.55) to (5.57) is the same as that of MSK (checked using MathCad).

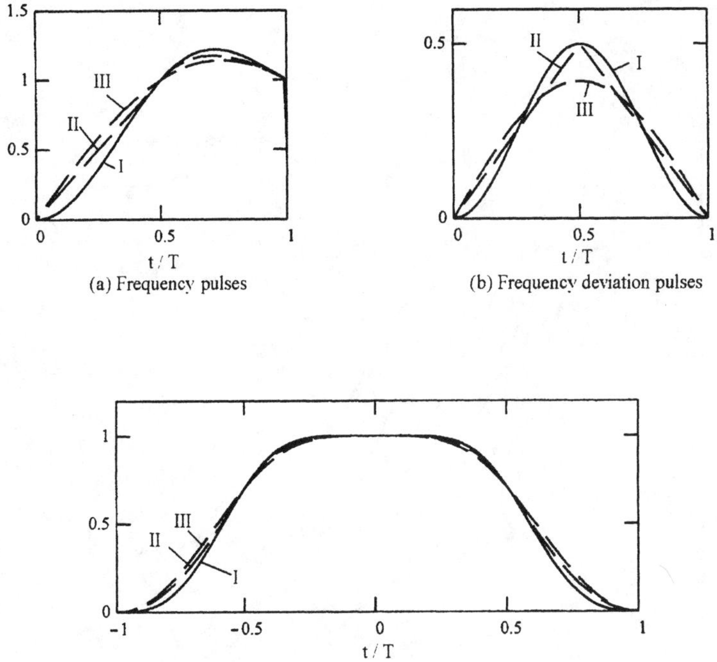

(a) Frequency pulses

(b) Frequency deviation pulses

(c) Amplitude pulses

Legend

I First type (SFSK)
II Second type (polynomial, n=1)
III Third type

Figure 5.23 Simon's frequency pulses (a), frequency deviation pulses (b), and amplitude pulses (c).

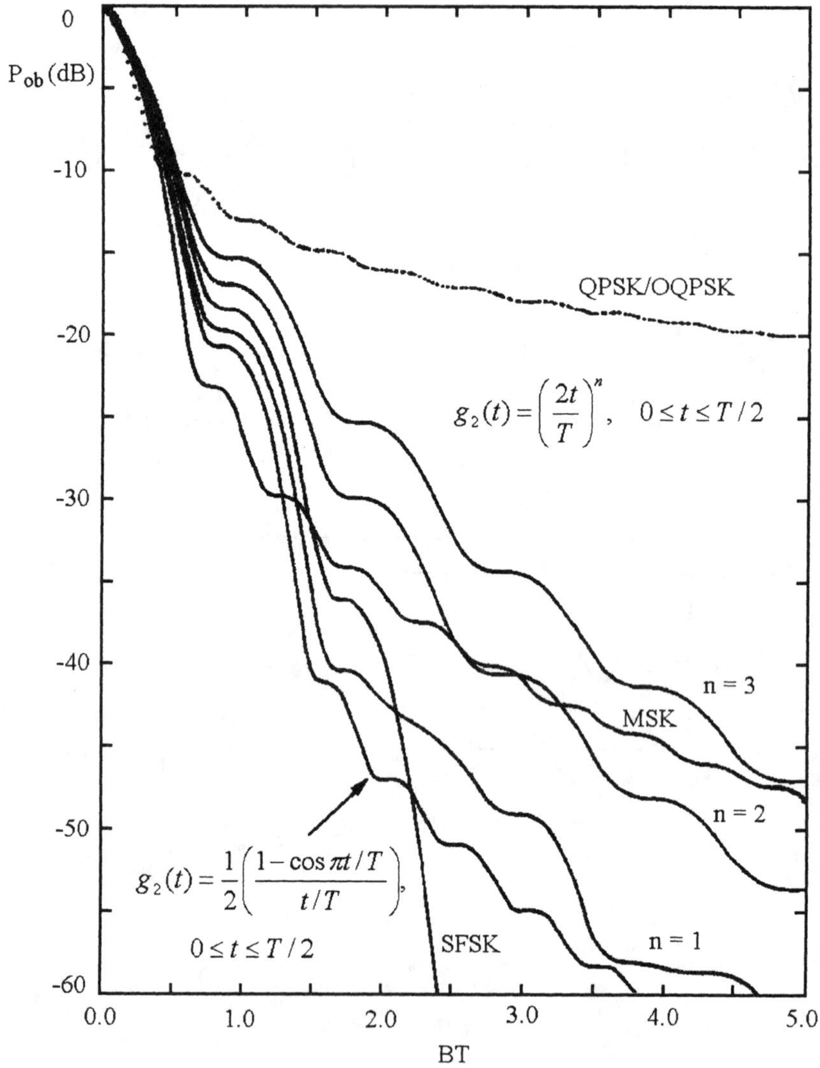

Figure 5.24 Fractional out-of-band behavior for several types of modulating pulses. From [12]. Copyright © 1976 IEEE.

5.11 RABZEL AND PASUPATHY'S SYMBOL-SHAPING PULSES

It is well known that if pulse $p(t)$ has $(N-1)$ continuous derivatives which are equal to zero at the leading and trailing edges of the pulse, then its Fourier transform decays asymptotically as $f^{-(N+1)}$ and the power spectrum as $f^{-(2N+2)}$ [14]. (Only positive f need be considered since spectra are symmetrical about $f=0$.) Equivalent conditions on $g(t)$, which is denoted as $g_2(t)$ in (5.53) and (5.54), are derived in [13] as

$$g^{(i)}(T)=(-1)^i\frac{i!}{T^i},\qquad i=0,1,2,\cdots,N-1$$

For an MSK-type signal, the above conditions are equivalent to

$$g^{(i-1)}(0)=0,\qquad i=0,1,2,\cdots,N-1 \tag{5.58}$$

A general class of pulses was proposed by Rabzel and Pasupathy [13] as

$$G(t,\alpha,M)=1-\frac{\sum_{i=1}^{M}K_i\left[\sin(\alpha\frac{2\pi t}{T})\right]^{2i-1}}{(\alpha\frac{2\pi t}{T})},\qquad 0\le t\le T \tag{5.59}$$

where

$$K_i=\frac{(2i-2)!}{2^{2i-2}[(i-1)!]^2(2i-1)}$$

and $\alpha=1,2,3,\cdots$. The K_is are the coefficients of the series expansion of the inverse sine function, that is,

$$\sin^{-1}(x)=\sum_{j=1}^{\infty}K_jx^{2j-1}$$

The first 10 K_is are 1, 0.167, 0.075, 0.045, 0.03, 0.022, 0.017, 0.014, 0.012, and 0.0097.

Note that an MSK-type signal using $G(t,\alpha,M)$ has a constant envelope since the signal is given in (5.40).

For finite M, it was shown in [13] that $G(t,\alpha,M)$ satisfies (5.46) and (5.58) with $N=2M+1$. Thus the power spectrum of an MSK-type signal employing this class of frequency shaping pulses will decay asymptotically as $f^{-(4M+4)}$. Some shaping pulses previously considered are special cases of (5.59). Specifically,

$$G(t,\alpha,0)=1$$

defines MSK, with an asymptotic decay rate of f^{-4} and

$$G(t,1,1) = 1 - \frac{\sin(2\pi t/T)}{(2\pi t/T)}$$

generates the SFSK with an asymptotic decay rate of f^{-8}. The shaping pulse defined in (5.57), which is not in the class, has an asymptotic decay rate of f^{-6}. These signals should be contrasted with the OQPSK signal which has an asymptotic decay rate of f^{-2} only.[7]

However, when $M = \infty$, the pulse is

$$G(t,1,\infty) = \begin{cases} 0, & 0 \le t \le T/4 \\ 2 - \frac{T}{2t}, & T/4 \le t \le 3T/4 \\ \frac{T}{t}, & 3T/4 \le t \le T \end{cases}$$

which has a discontinuous first derivative (just like MSK) and hence its power spectrum decays as f^{-4}.

The plots of this class of frequency-shaping pulses and in-phase symbol-shaping pulses are given in Figures 5.25 and 5.26. The power spectra (in the form of out-of-band power P_{ob}) of corresponding MSK-type signals are given in Figure 5.27. For fixed $\alpha = 1$, the higher the values of M (except for $M = \infty$), the smoother the pulse shapes and the lower the out-of-band power. The $P_{ob(\text{min})}$ in the figure refers to a lower bound on P_{ob} as calculated by Prabhu [15], using the optimum prolate spheroidal wave function for $p(t)$. The effect of α was also examined for fixed $M = 1$ case as shown in Figure 5.28. It was found that for large BT values the P_{ob} is smaller for $\alpha = 1$, and for small BT values the P_{ob} is smaller for larger α. However, using MSK at the small BT ranges is a better alternative since its P_{ob} is indistinguishable from that of SFSK but its system complexity is less.

The generation of the MSK-type signaling using $G(t, \alpha, M)$ is a straightforward extension of the scheme shown in Figure 5.20. The generator will use the frequency deviation function as an input to a VCO whose outputs are the I- and Q-channel symbol pulses $p(t)$ and $p(t-T)$. The generator is shown in Figure 5.29. The structure is based on the frequency deviation function.

$$f_d(t) = \frac{1}{4T}\frac{d}{dt}[tg(t)], \quad 0 \le t \le T$$

---

[7] The symbol shaping pulse for OQPSK is $p(t) = 1, 0 \le t \le T$, and 0 elsewhere, which is discontinuous. Its integration (or -1th derivative) is continuous. Thus $N = 0$, its power spectrum decays with an asymptotic decay rate of $|f|^{-(2N+2)} = |f|^{-2}$.

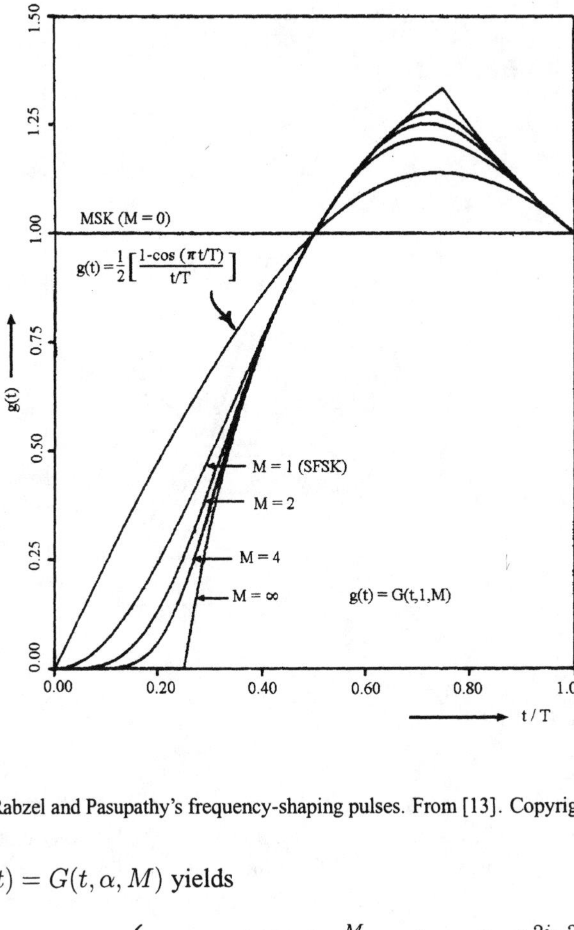

Figure 5.25 Rabzel and Pasupathy's frequency-shaping pulses. From [13]. Copyright © 1978 IEEE.

for which $g(t) = G(t, \alpha, M)$ yields

$$f_d(t) = \frac{1}{4T}\left\{1 - \cos\left(\frac{\alpha 2\pi t}{T}\right)\sum_{i=1}^{M} B_i \left[\sin\frac{\alpha 2\pi t}{T}\right]^{2i-2}\right\}$$

where

$$B_i = \frac{(2i-2)!}{2^{2i-2}[(i-1)!]^2}$$

The first four B_is are 1, 1/2, 3/8, and 5/16. Note that the output of the VCO in Figure 5.20 is directly the modulated signal since the input is a keyed sine signal, whereas the outputs of the VCO in Figure 5.29 are baseband symbol pulses. Keying is performed

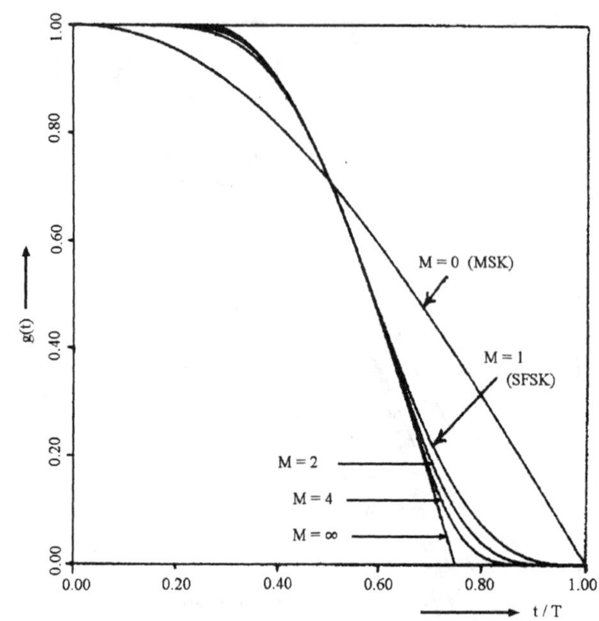

Figure 5.26 In-phase symbol weightings for $g(t) = G(t, 1, M)$. From [13]. Copyright © 1978 IEEE.

in the modulator.

Again the optimum demodulator is the same as that of MSK with the receiver matched filters matched over the $2T$ decision interval to the I- and Q-channel shaping pulses. The error performance is also the same since the bit energy of the signals using the pulse shapes given in (5.59) is the same as that of MSK (checked using MathCad).

5.12 BAZIN'S CLASS OF SYMBOL-SHAPING PULSES

A class of symbol-shaping pulses was proposed by Bazin [16] which includes the class of (5.59) as a subclass. Hence SFSK is also an element of the class. This class is defined to have up to Nth continuous derivatives so that its power spectrum decays asymptotically as f^{-2N-4} and to produce a constant envelope of the MSK-

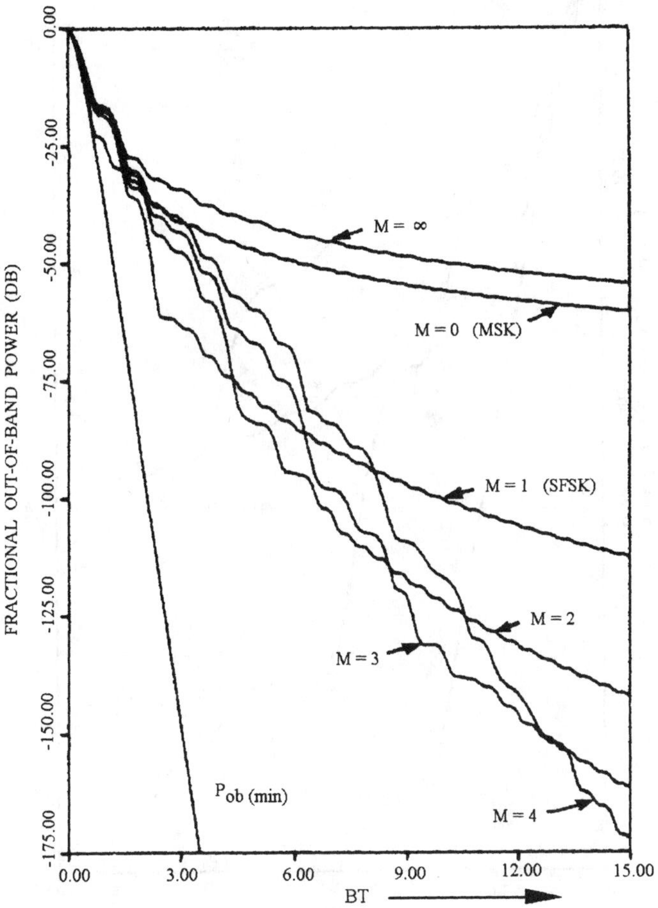

Figure 5.27 Fractional out-of-band powers for $G(t, 1, M)$. From [13]. Copyright © 1978 IEEE.

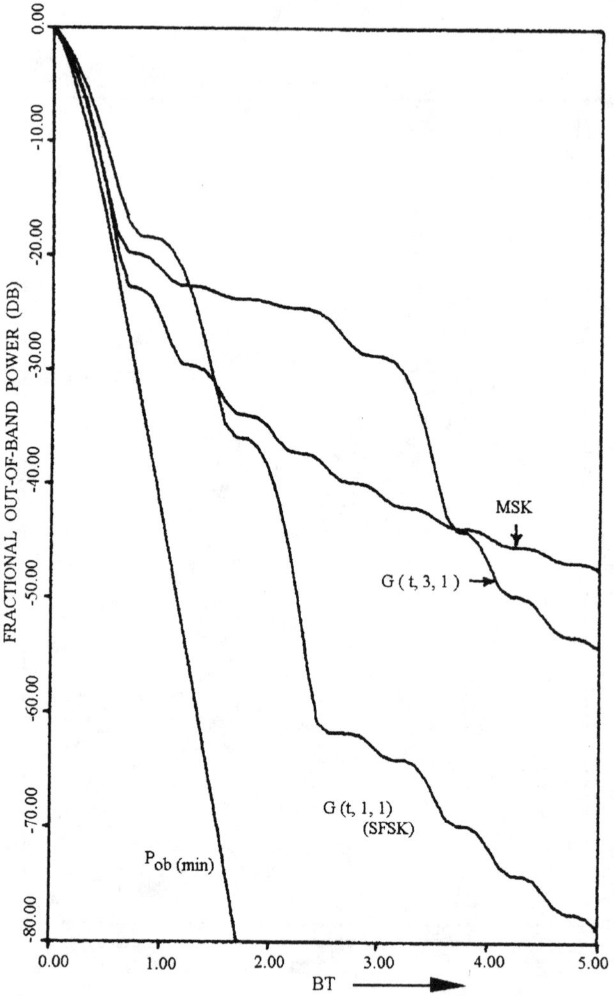

Figure 5.28 Effect of α in $G(t, \alpha, 1)$ on the fractional out-of-band power. From [13]. Copyright © 1978 IEEE.

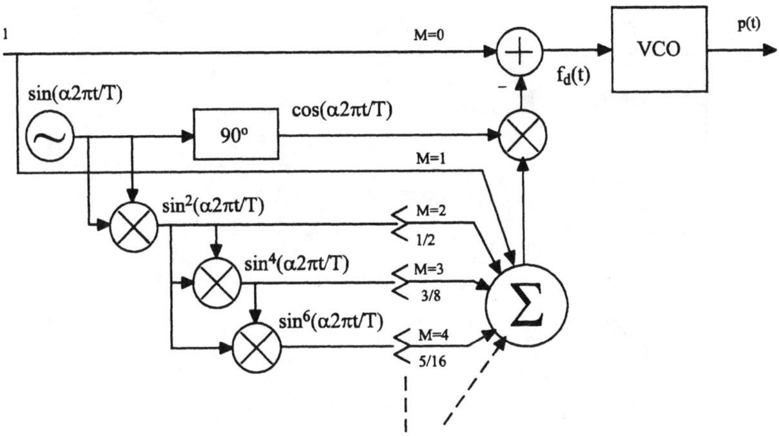

Figure 5.29 Generation of amplitude pulses. From [13]. Copyright © 1978 IEEE.

type which generates:

$$p(t) = \cos\left[\frac{\pi t}{2T} - \sum_{k=1}^{N'} A_k \sin\frac{2\pi kt}{T}\right] \quad \text{and } N' \geq \frac{N}{2} \tag{5.60}$$

The A_k coefficients are solutions of the linear system:

$$\frac{d_i p(\pm T)}{dt^i} = 0, \qquad i = 1, 2, \cdots, N$$

Some A_k coefficients may be zero. The magnitude of the vector sum of the in-phase and quadrature pulse envelope is invariant since $p(t) = p(-t)$, and $p^2(t) + p^2(t - T) = 1$. SFSK is apparently an element of this class with $k = 1$ and $N = 2$ and $A_1 = 0.25$, hence the power spectrum asymptotically decays as f^{-8}. Another example of this class, called double SFSK (DSFSK) is defined by choosing $N^{'} = 2$ and $N = 4$. The pulse is

$$p(t) = \cos\left[\frac{\pi t}{2T} - \frac{1}{3}\sin\frac{2\pi t}{T} + \frac{1}{24}\sin\frac{4\pi t}{T}\right]$$

Its power spectrum decays asymptotically as f^{-12}. It is interesting to compare this pulse to one of Rabzel's pulses which is close to it. By choosing $M = 2$ and $\alpha = 1$ in

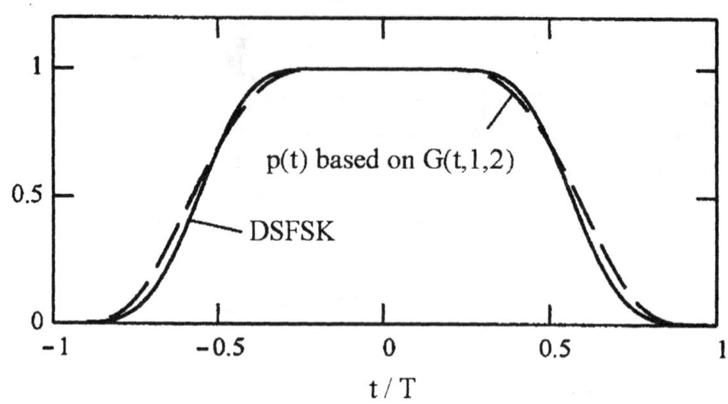

Figure 5.30 Bazin's pulse.

(5.59), Rabzel's frequency-shaping pulse leads to a symbol-shaping pulse as follows

$$p(t) = \cos\left[\frac{\pi t}{2T} - \frac{9}{32}\sin\frac{2\pi t}{T} + \frac{1}{96}\sin\frac{6\pi t}{T}\right] \tag{5.61}$$

The pulses are shown in Figure 5.30.

The comparison of power spectral densities in Figure 5.31 shows that the spectrum of DSFSK decays, on the average as f^{-12} beyond $f = 4.75/T$, and coincides with the SFSK spectrum near $f = 4/T$ and departs from the latter from $f = 1/T$ up to $f = 3.75/T$ where the power density is larger for DSFSK than for SFSK. The difference is maximum at $f = 3/T$ and equal to about 20 dB. The power spectrum of (5.61) is larger than SFSK and close to DSFSK before $f = 3.75/T$, and it approaches the SFSK spectrum beyond $f = 4.5/T$, but the effect of the asymptotic slope (as f^{-12}) does not happen before $f = 6/T$.

The error performance of above schemes is the same as MSK as long as the receiver uses matched filters which are matched to I- and Q-channel symbol-shaping pulses.

5.13 MSK-TYPE SIGNAL'S SPECTRAL MAIN LOBE

While above shaping techniques can reduce MSK spectral sidelobes considerably,

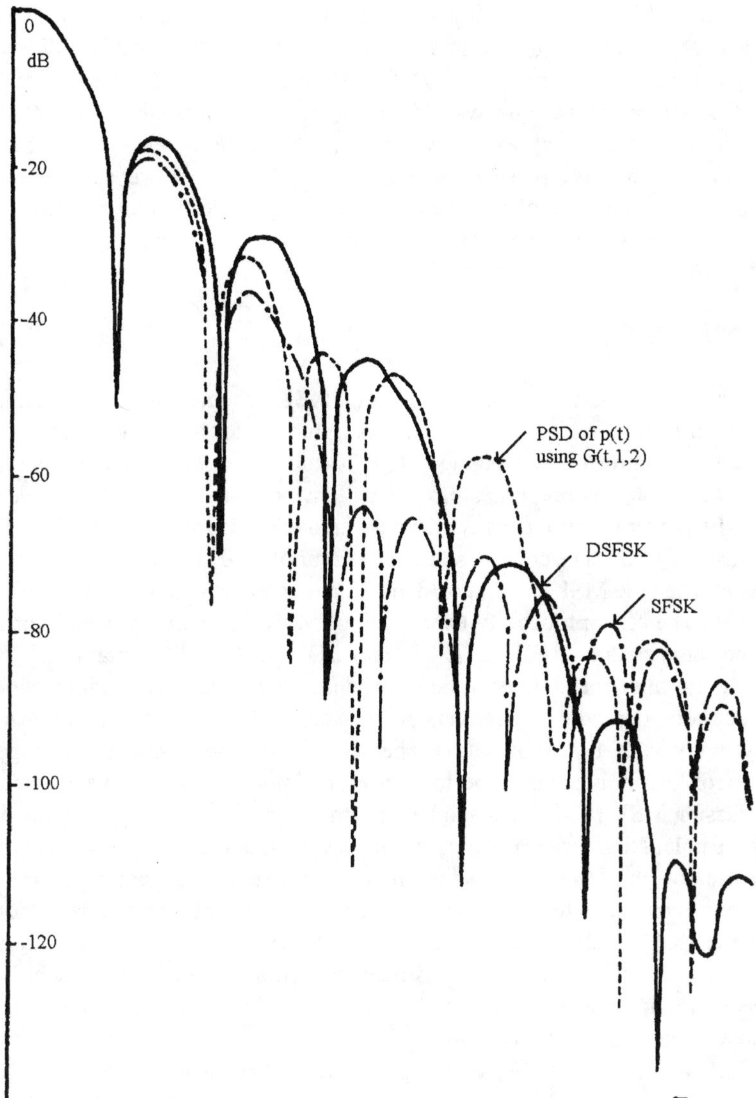

Figure 5.31 Power spectra. From [16]. Copyright © 1979 IEEE.

none of them can reduce the main lobe of MSK spectrum. A study of all these proposed symbol-shaping pulses $p(t)$ revealed the following common points: 1) All the proposed $p(t)$ are monotonic symmetrical pulses of finite duration $[-T, T]$, which is defined as follows: $p(t) = p(-t), p'(t) \leq 0$ for $0 < t < T$. 2) The spectral main lobes of all of these $p(t)$ are always wider than that of a square pulse having the same time duration. In fact, it has been shown that an MSK-type signal generated by using a monotonic symmetrical pulse of finite duration $[-T, T]$ will always have a spectral main lobe wider than that of a conventional PSK signal generated by using a polar NRZ pulse of the same duration [17].

5.14 SUMMARY

In this chapter we first described the original MSK scheme which is implemented in a parallel fashion with I- and Q-channels in modulator and demodulator. It was described as a sinusoidally weighted OQPSK and as a special case of continuous phase modulation. Its properties of constant envelope and continuous phase were proved. Its power spectral density was derived and compared with those of BPSK, QPSK, and OQPSK. The comparison was in favor of MSK in terms of sidelobe roll-off speed. Then the MSK modulator, demodulator, and synchronization circuit were described. The error probability expression of MSK was derived, which turns out to be the same as that of BPSK, QPSK, and OQPSK. Next the serial MSK, which is suitable for high-speed transmission, was described in detail, including its principle, modulator, demodulator, conversion and matched filters, and synchronization. Finally, many MSK-type modulation schemes, which generally have better spectral sidelobe roll-offs or better error performances, but more complexity, were discussed.

Gaussian MSK (GMSK) is another spectral compact MSK-type scheme which passes the polar NRZ data waveform through a Gaussian filter before sending it to an FM modulator [18]. It also suffers loss in error performance depending on the filter bandwidth. This leads to a trade-off between error performance and bandwidth. It was shown that 0.7 dB loss for a filter bandwidth $B = 0.25/T$ is a good trade-off for mobile radio channels [18]. It has been chosen as the modulation scheme for European mobile GSM system. It will be discussed in Chapter 6 in the context of continuous phase modulation (CPM).

Efforts to improve the MSK error performance were also reported [19,20]. Improvements are achieved by extending the observation interval from $2T$ for MSK to longer intervals. The improvements in E_b/N_o are on the order of 1 dB for $3T$ and 1.2 dB for $5T$ for an optimum value of $\Delta f = (f_+ - f_-) = 0.715/T$. For observation intervals longer than $5T$, improvements are minor. The complexity increase does not seem to favor these schemes over the simple yet efficient MSK.

We will revisit MSK and MSK-type schemes in the next chapter in the context of continuous phase modulation. In particular, one important MSK-type scheme, the GMSK, will be discussed in detail due to its application in the GSM system.

References

[1] Doelz, M. L., and E. H. Heald, "Minimum-shift data communication system," Collins Radio Co., U.S. Patent 2977 417, March 28, 1961.

[2] DeBuda, R., "Coherent demodulation of frequency-shift keying with low deviation ratio," *IEEE Trans. Commun.*, vol. 20, June 1972, pp. 429–435.

[3] Gronemeyer, S. A., and A. L. McBride, "MSK and offset QPSK modulation," *IEEE Trans. Commun.*, vol. 24, August 1976, pp. 809–820.

[4] Amoroso, F., and J. A. Kivett, "Simplified MSK signaling technique," *IEEE Trans. Commun.*, vol. 25, April 1977, pp. 433–441.

[5] *System Handbook of Advanced Communications Technology Satellite (ACTS)*, NASA, Lewis Research Center, Cleveland, Ohio, 1993.

[6] Steele, R., *Mobile Radio Communications*, New York: IEEE Press, 1995.

[7] Bhargava, V. K., et al., *Digital Communications by Satellite,* New York: John Wiley and Sons, 1981.

[8] Taylor, D. P., et al., "A high speed digital modem for experimental work on the communications technology satellite," *Canadian Electrical Engineering Journal*, vol. 2, no. 1, 1977, pp. 21–30.

[9] Pasupathy, S., "Minimum shift keying: a spectrally efficient modulation," *IEEE Communications Magazine*, July 1979, pp. 14–22.

[10] Ziemer, R. E., and C. R. Ryan, "Minimum-shift keyed modem implementations for high data rates," *IEEE Communications Magazine*, vol. 21, October 1983, pp. 28–37.

[11] Amoroso, F., "Pulse and spectrum manipulation in the minimum (frequency) shift keying (MSK) format," *IEEE Trans. Commun.*, vol. 24, March 1976, pp. 381–384.

[12] Simon, M. K., "A generalization of minimum-shift-keying (MSK)-type signaling based upon input data symbol pulse shaping," *IEEE Trans. Commun.*, vol. 24, August 1976, pp. 845–856.

[13] Rabzel, M., and S. Pasupathy, "Spectral shaping in MSK-type signals," *IEEE Trans. Commun.*, vol. 26, January 1978, pp. 189–195.

[14] Bennett, W. R., *Introduction of Signal Transmission,* New York: McGraw-Hill, 1970, p.17.

[15] Prabhu, V. K., "Spectral occupancy of digital angle-modulated signals," *Bell Syst. Tech. J.*, April 1976, pp. 429–453.

[16] Bazin, B., "A class of MSK baseband pulse formats with sharp spectral roll-off," *IEEE Trans. Commun.*, vol. 27, May 1979, pp. 826–829.

[17] Boutin, N., and S. Morissette, "Do all MSK-type signaling waveforms have wider spectra than those of PSK?" *IEEE Trans. Commun.*, vol. 29, July 1981, pp. 1071–1072.

[18] Murota, K., and K. Hirade, "GMSK modulation for digital mobile radio telephony," *IEEE Trans.*

Commun., vol. 29, July 1981, pp. 1044–1050.

[19] Osborn, W. P., and M. B. Luntz, "Coherent and noncoherent detection of CPFSK," *IEEE Trans. Commun.*, vol. 22, August 1974, pp. 1023–1036.

[20] De Buda, R., "About optimal properties of fast frequency-shift keying," *IEEE Trans. Commun.*, vol. 22, October 1974, pp. 1726–1727.

Selected Bibliography

- Haykin, S., *Digital Communications*, New York, John Wiley, 1988.
- Sklar, B., *Digital Communications: Fundamentals and Applications*, Englewood Cliffs, New Jersey: Prentice Hall, 1988.
- Svensson, A., and C.-E. Sundberg, "Serial MSK-type detection of partial response continuous phase modulation," *IEEE Trans. Commun.*, vol. 33, Jan. 1985, pp. 44–52.
- Ziemer, R. E., and R. L. Peterson, *Introduction to Digital Communication*, 4th ed., Boston: Houghton Mifflin, 1995.
- Ziemer, R. E., and W. H. Tranter, *Principles of Communications: Systems, Modulation, and Noise*, New York: Macmillan, 1992.

Chapter 6

Continuous Phase Modulation

From Chapter 5 we have learned that MSK signal has continuous phase. In fact MSK is just a special case of a large class of constant amplitude modulation schemes called continuous phase modulation (CPM). This class of modulation is jointly power and bandwidth efficient. With proper choice of pulse shapes and other parameters, CPM schemes may achieve higher bandwidth efficiency than QPSK and higher order MPSK schemes. Even though high-order QAM may outperform MPSK in terms of power or bandwidth efficiency (see Section 8.7 for comparison between QAM and MPSK), QAM's nonconstant envelope may hinder its use in channels with nonlinear power amplifiers. Therefore CPM has been getting a lot of attention in satellite channels and other channels. Some of the CPM schemes have been used in practical communication systems. For example, MSK has been used in NASA's Advanced Communication Technology Satellite (ACTS) system, GMSK (Gaussian MSK) has been used in the U.S. cellular digital packet data (CDPD) system and the European global system for mobile (GSM) system.

Significant contributions to CPM schemes, including signal design, spectral analysis, and error performance analysis were made by C-E. Sundberg, T. Aulin, A. Svensson, and J. Anderson, among other authors [1–9]. Excellent treatment of CPM up to 1986 can be found in the book by J. Anderson, T. Aulin, and C-E. Sundberg [9] or the article by C-E. Sundberg [1]. In this chapter we will cover all basic aspects of CPM and present research results up to date. The treatment here is limited to the AWGN channel as we did for previous chapters. The multiple index continuous phase modulation (MHPM) will be covered in the next chapter.

We define CPM signal and study its phase properties in Section 6.1. Its power spectral density is studied in Section 6.2. The error probability of CPM schemes is determined by the Euclidean distances between signals. In Section 6.3, we derive the distance expression for CPM signals and compare distances for different CPM schemes. CPM modulators and demodulators are presented in Sections 6.4 and 6.5, respectively. Section 6.6 is for synchronization (carrier and symbol) of CPM signals.

Since it is currently used in practical systems, a comprehensive treatment on GMSK is given in Section 6.7. Section 6.8 summarizes this chapter.

6.1 DESCRIPTION OF CPM

CPM signal is defined by

$$s(t) = A\cos(2\pi f_c t + \Phi(t, \mathbf{a})), \quad -\infty \le t \le \infty \tag{6.1}$$

The signal amplitude is constant. Unlike signals of previously defined modulation schemes such as FSK and PSK, where signals are usually defined on a symbol interval, this signal is defined on the entire time axis. This is due to the continuous, time-varying phase $\Phi(t, \mathbf{a})$, which usually is influenced by more than one symbol. The transmitted M-ary symbol sequence $\mathbf{a} = \{a_k\}$ is imbedded in the *excess phase*

$$\Phi(t, \mathbf{a}) = 2\pi h \sum_{k=-\infty}^{\infty} a_k q(t - kT) \tag{6.2}$$

with

$$q(t) = \int_{-\infty}^{t} g(\tau)d\tau \tag{6.3}$$

The M-ary data a_k may take any of the M values: $\pm 1, \pm 3, ..., \pm(M-1)$,where M usually is a power of 2. The phase is proportional to the parameter h which is called the *modulation index.*[1] *Phase function* $q(t)$, together with modulation index h and input symbols a_k, determine how the phase changes with time. The derivative of $q(t)$ is function $g(t)$, which is the *frequency shape pulse.* The function $g(t)$ usually has a smooth pulse shape over a finite time interval $0 \le t \le LT$, and is zero outside. When $L \le 1$, we have a full-response pulse shape since the entire pulse is in a symbol time T. When $L > 1$, we have a partial-response pulse shape since only part of the pulse is in a symbol time T.

The modulation index h can be any real number in principle. However, for development of practical maximum likelihood CPM detectors, h should be chosen as a rational number. Rational h makes the number of the phase states finite, thus maximum likelihood detectors using the Viterbi algorithm can be used. The Viterbi algorithm will be discussed when CPM demodulation is addressed.

[1] If the modulation index h varies cyclically from symbol to symbol, we have a modulation scheme called multi-h phase modulation (MHPM). Due to its complexity and importance, we will study MHPM in Chapter 7.

We have stated that the phase $\Phi(t, \mathbf{a})$ is continuous without proof. We will prove its continuousness in Section 6.1.2.

6.1.1 Various Modulation Pulse Shapes

By choosing different pulses $g(t)$ and varying the modulation index h and size of symbol alphabet M, a great variety of CPM schemes can be obtained. Some of the popular pulse shapes are listed in the following [1]. All pulse functions in the list have been normalized such that

$$\int_{-\infty}^{\infty} g(t)dt = 1/2$$

This makes the maximum phase change of the signal to be $(M-1)h\pi$ for the period of $g(t)$.

6.1.1.1 Rectangular (LREC), CPFSK, and MSK

LREC is the rectangular pulse with a length of L symbols. For example, 3REC has $L = 3$. LREC's $g(t)$ is defined by

$$g(t) = \begin{cases} \frac{1}{2LT}, & 0 \leq t \leq LT \\ 0, & \text{otherwise} \end{cases} \tag{6.4}$$

A special case is 1REC, which is most often referred to as CPFSK (continuous phase frequency shift keying). Further, if $M = 2$ and $h = 1/2$, 1REC becomes MSK. Substituting (6.4) with $L = 1$ into (6.3), we have

$$q(t) = \int_{-\infty}^{t} \frac{1}{2T} dt = \begin{cases} \int_0^t \frac{1}{2T} dt, & 0 < t < T \\ \int_0^T \frac{1}{2T} dt, & t > T \end{cases} = \begin{cases} \frac{t}{2T}, & 0 < t < T \\ \frac{1}{2}, & t > T \end{cases}$$

Then substituting this into (6.2) we have in the interval $[kT, (k+1)T]$

$$\begin{aligned} \Phi(t, \mathbf{a}) &= 2\pi h \left[\sum_{i=-\infty}^{k-1} a_i \frac{1}{2} + a_k \frac{1}{2T}(t - kT) \right] \\ &= \pi h \sum_{i=-\infty}^{k-1} a_i + h a_k \frac{\pi t}{T} - \pi h a_k k) \\ &= h a_k \frac{\pi t}{T} + \Phi_k, \quad kT \leq t \leq (k+1)T \end{aligned}$$

where

$$\Phi_k = \pi h \left(\sum_{i=-\infty}^{k-1} a_i - k a_k \right) (\bmod 2\pi)$$

Thus

$$s(t) = A\cos(2\pi f_c t + h a_k \frac{\pi t}{T} + \Phi_k), \quad kT \le t \le (k+1)T$$

This is the expression of the CPFSK signal which we have mentioned in Chapter 5 (see Section 5.1.2). With $h = 0.5$, the above expression becomes

$$s(t) = A\cos(2\pi f_c t + a_k \frac{\pi t}{2T} + \Phi_k), \quad kT \le t \le (k+1)T$$

which is the MSK signal if a_k is binary (see Section 5.1.2).

Another special case is binary 2REC, also called duobinary MSK (DMSK). Thus for DMSK

$$g(t) = \begin{cases} \frac{1}{4T}, & 0 \le t \le 2T \\ 0, & \text{otherwise} \end{cases}$$

6.1.1.2 Raised Cosine (LRC)

LRC is the raised cosine with a length of L symbols. For example, 3RC has $L = 3$. LRC's $g(t)$ is defined by

$$g(t) = \begin{cases} \frac{1}{2LT}\left[1 - \cos\left(\frac{2\pi t}{LT}\right)\right], & 0 \le t \le LT \\ 0, & \text{otherwise} \end{cases} \tag{6.5}$$

6.1.1.3 Spectrally Raised Cosine (LSRC)

LSRC is the spectrally raised cosine with length L. For example, 2SRC has $L = 2$. LSRC's $g(t)$ is defined by

$$g(t) = \frac{1}{LT} \frac{\sin\left(\frac{2\pi t}{LT}\right)}{\frac{2\pi t}{LT}} \frac{\cos\left(\beta \frac{2\pi t}{LT}\right)}{1 - \left(\frac{4\beta t}{LT}\right)^2}, \quad 0 \le \beta \le 1 \tag{6.6}$$

6.1.1.4 Tamed Frequency Modulation (TFM)

TFM is the tamed frequency modulation. TFM's $g(t)$ is defined by

$$\begin{aligned} g(t) &= \frac{1}{8}[ag_o(t-T) + bg_o(t) + ag_o(t+T)], \quad a=1, b=2 \qquad (6.7) \\ g_o(t) &\approx \sin(\frac{\pi t}{T})\left[\frac{1}{\pi t} - \frac{2 - \frac{2\pi t}{T}\cot\left(\frac{\pi t}{T}\right) - \left(\frac{\pi t}{T}\right)^2}{\frac{24\pi t^3}{T^2}}\right] \end{aligned}$$

6.1.1.5 Gaussian MSK (GMSK)

GMSK is the Gaussian minimum shift keying. GMSK's $g(t)$ is defined by

$$\begin{aligned} g(t) &= \frac{1}{2T}\left[Q\left(2\pi B_b\frac{t-\frac{T}{2}}{\sqrt{\ln 2}}\right) - Q\left(2\pi B_b\frac{t+\frac{T}{2}}{\sqrt{\ln 2}}\right)\right], \quad 0 \leq B_bT \leq 1 \\ & \qquad (6.8) \\ Q(t) &= \int_t^{\infty}\frac{1}{\sqrt{2\pi}}\exp(-\frac{\tau^2}{2})d\tau \end{aligned}$$

Due to the importance of GMSK, we will describe it in more detail in Section 6.7.

Figure 6.1 shows the $g(t)$ pulse shapes and $q(t)$ pulse shapes defined in the above list. The time axis is normalized to T. The familiar rectangular pulse is defined on $[0, LT]$ and its phase function is linear, reaching the maximum (0.5) at the end of the period. The raised-cosine pulse is defined in $[0, LT]$, the corresponding phase function $q(t)$ is a nonlinear yet smooth curve in $[0, LT]$. The $q(t)$ reaches its maximum at the end of its period. The tamed frequency modulation pulse is defined in $[-\infty, \infty]$, however its major energy is within $[-2T, +2T]$. Its phase function $q(t)$ changes smoothly in the interval and reaches the maximum at about $2T$. The spectrally raised cosine pulse has similar properties of the TFM. However, the phase function exhibits small oscillation around 0 and 0.5. The GMSK $g(t)$ is also defined in $[-\infty, \infty]$, but the main energy is in $[-T, +T]$ (for $B_bT = 0.25$). The phase function changes smoothly in the same interval, reaching its maximum at T. Since the length of frequency pulse functions of TFM, SRC, and GMSK is infinite, they must be truncated in time-domain implementation.

As will be seen shortly, $g(t)$ and h can be chosen to enable the CPM schemes to outperform MSK in terms of power efficiency and bandwidth efficiency. The reason that this can happen is that memory has been introduced into the CPM signal by means of the continuous phase. Further memory can be built into the CPM signal by choosing a partial response $g(t)$ with $L > 1$.

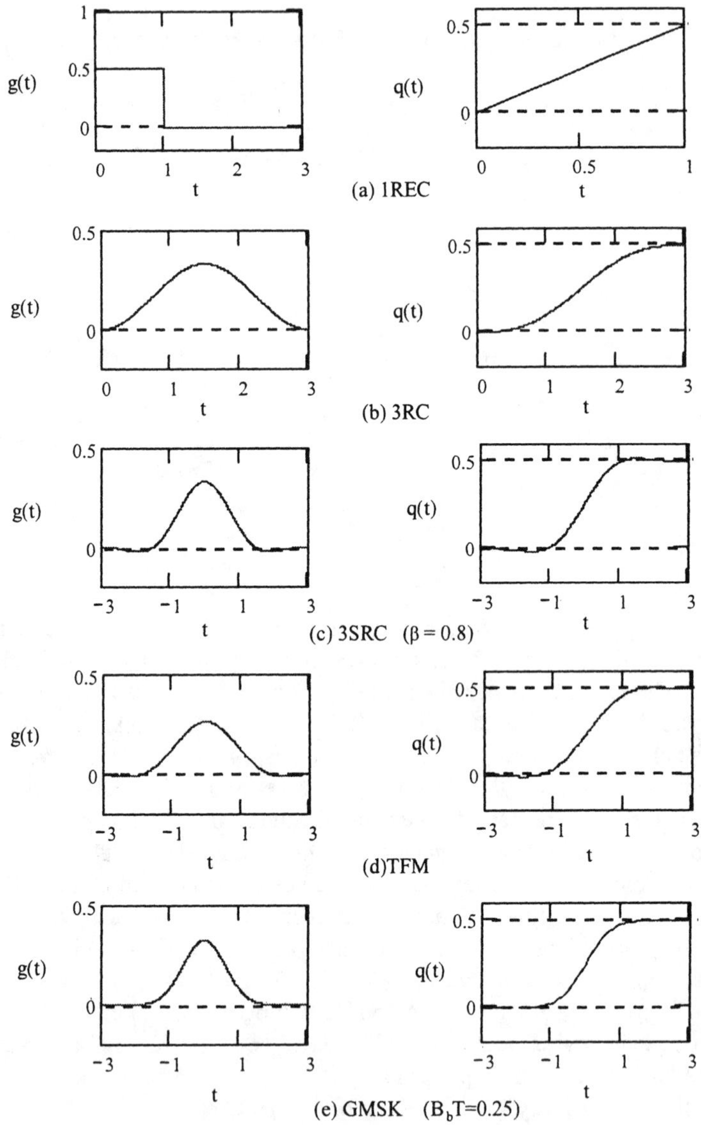

Figure 6.1 Frequency pulse shapes for CPM (a–e).

6.1.2 Phase and State of the CPM Signal

Since the information symbols are imbedded in the phase of the CPM signal, demodulation will be solely based on its phase. It is therefore very important that the phase behavior of CPM signals is fully understood.

First we examine an example shown in Figure 6.2 where phase $\Phi(t, \mathbf{a})$ of binary 3RC with $h = 2/3$ is shown for a particular data sequence. The unit of the phase is radian. We assume that the initial phase is zero, that is $\Phi(0, \mathbf{a}) = 0$. We have plotted the weighted $q(t - kT)$ for the first four data symbols and omitted the rest. The total phase $\Phi(t, \mathbf{a})$ (modulo 2π) for the sequence is shown on the bottom of the figure.

The continuousness of the phase $\Phi(t, \mathbf{a})$ can be easily proved with the help of Figure 6.2. Since $g(t)$ is a smooth function, its integral $q(t)$ is also a smooth function. The phase $\Phi(t, \mathbf{a})$ is a weighted sum of shifted versions of $q(t)$. Along the time axis, as t increases, a new weighted $q(t - kT)$ is added in at every symbol boundary. The value of $\Phi(t, \mathbf{a})$ will not change abruptly since $q(t)$ always starts from 0. Therefore the phase $\Phi(t, \mathbf{a})$ is continuous, even at the symbol boundaries.

From Figure 6.2 we can see that $q(t-kT)$ reaches its maximum at $t = (k+L)T$ and stays at the maximum for the rest of the time. Figure 6.2 is for 3RC which has a finite duration $L = 3$. For pulse shapes with infinite duration such as LSRC, TFM, and GMSK, the $q(t-kT)$ also reaches its maximum at $t = (k+L)T$, approximately (see Figure 6.1(c–e)). These maximum values are accumulated along the time axis. Thus we can separate the excess phase $\Phi(t, \mathbf{a})$ into two parts as follows.

The excess phase of a CPM signal during interval $kT < t < (k+1)T$ can be written as

$$\begin{aligned} \Phi(t, \mathbf{a}) &= 2\pi h \sum_{i=k-L+1}^{k} a_i q(t - iT) + \theta_k \\ &= \theta(t, \mathbf{a}_k) + \theta_k \end{aligned} \tag{6.9}$$

where

$$\theta(t, \mathbf{a}_k) = 2\pi h \sum_{i=k-L+1}^{k} a_i q(t - iT) \tag{6.10}$$

is the *instant phase*, which represents the changing part of the total excess phase in $[kT, (k+1)T]$, and

$$\theta_k = \left[h\pi \sum_{i=-\infty}^{k-L} a_i \right] \pmod{2\pi} \tag{6.11}$$

is the *cumulate phase,* which represents the constant part of the total excess phase in

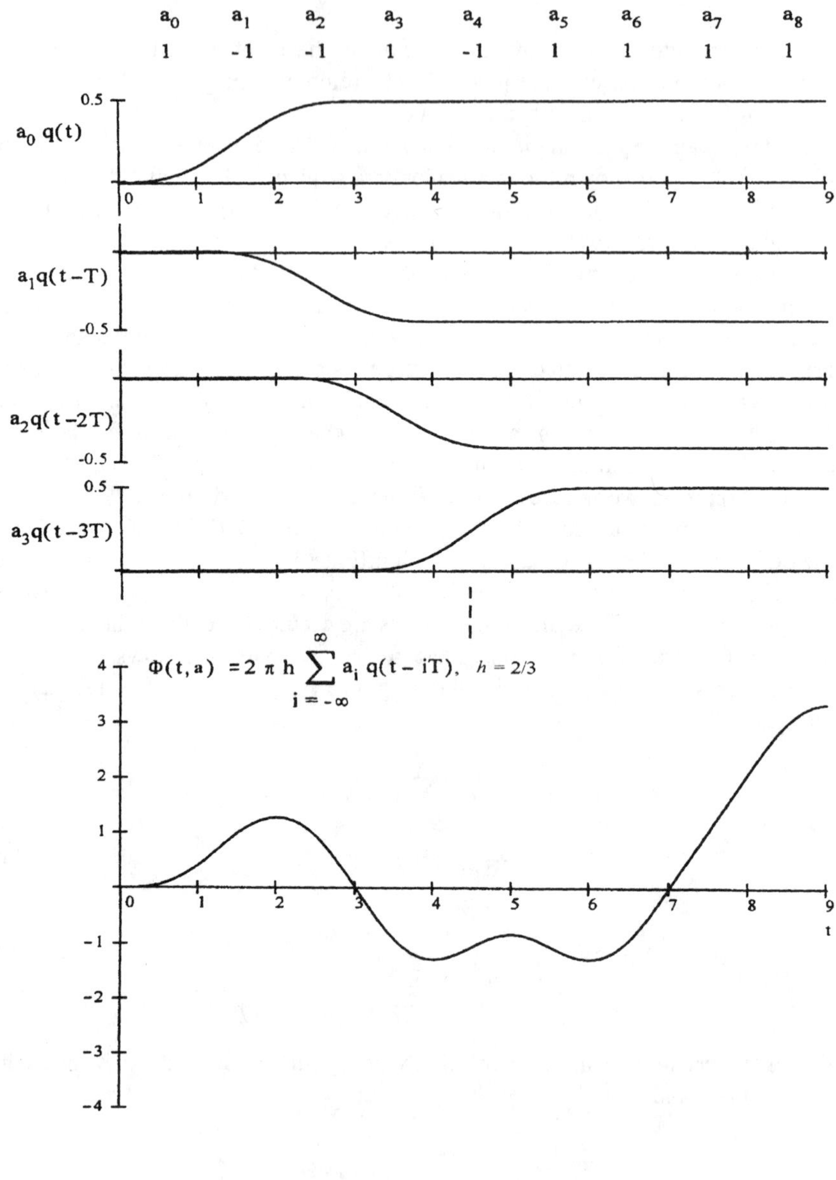

Figure 6.2 Phase $\Phi(t, \mathbf{a})$ of binary 3RC with $h = 2/3$.

$[kT, (k+1)T]$, and is equal to the sum of the maximum phase changes contributed by each symbol, accumulated along the time axis up to the $(k-L)$th symbol interval. It can be conveniently computed recursively as

$$\theta_{k+1} = \theta_k + h\pi a_{k-L+1} \tag{6.12}$$

The instant phase $\theta(t, \mathbf{a}_k)$ is determined by the data symbol a_k and previous $L-1$ symbols. For example, if $L = 1$, then

$$\theta(t, \mathbf{a}_k) = 2\pi h a_k q(t - kT), \quad kT \leq t \leq (k+1)T$$

which is a changing phase proportional to the current data and the phase function, where $q(t)$ is nonzero in $(0, T)$ and zero otherwise. If $L = 2$, then

$$\theta(t, \mathbf{a}_k) = 2\pi h[a_{k-1}q(t - kT + T) + a_k q(t - kT)], \quad kT \leq t \leq (k+1)T$$

where $q(t)$ is nonzero in $(0, 2T)$ and zero otherwise. $\theta(t, \mathbf{a}_k)$ is a changing phase proportional to the weighted sum of the current phase function and the previous phase function. The weighting factors are the current data symbol and the previous data symbol, respectively.

The cumulate phase θ_k is the accumulated phase due to the data up to $t = (k - L)T$, not including the phase accumulated by the carrier, and also not including the phase accumulated in the instant phase due to the previous $L-1$ symbols. Therefore it is in general not the initial phase at time kT (at first glance, it seems to be the initial phase). The initial phase at $t = kT$ is $2\pi f_c kT + \theta_k + \theta(kT, \mathbf{a}_k)$. However, if $L = 1$, then $\theta(kT, \mathbf{a}_k) = 0$, and in addition if f_c is an integer multiple of symbol rate, θ_k would become the initial phase of the kth symbol interval.

If h is rational, that is, $h = 2q/p$ where q and p have no common factors, the number of distinct values of θ_k is p.

Proof: Since $h = 2q/p$,

$$\theta_k = \frac{2q\pi}{p} \sum_{i=-\infty}^{k-L} a_i = \text{multiple of } \frac{2\pi}{p}$$

Thus the number of states is

$$2\pi/(2\pi/p) = p$$

For example, $h = 1/2 = 2/4$ for MSK, its number of phase states is four (see Figure 5.3). If h is a real number, the number of distinct values of θ_k is infinite.

We define a *state* of a CPM signal at $t = kT$ as the vector

$$\mathbf{s}_k = (\theta_k, a_{k-1}, a_{k-2}, ..., a_{k-L+1})$$

which consists of the value of the cumulate phase θ_k and the previous $L-1$ symbols. For rational h, since the number of distinct values of θ_k is p, the number of states is at most pM^{L-1}. However, for real h, the number of states is infinite. Each state corresponds to a specific function form of the excess phase $\Phi(t, \mathbf{a})$.

Using $3RC$ with $h = 2/3$ and binary symbols ($M = 2$) as an example, we have

$$\mathbf{s}_k = (\theta_k, a_{k-1}, a_{k-2}) \tag{6.13}$$

where

$$\theta_k = \left[\frac{2}{3}\pi \sum_{i=-\infty}^{k-L} a_i\right] (\text{mod } 2\pi) = \begin{cases} 0 \\ \frac{2}{3}\pi \\ \frac{4}{3}\pi \end{cases} \tag{6.14}$$

The number of θ_k is 3. Thus the binary $3RC$ has a total of 12 states. The recursive expression for θ_k is

$$\theta_{k+1} = \theta_k + \frac{2}{3}\pi a_{k-2} \tag{6.15}$$

The information bearing phase $\Phi(t, \mathbf{a})$ for the binary $3RC$ is

$$\begin{aligned} \Phi(t, \mathbf{a}) &= \frac{4\pi}{3}[a_{k-2}q(t - kT + 2T) + a_{k-1}q(t - kT + T) \\ &\quad + a_k q(t - kT)] + \theta_k, \\ kT &\le t \le (k+1) \end{aligned} \tag{6.16}$$

where

$$q(t) = \int_0^t \frac{1}{6T}\left[1 - \cos\left(\frac{2\pi t}{3T}\right)\right] dt$$

which is nonzero in $(0, 3T)$ and zero otherwise. At $t = kT$,

$$\Phi(kT, \mathbf{a}) = \frac{4\pi}{3}[a_{k-2}q(2T) + a_{k-1}q(T) + a_k q(0)] + \theta_k$$

where $q(0) = 0$, $q(T) = 0.098$, and $q(2T) = 0.402$. Thus

$$\Phi(kT, \mathbf{a}) = \frac{4\pi}{3}[0.402a_{k-2} + 0.098a_{k-1}] + \theta_k \tag{6.17}$$

Table 6.1 lists the 12 values of $\Phi(kT, \mathbf{a})$ corresponding to the 12 states. The fourth column contains the direct results from (6.17). The last column contains the results converted into $[0, 2\pi]$. These phase values can also be converted into $[-2\pi, 0]$, with the same array of absolute values. From the last column we can see that in fact there are only nine distinct values. Some of the phase values are produced by more

State	θ_k	a_{k-1}	a_{k-2}	$\Phi(kT, \mathbf{a})$ direct results from (6.17)	$\Phi(kT, \mathbf{a})$ converted into $[0, 2\pi]$
1	0	-1	-1	$-\frac{2}{3}\pi$	$\frac{4}{3}\pi$
2	0	-1	$+1$	$\frac{1.22}{3}\pi$	$\frac{1.22}{3}\pi$
3	0	$+1$	-1	$-\frac{1.22}{3}\pi$	$\frac{4.78}{3}\pi$
4	0	$+1$	$+1$	$\frac{2}{3}\pi$	$\frac{2}{3}\pi$
5	$\frac{2}{3}\pi$	-1	-1	0	0
6	$\frac{2}{3}\pi$	-1	$+1$	$\frac{3.22}{3}\pi$	$\frac{3.22}{3}\pi$
7	$\frac{2}{3}\pi$	$+1$	-1	$\frac{0.78}{3}\pi$	$\frac{0.78}{3}\pi$
8	$\frac{2}{3}\pi$	$+1$	$+1$	$\frac{4}{3}\pi$	$\frac{4}{3}\pi$
9	$\frac{4}{3}\pi$	-1	-1	$\frac{2}{3}\pi$	$\frac{2}{3}\pi$
10	$\frac{4}{3}\pi$	-1	$+1$	$\frac{5.22}{3}\pi$	$\frac{5.22}{3}\pi$
11	$\frac{4}{3}\pi$	$+1$	-1	$\frac{2.78}{3}\pi$	$\frac{2.78}{3}\pi$
12	$\frac{4}{3}\pi$	$+1$	$+1$	2π	2π

Table 6.1 3RC states and phase values.

than one state. $\frac{2}{3}\pi$ and $\frac{4}{3}\pi$ are repeated once, and 0 and 2π are the same. In this sense, these states are equivalent. That is, states 1 and 8, 4 and 9, and 5 and 12 are equivalent.

6.1.3 Phase Tree and Trellis and State Trellis

For an arbitrary data sequence, the phases of the CPM signals will follow a unique continuous phase trajectory (or path). The collection of all possible phase paths forms a phase tree. When h is rational the number of phase states is finite and the phase tree can be collapsed into a trellis.

In Chapter 5 we have seen the phase tree and trellis of MSK, which is a 1REC with $M = 2$ and $h = 1/2$ (see Figures 5.2 and 5.3). Here we examine an additional example with nonlinear yet smooth phase function $q(t)$: binary 3RC with $h = 2/3$. Its phase tree is drawn in Figure 6.3 using (6.16). We assume that the initial phase is zero, that is, $\Phi(0, \mathbf{a}) = 0$. We also assume that the two symbols before $t = 0$ are $+1$. The ±1 on a branch is the data symbol that, together with the previous two symbols and the phase θ_k, produces the phase trajectory (see (6.16)). It can be easily examined that all the phase values at $t = kT$ in the phase tree are included in Table 6.1.

As time increases, the tree will grow bigger. However, by using a modulo-2π operation, this phase tree can be collapsed into a trellis. Figure 6.4 shows the trellis. The trellis is drawn by using a modulo 2π operation first, then converting those phases in $[0, -2\pi]$ into $[0, 2\pi]$ by simply adding 2π to the values. The trellis becomes

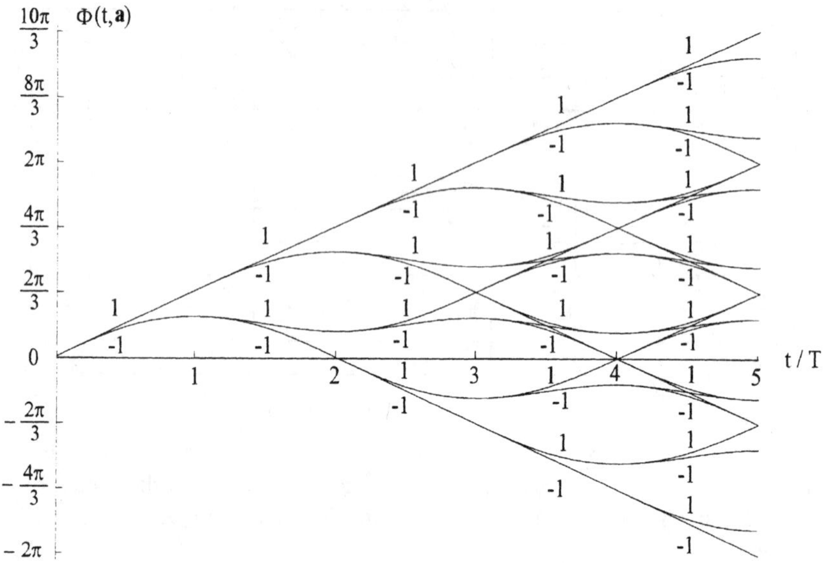

Figure 6.3 Phase tree of binary $3RC$ with $h = 2/3$. The phase starts with value zero at $t = 0$. After: [9].

fully developed only after $t = 4T$. This is because it takes $2T$ to have all possible θ_k (see (6.14)) and another $2T$ to have all possible instant phase values (see (6.16)). Now we examine the phase values at $t = 4T$ or $5T$. From the top to the bottom, those values correspond to states 12 (or 5), 10, 3, 1 (or 8), 6, 11, 4 (or 9), 2, 7, and 5 (or 12) in Table 6.1. At each node, there are two branches coming in and two branches going out except for those double-state nodes where four branches are coming in and going out. Each branch represents an input symbol (+1 or −1). Each branch is the phase trajectory from the previous node to the current node.

The phase tree and the phase trellis are very useful for understanding the phase behavior of the CPM signals. However, for demodulating CPM signals, a *state trellis* instead of phase trellis is more convenient. Figure 6.5 shows the state trellis of the 3RC example. The trellis is drawn using (6.13), (6.14), and (6.15). In the figure,

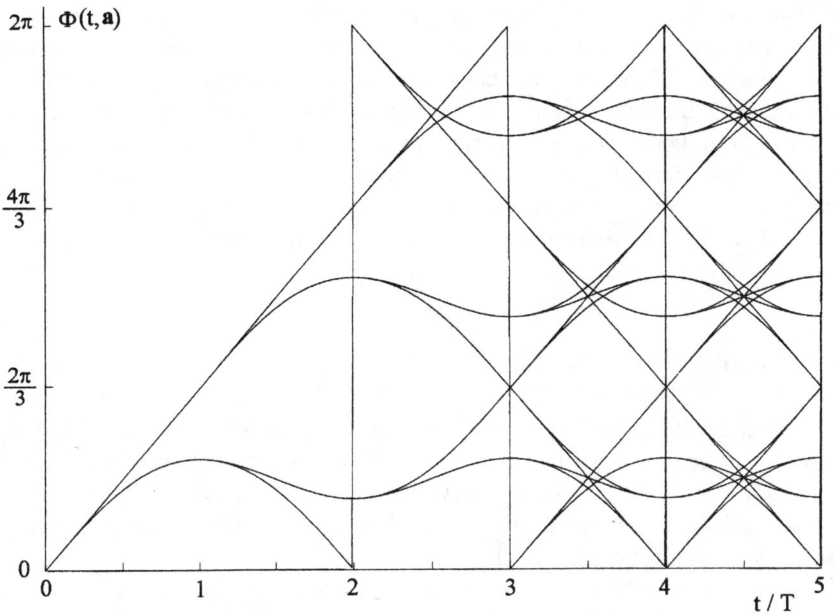

Figure 6.4 Phase trellis of binary 3RC with $h = 2/3$. The phase starts with value zero at $t = 0$.

each state is represented by a node, even for those states that produce the same phase values. In this way there are always two branches coming in and going out at a state. In general, there are always M branches in and out at a state for M-ary symbols. In addition, in the state trellis, the lines connecting states are not trajectories of the signal phases. They are there simply to show the transitions from one state to the other. Thus a state trellis is almost the same as a phase trellis except that the transition branches are not phase trajectories and the nodes are labeled by states instead of phases. However, we know that states can be mapped into phases even though the mapping may not be completely one-to-one (see Figures 6.4 and 6.5).

For full-response CPM, the state vector is just θ_k, thus the nodes of the state trellis are identical to the nodes of the phase trellis, but the branches are still different. For linear full-response CPM (CPFSK), even the branches of the phase trellis are straight lines, thus the appearance of the state trellis and the phase trellis are completely the same.

Another interesting fact is that the frequency pulse shape $g(t)$ does not affect the structure of the trellis as long as the convention $\int_{-\infty}^{\infty} g(t)dt = 1/2$ (or other constant) is observed. This is because the states are defined by $(\theta_k, a_{k-1}, a_{k-2}, ..., a_{k-L+1})$, where data have nothing to do with $g(t)$, and θ_k also is independent of $g(t)$ provided $\int_{-\infty}^{\infty} g(t)dt = 1/2$ (or other constant). This fact makes a trellis applicable to many schemes with different frequency pulse shapes as long as their L and h are the same.

If h is irrational, $\Phi(t, \mathbf{a})$ will have an infinite number of possible values, depending on the current symbol and all previous symbols. The phase tree still exists, but the number of branches at each node for a specific input symbol will no longer be M. Rather it grows exponentially with the depth of the tree (M^N, N is the depth). This tree cannot be collapsed into a trellis.

6.2 POWER SPECTRAL DENSITY

The methods of calculating power spectral densities of CPM signals, or other digitally modulated signals, for that matter, basically fall into three classes: direct method [10], Markov method [11], and correlation method [12]. Computer simulation can also be used. Finally, measuring is always a method of finding CPM spectra, and is also the ultimate verification of the calculated spectra.

In the direct method one takes the Fourier transform $S_N(f, \mathbf{a}, \phi_o)$ of a truncated deterministic CPM signal $s_N(t) = s_N(t, \mathbf{a}, \phi_o)$ and then forms the average over data $\mathbf{a}$ and initial phase ϕ_o which is distributed uniformly over $(0, 2\pi)$. That is

$$PSD = \lim_{N\to\infty} \frac{1}{NT} E\{|S_N(f, \mathbf{a}, \phi_o)|^2\}$$

where N is an integer. We have used this method to find out the PSD formulas of the baseband modulations in Appendix A. The results turned out to be quite simple. For CPM, however, the resultant equations are often complicated and two-dimensional numerical integrations are required.

In the Markov method, the random data are modeled as a Markov process characterized by a transition matrix. Then the autocorrelation function can be expressed in terms of this transition matrix and the matrix of the correlation between the basic baseband pulses. The Fourier transform of the autocorrelation function is the power spectral density of the modulated signal. We have used this method in Chapter 2 for finding the PSDs of Bi-Φ-M and the delay modulation line codes.

In the correlation method, the correlation function of a CPM signal is calculated first, then one takes the Fourier transform of the correlation function to get the PSD. The correlation is formed by first taking the average over data $\mathbf{a}$ of the product of

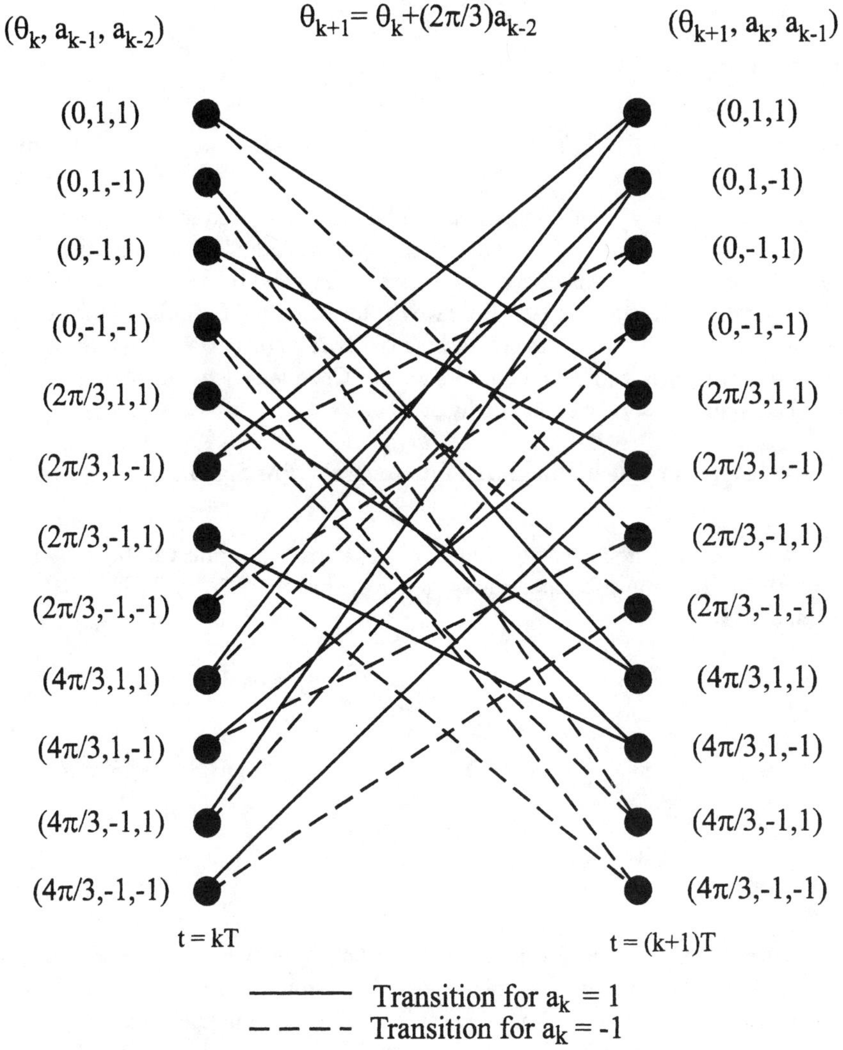

Figure 6.5 State trellis of binary 3RC with $h = 2/3$. After: [9].

the signal's complex envelopes $\widetilde{s}(t+\tau,\mathbf{a})$ and $\widetilde{s}(t,\mathbf{a})$:

$$R_{\widetilde{T}}(t+\tau,t) = E\{\widetilde{s}(t+\tau,\mathbf{a})\widetilde{s}(t,\mathbf{a})\} \tag{6.18}$$

and then taking the time average over a period of T:

$$R_{\widetilde{T}}(\tau) = \frac{1}{T}\int_0^T R_{\widetilde{T}}(t+\tau,t)dt \tag{6.19}$$

A fast, relatively simple, numerical method first appeared in [13]. It also can be found in [9, 14]. This method is a type of correlation method. The derivation of the method is presented in Appendix A together with methods of computing PSDs of other digitally modulated signals. Here we list the steps and formulas of the numerical calculation resulting from the two steps in (6.18) and (6.19). All PSD expressions are for CPM signals with a unit amplitude. For CPM signals with an amplitude A, their PSDs just need to be scaled up by A^2.

6.2.1 Steps for Calculating PSDs for General CPM Signals

1. Calculate the autocorrelation function $R_{\widetilde{s}}(\tau)$ over the interval $[0,(L+1)T]$. Note that the time difference τ is written in the form $\tau=\xi+mT$ with $0\leq\xi<T$ and $m=0,1,2,...$

$$\begin{aligned} R_{\widetilde{s}}(\tau) &= R_{\widetilde{s}}(\xi+mT) \\ &= \frac{1}{T}\int_0^T \prod_{k=1-L}^{m+1} \left\{ \sum_{\substack{n=-(M-1)\\ n \text{ odd}}}^{M-1} P_n \exp\{j2\pi hn[q(t+\xi-(k-m)T)\right. \\ &\quad -q(t-kT)]\}dt\} \end{aligned} \tag{6.20}$$

where P_n is the a priori probability of the nth symbol. The weighted (by P_n) sum is the result of averaging over M symbols. The product represents correlating over $L+m$ symbol periods. Finally, the integration is the result of averaging over the time for a symbol period.

2. Calculate

$$C_a = \sum_{\substack{n=-(M-1)\\ n \text{ odd}}}^{M-1} P_n \exp\left[j2\pi hnq(LT)\right] \tag{6.21}$$

The coefficient C_a will be used in the next step of computing. If h is not an integer, then $|C_a| < 1$, the PSD is purely continuous. If h is an integer, then $|C_a| = 1$, the PSD contains a continuous part as well as a discrete part (spectral lines).

3. IF $|C_a| < 1$, which is the common case when h is not an integer, usually rational, calculate the PSD using

$$\begin{aligned}\Psi_{\tilde{s}}(f) &= 2\,\mathrm{Re}\left\{\int_0^{LT} R_{\tilde{s}}(\tau)\, e^{-j2\pi f\tau} d\tau + \right. \\ &\left. \frac{e^{-j2\pi fLT}}{1 - C_a e^{-j2\pi fT}} \int_0^T R_{\tilde{s}}(\tau + LT)\, e^{-j2\pi f\tau} d\tau\right\} \qquad (6.22)\end{aligned}$$

The PSD is continuous. A special case is when the data symbols are equally likely, which is also the most common case, then $P_n = 1/M$, for $n = \pm1, \pm3, ..., \pm(M-1)$. For this case, the sum of the exponential functions in (6.20) becomes real-valued due to symmetry of the data symbols, we have

$$R_{\tilde{s}}(\tau) = \frac{1}{T}\int_0^T \prod_{k=1-L}^{\lceil \tau/T \rceil} \frac{1}{M}\frac{\sin 2\pi hM[q(t+\tau-kT) - q(t-kT)]}{\sin 2\pi h[q(t+\tau-kT) - q(t-kT)]} dt \qquad (6.23)$$

and the PSD is

$$\begin{aligned}\Psi_{\tilde{s}}(f) &= 2\int_0^{LT} R_{\tilde{s}}(\tau)\cos 2\pi f\tau d\tau + \frac{2}{1 + C_a^2 - 2C_a \cos 2\pi fT} \\ &\cdot \left\{(1 - C_a\cos 2\pi fT)\int_{LT}^{(L+1)T} R_{\tilde{s}}(\tau)\cos 2\pi f\tau d\tau \right. \\ &\left. -C_a \sin 2\pi fT \int_{LT}^{(L+1)T} R_{\tilde{s}}(\tau)\sin 2\pi f\tau d\tau\right\} \qquad (6.24)\end{aligned}$$

and (6.21) becomes

$$C_a = \frac{1}{M}\frac{\sin M\pi h}{\sin \pi h} \qquad (6.25)$$

4. IF $|C_a| = 1$, which is the rare case when h is an integer, the autocorrelation function contains an aperiodic part $R_{con}(\tau)$ and a periodical part $R_{dis}(\tau)$.

$$R_{\tilde{s}}(\tau) = R_{con}(\tau) + R_{dis}(\tau)$$

As a result, the PSD contains a continuous part and a discrete part.

$$\Psi_{\tilde{s}}(f) = 2\,\mathrm{Re}\left[\int_0^{LT} R_{con}(\tau)\, e^{-j2\pi f\tau} d\tau\right] + F_{dis}(f) \qquad (6.26)$$

where $F_{dis}(f)$ is the discrete PSD which is the Fourier series coefficient of $R_{dis}(\tau)$. It is shown in Appendix A that when h is even the period of $R_{dis}(\tau)$ is T, and the discrete frequency components appear at $f = \pm k/T$, $k = 0, 1, 2, \ldots$. When h is odd, $R_{dis}(\tau)$ is a periodic, *odd half-wave symmetrical* function with a period of $2T$. Its spectrum would only have odd harmonics at $f = \pm(2k+1)/2T$. $k = 0, 1, 2, \ldots$ [15, p. 103]. The property that a CPM signal with integer index has discrete frequency components can be used to recover the carrier and symbol timing in CPM receivers.

6.2.2 Effects of Pulse Shape, Modulation Index, and A Priori Distribution

Many numerical results are given in [1, 7–9]. The results include the PSDs for CPM schemes with (1) different frequency pulses ($g(t)$), (2) different modulation indexes (h), and (3) different a priori distributions (P_n). The results are often shown in comparison with the PSD of MSK since MSK is a bandwidth-efficient modulation scheme that appeared earlier than general CPM schemes. Here we quote some of the important results.

First we want to show the effect of the shape of $g(t)$. Figure 6.6 shows PSDs of some binary CPM schemes with different $g(t)$, with a fixed index $h = 1/2$, and a uniform a priori distribution. GMSK4 and 3SRC6 mean that the $g(t)$ is truncated symmetrically to a length of 4 and 6 symbols, respectively. All four CPM schemes have better PSDs than MSK in that their spectra fall faster with frequency. The PSD of 3RC, GMSK4 with $B_bT = 0.25$, and 3SRC6 with $\beta = 0.8$ are very similar. This is no surprise since their $g(t)s$ are similar (see Figure 6.1). Figure 6.7 shows PSDs for some quaternary CPM schemes for a fixed index $h = 1/3$. It is clear that 3RC is better than 2RC, which in turn is better than 1REC and MSK. The spectrum of TFM is not shown, however, it is reported that TFM has a PSD similar to binary 3.7RC or 3.7SRC or GMSK with $B_bT = 0.2$ [1]. The conclusion we can draw from Figures 6.6 and 6.7 is that the effect of pulse shape on PSDs is significant, a longer and smoother pulse $g(t)$ yields narrower power spectra for fixed h and M.

Next, the effect of modulation index difference is demonstrated in Figure 6.8 using binary 4RC as an example. It is seen that lower h values yield lower spectral side-lobes as we would expect, since h controls the frequency deviation from the carrier frequency. The effect of h on PSD is significant. However, as we will see shortly, h also affects bit error probability. So the choice of h is not simply based on

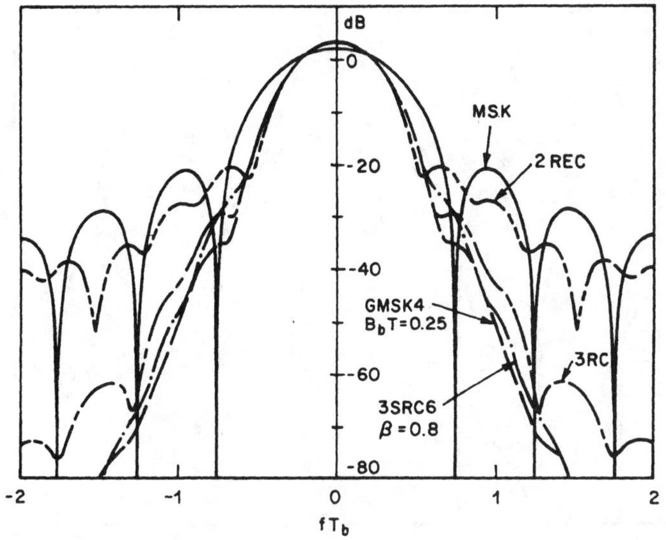

Figure 6.6 PSDs for some binary CPM schemes with $h = 1/2$, in comparison with MSK. From [1]. Copyright © 1986 IEEE.

PSD properties.

Finally, the effect of different a priori distributions (P_n) can be seen from Figure 6.9 where B_1, B_2, and B_3 are three a priori distributions (PDFs) defined as follows

$$(P_{-1}, P_{+1}) = \begin{cases} (\frac{1}{2}, \frac{1}{2}), & B_1 \\ (\frac{1}{4}, \frac{3}{4}), & B_2 \\ (\frac{1}{10}, \frac{9}{10}), & B_3 \end{cases}$$

From the figure we can see that the PSD of the symmetrical PDF is also symmetrical about $fT = 0$. The PSDs of the asymmetrical PDFs are also asymmetrical. However, all of them are very similar even for the very skewed distribution B_3. Thus the conclusion is that the a priori distributions (P_n) do not affect the PDFs significantly.

6.2.3 PSD of CPFSK

For M-ary 1REC (CPFSK), the PSD has a closed form expression [9, 10]. The ex-

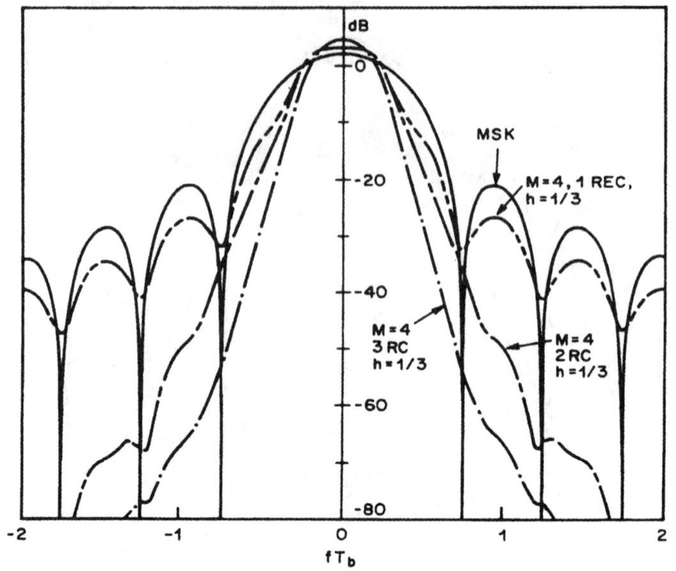

Figure 6.7 PSDs for some quaternary CPM schemes with $h = 1/3$, in comparison with MSK. From [1]. Copyright © 1986 IEEE.

pression is

$$\Psi_{\tilde{s}}(f) = \frac{A^2T}{M}\sum_{i=1}^{M}\left[\frac{1}{2}\frac{\sin^2\gamma_i}{\gamma_i^2} + \frac{1}{M}\sum_{j=1}^{M}A_{ij}\frac{\sin\gamma_i}{\gamma_i}\frac{\sin\gamma_j}{\gamma_j}\right] \tag{6.27}$$

where A is the signal amplitude. Other parameters are defined as

$$\gamma_i = (fT - (2i - M - 1)\frac{h}{2})\pi, \qquad i = 1, 2, ..., M$$

$$A_{ij} = \frac{\cos(\gamma_i + \gamma_j) - C_a\cos(\gamma_i + \gamma_j - 2\pi fT)}{1 + C_a^2 - 2C_a\cos 2\pi fT}$$

$$C_a = \frac{1}{M}\frac{\sin M\pi h}{\sin \pi h}$$

This group of expressions has been presented in Chapter 3 in the context of M-ary

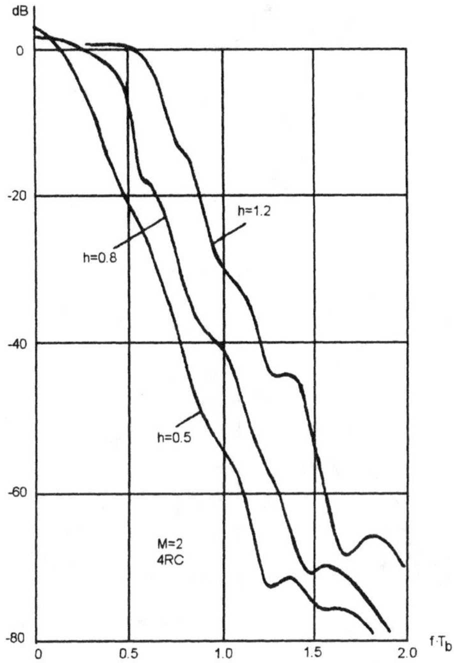

Figure 6.8 PSDs for binary 4RC for different h values. From [8]. Copyright © 1981 IEEE.

FSK.[2] PSD plots have also been given in Chapter 3 for $M = 2, 4$, and 8.

6.3 MLSD FOR CPM AND ERROR PROBABILITY

Unlike classic FSK and PSK modulation schemes, CPM signals have memory. The signal in one symbol duration is determined by the current input symbol and the state. The state vector consists of the previous $L - 1$ input symbols, and the phase θ_k, which is determined by all previous input symbols. Even when $g(t)$ has a finite length L, the length of phase function $q(t)$ is still infinite (see Figure 6.2). This is to say, CPM signals have an infinite long memory. It is for this reason that CPM signals can achieve better error performance than that of symbol-by-symbol modulation schemes. Since the memory length is infinite, for optimum detection, a receiver

[2] C_a has a different form in Chapter 3, but it can be reduced to the simpler form here.

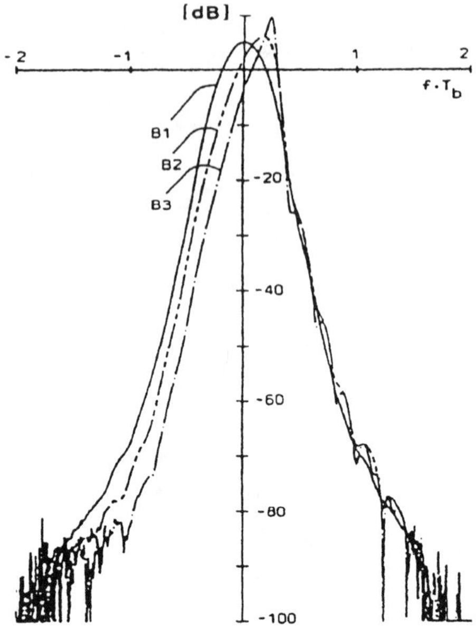

Figure 6.9 PSDs for binary 6RC with $h = 1/2$. PDFs are B_1, B_2, and B_3. From [9]. Copyright 1986 Plenum.

must observe a transmitted waveform with infinite length and chooses the infinitely long sequence $\{a_k\}$ which minimizes the error probability. This can be achieved by the maximum likelihood sequence detection (MLSD). Of course this type of receiver does not exist in practice. The practical receivers can only observe a waveform with a finite length (N). As a result, the receiver is suboptimum. When $N \to \infty$, the receiver becomes optimum.

In this section we will analyze the error performance of the MLSD for CPM signals. The practical demodulator may, or may not use MLSD, depending on the trade-off between transmission power and system complexity. But the error performance of the MLSD can always serve as a bench mark. We will discuss various types of CPM receivers, optimum or nonoptimum, coherent or noncoherent, in Section 6.5. We will see that the error performance of the MLSD receiver and other receivers is determined by the Euclidean distances, especially the minimum distance, between signals. Therefore we will study the Euclidean distance of CPM signals in this section in detail.

6.3.1 Error Probability and Euclidean Distance

Assuming the CPM receiver observes a sequence of N symbols. That is, the received signal is

$$r(t) = s_i(t) + n(t), \qquad 0 \le t \le NT, \qquad i = 1, 2, ..., M^N$$

where $s_i(t)$ is the signal determined by the ith data sequence $\mathbf{a}_i$. For M-ary symbols there are a total of M^N different sequences. Therefore there are a total of M^N different $s_i(t)$. Viewing each $s_i(t)$ as a member of the M^N-ary signals, the results of detection of M-ary signals in Appendix B can be applied (M-ary becomes M^N-ary here). Since all $s_i(t)$ have the same energy and are equally likely, from Appendix B we know that the optimum detection is the MLSD which also achieves minimum sequence detection error probability. The MLSD receiver correlates the received signal with all possible signals

$$l_i = \int_0^{NT} r(t)s_i(t)dt, \qquad i = 1, 2, ..., M^N$$

and chooses the signal sequence which maximizes l_i.

The error probability that $s_i(t)$ is detected as any other sequence is bounded by the union bound (Appendix B, (B.43))

$$P_e(i) \le \sum_{j \ne i} Q\left(\frac{D_{ij}}{\sqrt{2N_o}}\right) \tag{6.28}$$

where D_{ij} denotes the Euclidean distance between the two signals, which is defined as

$$D_{ij} = \left[\int_0^{NT} [s_i(t) - s_j(t)]^2 dt\right]^{\frac{1}{2}} \tag{6.29}$$

The squared distance is

$$\begin{aligned} D_{ij}^2 &= \int_0^{NT} [s_i(t) - s_j(t)]^2 dt \\ &= \int_0^{NT} [s_i^2(t) - 2s_i(t)s_j(t) + s_j^2(t)]dt \end{aligned}$$

$$
\begin{aligned}
&= 2NE_s - \int_0^{NT} 2s_i(t)s_j(t)dt \\
&= 2NE_s - \int_0^{NT} 2A^2 \cos[2\pi f_c t + \Phi(t, \mathbf{a}_i)] \cos[2\pi f_c t + \Phi(t, \mathbf{a}_j)]dt \\
&= 2NE_s - \int_0^{NT} \frac{2E_s}{T}\{\cos[\Phi(t, \mathbf{a}_i) - \Phi(t, \mathbf{a}_j)] \\
&\quad + \cos[4\pi f_c t + \Phi(t, \mathbf{a}_i) + \Phi(t, \mathbf{a}_j)]\}dt \\
&= 2NE_s - \int_0^{NT} \frac{2E_s}{T} \cos[\Phi(t, \mathbf{a}_i) - \Phi(t, \mathbf{a}_j)]dt + o(1/f_c)
\end{aligned}
$$

where E_s is the symbol energy, and $o(1/f_c)$ denotes a term of order $1/f_c$, which means that the term is proportional to a factor $1/f_c$. This term is negligible when $f_c \gg 1$, which is the usual case in CPM. Thus the distance in the limit of high frequency is

$$
D_{ij}^2 = 2NE_s - \int_0^{NT} \frac{2E_s}{T} \cos \Delta\Phi(t, \gamma_{ij})dt \tag{6.30}
$$

where

$$
\begin{aligned}
\Delta\Phi(t, \gamma_{ij}) &= \Phi(t, \mathbf{a}_i) - \Phi(t, \mathbf{a}_j) \\
&= 2\pi h \sum_{k=-\infty}^{\infty} (a_k^i - a_k^j) q(t - kT) \\
&= 2\pi h \sum_{k=-\infty}^{\infty} \gamma_k q(t - kT)
\end{aligned} \tag{6.31}
$$

The superscript denotes the ith or jth data sequence and γ_{ij} is the *difference data sequence* between $\mathbf{a}_i$ and $\mathbf{a}_j$.

Expression (6.30) is still not the most convenient one. The energy E_s is the symbol energy. In comparison modulation schemes, we must compare the error performance based on the same bit energy E_b. Since $E_s = (\log_2 M)E_b$, we normalize the distance to obtain

$$
\begin{aligned}
d_{ij}^2 &= D_{ij}^2/2E_b \\
&= N(\log_2 M) - \int_0^{NT} \frac{(\log_2 M)}{T} \cos \Delta\Phi(t, \gamma_{ij})dt \\
&= \frac{\log_2 M}{T} \int_0^{NT} [1 - \cos \Delta\Phi(t, \gamma_{ij})]dt
\end{aligned}
$$

$$= \frac{\log_2 M}{T} \sum_{k=0}^{N-1} \int_{kT}^{(k+1)T} [1 - \cos \Delta\Phi(t, \gamma_{ij})] dt \qquad (6.32)$$

This shows that d_{ij}^2 is a nondecreasing function of the observation length N for a fixed pair of phase paths.

Now let us return to (6.28). Using an upper bound of the Q function [9, p. 21]

$$Q(x) < \frac{1}{2} e^{-x^2/2}, \qquad x > 0 \qquad (6.33)$$

we can write

$$P_e(i) \leq \frac{1}{2} \sum_{j \neq i} \exp\{-d_{ij}^2 (E_b/2N_o)\} \qquad (6.34)$$

Each exponent contains the signal-to-noise ratio (SNR) E_b/N_o. For high SNR, the term with the smallest d_{ij}^2 will strongly dominate (6.34). Therefore at high SNR, we can approximate the bound as

$$P_e(i) \lesssim \frac{K_i}{2} \exp\{-\min_j [d_{ij}^2](E_b/2N_o)\} \qquad (6.35)$$

where K_i is the number of signals that have the minimum distance with $s_i(t)$. The total error probability can be bounded by

$$P_e \lesssim \frac{K}{2} \exp\{-\min_{\substack{i,j \\ i \neq j}} [d_{ij}^2](E_b/2N_o)\} \qquad (6.36)$$

where K is the number of signals that attain the minimum distances in all cases. A definition of minimum distance that applies to any signal set can be written as[3]

$$d_{\min}^2 \triangleq \frac{1}{2E_b} \min_{\substack{i,j \\ i \neq j}} [\int_0^{NT} [s_i(t) - s_j(t)]^2 dt] \qquad (6.37)$$

For CPM signals it becomes

$$d_{\min}^2 = \frac{1}{2E_b} \min_{\substack{i,j \\ i \neq j}} [\frac{\log_2 M}{T} \int_0^{NT} [1 - \cos \Delta\Phi(t, \gamma_{ij})] dt] \qquad (6.38)$$

---

[3] When the observation length N goes to infinity, the minimum distance is called *free distance*. That is

$$d_{free} = \lim_{N \to \infty} d_{\min}$$

Using d_{min} notation (6.36) becomes

$$P_e \lesssim \frac{K}{2} \exp\{-d^2_{\text{min}}(E_b/2N_o)\} \tag{6.39}$$

or using (6.33) we have

$$P_e \lesssim KQ\left(d_{\text{min}}\sqrt{E_b/N_o}\right) \tag{6.40}$$

Note P_e is the error probability of sequence detection, not the symbol or bit error probability. However, (6.39) and (6.40) are approximately equal to the symbol or bit error probability, since at high SNR a sequence error is most likely caused by one symbol error or one bit error. Even if a sequence error incurs multiple symbol or bit errors, the result is just an increase of the value of the factor K. This may cause a fraction of a decibel increase in the error probability graph.

Since the error performance is primarily determined by the minimum distance d_{min}, the error performance assessment of various CPM schemes can be replaced by assessment of d_{min} or d^2_{min}. In other words, d^2_{min} will become the indicator for error performance evaluation and comparison.

To calculate d^2_{min} for an observation length of N symbols, all pairs of phase paths in the phase tree (or trellis, for rational h) over N symbol intervals must be considered. The phase paths must not coincide over the first symbol interval. The Euclidean distance is calculated according to (6.32) for all these pairs, and the minimum is the d^2_{min} over the observation of N symbols. The phrase "over the observation of N symbols" implies that d^2_{min} varies with N. Of course d^2_{min} varies with h and $g(t)$ too since $\Delta\Phi(t, \gamma_{ij})$ in (6.32) is a function of h and $g(t)$. But for the discussion below, we are only interested in its dependance on N and denote it as $d^2_{\text{min}}(N)$.

A tight upper bound d^2_B on $d^2_{\text{min}}(N)$ is an important indicator of the error performance of the scheme since it gives the maximum achievable performance. To construct an upper bound on $d^2_{\text{min}}(N)$ we can choose any pair of phase paths of length N. The distance of this pair must equal or exceed $d^2_{\text{min}}(N)$. Since d^2_{min} is a nondecreasing function of N for a fixed pair of phase paths, $d^2_{\text{min}}(\infty)$ must equal or exceed $d^2_{\text{min}}(N)$ for any finite N. Thus we can choose any pair of phase paths of an infinite length to find an upper bound on $d^2_{\text{min}}(N)$ for any N. Good candidates for a tight bound are infinitely long pairs of phase paths that merge as soon as possible. Some merges occur only for some specific h values. Others occur independent of h, which are called *inevitable*. The first inevitable merge is usually used for upper-bound calculation. The first inevitable merge occurs in general at $t = (L+1)T$. The distances for all pairs of phase paths which give a first inevitable merger in the phase tree (for real h) or trellis (for rational h only) are calculated and compared.

The minimum of them is the upper bound d_B^2. Sometimes the second, third or even deeper inevitable mergers may provide even tighter upper bound. For details about upper-bound calculation, refer to [7,8].

6.3.2 Comparison of Minimum Distances

Efficient algorithms using the phase tree or trellis exist for computing $d_{\min}^2$ and its upper bound d_B^2 for different $g(t)$, L, h, and M [9,16]. We will not attempt to describe the algorithms here since they are quite complicated. However, for the simple case of CPFSK (i.e.,1REC), an exact expression of $d_{\min}^2$ (in fact, d_{free}^2) has been found. First the upper bound for binary 1REC has been shown in [9] as $d_B^2 = 2[1-\text{sinc}(2h)]$. Later, it was shown that this is in fact the exact value of $d_{\min}^2$ for $h < 1/2$ for binary 1REC [17]. Finally, this result was extended to more general cases in [18]. Specifically, $d_{\min}^2$ of M-ary CPFSK at any rational modulation index $h = p/q$, where p and q are relatively prime positive integers, is given by

$$d_{min}^2 = \begin{cases} \log_2 M, & \text{integer } h \\ \min\limits_{\gamma}\{2\log_2 M[1-\text{sinc}(\gamma h)]\}, & q \geq M \\ 2\log_2 M[1-\text{sinc}(2h)], & h < 0.3016773,\ q < M \\ \log_2 M, & h > 0.3016773,\ q < M \end{cases} \tag{6.41}$$

where $\gamma = 2, 4, 6, ..., 2(M-1)$. It is clear from the above expression that $d_{\min}^2$ is proportional to $\log_2 M$. In other words, larger M yields larger minimum distance.

We now present some results as examples in order to get an idea of the behavior of $d_{\min}^2$ and d_B^2 for more general cases. First we want to show how d_B^2 varies with h and how $d_{\min}^2$ varies with h and N. An example for a binary 3RC scheme is shown in Figure 6.10. The upper bound reaches its peak at an h value slightly smaller than one. The actual minimum distance $d_{\min}^2$ also generally increases with h for small h values and become oscillating for larger h values. For $N = 1$, $d_{\min}^2$ is very poor, but with $N = 2$, it already increases significantly. When $N = 4$, the upper bound is reached in the region $0 < h \lesssim 0.6$. For $h \approx 0.85$ the upper-bound value of 3.35 is reached with $N = 6$. Further increase of N does not increase $d_{\min}^2$ significantly. There are some weak points of h: $h = 2/3$ and $h = 1$, where the values of $d_{\min}^2$ are significantly lower. For practical design, we can approximate the optimum $h \approx 0.85$ by a rational number, for instance, $h = 7/8 = 0.875$. Then the scheme will have a state trellis and the MLSD can be performed by the Viterbi algorithm.

Figure 6.11 compares the d_B^2 for binary 1RC through 6RC. The linear vertical scale is the absolute distance and the dB scale is the relative distance normalized to that of MSK. The most important observation from this figure is that the distance increases with L for large h. For $h < 0.5$, the distance actually decreases with L.

But this comparison is not quite fair since the bandwidth also changes with L and h. A better comparison is given in Figure 6.12.

Figure 6.12 is a comparison of $d^2_{\rm min}$ between MSK and several schemes of CPM versus bandwidth. Each point represents a CPM scheme with its 99% double-sided normalized power bandwidth $2BT_b$ shown on the horizontal axis and the $d^2_{\rm min}$ difference relative to MSK shown on the vertical axis. If we draw a vertical line and a horizontal line through the MSK point, schemes to the left side of the vertical line are more bandwidth efficient than MSK and the ones to the right of the line; and the schemes above the horizontal line are more power efficient than MSK and those below the line. Those in the upper-left quadrant are both more bandwidth efficient and power efficient than MSK. It is evident that larger L and M yield more efficient schemes.

6.4 MODULATOR

A conceptual CPM modulator is shown in Figure 6.13. This is a direct implementation of (6.1). Data sequence $\{a_k\}$ passes through the filter and the multiplier to form the frequency pulse sequence $\{2\pi h a_k g(t-kT)\}$ which is used by the FM modulator to yield the required phase $\Phi(t,\mathbf{a})$. Note that an FM modulator instead of a phase modulation must be used since the input is $g(t)$, the frequency pulse, instead of $q(t)$, the phase pulse. If the FM modulator is a VCO (voltage-controlled oscillator), the VCO's control voltage is

$$v(t) = 2\pi h \sum_{k=-\infty}^{\infty} a_k g(t-kT)$$

But the VCO implementation is not practical since conventional free-running VCOs can not achieve either acceptable frequency stability or the linearity required for low distortion. Many practical solutions have been proposed. One solution is to use the quadrature structure proposed in [19] and also described in [9]. Other structures that use phase lock loop or bandpass filter and limiter are also described in [9]. Another method is to use all digital techniques, with the analog VCO replaced by a digital NCO (numerically controlled oscillator) [20]. We will describe these structures below.

6.4.1 Quadrature Modulator

The quadrature structure can be derived by rewriting the normalized CPM waveform

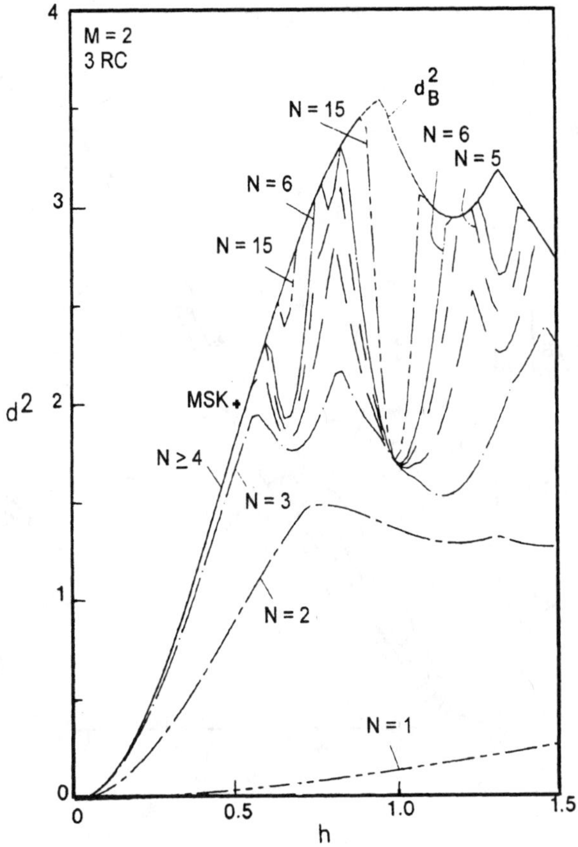

Figure 6.10 Minimum distance versus h for binary 3RC scheme. From [9]. Copyright © 1986 Plenum.

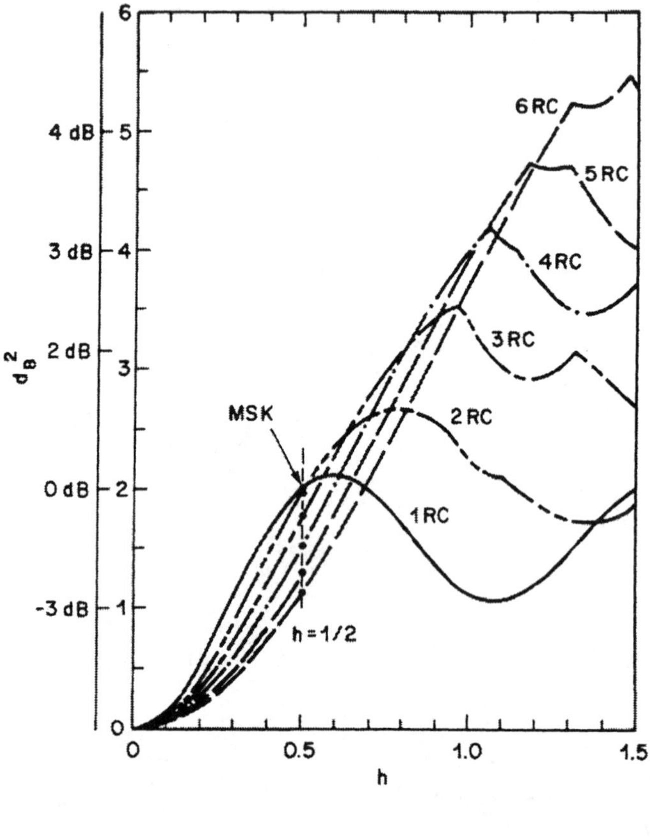

Figure 6.11 The upper bound d_B^2 as a function of h for LRC, $L = 1, 2, ..., 6$. From [1]. Copyright © 1986 IEEE.

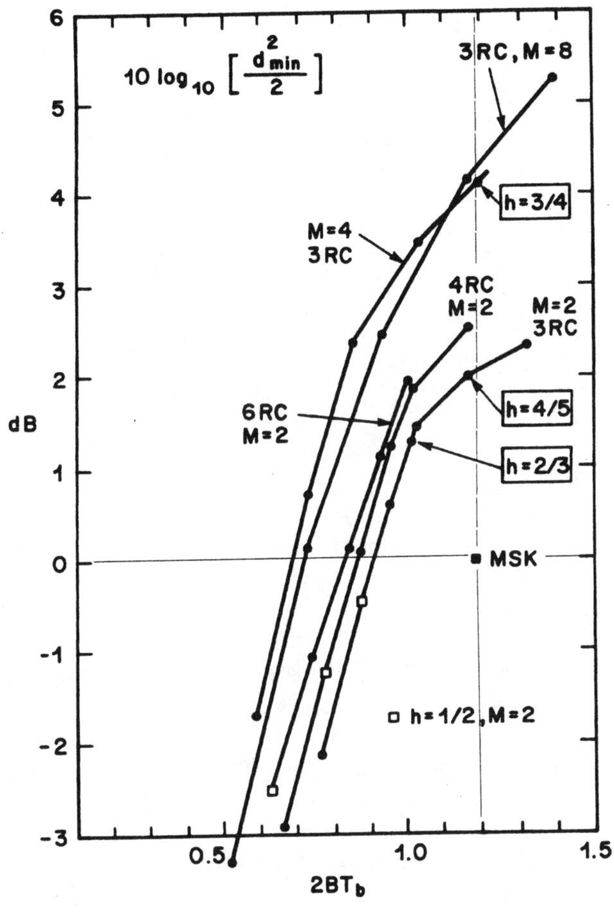

Figure 6.12 Power-bandwidth trade-off for CPM schemes using RC pulses. The bandwidth is defined with the 99%-power-in-band definition. The specific schemes are plotted as points and connected by straight lines. From [1]. Copyright © 1986 IEEE.

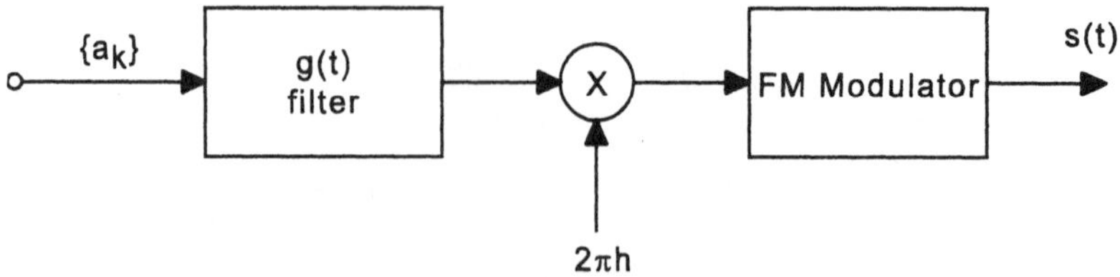

Figure 6.13 Conceptual CPM modulator [9]. Copyright © 1986 Plenum.

$s_o(t) = s(t)/A$ as:

$$s_o(t) = I(t)\cos(2\pi f_c t) - Q(t)\sin(2\pi f_c t), \tag{6.42}$$

where

$$I(t) = \cos\Phi(t, \mathbf{a}) = \cos[\theta(t, \mathbf{a}_k) + \theta_k] \tag{6.43}$$

and

$$Q(t) = \sin\Phi(t, \mathbf{a}) = \sin[\theta(t, \mathbf{a}_k) + \theta_k] \tag{6.44}$$

This expression is very similar to those of MPSK or MSK. The difference here lies in the generation of $I(t)$ and $Q(t)$. The most general and straightforward way is to use stored lookup tables for $I(t)$ and $Q(t)$. Figure 6.14 is the modulator based on (6.42) using ROM (read only memory) to store $I(t)$ and $Q(t)$ [7,8]. Here we assume a rational h. Thus the number of shapes of $\Phi(t, \mathbf{a})$ in a symbol interval is finite (e.g., see Figure 6.4). So are the number of shapes of $I(t)$ and $Q(t)$ in a symbol interval. Therefore all possible shapes of $I(t)$ and $Q(t)$ in a symbol interval can be stored in ROMs. Since the shape of $\Phi(t, \mathbf{a})$ in a symbol interval is determined by the state $\mathbf{s}_k = (\theta_k, \mathbf{a}_{k-1}) = (\theta_k, a_{k-1}, a_{k-2}, ..., a_{k-L+1})$ and the current input symbol a_k (again, refer to Figure 6.4), the shapes of $I(t)$ and $Q(t)$ in a symbol interval are addressed by the states $\mathbf{s}_k$ and the current input symbol a_k.

In Figure 6.14, the data symbols are received by the shift register of length L. The output of the shift register is a symbol vector $\mathbf{a}_k = (a_k, a_{k-1}, a_{k-2}, ..., a_{k-L+1})$. The phase states θ_k can be generated by a phase state ROM and a delay of T. At each symbol time, $\mathbf{a}_k$ and θ_{k-1} are used as input to the state ROM to obtain θ_k. Preferably a phase state serial number ν_k instead of θ_k can be used. By doing this the phase state generation can be implemented as an up/down counter. One way to relate ν_k to θ_k is to use the relation

$$\theta_k = (h\pi\nu_k + \phi_o) \bmod (2\pi) \tag{6.45}$$

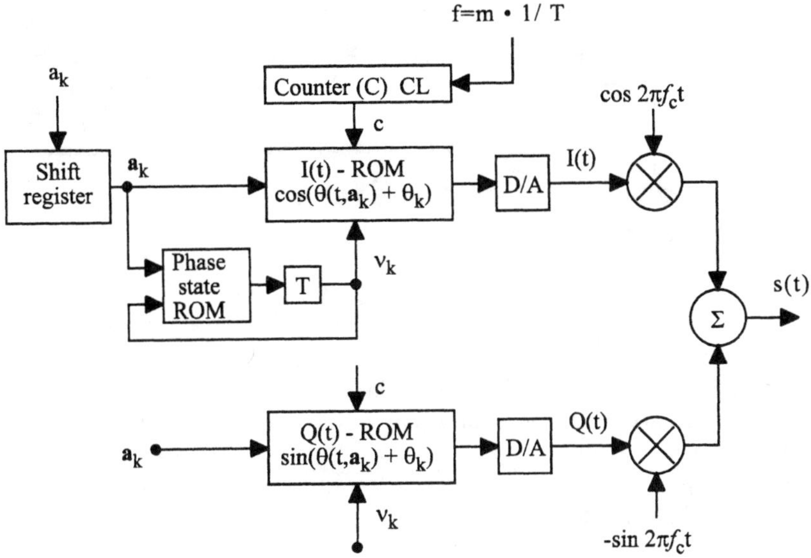

Figure 6.14 Quadrature modulator using phase state ROM. From [9]. Copyright © 1986 Plenum.

where ϕ_o is an arbitrary phase constant which is usually set to zero in analysis. For example, values of θ_k of the binary 3RC with $h = 2/3$ are $0, 2\pi/3$ and $4\pi/3$ (see (6.14)), which can be represented by $\nu_k = 0, 1, 2$.

The $I(t)$ and $Q(t)$ shapes are swept by a counter C at a speed m/T, where m is the number of stored samples per symbol interval. The symbol vector $\mathbf{a}_k$, the phase state serial number ν_k, and the counter output c are combined to form the addresses to obtain the samples of $I(t)$ and $Q(t)$ in the ROMs. These samples are converted to analog by the D/A converters. The rest of the modulator is the same as that of typical quadrature modulators.

The address field of the ROM is roughly $L \log_2 M + \lceil \log_2 p \rceil + 1$ bits. $L \log_2 M$ is for $\mathbf{a}_k$, and $\lceil \log_2 p \rceil + 1$ is for θ_k or ν_k, where $\lceil x \rceil$ denotes the smallest integer which is greater than x. Parameter p is the number of different θ_k and is related to index h by $h = 2q/p$ which we have stated before. The ROM size can be easily figured out. Since there are pM^{L-1} states, the current symbol has M different values, thus the number of $I(t)$ or $Q(t)$ shapes is $pM^{L-1}M = pM^L$. Each shape has m samples and each sample is quantized into m_q bits. Thus the ROM size is $pM^L mm_q$ bits. The above address length and ROM size are usually small. For example, for binary 3RC with $h = 2/3$, the ROM address length is only 5 bits and the ROM size is 1536 bits

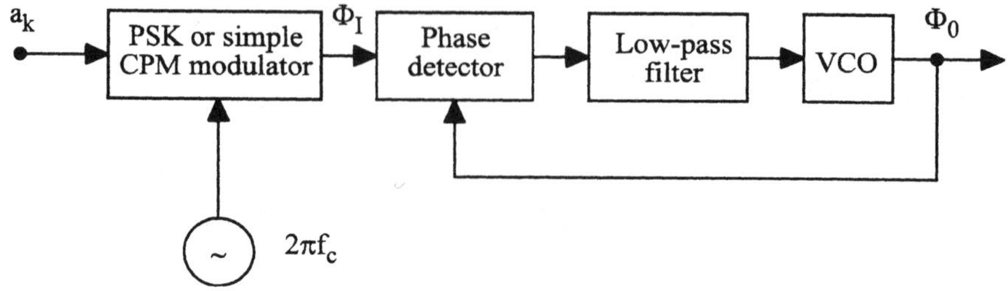

Figure 6.15 PLL implementation of CPM modulator. From [9]. Copyright 1986 © Plenum.

assuming $m = 8$ and $m_q = 8$. For larger values of $M, L, m,$ and m_q the ROM size is still very small in comparison to the capacity of ROMs that today's technology can achieve.

6.4.2 Serial Modulator

Similar to the technique used for serial MSK (SMSK) studied in Chapter 5, a CPM modulator may be implemented serially. Figure 6.15 is a CPM modulator consisting of a PSK or simple CPM modulator and a PLL (phase locked loop). By replacing the PLL with a bandpass filter and a hard limiter one can implement a CPM modulator as in Figure 6.16. Like the SMSK case, the basic function of the PLL or bandpass filter is to smooth the incoming waveform, thus creating a signal with correlation over several symbol intervals and a narrower spectrum. By properly choosing the PLL or the filter, it is possible to approximately generate a signal very close to the desired CPM signal (in the SMSK case, the generated signal is exactly the MSK signal). The limiter in Figure 6.16 is used to keep the signal amplitude constant in case there is any amplitude variation after filtering. In the following we show how CPM modulation can be achieved by these two structures.

6.4.2.1 PLL Modulator

Refer to Figure 6.15. Assuming the phase detector is used in its linear region, the input phase from the simple CPM modulator is

$$\Phi_I(t, \mathbf{a}) = \sum_k 2\pi h a_k q_I(t - kT)$$

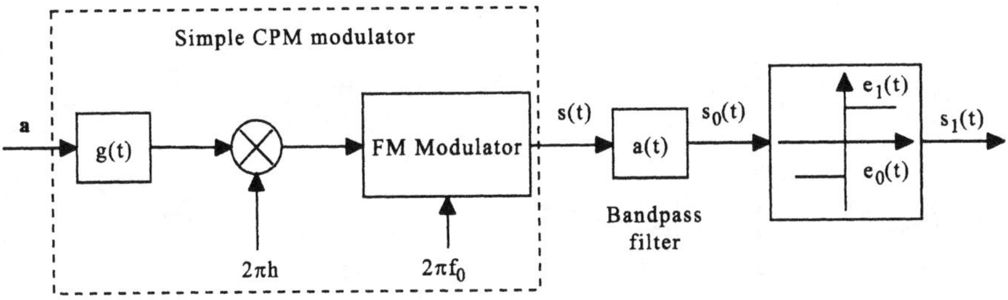

Figure 6.16 Bandpass filter and hard-limiter implementation of CPM modulator. From [9]. Copyright © 1986 Plenum.

Assuming the PLL impulse response is $h(t)$ then the output phase is

$$\Phi_o(t, \mathbf{a}) = \sum_k 2\pi h a_k q_I(t - kT) * h(t)$$

where * denotes convolution. Comparing this to the CPM phase expression, we have

$$q(t - kT) = q_I(t - kT) * h(t)$$

Thus the desired $q(t)$ can be deconvolved into two simpler functions, $q_I(t)$ and $h(t)$, and the CPM modulator complexity is reduced. The deconvolution may be done easier in frequency domain since

$$Q(f) = Q_I(f)H(f)$$

The problem with the PLL modulator is the limit on modulation index h since the maximum of the phase deviation $\Phi_o(t, \mathbf{a}) - \Phi_I(t, \mathbf{a})$ is limited by the linear region of the PLL operation characteristic. The limit also depends on the phase pulse length L and number of levels M. A rough estimation is that [9, p. 226]

$$h < \frac{2}{(L+1)(M-1)}$$

which is quite small for large L and M. For example, a binary 3RC CPM would have a maximum index of 0.5 and a quaternary 2RC a maximum index of 0.2.

6.4.2.2 BPF-Limiter Modulator

In Figure 6.16, the phase of the limiter output is the same as the input. That is

$$\Phi_1(t, \mathbf{a}) = \Phi_0(t, \mathbf{a}) = \arg\{\widetilde{s}_0(t) * \widetilde{a}(t)\}$$

where signal and filter are denoted by their complex envelopes, and $arg\{z\}$ denotes the argument of the complex quantity z. The above expression can be written as

$$\Phi_1(t, \mathbf{a}) = \arg\{\exp[j\Phi(t, \mathbf{a})] * \widetilde{a}(t)\}$$

where $\Phi(t, \mathbf{a})$ is the phase of the signal from the simple CPM stage which can be written as

$$\begin{aligned}\Phi(t, \mathbf{a}) &= 2\pi h \sum_{i=k-L+1}^{k} a_i q(t - iT) + \theta_k \\ &= \theta(t, \mathbf{a}_k) + \theta_k\end{aligned}$$

where we have assumed that $q(t)$ has a finite length L. For low modulation indexes, the time varying part of $\Phi(t, \mathbf{a})$ may be considered small enough to expand the exponential function into a Taylor series. That is

$$\theta_1(t, \mathbf{a}_k) = \arg\{[1 + j\theta(t, \mathbf{a}_k) - \frac{\theta^2(t, \mathbf{a}_k)}{2} + ...] * \widetilde{a}(t)$$

To a first-order approximation, we have

$$\theta_1(t, \mathbf{a}_k) = \arg\{[1 + j\theta(t, \mathbf{a}_k)] * \widetilde{a}(t)$$

Now we assume the filter $a(t)$ is real (this requires the bandpass filter have a symmetrical frequency response), we have

$$\begin{aligned}\theta_1(t, \mathbf{a}_k) &= \arg\{[1 + j\theta(t, \mathbf{a}_k)] * a(t) \\ &= \arctan\left[\frac{\theta(t, \mathbf{a}_k) * a(t)}{1 * a(t)}\right]\end{aligned}$$

The denominator is a constant, denoted by A, and for small h values the arctan may be omitted, thus

$$\theta_1(t, \mathbf{a}_k) \approx \theta(t, \mathbf{a}_k) * \frac{a(t)}{A}$$

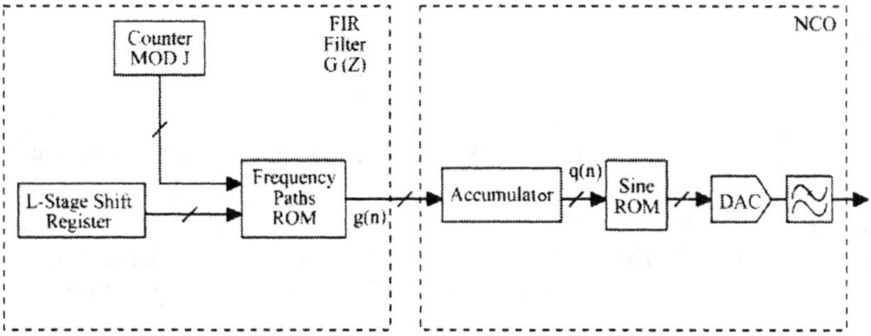

Figure 6.17 All digital CPM modulator. From [20]. Copyright © 1987 IEEE.

or

$$\theta_1(t, \mathbf{a}_k) \approx 2\pi h \sum_{i=k-L+1}^{k} a_i q(t - iT) * \frac{a(t)}{A}$$

When this expression is compared with the standard CPM phase, it is seen that the bandpass filter to a first-order approximation acts as a linear filter on the phase response. Thus one can synthesize the desired $q_1(t)$ from two simpler functions $q(t)$ and $a(t)$.

The problem of this approach is of course the imperfect linearity of the phase, which may cause some spectrum spread [9, p. 231].

The advantages of the two serial modulators in comparison with the quadrature modulator are faster speed and no need of balance between two channels.

6.4.3 All-Digital Modulator

A completely digital implementation of the CPM modulator is proposed in [20]. The structure is based on the conceptual CPM modulator in Figure 6.13, but all function blocks are implemented digitally. Filter $G(j\omega)$ (Fourier transform of $g(t)$) is replaced by a digital FIR filter $G(z)$, and the VCO by a numerically controlled oscillator (NCO). Figure 6.17 is the all-digital CPM modulator. Even though the ROM-based quadrature modulator (Figure 6.14) uses digital techniques to generate $I(t)$ and $Q(t)$, the modulation stages are still analog. In this all-digital implementation, the modulation stage is realized by the NCO.

The NCO consists of an accumulator, a sine ROM, a DAC, and a low-pass filter. The low-pass filter is to filter out any high-frequency harmonics after the DAC. The sine ROM stores samples of just *one* sinusoid so that the content of the nth memory location is

$$\langle n \rangle = [2^{K-1} \cos(2\pi n/N)]$$

where $[\cdot]$ means the nearest integer, and K is the number of bits per location in the ROM (i.e., the DAC converter resolution). N is an integer which is the discrete period of the sinusoid. The analog period or frequency is determined by N and the clock frequency f_s of the circuit. The accumulator is in fact a discrete integrator. At symbol interval j, it provides a number n_j as an address for the ROM. Assume the accumulator increases its output by 1 at each clock interval, the output of the sinusoid will repeat itself every N clock intervals. That is, the output signal will have a frequency of

$$K_{NCO} = \frac{f_s}{N}$$

which may be called the sensitivity of the NCO. K_{NCO} is usually an integer. Suppose now the accumulator accepts a number k at its input for each clock interval, the output of the accumulator in the jth interval of the clock is

$$n_j = \{k + n_{j-1}\} \operatorname{mod}(N)$$

Thus the frequency of the signal at the output of the DAC is

$$f_{NCO} = kK_{NCO}$$

The constraint on k is that $k \leq N/2$ due to the sampling theorem. By changing k, different frequencies can be obtained. The highest is $f_s/2$ and the lowest is $f_s/N = K_{NCO}$.

For CPFSK, the only thing that the FIR filter in Figure 6.17 needs to do is to put out different k values from the frequency paths ROM for different data symbols. In fact, M different values of k are needed for an M-ary CPFSK scheme. Since usually f_{NCO} is chosen as an integer multiple of the symbol rate $(1/T)$, the number of periods of the sinusoid in T is an integer. Thus the signal can start with zero phase and also can end with zero phase in every symbol interval. In this way, the phase continuity can be achieved.

For more complex CPM schemes, the FIR must provide frequency pulse-shaping processing. In principle, this processing can always be provided by an FIR filter, but in practice it is usually much simpler to use the structure shown in Figure 6.17. The L-stage shifter register remembers the last L symbols and uses them as the addresses for samples of one of M^L possible frequency paths. The frequency paths can be easily derived from the phase paths (see (6.9)):

$$f(t, \mathbf{a}) = \frac{d}{dt}\Phi(t, \mathbf{a}) = 2\pi h \sum_{i=j-L+1}^{j} a_i g(t - iT) \tag{6.46}$$

whose samples are stored in the frequency paths' ROM.

This all-digital CPM modulator is extremely stable, with very low phase noise, and very flexible. For instance, by increasing N, the resolution can be increased. For more design details the reader is referred to [20] where simulation and experimental results are also available.

6.5 DEMODULATOR

The demodulator is sometimes referred to as the receiver in the literature, even though their meanings are slightly different. We will use them interchangeably.

An important parameter which dictates the structure of the receiver is the modulation index h. If h is rational, as we demonstrated in Section 6.1, there is a state trellis for this CPM scheme. Each path in the trellis uniquely corresponds to a data sequence. Some search algorithms (e.g., the Viterbi algorithm) can be used to search through the trellis to find the transmitted data sequence under certain criterion. If h is irrational, there is no state trellis that exists. In this case a sequence detection based on the phase tree is theoretically possible, but it is impractical since the number of branches at a node in the phase tree grows exponentially with the length of the path. Thus there exists no trellis for irrational h.

We will consider rational h only.

6.5.1 Optimum ML Coherent Demodulator

An optimum MLSD coherent demodulator structure was first developed for binary CPFSK in [21] and extended for M-ary CPFSK in [22]. However, the results are in fact applicable to any CPM scheme as pointed out by [9, p. 233]. The receiver is to detect the first symbol using an observation of N symbols. As we have pointed out previously, the phase produced by the first symbol (any symbol, in fact) lasts forever. Therefore, only when $N \to \infty$, the receiver is optimum in the sense that all phase information of the first symbol contained in the entire signal is used. When N is finite, the receiver is only suboptimum in the sense that not all phase information is used. However, if only N symbols are observed, the best we can do is to perform MLSD based on the given observation. In this sense, it is optimum. That is why it is called an optimum ML demodulator.

We use a shorthand notation developed in [21] for the received signal

$$r(t) = s(t, a_1, A_k) + n(t), \quad 0 \leq t < NT$$

where $s(t, a_1, A_k)$ is a general CPM signal corresponding to the data symbol sequence $\{a_1, A_k\}$, a_1 is the first symbol to be detected, and $A_k = \{a_2, a_3, ..., a_N\}$, $k = 1, 2, ..., M^{N-1}$ are the all possible data sequences with $N - 1$ bits following a_1. For coherent detection, the initial phase of the signal is known, and hence will be assumed to be zero without loss in generality.

The optimum receiver is a first-symbol detector [7,21,22]. The receiver is only interested in finding an estimate of the first symbol a_1, not the entire sequence. Thus we can partition all M^N sequences into M groups:

$$\begin{cases} 1, a_2, a_3, ..., a_N \\ -1, a_2, a_3, ..., a_N \\ 3, a_2, a_3, ..., a_N \\ -3, a_2, a_3, ..., a_N \\ \vdots \\ (M-1), a_2, a_3, ..., a_N \\ -(M-1), a_2, a_3, ..., a_N \end{cases}$$

Note that each group contains M^{N-1} sequences since $\{a_2, a_3, ..., a_N\}$ has M^{N-1} possible combinations. The receiver must find which group of sequences maximizes the likelihood function, and take the first symbol of that group as the estimate. This problem is the composite hypothesis problem treated in [23,24] and other books. The main point about the composite hypothesis problem is that the observables pertaining to the symbol to be estimated (a_1, here) are distributed according to certain probability density, and to estimate the symbol, the likelihood of the symbol must be averaged over the density. In our case (AWGN channel with a noise spectral density N_o), the likelihood conditioned on a specific A_k is $\exp\left(\frac{2}{N_o}\int_0^{NT} r(t)s(t, a_1, A_k)dt\right)$ [23,24]. It must be averaged over the probability distribution density of A_k to obtain the unconditional likelihood of a_1. Since A_k can have $m = M^{N-1}$ different possibilities, the discrete probability density function (PDF) of A_k is $f(A_k) = 1/M^{N-1}$ for each possible A_k. Thus the likelihoods for all M possible a_1s are (ignoring factor $1/M^{N-1}$ since it is the same for all likelihoods)

$$l_1 = \sum_{j=1}^{m} \exp\left[\frac{2}{N_o}\int_0^{NT} r(t)s(t, 1, A_j)dt\right]$$

$$
\begin{aligned}
l_2 &= \sum_{j=1}^{m} \exp\left[\frac{2}{N_o}\int_0^{NT} r(t)s(t,-1,A_j)dt\right] \\
&\vdots \\
l_{M-1} &= \sum_{j=1}^{m} \exp\left[\frac{2}{N_o}\int_0^{NT} r(t)s(t,M-1,A_j)dt\right] \\
l_M &= \sum_{j=1}^{m} \exp\left[\frac{2}{N_o}\int_0^{NT} r(t)s(t,-(M-1),A_j)dt\right]
\end{aligned}
\tag{6.47}
$$

The receiver then chooses the data symbol a_1 that corresponds to the largest of l_1 to l_M. The receiver is shown in Figure 6.18. The correlators' outputs are denoted by $x_{\lambda j}, \lambda = 1, 2, ..., M, j = 1, 2, ..., m$ where

$$
x_{\lambda j} = \begin{cases} \int_0^{NT} r(t)s(t,\lambda,A_j)dt, & \lambda \text{ odd} \\ \int_0^{NT} r(t)s(t,-(\lambda-1),A_j)dt, & \lambda \text{ even} \end{cases}
$$

Then the M likelihoods can be written as

$$
l_\lambda = \sum_{j=1}^{m} \exp\left(\frac{2}{N_o} x_{\lambda j}\right), \quad \lambda = 1, 2, ..., M
$$

It is impossible to analyze the exact error performance of the optimum receiver. However, for high SNR the receiver can be simplified to a suboptimum receiver whose error performance can be analyzed by means of union bound.

For large SNR,

$$
l_\lambda = \sum_{j=1}^{m} \exp\left(\frac{2}{N_o} x_{\lambda j}\right) \approx \exp\left(\frac{2}{N_o} x_\Lambda\right)
$$

where x_Λ is the largest of $x_{\lambda j}$. Furthermore, since exp( ) is a monotonic function, x_Λ is an equivalent parameter for decision making. Thus a suboptimum receiver does not need the exp( ) blocks and the summers in Figure 6.18. The receiver directly checks the outputs of the correlators and makes the decision on a_1 that corresponds to the largest $x_{\lambda j}$. Its performance should be very close to that of the optimum receiver for high SNR.

The suboptimum receiver is in fact an ML sequence detector. As we pointed out before, if $N \to \infty$, the receiver is optimum. Therefore, the performance of the suboptimum receiver can be compensated for by increasing the observation length.

The error probability of the optimum first-symbol receiver is difficult to analyze.

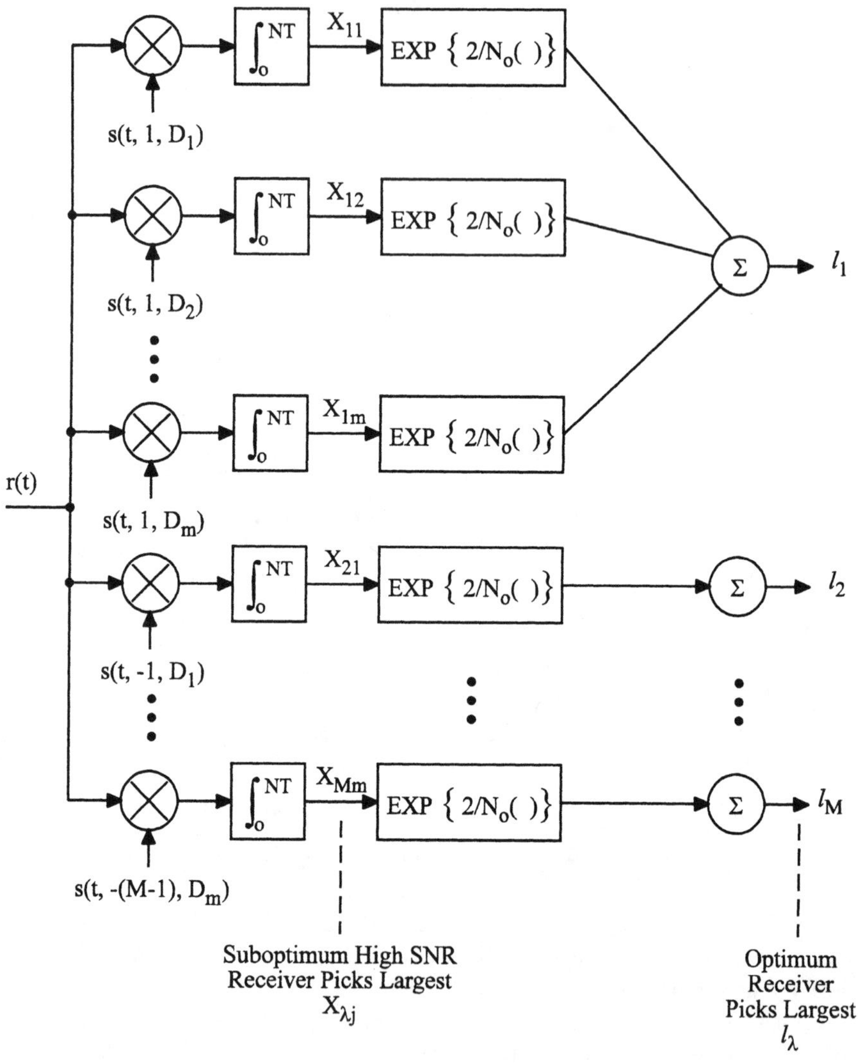

Figure 6.18 Optimum and high-SNR suboptimum CPM receivers. From [22]. Copyright © 1976 IEEE.

However, for high SNR, the suboptimum receiver is just as good as the optimum one. The error performance thus can be evaluated by using the suboptimum receiver which is the ML sequence detector. Its performance has been analyzed in Section 6.3, and is dominated by the minimum distance between all sequences. That is

$$P_e \lesssim KQ\left(d_{\min}\sqrt{E_b/N_o}\right)$$

where K is the number of sequences that attain the minimum distances in all cases. Using the definitions of the distance and x, they can be related as $d^2 = \frac{1}{E_b}[NE_s - x]$ and $d^2_{\min} = \frac{1}{E_b}[NE_s - x_\Lambda]$.

The error probability upper-bound curves for binary, quaternary, and octal CPFSK schemes with various h values are given in [22] (Figures 6.19 to 6.21). The h values used in the figures are the optimum values, that is, the values for which the upper bounds are the smallest. These values are found by trial-and-error numerically. The curves show that binary CPFSK $(h = 0.715)$ offers up to 1.1 dB improvement for five-symbol detection with respect to coherent BPSK. Quaternary CPFSK offers a 2.5 dB improvement for two-symbol detection $(h = 1.75)$ and a further 1 dB improvement for five-symbol detection $(h = 0.8)$ over coherent QPSK. Octal CPFSK $(h = 0.879)$ offers a 1.9 dB advantage over orthogonal signaling for two-symbol detection and a 2.6 dB improvement for three-symbol detection. The advantage over nonorthogonal coherent 8PSK is huge, in the neighborhood of 7.5 dB. Simulation results provided in [22] confirm the tightness of the upper bound at high SNR $(> 4$ dB$)$.

Note that the above comparison is based on the assumption that no signal distortion exists at the receiver. This implies the bandwidth of the communication system is, theoretically, infinite; and practically, wide enough to receive most of the signal spectrum. This in turn implies that the bandwidth requirement is generally higher for larger values of h (refer to Figures 3.12 to 3.14). Obviously this argument also applies to the noncoherent detection case in the next section.

6.5.2 Optimum ML Noncoherent Demodulator

The optimum ML noncoherent demodulator is also developed in [21,22]. For noncoherent detection, the carrier initial phase ϕ is unknown. We assume a uniform PDF $f(\phi)$ from 0 to 2π, that is

$$f(\phi) = \frac{1}{2\pi}, \quad 0 \le \phi \le 2\pi$$

Different from the coherent case, the $r(t)$ is observed for $(2n+1)$ symbols and the symbol to be detected is the middle symbol [21,22] (according to [21], the magnitude

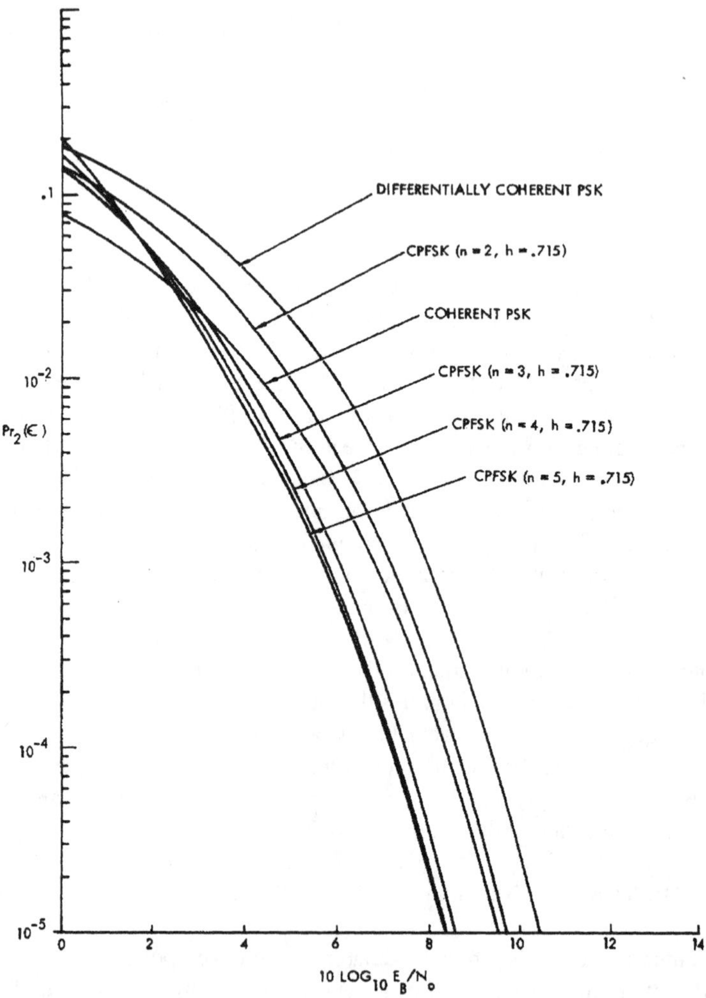

Figure 6.19 P_b for binary coherent CPFSK in comparison with coherent BPSK and differentially coherent BPSK. From [22]. Copyright © 1976 IEEE.

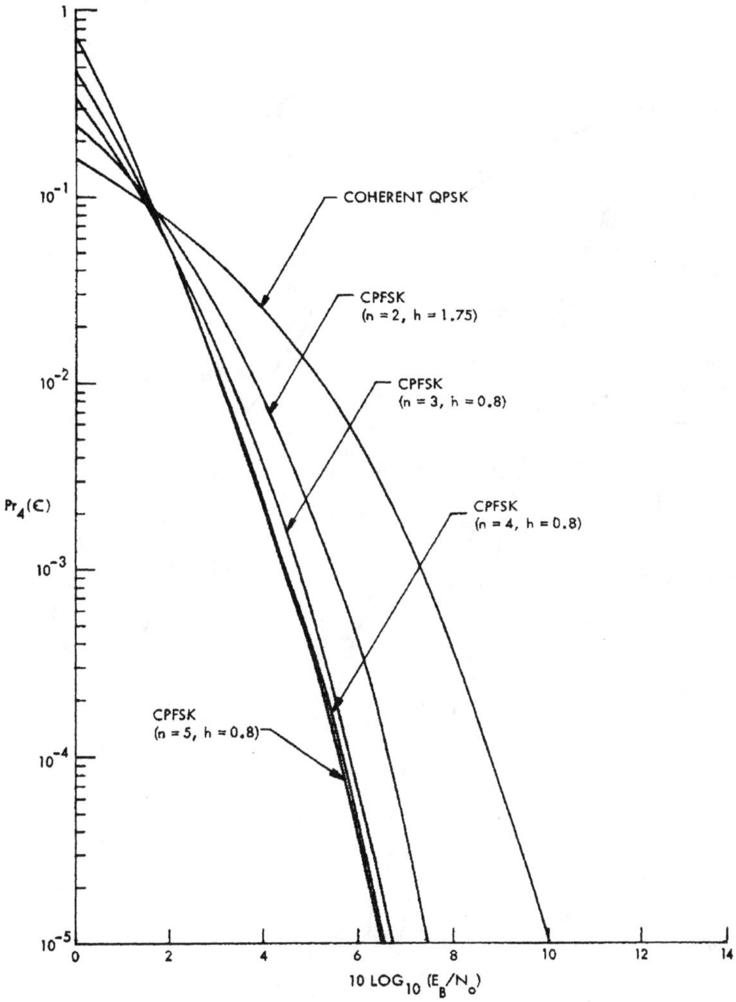

Figure 6.20 P_s for quaternary coherent CPFSK in comparison with coherent QPSK. From [22]. Copyright © 1976 IEEE.

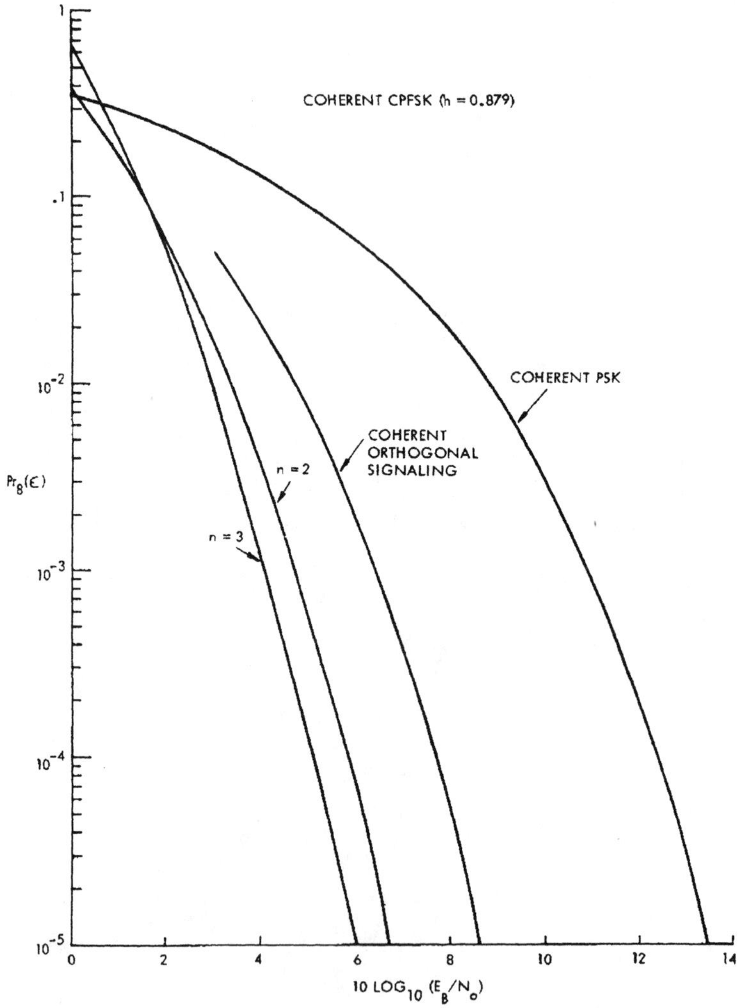

Figure 6.21 P_s for octal coherent CPFSK in comparison with coherent 8PSK and orthogonal signaling. From [22]. Copyright © 1976 IEEE.

of the complex correlation between two binary CPFSK waveforms corresponding to data sequences differing in only one bit is a minimum when the difference bit is in the middle). Using a shorthand notation for the signal we can write

$$r(t) = s(t, a_{n+1}, \Delta_k, \phi) + n(t), \quad 0 \le t \le (2n+1)T$$

where a_{n+1} is the middle symbol which is to be detected, Δ_k is a $2n$-tuple defined as

$$\Delta_k = \{a_1, a_2, ..., a_n, a_{n+2}, ..., a_{2n+1}\}$$

There are a total of $\mu = M^{2n}$ sequences of Δ_k. All of them are equally likely, that is, the PDF of Δ is $f(\Delta) = 1/\mu$. The conditional likelihood

$$\exp\left(\frac{2}{N_o}\int_0^{(2n+1)T} r(t)s(t, a_{n+1}, \Delta_k, \phi)dt\right)$$

must be averaged over Δ and ϕ. That is

$$\begin{aligned}
l_1 &= \frac{1}{\mu}\int_\phi \sum_{k=1}^{\mu} \exp\left[\frac{2}{N_o}\int_0^{(2n+1)T} r(t)s(t, 1, \Delta_k, \phi)dt\right] f(\phi)d\phi \\
l_2 &= \frac{1}{\mu}\int_\phi \sum_{k=1}^{\mu} \exp\left[\frac{2}{N_o}\int_0^{(2n+1)T} r(t)s(t, -1, \Delta_k, \phi)dt\right] f(\phi)d\phi \\
&\vdots \\
l_{M-1} &= \frac{1}{\mu}\int_\phi \sum_{k=1}^{\mu} \exp\left[\frac{2}{N_o}\int_0^{(2n+1)T} r(t)s(t, M-1, \Delta_k, \phi)dt\right] f(\phi)d\phi \\
l_M &= \frac{1}{\mu}\int_\phi \sum_{k=1}^{\mu} \exp\left[\frac{2}{N_o}\int_0^{(2n+1)T} r(t)s(t, -(M-1), \Delta_k, \phi)dt\right] f(\phi)d\phi
\end{aligned} \tag{6.48}$$

The average over the phase leads to a zeroth-order modified Bessel function (see Section B.3 of Appendix B, particularly (B.51) to (B.55)), so we have

$$\begin{aligned}
l_1 &= \frac{1}{\mu}\sum_{k=1}^{\mu} I_0\left(\frac{2}{N_o}z_{1k}\right) \\
l_2 &= \frac{1}{\mu}\sum_{k=1}^{\mu} I_0\left(\frac{2}{N_o}z_{2k}\right)
\end{aligned}$$

$$\vdots$$

$$l_M = \frac{1}{\mu}\sum_{k=1}^{\mu} I_0\left(\frac{2}{N_o}z_{Mk}\right) \tag{6.49}$$

where

$$z_{ik}^2 = x_{ik}^2 + y_{ik}^2 \tag{6.50}$$

$$x_{ik} = \begin{cases} \int r(t)s(t,i,\Delta_k,0)dt, & i \text{ odd} \\ \int r(t)s(t,-(i-1),\Delta_k,0)dt, & i \text{ even} \end{cases}$$

and

$$y_{ik} = \begin{cases} \int r(t)s(t,i,\Delta_k,\pi/2)dt, & i \text{ odd} \\ \int r(t)s(t,-(i-1),\Delta_k,\pi/2)dt, & i \text{ even} \end{cases}$$

Since x_{ik} and y_{ik} are Gaussian with nonzero mean, z_{ik} is a Rician statistic variable.

The optimum noncoherent receiver structure is shown in Figure 6.22 where all integration intervals are from 0 to $(2n+1)T$. It is optimum independent of SNR. Unfortunately, its error performance is difficult to analyze, as in the coherent case. However, we can derive a suboptimum receiver for high SNR and its error performance bound can be determined.

For high SNR, we have with good approximation

$$\sum_{k=1}^{\mu} I_0\left(\frac{2}{N_o}z_{ik}\right) \approx I_0\left(\frac{2}{N_o}z_{i\Lambda}\right)$$

where $z_{i\Lambda}$ is the largest of the z_{ik}. In addition, the Bessel function is a monotonic function, thus the suboptimum receiver needs only to examine all the z_{ik}, $i = 1, 2, \ldots, M$, and chooses the a_{n+1} corresponding to the largest. The suboptimum noncoherent receiver is also shown in Figure 6.22. The error performance of the suboptimum receiver can be analyzed using the union bound. The reader is referred to [22] for details.

The error probability upper-bound curves for noncoherent binary, quaternary, and octal CPFSK schemes with various h values are given in [22] (Figures 6.23 to 6.25). The h values used in the figure are the optimum h values that are similar to the coherent ones. The binary noncoherent CPFSK ($h = 0.715$) outperforms coherent BPSK up to 0.5 dB for $SNR > 7.5$ dB for five-symbol detection, while for three-symbol detection, it is inferior by about 1 dB at high SNR. Quaternary noncoherent CPFSK also achieves better performance than coherent QPSK, offers a 2 dB improvement for three-symbol detection ($h = 0.8$), and a further 0.8 dB improvement for five-symbol detection ($h = 0.8$). Noncoherent octal CPFSK ($h =$

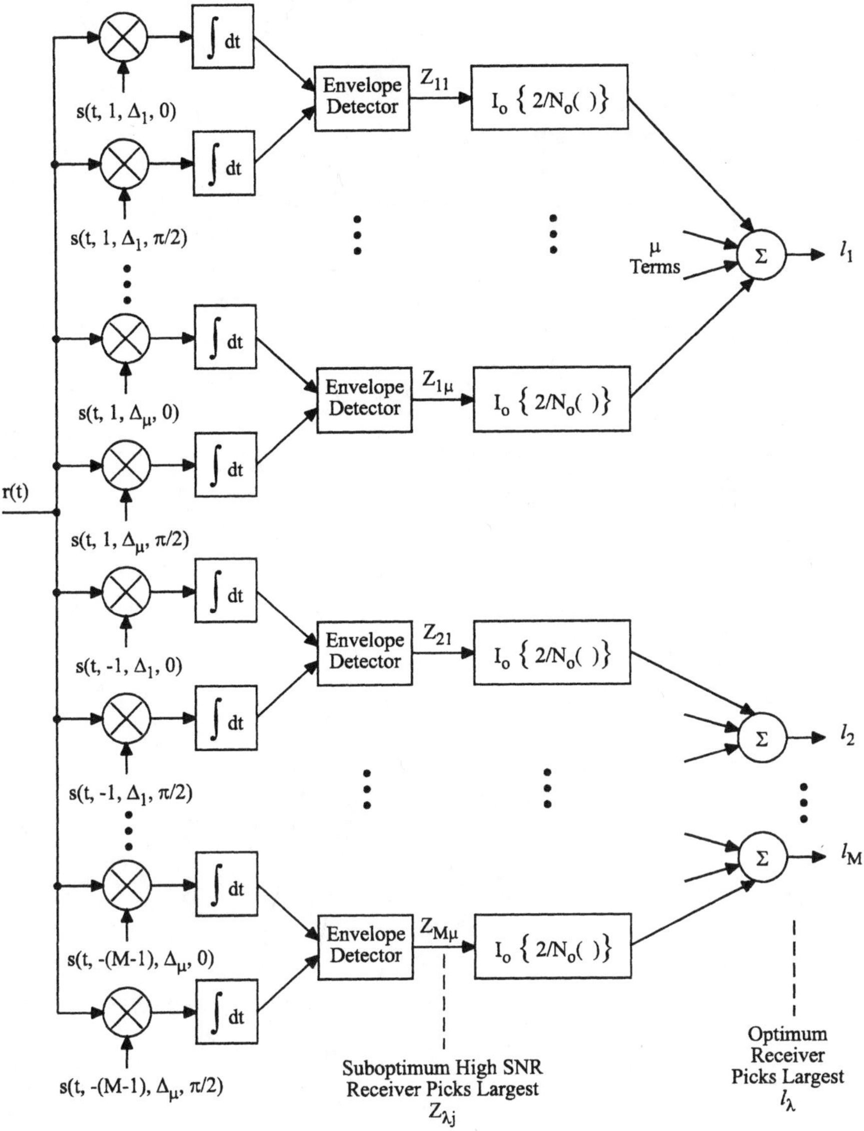

Figure 6.22 Optimum and high-SNR suboptimum CPM noncoherent receiver. From [22]. Copyright © 1976 IEEE.

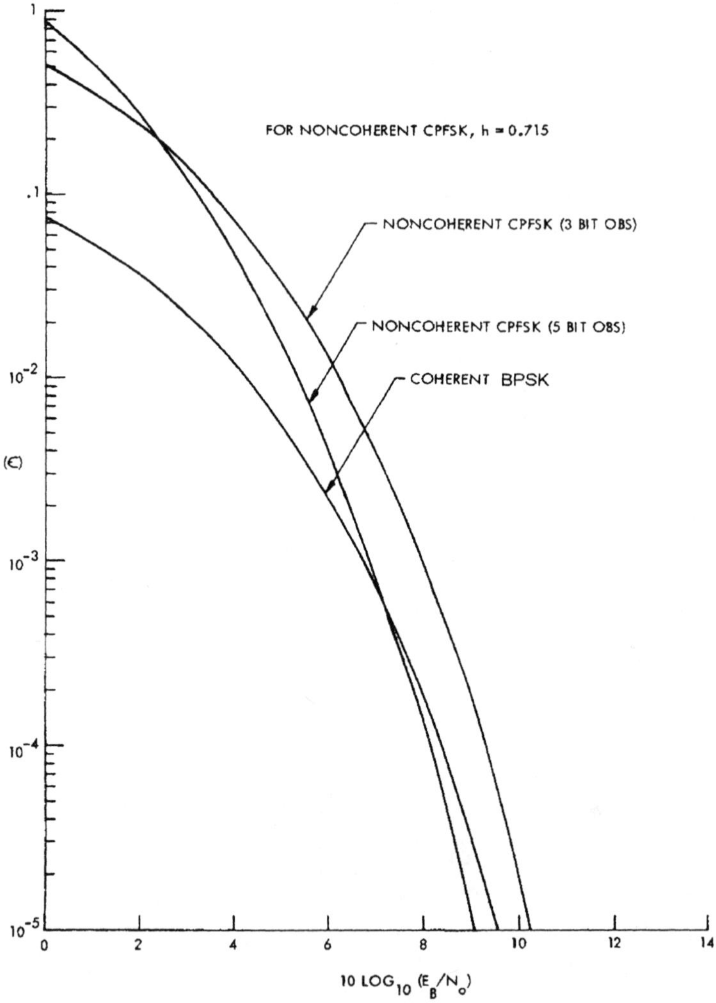

Figure 6.23 P_b of binary noncoherent CPFSK in comparison with coherent BPSK. From [22]. Copyright © 1976 IEEE.

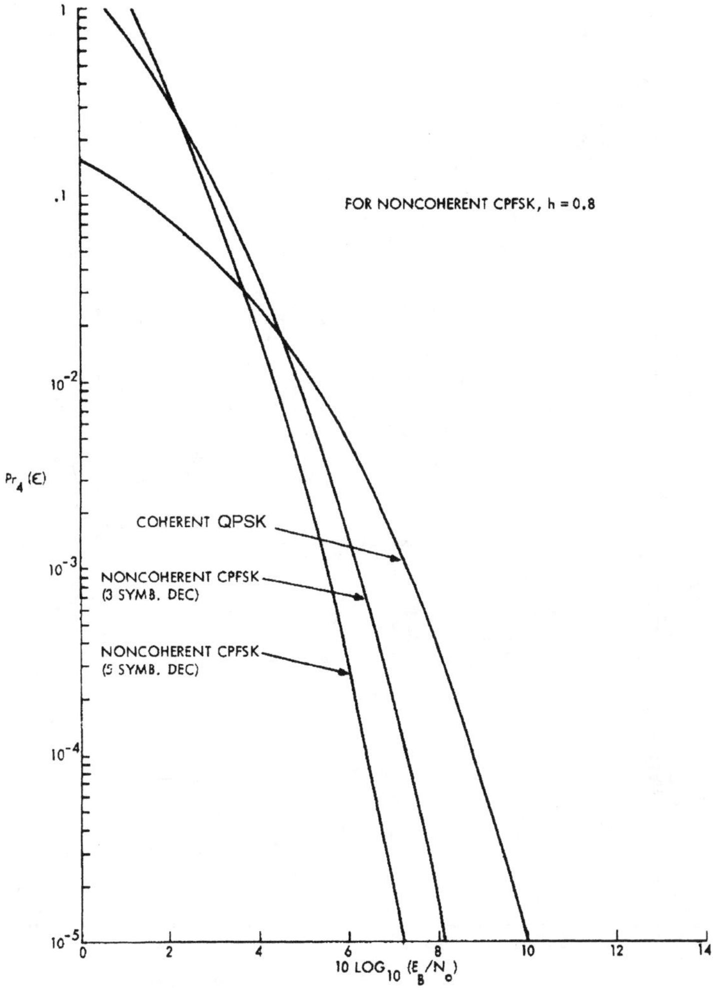

Figure 6.24 P_s of 4-ary noncoherent CPFSK in comparison with coherent QPSK. From [22]. Copyright © 1976 IEEE.

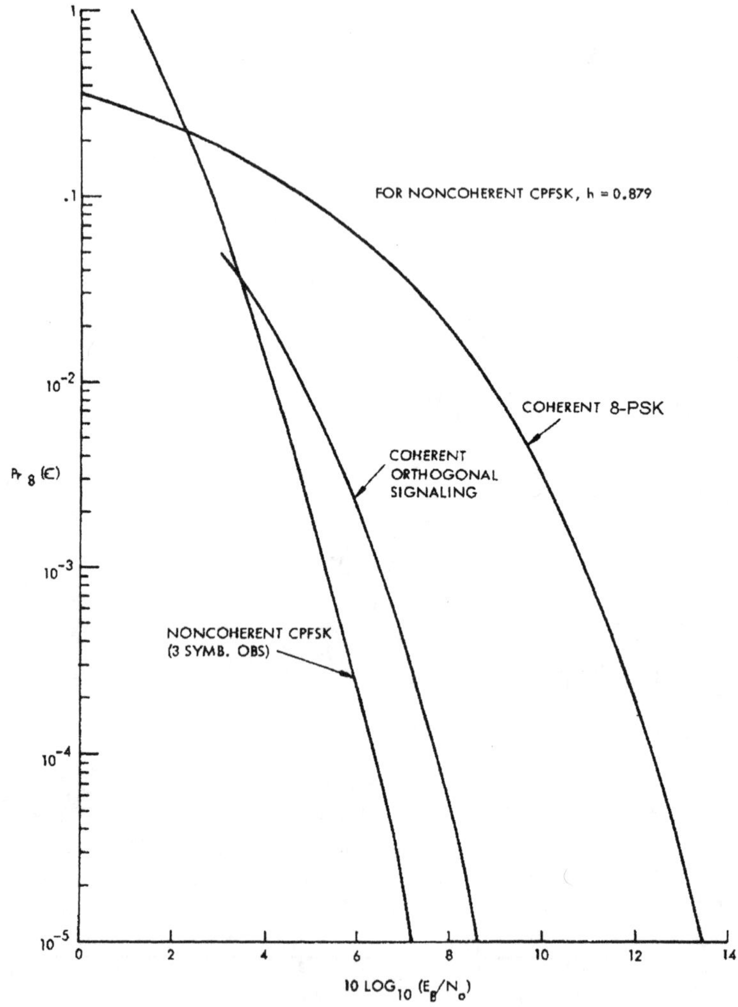

Figure 6.25 P_s of 8-ary noncoherent CPFSK in comparison with coherent 8PSK and coherent orthogonal signaling. From [22]. Copyright © 1976 IEEE.

0.879) offers a 1 dB advantage over orthogonal signaling for three-symbol detection and a significant advantage over nonorthogonal 8PSK, in the neighborhood of 6 dB. Simulation results provided in [22] confirm the tightness of the upper bound at high SNR (> 4 dB).

6.5.3 Viterbi Demodulator

Even though the above two receivers can be applied to general CPM, they become unreasonably complex. Also, it is not clear how successive symbol decisions can be made in a convenient way using the structures in Figures 6.18 and 6.22. An MLSD demodulator using the Viterbi processor was proposed in [8] for partial (and full) response CPM. We will study this structure in this section. We will see that this structure can provide successive real-time symbol decisions in a convenient manner.

This structure is based on the state trellis that we developed in Section 6.1. As pointed out before, for CPM schemes to have a state trellis, the modulation index h must be a rational number. Thus in this section, we exclusively deal with rational h.

We have studied the state trellis of CPM schemes in Section 6.1, and an example is given in Figure 6.5. Here we briefly repeat the state development in Section 6.1. A state at $t = kT$ is defined by the vector

$$\mathbf{s}_k = (\theta_k, a_{k-1}, a_{k-2}, ..., a_{k-L+1})$$

where L is the finite length of the frequency shape pulse $g(t)$. For rational $h = 2q/p$ (q, p integers), there are p different phase states with values $0, 2\pi/p, 2 \cdot 2\pi/p, ..., (p-1)2\pi/p$. Thus the number of states is

$$S = pM^{L-1}$$

In the state trellis, there are M branches coming into a state and M branches leaving from a state. So the total number of in-branches is pM^L and number of out-branches is also pM^L.

The receiver observes the signal $r(t) = s(t, \mathbf{a}) + n(t)$, where the noise $n(t)$ is Gaussian and white. Here we explicitly show the data sequence $\mathbf{a}$ as an argument of the signal. The MLSD receiver maximizes the log likelihood function, or equivalently, the correlation (see Appendix B, Detection of M-ary Signals)

$$l(\widetilde{\mathbf{a}}) = \int_{-\infty}^{\infty} r(t)s(t, \widetilde{\mathbf{a}})dt$$

where $\widetilde{\mathbf{a}}$ is any possible data sequence which might be transmitted, one of them is $\mathbf{a}$, the actual transmitted one. A practical receiver can only observe signals for a finite period. Thus we rewrite the above as

$$l_k(\widetilde{\mathbf{a}}) = \int_0^{(k+1)T} r(t)s(t,\widetilde{\mathbf{a}})dt$$

and it can be written in a recursive form

$$l_k(\widetilde{\mathbf{a}}) = l_{k-1}(\widetilde{\mathbf{a}}) + Z_k(\widetilde{\mathbf{a}})$$

where

$$\begin{aligned} Z_k(\widetilde{\mathbf{a}}) &= \int_{kT}^{(k+1)T} r(t)s(t,\widetilde{\mathbf{a}})dt \\ &= \int_{kT}^{(k+1)T} r(t)\cos[2\pi f_c t + \theta(t,\widetilde{\mathbf{a}}_k) + \widetilde{\theta}_k]dt \end{aligned} \quad (6.51)$$

which is called the *metric* of the branch corresponding to the signal $s(t,\widetilde{\mathbf{a}})$ in the kth symbol interval (the amplitude of $s(t,\widetilde{\mathbf{a}})$ is normalized to 1 since it is the same for all signals). This metric is the correlation between the received signal and $s(t,\widetilde{\mathbf{a}})$ over the kth symbol interval.

The Viterbi algorithm (VA) recursively accumulates the branch metrics up to the kth symbol interval for certain paths in the trellis and chooses those paths that have the maximum *path metrics*. Since the VA does not search all paths, the algorithm is much more efficient than the exhaustive search. Yet the VA can guarantee to find the maximum likelihood path. We will describe the VA in detail shortly.

Now we have to construct a receiver that can produce the branch metric $Z_k(\widetilde{\mathbf{a}})$ in an efficient way. Since the number of branches (in or out) in the trellis is pM^L, there are pM^L different values of $Z_k(\widetilde{\mathbf{a}})$. It is well known that a correlator can be implemented as a matched filter with its output sampled at the end of the integration period. Thus we want to realize (6.51) by matched filters. The signal can be written in quadrature form as (see (6.42) to (6.44))

$$s(t,\mathbf{a}) = AI(t,\mathbf{a})\cos(2\pi f_c t) - AQ(t,\mathbf{a})\sin(2\pi f_c t)$$

and the received noise is bandpass noise which can also be put in a quadrature form

$$n(t) = x(t)\cos(2\pi f_c t) - y(t)\sin(2\pi f_c t)$$

Thus $r(t)$ can be written as

$$\begin{aligned} r(t) &= s(t,\mathbf{a}) + n(t) \\ &= \widehat{I}(t,\mathbf{a})\cos(2\pi f_c t) - \widehat{Q}(t,\mathbf{a})\sin(2\pi f_c t) \end{aligned} \quad (6.52)$$

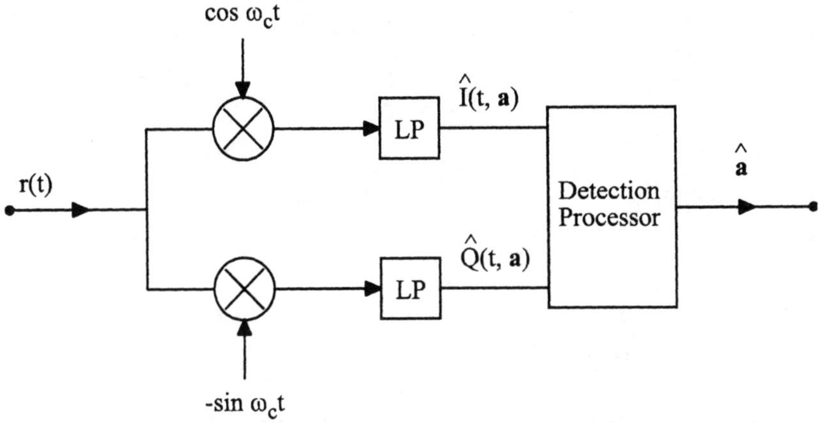

Figure 6.26 Basic quadrature receiver. From [9]. Copyright © 1986 Plenum.

where

$$\left.\begin{aligned} \widehat{I}(t,\mathbf{a}) &= AI(t,\mathbf{a}) + x(t) \\ \widehat{Q}(t,\mathbf{a}) &= AQ(t,\mathbf{a}) + y(t) \end{aligned}\right\}$$

These two quadrature components can be produced by the quadrature receiver in Figure 6.26. Using (6.52) in (6.51) and omitting double frequency terms and a constant of $1/2$, we have

$$\begin{aligned} Z_k(\widetilde{\mathbf{a}}_k, \widetilde{\theta}_k) \;=\; & \cos(\widetilde{\theta}_k) \int_{kT}^{(k+1)T} \widehat{I}(t,\mathbf{a}) \cos[\theta(t,\widetilde{\mathbf{a}}_k)]dt \\ & + \cos(\widetilde{\theta}_k) \int_{kT}^{(k+1)T} \widehat{Q}(t,\mathbf{a}) \sin[\theta(t,\widetilde{\mathbf{a}}_k)]dt \\ & + \sin(\widetilde{\theta}_k) \int_{kT}^{(k+1)T} \widehat{Q}(t,\mathbf{a}) \cos[\theta(t,\widetilde{\mathbf{a}}_k)]dt \\ & - \sin(\widetilde{\theta}_k) \int_{kT}^{(k+1)T} \widehat{I}(t,\mathbf{a}) \sin[\theta(t,\widetilde{\mathbf{a}}_k)]dt \end{aligned} \tag{6.53}$$

This can be implemented by $4M^L$ baseband filters with the impulse responses

$$h_{c,i}(t,\widetilde{\mathbf{a}}) \;=\; \begin{cases} \cos[\theta(T-t,\widetilde{\mathbf{a}}_k)] & \\ 0, & \text{for } t \text{ outside } [0,T] \end{cases}$$

$$= \begin{cases} \cos[2\pi h \sum_{j=-L+1}^{0} \tilde{a}_j q((1-j)T - t)] \\ 0 \qquad \text{for } t \text{ outside } [0,T] \end{cases} \tag{6.54}$$
$$(i = 1, 2, ..., 2M^L)$$

and

$$h_{s,i}(t, \widetilde{\mathbf{a}}) = \begin{cases} \sin[\theta(T-t, \widetilde{\mathbf{a}}_k)] \\ 0 \qquad \text{for } t \text{ outside } [0,T] \end{cases}$$
$$= \begin{cases} \sin[2\pi h \sum_{j=-L+1}^{0} \tilde{a}_j q((1-j)T - t)] \\ 0 \qquad \text{for } t \text{ outside } [0,T] \end{cases} \tag{6.55}$$
$$(i = 1, 2, ..., 2M^L)$$

There are M^L different $\widetilde{\mathbf{a}}_k$ sequences. For each sequence a pair of matched filters is needed for $\widehat{I}(t, \mathbf{a})$ and another pair is needed for $\widehat{Q}(t, \mathbf{a})$. Thus total $2M^L$ cosine matched filters and $2M^L$ sine matched filters are needed. Realizing that every $\widetilde{\mathbf{a}}_k$ sequence has a corresponding sequence with reversed sign, the number of matched filters can be reduced by a factor of 2. Figure 6.27, where $H = M^L$, is the optimum receiver based on (6.53) using $2M^L$ matched filters. The outputs of these filters are sampled at the end of the kth symbol interval, which produces the branch metrics $Z_k(\widetilde{\mathbf{a}}_k, \widetilde{\theta}_k)$. The processor in the figure is the Viterbi processor.

The Viterbi algorithm was discovered and analyzed by Viterbi [25] in 1967 for decoding convolutional codes. Later on it was extended to maximum likelihood sequence estimation (MLSE) for channels with intersymbol interference [26] and for partial response continuous phase modulation [8]. All of them have one thing in common: there is memory in the received signals. That is, the received signal values are determined by not only the current symbol, but also some or all previous symbols. Each of the possible combinations of the previous symbols constitutes a state. In the CPM case, the signal value is also affected by the cumulate phase θ_k (see Section 6.1.2). Assuming that $L-1$ is the memory length, the symbol is M-ary, and θ_k has p different values, then there are total $S = pM^{L-1}$ states. Each of the possible transmitted sequence is a path in the S-state trellis. The MLSE receiver searches through the trellis to find the path which best matches to the transmitted sequence in the maximum likelihood sense. The Viterbi algorithm is an efficient algorithm to implement the ML search. The brutal-force search (i.e., test each path) is too time consuming. It needs to test M^N paths, where N is the number of symbols in a frame of data. To get an idea of the number of paths, let us assume $M = 2$ (binary), $N = 64$ (8 bites), then $M^N = 1.89 \times 10^{19} = 1.89 \times 10^{10}$ billion paths. Searching this trellis is impossible in practice. However, with the Viterbi algorithm, as we will see shortly, the number of paths to be searched is only the number of states:

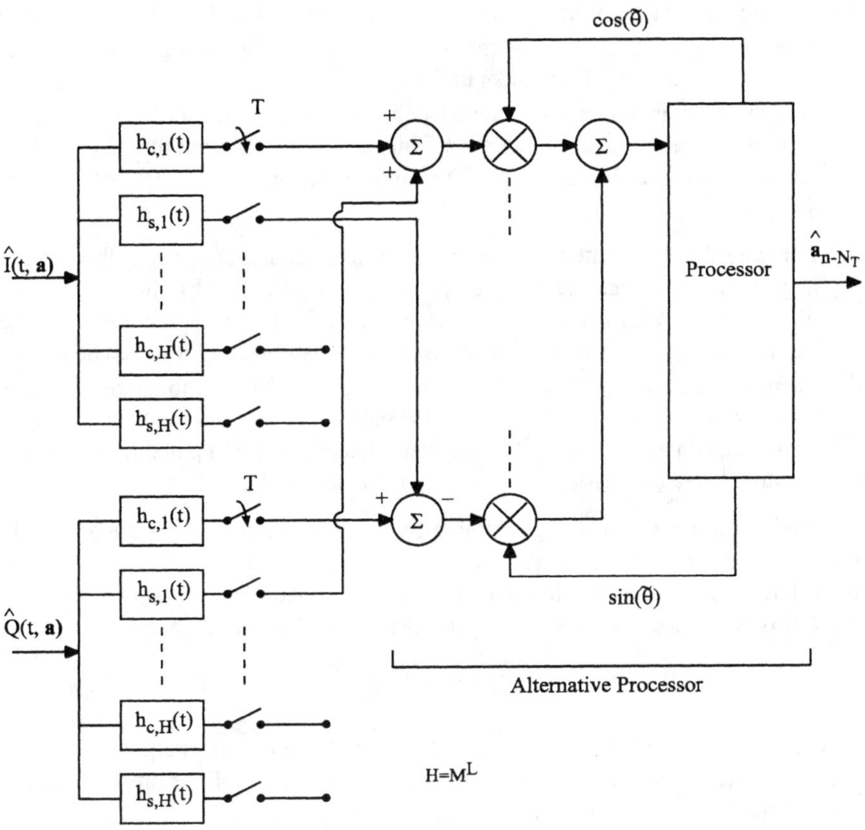

Figure 6.27 Matched filter bank. From [8]. Copyright © 1981 IEEE.

$S = pM^{L-1}$. Usually L and p are small integers. For example, if $L = 3$, $p = 3$, then $S = 12$ (Figure 6.5). The savings on the search time are tremendous, and the search time of the Viterbi algorithm is independent of the sequence length.

Now we describe the Viterbi algorithm for the partial response CPM case. Refer to Figure 6.5 while reading the following steps of the Viterbi algorithm.

1. At time $(k+1)T$, for any state node, compute the M ($M = 2$ in Figure 6.5) branch metrics $Z_k(\widetilde{\mathbf{a}}_k, \widetilde{\theta}_k)$ for the incoming M branches using (6.53). This is done by the matched filter banks in Figure 6.27.
2. Add the M branch metrics to the M path metrics $l_{k-1}(\widetilde{\mathbf{a}}_k)$ which are the path metrics for the M paths up to time kT and are connected with the M branches, respectively. Thus the path metrics are updated to time $(k+1)T$. Denote them as $l_k^{(i)}(\widetilde{\mathbf{a}}_{k+1})$, $i = 1, 2, ..., M$.
3. Compare the M updated path metric $l_k^{(i)}(\widetilde{\mathbf{a}}_{k+1})$ and choose the path with the largest path metric as the *survivor* path of the state node. Eliminate the other $M-1$ paths. When step one to three are done for all states, there is only one survivor path for each state. The survivor paths' data and metrics must be stored.
4. Then advance time by T and repeat steps one to three until demodulation reaches the end of the sequence. Then choose the path with the largest path metric as the demodulated path, which is the maximum likelihood (ML) path. By retrieving the stored data of the ML path, we obtain the demodulated data.

We have asserted that the path found by the Viterbi algorithm is the ML path. This assertion is not difficult to prove. Assume that the ML path is eliminated by the algorithm at time t_i. This implies that the partial path metric of the survivor (denoted as $l(t_i)$) is larger than that of the ML path (denoted as $l_{ML}(t_i)$). That is

$$l(t_i) > l_{ML}(t_i)$$

Now if the remaining (future) portion of the ML path (denote its metric as $l_{ML}(t > t_i)$) is appended on to the survivor at time t_i, the total metric will be greater than the total metric of the ML path. That is

$$l(t_i) + l_{ML}(t > t_i) > l_{ML}(t_i) + l_{ML}(t > t_i)$$

However, this is impossible since the right-hand side is the ML path metric which is supposed to be the largest. Hence the ML path cannot be eliminated by the VA.

References on the Viterbi algorithm are widely available, such as [25], [27], and [28] or any coding book.

Since CPM signal is normally not in a block structure, a modified VA is used for the CPM, much the same way as a modified VA is used for decoding convolutional codes [28]. It puts out one symbol at a time successively after a decoding delay that is the same for every symbol, and it accepts sequences of indefinite length. The surviving paths at each state of the trellis are saved only back to a certain point, a length called path memory N_T. The decision is made on the oldest symbol by certain criterion, such as taking the symbol that belongs to the majority of the surviving paths, or the symbol that belongs to the maximum likelihood path, or by simply choosing at random. The best criterion depends on conditions like SNR. The choice of N_T can be guided by the distance property of the CPM. If N_B is the length for the paths to reach the upper bound of the minimum distance, then the path memory N_T should be at least N_B, and experiment has shown that N_B is almost always enough [9].

The error probability of the perfect Viterbi demodulator, of course, is the one of the MLSD as given in Section 6.3.1, particularly, (6.39) and (6.40). The error probability of the modified Viterbi demodulator should be very close to the perfect one, especially at high SNR.

Simulation results of the error probability of the modified Viterbi receiver for some representative CPM schemes, such as 8-ary CPFSK and binary 3RC are shown in Figures 6.28 and 6.29 [9]. In Figure 6.28, the path memory has been set to 50 symbol intervals, which is more than enough for the minimum distance to occur. Twenty sample points per symbol interval have been used. Figure 6.28 shows two 8-ary CPFSK schemes, one with $h = 1/4$ and the other with $h = 5/11$. The 8-ary symbols are Gray coded. Figure 6.29 shows the results for binary 3RC with $h = 4/5$ when $N_T = 1, 2, ..., 20$ and $N_T = \infty$. By increasing N_T the upper bound is improved at high SNR, but it is still loose at low SNR. The asymptotic gain is about 2 dB over QPSK, which can be fairly accurately predicted by $d^2_{\min} = 3.17$ ($d^2_{\min} = 2$ for QPSK). This conclusion holds for many schemes. That is, $d_{\min}$ is sufficient for the characterization of performance in terms of symbol error probability.

More results have been obtained for a wide variety of CPM schemes in the literature, see for instance [29] and [30].

The complexity of the matched filter bank and the Viterbi algorithm limit the implementation of such an optimum receiver to small values of M and L. In the following sections, we will present some important research results on reduction of complexity of CPM receivers, sometimes at expenses of error performance.

6.5.4 Reduced-Complexity Viterbi Demodulator

A class of reduced-complexity Viterbi detectors for partial response CPM schemes has been proposed in [31]. The key concept is that the approximate receiver is based on a less complex CPM scheme than the transmitted scheme. The less complex scheme is a scheme with a shorter frequency pulse length and sometimes simpler frequency

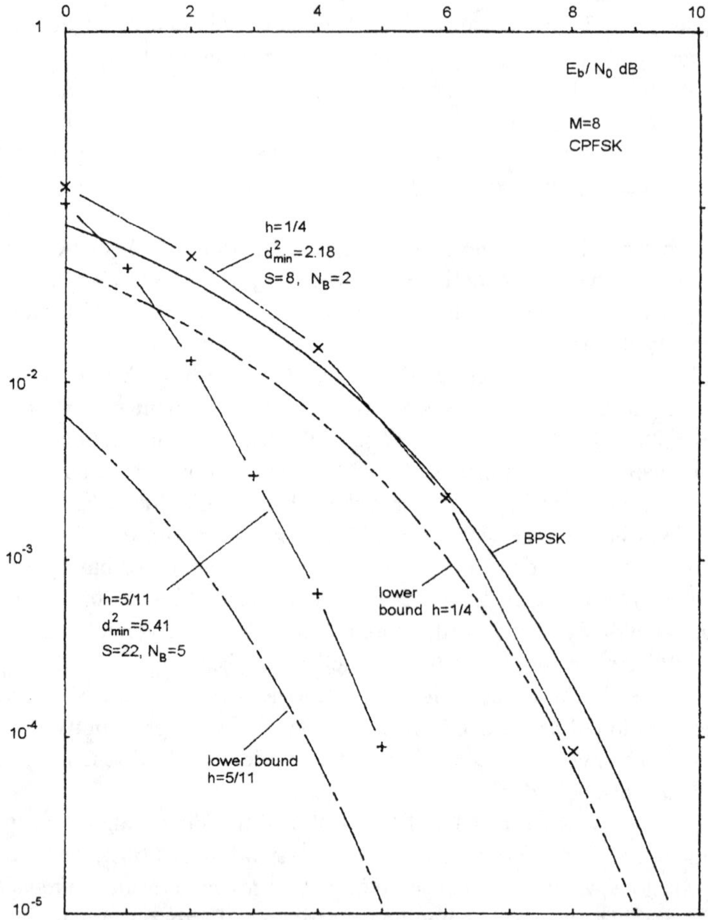

Figure 6.28 Simulated P_b for 8-ary CPFSK with comparisons with asymptotic lower bounds. Modified Viterbi receiver. From [9, p. 261]. Copyright © 1986 Plenum.

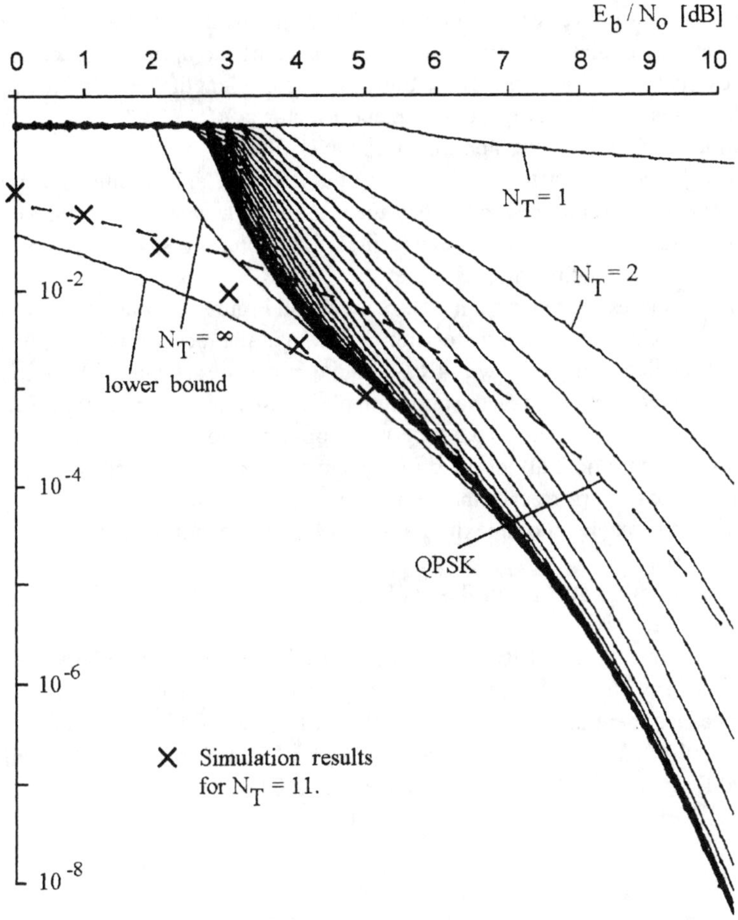

Figure 6.29 Simulated P_b for binary 3RC, $h = 4/5$, $N_T = 11$. Upper bounds for $N_T = 1, 2..., 20$ and a lower bound are also shown. Modified Viterbi receiver. From [8]. Copyright © 1981 IEEE.

pulse. Figure 6.30 shows the transmitter phase tree for 3RC and the receiver phase tree for 2REC which is used to approximate the transmitter phase tree. The 2REC phase tree is shown with a small phase offset ($T/2$ in this case). From the figure we can see that the two trees are very close. Since all the data information is carried by the phase tree, we have reason to believe that using the approximate phase tree for demodulation can almost achieve the performance of the optimum receiver which uses the transmitter phase tree. If the transmitter pulse length is L_T and the receiver pulse length is L_R, the complexity reduction factor is $M^{(L_T - L_R)}$ in terms of both the number of receiver states and that of receiver filters.

For a particular transmitter frequency pulse $g_T(t)$, an optimum receiver frequency pulse $g_R(t)$ can be found so that the minimum distance of the receiver trellis is maximized. In particular, a piecewise linear function for $0 < t < L_R T$ is defined in [31] and used for optimization.

Simulation results reported in [31] are shown in Figures 6.31 and 6.32. In Figure 6.31, 4RC-4RC2 means the transmission is 4RC and the receiver is based on a 4RC pulse truncated to a length of two. 4RC-2T0.5 denotes the scheme with a transmitter of 4RC and a receiver based on the optimum piecewise linear function with $L_R = 2T$ and an index $h = 0.5$. The 2T0.5 receiver is optimum to the 4RC transmitter. The notations in Figure 6.32 are defined in the same manner. From the figures we can see that the loss of error performance is very small, which is about 0.1 dB at high SNR, while the receiver complexity reduction factor is two to four.

6.5.5 Reduction of the Number of Filters for LREC CPM

It has been shown that for LREC CPM, the size of the matched filter bank need only increase linearly with L [32]. In Section 6.5.3, we stated that there are $2M^L$ matched filters because there are M^L different values of $\theta(t, \tilde{\mathbf{a}}_k)$. Here we can show that there are only $L(M-1)+1$ different $\theta(t, \tilde{\mathbf{a}}_k)$ for LREC CPM. Thus the number of matched filters is only $2L(M-1)+2$ for LREC CPM.

The LREC phase pulse is given by

$$q(t) = \begin{cases} 0, & t < 0 \\ \frac{t}{2LT}, & 0 \le t < T \\ 1/2, & t \ge LT \end{cases}$$

In the interval $[0, T]$, it can be easily verified using an example (say, 3REC),

$$a_0 q(t) + a_{-1} q(t+T) + \cdots + a_{-(L-1)} q(t + (L-1)T) = Aq(t) + B$$

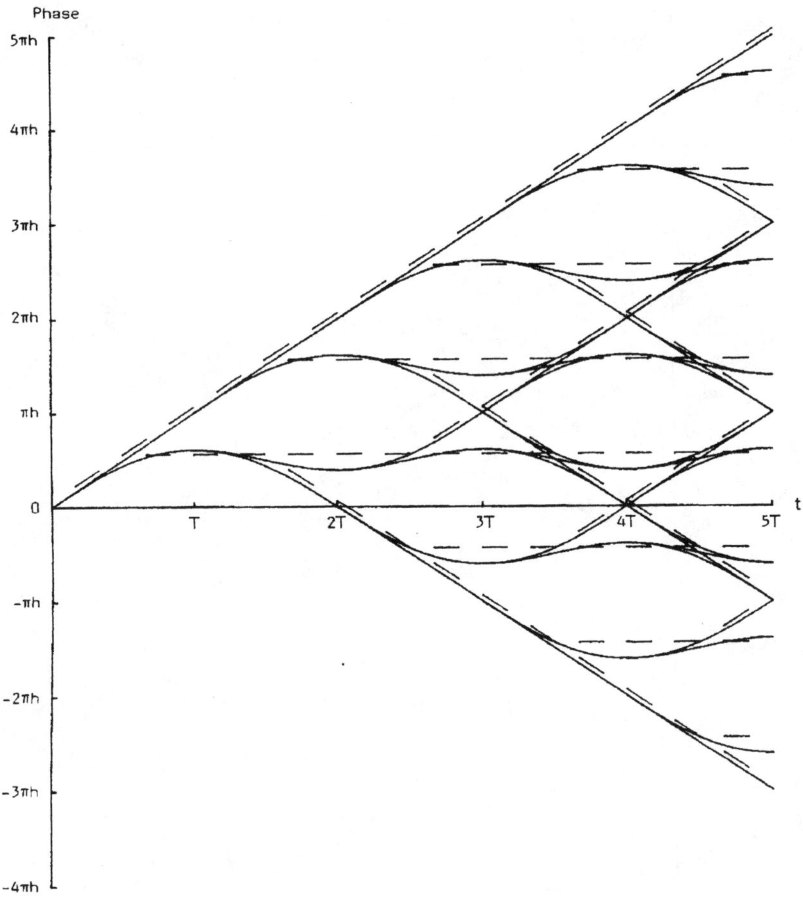

Figure 6.30 Transmitter phase tree (solid) for 3RC and receiver phase tree for 2REC. Correct timing, small phase offset (T/2). From [31]. Copyright © 1984 IEEE.

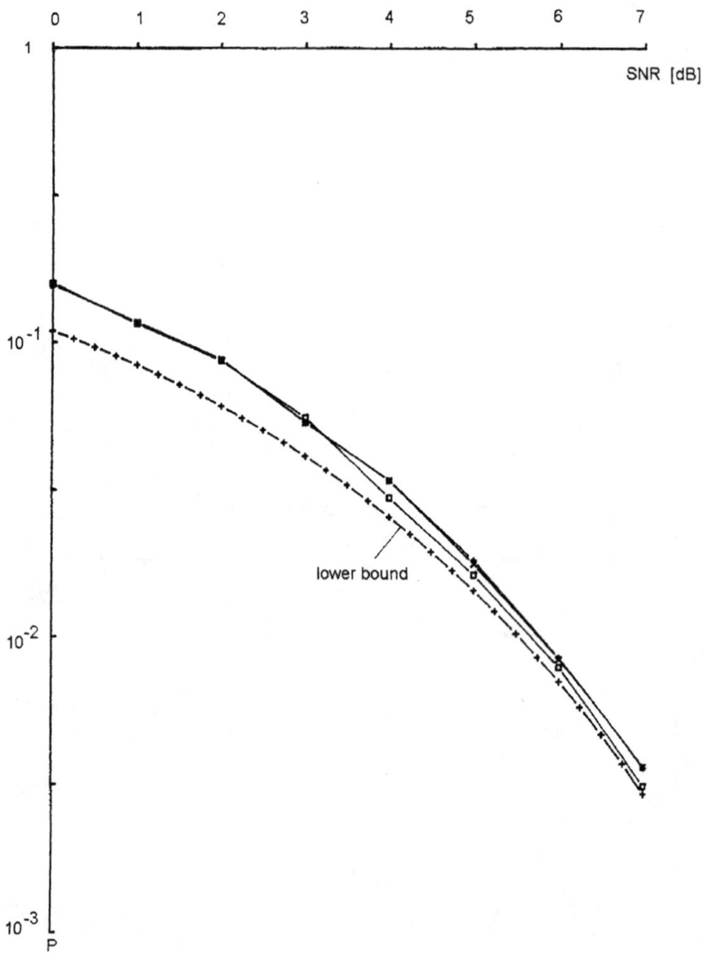

Figure 6.31 Simulated P_b for 4RC-4RC2 (+), 4RC-2T0.5 ($\times$), and optimum 4RC ($\square$) for $h = 1/2$. The lower bound for the optimum receiver is also shown. The results are based on 2500 errors. From [31]. Copyright © 1984 IEEE.

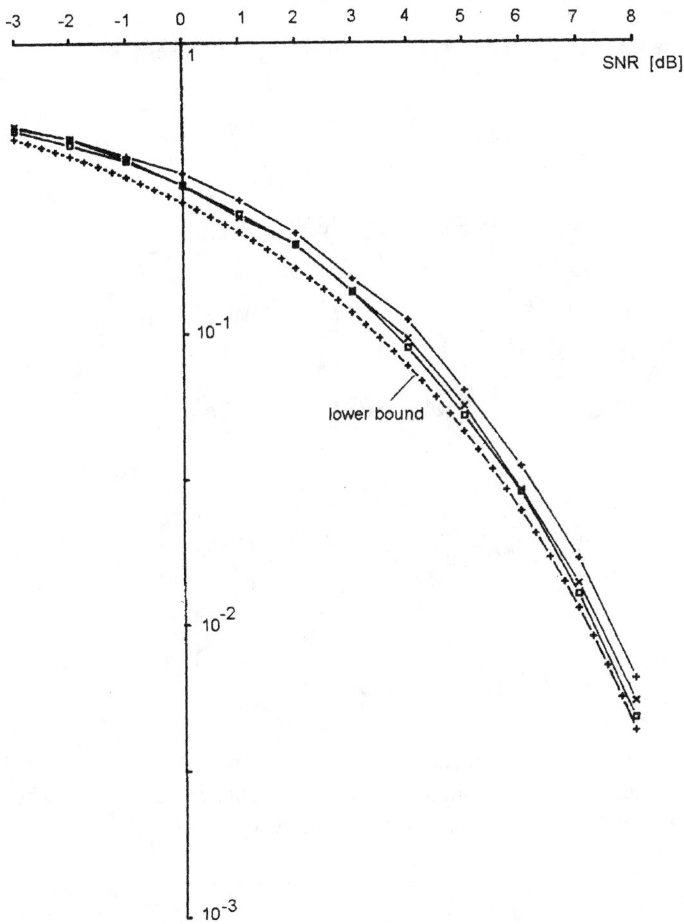

Figure 6.32 Simulated P_s for quaternary 2RC-2RC1 (+), 2RC-1T0.25 ($\times$), and optimum 2RC ($\square$) for $h = 1/4$. The lower bound for the optimum receiver is also shown. The results are based on 2500 errors. From [31]. Copyright © 1984 IEEE.

where

$$A = \sum_{i=0}^{L-1} a_{-i}, \qquad B = \sum_{i=0}^{L-1} a_{-i}(\frac{i}{2L})$$

During the interval $[kT, (k+1)T]$, the time-varying phase

$$\begin{aligned} \theta(t, \mathbf{a}_k) &= 2\pi h \sum_{i=k-L+1}^{k} a_i q(t - iT) \\ &= 2\pi h q(t - kT) \sum_{i=0}^{L-1} a_{k-i} + 2\pi h \sum_{i=0}^{L-1} a_{k-i}(\frac{i}{2L}) \end{aligned}$$

Where $a_j \in \{-(M-1), ..., -1, 1, ..., (M-1)\}$. We can write $a_j = 2u_j - (M-1)$ where $u_j \in \{0, 1, ..., (M-1)\}$. Then

$$\theta(t, \mathbf{a}_k) = \theta_t(t, \mathbf{a}_k) + \theta_c(\mathbf{a}_k)$$

where

$$\theta_t(t, \mathbf{a}_k) = 4\pi h q(t - kT) \sum_{i=0}^{L-1} [u_{k-i} - \frac{M-1}{2}]$$

$$\theta_c(\mathbf{a}_k) = 4\pi h \sum_{i=0}^{L-1} [u_{k-i} - \frac{M-1}{2}](\frac{i}{2L})$$

Since $\sum_{i=0}^{L-1} u_{k-i} \in \{0, 1, ..., L(M-1)\}$, there are only $L(M-1)+1$ different $\theta_t(t, \mathbf{a}_k)$. Thus the ML demodulator only needs $2L(M-1)+2$ matched filters to compute the metrics for all pM^L states as follows.

$$\begin{aligned} Z_k(\widetilde{\mathbf{a}}_k, \widetilde{\theta}_k) &= \cos[\widetilde{\theta}_k + \theta_c(\widetilde{\mathbf{a}}_k)] \int_{kT}^{(k+1)T} \widehat{I}(t) \cos[\theta_t(t, \widetilde{\mathbf{a}}_k)] dt \\ &+ \cos[\widetilde{\theta}_k + \theta_c(\widetilde{\mathbf{a}}_k)] \int_{kT}^{(k+1)T} \widehat{Q}(t) \sin[\theta_t(t, \widetilde{\mathbf{a}}_k)] dt \\ &+ \sin[\widetilde{\theta}_k + \theta_c(\widetilde{\mathbf{a}}_k)] \int_{kT}^{(k+1)T} \widehat{Q}(t) \cos[\theta_t(t, \widetilde{\mathbf{a}}_k)] dt \\ &- \sin[\widetilde{\theta}_k + \theta_c(\widetilde{\mathbf{a}}_k)] \int_{kT}^{(k+1)T} \widehat{I}(t) \sin[\theta_t(t, \widetilde{\mathbf{a}}_k)] dt \end{aligned}$$

The complexity of the Viterbi processor remains unchanged since the number of states is still pM^L.

This simplified receiver also can be applied to any CPM with a piecewise linear pulse. For smoother pulses one can use a piecewise linear approximation to the pulse [32].

6.5.6 ML Block Detection of Noncoherent CPM

Maximum likelihood block detection of noncoherent full-response CPM is proposed in [33]. The derivation of this receiver starts with the likelihood function conditioned on the received signal phase. Then averaging it over the random phase leads to the likelihood function which in turn gives the sufficient statistics for decision making.

Assume the received signal $r(t)$ is observed for an N-symbol period. Denoting the M-ary symbol set as $\{\Delta_i = -M + (2i-1); i = 1, 2, ..., M\}$, let $\mathbf{i} = (i_1, i_2, ..., i_N)$ be a sequence of indexes whose elements take on values from the set of integers $\{1, 2, ..., M\}$. The detection rule is as follows. At nth symbol time, for each particular input data vector $\Delta = (\Delta_{i_1}, \Delta_{i_2}, ..., \Delta_{i_N})$ compute

$$\beta_{\mathbf{i}} = \sum_{k=1}^{N} \Gamma_{i_k, N-k} C_k$$

where

$$\Gamma_{ij} = \int_{(n-j)T}^{(n-j+1)T} r(t) e^{-j2\pi h \Delta_i q(t-(n-j)T)} dt$$

and the C_is are complex constants defined recursively as follows

$$C_1 = 1; \qquad C_{k+1} = C_k e^{-j\pi h \Delta_{i_k}}; \qquad k = 1, 2, ..., N-1$$

Then

$$\text{choose } a_{n-N+1} = \Delta_{i_1^*}, a_{n-N+2} = \Delta_{i_2^*}, ..., a_n = \Delta_{i_N^*}, \text{ if } |\beta_{\mathbf{i}}|_{\max} = |\beta_{\mathbf{i}^*}|$$

where $\mathbf{i}^* = (i_1^*, i_2^*, ..., i_N^*)$ is a particular value of $\mathbf{i}$ and $\Delta^* = (\Delta_{i_1^*}, \Delta_{i_2^*}, ..., \Delta_{i_N^*})$ is the corresponding input data vector.

The implementation of the above rule is straight forward in complex form (Figure 6.33). For MSK (1REC, $h = 1/2$), the receiver can be implemented in a real I-Q form as shown in Figure 6.34. The receiver has a front end (correlators to produce Γ_{ij}) analogous to that for a coherent CPM receiver. What is different from the coherent ML receiver is that the number of the correlators is M, not M^L. The complexity of the processor that follows depends on the length of the observation. This

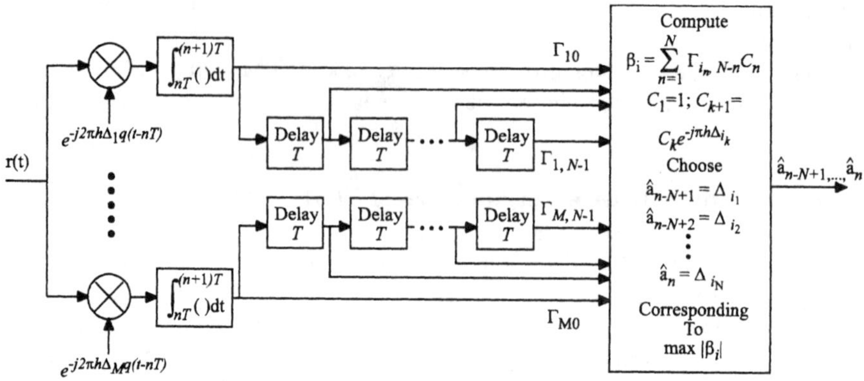

Figure 6.33 ML block estimator for noncoherent CPM based on an N-symbol observation. From [33]. Copyright © 1993 IEEE.

structure is much less complex than that of the coherent one.

The error performance analysis in [33] shows that large gain can be achieved by making block-by-block decisions in proportion to the length of the observation rather than making symbol-by-symbol decisions based on the same observation length. Figure 6.35 shows the P_b upper bounds for 1REC CPM with $h = 0.5$ (MSK) with N as a parameter. It is evident that dramatic improvement in P_b can be obtained with moderate values of N.

It should be pointed out that Figure 6.35 curves compare poorly with Figure 6.23 which is also for noncoherent CPM (not exactly MSK since $h = 0.715$). However, the noncoherent detector for Figure 6.23 detects only one symbol based on N (odd) observations, while the detector here detects a block of N symbols based on N observations. So the speed is N times faster here.

6.5.7 MSK-Type Demodulator

Recall that MSK is a member of the CPM class. It is a 1REC CPM with $h = 1/2$. Its demodulation can be performed by a parallel-type receiver (PMSK), or equivalently, by a serial-type receiver (SMSK), as studied in Chapter 5. Research shows that the MSK-type receiver works well for binary CPM schemes with $h = 1/2$ [9, 34–36]. Of course this receiver is not optimum in general, but the performance of this type of receiver is almost equal to the optimum Viterbi receiver for schemes with a moderate degree of smoothing, that is, overlapping frequency pulses of length L up to three to four symbol intervals, such as 3RC, 4RC, TFM, and some GMSK schemes.

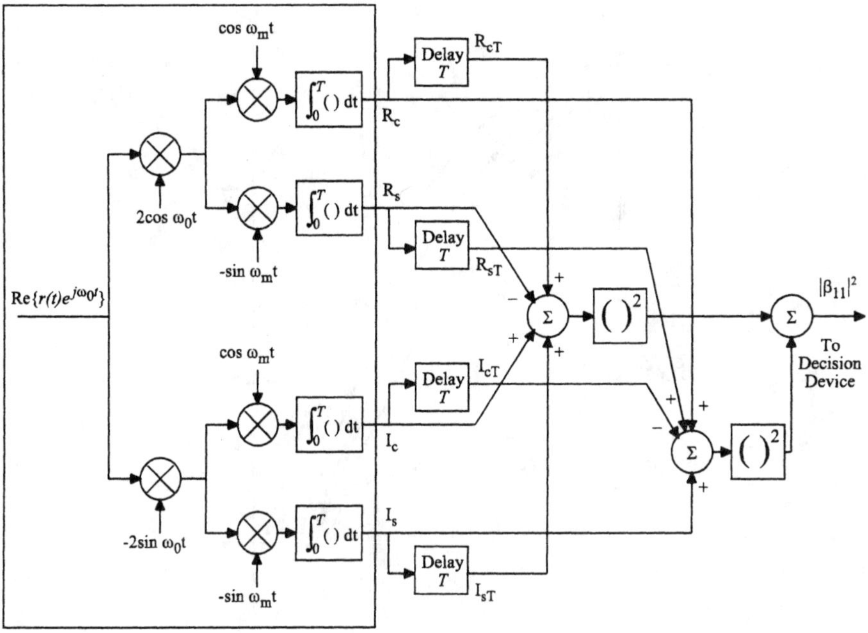

Figure 6.34 A real I-Q implementation of an ML block estimator for noncoherent MSK based on a two-symbol observation. From [33]. Copyright © 1993 IEEE.

Figure 6.36 is the structure of a parallel MSK-type receiver, where perfect carrier recovery and symbol timing are assumed, that is, the demodulation is coherent. This receiver works for binary CPM schemes with $h = 1/2$ because these schemes have $\cos[\Phi(t, \mathbf{a})]$ eye patterns that have maximum open at $t = (2n+1)T$, $n = 0, 1, 2, \ldots$. Figure 6.37 is the $\cos[\Phi(t, \mathbf{a})]$ eye pattern for 3RC, $h = 1/2$ CPM [36]. If the filter $a(t)$ is an ideal low-pass filter, the output of the upper arm of the receiver is exactly $\cos[\Phi(t, \mathbf{a})]$ and the output of the lower arm is exactly $\sin[\Phi(t, \mathbf{a})]$. The eye pattern of $\sin[\Phi(t, \mathbf{a})]$ is the same as $\cos[\Phi(t, \mathbf{a})]$ but shifted by T in time. Therefore the receiver samples the upper arm and lower arm at $t = (2n+1)T$ and $t = 2nT$, alternatively. The "decision logic" observes the sampled signals for a length of N_T symbol intervals and makes an optimum decision on one symbol [9, p. 297]. The decision logic also performs differential decoding (for differential encoded data) and multiplexing. The filter $a(t)$ can be optimized to minimize the average symbol error probability [34, 37].

As in the MSK case, the parallel receiver in Figure 6.36 can be replaced by an

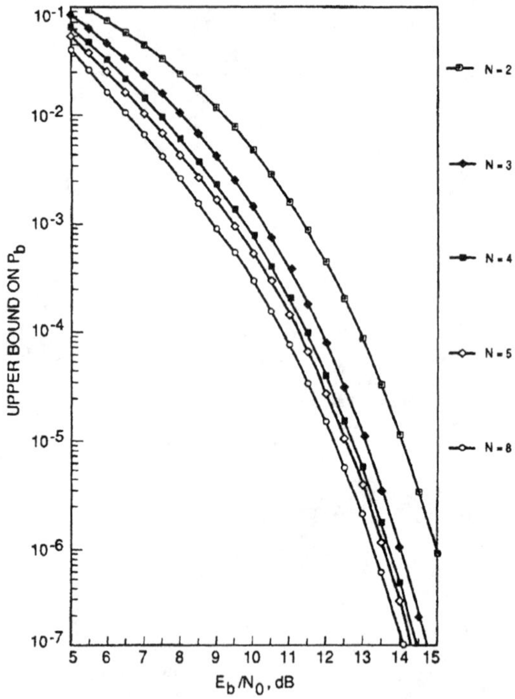

Figure 6.35 Upper bounds on P_b for MSK with multiple-symbol noncoherent detection. From [33]. Copyright © 1993 IEEE.

equivalent serial MSK-type receiver (Figure 6.38). For MSK, the filter $a(t)$ in the parallel receiver is [38]

$$a(t) = \begin{cases} \cos(\frac{\pi t}{2T}), & |t| \leq T \\ 0, & \text{otherwise} \end{cases}$$

and the corresponding serial filters are

$$h_1(t) = \begin{cases} \cos^2(\frac{\pi t}{2T}), & |t| \leq T \\ 0, & \text{otherwise} \end{cases}$$

$$h_2(t) = \begin{cases} -\frac{1}{2}\sin(\frac{\pi t}{T}), & |t| \leq T \\ 0, & \text{otherwise} \end{cases}$$

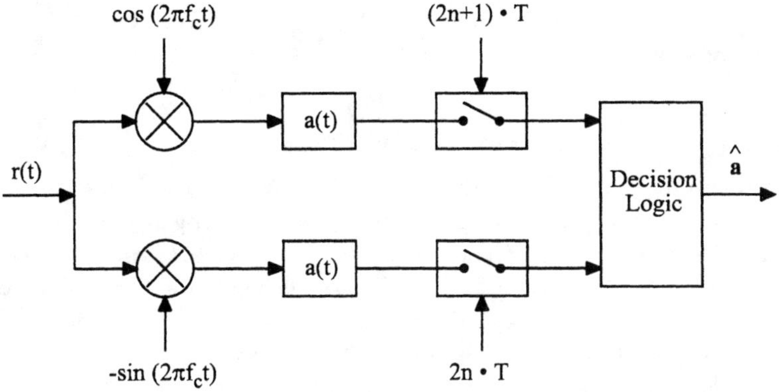

Figure 6.36 Parallel MSK-type receiver for binary CPM with $h = 1/2$. From [35]. Copyright © 1985 IEEE.

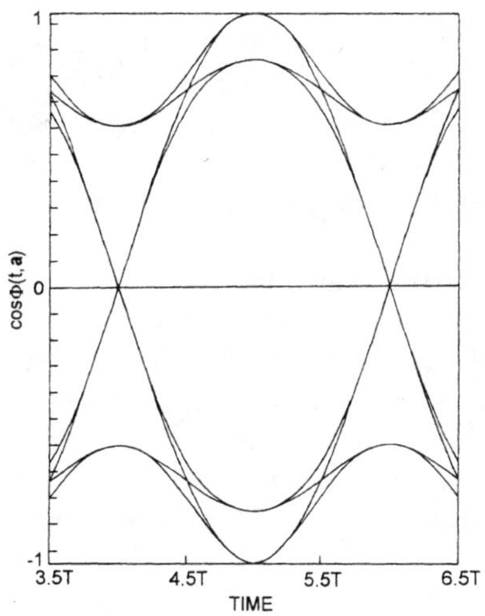

Figure 6.37 cos$\Phi(t, \mathbf{a})$ eye pattern for parallel MSK-type receiver for binary 3RC, $h = 1/2$, scheme. From [36]. Copyright © 1982. AT&T. All rights reserved. Reprinted with permission.

(These two filters are denoted as $h_{mI}(t)$ and $h_{mQ}(t)$ in Figure 5.15, Chapter 5. Their frequency responses are shown in Figure 5.17.) The above filters in fact satisfy the following relations

$$\begin{cases} h_1(t) = a(t)\cos(\frac{\pi t}{2T}) \\ h_2(t) = -a(t)\sin(\frac{\pi t}{2T}) \end{cases} \tag{6.56}$$

It is therefore reasonable to assume that the serial receiver could also be used for general $h = 1/2$ binary CPM with the filters defined according to (6.56), and $a(t)$ the corresponding filter for the parallel receiver. Note that the local reference frequency of the serial receiver is $f_c - 1/4T$. Thus the lowpass signals in the quadrature arms of the serial receiver are $\cos[\Phi(t, \mathbf{a}) + \frac{\pi t}{2T}]$ and $\sin[\Phi(t, \mathbf{a}) + \frac{\pi t}{2T}]$ before the filters. The eye pattern of the sum of these two signals is shown in Figure 6.39 for binary, $h = 1/2$, 3RC scheme [35]. It can be seen that the serial eye pattern has open eyes at every $t = nT$. The sum of these signals are then sampled at every $t = nT$ and sent to the decision logic for decision making and differential decoding if needed.

It was shown that the serial and parallel receivers have equal performance, assuming perfect phase and time synchronization [35]. Figure 6.40 shows the error probability P in estimating the phase node θ_n for some binary CPM schemes with optimum filters in the MSK-type (parallel or serial) receiver. The bit error probability is about $2P$ [9, p. 299].

Serial and parallel MSK-type detection with phase errors and timing errors were compared for partial response CPM. It was found that the serial MSK-type receiver is less sensitive to phase errors while the parallel MSK-type receiver is less sensitive to symbol timing errors. Assuming that it is easier to obtain an accurate timing than phase synchronization, the serial receiver has an advantage over the parallel one.

6.5.8 Differential and Discriminator Demodulator

Besides the optimum and suboptimum receivers described so far in this chapter, there exists simple noncoherent receivers. Figure 6.41 shows two such simple noncoherent receivers, differential and discriminator receivers for binary partial response CPM [9]. In the differential receiver the output of the filter $A(f)$ is a signal with time-varying amplitude $\mathcal{R}(t, \mathbf{a})$ and a distorted phase $\psi(t, \mathbf{a})$. The output of the differential detector would then become

$$y(t) \propto \mathcal{R}(t, \mathbf{a})\mathcal{R}(t - T, \mathbf{a}) \sin \Delta\psi(t, T, \mathbf{a})$$

where $\Delta\psi(t, T, \mathbf{a}) = \psi(t, \mathbf{a}) - \psi(t - T, \mathbf{a})$ is the phase difference between the current symbol and the previous symbol. The eye pattern of this signal has open eyes at every $t = kT$. This signal is sampled and a hard decision is made on the sample.

In the case of the discriminator, the output of the discriminator is the derivative of

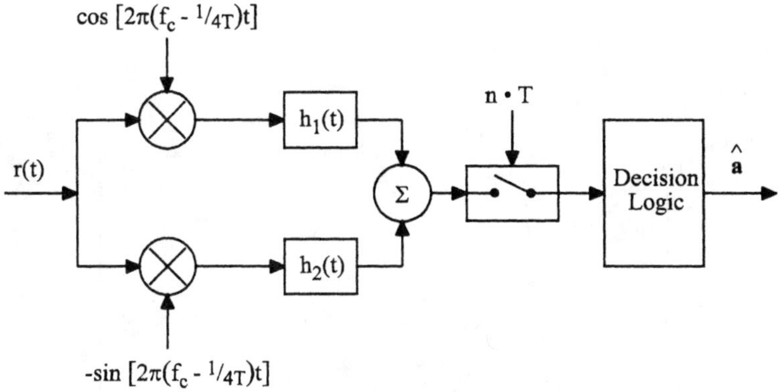

Figure 6.38 Serial MSK-type receiver for binary CPM with $h = 1/2$. From [35]. Copyright © 1985 IEEE.

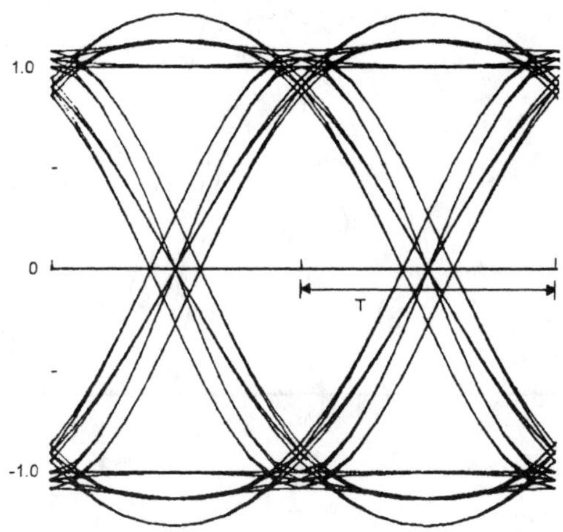

Figure 6.39 The eye pattern for serial MSK-type receiver for binary, $h = 1/2$, 3RC scheme. From [35]. Copyright © 1985 IEEE.

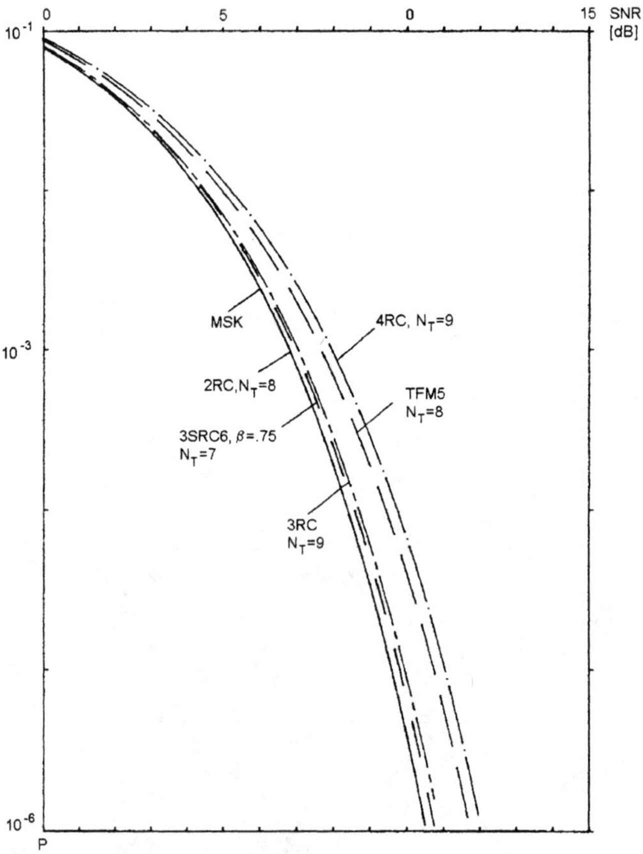

Figure 6.40 The error probability P for estimating phase node θ_n for some CPM schemes with optimum filters in the MSK-type (parallel or serial) receiver. The bit error probability is about $2P$. From [9, p. 320]. Copyright © 1986 Plenum.

the phase $\psi(t, \mathbf{a})$. An integrate-and-dump filter following the discriminator produces a phase difference $\Delta\Psi(t, T, \mathbf{a})$ which is slightly different from $\Delta\psi(t, T, \mathbf{a})$ [9, p. 269]. The hard decision detector decides $\widetilde{\mathrm{a}}_{n-1} = 1$ if $\Delta\Psi(t, T, \mathbf{a})$ >0, $\widetilde{\mathrm{a}}_{n-1} = -1$ otherwise.

The filter $A(f)$ is chosen as a raised cosine (LRC) type in [9, p. 265]. As to the error performance of these two receivers, it is very difficult to analyze due to the nonlinearity. Some numerical results are available in [9, section 7.5.3]. It is expected that the simple receivers would have losses in error performance. However, for good combinations of modulations and differential detectors or discriminators, the loss at error rate 10^{-6} is only about 2 dB compared to coherent MSK. Figures 6.42 and 6.43 show some numerical results for binary CPM schemes, where 1RC/2RC means that the CPM scheme is 1RC and the receiver filter $A(f)$'s impulse response is 2RC (for LRC filter, the longer the L, the narrower the bandwidth required). Differential detection is best suited for schemes with $h \approx 1/2$, so that the index in Figure 6.42 is $1/2$. It is seen from Figure 6.42 that 1RC/3RC is quite poor and 1RC/2RC is the best, with a loss of about 2.5 dB over coherent MSK at $P_b = 10^{-4}$. The error performance of discriminator detection of $h = 1/2$ is very close to that of the differential detector. Figure 6.43 shows some results for discriminator detection with $h = 0.62$. It turns out this index value makes the discriminator detection slightly better than differential detection: the loss of 1RC/2RC is about 2 dB over coherent MSK at $P_b = 10^{-4}$.

Sequence detection can also be applied to the differential and discriminator receivers since the memory in the continuous phase also can help in noncoherent reception. Receivers that use Viterbi processing following the discriminator [39–41,43] or differential detector [3,41–43] were proposed.

6.5.9 Other Types of Demodulators

In this section we present some less known, but maybe potentially useful CPM demodulators.

MSK and OQPSK are two well-known modulations that can be interpreted as a set of time/phase-shifted AM pulses. Laurent proved that any constant amplitude binary phase modulation can be expressed as a sum of a finite number of time limited amplitude modulated pulses (AMP) [44]. Based on this notion, Kaleh developed a coherent receiver using the Viterbi algorithm for binary partial response CPM [45]. The receiver's complexity is nearly the same as the old Viterbi receiver. However, the complexity can be reduced by using fewer matched filters and consequently a smaller number of VA states while maintaining a near optimum error performance. A linear filter receiver can also be derived from the AMP representation. An example given in [45] is the binary GMSK with $BT = 0.25$ and $h = 0.5$, for which the simplified VA receiver only needs two matched filters and a four-state Viterbi processor, while

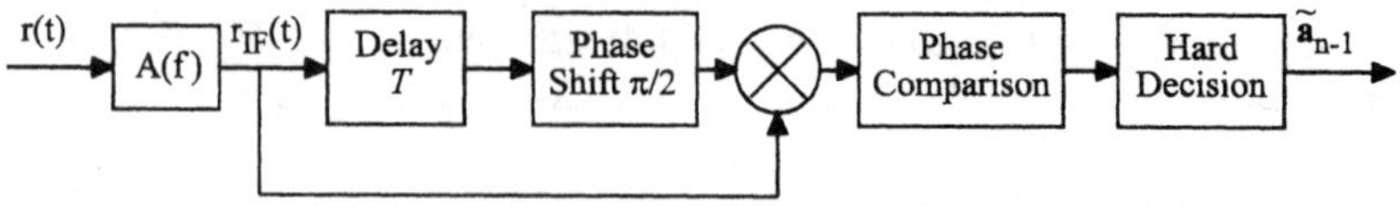

(a)

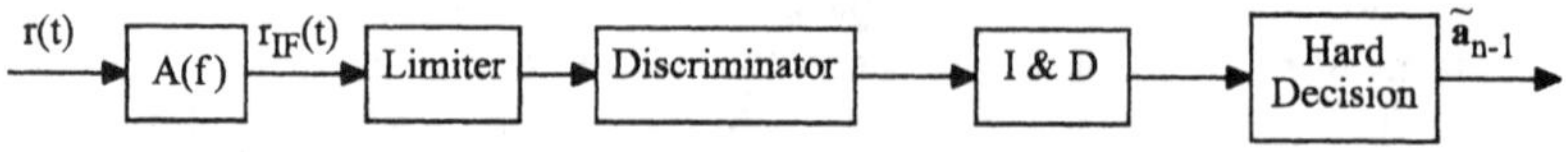

(b)

Figure 6.41 Differential (a) and discriminator (b) receiver for binary CPM schemes. From [9]. Copyright © 1986 Plenum.

the error performance loss is only about 0.24 dB. The error performance of the linear receiver is also only slightly worse than the simplified VA receiver.

The optimum ML coherent and noncoherent demodulators discussed in Sections 6.5.1 and 6.5.2 can be simplified. The idea is to use a matched filter that matches to the average of all signals having the same first symbol a_1. Thus the total number of filters (or correlators) is reduced by a factor of M^{N-1} where N is the length in symbols of the observation. This suboptimum receiver is called average matched filter (AMF) receiver and is given in [46]. However, the error performance of this type of AMF receiver is not good since the received signal is poorly matched. For example, the loss of CPFSK at $P_b = 10^{-4}$ is about 3 dB [21,46]. Another type of AMF receiver is proposed in [9] for partial response CPM. The impulse response of the filters is the average over all symbols except the prehistory and the decision symbol, where prehistory refers to the $L-1$ symbols before the decision symbol.

In Section 6.5.4 we have explored the reduced state Viterbi demodulator. On the other hand, complexity reduction of the ML demodulator can be achieved by using search algorithms other than the Viterbi algorithm. Sequential algorithms are a class of suboptimum algorithms, which search through the state tree or trellis along only one path. The best known sequential algorithms are the Fano algorithm and the stack algorithm [28]. Other algorithms compromise between the Viterbi algorithm and the sequential algorithms, thus maintaining most of the optimum error performance while reducing the computational complexity. Refer to [47] for a survey of search algorithms.

In the above discussion, either we assume that the demodulation is coherent (i.e.,

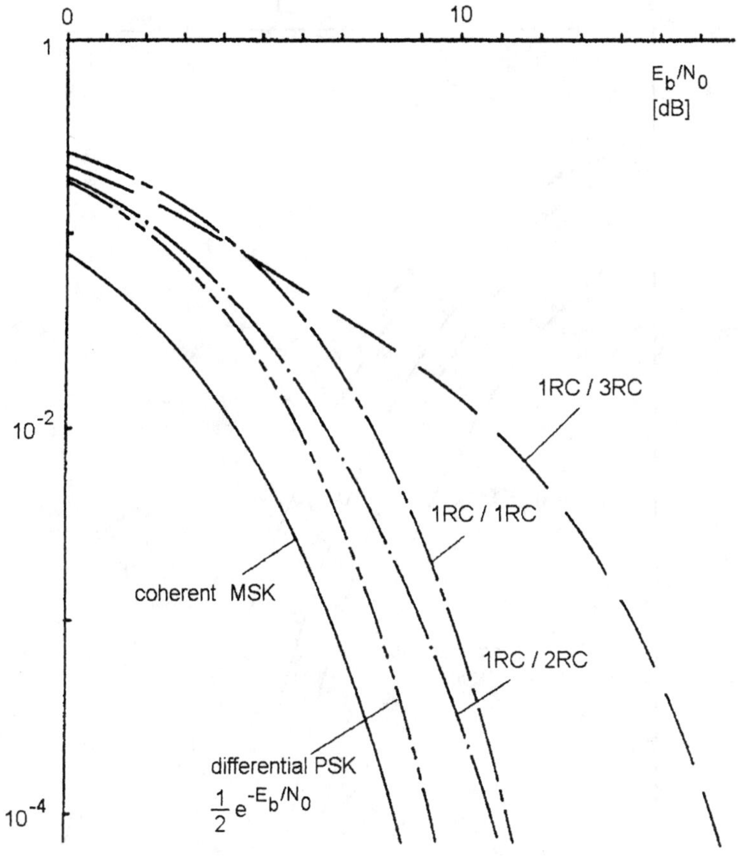

Figure 6.42 P_b for differential detection of 1RC with $h = 1/2$. From [9, p. 273]. Copyright © 1986 Plenum.

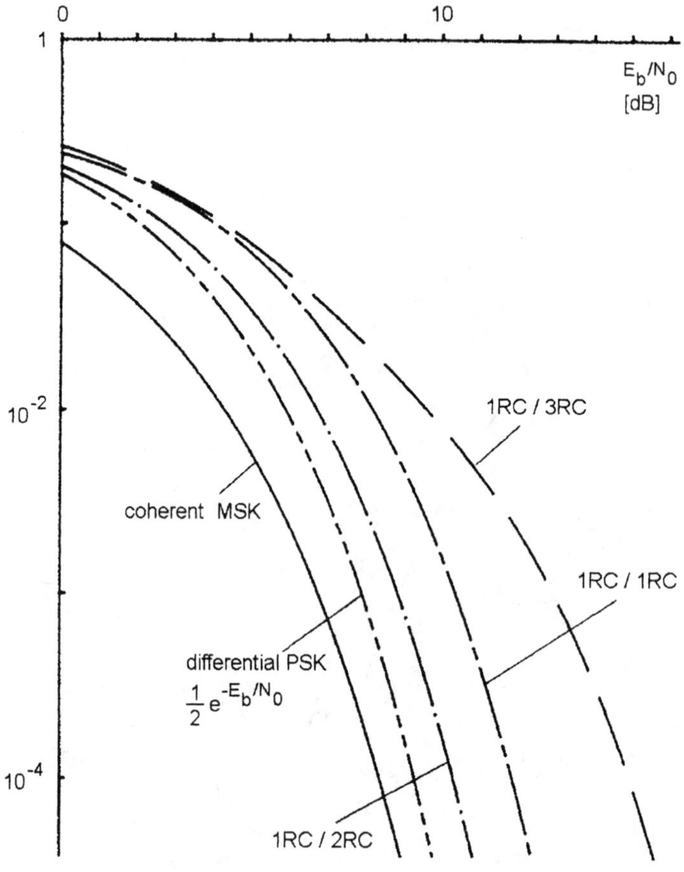

Figure 6.43 P_b for discriminator detection of 1RC with $h = 0.62$. From [9, p. 274]. Copyright © 1986 Plenum.

the received carrier phase is completely known (synchronized) to the receiver), or we assume that the demodulation is noncoherent (i.e., the received carrier phase is completely unknown (random) to the receiver). The complete randomness of the phase error in a noncoherent case is manifested by assuming it to have a uniform distribution in $[0, 2\pi]$. The practical situation of a coherent demodulator is in between. The best we can hope is that the carriers are synchronized with a phase error that fluctuates around a mean value zero. To deal with this situation, partial coherent receivers are considered in [9] where the phase error is assumed to have a PDF other than uniform, with various degrees of randomness.

Before we end this section, we would like to point out that demodulation for CPM has been, and still is, an active area of research. New ideas about the CPM demodulator are expected to appear in the literature from time to time in the future.

6.6 SYNCHRONIZATION

Perhaps synchronization is the most difficult part of the CPM technique. Synchronization techniques for special CPM cases, like MSK, are relatively mature and in practical use. In this section we will discuss synchronizers for general CPM schemes. Some are extended from the MSK synchronizer and others are based on new ideas. Research on this subject is still active. New results are expected to appear in the future.

6.6.1 MSK-Type Synchronizer

The synchronization technique for MSK [48] (Figure 5.11) was first generalized by Lee [49] to include M-ary data and any rational modulation index, but the frequency pulse is still 1REC. This synchronizer was further generalized by Sundberg [9] to be useful for any CPM schemes with rational index. Figure 6.44 is the generalized synchronizer. Now we explain how it works. The modulation index is assumed rational as $h = k_1/k_2$, where k_1, k_2 are nonzero integers. The nonlinearity in the synchronizer will raise the received signal to the power of k_2.

$$[r(t)]^{k_2} = [s(t) + n(t)]^{k_2}$$

and one of the resultant components is [50]

$$\frac{[2E_s/T]k_2/2}{2^{k_2-1}} \cos[2\pi k_2 f_c t + k_2 \Phi(t, \mathbf{a})]$$

The amplitude of this signal is not important in the following discussion. What is

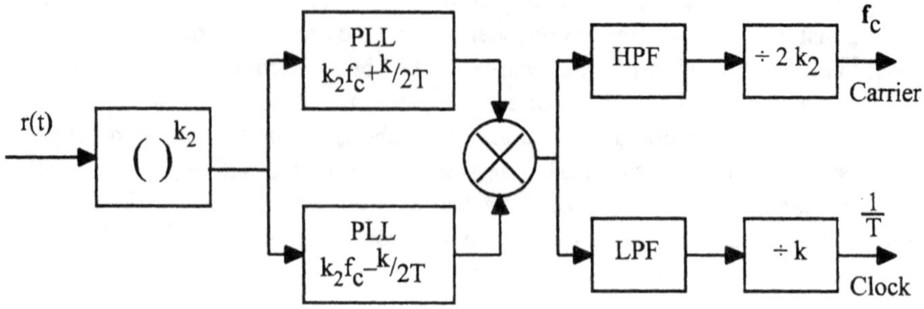

Figure 6.44 CPM synchronizer generalized from MSK synchronizer. After [9].

important is that this is a CPM signal with carrier frequency $k_2 f_c$ and phase

$$k_2\Phi(t, \mathbf{a}) = k_2 2\pi \frac{k_1}{k_2} \sum_i a_i q(t - iT) = 2\pi k_1 \sum_i a_i q(t - iT)$$

which has an integer index $h' = k_1$. Recall that in Section 6.2.1, we state that when h' is an integer the PSD of the CPM signal contains a discrete part. The discrete frequency components appear at

$$f = \begin{cases} k_2 f_c \pm 2k/2T, & k = 0, 1, 2, ..., \quad \text{for even } h' \\ k_2 f_c \pm (2k+1)/2T, & k = 0, 1, 2, ..., \quad \text{for odd } h' \end{cases} \tag{6.57}$$

Figure 6.45 illustrates such a scenario, where the spectrum is drawn in the baseband. Combining even and odd h' cases together, we can see

$$f = k_2 f_c \pm k/2T, \qquad k = 0, 1, 2, ...,$$

The two phase-lock loops (or narrow-band bandpass filters) in Figure 6.44 can select any pair of frequencies $k_2 f_c \pm k/2T$. The output of the mixer contains the sum and the difference of these two frequencies: $2k_2 f_c$ and k/T. Since in most cases the strongest components appear at $h'/2T$, thus the frequencies selected by the PLLs are usually $k_2 f_c \pm h'/2T$. The difference frequency would be $h'/T = k_1/T$. The high-pass filter and the low-pass filter pick up one of them. The frequency dividers divide two frequencies to produce f_c for the carrier and $1/T$ for the symbol timing.

The recovered carrier has a phase ambiguity of $2\pi/2k_2 = \pi/k_2$ because the received carrier frequency f_c has been multiplied by $2k_2$ in the recovery process. The symbol timing phase ambiguity is $2\pi/k_1$ for similar reasons. Recall that if $h = 2q/p$

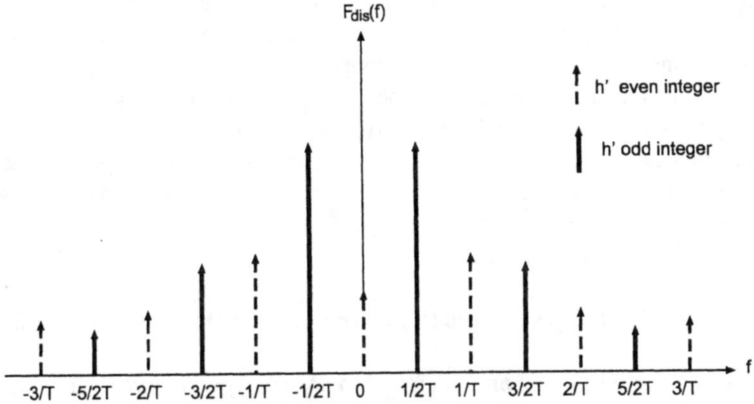

Figure 6.45 Discrete frequency components of CPM when h is an integer. From [9]. Copyright © 1986 Plenum.

where q and p are integers with no common factors, then there are p phase states in the signal. A phase shift of $2\pi/p$ merely renumbers the phase's states. The trellis remains the same. Thus the phase of the carrier only has to be known modulo $2\pi/p$ if the Viterbi receiver is used. This greatly reduces and in most cases eliminates the effect of carrier phase ambiguity. For example, in MSK, $h = 1/2 = 2/4$, $p = 4$, $k_1 = 1, k_2 = 2$, the carrier phase ambiguity is $\pi/k_2 = \pi/2$, but the Viterbi processor can tolerate a phase shift of $2\pi/p = \pi/2$, thus the effect of carrier phase ambiguity is completely eliminated. However, optimum demodulation of MSK need not be performed by the Viterbi processor, rather, it can be performed by the quadrature receiver depicted in Figure 5.10. The carrier and the symbol timing are recovered by the circuit in Figure 5.11. It has a phase ambiguity of π which can be resolved by differentially encoding the data. The symbol phase ambiguity must be resolved in some way if $k_1 \neq 1$. If $k_1 = 1$, like in MSK, there is no phase ambiguity. Performance analysis results of the synchronizer are available in [9,51].

There are some problems with the MSK-type synchronizer. First, both carrier phase and symbol timing are recovered from the same outputs of the two tracking loops. As a result, the equivalent bandwidths for both carrier recovery and symbol-timing recovery are the same. For carrier recovery, the bandwidth of the PLLs cannot be made arbitrarily small due to factors such as phase noise and fading. As a result, the PLL bandwidth may be too wide for symbol-timing recovery. The recovered clock may suffer from a large level of phase jitter. Furthermore, since the mean time between cycle slips decreases as the PLL bandwidth is increased, the recovered timing may have an undesirable rate of cycle slipping. The second problem of the MSK-type synchronizer is the difficulty of acquiring signals with a large Doppler shift with respect to the data rate. If the loops are to be swept, then particular care must be taken so that the two loops do not lock on to the same frequency. A signal dropout problem also exists in the MSK-type synchronizer. Certain data sequences may lead to the transmission of only one tone, so that one of the loops loses lock and returns to its rest frequency. This situation produces both carrier and clock tracking errors. Finally, the MSK-type synchronizer is not easy to be realized in VLSI, which reduces its practical significance.

6.6.2 Squaring Loop and Fourth-Power Loop Synchronizers

To overcome some of the shortcomings of the MSK-type synchronizer, a squaring loop and a fourth-power loop synchronizer were proposed by [52] for binary CPM schemes with $h = 1/2$ (Figure 6.46). These two synchronizers are for carrier recovery only. The timing recovery must be done separately. Thus its bandwidth can be selected independently of the carrier recovery bandwidth.

Figure 6.46(a) is the squaring loop synchronizer. Since we assume that the received CPM is binary with $h = 1/2 = k_1/k_2$, according to the argument in the previous subsection, the squaring operation generates a CPM signal at $2f_c$ with $h' = k_1 = 1$. This leads to discrete spectral lines at $2f_c \pm (2k+1)/2T$, k integer, with $2f_c \pm 1/2T$ the strongest. The lines are shifted to $2f_c$ by multiplication with $\cos[\pi(t - T_d)/T]$, where T_d is the symbol timing delay (error). It is easy to see that when $T_d \neq 0$, the shifted spectral lines cannot be at $2f_c$ exactly. Therefore the prior knowledge of the symbol timing is essential to this synchronizer. It is shown in [52] that the phase error variance of this synchronizer increases with the increase of T_d. Other than that, the squaring loop's jitter performance is essentially the same as the MSK-type synchronizer, but the latter's practical limitations are eliminated. Most importantly, the squaring loop leaves the symbol timing to be recovered separately. The timing recovery bandwidth can be made much smaller than the carrier recovery bandwidth. Consequently the rate of clock slips can be reduced. The phase ambiguity of the squaring loop is π due to the squaring operation, which can be eliminated by differentially encoding the data symbols.

Figure 6.46(b) is the fourth-power loop synchronizer. After the fourth-power nonlinearity, the signal contains a component $\cos[8\pi f_c t + 4\Phi(t, \mathbf{a})]$ which has an even index $h' = 2$. Thus the discrete lines are at $4f_c \pm k/T$, k integer. Setting $k = 0$, we obtain a line at $4f_c$ which is strong for most of the popular binary, $h = 1/2$ CPM schemes except MSK, for which the magnitude of this spectral line happens to be zero [52, Figure 4]. Thus this loop will not work for MSK. The advantage of the fourth-power loop is that it does not require prior knowledge of the symbol timing.

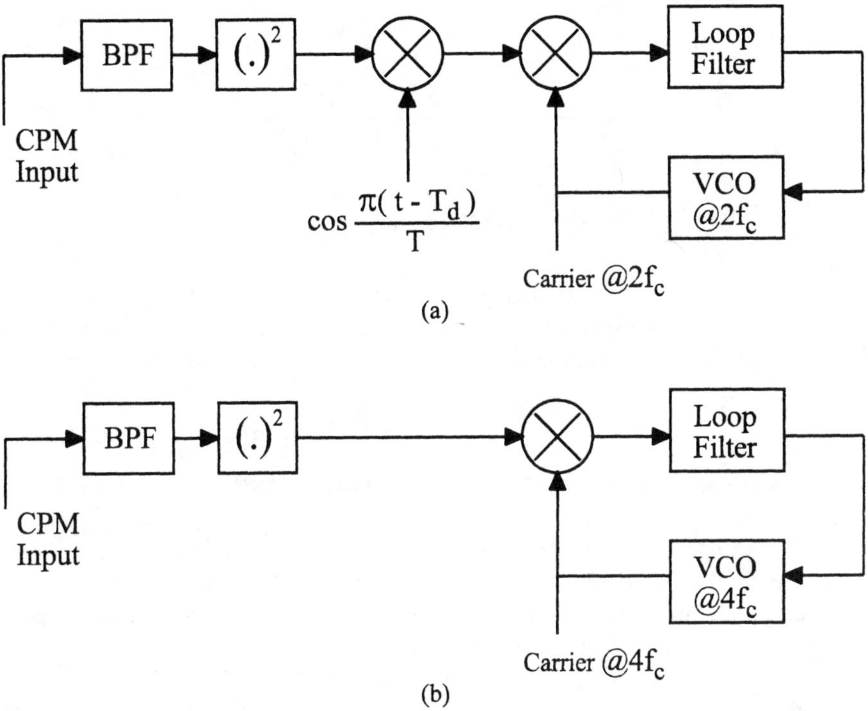

Figure 6.46 Squaring loop and fourth-power loop synchronizers for binary CPM with $h = 1/2$. From [52]. Copyright © 1989 IEEE.

As a result, it allows carrier phase and symbol timing to be recovered sequentially. A faster acquisition can be achieved.

The results in [52, Fig. 4] show that the fourth-power loop is inferior to the squaring loop at lower SNR and the situation is reversed at higher SNR. The two loops were evaluated with several popular binary $h = 1/2$ CPM schemes, including 2RC, 3RC, GMSK, TFM, and DMSK (duobinary MSK). It was found that the best performance (phase variance) is achieved with DMSK.

6.6.3 Other Types of Synchronizers

The synchronizers described above are relatively simple. More sophisticated synchronizers are reported in the literature. A synchronizer using the maximum a posteriori (MAP) technique to jointly estimate the carrier phase and symbol timing for

B_bT	90%	99%	99.9%	99.99%
0.2	0.52	0.79	0.99	1.22
0.25	0.57	0.86	1.09	1.37
0.5	0.69	1.04	1.33	2.08
MSK	0.78	1.20	2.76	6.00
TFM	0.52	0.79	1.02	1.37

Table 6.2 GMSK percentage bandwidth.

MSK is reported in [53]. A MAP symbol synchronizer for partial-response CPM is proposed in [54]. More synchronizers are mentioned in [9]. All these synchronizers are too complicated. More research is needed to make them easier to implement in practice.

6.7 GAUSSIAN MINIMUM SHIFT KEYING (GMSK)

GMSK was first proposed by Murota and Hirade for digital mobile radio telephony [3]. Currently GMSK is used in the U.S. cellular digital packet data (CDPD) system and European GSM system [55, p. 265]. The wide spread use is due to its compact power spectral density and excellent error performance.

GMSK, as its name suggests, is based on MSK and is developed to improve the spectral property of MSK by using a premodulation Gaussian filter. The transfer function of the filter is

$$H(f) = \exp\left\{-\left(\frac{f}{B_b}\right)^2 \frac{\ln 2}{2}\right\} \tag{6.58}$$

where B_b is the 3-dB bandwidth. We have defined the frequency pulse $g(t)$ of GMSK in (6.8). This $g(t)$ can be generated by passing a rectangular pulse $rec(t/T)$ through this filter [56, p. 183].

The power spectral density of GMSK is shown in Figure 6.47, where B_bT is a parameter. The spectrum of MSK ($B_bT = \infty$) is also shown for comparison. It is clear that the smaller the B_bT, the tighter the spectrum. However, the smaller the B_bT, the farther the GMSK is from the MSK. Then the degradation in error performance using an MSK demodulator will be larger. We will see this shortly. A fact that needs to be pointed out is that the spectrum of GMSK with $B_bT = 0.2$ is nearly equal to that of TFM. Table 6.2 shows the bandwidth (normalized to symbol rate) for the prescribed percentage of power within the bandwidth.

The modulator for GMSK currently used in CDPD and GSM systems is of the

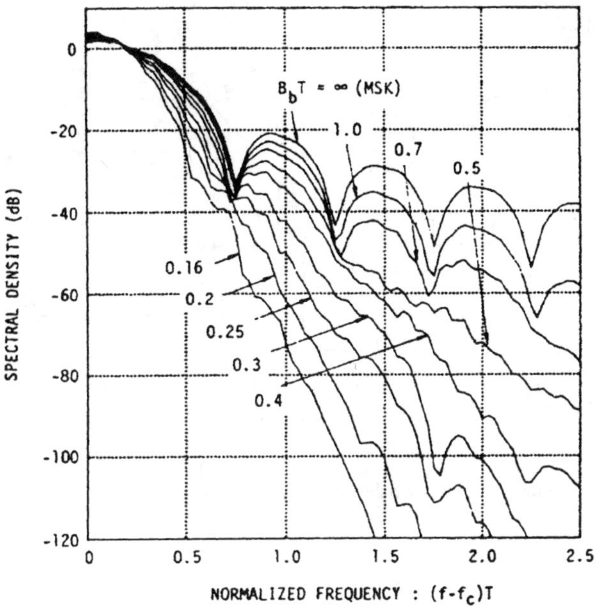

Figure 6.47 Power spectra of GMSK. From [3]. Copyright © 1981 IEEE.

type of Figure 6.13, where the filter must be a Gaussian filter and the FM modulator must be an MSK modulator (that is, a 1REC modulator with $h = 0.5$). Of course other types of CPM modulators can also be used for GMSK.

The demodulation for GMSK of course can be done using all types of demodulators that we described in this chapter. The demodulator suggested in [3] is the Costas loop type shown in Figure 6.48 where demodulation and carrier recovery are combined. Clock recovery is still separated. Figure 6.49 is the digital implementation of Figure 6.48. In Figure 6.49, two D flip-flops act as the quadrature product demodulators and both the exclusive-or logic circuits are used for the baseband multipliers. The mutually orthogonal reference carriers are generated by the use of two D flip-flops. The VCO center frequency is then set equal to four times carrier center frequency. This circuit is considered to be especially suitable for mobile radio units which must be simple, small, and economical.

The theoretical BER performance of the coherent GMSK is of course described by the expressions given in Section 6.3.1. Figure 6.50 shows some measured BER

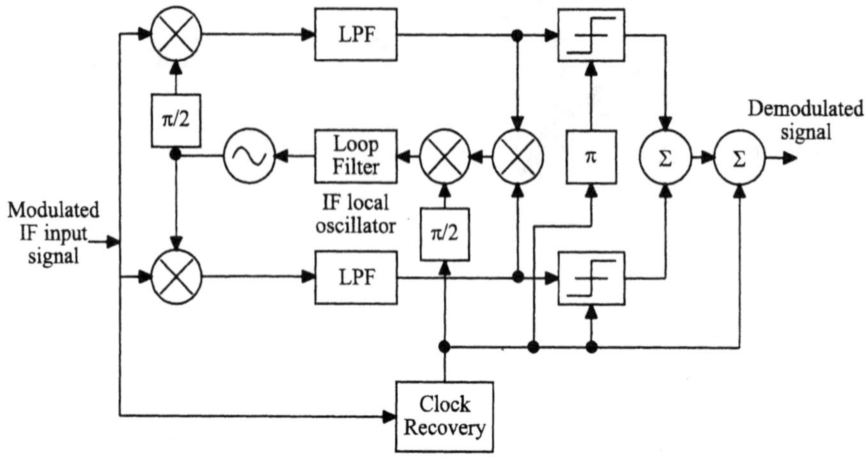

Figure 6.48 Block diagram of Costas loop demodulator for GMSK. From [3]. Copyright © 1981 IEEE.

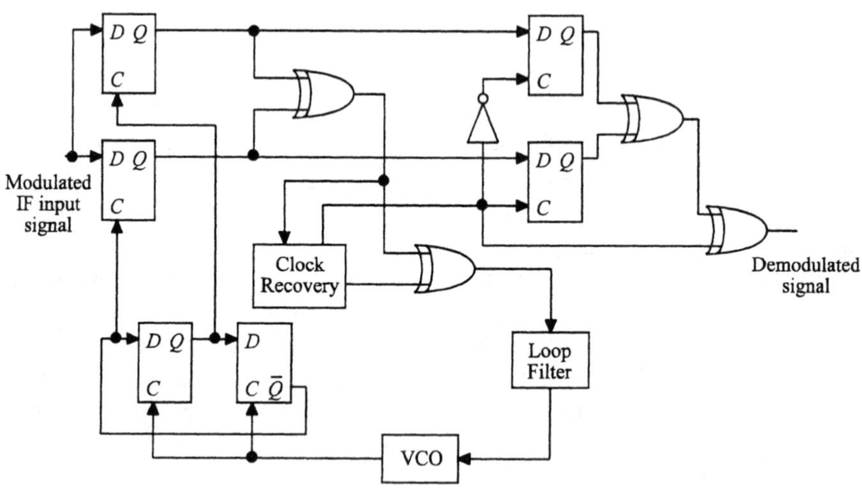

Figure 6.49 Digital circuit implementation of the Costas loop demodulator for GMSK. From [3]. Copyright © 1981 IEEE.

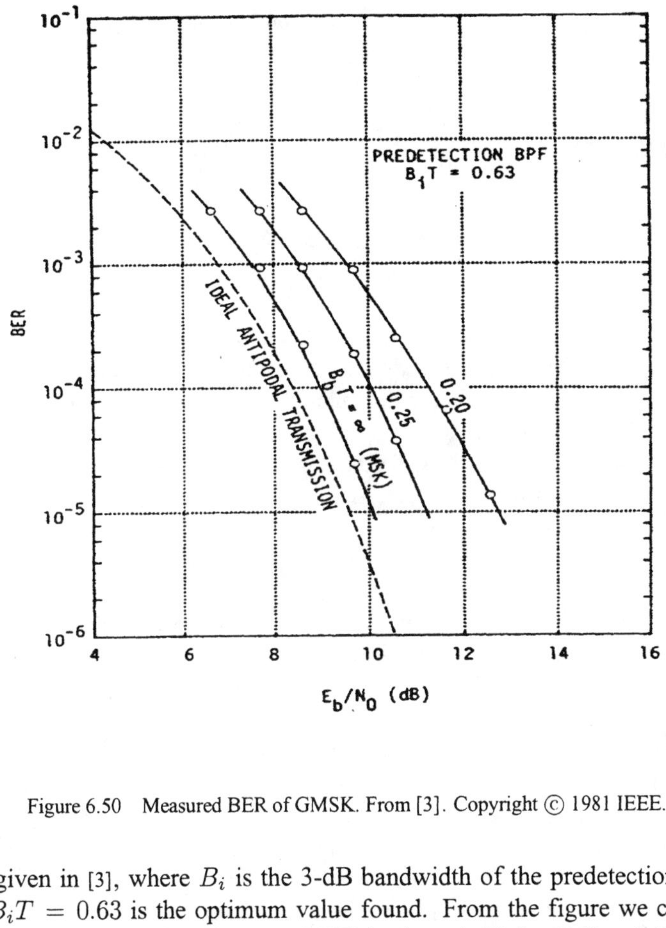

Figure 6.50 Measured BER of GMSK. From [3]. Copyright © 1981 IEEE.

results given in [3], where B_i is the 3-dB bandwidth of the predetection Gaussian filter. $B_iT = 0.63$ is the optimum value found. From the figure we can see that the degradation of GMSK relative to MSK is about 1 dB for $B_bT = 0.25$. Also the spectrum with $B_bT = 0.25$ is quite tight, so this bandwidth is considered a good choice. The measured BER curves can be approximated by the following equations

$$P_b \approx Q\left(\sqrt{\frac{2\alpha E_b}{N_o}}\right) \tag{6.59}$$

where

$$\alpha = \begin{cases} 0.68, & \text{for GMSK with } B_bT = 0.25 \\ 0.85, & \text{for simple MSK } (B_bT \to \infty) \end{cases}$$

6.8 SUMMARY

In this chapter we have studied the continuous phase modulation. We first defined the CPM signal. Then various important frequency pulses, including LREC, LRC, LSRC, TFM, and GMSK were presented. Both their mathematical expressions and waveforms were given. The phase, phase state, phase tree, phase trellis, state, and state trellis of the CPM signal were studied in great detail since the information symbols are imbedded in the phase, and demodulation is based on the phase tree or the state trellis. Steps to calculate power spectral density of CPM were given, but the derivation was left in Appendix A. PSDs of some of the important CPM schemes were presented. The influence of pulse shape, modulation index, and a priori probability distribution on the PSD was demonstrated using examples. It was found that unlike pulse shape and modulation index, a priori probability distribution has no significant impact on the spectrum. The formula of the distance of CPM was derived. The error performance was related to the distance, particularly, the minimum distance of the CPM signal. It was found that the error performance is mainly determined by the minimum distance. Thus the minimum distance is often used as an indicator of the error performance of a CPM scheme. In the section on modulators, we covered direct modulator, quadrature modulator, serial modulator, and all-digital direct modulator. The great challenge of CPM lies in the design of demodulators. We studied optimum ML coherent and noncoherent demodulators which detect one symbol based on observation of several symbols. We described the Viterbi demodulator which is an ML detector based on the state trellis. The Viterbi demodulator detects the signal in a convenient recursive fashion, thus it allows the data stream to be continuous without frame structure. Demodulators that simplify the Viterbi detector, either based on a simpler trellis or based on a smaller number of matched filters were described. MSK-type demodulators, either parallel or serial, were discussed. They are very practical in terms of their simple structures, even though they offer slightly inferior error performance and are mainly suitable for binary CPM schemes with $h = 1/2$. The still simpler noncoherent differential and discriminator demodulators were also studied. Of course the price paid is the degradation in error performance, or power efficiency. However, when receiver complexity is a more severe problem than transmission power, they are not a bad choice. The synchronization problem is also a big challenge of CPM techniques. We described the popular MSK-type synchronizer and newer squaring loop and fourth-power loop synchronizers. Other complex synchronizers were also mentioned. Finally, we discussed GMSK in detail in a dedicated section due to its importance in practical use.

Multi-h phase modulation is developed from single-h CPM that we have just studied. It offers even better performance, but with a greater complexity. We will study the multi-h phase modulation in the next chapter.

References

[1] Sundberg, C-E, "Continuous phase modulation: a class of jointly power and bandwidth efficient digital modulation schemes with constant amplitude," *IEEE Communications Magazine*, vol. 24, no. 4, April 1986, pp. 25–38.

[2] de Jager, F., and C. B. Dekker, "Tamed frequency modulation, a novel method to achieve spectrum economy in digital transmission," *IEEE Trans. on Comm.*, vol. 26, no. 5, May 1978, pp. 534–542.

[3] Murota, K., and K. Hirade, "GMSK modulation for digital mobile telephony," *IEEE Trans. on Comm.*, vol. 29, no. 7, July 1981, pp. 1044–1050.

[4] Muilwijk, D., "Correlative phase shift keying-a class of constant envelope modulation techniques," *IEEE Trans. on Comm.*, vol. 29, no. 3, March 1981, pp. 226–236.

[5] Deshpande, G. S. and P. H. Wittke, "Correlative encoded digital FM," *IEEE Trans. on Comm.*, vol. 29, no. 2, February 1981, pp. 156–162.

[6] Chung, K. S., "General tamed frequency modulation and its application for mobile radio communication," *IEEE Jour. on Selected Areas in Comm.*, vol. 2, no. 4, July 1984, pp. 487–497.

[7] Aulin, T., and C-E. Sundberg, "Continuous phase modulation—Part I: Full response signaling," *IEEE Trans. on Comm.*, vol. 29, no. 3, March 1981, pp. 196–206.

[8] Aulin, T., N. Rydbeck, and C-E. Sundberg, "Continuous phase modulation—Part II: Partial response signaling," *IEEE Trans. on Comm.*, vol. 29, no. 3, March 1981, pp. 210–225.

[9] Anderson, J., T. Aulin, and C-E. Sundberg, *Digital Phase Modulation,* New York: Plenum Publishing Company, 1986.

[10] Anderson, R. R., and J. Salz, "Spectra of digital FM," *Bell System Tech. J.*, vol. 44, 1965, pp. 1165–1189.

[11] Prabhu, V. K., and H. E. Rowe, "Spectra of digital phase modulation by matrix methods," *Bell System Tech. J.*, vol. 53, 1974, pp. 899–935.

[12] Greenstein, L. J., "Spectra of PSK signals with overlapping baseband pulses," *IEEE Trans. on Comm.*, vol. 25, 1977, pp. 523–530.

[13] Aulin, T., and C-E. Sundberg, "An easy way to calculate power spectra for digital FM," *IEE proceedings*, Part F, vol. 130, no. 6, 1983, pp. 519–526.

[14] Proakis, J. G., *Digital Communications,* 2nd ed., New York: McGraw-Hill, 1989.

[15] McGillem, C., and G. Cooper, *Continuous and Discrete Signal and System Analysis*, 3rd ed., Philadelphia: Saunders College Publishing, 1991.

[16] Wilson, S. G., and M. G. Mulligan, "An improved algorithm for evaluating trellis phase codes," *IEEE Trans. on Information Theory,* vol. 30, no. 6, November 1984, pp. 846–851.

[17] Rimoldi, B., "Exact formula for the minimum squared Euclidean distance of CPFSK," *IEEE Trans. on Comm.,* vol. 39, no. 9, September 1991, pp. 1280–1282.

[18] Ekanayake, N., and R. Liyanapathirana, "On the exact formula for minimum squared distance of CPFSK," *IEEE Trans. on Comm.,* vol. 42, no. 11, Nov. 1994, pp. 2917–2918.

[19] Aulin, T., N. Rydbeck, and C-E. Sundberg, "Transmitter and receiver structure for M-ary partial response FM," *IEEE Trans. on Comm.,* vol. 26, no. 3, May 1978, pp. 534–538.

[20] Kopta, A., S. Budisin, and V. Jovanovic, "New universal all-digital CPM modulator," *IEEE Trans. on Comm.,* vol. 35, no. 4, April 1987, pp. 458–462.

[21] Osborne, W. P., and M. B. Luntz, "Coherent and noncoherent detection of CPFSK," *IEEE Trans. on Comm.,* vol. 22, no. 8, August 1974, pp. 1023–1036.

[22] Schonhoff, T. A., "Symbol error probabilities for M-ary CPFSK: coherent and noncoherent detection," *IEEE Trans. on Comm.,* vol. 24, no. 6, June 1976, pp. 644–652.

[23] Van Trees, H. L., *Detection, Estimation, and Modulation Theory, Part I,* New York: John Wiley & Sons, Inc., 1968.

[24] Stiffler, J. J., *Theory of Synchronous Communications*, Englewood Cliffs, New Jersey: Prentice Hall, 1971.

[25] Viterbi, A. J., "Error bounds for convolutional codes and an asymptotically optimum decoding algorithm," *IEEE Trans. on Inform. Theory,* vol. 13, April 1967, pp. 260–269.

[26] Forney, G. D., Jr., "Maximum likelihood sequence estimation of digital sequences in the presence of intersymbol interference," *IEEE Trans. Inform. Theory,* vol. 18, May 1972, pp. 363–378.

[27] Forney, G. D., Jr., "The Viterbi algorithm," *Proceedings of IEEE,* vol. 61, 1973, pp. 268–278.

[28] Viterbi, A. J., and J. K. Omura, *Principles of Digital Communication and Coding*, New York: McGraw-Hill, 1979.

[29] Aulin, T., "Three papers on continuous phase modulation (CPM)," Thesis, University of Lund, November 1979.

[30] Aulin, T., "Symbol error probability bounds for coherently Viterbi detected continuous phase modulated signals," *IEEE Trans. on Comm.,* vol. 29, 1981, pp. 1707–1715.

[31] Svensson, A., C-E. Sundberg, and T. Aulin, "A class of reduced-complexity Viterbi detectors for partial response continuous phase modulation," *IEEE Trans. on Comm.,* vol. 32, no. 10, October 1984, pp. 1079–1087.

[32] Kramer, G. G., and E. Shwedyk, "On reducing the number of matched filters for optimum detection of LREC CPM," *Proceedings of Queen's University Symposium of Communications*, Canada, May 1992.

[33] Simon, M., and D. Divsalar, "Maximum-likelihood block detection of noncoherent continuous phase modulation," *IEEE Trans. on Comm.,* vol. 41, no. 1, January 1993, pp. 90–98.

[34] Svensson, A., and C-E. Sunderberg, "Optimum MSK-type receivers for CPM on Gaussian and Rayleigh fading channels," *IEE Proceedings*, Part F, vol. 131, no. 5, 1984, pp. 480–490.

[35] Svensson, A., and C-E. Sunderberg, "Serial MSK-type detection of partial response continuous phase modulation," *IEEE Trans. on Comm.*, vol. 33, no. 1, January 1985, pp. 44–52.

[36] Sundberg, C-E., "Error probability of partial response continuous phase modulation with coherent MSK-type receiver, diversity and slow Rayleigh fading in Gaussian noise," *The Bell System Technical Journal,* vol. 61, no. 8, 1982, pp. 1933–1963.

[37] Galko, P., "Generalized MSK," Ph.D. thesis, University of Toronto, Canada, August 1982.

[38] Ziemer, R., C. Ryan, and J. Stilwell, "Conversion and matched filter approximations for serial minimum-shift keyed modulation," *IEEE Trans. on Comm.,* vol. 30, no. 3, March 1982, pp. 495–509.

[39] Chung, K. S., and L. E. Zegers, "Generalized tamed frequency modulation, " *Proc. IEEE International Conf. on Acoustics, Speech and Signal Processing*, vol. 3, Paris, May 1982, pp. 1805–1808.

[40] Chung, K. S., "A noncoherent receiver for GTFM signals," *Proc. IEEE GLOBECOM'82*, Miami, December 1982, pp. B3.5.1–N3.5.5.

[41] Stjernvall, J-E., and J. Uddenfeldt, "Gaussian MSK with different demodulators and channel coding for mobile telephony," *Proc. ICC'84*, Amsterdam, the Netherlands, May 14–17, 1984, pp. 1219–1222.

[42] Hirade, K. M. et al., "Error-rate performance of digital FM with differential detection in mobile radio channels," *IEEE Trans. on Vehicular Technology,* vol. 28, 1979, pp. 204–212.

[43] Svensson, A., and C-E. Sundberg, "On error probability for several types of noncoherent detection of CPM," *Proc. IEEE GLOBECOM'84*, Atlanta, Georgia, November 1984, pp. 22.5.1–22.5.7.

[44] Laurent, P., "Exact and approximate construction of digital phase modulations by superposition of amplitude modulated pulses (AMP)," *IEEE Trans. Commun.*, vol. 34, no. 2, February 1986, pp. 150–160.

[45] Kaleh, G., "Simple coherent receivers for partial response continuous phase modulation," *IEEE Journal on Selected Areas in Commun.*, vol. 7, no. 9, December 1989, pp. 1427–1436.

[46] Hirt, W., and S. Pasupathy, "Suboptimal reception of binary CPSK signals," *IEE Proceedings*, Part F, vol. 128, no. 3, pp. 125–134, 1981.

[47] Anderson, J. B., and S. Mohan, "Sequential coding algorithms: a survey and cost analysis," *IEEE Trans. on Commun.,* vol. 32, 1984, pp. 169–176.

[48] DeBuda, R., "Coherent demodulation of frequency-shift keying with low deviation ratio," *IEEE Trans. Commun.*, vol. 20, June 1972, pp. 429–435.

[49] Lee, W. U., "Carrier synchronization of CPFSK signals," *Proc. NTC'77*, 1977, pp. 30.2.1–30.2.4.

[50] Aulin, T., and C-E. Sundberg, "Synchronization properties of continuous phase modulation," *Proc. IEEE GLOBECOM'82*, Miami, December 1982, pp. D7.1.1–D7.1.7.

[51] D'andrea, A., U. Mengali, and R. Reggiannini, "Carrier phase and clock recovery for continuous phase modulated signals," *IEEE Trans. Commun.*, vol. 35, no. 10, October 1972, pp.1095–1101.

[52] El-tanany, M., and S. Mahmoud, "Analysis of two loops for carrier recovery in CPM with index 1/2," *IEEE Trans. Commun.*, vol. 37, no. 2, February 1989, pp. 164–176.

[53] Booth, R., "An illustration of the MAP estimation for deriving closed-loop phase tracking topolo-

gies: the MSK signal structure," *IEEE Trans. Commun.*, vol. 28, no. 8, August 1980, pp. 1137–1142.

[54] Glisic, S., "Symbol synchronization in digital communication systems using partial response CPM signaling," *IEEE Trans. Commun.*, vol. 37, no. 3, March 1989, pp. 298–308.

[55] Rappaport, T., *Wireless Communications*, Upper Saddle River, New Jersey: Prentice Hall, 1996.

[56] Stüber, G., *Principles of Mobile Communication*, Boston: Kluwer Academic Publishers, 1996.

Chapter 7

Multi-*h* Continuous Phase Modulation

Multi-h continuous phase modulation (MHPM) is a special class of CPM where the modulation index h is cyclically changed for successive symbol intervals. Cyclically changing indexes leads to delayed merging of neighboring phase trellis paths, which in turn increases the minimum Euclidean distance and improves the error performance. Binary multi-h CPFSK schemes were first proposed by Miyakawa et al [1], and were generalized by Anderson, De Buda, and Taylor [2–4]. Recently, MHPM technique has received a great deal of attention for application in satellite communications due to its power and bandwidth efficiency which are even better than that of CPM [5].

In this chapter we will cover all basic aspects of MHPM in great detail. However, the treatment here is limited to an AWGN channel as we did for previous chapters. The application of MHPM in fading channels will be treated in Chapter 10. Combined MHPM and convolutional codes are beyond the scope of this book and are not covered.

We define MHPM signal and study its phase and phase trellis properties in Section 7.1. Its power spectral density is studied in Section 7.2. The Euclidean distances of MHPM signals and their error probabilities are discussed in Section 7.3. MHPM modulators are basically the same as those for CPM, which is briefly mentioned in Section 7.4. The MHPM demodulators and synchronizers are more complicated than those of CPM. They are presented in Section 7.5. New developments of MHPM are discussed in Section 7.6. Summary of this chapter is given in Section 7.7.

7.1 MHPM SIGNAL, PHASE TREE, AND TRELLIS

The expressions for MHPM are essentially the same as those for CPM ((6.1) to (6.3)) except that h is replaced by h_k. That is

$$s(t) = A\cos(2\pi f_c t + \Phi(t, \mathbf{a})), \quad -\infty < t < \infty \tag{7.1}$$

where the phase is

$$\Phi(t, \mathbf{a}) = 2\pi \sum_{k=-\infty}^{\infty} h_k a_k q(t - kT) \tag{7.2}$$

where T is the symbol period and

$$q(t) = \int_{-\infty}^{t} g(\tau)d\tau \tag{7.3}$$

The M-ary data a_k may take any of the M values: $\pm1, \pm3, ..., \pm(M-1)$. The index h_k cyclically changes from symbol to symbol with a period of K, but only one index is used during one symbol interval, that is, $h_1, h_2, ..., h_K, h_1, h_2, ..., h_K$, and so on. It is normally assumed that the set of modulation index is rational with a common denominator

$$H_K = (h_1, h_2, ..., h_K) = (\frac{p_1}{q}, \frac{p_2}{q}, ..., \frac{p_K}{q}) \tag{7.4}$$

where p_i, $i = 1, ..., K$ and q are small integers and all h_i values are multiples of $1/q$. Note that q in denominator of (7.4) has no relation to $q(t)$ in (7.3). For bandwidth considerations, h_i is usually chosen to be smaller than one. The index set H_K is also called a multi-h phase code since it controls the signal phase.

As in the case of single-h CPM, the frequency pulse $g(t)$ can be full response or partial-response. It can be any one of the smooth waveforms defined in Section 6.1.1. In addition, we still keep the convention that

$$\int_{-\infty}^{\infty} g(t)dt = 1/2 \tag{7.5}$$

This makes the maximum phase change of the signal to be $(M-1)h_i\pi$ for the period of $g(t)$. Usually $g(t)$ is a pulse defined in $[0, LT]$ and is zero outside this interval. Thus we have $q(LT) = \int_0^{LT} g(t)dt = 1/2$.

The definition of instant phase and cumulate phase of MHPM are similar to those of single-h CPM (see Section 6.1.2). During interval $kT < t < (k+1)T$ the instant phase of an MHPM signal with a $g(t)$ of length LT is

$$\theta(t, \mathbf{a}_k) = 2\pi \sum_{i=k-L+1}^{k} h_i a_i q(t - iT) \tag{7.6}$$

the cumulate phase is

$$\theta_k = \left[\pi \sum_{i=-\infty}^{k-L} h_i a_i \right] \pmod{2\pi} \tag{7.7}$$

and the total phase is the sum of the above two

$$\Phi(t, \mathbf{a}) = \theta(t, \mathbf{a}_k) + \theta_k \tag{7.8}$$

A simple but very important MHPM scheme is the 1REC MHPM. Since $g(t) = 1/2T$ for $[0, T]$ and zero elsewhere, $\theta(t, \mathbf{a}_k) = \pi h_k a_k (t - kT)/T$ and the signal in $[kT, (k+1)T]$ is

$$s(t) = A\cos(2\pi f_c t + \frac{a_k \pi h_k (t - kT)}{T} + \theta_k), \quad kT \le t \le (k+1)T \tag{7.9}$$

The phase change at the end of each interval is $a_k \pi h_k$.

As in the single-h CPM case, there exists a phase tree for MHPM whether the indexes are rational or irrational. If the indexes are rational, the phase tree can be collapsed into a periodic phase trellis. The difference of using multiple indexes lies in the phase changing slopes in the phase tree and delayed path merge in the trellis. Figure 7.1 compares the phase trees of two binary CPFSK schemes with single-h and $H_2 = (2/4, 3/4)$ side by side. The absolute value of the phase tree slope of the single-h CPFSK is constant, which is $h\pi/T$, whereas it is $\pi/2T$ and $3\pi/4T$ alternatively for the H_2 CPFSK. Figure 7.2 shows the phase trellises for $h = 1/2$ and $h = 3/4$ single-h binary CPFSK. The first merges happen at $t = 2T$ (see the bold line paths). Figure 7.3 shows the trellis of $H_2 = (2/4, 3/4)$ binary CPFSK scheme where the first merge occurs at $t = 3T$. It is this delayed merge that gives the MHPM larger minimum Euclidean distance between neighboring paths that in turn gives rise to a better error performance.

Recall that we defined the state of CPM as

$$\mathbf{s}_k = (\theta_k, a_{k-1}, a_{k-2}, ..., a_{k-L+1})$$

and the Viterbi demodulator relies on the state trellis (see, for example, Figure 6.5). This definition is also applicable to MHPM and a state trellis is generally needed for the Viterbi demodulator. However, as we state in Section 6.1.1, for full-response MHPM, the phase trellis is equivalent to the state trellis for explaining and designing the Viterbi demodulator. Therefore the phase trellis of full-response MHPM is often called state trellis in the literature. For partial-response MHPM, a state trellis like the one for single-h CPM is needed (see Figure 6.5).

Historically, multi-h CPFSK, which is full-response with a rectangular frequency

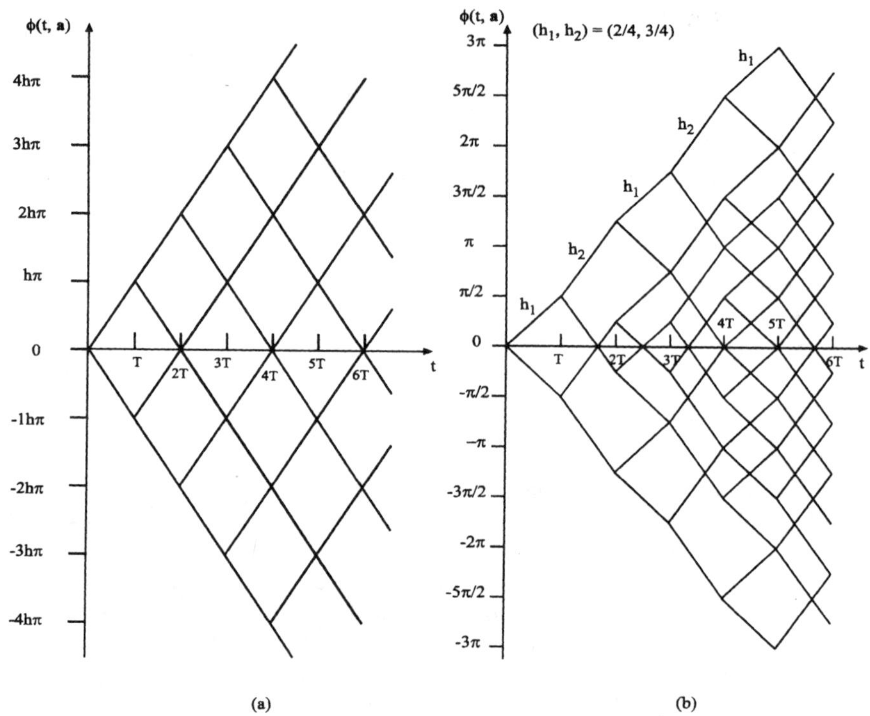

Figure 7.1 Phase trees for (a) single-h (1/2) binary CPFSK, and (b) multi-h (2/4,3/4) binary CPFSK. From [5]. Copyright © 1991 IEEE.

pulse, was first studied and the majority of the research results in the literature were about it. In the context of MHPM, we prefer to name it as 1REC MHPM. In the following, we will present the results for both 1REC MHPM and other full-response or partial-response MHPM with the emphasis on the 1REC MHPM.

For MHPM with θ_k given in (7.7) and K indexes given in (7.4), we have

$$\theta_k = \left[\frac{\pi}{q} \sum_{i=-\infty}^{k-L} p_i a_i \right] \pmod{2\pi}$$

It is clear that cumulate phase θ_k is always some multiple of π/q. The maximum number of distinct phase states θ_k in the trellis is therefore $N_s = 2\pi/(\pi/q) = 2q$. However, it was shown that the number of phase states at any $t = kT$ is q, even

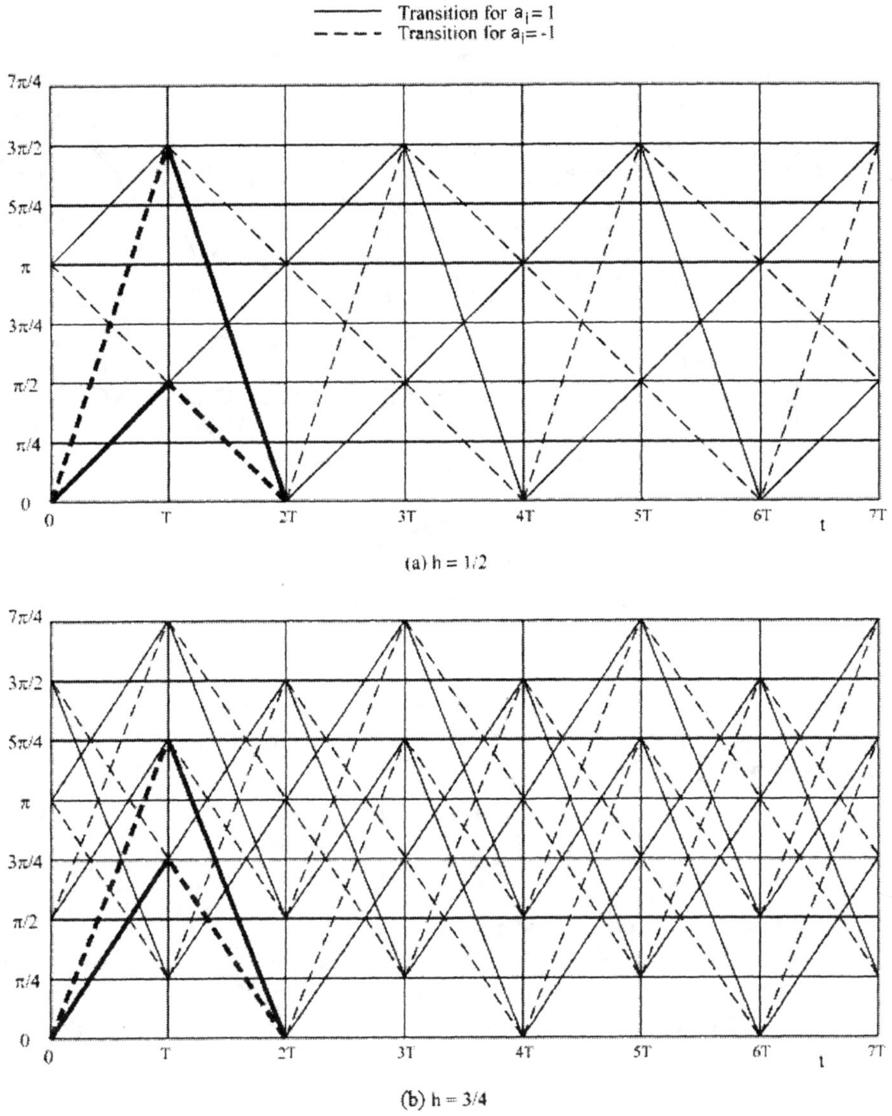

Figure 7.2 Phase trellis for (a) $h = 1/2$ binary CPFSK, and (b) $h = 3/4$ binary CPFSK. From [5]. Copyright © 1991 IEEE.

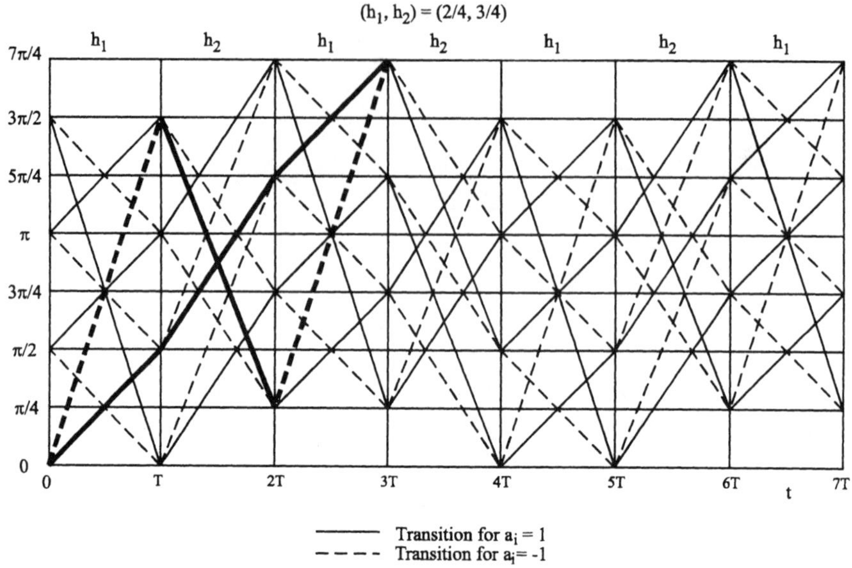

Figure 7.3 Phase trellis for $H_2 = (2/4, 3/4)$ binary CPFSK. From [5]. Copyright © 1991 IEEE.

though the total number of phase states may be q or $2q$ [6, 7]. Here a rigorous proof is given. To simplify the proof without loss of generality, we assume that the time index i starts at $i = 1$

$$\theta_k = \left[\frac{\pi}{q} \sum_{i=1}^{k-L} p_i a_i \right] (\text{mod}\, 2\pi)$$

where p_i could be an even or odd number.

First we prove that if $\sum_{i=1}^{k-L} p_i$ is odd, $\sum_{i=1}^{k-L} p_i a_i$ must be odd. This conclusion can be proved as follows. First this is apparently true for all $a_i = 1$ or all $a_i = -1$. In general, for any set of $a_i \in \{\pm 1, \pm 3, ..., \pm(M-1)\}$, if $\sum_{i=1}^{k-L} p_i$ is odd, then there must be an odd number (say, N_{odd}) of odd p_is and any number (odd or even, say, N_{even}) of even p_is in $\sum_{i=1}^{k-L} p_i$. In the sum $\sum_{i=1}^{k-L} p_i a_i$, it is obvious that a term generated by an even p_i is always even. A term generated by an odd p_i is always odd since p_i and a_i are both odd. Thus in the sum $\sum_{i=1}^{k-L} p_i a_i$, the subsum of those N_{even} terms generated by even p_is is even; the subsum of those N_{odd} terms generated by odd p_is is always odd since each term is odd and the number of such terms is also odd. Thus the total sum $\sum_{i=1}^{k-L} p_i a_i$ is odd.

Likewise, we prove that if $\sum_{i=1}^{k-L} p_i$ is even, $\sum_{i=1}^{k-L} p_i a_i$ must be even. In this case, the number of odd p_is (N_{odd}) must be even. The subsum of N_{odd} odd $p_i a_i$ terms is even since each term is odd but the number of terms is even. Thus the total sum $\sum_{i=1}^{k-L} p_i a_i$ is even.

When $\sum_{i=1}^{k-L} p_i a_i$ is odd, it takes values of $\pm 1, \pm 3, ...$ and $\theta_k = \pm\pi/q, \pm 3\pi/q$, ...; when $\sum_{i=1}^{k-L} p_i a_i$ is even, it takes values of $0, \pm 2, \pm 4, ...$ and $\theta_k = 0, \pm 2\pi/q$, $\pm 4\pi/q, ...$ We can convert the phases into $[0, 2\pi]$ and summarize the above results as follows.

$$\theta_k = \frac{\pi}{q}, \frac{3\pi}{q}, \frac{5\pi}{q}, ..., \frac{(2q-1)\pi}{q}, \quad \Lambda_k = \sum_{i=1}^{k-L} p_i \text{ odd} \tag{7.10}$$

$$\theta_k = 0, \frac{2\pi}{q}, \frac{4\pi}{q}, ..., \frac{2\pi(q-1)}{q}, \quad \Lambda_k = \sum_{i=1}^{k-L} p_i \text{ even} \tag{7.11}$$

In either case, the number of possible phases ($mod\ 2\pi$) is q with a distance of $2\pi/q$ between two adjacent phases. Thus at any $t = kT$, the number of possible phases is q.

The number of total phase states (N_s) in the phase trellis could be $2q$ or q. If not all p_is are even, Λ_k is odd at some times and even at other times, then the phase can take values in (7.10) and (7.11) (i.e., $N_s = 2q$). If all p_is are even, Λ_k is even all the time, then the phase can take values in (7.11) only, i.e., $N_s = q$. The first case is the usual case since usually q is chosen as a power of 2, thus not all p_is are even. Otherwise if all p_is are even, q can be reduced by a factor of two in defining H_K (see (7.4)), still resulting in the case that "not all p_is are even." The second case, which is not the usual case, happens when q is an odd number. We summarize the conclusions as follows.

$$\text{If not all } p_i s \text{ are even, } N_s = 2q \tag{7.12}$$

$$\text{If all } p_i s \text{ are even,} \quad N_s = q \tag{7.13}$$

Some examples which support the above theory are given here. An example for the first case (not all p_is are even) is $H_2 = (2/4, 3/4)$, starting from $k = L + 2$, $\Lambda_k = 5, 7, 10, 12, 15, 17, ...$ The values of Λ_k are odd or even. $N_s = 2q = 8$ (Figure 7.3). Another example for the first case is $H_3 = (1/8, 2/8, 3/8)$, starting from $k = L+3$, $\Lambda_k = 6, 7, 9, 12, 13, ...$ The values of Λ_k are odd or even, $N_s = 2q = 16$. The trellis is shown in Figure 7.4. Finally, an example for the second case (all p_is are even) is $H_2 = (2/5, 4/5)$. Starting from $k = L + 2$, $\Lambda_k = 6, 8, 12, 14, 18, ...$ All values of Λ_k are even in this case, thus $N_s = q = 5$. The trellis is in Figure 7.5.

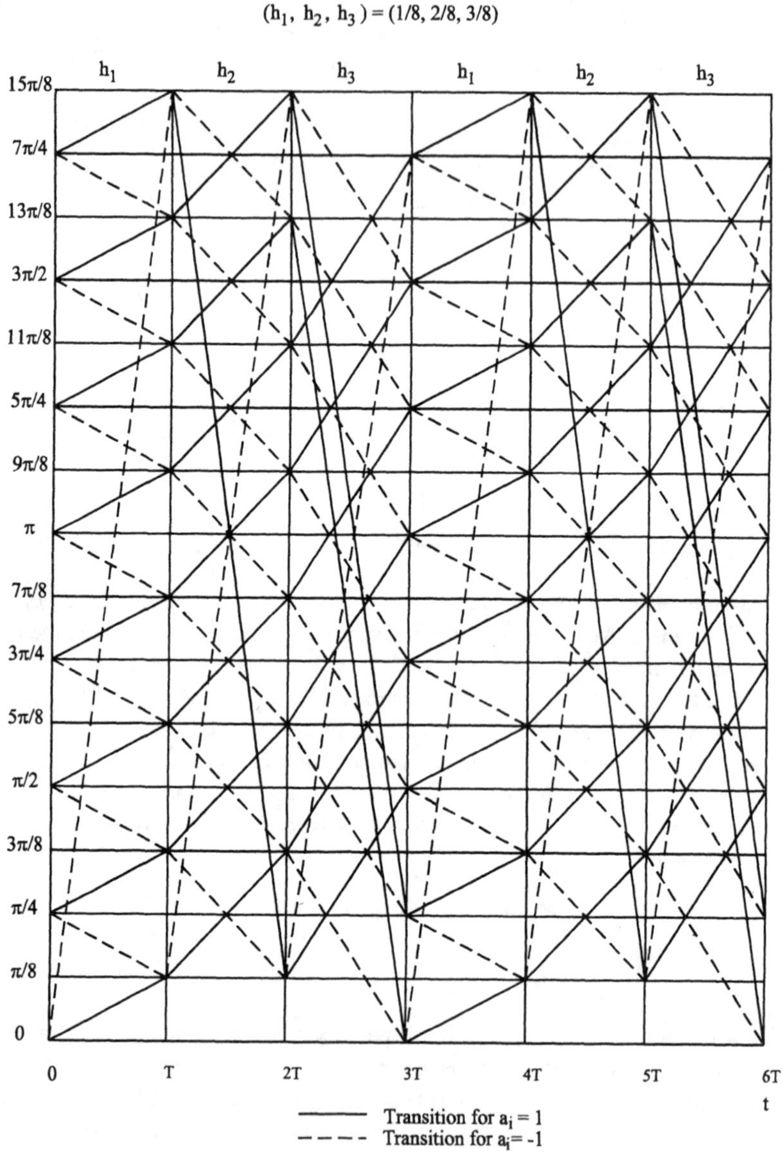

Figure 7.4 Phase trellis for $H_3 = (1/8, 2/8, 3/8)$ binary CPFSK.

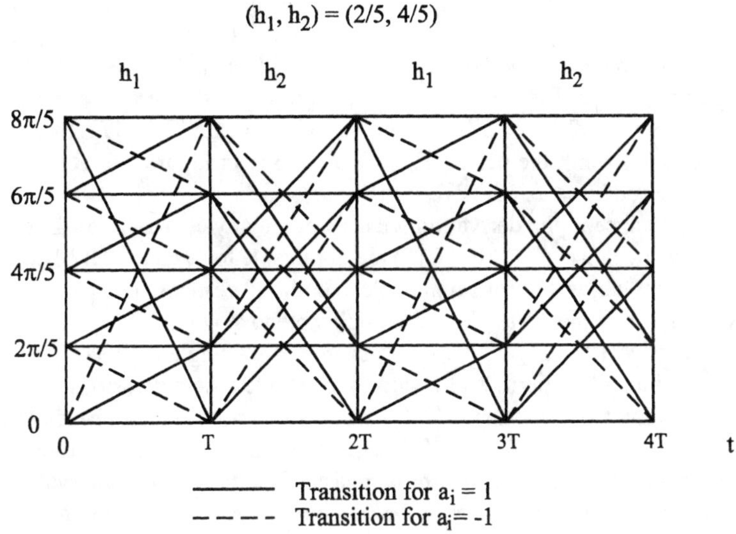

Figure 7.5 Phase trellis for $H_2 = (2/5, 4/5)$ binary CPFSK.

It is natural to think that the period of the trellis (T_p) is KT since K modulation indexes are cyclically rotated. In fact T_p could be KT or $2KT$, depending on a quantity related to the modulation indexes.

$$T_p = KT, \quad \Gamma = \sum_{i=1}^{K} p_i \text{ even} \tag{7.14}$$

$$T_p = 2KT, \quad \Gamma = \sum_{i=1}^{K} p_i \text{ odd} \tag{7.15}$$

Here is the proof. Assume the trellis starts from $k = 1$. When Γ is even, from (7.11) we can see that $\theta_K = 0, 2\pi/q, 4\pi/q, ..., 2\pi(q-1)/q$. Starting from this point, after K symbols, θ_K will increase to $\theta_{2K} = \theta_K + \sum_{i=1}^{K} p_i a_i$. Since Γ is even, from (7.11) we can see that the increment $\sum_{i=1}^{K} p_i a_i$ is one of the following values, $0, 2\pi/q, 4\pi/q, ..., 2\pi(q-1)/q$. The resulting θ_{2K} is also one of the same set of values $(mod\, 2\pi)$. Thus the period is KT. When Γ is odd, from (7.10) we can see that $\theta_K = \pi/q, 3\pi/q, 5\pi/q, ..., (2q-1)\pi/q$. Starting from this point, after K symbols, θ_K will increase to $\theta_{2K} = \theta_K + \sum_{i=1}^{K} p_i a_i$. Since Γ is odd, from (7.10) we can see that the increment $\sum_{i=1}^{K} p_i a_i$ is one of the following values, $\pi/q, 3\pi/q, 5\pi/q, ..., (2q - 1)\pi/q$. The resulting θ_{2K} is one of the values in (7.11) $(mod\ 2\pi)$. Only after next K symbols, the phases could become the values in (7.10). Thus the period is $2KT$.

Again we can use Figures 7.3, 7.4, and 7.5 to verify the periods of phase trellis. In Figure 7.3, $K = 2, \Gamma = 5$ (odd), the period $T_p = 2KT = 4T$. In Figure 7.4, $K = 3, \Gamma = 6$ (even), the period $T_p = KT = 3T$. In Figure 7.5, $K = 2, \Gamma = 6$ (even), the period $T_p = KT = 2T$.

Note that since in the derivation we did not put any constraint on L, the length of the frequency pulse, (7.11) to (7.15) are applicable to M-ary MHPM with any L values. In other words, they are applicable to both full-response and partial-response M-ary MHPM schemes. They are also applicable to single-h CPM since it is just a special case of MHPM. Figure 7.2 is two examples for single-h CPM. Their number of phases and length of periods also satisfy the above four expressions.

As we pointed out in Chapter 6, θ_k is in general not the total phase at time kT. However, if $L = 1$ and in addition f_c is an integer multiple of symbol rate, and θ_k is the total phase at $t = kT$, thus the initial phase of the kth symbol interval.

Another important parameter of the phase trellis is the *constraint length* which is the minimum number of symbol intervals over which any two paths remain unmerged. The constraint length determines the minimum Euclidean distance which in turn determines the error probability. A set H_K cannot have constraint length longer than $K+1$, since a merge is always possible over $K+1$ intervals. For instance, data $(K+1)$-tuples $(1, -1, ..., -1)$ and $(-1, ..., -1, 1)$, where $-1, ..., -1$ indicates K successive -1s and that corresponding modulation indexes are $(h_1, h_2, ..., h_K, h_1)$, represent two paths which diverge at the first symbol and merge at the last symbol. The first merger that occurs independently of the values of modulation indexes is called the first *inevitable* merger. It is clear from above that the first inevitable merger occurs at $t = (K+1)T$ for full-response MHPM schemes.

It was found that there exist sets H_K which can achieve the maximum attainable constraint length $(K+1)$ [4, 8]. A necessary and sufficient condition for the set to achieve the upper limit $K+1$ is that the weighted sum $(k_1 h_1 + k_2 h_2 + ... + k_K h_K)$ must not be integer-valued for any integer K-tuple $(k_1, k_2, ..., k_K)$ with $k_i \in \{0, 1, 2, ..., M-1\}$. Furthermore, if h_i are multiples of $1/q$, then H_K can achieve the maximum constraint length only if $q \geq M^K$. However, it has been shown that MHPM schemes exist that do not satisfy $q \geq M^K$, but still offer significant gain over the best CPFSK schemes with comparable bandwidth and number of receiver states [9].

For partial-response MHPM with K indexes and a $g(t)$ of L symbol duration, the first inevitable merger in the phase trellis occurs at $t = (L+K)T$ [10]. Thus the constraint length of such a scheme cannot be longer than $L+K$.

7.2 POWER SPECTRAL DENSITY

Similar to the calculation of the power spectral density of single-h CPM, there are three methods for computing the power spectral density of MHPM schemes: the direct method, the Markov chain method, and the autocorrelation method. See Section 6.2 for general descriptions of these methods. For MHPM they need modifications.

First we consider binary 1REC MHPM (binary multi-h CPFSK). The autocorrelation function has been analytically found in [11]. It was shown in [12, pp. 377–378] that the autocorrelation function of a phase-modulated signal is the same as the real part of the characteristic function of the probability density function of the *phase changing variable* $\Phi(t+\tau, \mathbf{a}) - \Phi(t, \mathbf{a})$. Using this property, the autocorrelation function of binary 1REC MHPM was found as

$$R(x) = \begin{cases} \frac{1}{K}\sum_{i=1}^{K}\left[\frac{x}{2}D(-\pi h_i x, \pi h_{i+1}x) + \frac{x}{2}D(\pi h_i x, \pi h_{i+1}x)\right. \\ \left. +(1-x)\cos\pi h_i x\right], & \text{for } n=0 \\ \frac{1}{K}\sum_{i=1}^{K}\left(\prod_{j=1}^{n-1}\cos h_{i+j}\right) \\ \left[\frac{x}{2}\cos h_{i+n}[D(-\pi h_i x, \pi h_{i+n+1}x) + \frac{x}{2}D(\pi h_i x, \pi h_{i+n+1}x)]\right. \\ +\frac{(1-x)}{2}[D(\pi h_i + \pi h_{i+n}x, \pi h_i x + \pi h_{i+n}) \\ \left. +D(\pi h_i - \pi h_{i+n}x, \pi h_i x - \pi h_{i+n})]\right], & \text{for } n = 1, 2, \ldots \end{cases} \tag{7.16}$$

where x is the normalized time variable

$$\begin{aligned} x &= \Delta\tau/T \\ \tau &= nT + \Delta\tau \text{ and } 0 \leq \Delta\tau \leq T \end{aligned}$$

and

$$D(x,y) = \begin{cases} \cos x, & x = y \\ (\sin y - \sin x)/(y-x), & x \neq y \end{cases}$$

A special case is the MSK $(K = 1, h = 1/2)$ for which (7.16) becomes

$$R(x) = \begin{cases} (1-\frac{x}{2})\cos(\frac{\pi}{2}x) + \frac{1}{\pi}\sin(\frac{\pi}{2}x), & \text{for } n = 0 \\ (\frac{x-1}{2})\sin(\frac{\pi}{2}x) + \frac{1}{\pi}\cos(\frac{\pi}{2}x), & \text{for } n = 1 \\ 0, & \text{for } n > 1 \end{cases}$$

which is the same as that derived in [13].

An analytical expression for the power spectral density can be found by taking the Fourier transform of (7.16). Note that since $R(\tau)$ will rapidly approach zero as n increases, it is sufficient to consider only a few symbol intervals. Figure 7.6 shows the autocorrelation function and power spectral density of the binary 1REC MHPM scheme with $H_3 = \{4/8, 5/8, 6/8\}$, obtained using the above method.

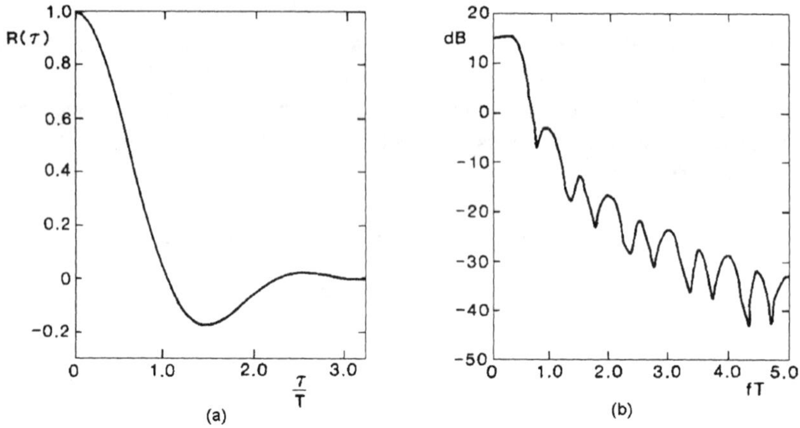

Figure 7.6 Autocorrelation function (a), and power spectral density (b), of the binary 1REC MHPM scheme with $H_3 = \{4/8, 5/8, 6/8\}$. From [11]. Copyright © 1985 IEEE.

The numerical autocorrelation method for single-h CPM described in Chapter 6 can also be generalized for MHPM. A guide line for this generalization is given in [14, p. 162]. However, no detailed results are available.

The direct method has been applied to binary 1REC MHPM [4] and M-ary full-response MHPM [8]. This method is general and has the ability to provide exact (via numerical integration) power spectral density. In this method, the power spectral density is given by

$$\Psi_{\tilde{s}}(f) = \lim_{N \to \infty} E\{|\Psi_{NT}(f)|^2/NT\}$$

where $\Psi_{NT}(f)$ is the direct Fourier transform of an NT second segment of the MHPM signal. The expectation is over the data modulation and the random initial phase angle.

The work of [8] for PSD of M-ary full-response MHPM is based on the work for PSD of FSK in [15] where it is assumed that the random modulation waveform is selected in an independent, identically distributed manner in each symbol interval. In attempting to apply the results of [15] to MHPM, one faces the difficulty that the selection of modulation is not identically distributed, though independent, in each interval. This is because the h_k index is cyclically chosen, and the allowable modulating signals form different sets for different time indexes. A way out of this difficulty is to view the modulation as having a basic signaling interval of $T' = KT$ seconds, thus the distribution on each super-interval is identical. Now there are M^K distinct waveforms to choose from. Based on this concept, the results of [15] can be applied directly to the M-ary full-response MHPM case. The one-side low-pass (or complex envelope) PSD is given by the following equations:

$$\Psi_{\tilde{s}}(f) = \frac{2}{T'}[P(f) + 2\,\mathrm{Re}\{F(f)F_b^*(f)\exp(-j2\pi fT') + F(f)F_b^*(f)\Lambda(f)\}] \tag{7.17}$$

where

$$\Lambda(f) = \exp(-j4\pi fT')\frac{C(T')}{1 - C(T')\exp(-j2\pi fT')} \tag{7.18}$$

To evaluate the above we define

$$b_n(t) = \pi \int_0^t \sum_{i=1}^{K} a_{in} h_i g(\tau - iT)d\tau, \quad n = 1, 2, ..., M^K \tag{7.19}$$

where $\{a_{in}, i = 1, 2, ..., K\}$ is one of the M^K distinct data vectors of length K. From $b_n(t)$ we obtain

$$B_n = b_n(T') \tag{7.20}$$

$$F_n(f) = \int_0^{T'} \exp(-j2\pi ft + jb_n(t))dt \tag{7.21}$$

and from B_n we have

$$C(T') = E\{\exp(jB_n)\}$$

and using $F_n(f)$ we obtain

$$P(f) = E\{|F_n(f)|^2\} = M^{-K}\sum_n |F_n(f)|^2 \tag{7.22}$$

$$F(f) = E\{F_n(f)\} = M^{-K}\sum_n F_n(f) \tag{7.23}$$

Finally using $F_n(f)$ and B_n we define

$$F_b^*(f) = E\{F_n^*(f)\exp(jB_n)\} = M^{-K}\sum_n F_n^*(f)\exp(jB_n) \tag{7.24}$$

In the above series of equations, $E\{\cdot\}$ denotes the expectation with respect to the random index $n = 1, 2, ..., M^K$.

The expression in (7.17) is valid only when $|C(T')| < 1$ [15]. This is also the condition for absence of discrete components in the PSD, and is always satisfied for those MHPM schemes of interest.

Computation of the spectrum of a particular frequency starts from (7.21) for all n. Then other quantities can be computed. The only difficult calculation is (7.21) where numerical integration must be performed. In the cases of LREC and LRC frequency pulses, analytical expressions may be obtained for (7.21), thus considerably simplifying the calculation. Programming efficiency may be increased by exploiting the symmetry in $b_n(t)$ and by utilizing the tree structure of the M^K signals. The entire computation effort is roughly proportional to M^K. For practical schemes K usually is a small number. The computation complexity should not be a problem.

As in the case of single-h CPM, when H_K is a set of integer indexes, $|C(T')| = 1$, line spectral components will appear in the PSD [8].

Figure 7.7 shows PSDs for several sets of indexes obtained using the above method. Figure 7.7(a) also shows the PSD of MSK for comparison. It is seen that the spectrum of the $\{2/4, 3/4\}$ binary 1REC scheme is largely the same as MSK, but it has an asymptotic coding gain over MSK of about 1.4 dB. The spectrum for $\{4/9, 6/9\}$ binary 1REC scheme is shown in Figure 7.7(b), which also is similar to the MSK spectrum. The asymptotic coding gain over MSK is 2.3 dB. Both multi-h spectra in Figures 7.7(a, b) are absent from nulls. Multiple indexes appear to smear the spectra and fill the nulls in the MSK spectra. The roll-off rate of the two spectra is proportional to f^{-4}. Figure 7.7(c) is the PSD for $\{4/8, 5/8, 6/8\}$ binary 1REC scheme. This spectrum is also shown in Figure 7.6 where the autocorrelation method is used. It can be seen that both methods produce the same PSD. Again the spectrum of this scheme is similar to those in Figure 7.7(a, b). This is not surprising since their minimum and maximum deviations (determined by the indexes) are equivalent or close. This scheme has a 2.8 dB asymptotic coding gain over MSK. PSD of the $\{4/9, 6/9\}$ binary RC scheme is shown in Figure 7.7(d). The roll-off rate is proportional to f^{-8}.

For phase modulation, it was shown that if the phase pulse $q(t)$ has $(N-1)$ continuous derivatives which are equal to zero at the leading and trailing edges, then the power spectrum decays asymptotically as $f^{-2(N+1)}$ [16].[1] For 1REC MHPM schemes, $N = 1$, thus their spectra decay as f^{-4} as shown in Figures 7.7 (a–c). For raised-cosine pulse, $N = 3$, and its spectrum decays as f^{-8} as shown in Figure 7.7(d).

There are two simple methods that approximate the spectrum of an MHPM

---

1 Similar results are stated in Chapter 5, Section 5.11 for symbol shaping pulse.

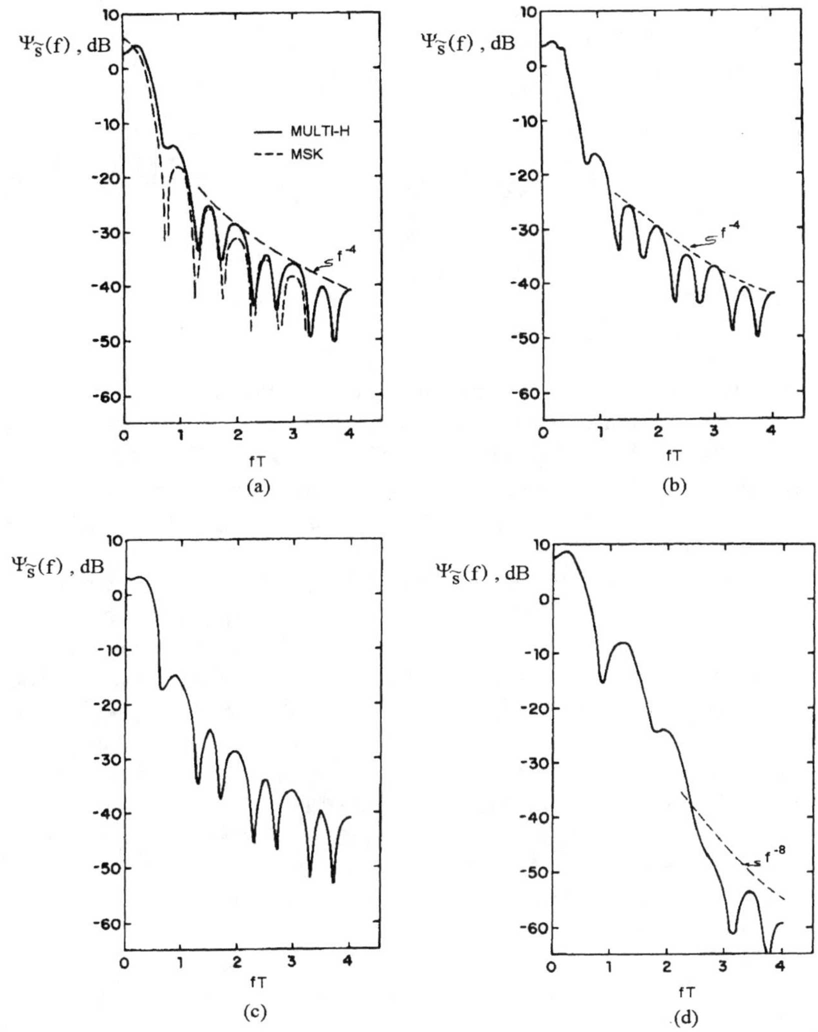

Figure 7.7 (a) PSD for $\{2/4, 3/4\}$ binary 1REC MHPM in comparison with MSK, (b) PSD for $\{4/9, 6/9\}$ binary 1REC MHPM, (c) PSD for $\{4/8, 5/8, 6/8\}$ binary 1REC MHPM, and (d) PSD for $\{2/4, 3/4\}$ binary RC MHPM. From [8]. Copyright © 1981 IEEE.

scheme [8]. Method one is to treat the spectrum as an average of constant-h spectra. That is

$$\Psi_{\tilde{s}}(f) \approx \frac{1}{K}\sum_{i=1}^{K}\Psi_{h_i}(f)$$

where $\Psi_{h_i}(f)$ is the spectrum corresponding to index h_i. Method two is to average the indexes first

$$\overline{h} = \frac{1}{K}\sum_{i=1}^{K}h_i$$

then calculate the spectrum for the single-h CPM with $h = \overline{h}$. It is intuitive that the closer the indexes become, the more accurate the approximation is. It was reported in [8] that with "good" multi-h schemes, method one is generally quite accurate (i.e., ± 2 dB error in the spectrum). Good schemes appear to be those with indexes grouped near one value. Method two also gives the general spectrum shape, but it gives true nulls rather than local minima, and is somewhat misleading in that regard. Both methods are quite accurate in terms of describing out-of-band power behavior.

The spectra of some MHPM schemes with partial-response frequency pulse were obtained by simulations [10]. Figure 7.8 shows PSDs for several binary 4S MHPM schemes with various indexes. A 4S frequency pulse is obtained by convolving 3S with 1REC, while 3S is obtained by convolving 2TRI (2S) with 1REC, where 2TRI is a triangular pulse over two-symbol intervals. Figure 7.9 shows the LS pulses and some other pulses. In Figure 7.8, the average indexes are 0.5 and 0.8. The spectra of the 0.8 group is wider than that of 0.5 group. In each group, the spreading between the two indexes determines the spectral width. Generally, the wider the spread, the wider the spectrum. Also it is clear from the figure that the average index can be used to calculate approximately the MHPM spectra, especially when the spread is narrow.

7.3 DISTANCE PROPERTIES AND ERROR PROBABILITY

The distance equations of single-h CPM are still applicable to MHPM. Particularly we rewrite the normalized Euclidean distance (6.32) here

$$d_{ij}^2 = \frac{\log_2 M}{T}\int_0^{NT}[1 - \cos\Delta\Phi(t,\gamma_{ij})]dt \qquad (7.25)$$

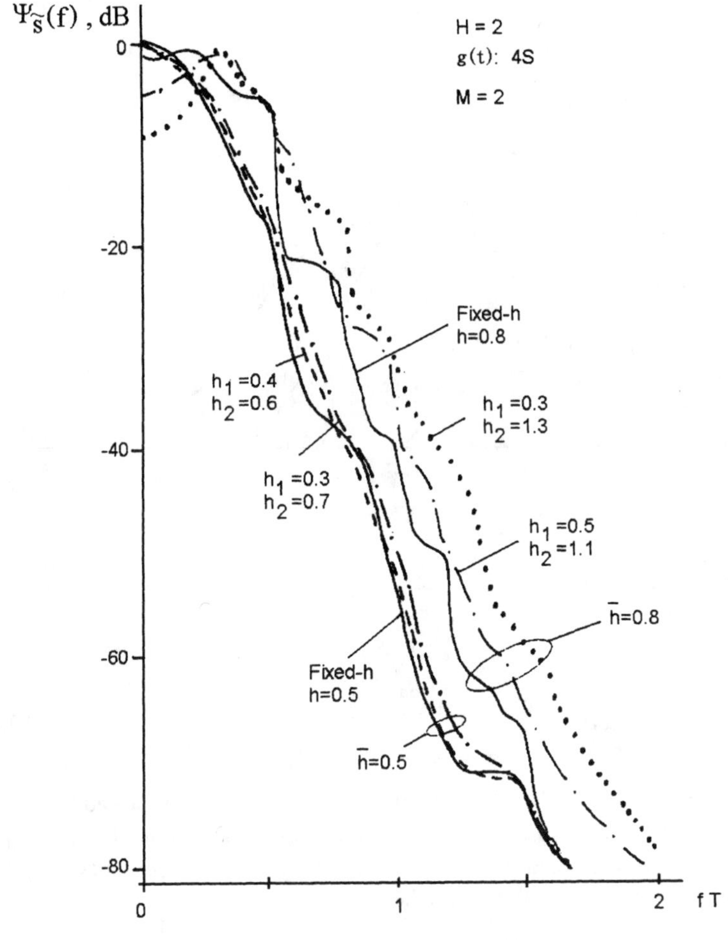

Figure 7.8 PSDs for several binary 4S MHPM schemes with various indexes. From [10]. Copyright © 1982 IEEE.

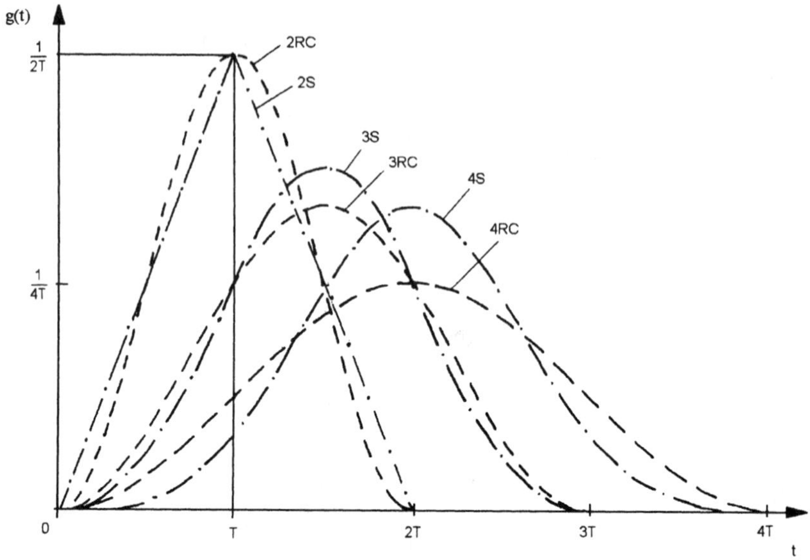

Figure 7.9 Some frequency pulse shapes. From [10]. Copyright © 1982 IEEE.

where

$$\Delta\Phi(t, \gamma_{ij}) = 2\pi \sum_{k=-\infty}^{\infty} h_k \gamma_k q(t - kT) \tag{7.26}$$

is slightly different from (6.31). The difference is that h has been replaced by h_k.

The error probability is again largely determined by the minimum distance $d_{\min}$ as in (6.39) and (6.40) when detected with the maximum likelihood sequence detector at high signal-to-noise ratios. We rewrite them here

$$P_e \lesssim \frac{K}{2} \exp\{-d_{\min}^2 (E_b/2N_o)\} \tag{7.27}$$

or

$$P_e \lesssim KQ\left(d_{\min}\sqrt{E_b/N_o}\right) \tag{7.28}$$

As we pointed out in Chapter 6, that even though P_e is the error probability of sequence detection, it is approximately equal to the symbol error probability at high

SNR.

Distance properties of M-ary 1REC (or linear phase) MHPM schemes have been analyzed in [17]. Upper bounds on the minimum distances were calculated for binary 1REC 2-h schemes, M-ary 1REC 2-h schemes, binary 1REC 3-h schemes, binary 1REC 4-h schemes, quaternary and octal 1REC MHPM schemes. An upper bound on the minimum distance is an important parameter of an MHPM scheme since it gives the maximum achievable minimum distance, hence the minimum achievable error probability of the scheme. The actual minimum distances were calculated for several different 2-h schemes and for various parameter combinations.

As we described in Chapter 6, upper bound on the $d^2_{\min}$ can be obtained by calculating the distances and choosing the minimum for all pairs of phase paths which give first, or deeper, inevitable mergers in the phase tree or trellis. As we already know from Chapter 6, upper bound d^2_B is a function of h for fixed frequency pulse $g(t)$. For MHPM it is a function of H_K for fixed frequency pulse $g(t)$. Logically the upper bounds for multi-h schemes should be calculated against various sets of $(h_1, h_2, ..., h_K)$. However, empirical results show that most of the good 2-h, 3-h, and 4-h schemes are those with low and moderate average $\overline{h}$ values and the individual h_i values are close together. Thus it is of special interest to study the upper bound for the case $h_i = h$. This of course degenerates the multi-h schemes back to single-h schemes whose upper-bound expressions are different and upper-bound values are smaller. Typically, first merges of single-h schemes occur earlier and the upper bound is smaller. However, we can still pretend that the scheme is multi-h and the merges remain unchanged, thus the expressions also remain unchanged. Then we let all $h_i = \overline{h}$, the resulting expressions would become the approximate upper bounds for schemes with h_i close to each other but still different. Strictly speaking, these upper bounds are not upper bounds for any scheme, single-h or multi-h, since for single-h, bound expressions are different; for multi-h, there exist no such schemes where all h_i are equal. Nonetheless the upper bound for $h_i = \overline{h}$ is representative for schemes with h_i close to each other but still different [17]. Figure 7.10 is a summary of upper bounds on $d^2_{\min}$ for various M and K versus $\overline{h}$ for multi-h schemes and versus h for single-h schemes, respectively. This figure is simple, clear, and representative due to the use of average index $\overline{h}$. Otherwise, if various index sets H_K were used, the figure would have been too complicated, and not representative.

Now let us examine Figure 7.10. For $M = 2$, the maximum increase of bound for $K \geq 2$ over the $K = 1$ case is 3 dB for $\overline{h} \leq 0.5$. For larger $\overline{h}$ values slightly better bounds are obtained with increasing values of K. Further increase in K leads to smaller gain in bounds. On the other hand, increasing M gives significant gain in distance bounds. For instance, $M = 4$ single-h case has the same bound as that for $M = 2$, $K = 2, 3, 4$ cases for $\overline{h} \leq 0.5$. And $M = 8$ single-h case has the higher bound than that for $M = 2$, $K = 2, 3, 4$ cases for $\overline{h} \leq 0.5$. The conclusion from Fig-

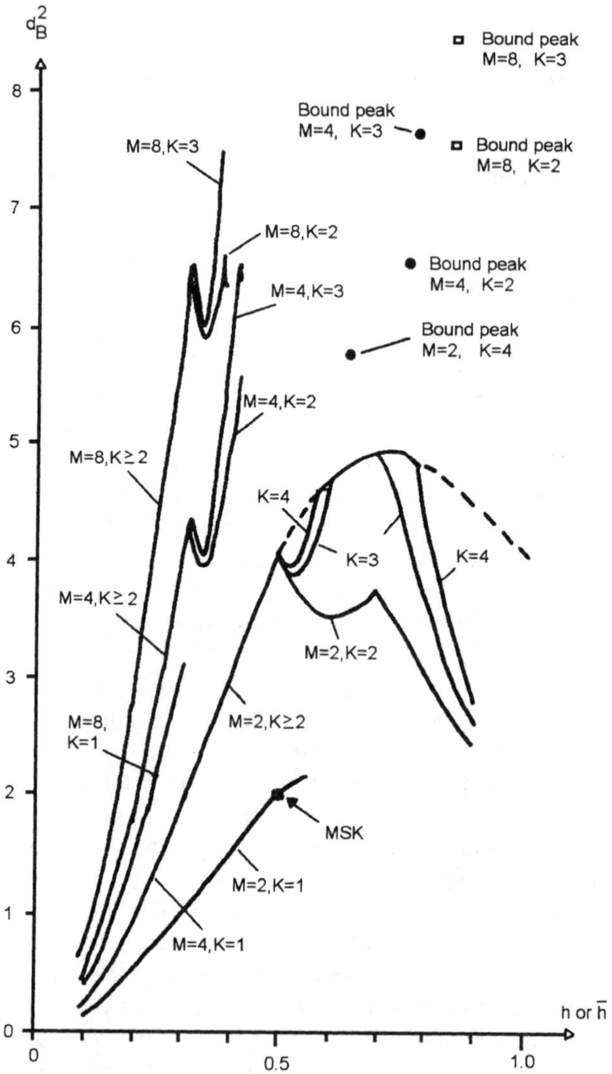

Figure 7.10 Summary of upper bounds d_B^2 on $d_{\min}^2$ for various M and K values. From [17]. Copyright © 1982 IEEE.

ure 7.10 is that to keep system complexity down while achieving fairly large distance gain, the good choices are binary 2-h schemes and quaternary single-h schemes for $\overline{h} \leq 0.5$. Further increasing K and/or M will not increase distance gain significantly while the system complexity will increase significantly. In addition, using $\overline{h} > 0.5$ is not preferable since the signal bandwidth will be increased.

Actual d^2_{min} has been calculated in [17] for various schemes. Figure 7.11 shows d^2_{min} for binary 1REC 2-h schemes where $h_1 = 0.5$ is fixed and h_2 is allowed to change. The path length is $N = 30$ for calculating d^2_{min}. The dashed line is the upper bound. From the figure, it can be seen that the upper bound is tight in this h region except for those weak or catastrophic values of the modulation indexes. For these specific values of modulation indexes, merges occur earlier in the phase tree, leading to reduced minimum distances. In Figure 7.11, the weak indexes are $(1/2, 1/4)$, $(1/2, 1/3)$, $(1/2, 1/2)$, $(1/2, 3/4)$, and $(1/2, 1)$. More results of d^2_{min} for binary and quaternary 1REC MHPM schemes are available in [17].

By using partial-response frequency pulses, further gain of minimum distance can be obtained [10]. Of course the price is the increased system complexity.

The trade-offs between distance and bandwidth for full-response 1REC-MHPM schemes and partial-response 3RC MHPM schemes are considered in [10] and the results are summarized in Figures 7.12 and 7.13, respectively.

In Figure 7.12, the bandwidth is the two-sided normalized 99% in-band power bandwidth and the distance is relative to MSK. The MSK's two-sided 99% bandwidth is $1.2R_b$ (see Section 5.2.2). Among single-h schemes, $M = 4$ schemes outperform $M = 2$ schemes considerably. Slight further improvement can be obtained by increasing M to 8, especially at large h values (large $2BT_b$). The figure also shows the equal-h bounds for $K = 2$ schemes with $M = 2, 4$, and 8. Again we can see that going from $M = 2$ to $M = 4$ improves significantly and going from $M = 4$ to $M = 8$ does not make much difference. Specific schemes are shown for the $K = 2, M = 2$ and $K = 2, M = 4$ cases. It can be seen that these schemes are slightly below corresponding bounds. For $K = 3$, bound peak points are shown. The $M = 2, K > 2$ equal-h bound is also shown, which coincides with the $M = 2, K = 2$ bound in the first part of the curve (also see Figure 7.10). An overall comparison reveals that binary multi-h schemes perform roughly the same as single-h $M = 4$ or $M = 8$ schemes. In most cases M-ary single-h schemes are better. Considerable improvements are obtained by using $K = 2, M = 4$ schemes. Only minor further improvement can be obtained by further increasing K and/or M.

Figure 7.13 shows power-bandwidth trade-off for M-ary partial-response 3RC MHPM schemes. The bandwidth is defined as the bit rate normalized width of the smooth power spectra at the -60 dB level, not the fractional out-of-band power bandwidth used in Figure 7.12. The figure shows four cases: $M = 2, 4$ and $K = 1, 2$.

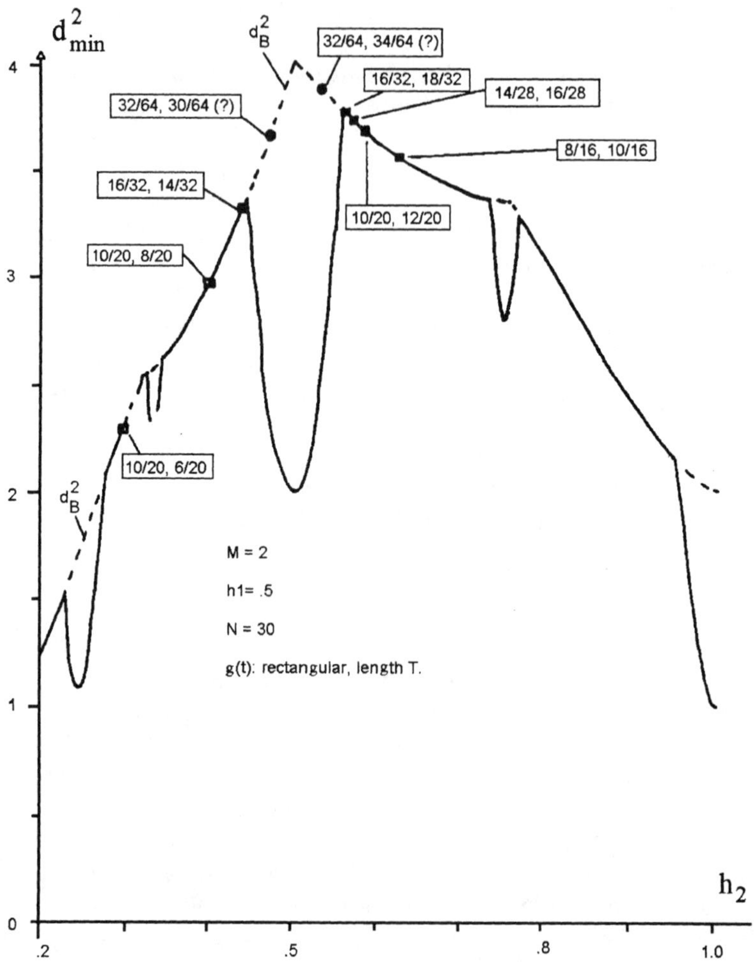

Figure 7.11 The minimum squared normalized Euclidean distance $d^2_{\min}$ for binary 1REC 2-h schemes with $h_1 = 0.5$ versus h_2 for $N = 30$. From [17]. Copyright © 1982 IEEE.

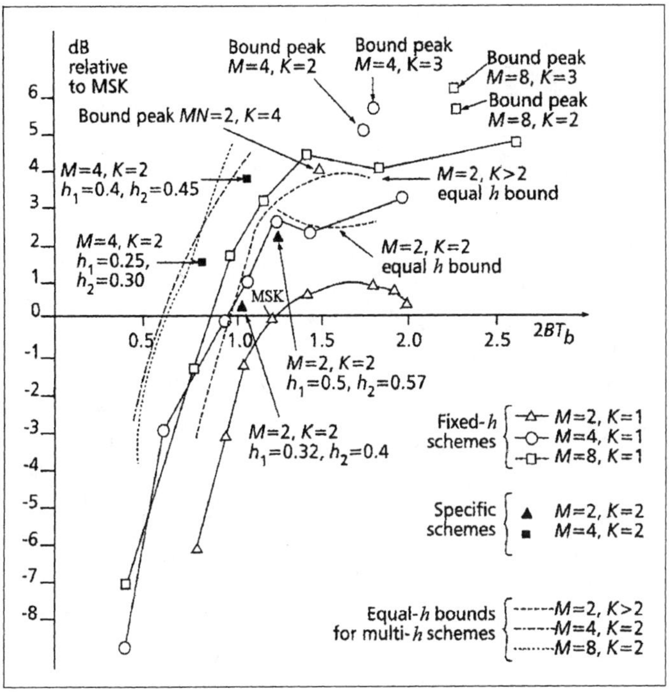

Figure 7.12 Power bandwidth trade-off for M-ary 1REC MHPM schemes. The bandwidth is the –20 dB (99%) out-of-band power bandwidth. All schemes are compared at equal bit rate. From [10]. Copyright © 1982 IEEE.

Again from this figure we can see that $M = 4$ schemes are considerably better than $M = 2$ schemes for the same h or H_K; and 2-h schemes are slightly better than single-h schemes for the same M. There are two 2-h 3RC schemes that are considerably better than MSK: $M = 4$, $H_2 = (0.35, 0.40)$, and $(0.40, 0.45)$. They have 1.81 dB and 3.14 dB gain in distance over MSK.

Lereim investigated the coding gain and spectral properties of the various binary 1REC MHPM signals extensively and the results are summarized in Tables 7.1, 7.2, and 7.3 [6]. The modulation indexes are expressed in the form of $\{q/p_1, p_2, ...p_K\}$. The decision depth is the depth in the trellis beyond which all unmerged paths have distance exceeding d_{free}, the free distance, and is a useful design parameter for the

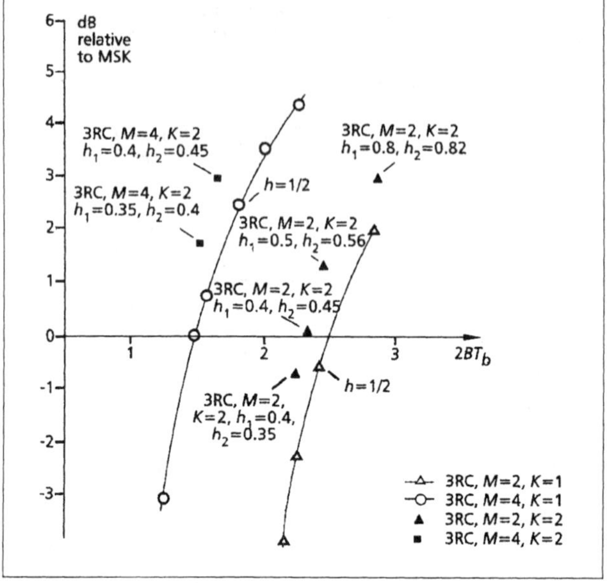

Figure 7.13 Power bandwidth trade-off for M-ary partial response 3RC MHPM schemes. The bandwidth is defined as the bit rate normalized width of the PSDs at the –60 dB level. From [10]. Copyright © 1982 IEEE.

practical Viterbi demodulator.[2] The coding gain is the gain in $d^2_{\min}$ over that of MSK. The bandwidths are the one-sided fractional out-of-band power bandwidths normalized to the bit rate. For MSK, we can easily find out (using (5.14) and (2.21)) $B_{95\%} \approx 0.45$, $B_{99\%} \approx 0.6$, and $B_{99.5\%} \approx 0.81$. These are benchmarks for bandwidth comparison when reading the tables. Table 7.1 lists the schemes with high coding gains. But these schemes also need higher bandwidth than MSK; and the schemes have at least three indexes, which translate into high system complexity. Table 7.2 lists the schemes with moderate coding gain and less bandwidth needs than MSK, and some schemes have only two indexes. Table 7.3 lists the schemes with some coding loss and very high bandwidth savings with respect to MSK.

[2] d_{free} is defined as the distance from a given ("transmitted") path to its nearest neighbor of any length. For linear codes, such as convolutional codes, d_{free} is independent of the transmitted path. For multi-h codes, d_{free} varies with the transmitted path. Thus there is a $d_{free,\min}$ which dominates the error performance. See [4] for more about d_{free}.

H_K	K	Coding gain (dB)	Bandwidth			Decision depth
			95.0%	99.0%	99.5%	
16/12 10 11 8	4	3.66	0.53	0.85	1.00	26-38
16/10 11 12	3	3.36	0.53	0.89	1.03	26-27
13/ 8 9 10	3	3.19	0.53	0.90	1.03	17-20
16/11 12 10 8	4	3.16	0.53	0.85	1.00	11-27
16/12 11 10 8	4	3.16	0.53	0.85	1.00	11-34
16/ 9 10 11	3	3.13	0.51	0.81	0.98	24-26
14/ 8 9 10	3	3.13	0.53	0.84	0.99	18-20
16/12 9 10 8	4	3.03	0.51	0.81	0.98	28
14/ 9 10 11	3	2.96	0.54	0.91	1.04	17-26
16/12 10 9 8	4	2.91	0.51	0.81	0.98	29-30
16/10 12 9 8	4	2.91	0.51	0.81	0.98	29
16/ 8 9 10	3	2.82	0.49	0.64	0.93	28
14/ 7 8 9	3	2.81	0.50	0.65	0.94	22
12/ 6 7 8	3	2.81	0.50	0.66	0.95	17
10/ 5 6 7	3	2.81	0.51	0.78	0.90	13
12/ 6 7 9	3	2.77	0.53	0.83	0.99	7-17
8/ 4 5 6	3	2.77	0.53	0.84	1.00	7-9
12/ 6 8 9	3	2.77	0.53	0.85	1.00	13-14
16/ 8 9	2	2.74	0.48	0.61	0.89	20
10/ 5 6 8	3	2.74	0.54	0.86	1.01	11-13

Table 7.1 Binary 1REC MHPM schemes with high coding gain over MSK. From [6].

H_K	K	Coding gain (dB)	Bandwidth			Decision depth
			95.0%	99.0%	99.5%	
16/ 7 8	2	2.19	0.45	0.59	0.68	21
16/ 6 7 8	3	2.10	0.43	0.59	0.66	20
14/ 6 7	2	2.07	0.44	0.59	0.68	19
14/ 5 6 7	3	1.96	0.43	0.58	0.65	16-18
10/ 4 5	2	1.72	0.44	0.59	0.66	11
15/ 6 7	2	1.67	0.43	0.58	0.65	20
16/ 6 4 7 8	4	1.65	0.40	0.56	0.64	7-20
15/ 5 6 7	3	1.53	0.40	0.56	0.64	20
12/ 4 5 6	3	1.47	0.41	0.58	0.65	13
8/ 3 4	2	1.45	0.43	0.59	0.66	8
16/ 4 6 7 8	4	1.45	0.40	0.58	0.65	20
13/ 5 6	2	1.44	0.41	0.58	0.65	15
14/ 5 7	2	1.28	0.43	0.59	0.66	11
16/ 6 7	2	1.28	0.40	0.56	0.64	21
16/ 4 7 6 8	4	1.21	0.40	0.59	0.65	11-20
11/ 4 5	2	1.12	0.40	0.58	0.64	11
16/ 5 6 7	3	1.07	0.39	0.55	0.63	20
14/ 5 6	2	0.97	0.40	0.56	0.63	16
13/ 4 5 6	3	0.89	0.39	0.56	0.64	14
16/ 5 8	2	0.87	0.40	0.60	0.68	8

Table 7.2 Bandwidth efficient binary 1REC MHPM schemes with moderate coding gain over MSK. From [6].

H_K	K	Coding gain (dB)	Bandwidth			Decision depth
			95.0%	99.0%	99.5%	
15/ 3 4 5	3	-2.40	0.29	0.46	0.54	11
14/ 3 4	2	-2.67	0.26	0.46	0.54	8
16/ 3 4 5	3	-2.92	0.26	0.46	0.53	11
16/ 3 5	2	-3.03	0.28	0.46	0.55	11
13/ 3 4	2	-2.08	0.29	0.48	0.56	8
15/ 3 5	2	-2.50	0.29	0.48	0.56	11
16/ 4 5	2	-1.57	0.30	0.49	0.56	12
11/ 2 3 4	3	-2.48	0.30	0.49	0.56	9-10
9/ 2 3	2	-2.54	0.30	0.49	0.58	6
15/ 4 5	2	-1.08	0.31	0.50	0.58	12
10/ 2 3 4	3	-1.70	0.33	0.50	0.56	9-10
14/ 3 4 5	3	-1.84	0.31	0.50	0.58	10
16/ 4 5 6	3	-0.68	0.33	0.51	0.58	15-16
13/ 3 4 5	3	-1.25	0.33	0.51	0.59	10
8/ 2 3	2	-1.62	0.33	0.51	0.59	6
14/ 3 5	2	-1.94	0.30	0.51	0.59	10
16/ 5 6	2	0.06	0.35	0.53	0.60	16
14/ 4 5	2	-0.56	0.34	0.53	0.59	12
11/ 3 4	2	-0.80	0.34	0.53	0.60	8
13/ 3 5	2	-1.35	0.33	0.53	0.61	10

Table 7.3 Bandwidth efficient binary 1REC MHPM schemes (with negative coding gain over MSK). From [6].

So far we have been focusing on $d^2_{\min}$ and its upper bound d^2_B. Even though $d^2_{\min}$ and d^2_B are good performance indicators, it would be desirable to have the results of the symbol or bit error probability directly. Over bounds to probability of error event are given in [4] for binary 1REC MHPM. Bit error probabilities for the same class of schemes were found by simulation and presented in [18].[3] Further improved results were reported in [19] where upper and lower bounds as well as simulation results for the bit error probabilities of the M-ary 1REC schemes were presented. Figures 7.14 to 7.16 are from [19]. Simulations were performed by the Viterbi demodulator.

Three schemes, binary $\{2/4, 3/4\}$, $\{4/8, 5/8, 6/8\}$, and 4-ary $\{3/16, 4/16\}$ were considered. The free distance calculations for these schemes project an asymptotic gain over coherent BPSK or QPSK by 1.4 dB, 2.8 dB, and –0.8 dB, respectively. The first two have spectra comparable to that of CPFSK with $h = 5/8$, and the 99% bandwidth is about 1.67 times the bit rate. The 4-ary scheme has a narrower bandwidth. Its 99% bandwidth is about 0.7 times the bit rate.

The upper bounds and lower bounds and simulated bit error probabilities for the binary $\{2/4, 3/4\}$ scheme are shown in Figure 7.14 where N is the number of symbols of the observation (or path memory length). The decision depth is four for this scheme. Thus we expect that $N = 4$ should be adequate as a practical path memory length. From Figure 7.14 we can see that this indeed is true. For $P_b < 10^{-4}$, the upper bound for the $N = 4$ case is only 0.2 dB or less higher than the unlimited path memory case. The simulation results for $N = 4$ and $N = \infty$ are very close. The upper and lower bounds for $N \geq 4$ are very tight at $E_b/N_o \geq 8$ dB. With $N = 2$, the performance is much poorer. It is essentially equal to that of BPSK (see the BPSK curve in Figure 4.4).

Figure 7.15 shows the bounds for the binary $\{4/8, 5/8, 6, 8\}$ scheme. Larger asymptotic coding gain with respect to coherent BPSK is observed. When compared with the $\{2/4, 3/4\}$ scheme, this stronger scheme needs longer path memory length to realize its potential. The decision depth for this scheme is nine. The upper bound for $N = 8$ is about 0.3 dB higher than the unlimited memory bound.

Figure 7.16 shows the bounds for the 4-ary $\{3/16, 4/16\}$ scheme. This scheme has a narrower bandwidth than the binary ones. However, the coding gain is negative with respect to coherent BPSK. The asymptotic loss is about 0.8 dB. Again the decision depth for this scheme is nine, the upper bound for $N = 8$ is about 0.3 dB higher than the unlimited memory bound.

---

[3] The binary data symbols are [0,1] in [4, 18] instead of [–1,1] which is the common assumption in the literature and this book. The [0,1] symbol assumption does not change the distance property of the phase trellis, hence the error performance. However, it does reduce the number of distinct phases in the trellis to half. That is, the number of distinct phases is q instead of $2q$ [4].

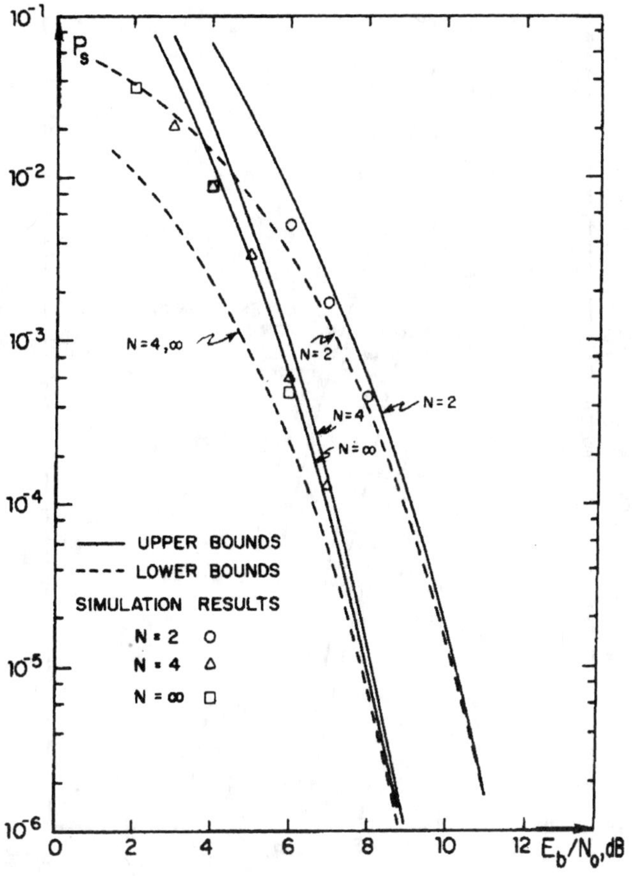

Figure 7.14 Bounds and simulated bit error probabilities for the binary 1REC $\{2/4, 3/4\}$ scheme. The decision depth of the Viterbi algorithm is four. From [19]. Copyright © 1982 IEEE.

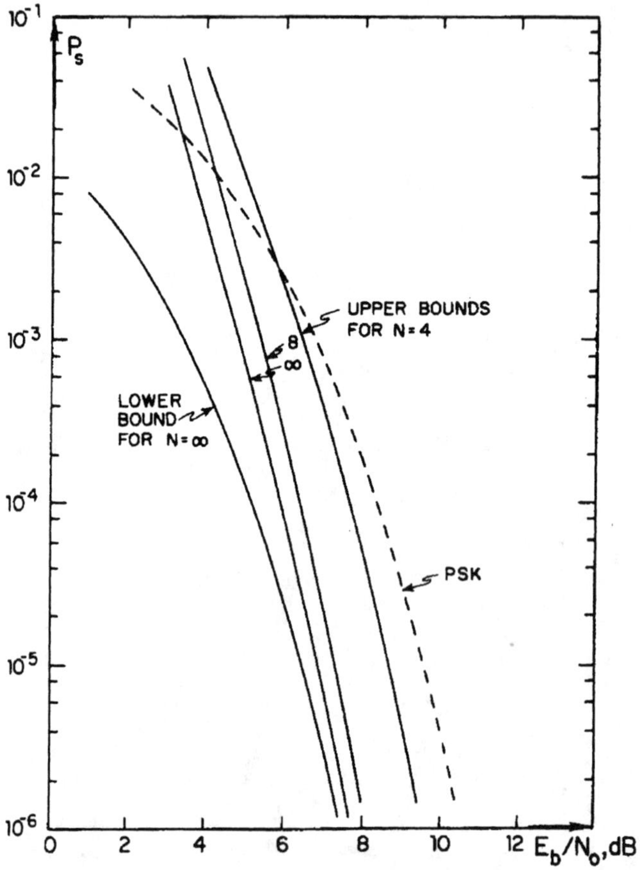

Figure 7.15 Bounds for the binary 1REC $\{4/8, 5/8, 6, 8\}$ scheme. The decision depth of the Viterbi algorithm is nine. From [19]. Copyright © 1982 IEEE.

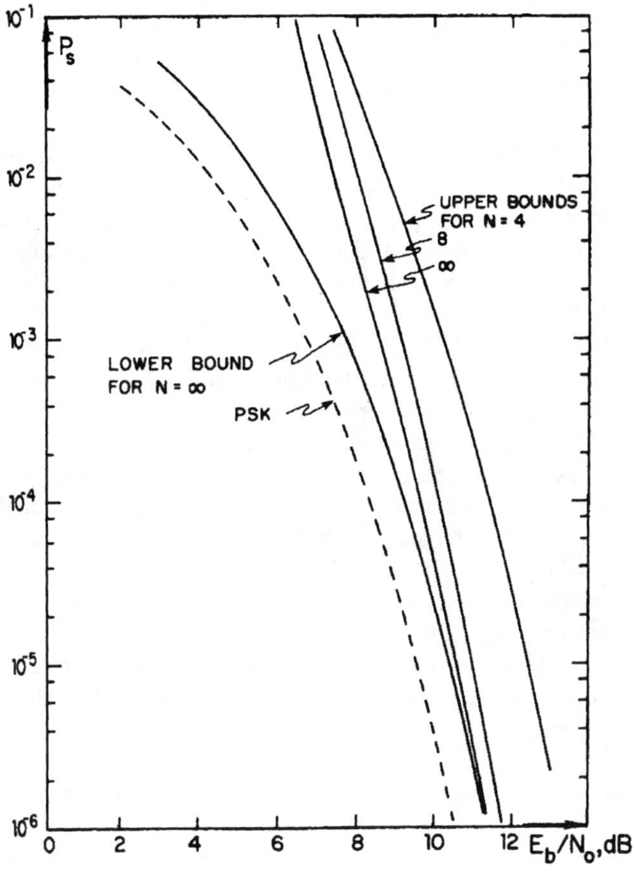

Figure 7.16 Bounds for the 4-ary 1REC $\{3/16, 4/16\}$ scheme. The decision depth of the Viterbi algorithm is nine. From [19]. Copyright © 1982 IEEE.

7.4 MODULATOR

MHPM modulators are almost identical to those of single-h CPM except that the index h must be cyclically rotated when modulating. Specifically, Figures 6.13 to 6.17 all can be adapted to MHPM by cyclically changing the modulation indexes.

7.5 DEMODULATOR AND SYNCHRONIZATION

MHPM demodulators are also similar to single-h CPM demodulators except that the demodulator must have a prior knowledge of the sequence of indexes h_k employed at the transmitter. Thus in principle we can adapt all demodulators in Chapter 6 to MHPM. However, since the MLSE demodulator using the Viterbi algorithm is the most promising one and is explored the most in the literature, we will only consider this type of MHPM demodulators.

For MHPM schemes with rational indexes, MLSE demodulation is based on their state trellises using the Viterbi algorithm. For MHPM schemes with rational indexes and full-response frequency pulse, the phase trellis is equivalent to the state trellis, thus MLSE demodulation is based on their phase trellises using the Viterbi algorithm. Difference between various MLSE demodulators lie in the "front end" of the demodulator (i.e., the method of obtaining the branch metrics in the trellis).

MHPM demodulators reported in the literature so far are for full-response binary or M-ary CPFSK (i.e., 1REC-MHPM) schemes [4, 7, 18, 20–23]. Demodulators for partial-response MHPM schemes and frequency pulses other than 1REC have not been reported in the literature. In the following, we will first discuss the demodulator for binary multi-h CPFSK, which has a very simple structure. Then we will discuss the more complex demodulators for M-ary CPFSK with carrier and/or symbol synchronization.

7.5.1 A Simple ML Demodulator for Multi-*h* Binary CPFSK

Figure 7.17 is the simple ML multi-h binary CPFSK demodulator which employs four bandpass correlators. The reference signals must be updated for each symbol interval by cyclically switching indexes h_k. The coefficients $\{A'_1, A'_2, A'_3, A'_4\}$ are sufficient statistics for the signal. They are used to compute the branch metrics of the phase trellis for the Viterbi processor which eventually demodulates the signal. Factors $C_{1,k}$, $C_{2,k}$, and D_k are related to indexes too and must be cyclically changed.

This demodulator was first proposed in [4] and later in [18] as a means of simulating error performance. It was later presented as a formal demodulator in [7].

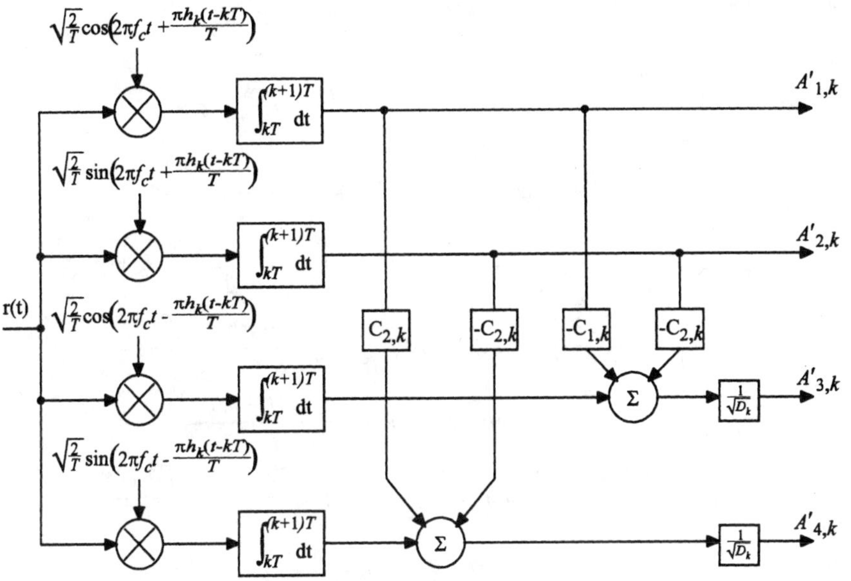

Figure 7.17 Multi-h binary CPFSK demodulation: determination of coefficients. From Bhargava et al., *Digital Communications by Satellite,* Copyright © 1981 John Wiley & Sons, Inc. Reprinted by permission of John Wiley & Sons, Inc.

Although this demodulator was proposed for multi-h CPFSK, we have no reason to believe that this demodulator cannot be used for single-h binary CPFSK. We need to simply replace h_k by h. We have presented the Viterbi demodulator for the general CPM case in Chapter 6, Section 6.5.3. It needs a basic quadrature receiver (Figure 6.26) and a bank of $2M^L$ baseband matched filters (Figure 6.27). When applied to binary CPFSK (single-h or multi-h), the number of the matched filters in Figure 6.27 is four which is the same as the number of correlators in Figure 7.17. The structure of Figure 7.17 looks a bit simpler since no quadrature receiver is needed. However, it requires four coherent reference signals with not only the exact carrier frequency but also a linear phase $\pi h_k t/T$ which depends on the modulation index of the symbol interval. The four coherent reference signals in Figure 7.17 must be provided by a separate carrier recovery circuit.

We now derive the demodulator structure of Figure 7.17. It is obtained by the Gram-Schmidt procedure (which is also used in the discussion of detection of binary and M-ary signals in Appendix B).

Using trigonometric identities, we can expand the CPFSK signal in (7.9) as (expressing the amplitude in terms of the signal energy $E : A = \sqrt{2E/T}$)

$$
\begin{aligned}
s(t) &= \sqrt{\frac{2E}{T}} \cos(2\pi f_c t + \frac{a_k \pi h_k (t-kT)}{T} + \theta_k) \\
&= \sqrt{\frac{2E}{T}} \left[\cos\theta_k \cos(2\pi f_c t + \frac{a_k \pi h_k (t-kT)}{T}) \right. \\
&\quad \left. - \sin\theta_k \sin(2\pi f_c t + \frac{a_k \pi h_k (t-kT)}{T}) \right] \\
&= \sqrt{\frac{2E}{T}} \left\{ \cos\theta_k \left[\cos(2\pi f_c t) \cos \frac{a_k \pi h_k (t-kT)}{T} \right. \right. \\
&\quad \left. - \sin(2\pi f_c t) \sin \frac{a_k \pi h_k (t-kT)}{T} \right] \\
&\quad - \sin\theta_k \left[\sin(2\pi f_c t) \cos \frac{a_k \pi h_k (t-kT)}{T} \right. \\
&\quad \left. \left. + \cos(2\pi f_c t) \sin \frac{a_k \pi h_k (t-kT)}{T} \right] \right\}
\end{aligned}
$$

Realizing that if $a = \pm 1$, then $\cos(ax) = \cos x$ and $\sin(ax) = a \sin x$, the above can be reduced to

$$
\begin{aligned}
s(t) &= \sqrt{\frac{2E}{T}} \left\{ \cos\theta_k \left[\cos(2\pi f_c t) \cos \frac{\pi h_k (t-kT)}{T} \right. \right. \\
&\quad \left. - a_k \sin(2\pi f_c t) \sin \frac{\pi h_k (t-kT)}{T} \right] \\
&\quad - \sin\theta_k \left[\sin(2\pi f_c t) \cos \frac{\pi h_k (t-kT)}{T} \right. \\
&\quad \left. \left. + a_k \cos(2\pi f_c t) \sin \frac{\pi h_k (t-kT)}{T} \right] \right\}
\end{aligned}
\tag{7.29}
$$

Using trigonometric identities for $\cos u \cos v$, $\cos u \sin v$, $\sin u \sin v$, $\sin u \cos v$, the above can be converted to

$$
\begin{aligned}
s(t) &= \sqrt{\frac{2E}{T}} (\cos\theta_k) \left(\frac{1+a_k}{2} \right) \cos\left(2\pi f_c t + \frac{\pi h_k (t-kT)}{T} \right) \\
&\quad - \sqrt{\frac{2E}{T}} (\sin\theta_k) \left(\frac{1+a_k}{2} \right) \sin\left(2\pi f_c t + \frac{\pi h_k (t-kT)}{T} \right)
\end{aligned}
$$

$$+\sqrt{\frac{2E}{T}}(\cos\theta_k)\left(\frac{1-a_k}{2}\right)\cos\left(2\pi f_c t-\frac{\pi h_k(t-kT)}{T}\right)$$
$$-\sqrt{\frac{2E}{T}}(\sin\theta_k)\left(\frac{1-a_k}{2}\right)\sin\left(2\pi f_c t-\frac{\pi h_k(t-kT)}{T}\right) \quad (7.30)$$

Although each pair of terms with $\pi h_k(t-kT)/T$ and $-\pi h_k(t-kT)/T$ is orthogonal, terms in one pair are not orthogonal to terms in the second pair. By using the Gram-Schmidt procedure, however, we can transform (7.30) into an orthogonal expansion based on the following four orthonormal base functions (see Appendix 7A at the end of this chapter)

$$\psi_1(t)=\sqrt{\frac{2}{T}}\cos\left(2\pi f_c t+\frac{\pi h_k(t-kT)}{T}\right) \quad (7.31)$$

$$\psi_2(t)=\sqrt{\frac{2}{T}}\sin\left(2\pi f_c t+\frac{\pi h_k(t-kT)}{T}\right) \quad (7.32)$$

$$\psi_3(t)=\frac{1}{\sqrt{D_k}}\left[\sqrt{\frac{2}{T}}\cos\left(2\pi f_c t-\frac{\pi h_k(t-kT)}{T}\right)-C_{1,k}\psi_1(t)-C_{2,k}\psi_2(t)\right] \quad (7.33)$$

$$\psi_4(t)=\frac{1}{\sqrt{D_k}}\left[\sqrt{\frac{2}{T}}\sin\left(2\pi f_c t-\frac{\pi h_k(t-kT)}{T}\right)+C_{2,k}\psi_1(t)-C_{1,k}\psi_2(t)\right] \quad (7.34)$$

where

$$C_{1,k}=\frac{\sin 2\pi h_k}{2\pi h_k}$$

$$C_{2,k}=\frac{1-\cos 2\pi h_k}{2\pi h_k}$$

$$D_k=1-C_{1,k}^2-C_{2,k}^2$$

Using these four functions, the signal $s(t)$ can be written as

$$s(t) = \sqrt{E}(\cos\theta_k)\left(\frac{1+a_k}{2}\right)\psi_1(t)$$
$$-\sqrt{E}(\sin\theta_k)\left(\frac{1+a_k}{2}\right)\psi_2(t)$$

$$+\sqrt{E}\,(\cos\theta_k)\left(\frac{1-a_k}{2}\right)\left[C_{1,k}\psi_1(t)+C_{2,k}\psi_2(t)+\sqrt{D_k}\psi_3(t)\right]$$
$$-\sqrt{E}\,(\sin\theta_k)\left(\frac{1-a_k}{2}\right)\left[-C_{2,k}\psi_1(t)+C_{1,k}\psi_2(t)+\sqrt{D_k}\psi_4(t)\right] \tag{7.35}$$

Rearranging terms in (7.35) gives

$$s(t)=\sqrt{E}\,[A_{1,k}\psi_1(t)+A_{2,k}\psi_2(t)+A_{3,k}\psi_3(t)+A_{4,k}\psi_4(t)] \tag{7.36}$$

where

$$A_{1,k}=(\cos\theta_k)\left(\frac{1+a_k}{2}\right)+C_{1,k}\,(\cos\theta_k)\left(\frac{1-a_k}{2}\right)+C_{2,k}\,(\sin\theta_k)\left(\frac{1-a_k}{2}\right) \tag{7.37}$$

$$A_{2,k}=-\,(\sin\theta_k)\left(\frac{1+a_k}{2}\right)+C_{2,k}\,(\cos\theta_k)\left(\frac{1-a_k}{2}\right)-C_{1,k}\,(\sin\theta_k)\left(\frac{1-a_k}{2}\right) \tag{7.38}$$

$$A_{3,k}=(\cos\theta_k)\left(\frac{1-a_k}{2}\right)\sqrt{D_k} \tag{7.39}$$

$$A_{4,k}=-\,(\sin\theta_k)\left(\frac{1-a_k}{2}\right)\sqrt{D_k} \tag{7.40}$$

The coefficients reflect the data, the cumulate phase, and the modulation index (contained in $C_{i,k}$ and D_k) in each symbol interval. They completely determine the signal $s(t)$. Due to the orthogonality of the base functions, these coefficients can be obtained by correlating $s(t)$ with each of the base functions

$$A_{i,k}=\frac{1}{\sqrt{E}}\int_{kT}^{(k+1)T}s(t)\psi_i(t)dt$$

which is equivalent to passing the signal $s(t)$ through the structure in Figure 7.17 (except for a factor $\sqrt{E}$). The reference signals for the correlators are not completely orthogonal, but with the additional multipliers and adders, it is equivalent to a structure with the four orthonormal base functions given in (7.31) to (7.34).

In the following we will show how the structure of Figure 7.17 is used in the Viterbi MLSE demodulator.

Recall the derivation of the Viterbi demodulator in Section 6.5.3 which is applicable to the multi-h case too. The branch metric of the Viterbi processor is

$$Z_k(\widetilde{\mathbf{a}}) = \int_{kT}^{(k+1)T} r(t)s(t,\widetilde{\mathbf{a}})dt \tag{7.41}$$

where $\widetilde{\mathbf{a}}$ is an estimate of the transmitted data sequence $\mathbf{a}$. For the estimate data sequence the signal in the kth interval is

$$s(t,\widetilde{\mathbf{a}}) = \sqrt{E}\left[\widetilde{A}_{1,k}\psi_1(t) + \widetilde{A}_{2,k}\psi_2(t) + \widetilde{A}_{3,k}\psi_3(t) + \widetilde{A}_{4,k}\psi_4(t)\right] \tag{7.42}$$

where

$$\widetilde{A}_{i,k} = \frac{1}{\sqrt{E}}\int_{kT}^{(k+1)T} s(t,\widetilde{\mathbf{a}})\psi_i(t)dt \tag{7.43}$$

are determined by the sequence $\widetilde{\mathbf{a}}$. These coefficients are computed using (7.37) to (7.40) given the sequence $\widetilde{\mathbf{a}}$.

Substitute (7.42) into (7.41) we have

$$\begin{aligned} Z_k(\widetilde{\mathbf{a}}) &= \int_{kT}^{(k+1)T} r(t)\sqrt{E}\left[\widetilde{A}_{1,k}\psi_1(t) + \widetilde{A}_{2,k}\psi_2(t)\right. \\ &\quad \left. + \widetilde{A}_{3,k}\psi_3(t) + \widetilde{A}_{4,k}\psi_4(t)\right] dt \\ &= \sum_{i=1}^{4}\int_{kT}^{(k+1)T} \sqrt{E}r(t)\widetilde{A}_{i,k}\psi_i(t)dt \\ &= \sqrt{E}\sum_{i=1}^{4}\widetilde{A}_{i,k}A'_{i,k} \end{aligned} \tag{7.44}$$

where

$$\begin{aligned} A'_{i,k} &= \int_{kT}^{(k+1)T} r(t)\psi_i(t)dt \\ &= \int_{kT}^{(k+1)T} \left[s(t,\mathbf{a}) + n(t)\right]\psi_i(t)dt \\ &= \int_{kT}^{(k+1)T} s(t,\mathbf{a})\psi_i(t)dt + \int_{kT}^{(k+1)T} n(t)\psi_i(t)dt \\ &= \sqrt{E}A_{i,k} + n_{i,k} \end{aligned} \tag{7.45}$$

are the coefficients obtained by passing $r(t)$ through the correlators in Figure 7.17. These are the coefficients of the transmitted signal and noise terms $n_{i,k}$ which are white Gaussian.

Equation (7.44) shows that the branch metric is equal to the inner product of the received coefficients $\{A'_{1,k}, A'_{2,k}, A'_{3,k}, A'_{4,k}\}$ obtained by the correlators of Figure 7.17 and the locally available coefficients $\{\widetilde{A}_{1,k}, \widetilde{A}_{2,k}, \widetilde{A}_{3,k}, \widetilde{A}_{4,k}\}$. These local coefficients are based on the estimate sequence $\widetilde{\mathbf{a}}$ and are updated for each symbol interval as the length of $\widetilde{\mathbf{a}}$ grows in the Viterbi demodulation process. Particularly, from (7.37) to (7.40) we can see that to compute $\widetilde{A}_{i,k}$, we need the symbol estimate $\widetilde{a}_k$, cumulate phase estimate $\widetilde{\theta}_k$, and the current modulation index h_k. The Viterbi processor is able to track the estimate sequence (path) $\widetilde{\mathbf{a}}$ and related phase sequence $\boldsymbol{\theta}$, and the modulation index h_k must be provided by a synchronizer.

7.5.2 Joint Demodulation and Carrier Synchronization of Multi-*h* CPFSK

Although the demodulator described above is simple, it requires separate circuits for carrier synchronization and symbol synchronization. A joint demodulation and carrier synchronization scheme was proposed by Mazur and Taylor [23] for binary multi-h CPFSK. The approach is to use the Viterbi algorithm for estimating the data and the carrier phase error. The estimated phase error is used in a closed phase-tracking loop to achieve carrier phase synchronization. The scheme is named decision-directed demodulator. The scheme is shown in Figure 7.18. The inner product calculator (IPC) is to produce the branch metrics needed in the Viterbi algorithm. As we discussed in Section 6.5.3 (see (6.51)), for the AWGN channel, the branch metrics are the correlations of the received signal $r(t)$ with the local carriers. For the jth signaling interval the metrics are

$$\begin{aligned} b(1, h_j, n) &= \int_{jT}^{(j+1)T} r(t) \cos\left[2\pi f_c t + \frac{\pi h_j}{T}(t - jT) + \frac{n\pi}{q} + \phi(0)\right] dt \\ b(-1, h_j, n) &= \int_{jT}^{(j+1)T} r(t) \cos\left[2\pi f_c t - \frac{\pi h_j}{T}(t - jT) + \frac{n\pi}{q} + \phi(0)\right] dt \\ n &= 0, 1, 2, ..., 2q - 1 \end{aligned} \tag{7.46}$$

for data symbol 1 and –1, respectively. Recall $h_i = p_i/q$ and there are $2q$ phase states in the trellis for an even q which is the usual case. Therefore the phase state at the beginning of the interval is $n\pi/q$, $n = 0, 1, 2, ..., 2q - 1$. The cosine functions are the local carriers, corresponding to 1 and –1, respectively. $\phi(0)$ is the initial phase of the local carriers. To have coherent detection, $\phi(0)$ must be estimated from $r(t)$, that is, the carriers must be synchronized with the received signal.

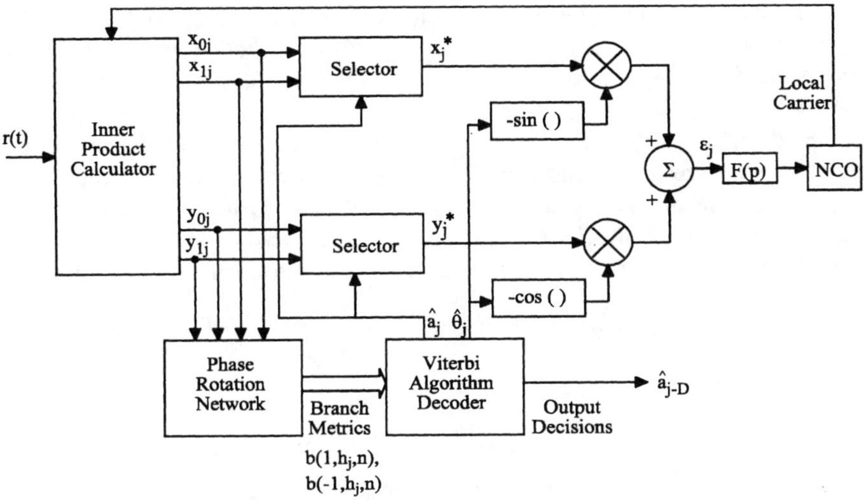

Figure 7.18 Receiver for binary multi-h CPFSK showing the decision-directed carrier recovery loop and the use of the Viterbi algorithm in phase error computation. From [23]. Copyright © 1981 IEEE.

The branch metrics can be rewritten as

$$\begin{aligned} b(1, h_j, n) &= X_{1j} \cos \frac{n\pi}{q} - Y_{1j} \sin \frac{n\pi}{q} \\ b(-1, h_j, n) &= X_{0j} \cos \frac{n\pi}{q} - Y_{0j} \sin \frac{n\pi}{q} \\ n &= 0, 1, 2, ..., 2q-1 \end{aligned} \tag{7.47}$$

where

$$\begin{aligned} X_{1j} &= \int_{jT}^{(j+1)T} r(t) \cos \left[2\pi f_c t + \frac{\pi h_j}{T}(t - jT) + \phi(0) \right] dt \\ Y_{1j} &= \int_{jT}^{(j+1)T} r(t) \sin \left[2\pi f_c t + \frac{\pi h_j}{T}(t - jT) + \phi(0) \right] dt \\ X_{0j} &= \int_{jT}^{(j+1)T} r(t) \cos \left[2\pi f_c t - \frac{\pi h_j}{T}(t - jT) + \phi(0) \right] dt \\ Y_{0j} &= \int_{jT}^{(j+1)T} r(t) \sin \left[2\pi f_c t - \frac{\pi h_j}{T}(t - jT) + \phi(0) \right] dt \end{aligned} \tag{7.48}$$

The benefit of converting (7.46) to (7.47) and (7.48) is that the local carriers in (7.48) no longer depend on $n\pi/q$, unlike the ones in (7.46). In addition, the computation of branch metrics is made more efficient by means of the two-step calculation depicted in (7.48) and (7.47).

Since $r(t) = s(t) + n(t)$, where $s(t)$ is the signal defined in (7.9) and $n(t)$ is the AWGN with zero mean and a single-sided power spectral density N_o, each of the above four correlations consists of a signal part and a noise part. In the following we will show how the phase error is extracted. For this purpose we assume that noise is absent. Under this assumption, we rewrite the received $r(t)$ in a different form

$$r(t) = s(t) = A\cos[2\pi f_c t + \theta(t, a_j) + \theta_j + \phi_j], \quad jT \le t \le (j+1)T$$

where

$$\begin{aligned} \theta(t, a_j) &= \pm\frac{\pi h_j}{T}(t - jT), \quad \text{for } a_j = \pm 1 \\ \theta_j &= \frac{n\pi}{q}, \quad n = 0, 1, 2, ..., 2q-1 \end{aligned}$$

and ϕ_j is the random phase introduced by the channel. Then when $a_j = 1$, using (7.48) we have

$$\begin{aligned} X_{1j} &= \int_{jT}^{(j+1)T} A\cos[2\pi f_c t + \theta(t, a_j) + \theta_j + \phi_j] \\ &\quad \cdot \cos[2\pi f_c t + \theta(t, a_j) + \phi(0)]dt \\ &= \frac{1}{2}\int_{jT}^{(j+1)T} A\cos[4\pi f_c t + 2\theta(t, a_j) + \theta_j + \phi_j + \phi(0)]dt \\ &\quad + \frac{1}{2}\int_{jT}^{(j+1)T} A\cos[\theta_j + \phi_j - \phi(0)]dt \\ &\approx \frac{1}{2}A\cos(\theta_j + \varepsilon_j), \quad a_j = 1 \end{aligned}$$

since the first integral is approximately zero for $f_c \gg 1/T$. In the above

$$\varepsilon_j = \phi_j - \phi(0)$$

is the difference between the signal random phase and the carrier initial phase which is the estimate of the signal random phase. Thus ε_j is the estimate error. Similarly we can find out others. In summary, we have

$$X_{1j} \approx \frac{1}{2}A\cos(\theta_j + \varepsilon_j), \quad a_j = 1 \tag{7.49}$$

$$Y_{1j} \approx -\frac{1}{2}A\sin(\theta_j+\varepsilon_j), \quad a_j=1 \tag{7.50}$$

$$X_{0j} \approx \frac{1}{2}A\cos(\theta_j+\varepsilon_j), \quad a_j=-1 \tag{7.51}$$

$$Y_{0j} \approx -\frac{1}{2}A\sin(\theta_j+\varepsilon_j), \quad a_j=-1 \tag{7.52}$$

The nonmatched cases: $(X_{1j}, a_j=-1)$, $(Y_{1j}, a_j=-1)$, $(X_{0j}, a_j=1)$, and $(Y_{0j}, a_j=1)$, are not listed above since they are irrelevant to the discussion that follows.

In Figure 7.18, the inner product calculator (IPC) is used to generate the primitive set of four correlations X_{1j}, Y_{1j}, X_{0j}, and Y_{0j} needed for each signaling interval, and a phase rotation network (PRN) uses these to realize the calculation in (7.47) (which is a phase rotation operation). Table lookup techniques may be used to implement the PRN. The outputs of the PRN are the branch metrics given in (7.47), which are used in the Viterbi algorithm detector. For bit interval j and state n, the PRN produces two branch metrics $b(1,h_j,n)$ and $b(-1,h_j,n)$ for $\widehat{a}_j=1$ and -1, respectively. Based on these metrics, the Viterbi detector makes a tentative decision $\widehat{a}_j$ on the data symbol for the current interval (choosing the one with a larger branch metric). Based on a sequence of $\widehat{a}_j$ the Viterbi detector makes the final decision $\widehat{a}_{j-D}$ on the symbol back by D intervals, where D is the decision depth, typically 4 to 8 constraint lengths (see Section 6.5.3). The Viterbi detector also makes a tentative decision $\widehat{\theta}_j$ on the cumulate phase θ_j. Assume that the data estimate $\widehat{a}_j$ and phase estimate $\widehat{\theta}_j$ are correct, $\widehat{a}_j$ and $\widehat{\theta}_j$ are used to estimate the phase error ε_j as follows

$$\text{If } \widehat{a}_j=1, \text{ choose } X_j^*=X_{1j}, Y_j^*=Y_{1j}$$

then in the absence of noise

$$\begin{aligned}
& -X_j^*\sin\widehat{\theta}_j - Y_j^*\cos\widehat{\theta}_j \\
= & -X_{1j}\sin\widehat{\theta}_j - Y_{1j}\cos\widehat{\theta}_j \\
= & -\frac{1}{2}A\cos(\theta_j+\varepsilon_j)\sin\widehat{\theta}_j + \frac{1}{2}A\sin(\theta_j+\varepsilon_j)\cos\widehat{\theta}_j \\
= & \frac{1}{2}A\sin(\theta_j+\varepsilon_j-\widehat{\theta}_j) \\
= & \frac{1}{2}A\sin\varepsilon_j \\
\approx & \frac{1}{2}A\varepsilon_j \propto \varepsilon_j, \text{ for small } \varepsilon_j
\end{aligned}$$

Similarly

$$\text{If } \widehat{a}_j = -1, \text{ choose } X_j^* = X_{0j}, Y_j^* = Y_{0j}$$

then

$$\begin{aligned} & -X_j^* \sin\widehat{\theta}_j - Y_j^* \cos\widehat{\theta}_j \\ = & -X_{0j} \sin\widehat{\theta}_j - Y_{0j} \cos\widehat{\theta}_j \\ = & \frac{1}{2} A \sin\varepsilon_j \\ \approx & \frac{1}{2} A\varepsilon_j \propto \varepsilon_j, \text{ for small } \varepsilon_j \end{aligned}$$

It is clear that if the Viterbi estimates $\widehat{a}_j$ and $\widehat{\theta}_j$ are correct, the operation $-X_j^* \sin\widehat{\theta}_j$ $-Y_j^* \cos\widehat{\theta}_j$ produces a quantity that is proportional to the phase estimation error (Figure 7.18). With noise present, the error signal is passed through the low-pass filter $F(p)$ in order to reduce the noise. The output is used to control the NCO (numerically controlled oscillator) to generate a carrier with a decreasing phase error as this phase lock loop converges.

A scheme of $(3/8, 4/8)$ and other schemes were simulated in [23]. It was found empirically that the minimum number of bits of quantization required for demodulation/decoding is given by $1 + \log_2 q$, and that rounding to the nearest quantization level, as opposed to truncating, leads to superior performance. It was reported that a digital word size of 5 to 6 bits appears sufficient and the expected coding gain of 1.45 dB for $(3/8, 4/8)$ was nearly reached. For the schemes tested, an rms phase error of $0.2(\pi/2q)$ rad is typically sufficient to achieve almost optimum (within 0.2 dB) error-rate performance. In order to achieve this phase error, a minimum loop SNR of 30 dB is required.

7.5.3 Joint Carrier Phase Tracking and Data Detection of Multi-h CPFSK

A joint demodulation and carrier phase tracking (not synchronization) receiver was proposed in [20] for binary multi-h CPFSK. In this receiver, the demodulator is of the correlator type shown in Figure 7.17. It is modified to take into consideration the carrier phase error. The demodulator is followed by the Viterbi algorithm, which maximizes the a posteriori path metric. Thus the receiver is called a MAP-VA receiver. The Viterbi algorithm estimates both data and phase error of the local carrier. The estimated phase error is not used to adjust the local carrier phase to achieve phase synchronization. Instead, it is used in the Viterbi algorithm to assist in demodulating the data. Because no carrier synchronization is really achieved, the error performance of this receiver appears inferior to the decision-directed receiver given in [23].

Based on the simulation in [20] we can make a comparison between this MAP-VA receiver with the decision-directed receiver. The rms phase error of $0.2(\pi/2q)$ rad mentioned above translates to a squared phase error of 0.16 rad^2 for the $(3/8, 4/8)$ scheme. According to the results given in [20], for this amount of phase error, the loss in error performance at $P_e = 10^{-3}$ is about 2.5 dB (Figure 4 in [20]), whereas the loss is only 0.2 dB in [23]. The advantage of this receiver is that its system complexity is simpler, no phase lock loop is utilized. The detail of this receiver is omitted here due to the highly mathematical derivation.

7.5.4 Joint Demodulation, Carrier Synchronization, and Symbol Synchronization of M-ary Multi-h CPFSK

A receiver which can jointly perform demodulation, carrier recovery, and symbol synchronization for 4-ary multi-h CPFSK was proposed by Premji and Taylor in [21]. The receiver is an extension of the work on binary multi-h CPFSK in [23], and again is based on maximum likelihood sequence estimation via the Viterbi algorithm.

The derivation of this receiver basically follows that in Section 7.5.2. Since carrier and symbol synchronization are considered, the signal model includes an unknown carrier phase θ_0 and an unknown symbol timing offset τ_0 as

$$s(t) = A\cos[2\pi f_c t + a_j \frac{\pi h_j}{T}(t - jT - \tau_0) + \theta_0 + \theta_j], \quad jT + \tau_0 \le t \le (j+1)T + \tau_0$$

where $\theta_j = n\pi/q, n = 0, 1, 2, ..., 2q - 1$ is the accumulated phase due to previous data. For the jth signaling interval the branch metrics, corresponding to the branches of the phase trellis extending from the nth trellis node at time $t = jT + \widehat{\tau}_0$, are the correlations between the $r(t)$ and the local carriers:

$$\begin{aligned} & b(a_k, h_j, n) \\ &= \int_{jT+\widehat{\tau}_0}^{(j+1)T+\widehat{\tau}_0} r(t)\cos\left[2\pi f_c t + a_k\frac{\pi h_j}{T}(t - jT - \widehat{\tau}_0) + \widehat{\theta}_0 + \frac{n\pi}{q}\right] dt \\ &= \cos(\frac{n\pi}{q})X_j - \sin(\frac{n\pi}{q})Y_j \\ n &= 0, 1, 2, ..., 2q-1, \quad k = 1, 2, 3, 4 \end{aligned} \tag{7.53}$$

where

$$\begin{aligned} X_j &= \int_{jT+\widehat{\tau}_0}^{(j+1)T+\widehat{\tau}_0} r(t)\cos\left[2\pi f_c t + a_k\frac{\pi h_j}{T}(t - jT - \widehat{\tau}_0) + \widehat{\theta}_0\right] dt \\ Y_j &= \int_{jT+\widehat{\tau}_0}^{(j+1)T+\widehat{\tau}_0} r(t)\sin\left[2\pi f_c t + a_k\frac{\pi h_j}{T}(t - jT - \widehat{\tau}_0) + \widehat{\theta}_0\right] dt \end{aligned}$$

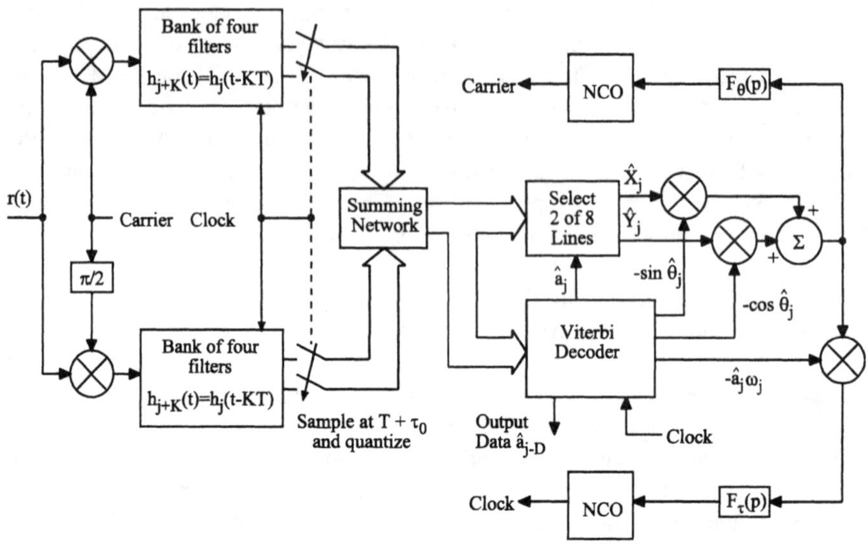

Figure 7.19 Joint demodulation, carrier synchronization, and symbol synchronization for 4-ary multi-h CPFSK. From [21]. Copyright © 1987 IEEE.

$$a_k = +1, -1, +3, \text{or} -3 \text{ for } k = 1, 2, 3, 4 \tag{7.54}$$

where $\widehat{\tau}_0$ and $\widehat{\theta}_0$ are estimates of τ_0 and θ_0, respectively. There are four a_ks, that is, $a_k = +1, -1, +3$, or -3 for $k = 1, 2, 3, 4$, corresponding to four branches of a node in the trellis. The local carriers in (7.54) no longer depend on $n\pi/q$. Since $a_k = \pm 1, \pm 3$, during the jth interval four correlation operations are needed for X_j and another four correlation operations are needed for Y_j. The total number of correlations is eight in each interval.

The receiver is shown in Figure 7.19. The correlations of (7.54) are implemented by the two carrier-multipliers, two cycled banks of matched filters, and the summing network. The Viterbi decoder calculates the branch metric using (7.53) and performs the MLSE based on the phase trellis. In the following, we explain the correlation operation first and then the Viterbi decoder.

The correlations of (7.54) are actually implemented by matched filtering based on the well-known fact that correlation can be realized by matched filtering sampled at the end of the correlation period. Further, the received signal is down-converted to the baseband by the multipliers before filtering is performed. The equivalence of the operation in Figure 7.19 to (7.54) can be shown as follows. First, define a shorthand notation $\alpha_j(t)$ for the linear phase term,

$$\alpha_j(t) = \frac{a_j \pi h_j}{T}(t - jT - \tau_0), \quad jT + \tau_0 \le t \le (j+1)T + \tau_0$$

Then the integrand of X_j can be written as

$$\begin{aligned} & \cos\left[2\pi f_c t + \alpha_j(t) + \theta_j + \theta_0\right] \cos\left[2\pi f_c t + \widehat{\alpha}_j(t) + \widehat{\theta}_0\right] \\ = & \ \frac{1}{2}\cos\left[4\pi f_c t + \alpha_j(t) + \theta_j + \theta_0 + \widehat{\alpha}_j(t) + \widehat{\theta}_0\right] \\ & + \frac{1}{2}\cos\left[\alpha_j(t) - \widehat{\alpha}_j(t) + \theta_j + \theta_0 - \widehat{\theta}_0\right] \end{aligned}$$

Since the double frequency term integrates to zero, only the low-pass term of the above contributes to X_j (a factor of 1/2 is ignored and set $\varepsilon_{j\theta} = \theta_0 - \widehat{\theta}_0$)

$$\begin{aligned} X_j &= \int_{jT+\widehat{\tau}_0}^{(j+1)T+\widehat{\tau}_0} \cos\left[\alpha_j(t) + \theta_j + \varepsilon_{j\theta} - \widehat{\alpha}_j(t)\right] dt \\ &= \int_{jT+\widehat{\tau}_0}^{(j+1)T+\widehat{\tau}_0} \cos\left[\alpha_j(t) + \theta_j + \varepsilon_{j\theta}\right] \cos\widehat{\alpha}_j(t) dt \\ &\quad + \int_{jT+\widehat{\tau}_0}^{(j+1)T+\widehat{\tau}_0} \sin\left[\alpha_j(t) + \theta_j + \varepsilon_{j\theta}\right] \sin\widehat{\alpha}_j(t) dt \end{aligned} \tag{7.55}$$

Similarly Y_j can be shown as

$$\begin{aligned} Y_j &= -\int_{jT+\widehat{\tau}_0}^{(j+1)T+\widehat{\tau}_0} \sin\left[\alpha_j(t) + \theta_j + \varepsilon_{j\theta} - \widehat{\alpha}_j(t)\right] dt \\ &= -\int_{jT+\widehat{\tau}_0}^{(j+1)T+\widehat{\tau}_0} \sin\left[\alpha_j(t) + \theta_j + \varepsilon_{j\theta}\right] \cos\widehat{\alpha}_j(t) dt \\ &\quad + \int_{jT+\widehat{\tau}_0}^{(j+1)T+\widehat{\tau}_0} \cos\left[\alpha_j(t) + \theta_j + \varepsilon_{j\theta}\right] \sin\widehat{\alpha}_j(t) dt \end{aligned} \tag{7.56}$$

Now assume that the carrier in Figure 7.19 is

$$C(t) = \cos(2\pi f_c t + \widehat{\theta}_0)$$

where the estimated phase $\widehat{\theta}_0$ is included. Then the I-channel (upper channel) multiplier output is

$$r(t)C(t) = \cos\left[2\pi f_c t + \alpha_j(t) + \theta_j + \theta_0\right] \cos(2\pi f_c t + \widehat{\theta}_0)$$

the low-pass term of it is (a factor of 1/2 is ignored)

$$x_j(t) = \cos[\alpha_j(t) + \theta_j + \varepsilon_{j\theta}] \tag{7.57}$$

Similarly, the low-pass term of the output of the Q-channel (lower channel) multiplier is

$$y_j(t) = \sin[\alpha_j(t) + \theta_j + \varepsilon_{j\theta}] \tag{7.58}$$

By inspecting the expressions for $x_j(t)$, $y_j(t)$, X_j, and Y_j, we can see that

$$\begin{aligned} X_j &= \int_{jT+\widehat{\tau}_0}^{(j+1)T+\widehat{\tau}_0} x_j(t)\cos\widehat{\alpha}_j(t)dt \\ &+ \int_{jT+\widehat{\tau}_0}^{(j+1)T+\widehat{\tau}_0} y_j(t)\sin\widehat{\alpha}_j(t)dt \end{aligned} \tag{7.59}$$

$$\begin{aligned} Y_j &= -\int_{jT+\widehat{\tau}_0}^{(j+1)T+\widehat{\tau}_0} y_j(t)\cos\widehat{\alpha}_j(t)dt \\ &+ \int_{jT+\widehat{\tau}_0}^{(j+1)T+\widehat{\tau}_0} x_j(t)\sin\widehat{\alpha}_j(t)dt \end{aligned} \tag{7.60}$$

All the four integrations in (7.59) and (7.60) are now correlations in baseband. These correlations can be implemented by matched filtering. The filter impulse responses must match to

$$\cos\widehat{\alpha}_j(t) = \cos\frac{a_j\pi h_j}{T}(t - jT - \widehat{\tau}_0)$$

and

$$\sin\widehat{\alpha}_j(t) = \sin\frac{a_j\pi h_j}{T}(t - jT - \widehat{\tau}_0)$$

Since $a_j = \pm1, \pm3$ and by considering $\pm\sin\widehat{\alpha}_j(t)$ as just one filter since the sign can be taken care of later when summing is done, the number of matched filter impulse responses is actually four. Recall that if the reference signal in the correlation is $s(t)$, then the matched filter must be $h(t) = ks(T-t)$, k constant. Thus in the 0th interval the four matched filter impulse responses are (assuming symbol timing offset $\widehat{\tau}_0 = 0$, since $\widehat{\tau}_0$ is random and cannot be anticipated when filters are determined).

$$\{h_0(t)\} = \{\cos\frac{\pi h_0(T-t)}{T}, \sin\frac{\pi h_0(T-t)}{T}, \cos\frac{3\pi h_0(T-t)}{T}, \sin\frac{3\pi h_0(T-t)}{T}\} \tag{7.61}$$

where h_0 is one of the K indexes, usually the first one. In the jth interval, the filter expressions can be obtained by replacing h_0 by h_j and t by $t - jT$ in the above expressions.

For a K-index CPFSK, there are K sets of four filters in the above forms, with each set using a different index. In the receiver these K sets are cyclically reused such that

$$\{h_{j+K}(t)\} = \{h_j(t - KT)\}$$

From (7.59) and (7.60) we also see that the I-channel alone cannot produce X_j and the Q-channel alone cannot produce Y_j. Thus a summing network is needed to perform the additions and subtractions in (7.59) and (7.60).

A selector in Figure 7.19 selects a pair of signals $(\widehat{X}_j, \widehat{Y}_j)$ out of four pairs of (X_j, Y_j) based on the data estimate $\widehat{a}_j$ provided by the Viterbi decoder. This $(\widehat{X}_j, \widehat{Y}_j)$ and the phase estimate $\widehat{\theta}_j$ are used to generate the phase error estimate $\widehat{\varepsilon}_{j\theta}$ and timing offset estimate $\widehat{\varepsilon}_{j\mathcal{T}}$ as follows

$$\begin{aligned}\widehat{\varepsilon}_{j\theta} &= -[\widehat{X}_j \sin\widehat{\theta}_j + \widehat{Y}_j \cos\widehat{\theta}_j] \\ \widehat{\varepsilon}_{j\mathcal{T}} &= -\widehat{a}_j \omega_j \widehat{\varepsilon}_{j\theta}\end{aligned}$$

where $\omega_j = \pi h_j / T$. These have been shown to be optimum under the condition that the tracking loop bandwidth is much smaller than the data rate [21]. These error signals are passed through loop filters and eventually are used to control two NCOs to produce a synchronous carrier and a synchronous symbol timing clock, respectively.

The performance analysis of this receiver is very complicated and is thus omitted here. The receiver was simulated for a $(12/16, 13/16)$ multi-h CPFSK. The simulation used a six-bit quantization and a 15-symbol decision depth in the Viterbi decoder. The results showed that a coding gain of approximately 3.8 dB (over MSK) is already realized at a BER of 10^{-3}, while the expected asymptotic gain is about 4.52 dB for an ideal receiver. At values of E_s/N_o above 13 dB, a reasonable acquisition time/steady-state jitter trade-off may be attained through a suitable choice of loop SNRs. At lower E_s/N_o, however, some form of gear shifting of the loop bandwidth becomes necessary to ensure both a reasonable acquisition time and low steady-state jitter subsequent to acquisition. At an input E_s/N_o of 13 dB, satisfactory receiver performance mandates loop SNR values of approximately 35 dB for the carrier recovery loop and 33 dB for the clock loop.

A simplified suboptimum version of the above receiver is given in [22]. The major change of the receiver is that a fixed bank of filters has replaced the cyclically changing matched filters. The impulse responses of the fixed filters have the average h of the indexes. Thus they still have the forms given in (7.61) except that h_i is replaced by h. By using fixed filters, a complexity reduction of $1/K$ is achieved while the penalty is a degradation in error-rate performance of 0.6 to 0.7 dB at $BER = 10^{-3}$ for the code $(15/20, 17/20)$ that was simulated.

7.5.5 Synchronization of MHPM

Three levels of synchronization are required in an MHPM demodulator: carrier phase, symbol timing, and superbaud timing. Carrier phase and symbol-timing synchronization are common for all coherent phase modulations. What is unique here is the superbaud timing. The superbaud is the period of cyclic rotation of the K modulation indexes. A superbaud period is KT. Since symbol timing and superbaud timing are harmonically related, either one can be derived from the other easily. For example, if superbaud timing is established, a frequency multiplier of K times is sufficient to produce the symbol timing.

For the receiver in Section 7.5.4, carrier synchronization and symbol-timing synchronization are incorporated in the receiver already. For the receiver in Section 7.5.3, carrier phase synchronization is not needed and only symbol-timing synchronization is needed. For the receiver in Section 7.5.2, carrier phase synchronization is incorporated in the receiver, thus only symbol-timing synchronization is needed. For the receiver in Section 7.5.1, both carrier synchronization and symbol-timing synchronization are needed.

Recall that in Section 6.6.1 we discussed the MSK-type carrier and symbol synchronizer for single-h CPM (Figure 6.44). This synchronizer can also be used for MHPM [23]. First the received signal is raised to the power of q (recall $h_i = p_i/q$) by the nonlinear device in the synchronizer.

$$[r(t)]^q = [s(t) + n(t)]^q$$

and one of the resultant components is (the amplitude is immaterial; it is set to 1)

$$\cos[2\pi q f_c t + q\Phi(t, \mathbf{a})]$$

This is an MHPM signal with carrier frequency qf_c and phase

$$q\Phi(t, \mathbf{a}) = 2\pi \sum_i a_i p_i q(t - iT)$$

which has an integer index p_i. Recall that in Section 6.2.1, we state that when the index is an integer the PSD of the CPM signal contains a continuous part and a discrete part. Similarly, it can be shown [24] that when the index is an integer the PSD of the MHPM signal contains a continuous part and a discrete part. The discrete frequency components appear at

$$f = \begin{cases} qf_c \pm \frac{2k}{2KT}, & k = 0, 1, 2, ..., \quad \text{if } \Gamma = \sum_{i=1}^{K} p_i \text{ is even} \\ qf_c \pm \frac{2k+1}{2KT}, & k = 0, 1, 2, ..., \quad \text{if } \Gamma = \sum_{i=1}^{K} p_i \text{ is odd} \end{cases} \tag{7.62}$$

which can be simplified as

$$f = qf_c \pm \frac{n}{2KT} \begin{cases} n \text{ is even if } \Gamma = \sum_{i=1}^{K} p_i \text{ is even} \\ n \text{ is odd if } \Gamma = \sum_{i=1}^{K} p_i \text{ is odd} \end{cases} \tag{7.63}$$

When $K = 1$, the above expression degenerates back to the expression for single-h CPM. Those spectral lines that contain significant power are few in number since their weights decay rapidly at higher harmonics (large n). For any given multi-h codes, however, there are always two or three adjacent lines near the peak (qf_c) of the spectrum which contain significant power. The spectral lines are separated by $1/KT$ Hz.

Based on (7.63), a synchronizer that is similar to the MSK-type carrier and symbol synchronizer for single-h CPM (Figure 6.44) can be constructed as shown in Figure 7.20. The two phase lock loops (or narrow-band bandpass filters) can select any pair of frequencies $qf_c \pm n/2KT$. The output of the mixer contains the sum and the difference of these two frequencies: $2qf_c$ and n/KT. The high-pass filter and the low-pass filter pick up one of them. The frequency dividers divide two frequencies to produce f_c for the carrier and $1/KT$ for the superbaud timing. The symbol timing is derived from the superbaud timing by a frequency multiplier.

If carrier recovery is already incorporated in the demodulator, like the one in Figure 7.18, the carrier recovery part of Figure 7.20 is not needed.

7.6 IMPROVED MHPM SCHEMES

Improvements on MHPM schemes have been made during the recent years. Techniques that combine MHPM with error control codes, such as convolutional codes, will be discussed in a later chapter. What we present here are techniques which make changes on the MHPM phase characteristic itself. These include MHPM schemes with asymmetrical modulation indexes, multi-T realization of multi-h, correlatively encoded MHPM, and nonlinear multi-h CPFSK. It should be pointed out that these improved MHPM schemes may require more complex receivers which have not been available in the published literature.

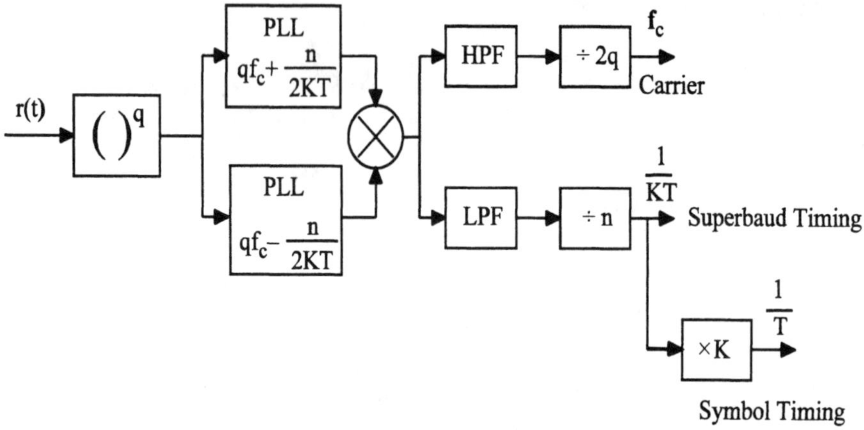

Figure 7.20 MHPM carrier, symbol, and superbaud synchronizer.

7.6.1 MHPM with Asymmetrical Modulation Indexes

Binary MHPM schemes with asymmetrical modulation indexes are proposed in [25, 26]. They use a set of modulation indexes h_{+k} for data $+1$ and h_{-k} for data -1 during any kth signaling interval. One simple way of doing this is to use the same set of the modulation indexes for both h_{+k} and h_{-k} and simply shift h_{-k} with respect to h_{+k}. For example, let $\{h_a, h_b, h_c\}$ be the modulation index set for a 3-h binary scheme. The asymmetric modulation index sets can be arranged as follows.

$k:$	1	2	3	4	5	6	...
$h_{+k}:$	h_a	h_b	h_c	h_a	h_b	h_c	...
$h_{-k}:$	h_b	h_c	h_a	h_b	h_c	h_a	...

That is, $h_{+(k+1)} = h_{-k}$. This makes the index for the ith interval in the phase trellis, h_{+k} or h_{-k}, not necessarily equal for different paths, whereas in conventional MHPM it is fixed. As a result, the phase values at $t = kT$ for asymmetric MHPM may be a multiple of π/q, instead of $2\pi/q$ as in the conventional MHPM. This gives more flexibility to make the minimum Euclidean distance larger. Improvements over the symmetrical ones in the range of 0.5 to 1 dB in the upper bounds of error probability are reported. The spectral properties of the asymmetrical schemes are almost the same of the symmetrical ones.

7.6.2 Multi-*T* Realization of Multi-*h* Phase Codes

Binary and M-ary linear phase multi-T phase codes are investigated in [27,28]. The modulation index is effectively changed by cyclically varying the symbol interval with a constant frequency deviation (excluding the effect of the data symbol). For every multi-h code $\{h_1, h_2, ..., h_K\}$ a set of symbol periods can be defined as

$$\{T_1, T_2, ..., T_K\} = \{\lambda_1 T, \lambda_2 T, ..., \lambda_K T\}$$

where T is the average period

$$T = (T_1 + T_2 + ... + T_K)/K$$

and

$$\lambda_i = h_i/h$$

h is the average index

$$h = (h_1 + h_2 + ... + h_K)/K$$

Imitating the signal form of the conventional CPFSK, this signal can be written as

$$s(t) = A\cos(2\pi f_c t + \frac{a_k \pi h(t - t_k)}{T} + \theta_k), \quad t_k \leq t \leq t_{k+1} \tag{7.64}$$

where

$$t_k = \sum_{i=-\infty}^{k-1} T_i$$

The phase change at the end of each interval is $a_k \pi h T_k/T$. Thus the effective index is $h_k = hT_k/T$ in terms of phase increment for each symbol interval. However, in terms of phase change slope, it is like a single-h CPM.

It is shown that multi-T codes are similar to multi-h codes in terms of minimum distance and spectral properties. However, the extraction of the synchronization information by means of a PLL seems to be easier for multi-T codes than multi-h codes since the spectrum for the qth power of the binary multi-T signal contains only two lines in the vicinity of qf_c, whereas there are more than two lines in the case of binary multi-h signals [28].

7.6.3 Correlatively Encoded Multi-*h* Signaling Technique

A correlatively encoded multi-h signaling technique was introduced by Fonseka et al. [29]. It was claimed that this scheme performed significantly better than partial-response MHPM. In this scheme, the MHPM signal is defined as

$$\begin{aligned} s(t) &= A\cos\left[2\pi f_c t + \frac{\pi h_k(a_k + a_{k-1} + \ldots + a_{k-L+1})}{LT}(t-kT) + \theta_k\right] \\ &= A\cos\left[2\pi f_c t + \Phi(t,\mathbf{a})\right], \quad kT \le t \le (k+1)T \end{aligned} \tag{7.65}$$

This signal is a multi-h CPFSK signal with a depth of correlation L, is abbreviated by LCE-MH. Recall that the total phase of the partial-response multi-h CPFSK (LREC-MH) signal is

$$\Phi(t,\mathbf{a}) = \frac{\pi(a_k h_k + a_{k-1}h_{k-1} + \ldots + a_{k-L+1}h_{k-L+1})}{LT}(t-kT) + \theta_k \tag{7.66}$$

Clearly, when h is constant, LCE-MH is identical to LREC-MH. LREC-MH signals are generated by first multiplying the information data by the modulation indexes and then correlative encoding, while LCE-MH signals are generated by first correlative encoding the information data and then multiplying by the modulation indexes. Thus the phase variation of LREC-MH during any signal interval is influenced not only by the modulation index used in that interval, but also by the modulation indexes used in the previous $(L-1)$ intervals. The phase variation of LCEC-MH signals during any signal interval is influenced by only the modulation index used in that interval.

In addition to correlative coding as depicted in (7.65), modulation indexes can be arranged in certain patterns instead of simply cycling them. This can further increase the constraint length and the minimum distance. Two good patterns used in (7.65) are 2-h slots of $2T$ pattern

$$\{h_1, h_1, h_2, h_2, h_1, h_1, h_2, h_2, \ldots\}$$

and 2-h slots of $3T$ pattern

$$\{h_1, h_1, h_1, h_2, h_2, h_2, h_1, h_1, h_1, h_2, h_2, h_2, \ldots\}$$

Some numerical results in [29] are listed in Table 7.4, where comparisons are made between the correlatively encoded/2-h signals with partial-response/2-h signals at fixed modulation indexes. In the table, N_R is the receiver path memory length and is equal to the minimum number of intervals required to ensure that the distance between any two paths is less than the minimum distance of the signals. From the table, it is seen that significant coding gain can be achieved by correlative encoding over the conventional partial-response MHPM.

h_1, h_2	signal type	index pattern	$d^2_{\min}$	N_R	p.r./reg. 2-h $d^2_{\min}$	p.r./reg. 2-h N_R	gain (dB)
0.5,0.7	2CE-MH	slots of 2T	3.28	36	2.518	47	1.152
0.5,0.7	2CE-MH	slots of 2T	3.97	16	3.226	8	0.876
0.5,0.7	3CE-MH	regular 2-h	2.39	22	1.226	7	2.890
0.8,1.0	3CE-MH	regular 2-h	4.01	19	3.000	31	1.263
19/24,5/4	3CE-MH	regular 2-h	5.00	40	2.217	7	3.535
19/24,7/6	3CE-MH	slots of 3T	5.13	76	2.487	7	3.140

Table 7.4 Comparison of correlatively encoded 2-*h* signals with partial response 2-*h* signals. From [29]. Copyright © IEEE 1993.

7.6.4 Nonlinear Multi-*h* CPFSK

A scheme called nonlinear multi-h CPFSK is proposed in [30]. In this scheme, the signals still have the same form of (7.9) except that the modulation index during any kth interval h_k is a function of signaling interval k and m previous symbols. The m previous symbols define 2^m states $S_i = 0, 1, 2, ..., 2^m - 1$. The scheme uses 2^m modulation index patterns, and a pattern

$$H_{S_i} = \{h_{S_i}(0), h_{S_i}(1), ..., h_{S_i}(K-1)\}$$

contains K distinct modulation indexes where H_{S_i} denotes the modulation index pattern selected for symbol state S_i. Once the modulation index pattern is selected, a modulation index is picked from it by cyclically shifting the modulation indexes with period K as in ordinary multi-h CPFSK. It is observed that the constraint length of this scheme is $m + K + 1$, whereas it is $K + 1$ for the ordinary multi-h CPFSK. By properly choosing the index pattern, increases in the minimum Euclidean distance of up to 2.7 dB over ordinary multi-h CPFSK and up to 1.8 dB over asymmetrical multi-h CPFSK have been observed.

7.7 SUMMARY

In this chapter we have studied the multi-h continuous phase modulation (MHPM). We first defined the MHPM signal. Then we studied the properties of its phase trellis, including the number of phase states, the period of the trellis. A direct method of computing the spectra of the MHPM schemes was described and spectra of some important MHPM schemes were presented. The calculation of the distances is similar to CPM. Many results of the minimum distance and its upper bound for various binary and M-ary MHPM schemes were presented. Three extensive tables compared many multi-h codes in terms of their coding gain, bandwidth, and decision depth. Those are valuable design references. The modulator for MHPM is almost the same as that of the CPM except that the index must be cyclically rotated. Thus it is not repeated in this chapter. The demodulators for MHPM are also similar to single-h CPM demodulators except that the demodulator must have a prior knowledge of the sequence of indexes h_k employed at the transmitter. Thus in principle we can modify all demodulators in Chapter 6 for MHPM. However, since the MLSE demodulator using the Viterbi algorithm is the most promising one, several demodulators of this type were presented. Some of these receivers incorporate carrier and symbol synchronization in the receiver. For those that do not, a separate synchronizer is needed, which was discussed. Finally, we briefly described several improved MHPM schemes in the end of the chapter. These schemes manipulate the modulation indexes in a variety of ways in an attempt to increase the minimum distance, while maintaining a similar spectrum. The receivers for many of these schemes have not yet appeared in the literature.

In the next chapter we will study quadrature amplitude modulation (QAM) which is a nonconstant envelope modulation scheme with very high bandwidth efficiency, and is widely used in systems where constant envelope is not required.

APPENDIX 7A ORTHONORMAL BASE FUNCTIONS

By inspecting (7.30), two of the orthonormal functions can be assigned as

$$\psi_1(t) = \sqrt{\frac{2}{T}} \cos\left(2\pi f_c t + \frac{\pi h_k (t - kT)}{T}\right) \tag{7.67}$$

$$\psi_2(t) = \sqrt{\frac{2}{T}} \sin\left(2\pi f_c t + \frac{\pi h_k (t - kT)}{T}\right) \tag{7.68}$$

The orthogonal relationship can be verified by integration:

$$\begin{aligned}
&\int_{kT}^{(k+1)T} \psi_1(t)\psi_2(t)dt \\
= \; & \frac{2}{T}\int_{kT}^{(k+1)T} \cos\left(2\pi f_c t + \frac{\pi h_k (t - kT)}{T}\right) \sin\left(2\pi f_c t + \frac{\pi h_k (t - kT)}{T}\right) dt \\
= \; & \frac{1}{T}\int_{kT}^{(k+1)T} \sin\left(4\pi f_c t + \frac{2\pi h_k (t - kT)}{T}\right) dt \\
= \; & \frac{-1}{T(4\pi f_c + 2\pi h_k/T)} \left[\cos\left(4\pi f_c t + \frac{2\pi h_k (t - kT)}{T}\right)\right]_{kT}^{(k+1)T}
\end{aligned}$$

$$= \frac{-1}{T(4\pi f_c + 2\pi h_k/T)} \left[\cos\left(4\pi f_c(k+1)T + 2\pi h_k\right) - \cos\left(4\pi f_c kT\right)\right]$$

In most systems of practical interest $f_c \gg 1/T$, this means $2\pi f_c T \geq 2\pi$. The value in the bracket is between -2 and $+2$, thus the above is approximately equal to zero. The norm of $\psi_1(t)$ is

$$\begin{aligned}
\int_{kT}^{(k+1)T} \psi_1^2(t)dt &= \frac{2}{T}\int_{kT}^{(k+1)T} \cos^2\left(2\pi f_c t + \frac{\pi h_k(t-kT)}{T}\right)dt \\
&= \frac{1}{T}\int_{kT}^{(k+1)T} \left[1 + \cos 2\left(2\pi f_c t + \frac{\pi h_k(t-kT)}{T}\right)\right]dt \\
&= 1 + \left.\frac{\sin 2\left(2\pi f_c t + \pi h_k(t-kT)/T\right)}{2T\left(2\pi f_c + \pi h_k/T\right)}\right|_{kT}^{(k+1)T} \\
&\approx 1
\end{aligned}$$

Similarly we can show $\int_{kT}^{(k+1)T} \psi_2^2(t)dt \approx 1$. The remaining orthonormal functions are found as follows. Define

$$s_3(t) = \sqrt{\frac{2}{T}}\cos\left(2\pi f_c t - \frac{\pi h_k(t-kT)}{T}\right) \tag{7.69}$$

$$\begin{aligned}
s_{31} &= \int_{kT}^{(k+1)T} s_3(t)\psi_1(t)dt \\
&= \frac{2}{T}\int_{kT}^{(k+1)T} \cos\left(2\pi f_c t - \frac{\pi h_k(t-kT)}{T}\right) \\
&\quad \cdot \cos\left(2\pi f_c t + \frac{\pi h_k(t-kT)}{T}\right)dt \\
&= \frac{1}{T}\int_{kT}^{(k+1)T} \left(\cos 4\pi f_c t + \cos\frac{2\pi h_k(t-kT)}{T}\right)dt \\
&= \frac{1}{T}\left(\frac{\sin 4\pi f_c t}{4\pi f_c} + \frac{T\sin 2\pi h_k(t-kT)/T}{2\pi h_k}\right)_{kT}^{(k+1)T} \\
&\approx \frac{\sin 2\pi h_k}{2\pi h_k} \triangleq C_{1,k}
\end{aligned} \tag{7.70}$$

Similarly

$$\begin{aligned} s_{32} &= \int_{kT}^{(k+1)T} s_3(t)\psi_2(t)dt \\ &= \frac{2}{T}\int_{kT}^{(k+1)T} \cos\left(2\pi f_c t - \frac{\pi h_k(t-kT)}{T}\right) \\ &\quad \cdot \sin\left(2\pi f_c t + \frac{\pi h_k(t-kT)}{T}\right) dt \\ &= \frac{1}{T}\int_{kT}^{(k+1)T} \left(\sin 4\pi f_c t + \sin\frac{2\pi h_k(t-kT)}{T}\right) dt \\ &= \frac{1}{T}\left(\frac{-\cos 4\pi f_c t}{4\pi f_c} - \frac{T\cos 2\pi h_k(t-kT)/T}{2\pi h_k}\right)_{kT}^{(k+1)T} \\ &\approx \frac{1-\cos 2\pi h_k}{2\pi h_k} \triangleq C_{2,k} \end{aligned} \tag{7.71}$$

According to the Gram-Schmidt procedure, the third orthonormal function is

$$\psi_3(t) = \frac{f_3(t)}{\sqrt{E_3}}$$

where

$$f_3(t) = s_3(t) - s_{31}\psi_1(t) - s_{32}\psi_2(t)$$

and

$$\begin{aligned} E_3 &= \int_{kT}^{(k+1)T} f_3^2(t)dt \\ &= \int_{kT}^{(k+1)T} \left[s_3(t) - s_{31}\psi_1(t) - s_{32}\psi_2(t)\right]^2 dt \\ &= 1 + s_{31}^2 + s_{32}^2 - 2s_{31}^2 - 2s_{32}^2 \\ &= 1 - C_{1,k}^2 - C_{2,k}^2 \triangleq D_k \end{aligned} \tag{7.72}$$

Thus

$$\begin{aligned} \psi_3(t) &= \frac{s_3(t) - C_{1,k}\psi_1(t) - C_{2,k}\psi_2(t)}{\sqrt{D_k}} \\ &= \frac{1}{\sqrt{D_k}}\left[\sqrt{\frac{2}{T}}\cos\left(2\pi f_c t - \frac{\pi h_k(t-kT)}{T}\right)\right. \\ &\quad \left. -C_{1,k}\psi_1(t) - C_{2,k}\psi_2(t)\right] \end{aligned} \tag{7.73}$$

Finally

$$\psi_4(t) = \frac{f_4(t)}{\sqrt{E_4}}$$

where

$$f_4(t) = s_4(t) - s_{41}\psi_1(t) - s_{42}\psi_2(t) - s_{43}\psi_3(t)$$

where

$$s_4(t) = \sqrt{\frac{2}{T}}\sin\left(2\pi f_c t - \frac{\pi h_k(t-kT)}{T}\right) \tag{7.74}$$

The coefficients s_{4i} are determined as follows:

$$\begin{aligned} s_{41} &= \int_{kT}^{(k+1)T} s_4(t)\psi_1(t)dt \\ &= \frac{2}{T}\int_{kT}^{(k+1)T} \sin\left(2\pi f_c t - \frac{\pi h_k(t-kT)}{T}\right) \\ &\quad \cdot\cos\left(2\pi f_c t + \frac{\pi h_k(t-kT)}{T}\right) dt \\ &= \frac{1}{T}\left(\frac{-\cos 4\pi f_c t}{4\pi f_c} + \frac{T\cos 2\pi h_k(t-kT)/T}{2\pi h_k}\right)_{kT}^{(k+1)T} \\ &\approx \frac{\cos 2\pi h_k - 1}{2\pi h_k} = -C_{2,k} \end{aligned} \tag{7.75}$$

$$\begin{aligned} s_{42} &= \int_{kT}^{(k+1)T} s_4(t)\psi_2(t)dt \\ &= \frac{2}{T}\int_{kT}^{(k+1)T} \sin\left(2\pi f_c t - \frac{\pi h_k(t-kT)}{T}\right) \\ &\quad \cdot\sin\left(2\pi f_c t + \frac{\pi h_k(t-kT)}{T}\right) dt \\ &= \frac{1}{T}\left(\frac{T\sin 2\pi h_k(t-kT)/T}{2\pi h_k} - \frac{\sin 4\pi f_c t}{4\pi f_c}\right)_{kT}^{(k+1)T} \\ &\approx \frac{\sin 2\pi h_k}{2\pi h_k} = C_{1,k} \end{aligned} \tag{7.76}$$

$$\begin{aligned}
s_{43} &= \frac{2}{T}\int_{kT}^{(k+1)T} \sin\left(2\pi f_c t - \frac{\pi h_k(t-kT)}{T}\right) \\
&\quad \cdot \cos\left(2\pi f_c t - \frac{\pi h_k(t-kT)}{T}\right) dt \\
&\approx 0
\end{aligned} \tag{7.77}$$

$$\begin{aligned}
E_4 &= \int_{kT}^{(k+1)T} f_4^2(t)dt \\
&= \int_{kT}^{(k+1)T} \left[s_4(t) - s_{41}\psi_1(t) - s_{42}\psi_2(t)\right]^2 dt \\
&= 1 - C_{1,k}^2 - C_{2,k}^2 = D_k
\end{aligned} \tag{7.78}$$

Thus

$$\begin{aligned}
\psi_4(t) &= \frac{s_4(t) + C_{2,k}\psi_1(t) - C_{1,k}\psi_2(t)}{\sqrt{D_k}} \\
&= \frac{1}{\sqrt{D_k}}\left[\sqrt{\frac{2}{T}}\sin\left(2\pi f_c t - \frac{\pi h_k(t-kT)}{T}\right)\right. \\
&\quad \left. + C_{2,k}\psi_1(t) - C_{1,k}\psi_2(t)\right]
\end{aligned} \tag{7.79}$$

References

[1] Miyakawa, H., H. Harashima, and Y. Tanaka, "A new digital modulation scheme, multi-code binary CPFSK," *Proc. 3rd Int. Conf. on Digital Satellite Comm.*, November 1975, pp. 105–112.

[2] Anderson, J., and R. DeBuda, "Better phase-modulation error performance using trellis phase code," *Electron. Lett.*, vol. 12, no. 22, October 1976, pp. 587–588.

[3] Anderson, J., and D. P. Taylor, "Trellis phase-modulation coding: minimum distance and spectral results," *Proc. Electron. Aerospace Syst. Conf.*, September 1977, pp. 29-1A–1G.

[4] Anderson, J., and D. P. Taylor, "A bandwidth efficient class of signal space codes," *IEEE Trans. Inform. Theory*, vol. 24, no. 6, November 1978, pp. 703–712.

[5] Sasase, I., and S. Mori, "Multi-h phase-coded modulation," *IEEE Communications Magazine*, vol. 29, no. 12, December 1991, pp. 46–56.

[6] Lereim, A., "Spectral properties of multi-h codes," M. Eng. Thesis, McMaster Univ. Techn. Report, no. CRL-57, Communications Research Laboratory, McMaster University, Hamilton, Ontario, Canada, July 1978.

[7] Bhargava, V. R., et al., *Digital Communications by Satellite*, New York: John Wiley & Sons, 1981.

[8] Wilson, S., and R. Gaus, "Power spectra of multi-h phase codes," *IEEE Trans. on Comm.*, vol. 29, no. 3, March 1981, pp. 250–256.

[9] Holubowicz, W., "Optimum parameter combinations for multi-h phase codes," *IEEE Trans. on Comm.*, vol. 38, no. 11, November 1990, pp. 1929–1931.

[10] Aulin, T., and C-E. Sundberg, "Minimum Euclidean distance and power spectrum for a class of smoothed phase codes with constant envelope," *IEEE Trans. on Comm.*, vol. 30, no. 7, July 1982, pp. 1721–1729.

[11] Maseng, T., "The autocorrelation function for multi-h coded signals," *IEEE Trans. on Comm.*, vol. 33, no. 5, May 1985, pp. 481–484.

[12] Papoulis, A., *Probability, Random Variables, and Stochastic Processes,* Tokyo, Japan: McGraw-Hill Kogakusha, 1965.

[13] Bennett, W., and S. Rice, "Spectral density and autocorrelation functions associated with binary frequency shift keying," *Bell Syst. Tech. J.*, vol. 42, 1963, pp. 2355–2385.

[14] Anderson, J., T. Aulin, and C-E. Sundberg, *Digital Phase Modulation,* New York: Plenum Publishing Company, 1986.

[15] Mazo, J., and J. Salz, "Spectra of frequency modulation with random waveforms," *Inform. Contr.*, vol. 42, 1966, pp. 414–422.

[16] Baker, T., "Asymptotic behavior of digital FM spectra," *IEEE Trans. on Comm.*, vol. 22, no. 10, October 1974, pp. 1585–1594.

[17] Aulin, T., and C-E Sundberg, "On the minimum Euclidean distance for a class of signal space codes," *IEEE Trans. Inform. Theory,* vol. 28, no. 1, January 1982, pp. 43–55.

[18] Anderson, J., "Simulated error performance of multi-h phase codes," *IEEE Trans. Inform. Theory,* vol. 27, no. 3, May 1981, pp. 357–362.

[19] Wilson, S., J. Highfill III, and C-D. Hsu, "Error bounds for multi-h phase codes," *IEEE Trans. Information Theory,* vol. 28, no. 4, July 1982, pp. 650–665.

[20] Liebetreu, J., "Joint carrier phase estimation and data detection algorithm for multi-h CPM data transmission," *IEEE Trans. on Comm.*, vol. 34, no. 9, September 1986, pp. 873–881.

[21] Premji, A., and D. Taylor, "Receiver structures for multi-h signaling formats," *IEEE Trans. on Comm.*, vol. 35, no. 4, April 1987, pp. 439–451.

[22] Premji, A., and D. Taylor, "A practical receiver structures for multi-h signals," *IEEE Trans. on Comm.*, vol. 35, no. 9, September 1987, pp. 901–908.

[23] Mazur, B., and D. Taylor, "Demodulation and carrier synchronization of multi-h phase codes," *IEEE Trans. on Comm.*, vol. 29, no. 3, March 1981, pp. 257–266.

[24] Taylor D., and B. Mazur, "Research on self-synchronization of phase codes," *McMaster Univ. Techn. Report*, no. CRL-63, Communications Research Laboratory, McMaster University, Hamilton, Ontario, Canada, April 1979.

[25] Hwang, H., L. Lee, and S. Chen, "Multi-h phase coded modulations with asymmetrical modulation indices," *IEEE Journal of Selected Areas in Communications*, vol. 7, no. 9, December 1989, pp. 1450–1461.

[26] Hwang, H., L. Lee, and S. Chen, "Asymmetrical modulation indexes in full/partial-response multi-h phase-coded modulation with different phase pulse functions," *IEE Proc.*, vol. 139, Part I, October 1992, pp. 508–514.

[27] Holubowicz, W., and P. Szulakiwicz, "Multi-T realization of multi-h phase codes," *IEEE Trans. Inform. Theory*, vol. 31, no. 4, July 1985, pp. 528–529.

[28] Szulakiwicz, P., "M-ary linear phase multi-T codes," *IEEE Trans. on Comm.*, vol. 37, no. 3, March 1989, pp. 197–199.

[29] Fonseka, J., and G. Davis, "Correlatively encoded multi-h signals," *IEEE Trans. on Comm.*, vol. 41, no. 3, March 1993, pp. 444–446.

[30] Mao, R., and J. Fonseka, "Nonlinear multi-h phase codes for CPFSK signaling," *IEEE Trans. on Comm.*, vol. 43, no. 8, August 1995, pp. 2350–2359.

Selected Bibliography

- Ziemer, R., and R. Peterson, *Introduction to Digital Communications*, New York: Macmillan, 1992.

Chapter 8

Amplitude Shift Keying

In Chapter 1 we stated that there are three basic passband digital modulations: amplitude shift keying (ASK), frequency shift keying (FSK), and phase shift keying (PSK). Their simplest form is the binary form and the waveforms are shown in Figure 1.4. PSK is more bandwidth efficient than FSK and more power efficient than ASK and FSK. Therefore PSK has gained more popularity. However, ASK and FSK are still used in some communication systems due to their implementation simplicity. We have studied binary and M-ary FSK and PSK in detail in Chapters 3 and 4. In this chapter, binary ASK and M-ary ASK (MASK) are studied. Furthermore, an understanding of M-ary ASK will serve as a background for understanding the popular quadrature amplitude modulation (QAM), which is a very power-efficient and bandwidth-efficient two-dimensional modulation scheme. QAM will be studied in the next chapter.

MASK can be considered as a subclass of a more general class of modulations, the pulse amplitude modulation (PAM)[1] [1,2]. PAM could be in baseband without a carrier or in passband with a carrier. MASK is a PAM scheme with a carrier. Thus to make the results in this chapter more general, we will discuss PAM first. We will derive the power spectral density (PSD) expressions of PAM and discuss its optimum detection and error probability. Then we will focus on MASK. Since it is a subclass of PAM, the results of PAM will be adapted for MASK.

We will first present PAM signal in its most general form in Section 8.1, where a general form of the PSD expression is given and optimum detection and error probability expression are derived. In Section 8.2, bipolar symmetrical MASK and its PSD, error probability, modulator, and demodulator are presented. Section 8.3 is for unipolar MASK, where its PSD expressions and error probabilities for coherent and noncoherent demodulation are derived. On-off keying (OOK), the simplest ASK, is

[1] Some authors use the term PAM for the sampling technique that converts an analog signal to a pulse-type signal where the amplitude of the pulse represents the amplitude of the analog signal at the sampling time instant [3,4].

treated as a special case of unipolar MASK in Section 8.4. Comparison of MASK with MPSK is made in Section 8.5. Finally, Section 8.6 summarizes the chapter.

8.1 PULSE AMPLITUDE MODULATION

The M-ary amplitude modulation signal alphabet can be written in a general form as

$$s_i(t) = s_i\phi(t), \quad 0 \le t \le T \tag{8.1}$$

for $i = 1, 2, ...M$. $\phi(t)$ is any unit energy function of duration T, that is $\int_0^T \phi^2(t)dt = 1$. Consequently $s_i^2 = E_i$ is the energy of $s_i(t)$. If $\phi(t)$ is a baseband pulse, then $s_i(t)$ is the baseband PAM. The NRZ and RZ line codes described in Chapter 2 are examples of baseband binary PAM schemes. If $\phi(t)$ is a high frequency sinusoidal carrier, then $s_i(t)$ is the passband PAM, or M-ary amplitude shift keying (MASK). OOK is a binary ASK with one of the s_i being zero. BPSK can be also viewed as a binary passband PAM scheme with two antipodal s_i.

8.1.1 Power Spectral Density

On the entire time axis, we can write PAM signal as

$$s(t) = \sum_{k=-\infty}^{\infty} s_k\phi(t - kT), \quad -\infty < t < \infty \tag{8.2}$$

where amplitude s_k is determined by the message data which are random. Thus s_k is a random variable. Equation (8.2) is in the form of (A.14) in Appendix A. Further assuming that data are uncorrelated, the PSD of $s(t)$ is given by (A.18), that is

$$\Psi_s(f) = \frac{\sigma_s^2|\Phi(f)|^2}{T} + \left(\frac{m_s}{T}\right)^2 \sum_{k=-\infty}^{\infty} |\Phi(\frac{k}{T})|^2\delta(f - \frac{k}{T}) \tag{8.3}$$

where $\Phi(f) = \mathcal{F}\{\phi(t)\}$, σ_s^2 is the variance of s_k and m_s is the mean value of s_k. The first part of (8.3) is the continuous spectrum and the second part is the discrete spectral lines. Note that $\Phi(f)$ could be a baseband spectrum or a passband spectrum, depending on whether $\phi(t)$ is a baseband or passband signal.

When the amplitudes are symmetrically distributed around zero, then $m_s = 0$, and

$$\Psi_s(f) = \frac{\sigma_s^2|\Phi(f)|^2}{T} \tag{8.4}$$

This tells us that the PSD of a symmetrical PAM is determined by the PSD of the basis function $\phi(t)$. If the amplitude distribution is not symmetrical around zero, then $m_s \neq 0$, discrete spectral lines will exist.

Note that (8.3) and (8.4) are applicable to both baseband and passband $\phi(t)$. However, for passband $\phi(t)$ with a carrier frequency f_c (that is, the signal is ASK), the spectrum $\Phi(f)$ has two parts, centered around f_c and $-f_c$. Thus (8.3) and (8.4) are not convenient to use. For passband random signal the PSD is completely determined by the PSD of its complex envelope $\widetilde{s}(t)$ (see (A.13)). Thus it suffices to have the complex envelope's PSD, $\Psi_{\widetilde{s}}(f)$, instead of $\Psi_s(f)$. We will derive the $\Psi_{\widetilde{s}}(f)$ expression for ASK signal in the next section.

8.1.2 Optimum Detection and Error Probability

In the M-ary PAM signal set, there is only one basis function which is the $\phi(t)$. Geometrically, each signal can be represented by its projection on the $\phi(t)$

$$s_i = \int_0^T s_i(t)\phi(t)dt$$

Thus for the AWGN channel, from (B.37), the optimum detector is a one-dimensional minimum distance coherent detector. For a coherent detector, the received signal can be modelled as

$$r(t) = s_i(t) + n(t)$$

Note that in a real-world system, the signal will have an initial phase, say θ, due to propagation delay and channel phase distortion. However, since we are assuming coherent detection, the reference carrier in the receiver has the same phase as the incoming signal; and we can assume $\theta = 0$ without loss of generality. $n(t)$ is a Gaussian random process with a two-sided spectral density of $N_o/2$.

The sufficient statistic is

$$r = \int_0^T r(t)\phi(t)dt \tag{8.5}$$

and the detector compares r to s_i and chooses the closest (minimum distance decision rule).

The error probability of the coherent detection for an M-ary PAM with *equal amplitude spacings* can be derived as follows. Assuming an AWGN channel with two-sided noise PSD of $N_o/2$,

$$r = \int_0^T r(t)\phi(t)dt = \int_0^T [s_i(t) + n(t)]\phi(t)dt = s_i + n_i$$

$$f(r/x) = \frac{1}{\sqrt{\pi \cdot N_o}} \exp\left[\frac{-(r-x)^2}{N_o}\right]$$

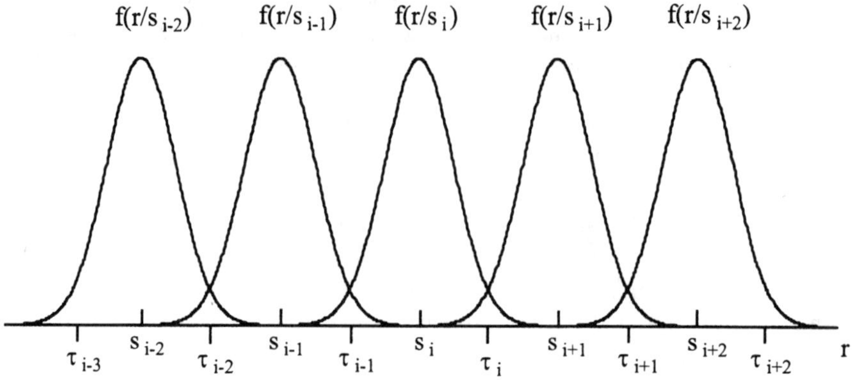

Figure 8.1 Thresholds and decision regions for M-ary PAM.

where n_i is Gaussian with zero mean and a variance of $N_o/2$. Thus r is Gaussian with mean s_i and variance $N_o/2$. Figure 8.1 shows the probability distribution densities of r conditioned on s_i, where τ_i are thresholds. This figure can help us derive the error probability.

Assuming s_i is transmitted, a symbol error occurs when the noise n_i exceeds in magnitude one-half of the distance between two adjacent levels. This probability is the same for each s_i except for the two outside levels, where an error can occur in one direction only. Assuming all amplitude levels are equally likely, the average symbol error probability is

$$P_s = \frac{M-1}{M} \Pr\left(|r - s_i| > \frac{\Delta}{2}\right)$$

where

$$\Delta = |s_i - s_{i-1}|$$

is the distance between adjacent signal levels, and also is the distance between adjacent thresholds. Thus

$$P_s = \frac{M-1}{M} \Pr\left(|r - s_i| > \frac{\Delta}{2}\right) = \frac{M-1}{M} \frac{2}{\sqrt{\pi N_o}} \int_{\Delta/2}^{\infty} e^{-x^2/N_o} dx$$

$$= \frac{M-1}{M}\frac{2}{\sqrt{2\pi}}\int_{\Delta/\sqrt{2N_o}}^{\infty} e^{-x^2/2}dx$$

That is

$$P_s = \frac{2(M-1)}{M}Q\left(\sqrt{\frac{\Delta^2}{2N_o}}\right) \tag{8.6}$$

Note that this expression is applicable to any uniformly spaced PAM, whether it is baseband or passband, bipolar or unipolar, symmetrical or asymmetrical.

8.2 BIPOLAR SYMMETRICAL MASK

Having derived general expressions for PAM in the last section, now we turn our attention to the passband PAM, that is, the M-ary amplitude shift keying (MASK).

In MASK, data bits are grouped into n-tuples, where $n = \log_2 M$. Each n-tuple is mapped into an amplitude A_i. Thus a symbol in MASK is a burst of carrier signal with an amplitude determined by the n-tuple and it represents n bits instead of only one bit as in OOK. As we will see later, the bandwidth of the MASK signal's PSD is determined by the symbol duration. By keeping the symbol duration the same, an MASK system and an OOK system will need the same bandwidth, but the bit rate of the MASK system will be n times that of the OOK system.

In the basic symbol period $[0, T]$, an MASK signal with a pulse shape $p(t)$ can be expressed as

$$s_i(t) = A_i p(t)\cos 2\pi f_c t, \quad 0 \le t \le T \tag{8.7}$$

for $i = 1, 2, ...M$. Note that T is the symbol duration and $T = nT_b$, where T_b is the bit duration.

The amplitudes of the symmetrical uniformly-spaced bipolar distribution is given by

$$A_i = (2i - 1 - M)A, \quad i = 1, 2, ..., M \tag{8.8}$$

where $A > 0$. The amplitude spacing is $2A$. For example, if $M = 4$, then $A_i \in (-3A, -A, A, 3A)$.

8.2.1 Power Spectral Density

From (A.13) we know that the PSD of a passband random signal is completely determined by the PSD of its complex envelope. That is

$$\Psi_s(f) = \frac{1}{2}[\Psi_{\tilde{s}}(f - f_c) + \Psi_{\tilde{s}}(-f - f_c)] \tag{8.9}$$

where $\Psi_{\tilde{s}}(f)$ is the PSD of the complex envelope.

From (8.7), the complex envelope of MASK on the entire time axis is

$$\tilde{s}(t) = \sum_{k=-\infty}^{\infty} A_k p(t - kT), \quad -\infty < t < \infty$$

which is in the form of (A.14). Thus for uncorrelated data, the PSD of MASK signal is

$$\Psi_{\tilde{s}}(f) = \frac{\sigma_A^2 |P(f)|^2}{T} + \left(\frac{m_A}{T}\right)^2 \sum_{k=-\infty}^{\infty} |P(\frac{k}{T})|^2 \delta(f - \frac{k}{T}) \tag{8.10}$$

where $P(f) = \mathcal{F}\{p(t)\}$ is the spectrum of $p(t)$, σ_A^2 is the variance of A_k, and m_A is the mean value of A_k. Note that $\Psi_{\tilde{s}}(f)$ in (8.10) is different from $\Psi_s(f)$ in (8.3). The former is the PSD of the complex envelope of the signal and the latter is the PSD of the signal itself.

Now let us assume $p(t)$ is rectangular with unit amplitude; then

$$|P(f)| = \left| T \frac{\sin(\pi f T)}{\pi f T} \right| \tag{8.11}$$

and

$$\left| P(\frac{k}{T}) \right| = \left| T \frac{\sin(\pi k)}{\pi k} \right| = \begin{cases} T, & k = 0 \\ 0, & k \neq 0 \end{cases}$$

then the PSD of the complex envelope of the passband MASK signal is

$$\Psi_{\tilde{s}}(f) = \sigma_A^2 T \left(\frac{\sin(\pi f T)}{\pi f T} \right)^2 + (m_A)^2 \delta(f) \tag{8.12}$$

The PSD of the complex envelope of the symmetrical bipolar MASK (not necessarily uniformly spaced) is obtained by setting $m_A = 0$ in (8.10):

$$\Psi_{\tilde{s}}(f) = \frac{\sigma_A^2 |P(f)|^2}{T} \tag{8.13}$$

The above shows that the PSD of symmetrical bipolar MASK is determined by the PSD of the pulse-shaping function $p(t)$. If the pulse is rectangular, from (8.12) we

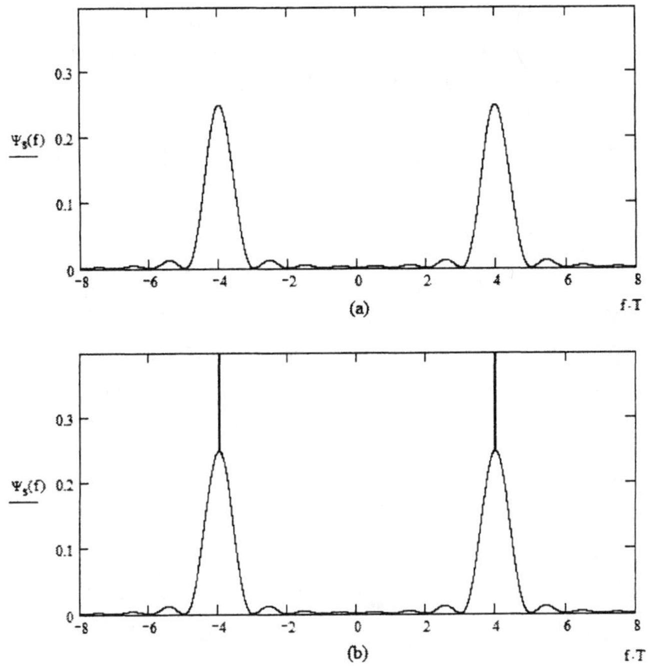

Figure 8.2 PSD of MASK with rectangular pulse and $\sigma_A^2 T = 1/2$, $f_c = 4/T$: (a) biplolar and (b) unipolar.

have

$$\Psi_{\tilde{s}}(f) = \sigma_A^2 T \left(\frac{\sin(\pi f T)}{\pi f T} \right)^2 \tag{8.14}$$

which has the same shape of the PSD of MPSK (see Figure 4.15). The passband PSD shape is shown in Figure 8.2(a).

8.2.2 Modulator and Demodulator

A bipolar MASK can only be demodulated coherently since noncoherent demodulation can not distinguish the sign of the signal. A unipolar MASK can be demodulated noncoherently as well as coherently. However it is usually used in a system where coherent demodulation is difficult thus noncoherent demodulation is preferred. If coherent demodulation is feasible, the symmetrical uniform-spacing bipolar MASK is preferred over the uniform-spacing unipolar MASK since the former has better BER performance. In the following, we will present the modulator and coherent demodulator for MASK. They are applicable to both bipolar and unipolar signals. The noncoherent demodulator will be described in Section 8.3.

MASK signal given in (8.7) can be easily written in the form of (8.1) by defining

$$\phi(t) = \sqrt{2/E_p}p(t)\cos 2\pi f_c t \tag{8.15}$$

where $E_p = \int_0^T p^2(t)dt$ is the energy of the pulse-shaping signal $p(t)$ in $[0, T]$. $E_p = T$ when $p(t)$ is a rectangular pulse with amplitude 1 in $[0, T]$. With the unit amplitude pulse, $\int_0^T \phi^2(t)dt = 1$. For an arbitrary pulse, $\int_0^T \phi^2(t)dt \cong 1$ for $f_c >> 1/T$. Thus for most practical cases, $\phi(t)$ is considered as a normalized basis function and $s_i(t) = s_i\phi(t)$ with

$$s_i = A_i\sqrt{E_p/2} \tag{8.16}$$

The modulator is shown in Figure 8.3. Figure 8.3(a) is a direct implementation of (8.7). The level generator takes $n = \log_2 M$ bits from the binary data stream and maps them into an amplitude level $A_k \in \{A_i\}$, where the subscript k indicates the kth symbol interval. The mapping is preferably Gray coding so that the n-tuples representing the adjacent amplitudes differ only by one bit. The functions of the rest blocks are self-explanatory. The equivalent implementation is shown in Figure 8.3(b). It is more practical for hardware implementation. The $p(t)$ multiplier is replaced with a filter with an impulse response $p(t)$. In order to generate a pulse $A_ip(t)$, the input to the filter must be an impulse $A_i\delta(t)$. In practice, this can be realized by a very narrow pulse with amplitude A_i.

The optimum receiver implementing the minimum distance rule is shown in Figure 8.4 where the last block is a threshold detector. Figure 8.4(a) is a direct implementation of (8.5) and the minimum distance rule. In Figure 8.4(b), the correlation with $p(t)$ is replaced by a matched filter and a sampler. Figure 8.4(b) is more practical for hardware implementation. The equivalence between parts (a) and (b) in Figure 8.4 can be shown as follows. For the signal part in the noise-corrupted received signal, the output of the down-convertor in both cases is

$$\begin{aligned} & A_ip(t)\cos 2\pi f_c t\sqrt{\frac{2}{E_p}}\cos 2\pi f_c t \\ = \ & \frac{1}{2}A_i\sqrt{\frac{2}{E_p}}p(t)[1 + \cos 4\pi f_c t] \end{aligned}$$

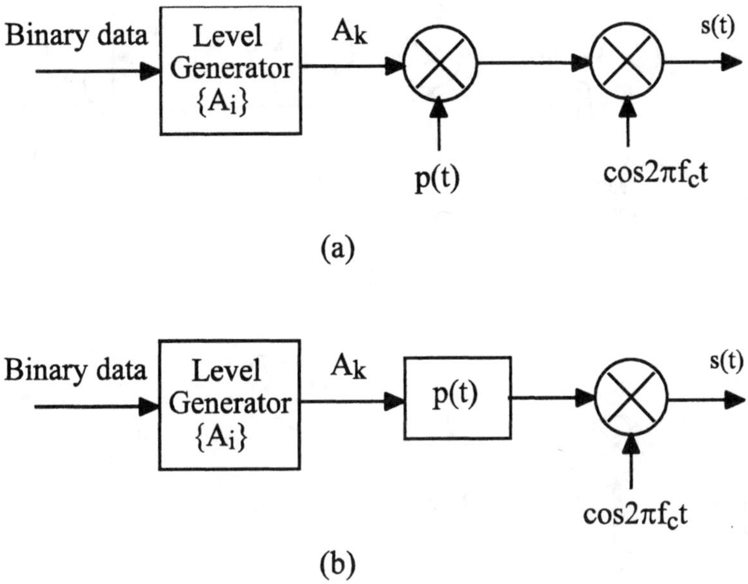

Figure 8.3 MASK modulators (a, b).

In Figure 8.4(a), the integrator output is

$$\frac{1}{2}A_i\sqrt{\frac{2}{E_p}}\int_0^T p^2(t)[1+\cos 4\pi f_c t]dt = A_i\sqrt{\frac{E_p}{2}} = s_i$$

for $f_c >> 1/T$. In Figure 8.4(b), the high-frequency term $\cos 4\pi f_c t$ is blocked by the low-pass filter $p(T-t)$. The output of the filter is

$$\frac{1}{2}A_i\sqrt{\frac{2}{E_p}}\int_0^t p(\tau)p(T-t+\tau)d\tau\bigg|_{t=T}$$

$$= \frac{1}{2}A_i\sqrt{\frac{2}{E_p}}\int_0^T p(\tau)p(\tau)d\tau = A_i\sqrt{\frac{E_p}{2}} = s_i$$

which is exactly the same as the output of the integrator in Figure 8.4(a). As to the noise in the received signal, parts (a) and (b) in Figure 8.4 are also equivalent. Assuming the noise in the received signal is $n(t)$, the output noise of the integrator

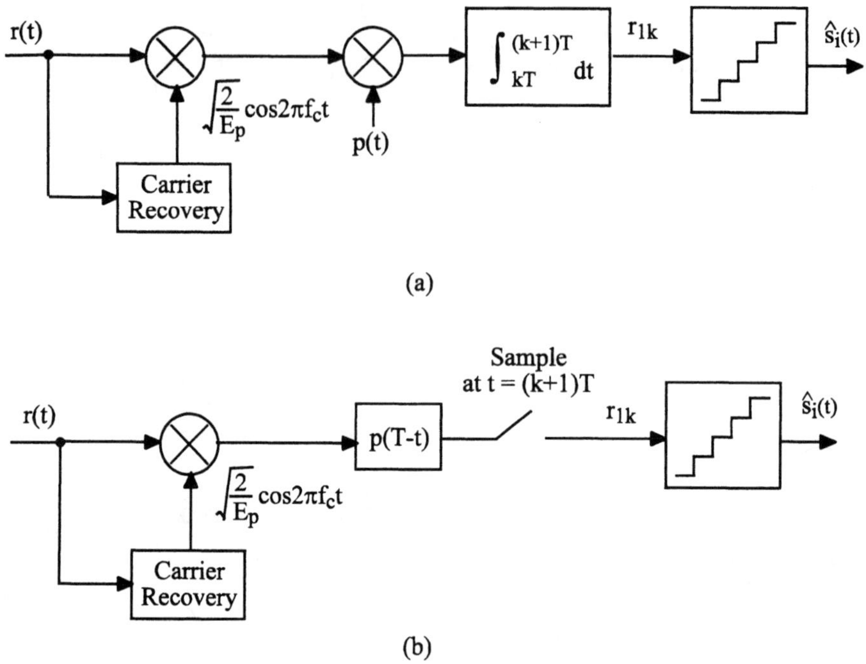

Figure 8.4 Optimum coherent demodulators for MASK (a, b).

in Figure 8.4(a) is $\int_0^T n(t)\phi(t)dt$. In Figure 8.4(b), the output noise of the down-convertor is

$$n_1(t) = n(t)\sqrt{\frac{2}{E_p}}\cos 2\pi f_c t = \frac{n(t)\phi(t)}{p(t)}$$

Then the output of the matched filter is

$$\begin{aligned}\left.\int_0^t n_1(\tau)p(T-t+\tau)d\tau\right|_{t=T} &= \left.\int_0^t \frac{n(\tau)\phi(\tau)}{p(\tau)}p(T-t+\tau)d\tau\right|_{t=T} \\ &= \int_0^T \frac{n(\tau)\phi(\tau)}{p(\tau)}p(\tau)d\tau = \int_0^T n(\tau)\phi(\tau)d\tau\end{aligned}$$

which is exactly the same as that of the correlator. Thus the equivalence between parts (a) and (b) in Figure 8.4 is established.

The threshold detector has $M-1$ thresholds, each placed in the middle of two signal points (refer to Figure 8.1). The correlator computes the r and the threshold detector is actually computing the distance of r to the amplitudes s_i, $i = 1, 2, ...M$ and chooses the smallest. Note that the reference signal can be any scaled version of $\phi(t)$ as long as the thresholds are scaled accordingly.

The coherent carrier in the coherent demodulator can be generated using the synchronization methods described in Chapter 4, such as a squaring loop (Figure 4.35 with $M = 2$). The π phase ambiguity associated with the squaring loop can be solved by differential coding. The Costas loop (Figure 4.36) can also be used to demodulate MASK. The symbol timing clock can be recovered by using the methods described in Chapter 4.

8.2.3 Error Probability

The error probability of bipolar symmetrical uniformly spaced MASK is obtained from (8.6), but we need to figure out what Δ is. For bipolar symmetrical uniformly spaced MASK, the amplitudes A_i are given in (8.8). In terms of signal projection s_i, from (8.16) we can write

$$\begin{aligned} s_i &= A_i\sqrt{E_p/2} = (2i-1-M)A\sqrt{E_p/2} \\ &= (2i-1-M)A_o, \quad i = 1, 2, ..., M \end{aligned} \tag{8.17}$$

where $A_o = A\sqrt{E_p/2}$. Then

$$\Delta = |s_i - s_{i-1}| = 2A_o$$

Thus from (8.6) we have

$$P_s = \frac{2(M-1)}{M}Q\left(\sqrt{\frac{2A_o^2}{N_o}}\right) \tag{8.18}$$

The symbol error probability can be expressed in terms of the average energy of the signals. The average symbol energy of the signals is

$$\begin{aligned} E_{avg} &= \frac{1}{M}\sum_{i=1}^{M} E_i = \frac{1}{M}\sum_{i=1}^{M} s_i^2 \\ &= \frac{1}{M}\sum_{i=1}^{M}(2i-1-M)^2 A_o^2 = \frac{1}{M}\frac{M(M^2-1)A_o^2}{3} \end{aligned}$$

$$= \frac{1}{3}(M^2 - 1)A_o^2 \tag{8.19}$$

As a result, (8.18) becomes

$$P_s = \frac{2(M-1)}{M} Q\left(\sqrt{\frac{6E_{avg}}{(M^2-1)N_o}}\right) \tag{8.20}$$

Since the average energy per bit is $E_b = E_{avg}/\log_2 M$, (8.20) can be written as

$$P_s = \frac{2(M-1)}{M} Q\left(\sqrt{\frac{6(\log_2 M)E_b}{(M^2-1)N_o}}\right) \tag{8.21}$$

Figure 8.5 shows the curves of the symbol error probability of the bipolar symmetrical uniformly spaced MASK versus E_b/N_o. We emphasize that (8.20) to (8.21) depend on only average energy of the signal symbols, not the shape of $p(t)$. In general, bit error probability is not simply related to symbol error probability. However, for Gray coded n-tuple to amplitude mapping, the adjacent symbols differ only by one bit. At high signal-to-noise ratios, the most likely errors are the adjacent symbols, so in this case

$$P_b \approx \frac{P_s}{\log_2 M} \tag{8.22}$$

8.3 UNIPOLAR M-ARY ASK

The amplitude distribution of the uniform-spacing unipolar distribution is given by

$$A_i = (i-1)A, \quad i = 1, 2, ..., M \tag{8.23}$$

The amplitude spacing is A. As an example, let $M = 4$, then $A_i \in (0, A, 2A, 3A)$. OOK is just a special case of the unipolar MASK when $M = 2$.

As pointed out in Section 8.2.2 a unipolar MASK can be demodulated noncoherently as well as coherently. Even though coherent unipolar MASK is not a usual choice since the symmetrical bipolar MASK is better in terms of BER, for the sake of completeness and also due to the ease of doing this, we will present the error probability expression. Then we will focus on the noncoherent unipolar MASK. But first we examine its power spectral density.

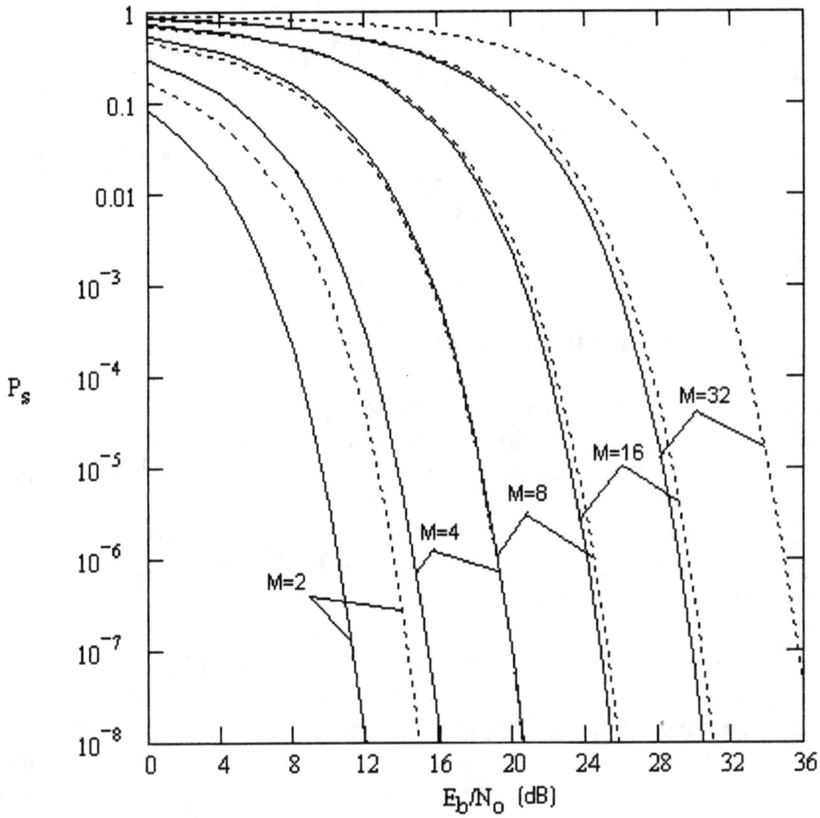

Figure 8.5 Symbol error probability of coherent bipolar symetrical uniformly spaced MASK (solid line) and coherent unipolar uniformly spaced MASK (dotted line). From [5, p. 123]. Copyright © 2005 John Wiley.

8.3.1 Power Spectral Density

In general, the PSD of unipolar MASK is given by (8.10). For uniformly spaced amplitudes and rectangular symbol pulse, the PSD of the complex envelope is given in (8.12). For the amplitudes given in (8.23), the mean value of the amplitudes is

$$m_A = \frac{1}{M}(0 + A + ... + (M-1)A) = \frac{(M-1)A}{2}$$

and then

$$\Psi_{\tilde{s}}(f) = \sigma_A^2 T \left(\frac{\sin \pi f T}{\pi f T} \right)^2 + \frac{(M-1)^2 A^2}{4} \delta(f) \tag{8.24}$$

The passband PSD shape is shown in Figure 8.2(b).

8.3.2 Modulator and Demodulator

The structure of the modulator for unipolar MASK is almost the same as that for bipolar MASK as shown in Figure 8.3. The difference is only in the settings of the level generator due to the signal's unipolarity. The signal levels are now only positive and equally spaced.

The structure of the coherent demodulator for unipolar MASK is almost the same as that for bipolar MASK as shown in Figure 8.4. The difference is only in the settings of the threshold detector due to the signal's unipolarity. The thresholds now are only positive and equally spaced.

The structure of the noncoherent demodulator for unipolar MASK is shown in Figure 8.6(a). The first stage is a bandpass filter (BPF) with a center frequency of f_c and adequate bandwidth. This BPF has the function of selecting the desired signal and rejecting the noise. The filtered signal then passes an envelope detector which extracts the envelope of the modulated carrier, (i.e., the noise-corrupted baseband pulses). Finally, the noise-corrupted baseband pulses pass the threshold detector to produce the binary data sequence.

The envelope detector can be implemented in the form of a quadrature detector as shown in Figure 8.6(b) (see also Figure B.10; however only detector arm is needed here as there is only one linearly independent signal in MASK symbol set). Each of the multipliers in Figure 8.6(b) may be replaced by a matched filter-sampler combination for implementation simplicity (see Figure B.11). It can also be approximately implemented in other forms. For example, the well-known AM demodulator consisting of a diode followed by a resistor and a capacitor in parallel is an envelope detector.

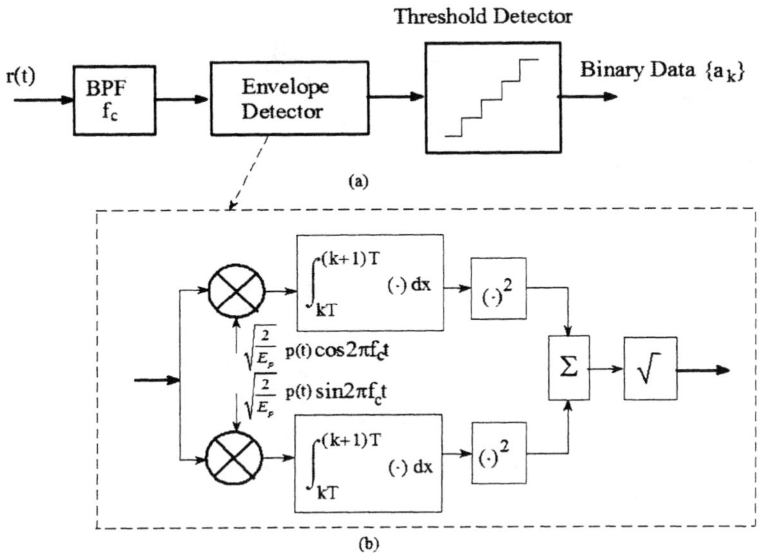

Figure 8.6 (a) Noncoherent demodulator for unipolar MASK, and (b) quadrature envelope detector.

8.3.3 Error Probability of Coherent Demodulation

The symbol error probability is again given by (8.6), but we need to figure out what Δ is. For unipolar uniformly spaced MASK, the amplitudes are given in (8.23). In terms of signal projection s_i, from (8.16) we can write

$$\begin{aligned} s_i &= A_i\sqrt{E_p/2} = (i-1)A\sqrt{E_p/2} \\ &= (i-1)A_o, \quad i = 1, 2, ..., M \end{aligned} \tag{8.25}$$

where $A_o = A\sqrt{E_p/2}$. Then

$$\Delta = |s_i - s_{i-1}| = A_o$$

Thus from (8.6) we have

$$P_s = \frac{2(M-1)}{M} Q\left(\sqrt{\frac{A_o^2}{2N_o}}\right) \tag{8.26}$$

The average symbol energy of the signals is

$$\begin{aligned} E_{avg} &= \frac{1}{M}\sum_{i=1}^{M} E_i = \frac{1}{M}\sum_{i=1}^{M} s_i^2 \\ &= \frac{1}{M}\sum_{i=1}^{M}(i-1)^2 A_o^2 = \frac{1}{M}\sum_{j=0}^{M-1} j^2 A_o^2 \\ &= \frac{1}{6}(2M^2 - 3M + 1)A_o^2 \end{aligned}$$

As a result, (8.26) becomes

$$P_s = \frac{2(M-1)}{M} Q\left(\sqrt{\frac{3E_{avg}}{(2M^2 - 3M + 1)N_o}}\right) \tag{8.27}$$

In terms of average energy per bit, (8.27) can be written as

$$P_s = \frac{2(M-1)}{M} Q\left(\sqrt{\frac{3(\log_2 M)E_b}{(2M^2 - 3M + 1)N_o}}\right) \tag{8.28}$$

This expression is plotted in Figure 8.5 (dotted line). The P_b at high SNRs can be estimated by (8.22).

Comparing (8.21) and (8.28) reveals that the ratio of the expression in the square root sign of (8.21) over that of (8.28) is

$$R(M) = \frac{2(2M^2 - 3M + 1)}{M^2 - 1}$$

Table 8.1 shows some values of $R(M)$. It shows that the bipolar uniformly spaced MASK is superior over the unipolar uniformly spaced MASK. The asymptotic ratio is 4, or 6 dB. This confirms the statement we said earlier that coherent unipolar MASK is not a usual choice since the symmetrical bipolar MASK is better in terms of BER.

M	2	4	8	16	32	64	∞
$R(M)$	2	2.8	3.333	3.647	3.818	3.908	4
$10\log R(M)$ (dB)	3.01	4.472	5.229	5.619	5.819	5.919	6.02

Table 8.1 Comparison between bipolar and unipolar uniformly spaced MASK.

8.3.4 Error Probability of Noncoherent Demodulation

When carrier synchronization is difficult or system simplicity is desired, noncoherent unipolar MASK can be used. The thresholds and symbol error probability for noncoherent demodulation of the unipolar uniformly spaced MASK in the AWGN channel have been derived in [6]. We summarize the derivation here and present the results.

The derivation is based on the envelope detector in Figure 8.6(b). Given the transmitted signal in (8.7), the received signal is

$$r(t) = A_i p(t)\cos(2\pi f_c t + \theta) + n(t)$$

where $n(t)$ is a zero-mean Gaussian process with a two-sided PSD of $N_o/2$, and θ reflects the phase difference between the incoming signal and the reference signals. Since the two reference signals ($\sqrt{2/E_p}p(t)\cos 2\pi f_c t$ and $\sqrt{2/E_p}p(t)\sin 2\pi f_c t$) of the envelope detector are normalized, the noise at the output of each integrator has a variance $\sigma^2 = N_o/2$ (see Section B.2). The signal component at the final output of the quadrature envelope detector is $\sqrt{E_i} = A_i\sqrt{E_p/2}$, where $E_i = A_i^2 E_p/2$ is the energy of signal with amplitude A_i. This can be shown as follows. The signal component of the output of the upper integrator is

$$\begin{aligned}
I_i &= \int_0^T A_i\sqrt{2/E_p}p^2(t)\cos(2\pi f_c t + \theta)\cos(2\pi f_c t)dt \\
&= A_i\sqrt{2/E_p}\int_0^T p^2(t)[\cos\theta\cos^2(2\pi f_c t) - \sin\theta\cos(2\pi f_c t)\sin(2\pi f_c t)]dt \\
&\cong A_i\sqrt{2/E_p}(\cos\theta)\int_0^T p^2(t)\cos^2(2\pi f_c t)dt \\
&= A_i\sqrt{2/E_p}(\cos\theta)\int_0^T p^2(t)\frac{1}{2}[1+\cos(4\pi f_c t)]dt \\
&\cong A_i\sqrt{2/E_p}(\cos\theta)\frac{1}{2}\int_0^T p^2(t)dt = A_i\sqrt{2/E_p}(\cos\theta)\frac{1}{2}E_p \\
&= A_i\sqrt{E_p/2}\cos\theta
\end{aligned}$$

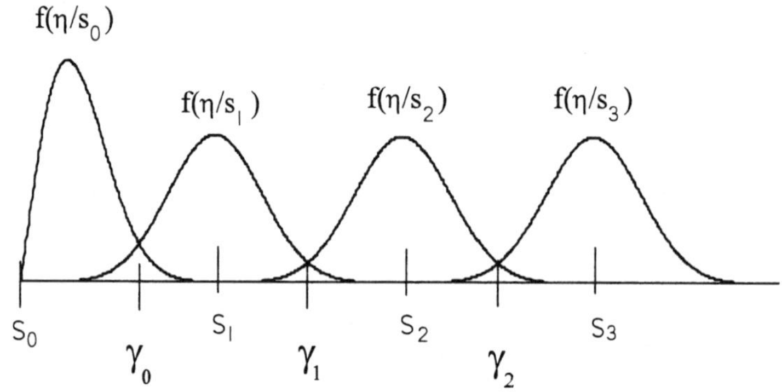

Figure 8.7 Threshold and decision regions for unipolar noncoherent MASK ($M = 4$). From [5, p. 125]. Copyright © 2005 John Wiley.

where the sign "$\cong$" holds for $f_c >> 1/T$. When $p(t)$ is the rectangular pulse, sign "$\cong$" is replaced by sign "$=$". Similarly it can be shown that the signal component of the output of the lower integrator is $Q_i = A_i\sqrt{E_p/2}\sin\theta$. Thus the final output of the envelope detector is

$$\sqrt{I_i^2 + Q_i^2} = A_i\sqrt{E_p/2} = \sqrt{E_i}$$

Recall that the unipolar uniformly spaced MASK signal set is $(0, A, 2A, ...(M-1)A)$. When the amplitude $A_0 = 0$ is transmitted, the output of the envelope detector, denoted as η, has Rayleigh distribution (the first probability density function (pdf) in Figure 8.7). An error will happen if the envelope η has a value greater than the threshold γ. The probability of error for this case will be

$$p(e|A_0) = \int_\gamma^\infty \frac{\eta}{\sigma^2}\exp\left(-\frac{\eta^2}{2\sigma^2}\right)d\eta$$

At high signal-to-noise ratio, the optimum threshold (γ_0) can be closely approximated by $\sqrt{E_{\min}}/2$ [7], where $E_{\min} = A^2E_p/2$ is the energy of the symbol with the minimum nonzero amplitude. Therefore, from above

$$p(e|A_0) = \exp\left(-\frac{E_{\min}}{4N_o}\right)$$

The other case that should be considered separately is the case when the symbol A_{M-1} is transmitted (see the last pdf in Figure 8.7). In this case, the probability density function of the envelope η is Rician and it is centered at $(M-1)\sqrt{E_{\min}}$. An error will happen if the value of the envelope η has value less than the threshold. At high signal-to-noise ratio, the Rician distribution is well approximated by a Gaussian distribution [7]. Using the Gaussian approximation, the probability of error for this case will be

$$p(e|A_{M-1}) \approx \int_{-\infty}^{\gamma_{M-2}} \frac{1}{\sigma\sqrt{2\pi}} \exp\left[-\frac{(\eta-(M-1)\sqrt{E_{\min}})^2}{2\sigma^2}\right] d\eta$$

Since we have used the Gaussian approximation, the threshold will be taken as the intersection of the adjacent probability density functions $p(\eta|A_{M-1})$ and $p(\eta|A_{M-2})$, (i.e., $\gamma_{M-2} = (M-1.5)\sqrt{E_{\min}}$), then

$$p(e|A_{M-1}) \approx Q\left(\sqrt{\frac{E_{\min}}{2N_o}}\right)$$

All other transmitted symbols have the same Rician distribution with different means. Due to the equal spacing of these signals and their thresholds, they have equal probability of error. Regardless of what symbol was transmitted, we can assume that the Rician pdf is centered at the zero point and an error will happen if η exceeds the upper or lower threshold (i.e., $\eta > \sqrt{E_{\min}}/2$ or $\eta < -\sqrt{E_{\min}}/2$). Since the Gaussian approximation has symmetrical pdf around its mean, the average probability of error can be expressed as

$$p(e|A_i) \approx 2\int_{\gamma}^{\infty} \frac{1}{\sigma\sqrt{2\pi}} \exp\left(-\frac{\eta^2}{2\sigma^2}\right) d\eta = 2Q\left(\sqrt{\frac{E_{\min}}{2N_o}}\right), \qquad i = 1, ..., M-2$$

Multiplying the three conditional probabilities by $(1/M)$, $(1/M)$, and $[(M-2)/M]$, respectively, and adding them will give

$$P_s = \frac{1}{M}\left[\exp\left(-\frac{E_{\min}}{4N_o}\right) + (2M-3)Q\left(\sqrt{\frac{E_{\min}}{2N_o}}\right)\right]$$

The average symbol energy E_{avg} is related to the minimum symbol energy by

$$\begin{aligned} E_{avg} &= \frac{1}{M}\sum_{i=1}^{M} E_i = \frac{1}{M}\sum_{i=1}^{M}(i-1)^2\left(A^2\frac{E_p}{2}\right) = \frac{1}{M}\sum_{j=0}^{M-1} j^2\left(A^2\frac{E_p}{2}\right) \\ &= \frac{1}{6}(2M^2-3M+1)\left(A^2\frac{E_p}{2}\right) = \frac{1}{6}(2M^2-3M+1)E_{\min} \end{aligned}$$

or

$$E_{\min} = \frac{6E_{avg}}{2M^2-3M+1} = 2B_M E_{avg}$$

where

$$B_M = \frac{3}{2M^2-3M+1} \tag{8.29}$$

This equation can be written in terms of E_{avg}/N_o using the following equation

$$P_s = \frac{1}{M}\left[\exp\left(-\frac{E_{avg}}{2N_o}B_M\right) + (2M-3)Q\left(\sqrt{\frac{E_{avg}}{N_o}B_M}\right)\right] \tag{8.30}$$

or in terms of E_b/N_o

$$P_s = \frac{1}{M}\left[\exp\left(-\frac{1}{2}(\log_2 M)B_M\frac{E_b}{N_o}\right) + (2M-3)Q\left(\sqrt{(\log_2 M)B_M\frac{E_b}{N_o}}\right)\right] \tag{8.31}$$

where E_b is the average energy per bit. Figure 8.8 shows the above symbol error probability for noncoherent demodulation (dotted lines), together with those for coherent demodulation (8.28). From the figure we can see that the noncoherent demodulation incurs a small performance degradation. When M is high, the degradation is negligible.

8.4 BINARY ASK (ON-OFF KEYING)

A special case of unipolar MASK is the binary on-off keying (OOK). It is the simplest ASK. The OOK signal set is

$$\begin{aligned} s_1(t) &= A\cos 2\pi f_c t, \quad \text{for } a=1, \quad 0 \le t \le T \\ s_2(t) &= 0, \qquad\qquad\quad \text{for } a=0, \quad 0 \le t \le T \end{aligned} \tag{8.32}$$

where a is the binary data which are assumed uncorrelated and equally likely. The

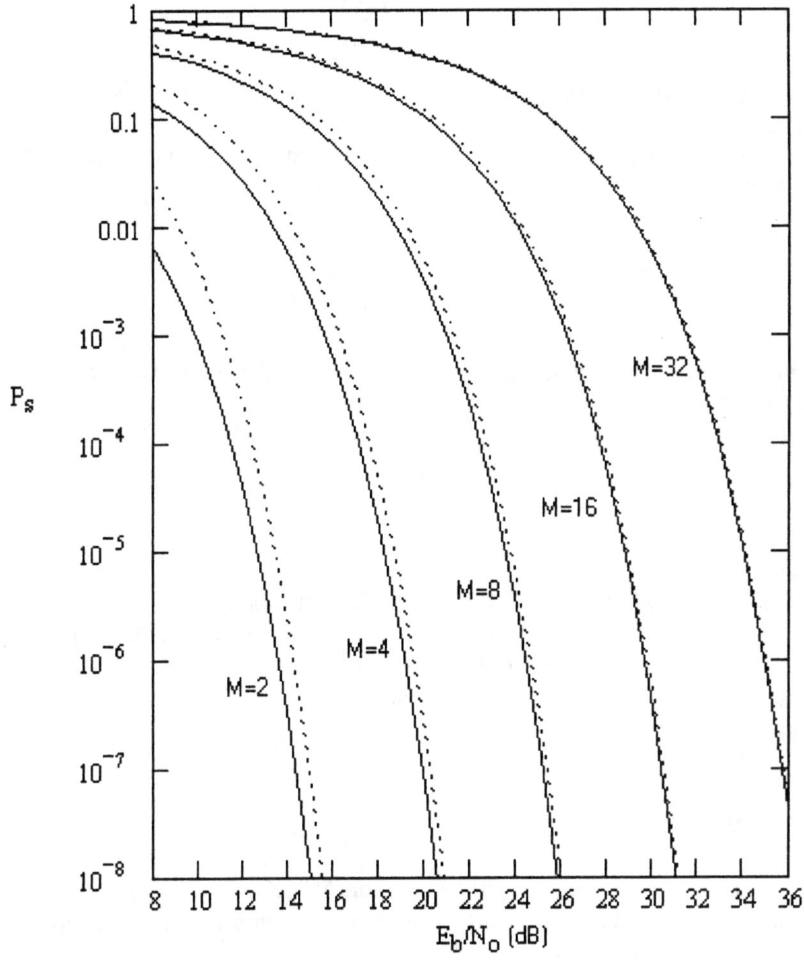

Figure 8.8 Symbol error probability of unipolar uniformly spaced MASK: solid line: coherently demodulated, dotted line: noncoherently demodulated. From [5, p. 126]. Copyright © 2005 John Wiley.

pulse shape $p(t)$ is a rectangular pulse with unit amplitude. From (8.24) we have the PSD for OOK as

$$\Psi_{\tilde{s}}(f) = \frac{A^2T}{4}\left(\frac{\sin\pi fT}{\pi fT}\right)^2 + \frac{A^2}{4}\delta(f) \tag{8.33}$$

Note this is exactly the same as the PSD of the unipolar NRZ line codes (see (2.25) and Figure 2.3(b)).

The symbol error probability for coherent demodulation of OOK can be obtained from (8.28) or from (B.32), the BER expression for binary signaling. That is

$$P_b = Q\left(\sqrt{\frac{E_b}{N_o}}\right) \tag{8.34}$$

The bit error probability for noncoherent demodulation of OOK can be obtained from (8.31) be setting $M = 2$:

$$P_b = \frac{1}{2}\exp\left(-\frac{E_b}{2N_o}\right) + \frac{1}{2}Q\left(\sqrt{\frac{E_b}{N_o}}\right) \tag{8.35}$$

which was also given in [8].

When compared with BPSK, the PSD is the same except that it has a spectral line at f_c, which can be locked on by a phase lock loop to recover the carrier, but the BER performance of coherently-demodulated OOK is 3 dB inferior to that of BPSK. OOK is not usually preferred against BPSK unless noncoherent detection is necessary.

8.5 COMPARING MASK WITH MPSK

For the same symbol rate and the pulse shape, the MPSK and MASK have the same PSD and hence the same bandwidth. Thus it makes sense to make a comparison between their BER performances.

In comparison to MPSK, starting from $M = 4$, the error probability of uniformly-spaced bipolar MASK is inferior to that of the MPSK. In fact we can easily compare them. From the P_s of the MPSK (4.24) and that of MASK (8.20), the ratio (MASK over MPSK) of the arguments inside the square root sign of the Q-function is

$$R_M = \frac{3}{(M^2-1)\sin^2\frac{\pi}{M}} \tag{8.36}$$

M	2	4	8	16	32	64	128
$R(M)$	1	0.4	0.325	0.309	0.305	0.304	0.304
$10 \log R(M)$ (dB)	0	-3.98	-4.88	-5.10	-5.16	-5.17	-5.17

Table 8.2 Power penalty of bipolar MASK over MPSK.

It is tabulated in Table 8.2. If $M \geq 32$ the degradation is constantly 5.17 dB. This is easily seen from (8.36). For large M, $\sin\pi/M \cong \pi/M$, and $R_M \cong 3/\pi^2 \cong 0.304$. Another fact is that in both MASK and MPSK, for the same error probability, the power increase is 6 dB for doubling M for large M. This can be seen from (4.24) and (8.20). This means in terms of increasing bandwidth efficiency with increased M, both schemes pay the same penalty in BER performance. However, MASK still has a fixed 5.17 dB disadvantage against MPSK. This shows MASK is inferior to MPSK in terms of error probability. However, we will see that going to two-dimensional MASK, namely, QAM, improves error rate performance significantly. As a result, QAM is superior to MPSK when $M > 4$. QAM will be studied in the next chapter.

8.6 SUMMARY

ASK belongs to a broader class of modulations, the PAM, including baseband and passband PAM. Thus we first discussed PAM signal, its PSD, and its optimal detection in Section 8.1. The results are applicable to both baseband and passband PAM. For uncorrelated data, the PSD of PAM is determined by the Fourier transform of the symbol pulse shape. The optimum detector is a minimum-distance coherent detector. Next we turn our attention to our main subject of the chapter, the passband PAM or ASK. ASK can be made bipolar or unipolar. The amplitude spacing is usually made uniform for convenience of design and analysis. In Section 8.2, we discussed in detail the symmetrical uniformly spaced bipolar M-ary ASK. We derived its PSD expressions, modulator and demodulator, and BER expressions, where the demodulator must be coherent. In Section 8.3, we discussed in detail the uniformly spaced unipolar M-ary ASK. Again, we derived its PSD which includes a delta function at the zero frequency due to the dc value in the average value of its data sequence. Its demodulator can be coherent or noncoherent. The BER expressions for both cases were derived. In Section 8.4, OOK, the binary unipolar ASK was treated as a special case of the unipolar M-ary ASK. Its PSD expression and BER expression were given. Finally, we made a comparison between MASK and MPSK. While they have the same bandwidth efficiency, the BER performance of MPSK is much better. Asymptotically, the difference approaches 5.17 dB. Because of this, coherent MASK is not usually a choice of modulation. Only the noncoherent unipolar MASK can be of some practical use in systems where circuit simplicity is of prime concern, and power efficiency is not. The discussion of bipolar MASK primarily serves as a prerequisite for understanding QAM in the next chapter.

References

[1] Proakis, J., *Digital Communications*, 3rd ed., New York: McGraw Hill, 1989.

[2] Lee, A., and D. Messerschmett, *Digital Communication*, 2nd ed., Boston: Kluwer Academic Publishers, 1994.

[3] Couch II, L. W., *Digital and Analog Communication Systems*, 3rd ed., New York: Macmillan Publishing Company, 1990.

[4] Haykin, S., *Digital Communications*, New York: John Wiley & Sons, 1988.

[5] Xiong, F., "Amplitude Shift Keying," *Encyclopedia of RF and Microwave Engineering*, vol. 1, pp. 117–126, edited by K. Chang, Hoboken, New Jersey: John Wiley, 2005.

[6] Al-Dweik, A. J. and F. Xiong, "Frequency-hopped multiple-access communication with noncoherent OFDM-MASK in AWGN channels," *International Conference on Military Communications (MILCOM)* 2001, McLean, Virginia, October 28–31, 2001.

[7] Ziemer, R. E., and W. H. Tranter, *Principles of Communications*, Boston: Houghton Mifflin Company, 1985.

[8] Stremler, F. G., *Introduction to Communication Systems*, 3rd ed., Reading Massachusetts: Addison-Wesley, 1990.

Selected Bibliography

- Haykin, S., *Communication Systems*, 3rd ed., New York: John Wiley, 1994.
- Simon, K. M., S. M. Hinedi, and W. C. Lindsey, *Digital Communication Techniques, Signal Design, and Detection*, Englewood Cliffs, New Jersey: Prentice Hall, 1995.
- Smith, D. R., *Digital Transmission Systems*, 2nd ed., New York: Van Nostrand Reinhold, 1993.
- Whalen, A. D., *Detection of Signals in Noise*, New York: Academic Press, 1971.
- Ziemer, R. E., and R. L. Peterson, *Introduction to Digital Communication*, New York: Macmillan, 1992.

Chapter 9

Quadrature Amplitude Modulation

The majority of the passband modulation schemes we have studied in previous chapters are constant envelope schemes. The constant envelope property of these schemes is especially important to systems with power amplifiers which must operate in the nonlinear region of the input-output characteristic for maximum power efficiency, like the satellite transponders. For some other communication systems, constant envelope may not be a crucial requirement, whereas bandwidth efficiency is more important. Quadrature amplitude modulation (QAM) is a class of nonconstant envelope schemes that can achieve higher bandwidth efficiency than MPSK with the same average signal power. QAM is widely used in modems designed for telephone channels. The CCITT telephone circuit modem standards V.29 to V.33 are all based on various QAM schemes ranging from uncoded 16QAM to trellis coded 128QAM. The research of QAM applications in satellite systems, point-to-point wireless systems, and mobile cellular telephone systems also has been very active.

In Chapter 8 we studied bipolar M-ary ASK. MASK is usually no longer a preferable choice due to its poor power efficiency. However, since QAM signal consists of two MASK components and they can be demodulated in two separate channels, the knowledge of MASK will serve as a basis for developing QAM in this chapter.

In this chapter, we first define QAM signal and constellation in Section 9.1. Various QAM constellations are introduced in Section 9.2, but only the square QAM constellations are described in detail. QAM's PSD, modulator, demodulator, error probability, synchronization, and differential coding are discussed in Sections 9.3–9.8. Section 9.9 summarizes the chapter.

9.1 QAM SIGNAL DESCRIPTION

Having studied MASK in Chapter 8, we are ready to discuss QAM. A single attribute of the carrier is used in MASK (amplitude) and MPSK (phase) to convey the information. Naturally, the next step is to consider using both amplitude and phase modulations in a scheme. This leads to the concept of QAM.

QAM signal can be written as

$$s_i(t) = A_i \cos(2\pi f_c t + \theta_i), \quad i = 1, 2, ...M \tag{9.1}$$

where A_i is the amplitude and θ_i is the phase of the ith signal in the M-ary signal set. Pulse shaping is usually used to improve the spectrum and for ISI control purpose in QAM. With pulse shaping, QAM signal is

$$s_i(t) = A_i p(t) \cos(2\pi f_c t + \theta_i), \quad i = 1, 2, ...M \tag{9.2}$$

where $p(t)$ is a smooth pulse defined on $[0, T]$.[1] Equation (9.2) can be written as

$$s_i(t) = A_{i1} p(t) \cos 2\pi f_c t - A_{i2} p(t) \sin 2\pi f_c t \tag{9.3}$$

where

$$A_{i1} = A_i \cos \theta_i \tag{9.4}$$

$$A_{i2} = A_i \sin \theta_i \tag{9.5}$$

and

$$A_i = \sqrt{A_{i1}^2 + A_{i2}^2} \tag{9.6}$$

Similar to MPSK, QAM signal can be expressed as a linear combination of two orthonormal functions. Equation (9.3) can be written as

$$s_i(t) = s_{i1}\phi_1(t) + s_{i2}\phi_2(t) \tag{9.7}$$

where

$$\phi_1(t) = \sqrt{\frac{2}{E_p}} p(t) \cos 2\pi f_c t, \quad 0 \le t \le T \tag{9.8}$$

[1] Even if pulse shaping is not desired, there is still inevitably pulse shaping due to the limited bandwidth of the system. In fact, deliberate pulse shaping is usually achieved through filtering. That is to make $P(f) = H_T(f)H_C(f)H_R(f)$ or equivalently $p(t) = h_T(t) * h_C(t) * h_R(t)$, where $h_T(t)$, $h_C(t)$ and $h_R(t)$ are the impulse responses of the transmitter filter, channel, and receiver filter, respectively. $H_T(f)$, $H_C(f)$, and $H_R(f)$ are their transfer functions. A common choice of $P(f)$ is the raised-cosine, whose time domain function $p(t)$ has zero values at sampling instants except for at $t = 0$. Thus $p(t)$ incurs no ISI. However, the raised-cosine response is noncausal, only an approximate delayed version is realizable. See [1, pp. 100–102].

$$\phi_2(t) = -\sqrt{\frac{2}{E_p}}p(t)\sin 2\pi f_c t, \quad 0 \le t \le T \tag{9.9}$$

and

$$s_{i1} = \sqrt{\frac{E_p}{2}}A_{i1} = \sqrt{\frac{E_p}{2}}A_i \cos\theta_i \tag{9.10}$$

$$s_{i2} = \sqrt{\frac{E_p}{2}}A_{i2} = \sqrt{\frac{E_p}{2}}A_i \sin\theta_i \tag{9.11}$$

where E_p is the energy of $p(t)$ in $[0, T]$. That is $E_p = \int_0^T p^2(t)dt$. The factor $\sqrt{2/E_p}$ is to normalize the basis functions $\phi_1(t)$ and $\phi_2(t)$.

It can be easily verified that the basis functions $\phi_1(t)$ and $\phi_2(t)$ are virtually orthonormal for $f_c >> 1/T$. When $f_c >> 1/T$, $p(t)$ is a slow-varying envelope. First they are virtually normalized since

$$\begin{aligned} \int_0^T \phi_1^2(t)dt &= \frac{2}{E_p}\int_0^T p^2(t)\cos^2 2\pi f_c t dt \\ &= \frac{1}{E_p}\int_0^T p^2(t)[1+\cos 4\pi f_c t]dt \\ &\cong 1, \quad \text{for } f_c >> 1/T \end{aligned}$$

The same is true for $\phi_2(t)$. Second, they are virtually orthogonal since

$$\begin{aligned} \int_0^T \phi_1(t)\phi_2(t)dt &= -\frac{2}{E_p}\int_0^T p^2(t)\cos 2\pi f_c t \sin 2\pi f_c t dt \\ &= -\frac{2}{E_p}\int_0^T p^2(t)\sin 4\pi f_c t dt \\ &\cong 0, \quad \text{for } f_c >> 1/T \end{aligned}$$

Thus for most practical cases, $\phi_1(t)$ and $\phi_2(t)$ are orthonormal. When there is no pulse shaping, that is, $p(t) = 1$ in $[0, T]$, $E_p = T$. Then (9.8) and (9.9) have the same forms of (4.2) and (4.3). They are precisely orthonormal.

The energy of the ith signal is

$$E_i = \int_0^T s_i^2(t)dt \cong \frac{1}{2}A_i^2 E_p \tag{9.12}$$

and the average signal energy is

$$E_{avg} = \frac{1}{2} E_p \cdot E\{A_i^2\} \tag{9.13}$$

The average power is

$$P_{avg} = \frac{E_{avg}}{T} \tag{9.14}$$

The average amplitude is

$$A_{avg} = \sqrt{2P_{avg}} \tag{9.15}$$

Similar to MPSK, a geometric representation called constellation is a very clear way of describing a QAM signal set. The horizontal axis of the constellation plane is $\phi_1(t)$ and the vertical axis is $\phi_2(t)$. A QAM signal is represented by a point (or vector, or phasor) with coordinates (s_{i1}, s_{i2}). Alternatively, the two axes can be simply chosen as $p(t)\cos 2\pi f_c t$ and $-p(t)\sin 2\pi f_c t$. Then the signal coordinates are (A_{i1}, A_{i2}). The two axes sometimes are simply labeled as I-axis and Q-axis, and sometimes are even left unlabeled. Figure 9.1 shows examples of three types of QAM constellations.

Now let us examine the properties of the QAM constellation. Assuming the axes are $\phi_1(t)$ and $\phi_2(t)$, then each signal is represented by the phasor

$$\mathbf{s}_i = (s_{i1}, s_{i2})$$

The magnitude of the phasor is

$$||\mathbf{s}_i|| = \sqrt{s_{i1}^2 + s_{i2}^2} = \sqrt{E_i} \tag{9.16}$$

which is related to the signal amplitude by (from (9.12) and (9.16))

$$A_i = \sqrt{\frac{2}{E_p}}||\mathbf{s}_i|| \tag{9.17}$$

The average energy is

$$E_{avg} = E\{E_i\} = E\{||\mathbf{s}_i||^2\} \tag{9.18}$$

The phase θ_i is the angle of the corresponding phasor

$$\theta_i = \tan^{-1}\frac{s_{i2}}{s_{i1}} \tag{9.19}$$

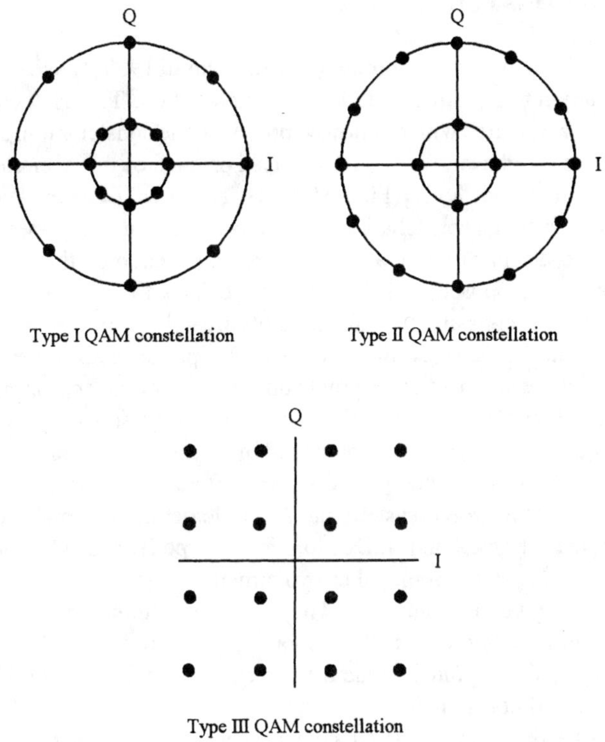

Figure 9.1 Examples of type I, II, and III QAM constellations. From [2]. Copyright © 1994 IEEE.

The distance between any pair of phasors is

$$\begin{aligned} d_{ij} &= \sqrt{|\mathbf{s}_i - \mathbf{s}_j|^2} \\ &= \sqrt{(s_{i1} - s_{j1})^2 + (s_{i2} - s_{j2})^2}, \qquad i, j = 1, 2, ...M \end{aligned} \tag{9.20}$$

Depending what values (s_{i1}, s_{i2}) or (A_i, θ_i) are assigned with, a variety of QAM constellations can be realized.

9.2 QAM CONSTELLATIONS

The first QAM scheme was proposed by C. R. Cahn in 1960 [3]. He simply extended phase modulation to a multiamplitude phase modulation. That is, there is more than one amplitude associated with an allowed phase. In the constellation, a fixed number of signal points (or phasors) are equally spaced on each of the N circles, where N is the number of amplitude levels (Figure 9.1(a)). This is called a type I constellation in the literature. In a type I constellation, the points on the inner ring are closest together in distance and are most vulnerable to errors. To overcome this problem, type II constellation was proposed by Hancock and Lucky a few months later [4] (Figure 9.1(b)). In a type II constellation, signal points are still on circles, but the number of points on the inner circle is less than the number of points on the outer circle, making the distance between two adjacent points on the inner circle approximately equal to that on the outer circle. The type III constellation is the square QAM constellation shown in Figure 9.1(c), which was proposed by Campopiano and Glazer in 1962 [5]. Their analysis showed that the type III system offered a very small improvement in performance over the type II system, but its implementation would be considerably simpler than that of type I and II. Due to this, the type III constellation has been the most widely used system. Some other two dimensional constellations considered in the literature are given in Figure 9.2. The circular constellations are denoted by the notation $(n_1, n_2, ...)$ where n_1 is the number of signal points on the inner circle, n_2 is the number of signal points on the next circle, and so on. Figure 9.2 contains the type II and type III constellations.

When designing a constellation, consideration must be given to:

1. The minimum Euclidean distance $d_{\min}$ among the phasors (signal points). It should be as large as possible under other constraints, since it determines the symbol error probability of the modulation scheme.
2. The phase differences among the phasors. It should be as large as possible under other constraints, since it determines the phase jitter immunity and hence the scheme's resilience against the carrier- and clock-recovery imperfections and channel phase rotations.
3. The average power of the phasors. It should be as small as possible under other constraints.
4. The ratio of the peak-to-average phasor power, which is a measure of robustness against nonlinear distortion caused by the power amplifier. It should be as close to unity as possible under other constraints.
5. The implementation complexity.
6. Other properties, such as resilience against fading.

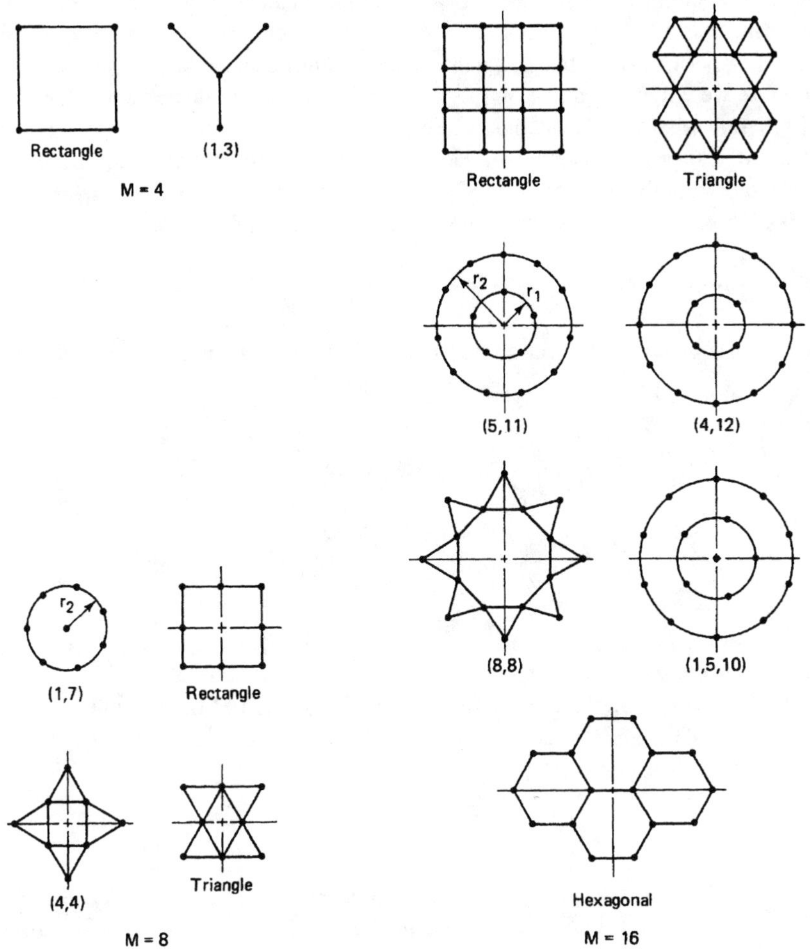

Figure 9.2 Various QAM constellations. From [6]. Copyright © 1974 IEEE.

Research results have shown that the square constellation (type III) is the most appropriate choice in AWGN channels. It can be easily generated as two MASK signals impressed on two phase-quadrature carriers. It can be easily demodulated to yield two quadrature components. Each component can be individually detected by comparing it to a set of thresholds. A few of the other constellations offer slightly better error performance, but with a much more complicated system implementation. Therefore we will concentrate on the square constellation in this chapter. The type I constellation (also called star constellation) is not optimum in terms of $d_{\min}$ under the constraint of average phasor power. However, it allows efficient differential encoding and decoding methods to be used. This makes it suitable for fading channels. Its application in fading channels will be covered in Chapter 11.

9.2.1 Square QAM

For M-ary square QAM signals, (9.3) and (9.7) can be written in the following form

$$\begin{aligned} s_i(t) &= I_i\sqrt{\frac{E_0}{E_p}}p(t)\cos 2\pi f_c t - Q_i\sqrt{\frac{E_0}{E_p}}p(t)\sin 2\pi f_c t \\ &= I_i\sqrt{\frac{E_0}{2}}\phi_1(t) + Q_i\sqrt{\frac{E_0}{2}}\phi_2(t) \end{aligned} \quad (9.21)$$

where E_0 is the energy of the signal with the lowest amplitude, and (I_i, Q_i) are a pair of independent integers which determine the location of the signal point in the constellation. The minimum values of (I_i, Q_i) are $(\pm 1, \pm 1)$. The pair (I_i, Q_i) is an element of the $L \times L$ matrix:

$$[I_i, Q_i] = \begin{bmatrix} (-L+1, L-1) & (-L+3, L-1) & \cdots & (L-1, L-1) \\ (-L+1, L-3) & (-L+3, L-3) & \cdots & (L-1, L-3) \\ \vdots & \vdots & & \vdots \\ (-L+1, -L+1) & (-L+3, -L+1) & \cdots & (L-1, -L+1) \end{bmatrix} \quad (9.22)$$

where

$$L = \sqrt{M}, \quad M = 4^n, \quad n = 1, 2, 3, ...$$

For example, for the 16QAM in Figure 9.3, where $L = 4$, the matrix is

$$[I_i, Q_i] = \begin{bmatrix} (-3,3) & (-1,3) & (1,3) & (3,3) \\ (-3,1) & (-1,1) & (1,1) & (3,1) \\ (-3,-1) & (-1,-1) & (1,-1) & (3,-1) \\ (-3,-3) & (-1,-3) & (1,-3) & (3,-3) \end{bmatrix} \quad (9.23)$$

When $M = 2^n$ but not 4^n, L is not an integer, we cannot use the matrix (9.22) directly to define the QAM. However, we may use a modified matrix to define the QAM. For example, the 32QAM can be defined by a 6×6 matrix without the four elements on the four corners.

The constellation can be conveniently expressed in terms of (I_i, Q_i). The phasors for the square QAM are

$$\mathbf{s}_i = (I_i\sqrt{\frac{E_0}{2}}, Q_i\sqrt{\frac{E_0}{2}}) \qquad i = 1, 2, ...M$$

The magnitude of a phasor is

$$||\mathbf{s}_i|| = \sqrt{\frac{E_0}{2}(I_i^2 + Q_i^2)}$$

The QAM signal in (9.21) can also be written as

$$s_i(t) = A_i \cos(2\pi f_c t + \theta_i)$$

where the amplitude is

$$A_i = \sqrt{\frac{E_0}{E_p}(I_i^2 + Q_i^2)} = \sqrt{\frac{2}{E_p}}||\mathbf{s}_i||$$

The phase θ_i is the angle of the corresponding phasor

$$\theta_i = \tan^{-1}\frac{Q_i}{I_i}$$

The distance between any pair of phasors is

$$\begin{aligned} d_{ij} &= \sqrt{|\mathbf{s}_i - \mathbf{s}_j|^2} \\ &= \sqrt{\frac{E_0}{2}[(I_i - I_j)^2 + (Q_i - Q_j)^2]}, \qquad i, j = 1, 2, ...M \end{aligned}$$

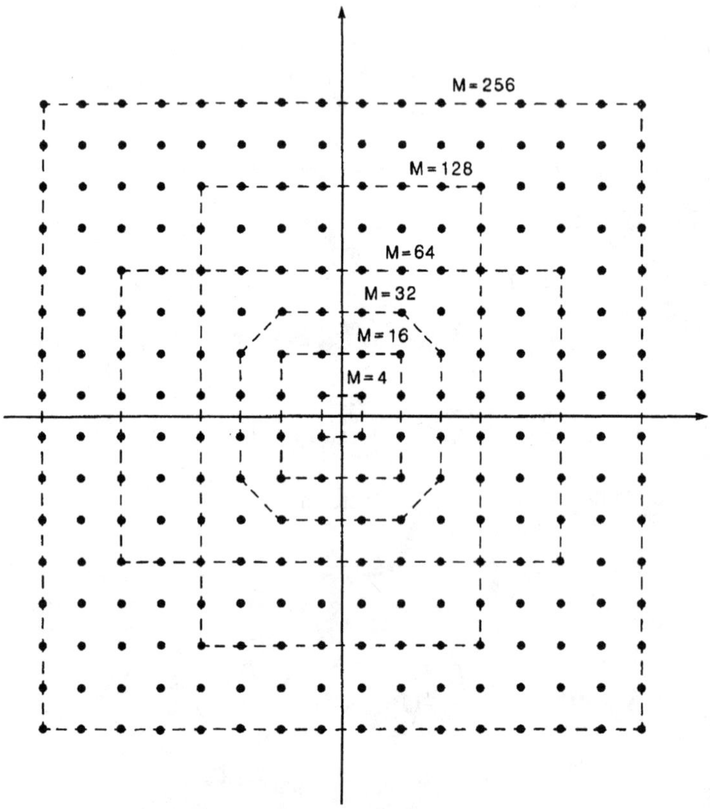

Figure 9.3 Square QAM constellations. From [7, p. 224]. Copyright © 1987 Prentice Hall.

The average energy is

$$E_{avg} = E\{\frac{E_0}{2}(I_i^2 + Q_i^2)\} = \frac{E_0}{2}[E\{I_i^2\} + E\{Q_i^2\}] = E_0 E\{I_i^2\}$$

and the average power is

$$P_{avg} = \frac{E_0}{T} E\{I_i^2\}$$

where for the strict square ($L \times L$) QAM

$$\begin{aligned} E\{I_i^2\} &= \frac{1}{L}[(-(L-1))^2 + (-(L-3))^2 + \cdots + (L-3)^2 + (L-1)^2] \\ &= \frac{2}{L}[1^2 + 3^2 + \cdots + (L-1)^2] \\ &= \frac{1}{3}(L^2 - 1) = \frac{1}{3}(M-1) \end{aligned} \quad (9.24)$$

Thus

$$P_{avg} = \frac{E_0}{3T}(M-1) = P_0 \frac{1}{3}(M-1) \quad (9.25)$$

where P_0 is the power of the smallest signal.

9.3 POWER SPECTRAL DENSITY

Using the PSD formula for quadrature modulation (see (A.21)), the PSD of the square QAM can be computed as follows.

In order to include the most general case, we consider QAM with pulse shaping.

$$s_i(t) = A_i p(t) \cos(2\pi f_c t + \theta_i), \quad i = 1, 2, ...M$$

On the entire time axis, the QAM signal can be written as

$$s(t) = \text{Re}\left\{\left[\sum_{k=-\infty}^{\infty} A_k \exp(j\theta_k) p(t - kT)\right] \exp(j2\pi f_c t)\right\}, \quad -\infty < t < \infty \quad (9.26)$$

The complex envelope of the QAM signal is

$$\widetilde{s}(t) = \sum_{k=-\infty}^{\infty} A_k \exp(j\theta_k) p(t - kT)$$

$$= \sum_{k=-\infty}^{\infty} A_k \cos\theta_k p(t-kT) + j \sum_{k=-\infty}^{\infty} A_k \sin\theta_k p(t-kT) \quad (9.27)$$

$$= \sum_{k=-\infty}^{\infty} A_{k1} p(t-kT) + j \sum_{k=-\infty}^{\infty} A_{k2} p(t-kT)$$

where

$$A_{k1} = A_k \cos\theta_k$$

$$A_{k2} = A_k \sin\theta_k$$

These are random variables with equal probability for each value. They have zero means for symmetrical constellations (Figures 9.1–9.3). The variances of them depend on the constellation shape.

$$\sigma_1^2 = E\{A_{k1}^2\}$$

$$\sigma_2^2 = E\{A_{k2}^2\}$$

From (9.18), the average power of the signal is

$$P_{avg} = \frac{1}{2T} E_p \cdot E\{A_i^2\} = \frac{1}{2T} E_p(\sigma_1^2 + \sigma_2^2)$$

Now we use the PSD formula (A.21) to compute the PSD of QAM. We rewrite it in the following

$$\Psi_{\tilde{s}}(f) = \frac{\sigma_x^2 |P(f)|^2}{T} + \frac{\sigma_y^2 |Q(f)|^2}{T}$$

where $P(f)$ and $Q(f)$ are the spectra of the I-channel signal and Q-channel signal pulse shape, respectively. This is a very general formula, applicable to any quadrature modulated signal. For QAM schemes $P(f) = Q(f)$, $\sigma_x^2 = \sigma_1^2$, and $\sigma_y^2 = \sigma_2^2$, we have

$$\Psi_{\tilde{s}}(f) = \frac{|P(f)|^2}{T}(\sigma_1^2 + \sigma_2^2) = \frac{2P_{avg}}{E_p}|P(f)|^2 \quad (9.28)$$

This equation tells us that the shape of the PSD of a QAM scheme is determined by the baseband pulse shape, and the magnitude of the PSD is determined by the average power (or average amplitude) of the QAM signal set. It is also worthwhile to point out that the shape of the PSD of a QAM scheme is independent of the constellation. In other words, no matter what the constellation is, be it square, circular, or others, the PSD shape is the same as long as the $p(t)$ is the same, the PSD magnitude is also the same as long as the average signal power is also the same.

Without particular pulse shaping, $p(t)$ is just a rectangular pulse with unit amplitude. Then $E_p = T$ and

$$|P(f)| = \left| T \frac{\sin \pi f T}{\pi f T} \right|$$

Therefore

$$\begin{aligned} \Psi_{\tilde{s}}(f) &= 2P_{avg}T \left(\frac{\sin \pi f T}{\pi f T} \right)^2 \\ &= A_{avg}^2 T \left(\frac{\sin \pi f T}{\pi f T} \right)^2 \\ &= A_{avg}^2 n T_b \left(\frac{\sin \pi f n T_b}{\pi f n T_b} \right)^2 \end{aligned} \tag{9.29}$$

where $n = \log_2 M$ and $T_b = T/n$ is the bit period. This PSD has the same shape of the PSD of MPSK (see (4.26)). The only difference lies in the magnitude. In the MPSK case, the PSD magnitude depends on the signal amplitude because there is only one signal amplitude. In the QAM case, the PSD magnitude depends on the average signal amplitude. Thus, the PSD curves for MPSK in Figure 4.15 are also applicable to QAM schemes as long as the average amplitude of the QAM is used. For example, for M-ary square QAM, from (9.29) and (9.25) we have

$$\begin{aligned} \Psi_{\tilde{s}}(f) &= \frac{2E_0}{3}(M-1) \left(\frac{\sin \pi f T}{\pi f T} \right)^2 \\ &= \frac{2E_0}{3}(M-1) \left(\frac{\sin \pi f n T_b}{\pi f n T_b} \right)^2 \end{aligned} \tag{9.30}$$

9.4 MODULATOR

The QAM modulator is almost identical to that of MPSK since both of them are quadrature schemes. We can write the QAM signal as

$$s(t) = s_1(t) \cos 2\pi f_c t - s_2(t) \sin 2\pi f_c t, \quad -\infty < t < \infty \tag{9.31}$$

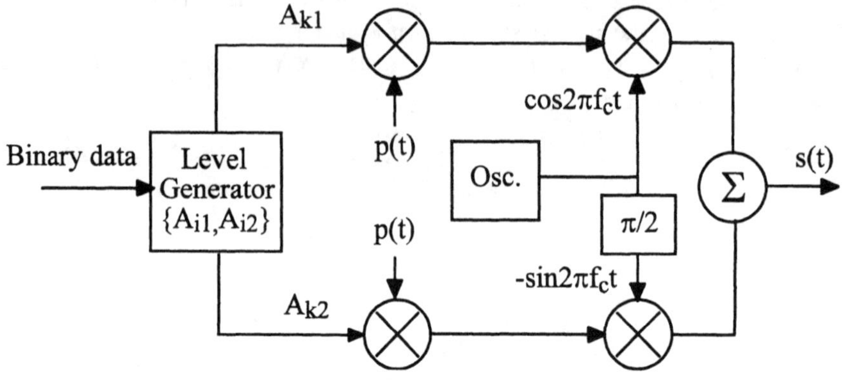

Figure 9.4 QAM modulator.

where

$$s_1(t) = \sum_{k=-\infty}^{\infty} A_{k1} p(t - kT)$$

$$s_2(t) = \sum_{k=-\infty}^{\infty} A_{k2} p(t - kT)$$

The modulator derived directly from (9.31) is shown in Figure 9.4. If pulse shaping is not desired, the $p(t)$ block will be absent. The data bit sequence is divided into n-tuples of n bits. There are $M = 2^n$ distinct n-tuples. Each n-tuple of the input bits is used to control the level generator. The level generator provides the I- and Q-channel the particular sign and level for a signal's horizontal and vertical coordinates (A_{k1}, A_{k2}), respectively. The mapping from n-tuples to QAM points are usually Gray coded for minimizing bit errors. For square QAM, perfect Gray coding is possible. Figure 9.5 is a Gray coded square 16QAM constellation. For some constellations, such as a circular QAM with four points on the inner ring and eight on the outer, it is not possible to have perfect Gray coding.

Digital synthesis techniques can be used to generate QAM signals. Each signal in the constellation can be stored as a set of samples and the data n-tuple is used as the address to obtain the samples. The samples are fed to a D/A converter whose output is the desired QAM signal.

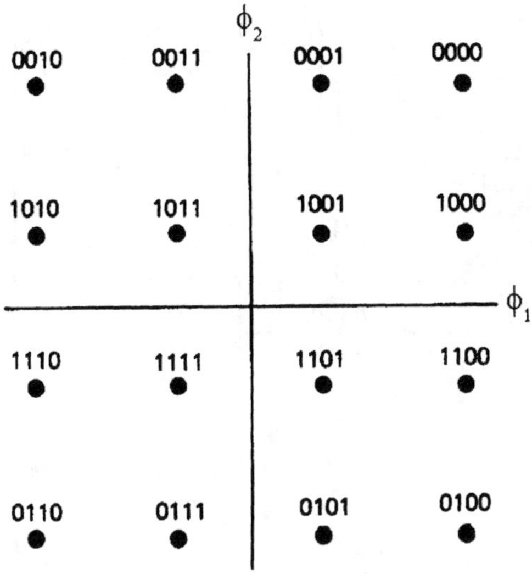

Figure 9.5 Gray coded square 16QAM constellation. The signal points are labeled with 4-bit Gray codes.

9.5 DEMODULATOR

Similar to MPSK, the coherent demodulation of QAM could be implemented by one of the coherent detectors for M-ary signals as described in Appendix B. Since the QAM signal set has only two basis functions, the simplest receiver is the one that uses two correlators (Figure B.8 with $N = 2$). Due to the special characteristic of the QAM signal, the general demodulator of Figure B.8 can be simplified.

The received signal is

$$r(t) = s_i(t) + n(t)$$

According to (B.37) in Appendix B, for QAM signal detection the sufficient statistic is the (squared) distance[2]

$$l_i = (r_1 - s_{i1})^2 + (r_2 - s_{i2})^2 \tag{9.32}$$

---

[2] Note that this is different from the sufficient statistic of MPSK. In the case of MPSK, each signal has the same energy, thus (B.38) with B_j dropped instead of (B.37) is used, which leads to a decision rule that compares only angles not distances.

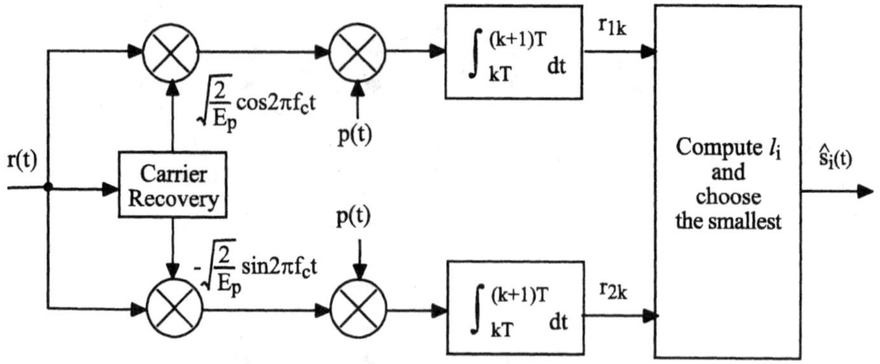

Figure 9.6 Coherent demodulator for QAM.

where

$$r_1 \triangleq \int_0^T r(t)\phi_1(t)dt = s_{i1} + n_1$$

$$r_2 \triangleq \int_0^T r(t)\phi_2(t)dt = s_{i2} + n_2$$

are independent Gaussian random variables with mean values s_{i1} and s_{i2}, respectively. Their variance is $N_o/2$. The pair (r_1, r_2) determines a point in the QAM constellation plane, representing the received noisy signal. The detector compares the distances from (r_1, r_2) to all pairs of (s_{i1}, s_{i2}) and chooses the closest one.

Figure 9.6 is the demodulator based on the above decision rule where subscript k indicates the kth symbol period. Note that the amplitude of the reference signals can be any value, which is $\sqrt{2/E_p}$ in the figure, as long as (s_{i1}, s_{i2}) are also scaled accordingly. As we have shown in Section 8.2.2, the integrators may be replaced by matched filters whose outputs are sampled at $t = (k+1)T$ (Figure 9.7). The filter impulse responses match to the shaping pulse $p(t)$. For the square QAM, the r_{1k} and r_{2k} can be detected separately by two multithreshold detectors to yield s_{i1} and s_{i2}, and then signal $s_i(t)$ can be determined (Figure 9.8).

9.6 ERROR PROBABILITY

For square QAM constellations with $M = 2^k$ where k is even, the QAM constella-

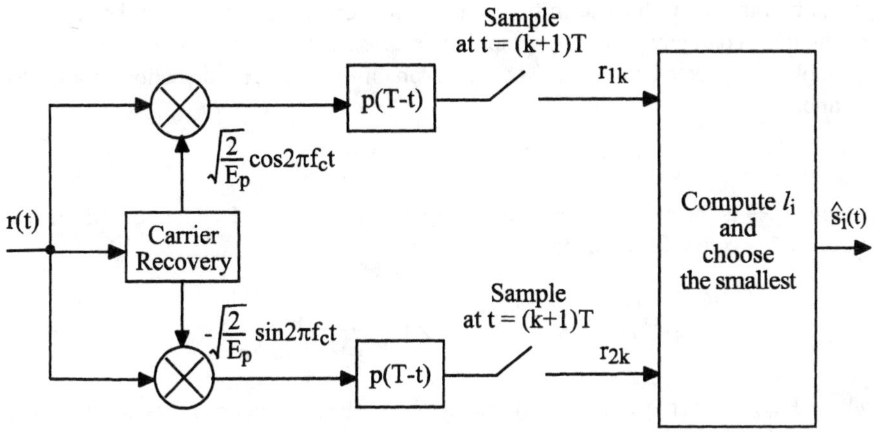

Figure 9.7 Coherent QAM demodulator using matched filters.

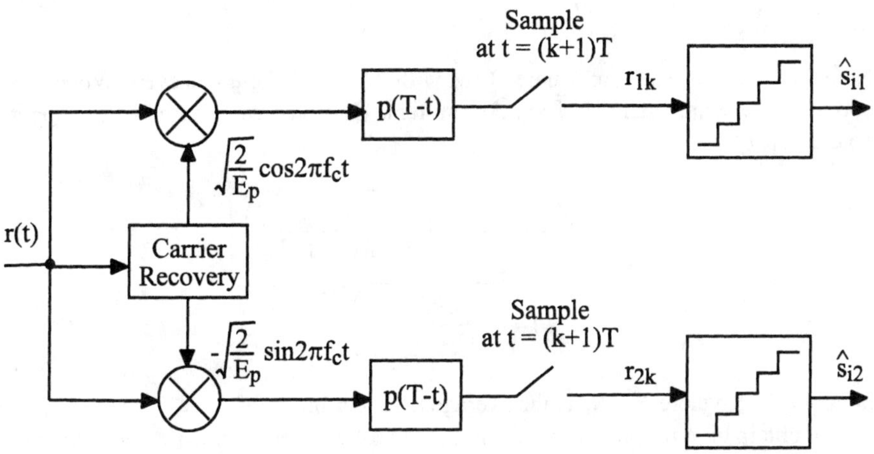

Figure 9.8 Coherent demodulator for square QAM using threshold detectors. I-channel and Q-channel are demodulated separately.

tion is equivalent to two MASK signals on quadrature carriers, each having $L = \sqrt{M}$ signal points. As we have seen from the last section, each MASK signal can be demodulated separately. A QAM symbol is detected correctly only when two MASK symbols are detected correctly. Thus the probability of correct detection of a QAM symbol is

$$P_c = (1 - P_{\sqrt{M}})^2$$

where $P_{\sqrt{M}}$ is the symbol error probability of a $\sqrt{M}$-ary AM with one-half the average power of the QAM signal. From (8.20) we have

$$P_{\sqrt{M}} = \frac{2(\sqrt{M}-1)}{\sqrt{M}} Q\left(\sqrt{\frac{3E_{avg}}{(M-1)N_o}}\right) \tag{9.33}$$

where E_{avg}/N_o is the average SNR per symbol. The symbol error probability of the square QAM is

$$P_s = 1 - (1 - P_{\sqrt{M}})^2 = 2P_{\sqrt{M}} - P_{\sqrt{M}}^2 \tag{9.34}$$

At high SNR,

$$P_s \cong 2P_{\sqrt{M}} = \frac{4(\sqrt{M}-1)}{\sqrt{M}} Q\left(\sqrt{\frac{3E_{avg}}{(M-1)N_o}}\right) \tag{9.35}$$

Note that (9.34) is exact for square QAM with $M = 2^k$ where k is even. When k is odd there is no equivalent $\sqrt{M}$-ary AM system. However, we can find a tight upper bound [8, p. 655]

$$\begin{aligned} P_s &\leq 1 - \left[1 - 2Q\left(\sqrt{\frac{3E_{avg}}{(M-1)N_o}}\right)\right]^2 \\ &\leq 4Q\left(\sqrt{\frac{3kE_b}{(M-1)N_o}}\right) \end{aligned} \tag{9.36}$$

for any $k \geq 1$, where E_b/N_o is the average SNR per bit.

To obtain bit error probability from the symbol error probability, we observe that square QAM can be perfectly Gray coded. That is, there is only one bit difference between adjacent symbols. Each symbol error most likely causes one bit error at large SNR. Thus

$$P_b \cong \frac{P_s}{\log_2 M} \tag{9.37}$$

M	R_M	$10\log R_M$ (dB)
4	1	0
8	1.463	1.65
16	2.627	4.20
32	5.036	7.02
64	9.889	9.95
128	19.61	12.92
256	39.06	15.92

Table 9.1 Power savings of QAM over MPSK.

A more accurate expression for direct computation of BER of Gray coded square QAM is given in [9] as

$$P_b \cong \frac{4}{\log_2 M}\left(1-\frac{1}{\sqrt{M}}\right)\sum_{i=1}^{\sqrt{M}/2} Q\left((2i-1)\sqrt{\frac{3E_b\log_2 M}{(M-1)N_o}}\right)$$

It was found by simulations that this expression is very accurate when the first two terms in the sum are included. If all terms are taken, this expression gives an upper bound. If one term is taken, it becomes the expression given by (9.35) and (9.37), which is in fact a less accurate lower bound.

Figure 9.9 shows the P_b curves for $M = 4, 8, 16, 32, 64, 128,$ and 256, using the accurate P_b expression above with 2 terms in the sum are taken for $M \geq 8$.

In the following we compare QAM with MPSK. From the P_s of MPSK (4.24) and (9.35) or (9.36), the ratio (QAM over MPSK) of the arguments inside the square root sign of the Q-function is

$$R_M = \frac{3}{2(M-1)\sin^2 \pi/M} \tag{9.38}$$

This reflects the ratio of the signal power. This ratio is tabulated in Table 9.1.

From the table we can see that for $M > 4$, the QAM is superior to MPSK. Furthermore, for large $M(\geq 32)$ the power savings increases 3 dB for doubling the number of signal points. This can be explained from (9.38) by observing that $R_M \cong 3M/2\pi^2$ for large M. This can also be looked at from another point of view. Examining (9.35) or (9.36) reveals that for large M, doubling M incurs a 3 dB penalty in signal power, whereas the penalty for MPSK and MASK is 6 dB as we pointed out before. That is why Table 9.1 shows a 3 dB increase in power savings for doubling M when QAM is compared with MPSK (and it is also true compared with MASK).

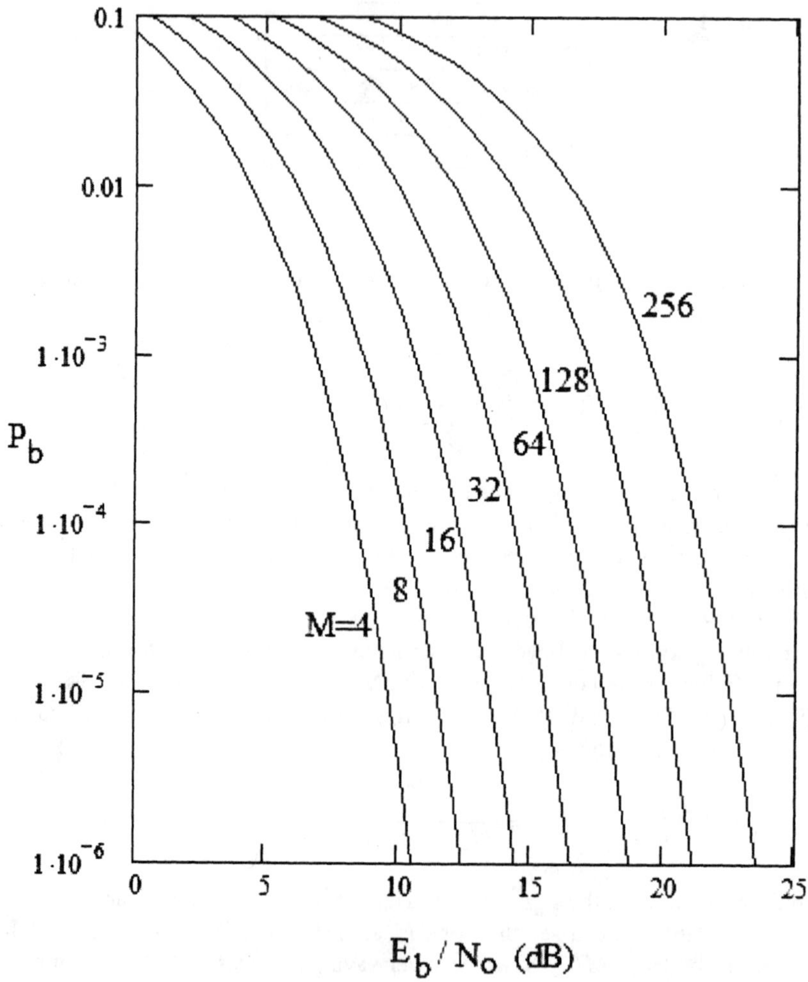

Figure 9.9 Square QAM bit error probability.

9.7 SYNCHRONIZATION

Clock synchronization for QAM is usually achieved by a clock recovery circuit which extracts the clock from the demodulated signal or uses the demodulated signal to control the local oscillator. Clock recovery techniques for QAM are the same as those for MPSK. The open-loop clock recovery circuits in Figure 4.37 and the early/late-gate clock recovery circuit in Figure 4.38 are all applicable to QAM. Refer to Chapter 4 for details.

Carrier synchronization is always necessary for square QAM constellations, even when differential coding is used. This is because differential coding for the square QAM is only for some of the bits in a symbol. Even though these bits can be determined by comparing two consecutive symbols (differential demodulation), the rest of the bits still must be determined by coherent demodulation. Thus the entire symbol might as well be coherently demodulated. For a description of differential coding for QAM, refer to the next section. Circular QAM constellations do not require carrier synchronization if differential encoding is used. We will discuss this in Chapter 11.

Carrier synchronization can be achieved by the pilot-tone technique or a separate synchronization channel, which requires extra bandwidth, as we mentioned in Chapter 4. Again we will not elaborate on this. Instead we focus on the carrier recovery techniques. There are two major types of carrier recovery techniques for QAM. One is the fourth-power loop (or times-four loop) and another is the decision-directed carrier recovery (DDCR). According to [2, p. 182], the decision-directed carrier recovery technique is one of the most popular carrier recovery schemes used in fixed-link QAM systems. The fourth-power loop carrier recovery has been suggested for digital radio in fading environment [10]. In the following we first discuss the fourth-power loop technique and then the decision-directed carrier recovery technique.

Recall that the Mth-power loop (Figure 4.35) is used for MPSK, but for M-ary QAM, the Mth-power loop is not necessary, only the fourth-power loop is needed. Unfortunately the squaring loop does not work for symmetrical QAM since the average energy at $2f_c$ is zero. It is also known that for M greater than four, the data pattern effects lead to an increased amount of carrier phase jitter in an Mth-power loop [10]. Thus the fourth-power loop is the right choice for QAM.

We show below that the squaring loop does not work for symmetrical QAM, but the fourth-power loop does. From Section 9.4 we see that A_{i1} and A_{i2} are generated by level generators using the input data which are random. Therefore A_{i1} and A_{i2} are random. We assume that their values are symmetrical about zero, that is, they have zero means. We further assume that the input data are ergodic random processes. Consequently A_{i1} and A_{i2} are ergodic random processes so that the time average is equal to the statistical average. All the above assumptions are the usual case in digital communications. We square the QAM signal given in (9.3) and take the average over all possible i to obtain

$$\begin{aligned} E\{s_i^2(t)\} &= E\{p^2(t)[A_{i1}^2\cos^2\omega_c t - 2A_{i1}A_{i2}\cos\omega_c t\sin\omega_c t + A_{i2}^2\sin^2\omega_c t]\} \\ &= E\{p^2(t)[A_{i1}^2\cos^2\omega_c t + A_{i2}^2\sin^2\omega_c t]\} \\ &= \frac{1}{2}p^2(t)\left[(E\{A_{i1}^2\} + E\{A_{i2}^2\}) + (E\{A_{i1}^2\} - E\{A_{i2}^2\})\cos 2\omega_c t\right] \end{aligned}$$

where $\omega_c = 2\pi f_c$ and the middle term in the first expression vanishes since A_{i1} and A_{i2} are independent zero-mean random processes so that

$$E\{A_{i1}A_{i2}\} = E\{A_{i1}\}E\{A_{i2}\} = 0$$

For symmetrical QAM, $E\{A_{i1}^2\} = E\{A_{i2}^2\}$, the above becomes

$$E\{s_i^2(t)\} = p^2(t)E\{A_{i1}^2\}$$

which has no periodical component at f_c or its multiples. We can intentionally unbalance the QAM constellation to make carrier recovery possible, but this leads to inefficient signal constellations. A fourth-power nonlinearity can produce a nonzero component at $4f_c$ even if the QAM is symmetrical. Using trigonometrical identities we can show that (see Appendix 9A)

$$\begin{aligned} E\{s_i^4(t)\} &= \frac{1}{8}p^4(t)[E\{A_{i1}^4\}(3 + 4\cos 2\omega_c t + \cos 4\omega_c t) \\ &\quad + E\{A_{i2}^4\}(3 - 4\cos 2\omega_c t + \cos 4\omega_c t)] \\ &\quad + \frac{3}{4}p^4(t)E\{A_{i1}^2\}E\{A_{i2}^2\}(1 - \cos 4\omega_c t) \end{aligned} \tag{9.39}$$

For symmetrical QAM, since $E\{A_{i1}^2\} = E\{A_{i2}^2\}$, we have

$$\begin{aligned} E\{s_i^4(t)\} &= \frac{3}{4}p^4(t)\left[E\{A_{i1}^4\} + (E\{A_{i1}^2\})^2\right] \\ &\quad + \frac{1}{4}p^4(t)\left[E\{A_{i1}^4\} - 3(E\{A_{i1}^2\})^2\right]\cos 4\omega_c t \end{aligned} \tag{9.40}$$

There is a nonzero component at $4f_c$.

For the square QAM, $A_{i1} = I_i\sqrt{E_0/E_p}$, $A_{i2} = Q_i\sqrt{E_0/E_p}$ (see (9.21)).

$$E\{A_{i1}^2\} = \frac{E_0}{E_p}\frac{1}{3}(M-1) \tag{9.41}$$

where the result of (9.24) is used, and

$$E\{A_{i1}^4\} = \left(\frac{E_0}{E_p}\right)^2 E\{I_i^4\}$$

where

$$\begin{aligned} E\{I_i^4\} &= \frac{1}{L}[(-(L-1))^4 + (-(L-3))^4 + \cdots + (L-3)^4 + (L-1)^4] \\ &= \frac{2}{L}[1^4 + 3^4 + \cdots + (L-1)^4] \\ &= \frac{1}{15}(3L^4 - 10L^2 + 7) = \frac{1}{15}(3M^2 - 10M + 7) \end{aligned} \quad (9.42)$$

Substitute (9.41) and (9.42) into (9.40), we have

$$E\{s_i^4(t)\} = \frac{1}{30}p^4(t)\left(\frac{E_0}{E_p}\right)^2 [(7M^2 - 20M + 13) - (M^2 - 1)\cos 4\omega_c t] \quad (9.43)$$

From (9.43) we can see that the $4f_c$ component always has a phase of π for the square QAM.

For the square 16QAM we have

$$E\{s_i^4(t)\} = \frac{1}{2}p^4(t)\left(\frac{E_0}{E_p}\right)^2 (99 - 17\cos 4\omega_c t)$$

If $p(t) = 1$, then $E_p = T$, then

$$E\{s_i^4(t)\} = \frac{1}{2}\left(\frac{E_0}{T}\right)^2 (99 - 17\cos 4\omega_c t)$$

This shows that the $4f_c$ component has a phase of π and an amplitude of $17/99 \cong 17.2\%$ of the dc component. In fact we can calculate this ratio for a series of M, as listed in Table 9.2.

$$R_M = \frac{M^2 - 1}{7M^2 - 20M + 13} \quad (9.44)$$

Note that the above results (Table 9.2) are for ideal fourth-power devices and square QAM. For other types of constellations and nonideal devices, the results would be different. For some types of constellations, the $4f_c$ component may be too weak to be useful. Then other carrier recovery techniques, such as the decision-directed carrier recovery technique may be needed.

M	r_M	$20 \log R_M$ (dB)
4	0.333	-9.54
16	0.172	-15.30
64	0.149	-16.51
256	0.144	-16.80

Table 9.2 Amplitude ratio.

The block diagram in Figure 4.35 can be used for the fourth-power loop by letting $M = 4$. A slightly different version was proposed by Rustako et al. [10] where the fourth-power loop was proposed for QAM digital radio receivers. Figure 9.10 is the diagram given in [10]. The VCO is working at f_c. A $\times 4$ block is inserted in the PLL loop to raise the VCO frequency to $4f_c$, which is then compared with the output of another $\times 4$ block whose input is the received signal. The $\times 4$ block is actually realized by a frequency-doubler whose nonlinearity generates a $4f_c$ component as a by-product at a level of about 20 dB below the $2f_c$ component. A detailed analysis of the times-four loop can be found in [10]. Rustako notes that in a fading environment, decision-directed carrier recovery suffers when occasional receiver outage destroys the accuracy of the data detections. This causes loss of carrier recovery and considerable time may be required to reacquire the carrier. He notes that times-four carrier recovery does not depend on data decision. For this and other reasons, times-four carrier recovery was proposed for digital radio receivers.

As mentioned, the decision-directed carrier recovery technique is one of the most popular carrier recovery schemes used in fixed-link QAM systems. The mechanism of DDCR is not based on nonlinearity. In this scheme, when a symbol is received, it is demodulated and a decision is made as to which was the most likely constellation point transmitted. Then it is assumed that the phase difference between the received symbol and the constellation point is due to carrier-recovery error, and the carrier recovery is updated accordingly. DDCR has the advantage that it can be used for all types of constellations. However, it exhibits a BER threshold. If the receiver BER is less than this threshold, it works extremely well since the "decisions" that assist the carrier recovery are correct almost all the time. If the receiver BER is higher than this threshold, the decisions on the constellation points are wrong more frequently. The update signal to the carrier recovery system therefore is erroneous more frequently, the cumulative effect can drive the carrier phase further and further away from the correct value. Whether DDCR is suitable for a particular application depends on this BER threshold.

Horikawa et al. designed a DDCR system for a 200 Mbps square 16QAM [11]. In their design, the square 16QAM constellation points are divided into two subgroups, called class I and class II phasors (Figure 9.11). The class II phasors are those on the

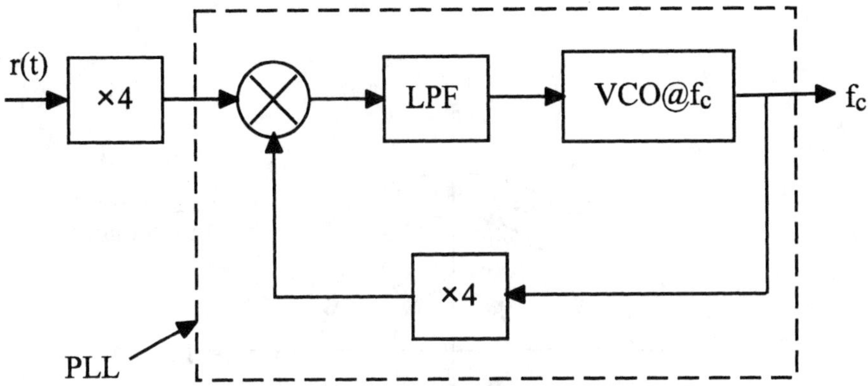

Figure 9.10 Times-four carrier recovery loop for QAM.

second circle with a normalized amplitude of $\sqrt{10}$. The rest are the class I phasors, four are on the outer circle, with an amplitude of $\sqrt{18}$; four are on the inner circle, with an amplitude of $\sqrt{2}$. The quadrature components of the class I phasors are equal in magnitude (i.e., $|I| = |Q|$), whereas the quadrature components of the class II phasors have a 3 to 1 ratio (i.e., $|I| = 3|Q|$ or $|Q| = 3|I|$).

Horikawa's DDCR system operates only on class I phasors. In other words, only class I phasors in the received signal are utilized by the DDCR system to recover the carrier. The DDCR system compares the values of the demodulated signal I and Q. If they satisfy the relationship

$$\frac{|I|}{2} < |Q| < 2|I| \tag{9.45}$$

then the system decides that a class I phasor has been sent. This rule is designed to effectively reject class II phasors while detecting class I phasors in a noisy environment. Once a class I phasor is detected, its I and Q values are used to determine polarity of the carrier drift and subsequently to adjust the VCO. The algorithm is as follows [2]. First compute the quantities

$$\begin{aligned} a &= pol(I) \oplus pol(Q) \\ b &= pol(I+Q) \\ c &= pol(I-Q) \\ d &= (a \oplus b) \oplus c \end{aligned} \tag{9.46}$$

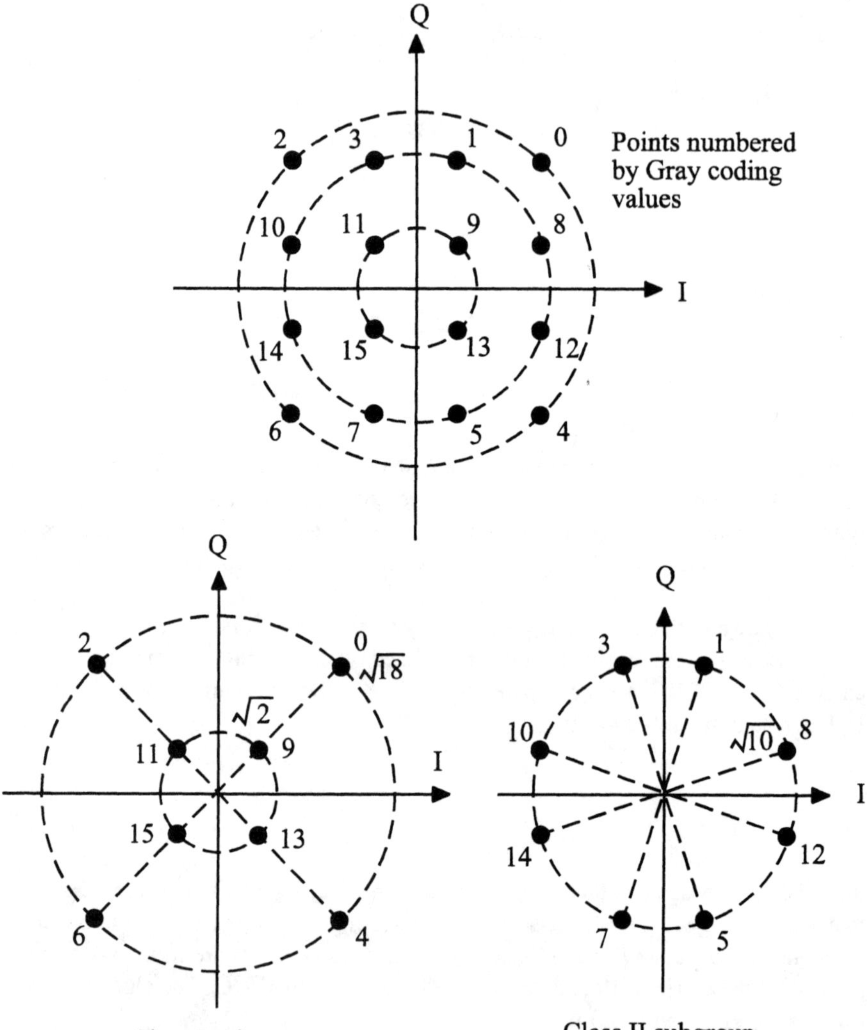

Figure 9.11 Square 16QAM subgroups.

where $\oplus$ denotes modulo-2 addition or XOR operation, and

$$pol(x) = \begin{cases} 1, & x > 0 \\ 0, & x < 0 \end{cases}$$

The d is used to control the VCO.

According to the above algorithm, the values of (abc) and d are shown in Figure 9.12 for each octant. As can be seen, the value of d changes between 0 and 1 alternately from one octant to the next. The quantity d is used to adjust the VCO frequency and phase. The demodulated signal is

$$r = LPF\{A_i \cos(\omega_c t + \theta_i) \cdot 2\cos(\omega_c t + \Delta\theta)\} = A_i \cos(\theta_i - \Delta\theta) \qquad (9.47)$$

where LPF denotes low-pass filtering. The VCO is assumed to have an amplitude of 2 and a phase error $\Delta\theta$ which is due to frequency error and initial phase error. From Figure 9.12, when $d = 0$, the angle of **r** is less than the transmitted phasor. From (9.47), this implies that the VCO has a positive phase error. Thus the VCO is instructed to run slower (reduce its frequency). When $d = 1$, the angle of **r** is greater than the transmitted phasor. From (9.47), this implies that the VCO has a negative phase error. Thus the VCO is instructed to run faster (increase its frequency). For example, assume $\mathbf{r} = (I, Q) = (2.24, 3.6)$ is received (see Figure 9.12). Then $a = pol(2.24) \oplus pol(3.6) = 1 \oplus 1 = 0$, $b = pol(5.84) = 1$, $c = pol(-1.36) = 0$, and $d = 0 \oplus 1 \oplus 0 = 1$. Thus the VCO is instructed to increase its frequency.

The DDCR system is shown in Figure 9.13. The received signal $r(t)$ is demodulated by the I- and Q-branches to produce I and Q. Then I and Q are fed into four comparators to generate a, b, and c, which are fed into XORs to generate d. At the same time, the I and Q are full-wave rectified to produce |I| and |Q|, which are passed through the other two comparators to check the relation (9.45). If they do, the selective gate will open, allowing signal d to pass the D flip-flop to control the VCO. Once synchronization is established, the I- and Q-channel outputs can be tapped by two threshold detectors to determine the transmitted symbol. Thus only two extra threshold detectors are needed to make Figure 9.13 a complete demodulator.

9.8 DIFFERENTIAL CODING IN QAM

In Chapter 4 we studied differential coding for MPSK signals. Similarly, differential coding is needed for QAM. It is needed for square QAM to resolve the phase ambiguity in carrier recovery. For circular QAM the use of differential coding can eliminate the need for carrier recovery, which is particularly attractive for fading channels. We will elaborate on this in Chapter 11. For now we focus on how differential coding is

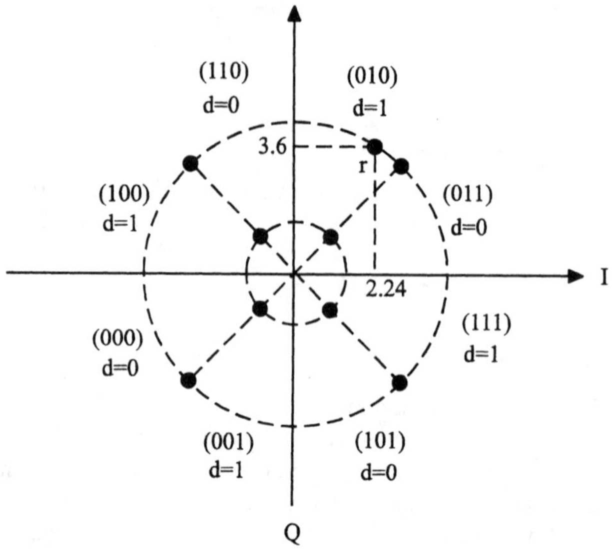

Figure 9.12 Distribution of the quantities in the DDCR algorithm.

constructed in square QAM.

Weber proposed a scheme for differentially encoding the signals for QAM and MPSK [12]. We follow his paper here. First, a signal set has an L-fold rotational symmetry if the signal set pattern remains unchanged after a rotation of $\pm I \cdot (2\pi/L)$ radians, where I and L are integers. For example, in Figure 9.1, the type I constellation has an 8-fold symmetry, the type II constellation has a 4-fold symmetry, and the type III constellation has also a 4-fold symmetry. As a matter of fact, all square QAM schemes are 4-fold symmetrical. It is also clear that all MPSK schemes are M-fold symmetrical. The L-fold rotational symmetry causes a phase ambiguity of $2\pi/L$, that is, any phase rotation of $2\pi/L$ or its multiples in the received signal will cause the signal to be demodulated as another signal. For example, for the square QAM, times-four loop is used for carrier recovery. Any phase rotation of $\pi/2$ in the received signal will not change the constellation, thus the recovered carrier, which is based on the average of signals over the constellation (over a long period of time in practice), will always have the same phase. Thus the signals in the first quadrant may be demodulated as the signals in the second quadrant, or third quadrant, or fourth quadrant, depending on the actual amount of phase rotation. Differential coding is an efficient way to resolve the phase ambiguity. Other methods include sending a separate synchronizing signal or periodic insertion of a synchronizing se-

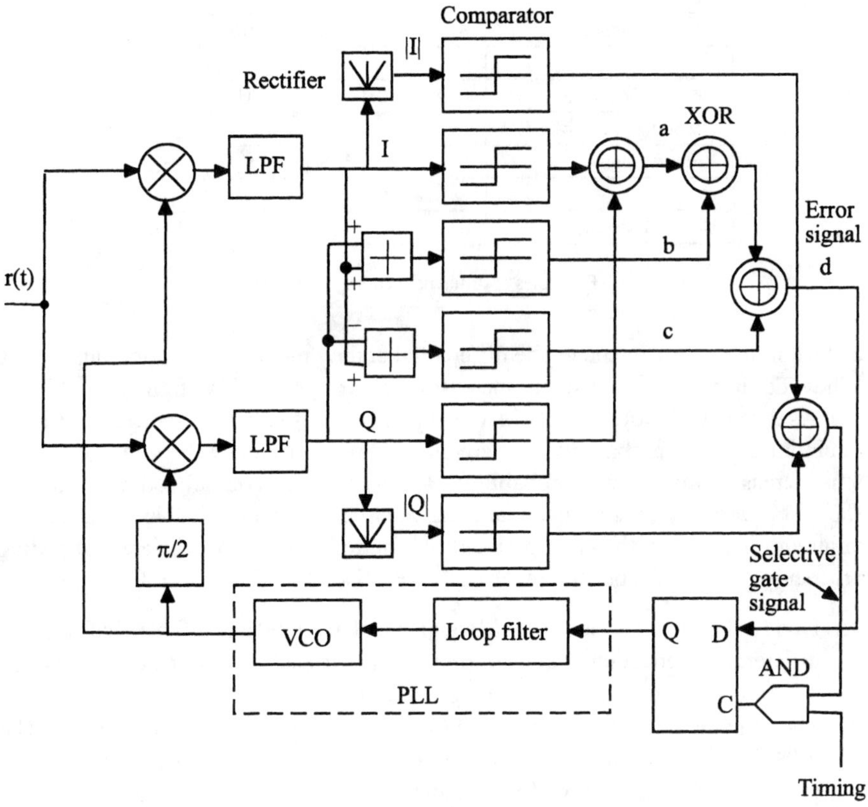

Figure 9.13 Implementation of a decision-directed carrier recovery system for square 16QAM. From [11]. Copyright © 1979 IEEE.

Encoding	ref.				
Message	0000	1011	0001	1100	0110
Δq		-1	0	+2	+1
q_i	q_1	q_4	q_4	q_2	q_3
Transmitted signal	$\mathbf{s}_1$	$\mathbf{s}_{16}$	$\mathbf{s}_{14}$	$\mathbf{s}_5$	$\mathbf{s}_{11}$
Decoding ($\Delta\phi = 0$)	ref.				
Received signal	$\mathbf{s}_1$	$\mathbf{s}_{16}$	$\mathbf{s}_{14}$	$\mathbf{s}_5$	$\mathbf{s}_{11}$
Δq		-1	0	+2	+1
Decoded Message	0000	1011	0001	1100	0110
Decoding ($\Delta\phi = \pi/2$)					
Received signal	$\mathbf{s}_5$	$\mathbf{s}_4$	$\mathbf{s}_2$	$\mathbf{s}_9$	$\mathbf{s}_{15}$
Δq		-1	0	+2	+1
Decoded Message	0000	1011	0001	1100	0110

Table 9.3 Examples of differential coding for square 16QAM.

quence in the data sequence. The differential coding method does not require extra bandwidth like the other two, but increases the error probability slightly.

As discussed before, the QAM signal points are usually Gray coded in order to minimize bit error probability. Differential coding will violate the Gray code rule, thus increasing the bit error probability. However, the differential coding procedure should be carefully chosen such that it can remove the symmetrical ambiguities at a minimum increase of the error probability. The following is the differential coding procedure given by Weber for an $M = 2^K$ signal set for which $L = 2^N$.

1. Divide the signal space into L equal pie-shaped sectors. These sectors are differentially encoded using the first N bits of each K bits of data. That is, the first N bits determine the *change in sector.*
2. The remaining $K - N$ bits determine the signal point within the sector. The $2^{(K-N)}$ signal points within the sector must be Gray coded to reduce the probability of error. (However, perfect Gray coding may not be possible.)

We will see how this is done through an example. Figure 9.14 is the differential encoding for the square 16QAM. In the figure, the signal space is divided by the I- and Q-axes into $L = 4$ pie-shaped sectors (quadrants in this case). Signal points are labeled with the last two bits in a 4-tuple. As far as these two bits are concerned, they are Gray coded in each sector. In addition they are arranged so that they are 4-fold symmetrical in terms of the last two bits. That is, a rotation of $\pi/2$ or its multiples of any signal will result in a signal having the same last two bits. The first two bits are used to determine the change of quadrant. Table 9.3 shows the encoding and decoding procedure for Figure 9.14 through an example. The first quadrant is

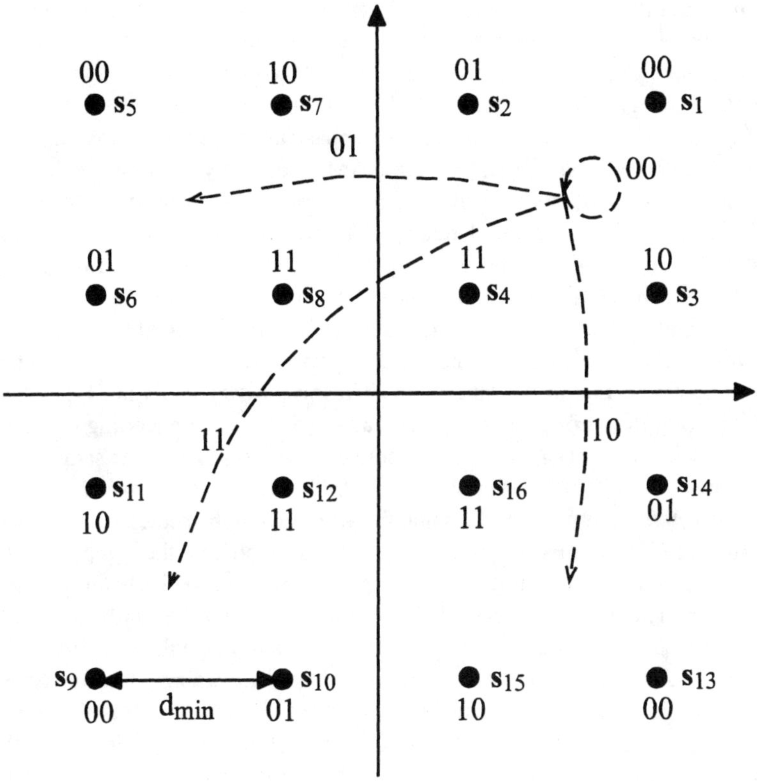

Figure 9.14 Differential encoding for square 16QAM. From [12]. Copyright © 1978 IEEE.

denoted as q_1, the second q_2, and so on. The change in quadrant (Δq) is denoted as 0 for no change, $+1$ for forwarding one quadrant, and so on. We assume the initial reference signal is s_1, corresponding to bits 0000. Take the second four bits 1011 as an example. The first two bits are 10, which means the quadrant change is –1 according to Figure 9.14. Since the current quadrant is q_1, the next will be q_4, that is, the next signal must be in q_4. Since the last two bits are 11, within q_4, the signal labelled with 11 is signal s_{16}. Thus the signal point to be transmitted is s_{16}. When s_{16} is received without phase rotation $(\Delta\phi = 0)$, which is compared with the previous signal (s_1) to determine the quadrant change (-1). This determines 10 as the first two bits. The last two bits then are determined as 11 by the fact that s_{16} is labeled 11. When s_{16} is received with phase rotation (say $\Delta\phi = \pi/2$), it becomes s_4. At the same time the reference signal is also changed to s_5 due to the rotation. Signal s_4 is compared with the previous signal (s_5) to determine the quadrant change (-1). This determines 10 as the first two bits. The last two bits then are determined as 11 by the fact that s_4 is labeled 11. From the table we can see that phase rotation does not affect decoding of the signals. Thus the phase ambiguity is removed.

For general K-bit (K even) square QAM, again only the first two bits need to be differentially encoded for resolving the quadrant ambiguity, while the remaining $K-2$ bits are Gray coded in each quadrant. For other constellations, such as circular QAM, the same encoding principles apply. The concrete encoding rules depend on the concrete constellation structure. See [12] for more examples.

In evaluating the error probability of the differentially coded QAM, a high SNR is assumed. This implies the errors are dominantly the errors made between points separated by the minimum distance d_{min}. Thus we only consider this type of error in approximating the error probability. The above differential encoding schemes produce two types of symbol errors: those between symbols within the same sector and those between symbols of different sectors. In the first type of error, the first N bits are not affected since they are the same for all points in the sector. Thus only one or more of the $(K-N)$ bits will be in error, depending on how the symbols in the sector have been Gray coded. If the symbols in the sector are perfectly Gray coded, like in the square 16QAM case, only one bit will be in error. If the symbols in the sector are not perfectly Gray coded, more than one bit on average will be in error. Then we say there is a Gray code penalty for the imperfect Gray coding. For this type of error, no error propagation exists since the differentially encoded bits are unaffected. In the second type of error, bit errors can arise from two sources. First, since the sector boundary has been crossed, a minimum of two bit errors will occur in the first N bits, one in each of two succeeding symbols (error propagation), due to the comparisons used in differential decoding. Additional bit errors may occur in the remaining $K-N$ bits depending on how these adjacent border symbols are encoded. While a perfect Gray code may be found for the symbols within a sector, a Gray code penalty may exist with symbols lying on the sector boundaries.

The relationship between the error probability of the differentially encoded system and that of the uncoded system can be given in the following general formula

$$P_{b,d} = F \cdot P_b$$

where $P_{b,d}$ is the bit error probability of the differentially coded system, and P_b is the bit error probability of the differentially uncoded system. The factor F is the penalty for differential encoding. For the differentially uncoded system,

$$P_b = \frac{g}{K} P_s$$

where P_s is the symbol error probability, $g \geq 1$ is the Gray code penalty, representing the average number of erroneous bits in a symbol error. For the perfect Gray coded system, $g = 1$. For the differentially encoded system,

$$P_{b,d} = \frac{f}{K} P_s$$

where f is the differentially encoded Gray code penalty, also representing the average number of erroneous bits in a symbol error. Thus the penalty for differential encoding is

$$F = \frac{f}{g} \tag{9.48}$$

The factor f can be found as follows.

1. Within a single sector draw lines between symbol point pairs separated by $d_{\min}$; denote the total number of these lines by N_1;
2. Next to each line write the Hamming distance (number of bits which differ) between the two signal points; denote the sum of all such Hamming distances in the sector by H_1;
3. Draw lines between points lying on *one* of the sector boundaries with points in the adjacent sectors separated by $d_{\min}$; denote the number of these lines by N_2 and denote the sum of their Hamming distances by H_2.

Then the differential encoding penalty is

$$f = \frac{H_1 + H_2 + 2N_2}{N_1 + N_2} \tag{9.49}$$

where the third term in the numerator is due to error propagation in the differentially encoded bits. The numerator is in fact the total number of bit errors when all possible symbol errors are considered, and the denominator is in fact the total number of possible symbol errors. Thus f is the average bit errors per symbol. Note that H_1/N_1 is the Gray code penalty within a sector and H_2/N_2 is the Gray code penalty across the boundary.

For the differentially encoded square 16QAM in Figure 9.14, $N_1 = 4$ and $H_1 = 4$ while $N_2 = 2$ and $H_2 = 2$. Thus $f = (4 + 2 + 4)/(4 + 2) = 10/6 = 1.67$. For general K-bit (K even) square QAM with the first two bits differentially encoded for resolving the quadrant ambiguity and the remaining $K - 2$ bits Gray coded, it can be shown that

$$\begin{aligned} N_1 &= 2^{K-1} - 2^{K/2} \\ H_1 &= 2^{K-1} - 2^{K/2} \end{aligned}$$

$$\begin{aligned} N_2 &= 2^{(K/2-1)} \\ H_2 &= \left(\frac{K}{2}-1\right)2^{(K/2-1)} \end{aligned} \tag{9.50}$$

For the above general case $g = 1$ so that from (9.50) and (9.49)

$$F = 1 + \frac{K/2}{2^{K/2}-1}, \quad \text{(square QAM, } K \text{ even)}$$

Thus F goes from two for $K = 2$ (4QAM or QPSK) to nearly one for very large K. This penalty of two or less is insignificant in term of increase in SNR, which is usually a fraction of a dB.

9.9 SUMMARY

We defined QAM signal and constellation in Section 9.1 where the orthogonality of the two components of the QAM was proved in the presence of the pulse-shaping function. Various QAM constellations were introduced in Section 9.2 in order to give the reader an overview of the QAM constellations. However, only the square QAM constellations were described in detail since they are among the most efficient yet their implementations are the simplest. In Section 9.3, QAM's PSD was derived. It turned out that the shape of PSD of QAM is solely determined by the pulse-shaping function. This property is the same as that of MASK and MPSK. In Sections 9.4 and 9.5 we presented the modulator and demodulator based on those of MASK. The modulator is almost identical to that of MPSK except that the level settings of the level generators are different. Gray coding is usually used for mapping from data n-tuples to QAM points for minimizing bit errors. The demodulator is also similar to the MPSK demodulator. The error probability of QAM was derived in Section 9.6 based on that of MASK. It was shown that QAM requires less signal-to-noise ratios than MPSK for achieving the same error performance. At the $M = 4$ level, they are the same since 4PSK is 4QAM. Above $M = 4$, the signal power savings range from 1.65 to 16 dB for $M = 8, 16, ..., 256$. The savings increase by approximately 3 dB for doubling the number of points in the constellation. This is what makes QAM very attractive. Synchronization for QAM was discussed in Section 9.7. The clock recovery of QAM is not a particular problem. The clock recovery techniques in Chapter 4 are applicable. The carrier recovery of QAM has its particular feature. It turned out that it does not require Mth-power nonlinearity as in the MPSK case. It requires a fourth-power loop, but a squaring loop does not work. We showed in detail why this is true. We also described a decision-directed carrier recovery system which does not rely on nonlinearity at all. Finally in Section 9.8 we discussed differential coding for QAM for the purpose of phase ambiguity elimination.

The application of QAM, particularly the star QAM, to fading channels will be discussed in Chapter 11.

APPENDIX 9A PROOF OF (9.39)

We form the fourth power of $s_i(t)$:

$$\begin{aligned} E\{s_i^4(t)\} &= E\{[A_{i1}p(t)\cos\omega_c t - A_{i2}p(t)\sin\omega_c t]^4\} \\ &= p^4(t)E\{[A_{i1}\cos\omega_c t - A_{i2}\sin\omega_c t]^4\} \end{aligned} \quad (9.51)$$

To simplify the derivation we denote

$$x = A_{i1}\cos\omega_c t$$

$$y = -A_{i2}\sin\omega_c t$$

Note that $E\{x\} = E\{y\} = 0$ since $E\{A_{i1}\} = E\{A_{i2}\} = 0$. Using x and y notations we can write (9.51) as

$$\begin{aligned} E\{s_i^4(t)\} &= p^4(t)E\{[x+y]^4\} \\ &= p^4(t)E\{x^4 + 4x^3y + 6x^2y^2 + 4xy^3 + y^4\} \\ &= p^4(t)E\{x^4 + 6x^2y^2 + y^4\} \end{aligned} \quad (9.52)$$

where $E\{4x^3y\} = 4E\{x^3\}E\{y\} = 0$ and $E\{4xy^3\} = 4E\{x\}E\{y^3\} = 0$. Using trigonometrical identities we obtain

$$\begin{aligned} x^4 &= A_{i1}^4\cos^4\omega_c t = A_{i1}^4[\cos^2\omega_c t]^2 = A_{i1}^4\left[\frac{1}{2}(1+\cos 2\omega_c t)\right]^2 \\ &= \frac{1}{4}A_{i1}^4(1 + 2\cos 2\omega_c t + \cos^2 2\omega_c t) \\ &= \frac{1}{4}A_{i1}^4\left[1 + 2\cos 2\omega_c t + \frac{1}{2}(1+\cos 4\omega_c t)\right] \\ &= \frac{1}{8}A_{i1}^4\left[2 + 4\cos 2\omega_c t + (1+\cos 4\omega_c t)\right] \\ &= \frac{1}{8}A_{i1}^4(3 + 4\cos 2\omega_c t + \cos 4\omega_c t) \end{aligned} \quad (9.53)$$

$$y^4 = A_{i2}^4\sin^4\omega_c t = A_{i2}^4[\sin^2\omega_c t]^2 = A_{i2}^4\left[\frac{1}{2}(1-\cos 2\omega_c t)\right]^2$$

$$
\begin{aligned}
&= \frac{1}{4}A_{i2}^4(1 - 2\cos 2\omega_c t + \cos^2 2\omega_c t) \\
&= \frac{1}{4}A_{i2}^4\left[1 - 2\cos 2\omega_c t + \frac{1}{2}(1 + \cos 4\omega_c t)\right] \\
&= \frac{1}{8}A_{i2}^4\left[2 - 4\cos 2\omega_c t + (1 + \cos 4\omega_c t)\right] \\
&= \frac{1}{8}A_{i2}^4(3 - 4\cos 2\omega_c t + \cos 4\omega_c t) \qquad (9.54)
\end{aligned}
$$

$$
\begin{aligned}
6x^2y^2 &= 6A_{i1}^2A_{i2}^2\cos^2\omega_c t\sin^2\omega_c t \\
&= 6A_{i1}^2A_{i2}^2\frac{1}{2}(1 + \cos 2\omega_c t)\frac{1}{2}(1 - \cos 2\omega_c t) \\
&= \frac{3}{2}A_{i1}^2A_{i2}^2(1 - \cos^2 2\omega_c t) = \frac{3}{2}A_{i1}^2A_{i2}^2\sin^2 2\omega_c t \\
&= \frac{3}{2}A_{i1}^2A_{i2}^2\frac{1}{2}(1 - \cos 4\omega_c t) = \frac{3}{4}A_{i1}^2A_{i2}^2(1 - \cos 4\omega_c t) \qquad (9.55)
\end{aligned}
$$

Substituting (9.53), (9.54), and (9.55) into (9.52), we obtain

$$
\begin{aligned}
E\{s_i^4(t)\} &= \frac{1}{8}p^4(t)[E\{A_{i1}^4\}(3 + 4\cos 2\omega_c t + \cos 4\omega_c t) \\
&\quad + E\{A_{i2}^4\}(3 - 4\cos 2\omega_c t + \cos 4\omega_c t)] \\
&\quad + \frac{3}{4}p^4(t)E\{A_{i1}^2\}E\{A_{i2}^2\}(1 - \cos 4\omega_c t)
\end{aligned}
$$

References

[1] Sklar, B., *Digital Communications: Fundamentals and Applications*, Englewood Cliffs, New Jersey: Prentice Hall, 1988.

[2] Webb, W. T., and L. Hanzo, *Modern Quadrature Amplitude Modulation*, New York: IEEE Press, and London: Pentech Press, 1994.

[3] Cahn, C. R.,"Combined digital phase and amplitude modulation communication system," *IRE Trans. Comms.,* vol. 8, September 1960, pp. 150–155.

[4] Hancock, J. C., and R. W. Lucky, "Performance of combined amplitude and phase modulated communications system," *IRE Trans. Comms.,* vol. 8, December 1960, pp. 232–237.

[5] Campopiano, C. N., and B. G. Glazer, "A coherent digital amplitude and phase modulation system," *IRE Trans. Comms.,* vol. 10, March 1962, pp. 90–95.

[6] Thomas, C. M., M. Y. Weidner, and S. H. Durrani, "Digital amplitude-phase keying with M-ary alphabets," *IEEE Trans. Comms.,* vol. 22, no. 2, February 1974, pp. 168–180.

[7] Benedetto, S., E. Biglieri, and V. Castellani, *Digital Transmission Theory*, Englewood Cliffs, New

Jersey: Prentice Hall, 1987.

[8] Proakis, J., and M. Salehi, *Communication Systems Engineering*, Englewood Cliffs, New Jersey: Prentice Hall, 1994.

[9] Lu, J., et al., "MPSK and MQAM BER computation using signal-space concepts," *IEEE Trans. Commun.*, vol. 47, no. 2, February 1999, pp. 181–184.

[10] Rustako, A., et al., "Using times four carrier recovery in MQAM digital radio receivers," *IEEE Journal on Selected Areas in Commun.*, vol. 5, no. 3, April 1987, pp. 524–533.

[11] Horikawa, I., T. Murase, and Y. Saito, "Design and performances of a 200 Mbit/s 16 QAM digital radio system," *IEEE Trans. Commun.*, vol. 27, no. 12, December 1979, pp. 1953–1958.

[12] Weber, W. J., "Differential encoding for multiple amplitude and phase shift keying systems," *IEEE Trans. Comms.*, vol. 26, no. 3, Mar. 1978, pp. 385–391.

Selected Bibliography

- Couch II, L. W., *Digital and Analog Communication Systems*, 3rd ed., New York: Macmillan, 1990.
- Haykin, S., *Communication Systems*, 3rd ed., New York: John Wiley, 1994.
- Simon, K. M., S. M. Hinedi, and W. C. Lindsey, *Digital Communication Techniques: Signal Design and Detection*, Englewood Cliffs, New Jersey: Prentice Hall, 1995.
- Smith, D. R., *Digital Transmission Systems*, 2nd ed., New York: Van Nostrand Reinhold, 1993.

Chapter 10

Nonconstant-Envelope Bandwidth-Efficient Modulations

The schemes we studied in Chapters 5, 6, and 7 are all bandwidth-efficient modulation schemes with constant envelopes. The next natural step is to explore the possibility of obtaining high bandwidth efficiency using nonconstant-envelope schemes. QAM is very bandwidth efficient, but its amplitude can vary considerably. This makes it unsuitable for transmitters with power amplifiers that must operate in a nonlinear region for maximum efficiency. A lot of effort has been devoted to finding bandwidth-efficient schemes without too much amplitude variations. This can be achieved by using pulse shaping in quadrature modulation or other means.

In this chapter modulation schemes with compact spectrum, low spectral spreading caused by nonlinear amplification, good error performance, and simple hardware implementation are presented. In describing these nonconstant-envelope schemes, the emphasis is on the pulse shape and spectral properties. The eye diagram of each scheme is also presented since it is critical in demodulation if a conventional OQPSK-type demodulator is used. Even though the OQPSK-type demodulator is not optimum for these schemes, it is often suggested due to its simplicity and small loss in error performance. The error performance loss with respect to MSK is evaluated.

Since the majority of the schemes presented in this chapter are two-symbol-time ($2T_s$) schemes, we have a general discussion of this type of scheme in Section 10.1. Particularly, an optimum receiver for the AWGN channel is developed in this section. Individual schemes are described in the sections that follow. Section 10.2 describes the quasi-bandlimited modulation (QBL) [1]. The quadrature overlapped raised-cosine modulation (QORC), and its staggered version (SQORC) proposed in [2] are described in Section 10.3. Later a scheme named modified QORC (MQORC) was proposed in [3]. MQORC uses a different pulse-shaping function which is similar to that of SFSK (sinusoidal frequency shift keying, see Section 5.9). This pulse improves the spectrum further, however the pulse involves Bessel functions and it is not easy to implement. We will not include this scheme here. But the quadrature overlapped squared raised-cosine (QOSRC) modulation [4] is simple to realize. It is also described in Section 10.3. The research group led by Dr. Kamilo Feher developed a family of power-efficient coherent nonconstant-envelope modems. Prior to a hard limiter, the envelope of the modulated signals is not constant, but a hard limiter inserted into the transmission channel does not significantly spread the processed signal spectrum. Thus, these techniques are suitable for nonlinearly amplified satellite channels in a densely packed ACI (adjacent channel interference) environment. This family of modulation schemes include IJF (intersymbol-interference/jitter-free) OQPSK [5], TSI (two-symbol-interval) OQPSK [6], SQAM (superposed-QAM) [7], and XPSK (crosscorrelated QPSK) [8]. Among them, the XPSK is the most complex one. XPSK involves 16 different cross-correlated signal combinations using 14 different signal patterns. It can achieve almost constant signal envelope. Its spectrum is similar to that of TFM (tamed frequency modulation, see Chapter 6 for TFM). However, due to its complex signal format, it may not be a preferable choice over other schemes. We will not include it here. The other Feher's schemes all use one pulse pattern. IJF-OQPSK and TSI-OQPSK are discussed in Section 10.4. SQAM is discussed in Section 10.5. Section 10.6 describes a new approach of achieving compact spectrum, other than pulse shaping. The scheme is called quadrature quadrature phase shift keying (Q^2PSK). Section 10.7 summarizes this chapter.

10.1 TWO-SYMBOL-PERIOD SCHEMES AND OPTIMUM DEMODULATOR

The majority of the schemes in this chapter are $2T_s$ schemes, they can be expressed as

$$s(t) = s_I(t)\cos 2\pi f_c t + s_Q(t)\sin 2\pi f_c t, \quad -\infty \leq t \leq \infty \tag{10.1}$$

where

$$s_I(t) = \sum_{k=-\infty}^{\infty} I_k p(t - 2kT_b) \tag{10.2}$$

$$s_Q(t) = \sum_{k=-\infty}^{\infty} Q_k p(t - 2kT_b - \tau) \tag{10.3}$$

where T_b is the bit time interval corresponding to the input data sequence $\{a_k \in (-1, +1)\}$ that has been demultiplexed into $\{I_k\}$ and $\{Q_k\}$. It is clear that each data symbol lasts for a duration of $T_s = 2T_b$ in I- and Q-channels. Each data is weighted by a pulse-shaping function $p(t)$ which has a duration of $2T_s = 4T_b$. If the delay

τ is zero we have a nonstaggered scheme, otherwise we have a staggered scheme. Usually, if a staggered scheme is desired, τ is set to be T_b. In this case, due to the staggering of the I- and Q-channels, the symbol duration of the modulated signal is T_b instead of $2T_b$ despite that the symbol durations are $2T_b$ for $s_I(t)$ and $s_Q(t)$. However, demodulation must be performed in a $2T_b$ duration.

The modulators for $2T_s$ schemes are the same as that of QPSK/OQPSK, except that a baseband signal processor (filter) is inserted in each channel prior to the carrier multiplier. The processor weights the data by the symbol pulse and overlaps them by T_s. This can be realized by using a filter with an impulse response of the pulse-shaping function ($p(t)$). In the following sections of $2T_s$ schemes, we will not repeat this statement again.

By a reasoning similar to that in Section 9.1 for QAM, the I- and Q-channel components in (10.1) are essentially orthogonal for $f_c >> 1/T_s$. Thus the receiver will have the same quadrature structure as in MSK (Figure 5.9). That is, an inphase carrier in the I-channel and a quadrature carrier in the Q-channel are used to separate the I- and Q-channel data streams. However, the post-separation processing is different, which is based on the possible composite waveforms in a $2T_b$-symbol duration in the I- or Q-channel data stream. In the detection interval $[0, 2T_b]$ (or any $[2kT_b, 2(k+1)T_b]$ interval), the I-channel baseband signal is (assuming $p(t)$ is defined on $[-2T_b, 2T_b]$)

$$s_I(t) = \pm p(t) \pm p(t - 2T_b), \quad 0 \le t \le 2T_b$$

Thus there are four possible waveforms of $s_I(t)$. But two are just negatives of the other two.

$$\begin{cases} f_1(t) = p(t) + p(t - 2T_b), & a_k = a_{k+1} = 1 \\ f_2(t) = p(t) - p(t - 2T_b), & a_k = 1, a_{k+1} = -1 \\ f_3(t) = -f_2(t), & a_k = -1, a_{k+1} = 1 \\ f_4(t) = -f_1(t), & a_k = a_{k+1} = -1 \end{cases} \tag{10.4}$$

The energy of each signal is

$$\mathcal{E}_i = \int_0^{2T_b} f_i^2(t)dt, \quad i = 1, 2, ..., 4 \tag{10.5}$$

The Q-channel waveforms are just the delayed-by-τ version of the I-channel waveforms. The following discussion is based on the I-channel. The results are obviously applicable to the Q-channel provided proper symbol timing is maintained (delay τ with respect to I-channel, $\tau = 0$ for nonstaggered schemes).

The postseparation demodulator thus can be based on these four waveforms. As usual we assume the noise $n(t)$ at the input of the entire demodulator is AWGN with zero mean and a two-sided PSD of $N_o/2$. The received signal at any moment is

$$r(t) = f_i(t)\cos 2\pi f_c t + f_j(t-\tau)\sin 2\pi f_c t + n(t), \quad i,j = 1,2,...,4$$

But as far as the detection of the I-channel signal is concerned, the Q-channel signal does not have any effect. Thus to the I-channel the received signal can be written as

$$r(t) = f_i(t)\cos 2\pi f_c t + n(t) = s_i(t) + n(t), \quad i = 1,2,...,4$$

where

$$s_i(t) = f_i(t)\cos 2\pi f_c t, \quad i = 1,2,...,4$$

The problem now becomes a 4-ary signal detection problem with different signal energies. From (B.38) the sufficient statistic is

$$l_i = r_i + c_i, \quad i = 1,2,3,4$$

where

$$r_i = \int_0^{2T_b} r(t)s_i(t)dt = \int_0^{2T_b} r_I(t)f_i(t)dt$$

$$r_I(t) = r(t)\cos 2\pi f_c t$$

is the post-separation signal which is obtained by a down-convertor with a reference signal of $\cos 2\pi f_c t$.

$$c_i = \frac{1}{2}\left(N_o \ln(P_i) - E_i\right)$$

where $P_i = 1/4$ for equally likely data.

$$E_i = \int_0^{2T_b} s_i^2(t)dt = \frac{1}{2}\mathcal{E}_i, \quad i = 1,2,...,4$$

Since the first term in c_i is the same for all four signals, it can be eliminated. The detector chooses the largest of l_i and the corresponding data pattern can be identified. From (10.4) we know that only two correlators for r_1 and r_2 are needed. That is

$$\begin{cases} l_1 = r_1 - E_1/2 \\ l_2 = r_2 - E_2/2 \\ l_3 = -r_2 - E_3/2 \\ l_4 = -r_1 - E_4/2 \end{cases}$$

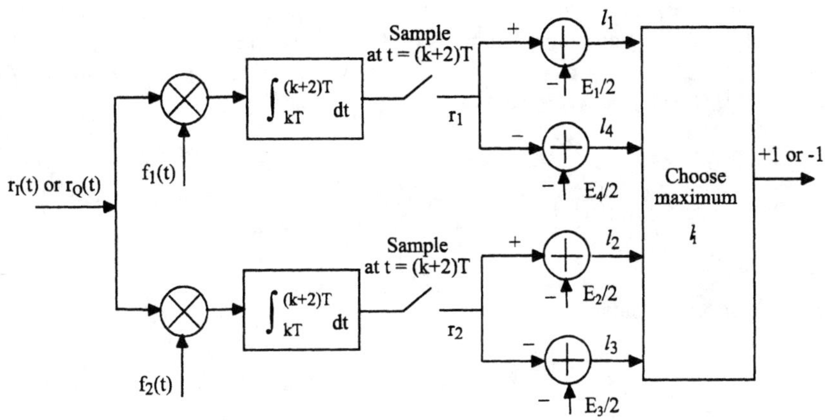

Figure 10.1 Correlation detector for $2T_s$ schemes with overlapping pulses.

Figure 10.1 shows the correlation detector which can be used for I- or Q-channel post-separation processing. The detector input is the post-separation signal $r_I(t)$ or $r_Q(t)$. The detector output is directly the first data symbol in the data pattern since once maximum l_i is determined the first data symbol is also determined.

The upper bound of the symbol error probability of this optimum receiver has been given in Appendix B (B.43), that is,

$$P_s \leq \sum_{j \neq i} Q\left(\sqrt{\frac{d_{ij}^2}{2N_o}}\right), \quad i,j = 1,2,...,4 \tag{10.6}$$

where

$$\begin{aligned} d_{ij}^2 &= \int_0^{2T_b} [s_i(t) - s_j(t)]^2 dt \\ &= \frac{1}{2}\int_0^{2T_b} [f_i(t) - f_j(t)]^2 dt \\ &= \frac{1}{2}\int_0^{2T_b} f_i^2(t)dt + \frac{1}{2}\int_0^{2T_b} f_j^2(t)dt - \int_0^{2T_b} f_i(t)f_j(t)dt \\ &= \frac{1}{2}\mathcal{E}_i + \frac{1}{2}\mathcal{E}_j - \int_0^{2T_b} f_i(t)f_j(t)dt, \quad i,j = 1,2,...,4 \end{aligned} \tag{10.7}$$

Usually the pulse $p(t)$ is symmetrical about $t=0$ (if $p(t)$ is defined in $[-2T_b, 2T_b]$). Then $f_1(t)$ and $f_4(t)$ are even functions about $t=T_b$ and $f_2(t)$ and $f_3(t)$ are odd functions about $t=T_b$. As a result, $f_1(t)$ is orthogonal to $f_2(t)$ and $f_3(t)$ in $[0, 2T_b]$. So is $f_4(t)$ to $f_2(t)$ and $f_3(t)$. That is

$$\int_0^{2T_b} f_i(t)f_j(t)dt = 0, \quad \text{for} \begin{cases} i=1,\ j=2 \\ i=1,\ j=3 \\ i=4,\ j=2 \\ i=4,\ j=3 \end{cases} \tag{10.8}$$

We check one example:

$$\begin{aligned} \int_0^{2T_b} f_1(t)f_2(t)dt &= \int_0^{2T_b} [p(t)+p(t-2T_b)][p(t)-p(t-2T_b)]dt \\ &= \int_0^{2T_b} p^2(t)dt - \int_0^{2T_b} p^2(t-2T_b)dt \\ &= \int_0^{2T_b} p^2(t)dt - \int_{-2T_b}^{0} p^2(\tau)d\tau = 0 \end{aligned}$$

The last step holds because $p(t)$ is symmetrical about $t=0$. The orthogonality of other pairs can be similarly verified. From (10.7) and (10.8), and noticing that $\mathcal{E}_4 = \mathcal{E}_1$ and $\mathcal{E}_3 = \mathcal{E}_2$, we have

$$[d_{ij}^2] = \frac{1}{2}\begin{bmatrix} 0 & \mathcal{E}_1+\mathcal{E}_2 & \mathcal{E}_1+\mathcal{E}_2 & 4\mathcal{E}_1 \\ \mathcal{E}_1+\mathcal{E}_2 & 0 & 4\mathcal{E}_2 & \mathcal{E}_1+\mathcal{E}_2 \\ \mathcal{E}_1+\mathcal{E}_2 & 4\mathcal{E}_2 & 0 & \mathcal{E}_1+\mathcal{E}_2 \\ 4\mathcal{E}_1 & \mathcal{E}_1+\mathcal{E}_2 & \mathcal{E}_1+\mathcal{E}_2 & 0 \end{bmatrix} \tag{10.9}$$

Since $\mathcal{E}_i$s are usually close, the distance $(\mathcal{E}_1+\mathcal{E}_2)/2$ is the minimum $(d_{\min})$. At high signal-to-noise ratio, the larger distance terms may be ignored. If that is the case, only the four $d_{\min}$ terms need be considered. They are d_{12}, d_{13}, d_{24}, and d_{34}. Their value is

$$d_{\min}^2 = (\mathcal{E}_1+\mathcal{E}_2)/2 = E_1 + E_2 = 2E_{avg} = 4E_b$$

where E_{avg} and E_b are average symbol and bit energy, respectively. Thus the symbol error probability is approximately

$$P_s \lessapprox 4Q\left(\sqrt{\frac{2E_b}{N_o}}\right)$$

Furthermore, among the four error events associated with the $d_{\min}$ terms, two events (associated with d_{12} and d_{34}) actually do not cause bit errors (see (10.4) where $f_1(t)$ and $f_2(t)$ have the same first bit, so do $f_3(t)$ and $f_4(t)$). Thus we have the bit error probability

$$P_b \lessapprox 2Q\left(\sqrt{\frac{2E_b}{N_o}}\right) \tag{10.10}$$

This is two times that of MSK. However, in terms of SNR required for the same P_b, this is only slightly inferior to MSK (about 0.3 dB at 10^{-5}). Sometimes the second smallest distance is too close to d_{min} to be ignored. Then the error probability expression has to include the term associated with the second smallest distance.

The synchronization schemes for $2T_s$ schemes are in general the same as those of QPSK/OQPSK since all of them are quadrature modulations. Thus the carrier and symbol synchronization techniques described in Chapter 4 can be applied. Recall MSK has a combined carrier and symbol recovery circuit (Figure 5.11) due to its unique spectral components. It is also possible to exploit the spectral properties of $2T_s$ schemes to obtain better synchronization schemes.

10.2 QUASI-BANDLIMITED MODULATION

Amoroso proposed the use of quasi-bandlimited pulses in MSK-type modulation to improve the spectrum for the near-center region ($f \leq 1/T_b$), including reducing the main lobe width [1]. The proposed pulses are

$$\begin{aligned} p(t) &= \begin{cases} \left[\dfrac{\sin\frac{\pi t}{2T_b}}{\frac{\pi t}{2T_b}}\right]^n, & -2T_b \leq t \leq 2T_b \\ 0, & \text{elsewhere} \end{cases} \\ n &= 1, 2, 3, ... \end{aligned} \tag{10.11}$$

These pulses will not maintain a constant envelope for the MSK-type signal which they generate. But a bandpass hard limiter will be introduced just before transmission to ensure envelope constancy. The term "quasi-bandlimited" refers to the fact that the pulse duration is relaxed to $4T_b$ from $2T_b$ of the MSK, thus the bandwidth is somewhat more limited, but not completely limited as in the case where the duration is allowed to extend to $\pm\infty$.

Figure 10.2(a) shows $p(t)$ for $n = 3$ (denoted as QBL-3) in comparison with the pulse of MSK. The QBL pulse has a duration of $4T_b$. From (10.2) and (10.3), the adjacent pulses in each channel are overlapped by a length of $2T_b$. However, if sampled at center of each pulse (i.e., at $t = 2kT_b$ for the I-channel and $t = (2k+1)T_b$ for the Q-channel), there is no intersymbol interference in the samples (Figure

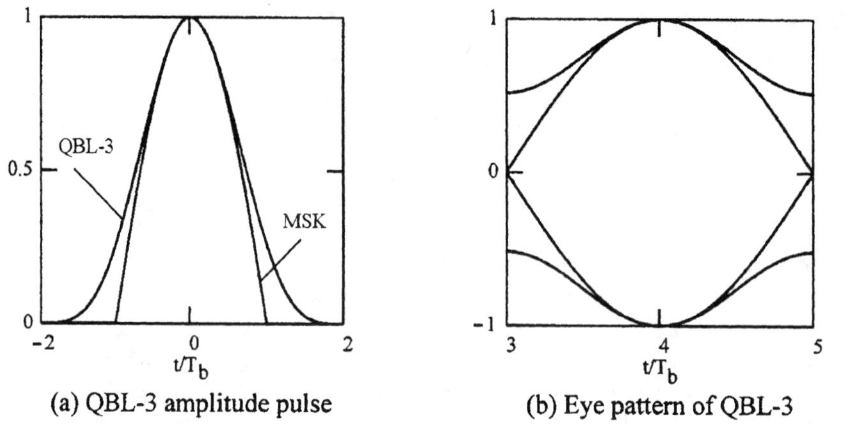

Figure 10.2 (a) QBL-3 amplitude pulse in comparison with MSK amplitude pulse, and (b) eye pattern of QBL-3 for I-channel.

10.2(b)).

The modulated QBL signal is given by (10.1) to (10.3). The pulse sequences in the I- and Q-channel are staggered by a length of T_b. Since the pulse occupies a duration of $4T_b$, in any period of T_b, there are four segments of pulses which affect the envelope of the modulated signal. We can use any period of T_b, say, the period $[0, T_b]$, to compute the amplitude distribution. In $[0, T_b]$, the four pulse segments are from $p(t)$, $p(t-2T_b)$, $p(t+T_b)$, and $p(t-T_b)$. The first two affect the I-channel signal, and the last two affect the Q-channel signal. Thus the amplitude of the QBL signal is

$$A(t) = \sqrt{[I_0 p(t) + I_1 p(t-2T_b)]^2 + [Q_{-1} p(t+T_b) + Q_0 p(t-T_b)]^2}$$

There are 16 different combinations of $[I_0, I_1, Q_{-1}, Q_0]$. However, one-half of them is just the negated version of another half. We only need to compute the amplitude for the first half. For $n = 3$, the possible envelopes found by numerical computation are shown in Figure 10.3. The minimum amplitude is $A_{\min} = 0.994$ and the maximum amplitude is $A_{\max} = 1.125$. The minimum to maximum ratio is

$$A_{\min}/A_{\max} = 0.884$$

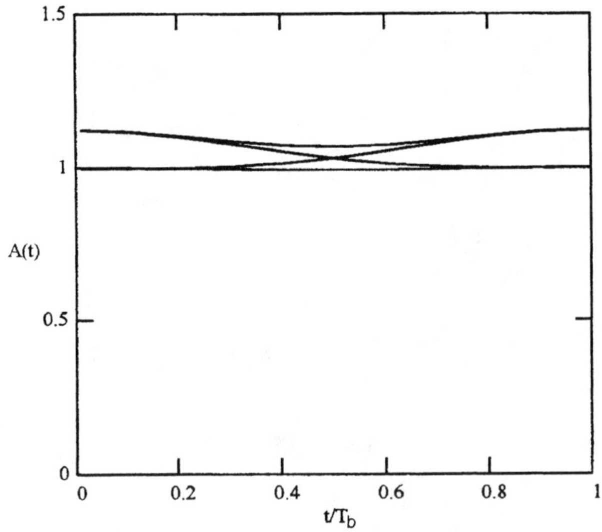

Figure 10.3 Possible envelopes of QBL-3 signal ($n = 3$).

Amoroso considered a hard limiter characterized by

$$y(t) = \frac{s(t)}{A(t)}$$

Then the amplitude of $y(t)$ is always one.

The spectral analysis is quite involved [1]. We only present the results here. Figure 10.4 shows the power spectral densities with hard limiting, for $n = 1, 2, 3, 4$. It is seen that $n = 3$ (QBL-3) is the best overall. Figure 10.5 compares the PSDs of QBL-3 with those of SFSK and MSK. It is seen that QBL-3 is much better in terms of having lower sidelobes. The hard limiting spreads the PSD slightly.

The implementation of QBL suggested in [1] is to use the serial MSK modulator and demodulator given in Figure 5.12. The only change is the conversion filter $H(f)$ in the transmitter, which must satisfy the following expression to produce the desired spectral shape

$$\Psi_s(f) = |H(f)|^2 \Psi_{BPSK}(f)$$

where $\Psi_s(f)$ is the QBL signal spectrum, $H(f)$ is the conversion filter transfer func-

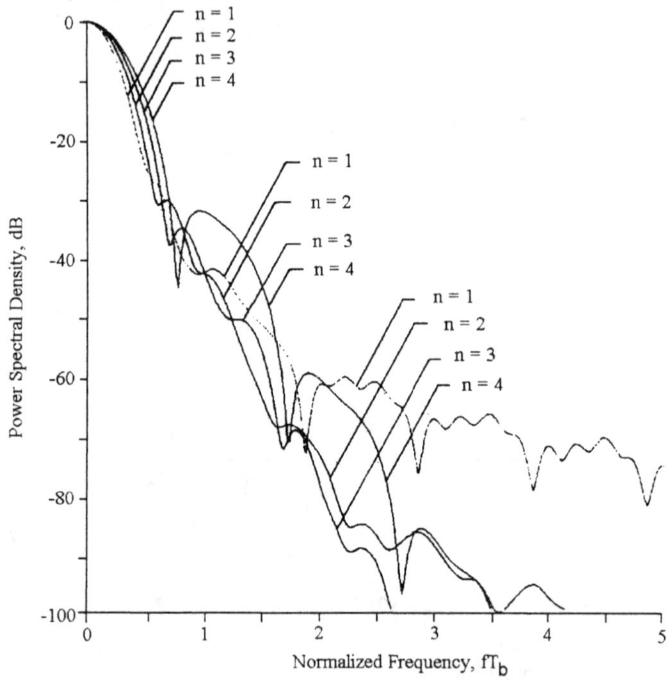

Figure 10.4 Power spectral densities of QBL-n with hard limiting, for $n = 1, 2, 3, 4$. From [1]. Copyright © 1979 IEEE.

n	Loss (dB) with hard limiting	Loss (dB) no limiting	Limiter loss (dB)	Eye Opening with hard limiting (%)	Eye Opening no limiting (%)
1	8.16	6.53	1.63	36	45
2	2.13	1.83	0.30	73	77
3	0.66	0.54	0.12	88	90
4	0.15	0.12	0.03	95	96

Table 10.1 Losses of power efficiency for QBL signals with respect to MSK.

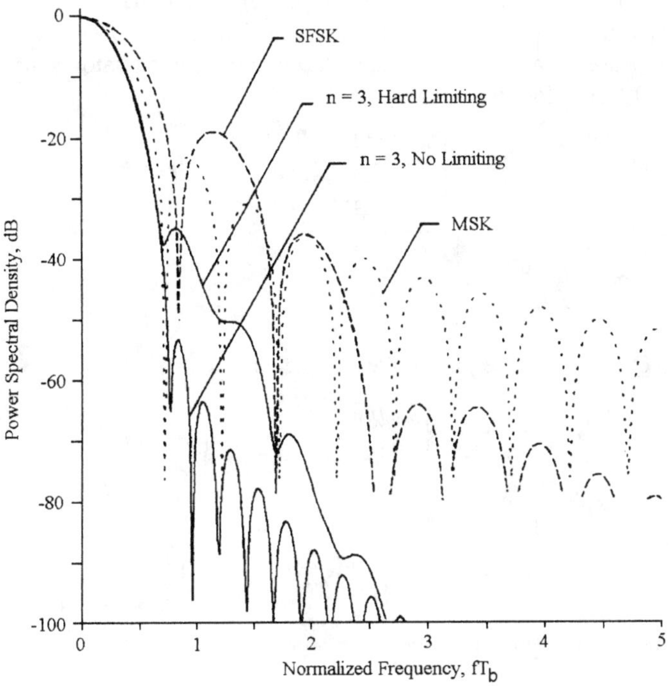

Figure 10.5 Power spectral densities for classical MSK, SFSK, and QBL-n for $n = 3$ with hard limiting, and $n = 3$ without limiting. From [1]. Copyright © 1979 IEEE.

tion, and $\Psi_{BPSK}(f)$ is the spectrum of the BPSK signal which enters the filter. At the receiver a filter matched to $p(t)$ is used. This receiver is not optimum since the matched filter does not match the envelopes of the received signal.

The loss in power efficiency in comparison with MSK is shown in Table 10.1. The eye opening is defined as the ratio of smallest detected voltage to mean detected voltage in I- or Q-channel baseband signal. A matched filter is assumed in the receiver, which introduces intersymbol interference. This accounts for the less than 100% eye openings even without hard limiting. The signal would always have 100% eye openings if there were no hard limiting and filtering (Figure 10.2(b)). The table shows that the losses range from a fraction of a decibel to many decibels. It is clear that the $n = 3$ case is the best choice since its loss is only a fraction of a decibel while its PSD is the most compact one. Its eye opening loss is also small, about

10%. According to [1], it appears that much of the loss is due to intersymbol interference of the commonly encountered linear type, suggesting that a moderate amount of transversal filtering could reduce the loss considerably.

If QBL is used in an AWGN channel, the optimum demodulator is the one given in Section 10.1. The four waveforms are

$$\begin{cases} f_1(t) = \left[\dfrac{\sin\frac{\pi t}{2T_b}}{\frac{\pi t}{2T_b}}\right]^n + \left[\dfrac{\sin\frac{\pi(t-2T_b)}{2T_b}}{\frac{\pi(t-2T_b)}{2T_b}}\right]^n, & a_k = a_{k+1} = 1 \\ f_2(t) = \left[\dfrac{\sin\frac{\pi t}{2T_b}}{\frac{\pi t}{2T_b}}\right]^n - \left[\dfrac{\sin\frac{\pi(t-2T_b)}{2T_b}}{\frac{\pi(t-2T_b)}{2T_b}}\right]^n, & a_k = 1, a_{k+1} = -1 \\ f_3(t) = -f_2(t), & a_k = -1, a_{k+1} = 1 \\ f_4(t) = -f_1(t), & a_k = a_{k+1} = -1 \end{cases} \tag{10.12}$$

For $n = 3$ (QBL-3), the energies of these signal are

$$\mathcal{E}_i = \begin{cases} 1.206T_b, & i = 1, 4 \\ 0.994T_b, & i = 2, 3 \end{cases} \tag{10.13}$$

The energies of $s_i(t)$ are

$$E_i = \frac{1}{2}\mathcal{E}_i = \begin{cases} 0.603T_b, & i = 1, 4 \\ 0.497T_b, & i = 2, 3 \end{cases} \tag{10.14}$$

and the distances of $s_i(t)$ are

$$[d_{ij}^2] = T_b \begin{bmatrix} 0 & 1.1 & 1.1 & 2.411 \\ 1.1 & 0 & 1.988 & 1.1 \\ 1.1 & 1.988 & 0 & 1.1 \\ 2.411 & 1.1 & 1.1 & 0 \end{bmatrix} \tag{10.15}$$

The upper bound on the symbol error probability is

$$P_s \leq 4Q\left(\sqrt{\frac{1.1T_b}{2N_o}}\right) + Q\left(\sqrt{\frac{1.988T_b}{2N_o}}\right) + Q\left(\sqrt{\frac{2.411T_b}{2N_o}}\right) \tag{10.16}$$

At high signal-to-noise ratio, since $d^2_{\text{min}} = 1.1T_b$ is much smaller (2.6 dB) than the next smallest distance, the P_b given by (10.10) is applicable to QBL signals. That is,

$$P_b \lessapprox 2Q\left(\sqrt{\frac{2E_b}{N_o}}\right) \tag{10.17}$$

This translates to a fraction of dB increase in SNR for the same P_b when compared with MSK.

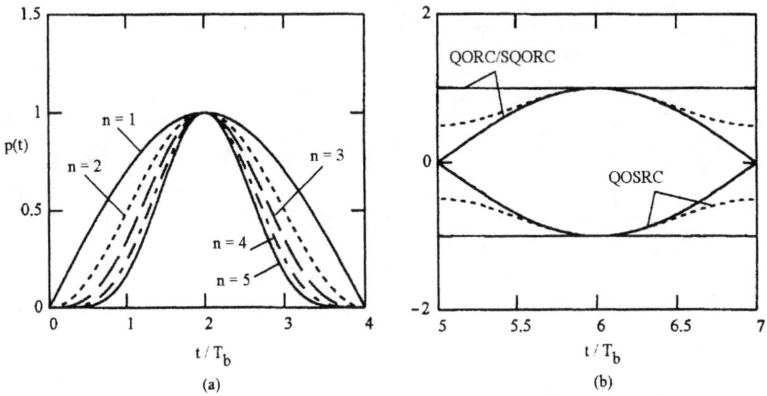

Figure 10.6 (a) Pulse shapes, and (b) eye patterns for QORC/SQORC (solid line) and QOSRC (dotted line).

10.3 QORC, SQORC, AND QOSRC

Quadrature overlapped raised-cosine modulation (QORC) and its staggered version (SQORC) were proposed in [2]. The amplitude pulse-shaping function (or baseband pulse) is the raised-cosine pulse given by

$$p(t) = \frac{1}{2}(1 - \cos\frac{\pi t}{2T_b}), \quad 0 \leq t \leq 4T_b \tag{10.18}$$

The QORC/SQORC signal's I- and Q-channel data streams take the form of overlapping raised-cosine pulse shapes, that is, $s_I(t)$ and $s_Q(t)$ are given by (10.2) and (10.3). If $\tau = 0$, the signal is QORC. If $\tau = T_b$, the signal becomes SQORC.

Later the pulse shape in (10.18) was generalized to [4]

$$p(t) = \left[\sin(\frac{\pi t}{4T_b})\right]^n, \quad 0 \leq t \leq 4T_b \tag{10.19}$$

Note that the $n = 2$ case is the pulse for QORC and that the $n = 4$ case is named as quadrature overlapped squared raised-cosine (QOSRC) modulation. The I- and Q-channel signals are staggered by T_b in QOSRC. The pulses for $n = 1, ..., 5$ are shown in Figure 10.6. Also shown are the eye patterns in $[4T_b, 8T_b]$ or any $[2kT_b, 2(k+1)T_b]$ interval. The inner eye of QOSRC coincides with that of QORC/SQORC. If sampled at $t = 2kT_b$, there is no ISI.

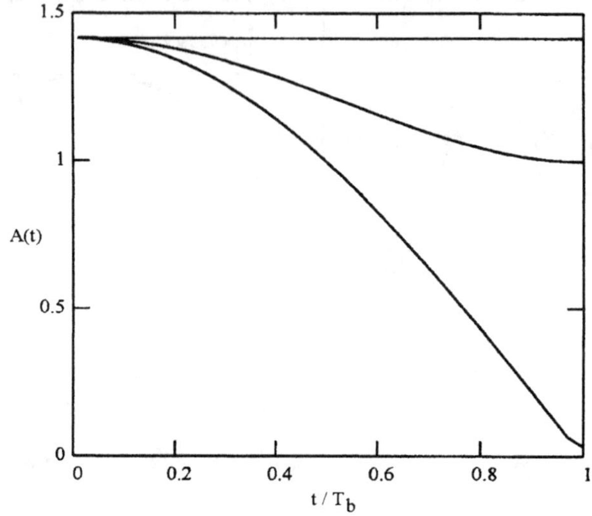

Figure 10.7 Possible envelopes for QORC.

Using the same technique for QBL, the amplitude variation of QORC, SQORC, and QOSRC can be found. For QORC, there is no staggering, the amplitude is given by

$$A(t) = \sqrt{[I_0 p(t) + I_{-1} p(t + 2T_b)]^2 + [Q_0 p(t) + Q_{-1} p(t + 2T_b)]^2}$$

It is found that $A_{\max} = \sqrt{2}$ and $A_{\min} = 0$. Figure 10.7 shows the three possible envelopes for QORC. It shows that the envelopes may descend down to zero at bit boundaries. For SQORC, there is staggering, the amplitude is

$$A(t) = \sqrt{[I_0 p(t) + I_{-1} p(t + 2T_b)]^2 + [Q_{-1} p(t + T_b) + Q_{-2} p(t + 3T_b)]^2} \tag{10.20}$$

It is found that $A_{\max} = \sqrt{2}$ and $A_{\min} = 1$. Thus

$$A_{\min}/A_{\max} \approx 0.7$$

For QOSRC, there is staggering too, and the amplitude formula is again (10.20). It

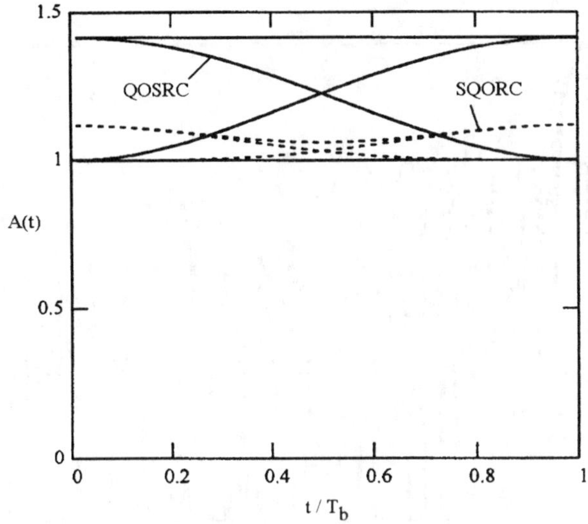

Figure 10.8 Possible envelopes for SQORC and QOSRC.

is found that $A_{\max} = 1.118$ and $A_{\min} = 1$. Thus

$$A_{\min}/A_{\max} \approx 0.9$$

Figure 10.8 shows the four possible envelopes for SQORC and QOSRC, respectively. The flat line $A(t) = 1$ is for both cases.

The power spectral density expression of QORC/SQORC derived in [2] is

$$\Psi_{\tilde{s}}(f) = P_c T_b \left[\frac{\sin(2\pi f T_b)}{2\pi f T_b} \right]^2 \left[\frac{\cos(2\pi f T_b)}{1 - (4 f T_b)^2} \right]^2 \tag{10.21}$$

where P_c is the power of the modulated signal. Comparing (10.21) with (4.38) and (5.14) reveals that except for a constant, the PSD of QORC/SQORC is the product of the PSDs of MSK and QPSK/OQPSK. The PSD is shown in Figure 10.9 in comparison with those of QPSK/OQPSK and MSK. The PSD of QORC/SQORC retains the same first null as QPSK/OQPSK. The remaining nulls occur twice as often as they do for QPSK/OQPSK. From the denominators of (10.21) we can see the side-lobe roll-off rate is proportional to f^{-6} which is the product of that of QPSK/OQPSK (f^{-2}) and that of MSK (f^{-4}).

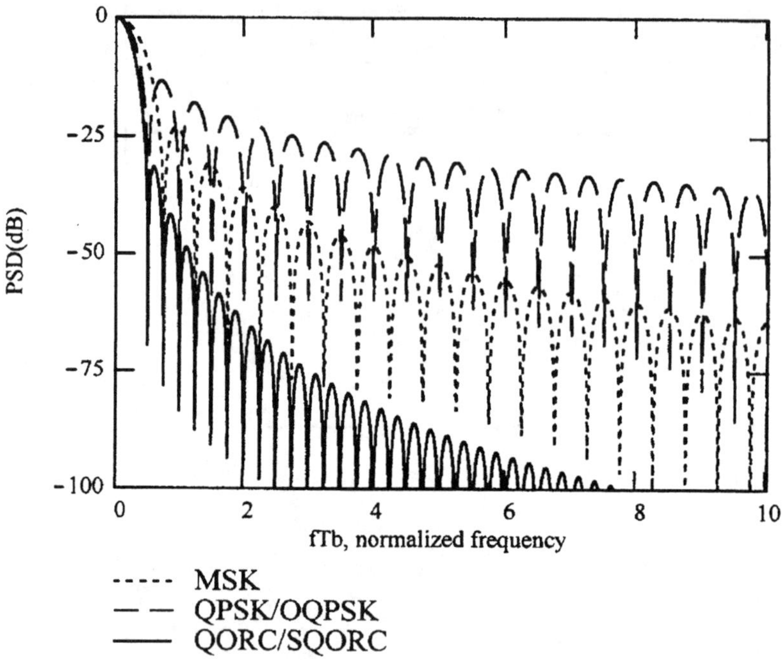

Figure 10.9 PSD of QORC/SQORC.

The PSDs for QOSRC and other pulses given in (10.19) were obtained by numerical Fourier transform in [4] and are shown in Figure 10.10. It is seen that all PSDs have almost the same characteristics in the region $0 \leq fT_b \leq 0.7$ while the sidelobes of the $n = 4$ case (QOSRC) drop much faster than others, including that of QORC/SQORC.

The error performance of QORC/SQORC in a nonlinear channel was evaluated through computer simulation in [2]. The channel consists of an input filter followed by a TWT (traveling wave tube) amplifier. The two types of input filter are a seven-pole Chebyshev design with a 3 dB RF bandwidth of 56 MHz and a 56 MHz phase equalized filter. The demodulator is the QPSK/OQPSK type demodulator with a third-order Butterworth filter following the carrier multiplier in I- and Q-channel, respectively. The BT products were optimized as the data rate was varied. Simulation results show that QORC and QPSK perform equally well for the entire bit rate range

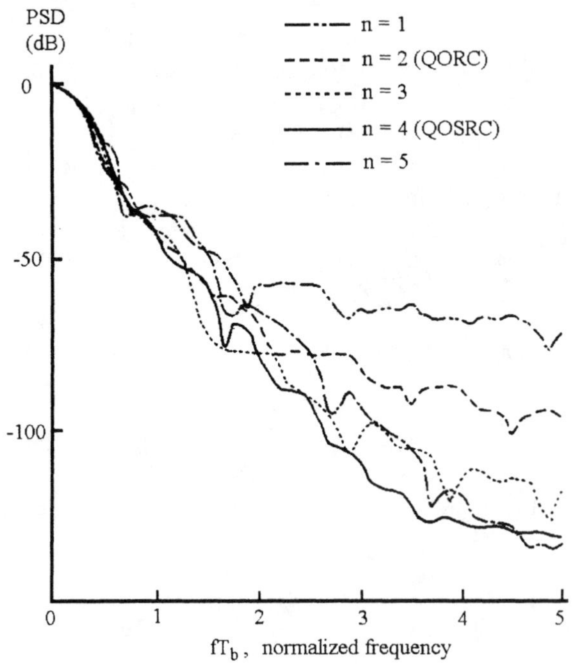

Figure 10.10 PSDs of modulated signals using the pulses in (10.19). From [4]. Copyright © 1985 IEEE.

(40 to120 Mbps). MSK performs about 0.7 dB better at lower rates and 0.5 to 1.0 dB worse at higher rates. SQPSK performs equally well with QORC at lower rates but about 1.0 dB worse at high rates. The error performance of QOSRC is considered on a hard-limited satellite channel when both uplink and downlink additive Gaussian noise and intersymbol interference caused by transmitter filtering are present [4]. It was found that the error performance of QOSRC is better than that of QORC.

If QORC/SQORC is used in an AWGN channel, the optimum demodulator given in Figure 10.1 can be used. The four waveforms in $[0, 2T_b]$ (or $[T_b, 3T_b]$ for Q-channel of SQORC) are

$$\begin{cases} f_1(t) = 1, & a_k = a_{k+1} = 1 \\ f_2(t) = \cos(\frac{\pi t}{2T_b}), & a_k = 1, a_{k+1} = -1 \\ f_3(t) = -f_2(t), & a_k = -1, a_{k+1} = 1 \\ f_4(t) = -f_1(t), & a_k = a_{k+1} = -1 \end{cases} \tag{10.22}$$

The energies of these signals are

$$\mathcal{E}_i = \begin{cases} 2T_b, & i = 1,4 \\ T_b, & i = 2,3 \end{cases} \tag{10.23}$$

The energies of $s_i(t)$ are

$$E_i = \frac{1}{2}\mathcal{E}_i = \begin{cases} T_b, & i = 1,4 \\ 0.5T_b, & i = 2,3 \end{cases} \tag{10.24}$$

and the distances of $s_i(t)$ are

$$[d_{ij}^2] = T_b \begin{bmatrix} 0 & 1.5 & 1.5 & 4 \\ 1.5 & 0 & 2 & 1.5 \\ 1.5 & 2 & 0 & 1.5 \\ 4 & 1.5 & 1.5 & 0 \end{bmatrix} \tag{10.25}$$

The upper bound on the symbol error probability is

$$P_s \leq 4Q\left(\sqrt{\frac{1.5T_b}{2N_o}}\right) + Q\left(\sqrt{\frac{2T_b}{2N_o}}\right) + Q\left(\sqrt{\frac{4T_b}{2N_o}}\right) \tag{10.26}$$

Since $d_{\min}^2 = 1.5T_b$ is only about 1.2 dB smaller than the second smallest distance, the P_b expression should include the second smallest distance. Among the four error events associated with the $d_{\min}$ terms, two events (events associated with d_{12} and d_{34}) actually do not cause bit errors. The error event associated with the second smallest distance (d_{23}) does make one bit error. From (10.24) we see that $E_b = E_{avg}/2 = 0.375T_b$, thus

$$P_b \lessapprox 2Q\left(\sqrt{\frac{2E_b}{N_o}}\right) + Q\left(\sqrt{\frac{2.67E_b}{N_o}}\right) \tag{10.27}$$

This is slightly higher than MSK and QBL.

For QOSRC we can find in $[0, 2T_b]$ for I-channel (or $[T_b, 3T_b]$ for Q-channel)

$$\begin{cases} f_1(t) = \left[\sin(\frac{\pi(t+2T_b)}{4T_b})\right]^4 + \left[\sin(\frac{\pi t}{4T_b})\right]^4, & a_k = a_{k+1} = 1 \\ f_2(t) = \left[\sin(\frac{\pi(t+2T_b)}{4T_b})\right]^4 - \left[\sin(\frac{\pi t}{4T_b})\right]^4, & a_k = 1, a_{k+1} = -1 \\ f_3(t) = -f_2(t), & a_k = -1, a_{k+1} = 1 \\ f_4(t) = -f_1(t), & a_k = a_{k+1} = -1 \end{cases} \tag{10.28}$$

The energy of each signal is

$$\mathcal{E}_i = \begin{cases} 1.188T_b, & i = 1, 4 \\ T_b, & i = 2, 3 \end{cases} \tag{10.29}$$

The energies of $s_i(t)$ are

$$E_i = \frac{1}{2}\mathcal{E}_i = \begin{cases} 0.594T_b, & i = 1, 4 \\ 0.5T_b, & i = 2, 3 \end{cases} \tag{10.30}$$

and the distances of $s_i(t)$ are

$$[d_{ij}^2] = T_b \begin{bmatrix} 0 & 1.094 & 1.094 & 2.375 \\ 1.094 & 0 & 2 & 1.094 \\ 1.094 & 2 & 0 & 1.094 \\ 2.375 & 1.094 & 1.094 & 0 \end{bmatrix} \tag{10.31}$$

The upper bound on the symbol error probability is

$$P_s \le 4Q\left(\sqrt{\frac{1.094T_b}{2N_o}}\right) + Q\left(\sqrt{\frac{2T_b}{2N_o}}\right) + Q\left(\sqrt{\frac{2.375T_b}{2N_o}}\right) \tag{10.32}$$

At high signal-to-noise ratio, since $d_{\min}^2 = 1.094T_b$ is much smaller (2.6 dB) than the next smallest distance, the P_b given by (10.10) is applicable to QOSRC signal. That is,

$$P_b \lessapprox 2Q\left(\sqrt{\frac{2E_b}{N_o}}\right) \tag{10.33}$$

Thus P_b of QOSRC is the same as that of QBL and is slightly higher than that of MSK.

10.4 IJF-OQPSK AND TSI-OQPSK

Intersymbol-interference/jitter-free OQPSK (IJF-OQPSK) was first proposed in [5]. An example of intersymbol-interference-free and jitter-free baseband signal and its generator are shown in Figure 10.10(a) and (b), respectively. When sampled at symbol boundaries, the sampled signal is clearly ISI-free. It is also jitter-free since there is no abrupt amplitude change at any sampling instant so that any symbol timing jitter essentially causes no errors in sampled signals. From the figure we can see that the filtered data stream consists of four pulse shapes: ± 1 and $\pm \sin(\pi t/T_s)$ defined in $[-T_s/2, T_s/2]$ where $T_s = 2T_b$ is the symbol duration.[1] Therefore we need only two pulse shapes to define an intersymbol-interference/jitter-free baseband signal. One is an even function $s_e(t)$ and the other is an odd function $s_o(t)$ with equal nonzero amplitude at the symbol boundaries, that is,

$$\begin{cases} s_e(t) = s_e(-t), & |t| \le T_s/2 \\ s_o(t) = -s_o(t), & |t| \le T_s/2 \\ s_e(t) = s_o(t) = 0, & |t| > T_s/2 \\ s_e(t) = s_o(t) \neq 0, & t = T_s/2 \end{cases} \tag{10.34}$$

The encoding of binary data stream $\{a_k\}$ into an IJF data stream $\{y_k(t)\}$ obeys the following rules

$$y_k(t) = \begin{cases} s_e(t - kT_s), & \text{if } a_k = a_{k-1} = 1 \\ -s_e(t - kT_s), & \text{if } a_k = a_{k-1} = -1 \\ s_o(t - kT_s), & \text{if } a_k = 1, a_{k-1} = -1 \\ -s_o(t - kT_s), & \text{if } a_k = -1, a_{k-1} = 1 \end{cases} \tag{10.35}$$

where $y_k(t)$ is the waveform in $[(k-1)T_s, kT_s]$. In Figure 10.11,

$$\begin{cases} s_e(t) = 1, & |t| \le T_s/2 \\ s_o(t) = \sin(\frac{\pi t}{T_s}), & |t| \le T_s/2 \end{cases} \tag{10.36}$$

and $s_e(t) = s_o(t) = 1$ for $t = T_s/2$. The encoding rules are clearly demonstrated by the example in Figure 10.11.

When I- and Q-channel data of an OQPSK modulator undergo the above data encoding and pulse shaping, we obtain the scheme named IJF-OQPSK in the literature.

This early IJF-OQPSK scheme was extended to a class of schemes called two-symbol-interval hard limited OQPSK (TSI-OQPSK) [6]. This class of two-symbol-

---

[1] In this section and the next section we use T_s instead of T_b as the basic time interval in order to be consistent with the literature that proposed the schemes. This enables the reader to refer to the literature without confusion caused by different basic time intervals.

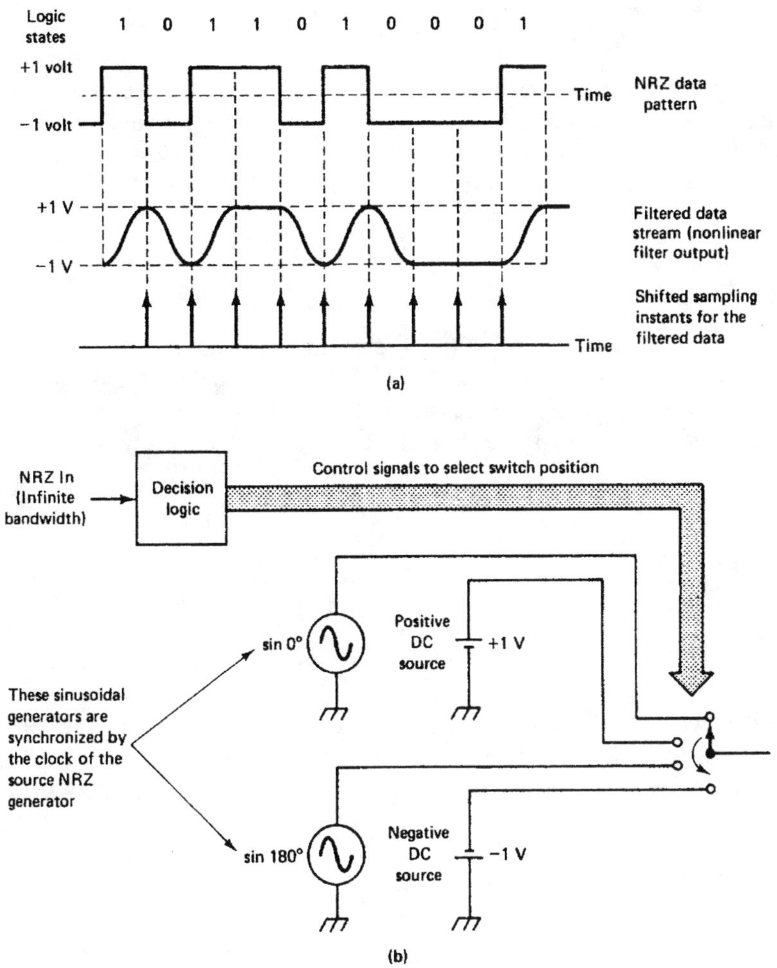

Figure 10.11 Intersymbol-interference/jitter-free signaling (a, b). From [9, p. 343]. Copyright © 1987 Kamilo Feher.

interval amplitude pulse-shaping functions is defined as

$$p^{(n)}(t) = \begin{cases} \frac{1}{2}\left[1 - \frac{\sin\{\frac{\pi}{nT_s}(|t|-T_s/2)\}}{\sin(\frac{\pi}{2n})}\right], & -T_s \le t \le T_s \\ 0, & \text{elsewhere} \quad n = 1, 2, \ldots. \end{cases} \tag{10.37}$$

The pulse lasts for two symbol durations in the I- and Q-channel. They are overlapped in the baseband signal streams. By choosing different values of n, a class of TSI-OQPSK schemes results. Figure 10.12(a) shows this class of pulses. When $n = 1$, (10.37) becomes a raised-cosine function

$$p^{(1)}(t) = \begin{cases} \frac{1}{2}\left[1 + \cos(\frac{\pi t}{T_s})\right], & -T_s \le t \le T_s \\ 0, & \text{elsewhere} \end{cases} \tag{10.38}$$

This is the same pulse defined in (10.18). Therefore TSI-OQPSK ($n = 1$) is SQORC and as will be shown next, it is also the IJF-OQPSK defined in (10.36).

To show that this class of pulses can realize an IJF scheme, we notice that in the interval of $[0, T_s]$, the composite waveform is one of the four waveforms: $\pm p^{(n)}(t) \pm p^{(n)}(t - T_s)$. Using (10.37) we have

$$p^{(n)}(t) + p^{(n)}(t - T_s) = 1, \quad 0 \le t \le T_s$$

and

$$-p^{(n)}(t) + p^{(n)}(t - T_s) = \frac{\sin\{\frac{\pi}{nT_s}(t - T_s/2)\}}{\sin(\frac{\pi}{2n})}, \quad 0 \le t \le T_s$$

The other two waveforms are just their negatives, respectively. These waveforms are just the shifted-by-$T_s/2$ version of the odd and even functions required by the IJF property. That is, the odd and even functions that satisfy the IJF conditions in (10.34) are

$$s_e^{(n)}(t) = p^{(n)}(t + \frac{T_s}{2}) + p^{(n)}(t - \frac{T_s}{2}) = \begin{cases} 1, & |t| \le T_s/2 \\ 0, & \text{elsewhere} \end{cases} \tag{10.39}$$

and

$$s_o^{(n)}(t) = -p^{(n)}(t + \frac{T_s}{2}) + p^{(n)}(t - \frac{T_s}{2}) = \begin{cases} \frac{\sin(\frac{\pi t}{nT_s})}{\sin(\frac{\pi}{2n})}, & |t| \le T_s/2 \\ 0, & \text{elsewhere} \end{cases} \tag{10.40}$$

Thus the waveforms of the baseband data stream are similar to Figure 10.11(a) with the curved parts obeying the functions $\pm s_o^{(n)}(t)$. Figure 10.12(b, c) shows the even and odd functions given in (10.39) and (10.40). Note that when $n = 1$, the odd and

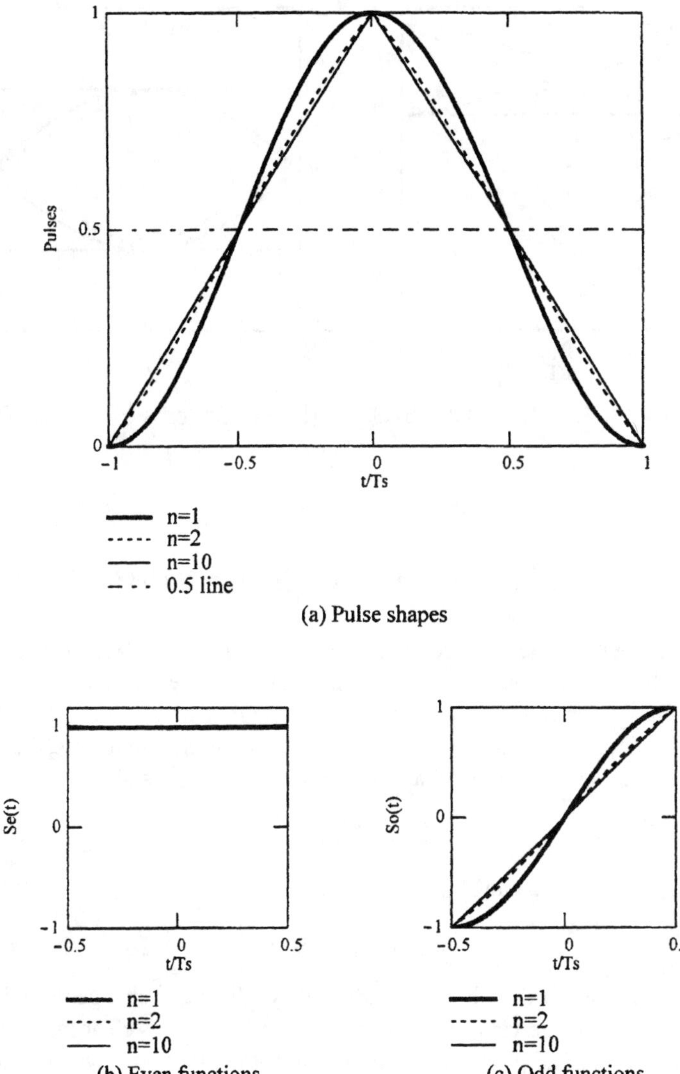

Figure 10.12 TSI pulses and odd and even functions (a–c).

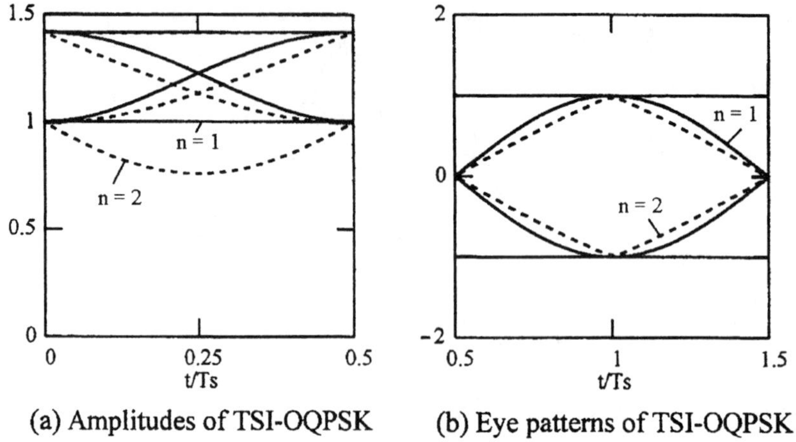

Figure 10.13 (a) Amplitudes and (b) eye patterns of TSI-OQPSK.

even pulses are the same as those in (10.36), that is, IJF-OQPSK is a special case of TSI-OQPSK. Its TSI pulse is a raised-cosine function given in (10.38).

This class of schemes is nonconstant envelope schemes. Figure 10.13 shows the possible amplitudes and the eye diagrams (the two flat lines in the eye diagrams are for any value of n) . It is found that for $n = 1$, $A_{\max} = \sqrt{2}$ and $A_{\min} = 1$. Thus

$$A_{\min}/A_{\max} \approx 0.7$$

For $n = 2$, $A_{\max} = \sqrt{2}$ and $A_{\min} = 0.765$.

$$A_{\min}/A_{\max} \approx 0.54$$

When n increases further the amplitude and eye diagrams change very little since the pulse shapes change very little. From the eye patterns it is clear that this class of schemes is ISI-free and jitter-free if sampled at the center of the eye diagram.

The PSDs of TSI-OQPSK baseband signals are shown in Figure 10.14. It is seen that $n = 1$ case (IJF-OQPSK) has the lowest sidelobes. However, TSI signals with $n = 2, 3, ...$ have a narrower main lobe than the one with $n = 1$. Figure 10.15 is the PSDs after nonlinear amplification. It is seen that the TSI's spectral components that cause significant ACI (adjacent channel interference) are about 10 dB lower than

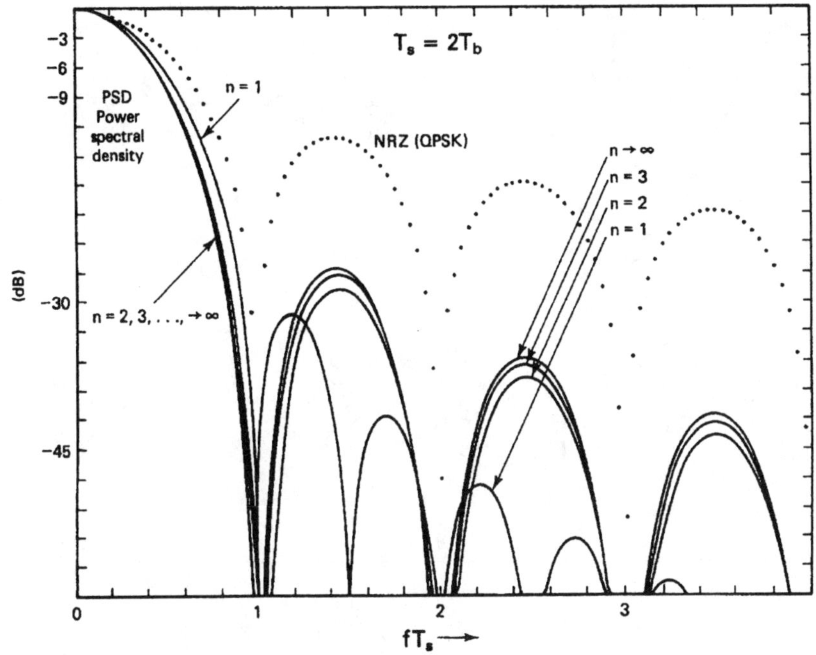

Figure 10.14 PSDs of TSI signals. From [6]. Copyright © 1983 IEEE.

QPSK and 5 dB lower than OQPSK.

Similar to QBL and QORC/SQORC, the optimum receiver of Section 10.1 can be used for TSI-OQPSK signals, including the IJF-OQPSK signal, in an AWGN channel. TSI-OQPSK ($n = 1$), IJF-OQPSK, and SQORC are identical. The bit error probability of SQORC has been proven to be essentially the same as that of MSK. Thus all of them have a bit error probability essentially equal to that of MSK.

For TSI-OQPSK ($n = 2$), substituting (10.37) into (10.4), we find in $[0, T_s]$ for I-channel (or $[T_s/2, 3T_s/2]$ for Q-channel)

$$\begin{cases} f_1(t) = 1, & a_k = a_{k+1} = 1 \\ f_2(t) = \sqrt{2}\sin\{\frac{\pi}{2T_s}(T_s/2 - t)\}, & a_k = 1, a_{k+1} = -1 \\ f_3(t) = -f_2(t), & a_k = -1, a_{k+1} = 1 \\ f_4(t) = -f_1(t), & a_k = a_{k+1} = -1 \end{cases} \tag{10.41}$$

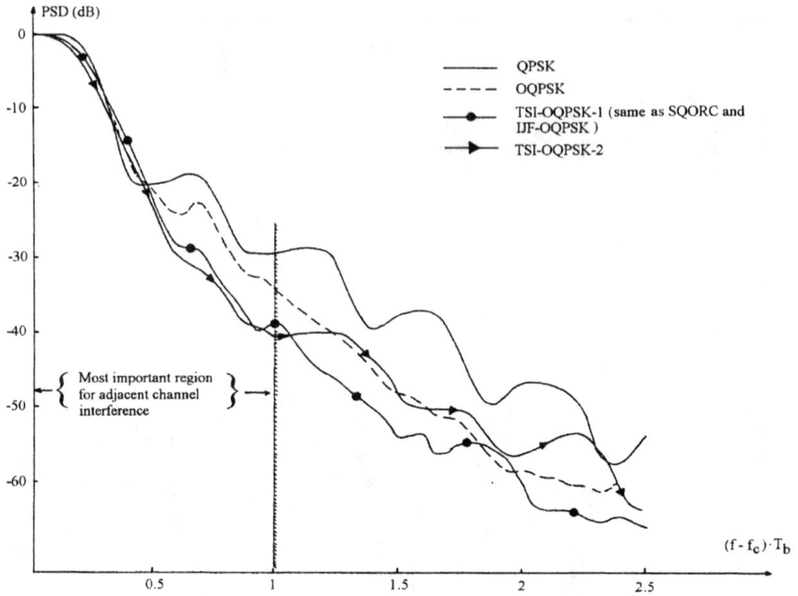

Figure 10.15 Nonlinearly amplified (saturated HPA with 0-dB input back off) PSDs of TSI signals. From [10]. Copyright © 1984 IEEE.

The energy of each signal is

$$\mathcal{E}_i = \begin{cases} 2T_b, & i = 1,4 \\ 0.726T_b, & i = 2,3 \end{cases} \tag{10.42}$$

Here we express $\mathcal{E}_i$ in terms of T_b in order that they can be readily compared with those of previously described schemes. The energies of $s_i(t)$ are

$$E_i = \frac{1}{2}\mathcal{E}_i = \begin{cases} T_b, & i = 1,4 \\ 0.363T_b, & i = 2,3 \end{cases} \tag{10.43}$$

and the distances of $s_i(t)$ are

$$[d_{ij}^2] = T_b \begin{bmatrix} 0 & 1.363 & 1.363 & 4 \\ 1.363 & 0 & 1.454 & 1.363 \\ 1.363 & 1.454 & 0 & 1.363 \\ 4 & 1.363 & 1.363 & 0 \end{bmatrix} \tag{10.44}$$

The upper bound on the symbol error probability is

$$P_s \leq 4Q\left(\sqrt{\frac{1.363T_b}{2N_o}}\right) + Q\left(\sqrt{\frac{1.454T_b}{2N_o}}\right) + Q\left(\sqrt{\frac{4T_b}{2N_o}}\right) \tag{10.45}$$

Since $d_{\min}^2 = 1.363T_b$ is only about 0.28 dB smaller than the second smallest distance, the P_b expression should include the second smallest distance. Among the four error events associated with the $d_{\min}$ terms, two events (events associated with d_{12} and d_{34}) actually do not cause bit errors. The error event associated with the second smallest distance ($d_{23} = 1.454T_b$) does make one bit error. From (10.43) we see that $E_b = E_{avg}/2 = 0.34075T_b$. The P_b is approximated as

$$P_b \lessapprox 2Q\left(\sqrt{\frac{2E_b}{N_o}}\right) + Q\left(\sqrt{\frac{2.133E_b}{N_o}}\right) \tag{10.46}$$

This is slightly higher than MSK and QBL.

The error performance of TSI-OQPSK signals was evaluated through simulation for satellite channels with saturated HPA (high power amplifier) or cascaded hard-limiter and HPA in an ACI environment with AWGN [10]. The demodulator in the simulation is just the ordinary OQPSK demodulator with proper filtering (see [10] for details). Due to the more compact PSDs of the TSI-OQPSK signals, it was found that in the ACI environment the degradation with respect to ideal channel is less for TSI-OQPSK signals than for conventional QPSK and OQPSK (Figures 10.16 to 10.18). In the figures, Δf is the carrier frequency spacing between two adjacent channels. The effect of the spectral advantages of TSI-OQPSK schemes becomes even more evident when the received modulated carrier power of the main channel is below that of the adjacent channels. This situation occurs in the case of an uplink fade of the desired channel in satellite communications. Figure 10.19 presents examples of E_b/N_o degradation as a function of the fade depth of the desired channel. As the main channel is attenuated, the ACI becomes predominant and for this reason the modulation technique which creates less ACI provides better performance. As an example, at a spacing of 92% of the bit rate, with a fade depth of 12 dB, the TSI-OQPSK-2 only has a degradation of 2 dB, whereas the degradation of QPSK is more than 7 dB (Figure 10.19(a)). With tighter spacing, the differences are even more

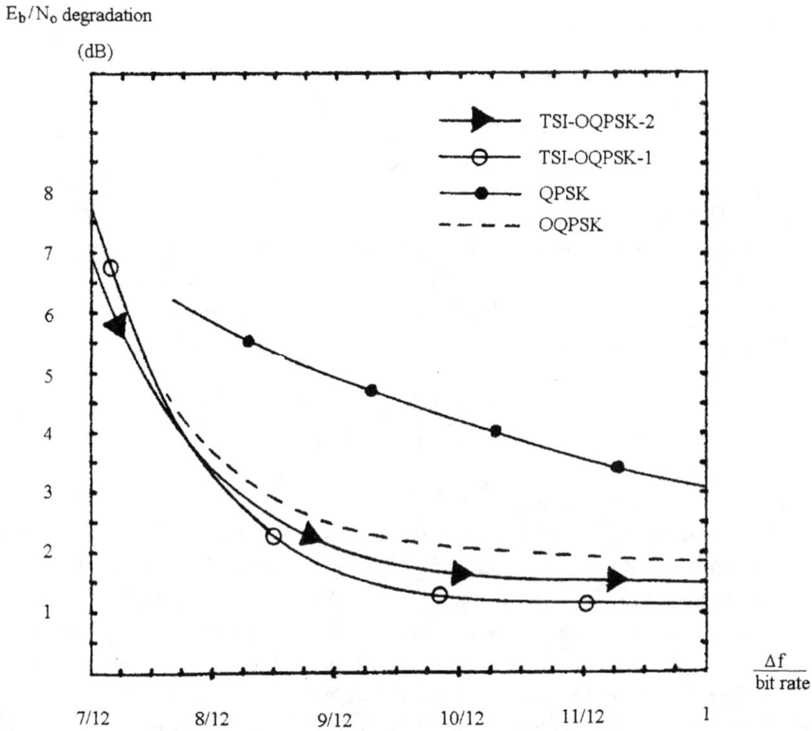

Figure 10.16 E_b/N_o degradation (at $P_b = 10^{-6}$) of ideal hard-limited QPSK, OQPSK, and TSI-OQPSK schemes in an ACI and AWGN environment. From [10]. Copyright © 1984 IEEE.

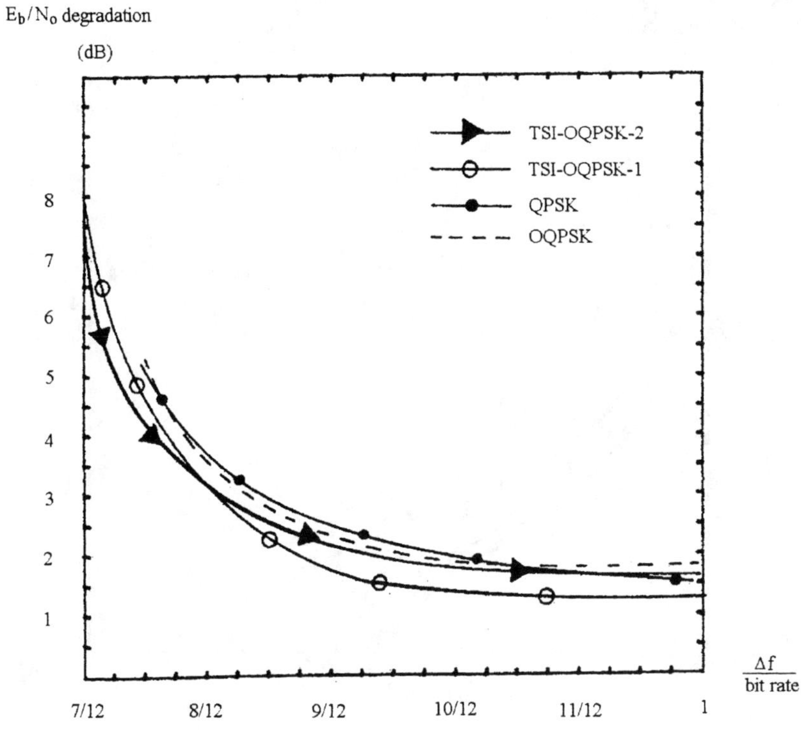

Figure 10.17 E_b/N_o degradation (at $P_b = 10^{-6}$) of saturated HPA-QPSK, OQPSK, and TSI-OQPSK schemes in an ACI and AWGN environment. From [10]. Copyright © 1984 IEEE.

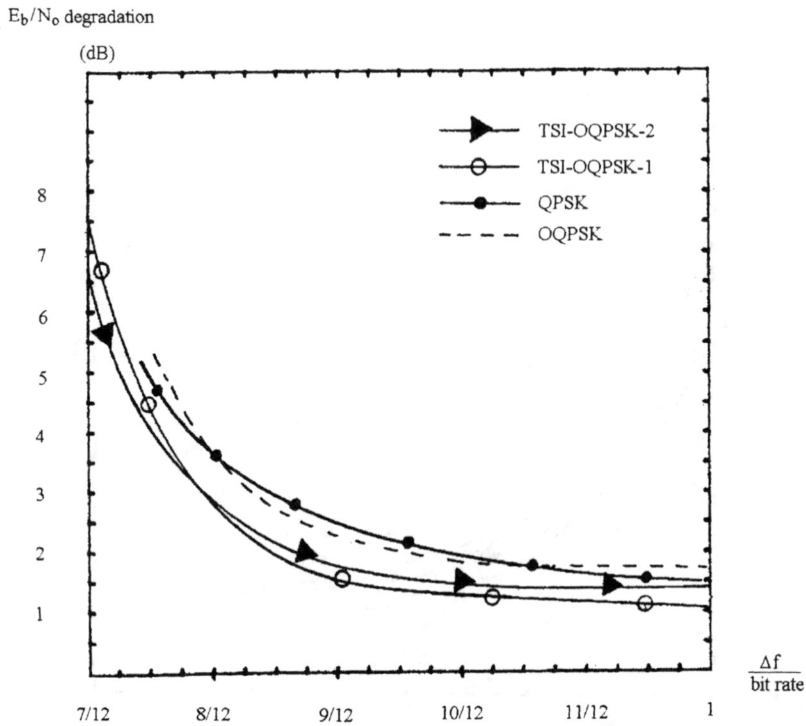

Figure 10.18 E_b/N_o degradation (at $P_b = 10^{-6}$) of a saturated HPA-QPSK, OQPSK, and cascaded hard-limiter-HPA TSI-OQPSK schemes in an ACI and AWGN environment. From [10]. Copyright © 1984 IEEE.

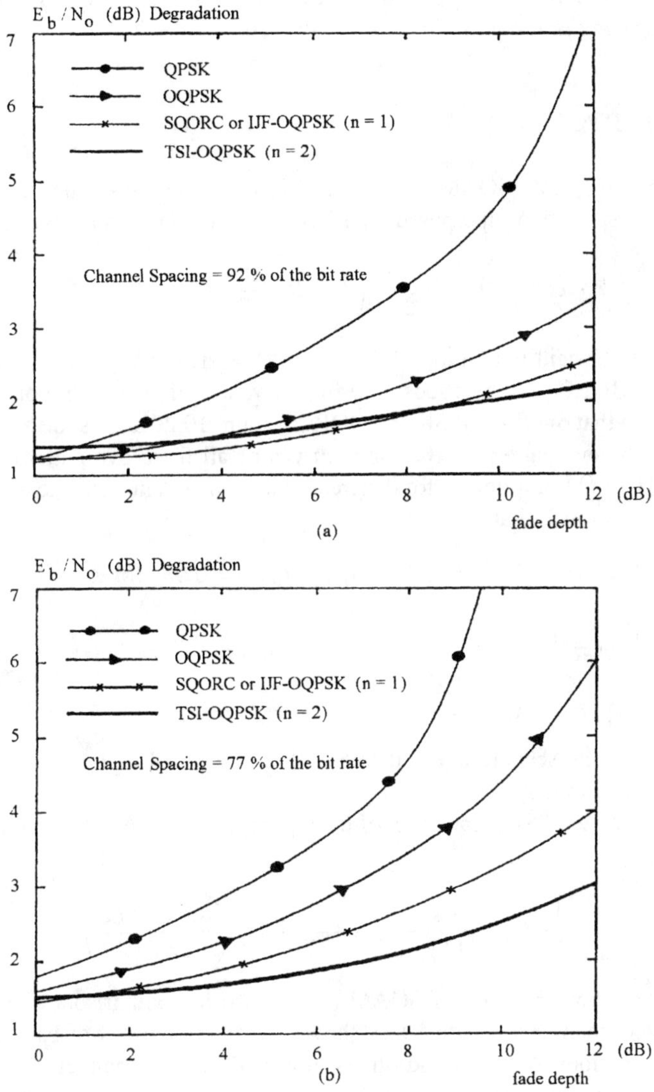

Figure 10.19 E_b/N_o degradation versus fade depth of the desired channel: (a) spacing of 92% of the bit rate, compared to $E_b/N_o = 8.4$ dB for $P_s = 10^{-4}$, (b) spacing of 77% of the bit rate, compared to $E_b/N_o = 8.4$ dB for $P_s = 10^{-4}$. From [10]. Copyright © 1984 IEEE.

evident and TSI-OQPSK-2 is better than the others at all fade depth (Figure 10.19(b) where the spacing is 77%).

10.5 SUPERPOSED-QAM

The superposed-QAM (SQAM) is proposed in [7]. The amplitude pulse-shaping function consists of two superposed raised-cosine functions and is given by

$$p(t) = \frac{1}{2}\left(1 + \cos\frac{\pi t}{T_s}\right) - \frac{1-A}{2}\left(1 - \cos\frac{2\pi t}{T_s}\right), \quad -T_s \le t \le T_s \tag{10.47}$$

where A is a constant within $[0.5, 1.5]$. When $A = 1$, the SQAM is equal to IJF-OQPSK or SQORC. But by choosing different values of A we can obtain a better spectrum than that of IJF-OQPSK or SQORC. Figure 10.20 shows the SQAM pulse shapes and the eye patterns. The inner curves of all three eye patterns coincide. Figure 10.21 shows amplitudes for different values of A. The flat line $A = 1$ is for both cases. It is found that

$$A_{\text{min}} = 1,\ A_{\text{max}} = 1.077,\ A_{\text{min}}/A_{\text{max}} = 0.93,\ \text{for } A = 0.7$$

$$A_{\text{min}} = 1,\ A_{\text{max}} = 1.166,\ A_{\text{min}}/A_{\text{max}} = 0.86,\ \text{for } A = 0.8$$

$$A_{\text{min}} = 1,\ A_{\text{max}} = 1.281,\ A_{\text{min}}/A_{\text{max}} = 0.78,\ \text{for } A = 0.9$$

When $A = 1$, the scheme is equal to SQORC, we recall that $A_{\text{max}} = \sqrt{2}$ and $A_{\text{min}} = 1$, the ratio is 0.7.

The normalized PSD expression of the equiprobable SQAM baseband signal is given by [7, 11]

$$\Psi_{\tilde{s}}(f) = \frac{1}{A^2}\left(\frac{1}{1-4T_s^2 f^2} + \frac{A-1}{1-T_s^2 f^2}\right)\left(\frac{\sin 2\pi f T_s}{2\pi f T_s}\right)^2 \tag{10.48}$$

Figure 10.22 shows the PSD of SQAM ($A = 0.8$) in comparison with others in a hard-limited channel. Figure 10.23 shows the fractional out-of-band power of SQAM with various A values and others in a hard-limited channel. From Figure 10.23 we can see that a decrease of A leads to a faster spectral roll-off at higher frequencies at the expense of a slightly wider main lobe. SQAM has spectral advantages over QPSK and MSK, and comparable spectral properties to TFM and QBL.

The error performance of SQAM was evaluated for a hard-limited channel by simulation in [7]. The demodulator in the simulation is the same as an OQPSK demodulator where a fourth-order Butterworth low-pass filter is used after the carrier

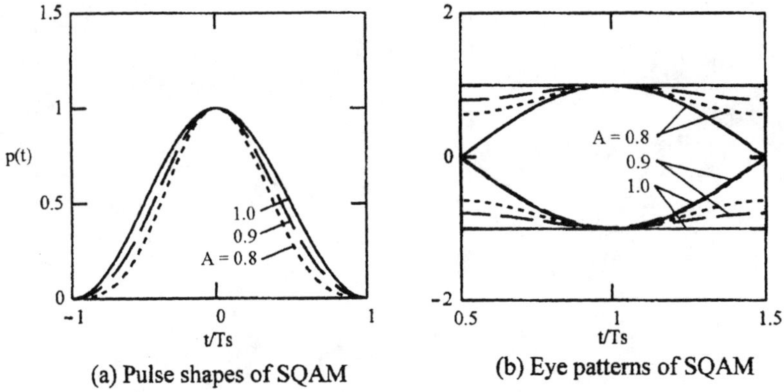

Figure 10.20 (a) Pulse shapes and (b) eye patterns of SQAM.

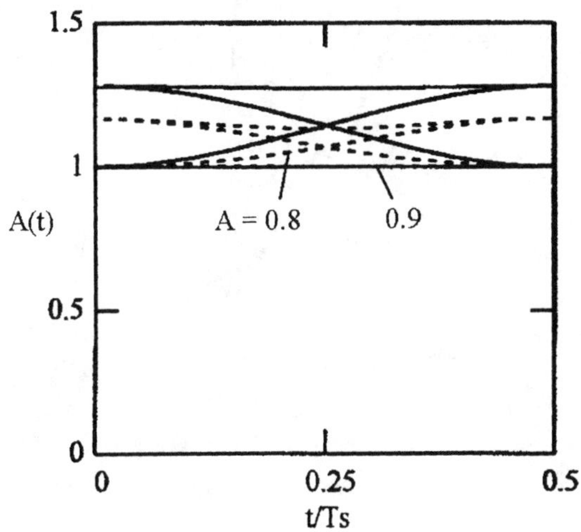

Figure 10.21 Amplitudes of SQAM.

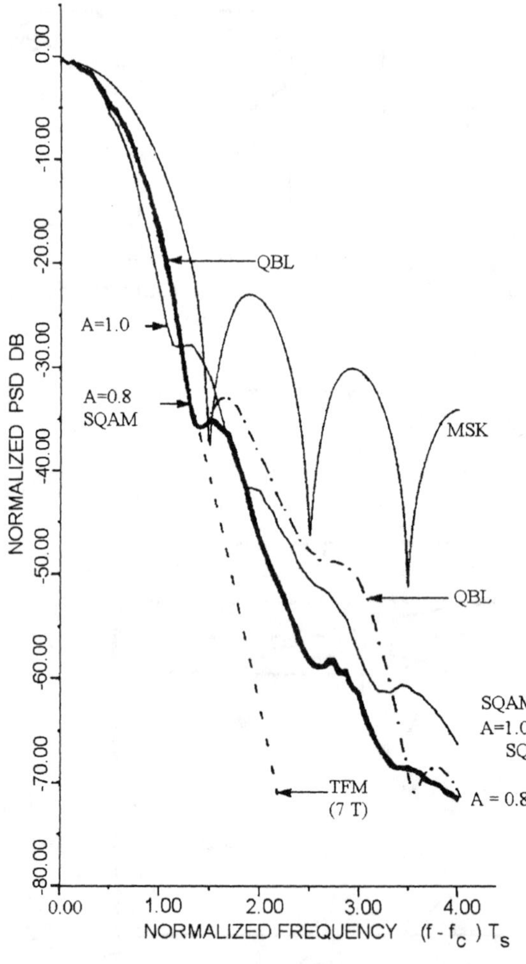

Figure 10.22 PSDs of SQAM, MSK, QBL, IJF-OQPSK (SQORC), and TFM in a nonlinear (hard-limited) channel. From [7]. Copyright © 1985 IEEE.

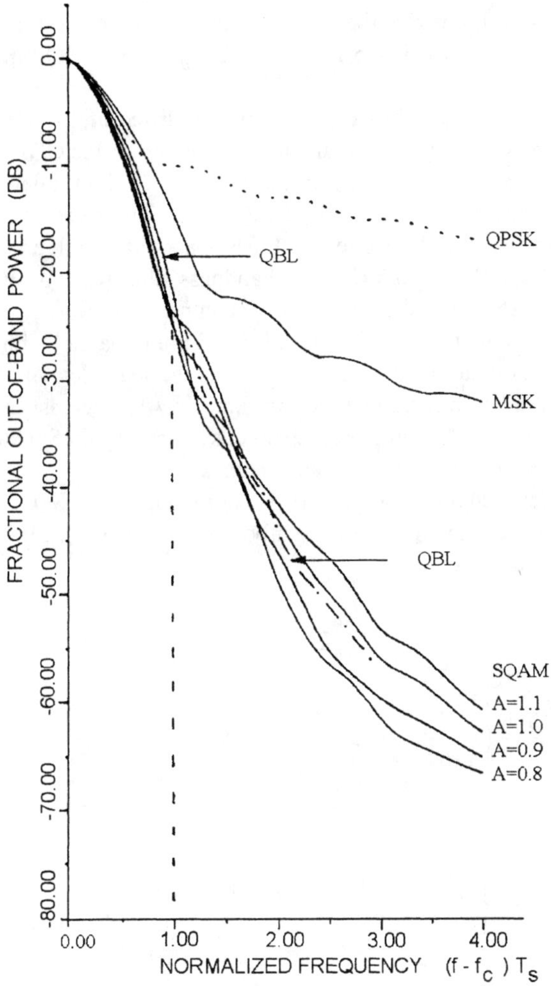

Figure 10.23 Out-of-band to total power ratio of SQAM, MSK, QBL, IJF-OQPSK (SQORC), and QPSK in a nonlinear (hard-limited) channel. From [7]. Copyright © 1985 IEEE.

multiplier. The filter's 3 dB bandwidth is $f_{3dB} = 1.1 f_N$ where f_N is the Nyquist bandwidth which in turn is one-half of the symbol rate. The simulation results are given in Figure 10.24. It can be seen that the performance is the best when $A = 0.8$. It was reported that $A = 0.8$ is also the best for a linear channel [7]. It is also shown that the performance of SQAM is 0.5 to 2 dB inferior to that of QPSK in a linear channel.

Error performance of SQAM in a nonlinearly amplified multichannel interference environment was studied in [11]. Again the demodulator is the OQPSK demodulator. In the simulation model, there is a transmitter filter (Tx filter) after the modulator and a receiver filter (Rx filter) before the demodulator. The Tx and Rx filters used in the simulation are fourth-order Butterworth bandpass filter with an equivalent 3 dB low-pass bandwidth $BT_s = 0.5$ (i.e., the bandpass filter has a 3 dB bandwidth of $1/T_s$) . Figure 10.25 shows the degradation (compared to $E_b/N_o = 8.4$ dB at $P_b = 10^{-4}$) against channel spacing for hard-limited multichannel with two equal-power ACIs. It is seen that SQAM ($A = 0.85$) has the least degradation. Figure 10.26 shows the degradation (compared to the same E_b/N_o) against fade-depth of the desired channel in a hard-limited multichannel system. Again SQAM (A=0.85) shows a significant improvement over other schemes.

The optimum demodulator in Figure 10.1 can be applied to SQAM. The four possible composite baseband signals in a symbol duration ($[0, T_s]$ for I-channel or $[T_s/2, 3T_s/2]$ for Q-channel) are

$$\begin{cases} f_1(t) = A + (1-A)\cos(2\pi t/T_s), & a_k = a_{k+1} = 1 \\ f_2(t) = -\cos(\pi t/T_s), & a_k = 1, a_{k+1} = -1 \\ f_3(t) = \cos(\pi t/T_s), & a_k = -1, a_{k+1} = 1 \\ f_4(t) = -A - (1-A)\cos(2\pi t/T_s), & a_k = a_{k+1} = -1 \end{cases} \tag{10.49}$$

The energy of each signal is

$$\mathcal{E}_i = \begin{cases} 2[A^2 + (1-A)^2/2]T_b, & i = 1, 4 \\ T_b, & i = 2, 3 \end{cases} \tag{10.50}$$

The energies of $s_i(t)$ are

$$E_i = \frac{1}{2}\mathcal{E}_i = \begin{cases} \left[A^2 + (1-A)^2/2\right] T_b, & i = 1, 4 \\ 0.5T_b, & i = 2, 3 \end{cases} \tag{10.51}$$

and the distances of $s_i(t)$ can be calculated by (10.9). The distances for $A = 0.8$ and

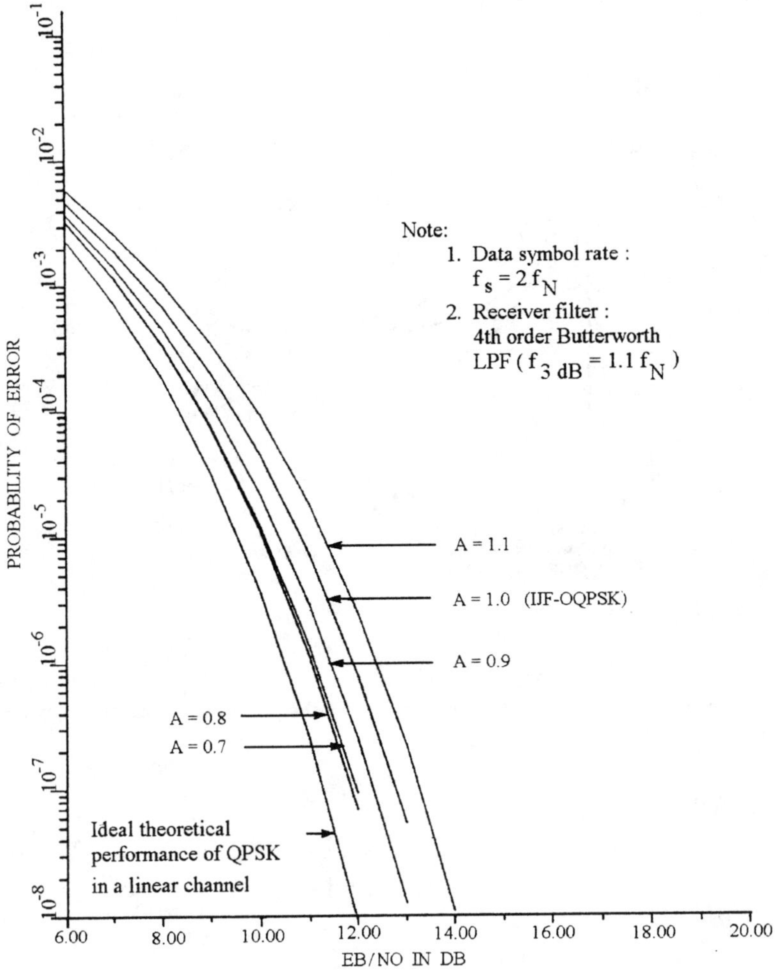

Figure 10.24 Error performance (P_b) of SQAM in a nonlinear (hard-limited) channel. From [7]. Copyright © 1985 IEEE.

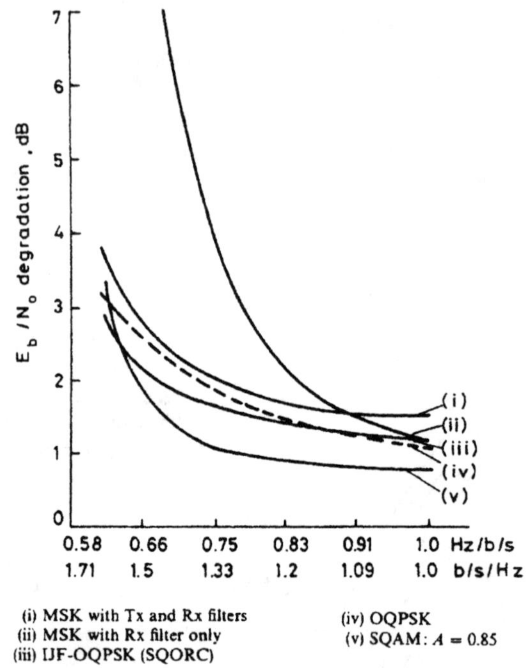

Figure 10.25 E_b/N_o degradation (compared to $E_b/N_o = 8.4$ dB at $P_b = 10^{-4}$) against channel spacing for a hard-limited multichannel system with two equal-power ACIs. From [11]. Copyright © 1985 IEE.

0.9 are as follows:

$$[d_{ij}^2] = T_b \begin{bmatrix} 0 & 1.16 & 1.16 & 2.64 \\ 1.16 & 0 & 2 & 1.16 \\ 1.16 & 2 & 0 & 1.16 \\ 2.64 & 1.16 & 1.16 & 0 \end{bmatrix}, \quad \text{for } A = 0.8 \tag{10.52}$$

$$[d_{ij}^2] = T_b \begin{bmatrix} 0 & 1.315 & 1.315 & 3.26 \\ 1.315 & 0 & 2 & 1.315 \\ 1.315 & 2 & 0 & 1.315 \\ 3.26 & 1.315 & 1.315 & 0 \end{bmatrix}, \quad \text{for } A = 0.9 \tag{10.53}$$

The symbol error probability upper bound can be found accordingly using (10.6).

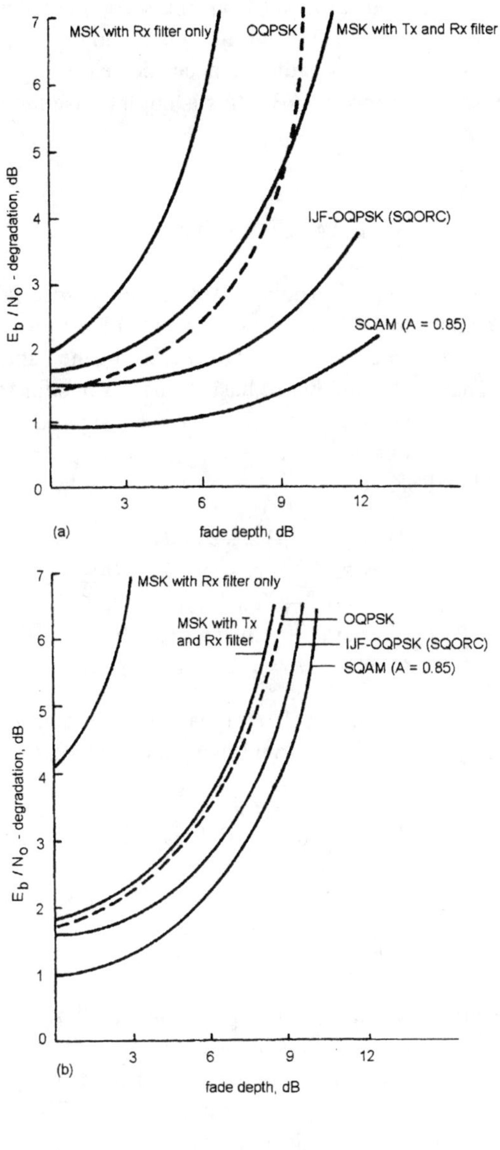

Figure 10.26 E_b/N_o degradation (compared to $E_b/N_o = 8.4$ dB at $P_b = 10^{-4}$) against fade depth of the desired channel in a hard-limited multichannel system. (a) $\Delta f = 0.83 f_b$, and (b)$\Delta f = 0.75 f_b$. From [11]. Copyright © 1985 IEE.

For both cases $d^2_{\min}$ is not too far away from the second smallest distance. This makes the error probability of SQAM slightly higher than that of MSK and QBL. It is clear that when A value decreases the distances decrease in general. Thus error performance degradation is expected. But this is not the case in an ACI channel as we have discussed already.

10.6 QUADRATURE QUADRATURE PSK

Q^2PSK is proposed in [12], which is a nonconstant-envelope spectrally efficient modulation scheme. We know that QPSK is more bandwidth efficient than BPSK. The increase of bandwidth efficiency is achieved by increasing the number of dimensions of the signal basis. That is, the number of basis signals is one for BPSK and two for QPSK. The Q^2PSK uses four basis signals. Four is considered to be the maximum number of dimensions achievable [12, Section III].

The signal set of Q^2PSK is

$$\begin{cases} s_1(t) = \cos(\frac{\pi t}{2T}) \cos 2\pi f_c t, & |t| \le T \\ s_2(t) = \sin(\frac{\pi t}{2T}) \cos 2\pi f_c t, & |t| \le T \\ s_3(t) = \cos(\frac{\pi t}{2T}) \sin 2\pi f_c t, & |t| \le T \\ s_4(t) = \sin(\frac{\pi t}{2T}) \sin 2\pi f_c t, & |t| \le T \\ s_i(t) = 0, \quad i = 1, 2, 3, 4, & |t| > T \end{cases} \tag{10.54}$$

where $2T$ is the duration of the signals. The signal set can be considered as consisting of two carriers $\cos 2\pi f_c t$ and $\sin 2\pi f_c t$ and two pulse-shaping functions:

$$p_1(t) = \begin{cases} \cos(\frac{\pi t}{2T}), & |t| \le T \\ 0, & |t| > T \end{cases} \tag{10.55}$$

and

$$p_2(t) = \begin{cases} \sin(\frac{\pi t}{2T}), & |t| \le T \\ 0, & |t| > T \end{cases} \tag{10.56}$$

The two carriers are orthogonal and the two pulse-shaping functions are also orthogonal. Note that between any two signals in the set $\{s_i(t)\}$, there is a common factor which is either a pulse-shaping function or a carrier; the remaining factor in one signal is in quadrature with the remaining factor in the other. This makes $\{s_i(t)\}$ a set of four equal-energy orthogonal signals under the restriction that

$$f_c = \frac{n}{4T}, \quad n = \text{integer} \ge 2$$

However, the signal set is not normalized. Each signal has an energy of 0.5 in

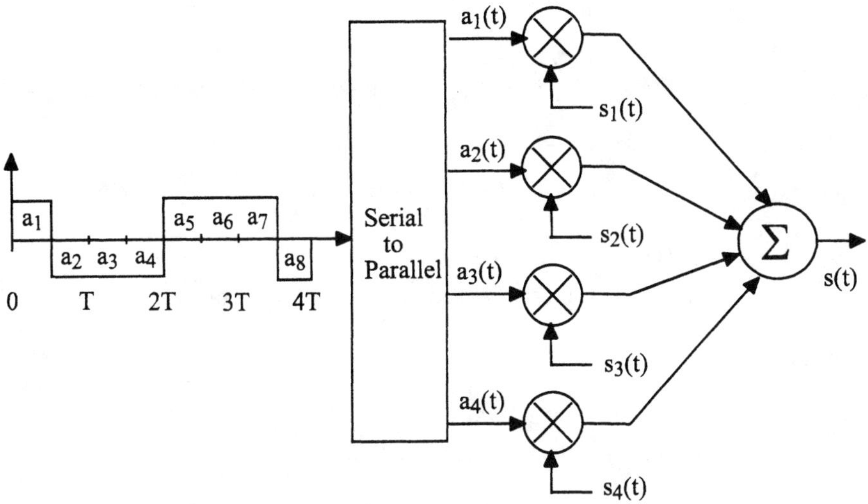

Figure 10.27 Q²PSK modulator.

$[-T, T]$. The orthogonality remains invariant with the shift of time origin by multiples of $2T$. This is to say that if $\{s_i(t)\}$ is defined by (10.54) for $|t| \leq \infty$ instead of $|t| \leq T$, then one will get orthogonality over every interval of $2T$ centered around $t = 2mT, m$ being an integer. This orthogonality suggests a modulation scheme shown in Figure 10.27. Data $\{a_i \in (\pm 1)\}$ from a binary source at a bit rate $R_b = 2/T$ is demultiplexed into four streams $\{a_i(t)\}$. Duration of each pulse in the streams is

$$T_s = 2T = 4T_b$$

which is four times the bit duration. If the symbol rate is the same, the bit rate of the Q^{2}PSK is twice that of QPSK and MSK. This is the fundamental reason why the bandwidth efficiency can be doubled with Q^{2}PSK with respect to MSK, as we will see shortly.

The modulated signal is

$$\begin{aligned} s(t) \quad = \quad & a_1(t)\cos(\frac{\pi t}{2T})\cos(2\pi f_c t) + a_2(t)\sin(\frac{\pi t}{2T})\cos(2\pi f_c t) \\ & + a_3(t)\cos(\frac{\pi t}{2T})\sin(2\pi f_c t) + a_4(t)\sin(\frac{\pi t}{2T})\sin(2\pi f_c t) \end{aligned}$$

$$\begin{aligned} &= \cos\left[2\pi\left(f_c + \frac{b_{14}(t)}{4T}\right)t + \phi_{14}(t)\right] \\ &\quad + \sin\left[2\pi\left(f_c + \frac{b_{23}(t)}{4T}\right)t + \phi_{23}(t)\right] \\ |t| &\le T \end{aligned} \tag{10.57}$$

where

$$\begin{aligned} b_{14}(t) &= -a_1(t)a_4(t) \\ \phi_{14}(t) &= 0 \text{ or } \pi \text{ depending on } a_1(t) = +1 \text{ or } -1 \end{aligned}$$

and

$$\begin{aligned} b_{23}(t) &= +a_2(t)a_3(t) \\ \phi_{23}(t) &= 0 \text{ or } \pi \text{ depending on } a_1(t) = +1 \text{ or } -1 \end{aligned}$$

At first glance, the two parts of the signal in (10.57) are like two MSK signals. However, there is a key difference between them and the MSK signal. In the MSK signal the I- and Q-channel signals are offset by T, which makes phase continuous at symbol boundaries, whereas there is no offset in Q^2PSK. Thus the Q^2PSK cannot be thought of as consisting of two MSK signals, rather it can be thought of as consisting of two FSK-BPSK signals whose phase is not always continuous at symbol boundaries as in MSK. Noting that there are two distinct frequencies and two distinct phases in each of the two components, the total number of distinct signals in Q^2PSK is 16.

This signal has a nonconstant envelope. Using trigonometrical identities and noting that $a_i = \pm 1$, (10.57) can easily be written as

$$s(t) = A(t)\cos(2\pi f_c t + \theta(t)) \tag{10.58}$$

where $\theta(t)$ is the carrier phase and $A(t)$ is the carrier amplitude given by

$$A(t) = \left[2 + (a_1 a_2 + a_3 a_4)\sin\frac{\pi t}{T}\right]^{1/2} \tag{10.59}$$

It is clear that the amplitude varies with time in general. Without any constraint on the data, $K = a_1a_2 + a_3a_4 = 0, \pm 2$. The possible amplitude variation in a symbol period ($T_s = 2T$) is shown in Figure 10.28. The amplitude can dip to zero at times. This is not a desired property for applications in nonlinear channels. However, when $K = 0$, the amplitude is constant. A coding scheme can be designed such that K is always zero, making the amplitude constant. We will discuss this shortly.

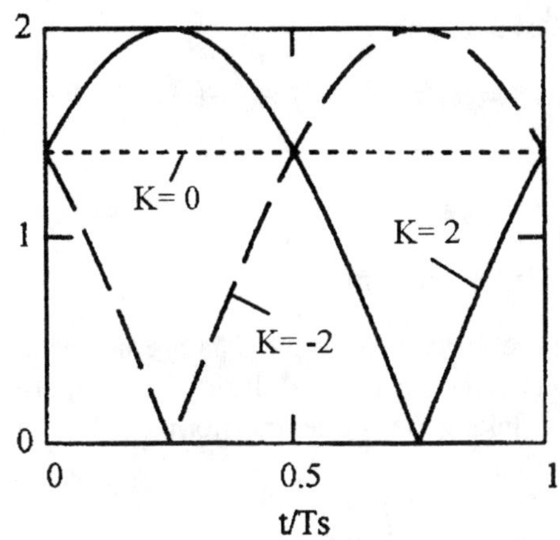

Figure 10.28 Possible amplitudes of Q^2PSK.

To find the PSD of the Q^2PSK signal we write (10.57) for $|t| \leq \infty$ as

$$s(t) = [x_1(t)\cos 2\pi f_c t + y_1(t)\sin 2\pi f_c t] + [x_2(t)\cos 2\pi f_c t + y_2(t)\sin 2\pi f_c t] \tag{10.60}$$

where

$$x_1(t) = \sum_{k=-\infty}^{\infty} a_{1k} p_1(t - kT_s)$$

$$x_2(t) = \sum_{k=-\infty}^{\infty} a_{2k} p_2(t - kT_s)$$

$$y_1(t) = \sum_{k=-\infty}^{\infty} a_{3k} p_1(t - kT_s)$$

$$y_2(t) = \sum_{k=-\infty}^{\infty} a_{4k} p_2(t - kT_s)$$

The complex envelope of the signal is

$$\widetilde{s}(t) = x_1(t) + x_2(t) - jy_1(t) - jy_2(t)$$

The autocorrelation is

$$\begin{aligned} R_{\widetilde{s}}(\tau) &= E\{\widetilde{s}(t)\widetilde{s}^*(t-\tau)\} = R_{x_1}(\tau) + R_{x_2}(\tau) + R_{y_1}(\tau) + R_{y_2}(\tau) \\ &= 2R_{x_1}(\tau) + 2R_{x_2}(\tau) \end{aligned}$$

where the cross terms are all zero since the data are independent and $R_{x_1}(\tau) = R_{y_1}(\tau)$, $R_{x_2}(\tau) = R_{y_2}(\tau)$ since their signals have the same pulse shape and the data have the same statistics. Taking the Fourier transform of $R_{\widetilde{s}}(\tau)$ and using (A.18) we obtain the PSD of Q^{2}PSK as

$$\Psi_{\widetilde{s}}(f) = \frac{2}{T_s}\left[|P_1(f)|^2 + |P_2(f)|^2\right]$$

where $P_1(f)$ and $P_2(f)$ are Fourier transforms of $p_1(t)$ and $p_2(t)$, respectively.

$$P_1(f) = \frac{4T}{\pi}\left(\frac{\cos 2\pi fT}{1 - 16f^2T^2}\right)$$

$$P_2(f) = \frac{-j16T}{\pi}\left(\frac{fT\cos 2\pi fT}{1 - 16f^2T^2}\right)$$

Thus

$$\Psi_{\widetilde{s}}(f) = \frac{16T}{\pi^2}(1 + 16f^2T^2)\left(\frac{\cos 2\pi fT}{1 - 16f^2T^2}\right)^2 \tag{10.61}$$

From this expression we can observe that the term in the squared parentheses is of the same form as that of MSK (see (5.14)) except that $T = 2T_b$. Thus the first null is at $f = 0.75/T = 0.375/T_b$, which is only one-half of that of MSK. Thus the bandwidth efficiency of Q^{2}PSK is twice as much as that of MSK as far as the main lobe is concerned. Figure 10.29 shows the PSD of Q^{2}PSK in comparison with those of OQPSK and MSK. Figure 10.30 is the fractional out-of-band power of the Q^{2}PSK. The main lobe of the Q^{2}PSK PSD is narrower than that of MSK (Figure 10.29), as a result, its roll-off within $2BT_b = 0.8$ is faster than that of MSK and QPSK/OQPSK (Figure 10.30(b)). Beyond $2BT_b = 0.8$ its roll-off rate is slower than that of MSK, and is the same as that of QPSK/OQPSK, however at a lower level (Figure 10.30(a)). The reason that the side lobes in Q^{2}PSK are higher than that of MSK is as follows.

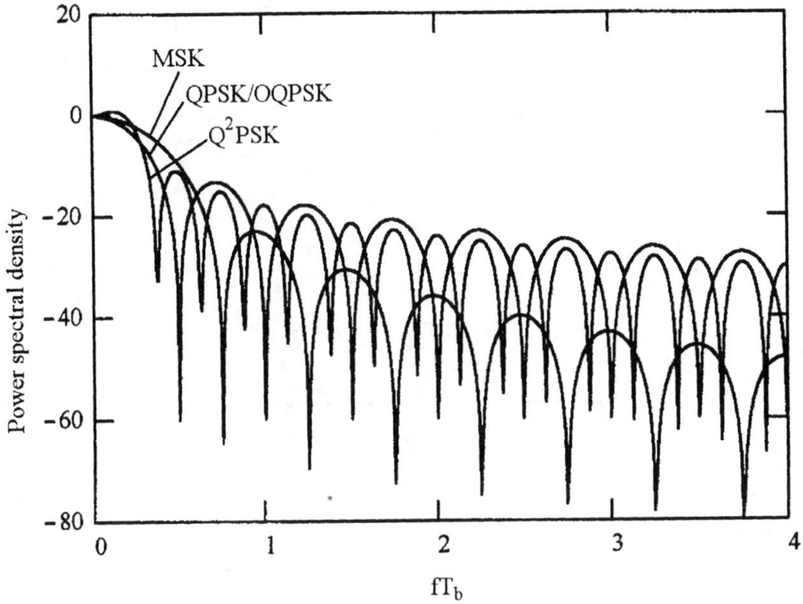

Figure 10.29 PSD of Q^2PSK in comparison with MSK, QPSK/OQPSK.

Q^2PSK uses two different pulses; one is $p_1(t)$ having a cosinusoidal shape as in MSK, the other is $p_2(t)$ having a sinusoidal shape. The shape of $p_1(t)$ is smoother than $p_2(t)$ in the sense that $p_2(t)$ has jumps at $t = \pm T$. As a result, for large f, the spectral roll-off associated with $p_2(t)$ is proportional to f^{-2}, while that with $p_1(t)$ varies as f^{-4}.

Since Q^2PSK signals are linear combinations of four orthogonal basis signals $\{s_i(t)\}$, the optimum demodulator is of the type given in Figures B.8 or B.9 with four correlators. Each signal is determined by a data four-tuple $\{a_1, a_2, a_3, a_4\}$. The total number of signals is $M = 2^4 = 16$. The bias terms $\{B_j\}_{j=1}^{16}$ can be eliminated since signals are equally likely and of the same energy. The weighting matrix consists of the 16 distinct data four-tuples. Each data four-tuple is a column of it.

However, we are not really interested in detecting any one of the 16 signals. Instead we are interested in detecting the data bits $\{a_1, a_2, a_3, a_4\}$ which are imbedded in the signal. Since each bit a_i is associated with only one of the four orthogonal carriers, each data bit can be detected independently. Thus the demodulator can be

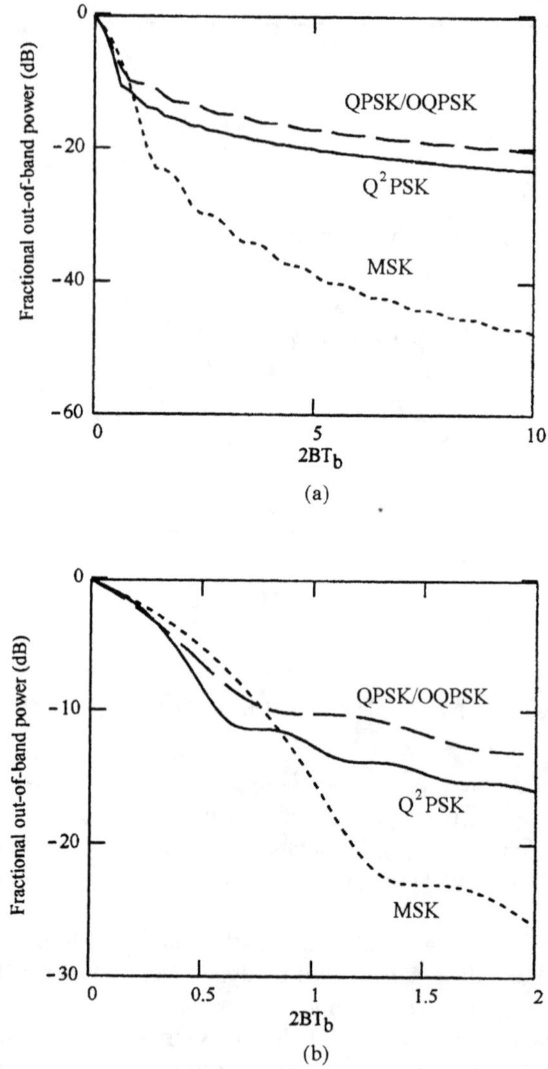

Figure 10.30 Fractional out-of-band power of Q^2PSK in comparison with MSK and QPSK/OQPSK (a, b).

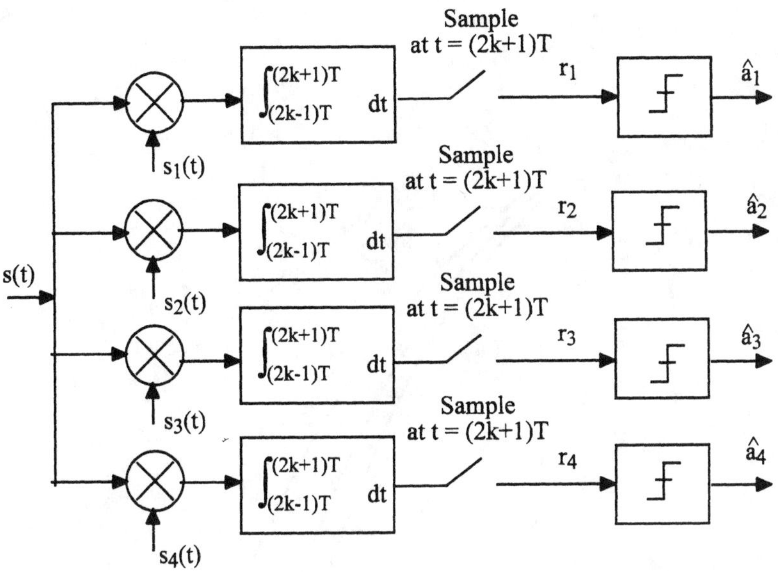

Figure 10.31 Q^{2}PSK demodulator.

simplified as shown in Figure 10.31. This is very much similar to the detection of I- and Q-channel bits in MSK. The bit error probability of this receiver is easy to determine. For any channel i in Figure 10.31, due to orthogonality, the correlator output caused by other three component signals other than $a_i s_i(t)$ is zero. As far as detection of $a_i s_i(t)$ is concerned, the other three signals do not have any effect. Thus the detection becomes just a detection of antipodal binary data a_i, modulated on a carrier $s_i(t)$. This detection problem is identical to that of I- or Q-channel detection of MSK (see Section 5.6). Therefore the bit error probability of Q^{2}PSK is the same as that of MSK. That is

$$P_b = Q\left(\sqrt{\frac{2E_b}{N_o}}\right) \tag{10.62}$$

It is interesting to note that the bit error probability is proved to be the same as in (10.62) if the symbols are detected using the demodulator depicted in Figure B.8 [12]. Thus the demodulator in Figure 10.31 is optimum as far as bit detection is concerned. The performance coincidence is due to the fact that each bit a_i is associated with only one of the four orthogonal carriers. We have seen similar phenomena in QPSK and

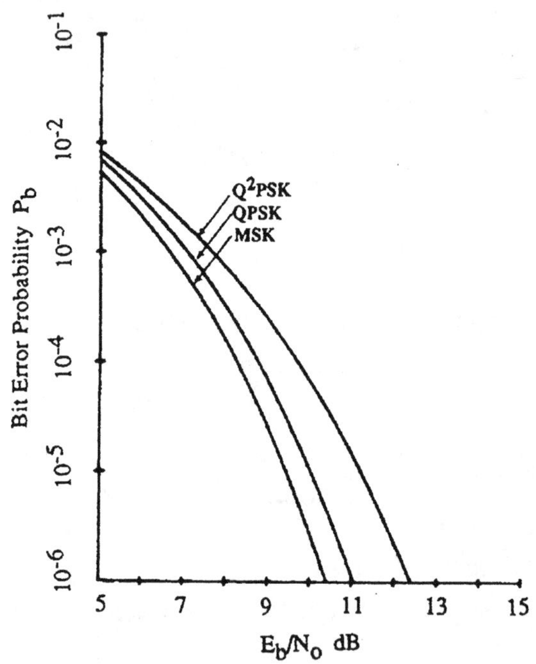

Figure 10.32 Bit error probability of Q^2PSK, QPSK , and MSK in a bandlimited channel. Bandlimiting filter is a sixth-order Butterworth with 3 dB bandwidth equal to 1.2/T. From [12]. Copyright © 1989 IEEE.

MSK.

While Q^2PSK has the same error performance as that of BPSK, QPSK, and MSK in an AWGN channel, for bandlimited channels, they behave differently due to their different spectra. Figure 10.32 is a comparison of their bit error probabilities. The performance evaluation assumed a sixth-order Butterworth filter with 3 dB bandwidth W equal to $1.2/T$. The E_b/N_o requirement of Q^2PSK is approximately 1.6 dB higher than that of MSK for $P_b = 10^{-5}$. The bit rate and bandwidth efficiency for MSK in this situation are $R_{b(MSK)} = 1/T$ and $R_{b(MSK)}/W = 0.83$. For Q^2PSK these values are $R_{b(Q^2PSK)} = 2/T$ and $R_{b(Q^2PSK)}/W = 1.66$ bits/s/Hz. Thus Q^2PSK achieves twice the bandwidth efficiency of MSK at the expense of a 1.6 dB increase in the average bit energy.

The synchronization scheme for Q^2PSK is derived as follows. If the Q^2PSK

signal passes through a squaring device, we get

$$
\begin{aligned}
s^2(t) &= 1+\frac{1}{2}(a_1a_2+a_3a_4)\sin(\frac{\pi t}{T}) \\
&+\frac{1}{2}(a_1a_2-a_3a_4)\sin(\frac{\pi t}{T})\cos(4\pi f_c t) \\
&+\cos(\theta_{12}-\theta_{34})\sin(4\pi f_c t) \\
&+\cos(\theta_{12}+\theta_{34})\cos(\frac{\pi t}{T})\sin(4\pi f_c t) \\
&+\sin(\theta_{12}+\theta_{34})\sin(\frac{\pi t}{T})\sin(4\pi f_c t)
\end{aligned}
\qquad (10.63)
$$

where

$$
\begin{aligned}
\theta_{12}(t) &= \tan^{-1}\left(\frac{a_2(t)}{a_1(t)}\right) \\
\theta_{34}(t) &= \tan^{-1}\left(\frac{a_4(t)}{a_3(t)}\right)
\end{aligned}
$$

There are five components in (10.63) which carry carrier- or symbol-timing information. The expected value of each component is zero. Filtering and further nonlinearity are needed in order to recover the carrier and symbol timing. By a low-pass and a bandpass filtering of the squared signal, we can obtain two signals

$$x_1(t)=\frac{1}{2}(a_1a_2+a_3a_4)\sin(\frac{\pi t}{T})$$

$$x_2(t)=\cos(\theta_{12}-\theta_{34})\sin(4\pi f_c t)$$

After squaring $x_1(t)$ and $x_2(t)$ and taking the expectation, we obtain

$$E\{x_1^2(t)\}=\frac{1}{4}(1-\cos\frac{2\pi t}{T})$$

$$E\{x_2^2(t)\}=\frac{1}{4}(1-\cos 8\pi f_c t)$$

Thus on the average $x_1^2(t)$ and $x_2^2(t)$ contain spectral lines at $1/T$ and $4f_c$. We can use these lines to lock phase-lock loops and carry out frequency division to recover the symbol-timing clock and carrier as

$$x_{c1}(t)=\cos(\pi t/T)$$

$$x_c(t)=\cos 2\pi f_c t$$

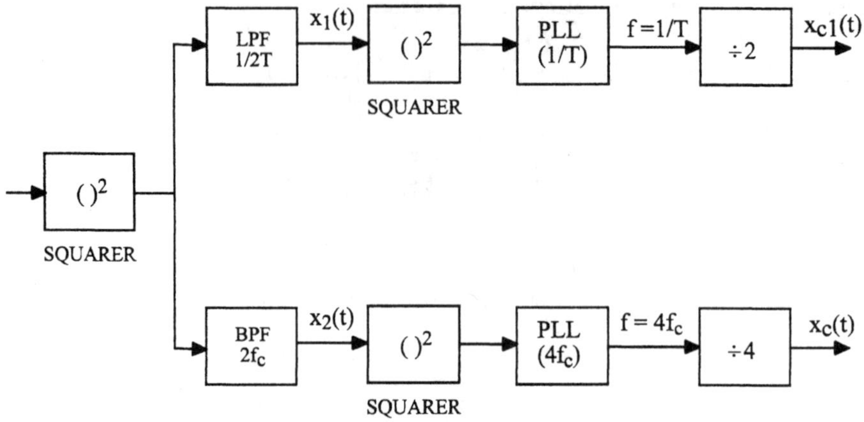

Figure 10.33 Synchronization scheme for coherent demodulation of Q^2PSK. From [12]. Copyright © 1989 IEEE.

Figure 10.33 is the block diagram of the synchronization scheme.

The realization of the modulator is shown in Figure 10.34. The basis signal set $\{s_i(t)\}$ is generated by a single oscillator of f_c and a bank of four bandpass filters and four adders.

The constant envelope version of Q^2PSK can be obtained by a simple coding scheme. From $K = a_1a_2 + a_3a_4 = 0$ we see that

$$a_4 = -\frac{a_1a_2}{a_3} \tag{10.64}$$

The encoder accepts serial data and for every three information bits $\{a_1, a_2, a_3\}$, it generates a codeword $\{a_1, a_2, a_3, a_4\}$ where a_4 satisfies (10.64). The rate of the code is $3/4$. The constant envelope feature is achieved at the expense of bandwidth efficiency. The information transmission rate is reduced from $R_b = 2/T$ to $3/2T$. But this is still 50% more than that of MSK ($R_b = 1/2T$ for MSK).

The modulator of the constant envelope Q^2PSK is the same as that of Q^2PSK except that an encoder must be added to the input to perform the coding given in (10.64).

Due to coding, the number of distinct signals is reduced from 16 to 8. Four of the eight possible codewords $\{C_i\}_{i=1}^{8}$ are as follows:

$$\begin{aligned} C_1 &= (+++-) \\ C_2 &= (++-+) \end{aligned}$$

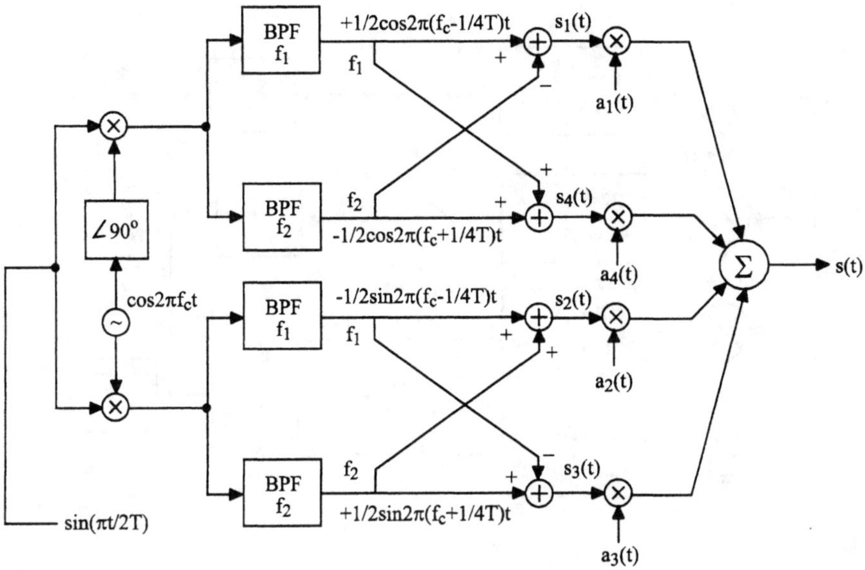

Figure 10.34 Realization of Q^{2}PSK modulator. From [12]. Copyright © 1989 IEEE.

$$\begin{aligned} C_3 &= (+-++) \\ C_4 &= (+---) \end{aligned}$$

where $+$ and $-$ represent $+1$ and -1, respectively. The remaining four codewords are just the negatives of these. This is a set of eight biorthogonal codewords with a minimum Hamming distance of two.[2] The code cannot be used for error correction, but the redundancy can be used to improve the error performance. The minimum Euclidean distance of the uncoded Q^{2}PSK is due to one different bit in the four-tuple, say a_j, then the squared minimum Euclidean distance is

$$\int_{-T}^{T}(s_j(t)+s_j(t))^2dt = 4\int_{-T}^{T} s_j^2(t)dt = 4E_b$$

The minimum Euclidean distance of the coded Q^{2}PSK is due to two different bits in

2 Hamming distance between two codewords is defined as the number of bits in which they differ.

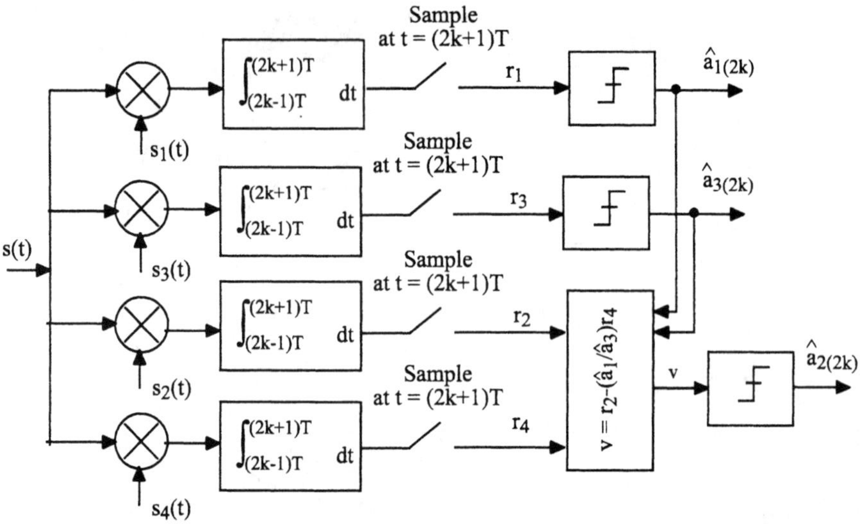

Figure 10.35 Constant envelope Q^{2}PSK demodulator. From [12]. Copyright © 1989 IEEE.

the four-tuple, say, a_i and a_j, then the squared minimum Euclidean distance is

$$\int_{-T}^{T} \left[(s_i(t) + s_i(t))^2 + (s_j(t) + s_j(t))^2\right] dt = 8 \int_{-T}^{T} s_i^2(t)dt = 6E_b$$

where $\int_{-T}^{T} s_i^2(t)dt = \frac{3}{4}E_b$ since a four-bit codeword represents three information bits. The coded Q^{2}PSK has a minimum Euclidean distance $\sqrt{1.5} = 1.22$ times greater than that of uncoded Q^{2}PSK. This provides a coding gain of 1.76 dB if the optimum demodulator (Figure B.8) is used.

From an implementation point of view, a nonoptimum demodulator based on hard decision may be of interest. Figure 10.35 is a nonoptimum demodulator for the coded Q^{2}PSK which is basically the same as Figure 10.31 except for the decoding part. According to [12], the pulses associated with $s_1(t)$ and $s_3(t)$ are relatively less distorted in a bandlimiting channel as compared to those associated with $s_2(t)$ and $s_4(t)$. Thus $\widehat{a}_1$ and $\widehat{a}_3$ are directly determined from the correlator outputs r_1 and r_3, while $\widehat{a}_2$ is determined from $\widehat{a}_1$, $\widehat{a}_3$, r_2, and r_4. The sufficient statistic for decision on $\widehat{a}_2$ is

$$V = r_2 - \frac{\widehat{a}_1}{\widehat{a}_3} r_4$$

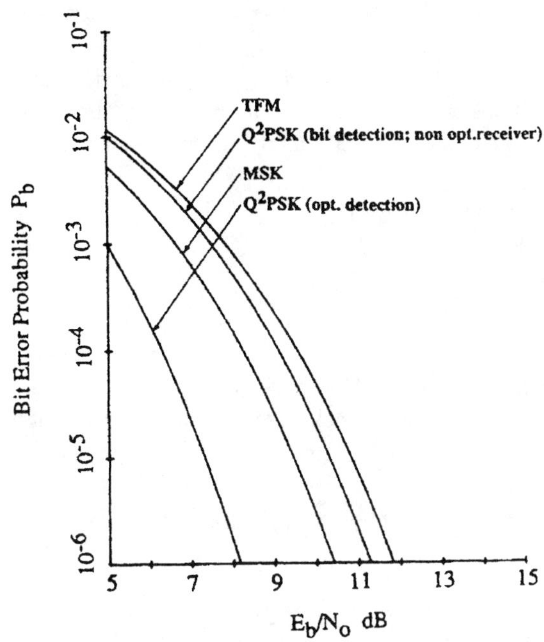

Figure 10.36 Bit error probabilities for constant envelope Q^2PSK, MSK, and TFM. From [12]. Copyright © 1989 IEEE.

The decoder decides $\widehat{a}_2$ as $+1$ or -1 according as $V \geq 0$ or $V < 0$. The correct decision on $\widehat{a}_2$ relies on the correct decision on $\widehat{a}_1$ and $\widehat{a}_3$. Figure 10.36 shows the performance of the constant envelope Q^2PSK in a bandlimited channel where a sixth-order Butterworth filter with 3 dB bandwidth of $1.2/T$ is used in both transmitter and receiver. It is seen that for $P_b = 10^{-5}$, the constant envelope Q^2PSK with nonoptimum demodulation requires an $E_b/N_o = 10.3$ dB while the MSK requires an $E_b/N_o = 9.6$ dB. Thus there is a 50% increase in bandwidth efficiency over MSK at a cost of a 0.7 dB increase in the average bit energy. The error performance of the optimum demodulator shown in the figure does not assume band limitation.

The synchronization system for the uncoded Q^2PSK is no longer applicable for the constant envelope Q^2PSK. Substituting (10.64) into (10.63), we get

$$s^2(t) \quad = \quad 1 + \frac{1}{2}(a_1 a_3 + a_1 a_2) \sin\left(2\pi t(2f_c + \frac{1}{2T})\right)$$

$$+\frac{1}{2}(a_1a_3 - a_1a_2)\sin\left(2\pi t(2f_c - \frac{1}{2T})\right) \tag{10.65}$$

There are two components in (10.65) which carry the carrier and clock information. But their expectations are zero. Further filtering and nonlinearity are needed to recover the carrier and clock information. By bandpass filtering we can get

$$x_1(t) = \frac{1}{2}(a_1a_3 + a_1a_2)\sin\left(2\pi t(2f_c + \frac{1}{2T})\right)$$

$$x_2(t) = \frac{1}{2}(a_1a_3 - a_1a_2)\sin\left(2\pi t(2f_c - \frac{1}{2T})\right)$$

It can be shown that $x_1^2(t)$ and $x_2^2(t)$, on the average, contain spectral lines at $4f_c \pm \frac{1}{T}$. We can use phase-locked loops to lock on these lines and carry out frequency division to get

$$x_3(t) = \cos\left(2\pi t(f_c + \frac{1}{4T})\right)$$

$$x_4(t) = \cos\left(2\pi t(f_c - \frac{1}{4T})\right)$$

Then a multiplication of these two followed by bandpass filtering and frequency division gives the desired carrier and clock signal

$$x_{c1}(t) = \cos(\frac{\pi t}{T})$$

$$x_c(t) = \cos 2\pi f_c t$$

The block diagram of this synchronization scheme is shown in Figure 10.37.

To further improve the bandwidth efficiency of Q^2PSK, pulse shapes other than the ones in (10.55) and (10.56) can be used [12]. A few transmitter baseband filter pairs that can achieve zero intersymbol interference and zero cross-correlation are given in Figure 10.38. These filters have a two-sided bandwidth of $1/T$, thus can achieve a bandwidth efficiency of 2 bits/s/Hz which is better than the 1.66 bits/s/Hz as previously demonstrated for (10.55) and (10.56). The filter pair in Figure 10.38(a) is a Hilbert transform pair. The rectangular frequency responses of this pair of filters are difficult to realize. Besides, a Hilbert transform pair has an additional problem because of the finite dc content. The pair in Figure 10.38(b) has no problem of dc content but it still has the problem of sharp cutoff at the band edge of $P_2(f)$. From the realization point of view, the pair in Figure 10.38(c) is especially convenient; one

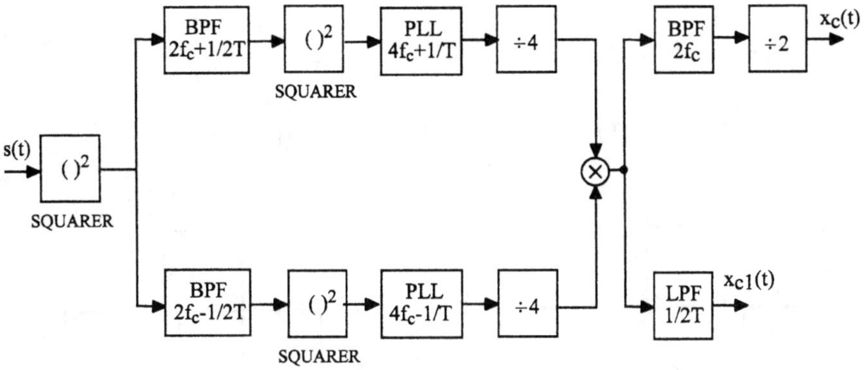

Figure 10.37 Synchronization scheme for coherent demodulation of constant envelope Q^{2}PSK. From [12]. Copyright © 1989 IEEE.

of the two filters is a cosine shape $P_1(f)$ and the other $P_2(t)$ is just a bandlimited differentiator. Note that $P_2(f)$ still has sharp cutoff edges at $f = \pm 1/2T$. However, if a Butterworth filter is incorporated with the ideal differentiator, the realization problem is greatly reduced. The bit error performance has been studied for $P_1(f) = \sqrt{2T}\cos \pi fT$ and $P_2(f) = jfB(f)$ where $B(f)$ is a Butterworth low-pass filter of second order with 3 dB bandwidth $W = 0.5R_b$. The corresponding transfer function is given by

$$P_2(s) = \frac{s}{s^2 + \sqrt{2}s + 1}$$

where s is the complex frequency (normalized with respect to 3 dB bandwidth). This transfer function is realizable. With this pair of filters, not only the bandwidth efficiency is improved to 2 bits/s/Hz, but also the energy efficiency is improved from 11.2 dB to 10.8 dB for a bit error probability of 10^{-5}.

A generalized Q^{2}PSK signaling format has been proposed for differential encoding and differential detection [13]. It is suitable for mobile and fixed radio links with multipath and fading problems. Differential detection of Q^{2}PSK requires approximately 3 dB extra E_b/N_o. However, the 3 dB loss can be fully or partially recovered if maximum likelihood decoding is used based on multiple (more than two) symbol observations.

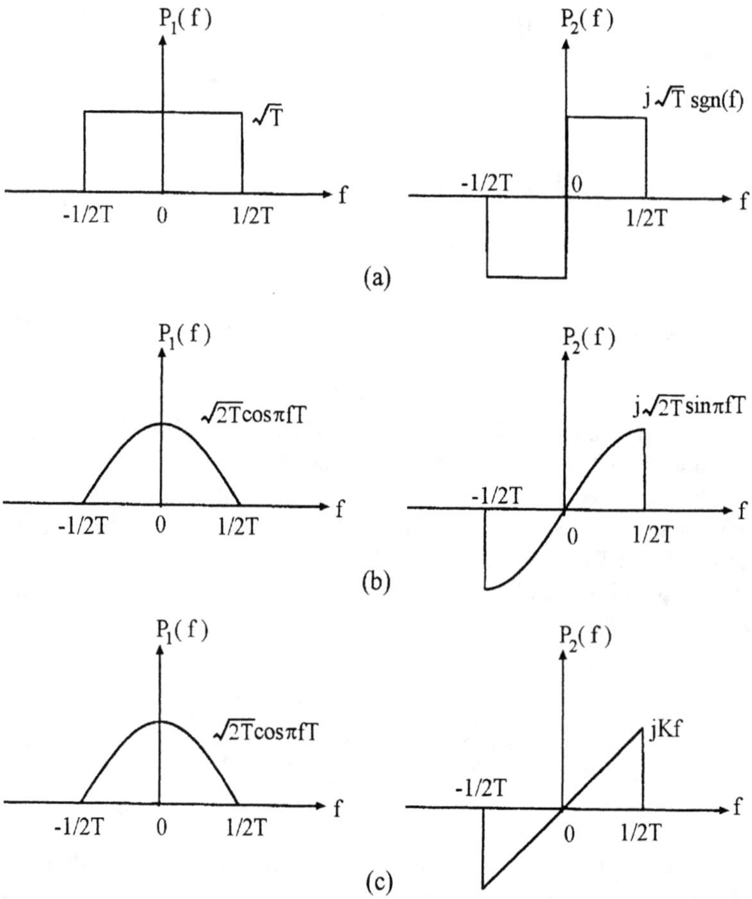

Figure 10.38 Transmitter filter pairs for Q^2PSK transmission with zero intersymbol interference and cross-correlation (a–c). From [12]. Copyright © 1989 IEEE.

10.7 SUMMARY

In this chapter we have covered the main research results on nonconstant-envelope bandwidth-efficient modulation schemes during the last twenty years. We have studied eight different schemes, namely, QBL, QORC, SQORC, QOSRC, IJF-OQPSK, TSI-OQPSK, SQAM, and Q^2PSK. The first seven are all similar in that each of them consists of two quadrature carriers weighted by pulses of a length of $2T_s$ or $4T_b$. The last one consists of four carriers weighted by sinusoidal or cosinusoidal pulses of a length of, again, $4T_b$.

By relaxing the constant envelope constraint, these proposed schemes are able to shrink the spectrum of the signal while maintaining a bit error probability close to MSK. The envelope fluctuations are not significant for staggered schemes. But the nonstaggered scheme QORC exhibits a big envelope variation. All of the eight schemes studied have spectral main lobes comparable to or smaller than that of QPSK and a faster spectral roll-off of the sidelobes. Therefore they provide low spectral spreading caused by nonlinear amplification.

Some schemes, namely the IJF-OQPSK, TSI-OQPSK, and SQAM have been examined against the ACI environment and found to have less degradation in error probability than QPSK/OQPSK. Other schemes have not been examined against the ACI environment and the study can be done in the future as research topics.

For the $2T_s$ schemes we have developed an optimum demodulator which uses two correlators in each of the I- and Q-channels. For all the $2T_s$ schemes, the bit error probability of the optimum demodulator is slightly higher than that of MSK. However, all these schemes can be demodulated by a conventional QPSK/OQPSK demodulator with small loss in error performance. For Q^2PSK, noncoherent demodulation is impossible and a coherent demodulator is developed. The Q^2PSK can also be made constant-envelope by a simple coding scheme. But the improvement of bandwidth efficiency over MSK is reduced from 100% to 50%.

References

[1] Amoroso, F., "The use of quasi-bandlimited pulses in MSK transmission," *IEEE Trans. Commun.*, vol. 27, no. 10, October 1979, pp. 1616–1624.

[2] Austin, M. C., and M. V. Chang, "Quadrature overlapped raised-cosine modulation," *IEEE Trans. Commun.*, vol. 29, no. 3, March 1981, pp. 237–249.

[3] Sasase, I., Y. Harada, and S. Mori, "Bandwidth efficient quadrature overlapped modulation," *IEEE Trans. Commun.*, vol. 32, no. 5, January 1984, pp. 638–640.

[4] Sasase, I., R. Nagayama, and S. Mori, "Bandwidth efficient quadrature overlapped squared raised-cosine modulation," *IEEE Trans. Commun.*, vol. 33, no. 1, January 1985, pp. 101–103.

[5] Le-Ngoc, T., K. Feher, and H. P. Van, "New modulation techniques for low-cost power and band-

width efficient satellite stations," *IEEE Trans. Commun.,* vol. 30, no. 1, July 1982, pp. 275–283.

[6] Van, H. P., and K. Feher, "A class of two symbol interval modems for nonlinear radio systems," *IEEE Trans. Commun.,* vol. 31, no. 3, March 1983, pp. 433–441.

[7] Seo, J., and K. Feher, "SQAM: a new superposed QAM modem technique," *IEEE Trans. Commun.,* vol. 33, no. 3, March 1985, pp. 296–300.

[8] Kato, S., and K. Feher, "XPSK: a new cross-correlated phase-shift-keying modulation technique," *IEEE Trans. Commun.,* vol. 31, no. 5, May 1983, pp. 701–706.

[9] Feher, K., et al., *Advanced Digital Communications: Systems and Signal Processing Techniques*, Englewood Cliffs, New Jersey: Prentice Hall, 1987.

[10] Van, H. P., and K. Feher, "TSI-OQPSK for multiple carrier satellite systems," *IEEE Trans. Commun.* vol. 32, no. 7, July 1984, pp. 818–825.

[11] Seo, J., and K. Feher, "Performance of SQAM systems in a nonlinearly amplified multichannel interference environment," *IEE Proceedings,* vol. 132. Pt. F. no. 3, June 1985, pp. 175–180.

[12] Saha, D., and T. G. Birdsall, "Quadrature-quadrature phase-shift keying," *IEEE Trans. Commun.,* vol. 37, no. 5, May 1989, pp. 437–448.

[13] El-Ghandour, O., and D. Saha, "Differential detection in quadrature-quadrature phase shift keying (Q^2PSK) systems," *IEEE Trans. Commun.,* vol. 39, no. 5, May 1991, pp. 703–712.

Chapter 11

Modulations in Fading Channels, Equalization, and Diversity

In wireless radio channels, a signal from the transmitter may arrive at the receiver's antenna through several different paths. The transmitted electromagnetic wave may be reflected, diffracted, and scattered by surrounding buildings and the terrain in the case of mobile radio communications, or by troposphere and ionosphere in the case of long-distance radio communications. As a result the signal picked up by the receiver's antenna is a composite signal consisting of these *multipath signals*. Sometimes a line-of-sight (LOS) signal may exist. The multipath signals arrive at the receiver at slightly different delays and have different amplitudes. The different delays translate to different phases. This results in a composite signal which can vary widely and rapidly in amplitude and phase. This phenomena is called *fading*. Variations in the property of the propagation medium, such as the occurrence of rain or snow, also can cause fading. However this type of fading is long-term fading, which we will not consider here. Multipath also causes intersymbol interference for digital signals. For mobile radio channels (ground and satellite), there is also Doppler frequency shift. Doppler shift causes carrier frequency drift and signal bandwidth spread. All these adversaries cause degradation in performance of modulation schemes in comparison with that in AWGN channels. In this chapter we study performances of modulation schemes in fading channels and remedial methods against fading, including equalization and diversity techniques. To do that we need to briefly study the characteristics of fading channels (Section 11.1). After that we first study flat-fading-channel performances of common binary and quaternary schemes, namely, BFSK, BPSK, DBPSK, QPSK, OQPSK, and MSK including GMSK in Section 11.2. In Section 11.3, their performances in frequency-selective channels are studied. The fading-channel performance of $\pi/4$-DQPSK, which is especially important since it is the standard of the U.S. and Japanese cellular telephone systems, is covered in great detail in Section 11.4. Then we move on to cover 1REC-MHPM or multi-h CPFSK in Section 11.5 and QAM in Section 11.6. Section 11.7

provides an overview of remedial measures against fading. In-depth discussions of remedial measures are provided in Sections 11.8 (channel estimation and correction), 11.9 (equalization), and 11.10 (diversity receptions). In Section 11.11, we introduce MIMO (multi-input-multi-output) wireless link and its companion technique: the space-time codes. MIMO, the breakthrough in communications in recent years, exploits instead of avoids multipath propagation in the wireless link so that the channel capacity is improved with the use of multiple antennas and space-time signal processing. Section 11.12 is a brief summary of this chapter.

11.1 FADING CHANNEL CHARACTERISTICS

11.1.1 Channel Characteristics

A fading-multipath channel is characterized by several parameters:

- *Delay spread.* In a multipath channel, the signal power at the receiver spreads over a certain range of time. The delay of the ith signal component in excess of the delay of the first arriving component is called excess delay, denoted as τ_i. Since τ is a random variable, the average $\overline{\tau}$ is the *mean excess delay*, the square root of the variance σ_τ is called *rms excess delay (or rms delay spread)*, and *excess delay spread* (X dB) is defined as the longest time delay during which multipath energy falls to X dB below the maximum. In other words, the maximum excess delay is defined as $\tau_X - \tau_0$, where τ_0 is the delay of the first arriving signal and τ_X is the maximum delay at which a multipath component is within X dB of the strongest signal which is not necessarily the first arriving signal. Figure 11.1 is an example of an indoor power delay profile, showing the definitions of delay spread parameters. The typical σ_τ values are on the order of microseconds (μs) in outdoor mobile radio channels and on the order of nanoseconds (ns) in indoor channels.
- *Coherence bandwidth.* The coherence bandwidth B_c is defined as the range of frequencies over which the channel can be considered "flat," meaning the channel passes all spectral components with approximately equal gain and linear phase. Frequency components in this bandwidth have a strong correlation in amplitude, hence the name "coherence bandwidth." On the other hand, two sinusoidal signals with frequency separation greater than B_c are affected quite differently. To derive a value of B_c, we define an envelope correlation

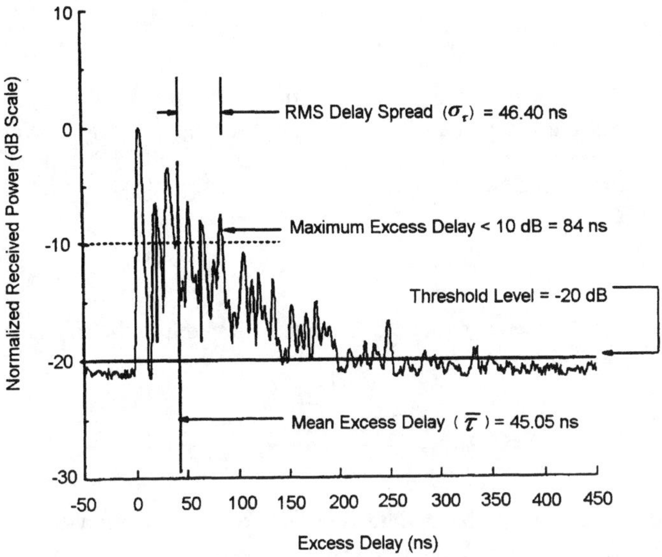

Figure 11.1 Delay spread parameters. From [1]. Copyright © 1996 Prentice Hall.

coefficient between two signals as

$$\rho(\Delta f, \Delta t) = \frac{E\{a_1 a_2\} - E\{a_1\}E\{a_2\}}{\sqrt{(E\{a_1^2\} - (E\{a_1\})^2)\,(E\{a_2^2\} - (E\{a_2\})^2)}}$$

where $E\{\cdot\}$ denotes the ensemble average and a_1 and a_2 represent the amplitudes of signals at frequencies f_1 and f_2, respectively, and at times t_1 and t_2, respectively, where $|f_2 - f_1| = \Delta f$ and $|t_2 - t_1| = \Delta t$. Next we employ the approximation that in the mobile radio environment the amplitude of each received signal is unity and that the joint probability density function (PDF) of receiving a signal with delay τ and angle θ is given by

$$p(\theta, \tau) = \frac{1}{2\pi\sigma_\tau} \exp(\frac{-\tau}{\sigma_\tau}) \tag{11.1}$$

where the PDF of θ is uniform in $[0, 2\pi)$ [3,4]. Then it can be shown [2,4] that the envelope correlation coefficient for two signals separated by Δf Hz and Δt

seconds is equal to

$$\rho(\Delta f, \Delta t) = \frac{J_0^2(2\pi f_M \Delta t)}{1 + (2\pi \Delta f)^2 \sigma_\tau^2} \tag{11.2}$$

where $J_0(\cdot)$ is the 0th order Bessel function of the first kind and $f_M = v f_c / c$ is the maximum Doppler shift for a carrier frequency of f_c and a velocity of v, with c representing the velocity of light. If $\Delta t = 0$, the envelope correlation coefficient becomes the frequency correlation coefficient

$$\rho(\Delta f, 0) = \frac{1}{1 + (2\pi \Delta f)^2 \sigma_\tau^2} \tag{11.3}$$

Now from (11.3) we can relate coherence bandwidth to the rms delay spread inversely as follows

$$B_c = \frac{1}{2\pi \sigma_\tau} \sqrt{\frac{1 - \rho(B_c, 0)}{\rho(B_c, 0)}} \tag{11.4}$$

Usually the coherence bandwidth is defined as the frequency separation at which $\rho(B_c, 0) = 0.5$. Thus from (11.4) we have

$$B_c = \frac{1}{2\pi \sigma_\tau} \tag{11.5}$$

However we can determine a coherence bandwidth for any value of frequency correlation $\rho(B_c, 0)$. For example if $\rho(B_c, 0) = 0.9$, we have

$$B_c = \frac{1}{6\pi \sigma_\tau} \tag{11.6}$$

Note that the relations given in (11.4)–(11.6) are based on (11.1) which is an approximation. Thus the above relations between B_c and σ_τ are just coarse estimates.

- *Doppler spread.* Doppler spread B_D is a measure of the spectral broadening caused by the relative movement of the mobile and base station, or by movement of objects in the channel. It is obvious that the Doppler spread is equal to the maximum Doppler frequency (i.e., $B_D = f_M$). The total bandwidth of the received signal is determined by the bandwidth of the baseband signal and the Doppler spread. If the baseband bandwidth is much greater than B_D, the effects of Doppler spread are negligible at the receiver.
- *Coherence time.* Similar to the definition of the coherence bandwidth, we can define a coherence time which is an estimate of at what transmitted signal

duration distortion becomes noticeable. The coherence time T_c is defined as the time difference between two signals with the same frequency, whose envelope correlation is 0.5. That is

$$\rho(0, T_c) = 0.5$$

which implies from (11.2)

$$\rho(0, T_c) = J_0^2(2\pi f_M T_c) = 0.5$$

This makes

$$T_c \approx \frac{9}{16\pi f_M} = \frac{9}{16\pi B_D} \tag{11.7}$$

Some authors simply use a more coarse estimate, that is,

$$T_c \approx \frac{1}{B_D} \tag{11.8}$$

This actually corresponds to $\rho(0, T_c) \approx J_0^2(2\pi) = 0.048$. Equation (11.8) is a much less restrictive definition than (11.7). According to [1], a popular rule of thumb for modern digital communications is to define the coherence time as the geometric mean of (11.7) and (11.8). That is

$$T_c = \sqrt{\frac{9}{16\pi B_D^2}} = \frac{0.423}{B_D} \tag{11.9}$$

11.1.2 Channel Classification

We use the above parameters to classify fading channels.

- *Flat fading.* Flat fading is also called frequency nonselective fading. If a wireless channel has a constant gain and linear phase response over a bandwidth which is greater than the signal bandwidth, then the signal will undergo flat or frequency nonselective fading. This type of fading is historically the most common fading model used in the literature. In flat fading, the multipath structure is such that the spectral characteristics of the transmitted signal are preserved at the receiver. However the strength of the signal changes with time, due to the variation of the gain of the channel caused by multipath. In terms of the parameters we just defined, the flat fading channel is characterized by

$$B_c \gg B_s$$

or equivalently

$$\sigma_\tau \ll T_s$$

where B_s is the signal bandwidth and T_s is the symbol period of the signal. *That is, a fading channel is flat or frequency nonselective if the channel coherence bandwidth is much greater than the signal bandwidth, or equivalently, the rms delay spread is much smaller than the signal symbol period.* Since the frequency response is flat, the impulse response of a flat fading channel can be modeled as a delta function without delay. The strength of the delta function changes with time.

- *Frequency selective fading.* If the channel has a constant gain and a linear phase response over a bandwidth which is smaller than the signal bandwidth, then the signal undergoes frequency selective fading. This is caused by such a multipath structure that the received signal contains multiple versions of the transmitted signal with different attenuations and time delays. Due to the time dispersion of the transmitted signal, the received signal has intersymbol interference. Thus the received signal is distorted. Viewed in the frequency domain, some frequency components have greater gains than others. Frequency selective fading channels are much more difficult to model than flat fading channels. Each multipath signal must be modeled and the channel is considered as a linear filter. Models are usually developed based on wideband measurement. In terms of the parameters, the frequency selective fading channel is characterized by

$$B_c < B_s$$

or equivalently

$$\sigma_\tau > T_s$$

That is, a fading channel is frequency selective if the channel coherence bandwidth is smaller than the signal bandwidth, or equivalently, the rms delay spread is greater than the signal symbol period. This rule may be too stringent for identifying a channel as frequency selective. A common rule of thumb is that a channel is frequency selective if $\sigma_\tau > 0.1T_s$, although this depends on the specific type of modulation used [1].

- *Fast fading.* If the channel impulse response changes rapidly within a signal symbol duration, the channel is classified as a fast fading channel; otherwise, it is classified as a slow fading channel. The fast change of the channel impulse response is caused by the motion, or equivalently, the Doppler spreading. *Quantitatively, when the channel coherence time is smaller than the symbol duration, or equivalently, the Doppler spreading is greater than the signal*

Based on multipath time delay spread	
Flat Fading	Frequency Selective Fading
$B_c \gg B_s$	$B_c < B_s$
$\sigma_\tau \ll T_s$	$\sigma_\tau > T_s$
Based on Doppler frequency spread	
Fast Fading	Slow Fading
$B_D > B_s$	$B_D \ll B_s$
$T_c < T_s$	$T_c \gg T_s$

Table 11.1 Fading channel classification.

bandwidth, a signal undergoes fast fading. That is, if

$$T_s > T_c$$

or equivalently

$$B_s < B_D$$

the channel is a fast fading channel.

- *Slow fading.* In a slow fading channel, the channel impulse response changes at a much slower rate than the symbol rate. *The channel coherence time is much greater than the symbol duration, or equivalently, the Doppler spreading is much smaller than the signal bandwidth.* That is, a signal undergoes slow fading if

$$T_s \ll T_c$$

or equivalently

$$B_s \gg B_D$$

In a slow fading channel, the channel impulse response can be considered static within one or several symbol durations.

Table 11.1 is a summary of the channel classifications.

A combination of low data rate (i.e., T_s is large) and high speed of the mobile unit corresponds to fast fading since the mobile experiences rapid electromagnetic field changes in a very short time (i.e., T_c is small). On the contrary a combination of high data rate and low speed corresponds to slow fading. Of course, when the mobile unit is not moving, the channel is a slow fading channel regardless of the data rate.

It should be pointed out that when a channel is classified as a fast or slow fading channel, it does not classify whether the channel is flat fading or frequency selective fading. A *flat fading, fast fading* channel is a channel whose impulse response is a delta function with a strength that varies faster than the symbol rate. In a *frequency selective, fast fading* channel, the amplitude, phase, and time delay of any of the multipath components vary faster than the symbol rate. The other two combinations are the *flat, slow fading* channel and *frequency selective, slow fading* channel. As a matter of fact, these two are the common models in practice since fast fading only occurs for very low data rates at which the mobile can move a long distance and experiences a wide range of signal strength change in a symbol period. We will focus on the *flat, slow fading* channel and *frequency selective, slow fading* channel in the rest of this chapter.

11.1.3 Fading Envelope Distributions

In a multipath channel, the received signal consists of a large number of plane waves, whose complex low-pass signal

$$\widetilde{r}(t) = r_I(t) + jr_Q(t)$$

can be modeled as a Gaussian random process. If none of the multipath components is dominant, $r_I(t)$ and $r_Q(t)$ are Gaussian processes with zero mean and a variance of σ^2. The envelope $z(t) = |\widetilde{r}(t)|$ obeys the *Rayleigh distribution*. The probability density function (PDF) of the Rayleigh envelope is given by

$$p(z) = \frac{z}{\sigma^2} \exp\left(-\frac{z^2}{2\sigma^2}\right), \quad z \geq 0 \tag{11.10}$$

and $p(z) = 0$ for $z < 0$. Rayleigh fading agrees very well with empirical observations for macrocellular applications. The plot of $p(z)$ is given in Figure 11.2. The maximum $p(z)$ occurs at $z = \sigma$.

The probability that the envelope does not exceed a specific value Z is given by the cumulative distribution function (CDF)

$$P(Z) = \Pr(z \leq Z) = \int_0^Z p(z)dz = 1 - \exp\left(-\frac{Z^2}{2\sigma^2}\right)$$

The mean value of the Rayleigh envelope is

$$E\{z\} = \int_0^\infty zp(z)dz = \sqrt{\frac{\pi}{2}}\sigma = 1.2533\sigma$$

The average power of the Rayleigh envelope is

$$E\{z^2\} = \Omega = 2\sigma^2$$

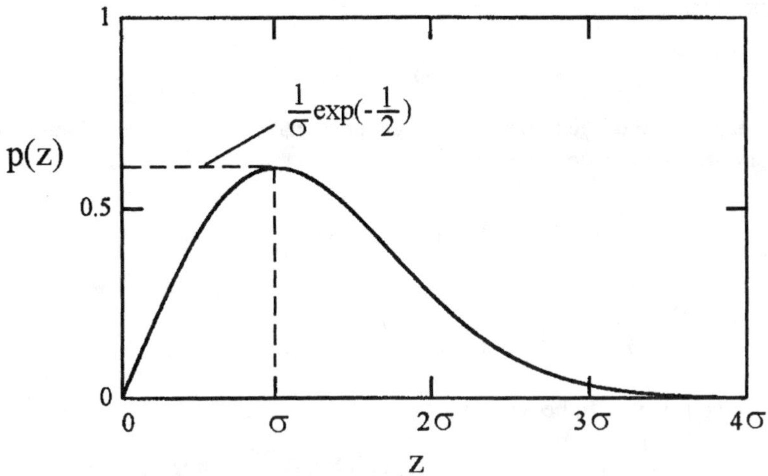

Figure 11.2 Rayleigh distribution density.

The ac power of the envelope is the variance

$$\begin{aligned} \sigma_z^2 &= E\{z^2\} - (E\{z\})^2 = 2\sigma^2 - \left(\sqrt{\frac{\pi}{2}}\sigma\right)^2 \\ &= (2 - \frac{\pi}{2})\sigma^2 = 0.4292\sigma^2 \end{aligned}$$

The Rayleigh density can be written in terms of the average power Ω as

$$p(z) = \frac{2z}{\Omega} \exp\left(-\frac{z^2}{\Omega}\right), \quad z \geq 0 \tag{11.11}$$

If a dominant component, such as the line-of-sight (LOS) or specular component, is present in the multipath channel, $r_I(t)$ and $r_Q(t)$ have nonzero mean and the envelope has a *Rician distribution* [5]

$$p(z) = \frac{z}{\sigma^2} \exp\left(-\frac{z^2 + A^2}{2\sigma^2}\right) I_0(\frac{Az}{\sigma^2}), \quad z \geq 0 \tag{11.12}$$

and $p(z) = 0$ for $z < 0$, where A is the peak amplitude of the dominant signal and $I_0()$ is the zeroth modified Bessel function of the first kind. Rician fading is very often observed in microcellular applications. A parameter K is often defined for a

Rician channel

$$K = \frac{A^2}{2\sigma^2} \tag{11.13}$$

which is the ratio of the power of the specular signal over the power of the scattered components. In terms of decibels, we have

$$K(dB) = 10 \log \frac{A^2}{2\sigma^2}$$

The average power

$$E\{z^2\} = \Omega = A^2 + 2\sigma^2 = 2\sigma^2(1+K)$$

The Rician density can be written in terms of K and Ω,

$$p(z) = \frac{2z(K+1)}{\Omega} \exp\left(-K - \frac{(K+1)z^2}{\Omega}\right) I_0\left(2z\sqrt{\frac{K(K+1)}{\Omega}}\right), \quad z \geq 0 \tag{11.14}$$

Figure 11.3 shows the Rician distribution density for some values of K. When $K = 0$, the Rician distribution becomes the Rayleigh distribution. When $K = \infty$ the channel does not exhibit any fading. The curves are for a fixed average power $\Omega = 2$, so that when $K = 0$, $\Omega = 2\sigma^2$ and $p(z)$ is identical to that given in (11.11).

Another model of envelope distribution is the Nakagami distribution which was selected to fit empirical data for long distance HF channels. The Nakagami distribution is given by [6]

$$p(z) = \frac{2m^m z^{2m-1}}{\Gamma(m)\Omega^m} \exp\left\{-\frac{mz^2}{\Omega}\right\}, \quad m \geq \frac{1}{2} \tag{11.15}$$

where $\Omega = E\{z^2\}$. When $m = 1$, the Nakagami distribution becomes the Rayleigh distribution. It is claimed to provide a closer match to some experimental data than Rayleigh or Rician distributions. However the claims were based on fitting envelope statistics around the mean or median, rather than the near-zero region which is fundamental to system performance over fading channels [7]. Therefore, the results based on a Nakagami distribution appear questionable for practical use, and they will not be discussed further in this chapter.

11.2 DIGITAL MODULATION IN SLOW, FLAT FADING CHANNELS

In a flat fading channel, the signal undergoes a multiplicative variation. In general,

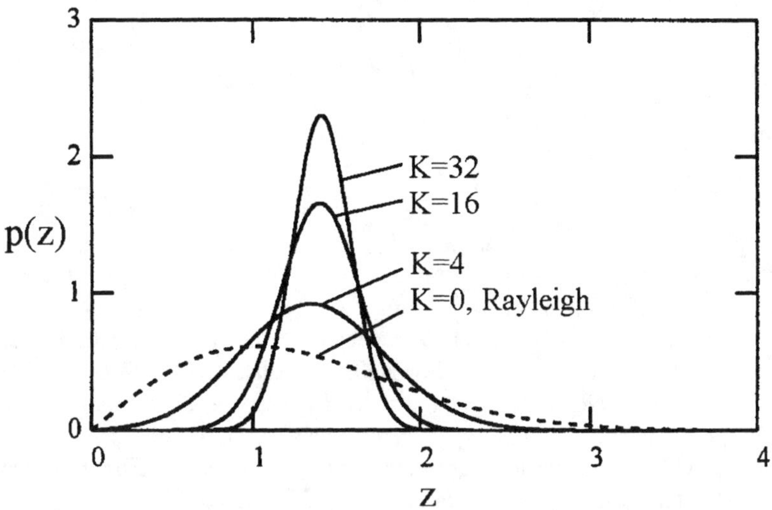

Figure 11.3 Rician distribution density curves for a fixed average power $\Omega = 2$ so that when $K = 0$, the Rayleigh curve is exactly the same as in Figure 11.2.

this multiplicative factor is complex, that is, the signal amplitude as well as phase are affected. If we further assume that the fading is slow, then the amplitude attenuation and phase shift of the received signal can be considered constant over at least a symbol duration. Therefore, if the transmitted equivalent low-pass complex signal is $\tilde{s}(t)$, the received equivalent lowpass complex signal can be written as

$$\tilde{r}(t) = ze^{-j\phi}\tilde{s}(t) + \tilde{n}(t) \tag{11.16}$$

where z is the amplitude of the signal (assuming $\tilde{s}(t)$ has a unit amplitude), ϕ is the phase shift of the signal caused by the channel, and $\tilde{n}(t)$ is the equivalent lowpass complex additive Gaussian noise.

In the following, we assume that the demodulators that we use for the fading channels are originally designed for the AWGN channel, which are not optimal for fading channels. The optimal receivers for fading channels can be found in [5]. The received signal may be coherently detected or noncoherently detected, depending on whether it is possible to accurately estimate the phase shift ϕ. In either case, the average error probability can be evaluated by averaging the error probability for a fixed amplitude z over the entire range of z. That is

$$P_e = \int_0^\infty P_e(\gamma_b)p(\gamma_b)d\gamma_b \tag{11.17}$$

where

$$\gamma_b = z^2 E_b/N_o$$

is the signal-to-noise ratio with fading for a particular value of z, $P_e(\gamma_b)$ is the symbol or bit error probability conditioned on a fixed γ_b, $p(\gamma_b)$ is the probability density function of γ_b, and P_e is the average symbol or bit error probability.

Note that to use (11.17) to evaluate the demodulator's performance, the demodulator must satisfy some conditions. For binary modulations like BPSK the detector has a zero threshold. No matter what the signal amplitude is, the detector for AWGN channel is the same; that is, a binary detector with a zero threshold. This situation is the same for QPSK since it can be detected by two binary zero-threshold detectors in the I- and Q-channels, respectively. In these cases, $\widetilde{r}(t)$ can be directly sent to the demodulator and the detection is optimal for that instantaneous γ_b. The average P_e can be evaluated by (11.17). However, for modulations that need detectors with multiple thresholds, such as MASK and QAM, $\widetilde{r}(t)$ cannot be directly sent to the demodulator designed for fixed signal levels; otherwise the BER caused by wrong thresholds will overwhelm the BER caused by the noise. This is because the detector's thresholds are designed for a set of fixed signal constellation $\{\widetilde{s}_i(t)\}$. With each value of the channel gain z, the detector thresholds need to be redesigned so that $P_e(\gamma_b)$ can be obtained using the AWGN formula and average P_e can be evaluated by (11.17). In practice, instead of adapting the threshold set, it is easier to correct the channel effect by dividing the received signal by the channel gain

$$\widetilde{r}_1(t) = \frac{\widetilde{r}(t)}{ze^{-j\phi}} = \widetilde{s}(t) + \frac{\widetilde{n}(t)}{ze^{-j\phi}}$$

Then the $\widetilde{r}_1(t)$ can be sent to the demodulator designed for $\{\widetilde{s}_i(t)\}$. $\widetilde{r}_1(t)$ still has an SNR of γ_b; and (11.17) can be used for the average P_e. Channel estimation is needed to estimate $ze^{-j\phi}$.

11.2.1 Rayleigh Fading Channel

For Rayleigh fading channel, z has a Rayleigh distribution, thus z^2 and γ_b have a chi-square distribution with two degrees of freedom. That is

$$p(\gamma_b) = \frac{1}{\Gamma}\exp(-\frac{\gamma_b}{\Gamma}), \quad \Gamma \geq 0 \tag{11.18}$$

where

$$\Gamma = E\{z^2\}\frac{E_b}{N_o}$$

is the average value of the signal to noise ratio.[1]

Substituting the error probability expression of a specific modulation in the AWGN channel and (11.18) into (11.17), we can obtain the error probability expression of the modulation in a slow, flat, Rayleigh fading channel. For many important modulation schemes, P_e expression in the AWGN channel is in the form of Q function or exponential function. Fortunately, for these two function forms, the P_e expressions in the fading channel are in closed forms. For other schemes numerical calculation is needed to obtain the error probabilities in the fading channel.

For many schemes, the symbol or bit error rate in the AWGN channel can be expressed in one of two general forms as

$$P_e = CQ\left(\sqrt{\frac{\delta E_b}{N_o}}\right)$$

$$P_e = C\exp\left(-\frac{\delta E_b}{N_o}\right)$$

where C and δ are constants (see Table 4.7 for many examples). In the fading channel, the signal-to-noise ratio E_b/N_o becomes $\gamma_b = z^2 E_b/N_o$. Correspondingly the conditional error probabilities are

$$P_e(\gamma_b) = CQ\left(\sqrt{\delta\gamma_b}\right) \tag{11.19}$$

$$P_e(\gamma_b) = C\exp(-\delta\gamma_b) \tag{11.20}$$

Substituting (11.19) or (11.20) and (11.18) into (11.17), we can obtain the corresponding symbol or bit error probabilities.

For the exponential function $P_e(\gamma_b)$, (11.17) can be evaluated easily. The result is

$$P_e = \frac{C}{1+\delta\Gamma} \tag{11.21}$$

---

1 From Section 11.1.3 we know that $z^2(t) = r_I^2(t) + r_Q^2(t)$ where $r_I(t)$ and $r_Q(t)$ are Gaussian random processes. It is well known that the sum of n Gaussian random variables obeys a distribution called chi-square (χ^2) distribution with n degrees of freedom. See [5, p. 109].

For the Q function $P_e(\gamma_b)$, (11.17) is evaluated using the following two formulas. The first is from the integral table [8]

$$\int_0^\infty [1 - \mathrm{erf}(\beta x)] e^{-\mu x^2} \cdot x dx = \frac{1}{2\mu}\left(1 - \frac{\beta}{\sqrt{\mu + \beta^2}}\right) \tag{11.22}$$

and the second is the well-known relation between the error function and the Q function

$$1 - \mathrm{erf}(x) = 2Q(\sqrt{2}x) \tag{11.23}$$

which is used to convert the Q function into the error function in order to use (11.22). Substituting (11.19) and (11.18) into (11.17), using (11.23), and making a variable change $\gamma_b = x^2$, we have

$$P_e = \int_0^\infty \frac{C}{\Gamma}[1 - \mathrm{erf}(\sqrt{\frac{\delta}{2}}x)]e^{-\frac{x^2}{\Gamma}} \cdot x dx$$

Recognizing $\sqrt{\delta/2} = \beta$ and $1/\Gamma = \mu$ in (11.22), the above is equal to

$$P_e = \frac{C}{2}\left[1 - \sqrt{\frac{\delta\Gamma}{2 + \delta\Gamma}}\right] \tag{11.24}$$

Using general expressions (11.21) and (11.24), error rates of many modulation schemes in the slow, flat, Rayleigh channel can be easily found out.

For *coherent BPSK, QPSK, OQPSK, and MSK*, their P_b expressions are the same in the AWGN channel. That is $P_b = Q\left(\sqrt{2E_b/N_o}\right)$. This means $C = 1$ and $\delta = 2$ in (11.24). The result is

$$P_b = \frac{1}{2}\left[1 - \sqrt{\frac{\Gamma}{1+\Gamma}}\right], \quad \text{(coherent BPSK, QPSK, OQPSK, and MSK)} \tag{11.25}$$

For *optimum differential BPSK* (Figure 4.7 and (4.10)), $P_b = \frac{1}{2}\exp(-\frac{E_b}{N_o})$ in the AWGN channel, which makes $C = 1/2$ and $\delta = 1$. From (11.21) we have

$$P_b = \frac{1}{2(1+\Gamma)} \quad \text{(optimum DBPSK)} \tag{11.26}$$

For *coherent binary FSK*, $P_b = Q\left(\sqrt{E_b/N_o}\right)$ in the AWGN channel. This implies $C = 1$ and $\delta = 1$. From (11.24) we obtain

$$P_b = \frac{1}{2}\left[1 - \sqrt{\frac{\Gamma}{2+\Gamma}}\right], \quad \text{(coherent BFSK)} \tag{11.27}$$

For *noncoherent orthogonal BFSK*, $P_b = \frac{1}{2}\exp(-\frac{E_b}{2N_o})$ in the AWGN channel. This leads to $C = 1/2$ and $\delta = 1/2$, thus

$$P_b = \frac{1}{2+\Gamma}, \quad \text{(noncoherent orthogonal BFSK)} \tag{11.28}$$

For large signal-to-noise ratios, the above error probability expressions can be approximated as

$$P_b \approx \frac{1}{4\Gamma}, \quad \text{(coherent BPSK, QPSK/OQPSK, and MSK)}$$

$$P_b \approx \frac{1}{2\Gamma}, \quad \text{(coherent BFSK and optimum DBPSK)}$$

$$P_b \approx \frac{1}{\Gamma}, \quad \text{(noncoherent orthogonal BFSK)}$$

For *GMSK*, the AWGN channel bit error rate is $P_b = Q\left(\sqrt{2\varepsilon E_b/N_o}\right)$. Thus $C = 1$, $\delta = 2\varepsilon$ and

$$P_b = \frac{1}{2}\left[1 - \sqrt{\frac{\varepsilon\Gamma}{1+\varepsilon\Gamma}}\right] \approx \frac{1}{4\varepsilon\Gamma}, \quad \text{(coherent GMSK)} \tag{11.29}$$

where

$$\varepsilon = \begin{cases} 0.68, & \text{for } BT_s = 0.25 \\ 0.85, & \text{for } BT_s = \infty \end{cases}$$

For $\pi/4$–*DQPSK*, the AWGN channel bit error rate is

$$P_b \approx Q\left(\sqrt{4E_b/N_o}\sin(\pi/4\sqrt{2})\right) = Q\left(\sqrt{1.112E_b/N_o}\right)$$

(see Table 4.7). Thus $C = 1$ and $\delta = 1.112$. From (11.24) we have

$$P_b = \frac{1}{2}\left[1 - \sqrt{\frac{0.556\Gamma}{1+0.556\Gamma}}\right] \approx \frac{1}{2.224\Gamma}, \quad \text{(optimum } \pi/4\text{-DQPSK)} \tag{11.30}$$

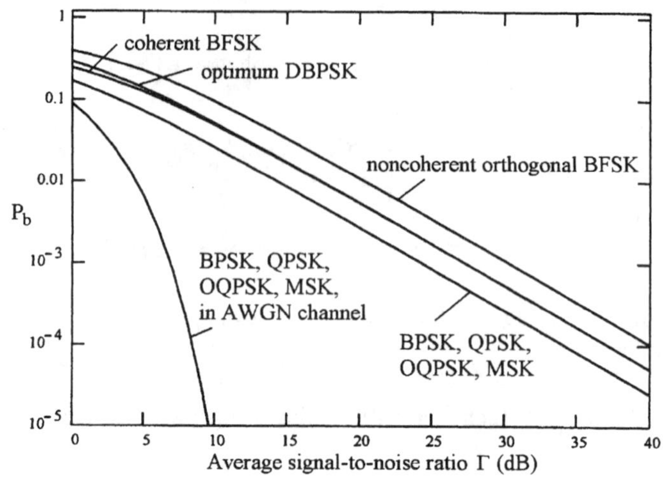

Figure 11.4 Bit error rates of several common modulation schemes in a slow, flat, Rayleigh fading channel.

The list can go on for many other schemes whose error rate expressions have the forms in (11.19) or (11.20).

Figures 11.4 and 11.5 illustrate the error performance of the above schemes in the slow, flat, Rayleigh channel. It is seen that Rayleigh fading causes a significant loss in signal-to-noise ratio for the same bit error probability, in comparison with the AWGN channel. For example at $P_b = 10^{-3}$, which is considered to be adequate for voice communications, the increase in the average signal-to-noise ratio is about 15 to 20 dB, depending on the particular modulation scheme used. The losses can be significantly reduced by using diversity techniques and error control coding techniques.

11.2.2 Rician Fading Channel

For the Rician fading channel the amplitude z of the received signal $r(t)$ in (11.16) has a density function given in (11.12). Closed-form expressions for the error probabilities of DPSK and orthogonal FSK in Rician fading channels are found in [9]. First the Rician density in (11.12) is transformed to an expression where the random variable is the instantaneous signal to noise ratio $\gamma_b = z^2 E_b/N_o$ instead of the signal

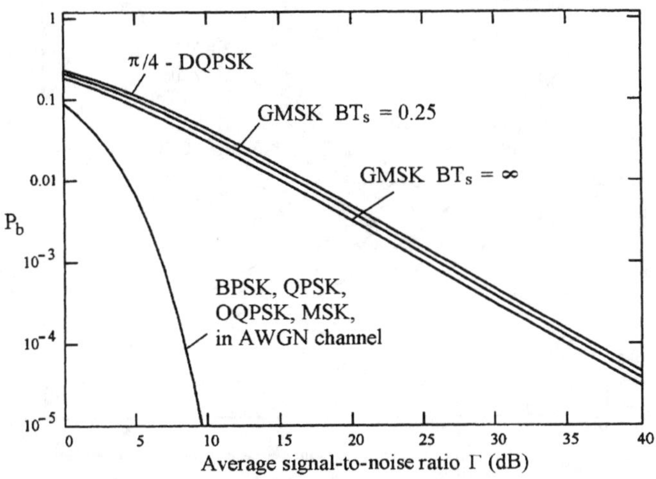

Figure 11.5 Bit error rates of GMSK and $\pi/4$-DQPSK in a slow, flat, Rayleigh fading channel.

amplitude z.

$$p(\gamma_b) = \frac{K+1}{\Gamma} \exp\left(-\frac{\gamma_b(K+1)+K\Gamma}{\Gamma}\right) I_0\left(\sqrt{\frac{4(K+1)K\gamma_b}{\Gamma}}\right) \tag{11.31}$$

where K is the ratio of the power of the specular signal over the power of the scattered components defined in (11.13). Using (11.31) and appropriate $P_e(\gamma_b)$ in (11.17) and again using the appropriate entry in the integral table [8], we obtain

$$P_b = \frac{K+1}{2(K+1+\Gamma)} \exp\left(-\frac{K\Gamma}{K+1+\Gamma}\right), \quad \text{(optimum DBPSK)} \tag{11.32}$$

and

$$P_b = \frac{K+1}{2(K+1)+\Gamma} \exp\left(-\frac{K\Gamma}{2(K+1)+\Gamma}\right), \quad \text{(noncoherent BFSK)} \tag{11.33}$$

Figures 11.6 and 11.7 show the plots from (11.32) and (11.33).

For other modulation schemes, we can substitute (11.31) and the appropriate AWGN error rate expressions into (11.17) and numerically evaluate their error performance in the slow, flat, Rician channel. Figures 11.8 to 11.12 are the numerical results. Figure 11.8 is for BPSK, QPSK/OQPSK, and MSK. Figure 11.9 is for BFSK.

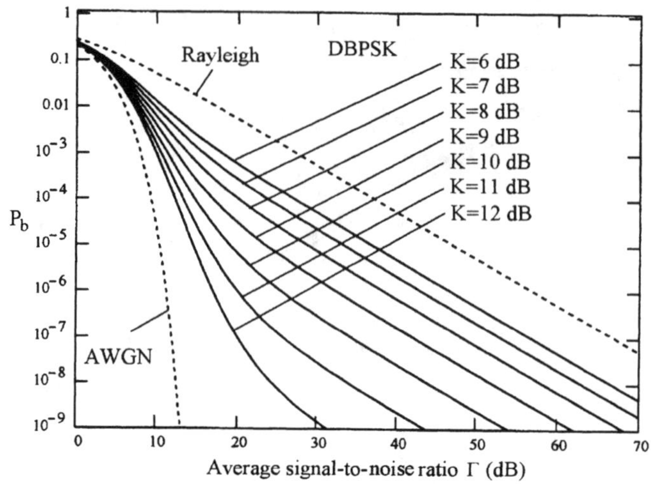

Figure 11.6 Bit error rate of optimum DBPSK in Rician fading channel.

Figure 11.10 is for GMSK ($BT = \infty$) and Figure 11.11 is for GMSK ($BT = 0.25$). Figure 11.12 is for $\pi/4$-DQPSK. It is seen from Figures 11.8–11.12 that the curves are more compact and more evenly spaced than those in Figures 11.6 and 11.7.

11.3 DIGITAL MODULATION IN FREQUENCY SELECTIVE CHANNELS

In a frequency selective fading channel, the received signal contains multiple delayed versions of the transmitted signal. The multipath signals cause intersymbol interference (ISI) which results in an irreducible BER floor. The BER falls initially with the increase of the signal-to-noise ratio (E_b/N_o), and stops falling when the signal-to-noise ratio is sufficiently high at which the errors are almost exclusively caused by the ISI. From the cause of the error floor, it is clear that the BER floor is directly related to the delays of the multipath components.

Simulation is the main tool of studying BER behavior in frequency selective fading channels. Chuang [10] simulated the error performance of unfiltered BPSK, QPSK, OQPSK, and MSK schemes in frequency selective fading channels. Chuang also simulated the error performance of GMSK with various BT_b of the premodula-

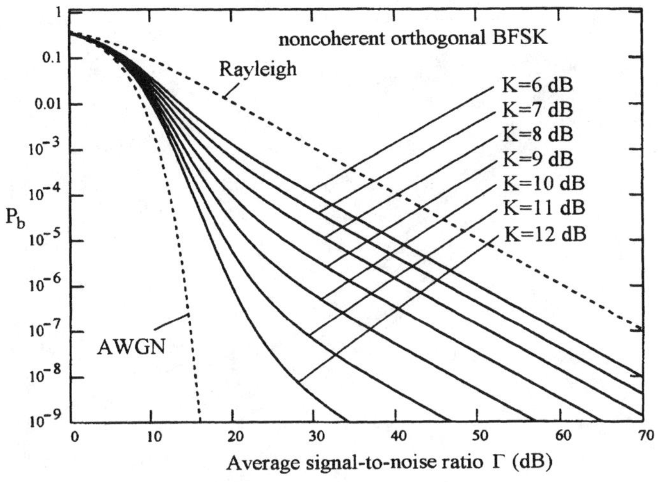

Figure 11.7 Bit error rate of noncoherent orthogonal BFSK in Rician fading channel.

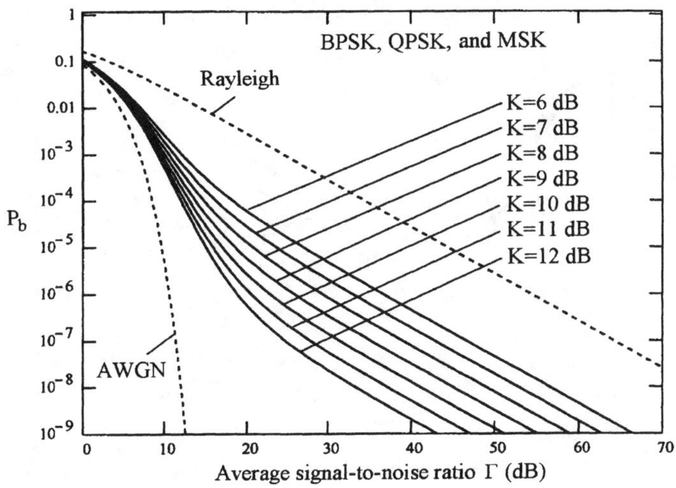

Figure 11.8 Bit error rate of BPSK, QPSK/OQPSK, and MSK in Rician fading channel.

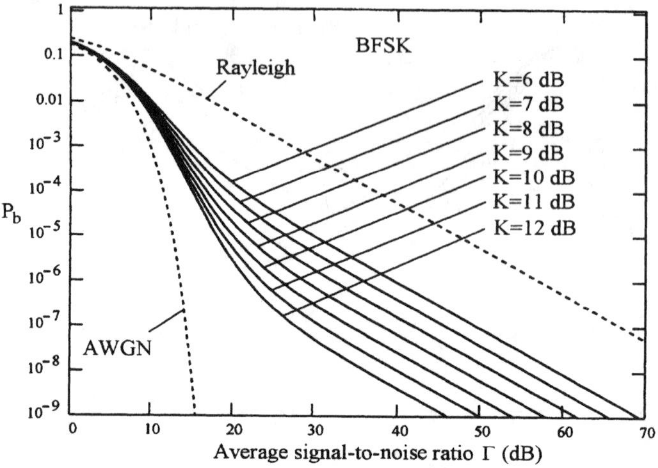

Figure 11.9 BER of BFSK in Rician fading channel.

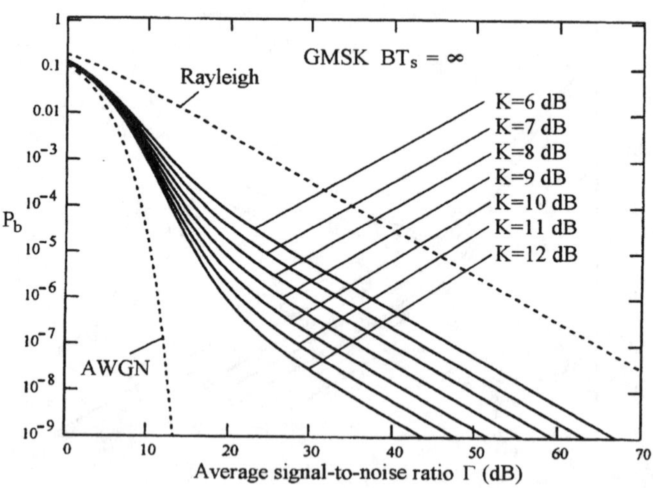

Figure 11.10 BER of GMSK $(BT = \infty)$ in slow, flat, Rician fading channel.

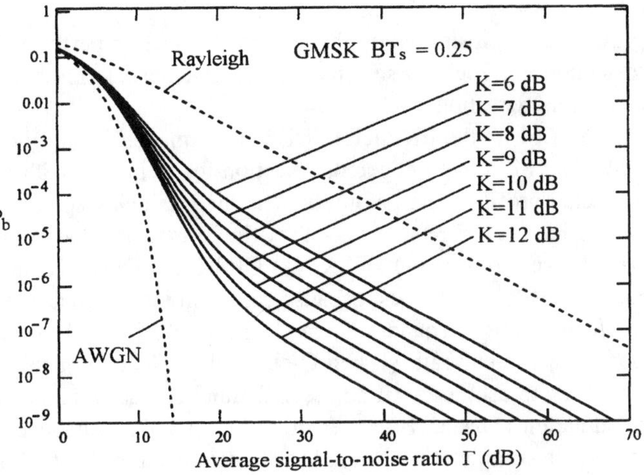

Figure 11.11 BER of GMSK ($BT = 0.25$) in slow, flat, Rician fading channel.

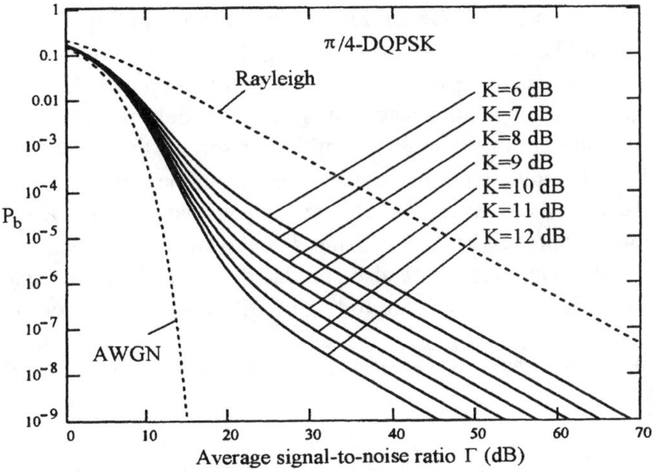

Figure 11.12 BER of $\pi/4$-DQPSK in slow, flat, Rician fading channel.

tion filter and QPSK with a raised-cosine Nyquist pulse (RC-QPSK)[2] using various roll-off factors α.

Chuang found through simulation that coherent detection performs better than differential detection in frequency selective fading channels; so his study was concentrated on coherent detection.

Figure 11.13 presents the irreducible BER performance for unfiltered BPSK, QPSK, OQPSK, and MSK with coherent detection for a channel with a Gaussian-shaped power delay profile. The parameter d is the rms delay spread normalized by the symbol period (i.e., $d = \sigma_\tau / T_s$). This figure indicates the performance is ranked in the following order: (1) BPSK, (2) QPSK, (3) OQPSK, and (4) MSK. The performance of BPSK is the best because cross-rail interference does not exist. Cross-rail interference exists in quadrature modulations (QPSK, OQPSK, and MSK), where the modulator has two rails (I- and Q-channels). The two bit sequences will have no interference to each other if an ideal channel condition is maintained and coherent demodulation is used. However in multipath fading channels, these two sequences can not be completely separated at the demodulator due to the multipath fading condition. Both OQPSK and MSK have a $T_s/2$ timing offset between two bit sequences, hence the cross-rail ISI is more severe. Therefore their performances are inferior to that of QPSK.

The normalization factor for parameter d in Figure 11.13 is the symbol period T_s, during which two bits are transmitted in QPSK, OQPSK, and MSK, whereas only one bit is transmitted in BPSK. In other words, the comparison of error floor in Figure 11.13 is based on different bit rate or information throughput. A fairer comparison should be based on $d' = \sigma_\tau / T_b$, that is, on the same bit rate. Figure 11.14 is the same set of curves as in Figure 11.13 plotted against d'. It is clear from the figure that 4-level modulations are more resistant to delay spread than BPSK for the same information throughput. The intuitive reason for this is that for the same bit rate the 4-level modulations have twice the length of the symbol period of that of BPSK; therefore for the same d', the relative delay spread d is only one half of that of BPSK for 4-level modulations. It is d, not d' that influences the detection of a bit.

Higher level modulations were also simulated. Figure 11.15 indicates that the performance of 8-PSK is not superior to that of QPSK as SNR approaches infinity,

---

2 The pulse shape of RC-QPSK is obtained by filtering the baseband NRZ pulse with a raised-cosine frequency response

$$H(f) = \begin{cases} 1, & \text{for } |f| < 2W_0 - W \\ \cos^2\left(\frac{\pi}{4}\frac{|f|+W-2W_0}{W-W_0}\right), & \text{for } 2W_0 - W < |f| < W \\ 0, & \text{for } |f| > W \end{cases}$$

where W is the absolute bandwidth of the filter and $W_0 = 1/2T_s$ is the Nyquist bandwidth. The roll-off factor is defined as $\alpha = (W - W_0)/W_0$. The larger the α, the larger the bandwidth W.

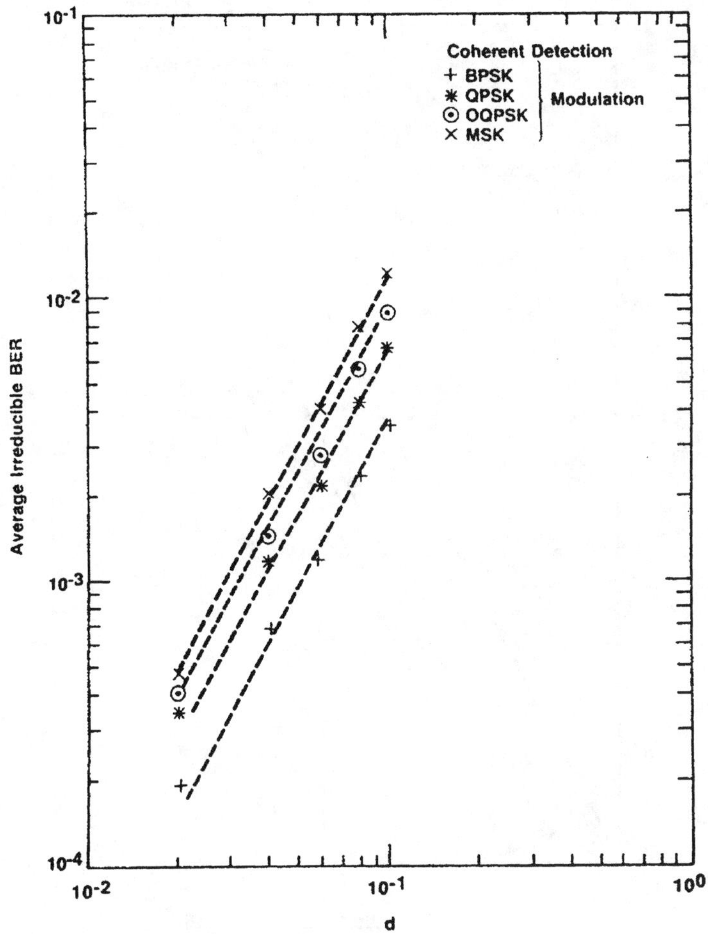

Figure 11.13 The irreducible BER performance for different modulations with coherent detection for a channel with a Gaussian shaped power delay profile. The parameter d is the rms delay spread normalized by the symbol period. From [10]. Copyright © 1987 IEEE.

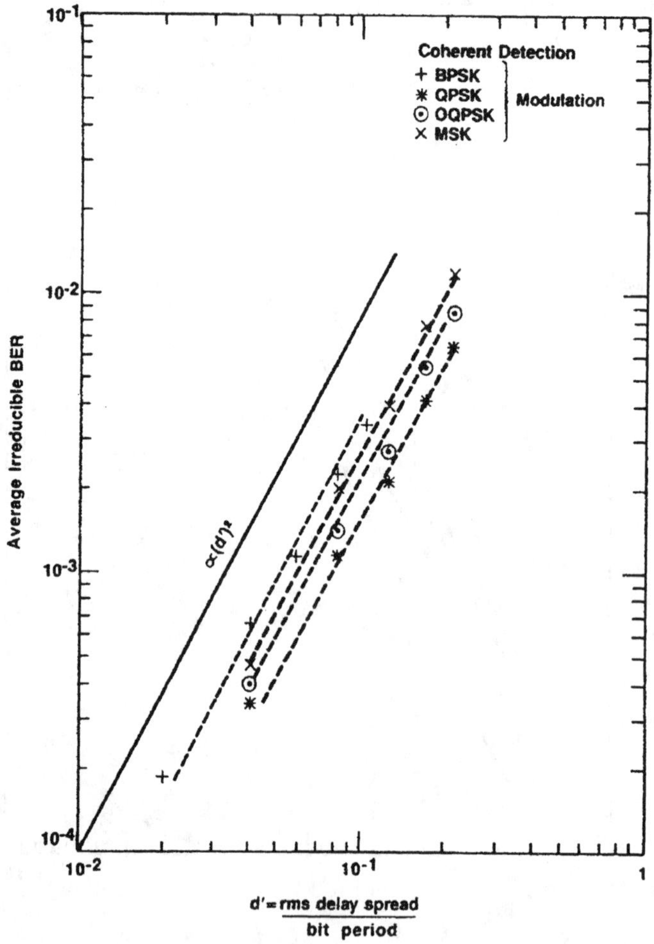

Figure 11.14 The same set of curves as in Figure 10.13, plotted against rms delay spread normalized by the bit period instead of symbol period. From [10]. Copyright © 1987 IEEE.

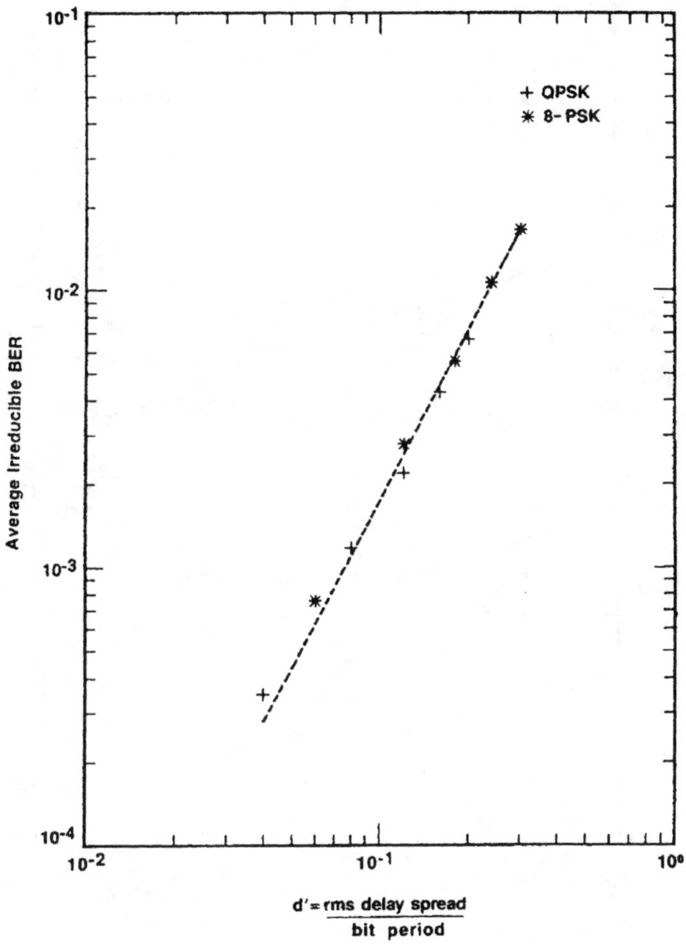

Figure 11.15 The irreducible BER performance for QPSK and 8-PSK. The parameter d' is the rms delay spread normalized by the bit period. From [10]. Copyright © 1987 IEEE.

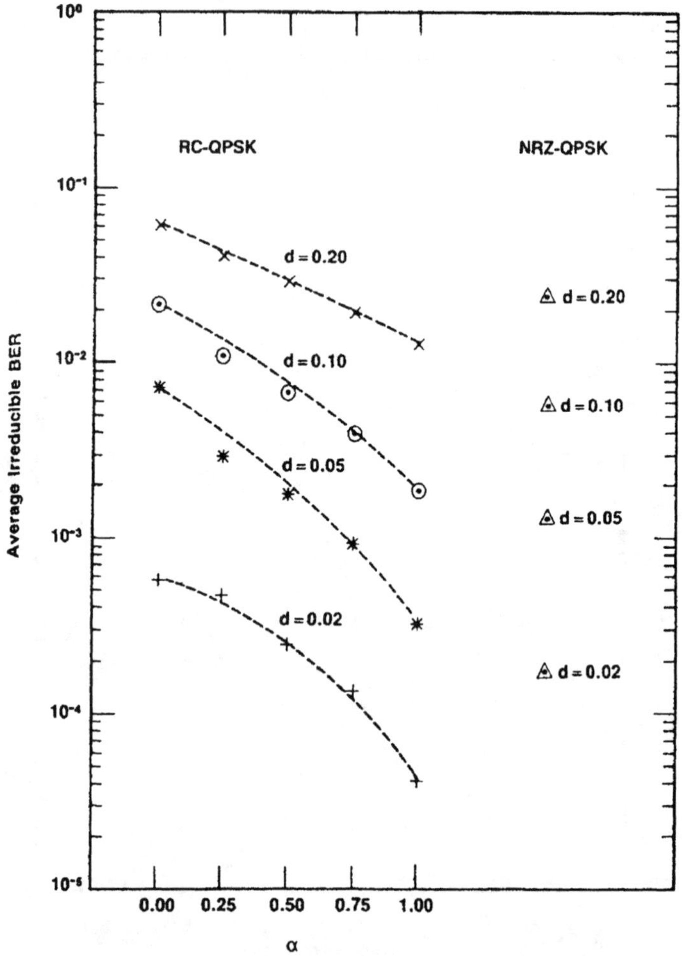

Figure 11.16 The irreducible BER performance for RC-QPSK with coherent detection for a channel with a measured power delay profile in Figure 10.18. Results for the unfiltered QPSK are also shown for comparison. The parameter α is the roll-off factor in raised-cosine filter. The parameter d is the rms delay spread normalized by the symbol period. From [10]. Copyright © 1987 IEEE.

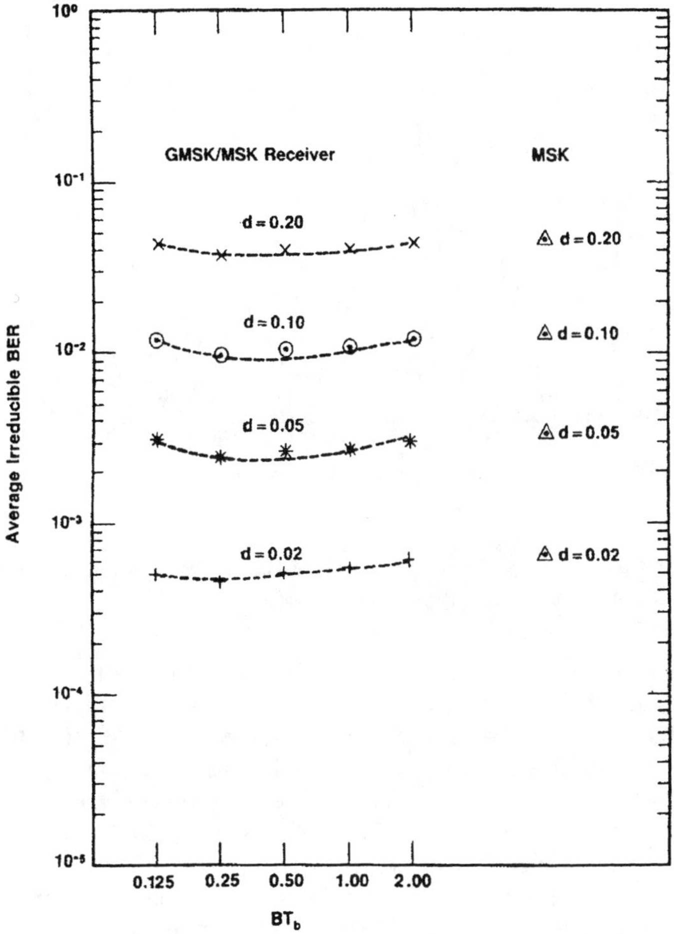

Figure 11.17 The irreducible BER performance for GMSK with coherent detection for a channel with a measured power delay profile as shown in Figure 10.18. Results for the unfiltered MSK are also shown for comparison. The parameter BT_b is the 3 dB bandwidth of the premodulation Gaussian filter normalized by bit rate. The parameter d is the rms delay spread normalized by the symbol period. From [10]. Copyright © 1987 IEEE.

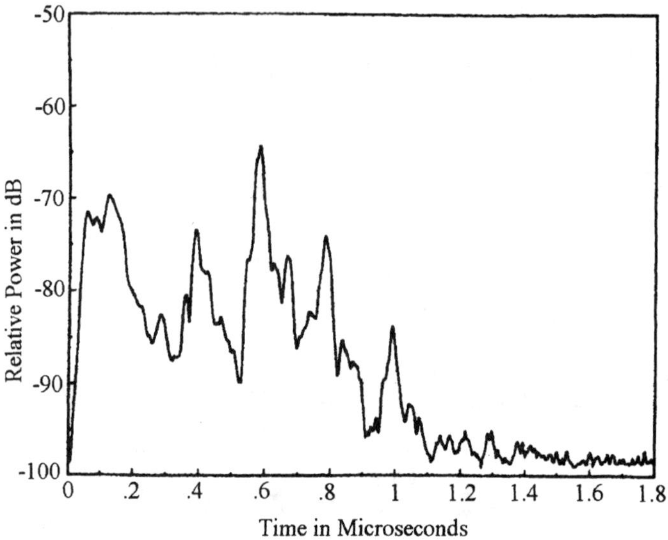

Figure 11.18 A power delay profile obtained from measurements done in an office building; the rms delay spread is approximately 250 ns. From [10]. Copyright © 1987 IEEE.

even though it has 3 bits per symbol. This is not surprising since the BER of 8-PSK is much higher than that of QPSK in the AWGN channel (see Figure 4.13 where the symbol error rate is shown, however it can be easily converted to BER using (4.25)). The advantage of smaller d' is offset by this disadvantage. This is why 4-level modulations are chosen for all third-generation wireless standards.

It is also interesting to note that all curves in Figures 11.13, 11.14, and 11.15 are nearly parallel to a straight line of slope 2. That is, an order of magnitude increase in delay spread results in about two orders of magnitude increase in the irreducible BER within the range of d simulated. These simulation results are consistent with an earlier theoretical finding given in [11].

Figure 11.16 shows the irreducible BER of RC-QPSK as a function of α, the roll-off factor in a raised-cosine filter, for different values of rms delay spread. The delay spread profile is a measured one shown in Figure 11.18. Results for the unfiltered QPSK are also shown for comparison. As α increases, the irreducible BER decreases monotonically due to a decrease in ISI. However, the spectral occupancy also increases. It is seen from the figure that the RC-QPSK with $\alpha \geq 0.75$ has lower irreducible BER than the unfiltered NRZ-QPSK.

Figure 11.17 shows the irreducible BER of GMSK as a function of BT_b, the 3 dB bandwidth of the premodulation Gaussian filter normalized by bit rate. The delay spread profile of the channel is again the one shown in Figure 11.18. Results for the unfiltered MSK are also shown for comparison. The GMSK is demodulated by an MSK receiver which is not matched to the GMSK signal. It is seen from the figure that the best BER performance is obtained by choosing $BT_b = 0.25$; however, the performance is not very sensitive to BT_b until BT_b is too small (< 0.25).

11.4 π/4-DQPSK IN FADING CHANNELS

$\pi/4$-DQPSK is the modulation of the third generation mobile telephone systems in the United States and Japan. Its performance in the mobile fading channel is of great interest. Its performance in a slow, flat, Rayleigh or Rician fading channel has been discussed in Section 11.2. The research groups led by Dr. Feher and Dr. Rappaport investigated its BER performance for other important channel conditions. Results were published in references [12–14].

In [13], the channel model is the flat fading indoor radio channel model based on measurements. The BER results of the simulation are based on the channel characteristics of 50 simulated measurement locations and narrow-band flat fading characteristics seen by a mobile at each location. Twenty-five of those 50 simulated channels are in LOS (line of sight) topography and 25 of them are in OBS (obstructed) topography. Figure 11.19 shows the results for $\pi/4$-DQPSK as well as BPSK and FSK. BPSK has a 3 dB advantage over $\pi/4$-DQPSK at low E_b/N_o values, and a 2.8 dB advantage at higher signal levels. However, this is offset by the 3 dB advantage on the spectral efficiency offered by $\pi/4$-DQPSK. This suggests that among the three schemes compared, $\pi/4$-DQPSK is the most appropriate choice for indoor flat fading radio channels.

The error performance of $\pi/4$-DQPSK is analyzed for a two-ray Rayleigh fading mobile channel with a cochannel interference (CCI) in [12]. This is considered to be the model for the cellular mobile communication. The channel is modeled by the following expression.

$$\widetilde{r}(t) = R_1(t)e^{-j\phi_1(t)}\widetilde{s}_T(t) + R_2(t)e^{-j\phi_2(t)}\widetilde{s}_T(t-\tau) + R_3(t)e^{-j\phi_3(t)}\widetilde{s}_C(t) + \widetilde{n}(t) \tag{11.34}$$

where $\widetilde{r}(t)$ is the equivalent low-pass received signal. $\widetilde{s}_T(t)$ is the transmitted low-pass signal and $\widetilde{s}_T(t-\tau)$ is the second ray with a delay τ. $\widetilde{s}_C(t)$ is the lowpass co-channel interference. $\widetilde{n}(t)$ is the low-pass AWGN. $R_i(t), i = 1, 2, 3$, are Rayleigh envelopes and $\phi_i(t), i = 1, 2, 3$, are random phases with uniform distribution. The combination of $s_T(t)$ and $s_T(t-\tau)$ accounts for the frequency-selective fading. If

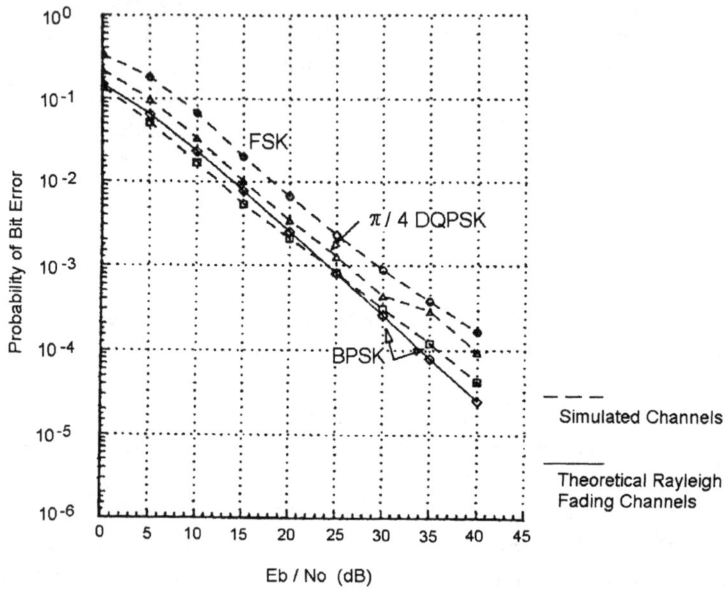

Figure 11.19 Comparison of simulated BER performance for various modulation schemes in the 50 combined simulated channels. From [13]. Copyright © 1991 IEEE.

$R_2(t) = 0$, the channel is not frequency-selective fading, it is flat fading. If $R_3(t) = 0$, there is no CCI. The channel could be slow fading or fast fading, depending on the speed of the mobile and the date rate. In terms of the received signal, this is reflected by the Doppler spread, or equivalently, the correlation of the signal samples.

In the analysis, to avoid ISI, a raised-cosine filter with roll-off factor α is used in the transmitter and a brick-wall filter with bandwidth (baseband) equal to $(1+\alpha)f_N$ $+f_{D,\max}$ is used in the receiver, where f_N is the Nyquist bandwidth ($f_N = 1/(2T_s)$ in baseband, see Section 1.4.2), and $f_{D,\max}$ is maximum Doppler spread. The receiver passes the received signal without distortion. The system is ISI-free. This causes 0.57 dB (= $10\log(1+\alpha)(1-0.25\alpha)$)[3] E_b/N_o degradation for

[3] This is the ratio between the equivalent noise bandwidths of the simulated system and the matched filtering system where a Nyquist filter is used in the transmitter and the receiver. The Nyquist filter is the most narrow-banded filter without ISI. But a Nyquist filter is susceptible to timing jitters and more difficult to implement. It can be easily shown that the equivalent noise bandwidth of a raised-cosine filter is $f_N(1-0.25\alpha)$. Thus the total equivalent noise bandwidth of the simulated system is

$\alpha = 0.2$. However when CCI is the dominant interference as in cellular systems, the performance is the same as the matched filtering system.

The analytical BER expressions for DBPSK, DQPSK, and $\pi/4$-DQPSK are derived in [12]. However these expressions are quite complicated. We would rather not list them here, instead, some representative numerical results for several sets of important parameters are presented. First we define the parameters.

1. C/N denotes the average carrier-to-noise power ratio with noise power measured in receiver bandwidth $(2(1+\alpha)f_N)$, if $f_{D,\max} \ll f_N$. Recall we usually draw BER curves against E_b/N_o instead of C/N. The relation between them is simple. We know that $C = E_b f_b$ and $N = N_o W_n$, where f_b is the data bit rate and W_n is the equivalent noise bandwidth, then

$$\frac{C}{N} = \frac{f_b}{W_n} \cdot \frac{E_b}{N_o} \tag{11.35}$$

   For $\pi/4$-DQPSK using the Nyquist filter, at the carrier band $W_n = 2f_N = f_s = f_b/2$. Thus $C/N = 2E_b/N_o$. *That is, C/N is greater than E_b/N_o by 3 dB* in a $\pi/4$-DQPSK system with matched Nyquist filters in the transmitter and the receiver. Since E_b/N_o and C/N are proportional to each other, the degradation caused by the raised-cosine filter is the same in C/N and E_b/N_o.
2. C/I denotes the average carrier-to-interference power ratio.
3. C/D denotes the average power ratio between the main-path signal and the delayed-path signal.
4. τ/T_s denotes the delay of the secondary path normalized to the symbol period.

Figures 11.20–11.23 show some representative numerical results. Figure 11.20 shows the BER ($P(e)$) versus C/N of $\pi/4$-DQPSK in a flat, slow, Rayleigh fading channel corrupted by AWGN and co-channel interference. No Doppler spread and time dispersion are assumed. That is, the mobile speed is zero and there is no second ray of signal. The filter roll-off factor $\alpha = 0.2$. The numbered curves are for different values of carrier-to-interference ratio.[4] Inspecting these curves carefully, we can see that when C/N is approximately equal to C/I, the effect of the CCI begins to become dominant. When C/N is much greater than C/I, the effect of the CCI is an irreducible error floor. This is no surprise since the cochannel signal is just another noise to the main channel signal.

Figure 11.21 shows the effect of mobile speed, or equivalently, the Doppler

---

$(1+\alpha)(1-0.25\alpha)f_N$.

[4] Curve (5) corresponds to a flat, slow, Rayleigh fading channel without CCI. So this curve should be equivalent to that in Figure 10.5. Recall that we just argued that C/N is greater than E_b/N_o by 3 dB and there is also a 0.57 dB degradation due to the filtering, then the C/N curve should be shifted to the right by 3.57 dB compared with the E_b/N_o curve. By comparing them we can see this is true.

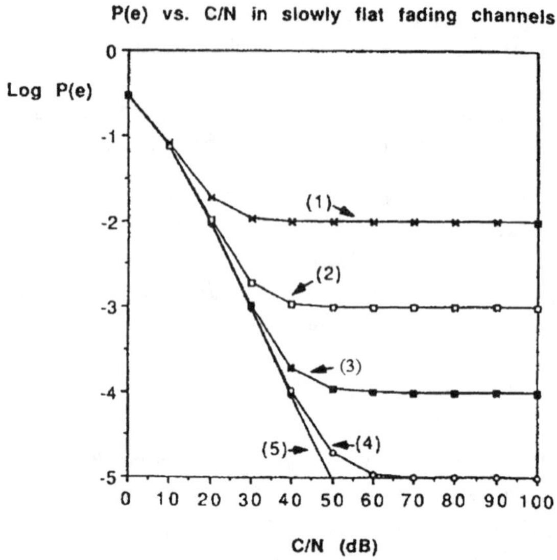

Figure 11.20 BER ($P(e)$) versus C/N of $\pi/4$-DQPSK in a flat slow Rayleigh fading channel corrupted by AWGN and cochannel interference. $f_c = 850$ MHz, $f_s = 24$ kbps, $\alpha = 0.2$. No Doppler spread and time dispersion. $C/I =$ (1) 20, (2) 30, (3) 40, (4) 50, (5) ∞ dB. From [12]. Copyright © 1991 IEEE.

spread. No CCI and time dispersion are assumed. Again $\alpha = 0.2$. The curves are for speed from 0 to 75 miles per hour (mph). It is seen that even when there are no CCI and no noise ($C/N \rightarrow \infty$), there are still errors as long as the mobile speed is not zero. The errors are caused by the random phase modulation of the fast channel (due to motion of the mobile).

Figure 11.22 illustrates the BER of $\pi/4$-DQPSK in a frequency-selective slow fading channel. There are no AWGN, CCI, and Doppler spread. The curves are drawn versus C/D, using τ/T_s as a parameter. Comparing Figures 11.22 and 11.21, it is seen that for small τ/T_s, the system can tolerate stronger delayed signal than CCI. For τ that approaches T_s, the BER caused by delayed signal approaches that caused by a CCI as we expect since the information source is assumed uncorrelated.

Figure 11.23 illustrates the BER of $\pi/4$-DQPSK in a frequency-selective fast fading channel. No AWGN or CCI is present in the channel. This figure shows that when the delayed signal is comparable to the main-path signal (C/D close to 0 dB), the error rate is controlled by the frequency selectivity (τ/T_s). When delayed signal power is small, BER is controlled by fast fading (speed v).

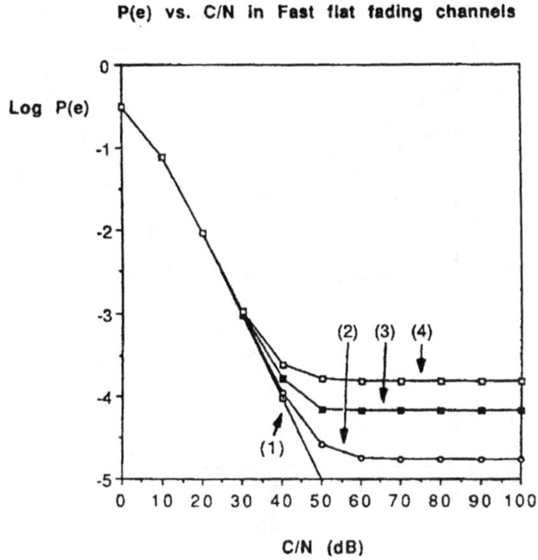

Figure 11.21 BER versus C/N of $\pi/4$-DQPSK in a flat fast Rayleigh fading channel corrupted by AWGN. No CCI and time dispersion. $f_c = 850$ MHz, $f_s = 24$ kbps, $\alpha = 0.2$. $v =$ (1) 0, (2) 25, (3) 50, (4) 75 mph. From [12]. Copyright © 1991 IEEE.

The simulation in [14] is for a two-ray Rayleigh fading model as well as measurement based models. The two-ray model is the same as that given in (11.34) and the filtering strategy is also the same as in [12]. The simulation results confirm the analytical results by Liu and Feher [12].

11.5 MHPM IN FADING CHANNELS

The performance of the rectangular frequency pulse multi-h modulation scheme (1REC-MHPM, also known as multi-h CPFSK) is evaluated for slow and flat or frequency selective Rician and Rayleigh channels with Doppler frequency shift [15]. The evaluation is performed with a method combining analysis and simulation. Performance degradations are evaluated for various direct-to-reflected signal ratio, Doppler shifts, and relative time delays in Rician fading channels. The channel model is a two-ray model consisting of a direct signal without fading, and a reflected signal with Rayleigh fading. AWGN is also included. The satellite-to-ground vehicle chan-

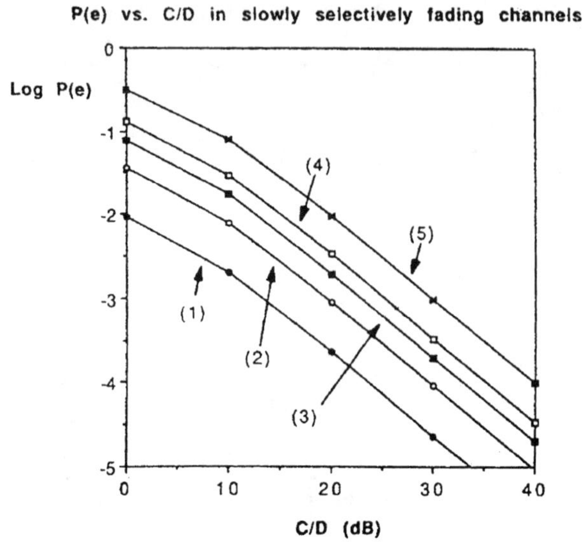

Figure 11.22 BER versus C/D of $\pi/4$-DQPSK in a frequency-selective slowly fading channel. No AWGN, CCI, and Doppler spread. $f_c = 850$ MHz, $f_s = 24$ kbps, $\alpha = 0.2$. $\tau/T_s =$ (1) 0.1, (2) 0.2, (3) 0.3, (4) 0.4, (5) 1.0. From [12]. Copyright © 1991 IEEE.

nel and the airplane-to-ground channel, where MHPM schemes are likely to be used, are of this type. The received signal is

$$r(t) = r_d(t) + r_r(t) + n(t)$$

where

$$r_d(t) = \sqrt{\frac{2E}{T_s}} \cos\left(\omega_c t + \frac{a_i \pi h_i t}{T_s} + \phi_i\right); \; 0 \le t \le T_s$$

is the direct component,

$$r_r(t) = v_i \sqrt{\frac{2E}{T_s}} \cos\left((\omega_c + \Delta\omega)(t - \tau) + \frac{a_i \pi h_i (t - \tau)}{T_s} + (\phi_i + \varphi_i)\right)$$

is the reflected component, and $n(t)$ is the AWGN. The parameters are defined as follows.

E = symbol energy.

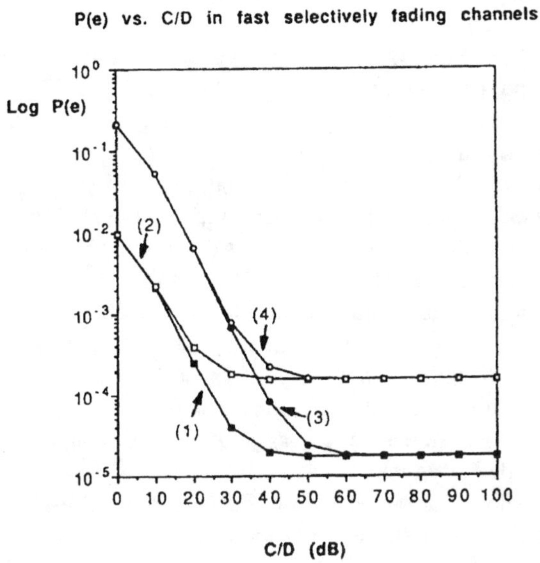

Figure 11.23 BER versus C/D of $\pi/4$-DQPSK in a frequency-selective fast fading channel. No AWGN, CCI. $f_c = 850$ MHz, $f_s = 24$ kbps, $\alpha = 0.2$. (1) $\tau/T_s = 0.1, v = 25$ mph, (2) $\tau/T_s = 0.1, v = 75$ mph, (3) $\tau/T_s = 0.5, v = 25$ mph, (4) $\tau/T_s = 0.5, v = 75$ mph. From [12]. Copyright © 1991 IEEE.

$\omega_c =$ carrier angular frequency.
$a_i =$ information data, $a_i \in [+1, -1]$.
$M =$ even constant and a power of 2.
$h_i =$ one of a set of modulation indices $(h_1, h_2, ...h_K)$.
$\phi_i =$ the phase at the beginning of the ith interval.
$v_i =$ Rayleigh envelope.
$\varphi_i =$ uniformly distributed phase in $(0, 2\pi)$.
$\tau =$ delay of the Rayleigh signal with respect to the direct signal.

The direct component of signal experiences a Doppler shift $\Delta\omega$ due to the relative motion between the transmitter and the receiver. For the coherent detection of 1REC-MHPM, the carrier needs to be recovered at the receiver. Since the direct signal is usually dominant, the recovered local carrier frequency is the same as the direct signal frequency, which is denoted as ω_c. Note that it is different from the transmitted frequency by $\Delta\omega$. The Doppler shift is not explicitly included in the Rayleigh fading signal model since its effect is modeled by the Rayleigh envelope and the random phase. Therefore the Rayleigh signal has the same carrier frequency as the transmission, but it is different from ω_c by $\Delta\omega$. There is also a delay τ in $r_r(t)$ with respect to the direct component $r_d(t)$.

Recall that 1REC-MHPM can be demodulated by an MLSE demodulator using four correlators and the Viterbi algorithm (see Section 7.5.1). The simulation in [15] is based on the analytical results of the correlators' outputs and the Viterbi algorithm. A similar method was used for the AWGN channel [16].

In the simulation the two independent Gaussian processes which produce the Rayleigh fading envelopes are bandlimited by a sixth-order Butterworth low-pass filter. It has a bandwidth of the fading process determined by the Doppler spread of the channel which, in turn, is determined by the mobile speed.

In the simulation it is found that P_b increases with the increase of Doppler shift f_d slightly, where $f_d = \Delta f T_s$ is the Doppler shift normalized to the symbol rate. However the effect of f_d is not as significant as that of other parameters. Therefore when we examine the effect of other parameters, we set $f_d = 0.2$ which gives an average value of P_b in the practical range of f_d. The simulation results are summarized in the following figures.

Figure 11.24 shows various P_b plots of Rician fading for different K, the direct-to-reflected signal power ratio, and normalized delay $t_d = \tau/T_s$ for the 1REC-MHPM scheme with $H_2 = (2/4, 3/4)$. When $K = 0$ we arrive at Rayleigh fading. It is seen from the plot of Rayleigh fading that the bit error rate P_b is very high $(0.25 - 0.26)$. This is due to the fact that the simulated receiver is not synchronized to the reflected component.[5] When the content of P_d is increased (i.e., when a carrier is recovered at the receiver from the direct signal component), a significant improvement in P_b is observed (see plots for $K = 0, 5, 10, 20$ dB). The P_b at $K = 20$ dB is comparable to that in the presence of only AWGN. For $P_b = 10^{-6}$, the performance degradation at K =10 dB and $t_d = 0$, with respect to the AWGN channel, is about 2 dB. Due to the Rayleigh component in the signal these plots show error floors. Figure 11.24 also shows the effect of t_d on the P_b. It is seen that when t_d increases P_b increases steadily, especially when $t_d \geq 0.6$. This is due to the fact that when t_d is much larger than 0, the channel becomes frequency-selective, which causes additional distortion of the signal on top of fading and AWGN. This in turn increases the P_b. However, the effect of t_d can be largely removed if a channel state tracking mechanism is incorporated.

Figure 11.25 is similar to Figure 11.24 except that it is for $H_3 = (4/8, 5/8, 6/8)$.

---

5 In the simulation, the Rayleigh channel is obtained as a special case of the Rician channel where carrier synchronization is established for the direct signal. When the power of the direct signal is reduced to zero, the channel becomes pure Rayleigh without carrier synchronization. If synchronization was established for the Rayleigh signal, the BER performance would be improved and the BER curve would be similar to that of BPSK or QPSK (see Figure 11.4) since the BER expression of MHPM is also of the form of a Q function (see (7.28)).

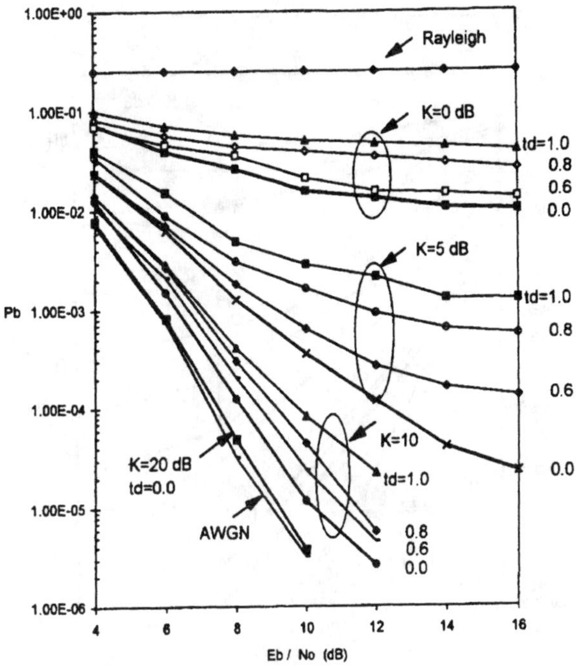

Figure 11.24 Performance of 1REC-MHPM $H_2 = (2/4, 3/4)$ in Rayleigh, Rician, and AWGN channels for various values of K and t_d ($f_d = 0.2$). Rayleigh channel is approximated by $K = -200$ dB, and no synchronization is done for the Rayleigh fading signal. From [15]. Copyright © 1997 IEEE.

Again we see that the direct to reflected signal ratio, K and t_d have significant impact on P_b.

The performance comparison of MSK and 1REC-MHPM schemes in Rician fading is plotted in Figure 11.26, for $K = 10$ dB and $t_d = 0$ and 1.0. This figure compares MSK, $H_2 = (2/4, 3/4)$, and $H_3 = (4/8, 5/8, 6/8)$. The solid lines are for $t_d = 0$, and the dotted lines are for $t_d = 1.0$. The performance of $H_2 = (2/4, 3/4)$ is found to be superior to MSK by about 2 dB for $t_d = 0$ throughout the E_b/N_o range. This coding gain is slightly larger than that in the AWGN channel (1.4 dB) [17]. Comparing $H_3 = (4/8, 5/8, 6/8)$ with MSK, the coding gain is around 3 dB for $t_d = 0$ which is slightly larger than that of the AWGN channel (2.8 dB). For $t_d = 1.0$, MSK suffers from more loss in P_b than MHPM schemes. This indicates that these

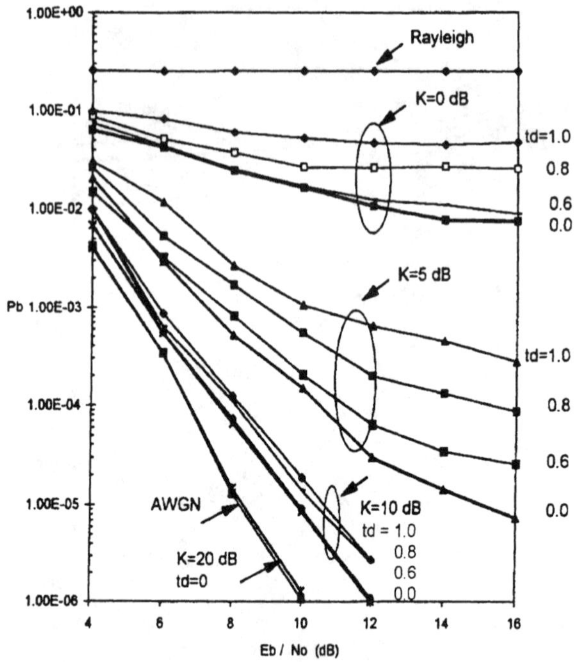

Figure 11.25 Performance of 1REC-MHPM $H_3 = (4/8, 5/8, 6/8)$ in Rayleigh, Rician, and AWGN channels for various values of K and t_d ($f_d = 0.2$). Rayleigh channel is approximated by $K = -200$ dB, and no synchronization is done for the Rayleigh fading signal. From [15]. Copyright © 1997 IEEE.

1REC-MHPM schemes have retained their coding gain over MSK in fading channels and are subject to less loss in error performance when delay spread increases.

11.6 QAM IN FADING CHANNELS

Because of its great bandwidth efficiency, applications of QAM in fading channels have always been attracting interest. In this section, we first evaluate the error performance of the square M-ary QAM in a slow flat Rayleigh or Rician fading channel, using the technique developed in Section 11.2. Then we present the star-QAM which is particularly suitable for fading channels.

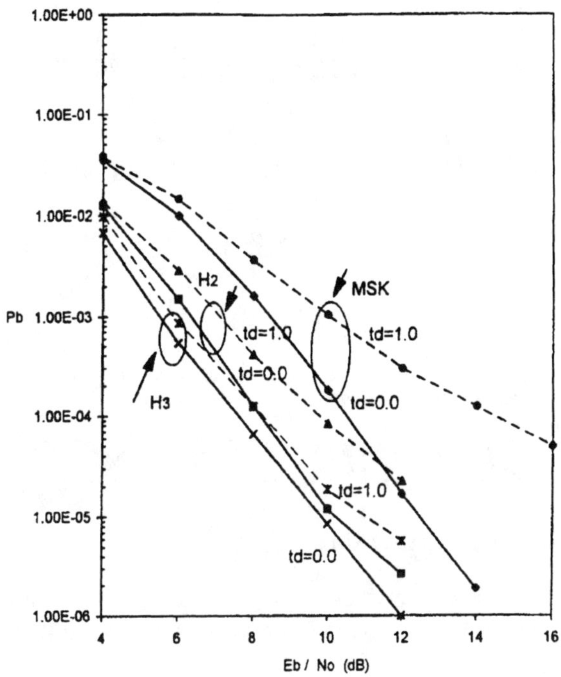

Figure 11.26 Comparison of MSK, $H_2 = (2/4, 3/4)$, and $H_3 = (4/8, 5/8, 6/8)$ in Rician channel ($K = 10$ dB, $f_d = 0.2$). From [15]. Copyright © 1997 IEEE.

11.6.1 Square QAM

Substituting the P_s expression for square M-ary QAM and corresponding fading distributions into (11.17), we can easily evaluate the symbol error probability of the square M-ary QAM in fading channels. If the bit-symbol mapping is Gray coded, as usually it is, the bit error probability can be found using the P_s-P_b relation expression (9.37).

At high SNRs, the P_s expression is (9.35) which is in Q-function form. We can directly use (11.24) to obtain the result. Note that in (9.35) E_{avg} is the energy per symbol, (that is $E_{avg} = E_b \log_2 M$), then in fading channel (9.35) can be written as

$$P_s \cong \frac{4(\sqrt{M}-1)}{\sqrt{M}} Q\left(\sqrt{\frac{3\log_2 M}{M-1}\gamma_b}\right) \tag{11.36}$$

Comparing (11.36) with (11.19) we can recognize

$$C = \frac{4(\sqrt{M}-1)}{\sqrt{M}}$$

and

$$\delta = \frac{3\log_2 M}{M-1}$$

Substituting them into (11.24) we have

$$P_s = \frac{2(\sqrt{M}-1)}{\sqrt{M}}\left[1-\sqrt{\frac{3\Gamma\log_2 M}{2(M-1)+3\Gamma\log_2 M}}\right] \tag{11.37}$$

(for square QAM at high SNR)

Note that when $M = 4$, square QAM degenerates to QPSK. In this case, we can easily check that (11.37) reduces to (11.25) except for a factor of 2, since one is P_s and the other is P_b.

At low SNRs, (11.37) can induce big inaccuracy just as (9.35) would for the AWGN channel. Therefore we need to use (9.33) and (9.34) in conjunction with (11.18) in (11.17) to obtain the results numerically. Equation (9.33) must be rewritten in terms of γ_b as

$$P_{\sqrt{M}} \cong \frac{2(\sqrt{M}-1)}{\sqrt{M}} Q\left(\sqrt{\frac{3\log_2 M}{M-1}\gamma_b}\right) \tag{11.38}$$

Figure 11.27 shows the numerical results which are accurate at both low and high SNRs.

For a slow, flat, Rician fading channel, we can numerically find out the error performance using the same set of formulas except that the distribution density is (11.31) instead of (11.18). Figures 11.28 to 11.30 show the numerical results for $M = 16, 64, 256$.

11.6.2 Star QAM

In Chapter 8 we have stated that star QAM is not optimum in terms of $d_{\min}$ under

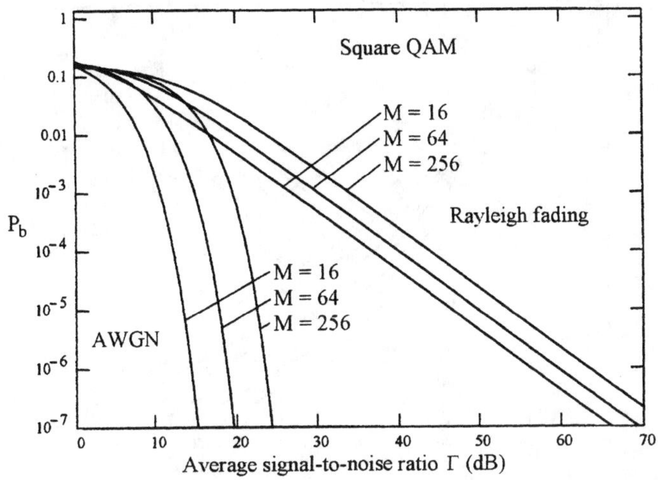

Figure 11.27 Bit error rates of square QAM in a slow, flat, Rayleigh fading channel.

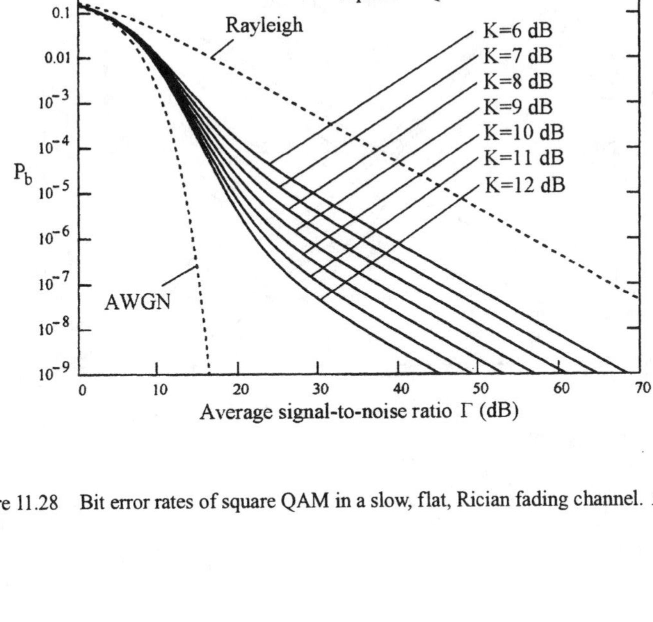

Figure 11.28 Bit error rates of square QAM in a slow, flat, Rician fading channel. $M = 16$.

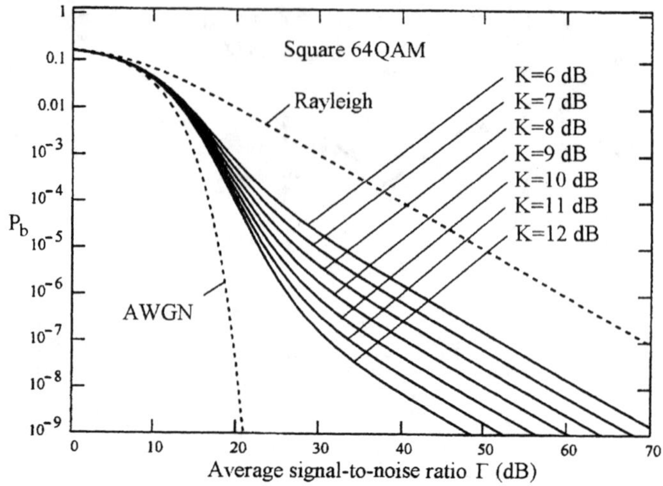

Figure 11.29 Bit error rates of square QAM in a slow, flat, Rician fading channel. $M = 64$.

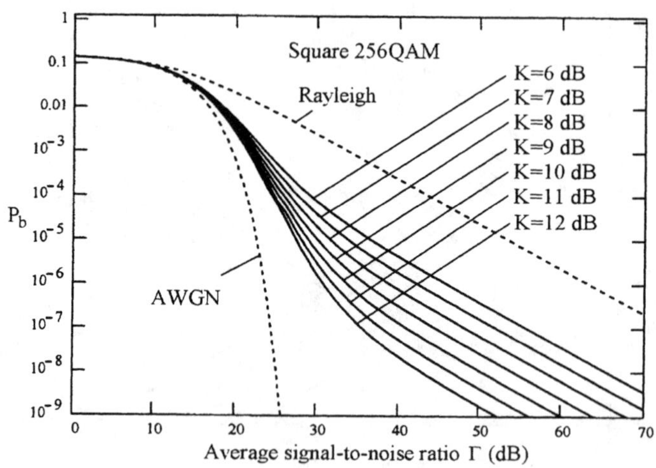

Figure 11.30 Bit error rates of square QAM in a slow, flat, Rician fading channel. $M = 256$.

the constraint of average phasor power. Therefore it is not a preferable choice in an AWGN channel. However it allows efficient differential encoding and decoding, which makes it suitable for fading channels. Recall that we have discussed differential coding for square QAM in Chapter 8. That is for the purpose of resolving phase ambiguity in carrier recovery. The method of differential coding for star QAM is different and the purpose is different too: to avoid carrier recovery and enable differential detection of signals.

For the M-ary star constellations, such as the one in Figure 9.1 Type I, assume $M = 2^n$, there are $M_1 = 2^{n_1}$ amplitudes (cycles) and there are $M_2 = 2^{n_2}$ phases (points on a cycle), where $n_1 + n_2 = n$ and $M = M_1 M_2$. For example, when $n_1 = 1, n_2 = 3$, we have the 16 star QAM in Figure 9.1 Type I.

For star QAM, since there are M_1 amplitudes, the average power is

$$P_{avg} = E\{A_i^2/2\} = \frac{1}{2M_1}\sum_{i=1}^{M_1} A_i^2 \tag{11.39}$$

where A_i is the amplitude of the ith phasor in the constellation. From (11.39) and (9.29) we have the PSD of star QAM as

$$\begin{aligned}\Psi_{\tilde{s}}(f) &= \frac{T_s}{M_1}\left(\sum_{i=1}^{M_1} A_i^2\right)\left(\frac{\sin \pi f T_s}{\pi f T_s}\right)^2 \\ &= \frac{nT_b}{M_1}\left(\sum_{i=1}^{M_1} A_i^2\right)\left(\frac{\sin \pi f n T_b}{\pi f n T_b}\right)^2 \end{aligned} \tag{11.40}$$

The signal points in star QAM can be easily encoded differentially [18, p. 325]. Use the star 16QAM as an example. Of the four bits in each symbol, b_1, b_2, b_3, and b_4, the first bit is used to encode the amplitude change. For example, a 1 denotes the signal amplitude is changed and a 0 denotes no change. The other three bits are Gray coded to denote the phase changes. For example we can code 000 as no phase change, 001 as $\pi/4$ change, and so forth. Table 11.2 shows the coding scheme. The Hamming distance between two adjacent bit patterns (including no. 1 and no. 8) is one and the corresponding phase changes differ by $\pi/4$ which is the smallest increment. These patterns are not mapped into transmitted symbols; instead they are mapped into the phase differences of two consecutive transmitted symbols. When two symbols are received, first their amplitude is detected and b_1 is determined. Next their phases are compared to determine what is the $\Delta\theta_i$. From $\Delta\theta_i$, using a look-up table like Table 11.2, bit pattern $b_2 b_3 b_4$ can be determined. When $\Delta\theta_i$ is corrupted by noise and fading, and a detection error is made, most likely the error is made by taking one of the adjacent bit patterns. This leads to only one bit error because of

Pattern no.	Bit pattern ($b_2b_3b_4$)	Phase change ($\Delta\theta_i$)
1	000	0
2	001	$\pi/4$
3	011	$2\pi/4$
4	111	$3\pi/4$
5	011	$4\pi/4$
6	010	$5\pi/4$
7	110	$6\pi/4$
8	100	$7\pi/4$

Table 11.2 Gray coded phase changes.

Gray coding.

In the fading channel, the signal amplitude varies. Thus the threshold for amplitude detection is made adaptive. Let A_1 and A_2 denote the amplitudes of the two rings in the star 16QAM constellation. Assume the received phasor amplitudes are Z_k and Z_{k+1} at time $t = kT_s$ and $(k+1)T_s$. The algorithm used in [18, p. 325] is as follows. If

$$\frac{Z_{k+1}}{Z_k} \geq \frac{A_1 + A_2}{2}$$

or if

$$\frac{Z_{k+1}}{Z_k} < \frac{2}{A_1 + A_2}$$

then a significant change in amplitude is deemed to have occurred and bit b_1 is set to logic 1 at time $(k+1)T_s$. This detection rule does not rely on the absolute value of the received signal amplitude, rather it depends on the relative amplitudes of the two consecutive signal amplitudes. This is exactly what is needed in a fading channel.

Figure 11.31 shows some simulation results for various 16QAM schemes. The simulation parameters are as follows. The carrier frequency is 1.9 GHz, the symbol rate is 16 k symbol/s, the mobile's speed is 30 mph, and the channel exhibits Rayleigh fading with AWGN. Both transmitter and receiver use a fourth-order Butterworth low-pass filter with a 3 dB point at 1.5 times that Baud rate. Curve (a) is the square 16QAM performance, which is worse than that given in Figure 11.27 where no filtering is assumed. Curve (b) is the performance of the differential star 16QAM, which is largely enhanced compared with square 16QAM. Curve (c) is again for star 16QAM, but an oversampling technique is used. By this technique, $n = 8$ samples equally spaced in time are made per symbol period. The variations observed in these samples are used to modify Z_k and Z_{k+1}, hence improve the amplitude detection (see [18, p. 325] for details). Curve (d) is for star 16QAM with two-antenna switched

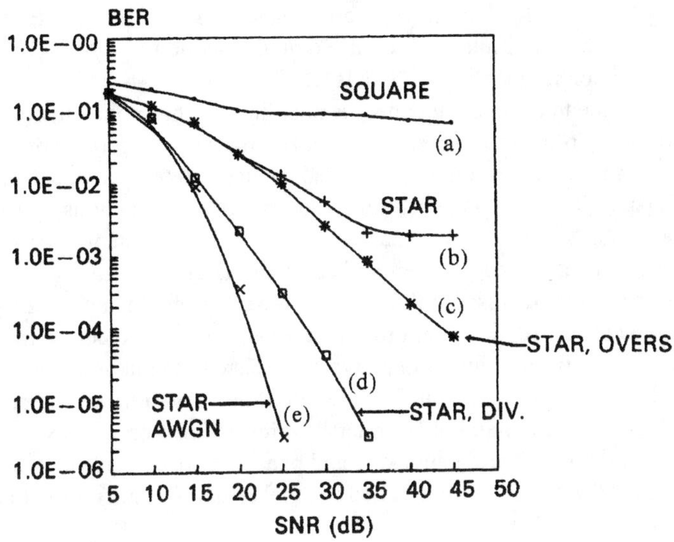

Figure 11.31 BER for various 16-QAM schemes using different demodulations. From [18]. Copyright © 1994 IEEE.

diversity where the larger of the two signals received is selected. Curve (e) is the performance of star 16QAM in an AWGN channel. From this figure it is seen that differential star QAM improves BER performance in fading channels significantly.

11.7 OVERVIEW OF REMEDIAL MEASURES AGAINST CHANNEL IMPAIRMENTS

From the above discussion we can see that fading, especially Rayleigh fading can reduce the BER performance of a modulation scheme significantly. Remedial measures against fading must be adopted in order to preserve or at least partially preserve the BER performance.

The first, also simplest, measure against fading is to use a differential modulation scheme where carrier phase synchronization is not required. Up to a 3 dB loss in error performance is expected. In the above discussion, the performance of DBPSK and $\pi/4$-DQPSK in fading channels has been addressed. Additional references exist for DBPSK and differential MSK in a Rician fast fading environment [19], differential GMSK in a land mobile channel [20], narrowband $\pi/4$-DQPSK in a Rayleigh fading channel [21], and multiple-symbol differential detection of $\pi/4$-DQPSK in land mobile satellite communications channels [22].

If the loss due to differential coding is not allowed, then coherent demodulation is required. To help the receiver establish carrier phase synchronization in a fading environment, a pilot tone or pilot symbols can be used. A technique called transparent tone in band (TTIB) is proposed for various mobile applications by McGeehan, Bateman et al. [23,24], and is also described in [18]. In TTIB systems for mobile radio applications, a spectral gap of a certain bandwidth is allocated in the center of the signal spectrum to allow insertion of the pilot tone. The frequency of the pilot tone is usually the carrier frequency in order that it can assist the receiver in carrier recovery. An alternative to TTIB is pilot symbol assisted modulation (PSAM) [25–29]. In this method, known phasors are inserted into the data stream. The receiver extracts channel attenuation and phase rotation information from the received known symbols and uses it for fading envelope and phase compensation (that is, eliminating $ze^{-j\phi}$ in (11.16)). This is often called channel estimation and correction (Section 11.8).

For frequency-selective fading channels, the existence of multipath signals causes extra terms (intersymbol interference or ISI) in (11.16) which can not be simply "corrected" as in the frequency-flat case. In this case, a class of techniques called equalization is used to eliminate or reduce the multipath terms. Equalization is not only needed for fading channels. It is necessary even for AWGN channels when there is ISI. This happens when the signal bandwidth is wider than the channel bandwidth. For example, if a square pulse train is transmitted through a telephone wire, the limited bandwidth will cause ISI in the pulse train at the receiver end. Equalization techniques for both cases are the same (Section 11.9).

Another class of remedial measures against fading are diversity techniques, which collect and combine the signals received from multiple antennas, or at different times, to form a composite signal. Due to the fact that fading is random and chances of all signals being in deep fades simultaneously are very small, the composite signal can stay strong with high probability (Section 11.10).

Recent developments of remedial measures against fading include orthogonal frequency division multiplexing (OFDM) and multi-input-multi-output (MIMO) links that use space-time codes and spatial multiplexing. OFDM is an important and mature modulation scheme and will be covered in Chapter 12 and subsequent chapters. MIMO is relatively new and is evolving fast. Its basics will be introduced in this chapter (Section 11.11).

11.8 CHANNEL ESTIMATION AND CORRECTION

Channel estimation is basically classified into three categories: pilot tone assisted, pilot symbol assisted, and decision feedback estimation.

11.8.1 Pilot Tone Methods

A pilot tone can be transmitted together with signals to assist demodulation in the receiver. Early systems positioned the pilot spectral line above or in the band of the SSB signal spectrum. The pilot subjects interference from signal frequency components. Later, a better scheme, transparent-tone-in-band (or TTIB), was proposed in [23] and researched in [24, 30–32]. The idea of TTIB is to insert the pilot spectral line in a "notch" artificially created in the signal spectral band so that the pilot will be easily extracted by filtering.

For example, the voice signal spectrum typically extends from 300 Hz to 3 kHz in mobile communications. The spectral band is split into two segments: 300 Hz–1.7 kHz and 1.7–3 kHz. Then by suitable mixing and filtering, the upper band is translated up in frequency by an amount equal to the required "notch" width or band separation. For example, the notch may be 1.2 kHz for the speech signal. Thus after spliting and shifting, the spectrum now extends from 300 Hz–1.7 kHz and 2.9–4.2 kHz. The low-level reference pilot tone then is inserted into the center of the notch, which for our particular example would be 2.3 kHz. This composite signal is transmitted using conventional SSB (single side band) techniques. In the receiver, the pilot tone will be extracted by a narrow-band filter or phase locked loop and is used as a reference signal for coherent demodulation. A reverse process of mixing and filtering will remove the notch and the original 300 Hz to 3 kHz band will be restored.

If the local oscillation frequency in the TTIB receiver is locked to the carrier frequency of the transmitter by the use of a phase locked loop, the system is called locked TTIB. Otherwise, the system is called an unlocked TTIB. It was found through experiments that unlocked TTIB works fine with speech signals, but fails for digital modulated signals (noncoherent FSK and DPSK) [23]. However, the locked TTIB works satisfactorily for digital modulations in AWGN and Rayleigh channels.

The disadvantages of TTIB are the relatively complex signal processing and increased peak-to-average power ratio. The scheme described next can avoid these problems.

11.8.2 Pilot Symbol Assisted Modulation (PSAM)

Pilot symbol assisted modulation (PSAM) is described in [25–29]. While we only briefly reviewed the TTIB, we will describe PSAM in detail since it is a better scheme and is used widely in digital modulations. We will present the results of [28] here since it addresses the most typical fading channel in mobile wireless communications, the fast Rayleigh channel.

The complex envelope of the transmitted signal is given by

$$s(t) = A \sum_{k=-\infty}^{\infty} b(k)p(t - kT)$$

where A is the amplitude, $b(k)$ is the kth symbol for BPSK, QPSK or a general QAM, and T is the symbol duration. $p(t)$ is a unit energy pulse

$$\int_{-\infty}^{\infty} |p(t)|^2 dt = 1$$

The symbols are divided into frames of M symbols and the pilot symbols are inserted at times $lM, l = 0, 1, ...$ with a known value $\widetilde{b}$.

The fading channel output is

$$r_c(t) = c(t)s(t) + n_c(t)$$

where $n_c(t)$ is complex AWGN with power spectral density N_o in both real and imaginary parts

$$c(t) = \exp(j2\pi f_0 t)g(t) \tag{11.41}$$

which includes residual frequency offset f_0 after AFC (automatic frequency control). $g(t)$ is the complex Gaussian fading process with variance σ_g^2 and Doppler spread f_D. Its autocorrelation is given as

$$R_c(\tau) = \sigma_g^2 \widetilde{R}_c(\tau)$$

where $\widetilde{R}_c(\tau)$ is the normalized autocorrelation.

At this point we need to introduce SNR measures which will be used later. Define the ratio q of pilot power to data power as

$$q = \frac{|\widetilde{b}|^2}{(M-1)E\{|b|^2\}} \tag{11.42}$$

where the expectation is over data symbols only. The expected total energy received in a frame is

$$E_f = \sigma_g^2 A^2 (|\widetilde{b}|^2 + (M-1)E\{|b|^2\}) \tag{11.43}$$

With n bits in a symbol, $M - 1$ data symbols in a frame, the received energy per data bit is

$$E_b = \frac{\sigma_g^2 A^2}{n(M-1)} |\widetilde{b}|^2 \left(\frac{1+q}{q}\right) \tag{11.44}$$

In the receiver, the signal is passed through a matched-filter and the symbol-spaced samples can be written as

$$r(k) = \frac{Ac(kT)}{\sqrt{N_o}} b(k) + n(k) = u(k)b(k) + n(k)$$

where $n(k)$ are white with unit variance. Assuming $b(0)$ is the pilot symbol in the frame and all symbols $b(k)$ in the frame are indexed such that $-\lfloor M/2 \rfloor \leq k \leq \lfloor (M-1)/2 \rfloor$. The autocorrelation of symbol gain $u(k)$ is

$$\begin{aligned} R_u(i-k) &= \frac{1}{2}E[u(i)u^*(k)] = \frac{A^2}{N_o}\frac{1}{2}E[c(iT)c^*(kT)] \\ &= \frac{A^2}{N_o}R_c((i-k)MT) = \frac{A^2}{N_o}\sigma_g^2 \widetilde{R}_c((i-k)MT) \\ &= \sigma_u^2 \widetilde{R}_c((i-k)MT) \end{aligned} \tag{11.45}$$

where σ_u^2 is the variance of symbol gain $u(k)$:

$$\sigma_u^2 = \frac{A^2}{N_o}\sigma_g^2 = R_u(0) \tag{11.46}$$

Using (11.44) to obtain σ_g^2 and substitute it into (11.46) we have

$$\sigma_u^2 = \gamma_b \frac{n(M-1)}{|\widetilde{b}|^2} \frac{q}{1+q} \tag{11.47}$$

where $\gamma_b = E_b/N_o$ is the signal-to-noise ratio.

To detect $b(k)$, a linear channel state estimator prepares an estimate of the symbol gain $u(k)$ using the K nearest pilot samples:

$$v(k) = \sum_{i=-\lfloor K/2 \rfloor}^{\lfloor K/2 \rfloor} h^*(i,k) r(iM)$$

Note that the coefficients $h(i,k)$ depend on the position index k within the frame. Denote the set of pilot samples $r(iM)$, $-\lfloor K/2 \rfloor \leq i \leq \lfloor K/2 \rfloor$ by a length K vector $\mathbf{r}$, and the corresponding set of coefficients for the kth position in the frame by $\mathbf{h}(k)$, then the estimated symbol gain can be written as

$$v(k) = \mathbf{h}^H(k)\mathbf{r} = [\mathbf{r}^H\mathbf{h}(k)]^* \tag{11.48}$$

where H denotes conjugate transpose (or Hermitian transpose). According to [33], the linear minimum mean square error estimator (LMMSE) requires that $\mathbf{h}(k)$ satisfy the orthogonal condition (or normal equation)

$$E\{\mathbf{r}[v(k) - u(k)]^*\} = E\{\mathbf{r}[(\mathbf{r}^H\mathbf{h}(k))^* - u(k)]^*\} = \mathbf{0}$$

This leads to

$$\mathbf{R}\mathbf{h}(k) = \mathbf{w}(k)$$

or

$$\mathbf{h}(k) = \mathbf{R}^{-1}\mathbf{w}(k) \tag{11.49}$$

where

$$\mathbf{R} = \frac{1}{2}E[\mathbf{r}\mathbf{r}^H] \tag{11.50}$$

is the $K \times K$ autocorrelation matrix of the pilot samples, and

$$\mathbf{w}(k) = \frac{1}{2}E[u^*(k)\mathbf{r}] \tag{11.51}$$

is the length K cross-correlation vector between $u(k)$ and pilot samples. The $\mathbf{h}(k)$ given by (11.49) is the Wiener filter coefficients vector.

Now we can relate $\mathbf{R}$ and $\mathbf{w}(k)$ to channel correlation function $\widetilde{R}_c(\tau)$ and signal to noise ratio $\gamma_b = E_b/N_o$. Using (11.45) and (11.47), the ikth element of $\mathbf{R}$ is shown as

$$\begin{aligned} R_{ik} &= \frac{1}{2}E\{r(i)r^*(k)\} = \frac{1}{2}E\{u(i)\widetilde{b}(i)u^*(k)\widetilde{b}^*(k) + n(i)n(k)\} \\ &= |\widetilde{b}|^2\frac{1}{2}E\{u(i)u^*(k)\} + \delta_{ik} = |\widetilde{b}|^2 R_u(i-k) + \delta_{ik} \\ &= \gamma_b n(M-1)\frac{q}{1+q}\widetilde{R}_c((i-k)MT) + \delta_{ik} \end{aligned} \tag{11.52}$$

where δ_{ik} is the Kronecker delta. Similarly, from (11.51), (11.45), and (11.47), we have

$$\begin{aligned} w_i(k) &= \frac{1}{2}E[u^*(k)r(iM)] = \frac{1}{2}E[u^*(k)\left(u(iM)\widetilde{b}(iM) + n(iM)\right)] \\ &= \frac{1}{2}\widetilde{b}E[u^*(k)u(iM)] + \frac{1}{2}E[u^*(k)n(iM)] \end{aligned}$$

$$= \widetilde{b}R_u(k-iM) = \gamma_b \frac{n(M-1)}{\widetilde{b}^*} \frac{q}{1+q} \widetilde{R}_c((k-iM)T) \qquad (11.53)$$

The normalized channel autocorrelation $\widetilde{R}_c(\tau)$ depends on channel characteristics. Assuming isotropic scattering radio channel (Clarke's or Jakes' model [3, 4]), with frequency offset as in (11.41), the autocorrelation is

$$\widetilde{R}_c(\tau) = \exp\{j2\pi f_0 \tau\} J_0(2\pi f_D \tau) \qquad (11.54)$$

where f_D is the maximum Doppler shift.

The correction to the received signal is accomplished by multiplying $r(k)$ by $v^*(k)$

$$r(k)v^*(k) = v^*(k)u(k)b(k) + v^*(k)n(k)$$

where the signal amplitude $v^*(k)u(k)$ will be very close to a real value so that the phase rotation introduced by the channel is approximately corrected.

For BPSK, the data symbols are $b(k) = \pm 1$. Set the pilot as $\widetilde{b} = 1$. The decision rule of the receiver is to compare the corrected signal to a threshold 0:

$$\text{Re}\{r(k)v^*(k)\} : 0 \quad \text{(BPSK)}$$

If $b(k) = 1$ and $\text{Re}\{r(k)v^*(k)\} < 0$, an error is made. The bit error probability is [28]

$$P_b(k) = \frac{1}{2}\left[1 - \sqrt{\frac{\text{Re}[\rho(k)]^2}{1 - \text{Im}[\rho(k)]^2}}\right] \qquad (11.55)$$

where $\rho(k)$ is the correlation coefficient between $r(k)$ and $v(k)$ and is defined as

$$\rho(k) = \frac{\sigma_{rv}^2(k)}{\sigma_r \sigma_v(k)}$$

Since $b(k) = 1$, we have $\sigma_{rv}^2(k) = \sigma_{uv}^2(k)$ and $\sigma_r^2 = \sigma_u^2 + 1$. For an arbitrarily selected filter $\mathbf{h}(k)$,

$$\rho(k) = \frac{\mathbf{w}^H(k)\mathbf{h}(k)}{\sqrt{(\sigma_u^2+1)\mathbf{h}^H(k)\mathbf{R}\mathbf{h}(k)}}, \quad \text{(BPSK, arbitrary } h(k)\text{)} \qquad (11.56)$$

For the optimum (Wiener) filter (11.49)

$$\rho(k) = \frac{\mathbf{w}^H(k)\mathbf{R}^{-1}\mathbf{w}(k)}{\sqrt{(\sigma_u^2+1)\mathbf{w}^H(k)\mathbf{R}^{-1}\mathbf{w}(k)}}, \quad \text{(BPSK, optimum } h(k)\text{)} \qquad (11.57)$$

and since $\rho(k)$ is real in this case, we have

$$P_b(k) = \frac{1}{2}\left[1 - \rho(k)\right], \quad \text{(BPSK, optimum } h(k)) \tag{11.58}$$

For QPSK, assume data symbols are $b(k) = \pm 1 \pm j$. Although the pilot symbol would be chosen from this set in practice, the resulting $\pi/4$ rotation must be corrected before use. To simplify the notation, we consider the pilot symbols to be real: $\widetilde{b} = \sqrt{2}$. The decision rules are then

$$\mathrm{Re}\{r(k)v^*(k)\} : 0, \ \mathrm{Im}\{r(k)v^*(k)\} : 0 \quad \text{(QPSK)}$$

The BER calculation is similar to that of BPSK and the resulting correlation coefficient is

$$\rho(k) = \frac{\mathbf{w}^H(k)\mathbf{h}(k)b(k)}{\sqrt{(2\sigma_u^2 + 1)\mathbf{h}^H(k)\mathbf{R}\mathbf{h}(k)}}, \quad \text{(QPSK, arbitrary } h(k)) \tag{11.59}$$

which is complex. Using (11.59) in (11.55), the BER of QPSK can be calculated.

Numerical simulations for BPSK in [28] show the following results.

1. BER is virtually independent of interpolation position k for $K > 7$, where K is the number of filter coefficients. Subsequent results are for $k = 1$, that is, for the symbol immediately following the pilot.
2. Pilot spacing M should satisfy the Nyquist sampling rate, that is,

$$M \leq \frac{1}{2f_D T} \tag{11.60}$$

   This is because when the maximum Doppler spread is f_D, the channel variation bandwidth in frequency domain is $2f_D$. Thus according to sampling theorem, the sampling interval in time domain should be $T_{sample} \leq 1/(2f_D)$, or $MT \leq 1/(2f_D)$. Subsequent results are for a benchmark spacing $M = 7$, which can handle Doppler shifts up to 5% of the symbol rate. In fact, for $f_D T = 5\%$, $M \leq 10$. That is, M can be a little bigger in theory; however simulations in [28] show that $M = 9$ is the maximum spacing without introducing significant BER degradation for $\gamma_b = 30$ dB and $K = 63$.
3. The number of coefficients K cannot be too small. However, improvement beyond 5 for $\gamma_b = 30$ dB, or 10 for $\gamma_b = 20$ dB, is very slight. K is chosen as 11 for the results shown in the following.

Results for $M = 7$ and $K = 11$ are shown in Figures 11.32 and 11.33 for optimized PSAM and mismatched PSAM, respectively. In Figure 11.32, the coefficients are optimized at every point. It can be seen that for 0 Hz Doppler spread,

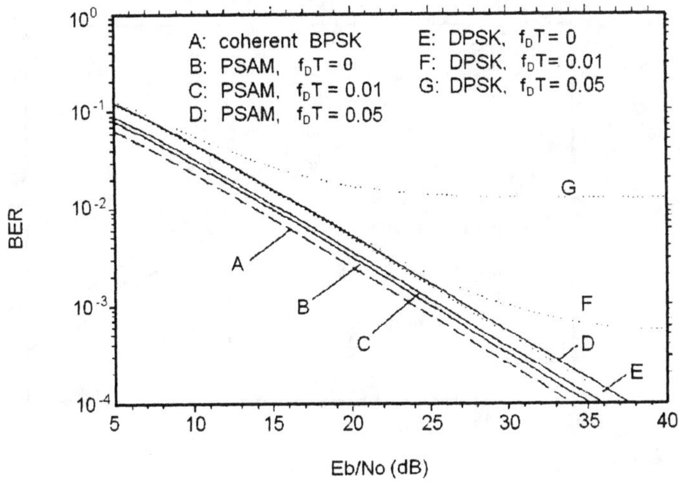

Figure 11.32 BER performance of optimized PSAM for BPSK. From [28]. Copyright © 1991 IEEE.

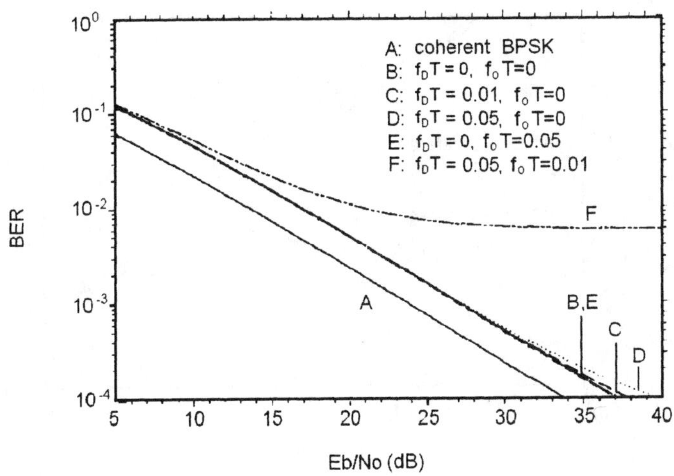

Figure 11.33 PSAM BPSK: Mismatched with coefficients optimized for $\gamma_b = 20$ dB, $f_oT = 0$, $f_DT = 0.05$. From [28]. Copyright © 1991 IEEE.

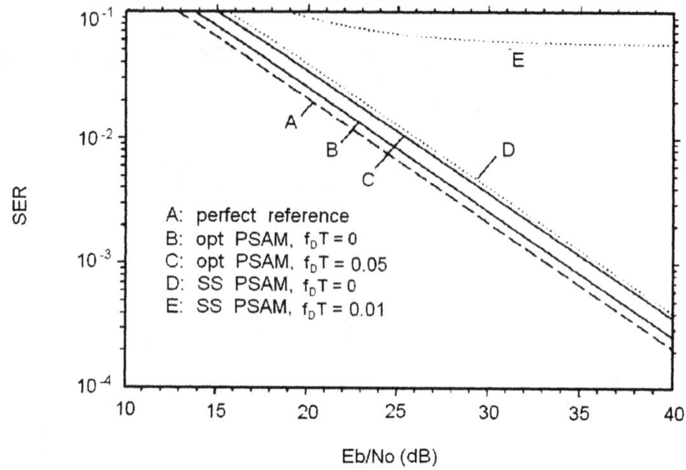

Figure 11.34 SER performance of optimized PSAM for 16QAM. From [28]. Copyright © 1991 IEEE.

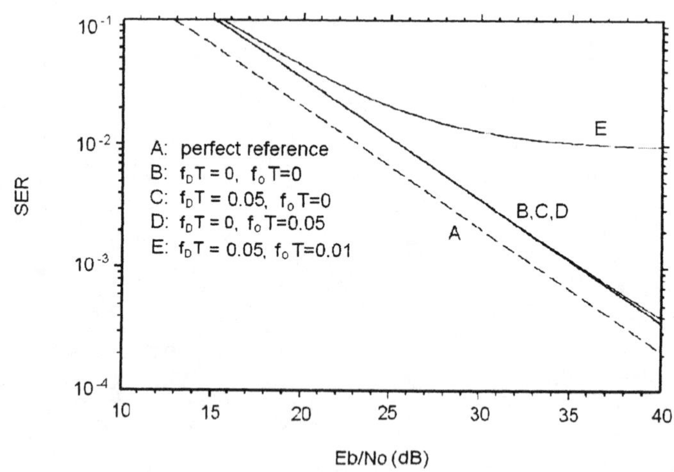

Figure 11.35 PSAM 16QAM: Mismatched with coefficients optimized for $\gamma_b = 30$ dB, $f_oT = 0$, $f_DT = 0.05$. From [28]. Copyright © 1991 IEEE.

PSAM is only 1 dB poorer than the unachievable coherent BPSK. Of the loss, 0.7 dB ($|\log_{10}(6/7)|$) is due to the energy spent on the pilot. For low Doppler spread case, this loss can be reduced by increasing the spacing. Even at 5% Doppler spread, the loss is only 3.5 dB; and there is no error floor, at least in the useful SNR range. When compared with DPSK, as calculated from expressions given in [34], PSAM is better for all values of SNR and Doppler spreads.. Moreover, when Doppler spread increases, DPSK develops a significant error floor, which makes it unusable. Figure 11.33 illustrates a trial of a mismatched PSAM, where the coefficients are optimized for $\gamma_b = 20$ dB, 5% Doppler, and no frequency offset. It is seen that Doppler spreads and frequency offsets alone do not have significant effects on the performance. However, the combined effect of 5% Doppler and 1% frequency offset causes substantial error floor, due to the fact that total bandwidth exceeds that of the interpolation filter.

Results for QPSK are very close to those of BPSK, within 2% of that of BPSK for error rates of 10^{-2} or less.

16QAM PSAM is also analyzed in [28] and a tight upper bound was developed (interested readers may refer to the reference). Numerical results for 16QAM PSAM are shown in Figures 11.34 and 11.35, respectively. Again, $M = 7$ and $K = 11$ were chosen. The results in Figure 11.34 were obtained with the filter coefficients optimized for every SNR. Also shown in the figure are the results obtained by a three-point interpolation given in [27] (labeled ss), which are unacceptable at 1% Doppler. Figure 11.35 shows the results for mismatched cases, where the coefficients are optimized for $\gamma_b = 40$ dB, $f_oT = 0$, $f_DT = 0.05$. Again, like in the BPSK PSAM case, the combined Doppler spread and frequency offset cause significant error floor.

11.8.3 Decision Feedback Channel Estimation (DFCE)

To save signal power and bandwidth spent on the pilot symbols, an alternative channel estimation method is decision feedback channel estimation [35–38]. DFCE has been applied to many cases: BPSK, QPSK, MPSK, 16QAM, and 16-star QAM; for AWGN, or frequency-nonselective Rayleigh slow fading, or frequency-selective slow Rayleigh fading; for coherent or differential detection. According to the analysis given in [36], the BER performance of BPSK, QPSK, and 16QAM using DFCE in a frequency-selective slow Rayleigh fading channel can be very close to the ideal coherent detection. We will present the results of [36] that are for the most popular digital modulation schemes and in the most common fading channel.

The DFCE system model is shown in Figure 11.36. Assume the transmitted

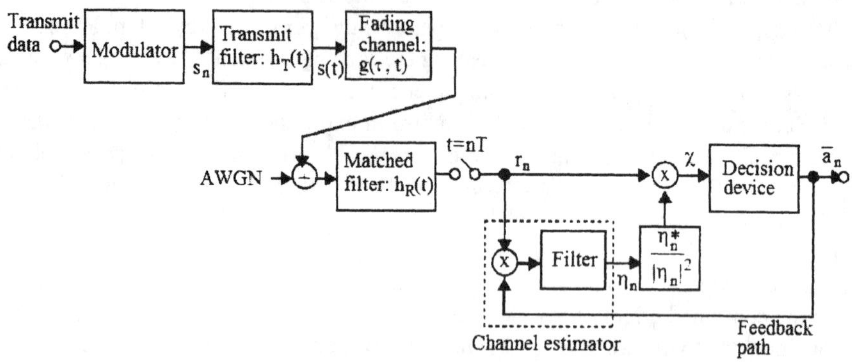

Figure 11.36 DFCE system model. From [36]. Copyright © 1999 IEEE.

signal is

$$s(t) = \sqrt{\frac{2E_s}{T}} \sum_{n=-\infty}^{\infty} a_n h_T(t - nT) \tag{11.61}$$

where E_s is the average signal energy per symbol, T is the symbol duration, a_n is the data symbol, $|a_n|^2$ is normalized to 1, and $h_T(t)$ is the transmitter filter. The fading channel $g(\tau, t)$ is assumed as the common WSSUS (wide-sense stationary uncorrelated scattering) channel, which is frequency-selective slow Rayleigh fading. The receiver has a matched filter $h_R(t) = (1/T)\, h_T(-t)$. The output of the matched filter sampled at $t = nT$ for detecting the symbol a_n is

$$r_n = \sqrt{\frac{2E_s}{T}} \sum_{m=-\infty}^{\infty} a_{n-m} d_m + w_n \tag{11.62}$$

where w_n is the filtered noise with zero mean and variance $2N_o/T$, where N_o is the one-sided power spectrum density of the AWGN. d_m is the composite channel impulse response,

$$d_m = \int_{-\infty}^{\infty} h(mT - \tau) g(\tau, t = nT) d\tau \tag{11.63}$$

where $h(t) = h_T(t) \otimes h_R(t)$, where $\otimes$ denotes convolution operation.

As shown in Figure 11.36, the decision feedback channel estimator feeds back the past detected symbols and remodulates the received signal samples with L feedback symbols to obtain the channel estimate η_n at $t = nT$ as

$$\eta_n = \frac{\sum_{l=1}^{L} r_{n-l}\overline{a}_{n-l}^*}{\sum_{l=1}^{L} |\overline{a}_{n-l}|^2} \tag{11.64}$$

where $\overline{a}_{n-l}$ is the detected symbol at $t = (n-l)T$, and * denotes the conjugate. The reasoning behind this estimate is as follows. If there were no ISI, each term in the numerator would be the product of channel gain η_{n-l} with the $|\overline{a}_{n-l}|^2$. If further we assume the channel is not changing at all, then the result of (11.64) would be exactly the channel gain η_n. In fact if this was the case, only one term would be needed in (11.64). In reality, there are ISI in the numerator and the channel is changing. However the summation operation in the numerator cancels some of the ISI and channel variation, and produces an approximation of $\eta_{n-1}\sum_{l=1}^{L}|\overline{a}_{n-l}|^2$, which in turn is divided by $\sum_{l=1}^{L}|\overline{a}_{n-l}|^2$ to produce the channel estimate η_n.

The channel correction is done by dividing r_n by η_n, or equivalently by multiplying r_n by $\eta_n^*/|\eta_n|^2$, to produce the decision variable

$$\chi = \frac{r_n}{\eta_n} = r_n\frac{\eta_n^*}{|\eta_n|^2} \tag{11.65}$$

The BER analysis of the DFCE given in [36] is quite involved and the results are not in closed form. Monte Carlo simulation was used to obtain the final results. We will omit the analysis and only present some simulation results. The simulation results are for a double-spike power delay profile

$$\xi(\tau, 0) = \frac{1}{2}\delta(\tau + \tau_{rms}) + \frac{1}{2}\delta(\tau - \tau_{rms}) \tag{11.66}$$

where τ_{rms} is the rms delay spread, and square-root-raised cosine transmitting/receiving filters whose impulse response is given as

$$h(t) = \frac{\sin(\pi t/T)}{\pi t/T}\frac{\cos(\nu\pi t/T)}{1-(2\nu t/T)^2} \tag{11.67}$$

Figure 11.37 shows BER performance of DFCE of 2PSK(BPSK), 4PSK(QPSK), and 16QAM in a Rayleigh channel with the double-spike power delay profile with various delay spreads and a roll-off factor of 0.5 of the $h(t)$. It is clearly seen from Figure 11.37(b, c) that when $\tau_{rms}/T_b = 0.05$ and 0.2, as the value of E_b/N_o increases the BER decreases and approaches the BER error floors due to the effect of delay spread. When τ_{rms}/T_b, where T_b is the bit interval, is negligible, the 16QAM is not as good as 2PSK or 4PSK (Figure 11.37(a)). However, when τ_{rms}/T_b is not negligible, 16QAM provides almost the same BER as 4PSK, and 2PSK provides the

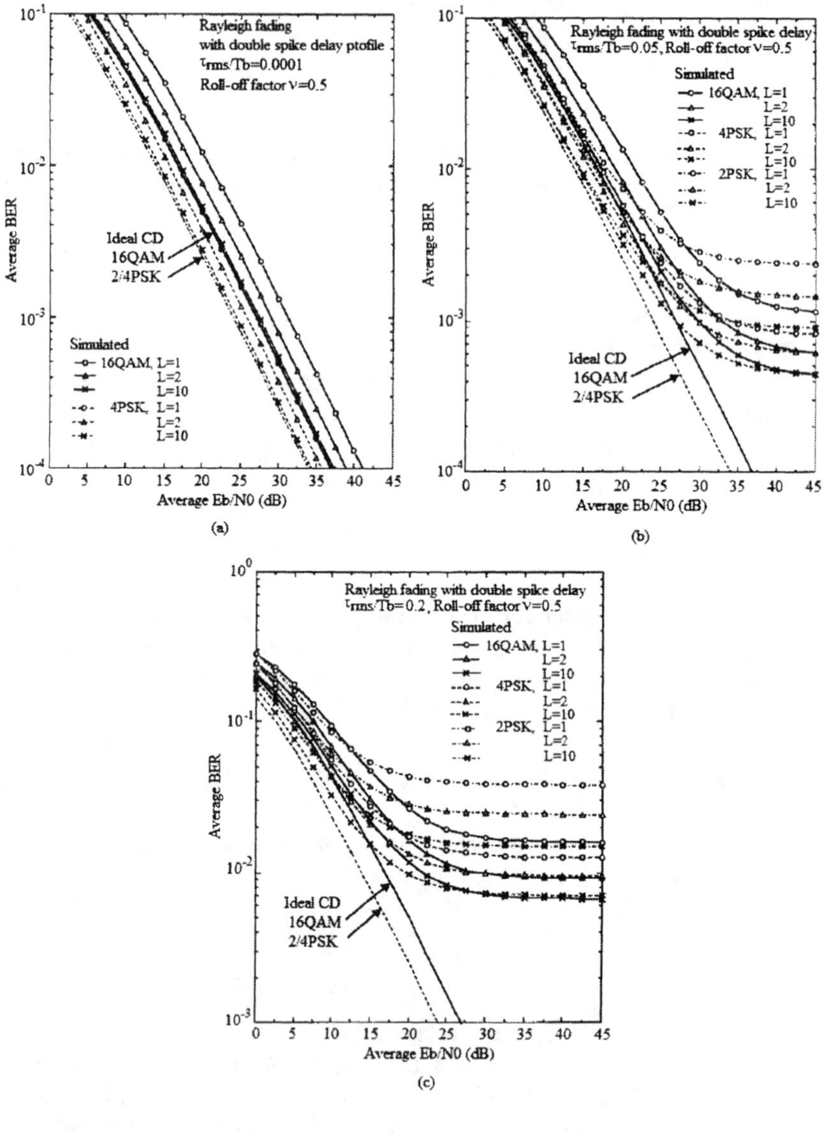

Figure 11.37 Performance comparison of 2PSK, 4PSK, and 16QAM. L = 1, 2, and 10. Double-spike delay profile with: (a) $\tau_{rms}/T_b = 0.0001$, (b) $\tau_{rms}/T_b = 0.05$, and (c) $\tau_{rms}/T_b = 0.2$. CD stands for coherent detection. From [36]. Copyright © 1999 IEEE.

worst. One possible reason is that 16QAM symbols have three different amplitudes and smaller amplitudes incur smaller ISI, contributing to smaller irreducible BER, while 2PSK and 4PSK symbols have the same amplitude and always incur the same amount of ISI.

11.9 EQUALIZATION

Equalization is an important remedial measure against ISI caused by multipath propagation or bandlimitation of the channel.

Consider equalization analysis for linear modulations, where the complex digital signal to be transmitted can be modelled as

$$\widetilde{x}(t) = \sum_{n=-\infty}^{\infty} \widetilde{x}_n \delta(t - nT)$$

where $\{\widetilde{x}_n\}$ is the complex symbol sequence. T is the symbol period. We represent the transmitter filter, the physical channel, and the receiver filter by a single complex impulse response $c(t)$, which we refer to as effective channel. Then we can write the output of the effective channel, sampled at the end of the symbol interval, $t = nT$, as

$$\begin{aligned}
\widetilde{r}(t)|_{t=nT} &= [\widetilde{x}(t) * c(t) + \widetilde{n}(t)]|_{t=nT} = [\sum_{l=-\infty}^{\infty} \widetilde{x}_l \delta(t - lT) * c(t) + \widetilde{n}(t)]|_{t=nT} \\
&= [\sum_{l=-\infty}^{\infty} \widetilde{x}_l c(t - lT) + \widetilde{n}(t)]|_{t=nT} = \sum_{l=-\infty}^{\infty} \widetilde{x}_l c(nT - lT) + \widetilde{n}(nT)
\end{aligned}$$

Using discrete-time notation, we have

$$\begin{aligned}
\widetilde{r}_n &= \sum_{l=-\infty}^{\infty} \widetilde{x}_l c_{n-l} + \widetilde{n}_n = \widetilde{x}_l * c_n + \widetilde{n}_n \\
&= c_0 \widetilde{x}_n + \sum_{\substack{l=-\infty \\ l \neq n}}^{\infty} \widetilde{x}_l c_{n-l} + \widetilde{n}_n \qquad (11.68)
\end{aligned}$$

In the discussion that follows, complex signal is to be replaced by real signal for notational simplicity. In fact, equalization of complex signal is realized by two real

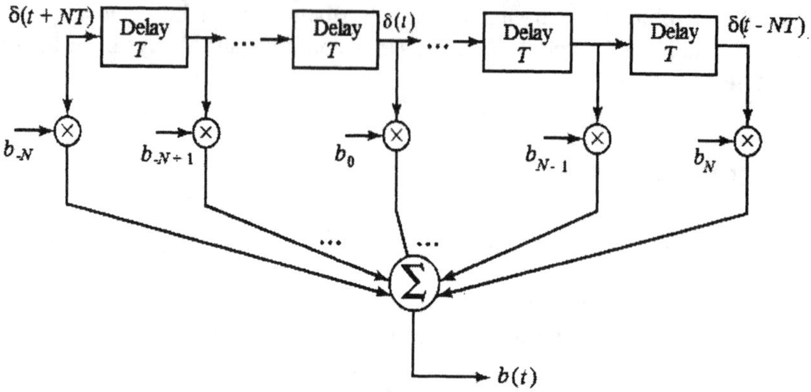

Figure 11.38 Structure of a linear equalizer. From [39]. Copyright © 2003. Adapted by permission of Pearson Education.

signal equalizations (I,Q channel equalization). Thus we have from (11.68)

$$r_n = c_0 x_n + \sum_{\substack{l=-\infty \\ l \neq n}}^{\infty} x_l c_{n-l} + n_n \qquad (11.69)$$

The task of an equalizer is to eliminate or reduce the second term (ISI) in the above expression. Popular equalizers include a zero-forcing linear equalizer, MMSE linear equalizer, decision-feedback equalizer, and blind equalizer. We will discuss them next.

11.9.1 Linear Equalizers (LE)

Linear equalizers are taped-delay-line filters having a general form of

$$b(t) = \sum_{k=-N}^{N} b_k \delta(t - kT)$$

This general form is a noncausal filter with $(2N + 1)$ taps. In many practical cases, $b(t)$ may be causal with only taps extending from 0 to N. The equalizer is shown in Figure 11.38.

The z-transform of $b(t)$ is

$$B(z) = \sum_{k=-N}^{N} b_k z^{-k}$$

and the z-transform of $\{r_n\}$ is

$$R(z) = \sum_{n=-\infty}^{\infty} r_n z^{-n}$$

and the z-transform of the equalizer output sequence is

$$\begin{aligned} Y(z) &= R(z)B(z) = \sum_{n=-\infty}^{\infty} r_n z^{-n} \sum_{k=-N}^{N} b_k z^{-k} \\ &= \sum_{n=-\infty}^{\infty} \sum_{k=-N}^{N} b_k r_n z^{-(k+n)} \end{aligned}$$

Letting $l = k + n$, we have

$$Y(z) = \sum_{n=-\infty}^{\infty} \sum_{k=-N}^{N} b_k r_{l-k} z^{-l} = \sum_{n=-\infty}^{\infty} y_l z^{-l}$$

where

$$y_l = \sum_{k=-N}^{N} b_k r_{l-k}$$

11.9.1.1 Zero-Forcing LE

In the absence of noise, the output of the effective channel is $R(z) = C(z)X(z)$, where $C(z) = \sum_{n=-\infty}^{\infty} r_n z^{-n}$ and $X(z) = \sum_{n=-\infty}^{\infty} x_n z^{-n}$ are the z-transforms of the effective channel and its input signals, respectively. Thus the output of the equalizer is

$$Y(z) = R(z)B(z) = C(z)X(z)B(z)$$

The transfer function of the equalized system is

$$H(z) = \frac{Y(z)}{X(x)} = B(z)C(z)$$

In the time domain, the discrete impulse response of the equalized system is

$$h_n = b_n * c_n = \sum_{k=-N}^{N} b_k c_{n-k} \tag{11.70}$$

The desired ISI-free output requires that

$$h_n = \begin{cases} 1, & n = 0 \\ 0, & n \neq 0 \end{cases} \tag{11.71}$$

This is called the Nyquist's first criterion. Note this only requires that $h(nT) = 0$, $n \neq 0$ (i.e., $h(t) = 0$ at sampling instants nT, $n \neq 0$ only). The equalizer satisfying (11.71) is called the "zero-forcing" linear equalizer. In the frequency domain, the desired ISI-free output requires that $H(z) = 1$ or

$$B(z) = \frac{1}{C(z)}$$

(that is, the transfer function of the equalizer must be the reciprocal of the effective channel impulse response). The equalizer is to compensate for the distortion on the transmitted signal introduced by the effective channel such that the equalized system has a constant (equal) transfer function in the frequency domain, hence the name "equalizer."

As an example, consider a channel with $c(t) = 1 + 0.2\delta(t - T)$. The transfer function is $C(z) = 1 + 0.2z^{-1}$. And the ZF equalizer transfer function is

$$B(z) = \frac{1}{1 + 0.2z^{-1}} = 1 - 0.2z^{-1} + (-0.2)^2 z^{-2} + ... + (-0.2)^k z^{-k} + ... \tag{11.72}$$

As shown in (11.72), an ideal ZF equalizer often requires an infinite number of taps. Practical filters only can have a finite number of taps. For example, (11.72) can be approximated by a large number of taps. In general, with a finite number of taps, we can only guarantee

$$\overline{h}_n = \begin{cases} 1, & n = 0 \\ 0, & n = \pm 1, \pm 2, ..., \pm N \end{cases} \tag{11.73}$$

by choosing the tap coefficients to satisfy

$$\begin{bmatrix} c_0 & \cdots & c_{-N+1} & c_{-N} & c_{-N-1} & \cdots & c_{-2N} \\ \vdots & & \vdots & \vdots & \vdots & & \vdots \\ c_{N-1} & \cdots & c_0 & c_{-1} & c_{-2} & \cdots & c_{-N-1} \\ c_N & \cdots & c_1 & c_0 & c_{-1} & \cdots & c_{-N} \\ c_{N+1} & \cdots & c_2 & c_1 & c_0 & \cdots & c_{-N+1} \\ \vdots & & \vdots & \vdots & \vdots & & \vdots \\ c_{2N} & \cdots & c_{N+1} & c_N & c_{N-1} & \cdots & c_0 \end{bmatrix} \begin{bmatrix} b_{-N} \\ \vdots \\ b_{-1} \\ b_0 \\ b_1 \\ \vdots \\ b_N \end{bmatrix} = \begin{bmatrix} 0 \\ \vdots \\ 0 \\ 1 \\ 0 \\ \vdots \\ 0 \end{bmatrix}$$

When $N \to \infty$, $\overline{h}_n \to h_n$. Equation (11.73) implies that there may be residual ISI when a ZF equalizer has a finite number of taps.

The disadvantage of the ZF linear equalizer is noise enhancement. The output noise of the equalizer is given by

$$v_n = n_n * b_n = \sum_{k=-N}^{N} b_k n_{n-k}$$

and the noise power is

$$\sigma_v^2 = \sigma_n^2 \sum_{k=-N}^{N} b_k^2$$

where σ_n^2 is the noise power at the input of the equalizer. The noise power is increased by the factor of $\sum_{k=-N}^{N} b_k^2$.

For the above example in (11.72), the noise power at the equalizer output is

$$\sigma_v^2 = \sigma_n^2 \sum_{k=0}^{\infty} (-0.2)^{2k} = \sigma_n^2 \frac{1}{1-0.2^2} = \frac{25}{24}\sigma_n^2$$

which is enhanced. When ISI is more severe, $\sum_{k=-N}^{N} b_k^2$ will be larger and the noise enhancement is more pronounced.

11.9.1.2 Minimum Mean Squared Error LE

While the ZF linear equalizer (ZF-LE) focuses on eliminating ISI and incurs noise enhancement, the minimum mean squared error linear equalizer (MMSE-LE) is to choose the equalizer coefficients such that the mean squared error (MSE), which is the expected sum of the squares of all the ISI terms plus the noise power at the output of the equalizer, is minimized. The MMSE linear equalizer provides a balanced solution to the problem of reducing the effects of both noise and ISI. For a prescribed computational complexity, the MMSE linear equalizer always performs as well as, and often better than its ZF counterpart. In addition, the MMSE linear equalizer is more robust and has better convergence properties for adjusting tap gain settings if an adaptive scheme is incorporated. The MMSE-LE has the same structure of ZF-LE except that the tap gains are chosen according to MMSE criterion, so it still incurs noise enhancement.

The MSE of the linear estimator is

$$\begin{aligned} \varepsilon &= E\{\varepsilon_n^2\} = E\{(x_n - y_n)^2\} = E\{(x_n - \sum_{k=-N}^{N} b_k r_{n-k})^2\} \\ &= \sum_{l=-N}^{N}\sum_{k=-N}^{N} b_k b_l R_r(k-l) - 2\sum_{k=-N}^{N} b_k R_{xr}(k) + E\{x_n^2\} \end{aligned}$$

where

$$R_r(k-l) = E\{r_{n-l} r_{n-k}\}$$

$$R_{xr}(k) = E\{x_n r_{n-k}\}$$

are the autocorrelation of r_n and cross-correlation functions for x_n and r_n. The minimum MSE solution is obtained by the usual "setting derivative to zero" method. Set

$$\frac{\partial \varepsilon}{\partial b_k} = 0, \quad k = 0, \pm 1, \pm 2, ..., \pm N$$

we obtain

$$\sum_{l=-N}^{N} b_l R_r(k-l) = R_{xy}(k), \quad k = 0, \pm 1, \pm 2, ..., \pm N$$

In matrix form, we have

$$\mathbf{R}_r \mathbf{b} = \mathbf{R}_{xr} \quad \text{or} \quad \mathbf{b} = \mathbf{R}_r^{-1}\mathbf{R}_{xr}$$

where

$$\mathbf{R}_r = \begin{bmatrix} R_r(0) & \cdots & R_r(N) & \cdots & R_r(2N) \\ R_r(-1) & \cdots & R_r(N-1) & \cdots & R_r(2N-1) \\ \cdots & \cdots & \cdots & \cdots & \cdots \\ R_r(-2N+1) & \cdots & R_r(-N+1) & \cdots & R_r(1) \\ R_r(-2N) & \cdots & R_r(-N) & \cdots & R_r(0) \end{bmatrix}^T$$

$$\mathbf{R}_{xr} = [R_{xr}(-N), \cdots, R_{xr}(-1), R_{xr}(0), R_{xr}(1), \cdots, R_{xr}(N)]^T$$

$$\mathbf{b} = [b_{-N} \cdots, b_{-1}, b_0, b_1, \cdots, b_N]^T$$

To find $\mathbf{b}$, we need $\mathbf{R}_r$ and $\mathbf{R}_{xr}$ which are unknown a priori. They can be estimated using a test signal over the channel and using the time-average

$$\widehat{R}_r(k-l) = \frac{1}{\widetilde{N}} \sum_{n=1}^{\widetilde{N}} r_{n-l} r_{n-k}$$

$$\widehat{R}_{xr}(k) = \frac{1}{\widetilde{N}} \sum_{n=1}^{\widetilde{N}} x_n r_{n-k}$$

where $\widetilde{N}$ is the number of data samples in the test signal.

For time-variant fading channels, the linear equalizers must be made adaptive. The tap coefficients of the linear equalizer need to be updated from symbol to symbol using the iteration:

$$\mathbf{b}(n+1) = \mathbf{b}(n) + \Delta\mathbf{b}(n)$$

where $\Delta\mathbf{b}(n)$ is an updating quantity. A hill-climbing method of updating $\mathbf{b}(n)$ is described in [39].

11.9.2 Decision-Feedback Equalizers (DFE)

The decision feedback equalizer (DFE) is a nonlinear equalizer widely used in wireless communication. DFE performs better than linear equalizers in channels with severe amplitude distortion. This is because linear equalizers have the disadvantage of noise enhancement.

The transfer function of the ZF-LE is

$$B(z) = 1/C(z) \tag{11.74}$$

For an MMSE-LE with infinite number of taps, the transfer function can be shown as [6],

$$B(z) = \frac{1}{C(z) + 2\sigma_n^2} \tag{11.75}$$

From (11.74) and (11.75) we can see that if $C(z)$ has frequency nulls, $B(z)$ will be very big and the noise variance after the equalizer will be very big since

$$\sigma_v^2 = \sigma_n^2 \sum_{k=-N}^{N} b_k^2$$

where b_k are the tap coefficients. Wireless channels often have frequency nulls due to severe fading. Thus linear equalizers are only effective when ISI is not severe and not effective for channels with severe ISI.

To derive the DFE, we rewrite the effective channel output in (11.69) as

$$r_n = c_o x_n + \sum_{l=-\infty}^{-1} c_l x_{n-l} + \sum_{l=1}^{\infty} c_l x_{n-l} + n_n \tag{11.76}$$

The first term is the current signal with gain c_o. The second term is the ISI due to precursors of the effective channel impulse response (c_l, $l = -\infty, ..., -1$), which is caused by future symbols. In a causal channel (the practical channel is usually causal), this term does not exist. The third term is the ISI due to post-cursors (c_l, $l = 1, ..., \infty$), which is caused by previously transmitted symbols. This part of ISI can be subtracted from r because past symbols have been detected. Assuming a very low detection error rate, the subtraction will reduce ISI in r_n. This observation prompts the idea of a decision feedback equalizer.

The structure of DFE is shown in Figure 11.39. The feedforward filter is like a linear equalizer which uses current and future (if any) noisy symbols (r_l, $l = n, ..., n + N$) to produce a quantity $\widetilde{r}_n$ according to one of the aforementioned criteria (zero-forcing, for example). The feedback filter uses estimates of past symbols ($\widehat{x}_{n-l}, l = 1, ..., M$) to estimate the ISI due to the post-cursors (the third term in (11.76)). Then this estimated ISI is subtracted from $\widetilde{r}_n$. The resulting quantity is denoted as y_n, which is sent to the decision device to produce an estimate of x_n. The decision device is determined by modulation type; for example, for BPSK, it is a threshold detector with zero as threshold.

Coefficients of the feed-forward filter $\{b_k\}$ and the feedback filter $\{d_k\}$ can be determined by different design criteria: zero-forcing or minimum mean squared error.

11.9.2.1 Zero-Forcing DFE

The transfer function of the channel is

$$C(z) = \sum_{k=-\infty}^{\infty} c_k z^{-k}$$

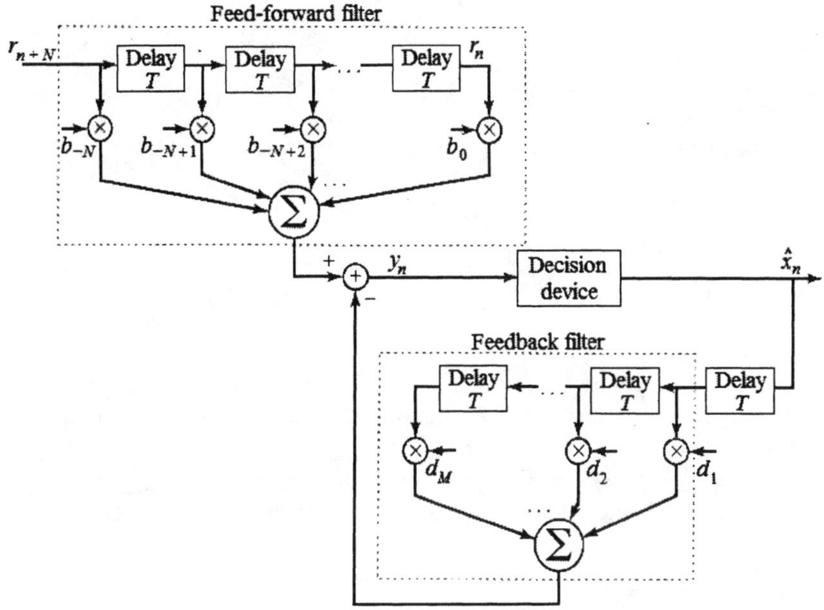

Figure 11.39 Structure of a decision-feedback equalizer. From [39]. Copyright © 2003. Adapted by permission of Pearson Education.

The transfer function of the feed-forward linear filter is

$$B(z) = \sum_{k=-N}^{0} b_k z^{-k}$$

The combined transfer function is

$$C(z)B(z) = \sum_{n=-\infty}^{\infty} h_n z^{-n}$$

where

$$h_n = \sum_{k=-N}^{0} b_k c_{n-k} \tag{11.77}$$

For the zero-forcing criterion, we want

$$h_n = \begin{cases} 1, & n = 0 \\ 0, & n = -1, -2, ..., -N \end{cases} \tag{11.78}$$

That is, zero-forcing for precursors. From (11.77) and (11.78) we have

$$\begin{bmatrix} c_0 & c_{-1} & \cdots & c_{-N} \\ c_1 & c_0 & \cdots & c_{-N+1} \\ \vdots & \vdots & \vdots & \vdots \\ c_N & c_{N-1} & \cdots & c_0 \end{bmatrix} \begin{bmatrix} b_{-N} \\ b_{-N+1} \\ \vdots \\ b_0 \end{bmatrix} = \begin{bmatrix} 0 \\ 0 \\ \vdots \\ 1 \end{bmatrix} \begin{matrix} \leftarrow h_{-N} \\ \\ \\ \leftarrow h_0 \end{matrix} \tag{11.79}$$

This is used to determine the coefficients of the feed-forward linear filter. Note that zero forcing for precursors of $\{h_n\}$ is equivalent to eliminating ISI due to the future symbols.

Once $\{b_k\}_{k=-N}^{0}$ is determined, we can use it to determine for $h_n = \sum_{k=-N}^{0} b_k c_{n-k}$ for $n = 1, 2, ..., M$, and set the feedback filter coefficients

$$d_j = h_j, j = 1, 2, ..., M$$

so that the effect of the postcursors (or effect of the past symbols) is removed. Mathematically, the input of the decision device is

$$y_n = \sum_{k=-N}^{0} b_k r_{n-k} - \sum_{j=1}^{M} h_j \widehat{x}_{n-j}$$

where the feed-forward filter output is

$$\sum_{k=-N}^{0} b_k r_{n-k} = x_n + \sum_{j=1}^{M} h_j x_{n-j} + \sum_{k=-N}^{0} b_k n_{n-k}$$

Due to the zero-forcing effect of the feed-forward filter the ISI term due to precursors (or future symbols) is eliminated in the above expression. However the ISI due to postcursors (or past symbols) still exists (the second term), which will be canceled by the output of the feedback filter. Further assuming the detections of previous symbols are correct, then from above we have $y_n = x_n + v_n$, $v_n = \sum_{k=-N}^{0} b_k n_{n-k}$. The equalization error at the input to the decision device is $\varepsilon_n = y_n - x_n = v_n$, which is due to channel noise only. Variance of v_n is $\sigma_n^2 \sum_{k=-N}^{0} b_k^2$, which is usually smaller than that of ZF-LE since the number of taps is smaller.

11.9.2.2 Minimum Mean Square Error DFE

The criterion of MMSE-DFE is to minimize the squared error at the input of the decision device. Assuming past symbols are detected correctly, we have

$$y_n = \sum_{k=-N}^{0} b_k r_{n-k} - \sum_{j=1}^{M} d_j x_{n-j}$$

$$\varepsilon = E\{\varepsilon_n^2\} = E\left\{\left(x_n - \sum_{k=-N}^{0} b_k r_{n-k} + \sum_{j=1}^{M} d_j x_{n-j}\right)^2\right\}$$

The mean-square error ε is minimum when the tap coefficient b_l of the feedforward filter, $l = -N, -N+1, ..., 0$ is chosen in such a way that

$$\begin{aligned}\frac{\partial \varepsilon}{\partial b_l} &= 2E\left\{-r_{n-l}\left(x_n - \sum_{k=-N}^{0} b_k r_{n-k} + \sum_{j=1}^{M} d_j x_{n-j}\right)\right\} \\ &= -2E[r_{n-l}\varepsilon_n] = 0\end{aligned}$$

and when the tap coefficient d_j of the feedback filter, $j = 1, 2, ..., M$, is chosen in such a way that

$$\begin{aligned}\frac{\partial \varepsilon}{\partial d_j} &= 2E\left\{x_{n-j}\left(x_n - \sum_{k=-N}^{0} b_k r_{n-k} + \sum_{j=1}^{M} d_j x_{n-j}\right)\right\} \\ &= 2E[x_{n-j}\varepsilon_n] = 0\end{aligned}$$

Under the assumption that $E\{x_n x_{n-j}\} = \sigma_x^2 \delta_j$, and noise and signals are independent from each other, representing r_n by $r_n = \sum_{l=-\infty}^{\infty} c_l x_{n-l} + n_n$, and working out the correlations, the above two equations become (see Appendix 11A)

$$\begin{aligned}\sum_{k=-N}^{0}\left[b_k R_c(k-l) + \frac{R_n(k-l)}{\sigma_x^2}\right] &= \sum_{j=1}^{M} d_j c_{j-l} + c_{-l}, \text{ for } l = -N, ..., 0 \\ \sum_{k=-N}^{0} b_k c_{j-k} &= d_j, \text{ for } j = 1, 2, ..., M \end{aligned} \quad (11.80)$$

where $R_n(k-l) = E\{n_{n-l} n_{n-k}\}$ is the noise correlation function and $R_c(k-l) = \sum_{m=-\infty}^{\infty} c_{m-l} c_{m-k}$. The above equations determine the MMSE-DFE taps.

DFE can be made adaptive for time-varying wireless channels; see [39] for details.

11.10 DIVERSITY RECEPTION

Another efficient remedial measure is the diversity technique [1, 6]. The concept of diversity is to enable the receiver to receive more than one signal at any moment. It is very unlikely that all signals are in deep fade. Therefore the receiver can usually receive adequate signal power by choosing the strongest signal, or combining them together.

11.10.1 Diversity Techniques

There are five major categories of diversity techniques.

1. *Space diversity*. It is usually achieved by using more than one receiver antenna. The spacing between antennas is choosing so that the received signals experience uncorrelated fading. A spatial separation of a half-wavelength is usually enough for two dimensional isotropic scattering. This is called microscopic diversity. Another type of space diversity is achieved by selecting different base stations. Shadowing due to variations in terrain and the nature of surroundings causes large-scale fading. By selecting a base station which is not shadowed, the mobile can improve its signal-to-noise ratio in the forward link significantly. This is called macroscopic diversity. Macroscopic diversity can also be used in the base station receiver. By using multiple base station antennas, the base station receiver can choose the strongest signal from the mobile.
2. *Polarization diversity*. A scattering environment tends to depolarize a signal. Receiver antennas having different polarizations can be used to obtain diversity. An example is the use of vertical monopole and patch antennas in cellular telephone user units. Another application of polarization diversity is at the base station. At base station space diversity is less practical because the narrow angle of incident fields requires large antenna separation. However orthogonal polarization diversity is possible since many users are using portable units and the signal polarizations are no longer pure vertical due to hand-tilting.
3. *Frequency diversity*. This is achieved by transmitting the same signal on multiple carriers. These carrier frequencies are separated by at least the coherent bandwidth of the channel. Thus each signal does not experience the same fading. Frequency diversity is often used in microwave line-of-sight links using frequency division multiplexing (FDM). In these links deep fading may occur

due to tropospheric propagation and resulting refraction. A spare frequency channel is usually provided for several frequency channels, each carrying an independent traffic. If any one of the frequency channel is experiencing deep fading, its traffic can be switched to the spare channel.

4. *Time diversity.* Signals are transmitted in multiple time slots separated by at least the coherence time of the channel. For example, repetition codes transmit a symbol in several time slots. Other sophisticated error control codes encode a message sequence into a longer coded sequence. This is also a type of time diversity because the information is distributed on the entire code so if a part of it is erased the information can still be recovered. Interleaving is usually used in combination with error control codes. When an interleaved sequence is deinterleaved at the receiver, bursty errors are spread out and become isolated errors which can be easily corrected by the error control code. Interleaving effectively separates the symbols corresponding to a message by more than the channel coherence time.
5. *Multipath diversity.* Using a tapped delay line, the energy in the multipath components can be collected and their weighted sum becomes a stronger signal. This type of receiver is called a RAKE receiver since the block diagram is just like a garden rake. Multipath diversity is also sometimes considered as time diversity since signal components at different times are collected.

There are also other diversity techniques. The *angle diversity* or *direction diversity* uses several directional antennas. Each antenna receives plane waves arriving from a narrow range of angles. Signals received by different antennas are thus uncorrelated. A new technique is called *diversity transform* [40] which transforms the message sequence into another sequence for transmission. However unlike error control coding, it does not increase the length of the sequence.

11.10.2 Combining Techniques

It is not enough to just have diverse signal sources (called diversity branches) for the receiver. It is important how the receiver processes these signals. The processing techniques are called *diversity combining* techniques. The *predetection selective combining* technique selects the signal with the highest signal-to-noise ratio. The *predetection maximal ratio combining* technique combines weighted signals from all diversity branches. Each signal is weighted by its complex fading gain which must be estimated at the receiver. This technique gives the best possible performance among the diversity combining techniques, but it requires channel estimation at the receiver. If the signals are not weighted, then the process is called *predetection equal gain combining*. The *predetection switched combining* technique uses a scanner to scan through the diversity branches until it finds one that has a signal-to-noise ratio that is above the specified threshold. This branch is used until the signal-to-noise ratio drops below the threshold, then the scan process starts again. Combining can also be done after detection, that is, *postdetection combining techniques* exist too [1, 6].

To illustrate the effectiveness of combining techniques, we present a detailed analysis of the most effective combining technique, the predetection maximal ratio combining, in the following [39].

Assuming there are L branches, the received lth diversity signal is

$$r_l(t) = \alpha_l(t)\exp[j\theta_l(t)]x(t) + w_l(t), \quad l = 1, 2, ..., L$$

where $\alpha_l(t)$, $\theta_l(t)$, and $w_l(t)$ are the random amplitude gain, phase, and AWGN of the channel. We make the following assumption. All random processes, $\alpha_l(t)$, $\theta_l(t)$, and $w_l(t)$ are statistically independent to each other and independent between different diversity channels; due to flat fading, there is no ISI; due to slow fading, $\alpha_l(t)$ and $\theta_l(t)$ are constant in a symbol period. The output of the coherent demodulator of the lth branch at the end of the kth symbol interval is $\alpha_{lk}\exp(j\theta_{lk})\mathbf{x}_k + \mathbf{n}_{lk}$, where $\alpha_{lk}\exp(j\theta_{lk})$ is the complex gain, $\mathbf{x}_k = (x_{k1}, x_{k2}, ..., x_{kN})$ is the signal vector, where N is the number of dimensions of the signal space (for example, if the modulation is QPSK, there are two orthogonal signals in the constellation, then, $N = 2$, see Chapter 4). $\mathbf{n}_{lk} = (n_{lk,1}, n_{lk,2}, ..., n_{lk,N})$ is the noise vector. The signal energy is $||\mathbf{x}_k||^2 = E_s$. The noise vector $\mathbf{n}_{lk}$ is a real Gaussian random vector with zero mean and variance $N_o/2$ (assuming $w_l(t)$ has a two-sided PSD of $N_o/2$ and the coherent demodulator is standard, that is, its reference signal is orthonormal).

The maximum ratio combining output is

$$\begin{aligned}
\mathbf{r}_k &= \sum_{l=1}^{L}[\alpha_{lk}\exp(-j\theta_{lk})][\alpha_{lk}\exp(j\theta_{lk})\mathbf{x}_k + \mathbf{n}_{lk}] \\
&= \left[\sum_{l=1}^{L}\alpha_{lk}^2\right]\mathbf{x}_k + \sum_{l=1}^{L}\alpha_{lk}\exp(-j\theta_{lk})\mathbf{n}_{lk} \\
&= g_k\mathbf{x}_k + \mathbf{n}_k
\end{aligned}$$

where

$$g_k = \sum_{l=1}^{L}\alpha_{lk}^2$$

is real and

$$\mathbf{n}_k = \sum_{l=1}^{L} \alpha_{lk} \exp(-j\theta_{lk})\mathbf{n}_{lk} = (n_{k,1}, n_{k,2}, ..., n_{k,N})$$

is complex. The component

$$n_{k,i} = \sum_{l=1}^{L} \alpha_{lk} \exp(-j\theta_{lk}) n_{lk,i}, \quad i = 1, 2, ..., N$$

is complex Gaussian with zero mean and variance $\sigma_{k,i}^2 = \frac{1}{2} N_o \sum_{l=1}^{L} \alpha_{lk}^2$.

To see the SNR improvement brought by the maximum ratio combining, consider BPSK. For BPSK, $N = 1$, $r_k = g_k x_k + n_k$, $x_k = \sqrt{E_b}$ for logic "1", $x_k = -\sqrt{E_b}$ for logic "0". SNR at the output of the combiner is

$$\gamma_k = \frac{[g_k x_k]^2}{2\sigma_{k,n}^2} = \frac{[\sum_{l=1}^{L} \alpha_{lk}^2]^2 (x_k)^2}{N_o \sum_{l=1}^{L} \alpha_{lk}^2} = \frac{E_b}{N_o} \sum_{l=1}^{L} \alpha_{lk}^2 \tag{11.81}$$

For a Rayleigh fading channel, α_{lk} are i.i.d. Rayleigh random variables with parameter σ_α^2, then γ_k has a chi-square distribution with 2L degrees of freedom. That is

$$f_\gamma(x) = \frac{x^{L-1} \exp\{-x/\Gamma_c\}}{(L-1)!\Gamma_c^L}$$

where

$$\Gamma_c = 2\sigma_\alpha^2 E_b / N_o$$

is the average SNR per bit at each individual branch. From (11.81), the mean SNR per bit after combining is

$$\Gamma_b = E\{\gamma_k\} = \frac{E_b}{N_o} \sum_{l=1}^{L} E\{\alpha_{lk}^2\} = \frac{E_b}{N_o} L(2\sigma_\alpha^2) = L\Gamma_c \tag{11.82}$$

This shows that the combined SNR Γ_b increases with L linearly.

The bit error probability of BPSK with maximal ratio combining is

$$P_b = \int_0^\infty P_{b/\gamma}(x) f_\gamma(x) dx$$

where $P_{b/\gamma}(x) = Q(\sqrt{2x})$ for BPSK. It can be derived that [34,39]

$$\begin{aligned} P_b &= \int_0^{\infty} Q(\sqrt{2x}) \frac{x^{L-1} \exp\{-x/\Gamma_c\}}{(L-1)!\Gamma_c^L} dx \\ &= \left[[0.5(1-\mu)]^L \sum_{l=0}^{L-1} \binom{L-1+l}{l} [0.5(1+\mu)]^l \right] \end{aligned} \tag{11.83}$$

where

$$\mu = \sqrt{\frac{\Gamma_c}{1+\Gamma_c}}$$

For $\Gamma_c >> 1$, we have $0.5(1+\mu) \approx 1, 0.5(1-\mu) \approx 1/(4\Gamma_c)$. Furthermore, $\sum_{l=0}^{L-1} \binom{L-1+l}{l} = \binom{2L-1}{L}$, then

$$P_b \approx \left(\frac{1}{4\Gamma_c}\right) \left[\begin{array}{c} 2L-1 \\ L \end{array} \right] \tag{11.84}$$

Figure 11.40 shows improvement of BER using diversity with maximal ratio combining.

11.11 MIMO WIRELESS LINK

The multiple-input-multiple-output (MIMO) wireless link has recently emerged as one of the most significant breakthroughs in modern digital communications. MIMO systems are described in two tutorial papers by Gesbert et al. [41] and Paulraji et al. [42] and other papers.

A MIMO wireless link consists of multiple transmitting antennas and multiple receiving antennas. Such a setup is illustrated in Figure 11.41. Each receiving antenna receives signals from all transmitting antennas. When these signals are "combined" in the receiver by proper signal processing, the BER and/or the data rate will be improved. Note that the word "combined" is used in a wide sense which refers to any type of signal processing, including decoding, imposed on the signals from receiving antennas.

In a MIMO system, supposing the number of transmitting antennas is M_T, and the number of receiving antennas is M_R, then the most general channel model, the

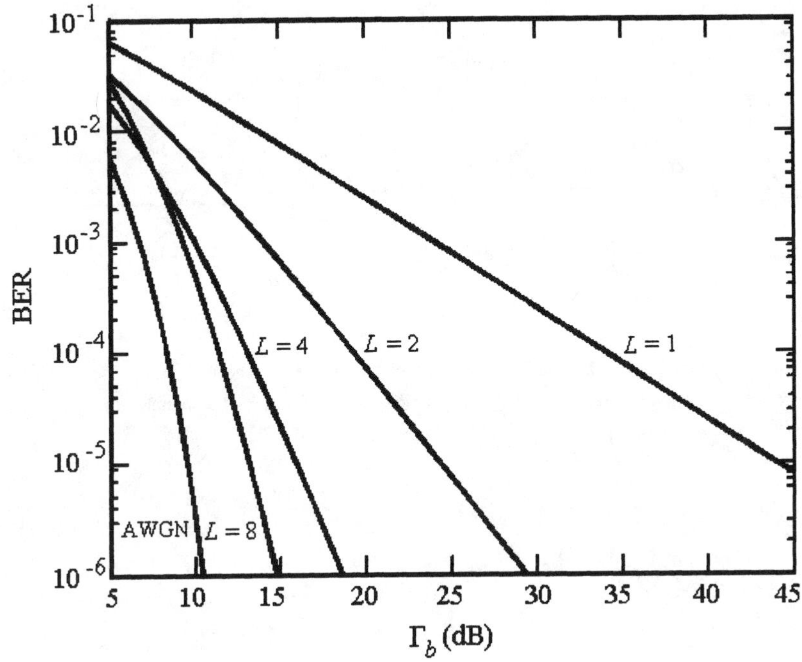

Figure 11.40 BER performance improvement using diversity with maximal ratio combining. From [39]. Copyright © 2003. Adapted by permission of Pearson Education.

time-varying channel model can be described by a $M_R \times M_T$ matrix $\mathbf{H}(\tau, t)$ [42]

$$\mathbf{H}(\tau,t)= \begin{bmatrix} h_{1,1}(\tau,t) & h_{1,2}(\tau,t) & \cdots & h_{1,M_T}(\tau,t) \\ h_{2,1}(\tau,t) & h_{2,2}(\tau,t) & \cdots & h_{2,M_T}(\tau,t) \\ \vdots & \vdots & \ddots & \vdots \\ h_{M_R,1}(\tau,t) & h_{M_R,2}(\tau,t) & \cdots & h_{M_R,M_T}(\tau,t) \end{bmatrix} \tag{11.85}$$

where $h_{i,j}(\tau, t)$ is the time-varying channel impulse response between the jth transmitting antenna and the ith receiving antenna. The column vector $[h_{1,j}(\tau, t), h_{2,j}(\tau, t), \cdots, h_{M_R,j}(\tau, t)]^T$ is called the spatio-temporal signature induced by the jth transmitting antenna across the receiving antenna array. On the receiving side, the signal received by the ith antenna is determined by the ith row vector $[h_{i,1}(\tau, t), h_{i,2}(\tau, t),$

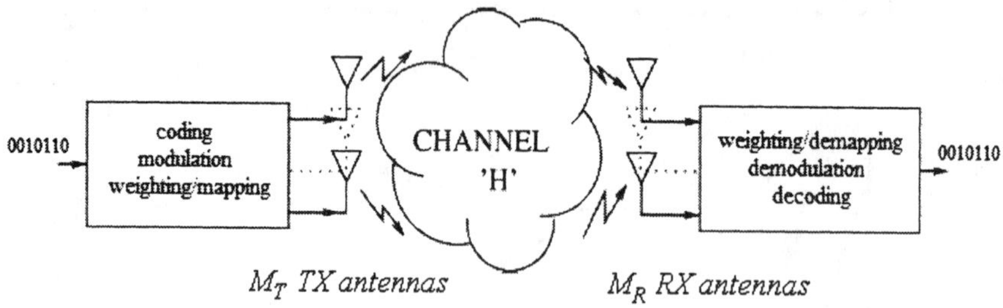

Figure 11.41 Diagram of a MIMO wireless transmission system. The transmitter and receiver are equipped with multiple antennas. Coding, modulation, and mapping of the signals onto the antennas may be realized jointly or separately. From [41]. Copyright © 2003 IEEE.

$\cdots$, $h_{i,M_T}(\tau, t)]$ as

$$y_i(t) = \sum_{j=1}^{M_T} h_{i,j}(\tau, t) * s_j(t) + n_i(t), \quad i = 1, 2, ..., M_R \tag{11.86}$$

where $*$ denotes convolution and $s_j(t)$ is the signal transmitted by the jth antenna, $n_i(t)$ is the AWGN at the ith receiving antenna.

11.11.1 Capacity of MIMO Channel

In order to see the benefit of a MIMO link, we examine the capacity of MIMO channels. First we assume that the channel is frequency-flat fading (the capacity of the frequency-selective fading MIMO channel will be discussed later). The discrete (sampled) input-output relation over a symbol period can be written as

$$\mathbf{y} = \sqrt{\frac{E_s}{M_T}}\mathbf{H}\mathbf{s} + \mathbf{n} \tag{11.87}$$

where $\mathbf{y}$ is the $M_R \times 1$ complex received signal vector (at the output of the demodulator), s with $E\{\mathbf{s}\} = \mathbf{0}$ is the $M_T \times 1$ complex transmitted signal vector (at the input of the modulator), $\mathbf{H}$ is the $M_R \times M_T$ complex MIMO channel matrix, $\mathbf{n}$ is additive temporally white complex Gaussian noise with $E\{\mathbf{n}\mathbf{n}^H\} = N_o\mathbf{I}_{M_R}$, where $\mathbf{n}^H$ denotes a Hermitian transpose of $\mathbf{n}$, and E_s is the total average energy available at the transmitter over a symbol period. The total average transmitted power over a symbol period is constrained to M_T, which means that the covariance matrix of s, $\mathbf{R_{ss}} = E\{\mathrm{ss}^H\}$, satisfies $\mathrm{Tr}(\mathbf{R_{ss}}) = M_T$.

11.11.1.1 Deterministic MIMO Channel

Even though **H** is random, we will first consider channel capacity conditioned on a specific realization of **H** (that is, it is deterministic). Then we will take the statistical average of the conditional capacity to obtain the averaged capacity.

If **H** is known at the transmitter, the mutual information for s having covariance matrix $\mathbf{R_{ss}}$ is [42,43]

$$I = \log_2 \det\left(\mathbf{I}_{M_R} + \frac{E_s}{M_T N_o}\mathbf{H}\mathbf{R_{ss}}\mathbf{H}^H\right) \quad b/s/Hz \tag{11.88}$$

and the channel capacity is

$$C = \max_{\mathbf{R_{ss}}} \log_2 \det\left(\mathbf{I}_{M_R} + \frac{E_s}{M_T N_o}\mathbf{H}\mathbf{R_{ss}}\mathbf{H}^H\right) \quad b/s/Hz \tag{11.89}$$

where the maximization is performed over all possible input covariance matrices satisfying $\mathrm{Tr}(\mathbf{R_{ss}}) = M_T$. However, **H** is very difficult to acquire at the transmitter. In the absence of the knowledge of **H**, it is reasonable to assume that $\mathbf{R_{ss}} = \mathbf{I}_{M_T}$, which means that the signals from all antennas are independent and with equal power. Then the channel capacity (or spectral efficiency) achieved with this covariance matrix is [41–44]

$$C_{EP} = \log_2 \det\left(\mathbf{I}_{M_R} + \frac{E_s}{M_T N_o}\mathbf{H}\mathbf{H}^H\right), \quad b/s/Hz \tag{11.90}$$

which may be decomposed as

$$C_{EP} = \sum_{i=1}^{r} \log_2\left(1 + \frac{E_s}{M_T N_o}\lambda_i\right), \quad b/s/Hz \tag{11.91}$$

where r is the rank of **H** and λ_i $(i = 1, 2, ..., r)$ denotes the positive eigenvalues of $\mathbf{H}\mathbf{H}^H$. Obviously $C_{EP} \leq C$. The subscript EP stands for "equal power."

Three special cases are as follows [41].

For a single-input-single-output (SISO) system: $M_T = M_R = 1$, $\mathbf{H} = h$, the channel capacity is

$$C_{SISO} = \log_2\left(1 + \frac{E_s}{N_o}|h|^2\right), \quad b/s/Hz \tag{11.92}$$

Comparing this to (11.91) shows that the spectral efficiency of the MIMO channel is the sum of the capacities of r SISO channels with corresponding channel gains of $\sqrt{\lambda_i}$ $(i = 1, 2, ..., r)$ and transmitted energy E_s/M_T.

For a single-input-multi-output (SIMO) system: $M_T = 1$, $M_R \neq 1$, $\mathbf{H} = [h_1, h_2, ..., h_{M_R}]^T$, where h_i is the gain of the ith receiving antenna, the channel capacity is

$$C_{SIMO} = \log_2\left(1 + \frac{E_s}{N_o}\sum_{i=1}^{M_R}|h_i|^2\right), \quad b/s/Hz \tag{11.93}$$

This expression shows that the spectral efficiency of the SIMO channel increases with the logarithm of the number of receiving antennas.

For a multi-input-single-output (MISO) system: $M_T \neq 1$, $M_R = 1$, $\mathbf{H} = [h_1, h_2, ..., h_{M_T}]$, where h_i is the gain of the ith transmitting antenna, channel capacity is

$$C_{MISO} = \log_2\left(1 + \frac{E_s}{M_T N_o}\sum_{i=1}^{M_T}|h_i|^2\right), \quad b/s/Hz \tag{11.94}$$

Letting all antennas have the same gain, then the above will become (11.92), that is, there is no array gain. This is because signals can not be added coherently in this case, whereas they are added coherently in the SIMO case.

11.11.1.2 Frequency-Flat Slow Fading MIMO Channel

We now consider the capacity of frequency-flat slow fading MIMO channels. The following results are based on three assumptions. First, the channel is the classical i.i.d. frequency-flat Rayleigh fading channel. That is, the elements of $\mathbf{H}$ are all i.i.d. standard complex Gaussian, denoted as $\mathcal{CN}(0, 1)$. Second, $\mathbf{H}$ is known at the receiver and no channel information is available at the transmitter. Third, the channel is assumed quasi-static, meaning the channel is constant for a block of symbols but varies from block to block.

There are two types of channel capacity for a fading MIMO channel: ergodic capacity and outage capacity. If the transmitted codewords span an infinite number of independently fading blocks, the Shannon capacity or *ergodic capacity* is given by [42,43,45]

$$C = E\{C_{EP}\} \tag{11.95}$$

where the expectation is with respect to random channel. At high SNR, it has been shown that [42,43,46]

$$C = \min(M_R, M_T)\log_2 \rho + O(1) \tag{11.96}$$

where $\rho = E_s/N_o$ is the average SNR at each receiving antenna. This expression shows that capacity increases linearly with the minimum of the numbers of transmitting and receiving antennas. The *outage capacity* $C_{out,q}$ is defined as the information rate that is guaranteed for $(100 - q)\%$ of the channel realization [47,48]—that is

$$P(C_{EP} \leq C_{out,q}) = q\%$$

Ergodic capacity and outage capacity are plotted in Figures 11.42 and 11.43, respectively. They behave similarly with respect to SNR: Capacities increase with $log_2(SNR)$. However, they behave differently with the number of transmitting and receiving antennas. The ergodic capacity is basically linearly proportional to $\min(M_R, M_T)$, while both M_R and M_T have considerable effect on the outage capacity.

11.11.1.3 Frequency-Selective Slow Fading MIMO Channel

The capacity of a frequency-selective fading slow fading MIMO channel can be obtained by dividing the frequency band of interest into N equal subbands. When N is sufficiently large, each subchannel is considered frequency-flat. Each subchannel is an $M_R \times M_T$ MIMO channel with a channel matrix H_i $(i = 0, 1, ..., N)$. Assuming that transmitted power is allocated uniformly across space and frequency, then the mutual information associated with a particular channel realization is given by

$$I_{FS} = \frac{1}{N}\sum_{i=1}^{N} \log_2 \det\left(\mathbf{I}_{M_R} + \frac{E_s}{M_T N_o}\mathbf{H}_i\mathbf{H}_i^H\right) \quad b/s/Hz \tag{11.97}$$

where E_s is energy allocated to the ith subchannel, and the ergodic capacity is

$$C_{FS} = E\{I_{FS}\} \tag{11.98}$$

and the outage capacity is defined in the same manner as in the frequency-flat case.

11.11.2 MIMO Signaling: Space-Time Coding and Spatial Multiplexing

We all know that channel capacity is an upper bound of the error-free transmission rate which can only be achieved by algorithms or codes with boundless complexity and latency. However it provides a goal to be neared by practical algorithms or codes.

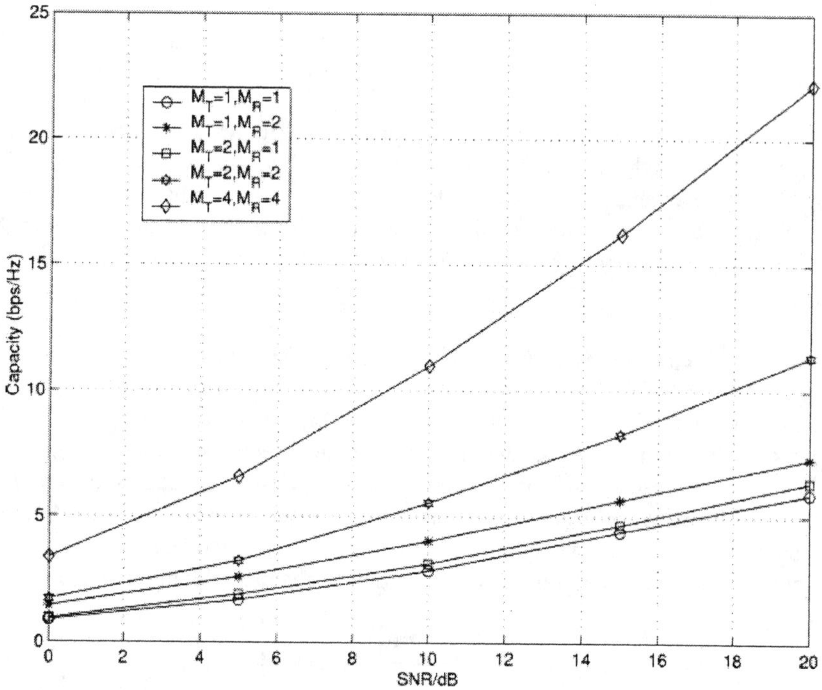

Figure 11.42 Ergodic capacity for different MIMO antenna configurations. Note that the SIMO channel has a higher ergodic capacity than the MISO channel. From [42]. Copyright © 2004 IEEE.

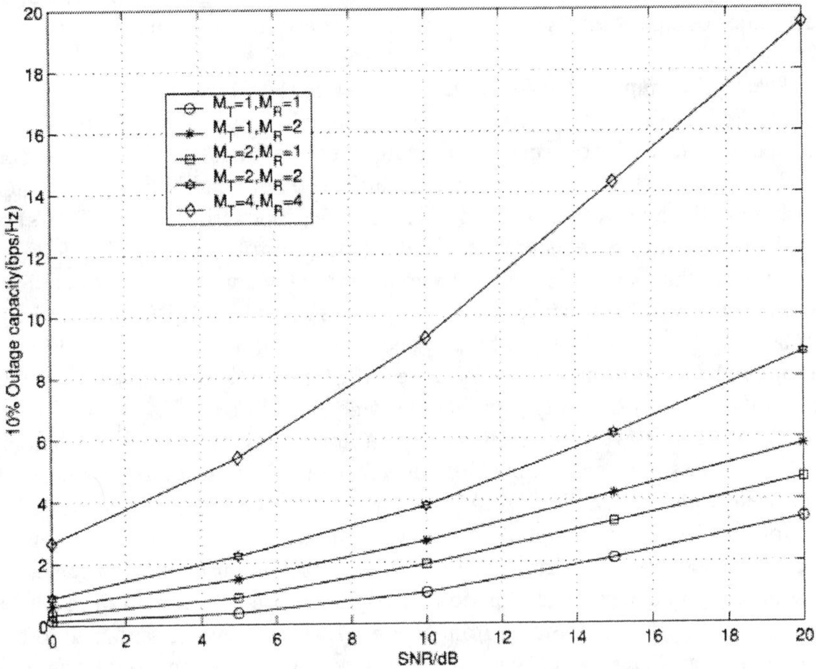

Figure 11.43 Ten-percent outage capacity for different MIMO configurations. MIMO yields significant improvements in terms of outage capacity. From [42]. Copyright © 2004 IEEE.

The key techniques to exploit the high channel capacity of the MIMO wireless link are *space-time diversity coding* and *spatial multiplexing*.

11.11.2.1 Space-Time Coding

The spatial separation of multiple antennas of the MIMO system provides space diversity. Combined with time diversity provided by error-control coding, space-time diversity is realized, which can provide excellent diversity gain. The objective of space-time coding is to extract the total available spatial diversity in the MIMO channel.

The first attempt to develop a space-time code (STC) was reported in [49] and was inspired by the delay diversity scheme reported in [50]. However, the key development of the STC concept was originally presented in [51] in the form of trellis coded modulation (trellis codes for short), which requires a multidimensional Viterbi algorithm at the receiver for decoding. These space-time trellis codes (STTCs) can provide a diversity gain equal to the number of transmitting antennas in addition to a coding gain that depends on the complexity of the code without any loss in bandwidth efficiency. The discovery of space-time block codes (STBCs) increased the popularity of STCs. STBCs can be decoded by simple linear processing instead of the complicated Viterbi algorithm. However, STBCs provide zero or minimal coding gains, although their diversity gains are the same as STTCs.

As an example, we will briefly describe Alamouti's STBC here [52]. Alamouti's STBC is a very simple space-time block code which only requires linear processing. It is currently part of both W-CDMA and CDMA-2000 standards. Alamouti's STBC is as follows. At a given symbol period, two symbols $\{s_0, s_1\}$ from the modulator are transmitted by two antennas, respectively. Antenna 0 transmits s_0, and antenna 1 transmits s_1. In the next symbol period, the signal $-s_1^*$ is transmitted from antenna 0 and the signal s_0^* is transmitted from antenna 1. The receiver is assumed to have only one antenna. The channel gains from the first and second transmitting antennas to the receiving antenna are denoted as h_0 and h_1, respectively. In addition, h_0 and h_1 are assumed to be scalar (no multipath signals) and constant over two consecutive symbol periods (T_s):

$$\begin{aligned} h_0(t) &= h_0(t+T_s) = h_0 = \alpha_0 e^{j\theta_0} \\ h_1(t) &= h_1(t+T_s) = h_1 = \alpha_1 e^{j\theta_1} \end{aligned} \tag{11.99}$$

Denoting the received signals over two consecutive periods as r_0 and r_1, they can be written as

$$r_0 = r(t) = h_0 s_0 + h_1 s_1 + n_0$$

$$r_1 = r(t+T) = -h_0 s_1^* + h_1 s_0^* + n_1 \quad (11.100)$$

where n_0 and n_1 are AWGN and are modeled as i.i.d. complex Gaussian random variables with zero mean and power spectral density $N_o/2$ per dimension. The received signals in the two consecutive periods are combined to build the following two signals

$$\begin{aligned} \widetilde{s}_0 &= h_0^* r_0 + h_1 r_1^* \\ \widetilde{s}_1 &= h_1^* r_0 - h_0 r_1^* \end{aligned} \quad (11.101)$$

Substituting (11.100) into (11.101) we have

$$\begin{aligned} \widetilde{s}_0 &= (\alpha_0^2 + \alpha_1^2) s_0 + h_0^* n_0 + h_1 n_1^* \\ \widetilde{s}_1 &= (\alpha_0^2 + \alpha_1^2) s_1 - h_0 n_1^* + h_1^* n_0 \end{aligned} \quad (11.102)$$

Each of these two signals is equivalent to that obtained by a maximum ratio combining diversity technique except that the noise components are rotated by different phases, which do not affect signal-to-noise power ratio [52]. Note that the first signal is only related to s_0 and the second signal is only related to s_1. Thus they can be detected separately. Each of the combined signals is sent to a maximum likelihood detector and the decision rule is to choose s_i if and only if [52]

$$(\alpha_0^2 + \alpha_1^2 - 1)|s_i|^2 + d^2(\widetilde{s}_0, s_i) \le (\alpha_0^2 + \alpha_1^2 - 1)|s_k|^2 + d^2(\widetilde{s}_0, s_k), \qquad \forall i \neq k$$

where $\widetilde{s}_0$ is replaced by $\widetilde{s}_1$ when detecting $\widetilde{s}_1$. $d(\widetilde{s}_0, s_i)$ is the Euclidean distance.

The SNR improvement in this scheme is $(\alpha_0^2 + \alpha_1^2)$, and the diversity order is two. This scheme can be extended to a scheme with two transmitting antennas and M receiving antennas. The encoding scheme is the same as before and the diversity order achieved is $2M$ [52].

The decoding of STBC requires knowledge of the channel. The channel state information can be estimated by sending training or pilot symbols.

Both STTC and STBC codes were first designed for narrow-band radio channels, that is, frequency-flat fading channels. If they are to be used in frequency-selective channels, channel equalization is needed since the channel causes ISI. However, using classical equalizers with STC is a challenging problem [41]. Another way to solve this problem is to use orthogonal frequency division multiplexing (OFDM) (see Chapter 12), which converts a wide-band channel to multiple narrow-band channels, and equalization is avoided.

11.11.2.2 Spatial Multiplexing

The objective of spatial multiplexing is to maximize the transmission data rate of the

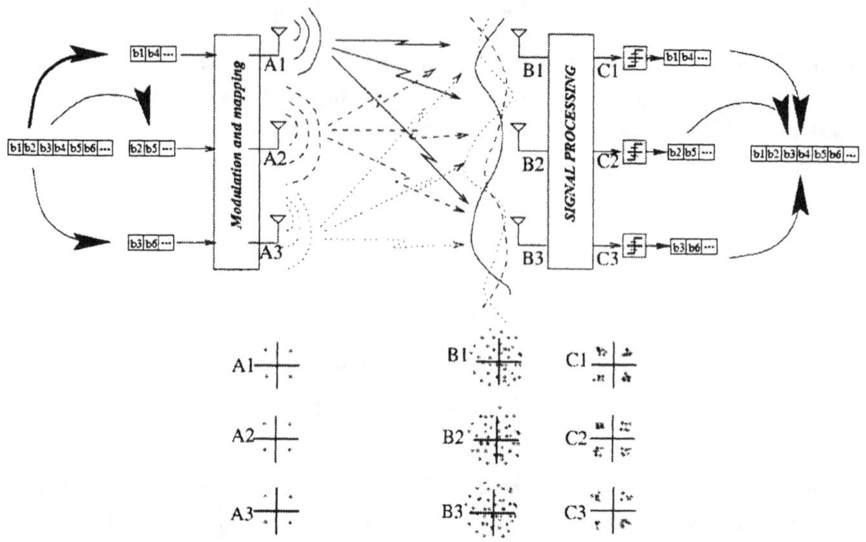

Figure 11.44 Vertical spacial-multiplexing MIMO system with three TX and RX antennas yielding three-fold improvement in spectral efficiency. Ai, Bi, and Ci represent symbol constellations (using QPSK as an example) for the three inputs at the various stages of transmission and reception. From [41]. Copyright © 2003 IEEE.

MIMO channel. Spatial multiplexing can be considered as a special class of STBCs. Two basic types of spatial multiplexing are the horizontal encoding (HE) and vertical encoding (VE). A combination of these two is called diagonal encoding (DE).

Figure 11.44 shows an HE system. With HE, the bit stream is first demultiplexed into M_T separate data streams. Each of them undergoes independent temporal encoding, symbol mapping, and interleaving and is then transmitted from a corresponding antenna. One antenna is for one data stream only and the association between the data stream and the antenna is fixed over time. The spatial rate is M_T, which increases the transmission rate by a factor of M_T. The diversity order is M_R, the number of receiving antennas since any information bit is transmitted from one antenna and received by M_R antennas.

With VE, the bit stream first undergoes temporal encoding, symbol mapping, and interleaving and is then demultiplexed into M_T separate data streams, and transmitted from different antennas. This scheme can achieve full diversity gain (M_TM_R-order) with proper temporal coding, since each information bit can be spread across all transmitting antennas.

With DE, the data stream first undergoes HE after which the antenna-stream association is rotated in a round-robin fashion. The codewords are made long enough so that each codeword is transmitted form all M_T antennas. In this way, the system can achieve the full ($M_T M_R$-order) diversity gain.

Examples of VE and DE are the Bell Labs's V-BLAST and D-BLAST reported in [53], where BLAST stands for Bell Labs Layered Space Time Architecture.

11.12 SUMMARY

In this chapter we studied the error performance of modulation schemes in fading channels. First we briefly studied the characteristics of fading channels in terms of four important parameters: delay spread, coherence bandwidth, Doppler shift, and coherence time. Using these parameters, we classified fading channels into slow or fast, frequency-nonselective (flat) or frequency-selective. We observed that slow channels are more common than fast channels which occur only when the data rate is very low. We described the major fading envelope distributions: Rayleigh and Rician. The flat-fading-channel (Rayleigh and Rician) performances of common binary and quaternary schemes, namely, BFSK, BPSK, DBPSK, QPSK, OQPSK, $\pi/4$-DQPSK, and MSK including GMSK were treated in great detail. Error probability expressions were derived and plots were given. It was observed that Rayleigh fading and low-K Rician fading cause substantial degradation in error performance. The rapid exponential roll-offs of the AWGN channel performance curves become slow reciprocal roll-off in fading channels. The performances of the above schemes in frequency-selective channel were studied through simulation and the results were presented in Section 11.3. Irreducible error floors are a feature of the performances in frequency-selective channels. The error floor increases with increasing delay spread, decreasing filter bandwidth. The fading-channel performance of $\pi/4$-DQPSK was covered in great detail in Section 11.4. What was covered includes the performance in a flat fading indoor channel, and performance in a two-ray Rayleigh fading channel (slow or fast) with CCI and Doppler shift. As expected, performance degrades with an increase in CCI and Doppler shift. The plots given can be valuable reference resources for design engineers. Simulation results of the performance of 1REC-MHPM or multi-h CPFSK in a slow, flat or frequency-selective Rician and Rayleigh channel were presented in Section 11.5. The fading channel performance of square QAM was discussed in Section 11.6. We also introduced star QAM in Section 11.6. The star constellation can be easily differentially coded and decoded, releasing the requirement of carrier phase synchronization.

Starting from Section 11.7 we turned our attention to remedial measures against fading. We first gave an overview of remedial measures against fading in Section 11.7. In Section 11.8 we studied channel estimation and correction, including mainly the pilot symbol assisted modulation (PSAM) and decision feedback channel estimation (DFCE); in comparison with the former, the latter saves bandwidth but may not perform well in severely fading environments. In Section 11.9 we discussed equalization, including linear equalizers and decision feedback equalizers with zero-forcing criterion or minimum mean squared error criterion. Decision feedback equalizers generally perform better than the linear equalizers since they suffer less from noise enhancement but they may be subject to decision error propagation when the channel is very noisy. In Section 11.10 diversity reception and combining techniques are presented. Multiple antenna diversity is an usual and effective option and maximal ratio combining is the most effective in comparison to the simpler selective combining and equal-gain combining. However it requires channel estimation, which can be difficult in severe fading channels. In Section 11.11, we introduce a new development in diversity techniques: the combination of the MIMO wireless link and space-time codes, which exploits rather than avoids multipath signals and increases channel capacity.

APPENDIX 11A DERIVATION OF (11.80)

We derive (11.80) here. The first equation in (11.80) is from

$$\frac{\partial \varepsilon}{\partial b_l} = -2E[r_{n-l}\varepsilon_n] = 0, \quad l = -N, -N+1, ..., 0 \tag{11.103}$$

and the second equation in (11.80) is from

$$\frac{\partial \varepsilon}{\partial d_j} = 2E[x_{n-j}\varepsilon_n] = 0, \quad j = 1, 2, ..., M \tag{11.104}$$

We derive the first equation in (11.80) first. The correlation in (11.103) is

$$\begin{aligned} E[r_{n-l}\varepsilon_n] &= E\left[r_{n-l}\left(x_n - \sum_{k=-N}^{0} b_k r_{n-k} + \sum_{j=1}^{M} d_j x_{n-j}\right)\right] \\ &= E[r_{n-l}x_n] - \sum_{k=-N}^{0} b_k E[r_{n-l}r_{n-k}] - \sum_{j=1}^{M} d_j E[r_{n-l}x_{n-j}] \end{aligned} \tag{11.105}$$

Using the fact that

$$r_n = \sum_{l=-\infty}^{\infty} c_l x_{n-l} + n_n \tag{11.106}$$

and under the assumption that $E\{x_n x_{n-j}\} = \sigma_x^2 \delta_j$, and noise and signals are independent from each other, the first term in (11.105) yields

$$E[r_{n-l} x_n] = \sum_{i=-\infty}^{\infty} c_i E[x_{n-l-i} x_n] + E[x_n n_n] = c_{-l}\sigma_x^2 \tag{11.107}$$

The second term in (11.105) yields

$$\begin{aligned}
&\sum_{k=-N}^{0} b_k E[r_{n-l} r_{n-k}] \\
&= \sum_{k=-N}^{0} b_k E\left[\left(\sum_{i=-\infty}^{\infty} c_i x_{n-l-i} + n_{n-l}\right)\left(\sum_{j=-\infty}^{\infty} c_j x_{n-k-j} + n_{n-k}\right)\right] \\
&= \sum_{k=-N}^{0} b_k E\left[\left(\sum_{i=-\infty}^{\infty} c_i x_{n-l-i}\right)\left(\sum_{j=-\infty}^{\infty} c_j x_{n-k-j}\right)\right] \\
&\quad + \sum_{k=-N}^{0} b_k E\left[n_{n-l} n_{n-k}\right]
\end{aligned}$$

In the first summation of above, only when $n - l - i = n - k - j$ or $j = i - k + l$, the correlation is nonzero.

Thus

$$\begin{aligned}
&\sum_{k=-N}^{0} b_k E[r_{n-l} r_{n-k}] \\
&= \sum_{k=-N}^{0} b_k E[c_i c_{i-(k-l)}] E[x_{n-l-i} x_{n-l-i}] + \sum_{k=-N}^{0} b_k E\left[n_{n-l} n_{n-k}\right] \\
&= \sum_{k=-N}^{0} b_k R_c(k-l)\sigma_x^2 + \sum_{k=-N}^{0} b_k R_n(k-l)
\end{aligned} \tag{11.108}$$

The third term in (11.105) yields

$$\begin{aligned}\sum_{j=1}^{M} d_j E[r_{n-l}x_{n-j}] &= \sum_{j=1}^{M} d_j E\left[\left(\sum_{i=-\infty}^{\infty} c_i x_{n-l-i} + n_n\right) x_{n-j}\right] \\ &= \sum_{j=1}^{M} d_j \sum_{i=-\infty}^{\infty} c_i E[x_{n-l-i}x_{n-j}]\end{aligned}$$

where only when $i = j - l$ the correlation is nonzero. Thus

$$\begin{aligned}\sum_{j=1}^{M} d_j E[r_{n-l}x_{n-j}] &= \sum_{j=1}^{M} d_j \sum_{i=-\infty}^{\infty} c_{j-l} E[x_{n-j}x_{n-j}] \\ &= \sum_{j=1}^{M} d_j c_{j-l}\sigma_x^2 \qquad (11.109)\end{aligned}$$

Combining (11.107), (11.108), and (11.109) we have

$$c_{-l}\sigma_x^2 - \sum_{k=-N}^{0} b_k R_c(k-l)\sigma_x^2 - \sum_{k=-N}^{0} b_k R_n(k-l) + \sum_{j=1}^{M} d_j c_{j-l}\sigma_x^2 = 0$$

or

$$\begin{aligned}&\sum_{k=-N}^{0}\left[b_k R_c(k-l) + b_k \frac{R_n(k-l)}{\sigma_x^2}\right] \\ &= \sum_{j=1}^{M} d_j c_{j-l} + c_{-l}, \quad l = -N, -N+1, ..., 0 \qquad (11.110)\end{aligned}$$

We derive the second equation in (11.80) now. For $j = 1, ..., M$,

$$\begin{aligned}E[x_{n-j}\varepsilon_n] &= E\left[x_{n-j}\left(x_n - \sum_{k=-N}^{0} b_k r_{n-k} + \sum_{i=1}^{M} d_j x_{n-j}\right)\right] \\ &= E[x_{n-j}x_n] - E\left[x_{n-j}\sum_{k=-N}^{0} b_k r_{n-k}\right] + E\left[x_{n-j}\sum_{i=1}^{M} d_i x_{n-i}\right]\end{aligned}$$

The first correlation is zero. The second correlation is

$$\begin{aligned} & E\left[x_{n-j}\sum_{k=-N}^{0} b_k r_{n-k}\right] \\ &= \sum_{k=-N}^{0} b_k E\left[x_{n-j}\left(\sum_{l=-\infty}^{\infty} c_l x_{n-k-l} + n_{n-k}\right)\right] \\ &= \sum_{k=-N}^{0} b_k \sum_{l=-\infty}^{\infty} c_l E(x_{n-j}x_{n-k-l}) \\ &= \sum_{k=-N}^{0} b_k c_{j-k} E(x_{n-j}x_{n-j}) = \sum_{k=-N}^{0} b_k c_{j-k}\sigma_x^2 \end{aligned} \tag{11.111}$$

The third correlation is

$$E\left[x_{n-j}\sum_{i=1}^{M} d_i x_{n-i}\right] = E\left[d_j x_{n-j}x_{n-j}\right] = d_j\sigma_x^2 \tag{11.112}$$

Combing (11.111) and (11.112) we have

$$\sum_{k=-N}^{0} b_k c_{j-k} = d_j, \quad j = 1, 2, ..., M \tag{11.113}$$

References

[1] Rappaport, T. S., *Wireless Communications: Principles and Practice*, Upper Saddle River, New Jersey: Prentice Hall, 1996.

[2] Lee, W. C. Y., *Mobile Communications Engineering*, New York: MacGraw Hill, 1982.

[3] Clarke, R., "A statistical theory of mobile radio reception," *Bell System Tech. J.*, vol. 47, 1968, pp. 957–1000.

[4] Jakes, W. C., ed., *Microwave Mobile Communications*, New York: Wiley, 1974.

[5] Whalen, A. D., *Detection of Signals in Noise*, New York and London: Academic Press, 1971.

[6] Stüber, G. L., *Principles of Mobile Communication*, Boston: Kluwer Academic Publishers, 1996.

[7] Stein, S., "Fading channel issues in system engineering," *IEEE Journal on Selected Areas in Communications*, vol. 5, no. 2, February 1987, pp. 68–89.

[8] Gradshteyn, I. S., and I. M. Rynzhik, *Table of Integrals, Series and Products*, New York and London: Academic Press, 1980, p. 649.

[9] Roberts, J., and J. Bargallo, "DPSK performance for indoor wireless Rician fading channels," *IEEE Trans. Communications*, vol. 42, no. 2/3/4, February/March/April 1994, pp. 592–596.

[10] Chuang, J., "The effects of time delay spread on the portable radio communications channels with digital modulation," *IEEE Journal on Selected Areas in Communications*, vol. 5, no. 5, June 1987, pp. 879–889.

[11] Anderson, J. B., and S. L. Lauritzen, and C. Thommesen, "Statistics of phase derivatives in mobile communications," *Proc. 1986 IEEE VT Conference,* May 20–22, 1986.

[12] Liu, C., and K. Feher, "Bit error rate performance of $\pi/4$-DQPSK in a frequency-selective fast Rayleigh fading channel," *IEEE Trans. on Vehicular Technology*, vol. 40, no. 3, August 1991, pp. 558–568.

[13] Rappaport, T., and V. Fung, "Simulation of bit error performance of FSK, BPSK, and $\pi/4$-DQPSK in flat fading indoor radio channels using a measurement-based channel model," *IEEE Trans. on Vehicular Technology*, vol. 40, no. 4, November 1991, pp. 731–740.

[14] Fung, V., T. Rappaport, and B. Thoma, "Bit error simulation for $\pi/4$-DQPSK mobile radio communications using two-ray and measurement-based impulse response channel models," *IEEE Journal on Selected Areas in Communications*, vol. 11, no. 3, April 1993, pp. 393–405.

[15] Xiong, F., and S. Bhatmuley, "Performance of MHPM in Rician and Rayleigh fading mobile channels," *IEEE Trans. Communications*, vol. 45, no. 3, March 1997, pp. 279–282.

[16] Anderson, J. B., "Simulated Error Performance of Multi-h Phase Codes," *IEEE Trans. Inform. Theory,* vol. IT-27, no. 3, May 1981, pp. 357–362.

[17] Wilson, S. G., J. H. Highfill, III, and C-D, Hsu, "Error bounds for multi-h phase codes," *IEEE Trans. Inform. Theory,* vol. IT- 28, no. 4, pp. 650–665, July 1982.

[18] Webb, W. T., and L. Hanzo, *Modern Quadrature Amplitude Modulation*, New York: IEEE Press, and London: Pentech Press, 1994.

[19] Mason, L.,"Error probability evaluation for systems employing differential detection in a Rician fast fading environment and Gaussian noise," *IEEE Trans. Communications*, vol. 35, no. 1, January 1987, pp. 39–46.

[20] Varshney, P., J. Salt, and S. Kumar, "BER analysis of GMSK with differential detection in a land mobile channel," *IEEE Trans. on Vehicular Technology*, vol. 42, no. 4, November 1993, pp. 683–689.

[21] Ng, C., et al., "On the error rates of differentially detected narrowband $\pi/4$-DQPSK in Rayleigh fading and Gaussian noise," *IEEE Trans. on Vehicular Technology*, vol. 42, no. 3, August 1993, pp. 259–265.

[22] Xiong, F., and D. Wu, "Multiple-symbol differential detection of $\pi/4$-DQPSK in land mobile satellite communications channels," *IEE Proc. Communications*, vol. 147, no. 3, June 2000, pp. 163–168.

[23] McGeehan, J., and A. Bateman, "Phase-locked transparent tone in band (TTIB): A new spectrum configuration particularly suited to the transmission of data over SSB mobile radio networks," *IEEE Trans. on Communications*, vol. 32, 1984, pp. 81–87.

[24] McGeehan, J., and A. Bateman, "Theoretical and experimental investigation of feedforward signal regeneration," *IEEE Trans. on Vehicular Technology*, vol. 32, 1983, pp. 106–120.

[25] Lodge, J., and M. Moher, "Time diversity for mobile satellite channels using trellis coded modulations," *IEEE Global Telecommun. Conf.*, Tokyo, 1987.

[26] Moher, M., and J. Lodge, "TCMP—A modulation and coding strategy for Rician fading channels," *IEEE J. Select. Areas Commun.*, vol. 7, December 1989, pp. 1347–1355.

[27] Sampei, S., and T. Sunaga,"Rayleigh fading compensation method for 16QAM in digital land mobile radio channels," *Proc. IEEE Veh. Technol. Conf.*, San Francisco, California, May 1989, pp. 640–646.

[28] Cavers, J., "An analysis of pilot symbol assisted modulation for Rayleigh fading channels," *IEEE Trans. on Vehicular Technology*, vol. 40, no. 4, November 1991, pp. 686–693.

[29] Hanzo, L., et al., "Transmission of digitally encoded speech at 1.2 K Baud for PCN," *IEE-Proc., Part-L* vol. 139, no. 4, August 1992, pp. 437–447.

[30] Bateman, A. J., et al., "Speech and data communications over 942 MHz TAB and TTIB single sideband mobile radio systems incorporating feed-forward signal regeneration," *IEEE Trans. on Vehicular Technology*, vol. 34, February 1985, pp. 13–21.

[31] Davarian, F., "Mobile digital communications via tone calibration," *IEEE Trans. on Vehicular Technology*, vol. 36, May 1987, pp. 55–62.

[32] Martin, P. M., et al., "The implementation of a 16QAM mobile data system using TTIB-based fading correction techniques," *Proc. IEEE Veh. Technol. Conf.*, Philadelphia, Pennsylvania, 1988, pp. 71–76.

[33] Scharf, L. L., *Statistical Signal Processing: Detection, Estimation, and Time Series Analysis*, Reading MA: Addison-Wesley, 1991.

[34] Proakis, J., *Digital Communications*, 2nd ed., New York: McGraw-Hill, 1989.

[35] Adachi, F., "Decision feedback differential detection of differentially encoded 16APSK signals," *IEEE Trans. Commun.*, vol. 44, April 1996, pp. 416–418.

[36] Adachi, F., "BER analysis of 2PSK, 4PSK, and 16QAM with decision feedback channel estimation in frequency-selective slow Rayleigh fading," *IEEE Trans. on Vehicular Technology*, vol. 48, no. 5, September 1999, pp. 1563–1571.

[37] Adachi, F., and M. Sawahashi, "Decision feedback multiple-symbol differential detection for M-ary DPSK," *Electronics Letters*, vol. 29, July 1993, pp. 1385–1387.

[38] Adachi, F., and M. Sawahashi, "Decision feedback differential detection of 16DAPSK signals, *Electronics Letters*, vol. 29, August 1993, pp. 1455–1457.

[39] Mark, J., and W. Zhuang, *Wireless Communications and Networking*, Upper Saddle River, New Jersey: Prentice Hall, 2003.

[40] Rainish, D., "Diversity transform for fading channels," *IEEE Trans. on Communications*, vol. 44, no. 12, December 1996, pp. 1653–1661.

[41] Gesbert, D., et al., "From theory to practice: an overview of MIMO space-time coded wireless systems," *IEEE Journal on Selected Areas in Communications*, vol. 21, no. 3, April 2003, pp. 281–301.

[42] Paulraji, A., et al., "An overview of MIMO communications—a key to gigabit wireless," *Proceedings of IEEE*, vol. 92, no. 2, February 2004, pp. 198–218.

[43] Telatar, I., "Capacity of multi-antenna Gaussian channels," *Eur. Trans. Tel.*, vol. 10, no. 6, November/December 1999, pp. 585–595.

[44] Foschini, G., and M. Gans, "On limits of wireless communications in a fading environment when using multiple antennas," *Wireless Pers. Commun.*, vol. 6, no. 3, March 1998, pp. 311–335.

[45] Marzetta, T., and B. Hochwald, "Capacity of a mobile multiple antenna communication link in Rayleigh flat fading," *IEEE Trans. Inform. Theory*, vol. 45, January 1999, pp. 139–157.

[46] Oyman, Ö., et al., "Characterizing the statistical properties of mutual information in MIMO channels," *IEEE Trans. Signal Processing,* vol. 51, November 2003, pp. 2784–2795.

[47] Ozarow, L., S. Shamai, and A. Wyner, "Information theoretic considerations for cellular mobile radio," *IEEE Trans. Veh. Technol.,* vol. 43, May 1994, pp. 359–378.

[48] Biglieri, E., J. Proakis, and S. Shamai, "Fading channels: Information-theoretic and communications aspects," *IEEE Trans. Inform. Theory,* vol. 44, October 1998, pp. 2619–2692.

[49] Seshadri, N., and J. H. Winters, "Two schemes for improving the performance of frequency-division duplex (FDD) transmission systems using transmitter antenna diversity," *Int. J. Wireless Information Networks*, vol. 1, January 1994, pp. 49–60.

[50] Wittneben, A., "A new bandwidth efficient transmit antenna modulation diversity scheme for linear digital modulation," *Proc. IEEE ICC93,* vol. 3, Geneva, Switzerland, 1993, pp. 1630–1634.

[51] Tarokh, V., N. Seshadri, and A. R. Calderbank, "Space-time codes for high data rate wireless communication: performance criterion and code construction," *IEEE Trans. Inform. Theory*, vol. 44, March 1998, pp. 744–765.

[52] Alamouti, S., "A simple transmitter diversity technique for wireless communications," *IEEE J. Select. Areas. Commun.,* vol. 16, no. 8, October 1998, pp. 1451–1458.

[53] Foschini, G. J., "Layered spacetime architecture for wireless communication in a fading environment when using multielement antennas," *Bell Labs Tech. J.*, Autumn 1996, pp. 41–59.

Selected Bibliography

- Lee, W. C. Y., *Mobile Cellular Telecommunications: Analog and Digital Systems,* 2nd ed., New York: McGraw-Hill, 1995.
- Sklar, B., *Digital Communications: Fundamentals and Applications*, Englewood Cliffs, New Jersey: Prentice Hall, 1988.
- Steele, R., *Mobile Radio Communications*, London: Pentech Press Publishers, and New York: IEEE Press, 1992.

Chapter 12

Orthogonal Frequency Division Multiplexing

All bandpass modulation schemes covered in previous chapters involve only a single carrier. Recently multicarrier modulations (MCM) are getting more and more attention and are used in many applications because of their many advantages. One obvious advantage is that transmitting N data symbols on N carriers simultaneously reduces the symbol rate to one Nth of the original symbol rate of the serial data, or increases the symbol duration by N times. Thus the effect of intersymbol interference due to time dispersion of the channel will be reduced and equalization in the receiver will be easier or even unnecessary.

The multiple carriers in MCM are called subcarriers. The frequency band occupied by the signal carried by a subcarrier is called a subband. To separate the signals of subbands at receiver, the earliest method, which is borrowed from "frequency division multiplexing (FDM)," is to space the subcarrier center frequencies far apart so that the spectra of N subbands are virtually nonoverlapped, and N bandpass filters are used in the receiver to separate the subbands. This method requires each bandpass filter to have a very sharp frequency response.

The second method is to allow overlapping of the adjacent subbands; and the separability of the signals at the receiver is achieved by spacing two adjacent subcarriers by $1/T$, where T is the symbol period, so that all subcarriers are orthogonal to each other and can be separated by correlators in the receiver. Even though spectral overlapping exists in this method, it does not require any filtering for each subcarrier. This method is called orthogonal frequency division multiplexing (OFDM), a not-so-accurate name derived from the historical term FDM. FDM refers to a technique where signals from each user or each group of users are modulated on a carrier with a different frequency. It is clear that FDM uses multicarrier modulation; however, multicarrier modulations are not necessarily used for multiple users.

Compared to the first method, OFDM uses less bandwidth and requires no bandpass filter for each subband. However, it is interesting to note that while both methods are meant for digital signals, the first method can be used to transmit analog signals as well.

OFDM offers many advantages over single-carrier modulations:

1. The maximum signaling rate (Nyquist rate) for a given channel can be approached without the use of sharp cutoff filters.
2. It elongates the symbol period so that the signal is more robust against intersymbol interference caused by channel dispersions and multipath interference.
3. It divides the entire frequency band into narrow bands so that it is less sensitive to wide-band impulse noise and fast channel fades.
4. The effect of a slow frequency-selective fade is a separate complex gain on each subband signal and it can be removed by simply multiplying the signal by the conjugate of the complex gain—that is, equalization can be easily done by a one-tap equalizer.
5. Different modulation formats and data rates can be used on different subcarriers depending on the noise level of individual subbands (the symbol periods are kept the same). In serial transmission, certain types of noise (such as time-varying tone interference) may cause an entire system to fail; the parallel OFDM system can avoid this problem by adaptively reducing the data rate of the affected subbands or dropping them.
6. OFDM can be implemented digitally using an inverse discrete Fourier transform and discrete Fourier transform (IDFT/DFT) pair (via the efficient fast algorithm IFFT/FFT pair), which greatly reduces the system complexity.

OFDM was first proposed in the 1960s [1–5] and further developed in the 1970s and 1980s [6–9]. Despite the above advantages, implementation complexity has impaired applications of OFDM in the past. However, it is gaining popularity recently due to the advances in speed and processing power of digital signal processing devices. OFDM is used for asymmetrical digital subscriber line (ADSL) services and wireless local area networks (IEEE 802.11a in the United States and HIPERLAN/2 in Europe). OFDM is also proposed for many other applications, including digital audio broadcasting, digital television, high definition digital television (HDTV) broadcasting, optical communications, and mobile communications.

In this chapter we will study the principle of the OFDM scheme and address related practical issues. This chapter is organized as follows: OFDM signal and spectrum will be described in Section 12.1. The OFDM modulator and demodulator (modem), including the IDFT/DFT-based OFDM implementation is discussed in Section 12.2. Section 12.3 is for real-output DFT for real-value OFDM signals. Sections 12.4 and 12.5 describe FFT and partial FFT algorithms which are cost-efficient for OFDM in asymmetrical communication links. Section 12.6 describes intersymbol interference (ISI) and interchannel interference (ICI) in OFDM when the channel is dispersive and the cyclic extension of the OFDM signal as a measure of eliminating them. Spectral shaping is discussed in Section 12.7. Finally, the chapter is concluded with a summary in Section 12.8.

The high peak-to-average power ratio (PAPR) of OFDM signals often puts big stress on power amplifiers; a great number of methods to reduce the PAPR are reported in the literature and will be presented in a separate chapter (Chapter 13). Synchronization of OFDM signals is another important issue for which there exists rich research results and it deserves a separate chapter too (Chapter 14). The performance of OFDM in an AWGN channel (that is, no other impairments such as ISI and ICI) is the same as that of its modulation schemes on the subcarriers in an AWGN channel. However, the performance of OFDM in fading channels degrades differently from single-carrier systems and its degradation and remedial measures will be described in Chapter 15. To improve the performance of OFDM, time-domain code, frequency-domain code, and spatial code can be used; these result in coded OFDM, vector OFDM, and MIMO-OFDM which will be briefly discussed in Chapter 15 too. And finally, some newly emerging non-FFT-based OFDMs (that is wavelet OFDM and MASK-OFDM) will be introduced in Chapter 16. All of these chapters cover the essentials of OFDM. Various applications of OFDM, such as the multicarrier (MC)-CDMA which is a combination of OFDM with code division multiple access, will not be covered in this book due to page limitations. However, readers should have no problem reading the descriptions of OFDM applications in the literature after reading the material covered in this chapter and Chapters 13–16.

12.1 OFDM SIGNAL AND SPECTRUM

Depending on types of applications, the final form of an OFDM signal can be a baseband signal or a bandpass signal. For wired systems, due to the limited bandwidth of the wires, baseband OFDM signals are transmitted. For wireless communications, such as wireless local area network (WLAN) specified by IEEE 802.11a Standard [10], OFDM signal frequency bands are allocated in the RF frequency band. In this case, OFDM signals are generated in baseband first and up-converted to the RF band for transmission. We describe baseband OFDM first and then bandpass OFDM.

Many OFDM papers or books start describing an OFDM signal from its IDFT generation expression. While the IDFT generation is currently the most efficient method for generating an OFDM signal, it is not the only method. Therefore, we take a different approach by describing the OFDM signal in its form at the transmitter output first, then describing its generation methods later. In addition, after having understood thoroughly the signal format, its parameters, and its properties (e.g., spectrum) first, a description of generation methods that can meet these requirements will be much easier to understand.

12.1.1 Baseband OFDM Signal

A general form of the baseband OFDM signal in the first symbol period can be written as

$$s(t) = \sum_{i=0}^{N-1} s_i(t) = \sum_{i=0}^{N-1} A_i \cos(2\pi f_i t + \phi_i), \quad 0 \le t \le T \tag{12.1}$$

where

$$s_i(t) = A_i \cos(2\pi f_i t + \phi_i), \quad 0 \le t \le T \tag{12.2}$$

where A_i, f_i, and ϕ_i are the amplitude, frequency, and phase of the ith subcarrier. N is the number of subcarriers. T is the symbol period of the data. If the signal is amplitude shift keyed (ASK), A_i is determined by the data, and ϕ_i is an initial phase which is usually assumed zero. If the signal is phase shift keyed (PSK), A_i is constant, and ϕ_i is determined by the data. If the signal is quadrature amplitude modulated (QAM), both A_i and ϕ_i are determined by the data. Therefore the above is a general form of the OFDM signal in the baseband. It is called the baseband OFDM signal because this signal is usually modulated on a higher frequency carrier before transmission. However, this signal can also be transmitted directly, without further frequency conversion (e.g., the system depicted in [6]).

In order for the subcarriers to be orthogonal to each other, f_i must be integer multiples of $1/2T$ (see Sections 3.1 and 3.3), and minimum frequency separation between subcarriers must be $1/T$. Usually f_i are chosen as integer multiples of the symbol rate $R_s = 1/T$ and are spaced in frequency by $R_s = 1/T$. However it can be easily verified that the spacing can be reduced to $1/2T$ if the modulation is ASK. Throughout this chapter, we will assume the subcarrier frequency spacing is $1/T$, which makes the subcarriers orthogonal regardless of the modulation format. Assuming the lowest subcarrier frequency is f_0, then the subcarrier frequencies are $f_0, f_0 + R_s, f_0 + 2R_s, ..., f_0 + (N-1)R_s$. Even though to minimize bandwidth requirement, one would let f_0 be zero, in a practical OFDM system, to avoid problems created by the DC offset of the D/A convertor, f_0 is usually not zero. However, for simplicity f_0 is often set to zero in theoretical study. This will not affect the validity of the theoretical results since when applying the theoretical results, one can simply leave the zero frequency subcarrier unused (see IEEE Standard 802.11a [10]).

Therefore, in this chapter we assume that subcarrier frequencies are as follows

$$f_i = iR_s = \frac{i}{T}, \quad i = 0, 1, ..., N-1 \tag{12.3}$$

With these subcarrier frequencies, it can be easily verified that

$$\begin{aligned} \int_0^T s_i(t)s_j(t)dt &= \int_0^T A_i \cos(2\pi f_i t + \phi_i) A_j \cos(2\pi f_j t + \phi_j) dt \\ &= \begin{cases} A_0^2 T \cos^2 \phi_0, & i = j = 0 \\ \frac{1}{2} A_i^2 T, & i = j \neq 0 \\ 0, & i \neq j \end{cases} \end{aligned} \tag{12.4}$$

This shows the orthogonality between the subcarriers. Note that the orthogonality holds for any values of A_i, A_j and ϕ_i, ϕ_j for the frequency allocation in (12.3).

Figure 12.1 shows two examples of OFDM waveforms.. Figure 12.1(a) is for a 4-channel QPSK-OFDM signal where each subcarrier is modulated by one of four possible symbols in the QPSK constellation with amplitude $\sqrt{2}$ and phases of $\pi/4$, $3\pi/4$, $5\pi/4$, and $7\pi/4$, respectively. Figure 12.1(b) is for a 16-channel QAM-OFDM signal where each subcarrier is modulated by one of 16 possible symbols in the square 16QAM constellation located on $(-3, -1, +1, +3) \times (-3, -1, +1, +3)$ grid.

Assume data are random and wide-sense stationary; the frequency spectrum of the OFDM signal is characterized by its power spectral density (PSD). Further assume data on each subcarrier are independent from others; then the total PSD is just a superposition of the PSDs of all subband signals. That is

$$S(f) = \sum_{i=0}^{N-1} S_i(f) \tag{12.5}$$

where $S_i(f)$ is the PSD of the ith subband signal $s_i(t)$.

For $i \neq 0$, $s_i(t)$ is just a linearly modulated signal whose PSD has been found in previous chapters. If the modulation is MASK or QAM with symmetrical constellation, the PSD of the ith subband signal is (see (8.14), (9.29), and (A.13))

$$S_i(f) = \frac{1}{2} A_{avg}^2 T \left[\left(\frac{\sin \pi(f - f_i)T}{\pi(f - f_i)T} \right)^2 + \left(\frac{\sin \pi(-f - f_i)T}{\pi(-f - f_i)T} \right)^2 \right], \quad i \neq 0 \tag{12.6}$$

where $A_{avg}^2 = 2\sigma_a^2$ for QAM and $A_{avg}^2 = \sigma_a^2$ for MASK, and σ_a^2 is the amplitude variance of the I- or Q-channel signal of QAM (only I-channel for MASK) over their signal constellation. If the modulation is MPSK, the PSD function is the same except that A_{avg} is just equal to the PSK signal amplitude (see (4.26)). If the modulation is

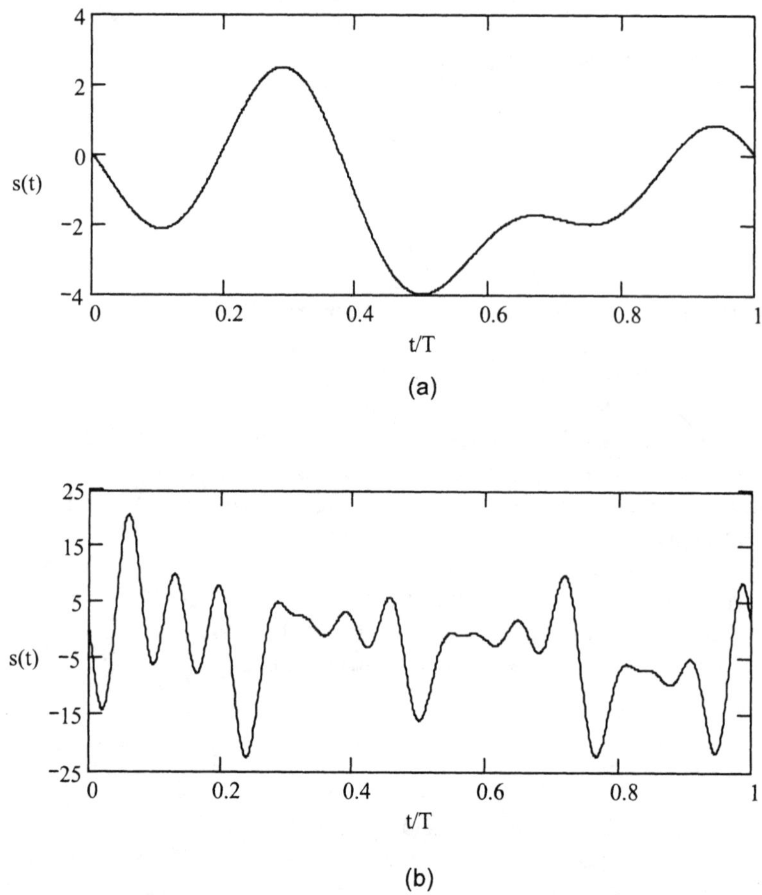

Figure 12.1 Examples of baseband OFDM signal waveform. (a) 4-channel QPSK-OFDM signal with each subcarrier modulated by one of four possible symbols in the QPSK constellation with amplitude $\sqrt{2}$ and phases of $\pi/4$, $3\pi/4$, $5\pi/4$, and $7\pi/4$, respectively. (b) 16-channel QAM-OFDM signal with each subcarrier modulated by one of 16 possible symbols in the square 16QAM constellation located on $(-3, -1, +1, +3) \times (-3, -1, +1, +3)$ grid.

unipolar MASK, the PSD will have a continuous part and a discrete part. Its continuous part is the same as in (12.6) and its discrete part is a spectral line at $\pm f_i$ with strength of $(m_A)^2$ where m_A is the mean of the amplitudes (see (8.13)).

For $i = 0$, $s_i(t) = A_i \cos \phi_i$ is a baseband signal, which is just equal to the I-channel data of QAM or PSK. From (A.18), its PSD is given by

$$S_0(f) = \sigma_a^2 T \left(\frac{\sin \pi (f) T}{\pi (f) T} \right)^2 = \frac{1}{2} A_{avg}^2 T \left(\frac{\sin \pi (f) T}{\pi (f) T} \right)^2 \tag{12.7}$$

Combining (12.6), (12.7), and (12.5), normalizing the PSD by its maximum, and showing only the positive frequency part, we have the final PSD expression

$$S(f) = \sum_{i=0}^{N-1} \left(\frac{\sin \pi (f - f_i) T}{\pi (f - f_i) T} \right)^2, \quad f \geq 0 \tag{12.8}$$

Each member of the PSD in (12.8) has the shape of a squared sinc function. The first PSD's first null is at $(f - f_i) = 1/T$. Since the subcarrier frequency separation is $1/T$, the first null point just coincides with the peak of the PSDs of adjacent subband signals. Figure 12.2 shows (a) spectra of four modulated subcarriers with frequencies 0, $1/T$, $2/T$, and $3/T$ and (b) their composite spectrum (setting $\frac{1}{2} A_{avg}^2 T = 1$). Note that the spectra in Figure 12.2 are one-sided spectra, mirror negative frequency components are not shown. From Figure 12.2(b) we see that the null bandwidth of the composite PSD of an N-subcarrier baseband OFDM signal is

$$B_{null} = \frac{N}{T} = N R_s \tag{12.9}$$

and the null-to-null bandwidth is

$$B_{null-to-null} = \frac{2N}{T} = 2 N R_s \tag{12.10}$$

It is interesting to see how the PSD of an OFDM signal evolves with the increase in the number of subcarriers. Figure 12.3 shows PSDs given in (12.8) for $N = 4, 16, 64$. The frequency axis is normalized to the total symbol rate of the OFDM signal, NR_s. First, it can be seen that the transition bands are getting sharper and sidelobes are getting lower as N increases. Second, it can be seen that the normalized zero-to-null frequency bandwidth approaches 1 when N approaches infinity.

12.1.2 Bandpass OFDM Signal

When an RF bandwidth is allocated for an OFDM signal, the subcarriers' frequencies are typically chosen to be symmetrically distributed around a nominal

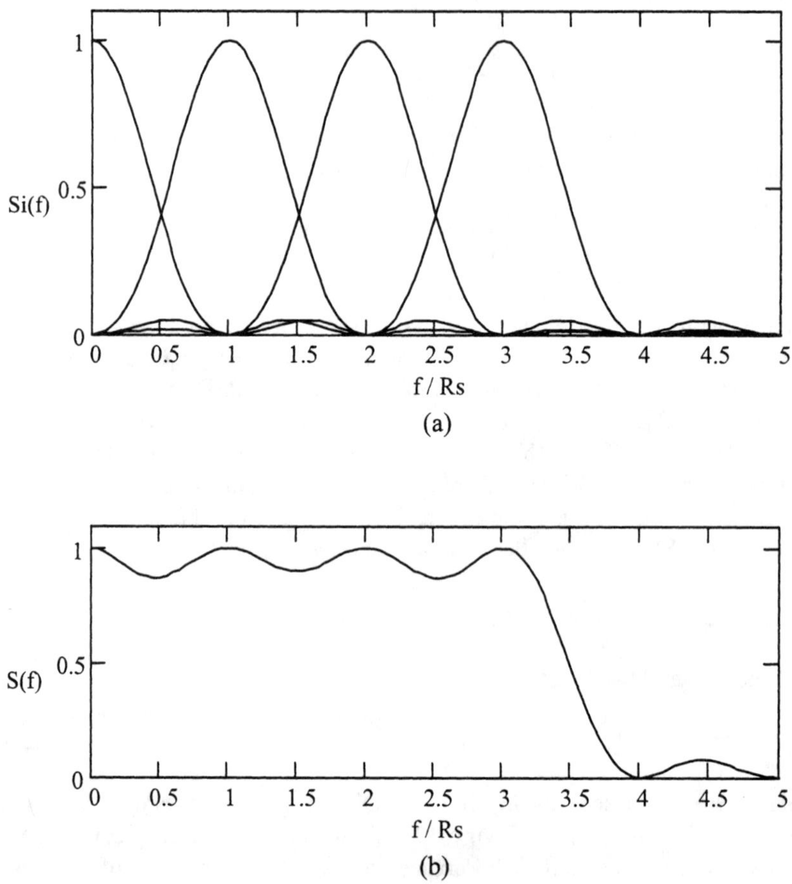

Figure 12.2 PSDs of a baseband OFDM signal with four subcarriers: (a) individual PSDs, and (b) composite PSD.

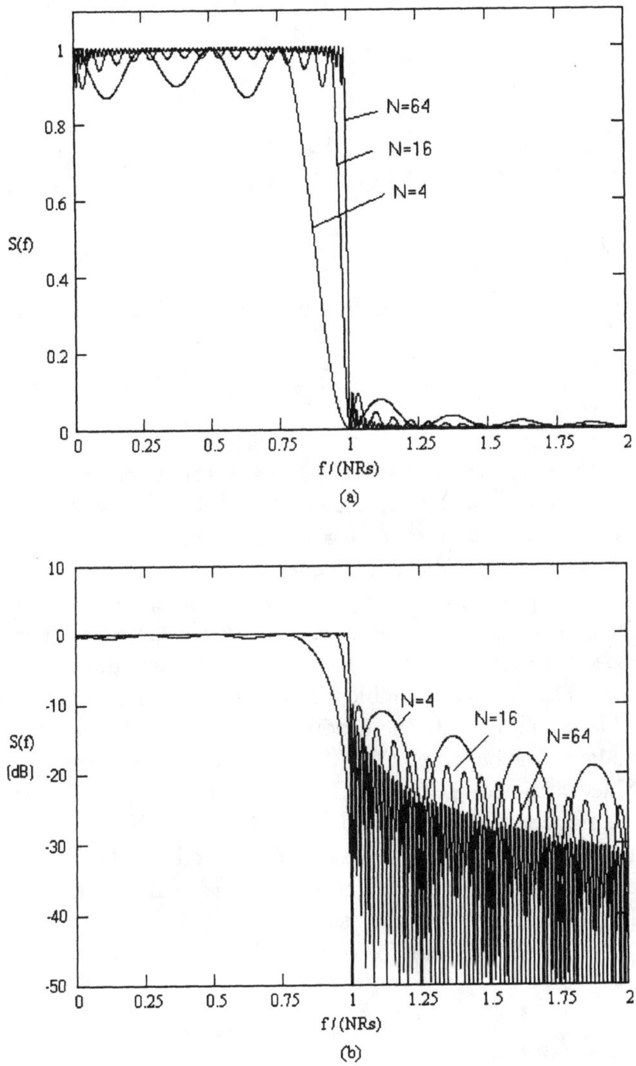

Figure 12.3 PSDs of baseband OFDM signals for $N = 4, 16, 64$. The transition band of the PSD of OFDM sharpens with the increase of number of subcarriers, N, when the total data rate is fixed. The f-axis is normalized to NR_s, the total data rate. The vertical axis is in linear scale (a) or logarithmic scale (b).

center frequency, f_c, of the band. That is, the subcarrier frequencies are

$$f_i = f_c - \frac{N-1}{2T} + \frac{i}{T}, \quad i = 0, 1, ..., N-1 \tag{12.11}$$

Thus the subcarrier frequencies range from $f_c - \frac{N-1}{2T}$ to $f_c + \frac{N-1}{2T}$. For example, if $N = 4$, the subcarrier frequencies will be

$$f_k = f_c - \frac{3}{2T}, \quad f_c - \frac{1}{2T}, \quad f_c + \frac{1}{2T}, \quad f_c + \frac{3}{2T}$$

which are symmetrical about f_c. Now, the bandpass OFDM signal can be written as

$$s(t) = \sum_{i=0}^{N-1} A_i \cos\left[2\pi\left(f_c - \frac{N-1}{2T} + \frac{i}{T}\right)t + \phi_i\right], \quad 0 \le t \le T \tag{12.12}$$

Note that it is not required for subcarriers at the RF band to be orthogonal to each other since demodulation is not performed at the RF band. Instead, the RF band OFDM signal will be down-converted back to the baseband and then demodulated. As stated above, at baseband, subcarrier frequencies f_i must be a multiple of $1/2T$; no such constraint is imposed on f_c. Figure 12.4 shows examples of bandpass OFDM waveforms. The OFDM signal parameters are the same as those for Figure 12.1 and the central frequency of the RF band is $f_c = 50/T$. Note that the envelopes of the waveforms in this figure do not resemble the baseband waveforms shown in Figure 12.1. This is due to the fact that the bandpass OFDM signal is not generated by amplitude-modulating the baseband OFDM signal onto the RF carrier. Instead, it is a frequency-shifted version of the baseband signal. By comparing (12.1) and (12.12), it is seen that the frequency shift is $f_c' = f_c - \frac{N-1}{2T}$.

When data on each subcarrier are independent of others, the OFDM signal in (12.12) is just a sum of independent signals. Thus the total PSD is only a sum of individual PSDs of each subcarrier. Each subcarrier's PSD is a squared sinc function centered at $f_c - \frac{N-1}{2T} + \frac{i}{T}$. Thus for positive frequency we have the total PSD expression as

$$S(f) = \sum_{i=0}^{N-1} \left(\frac{\sin\left[\pi\left(f - \left(f_c - \frac{N-1}{2T} + \frac{i}{T}\right)\right)T\right]}{\pi\left(f - \left(f_c - \frac{N-1}{2T} + \frac{i}{T}\right)\right)T} \right)^2, \quad f \ge 0 \tag{12.13}$$

This PSD is just a shifted version of the baseband PSD given in (12.8). Comparing (12.8) and (12.13) reveals that the frequency shift is

$$f_c' = f_c - \frac{N-1}{2T} \tag{12.14}$$

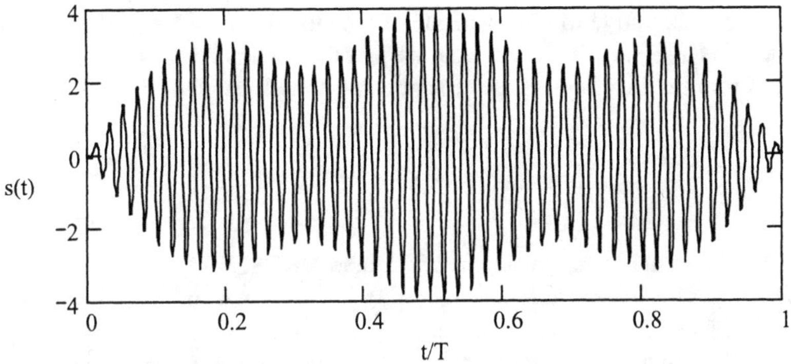

(a)

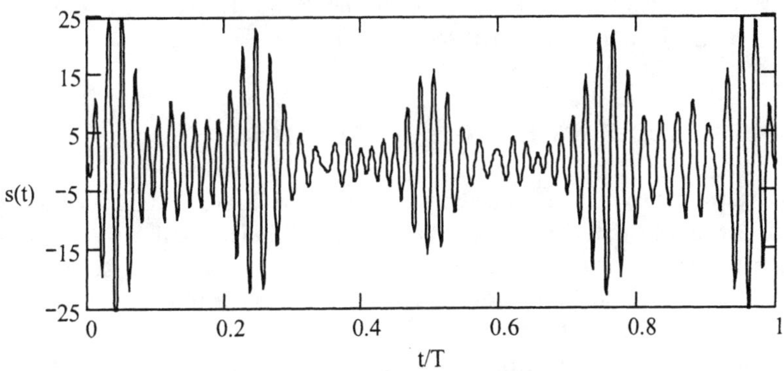

(b)

Figure 12.4 Examples of bandpass OFDM signal waveform: (a) QPSK, and (b) 16QAM. $f_c = 50/T$.

This frequency shift can be obtained by mixing the baseband OFDM signal with an RF signal of frequency f_c' and only keeping the upper sideband and filtering out the lower sideband, as will be described in the next section. It is worthwhile to note that the RF mixing frequency is not f_c, the nominal carrier frequency; rather, the frequency is offset by $(N-1)/(2T)$ from it.

It will be illustrative to observe PSD of a bandpass OFDM signal. Using the same examples in Figures 12.2 and 12.3, we plot the bandpass PSDs in Figures 12.5 and 12.6. From Figure 12.5 it is seen that the null-to-null bandwidth is

$$B_{null-to-null} = \frac{N+1}{T} = (N+1)R_s$$

Figure 12.6 shows that transition bands of the bandpass PSDs are getting sharper and sidelobes are getting lower as N increases, and the normalized null-to-null bandwidth approaches 1 when N approaches infinity.

It is well-known that the Nyquist rate of a channel with a rectangular frequency response is equal to the bandpass bandwidth of the channel. It is seen here that OFDM has achieved the Nyquist rate since its normalized bandpass frequency bandwidth is 1. We thus have seen the first advantage of OFDM as mentioned in the very beginning of this chapter, that is, "the maximum signaling rate (Nyquist rate) for a given channel can be approached without the use of sharp cutoff filters."

12.2 OFDM MODULATOR AND DEMODULATOR

Historically, two types of implementations of the OFDM modulator and demodulator (modem) were suggested. One uses oscillators and multipliers for the modulator and correlators for the demodulator [1–4]. Another uses discrete Fourier transform (DFT) [5–7]. The first implementation follows the single-carrier system and is derived from (12.1). For a large number of subcarriers, this method makes the system prohibitively complex and is not practical. The second implementation is based on the complex envelope of the OFDM signal as we will develop shortly. It is efficient for a large number of subcarriers. Details of these methods are given below.

In the discussion that follows, for the sake of clarity, only the bare-bone modems, which are optimum for the AWGN channel, are presented. Synchronization blocks are not included. However, synchronous carrier frequency, carrier phase, and symbol timing are assumed. Synchronization for OFDM signals will be discussed in a later chapter. Additional blocks needed for fading channels, the channel estimation block and the cyclic extension block, are also not included. They will be discussed separately too.

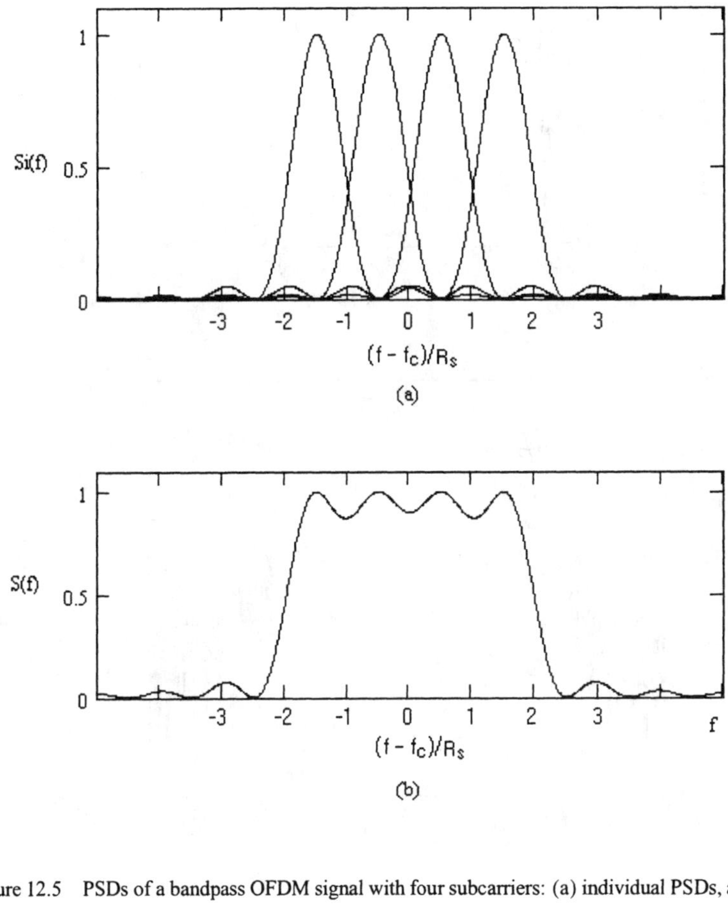

Figure 12.5 PSDs of a bandpass OFDM signal with four subcarriers: (a) individual PSDs, and (b) composite PSD.

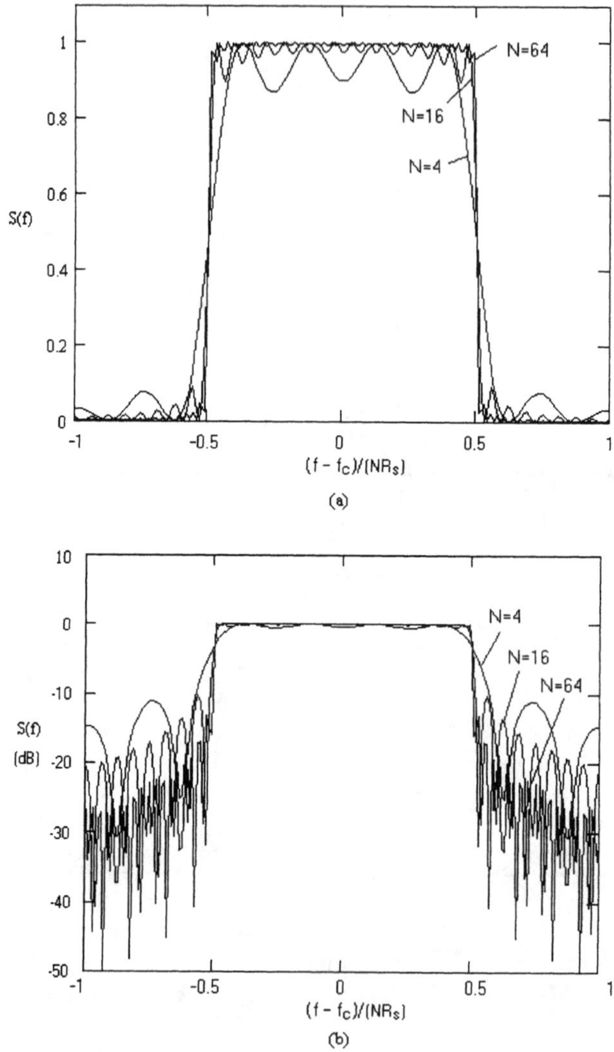

Figure 12.6 PSDs of bandpass OFDM signals for $N = 4, 16, 64$. The transition band of the PSD of OFDM sharpens with the increase of number of subcarriers, N, when the total data rate is fixed. The f-axis is normalized to NR_s, the total data rate. The vertical axis is in linear scale (a) or logarithmic scale (b).

12.2.1 Analog OFDM Modem

Since each modulated subcarrier in the OFDM signal in (12.1) is an MPSK, or MASK, or QAM signal, each of them can be generated using the modulators of MPSK, or MASK, or QAM, described in Chapters 4, 8, and 9, respectively. The sum of them is the baseband OFDM signal. Figure 12.7(a) is the analog OFDM modulator. Serial data bits $\{a_k\}$ are converted by the 1:N S/P (serial-to-parallel converter) to N data streams. In the mapper block, bits in each data stream are grouped into k-tuples and each k-tuple is mapped to a symbol which is denoted in general by a complex number

$$d_i = A_i \exp(j\phi_i) = I_i + jQ_i, \quad i = 0, 1, ..., N-1 \tag{12.15}$$

where

$$I_i = A_i \cos \phi_i$$

$$Q_i = A_i \sin \phi_i$$

are the inphase and quadrature components of the symbol, respectively (for MASK, only I_i exists). d_i is modulated onto the ith subcarrier in the modulator. Each subchannel can have a different modulation. There are total N modulators at N subcarrier frequencies: $f_0, f_1, ..., f_{N-1}$. Each of them is a complete modulator. Its construction depends on the chosen modulation (see Figure 4.10 for MPSK, Figure 8.3 for MASK, and Figure 9.4 for QAM, for example). It typically consists of a subcarrier oscillator, one (MASK) or two (QAM and MPSK) multipliers, and an adder. A $\pi/2$ phase shifter is also needed for quadrature modulations (QAM and MPSK). The adder of each modulator can be omitted since its task can be effectively carried out by the final adder. If a bandpass OFDM signal is desired, the adder in the modulator will be followed by a cascade of an RF mixer with a reference signal of frequency $f_c' = f_c - \frac{N-1}{2T}$ and a bandpass filter whose frequency response rejects the lower sideband of the mixer output. The output of the filter will be a bandpass OFDM signal with a spectrum centered at f_c. Up conversion may be conducted in more than one step: first to an IF frequency and finally to the RF frequency.

In the receiver, a down converter with a reference of f_c' will be used to translate the signal to the baseband before the demodulation is performed in the baseband demodulator (Figure 12.7(b)). The baseband OFDM demodulator is basically a bank of N demodulators at N subcarrier frequencies. Each of them is a complete demodulator whose construction depends on the modulation format (see Figure 4.11 for MPSK, Figure 8.4 for MASK, and Figure 9.6 for QAM, for example). It typically consists of local oscillators, multipliers, integrators, and threshold detectors. When

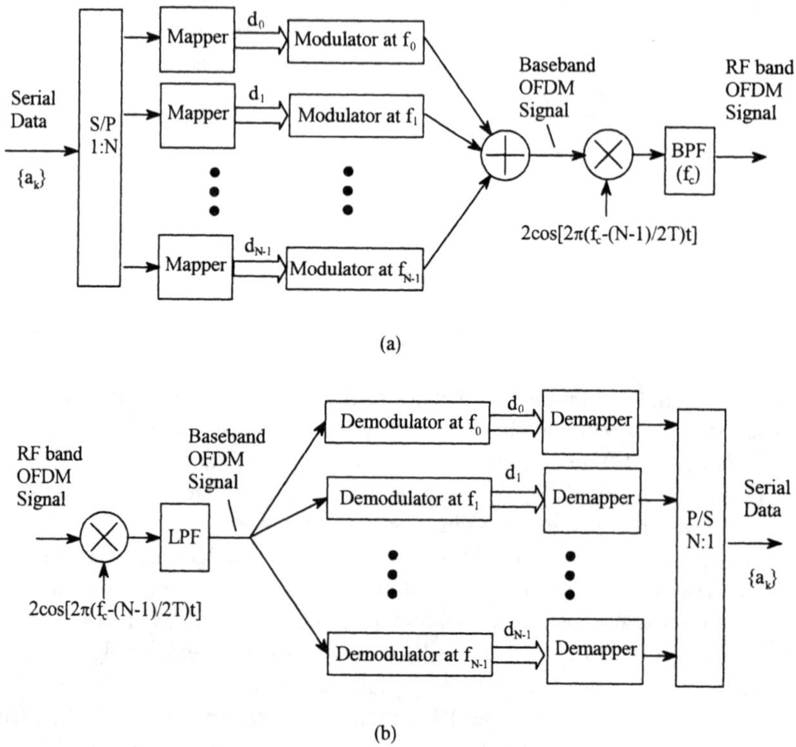

Figure 12.7 Basic OFDM modulator (a) and demodulator (b).

the number of channels, N, is big, the number of the components for the modulator and the demodulator is also big, which can be impractical.

Under ideal channel conditions, due to the orthogonality between subcarriers, each demodulator is independent of others. Thus the bit error probability of each channel is the same as that of a single-carrier system, as described in previous chapters.

Practical implementations were described in [1–4]; the predominant practical issue is that filters are included in the modulator and the demodulator. To maintain the orthogonality, certain requirements must be imposed on the filters' frequency responses.

Due to its complexity, this basic implementation using oscillators, multipliers, and correlators has become obsolete.

12.2.2 DFT-Based Digital OFDM Modem

The bandpass OFDM signal in (12.12) can be written as

$$s(t) = \text{Re}\left\{\left(\sum_{i=0}^{N-1} d_i \exp(j2\pi\frac{i}{T}t)\right)\exp\left[j2\pi\left(f_c - \frac{N-1}{2T}\right)t\right]\right\}, \quad 0 \le t \le T \tag{12.16}$$

where d_i is the complex data symbol already defined earlier in (12.15).

With respect to lower carrier frequency $f_c - \frac{N-1}{2T}$, the complex envelope of the bandpass OFDM signal is

$$\tilde{s}(t) = \sum_{i=0}^{N-1} d_i \exp(j2\pi\frac{i}{T}t), \quad 0 \le t \le T \tag{12.17}$$

The baseband OFDM signal in (12.1) can be written as

$$s(t) = \text{Re}\left[\sum_{i=0}^{N-1} d_i \exp(j2\pi\frac{i}{T}t)\right] = \text{Re}\left[\tilde{s}(t)\right], \quad 0 \le t \le T \tag{12.18}$$

That is, the baseband OFDM signal is the real part of the complex envelope of the bandpass OFDM signal.

If we sample the complex envelope with a sampling period of $\Delta t = T/N$, and add a normalizing factor $1/N$, we obtain

$$s_n = \frac{1}{N}\sum_{i=0}^{N-1} d_i \exp(j2\pi\frac{in}{N}), \quad n = 0, 1, ..., N-1 \tag{12.19}$$

This expression is the inverse discrete Fourier transform (IDFT). This fact implies that samples of the complex envelope of an OFDM signal can be generated by IDFT. The input complex data to the IDFT are in the frequency domain and the output samples, also complex in general, are in time domain. If the receiver receives these samples without any distortion, the data symbols, d_i, can be recovered by the discrete Fourier transform (DFT) given by

$$d_i = \sum_{n=0}^{N-1} s_n \exp(-j2\pi \frac{in}{N}), \quad i = 0, 1, ..., N-1 \tag{12.20}$$

This is to say that OFDM modulation and demodulation can be implemented by a pair of IDFT and DFT. This greatly simplifies OFDM modulation and demodulation. Further more IDFT and DFT can be calculated by their fast algorithms IFFT (inverse fast Fourier transform) and FFT (fast Fourier transform). Present day implementations of OFDM are almost exclusively by means of IFFT and FFT.

In a practical OFDM system, the complex signal samples $\{s_n\}$ in (12.19) must be separated into a real component (I-channel) and an imaginary component (Q-channel) first; then each of them must be converted to an analog signal before they are transmitted (for baseband transmission) or further modulated onto high frequency carriers (for RF band transmission). Now we show that distortionless I- and Q-channel analog signals cannot be recovered from these signal samples through the D/A converter due to inadequate sampling frequency. From (12.19) and (12.15), it is easy to see that the I- and Q-channel components of $\{s_n\}$ are

$$\begin{aligned} I_n &= \frac{1}{N} \sum_{i=0}^{N-1} A_i \cos(2\pi \frac{in}{N} + \phi_i), \quad n = 0, 1, ..., N-1 \\ Q_n &= \frac{1}{N} \sum_{i=0}^{N-1} A_i \sin(2\pi \frac{in}{N} + \phi_i), \quad n = 0, 1, ..., N-1 \end{aligned} \tag{12.21}$$

Obviously, they are samples of these two analog signals:

$$\begin{aligned} I(t) &= \frac{1}{N} \sum_{i=0}^{N-1} A_i \cos(2\pi \frac{i}{T} t + \phi_i), \quad 0 \le t \le T \\ Q(t) &= \frac{1}{N} \sum_{i=0}^{N-1} A_i \sin(2\pi \frac{i}{T} t + \phi_i), \quad 0 \le t \le T \end{aligned} \tag{12.22}$$

These are the baseband OFDM signals given in (12.1) (within a factor $1/N$). They have a bandwidth (null-to-null) of $2N/T$ (see (12.10)). Thus the sampling frequency should be at least $2N/T$ to essentially avoid aliasing [7]. However, the sampling frequency of $\{s_n\}$ is only N/T since there are N samples in a symbol period T.

To overcome this problem, an obvious measure is to increase the number of samples to $2N$. This was first observed by Hirosaki in his paper of baseband O(offset) QAM-OFDM in [7].[1] It requires appending N zeros in the data set $\{d_i\}$ to have a new data set of size $2N$:

$$d_0, d_1, ..., d_{N-1}, \underbrace{0, 0, ...0}_{N \text{ zeros}}$$

and a $2N$-point IDFT to generate the OFDM signal and an N-point DFT to demodulate the signal. Those subcarriers with zero data are called *dummy* or *virtual* subcarriers in the literature. Then the PSD of the signal samples will have no significant aliasing as shown by the example in Figure 12.8.

Considering using a $2N$-point DFT pair (we use the term DFT pair to refer to an IDFT in transmitter and a DFT in receiver) for an N-point data set as a defect, Hirosaki devised a scheme to use only an N-point DFT pair. In the modulator, the scheme requires preprocessing the complex data and causing the preprocessed data to pass through an N-point IDFT twice to obtain the real and imaginary part of the OFDM signal, respectively. In the receiver, demodulation is also performed for the real part and imaginary part in two uses of an N-point DFT processor, respectively. Thus the DFT processor pair is only N-point, but the speed of the processors is double that of the symbol rate. This method was actually realized experimentally and proved to be successful [7, 8]. Later, an improved scheme was proposed in [11] for bandpass OQAM-OFDM, which doubles the efficiency of the scheme in [7]. It is based on an N-point DFT pair and N-branch PPN (polyphase network) at the rate F, for the transmission of $2N$ complex symbols at a rate of $F/2$. This is equivalent to transmitting N complex symbols with an N-point DFT pair at the symbol rate. This efficiency can be achieved using the method described below with much less system complexity.

This method is to shift the DFT pair index range from $[0, N-1]$ to $[-N/2, N/2-1]$. By doing this, an N-point DFT pair will satisfy the Nyquist sampling rate.[2] The IDFT/DFT pair in (12.19) and (12.20) can be replaced by the IDFT/DFT pair

---

1 If the reader is going to read his paper, note that our N is L in [7] and his $N \geq 2L$.

2 Many OFDM papers in the literature use the 0 to (N-1) IDFT for the baseband signal generation and claim that the sampling rate needed is only N/T, which is not correct. However most results based on this model are probably still valid since $-N/2$ to $(N/2-1)$ DFT pair is equivalent to the 0 to $(N-1)$ DFT pair in terms of an end-to-end signal path and the time-domain signal generated by $-N/2$ to $(N/2-1)$ IDFT is only different by a factor of $(-1)^n$ from the signal generated by 0 to $(N-1)$ IDFT.

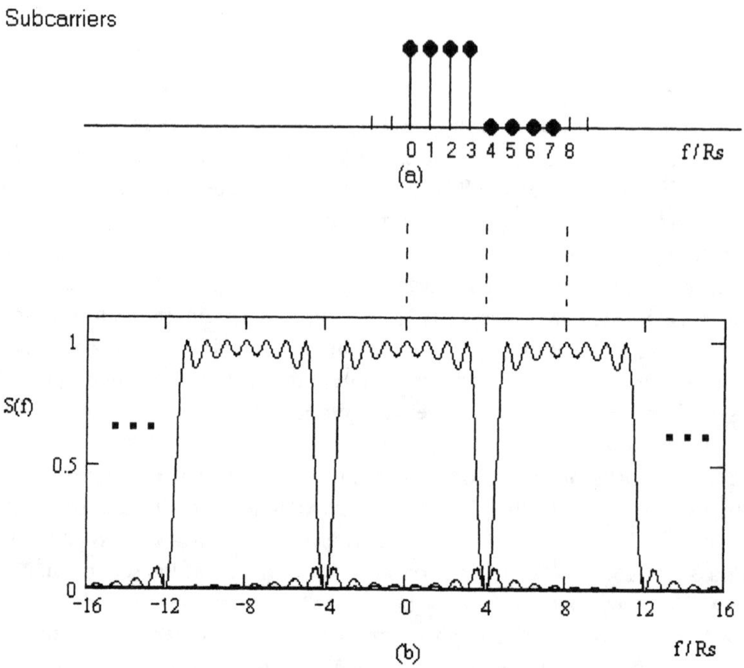

Figure 12.8 PSD of four-subcarrier OFDM signal samples generated by 0 to $(2N-1)$ IDFT: (a) four subcarriers and four virtual subcarriers, (b) replicas of the baseband PSD due to sampling (three replicas shown). Note that it is not yet the composite PSD of the sampled signal, but the composite PSD which is a sum of all replicas will have a very similar shape in the frequency range from $f/R_s = -12$ to 12.

defined by

$$s'_n = \frac{1}{N} \sum_{i=-\frac{N}{2}}^{\frac{N}{2}-1} d_i \exp(j2\pi \frac{in}{N}), \quad n = -\frac{N}{2}, -\frac{N}{2}+1, ..., \frac{N}{2}-1 \tag{12.23}$$

and

$$d_i = \sum_{n=-\frac{N}{2}}^{\frac{N}{2}-1} s'_n \exp(-j2\pi \frac{in}{N}), \quad i = -\frac{N}{2}, -\frac{N}{2}+1, ..., \frac{N}{2}-1 \tag{12.24}$$

This pair can be easily verified (see Appendix 12A). Note that in (12.23), data d_i remain the same except for an index shift and they are recovered without any difference from that of the previous IDFT/DFT pair, but the time domain signal samples are different, that is, $s'_n \neq s_n$; in fact (see Appendix 12A),

$$s'_n = \begin{cases} (-1)^n s_{n+N}, & n = -\frac{N}{2}, -\frac{N}{2}+1..., -1 \\ (-1)^n s_n, & n = 0, 1, ..., \frac{N}{2}-1 \end{cases} \tag{12.25}$$

From (12.23), it is easy to see that the I- and Q-channel components are

$$\begin{aligned} I_n &= \frac{1}{N} \sum_{i=-\frac{N}{2}}^{\frac{N}{2}-1} A_i \cos(2\pi \frac{in}{N} + \phi_i), \quad n = -\frac{N}{2}, -\frac{N}{2}+1, ..., \frac{N}{2}-1 \\ Q_n &= \frac{1}{N} \sum_{i=-\frac{N}{2}}^{\frac{N}{2}-1} A_i \sin(2\pi \frac{in}{N} + \phi_i), \quad n = -\frac{N}{2}, -\frac{N}{2}+1, ..., \frac{N}{2}-1 \end{aligned} \tag{12.26}$$

and corresponding analog signals are

$$\begin{aligned} I(t) &= \frac{1}{N} \sum_{i=-\frac{N}{2}}^{\frac{N}{2}-1} A_i \cos(2\pi \frac{i}{T} t + \phi_i), \quad -\frac{T}{2} \leq t \leq \frac{T}{2} \\ Q(t) &= \frac{1}{N} \sum_{i=-\frac{N}{2}}^{\frac{N}{2}-1} A_i \sin(2\pi \frac{i}{T} t + \phi_i), \quad -\frac{T}{2} \leq t \leq \frac{T}{2} \end{aligned} \tag{12.27}$$

Note that the period of the signals is from $-T/2$ to $T/2$. It seems that the signals are not causal. However, it is just a matter of choice of time reference. It means, after the D/A converter, the starting time of the waveforms obtained is designated as

$t = -T/2$ and the ending time is $t = T/2$. These baseband OFDM signals have subcarrier frequencies between $-N/(2T)$ to $(N-2)/(2T)$, the null-to-null two-sided bandwidth is $(N+2)/T$, which approaches N/T when $N \to \infty$. Thus as far as the main lobes of the PSD are concerned, the sampling frequency of (12.26) asymptotically satisfies the Nyquist rate since the sampling rate is N/T for a signal with a two-sided bandwidth of N/T. The sampling rate N/T is called *critical* sampling rate in the literature.

To further reduce aliasing, oversampling is often used. This is accomplished by padding zeros in the data set. Assuming L (an even integer) zeros are appended, then $L/2$ zeros are appended before and after the data sequence. The appended sequence becomes

$$\underbrace{0, 0, ...0}_{L/2 \text{ zeros}}, d_{-\frac{N}{2}}, d_{-\frac{N}{2}+1}, ..., d_{\frac{N}{2}-1}, \underbrace{0, 0, ...0}_{L/2 \text{ zeros}} \tag{12.28}$$

Figure 12.9 shows some examples of PSDs of the OFDM signal samples generated by the $-N/2$ to $N/2-1$ IDFT. Figure 12.9(a) shows PSD of four-subcarrier OFDM samples without virtual subcarriers. Some aliasing exists. When N increases, the aliasing reduces as shown in Figure 12.9(b) where $N = 52$ (the number of subcarriers for data and pilots in IEEE 802.11a Standard). However, there is almost no gap between the replicas of the baseband PSDs. An ideal low-pass filter is required in the D/A converter to obtain a distortionless analog OFDM signal. Figure 12.9(c) shows the PSD for the signal samples generated by a 64 DFT, where 12 are virtual subcarriers (IEEE 802.11a Standard). The spectral gaps allow nonideal low-pass filters.

In the literature, the DFT's properties and fast algorithms (FFT) are all based on a DFT defined on $[0, N-1]$ [13,14]. Therefore, one can perform the DFT on $[0, N-1]$ first, and then convert $\{s_n\}_{n=0}^{N-1}$ to $\{s'_n\}_{n=-N/2}^{N/2-1}$ using the relations in (12.25).

Figure 12.10 shows the OFDM modem based on the DFT pair given in (12.23) and (12.24). In the modulator (Figure 12.10(a)), serial data $\{a_k\}$ are converted to symbols $\{d_i\}_{i=-N/2}^{N/2-1}$ through a serial-to-parallel converter (S/P) and bits-to-symbol mappers. Then zeros for virtual carriers are added to form the data set as in (12.28). The new data set size is N_s. The IFFT block transforms $\{d_i\}_{i=-N_s/2}^{N_s/2-1}$ into complex time-domain signal samples $\{s'_n\}_{n=-N_s/2}^{N_s/2-1}$. A parallel-to-serial block (P/S) separates these complex samples into real and imaginary parts and multiplexes them into two data sample streams: the real stream $\{I_n\}_{n=-N_s/2}^{N_s/2-1}$ and the imaginary stream $\{Q_n\}_{n=-N_s/2}^{N_s/2-1}$, given in (12.26). Before they are sent to the D/A converter, a cyclic extension is added to the I- and Q-channel samples as a prefix in order to avoid

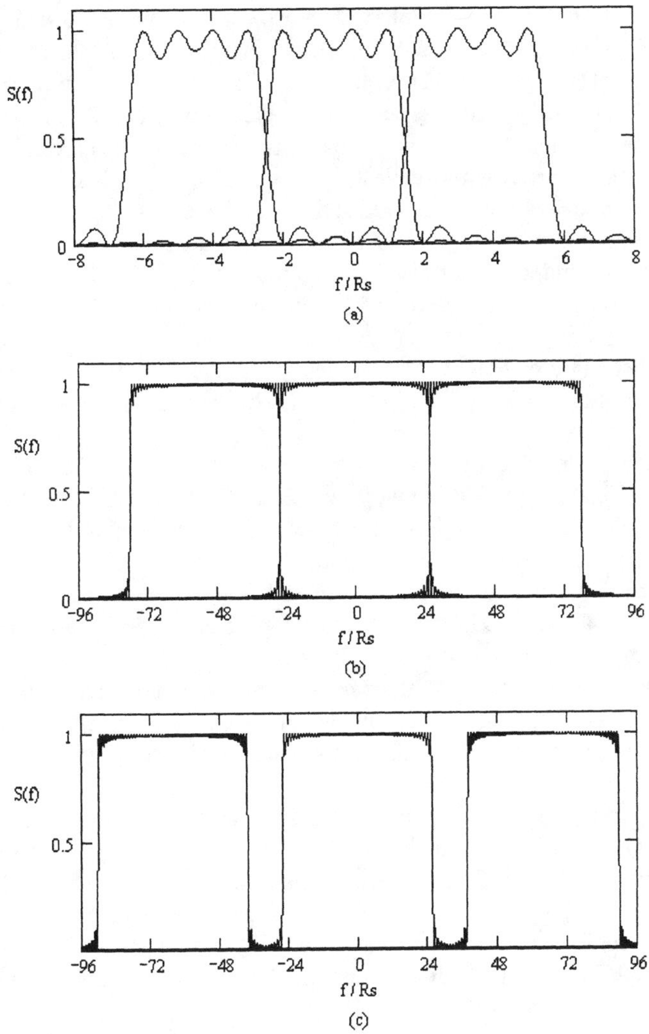

Figure 12.9 PSD of OFDM signal samples generated by $-N/2$ to $(N/2-1)$ IDFT: (a) four subcarriers without virtual carriers, (b) 52 subcarriers without virtual subcarriers, (c) 52 subcarriers plus six virtual subcarriers on each side (64 in total) (IEEE 802.11a). Note: three replicas shown, to illustrate aliasing. Also note that the PSDs are not centered at 0 due to the fact that the indexes $-N/2$ to $N/2-1$ are not centered at 0. This clearly shows when $N=4$, but is not apparent when N is very big as in (b) and (c).

intersymbol interference; and a time-domain windowing function is applied to the samples in order to reduce the sidelobes of the spectrum. We will discuss cyclic extension and windowing in later sections. Assuming perfect D/A conversion, the outputs of I-channel and Q-channel D/A blocks are just the continuous-time originals of the above two sampled signals, respectively; they are given in (12.27).

To move the baseband signal to an RF band, first $I(t)$ and $Q(t)$ are further I-Q modulated onto an IF carrier with frequency $f'_{IF} = f_{IF} + \frac{1}{2T}$. This shifts the baseband OFDM signal to the IF band centered at f_{IF} (see Appendix 12B for derivations of signal expressions at each stage of the modem). Second, an up-conversion stage (a multiplier plus a bandpass filter at f_c) is used to shift the IF OFDM signal to the RF band centered at f_c. The bandpass filter (BPF) must have a center frequency of f_c and a bandwidth of $(N+1)/T$ to allow the desired first term to pass and to reject the undesired second term whose center frequency is $f_c - 2f_{IF}$. The rejection is effective only when $f_{IF} \geq (N+1)/(2T)$. This must be satisfied when f_{IF} is chosen (refer to (12.65) in Appendix 12B).

In the demodulator, basically the reverse process happens. Figure 12.10(b) shows the demodulator. The RF OFDM signal is first down converted to an IF band and the IF OFDM signal then is demodulated to baseband OFDM signals $I(t)$ and $Q(t)$ (when noise is present, they will have distortion). The rest of the blocks are to sample, remove cyclic extension, perform N_s-point FFT, remove data on virtual carriers, and demap the signals back to binary data stream. The demapper should include a threshold detector that removes the effect of noise.

Comparing to the analog OFDM modem, the complexity of the IDFT/DFT based OFDM modem is greatly reduced. N analog subcarrier modems are replaced by the IFFT/FFT devices, only one I-Q modulator and demodulator at IF frequency is needed.

Note that even though the N_s-point DFT pair is actually performed in the modem, but all signal expressions (including $I(n)$, $I(t)$, $Q(n)$, and $Q(t)$) stay the same as those of the N-point DFT pair since data on the subcarriers beyond range $[-N/2, N/2-1]$ are zeros. In other words, as far as signal expressions are concerned, the N-point DFT pair is equivalent to the N_s-point DFT pair. Thus we will use an N-point DFT pair in many discussions and derivations throughout the rest of this chapter, unless otherwise noted.

12.3 REAL-OUTPUT DFT

For some types of channels, such as wireline channels, a baseband OFDM signal is transmitted without further modulation. However, it is not possible to transmit the complex samples generated by IDFT. Real samples are needed. There are two

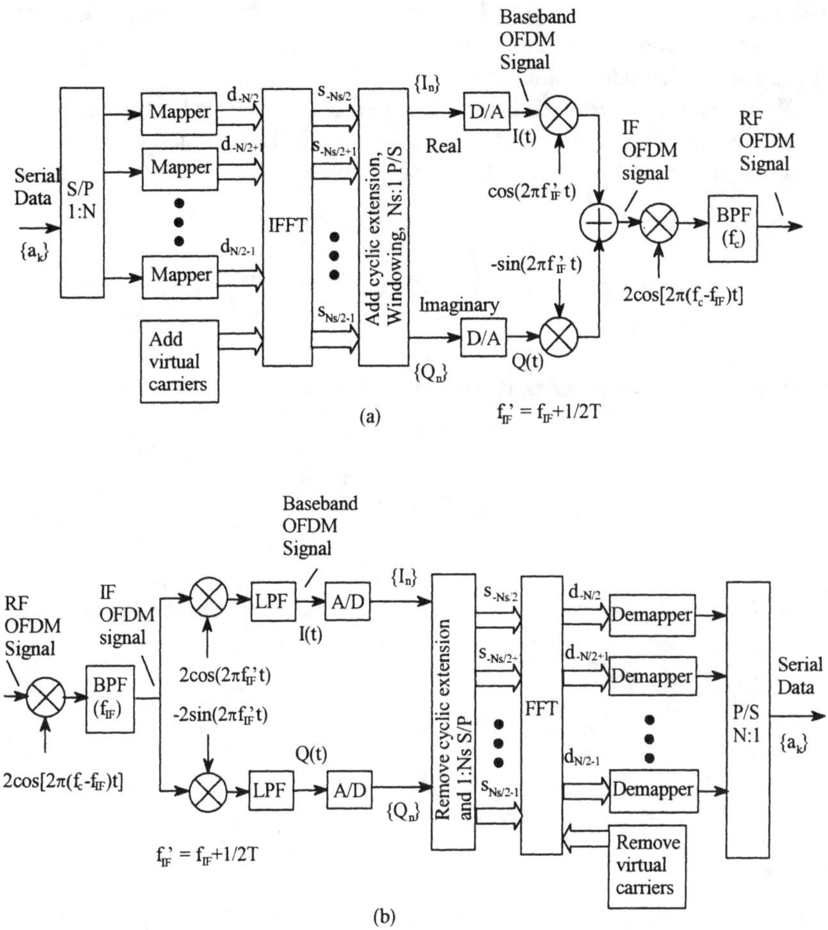

Figure 12.10 IFFT/FFT based OFDM modem: (a) IFFT based modulator and (b) FFT based demodulator.

methods to deal with this situation. One method is suggested by Weinstein and Ebert [6] where the real part of s_n is extracted and sent; at the receiver, signals are sampled at double speed and $2N$ samples are used for DFT to recover the signal. Another method is to have $2N$ data symbols when the OFDM signal is generated, where N symbols are the conjugates of the other N symbols [12]. At the receiver, the $2N$-point DFT is used to demodulate the signal.

Weinstein and Ebert's method of generating a real OFDM signal is to simply extract the real part of the OFDM signal from the IDFT output and pass it through a D/A converter. Thus the real OFDM signal is given by the first part of (12.27) as

$$I(t) = \frac{1}{N} \sum_{i=-\frac{N}{2}}^{\frac{N}{2}-1} \left(I_i \cos 2\pi \frac{i}{T} t - Q_i \sin 2\pi \frac{i}{T} t \right)$$

In the receiver, to recover data, it is sampled at intervals $\Delta t/2 = T/2N$,

$$I(k) = \frac{1}{N} \sum_{i=-\frac{N}{2}}^{\frac{N}{2}-1} \left(I_i \cos \frac{\pi i k}{N} - Q_i \sin \frac{\pi i k}{N} \right), \quad k = -N, ..., N-1$$

Using $\{I(k)\}_{k=-N}^{N-1}$ as input to a $2N$-point DFT, it can be shown (see Appendix 12C)[3]

$$\begin{aligned} s_n &= \sum_{k=-N}^{N-1} I(k) \exp(-j\frac{\pi n k}{N}) \\ &= \begin{cases} 2I_0, & n = 0 \\ I_n + jQ_n, & n = -N/2, ..., -1, 1, ..., N/2-1 \\ \text{irrelevant}, & n > N/2-1 \text{ or } n < -N/2 \end{cases} \end{aligned}$$

Thus $d_i = I_i + jQ_i$ can be recovered except for Q_0. Therefore, in this method, data for channel 0 must be real.

The second method of generating real OFDM is given in [12], where data set $\{d_i\}_{i=-N/2}^{N/2-1}$ is augmented with conjugates to form a new set $\{D_i\}_{i=-N}^{N-1}$ in the

[3] In [6], transmitted samples are produced by extracting the real parts of the output vector of an N-point DFT (not IDFT) of vector $\{2d_i\}$ (not $\{d_i\}$) and demodulation is performed by a $2N$-point DFT (but with a factor $1/(2N)$). The method here is essentially the same and the results are the same.

following manner

$$D_i = \begin{cases} \mathrm{Re}\{d_{-N/2}\}, & i=-N \\ d_{N/2+i}, & i=-1,\ -2,...,-(N-1). \\ \mathrm{Im}\{d_{-N/2}\}, & i=0 \\ d^*_{N/2-i}, & i=1,2,...,N-1 \end{cases} \tag{12.29}$$

The $2N$-point IDFT output of this new set of data are the $2N$ real-number samples of the OFDM signal (see Appendix 12D for derivations):

$$\begin{aligned} s_n &= \frac{1}{2N}\left[2\,\mathrm{Re}\left\{\sum_{i=1}^{N-1} d_{\frac{N}{2}-i}\exp(-j\frac{\pi ni}{N})\right\} + \mathrm{Re}\{d_{-\frac{N}{2}}\}(-1)^n + \mathrm{Im}\{d_{-\frac{N}{2}}\}\right] \\ n &= -N,...,N-1 \end{aligned}$$

Note that from (12.29) we see that $D_i = D^*_{-i}$, $i = 1,2,...,N-1$. It is this symmetrical relation that makes the imaginary parts cancelled. However there are no symmetrical terms of D_0 and D_{-N}, thus they must be real. Thus we can assign the real part of $d_{-N/2}$ to D_{-N} and imaginary part of $d_{-N/2}$ to D_0, as given in (12.29). This makes original data on every subcarrier complex (as required in quadrature modulations). This method is slightly better than that of Weinstein and Ebert where one channel's (channel 0) data cannot be complex.

In the above two methods, the index range of the $2N$-point IDFT must be $-N$ to $(N-1)$ to ensure no aliasing in the sampled signal's spectrum.

12.4 FFT ALGORITHMS

The fast algorithms of computing DFTs, collectively called fast Fourier transforms (FFTs), are described in many books [13,14]. In this section we briefly review the FFT. The inverse FFT can be easily obtained by modifying FFT since the only difference between DFT and IDFT is the sign of the exponent of the exponential kernel function and a scaling factor. The reader is assumed to know basic FFT already. Otherwise it is necessary to refer to other books containing detailed FFT description [13,14].

Recall that we need a $-N/2$ to $N/2-1$ DFT pair in the OFDM modem. However, the output samples of a 0 to $N-1$ IDFT are just as good since they can be converted to the output samples of the $-N/2$ to $N/2-1$ IDFT. Thus our description of FFTs and IFFTs is for 0 to $N-1$, which is what is commonly available in the literature. In describing FFT, we use conventional notations for the input and output data sets: $x(n)$ for the time-domain data (or points) and $X(k)$ for the frequency-domain data (or points). Note that when the FFT or IFFT is applied to OFDM, $X(k)$

correspond to data d_k, and $x(n)$ correspond to OFDM signal samples s_n.

The FFT is based on the strategy of decomposing the computation into successively smaller DFT computations. The decimation-in-time algorithm which is based on decomposing the time-domain sequence $x(n)$ into smaller sequences, is to be described in the following. The DFT can be written as

$$X(k) = \sum_{n=0}^{N-1} x(n)W_N^{kn}, \quad k = 0, 1, ..., N-1 \tag{12.30}$$

where for simplicity a short-hand notation is introduced:

$$W_N = \exp(-j\frac{2\pi}{N}) \tag{12.31}$$

The direct computation of (12.30) requires N^2 complex multiplications and $N(N-1)$ complex additions. To reduce the number of computations, fast algorithms are proposed. The first was by Cooley and Tukey [15].

For convenience in describing the FFT, we assume that N is an integer power of 2 (which is the most common case). Separating $x(n)$ into its even- and odd-numbered points, we have

$$\begin{aligned} X(k) &= \sum_{r=0}^{N/2-1} x(2r)W_N^{2rk} + \sum_{r=0}^{N/2-1} x(2r+1)W_N^{(2r+1)k} \\ &= \sum_{r=0}^{N/2-1} x(2r)(W_N^2)^{rk} + W_N^k \sum_{r=0}^{N/2-1} x(2r+1)(W_N^2)^{rk} \end{aligned}$$

However

$$W_N^2 = \exp(-j\frac{4\pi}{N}) = \exp(-j\frac{2\pi}{(N/2)}) = W_{N/2}$$

Thus

$$\begin{aligned} X(k) &= \sum_{r=0}^{N/2-1} x(2r)(W_{N/2})^{rk} + W_N^k \sum_{r=0}^{N/2-1} x(2r+1)(W_{N/2})^{rk} \\ &= G(k) + W_N^k H(k) \end{aligned} \tag{12.32}$$

where $G(k)$ is the $N/2$-point DFT of even-numbered data and $H(k)$ is the $N/2$-point DFT of the odd-numbered data. Although the index of k ranges from 0 to $N-1$, $G(k)$ and $H(k)$ only have to be computed for k from 0 to $N/2-1$, since they are periodic in k with period $N/2$. Thus (12.32) shows that an N-point DFT can be

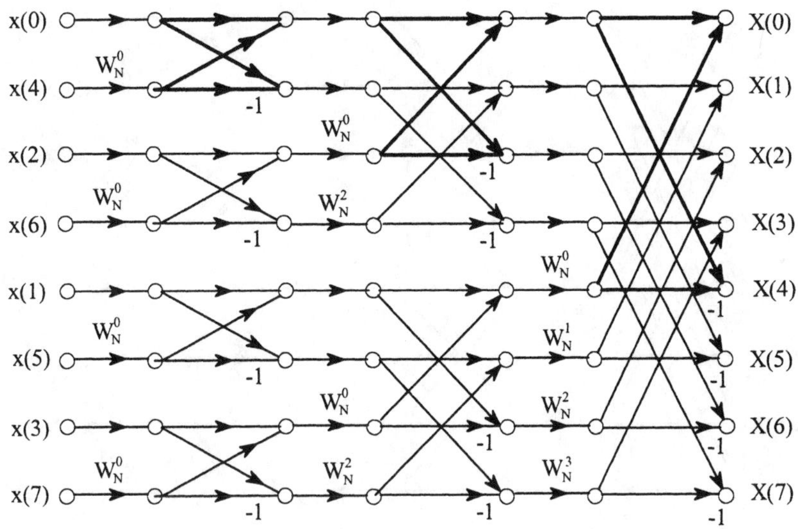

Figure 12.11 An 8-point decimation-in-time FFT.

obtained by combining two $N/2$-point DFTs. Continuing this way, a DFT eventually can be decomposed into a weighted sum of 2-point DFTs. This decomposing process can be conveniently depicted by a signal flow graph and based upon which the fast Fourier transform (FFT) algorithm is constructed. Figure 12.11 shows the flow graph of an 8-point decimation-in-time FFT. Data points are first divided into two groups $\{x(0), x(2), x(4), x(6)\}$ and $\{x(1), x(3), x(5), x(7)\}$. Each of them is further divided into two groups: $\{x(0), x(4)\}$ and $\{x(2), x(6)\}$, $\{x(1), x(5)\}$ and $\{x(3), x(7)\}$. These are listed on the left. By repeated use of (12.32), the flow graph can be drawn.

FFT also can be based on a flow graph of decimation-in-frequency DFT, where $X(k)$ is divided into smaller and smaller subsequences in the same manner. Figure 12.12 shows the flow graph of an 8-point decimation-in-frequency FFT. For details of decimation-in-frequency FFTs, refer to [13, 14]. Flow graphs for FFT can be easily modified to be used for IFFTs simply by replacing all W_N by W_N^{-1} and multiplying the results by a factor $1/N$.

The number of complex multiplications and additions of an N-point FFT (or

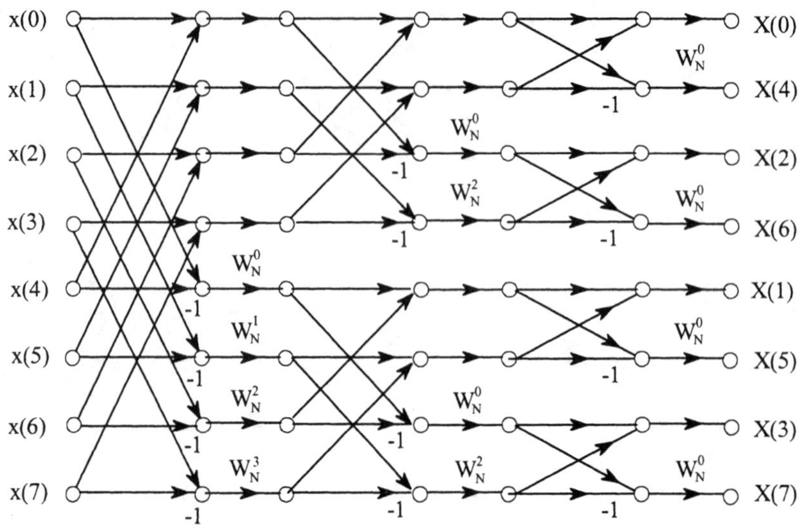

Figure 12.12 An 8-point decimation-in-frequency FFT.

IFFT) shown in Figure 12.11 or Figure 12.12 can be easily figured out. From Figure 12.11 we can see that at each stage of the flow graph, always two inputs are needed to calculate two outputs (heavy lines). That is, the entire flow graph is constructed by a number of this two-input-two-output sub-graph, which in fact is a two-point DFT and is commonly called a radix-2 *butterfly*. From Figure 12.11 we can see that there are $N/2$ radix-2 butterflies at each stage and there are total $\log_2 N$ stages; thus there are $\frac{N}{2}\log_2 N$ butterflies. Each butterfly requires one complex multiplication (by *twiddle* factors W_N^r, $r = 0, 1, 2, 3$ in Figure 12.11) and two complex additions. The multiplication by -1 in the radix-2 butterfly is just a change of sign and is not considered a multiplication. Further, multiplying by $W_N^0 = 1$ is also not a multiplication so that it can be excluded. There are $N/2$ of W_N^0 at the first stage and fewer at other stages, which will be ignored. Thus the total complex multiplications and complex additions are $\frac{N}{2}\log_2 N - \frac{N}{2}$ and $N\log_2 N$, respectively. A complex multiplication requires four real multiplications and two real additions and a complex addition requires two real additions. Thus the total number of real multiplications and real additions are

$$\#MUL_{COOLEY-TUKEY} = 2N \log_2 N - 2N \tag{12.33}$$

$$\#ADD_{COOLEY-TUKEY} = 3N \log_2 N - N \tag{12.34}$$

Since a multiplication takes much more resources than an addition, efforts have been directed toward reducing the number of multiplications. Two algorithms are worth mentioning. One of the algorithms is the radix-4 FFT which is based on decomposing DFT into a number of four-point DFT or radix-4 butterflies in the flow graph [16]. Radix-4 FFT requires the same number of additions and a reduced number of complex multiplications: $\frac{3}{8}N \log_2 N - \frac{3}{4}N$. Its number of multiplications is approximately $1/8$ less than the regular FFT. Another algorithm is the split-radix FFT (SRFFT) which is much more efficient [17, 18]. In this algorithm, a N-point DFT is first decomposed into one $N/2$-point DFT and two $N/4$-point DFTs. From the definition of DFT in (12.30), we have

$$X(2k) = \sum_{n=0}^{N/2-1} [x(n) + x(n + N/2)] W_N^{2nk} \tag{12.35}$$

for the even index terms, and

$$X(4k+1) = \sum_{n=0}^{N/4-1} [(x(n) - x(n+N/2)) - j\,(x(n+N/4) - x(n+3N/4))] W_N^n W_N^{4nk} \tag{12.36}$$

and

$$X(4k+3) = \sum_{n=0}^{N/4-1} [(x(n) - x(n+N/2)) + j\,(x(n+N/4) - x(n+3N/4))] W_N^{3n} W_N^{4nk} \tag{12.37}$$

for the odd index terms. This result shows that an N-point DFT can be decomposed into one $N/2$-point DFT (12.35) and two $N/4$-point DFTs ((12.36) and (12.37)). This process is repeated for the $N/2$-point DFT and $N/4$-point DFTs until scalars result. This process is very similar to the decimation-in-frequency Cooley Tukey algorithm.

The simplest SRFFT is the 4-point SRFFT. Its flow graph is shown in Figure 12.13(a) which can be easily verified using (12.35)–(12.37). Figure 12.13(b) shows an 8-point SRFFT [19]. It is seen that the flow graph of SRFFT looks like a radix-2 FFT except for the location of the twiddle factors. As a matter of fact, it is the location of the twiddle factors that makes this algorithm different.

It has been shown that the split-radix FFT with the best programming (three-butterfly program [24]) requires

$$\#MUL_{SPLT-RADIX} = N\log_2 N - 3N + 4 \tag{12.38}$$

real multiplications and

$$\#MUL_{SPLT-RADIX} = 3N\log_2 N - 3N + 4 \tag{12.39}$$

real additions. Compared with Cooley-Tukey FFT, the number of additions remains almost the same and the number of multiplications is reduced to less than half. More FFT algorithms that reduce the number of multiplications exist [13, Chapter 5]; however, they are not as efficient as the split-radix FFT.

12.5 PARTIAL FFT ALGORITHMS

In some systems, partial FFT-based OFDM may be a cost-effective solution. For example, in a multiple-access system, transmitters and receivers may have different capacity requirements. In the downlink, one transmitter sends the same composite signal to all receivers, but each receiver may be interested in a fraction of the data. In this case, the transmitter needs high capacity and each receiver needs small capacity. Alternately, in the uplink, multiple transmitters may simultaneously send separated signals to one receiver. In this scenario, the receiver needs high capacity and each transmitter needs small capacity. In both cases, if the transmitter and the receiver are designed to have the same capacity, the cost will be unnecessarily high. Digital broadcasting is an example of the downlink scenario, where the transmitter needs a high capacity, and can be equipped with a full-size FFT-based OFDM, say, N-point FFT OFDM. Each receiver needs different capacity, and can be equipped with a partial FFT-based OFDM. A partial FFT (or IFFT) computes only L outputs from all N inputs, where $L < N$; or it computes all N outputs from only L inputs. The aforementioned downlink case is an example of the first case, and the case where zeros are padded in the input for the FFT-based OFDM is the second case. Therefore, a partial FFT not only can be applied to the downlink but also to the uplink in the digital broadcasting scenario.

Two major efficient methods of computing partial DFT are the pruning method

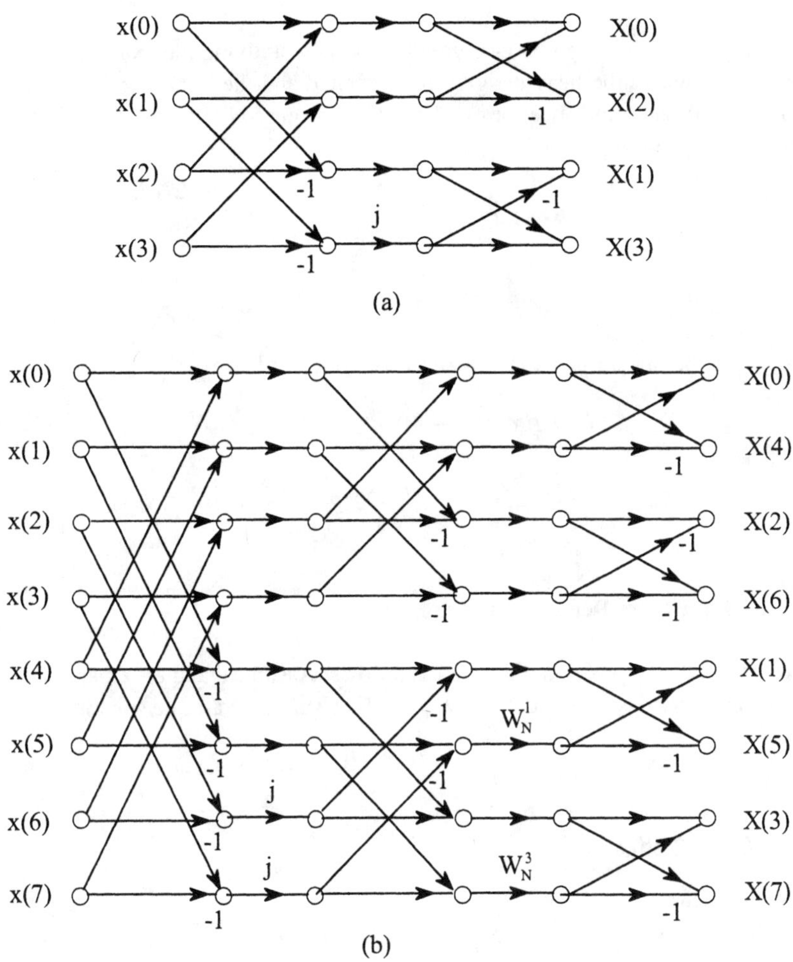

Figure 12.13 SRFFT flow graphs for (a) $N = 4$, and (b) $N = 8$. After [19].

[20–23] and the transform decomposition (TD) method [24].

12.5.1 The Pruned Partial FFT

The pruned partial FFT can be easily understood by studying the example given in Figure 12.14, where the heavy edges are computed and the light edges are pruned. The number of real multiplications and additions required by the pruning algorithm is [24]

$$\#MUL_{PRUNE} = 2N \lfloor \log_2 L \rfloor + 2N - 4L + \frac{2NL}{2^{\lfloor \log_2 L \rfloor}}$$

$$\#MUL_{PRUNE} = 3N \lfloor \log_2 L \rfloor + 3N - 6L + \frac{3NL}{2^{\lfloor \log_2 L \rfloor}}$$

where $\lfloor x \rfloor$ is the integer part of x. When L is a power of two,

$$\#MUL_{PRUNE} = 2N \log_2 L - 4L + 4N \tag{12.40}$$

$$\#ADD_{PRUNE} = 3N \log_2 L - 6L + 6N \tag{12.41}$$

12.5.2 Transform Decomposition

The transform decomposition is derived as follows. Assume that there exists a integer P such that P divides N and define $Q = N/P$. Using the variable substitution

$$n = Qn_1 + n_2, \quad \begin{array}{l} n_1 = 0, 1, ..., P-1 \\ n_2 = 0, 1, ..., Q-1 \end{array}$$

we can write (12.30) as

$$\begin{aligned} X(k) &= \sum_{n_2=0}^{Q-1} \sum_{n_1=0}^{P-1} x(n_1 Q + n_2) W_N^{(n_1 Q + n_2)k} \\ &= \sum_{n_2=0}^{Q-1} \left[\sum_{n_1=0}^{P-1} x(n_1 Q + n_2) W_N^{(n_1 Q)k} \right] W_N^{n_2 k} \end{aligned}$$

where

$$W_N^{(n_1 Q)k} = \exp(-j\frac{2\pi}{N} n_1 Qk) = \exp(-j\frac{2\pi}{N} n_1 \frac{N}{P} k) = W_P^{n_1 <k>_P}$$

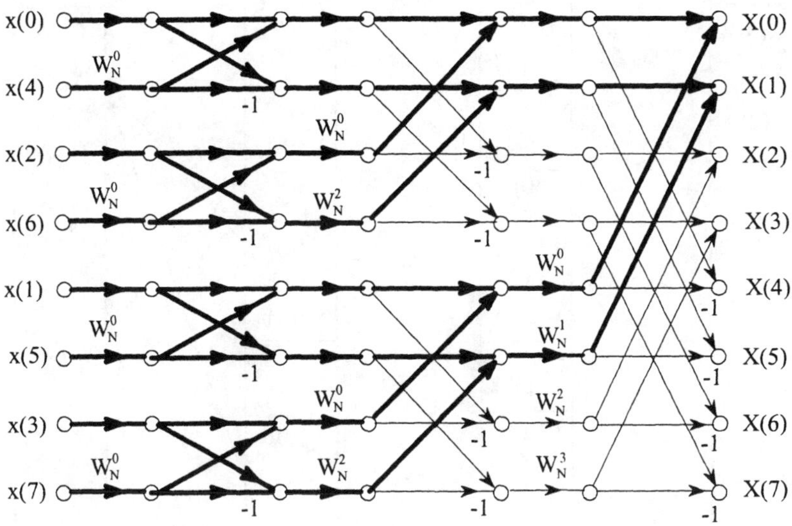

Figure 12.14 An 8-point pruned FFT for a subset of two output points.

where $< k >_P$ denotes reduction modulo-P, that is if $k = lP+k'$, then $< k >_P= k'$. Thus

$$\begin{aligned} X(k) &= \sum_{n_2=0}^{Q-1}\left[\sum_{n_1=0}^{P-1} x(n_1Q+n_2)W_P^{n_1<k>_P}\right]W_N^{n_2k} \\ &= \sum_{n_2=0}^{Q-1} X_{n_2}(< k >_P)W_N^{n_2k} \end{aligned} \tag{12.42}$$

where

$$\begin{aligned} X_{n_2}(i) &= \sum_{n_1=0}^{P-1} x(n_1Q+n_2)W_P^{n_1i} \\ &= \sum_{n_1=0}^{P-1} x_{n_2}(n_1)W_P^{n_1i}, \quad i=0,1,...,P-1 \end{aligned} \tag{12.43}$$

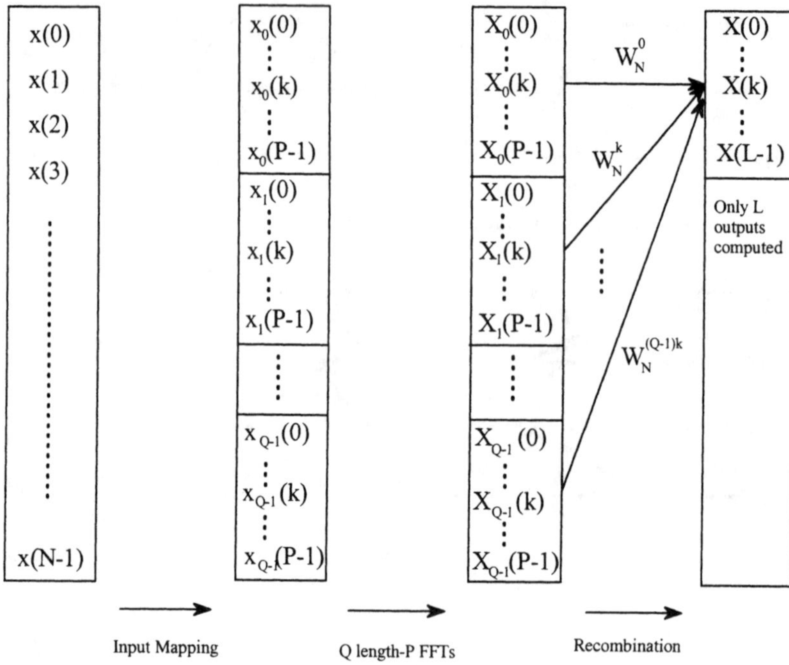

Figure 12.15 Block diagam of computing the transform decomposition FFT. After [24].

where

$$x_{n_2}(n_1) = x(n_1 Q + n_2) \tag{12.44}$$

Inspecting the above three equations, we can see the TD algorithm is as follows: The input sequence is decomposed into subsequences according to (12.44) $\{x_0(n_1)\}_{n_1=0}^{P-1}$, $\{x_1(n_1)\}_{n_1=0}^{P-1}$,..., $\{x_{Q-1}(n_1)\}_{n_1=0}^{P-1}$ (also refer to Figure 12.15 for a block diagram of TDFFT [24]). Each subsequence is transformed using a P-point DFT according to (12.43). After all Q sequences are transformed; the results are $\{X_0(i)\}_{i=0}^{P-1}$, $\{X_1(i)\}_{i=0}^{P-1}$,..., $\{X_{Q-1}(i)\}_{i=0}^{P-1}$. The kth element of each of them will be used to compute $X(k)$ for any k using (12.42).

Since the P-point DFT can be computed using any efficient FFT, such as the split-radix algorithm, this transform decomposition method outperforms the pruning

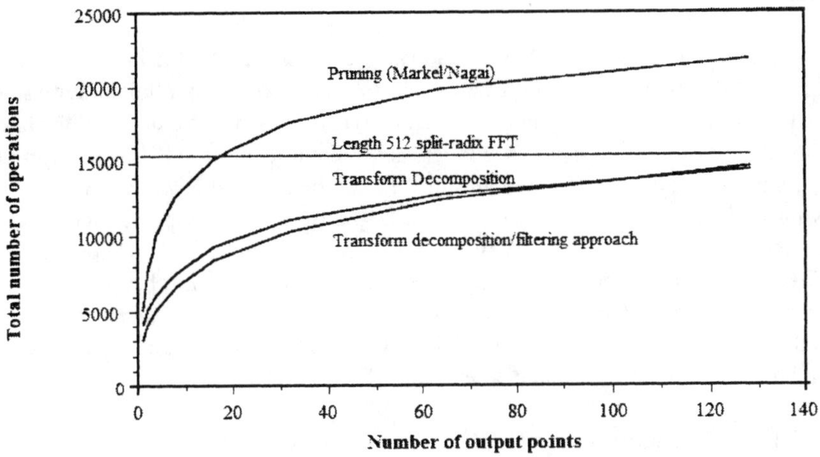

Figure 12.16 Computational complexities of FFT algorithms. From [24]. Copyright © 1993 IEEE.

method which must use a simple one-butterfly radix-2 FFT. If N is a power of two (and hence P is a power of two), the P-point FFT is the three-butterfly SRFFT whose computational complexity is given by (12.38) and (12.39), and the final computation of $X(k)$ for a small L value of k is a direct computation of (12.42), then the number of real multiplications and additions is [24]

$$\#MUL_{TD} = N \log_2 P - 3N + 4(L+1)\frac{N}{P} - 4L$$

$$\#ADD_{TD} = 3N \log_2 P - 3N + 4(L+1)\frac{N}{P} - 4L$$

Figure 12.16 compares the total number of real computations (multiplications and additions) of FFT algorithms for various numbers of output points. The lowest curve is for the TD-FFT with filtering which is one of the four variations of TD-FFTs given in [24]. It is only slightly better than the standard TD-FFT, and we will not elaborate on it here. It is seen that TD-FFTs are much better than the pruned FFTs for any L. When $L \geq N/2$, it is more efficient to compute all N output points using the split-radix FFT algorithm.

12.6 CYCLIC EXTENSION

As we mentioned earlier, one of the advantages of using OFDM instead of single-carrier modulation is that an OFDM signal has a much longer symbol duration so that the relative length of intersymbol interference (ISI) caused by a finite-length channel impulse response[4] $h(t)$ is much smaller; hence the effect of the ISI is much smaller. To remove the effect of ISI, a guard time T_g can be inserted between OFDM signal symbols.[5] T_g should be chosen so that it is greater than the expected length of the channel impulse response or delay spread. If the actual length of $h(t)$ or the delays of the multipath components is smaller than T_g, the ISI does not happen. However, interchannel interference (ICI) still exists. To avoid ICI, a cyclic extension is added to the signal in the guard interval. Using cyclic extension to avoid ICI can be done for both continuous-time and discrete-time OFDM signals.

12.6.1 Continuous-Time OFDM

We first study the continuous-time signal case. Figure 12.17 shows how cyclic extension is added as a prefix. The heavy-line waveform is $\cos(2\pi\frac{2}{T}t), 0 \leq t \leq T$ with $T = 1$. The segment in $(-T_g, 0)$ is the cyclic extension of the segment framed in the heavy dashed lines. With the cyclic extension, the new waveform is continuous at the connecting point $t = 0$, so that the new waveform expression is the same but its starting point is earlier at $t = -T_g$. That is, $\cos(2\pi\frac{2}{T}t), -T_g \leq t \leq T$. The light line is a multipath signal with a delay $\tau < T_g$. We can see that both the main waveform and the delayed multipath waveform have integer periods in $[0, T]$. It is this property that ensures the orthogonality between the multipath components of an OFDM signal, provided that the delay spread is less than T_g.

Mathematically, this can be shown as follows. To illustrate the concept, we assume a simple time-invariant multipath channel[6] whose complex impulse response

---

[4] An ideal channel would have an impulse response of zero length—that is, an impulse or delta function. A practical channel always has an impulse response with a finite length. Multipath channel is one of them, whose impulse response is a sum of several delta functions. Each has a different strength, different phase, and different delay, and they are usually time-varying. The length of the impulse response of the multipath channel is the length of the delay spread.

[5] Assuming the original OFDM symbol duration is T, then with a guard interval T_g, the total symbol duration is $T + T_g$. However, the subcarrier spacing of OFDM is still $1/T$, not $1/(T + T_g)$. This is because the integration or DFT for demodulation is still performed for a time length T. To ensure orthogonality between subcarriers, the spacing must be $1/T$.

[6] We assume that the channel is time invariant for at least a symbol period, which is called a quasi-stationary channel. A more realistic channel model is the time-varying multipath channel. Performance of OFDM in various fading channels will be studied in Chapter 15.

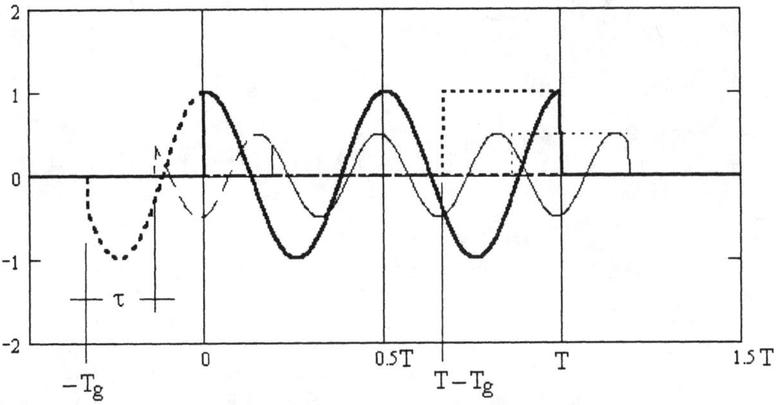

Figure 12.17 Cyclic extension.

is [25]

$$h(t) = \sum_{k=0}^{K} \alpha_k e^{-j\beta_k} \delta(t - \tau_k) \tag{12.45}$$

where α_k and β_k are the random amplitude and phase introduced by the channel. $\tau_k \leq T_g$ are the delays. Then the complex input of the correlator of the demodulator (Figure 12.7) is

$$\widetilde{x}(t) = \widetilde{s}(t) \otimes h(t) = \sum_{k=0}^{K} \alpha_k e^{-j\beta_k} \widetilde{s}(t - \tau_k)$$

where $\widetilde{s}(t)$ is the complex OFDM signal given in (12.17) from which we have

$$\widetilde{x}(t) = \sum_{k=0}^{K} \alpha_k e^{-j\beta_k} \sum_{j=0}^{N-1} d_j \exp\left[j2\pi f_j(t - \tau_j)\right]$$

Now we examine the output of the correlator of the demodulator. Take the I-channel as an example (similar in the Q-channel). The real part of $\widetilde{x}(t)$ is present at the input of the I-channel correlator. The ith correlator output is

$$I_i(t) = \int_0^T s_i(t)x(t)dt$$

$$= \int_0^T s_i(t) \sum_{k=0}^{K} \sum_{j=0}^{N-1} \alpha_k A_j \cos\left[2\pi f_j(t-\tau_j)+\phi_j+\beta_k\right] dt$$

For a particular j,

$$\begin{aligned} I_{ij}(t) &= \sum_{k=0}^{K} \alpha_k \int_0^T A_i \cos(2\pi f_i t + \phi_i) A_j \cos(2\pi f_j(t-\tau)+\phi_j+\beta_k) dt \\ &= \begin{cases} \sum_{k=0}^{K} \alpha_k A_0^2 T \cos\phi_0 \cos(\phi_0+\beta_k), & i=j=0 \\ \frac{1}{2}\sum_{k=0}^{K} \alpha_k A_i^2 T \cos(2\pi f_j \tau - \beta_k), & i=j\neq 0 \\ 0, \quad i\neq j \end{cases} \end{aligned} \quad (12.46)$$

where we assume that $f_i = \frac{i}{T}$. Thus orthogonality between $s_i(t)$ and $s_j(t-\tau)$ is preserved. Note that without the cyclic extension (assume that guard time still exists), the integration interval would be $[\tau, T]$ which will not ensure orthogonality. Now, although the orthogonality is preserved, the correlator output is $\sum_{k=0}^{K} \frac{1}{2}\alpha_k A_i^2 T \cos(2\pi f_j \tau - \beta_k)$ when $i = j \neq 0$, which is a function of the delay, random amplitude, and phase. When $(2\pi f_j \tau - \beta_k)$ is equal to or close to $k\pi/2$, k integer, the correlator output will be zero or near zero. This is what we do not want to see. Therefore channel estimation is needed and correction (equalization) can be made if the channel characteristic has been estimated. Channel estimation and equalization for OFDM signals will be discussed later in Chapter 15.

12.6.2 Discrete-Time OFDM

For discrete-time OFDM, the derivation and results are somewhat different. This derivation follows that in [26]. Assume that the discrete channel response is

$$\mathbf{h} = \{h_0,\ h_1, ..., h_{M-1}\} \quad (12.47)$$

which in general is complex.

If $\mathbf{h}$ has a length of $M \leq N$, then the cyclic extension adds $M-1$ samples in front of the original symbol. That is

$$\begin{aligned} \text{Original symbol} &: \quad \mathbf{s} = \{s_0, s_1, ..., s_{N-1}\} \\ \text{Extended symbol} &: \quad \mathbf{s}' = \{s_{N-M+1}, s_{N-M+2}, ..., s_{N-1}, s_0, s_1, ..., s_{N-1}\} \end{aligned}$$

For example $M = 3$, $N = 6$, we have

$$\begin{aligned} \text{Original symbol} &: \quad \mathbf{s} = \{s_0, s_1, s_2, s_3, s_4, s_5\} \\ \text{Extended symbol} &: \quad \mathbf{s}' = \{s_4, s_5, s_0, s_1, s_2, s_3, s_4, s_5\} \end{aligned}$$

$$\text{Impulse Response} \quad : \quad \mathbf{h} = \{h_0, h_1, h_2\}$$

The output of the channel is the linear convolution of $\mathbf{s}'$ with $\mathbf{h}$. Assuming the signal input to the channel starts at time $n = 0$, we have

$$x_0 = s_0 h_0 + s_5 h_1 + s_4 h_2$$

where the samples before $n = 0$, s_5 and s_4, have an effect on the current output due to the length of h_n. Similarly, at $n = 1$,

$$x_1 = s_0 h_1 + s_1 h_0 + s_5 h_2$$

where x_1 contains the effect of circular extension terms s_5. To express all samples of the output, we can write the samples of the channel output (received signal) as [26]

$$\mathbf{x} = \mathbf{s} \begin{bmatrix} h_0 & h_1 & h_2 & 0 & 0 & 0 \\ 0 & h_0 & h_1 & h_2 & 0 & 0 \\ 0 & 0 & h_0 & h_1 & h_2 & 0 \\ 0 & 0 & 0 & h_0 & h_1 & h_2 \\ h_2 & 0 & 0 & 0 & h_0 & h_1 \\ h_1 & h_2 & 0 & 0 & 0 & h_0 \end{bmatrix} \tag{12.48}$$

From above, the samples of output $\mathbf{x}$ are determined only by the current symbol $\mathbf{s}$. That is, there is no ISI. Equation (12.48) is in fact equivalent to the circular convolution of $\mathbf{s}$ and the response $\mathbf{h}$:

$$\mathbf{x} = \mathbf{s} \otimes \mathbf{h}$$

where $\otimes$ denotes circular convolution, or explicitly

$$x_k = \sum_{n=0}^{N-1} s^c_{k-n} h^c_n, \qquad k = 0, 1, ...N-1 \tag{12.49}$$

where $\{s^c_n\}$ and $\{h^c_n\}$ are the circular extensions of $\{s_n\}$ and $\{h_n\}$, respectively. And $\{s^c_{k-n}\}$ is the time reversed and shifted version of $\{s^c_n\}$. For the above example, we have

$$\{s^c_n\} = \{\cdots\cdots|s_0, s_1, ..., s_4, s_5|s_0, s_1, ..., s_4, s_5|s_0, s_1, ..., s_4, s_5|\cdots\cdots\}$$

$$\{s^c_{-n}\} = \{\cdots\cdots|s_5, s_4, ..., s_1, s_0|s_5, s_4, ..., s_1, s_0|s_5, s_4, ..., s_1, s_0|\cdots\cdots$$

$$\{h^c_n\} = \{\cdots\cdots|h_0, h_1, h_2, 0, 0, 0|h_0, h_1, h_2, 0, 0, 0|h_0, h_1, h_2, 0, 0, 0|\cdots\cdots\}$$

Note that $h_3 = h_4 = h_5 = 0$. So for this example, the summation in (12.49) only needs to calculate three terms instead of six terms.

Equation (12.49) is an N-point circular convolution. According to DFT theorem, we have the following pair

$$x_n = s_n \otimes h_n \Longleftrightarrow H_k S_k = X_k \tag{12.50}$$

where S_k and H_k are the N-point DFTs of s and $\{h_n\}$, respectively. After demodulation (DFT), the kth channel output is $X_k = H_k S_k$, thus

$$S_k = \frac{X_k}{H_k} \tag{12.51}$$

and the effect of the channel is removed. This is what a guard time and cyclic extension can achieve when ISI is present. Without guard time, x_0 and x_1 would have effects from the previous symbol which cannot be removed after DFT. Without cyclic extension, x_n would not be a circular convolution of s_n and h_n, and (12.50) would not hold; S_k cannot be recovered using (12.51). The estimation of H_k will be discussed in Chapter 15.

12.7 SPECTRUM SHAPING

The spectrum of the transmitted OFDM signal must satisfy a certain spectral mask to avoid interfering with other signals in adjacent communications bands. For example, Figure 12.18 shows the requirement of the IEEE 802.11a standard for an OFDM with $N = 64$ [10], where the horizontal axis is $f - f_c$. From the figure it is seen that the one-sided 3-dB bandwidth is 9 MHz, and at 30 MHz the attenuation must be 40 dB. This requires a 40 dB attenuation at a normalized frequency of $30/9 = 3.33$. This requirement cannot be met with the original OFDM signal with a rectangular pulse shape.

To reduce the sidelobes of the spectrum of the OFDM signal, or to shape its spectrum in general, three approaches can be considered. One approach is to shape the OFDM signal symbol by multiplying the symbol with a time-domain window function. This is commonly referred as "windowing," which is particularly convenient for discrete OFDM samples. Another approach is to pass the data through a pulse-shaping filter before modulating them onto subcarriers. This method can be referred to as "pulse shaping," even though windowing is also a type of pulse shaping. The third approach is to filter the OFDM signal after it is generated. We refer to this method as "filtering." For the OFDM signal generated by the analog method, which is rare, either one of the three approaches can be used, with the pulse-shaping method

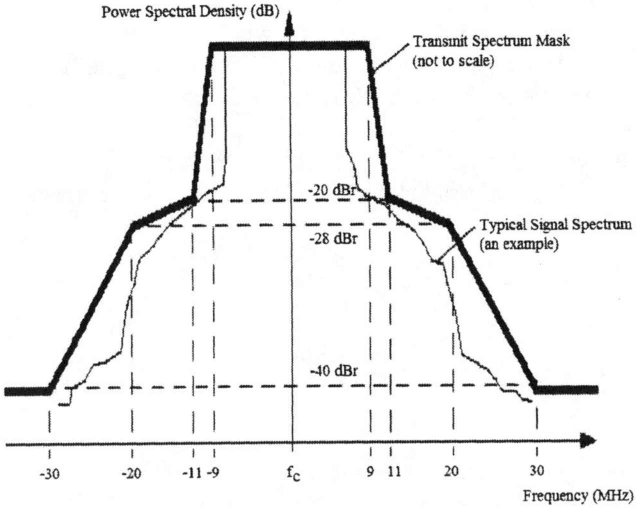

Figure 12.18 IEEE802.11a spectral mask. From [10]. Copyright © 1999 IEEE.

being preferable since it can be easily done on the data. In the DFT-based OFDM, which is common, the IDFT is followed by a digital-to-analog convertor that includes an analog low-pass filter (see Figure 12.10). The filter is to remove images of the spectrum of the analog OFDM signal (see Figure 12.9). In this case, since filtering is unavoidable, the filter could be designed by taking into consideration the sidelobe reduction as well, making the windowing or pulse-shaping unnecessary. However, windowing is commonly used in DFT-based OFDM systems despite the existence of the low-pass filter [10, 12, 27]. One of the reasons behind this is that windowing only involves a few samples on the edges of the OFDM symbol, which can be easily done. This greatly relieves the burden on the requirements to the analog low-pass filter. Another reason is that to preserve perfect orthogonality between subcarriers, the samples within the DFT interval must not be altered during spectral shaping; this can be easily controlled by windowing, but not by filtering. Thus the OFDM signal's spectral shape should be (and in practice is essentially) determined by the window function rather than the low-pass filter in the DAC.

In this section we will describe windowing in detail and provide only references for filtering of OFDM signals. Windowing is used also for other purposes in addition to reducing the out-of-band spectral energy, such as controlling ICI caused by frequency offset in the OFDM carrier. We will also present window functions for those cases.

In the following we will discuss several window functions. Even though windowing can be used in both analog and digital OFDM signal generations, the window functions will be presented in the continuous-time form for convenience. At the end of this section, we will discuss the effect of using a discrete window function when implementing OFDM using IDFT.

A common choice of window function is the raised cosine function given as

$$w(t) = \begin{cases} 0.5\left[1 - \cos\left[\frac{(2t+T_s+\beta T_s)\pi}{2\beta T_s}\right]\right], & \frac{-(1+\beta)T_s}{2} \leq t \leq \frac{-(1-\beta)T_s}{2} \\ 1.0, & \frac{-(1-\beta)T_s}{2} \leq t \leq \frac{(1-\beta)T_s}{2} \\ 0.5\left[1 + \cos\left[\frac{(2t-T_s+\beta T_s)\pi}{2\beta T_s}\right]\right], & \frac{(1-\beta)T_s}{2} \leq t \leq \frac{(1+\beta)T_s}{2} \\ 0, & \text{otherwise} \end{cases} \tag{12.52}$$

where β is the roll-off factor which represents the percentage of the roll-off segment on each side of the window (or pulse-shaping) function, and $T_s = T + T_g$ is the symbol duration.

The total length of the pulse, $(1+\beta)T_s$, is longer than the symbol duration. This has two consequences. First, overlap between adjacent symbols must be allowed to keep the original symbol rate. Second, copying the last length-T_g portion of the symbol and adding it as the prefix of the symbol only extends the symbol length to T_s, not the required $(1+\beta)T_s$. The solution is to copy the first length-βT_s portion of the symbol and append it as a postfix [27] (Figure 12.19). In this way the symbol is cyclically extended in both directions. For perfect demodulation, the orthogonality between subcarriers must be maintained. This requires that only the signal parts with constant envelop be used in demodulation. From Figure 12.19 we can see that the postfix cannot be used, and the first length-βT_s segment of the prefix cannot be used either. The usable segment now has the length of $T + T_g - \beta T_s$. This means the guard time is effectively reduced by βT_s.

With prefix and postfix and window function defined above, from (12.1), the baseband OFDM signal can be expressed as

$$s(t) = \sum_{i=0}^{N-1} A_i \cos(2\pi f_i t + \phi_i) w(t - \frac{T_s}{2}), \quad 0 \leq t \leq T_s \tag{12.53}$$

or in complex envelope notation, from (12.17), as

$$\tilde{s}(t) = \sum_{i=0}^{N-1} d_i \exp(j2\pi \frac{i}{T} t) w(t - \frac{T_s}{2}), \quad 0 \leq t \leq T_s \tag{12.54}$$

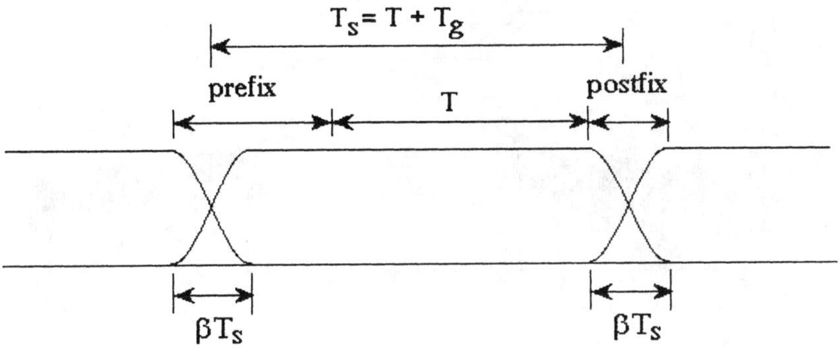

Figure 12.19 OFDM symbol sequence with cyclic extension and windowing.

where $w(t)$ is the window function.

It can be shown that the Fourier transform of $w(t)$ is [25, p. 536]

$$W(f) = \frac{T_s \cos(\pi f \beta T_s)}{1 - (2f\beta T_s)^2} \frac{\sin(\pi f T_s)}{\pi f T_s} \tag{12.55}$$

Following the argument in deriving the power spectral density of the unwindowed baseband OFDM in Section 12.1.1, the PSD of the windowed OFDM signal is just a sum of PSDs of all subcarriers. That is,

$$S_w(f) = \sum_{i=0}^{N-1} |W(f - f_i)|^2 \tag{12.56}$$

Using the raised-cosine window with $\beta = 0.025$, the PSD of the 64-channel QAM-OFDM of the IEEE 802.11a standard is shown in Figure 12.20. The OFDM of IEEE 802.11a is implemented by a 64-point IDFT, but only 52 subcarriers are used. The channel separation is $1/3.2\ \mu s = 0.3125$ MHz. The other 12 carriers are just null carriers. The unwindowed PSD (the wider one) is also shown for comparison. The difference is obvious. Without windowing, the 40-dB attenuation requirement at 30 MHz from the carrier frequency cannot be met. With windowing, this requirement is considerably exceeded. As discussed above, this is achieved without sacrificing the orthogonality between subcarriers and without decreasing the symbol rate. However, the guard time interval is shrunk by βT_s, which is only 2.5% of the total

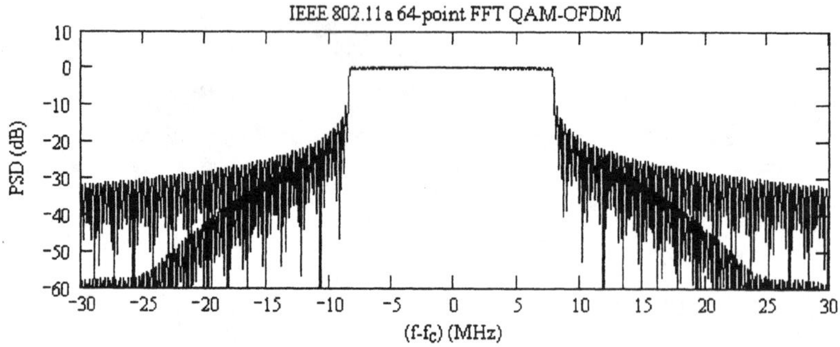

Figure 12.20 PSD of 64-point QAM-OFDM of IEEE 802.11a standard in comparison with the unwindowed PSD.

symbol interval in the IEEE 802.11a case.

Carrier frequency offset or Doppler frequency shift causes ICI in the OFDM system. Many authors tried to deal with the ICI problem by using new window functions or pulse-shaping filters. A new simple window function called "better than raised-cosine pulse" (btrc) is proposed in [28] to reduce ICI due to carrier frequency offset, which is given as

$$w_{btrc}(t) = \begin{cases} \frac{1}{T}, & 0 \leq |t| \leq \frac{(1-\beta)T_s}{2} \\ \frac{1}{T}\exp[\frac{-2\ln 2}{\beta T_s}][|t| - \frac{(1-\beta)T_s}{2}], & \frac{(1-\beta)T_s}{2} \leq |t| \leq \frac{T_s}{2} \\ \frac{1}{T}\left\{1 - \exp[\frac{-2\ln 2}{\beta T_s}][\frac{(1+\beta)T_s}{2} - |t|]\right\}, & \frac{T_s}{2} \leq |t| \leq \frac{(1+\beta)T_s}{2} \\ 0, & \text{otherwise} \end{cases} \tag{12.57}$$

In [28] it is shown that in comparison with the raised-cosine window this window function reduces ICI power to various extents, depending on the value of β and other system parameters.

Some other more sophisticated window functions or pulse-shaping filters are reported in [29–36], and a post-IDFT filtering method is proposed in [37].

For a DFT-based OFDM signal, the spectrum of the OFDM signal is determined by the pulse shape and the low-pass filter. If the low-pass filter is ideal, the main lobe will remain intact, and the sidelobes of the original spectrum of the continuous-time signal are lost. If the low-pass filter is not ideal, like any practical filter, not only are the sidelobes lost, the main lobe will also be distorted. This means the spectrum of the OFDM signal generated by the analog method and the spectrum of the DFT-based OFDM signal are different in general for the same pulse shape. This was noticed in [38], and doubts have arisen regarding the validity of the existing results on analysis of windowing or filtering based on the analog representation of the window functions or filter responses in the literature. However, we can see that with a proper design of the low-pass filter that ensures minimal mainlobe distortion, what is lost or reduced are only the sidelobes. This makes the sidelobes smaller than the ones of the analog OFDM signal (assuming the same pulse shape). Thus, if a window function or filter response improves roll-off property of the spectrum based on the analog representation, it also improves it for DFT-based OFDM signals. However, some other results that were obtained with the analog representation of the window function, such as the effect of window functions or filter responses on the ICI caused by the frequency offset, might not be valid [38].

Finally, before ending this section, we need to point out that not all OFDM systems need spectrum shaping. Earlier in this chapter, we have seen from Figure 12.3 that the transition bands are getting sharper and sidelobes are getting lower as N increases. When the number of channels is very big, like in the case of broadcasting systems where N could be in the hundreds or more [39], the roll-off of the spectrum can satisfy the spectral mask without spectrum shaping.

12.8 SUMMARY

In this chapter we have studied OFDM basics. OFDM is a type of multicarrier modulation where adjacent subcarriers are separated in frequency by the symbol rate for orthogonality between subcarriers. By transmitting data in parallel on multiple carriers, OFDM saves bandwidth and reduces symbol rate so that it can tolerate longer time dispersion and frequency-selective fading in the channel. We have studied OFDM signal's PSD. When N approaches infinity, the bandwidth normalized to the symbol rate approaches 1. That is, OFDM is asymptotically a Nyquist-rate (maximum rate) transmission technique. The OFDM signals can be generated by a bank of traditional multiplier-filter modulators and demodulated by a bank of traditional correlators. However, it can be generated more efficiently by an N-point IFFT and demodulated by an N-point FFT. It is because of this that OFDM has found widespread uses in all types of communication systems. We noted that IFFTs with indexes ranging from 0 to $N-1$ cannot satisfy the sampling theorem which is required by the A/D convertors in the modulator to produce OFDM signal without spectral aliasing. Instead, IFFTs with indexes ranging from $-N/2$ to $N/2-1$ can satisfy the sampling theorem. To further reduce aliasing, oversampling is often used, which is accomplished by padding zeros (virtual subcarriers) on the two sides of the data sequence.

We have discussed in great detail OFDM's FFT implementations, including the normal complex-output and real-output algorithms, various FFT algorithms, and partial FFT algorithms.

To reduce the effect of intersymbol interference (ISI), a guard interval is inserted between OFDM symbols. If the guard interval is greater than the length of the channel delay spread, ISI can be completely avoided. However, inter-subcarrier interference (ICI) still exists. A cyclic extension placed in the guard interval can solve the problem. With the cyclic extension, the received signal samples are the result of the circular convolution of the transmitted samples with the discrete-time channel impulse response, so that the FFT demodulator output sample is the product of the data symbol with a multiplicative factor which can be estimated and removed. Even though the edges of the OFDM spectrum are quite sharp, extra spectral shaping is often required to comply with the spectral mask set up by the regulatory authorities. Spectral shaping is done by time-domain windowing or filtering, with windowing being more popular. We have discussed the common raised cosine windowing function and other windowing functions.

The other issues and critical subsystems in OFDM systems will be discussed in later chapters, as we have already stated in the beginning of this chapter.

APPENDIX 12A DERIVATION OF (12.23), (12.24), AND (12.25)

This appendix establishes the DFT pair with indexes from $-N/2$ to $N/2-1$ and derives the relationship between values of IDFTs with indexes from 0 to $N-1$ and from $-N/2$ to $N/2-1$ as given in (12.25).

The following identity will be used in our proof. For even-integer N,

$$\frac{1}{N}\sum_{k=0}^{N-1}\exp(j\frac{2\pi ik}{N})=\begin{cases}1, & i=0,\ \pm N,\ \pm 2N,...\\ 0, & \text{otherwise}\end{cases} \tag{12.58}$$

This can be easily proved. For the cases where $i=0, \pm N, \pm 2N, ...$, $\exp(j\frac{2\pi ik}{N})=1$, so the sum is N. For other i values, $\exp(j\frac{2\pi ik}{N})$ is a set of evenly spaced phasors on the unit circle in the complex number plane. These phasors cancel each other in the sum due to the symmetry. So the sum is 0.

Let $l=k-N/2$, and (12.58) becomes

$$\frac{1}{N}\sum_{k=0}^{N-1}\exp(j\frac{2\pi ik}{N}) = \frac{1}{N}\sum_{l=-\frac{N}{2}}^{\frac{N}{2}-1}\exp(j\frac{2\pi il}{N})\exp(j\pi i)$$

$$= (-1)^i \frac{1}{N} \sum_{l=-\frac{N}{2}}^{\frac{N}{2}-1} \exp(j\frac{2\pi i l}{N})$$

$$= \begin{cases} 1, & i = 0, \ \pm N, \ \pm 2N, ... \\ 0, & \text{otherwise} \end{cases}$$

Thus we have an alternate form of the identity as

$$\frac{1}{N} \sum_{i=-\frac{N}{2}}^{\frac{N}{2}-1} \exp(j\frac{2\pi i k}{N}) = \begin{cases} 1, & i = 0, \ \pm N, \ \pm 2N, ... \\ 0, & \text{otherwise} \end{cases} \tag{12.59}$$

The DFT pair with indexes between $-N/2$ to $N/2 - 1$ is

$$s'_n = \frac{1}{N} \sum_{i=-\frac{N}{2}}^{\frac{N}{2}-1} d_i \exp(j2\pi \frac{in}{N}), \quad n = -\frac{N}{2}, -\frac{N}{2}+1, ..., \frac{N}{2}-1 \tag{12.60}$$

and

$$d_i = \sum_{n=-\frac{N}{2}}^{\frac{N}{2}-1} s'_n \exp(-j2\pi \frac{in}{N}), \quad i = -\frac{N}{2}, -\frac{N}{2}+1, ..., \frac{N}{2}-1 \tag{12.61}$$

By substituting the s'_n expression into (12.61) and using (12.59), we verify the new DFT pair.

The relationship between s'_n and s_n can be derived as follows. Let $i = k + N/2$ in (12.19) and we have

$$\begin{aligned} s_n &= \frac{1}{N} \sum_{k=-\frac{N}{2}}^{\frac{N}{2}-1} d_{k+\frac{N}{2}} \exp(j2\pi \frac{(k+\frac{N}{2})n}{N}), \quad n = 0, 1, ..., N-1 \\ &= \frac{1}{N} \sum_{k=-\frac{N}{2}}^{\frac{N}{2}-1} d_{k+\frac{N}{2}} \exp(j2\pi \frac{kn}{N}) \exp(j\pi n), \quad n = 0, 1, ..., N-1 \\ &= (-1)^n \left[\frac{1}{N} \sum_{k=-\frac{N}{2}}^{\frac{N}{2}-1} d_{k+\frac{N}{2}} \exp(j2\pi \frac{kn}{N}) \right], \quad n = 0, 1, ..., N-1 \end{aligned}$$

The expression in the brackets is the s'_n for data set $\{d_i\}_{i=0}^{N-1}$. However, s'_n does not exist for $n = N/2, ..., N-1$. But we at least can have

$$s_n = (-1)^n s'_n, \quad n = 0, 1, ..., \frac{N}{2} - 1 \tag{12.62}$$

For the other half, we have

$$s_n = (-1)^n \left[\frac{1}{N} \sum_{k=-\frac{N}{2}}^{\frac{N}{2}-1} d_{k+\frac{N}{2}} \exp(j2\pi \frac{kn}{N}) \right], \quad n = N/2, ..., N-1$$

Let $n = l + N$

$$\begin{aligned}
s_{l+N} &= (-1)^{l+N} \left[\frac{1}{N} \sum_{k=-\frac{N}{2}}^{\frac{N}{2}-1} d_{k+\frac{N}{2}} \exp(j2\pi \frac{k(l+N)}{N}) \right], \quad l = -\frac{N}{2}, ..., -1 \\
&= (-1)^l \left[\frac{1}{N} \sum_{k=-\frac{N}{2}}^{\frac{N}{2}-1} d_{k+\frac{N}{2}} \exp(j2\pi \frac{kl}{N}) \exp(j2\pi k) \right], \quad l = -\frac{N}{2}, ..., -1 \\
&= (-1)^l \left[\frac{1}{N} \sum_{k=-\frac{N}{2}}^{\frac{N}{2}-1} d_{k+\frac{N}{2}} \exp(j2\pi \frac{kl}{N}) \right], \quad l = -\frac{N}{2}, ..., -1 \\
&= (-1)^l s'_l, \quad l = -\frac{N}{2}, ..., -1
\end{aligned} \tag{12.63}$$

Thus from (12.62) and (12.63) we see

$$s'_n = \begin{cases} (-1)^n s_{n+N}, & n = -\frac{N}{2}, ..., -1 \\ (-1)^n s_n, & n = 0, 1, ..., \frac{N}{2} - 1 \end{cases} \tag{12.64}$$

APPENDIX 12B DERIVATION OF DFT-BASED OFDM MODEM

This appendix shows the mathematics of the DFT-based OFDM modem in Figure 12.10. The starting point is the I- and Q-channel signals after the D/A converter given by (12.27). For simplicity, time limit $-\frac{T}{2} \le t \le \frac{T}{2}$ associated with the OFDM signal is omitted in the derivation steps.

In the modulator (Figure 12.10(a)), let $f'_{IF} = f_{IF} + \frac{1}{2T}$; the IF OFDM signal after the I-Q modulator is

$$s_{IF}(t) = I(t)\cos(2\pi f'_{IF} t) - Q(t)\sin(2\pi f'_{IF} t)$$

$$= \sum_{i=-\frac{N}{2}}^{\frac{N}{2}-1} A_i \cos(2\pi \frac{i}{T} t + \phi_i) \cos(2\pi f'_{IF})$$

$$- \sum_{i=-\frac{N}{2}}^{\frac{N}{2}-1} A_i \sin(2\pi \frac{i}{T} t + \phi_i) \sin(2\pi f'_{IF})$$

$$= \sum_{i=-\frac{N}{2}}^{\frac{N}{2}-1} A_i \cos\left[2\pi (f'_{IF} + \frac{i}{T}) t + \phi_i\right]$$

$$= \sum_{i=-\frac{N}{2}}^{\frac{N}{2}-1} A_i \cos\left[2\pi (f_{IF} + \frac{1}{2T} + \frac{i}{T}) t + \phi_i\right]$$

The RF OFDM signal after the up-converter (the multiplier plus the BPF at f_c) is

$$2 s_{IF}(t) \cos[2\pi (f_c - f_{IF}) t]$$

$$= \sum_{i=-\frac{N}{2}}^{\frac{N}{2}-1} 2A_i \cos[2\pi (f_c - f_{IF}) t] \cos\left[2\pi (f_{IF} + \frac{1}{2T} + \frac{i}{T}) t + \phi_i\right]$$

$$= \sum_{i=-\frac{N}{2}}^{\frac{N}{2}-1} A_i \cos\left[2\pi (f_c + \frac{1}{2T} + \frac{i}{T}) t + \phi_i\right] + \underbrace{\sum_{i=-\frac{N}{2}}^{\frac{N}{2}-1} A_i \cos\left[2\pi (f_c - 2f_{IF} - \frac{1}{2T} - \frac{i}{T}) t + \phi_i\right]}_{\text{Rejected by the BPF at } f_c}$$

$$= \sum_{i=-\frac{N}{2}}^{\frac{N}{2}-1} A_i \cos\left[2\pi (f_c + \frac{1}{2T} + \frac{i}{T}) t + \phi_i\right] \triangleq s(t) \qquad (12.65)$$

In the demodulator (Figure 12.10(b)), after the down-converter (the first multiplier plus the bandpass filter at f_{IF}), the signal is

$$2 s(t) \cos[2\pi (f_c - f_{IF}) t]$$

$$\begin{aligned}
&= \sum_{i=-\frac{N}{2}}^{\frac{N}{2}-1} 2A_i \cos\left[2\pi(f_c + \frac{1}{2T} + \frac{i}{T})t + \phi_i\right] \cos\left[2\pi(f_c - f_{IF})t\right] \\
&= \underbrace{\sum_{i=-\frac{N}{2}}^{\frac{N}{2}-1} A_i \cos\left[2\pi(2f_c - f_{IF} + \frac{1}{2T} + \frac{i}{T})t + \phi_i\right]}_{\text{Rejected by the BPF at } f_{IF}} + \\
&\quad \sum_{i=-\frac{N}{2}}^{\frac{N}{2}-1} A_i \cos\left[2\pi(f_{IF} + \frac{1}{2T} + \frac{i}{T})t + \phi_i\right] \\
&\Rightarrow \sum_{i=-\frac{N}{2}}^{\frac{N}{2}-1} A_i \cos\left[2\pi(f_{IF} + \frac{1}{2T} + \frac{i}{T})t + \phi_i\right] \\
&= \sum_{i=-\frac{N}{2}}^{\frac{N}{2}-1} A_i \cos\left[2\pi(f'_{IF} + \frac{i}{T})t + \phi_i\right] = s_{IF}(t)
\end{aligned}$$

After the I-channel demodulator, the signal is

$$\begin{aligned}
&2s_{IF}(t)\cos(2\pi f'_{IF}t) \\
&= \sum_{i=-\frac{N}{2}}^{\frac{N}{2}-1} 2A_i \cos\left[2\pi(f'_{IF} + \frac{i}{T})t + \phi_i\right] \cos(2\pi f'_{IF}t) \\
&= \underbrace{\sum_{i=-\frac{N}{2}}^{\frac{N}{2}-1} A_i \cos\left[2\pi(2f'_{IF} + \frac{i}{T})t + \phi_i\right]}_{\text{Rejected by the LPF}} + \sum_{i=0}^{N-1} A_i \cos(2\pi\frac{i}{T}t + \phi_i) \\
&\Rightarrow \sum_{i=-\frac{N}{2}}^{\frac{N}{2}-1} A_i \cos(2\pi\frac{i}{T}t + \phi_i) = I(t)
\end{aligned}$$

After the Q-channel demodulator, the signal is

$$\begin{aligned}
&-2s_{IF}(t)\sin(2\pi f'_{IF}t) \\
&= -\sum_{i=-\frac{N}{2}}^{\frac{N}{2}-1} 2A_i \cos\left[2\pi(f'_{IF} + \frac{i}{T})t + \phi_i\right] \sin(2\pi f'_{IF}t)
\end{aligned}$$

$$= \underbrace{-\sum_{i=-\frac{N}{2}}^{\frac{N}{2}-1} A_i \sin\left[2\pi(2f'_{IF}+\frac{i}{T})t+\phi_i\right]}_{\text{Rejected by the LPF}} + \sum_{i=-\frac{N}{2}}^{\frac{N}{2}-1} A_i \sin(2\pi\frac{i}{T}t+\phi_i)$$

$$\Rightarrow \sum_{i=-\frac{N}{2}}^{\frac{N}{2}-1} A_i \sin(2\pi\frac{i}{T}t+\phi_i) = Q(t)$$

APPENDIX 12C RECOVERING DATA FROM REAL OFDM SIGNAL

This appendix proves Weinstein and Ebert's method of recovering data from the real part of the OFDM signal. Weinstein and Ebert used DFT to generate the OFDM signal as well as to demodulate the OFDM signal, while in our treatment here IDFT is used for signal generation and DFT is used for demodulation. However, their method is still applicable and the result here is only slightly different.

From (12.27) the real part of the OFDM signal after the D/A converter can be written as

$$I(t) = \frac{1}{N}\sum_{i=-\frac{N}{2}}^{\frac{N}{2}-1}\left(I_i \cos 2\pi\frac{i}{T}t - Q_i \sin 2\pi\frac{i}{T}t\right)$$

Sampled at intervals $\Delta t/2 = T/2N$,

$$I(k) = \frac{1}{N}\sum_{i=-\frac{N}{2}}^{\frac{N}{2}-1}\left(I_i \cos\frac{\pi i k}{N} - Q_i \sin\frac{\pi i k}{N}\right), \quad k=-N,...,N-1$$

Using $\{I(k)\}_{k=-N}^{N-1}$ as input to a $2N$-point DFT, we have

$$\begin{aligned} s_n &= \sum_{k=-N}^{N-1} I(k)\exp(-j\frac{\pi n k}{N}) \\ &= \frac{1}{N}\sum_{k=-N}^{N-1}\sum_{i=-\frac{N}{2}}^{\frac{N}{2}-1}\left[\left(I_i\cos\frac{\pi i k}{N} - Q_i \sin\frac{\pi i k}{N}\right)\exp(-j\frac{\pi n k}{N})\right], \\ n &= -N,...,N-1 \end{aligned}$$

When $n = 0$, we have

$$\begin{aligned} s_0 &= \frac{1}{N}\sum_{k=-N}^{N-1}\sum_{i=-\frac{N}{2}}^{\frac{N}{2}-1}\left(I_i\cos\frac{\pi ik}{N} - Q_i\sin\frac{\pi ik}{N}\right) \\ &= \frac{1}{N}\left[\sum_{i=-\frac{N}{2}}^{\frac{N}{2}-1} I_i\sum_{k=-N}^{N-1}\cos\frac{\pi ik}{N} - \sum_{i=-\frac{N}{2}}^{\frac{N}{2}-1} Q_i\sum_{k=-N}^{N-1}\sin\frac{\pi ik}{N}\right] \end{aligned}$$

where

$$\begin{aligned} &\frac{1}{N}\sum_{k=-N}^{N-1}\cos\frac{\pi ik}{N} = \frac{1}{2}\frac{1}{N}\sum_{k=-N}^{N-1}\left[\exp(j\frac{\pi ik}{N}) + \exp(-j\frac{\pi ik}{N})\right] \\ = &\left\{\begin{array}{ll} 2, & i = 0 \\ 0, & \text{otherwise} \end{array}\right. \end{aligned}$$

and

$$\begin{aligned} &\frac{1}{N}\sum_{k=-N}^{N-1}\sin\frac{\pi ik}{N} = \frac{1}{2j}\frac{1}{N}\sum_{k=-N}^{N-1}\left[\exp(j\frac{\pi ik}{N}) - \exp(-j\frac{\pi ik}{N})\right] \\ = &\ 0, \quad \text{for any } i \end{aligned}$$

where identity

$$\frac{1}{2N}\sum_{k=-N}^{N-1}\exp(j\frac{\pi ik}{N}) = \left\{\begin{array}{ll} 1, & i = 0,\ \pm 2N,\ \pm 4N, ... \\ 0, & \text{otherwise} \end{array}\right. \tag{12.66}$$

is used. Thus

$$s_0 = 2I_0 \tag{12.67}$$

When $n \neq 0$,

$$\begin{aligned} s_n &= \frac{1}{N}\sum_{k=-N}^{N-1}\sum_{i=-\frac{N}{2}}^{\frac{N}{2}-1}\left[\left(I_i\cos\frac{\pi ik}{N} - Q_i\sin\frac{\pi ik}{N}\right)\exp(-j\frac{\pi nk}{N})\right] \\ &= \frac{1}{N}\sum_{i=-\frac{N}{2}}^{\frac{N}{2}-1}\sum_{k=-N}^{N-1}\left[\left[\begin{array}{l} \frac{1}{2}I_i\left(\exp(j\frac{\pi ik}{N}) + \exp(-j\frac{\pi ik}{N})\right) \\ -\frac{1}{2j}Q_i\left(\exp(j\frac{\pi ik}{N}) - \exp(-j\frac{\pi ik}{N})\right) \end{array}\right]\exp(-j\frac{\pi nk}{N})\right] \end{aligned}$$

$$
\begin{aligned}
&= \frac{1}{N}\sum_{i=-\frac{N}{2}}^{\frac{N}{2}-1}\sum_{k=-N}^{N-1}\left[\begin{array}{c}\frac{1}{2}I_i\left(\exp(j\frac{\pi(i-n)k}{N})+\exp(-j\frac{\pi(i+n)k}{N})\right)\\ -\frac{1}{2j}Q_i\left(\exp(j\frac{\pi(i-n)k}{N})-\exp(-j\frac{\pi(i+n)k}{N})\right)\end{array}\right]\\
&= I_n+jQ_n
\end{aligned}
$$

where (12.66) is used again. Note that in deriving the last step, the summation of the second term in both brackets is zero. Combining the above with (12.67), we have

$$
s_n=\begin{cases}2I_0, & n=0\\ I_n+jQ_n, & n=-N/2,...,-1,1,...,N/2-1\\ \text{irrelevant}, & n>N/2-1 \text{ or } n<-N/2\end{cases}
$$

APPENDIX 12D METHOD OF GENERATING REAL OFDM SIGNAL

This appendix verifies the second method of generating real OFDM. From (12.29), the $2N$-point IDFT of this new set of data is

$$
\begin{aligned}
s_n &= \frac{1}{2N}\sum_{i=-N}^{N-1} D_i\exp(j\frac{\pi ni}{N})\\
&= \frac{1}{2N}\left[\sum_{i=-1}^{-(N-1)} d_{N/2+i}\exp(j\frac{\pi ni}{N})+\sum_{i=1}^{N-1} d^*_{N/2-i}\exp(j\frac{\pi ni}{N})\right.\\
&\quad \left.+\operatorname{Re}\{d_{-N/2}\}\exp(-j\pi n)+\operatorname{Im}\{d_{-N/2}\}\right]\\
n &= -N,...,N-1
\end{aligned}
$$

where the first sum can be proved as the conjugate of the second sum (change i to $-i$):

$$
\begin{aligned}
\sum_{i=-1}^{-(N-1)} d_{N/2+i}\exp(j\frac{\pi ni}{N}) &= \sum_{i=1}^{N-1} d_{N/2-i}\exp(-j\frac{\pi ni}{N})\\
&= \left[\sum_{i=1}^{N-1} d^*_{N/2-i}\exp(j\frac{\pi ni}{N})\right]^*
\end{aligned}
$$

Thus

$$
s_n = \frac{1}{2N}[2\operatorname{Re}\{\sum_{i=1}^{N-1} d_{N/2-i}\exp(-j\frac{\pi ni}{N})\}
$$

$$
\begin{aligned}
& + \operatorname{Re}\{d_{-N/2}\}(-1)^n + \operatorname{Im}\{d_{-N/2}\}], \\
n &= -N, ..., N-1
\end{aligned}
$$

References

[1] Chang, R. W., "Synthesis of band-limited orthogonal signals for multichannel data transmission," *Bell Syst. Tech. J*, vol. 45, December 1966, pp. 1775–1796.

[2] Chang, R. W., "A theoretical study of performance of an orthogonal multiplexing data transmission scheme," *IEEE Trans. Comm. Technol.*, vol. 16, August 1967, pp. 529–539.

[3] Chang, R. W., "Orthogonal frequency division multiplexing," U.S. Patent 3,488,445, filed November 14, 1966, issued January 6, 1970.

[4] Saltzberg, B. R., "Performance of an efficient parallel data transmission system," *IEEE Trans. Comm. Technol.*, vol. 15, December 1967.

[5] Zimmerman, M. S., and A. L. Kirsch, "The ANIGSC-10 (KATHRYN) variable rate data modem for HF radio," *IEEE Trans. Comm. Technol.*, vol. 15, April 1967, pp. 197–205.

[6] Weinstein, S. B., and P. W. Ebert, "Data transmission by frequency division multiplexing using the discrete Fourier transform," *IEEE Trans. Comm. Technol.*, vol. 19, no. 5, October 1971.

[7] Hirosaki, B.,"An orthogonally multiplexed QAM system using the discrete Fourier transform," *IEEE Trans. Commun.*, vol. 29, July 1981, pp. 982–989.

[8] Hirosaki, B., S. Hasegawa, and A. Sabato, "Advanced groupband data modem using orthogonally multiplexed QAM technique," *IEEE Trans. Commun.*, vol. 34. no. 6, June 1986, pp. 587–592.

[9] Hirosaki, B., et al., "A 19.2 kbits voiceband data modem based on orthogonally multiplexed QAM techniques," *Proc. of MEE ICCTS*, 1985, pp. 21.1.1–5.

[10] IEEE Standard 802.11a, 1999.

[11] Cariolaro, G., and F. C. Vagliani, "An OFDM scheme with a half complexity," *IEEE J. Select. Areas Commun.*, vol. 13, no. 9, December 1995, pp. 1586–1599.

[12] Bahai, A. R. S., and B. R. Saltzberg, *Multi-Carrier Digital Communications: Theory and Applications of OFDM*, New York: Kluwer Academic, 1999.

[13] Elliott, D. F., and K. R. Rao, *Fast Transforms: Algorithms, Analysis, Applications*, New York: Academic Press, 1982.

[14] Oppenheim, A., and R. W. Schafer, *Discrete-Time Signal Processing*, Englewood Cliffs, New Jersey: Prentice Hall, 1989.

[15] Cooley, J. W., and J. W. Turkey, "An algorithm for the machine calculation of complex Fourier series," *Math. Computation*, vol. 19, no. 2, April 1965, pp. 297–301.

[16] Blahut, R. E., *Fast Algorithms for Digital Signal Processing*, Reading, MA: Addison-Wesley, 1985.

[17] Duhamel, P., and H. Hollmann, "Split radix FFT algorithm," *Electron. Lett.*, vol. 20, January 5, 1984, pp. 14–16.

[18] Duhamel, P., "Implementation of 'split-radix' FFT algorithms for complex, real, and real-symmetric data," *IEEE Trans. Acoust., Speech, Signal Processing*, vol. 34, no. 1, February 1986, pp. 285–295.

[19] Sorensen, H. V., M. T. Heideman, and C. S. Burrus, "On computing the split-radix FFT," *IEEE Trans. Acoust., Speech, Signal Processing*, vol. 34, no. 1, February 1986, pp. 152–156.

[20] Markel, J. D., "FFT pruning," *IEEE Trans. Audio Electroacoust.*, vol. 19, no. 4, December 1971, pp. 305–311.

[21] Skinner, D. P., "Pruning the decimation-in-time FFT algorithm," *IEEE Trans. Acoust., Speech, Signal Processing*, vol. 24, no. 2, April 1976, pp. 193–194.

[22] Sreenivs, T. V., and P. V. S. Rao, "High resolution narrow-band spectra by FFT pruning," *IEEE Trans. Acoust., Speech, Signal Processing*, vol. 28, no. 2, April 1980, pp. 254–257.

[23] Nagai, K., "Pruning the decimation-in-time FFT algorithm with frequency shift," *IEEE Trans. Acoust., Speech, Signal Processing*, vol. 34, no. 4, August 1986, pp. 1008–1010.

[24] Sorensen, H. V., and C. S. Burrus, "Efficient computation of DFT with only a subset of input or output points," *IEEE Trans. Signal Processing*, vol. 41, no. 3, March 1993, pp. 1184–1199.

[25] Proakis, J., *Digital Communications*, 2nd ed., New York: McGraw-Hill, 1989.

[26] Ruiz, A., J. M. Cioffi, and S. Kasturia, "Discrete multiple tone modulation with coset coding for the spectrally shaped channel," *IEEE Trans. Commun.* vol. 40, no., 6, June 1992, pp. 1012–1029.

[27] van Nee, R., and R. Prasad, *OFDM for Wireless Multimedia Communications*, Norwood, Massachusetts: Artech House, 2000.

[28] Tan, P., and N. Beaulieu, "Reduced ICI in OFDM systems using the 'better than' raised-cosine pulse," *IEEE Communications Letters*, vol. 8, no. 3, March 2004, pp. 135–137.

[29] Bhatoolaul, D., and G. Wade, "Spectrum shaping in N-channel QPSK-OFDM systems," *IEE Proc. Vis. Image Signal Process*, vol. 142, no. 5, October 1995, pp. 333-338.

[30] Vahlin, A., and N. Holte, "Optimal finite duration pulses for OFDM," *IEEE Trans. Commun.*, vol. 44, no. 1, January 1996, pp. 10–14.

[31] Gudmundson, M., and P. O. Anderson, "An analysis of the effect of frequency errors in OFDM has Adjacent channel interference in an OFDM system," *Proc. IEEE 46th Vehicular Technology Conf.*, Atlanta, Georgia, April 1996, pp. 918–922.

[32] Muschallik, C., "Improving an OFDM reception using an adaptive Nyquist windowing," *IEEE Trans. Consumer Electron.*, vol. 42, no. 3, August 1996, pp. 259-269.

[33] Armstrong, J., "Analysis of new and existing methods of reducing intercarrier interference due to carrier frequency offset in OFDM," *IEEE Trans. Commun.*, vol. 47, no. 3, March 1999, pp. 365–369.

[34] Weinfurtner, S. H., "Optimum Nyquist windowing in OFDM receivers," *IEEE Trans. Commun.*, vol. 49, no. 3, March 2001, pp. 417–420.

[35] Redfern, A. J., "Receiver window design for multicarrier communication systems," *IEEE J. Select. Areas Commun.*, vol. 20, no. 5, June 2002, pp. 1029–1036.

[36] Strohmer, T., and S. Beaver, "Optimal OFDM Design for Time Frequency Dispersive Channels,"

IEEE Trans. Commun., vol. 51, no. 7, July 2003, pp. 1111–1122.

[37] Faulkner, M., "The effect of filtering on the performance of OFDM systems," *IEEE Trans. Vehicular Technology*, vol. 49, no. 5, September 2000, pp. 1877–1884.

[38] Lin, Y. P., and S. M. Phoong, "OFDM transmitters: analog representation and DFT-based implementation," *IEEE Trans. Signal Processing,* vol. 51, no. 9, September 2003, pp. 2450–2453.

[39] Zou, W. Y., and Y. Wu, "COFDM: an overview," *IEEE Trans. Broadcasting*, vol. 41, no. 1, March 1995, pp. 1–8.

Selected Bibliography

- Bahai, A. S. R., and B. R. Saltzberg, *Multi-Carrier Digital Communications: Theory and Applications of OFDM,* New York: Kluwer Academic/Plenum Publishers, 1999.
- Cimini, L., and L. Greenstein, "Orthogonal frequency-division multiplexing," *Wiley Encyclopedia of Telecommunications*, J. G. Proakis, ed., New York: John Wiley & Sons, 2003.
- Litwin, L., "An Introduction to multicarrier modulation," *IEEE Potentials*, April/May, 2000, pp. 36–38.
- Peled, A., and A. Ruiz, "Frequency domain data transmission using reduced computational complexity algorithms," *IEEE Int. Conf. Acoust., Speech, Signal Processing*, Denver, Colorado, April 1980, pp. 964–967.

Chapter 13

Peak-to-Average Power Ratio Reduction

In comparison with single-carrier modulation, a big disadvantage of OFDM is the high peak-to-average power ratio (PAPR). This is because the OFDM signal is a sum of N modulated subcarriers with different frequencies. The envelope of the sum varies with time. At some time points, the components add to a large envelope value, while at other points, they add to a small envelope value. In fact, we have seen this phenomenon in Figure 12.1, where (a) is a 4-channel QPSK-OFDM and (b) is a 16-channel QAM-OFDM. As we will see shortly, in the extreme case when all N subcarriers are cophasal at one point, the peak power at that point is N times the average power for PSK-OFDM signals and more than N times for QAM-OFDM signals. Big PAPR requires the DAC to have a large number of bits to accommodate the signal range. Even more seriously, it puts tremendous stress on the high-power amplifier (HPA) of the transmitter. To include the high peak power in its amplification range, the power amplifier must have a big back-off and operate at a very low average power level and thus is very inefficient. Thus techniques to reduce the PAPR of OFDM signals are widely employed in OFDM systems.[1]

In this chapter we first derive the maximum PAPRs for QAM-OFDM and PSK-OFDM (Section 13.1). Maximum PAPRs occur for the extreme cases which do not happen for every OFDM symbol. In fact, they rarely happen in a practical system. For each OFDM symbol, there is a peak power value, hence a PAPR. Thus PAPR is a random variable. In Section 13.2 we derive the statistical distribution of the instantaneous envelope power and PAPR, upon which PAPR reduction techniques are based. After that we present PAPR reduction techniques in the remaining sections of the chapter. Section 13.3 is a brief introduction to the PAPR reduction techniques. Section 13.4 describes peak clipping and methods for mitigating the clipping noise.

---

[1] Even when the PAPR reduction technique is used, the OFDM signal still faces power amplifier nonlinearity to various extents, depending on different tolerable performance degradations and desired amplifier efficiencies. However, this problem is not unique to OFDM; single-carrier signals have the same problem. Solutions for power amplifier nonlinearity, such as back-off and predistortion, belong to another big area of research and engineering, which is beyond the scope of this book.

Section 13.5 deals with the schemes which alter the amplitude of OFDM signal samples other than clipping, such as companding. Section 13.6 is for the third group of methods which alter the data, such as phase rotation, before performing IFFT. Section 13.7 presents the fourth group that exploits the PAPR reduction capability of certain error-control coding techniques.

13.1 MAXIMUM PEAK-TO-AVERAGE POWER RATIO

The peak power is defined as the power of a sine wave that has an amplitude equal to the maximum of the signal's envelope value. Thus a unmodulated carrier has a PAPR of 1, or 0 dB. A single-carrier constant-envelope modulated signal, such as PSK or MSK, also has a PAPR of 1. Another measure of the envelope variation that appears in the literature is the crest factor. Some authors define it as the ratio of maximum signal value over the rms signal value [1,2]. The crest factor for a unmodulated carrier or constant-envelope modulated carrier is $\sqrt{2}$, or 3 dB. This 3-dB difference holds for other narrow-band modulated signals. In [3], the crest factor is defined as the square root of the PAPR. Apparently, this definition is more convenient, because there is no difference in decibels between the crest factor and the PAPR.

In the following, we derive the maximum PAPR for QAM-OFDM and MPSK-OFDM signals.

The power of a cosine signal in any modulation format on a subcarrier is

$$P_i = \frac{1}{T}\int_0^T \left[A_i \cos(\omega_i t + \phi_i)\right]^2 dt = \frac{1}{2}A_i^2 \tag{13.1}$$

The OFDM signal at the transmitter power amplifier is a bandpass signal with a higher frequency carrier ω_c, and the total power is

$$\begin{aligned}
& P_{total} \\
= \; & \frac{1}{T}\int_0^T \left[\sum_{i=0}^{N-1} A_i \cos(\omega_i t + \phi_i)\cos\omega_c t\right]^2 dt \\
= \; & \frac{1}{2T}\int_0^T \left[\sum_{i=0}^{N-1} \underbrace{A_i \cos[(\omega_c + \omega_i)t + \phi_i]}_{f_i(t)} + \sum_{i=0}^{N-1} \underbrace{A_i \cos[(\omega_c - \omega_i)t - \phi_i]}_{g_i(t)}\right]^2 dt \\
= \; & \frac{1}{2T}\int_0^T \left[\begin{array}{c} \sum_{i=0}^{N-1}[f_i^2(t) + g_i^2(t)] + 2\sum_{i,j=0,\, i\neq j}^{N-1}[f_i(t)f_j(t) + g_i(t)g_j(t)] \\ +2\sum_{i,j=0}^{N-1} f_i(t)g_j(t) \end{array}\right] dt
\end{aligned}$$

The frequency separation between any $f_i(t)$ and $g_i(t)$ is $f_i + f_j$, which is an integer multiple of $1/T$. These terms are orthogonal. Also $f_i(t)$ and $f_j(t), i \neq j$, and $g_i(t)$ and $g_j(t), i \neq j$ are orthogonal by the same reasons. Therefore the integrals of the cross terms in the brackets are all zero. Thus the total power is

$$P_{total} = \frac{1}{2T}\sum_{i=0}^{N-1}\int_0^T [f_i^2(t) + g_i^2(t)]dt = \sum_{i=0}^{N-1}\frac{1}{2}A_i^2 = \sum_{i=0}^{N-1} P_i$$

That is, the total power of an OFDM signal is equal to the sum of the powers of all subcarriers. From this we can infer that the total average power is also equal to the sum of average powers of all subcarriers since

$$P_{avg(OFDM)} = E\{P_{total}\} = \sum_{i=0}^{N-1} E\{P_i\} = \sum_{i=0}^{N-1} P_{avg}$$

where $E\{x\}$ denotes the expectation of x. This conclusion will be used in the following.

For the square QAM, the maximum amplitude is $A_{\max(QAM)} = \sqrt{2}(\sqrt{M} - 1)$ (the outermost point in the constellation) (see (9.22)) and the maximum OFDM envelope is

$$A_{peak(QO)} = NA_{\max(QAM)} = \sqrt{2}N(\sqrt{M} - 1)$$

so the peak power is

$$P_{peak(QO)} = N^2(\sqrt{M} - 1)^2$$

The average power of the square QAM signal on a single subcarrier is (see (9.25))

$$P_{avg} = \frac{1}{3}(M - 1)P_o$$

where P_o is the power of the smallest signal. For the amplitude assignment we have assumed in (9.22), the minimum amplitude is $\sqrt{2}$, and from (13.1)

$$P_o = \frac{1}{2}(\sqrt{2})^2 = 1$$

and the average power of the OFDM signal on N subcarriers is

$$P_{avg(QO)} = \frac{1}{3}N(M - 1)$$

Order	$\text{PAPR}_{(PO)}$	$\text{PAPR}_{(QO)}$	Increase: QO/PO	
M	N	$3N\frac{\sqrt{M}-1}{\sqrt{M}+1}$	$\frac{3(\sqrt{M}-1)}{\sqrt{M}+1}$	dB
4	N	N	1	0
16	N	$1.8N$	1.8	2.553
64	N	$2.333N$	2.333	3.679
256	N	$2.647N$	2.647	4.228

Table 13.1 PAPR comparison of QAM-OFDM versus MPSK-OFDM.

Thus, the PAPR is

$$PAPR_{(QO)} = \frac{P_{peak(QO)}}{P_{avg(QO)}} = \frac{3N(\sqrt{M}-1)}{\sqrt{M}+1} \tag{13.2}$$

For the MPSK, all the amplitudes are the same, assuming it is A_{MPSK}. The maximum OFDM envelope is

$$A_{peak(PO)} = NA_{MPSK}$$

and the peak power is

$$P_{peak(PO)} = \frac{1}{2}N^2 A_{MPSK}^2$$

The average power of the MPSK signal is equal to the power of each individual signal and the average power of the MPSK-OFDM signal on N subcarriers is

$$P_{avg(PO)} = \frac{1}{2}NA_{MPSK}^2$$

Thus, the PAPR is

$$PAPR_{(PO)} = \frac{P_{peak(PO)}}{P_{avg(PO)}} = N \tag{13.3}$$

Comparing (13.2) and (13.3), it is seen that the PAPR of the QAM-OFDM is increased by a factor of $3(\sqrt{M}-1)/(\sqrt{M}+1)$ against PSK. Table 13.1 compares PAPRs.

13.2 ENVELOPE POWER AND PAPR DISTRIBUTION

While maximum PAPR tells us the extreme value of the ratio of the instantaneous signal power over the average power of the passband OFDM signal, the statistical distribution of the instantaneous power (or, equivalently, PAPR) is more important for system design. Since the signal envelope of a bandpass signal is the same as the magnitude of its baseband complex envelope (this can be easily verified or see [4, p. 730]), we examine the complex envelope of OFDM given in (12.17). We are interested in its probability distribution, not the absolute value; thus we rewrite it in a normalized form,

$$\widetilde{s}(t) = \frac{1}{\sqrt{N}} \sum_{i=0}^{N-1} d_i \exp(j2\pi \frac{i}{T} t), \quad 0 \le t \le T \tag{13.4}$$

Since $d_i = I_i + jQ_i$, we have

$$\widetilde{s}(t) = \frac{1}{\sqrt{N}} \sum_{i=0}^{N-1} [x_i(t) + jy_i(t)] = x(t) + jy(t)$$

where

$$x_i(t) = I_i \cos(2\pi \frac{i}{T} t) - Q_i \sin(2\pi \frac{i}{T} t)$$

$$y_i(t) = Q_i \cos(2\pi \frac{i}{T} t) + I_i \sin(2\pi \frac{i}{T} t)$$

Since complex data symbols $\{I_i, Q_i\}$ are independently identically distributed (i.i.d.) random variables, each of $x(t)$ and $y(t)$ is a sum of N i.i.d. random signals. When N is large, according to the central limit theorem, $x(t)$ and $y(t)$ are approximately Gaussian distributed. And $\widetilde{s}(t)$ is approximately a complex Gaussian random process.

We assume the modulation scheme has a symmetrical constellation, such as square QAM and MPSK. Then $E\{I_i\} = E\{Q_i\} = 0$. Consequently the mean values of $x(t)$ and $y(t)$ are zero, and the mean value of $\widetilde{s}(t)$ is zero.

As $x(t)$ and $y(t)$ both are a function of $\{I_i, Q_i\}$, we need to check if they are correlated. The covariance

$$\text{cov}(x(t), y(t)) = E\{x(t)y(t+\tau)\} = \frac{1}{N} \sum_{i=0}^{N-1} \sum_{k=0}^{N-1} E\{x_i(t) y_k(t+\tau)\} \tag{13.5}$$

where $E\{x_i(t)y_k(t+\tau)\} = 0$ for $i \neq k$, and for $i = k$,

$$\begin{aligned} & E\{x_i(t)y_i(t+\tau) \\ = \; & E\{I_i^2\} \cos(2\pi \frac{i}{T} t) \sin(2\pi \frac{i}{T}(t+\tau)) - E\{Q_i^2\} \sin(2\pi \frac{i}{T} t) \cos(2\pi \frac{i}{T}(t+\tau)) \end{aligned}$$

For a symmetrical constellation, such as square QAM and MPSK,[2]

$$E\{I_i^2\} = E\{Q_i^2\} \triangleq \sigma^2$$

then

$$E\{x_i(t)y_i(t+\tau)\} = \sigma^2 \sin(2\pi \frac{i}{T}\tau)$$

Thus

$$\text{cov}(x(t), y(t)) = \frac{\sigma^2}{N} \sum_{i=0}^{N-1} \sin(2\pi \frac{i}{T}\tau) \tag{13.8}$$

This expression indicates that the real part and the imaginary part of the complex baseband OFDM signal are correlated in general. However, the samples with sampling frequency of N/T (Nyquist), which usually are obtained by the N-point IDFT, are uncorrelated since in this case

$$\text{cov}(x(l), y(m)) = \frac{\sigma^2}{N} \sum_{i=0}^{N-1} \sin(2\pi \frac{i}{N}(l-m)) = 0, \qquad \text{for any integers } l, m \tag{13.9}$$

In the oversampling case, assuming the sampling frequency is nN/T, n is an integer, then

$$\text{cov}(x(l), y(m)) = \frac{\sigma^2}{N} \sum_{i=0}^{N-1} \sin(2\pi \frac{i}{nN}(l-m)) = \begin{cases} 0, & \text{for } \frac{(l-m)}{n} = \text{integer} \\ \text{nonzero}, & \text{otherwise} \end{cases} \tag{13.10}$$

This indicates that in general the real part and the imaginary part of the oversampled complex baseband OFDM signal samples are correlated.

---

2 For MPSK with $|d_i| = 1$, the variance

$$\sigma^2 = \frac{1}{2}E\{|d_i|^2\} = \frac{1}{2} \tag{13.6}$$

For square QAM, $|d_i|^2 = I_i^2 + Q_i^2$, where (I_i, Q_i) are defined in (9.22). From (9.24), we have

$$E\{|d_i|^2\} = E\{I_i^2\} + E\{Q_i^2\} = 2\sigma^2 = \frac{2}{3}(M-1)$$

Thus for square QAM defined by (9.22)

$$\sigma^2 = \frac{1}{3}(M-1) \tag{13.7}$$

When $M = 4$, QAM becomes QPSK with $|d_i| = \sqrt{2}$ and $\sigma^2 = 1$ from (13.6), which is consistent with (13.7).

We also need to check the autocovariance function of $x(t)$ and $y(t)$. Following the similar steps above we find

$$\text{cov}(x(t), x(t)) = E\{x(t)x(t+\tau)\} = \frac{\sigma^2}{N}\sum_{i=0}^{N-1}\cos(2\pi\frac{i}{T}\tau) \tag{13.11}$$

This indicates that values of $x(t)$ at different time instants are generally correlated. When $\tau = 0$, the autocovariance becomes the variance, which is

$$\text{var}(x(t)) = E\{x^2(t)\} = \sigma^2 \tag{13.12}$$

For the Nyquist samples, (13.11) becomes

$$\text{cov}(x(l), x(m)) = \frac{\sigma^2}{N}\sum_{i=0}^{N-1}\cos(2\pi\frac{i}{N}(l-m)) = \begin{cases} 0, & l \neq m \\ \sigma^2, & l = m \end{cases} \tag{13.13}$$

This means the Nyquist samples are uncorrelated and the variance of each sample is σ^2.

In the oversampling case, results similar to (13.10) are obtained:

$$\text{cov}(x(l), x(m)) = \frac{\sigma^2}{N}\sum_{i=0}^{N-1}\cos(2\pi\frac{i}{nN}(l-m)) = \begin{cases} 0, & \text{for } \frac{(l-m)}{n} = \text{integer} \\ \text{nonzero}, & \text{otherwise} \end{cases} \tag{13.14}$$

The same results of autocovariance are obtained for $y(t)$.

The instantaneous envelope of the OFDM signal is

$$r(t) = |\tilde{s}(t)| = \sqrt{x^2(t) + y^2(t)} \tag{13.15}$$

and the instantaneous envelope power of the OFDM signal with an RF carrier frequency f_c can be approximated as[3]

$$\begin{aligned} \xi(t) &= \frac{1}{T_c}\int_t^{t+T_c} \left[x(t)\cos(2\pi f_c\tau) + y(t)\sin(2\pi f_c\tau)\right]^2 d\tau \\ &= \frac{1}{2}\left[x^2(t) + y^2(t)\right] \end{aligned} \tag{13.16}$$

where $T_c = 1/f_c$. The statistical average of the instantaneous envelope power is

$$E\{\xi(t)\} = E\left\{\frac{1}{2}\left[x^2(t) + y^2(t)\right]\right\} = \sigma^2$$

3 Assume the envelope does not change appreciably over one RF cycle.

which is equal at any time instant. Thus the average bandpass OFDM signal power as also equal to σ^2. That is

$$P_{avg(OFDM)} = \sigma^2$$

(Note that the average baseband OFDM signal power is $E\left\{\left[x^2(t) + y^2(t)\right]\right\} = 2\sigma^2$.)

It is well-known that the probabilistic distribution of the envelope of a bandpass Gaussian random signal is Rayleigh (see Section 11.1.3). The probability density function of $r(t)$ at any time t is given by (from (11.10))

$$p_r(r) = \frac{r}{\sigma^2} e^{-\frac{r^2}{2\sigma^2}} U(r) \tag{13.17}$$

where $U(r)$ is the unit step function.

The square of a Rayleigh random variable has a central chi-square distribution with two degrees of freedom [5, p. 29]. Then distribution of the instantaneous envelope power is

$$p_\xi(\xi) = \frac{1}{2\sigma^2} e^{-\frac{\xi}{2\sigma^2}} U(\xi) \tag{13.18}$$

which in fact degenerates to an exponential distribution. The cumulative density function (cdf) of the instantaneous envelope power is

$$F_\xi(\xi) = \Pr(x \leq \xi) = \int_0^\xi p_\xi(x)dx = \int_0^\xi \frac{1}{2\sigma^2} e^{-\frac{x}{2\sigma^2}} dx = (1 - e^{-\frac{\xi}{2\sigma^2}})$$

Since σ^2 is the average power of the bandpass signal, $\xi(t)/\sigma^2$ is the instantaneous PAPR[4] of the bandpass signal. For convenience, the instantaneous PAPR of the baseband signal

$$\lambda(t) \triangleq \frac{\xi(t)}{2\sigma^2}$$

is usually used. Thus from (13.18)

$$p_\lambda(\lambda) = e^{-\lambda} U(\lambda) \tag{13.19}$$

and

---

4 This term is not quite appropriate in the strict sense of PAPR as it refers to the ratio of the statistical maximum (or peak) power over the average power. However, it is a convenient term and is used in the literature. It is correct in a wide sense as it is the ratio of the instantaneous peak power over the average power.

$$F_\lambda(\lambda) = 1 - e^{-\lambda} \tag{13.20}$$

Because the Nyquist samples are uncorrelated Gaussian random variables, they are independent. The probability that the PAPRs of all N samples are less than some threshold is given by

$$F_N(\lambda) = \Pr(PAPR \leq \lambda) = [F_\lambda(\lambda)]^N = (1 - e^{-\lambda})^N \tag{13.21}$$

which is the joint cdf of the PAPRs of the N samples.

The envelope distribution (in a slightly different form) is first derived in [6] for sampled OFDM signals. The fact that oversampled OFDM samples are correlated was noted in [1]. Here the derivation starts with the continuous-time OFDM signal, and the sampled OFDM signal is treated just as a special case. In this way, it is clearly seen that not only oversampled signal samples are correlated, but also the analog OFDM signal is in general correlated, only the samples of Nyquist rate are uncorrelated.

The Gaussian approximation of the OFDM signal and the above cdfs are not accurate when N is not too big. Simulation results given in [1] show that (13.21) is very accurate for $N = 1024$, and not accurate for $N \leq 256$. An empirical modifying factor is suggested to give

$$F_N(\lambda) \approx (1 - e^{-\lambda})^{\alpha N} \tag{13.22}$$

with $\alpha = 2.8$. This expression matches simulation curves quite accurately down to $N = 128$.

A more accurate distribution is derived in [3] and the N-sample cdf for high λ is approximated by

$$F_N(\lambda) \approx \exp\left[-\sqrt{\frac{\pi}{3}}\lambda N e^{-\lambda}\right] \tag{13.23}$$

Simulation results of 16QAM-OFDM signal in [3] show that (13.23) is more accurate than (13.22) for $N \geq 500$ and $\lambda \geq 4$. The difference between the results by simulation and calculation using (13.23) is indiscernible, while the difference between the results by simulation and calculation using (13.22) is noticeable. Figure 13.1 shows the complementary cdf of the PAPR given by (13.23), which is the probability that the PAPR is greater than the abscissa. From the figure it can be seen that the probability of large PAPR increases as the number of channel increases. The probability that PAPR is greater than certain threshold decreases rapidly at moderate λ level. For example, it is less than 10^{-5} for a threshold of 12 dB, less than 10^{-6} for a threshold of 13 dB, and so forth.

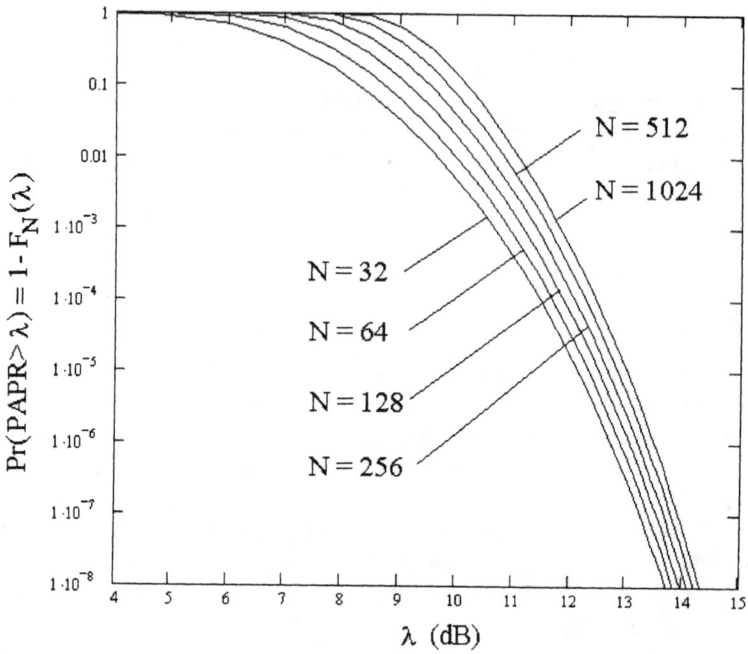

Figure 13.1 Probability that PAPR is greater than the abscissa.

13.3 INTRODUCTION TO PAPR REDUCTION TECHNIQUES

There are literally hundreds of papers for all kinds of PAPR reduction techniques. It is impossible and also not necessary to describe each of them here. In the sections that follow, we will present selected schemes for PAPR reduction. The criteria for selection are the efficiency and complexity as well as representativeness. Schemes selected include those having small or moderate gains but are simple to implement and those having high gains but high complexities.

PAPR reduction schemes can be classified into four groups. The schemes in the first group clip the peak samples and use a variety of methods to mitigate the clipping noise that resulted, such as oversampling and filtering. The schemes in the second group alter the amplitude of OFDM signal samples other than clipping, such as companding. The third group of methods alter the data before performing IFFT. A typical method used in this group is the phase rotation. The fourth group uses error-control coding techniques, which suppress PAPR as well as bit error rate at the expense of certain bandwidth expansion and system complexity. Some codes for PAPR reduction are very simple with small gains and others are more sophisticated with higher gains.

In the following, we present these four groups of PAPR reduction techniques one after another. We will largely stick to the notations used in the references, even though some of them may not be consistent with what are used previously in this book, so that the reader can easily refer to the original papers for further details.

13.4 CLIPPING AND CLIPPING NOISE MITIGATION

Clipping is done by limiting the amplitude of the OFDM samples to some desired threshold. Clipping is probably the first thing that comes to one's mind when thinking of PAPR reduction. From Figure 13.1 we see that large peaks are rare. Hopefully simply clipping the high peaks before sending the samples to the DAC would not introduce too much performance degradation. Indeed clipping is the simplest and most widely used method among all proposed PAPR reduction methods.

Results of performance analysis or simulation with clipping and filtering were presented in many articles for various system conditions [7–14]. There are two problems associated with clipping. First, the OFDM signal is distorted so that orthogonality between subcarriers is impaired. This leads to an increase in bit error probability in the receiver. Second, clipping leads to out-of-band spectrum regrowth. This can be intuitively understood as follows. Clipping is equivalent to multiplying the OFDM signal by a rectangular time window with rectangular narrow notches. The window function is one when samples are below the threshold and less than one when samples are above the threshold. The clipped OFDM spectrum is the result of convolution of this window's spectrum with the unclipped OFDM spectrum. The narrow notches in the clipping window will produce out-of-band sidelobes in the spectrum. This causes the clipped OFDM spectrum to regrow a certain amount of out-of-band sidelobes. Moreover, the low-pass filtering process of the clipped OFDM signal samples leads to considerable regrowth of peak power [16].

To mitigate the problems associated with clipping, several approaches were proposed in the literature [15–23]. In the following we describe some important methods in a concise way, emphasizing the concepts.

13.4.1 Decision-Aided Reconstruction

Kim et al. proposed a method called decision-aided reconstruction (DAR) to mitigate the clipping noise [15]. The method is based on the phenomenon that when the symbol decisions of the demodulator are converted back to time domain, the OFDM signal samples that had a large amplitude and were clipped in the transmitter will actually regrow. Although these samples will still be somewhat distorted due to clipping, they are now much less affected by the clipping noise. Consequently, an improvement can be made by replacing the clipped samples with the decisions and redoing the demodulation (FFT).

The targeted channel is a static time-dispersive channel with AWGN. The synchronization is assumed perfect. The guard interval is large enough so that there is no ISI in the received signal. Under these assumptions, the demodulator (FFT) output is

$$Z_n = \alpha_n \eta_n X_n + Q_n, \quad 0 \le n \le N-1 \tag{13.24}$$

where X_n is the nth transmitted symbol, α_n is a multiplicative distortion factor from the clipping operation, η_n is the complex channel gain, and Q_n is the sum of the clipping noise and AWGN. Typically, for a large clipping ratio, $\alpha_n \approx 1$. Since α_n changes randomly from block to block depending on which samples are clipped, estimating is difficult and may not be feasible at all. On the other hand, η_n can be estimated with very good accuracy [15, Section IV-B].

The receiver is assumed to know the clipping threshold A. The method is iterative and is quoted from [15, Section III] as follows.

1. By using the $\{Z_n\}$ in (13.24), an estimate of the clipped signal, $\{\widehat{y}_k\}$, is obtained and stored in memory by performing IFFT on $\{\widehat{Y}_n\}$, where

$$\widehat{Y}_n = Z_n / \eta_n, \quad 0 \le n \le N-1 \tag{13.25}$$

2. Decisions on the transmitted symbols are made in the frequency domain as

$$\widehat{X}_n^{(I)} = \min_{\{X\}} |Z_n^{(I)} - \eta_n X|, \quad 0 \le n \le N-1 \tag{13.26}$$

   where $\{X\}$ is the signal constellation. I represents an iteration number with an initial value of $I = 0$, and $Z_n^{(I)} = Z_n$.
3. The decisions in step 2 are converted back to the time domain using an IFFT, yielding $\{\widehat{x}_k^{(I)}\}$.
4. The clipped samples are detected and new sequence, $\{\widehat{y}_k^{(I)}\}$, is generated as

$$\widehat{y}_k^{(I)} = \begin{cases} \widehat{y}_k, & |\widehat{x}_k^{(I)}| \le A \\ \widehat{x}_k^{(I)}, & |\widehat{x}_k^{(I)}| > A \end{cases}, \quad 0 \le k \le N-1 \tag{13.27}$$

   where the sequence $\{\widehat{y}_k\}$ used in (13.27) is always retrieved from the memory in step 1. This step reconstructs the time-domain signal samples with the aid from $\{\widehat{x}_k^{(I)}\}$.

5. The sequence $\{\widehat{y}_k^{(I)}\}$ is converted to the frequency domain yielding $\{\widehat{Y}_n^{(I)}\}$.
6. Increment the index number $I \leftarrow I + 1$, and determine $\{Z_n^{(I)}\}$ by

$$Z_n^{(I)} = \eta_n \widehat{Y}_n^{(I-1)}, \quad 0 \leq n \leq N-1 \tag{13.28}$$

7. Go back to step 2. Decisions are made yielding $\{\widehat{X}_n^{(I)}\}$.

This completes the Ith iteration of the DAR algorithm. To continue iterations, repeat steps 3–7.

Numerical results in [15] show that the symbol error probability of 64QAM-OFDM is significantly improved with the DAR algorithm (Figures 13.2 and 13.3). Figure 13.2 also shows the influence of block size on the algorithm's effectiveness. It was found that this DAR scheme works better with large block sizes than small block sizes. Figure 13.3 shows the influence of iteration numbers. It is seen that with $I = 3$ the symbol error rate almost reaches the lower bound.

13.4.2 Oversampling and Frequency-Domain Filtering

The work in [16] considered the digital clipping performed on the OFDM signals sampled at the Nyquist rate, followed by the ideal low-pass filter (LPF). The results of [16] show that the significant PAPR reduction after LPF cannot be achieved by the Nyquist-rate clipping. Simulation results have shown, however, that the PAPR reduction capability can be considerably enhanced if the oversampled OFDM signal is digitally clipped and bandpass filtered [7,11]before low-pass filtering. Furthermore, the performance degradation is also alleviated [7, 11].

The scheme of oversampling and frequency-domain filtering was proposed by several authors [17, 18, 20, 24]. The frequency-domain filtering is implemented by an FFT pair. The minor difference between these methods is that [18] uses clipping and filtering repeatedly to reach a desired amplitude threshold, while others use clipping and filtering just once. Later, an algorithm that can replace the FFT pair for the frequency-domain filtering of the clipped signal was proposed by Leung et al. [19]. The complexity of Leung's algorithm is much less than that of the FFT-based original scheme.

The work by Ochiai et al. [17] not only presented the scheme, but also provided analysis of the PAPR reduction capability and channel capacity with the oversampling and frequency-domain filtering scheme. The presentation here follows that of [17].

The OFDM system considered in [17] is realized by a JN-point IDFT followed

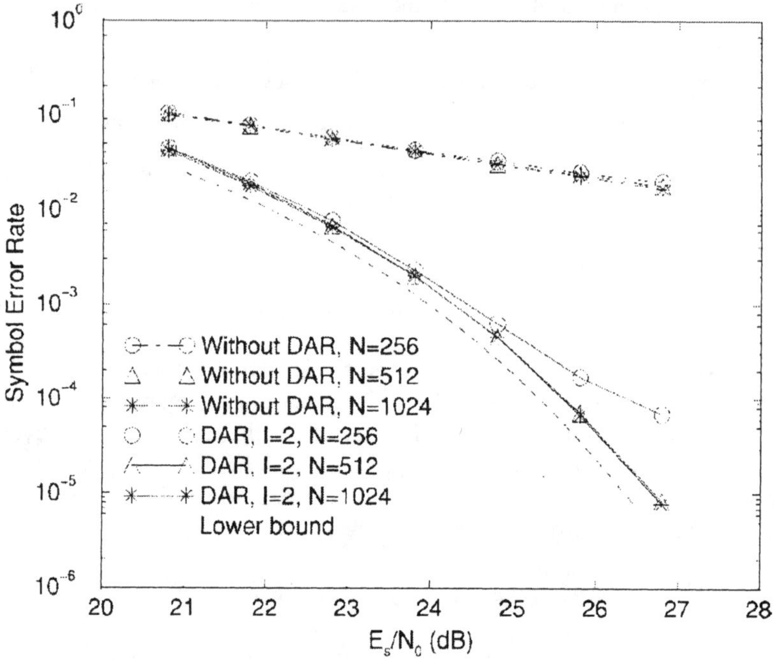

Figure 13.2 On AWGN chnnel with 64-QAM; $CR = 5$ dB. The lower bound is achieved when clipping is not used. From [15]. Copyright © 1999 IEEE.

by clipping and filtering, where N is the number of samples of a OFDM symbol for the Nyquist sampling rate and J is an oversampling factor. J could be an integer or noninteger, as long as JN is an integer. Figure 13.4 shows the system proposed in [17].

The data sequence is

$$\mathbf{A} = \{A_0, A_1, ..., A_{N-1}\}$$

After zero padding it becomes

$$\mathbf{A}' = \{A_0, A_1, ..., A_{N-1}, \underbrace{0, 0, ..., 0}_{(J-1)N \quad 0's}\}$$

Other symbols in the figure are self-explanatory except that the function of the offset block needs explanation. The frequency offset introduced here is mainly for analyti-

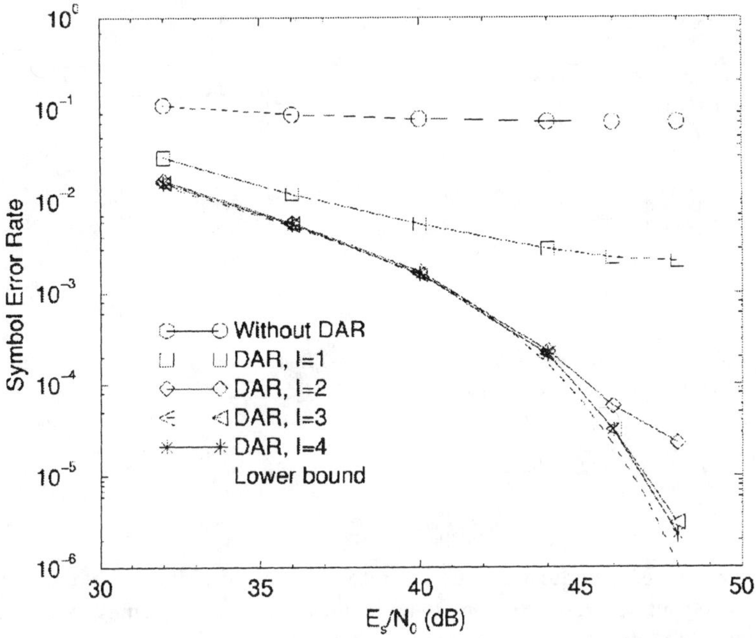

Figure 13.3 On static ISI channel with 64-QAM; $CR = 4$ dB, $N = 1024$. From [15]. Copyright © 1999 IEEE.

cal purposes; the offset guarantees the power spectrum of the input baseband signal to be conjugate symmetric, which will simplify the derivation of the signal-to-distortion ratio (SDR) of the clipped OFDM signal [17]. The offset is immaterial to analysis of peak power, since it does not change the amplitude of the complex envelope.

To remove the out-of-band frequency components due to clipping, a novel low-pass equivalent BPF (Figure 13.4(b)) is inserted before the LPF (in the DAC). This BPF actually is an LPF, but it is referred to as a BPF in distinction from the conventional LPF in the DAC. This BPF is implemented by a JN-point DFT/IDFT pair. First the clipped OFDM samples are converted to frequency domain using a JN-point DFT:

$$\widetilde{\mathbf{A}}' = \{\widetilde{A}_0, \widetilde{A}_1, ..., \widetilde{A}_{N-1}, \underbrace{\widetilde{A}_N, ..., \widetilde{A}_{JN-1}}_{\text{out-of-band}}\}$$

Then the out-of-band components are removed to obtain the distorted version of the

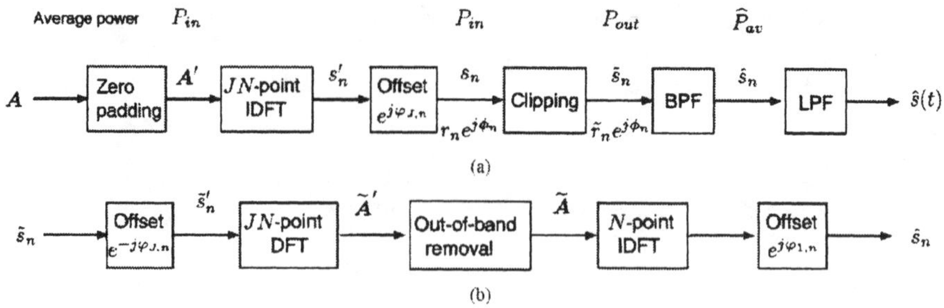

Figure 13.4 OFDM system with soft envelope limiter. (a) Transmistter, and (b) the low-pass equivalent BPF. From [17]. Copyright © 2000 IEEE.

original data sequence of length N

$$\widetilde{\mathbf{A}} = \{\widetilde{A}_0, \widetilde{A}_1, ..., \widetilde{A}_{N-1}\}$$

This sequence is remodulated back again to time domain by an N-point IDFT. In this way, the out-of-band radiation caused by the clipping process may be completely eliminated regardless of the oversampling factor J.

A parameter called *clipping ratio* γ is defined as

$$\gamma \triangleq \frac{A_{\max}}{\sqrt{P_{in}}} \tag{13.29}$$

where $A_{\max}$ is the maximum permissible amplitude over which the signal is clipped and $P_{in} = 2\sigma^2$ is the average input power of the OFDM signal before clipping. Two extremes of γ are $\gamma = 0$ and $\gamma = \infty$. The case of $\gamma = \infty$ corresponds to an ideal system without clipping. The case of $\gamma = 0$ corresponds to a hard (constant) envelope limiter. The case of $\gamma = 0$ seems impossible from (13.29); but it can be defined as a limit. When γ approaches zero, it can be shown that the output signal amplitude to rms output signal power ratio is 1, for all samples [17]; this justifies the claim that the limiter becomes a hard limiter when $\gamma \to 0$.

The following simulation results are based on several other assumptions: The LPF in the DAC of the transmitter is ideal, the amplification is linear after clipping and filtering, the channel is a slow Rayleigh fading channel, and timing synchronization and carrier recovery are perfect.

PAPR of clipped and bandlimited OFDM signals is studied under the above conditions. Figure 13.5 shows the complementary cdf of the PAPR, which is in fact the probability that the PAPR is greater than the abscissa, for different oversampling fac-

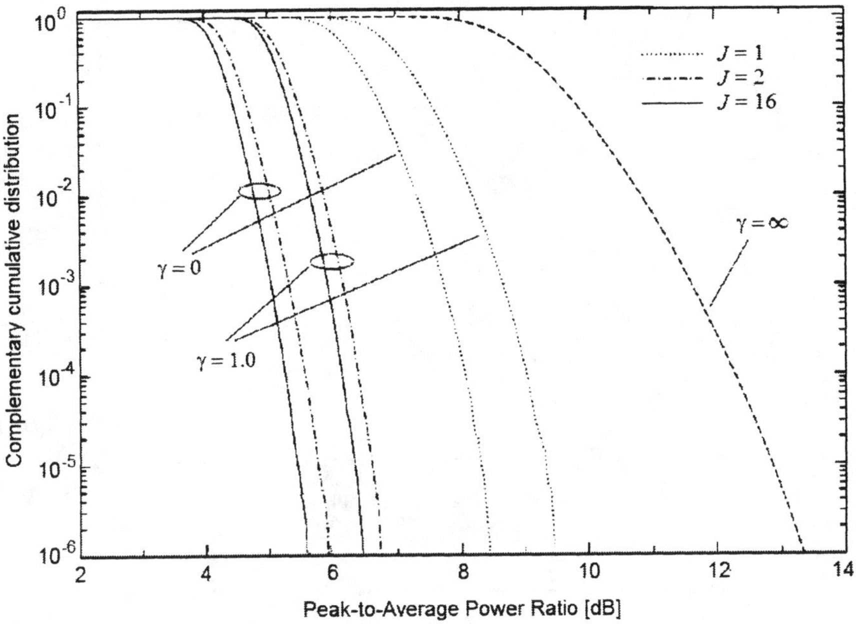

Figure 13.5 Comparison of the complementary cdf of the PAPR of the clipped and bandlimited OFDM signals with oversampling factor $J = 1, 2$, and 16 ($N = 512$, QPSK). From [17]. Copyright © 2000 IEEE.

tors J and clipping ratios γ. The curve of $J = \infty$ is plotted using (13.23). The other curves are based on simulations. We can see that even though the Nyquist-rate clipping considerably reduces the PAPR compared to that without clipping, the oversampling yields significantly lower PAPR. It also can be seen that oversampling by a factor of 2 achieves most of the reduction due to oversampling already; further increasing the sampling factor has no significant effect. Therefore $J = 2$ is a good choice in a practical system, which can reduce the PAPR by 2–3 dB from the $J = 1$ case, and 6–7 dB from the unclipping case.

Channel capacity is studied under the above conditions. Figures 13.6 and 13.7 show channel capacity of the clipped OFDM signal with QPSK input, and 16QAM input, respectively. In the figures, perfect CSI (channel state information) is assumed, which means the Rayleigh channel's fading gain is known to the receiver. In practice, it is estimated. As expected, without clipping ($\gamma = \infty$), QPSK achieves a capacity of 2 bits/subcarrier and 16QAM 4 bits/subcarrier, asymptotically. As γ decreases,

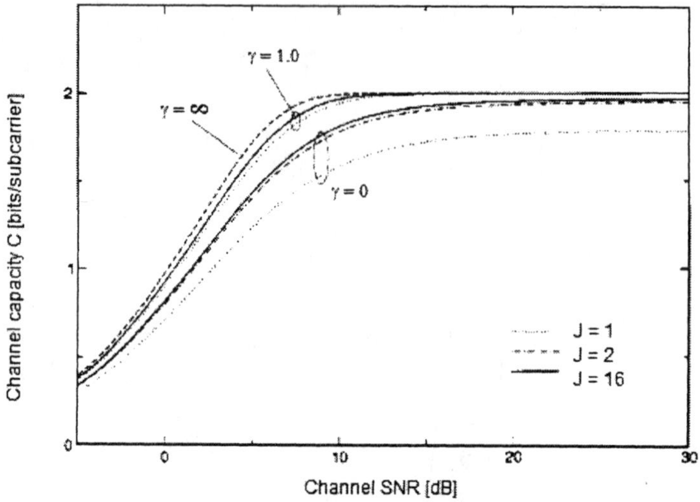

(a)

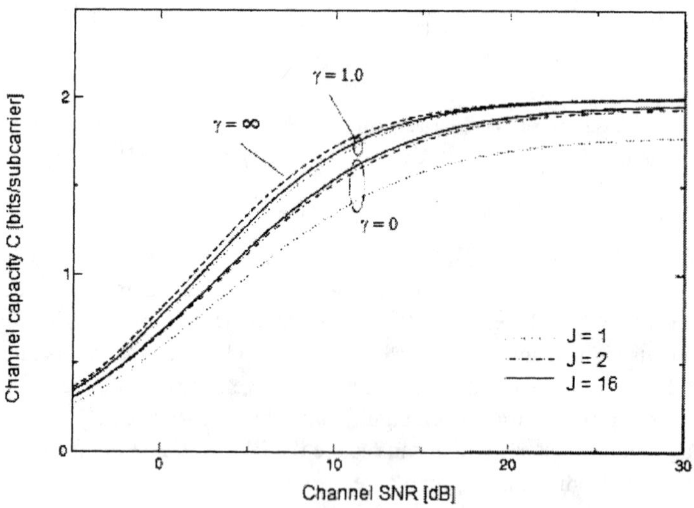

(b)

Figure 13.6 Channel capacity of the clipped OFDM signal with QPSK input. (a) AWGN channel. (b) Rayleigh fading channel (perfect CSI). From [17]. Copyright © 2000 IEEE.

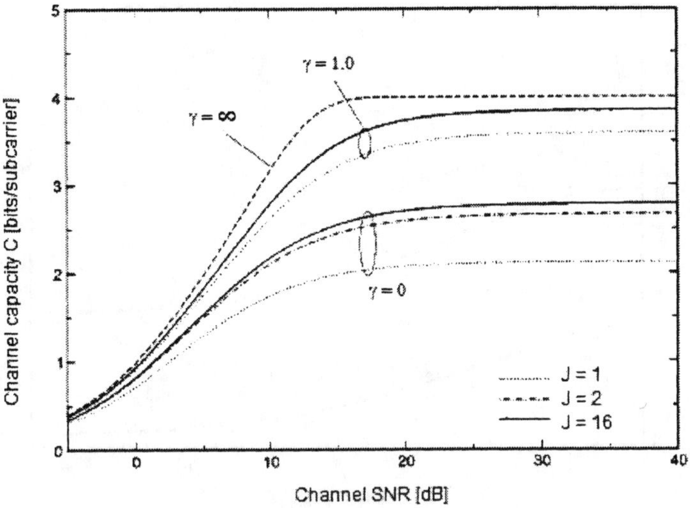

(a)

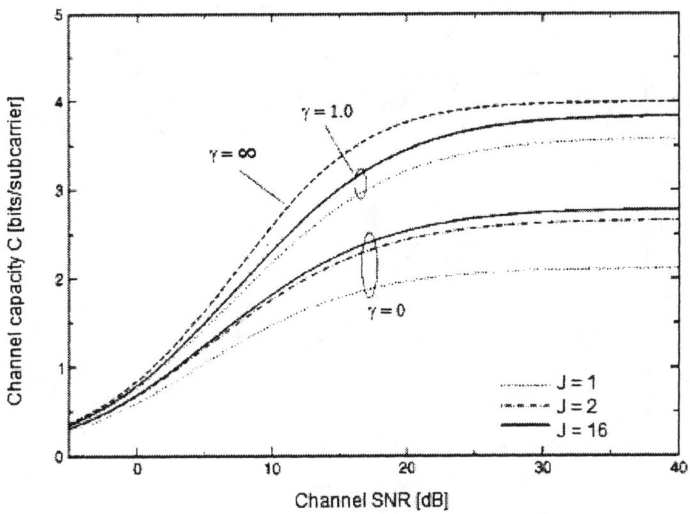

(b)

Figure 13.7 Channel capacity of the clipped OFDM signal with 16QAM input. (a) AWGN channel. (b) Rayleigh fading channel (perfect CSI). From [17]. Copyright © 2000 IEEE.

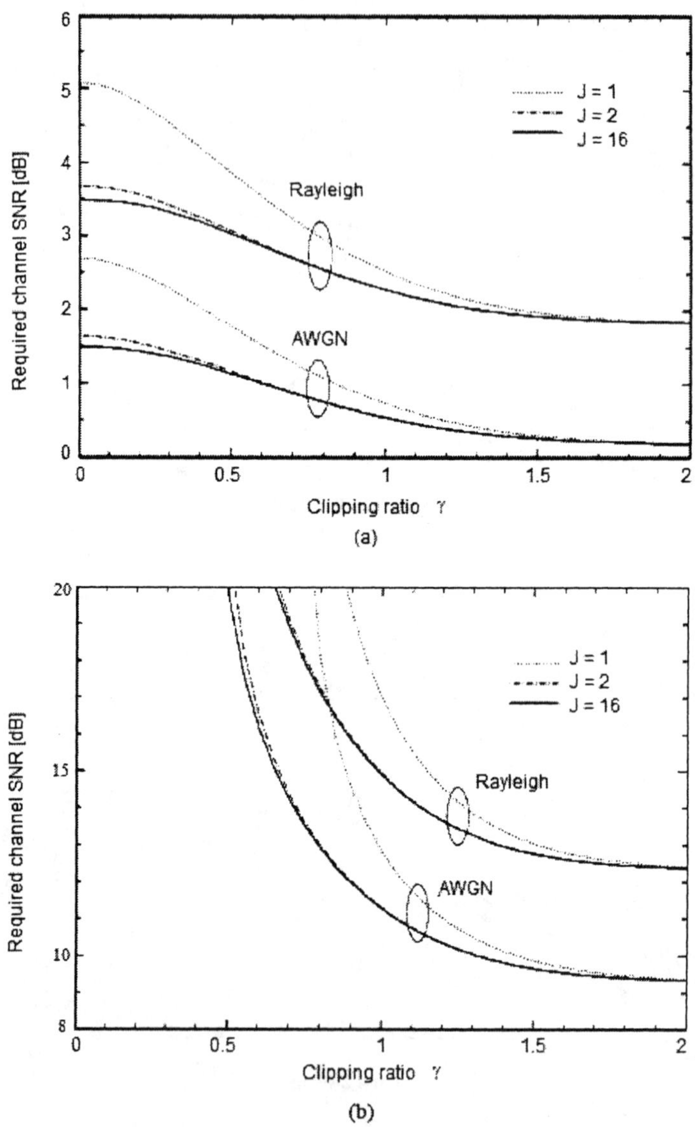

Figure 13.8 Required channel SNR for achieving channel capacity. (a) QPSK with 1 bit/subcarrier. (b) 16QAM with 3 bit/subcarrier. From [17]. Copyright © 2000 IEEE.

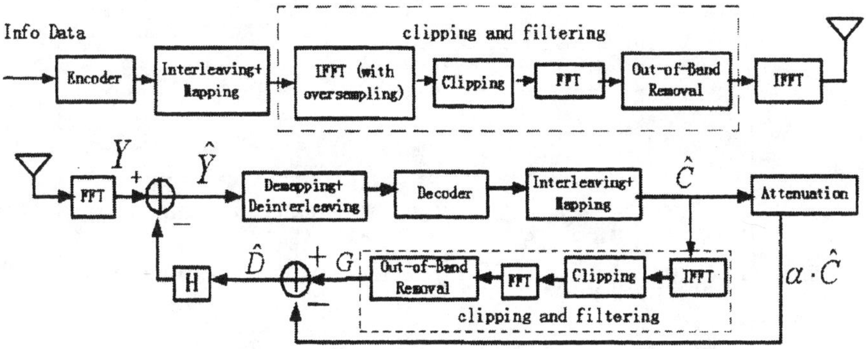

Figure 13.9 OFDM transmitter with (top) clipping and filtering and (bottom) receiver with iterative distortion cancellation. From [23]. Copyright © 2003 IEEE.

achievable capacity decreases. In particular for $\gamma = 0$, it is impossible to achieve error-free performance for 2 bits/subcarrier, no matter how much one may increase the SNR.

For a given targeted channel capacity, the required SNR is shown in Figure 13.8 as a function of the clipping ratio. The lower the clipping ratio (more clipping), the higher the SNR required.

13.4.3 Iterative Estimation and Canceling

An iterative method of estimating and cancelling the clipping noise is proposed in [23]. The system is shown in Figure 13.9. The transmitter part is the same as in Figure 13.4. The difference is in the receiver, which has an additional feedback-cancellation subsystem. This subsystem re-creates the clipping-filtering process of the transmitter using the detected data symbol $\widehat{C}$, then estimates and cancels the frequency domain clipping noise caused by it. Assuming perfect synchronization, the output of the FFT in the receiver is

$$Y_k = H_k(\alpha C_k + D_k) + Z_k, \quad k = 0, 1, ..., N-1$$

where H_k is the complex channel gain. α is the attenuation factor due to clipping, which is a function of clipping ratio. D_k represents clipping noise in frequency domain. Z_k is the AWGN in frequency domain.

The receiver works in an iteration fashion as described below (refer to

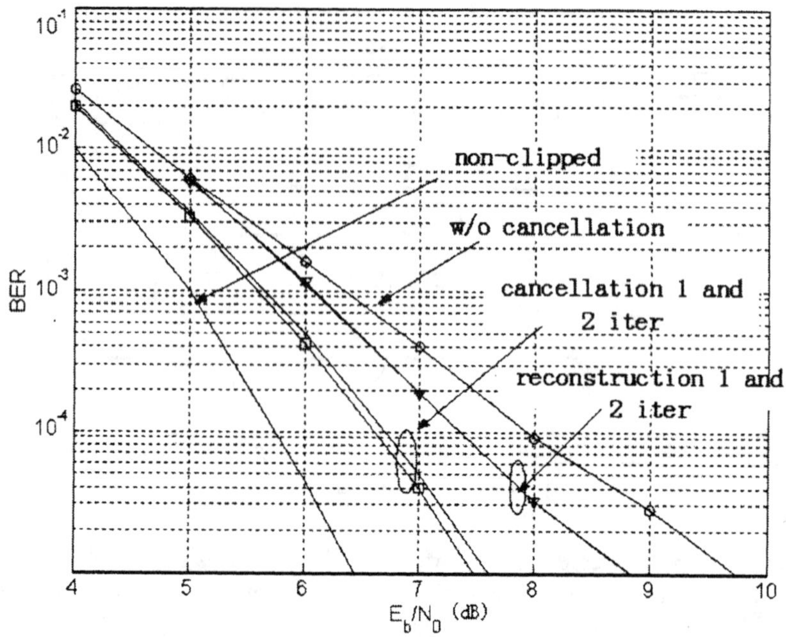

Figure 13.10 BER performance of the proposed receiver over AWGN channel and comparison with "signal reconstruction" approach. From [23]. Copyright © 2003 IEEE.

Figure 13.9):

(a) Channel observations $\{Y_k\}_{k=0}^{N-1}$ are decoded, detected, and recoded to create $\{\widehat{C}_k\}_{k=0}^{N-1}$.

(b) Sequence $\{\widehat{C}_k\}_{k=0}^{N-1}$ is sent to the clipping-filtering branch to obtain an estimate of the output of the clipping-filtering block in the transmitter, that is

$$G_k = \alpha\widehat{C}_k + \widehat{D}_k, \quad k = 0, 1, ..., N-1$$

$\widehat{C}_k$ also passes an attenuation block to form $\alpha\widehat{C}_k$which is subtracted from G_k to obtain an estimate of clipping noise as

$$\widehat{D}_k = G_k - \alpha\widehat{C}_k, \quad k = 0, 1, ..., N-1$$

(c) The estimated clipping noise is subtracted from the Y_k to obtain the refined

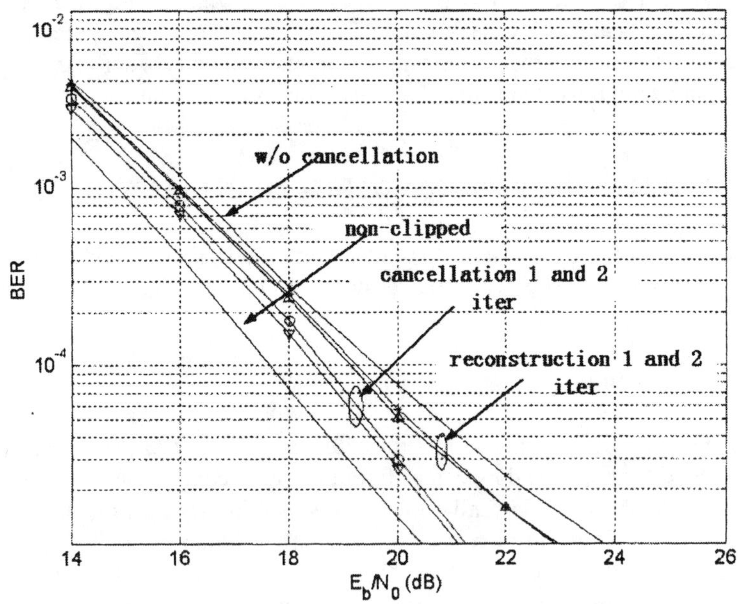

Figure 13.11 BER performance of the proposed receiver over Rayleigh fading channel and comparison with the "signal reconstruction" approach. From [23]. Copyright © 2003 IEEE.

complex symbol

$$\widehat{Y}_k = Y_k - H_k\widehat{D}_k = \alpha H_k C_k + H_k(D_k - \widehat{D}_k) + Z_k, \quad k = 0, 1, ..., N-1$$

where $(D_k - \widehat{D}_k)$ is the residual clipping noise.

(d) Go back to step (a) and replace $\{Y_k\}_{k=0}^{N-1}$ with $\{\widehat{Y}_k\}_{k=0}^{N-1}$.

The loop (a)–(d) continues for a few iterations. As the iteration proceeds, the estimation of the clipping noise is found to become increasingly accurate and the receiver performance is improved.

Simulation results shown in Figures 13.10 and 13.11 are for IEEE Standard 802.11a with the following parameters. The number of subcarriers is $N = 64$. Modulation is 16QAM. Channel code is a rate 1/2 convolutional code with generator polynomials $g_0 = 133_8$ and $g_1 = 171_8$, where the subscript 8 indicates the number is octal. Decoding is carried out using a soft Viterbi algorithm. The clipping ratio $\gamma = 0$ dB. E_b/N_0 is measured after signal clipping and filtering. From the figures we can see that in AWGN and Rayleigh fading channels the proposed method shows about

2.5 dB and 3 dB gains at $BER = 10^{-5}$ versus the case without cancellation, respectively. The results also show that no more than two iterations are required to secure the most of the gain, and the performance with this method is better than that of the method that attempts to reconstruct the signal to its original value.

13.5 AMPLITUDE ALTERATION OTHER THAN CLIPPING

This group of PAPR reduction methods alter the amplitude of the time-domain OFDM samples other than clipping. They include companding and complementary clipping transform.

13.5.1 Companding

Using the companding technique to reduce OFDM signals' PAPR was proposed in [24,25]. The idea of companding comes from the use of companding for speech signals. Since OFDM signal is similar to speech signal in the sense that large signals only occur very infrequently, the same companding technique might be effective for improving the PAPR of the OFDM signals.

In [25], the compression in the transmitter is governed by the μ-law compression equation

$$s_c(n) = \frac{A\mathrm{Sgn}\langle s(n)\rangle \ln\left[1 + \mu\left|\frac{s(n)}{A}\right|\right]}{\ln(1+\mu)}$$

where A is a normalization constant such that $0 \leq \left|\frac{s(n)}{A}\right| \leq 1$, $s(n)$ is the original OFDM signal, and $s_c(n)$ is the compressed signal. The expansion in the receiver is governed by the μ-law expanding equation

$$s'(n) = \frac{1}{\mathrm{Sgn}\langle s_c(n)\rangle \mu}\left[A\exp\left\{\frac{s_c(n) + q(n) + w(n)}{A\mathrm{Sgn}\langle s_c(n)\rangle}\ln(1+\mu)\right\} - A\right]$$

where $q(n)$ is the ADC error and $w(n)$ is the AWGN. The analysis shows that in the sense of a minimum quantization error, the optimal μ is

$$\mu = \sqrt{N-2}$$

where N is number of subcarriers of the OFDM signal. Symbol error probability is improved with companding (Figure 13.12). The improvement is the largest when $\mu = \sqrt{N-2}$ according to the analysis in [25]. This can be seen in Figure 13.12 where curve 3 is the best at high SNRs and it is for $\mu = 16$, a value near $\sqrt{N-2}$.

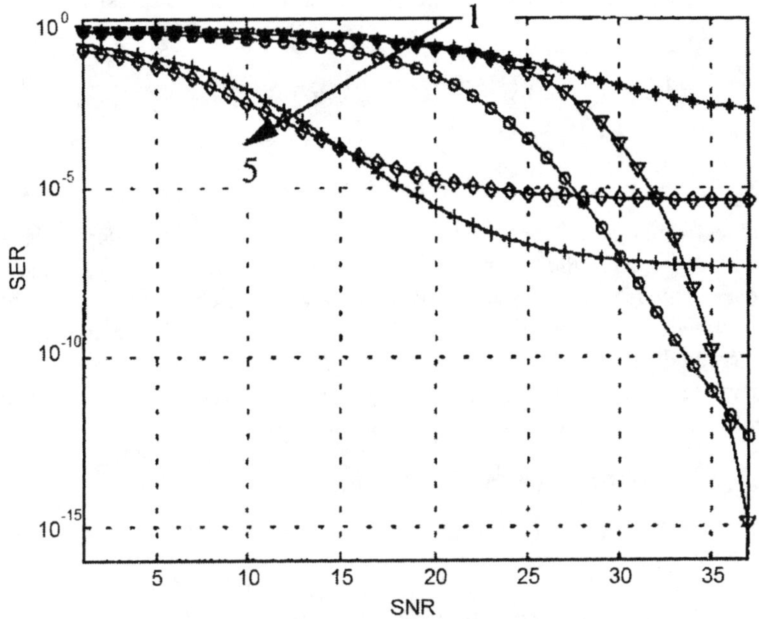

Figure 13.12 Error probability performance of the 16-QAM OFDM system with $N = 256$. The companding coefficients for curves 1, 3, 4, and 5 are 0, 16, 100, and 255, respectively. The word length is assumed to be 10 bits. Curve 1 is for the case without companding but with quantization noise. Curve 2 is for the case without companding but quantization noise is omitted. From [25]. Copyright © 1999 IEEE.

However, at lower SNRs curve 4 is the best which is for $\mu = 100$.

13.5.2 Complementary Clipping Transform

The idea of complementary clipping transform proposed in [26] is to replace large OFDM samples $s(n)$ by their complementary values which would be clipped otherwise. However, this method is only applicable to real-output OFDM signals, which can be generated using the methods in Section 12.3.

Given samples of a real OFDM symbol $\{s(n)\}_{n=0}^{N-1}$, the complementary clipping transform is defined as follows

$$z(n) = \begin{cases} s(n), & |s(n)| \leq A/2 \\ jy(n), & |s(n)| > A/2 \end{cases} \tag{13.30}$$

where $j = \sqrt{-1}$ and $y(n)$ is the complementary transform (CT) of $s(n)$ defined as

$$y(n) = CT(s(n)) = (A - |s(n)|) \cdot \text{sgn}(s(n)), \quad n = 0, 1, ..., N-1 \qquad (13.31)$$

where

$$A = \max\{|s(n)|\}$$

A changes from symbol to symbol. This transform reduces the amplitude of the samples but preserves their sign. The purpose of applying j to $y(n)$ to make it an imaginary value is to label transformed samples so that they can be easily identified in the receiver. In other words, the transformed samples form the imaginary part of the transmitted signal.

In the receiver the original OFDM signal samples can be recovered from the received samples $r(n)$ by the inverse transform as

$$s'(n) = CT^{-1}(r(n)) = (A' - |r(n)|) \cdot \text{sgn}(r(n)), \quad n = 0, 1, ..., N-1 \qquad (13.32)$$

where $A' = \max\{|r(n)|\}$. A' also changes among symbols. In a noise-free channel, $r(n) = y(n)$ and

$$A' = \max\{|y(n)|\} = A - \min\{|s(n)|\} = A - B$$

where

$$B = \min\{|s(n)|\}$$

Thus from (13.32) we can derive

$$s'(n) = \begin{cases} A' - y(n), & y(n) \geq 0 \\ -(A' + y(n)), & y(n) < 0 \end{cases} = \begin{cases} s(n) - B, & s(n) \geq 0 \\ s(n) + B, & s(n) < 0 \end{cases} \qquad (13.33)$$

From (13.33) we see that transformed samples are not completely recovered even in a noise-free channel unless B is zero. However, this is better than clipping since B is usually very small if not zero, much less than most clipped portions of samples. Thus, we can expect this method to have a better error performance. The simulation results for a 1,024-subcarrier QPSK-OFDM in [26] show that it outperforms the clipping system (without any mitigation) by 5 dB at $BER = 10^{-4}$ and $CR = 1.2$ (CR = clipping ratio).

13.6 PRE-IFFT DATA ALTERATION

This group of PAPR reduction techniques alters the pre-IFFT data to control post-

IFFT signal waveforms so that the PAPR is lowered. This group includes selective mapping (SLM), repeated random phase transform, selective scrambling, partial transmit sequence (PTS), and dummy sequence insertion (DSI).

13.6.1 Selective Mapping

The selective mapping (SLM) proposed in [27] generates N independent OFDM frames (symbol vectors used as inputs to the IFFT) representing the same information, then selects the one with the lowest PAPR to transmit. One possible promising method of generating them is to multiply the same data frame by different phase vectors. Define N distinct phase vectors $\mathbf{P}^{(n)} = [P_1^{(n)}, P_2^{(n)}, ..., P_D^{(n)}]$, where D is the dimension of the OFDM (i.e., the number of subcarriers) and $P_\mu^{(n)} = e^{j\varphi_\mu^{(n)}}$, $\varphi_\mu^{(n)} \in [0, 2\pi)$, $\mu = 1, 2, ..., D$, $n = 1, 2, ..., N$. This vector is used to multiply the information symbols, $V[\mu]$, as

$$V^{(n)}[\mu] = V[\mu]e^{j\varphi_\mu^{(n)}}, \quad \mu = 1, 2, ..., D,\ n = 1, 2, ..., N \tag{13.34}$$

Then all N frames are transformed into time domain and the one with the lowest PAPR is selected for transmission.

Recall that the probability that a sample is less than a threshold Y is $1 - e^{-Y}$ (see (13.20)). Then the probability that at least one sample is above the threshold is

$$\Pr\{PAPR > Y\} = 1 - (1 - e^{-Y})^D$$

The probability that the lowest PAPR in all N independent frames is above the threshold is

$$\Pr\{PAPR_{low} > Y\} = [\Pr\{PAPR > Y\}]^N = [1 - (1 - e^{-Y})^D]^N \tag{13.35}$$

Figure 13.13 shows some results from (13.35) for different values of N. From the figure it is seen that the probability of high peaks is greatly reduced with big N. For example, at $\Pr\{PAPR_{low} > Y\} = 10^{-6}$, the clipping happens in one out of 10^6 frames, which can be neglected in practice; even for $N = 4$, the gain is more than 3 dB. Simulation results in [27] also confirmed the results.

To recover data, the receiver must know which phase vector $\mathbf{P}^{(n)}$ is used. Thus, the side information of the number n needs to be transmitted to the receiver. This information is the most important so it must be protected by coding. As indicated in [27], even for simple block codes and moderate codeword lengths, erroneous detection of $\mathbf{P}^{(n)}$ can be neglected.

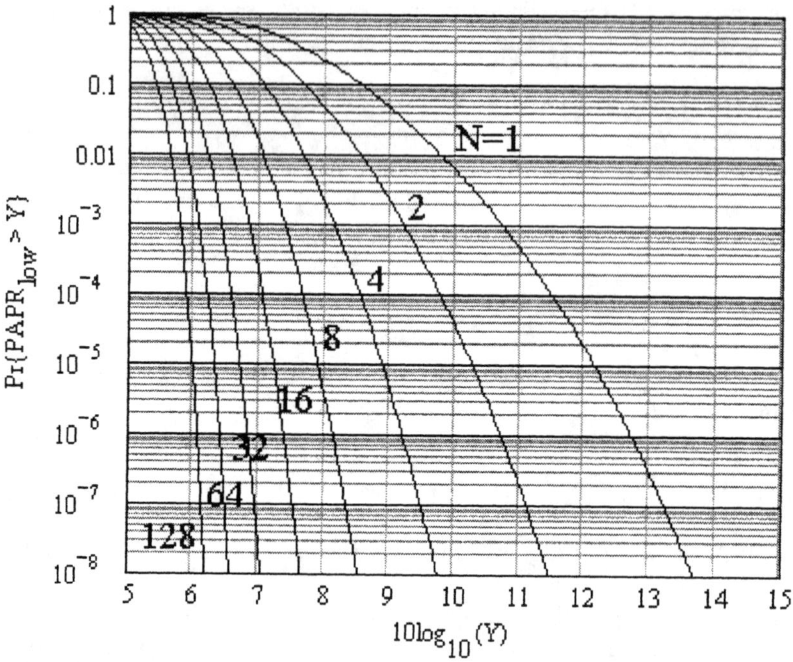

Figure 13.13 Complementary cumulative distribution function of PAPR, using selective mapping (SLM), if the frame with the lowest PAPR is selected out of N statistically independent frames. From [27]. Copyright © 1996 IEE.

13.6.2 Repeated Random Phasor Transform

Mestdagh et al. proposed a PAPR reduction method which is somewhat similar to selective mapping (SLM), but it is implemented serially [28]. The method is to reduce the probability of clipping by repassing randomly phasor-transformed signals through IFFT. A peak detector is used in the transmitter to detect whether the samples of OFDM symbols generated by the IFFT exceeds a threshold A_{clip}, which is chosen so that the probability of amplitude clipping is lower than a specified value. If the amplitude of the OFDM symbol never exceeds A_{clip} (Case a), then the symbol is sent to the transmitter front-end (including DAC, up-converter, HPA, and so forth) without any change. If the generated OFDM symbol has at least one sample whose amplitude exceeds A_{clip} (Case b), then it is not passed directly to the transmitter front end. Instead, the phasor of each QAM-modulated carrier is changed by means of a fixed phasor-transformation and a new OFDM symbol is generated by the IFFT. By careful selection of the phasor-transformation, the probability of clipping this new symbol (second pass) will be very low. The receiver is informed about whether case a or case b happened. Only one bit per OFDM symbol is required to carry this extra information. This bit could be provided by modulating the pilot tone that otherwise carries no information and that is used to maintain synchronization. Forward error correction coding and/or duplication of this information over another or several tones can be used to improve the reliability of this data recovery. The phasor-transformation can be in many forms. An easy-to-implement phasor-transformation is the fixed random phasor-transformation (known to the receiver).

It was easily shown that for large N and random phasor-transformation, the total probability of clipping after p passes is

$$P_{clip/total} = P^{p}_{clip/1} = \left[1 - \text{erf}\ ^{2N}(\frac{\mu}{\sqrt{2}})\right]^{p} \tag{13.36}$$

where $\mu = A_{clip}/\sigma$, and σ^2 is the total OFDM signal power; $\text{erf}(x)$ is defined in Section 4.3. When $P_{clip/1}$ is small, $P_{clip/total}$ will be very small. Figure 13.14 shows the probability of clipping the transformed OFDM signal as a function of μ for $p = 1, 2, 3$. We can see that after only two to three passes, the clipping probability is greatly reduced.

In addition, performance in terms of SNR (noise sources include clipping noise, DAC quantization noise, AWGN, and ADC quantization noise) was analyzed and significant improvements in SNR of about 3 dB up to 8 dB as compared to the case of brute-force clipping were observed. This method is simple and efficient, without the use of any precoding and hence any increase in data rate.

13.6.3 Selective Scrambling

This method is proposed in [29] for QPSK-OFDM. The message bit sequence is scrambled with each of four fixed, cyclically inequivalent m-sequences to form four scrambled bit streams (see [5, pp. 396–399] for m-sequence). Each of them is identified by one of four identifier bit pairs: 00, 01, 10, and 11, which is added to the head of the sequence. The identifier, which forms one QPSK symbol, will be used by the receiver to select the appropriate descrambler. One of the scrambled sequences is then chosen to be the one for transmission. The criterion of choosing the sequence should be the minimum PAPR of the time-domain OFDM signal generated by modulating the sequence onto the subcarriers. However a simpler, even though maybe not so accurate, criterion is developed in [29]. It is a heuristic *selection function* (SF), which

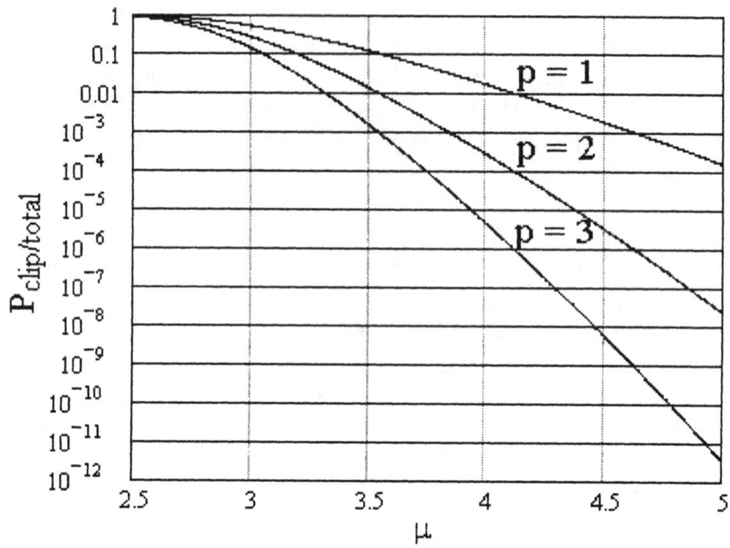

Figure 13.14 Probability of clipping the transformed OFDM signal as a function of μ for $p = 1, 2, 3$.

is developed as follows.

When a scrambled binary sequence with a high proportion of 1s or 0s is mapped into QPSK symbols and then inputted to an N-point IFFT modulator, it will generate subcarrier phasors that tend to align at some point in time, thereby giving high PAPR. On the other hand, a binary sequence of length $2N$ with Hamming weight W_H close to N often generates low PAPR. However, this is not a sufficient condition since highly structured sequences similar to $\{10101010...\}$ or $\{11001100...\}$, for example, will also have $W_H \approx N$, yet will generate high PAPR in QPSK-OFDM. Fortunately, structured sequences can be detected via R_1, which is defined as

$$R_1 = \frac{c_0}{c_1} + \frac{c_1}{c_2} + \frac{c_2}{c_3} + ... + \frac{c_{N-2}}{c_{N-1}} \tag{13.37}$$

where $c_k \in \{\frac{1}{\sqrt{2}}(\pm 1, \pm j)\}$ are QPSK symbols. For example, if we map 11 to $-1/\sqrt{2}$ and 00 to $1/\sqrt{2}$, then $R_1 = 1 - N$ for the sequence $\{11001100...\}$ of length $2N$. If a sequence is not structured, the value of R_1 will be small. A heuristic but practical selection function therefore uses both Hamming weight and R_1 and can be expressed as

$$SF = \frac{|W_H - N|^2 + |R_1|^2}{2N^2} \tag{13.38}$$

where $2N^2$ is just a normalization factor. The scrambled sequence with the lowest SF is chosen for IFFT.

Simulations for $N = 256$ show that the PAPR of the selected scrambled sequence is typically $<10\%$ of that of unscrambled ones, and is about 2% of the maximum value [29].

13.6.4 Partial Transmit Sequences

Müller et al proposed a PAPR reduction method called partial transmit sequences (PTS) [30]. This method is to partition the data sequence into several disjoint subsequences and linearly combine them to form a final sequence for IDFT. The combining coefficients are determined by an optimization process which minimizes the PAPR. This method is as follows.

First some notations are described. The subcarrier ν of the OFDM symbol at discrete time μ is modulated with the complex amplitude $A_{\mu,\eta}$. $\mathbf{A}_\mu$ is the frequency-domain vector consisting of all $A_{\mu,\eta}, 0 \leq \eta \leq D$. $\mathbf{a}_\mu = IDFT\{\mathbf{A}_\mu\}$ is the time-domain vector comprising the samples $a_{\mu,\rho}, 0 \leq \rho \leq D$. The data block $\mathbf{A}_\mu$ is partitioned into disjoint subblocks $\mathbf{A}_\mu^{(\nu)}, \nu = 1, ..., V$. Each is still a D-dimensional vector, but only part of the D positions are nonzero; other positions which are already represented in another subblock are set to zero. Thus

$$\mathbf{A}_\mu = \sum_{\nu=1}^{V} \mathbf{A}_\mu^{(\nu)}$$

For each subblock ν, a rotation factor $b_\mu^{(\nu)} = e^{j\phi_\mu^{(\nu)}}, \phi_\mu^{(\nu)} \in [0, 2\pi)$ is used to modify the subblock. The modified subcarrier block is

$$\breve{\mathbf{A}}_\mu = \sum_{\nu=1}^{V} b_\mu^{(\nu)} \mathbf{A}_\mu^{(\nu)}$$

$\breve{\mathbf{A}}_\mu$ represents the same information as $\mathbf{A}_\mu$, as long as the rotation vector $\{b_\mu^{(1)}, b_\mu^{(2)}, ..., b_\mu^{(V)}\}$, the so-called side information, is known.

Due to the linearity of IDFT, the subblocks are transformed by V separate IDFTs, yielding

$$\breve{\mathbf{a}}_\mu = \sum_{\nu=1}^{V} b_\mu^{(\nu)} IDFT\{\mathbf{A}_\mu^{(\nu)}\} = \sum_{\nu=1}^{V} b_\mu^{(\nu)} \mathbf{a}_\mu^{(\nu)} \tag{13.39}$$

where $\mathbf{a}_\mu^{(\nu)} = IDFT\{\mathbf{A}_\mu^{(\nu)}\}$ is called a *partial transmit sequence* (PTS). Note that the IDFT is a D-point IDFT. It is seen from (13.39) that each time-domain OFDM sample is a linear combination of V samples from V partial transmit sequences. Thus optimization can be performed to minimize the peak value of $\breve{\mathbf{a}}_\mu$ by suitably choosing the coefficients $b_\mu^{(\nu)}$. The optimum coefficients for the OFDM symbol μ are given by

$$\{\widetilde{b}_\mu^{(1)}, \widetilde{b}_\mu^{(2)}, ..., \widetilde{b}_\mu^{(V)}\} = \underset{\{\widetilde{b}_\mu^{(1)}, \widetilde{b}_\mu^{(2)}, ..., \widetilde{b}_\mu^{(V)}\}}{\arg\min} \left(\max_{0 \le \rho < D} \left| \sum_{\nu=1}^{V} b_\mu^{(\nu)} a_{\mu,\rho}^{(\nu)} \right| \right) \tag{13.40}$$

where $\arg\min(\cdot)$ yields the argument for which the given expression achieves the global minimum. Thus the optimized OFDM samples for transmission are

$$\widetilde{\mathbf{a}}_\mu = \sum_{\nu=1}^{V} \widetilde{b}_\mu^{(\nu)} \mathbf{a}_\mu^{(\nu)} \tag{13.41}$$

which have the lowest PAPR yielded by this PTS method.

One rotation factor can be fixed without any performance loss; for example, we can let $b_\mu^{(1)} = 1$. Simulations in [30] indicate that a restriction to four rotation angles already achieved a significant reduction in peak power. These four angles can be set to $b_\mu^{(\nu)} \in \{\pm 1, \pm j\}$, which can conveniently rotate the data without actual multiplications.

The receiver has to know $\{\widetilde{b}_\mu^{(1)}, \widetilde{b}_\mu^{(2)}, ..., \widetilde{b}_\mu^{(V)}\}$ in order to recover the original data. Hence it must be transmitted to the receiver without errors. The bits needed for this side information represent the redundancy associated with the PTS peak power reduction method.

Figure 13.15 shows the simulation results of the complementary cumulative distribution function of the crest factor ζ_0 of PTS-OFDM with QPSK, $b_\mu^{(\nu)} \in \{\pm 1, \pm j\}$, and $D = 512$. It can be seen that with $V = 4$, the gain is about 4 dB at $P_\zeta(\zeta_0) = 10^{-5}$ in comparison with $V = 1$, corresponding to the original OFDM. The redundancy needed is 6 bits for $V = 4$ since there are three free rotation factors and four (2 bits) allowed angles for each. For 512 QPSK subcarriers where there are usually some null subcarriers, these 6 bits represent a redundancy well below 1%.

Figure 13.15 also shows that PTS is better than selective mapping (SLM). Furthermore, as pointed out by [30], the method in [28] will achieve the same results as

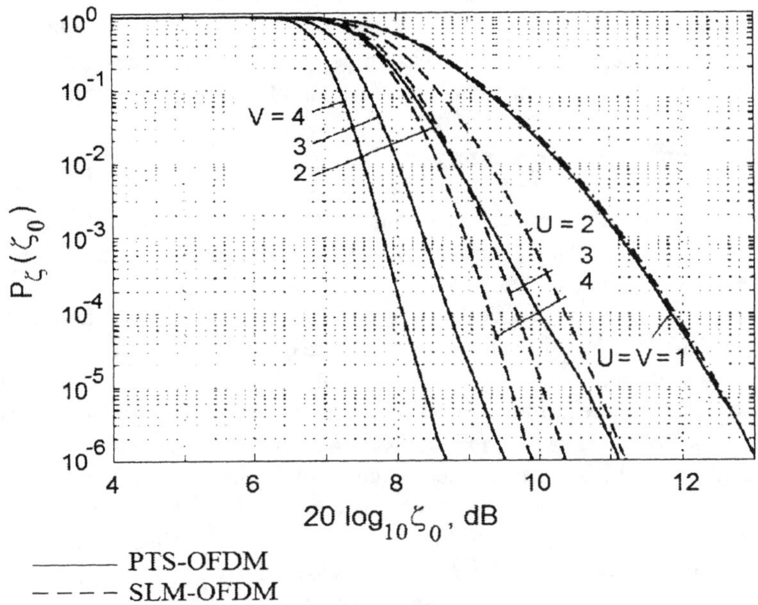

Figure 13.15 Probability $P_\zeta(\zeta_0)$ that OFDM symbol with $D = 512$ exceeds crest factor ζ_0 if partial transmit sequences (PTS) or selective mapping (SLM) is used with $V, U \in \{1, 2, 3, 4\}$. From [30]. Copyright © 1997 IEE.

SLM, if the therein introduced threshold is very low; and if the heuristic selection function in [29] works perfectly, the approach given there will achieve nearly the same results as SLM. This is to say that PTS is better than SLM and the methods in [28] and [29].

PTS works with little additional redundancy, as seen already, and a moderate increase in transmitter complexity, which includes $V = 2, ..., 4$ parallel IDFTs and an optimization step to combine the PTSs.

The optimization step was later simplified by simply flipping the signs of data in a progressive fashion [31]. Consider binary (± 1) weighting factors. As a first step, assume $b_\mu^{(\nu)} = 1$ for all ν and combine the PTSs and compute the PAPR. Next, invert the first phase factor (let $b_\mu^{(1)} = -1$) and compute PAPR again. If the new PAPR is reduced, retain $b_\mu^{(1)} = -1$ as part of the final phase factor sequence; otherwise revert it to its previous value. The algorithm continues this way until all factors are tried. The results are suboptimal in that the PAPR reduction is slightly less than that of the original PTS, but the number of trials for low PAPR is greatly reduced. The effects of using a larger set of $b_{\mu}^{(\nu)}$ (for example, four phases) are reported in [31], but improvements are small from the results of the simple flipping algorithm.

13.6.5 Dummy Sequence Insertion

Ryu et al. proposed a method called dummy sequence insertion (DSI) for PAPR reduction [32]. This method is to insert a dummy sequence in the data sequence for the sole purpose of PAPR reduction.

Four types of dummy sequences were proposed. The following description is from [32], which outlines these methods but not in detail. Method 1 inserts the complementary sequence. At first, a certain complementary sequence is inserted and PAPR is checked. If the PAPR is lower than the threshold, the sequence is transmitted. Otherwise, a feedback mechanism is activated and another complementary sequence is selected and PAPR is again checked. A common technique used in methods 2, 3, and 4 is the bit flipping technique which sequentially flips the bits of the dummy sequence one at a time, aiming to lower PAPR below the threshold. In method 2, data sequence of L bits is partitioned into M subblocks with equal length of L/M. The correlation sequence inversely corresponding to the first bits of partitioned subblocks is found and inserted as a dummy sequence. In method 3, the initial dummy sequence is an all-zero sequence. In method 4, the initial sequence is an all-one sequence. Simulation results show that method 1 is better than the others (Figure 13.16).

In the transmitter, the dummy sequence is simply amended to the data sequence before IFFT. It is discarded in the receiver after DFT. In comparison with other PAPR reduction methods, DSI is not better than PTS and block coding (see next section). However, unlike PTS, no side information is transmitted. Thus BER performance of DSI is better than PTS which suffers from BER degradation due to errors in side information. Simulation results in [32] support this claim. Figures 13.17 through 13.19 are the results, where "Theory" is the BER in the purely linear AWGN system without HPA, "without DSI" is the BER with HPA nonlinear distortion and without DSI method, "IBO = 0 dB" is the case without input back off in the power amplifier and DSI is used. The other cases are the cases where various IBO is applied and also DSI is used. SNR is equal to E_b/N_o. In comparison to "without DSI", the case "IBO = 0 dB" has about a 2-dB improvement over the "Theory" case for TWTA (traveling wave tube amplifier) and 3 dB for SSPA (solid state power amplifier).

In comparison with coding, DSI has no error detection or correction capability, however, it is claimed that DSI incurs less bandwidth expansion [32]. There are no simulation results in [32] to support this claim though.

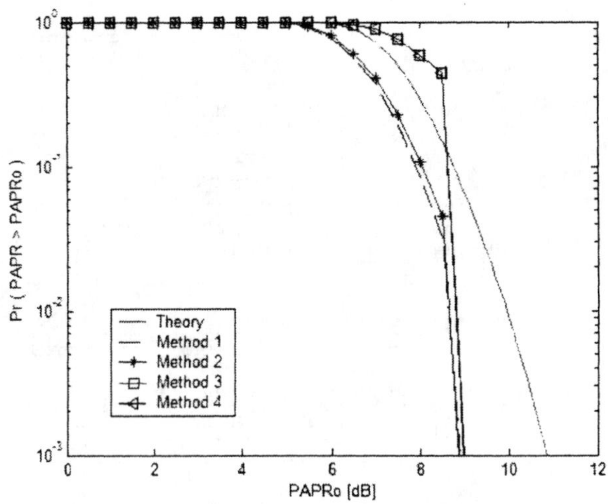

Figure 13.16 Complementary cumulative distribution function (CCDF) when $L = 48$, $M = 16$. "Theory" is CCDF of original OFDM in which DSI is not applied. From [32]. Copyright © 2004 IEEE.

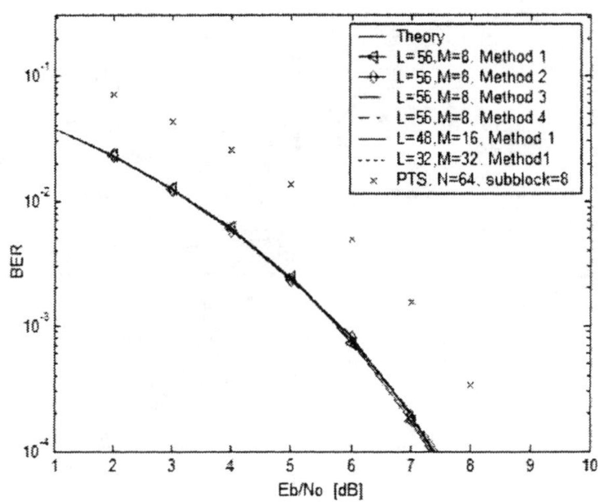

Figure 13.17 BER without HPA. From [32]. Copyright © 2004 IEEE.

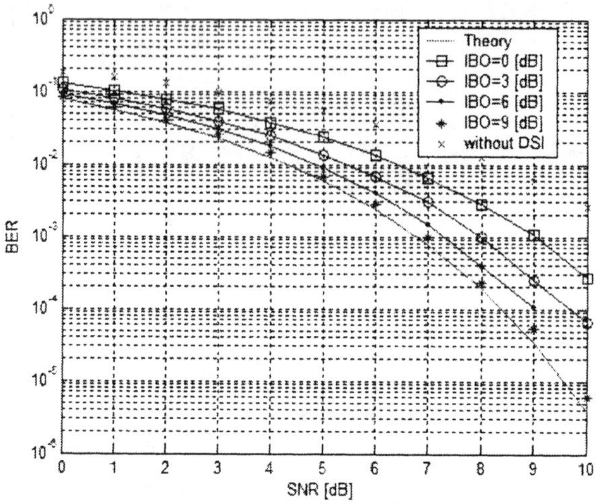

Figure 13.18 BER in nonlinear TWTA. From [32]. Copyright © 2004 IEEE.

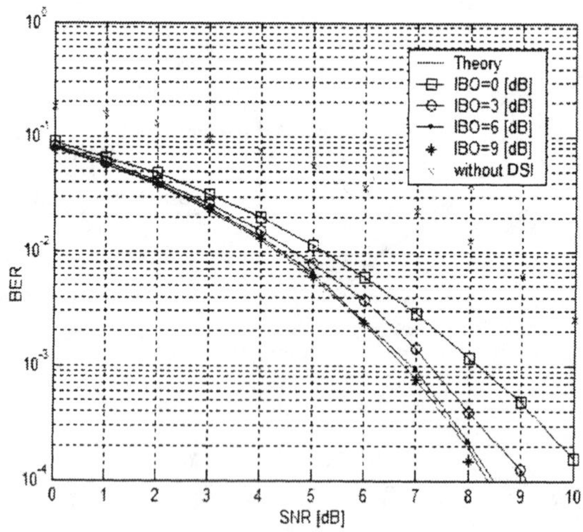

Figure 13.19 BER in nonlinear SSPA. From [32]. Copyright © 2004 IEEE.

13.7 CODING

This group of PAPR reduction techniques use channel codes as not only a means of error control and but also simultaneously as a sequence restrictor so that transmitted sequences are so chosen that the high signal peaks are avoided. Techniques in this group range from simple codes proposed in [33–35] to complicated codes proposed in [36,37]. We start with simple codes and progress to more sophisticated codes later.

13.7.1 Parity-Check Coding

An early paper by Jones et al proposed a simple parity-check code for PAPR reduction for a four carrier BPSK-OFDM system [33]. Assuming each subcarrier is normalized to have a power of 1W, the maximum PEP is 16W. The peak envelope powers (PEP) during symbol period for all possible data words are listed in Table 13.2. It is noticed that if four data bits (d_1, d_2, d_3, d_4) is an odd parity codeword, the PEP is 7.07W, the corresponding PAPR is 2.48 dB (10log(7.07/4)), which is 3.54 dB lower than the highest PAPR of 6.02 dB (10log16/4). Note that from the table, the first three bits of those 7.07W codewords exhaust all possible three tuples. Thus we can use first three bits from those data words to encode to four-bit parity codewords, with the fourth bit d_4 being the parity bit. In other words, we have a rate 3/4 parity code which reduces the PAPR by 3.54 dB (from 6.02 dB to 2.48 dB).

Wulich later proposed a method called cyclic coding for PAPR reduction for BPSK-OFDM signals [35]. The coding rule is: the phase of every fourth carrier is computed by

$$\begin{aligned} \varphi_{j+3} &= -\varphi_j + \varphi_{j+1} + \varphi_{j+2} + \pi \pmod{2\pi} \\ j &= 1, 5, ..., N-3 \end{aligned} \tag{13.42}$$

where $N = 4L$. L is any integer. Wulich pointed out that when $N = 4$, this method gives exactly the same results as [33]. This can be easily seen by listing the phases for $N = 4$, as shown in Table 13.3. Assume that 0 phase represents binary 0 and π represents binary 1, the code in Table 13.3 is obviously an odd parity-check code. For other $N = 4L$ values, this pattern repeats L times. Thus we can see that the cyclic coding method is in fact exactly the same method of [33] for $N = 4$ and is an extension of [33] for other $N = 4L$. The only difference is that one is encoded in binary form and another is encoded in carrier phases. The PAPR reductions for several N values were found through numerical computation in [35], which are listed in Table 13.4.

d_{dec}	d_1	d_2	d_3	d_4	PEP(W)
0	0	0	0	0	16.00
1	1	0	0	0	7.07
2	0	1	0	0	7.07
3	1	1	0	0	9.45
4	0	0	1	0	7.07
5	1	0	1	0	16.00
6	0	1	1	0	9.45
7	1	1	1	0	7.07
8	0	0	0	1	7.07
9	1	0	0	1	9.45
10	0	1	0	1	16.00
11	1	1	0	1	7.07
12	0	0	1	1	9.45
13	1	0	1	1	7.07
14	0	1	1	1	7.07
15	1	1	1	1	16.00

Table 13.2 PEP during symbol period for all possible 4-bit data words.

φ_j	φ_{j+1}	φ_{j+2}	φ_{j+3}
0	0	0	π
0	0	π	0
0	π	0	0
0	π	π	π
π	0	0	0
π	0	π	π
π	π	0	π
π	π	π	0

Table 13.3 Cyclic phase code.

N	$PAPR_{\max}$, dB	$10\log_{10} N - PAPR_{\max}$, dB
4	2.47	3.55
8	5.36	3.66
12	7.05	3.73
16	8.30	3.74

Table 13.4 Largest PAPR and PAPR reduction.

13.7.2 Rudin-Shapiro Coding

Shepherd et al. presented a simple rate 1/2 coding scheme based on Rudin-Shapiro polynomials for PAPR reduction [34]. This idea appeared earlier in [2] for multitone signals, but not specifically for OFDM signals.

The coding scheme is as follows. Consider a message space of m bits corresponding to $N = m/2$ carriers.

1. Break the m-bit message into 4-bit blocks. Designate the four bits a, b, c, d.
2. Append coding bits $\alpha, \beta, \gamma, \delta$ to each 4-bit message according to Table 13.5 (where bin = binary, dec = decimal). The four coding bits complete a Rudin-Shapiro polynomial of order 8 [38]. These polynomials are known to have good (but not optimum) peak factors [38].
3. Concatenate the message and the coding bits alternately to produce the final coded message of length $2m$ corresponding to $N = m$ subcarriers (for QPSK).

The logic function for the coding is

$$\begin{aligned} \text{If } a \oplus b \oplus c \oplus d &= 0 \\ \text{Then } \alpha &= a \text{ and } \beta = b \text{ and } \gamma = \overline{c} \text{ and } \delta = \overline{d} \\ \text{Else } \alpha &= \overline{a} \text{ and } \beta = \overline{b} \text{ and } \gamma = c \text{ and } \delta = d \end{aligned} \tag{13.43}$$

This logic can be easily implemented by hardware combinatorial circuits [34].

Monte Carlo simulation was performed for 200 random 32-bit messages, including all-zero and all-one sequences (16 carriers uncoded, 32 carriers coded). The results for each case show that the peak factor (or PAPR) never exceeds 4, one quarter of the worst case for the uncoded system. This means that this scheme reduces PAPR by more than 6 dB by using a rate 1/2 Rudin-Shapiro code in 200 runs of simulations. However, there are $2^{32} \approx 4 \times 10^9$ different 32-bit messages. To see whether this scheme can always reduce PAPR by 6 dB, more simulations are needed.

Message (bin) abcd	Message (dec)	Code (bin) $\alpha\beta\gamma\delta$	Code (dec)
0000	0	0011	3
0001	1	1101	13
0010	2	1110	14
0011	3	0000	0
0100	4	1000	8
0101	5	0110	6
0110	6	0101	5
0111	7	1011	11
1000	8	0100	4
1001	9	1010	10
1010	10	1001	9
1011	11	0111	7
1100	12	1111	15
1101	13	0001	1
1110	14	0010	2
1111	15	1100	12

Table 13.5 Rudin-Shapiro coding map.

13.7.3 Golay Complementary Sequences and Reed-Muller Codes

The two PAPR reduction codes above are very simple but not very efficient in terms of PAPR reduction and they do not have any error correction capability except that the parity check code has the ability to detect odd number errors. Besides, they are found in an ad hoc manner. We need a systematic way to construct codes that reduce PAPR and simultaneously correct bit errors efficiently.

Golay complementary sequences are found to have at most a PAPR of 2 [39]. This is much better than the above coding schemes in that it puts a very low upper bound (3 dB) on the PAPR and is independent of the number of subcarriers. However, Golay complementary sequences are polyphase (or multiphase) sequences; all elements in the sequence have the same magnitude. Therefore, they can be used to represent symbols of BPSK, QPSK, and higher-order PSK modulations, but not for QAM and other modulations.

In the following we first derive the PAPR expression of MPSK-OFDM signal. Then we show that the PAPR of MPSK-OFDM with Golay complementary sequences is at most 2. The argument follows that of [37] closely.

Let $\mathbb{Z}_M$ be a set of M integers: $(0, 1, 2, ..., M-1)$. For M-ary PSK-OFDM signal, denote the symbol value assigned to the ith subcarrier during a symbol period as an element a_i of $\mathbb{Z}_M$ for some $M \geq 2$, where $i = 0, 2, ..., N-1$. In each symbol period, the M-ary sequence $\mathbf{a} = (a_0, a_1, ..., a_{N-1})$ across the N subcarriers forms a codeword. Codewords in successive symbol periods belong to a code whose alphabets is $\mathbb{Z}_M$. For $\mathbb{Z}_2 = (0, 1)$, the code is binary; for $\mathbb{Z}_4 = (0, 1, 2, 3)$, the code is quaternary, and so forth. In an OFDM system, $\{a_i\}_{i=0}^{N-1}$ are mapped into a complex sequence $\boldsymbol{\xi}(\mathbf{a}) = (\xi^{a_0}, \xi^{a_1}, ..., \xi^{a_{N-1}})$, where $\xi = e^{j2\pi/M}$. Then $(\xi^{a_0}, \xi^{a_1}, ..., \xi^{a_{N-1}})$ can be modulated on to N subcarriers to form the OFDM signal. The modulation is M-ary phase shift keying. For example, when $M = 4$, we have

$$\boldsymbol{\xi}(\mathbf{a}) = (\xi^{a_0}, \xi^{a_1}, \xi^{a_2}, \xi^{a_3}) = (1, e^{j\pi/2}, e^{j\pi}, e^{j3\pi/2}) = (1, j, -1, -j)$$

which gives QPSK.

Using ξ notation, the complex envelope of the OFDM signal can be written as

$$s(t) = \sum_{n=0}^{N-1} \xi^{a_n} \exp(j2\pi f_n t) = \sum_{n=0}^{N-1} \xi^{a_n + M f_n t} \tag{13.44}$$

where

$$f_n = f_0 + n\Delta f \tag{13.45}$$

is the frequency of the ith subcarrier, where Δf is an integer multiple of the symbol rate $1/T$. The instantaneous envelope power is $P(t) = |s(t)|^2 = s(t)s^*(t)$, and from (13.44) and (13.45) we have

$$P(t) = \sum_{m=0}^{N-1} \sum_{n=0}^{N-1} \xi^{a_m - a_n + M(m-n)\Delta f t}$$

For the sequence $\mathbf{a} = (a_0, a_1, ..., a_{N-1})$, letting $n = m + u$, from the above, the envelope power is

$$P_a(t) = N + \sum_{u \neq 0} \sum_{m=0}^{N-1} \xi^{a_m - a_{m+u} - Mu\Delta f t} \tag{13.46}$$

where the summation is over only those integer values for which m and $m+u$ both lie within $\{0, 1, ..., N-1\}$. The aperiodic autocorrelation of $\mathbf{a}$ at displacement u is by definition[5]

$$C_a(u) = \sum_{m=0}^{N-1} \xi^{a_m - a_{m+u}} \tag{13.47}$$

Then (13.46) can be written as

$$P_a(t) = N + \sum_{u \neq 0} C_a(u) \xi^{-Mu\Delta f t} \tag{13.48}$$

The peak envelope power is the maximum over the symbol period of $P_a(t)$. From above we have

$$\begin{aligned} P_a(t) &\leq N + \sum_{u \neq 0} |C_a(u) \xi^{-Mu\Delta f t}| \\ &= N + \sum_{u \neq 0} |C_a(u)| \leq N + \sum_{u \neq 0} \sum_{m=0}^{N-1} |\xi^{a_m - a_{m+u}}| \\ &= N + \sum_{u \neq 0} N = N + N(N-1) = N^2 \end{aligned} \tag{13.49}$$

The $P_a(t)$ reaches its peak value when $\mathbf{a}$ has identical values on all subcarriers and when $t = 0$. In this case, $C_a(u) = \sum_{m=0}^{N-1} 1 = N$ and $\xi^{-Mu\Delta f t} = 1$, and the equality holds in the above expression.

---

5 This is not the usual definition of autocorrelation, where statistic average or time average is involved, depending on whether statistic or time autocorrelation is computed.

The average (or mean) envelope power is

$$\frac{1}{T}\int_0^T P_a(t)dt = N$$

since $\frac{1}{T}\int_0^T \xi^{a_m - a_{m+u}}dt = 0$ for any $u \neq 0$. Thus the maximum PAPR is $N^2/N = N$ for uncoded data. In fact we have proved this in an earlier section using trigonometrical representations instead of the complex envelope of the OFDM signal (Section 13.1). But here the envelope power $P_a(t)$ is related to the autocorrelation function of the data vector, which will enable the minimization of the PAPR by encoding the data to have a proper autocorrelation.

The Golay complementary pair is a pair of sequences $\mathbf{a}$ and $\mathbf{b}$ of length N over $\mathbb{Z}_M$, such that

$$C_a(u) + C_b(u) = 0, \quad \text{for } u \neq 0$$

Any sequence which is a member of a Golay complementary pair is called a Golay sequence. With sequences $\mathbf{a}$ and $\mathbf{b}$ in a Golay complementary pair, we have

$$P_a(t) + P_b(t) = 2N + \sum_{u \neq 0} [C_a(u) + C_b(u)]\, \xi^{-Mu\Delta ft} = 2N$$

and since $P_b(t) = |s_b(t)|^2 \geq 0$, we deduce $P_a(t) \leq 2N$ and $PAPR \leq 2$. This is an important conclusion which indicates that the PAPR of an OFDM signal modulated in M-ary PSK with any Golay sequence is upper bounded by 2 or 3 dB. Recall that the unmodulated carrier has a PAPR of 0 dB. This means Golay sequence only incurs a 3-dB increase in PAPR in comparison with unmodulated carrier, regardless of the number of subcarriers.

Now the problem that remains is how to encode data into Golay sequences. The binary Golay sequences were considered by Golay in [40]. Nonbinary (also called multiphase or polyphase) Golay sequences were studied by [41, 42]. However, the generation methods of Golay sequences described in [40–42] are by recursive expansion or computer search. More importantly, they are not encoding methods. We need a systematical method which encodes arbitrary data sequences to Golay complementary sequences. Davis and Jedwab solved this problem [36,37]. Their solution imposes the restriction $N = 2^m$ so that the sampled OFDM signal can be generated by IFFT. They also assumed that $M = 2^h$ so that in each symbol period the OFDM signal contains exactly h bits per subcarrier. Under these conditions, which are common for communication systems, Davis and Jedwab discovered that Golay complementary sequences can be generated by first-order Reed-Muller codes. This discovery solved the crucial problem of encoding arbitrary data sequences into Golay complementary sequences in the transmitter and decoding them back to data in the receiver.

In the following, we define the Reed-Muller codes first and describe the generation of Golay complementary sequences right after that.

Reed-Muller (RM) codes are most easily described by the use of Boolean functions. A Boolean function in m variables $f(x_1, x_2, ..., x_m)$ is defined as a mapping from

$$\mathbb{Z}_2^m = \{(x_1, x_2, ..., x_m) | x_i \in \{0, 1\}\}$$

to $\mathbb{Z}_2 = \{0, 1\}$. Any Boolean function can be expressed as a linear combination over $\mathbb{Z}_2$ of 2^m monomials

$$1, x_1, x_2, ..., x_m, x_1x_2, x_1x_3, ..., x_{m-1}x_m, ..., x_1x_2 \cdots x_m$$

That is

$$f = a_0 1 + a_1x_1 + ... + a_mx_m + a_{12}x_1x_2 + ... + a_{m-1m}x_{m-1}x_m + ... + a_{1...m}x_1 \cdots x_m$$

where the coefficients $a_i \in \{0, 1\}$ and the multiplications and summations are modulo-2. For example for $m = 2$, we can have $f_1 = 1 + x_1 + x_2$, $f_2 = x_1 + x_1x_2$, and so forth. For $(x_1, x_2) = (1, 1)$, we have $f_1 = 1$ and $f_2 = 0$.

The first-order binary Reed-Muller code $RM(1, m)$ of length 2^m is generated by a generator matrix consisting of the following $m + 1$ rows. The first (0th) row is an all-ones sequence. The ith row, $i = 1, 2, ..., m$ is a concatenation of 2^{i-1} copies of the sequence comprising 2^{m-i} 0s and followed by 2^{m-i} 1s.

The second-order binary Reed-Muller code $RM(2, m)$ of length 2^m is generated by a generator matrix consisting of all rows from the $RM(1, m)$ generator matrix plus rows formed by component-wise products $x_ix_j, 1 \leq i < j \leq m$. For example, the generator matrix of $R(2, 4)$ is

$$G = \begin{bmatrix}
1 & 1 & 1 & 1 & 1 & 1 & 1 & 1 & 1 & 1 & 1 & 1 & 1 & 1 & 1 & 1 \\
0 & 0 & 0 & 0 & 0 & 0 & 0 & 0 & 1 & 1 & 1 & 1 & 1 & 1 & 1 & 1 \\
0 & 0 & 0 & 0 & 1 & 1 & 1 & 1 & 0 & 0 & 0 & 0 & 1 & 1 & 1 & 1 \\
0 & 0 & 1 & 1 & 0 & 0 & 1 & 1 & 0 & 0 & 1 & 1 & 0 & 0 & 1 & 1 \\
0 & 1 & 0 & 1 & 0 & 1 & 0 & 1 & 0 & 1 & 0 & 1 & 0 & 1 & 0 & 1 \\
0 & 0 & 0 & 0 & 0 & 0 & 0 & 0 & 0 & 0 & 0 & 0 & 1 & 1 & 1 & 1 \\
0 & 0 & 0 & 0 & 0 & 0 & 0 & 0 & 0 & 0 & 1 & 1 & 0 & 0 & 1 & 1 \\
0 & 0 & 0 & 0 & 0 & 0 & 0 & 0 & 0 & 1 & 0 & 1 & 0 & 1 & 0 & 1 \\
0 & 0 & 0 & 0 & 0 & 0 & 1 & 1 & 0 & 0 & 0 & 0 & 0 & 0 & 1 & 1 \\
0 & 0 & 0 & 0 & 0 & 1 & 0 & 1 & 0 & 0 & 0 & 0 & 0 & 1 & 0 & 1 \\
0 & 0 & 0 & 1 & 0 & 0 & 0 & 1 & 0 & 0 & 0 & 1 & 0 & 0 & 0 & 1
\end{bmatrix}
\begin{matrix}
\mathbf{1} \\ \mathbf{x_1} \\ \mathbf{x_2} \\ \mathbf{x_3} \\ \mathbf{x_4} \\ \mathbf{x_1x_2} \\ \mathbf{x_1x_3} \\ \mathbf{x_1x_4} \\ \mathbf{x_2x_3} \\ \mathbf{x_2x_4} \\ \mathbf{x_3x_4}
\end{matrix} \tag{13.50}$$

The first five rows constitute the generator matrix of $R(1, 4)$.

For an $RM(r, m)$ code, the number of rows of the generator matrix is

$$k = 1 + \binom{m}{1} + \binom{m}{2} + ... + \binom{m}{r} \tag{13.51}$$

and the codeword length is 2^m. The code rate is $k/2^m$. Encoding is simply accomplished by multiplying the data vector with the generator matrix. Assuming $\mathbf{d} = (d_0, d_1, ..., d_{k-1})$ is the data vector and $\mathbf{c} = (c_0, c_1, ..., c_{2^m-1})$ is the codeword vector, then

$$\mathbf{c} = \mathbf{d}G = d_0\mathbf{1} + d_1\mathbf{x}_1 + ... + d_m\mathbf{x}_m + d_{m+1}\mathbf{x}_1\mathbf{x}_2 + d_{m+2}\mathbf{x}_1\mathbf{x}_3 + ... + d_{k-1}\mathbf{x}_{m-1}\mathbf{x}_m \tag{13.52}$$

which is a Boolean function of the monomials.

The minimum distance of an $RM(r, m)$ code is 2^{m-r}. Thus it can correct all error patterns of length $t \leq \left\lfloor \frac{2^{m-r}-1}{2} \right\rfloor$, where $\lfloor x \rfloor$ denotes the integer floor of x. For details of Reed-Muller codes, including decoding methods, the reader is referred to [43] or other coding theory books.

Davis and Jedwab made a crucial connection between Golay sequences and second-order Reed-Muller codes by the following theorem [36, 37].

Theorem 1 *The codeword of an $RM(2, m)$ code with each code bit given by*

$$\sum_{i=1}^{m-1} x_{\pi(i)} x_{\pi(i+1)} + \sum_{i=0}^{m} d_i x_i$$

is a binary Golay sequence of length 2^m for any permutation π of $\{1, 2, ..., m\}$ and for any data $d_i \in \{0, 1\}$.

The proof of this theorem is given in [37, Theorem 3 and Corollary 6]. Some explanations of this theorem follow.

Let π be a permutation of the numbers $\{1, 2, ..., m\}$; then $\pi(i)$ is the ith number of π and $\pi(i+1)$ is the $(i+1)$th number of π. For example, for $m = 4$, if $\pi = \{1, 3, 2, 4\}$, then $\sum_{i=1}^{m-1} x_{\pi(i)} x_{\pi(i+1)} = x_1 x_3 + x_3 x_2 + x_2 x_4$. It is well known that the number of permutations of $\{1, 2, ..., m\}$ is $m!$. However, $\sum_{i=1}^{m-1} x_{\pi(i)} x_{\pi(i+1)}$ is invariant under the mapping $\pi \longmapsto \pi'$, where $\pi'(k) = \pi(m+1-k)$ (i.e., a reversed π). For example, $\pi' = \{4, 2, 3, 1\}$ is a reversed $\pi = \{1, 3, 2, 4\}$, and it would result in the same sum. Thus the number of distinct sums $\sum_{i=1}^{m-1} x_{\pi(i)} x_{\pi(i+1)}$ is $m!/2$.

The number of terms in the sum $\sum_{i=1}^{m-1} x_{\pi(i)} x_{\pi(i+1)}$ is only $m - 1$, while the total number of second-order rows in G is $\binom{m}{2}$. Thus the codewords described in Theorem 1 are produced by those data sequences in which those bits corresponding to those rows included in the second-order sum must be 1 and those bits corresponding to those rows not included in the second-order sum must be zero. In other words, the codewords described in Theorem 1 are only a part of the $RM(2, m)$ code.

The number of the first-order sums $\sum_{i=0}^{m} d_i x_i$ is 2^{m+1}. The number of binary Golay sequences generated by Theorem 1 is $(m!/2) \times 2^{m+1} = 2^m m!$. The sums $\sum_{i=0}^{m} d_i x_i$ are codewords of an $RM(1, m)$. Thus we can see that the $2^m m!$ binary Golay sequences given by Golay's recursive and interleaving construction [40] can be explicitly represented by $m!/2$ cosets of $RM(1, m)$, each containing 2^{m+1} codewords. $\sum_{i=1}^{m-1} x_{\pi(i)} x_{\pi(i+1)}$ are called coset representatives.

Now it is time to show how an $RM(2, m)$ can be used to generate Golay sequences. There are two cases. Case 1 is to use only one coset representative [36, 44]. Case 2 is to use more than one coset representative [36, 37].

Case 1. One permutation, hence one coset representative, is selected. Note that for the chosen permutation, $\sum_{i=1}^{m-1} x_{\pi(i)} x_{\pi(i+1)}$ has different values for different columns in G. That is, the coset representative is a vector determined by the chosen permutation and the $RM(2, m)$ generator G. We denote it as $\mathbf{r}$. It has nothing to do with data so that it can be computed before encoding the data. Data sequence $\mathbf{d}$ is encoded as an $RM(1, m)$ codeword $\mathbf{c}'$. The coset representative $\mathbf{r}$ is added to each codeword of the $RM(1, m)$, resulting in an $RM(2, m)$ codeword $\mathbf{c} = \mathbf{r} + \mathbf{c}'$. $\mathbf{c}$ is a binary Golay sequence which guarantees that $PAPR \leq 2$ for a BPSK-OFDM. In the receiver, since the coset representative $\mathbf{r}$ is known, it is removed before decoding. The remaining codeword is a corrupted $\mathbf{c}'$, which is decoded by the simple decoding method for the $RM(1, m)$ code, the "Green machine" using the Hadamard transform method [43]. The code rate is $(m + 1)/2^m$. For example, if $m = 4$, the code rate is $5/16$ in this case. The minimum Hamming distance is 8, and the error correction capability is 3.

Case 2. Multiple permutations, hence multiple coset representatives, are selected. Up to $\lfloor \log_2(m!/2) \rfloor$ data bits can be encoded as the choice of coset representatives (for example, using a look-up table), where $\lfloor x \rfloor$ denotes the integer floor of x. Another $m+1$ bits are encoded as the $R(1, m)$ part. A total of $\lfloor \log_2(m!/2) \rfloor + m + 1$ bits are encoded as an $RM(2, m)$ code. In the receiver, an $RM(2, m)$ decoder, such as the majority-logic decoder [43], can be used to decode the codewords. The maximum code rate in this case is $(\lfloor \log_2(m!/2) \rfloor + m + 1) / 2^m$. For example, if $m = 4$, the code rate is $8/16$ in this case. The minimum Hamming distance is 4, and the error correction capability is 1. Compared with case 1, case 2 has a higher code rate but less capability for error correction.

If the phase modulation is nonbinary, such as QPSK and 8PSK, data sequences must be encoded into polyphase Golay sequences in order to keep the $PAPR \leq 2$. To do this, two new codes are defined [37].

$RM_{2^h}(r, m)$ code: for $h \geq 1$ and $0 \leq r \leq m$, the rth-order linear code

$RM_{2^h}(r,m)$ over $\mathbb{Z}_{2^h} = \{0, 1, ..., 2^h - 1\}$ of length 2^m is generated by the monomials in the x_i of degree at most r. $RM_{2^h}(r,m)$ is generalized from the binary $RM(r,m)$ code.

$ZRM_{2^h}(r,m)$ code: for $h > 1$ and $0 \le r \le m+1$, the rth-order linear code $ZRM_{2^h}(r,m)$ over $\mathbb{Z}_{2^h}$ of length 2^m is generated by the monomials in the x_i of degree at most $r-1$ together with two times the monomials in the x_i of degree r (with the convention that the monomials of degree -1 and $m+1$ are equal to zero). $ZRM_{2^h}(r,m)$ is generalized from the quaternary Reed-Muller code $ZRM(r,m)$ defined in [45] from the alphabet $\mathbb{Z}_4$ to the alphabet $\mathbb{Z}_{2^h}$.

In both cases, the formal generator matrix does not change as h varies. But the alphabet of the codeword symbols ($\mathbb{Z}_{2^h}$) are changed. That is, for $h = 1$, the codeword is binary, for $h = 2$, the codeword is quaternary, and so forth.

We are particularly interested in $RM_{2^h}(1,m)$ and $ZRM_{2^h}(2,m)$ for the reason that will be seen shortly. As examples, we show the generators for $RM_{2^h}(1,4)$ and $ZRM_{2^h}(2,4)$ in the following.

$$G_{RM} = \left[\begin{array}{cccccccccccccccc}
1&1&1&1&1&1&1&1&1&1&1&1&1&1&1&1\\
0&0&0&0&0&0&0&0&1&1&1&1&1&1&1&1\\
0&0&0&0&1&1&1&1&0&0&0&0&1&1&1&1\\
0&0&1&1&0&0&1&1&0&0&1&1&0&0&1&1\\
0&1&0&1&0&1&0&1&0&1&0&1&0&1&0&1
\end{array}\right]
\begin{array}{l}
\mathbf{1}\\ x_1\\ x_2\\ x_3\\ x_4
\end{array}
\tag{13.53}$$

$$G_{ZRM} = \left[\begin{array}{cccccccccccccccc}
1&1&1&1&1&1&1&1&1&1&1&1&1&1&1&1\\
0&0&0&0&0&0&0&0&1&1&1&1&1&1&1&1\\
0&0&0&0&1&1&1&1&0&0&0&0&1&1&1&1\\
0&0&1&1&0&0&1&1&0&0&1&1&0&0&1&1\\
0&1&0&1&0&1&0&1&0&1&0&1&0&1&0&1\\
0&0&0&0&0&0&0&0&0&0&0&0&2&2&2&2\\
0&0&0&0&0&0&0&0&0&0&2&2&0&0&2&2\\
0&0&0&0&0&0&0&0&0&2&0&2&0&2&0&2\\
0&0&0&0&0&0&2&2&0&0&0&0&0&0&2&2\\
0&0&0&0&0&2&0&2&0&0&0&0&0&2&0&2\\
0&0&0&2&0&0&0&2&0&0&0&2&0&0&0&2
\end{array}\right]
\begin{array}{l}
\mathbf{1}\\ x_1\\ x_2\\ x_3\\ x_4\\ 2x_1x_2\\ 2x_1x_3\\ 2x_1x_4\\ 2x_2x_3\\ 2x_2x_4\\ 2x_3x_4
\end{array}
\tag{13.54}$$

Based on $RM_{2^h}(1,m)$ and $ZRM_{2^h}(2,m)$, the following theorem gives the coding rules of generating polyphase Golay sequences [36, 37].

Theorem 2 *The codeword of $ZRM_{2^h}(2, m)$ with each code symbol given by*

$$2^{h-1}\sum_{i=1}^{m-1} x_{\pi(i)}x_{\pi(i+1)} + \sum_{i=0}^{m} d_i x_i$$

is a 2^h-phase Golay sequence of length 2^m for any permutation π of $\{1, 2, ..., m\}$ and for any data $d_i \in \mathbb{Z}_{2^h} = \{0, 1, ..., 2^h - 1\}$.

Similar to the binary case, the $2^{h(m+1)}$ first-order sums $\sum_{i=0}^{m} d_i x_i$ form an $RM_{2^h}(1, m)$ code. Each of $m!/2$ second-order sum $2^{h-1}\sum_{i=1}^{m-1} x_{\pi(i)}x_{\pi(i+1)}$ is a coset representative in $ZRM_{2^h}(2, m)$. The number of polyphase Golay sequences generated by Theorem 2 is $(m!/2) \times 2^{h(m+1)} = 2^{h(m+1)-1}m!$.

The encoding is also similar to binary case except that the data and codeword both are 2^h-ary. That is, encoding is simply accomplished by multiplying the data vector by the generator matrix. However, the multiplications and additions must be modulo-2^h. The 2^h-ary Golay codewords are modulated onto the subcarriers using (13.44) (where a_i are the codeword symbols). A decoding algorithm for $RM_{2^h}(1, m)$ is given in [37], which can be used in the case where only one coset representative is chosen so that the coset representative can be removed before decoding the $RM_{2^h}(1, m)$ code. In particular, a fast decoding algorithm is given in [44] for Q-ary modulated Reed-Muller codes.

13.8 SUMMARY

In this chapter we derived maximum PAPRs for MPSK-OFDM and QAM-OFDM signals. We have shown that the maximum PAPR for MPSK-OFDM is N, the number of subcarriers, while the maximum PAPR is $3N(\sqrt{M} - 1)/(\sqrt{M} + 1)$ for M-ary QAM-OFDM. The PAPR distributions were derived under the assumption that OFDM signal amplitude is Gaussian distributed for big N. For small N, a modified distribution is presented. It is found that Nyquist-rate samples are uncorrelated and oversampled samples are correlated. Major PAPR reduction techniques are presented. They are classified into four categories: clipping with or without clipping noise mitigation, post-IFFT amplitude alteration other than clipping, pre-IFFT data alteration, and block coding for PAPR reduction and simultaneously for error control. It is difficult to compare all these PAPR reduction methods since each of them may have advantages over others in certain situations. Users should select the method which suits one's circumstances the best, including BER performance requirements, bandwidth limits, allowable power level, system complexity, and so forth. Generally speaking, clipping without clipping noise reduction is the simplest but with the poorest BER performance, and coding, particularly the Reed Muller coding, is the most complex and requires bandwidth expansion, but with the biggest PAPR reduction and the best BER performance. The other methods are in between in terms of system complexity, bandwidth requirements, and BER performance.

References

[1] van Nee, R., and R. Prasad, *OFDM for Wireless Multimedia Communications*, Norwood, MA: Artech House, 2000.

[2] Boyd, S., "Multitone signals with low crest factor," *IEEE Trans. on Circuits and Systems*, vol. 33, no. 10, October 1986, pp. 1018–1021.

[3] Ochiai, H., and H. Imai, "On the distribution of the peak-to-average power ratio in OFDM signals," *IEEE Trans. on Communications*, vol. 49, no. 2, February 2001, pp. 282–289.

[4] Haykin, S., *Communication Systems*, 4th ed., New York: Wiley, 2001.

[5] Proakis, J., *Digital Communications*, 2nd ed., New York: McGraw-Hill, 1989.

[6] Shepherd, S., J. Orriss, and S. Barton, "Asymptotic limits in peak envelope power reduction by redundant coding in orthogonal frequency-division multiplex modulation," *IEEE Trans. on Communications*, vol. 46, no. 1, January 1998, pp. 5–10.

[7] O'Neill, R., and L. B. Lopes, "Envelope variations and spectral splatter in clipped multicarrier signals," *Proc. PIMRC'95*, Toronto, Canada, September 1995, pp. 71-75.

[8] Li, X., and L. J. Gimini, Jr., "Effects of clipping and filtering on the performance of OFDM," *Proc. VTC'97*, Phoenix, Arizona, May 1997, pp. 1634–1638.

[9] Merchan, S., A. G. Armada, and J. L. Garcia, "OFDM performance in amplifier nonlinearity," *IEEE Trans. on Broadcasting*, vol. 44, no. 1, March 1998, pp. 106–114.

[10] Chini, A., et al., "Hardware nonlinearities in digital TV broadcasting using OFDM modulation," *IEEE Trans. on Broadcasting*, vol. 44 , no. 1, March 1998, pp. 12–21.

[11] Dinis, R., and A. Gmmão, "On the performance evaluation of OFDM transmission using clipping techniques," *Proc. VTC'99 Fall*, Amsterdam, the Netherlands, September 1999, pp. 2923–2928.

[12] Wulich, D., N. Dinur, and A. Glinowiecki, "Level clipped high-order OFDM," *IEEE Trans. on Communications*, vol. 48, no. 6, June 2000, pp. 928–930.

[13] Bahai, A. R. S., et al., "A new approach for evaluating clipping distortion in multicarrier systems," *IEEE Journal on Selected Areas in Communications*, vol. 20, no. 5, June 2002, pp. 1037–1046.

[14] Panta, K. R., and J. Armstrong, "Effects of clipping on the error performance of OFDM in frequency selective fading channels," *IEEE Trans. on Wireless Communications*, vol. 3, no. 2, March 2004, pp. 668–671.

[15] Kim, D., and G. L. Stuber, "Clipping noise mitigation for OFDM by decision-aided reconstruction," *IEEE Communications Letters*, vol. 3, no. 1, January 1999, pp. 4–6.

[16] Ochiai, H., and H. Imai, "Performance of the deliberate clipping with adaptive symbol selection for strictly bandlimited OFDM systems," *IEEE J. Select. Areas Commrm.*, vol. 18, no. 11, November 2000, pp. 2270–2277.

[17] Ochiai, H., and H. Imai, "Performance analysis of deliberately clipped OFDM signals," *IEEE Trans. on Communications*, vol. 50, no. 1, January 2002, pp. 89–101.

[18] Armstrong, J., "Peak-to-average power reduction for OFDM by repeated clipping and frequency domain filtering," *IEE Electronics Letters*, vol. 38, no. 5, February 28, 2002, pp. 246–247.

[19] Leung, S. H., S. M. Ju, and G. G. Bi, "Algorithm for repeated clipping and filtering in peak-to-average power reduction for OFDM," *IEE Electronics Letters*, vol. 38, no. 25, December 5, 2002, pp. 1726–1727.

[20] Ryu, H., B. Jin, and I. Kim, "PAPR reduction using soft clipping and ACI rejection in OFDM system," *IEEE Trans. on Consumer Electronics*, vol. 48, no. 1, February 2002, pp. 17–22.

[21] Saeedi, H., M. Sharif, and F. Marvasti, "Clipping noise cancellation in OFDM systems using oversampled signal reconstruction," *IEEE Communications Letters*, vol. 6, no. 2, February 2002, pp. 73–75.

[22] Ju, S. M., and S. H. Leung, "Clipping on COFDM with phase on demand," *IEEE Communications Letters*, vol. 7, no. 2, February 2003, pp. 49–51.

[23] Chen, H., and A. M. Haimovich, "Iterative estimation and cancellation of clipping noise for OFDM signals," *IEEE Communications Letters*, vol. 7, no. 7, July 2003, pp. 305–307.

[24] Huang, X., et al., "Reduction of peak-to-average power ratio of OFDM signals with companding transform," *IEE Electronics Letters*, vol. 37, no. 8, April 2001, pp. 506–507.

[25] Wang, X., T. T. Tjhung, and C. S. Ng, "Reduction of peak-to-average power ratio of OFDM system using a companding technique," *IEEE Transactions on Broadcasting*, vol. 45, no. 3, September 1999, pp. 303–307.

[26] Ren, G., H. Zhang, and Y. Chang, "A complementary clipping transform technique for the reduction of peak-to-average power ratio of OFDM system," *IEEE Trans. on Consumer Electronics*, vol. 49, no. 4, November 2003, pp. 922–926.

[27] Bäuml, R., R. Fischer, and J. Huber, "Reducing the peak-to-average power ratio of multicarrier modulation by selected mapping," *Electronics Letters*, vol. 32, no. 22, October 1996, pp. 2056–2057.

[28] Mestdagh, D. J. G., and P. M. P. Spruyt, "A method to reduce the probability of clipping in DMT-based transceivers," *IEEE Trans. on Communications*, vol. 44, no. 10, October 1996, pp. 1234–1238.

[29] van Eetvelt, P., G. Wade, and M. Tomlinson, "Peak to average power reduction for OFDM schemes by selective scrambling," *Electronics Letters*, vol. 32, October 1996, pp. 1963–1964.

[30] Müller, S., and J. Huber, "OFDM with reduced peak-to-average power ratio by optimum combination of partial transmit sequences," *Electronics Letters*, vol. 33, February 1997, pp. 368–369.

[31] Cimini, L. J., and N. R. Sollenberger, "Peak-to-average power ratio reduction of an OFDM signal using partial transmit sequences," *IEEE Communications Letters*, vol. 4, no. 3, March 2000, pp. 86–88.

[32] Ryu, H. G., J. E. Lee, and J. S. Park, "Dummy sequence insertion (DSI) for PAPR reduction in the OFDM communication system," *IEEE Trans. on Consumer Electronics*, vol. 50, no. 1, February 2004, pp. 89–94.

[33] Jones, A., T. Wilkinson, and S. Barton, "Block coding scheme for the reduction of peak to mean envelope power ratio of multicarrier transmission schemes," *Electronics Letters*, vol. 30, no. 25, December 1994, pp. 2098–2099.

[34] Shepherd, S. J., et al., "Simple coding scheme to reduce peak factor in QPSK multicarrier modulation," *Electronics Letters*, vol. 31, no. 14, July 6, 1995, pp. 1131–1132.

[35] Wulich, D., "Reduction of peak to mean ratio of multicarrier modulation using cyclic coding," *Electronics Letters*, vol. 32, no. 5, 1996, pp. 432–433.

[36] Davis, J., and J. Jedwab, "Peak-to-mean power control and error correction for OFDM transmission using Golay sequences and Reed-Muller codes," *Electronics Letters*, vol. 33, no. 4, February 1997, pp. 267–268.

[37] Davis, J., and J. Jedwab, "Peak-to-mean power control in OFDM, Golay complementary sequences, and Reed-Muller codes," *IEEE Trans. on Information Theory*, vol. 45, no. 7, November 1999, pp. 2397–2417.

[38] Schroeder, M. R., *Number Theory in Science and Communication*, Berlin: Springer Verlag, 1984.

[39] Popović, B. M., "Synthesis of power efficient multitone signals with flat amplitude spectrum," *IEEE Trans. on Communications*, vol. 39, no. 7, July 1991, pp. 1031–1033.

[40] Golay, M. J. E., "Complementary series," *IRE Trans. on Information Theory*, vol. 7, no. 2, April 1961, pp. 82–87.

[41] Sivaswamy, R., "Mutiphase complementary codes," *IEEE Trans. on Information Theory*, vol. 24, no. 5, September 1978, pp. 546–552.

[42] Frank, R. L., "Polyphase complementary codes," *IEEE Trans. on Information Theory*, vol. 26, no. 6, November 1980, pp. 641–647.

[43] Wicker, S. B., *Error Control Systems for Digital Communication and Storage*, Upper Saddle River, New Jersey: Prentice-Hall, 1995.

[44] Grant, A., and R. van Nee, "Efficient maximum likelihood decoding of Q-ary modulated Reed-Muller codes," *IEEE Communications Letters*, vol. 2, no. 5, May 1998, pp. 134–136.

[45] Hammons, A. R. Jr., P. V. Kumar, A. R. Calderbank, N. J. A. Sloane, and P. Solé, "The $\mathbb{Z}_4$-linearity of Kerdock, Preparata, Goethals, and related codes, " *IEEE Trans. on Information Theory*, vol. 40, 1994, pp. 301–319.

Selected Bibliography

- Bahai, A. S. R., and B. R. Saltzberg, *Multi-Carrier Digital Communications: Theory and Applications of OFDM*, New York: Kluwer Academic/Plenum Publishers, 1999.
- Jones, A., and T. Wilkinson, "Performance of Reed-Muller codes and a maximum-likelihood decoding algorithm for OFDM," *IEEE Trans. on Communications*, vol. 47, no. 7, July 1999, pp. 949–952.
- Litwin, L., "An Introduction to multicarrier modulation," *IEEE Potentials*, April/May 2000, pp. 36–38.

Chapter 14

Synchronization for OFDM

Synchronization is important to OFDM systems, just as it is to single-carrier systems. Moreover, because of the existence of multiple subcarriers, frequency offset and phase errors between the transmitter carrier and the receiver carrier cause intercarrier (or interchannel) interference (ICI); hence, there is more severe performance degradation on OFDM than on single-carrier systems. Symbol timing errors can possibly destroy the orthogonality among subcarriers and hence incur severe BER increases, or at least introduce a phase rotation to the signal symbol, depending on the length and sign of the offset. Sample timing offset causes phase rotations in the signal, which results in BER increases. Sampling frequency error causes time-varying phase rotation and ICI.

In this chapter we will first study the effect of carrier frequency offset and phase noise. Then we will briefly discuss the effect of timing error. After that we discuss in detail the algorithms of carrier frequency and timing synchronization for OFDM, including important algorithms that use cyclic extensions or pilot symbols. Doppler frequency shift is also a type of carrier frequency error, whose effect and correctional methods will be discussed in Chapter 15 when OFDM performance in multipath fading channels is studied.

14.1 EFFECT OF CARRIER FREQUENCY OFFSET AND PHASE NOISE

Performance degradation of coherent OFDM modems due to frequency offset of local oscillators in the receiver is analyzed in [1,2]. In both papers, the OFDM signals are assumed to be generated by the inverse discrete Fourier transform (IDFT) and demodulated by the discrete Fourier transform (DFT).

Analysis and simulation of effect of carrier frequency offset to the signal-to-noise ratio (SNR) are reported in [1]. The channel is modeled by a complex transfer function H_k for the kth subcarrier and a frequency offset ε for all subcarriers, where

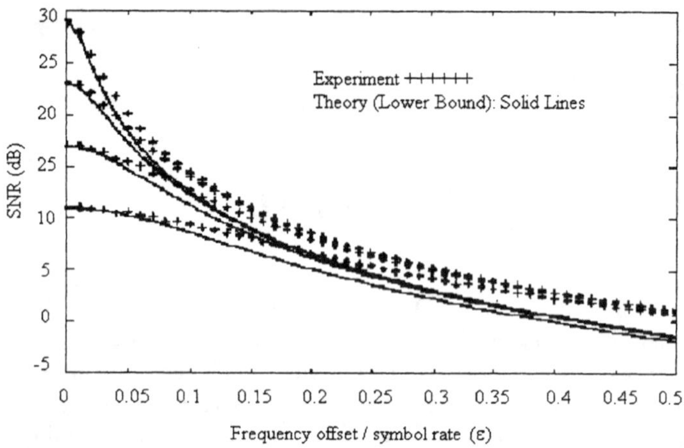

Figure 14.1 SNR versus frequency offset ε. From [1]. Copyright © 1994 IEEE.

ε is relative to the data symbol rate $R_s = 1/T$ (AWGN is also present). It is also assumed that H_k is not time-variant in one symbol period (this corresponds to a "slow-fading" channel, with the AWGN channel as its special case).

Based on this model, the signal-to-noise ratio (SNR) at the output of the DFT at the OFDM receiver is found in [1] to be lower bounded as follows

$$SNR \geq \frac{E_c}{N_o}\left(\frac{\sin \pi\varepsilon}{\pi\varepsilon}\right)^2 \frac{1}{1+0.5947\frac{E_c}{N_o}(\sin \pi\varepsilon)^2}, \quad |\varepsilon| \leq 0.5 \tag{14.1}$$

where E_c is the average received symbol energy of the individual subcarriers, which includes the channel gain. $N_o/2$ is the AWGN power spectral density in the bandpass transmission channel. Note that both SNR and E_c/N_o are in dB in (14.1).

Equation (14.1) is plotted in Figure 14.1 for values of $E_c/N_o = 11, 17, 23,$ and 29 dB. Simulation results for an 8-PSK OFDM with 193 subcarriers and 63 virtual subcarriers, implemented by 256-point IDFT/DFT pair, are also included for validation. It can be seen that the bound of (14.1) is quite tight for small values of ε.

At $BER = 10^{-6}$, the $E_c/N_o = 10.5, 13.5, 18.71,$ and 20.4 dB for BPSK, QPSK, 8PSK, and 16QAM. Lower bounds of degradations versus ε computed from (14.1) for these SNRs are plotted in Figure 14.2.

Similar results are reported in [2], where the channel is modeled by a time-varying phase $\theta(t)$ (no nonunity amplitude gain is assumed like in [1]) caused by either a

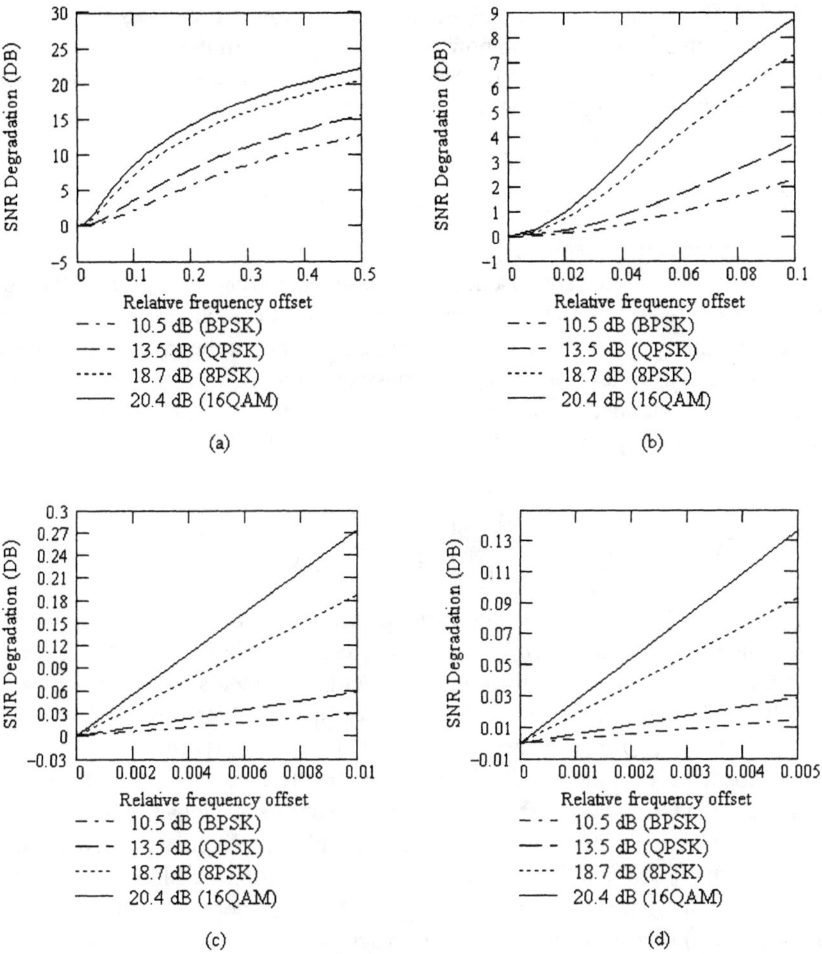

Figure 14.2 SNR degradation versus relative frequency offset ε (a–d). The original SNRs are the values at which $BER = 10^{-6}$: 20.4 dB for 16QAM, 18.7 dB for 8PSK, 13.5 dB for QPSK, and 10.5 dB for BPSK.

carrier offset between the receiver and transmitter carrier, or the phase noise of these carriers. In the first case, $\theta(t)$ is deterministic and equals $2\pi\Delta Ft + \theta_0$, where ΔF is the carrier offset. In the latter case, $\theta(t)$ is modeled as a Wiener process for which $E[\theta(t)] = 0$ and $E[\theta(t+t_0) - \theta(t_0)]^2 = 4\pi\beta|t|$, where β denotes the one-sided 3 dB linewidth of the Lorentzian power spectrum density of the free-running carrier generator. Lorentzian PSD has been verified by measurements and the double-sided Lorentzian PSD is given by [3, p. 74]

$$\Psi_s(f) = \frac{1}{\pi\beta[1 + |f - f_c|^2/\beta^2]}$$

where f_c is the carrier frequency.

Based on this model, and assuming the phase rotation caused by $\theta(t)$ of the signal on the considered subcarrier at the output of the FFT demodulator can be perfectly estimated and corrected, the degradation D (dB) of SNR at the input of the decision device is found as follows. For a fixed total symbol rate R, with $R = N/T = NR_s$, where N is the number of subcarriers, for OFDM, and $R = 1/T = R_s$ for single channel (SC), the degradation (dB) due to frequency offset is

$$D_f \cong \begin{cases} \frac{10}{\ln 10}\frac{1}{3}\left(\pi N\frac{\Delta F}{R}\right)^2 \frac{E_s}{N_o}, & \text{OFDM} \\ \frac{10}{\ln 10}\frac{1}{3}\left(\pi\frac{\Delta F}{R}\right)^2, & \text{SC} \end{cases} \tag{14.2}$$

where $E_s = E\{|a_k|^2\}$ is the average symbol energy for subcarrier k (the expectation is over the symbol alphabet $\{a_m\}$). It is assumed the same for each subcarrier. N_o is the thermal noise energy at the output of the FFT demodulator for any subcarrier. This N_o is the same as the one-sided noise power spectral density of the channel, N_o (see Section B.2.1.1 and consider that the FFT demodulator is equivalent to a complex orthonormal demodulator for each subcarrier). Therefore E_s/N_o is the same as E_c/N_o used in (14.1) except that E_c includes nonunity channel amplitude gain.

For OFDM, the degradation is proportional with E_s/N_o, but it is not the case for SC. For both OFDM and SC, the degradation is proportional with the square of the frequency offset. For OFDM, the degradation is also proportional with the E_s/N_o, and with the square of the number of subcarriers.

The degradation (dB) due to phase noise is

$$D_p \cong \begin{cases} \frac{10}{\ln 10}\frac{11}{60}\left(4\pi N\frac{\beta}{R}\right)\frac{E_s}{N_o}, & \text{OFDM} \\ \frac{10}{\ln 10}\frac{1}{60}\left(4\pi\frac{\beta}{R}\right)\frac{E_s}{N_o}, & \text{SC} \end{cases} \tag{14.3}$$

From (14.2) and (14.3) it is seen that for both OFDM and SC, the degradation (in dB) is proportional with E_s/N_o, and the linewidth β. We also see that the degradations of OFDM due to frequency offsets and phase noise is N^2E_s/N_o and N times greater than that of SC, respectively.

Using the subcarrier spacing (R_s), the OFDM parts of (14.2) and (14.3) can be written as

$$D_f \approx \left\{ \frac{10}{\ln 10}\frac{1}{3}(\pi\frac{\Delta F}{R_s})^2\frac{E_s}{N_o}, \quad \text{OFDM} \right. \tag{14.4}$$

and

$$D_p \approx \left\{ \frac{10}{\ln 10}\frac{11}{60}(4\pi\frac{\beta}{R_s})\frac{E_s}{N_o}, \quad \text{OFDM} \right. \tag{14.5}$$

respectively. In the expressions, $\Delta F/R_s$ and β/R_s are relative frequency offset and linewidth versus the subcarrier spacing, respectively.

Some numerical results for (14.4) and (14.5) are plotted in Figure 14.3. For example, for QPSK, from the figure it can be seen that the degradation due to Doppler shift $\varepsilon = \frac{\Delta F}{R_s} \cong 5 \times 10^{-3}$ ($\log \varepsilon = -2.3$) is about 0.01 dB. (This is similar to the results in Figure 14.2(d), which is 0.03 dB.) However, if the relative phase noise linewidth is also 5×10^{-3}, then the degradation is much larger: about 1.0 dB (Figure 14.3(b)). This means the degradation is more sensitive to the phase noise. This can be deduced from (14.4) and (14.5), where degradation is proportional to the relative phase noise linewidth while it is proportional to the square of the frequency offset. Since the relative frequency offset and phase noise linewidth are typically less than one, the squared frequency offset produces much less degradation than nonsquared phase noise linewidth.

SNR degradation due to carrier frequency offset for OFDM in shadowed multipath channels is derived in [4]. The channel is modeled as the shadowed tapped delay line channel plus AWGN with two-sided power spectral density $N_o/2$. The approximate SNR degradation (dB) due to relative frequency offset ΔF is found as

$$D_f \cong \frac{10}{\ln 10}\frac{1}{3}\left(\pi N\frac{\Delta F}{R}\right)^2 \frac{E[\alpha^2\gamma^2] + \frac{1}{2E_s/N_o}}{\text{Var}[\alpha\gamma] + \frac{1}{2E_s/N_o}} \tag{14.6}$$

where α is a factor reflecting the shadowing whose square is a lognormal random variable, and γ is related to channel impulse response, is the amplitude of the useful signal at the FFT output of the receiver, and is a Rician random variable [4]. Expression (14.6) is very similar to (14.2) when AWGN is dominant. However, when E_s/N_o is large and multipath effect becomes dominant, it is approximated further as

$$D_f \cong \frac{10}{\ln 10}\frac{1}{3}\left(\pi N\frac{\Delta F}{R}\right)^2 \frac{E[\alpha^2\gamma^2]}{\text{Var}[\alpha\gamma]} \tag{14.7}$$

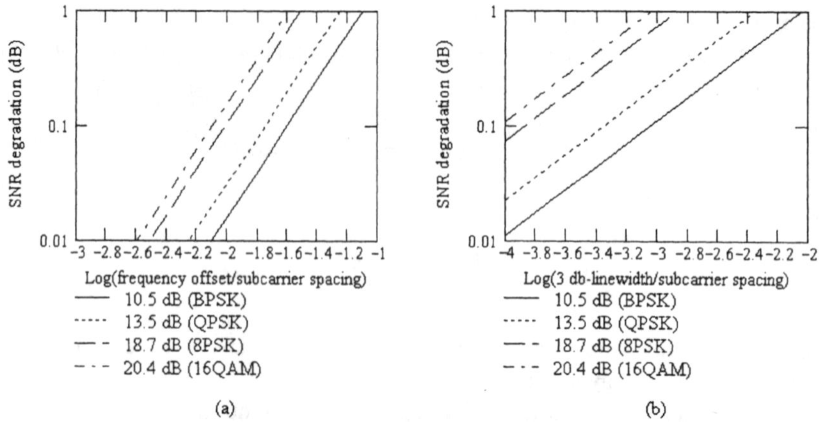

Figure 14.3 SNR degradation in dB versus the normalized frequency offset (a) and the normalized phase noise linewidth (b). The original SNRs are the values at which $BER = 10^{-6}$: 20.4 dB for 16QAM, 18.7 dB for 8PSK, 13.5 dB for QPSK, and 10.5 dB for BPSK.

which indicates the insensitivity to large E_s/N_o.

14.2 EFFECT OF TIMING ERRORS

In Section 12.6, we discussed the role of cyclic extension. We have shown that intersymbol interference (ISI) cannot destroy the orthogonality among subcarriers due to the cyclic extension, as long as the length of the ISI is not longer than the guard interval. Similarly here we can show that the cyclic extension can preserve the orthogonality as long as the timing offset is in advance with respect to the received signal and is shorter than the difference between the guard interval and the maximum delay spread of the channel [5].

Denote the sample indexes of a perfectly synchronized OFDM symbol as $\{-N_g, ..., -1, 0, 1, ..., N-1\}$, where N_g is the guard interval. Denote the timing offset as τ (in number of samples) and the maximum channel delay spread as $\tau_{\max}$ (in number of samples). The maximum length of ISI of the previous symbol into the guard interval of the current symbol is $\tau_{\max}$, we assume that $\tau_{\max} < N_g$, thus the ISI-free part of the guard interval has a minimum index range of $\{-N_g + \tau_{\max}, ..., -1\}$. We denote the received signal time index as t (integer) and the receiver timing index as $t + \tau$. Thus, a negative value of τ indicate a delay, and a positive value of τ indicates

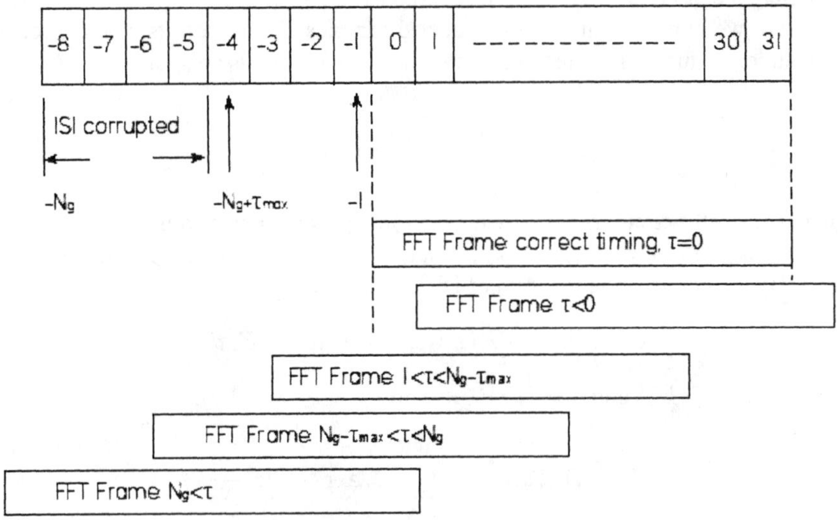

Figure 14.4 Cases of timing offsets. $N = 32, N_g = 8, \tau_{\max} = 4$.

an advance in the receiver timing with respect to the received signal, respectively. Figure 14.4 shows all possible cases of τ for an example of $N = 32, N_g = 8$, and $\tau_{\max} = 4$.

We consider first the case: $\tau \notin \{1, ..., N_g - \tau_{\max}\}$, which includes two subcases: $\tau < 0$ and $\tau > N_g - \tau_{\max}$. In these subcases, the DFT frame will include undesired samples that either belong to other symbols ($\tau < 0$ and $\tau > N_g$) or ISI-corrupted own cyclic extension samples (when $N_g - \tau_{\max} < \tau \leq N_g$). The orthogonality among subcarriers will be destroyed and severe symbol errors will happen.

The second case is that $\tau \in \{1, ..., N_g - \tau_{\max}\}$. In this case, the DFT frame will include τ samples of the cyclic extension of the current symbol and miss τ samples in the end of the current symbol. The orthogonality among subcarriers will be preserved and data symbols will be recovered with a phase rotation caused by the timing offset and a scaling factor caused by the channel transfer function. We will show this next.

The transmitted OFDM signal samples are

$$s_n = \frac{1}{N} \sum_{i=0}^{N-1} d_i \exp\left[j2\pi \frac{in}{N}\right], \qquad n = 0, ..., N-1$$

Assuming the discrete channel impulse response is $\{h_m\} = \{h_0, ..., h_{M-1}\}$, in Section 12.6, we have shown that the received samples are the results of the circular convolution of the transmitted signal and the channel impulse response:

$$x_n = \sum_{m=0}^{N-1} h_m^c s_{n-m}^c$$

where $\{s_n^c\}$ is the circularly extended $\{s_n\}$. $\{h_m^c\}$ is obtained by amending $(N-M)$ zeros to $\{h_m\}$ to form $\{h_m\} = \{h_0, ..., h_{M-1}, 0, ..., 0\}$ and then circularly extending it. By the convolution theorem of DFT,

$$\text{DFT}_k(\mathbf{x}) = \text{DFT}_k(\mathbf{h}) \times \text{DFT}_k(\mathbf{s}) = H_k d_k$$

where $\mathbf{x} = \{x_n\}_{n=0}^{N-1}$ and $\mathbf{s} = \{s_n\}_{n=0}^{N-1}$, and

$$H_k = \text{DFT}_k(\mathbf{h}) = \sum_{m=0}^{M-1} h_m \exp(-j2\pi \frac{km}{N}) \tag{14.8}$$

Now assuming timing offset $-\tau \in \{-N_g + \tau_{\max}, ..., -1\}$, then the samples in the DFT frame are $\mathbf{x}' = \{x_{n-\tau}\}_{n=0}^{N-1}$, where

$$x_{n-\tau} = \sum_{m=0}^{N-1} h_m^c s_{n-\tau-m}^c$$

From the DFT circular convolution theorem we have

$$\widehat{d_k} = \text{DFT}_k(\mathbf{x}') = \text{DFT}_k(\mathbf{h}) \times \text{DFT}_k(\mathbf{s}') = H_k d_k'$$

where $\mathbf{s} = \{s_{n-\tau}\}_{n=0}^{N-1}$ and

$$\begin{aligned} d_k' &= \text{DFT}_k(\mathbf{s}') = \sum_{n=0}^{N-1} s_{n-\tau} \exp(-j2\pi \frac{kn}{N}) = \sum_{l=-\tau}^{N-1-\tau} s_l \exp(-j2\pi \frac{k(l+\tau)}{N}) \\ &= \sum_{l=-\tau}^{-1} s_l \exp(-j2\pi \frac{k(l+\tau)}{N}) + \sum_{l=0}^{N-1-\tau} s_l \exp(-j2\pi \frac{k(l+\tau)}{N}) \end{aligned} \tag{14.9}$$

In the guard interval, due to the cyclic extension, we have $s_l = s_{N+l}$. Thus the first term above is

$$\sum_{l=-\tau}^{-1} s_{N+l} \exp(-j2\pi \frac{k(l+\tau)}{N}) = \sum_{v=N-\tau}^{N-1} s_v \exp(-j2\pi \frac{k(v-N+\tau)}{N})$$

Combining the above with the second term in (14.9), we obtain

$$\begin{aligned} d'_k &= \sum_{v=N-\tau}^{N-1} s_v \exp(-j2\pi \frac{k(v+\tau)}{N}) + \sum_{l=0}^{N-1-\tau} s_l \exp(-j2\pi \frac{k(l+\tau)}{N}) \\ &= \sum_{l=0}^{N-1} s_l \exp(-j2\pi \frac{k(l+\tau)}{N}) = \exp(-j2\pi \frac{k\tau}{N}) \sum_{l=0}^{N-1} s_l \exp(-j2\pi \frac{kl}{N}) \\ &= \exp(-j2\pi \frac{k\tau}{N}) d_k \end{aligned}$$

Thus the demodulator output in the absence of other noise is

$$Y_k = \exp(-j2\pi \frac{k\tau}{N}) H_k d_k \tag{14.10}$$

We conclude that the effect of timing offset within the range of the ISI-free part of the cyclic extension is to cause a phase rotation of $\exp(-j2\pi \frac{k\tau}{N})$ on the kth subcarrier. The phase rotation is linearly proportional to the subcarrier frequency and must be corrected. The phase rotation can be corrected by channel estimation (Chapter 15). Timing offset beyond this range destroys orthogonality and causes severe symbol demodulation errors.

Next we consider effects of small sample timing error and sampling frequency error. With sample timing error τ, $|\tau| < 1$, the received signal samples (ignoring any noise and distortion) are

$$s_{n+\tau} = \frac{1}{N} \sum_{i=0}^{N-1} d_i \exp\left[j2\pi \frac{i(n+\tau)}{N} \right], \qquad n = -\frac{N}{2}, \ldots, \frac{N}{2} - 1 \tag{14.11}$$

and the DFT demodulator outputs are

$$\begin{aligned} \widehat{d_k} &= \sum_{i=0}^{N-1} s_{n+\tau} \exp(-j2\pi \frac{kn}{N}) \\ &= \frac{1}{N} \sum_{i=0}^{N-1} d_i \exp(j2\pi \frac{i}{N} \tau) \sum_{i=0}^{N-1} \exp\left[j2\pi \frac{n}{N} (i-k) \right] \end{aligned}$$

The second summation is equal to 0 for $i \neq k$ and equal to N for $i = k$. Thus from the above we have

$$\widehat{d_k} = d_k \exp(j2\pi \frac{k}{N} \tau), \qquad k = -\frac{N}{2}, \ldots, \frac{N}{2} - 1 \tag{14.12}$$

This indicates that the sample timing error does not destroy the orthogonality between subcarriers. However, it does introduce a phase rotation on the data of each subcarrier. The phase rotation is linearly proportional to the subcarrier frequency. For example, supposing an OFDM system has QPSK as its modulation and the sample timing error is equal to $1/16$ of the OFDM interval, then the phase rotation of the data of the kth subcarrier is $2\pi k/16$. That is, each FFT output will appear at one of the 16, instead of 4, equally spaced points on a circle in the signal constellation diagram. This phase rotation must be estimated and corrected. We discussed estimation of phase rotation caused by sampling timing error together with multipath delay in Chapter 15.

An error in sampling frequency has two effects. First, it causes a time-varying timing offset, hence a time-varying phase rotation of the signal, which must be tracked by the receiver. Second, it results in ICI, since an error in sampling frequency means an error in DFT interval, which causes the loss of orthogonality of the subcarriers. Fortunately, as shown in [6], for practically achievable sampling offsets of 10 ppm, the amount of ICI is rather small, about 0.01 dB at an $E_s/N_o = 20$ dB.

14.3 ARCHITECTURE OF SYNCHRONIZATION SYSTEM

Figure 14.5 shows the architecture of the OFDM synchronization system. In a TDMA based OFDM system, first, the coarse timing of the OFDM signal frame (called variably a slot or packet in different applications) and the symbol are recovered. Then the frequency offset is estimated and corrected before FFT. Since small timing error does not affect orthogonality between subcarriers, its effect is eliminated by fine timing correction after FFT. The effect of residual frequency error is also eliminated by fine frequency correction after FFT. Note that phase noise of the local oscillator always exists, even in the case of an oscillator in a phase-locked loop. The phase noise cannot be reduced or eliminated through synchronization. It can only be reduced by improving the quality of the oscillator.

14.4 SYNCHRONIZATION METHODS OVERVIEW

As discussed in Section 4.10, in a single-carrier system, symbol timing is usually recovered by fairly simple methods such as differentiator-rectifier combination, the early-late gate method, or Gardner's zero-crossing detection method. All these methods are based on the fact that the shapes of baseband signals of the modulations are near rectangular (e.g., MPSK or QAM) or other simple waveforms (e.g., sinusoidal in MSK). Waveforms of baseband OFDM signals are very irregular; thus symbol tim-

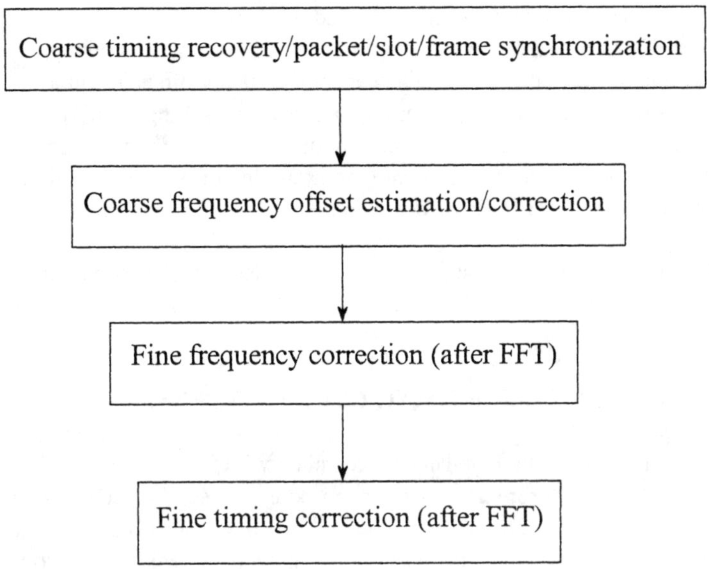

Figure 14.5 Synchronization sequence in OFDM system. From [7]. Copyright © 1999. With kind permission of Springer Science and Business Media.

ing methods for single-carrier systems are not applicable for OFDM signals. Most proposed timing recovery methods for OFDM found in literature or used in practical systems rely on correlation. It is either between a received pilot symbol and local coefficients which are conjugates of the samples of the transmitted pilot symbol, or between two identical parts of the pilot symbol, or between the cyclic extension of the OFDM symbol and its replica in the last part of the symbol.

In a single-carrier system, carrier frequency synchronization is usually accomplished by using a carrier frequency extracting circuit with a nonlinear device; or it can be accomplished by sending a pilot tone. For OFDM, due to the need for timing recovery, the correlation method must be used, which at the same time can estimate the carrier frequency offset. Thus there is no need to use another nonlinear circuit to synchronize the carrier frequency. Because of this reason, the frequency and timing synchronizations are achieved by the correlation method simultaneously.

Besides correlation methods, other schemes were proposed for OFDM synchronization. A null symbol is used as the first OFDM symbol in the time frame in the pan-European digital video broadcasting (DVB) system [8, 9]. No signal is transmitted in this null symbol period and it is detected by monitoring the symbol energy by the receiver. Using a null symbol to flag the start of a frame was also proposed for mobile radio communications [10]. The carrier frequency offset is found in the frequency domain after applying a Hamming window and taking FFT. While the null symbol method may be feasible for a continuous transmission system, it cannot be used in a packet-switched environment (for example, TDMA or ALOHA system) since there would be no difference between the null symbol and the idle period between signal bursts.

In the following, we will discuss the cyclic extension correlation method and the pilot symbol correlation method in detail.

14.5 SYNCHRONIZATION USING CYCLIC EXTENSION

The method of using cyclic extension to correlate with its replica in the end of the OFDM symbol for synchronization was proposed in [11–16]. The idea of using cyclic extension for symbol synchronization appeared first in [11] (1993) and later a scheme was devised in [13] (1995). The proposal of using cyclic extension for estimation of frequency offset was first given in [12] (1995). Based on previous results, a scheme of using cyclic extension for both symbol synchronization and estimation of frequency offset was proposed in [14] (1997). The presentation here closely follows that in [14], which is an ML estimator for both timing offset and carrier frequency offset.

Assuming the carrier frequency offset normalized to subcarrier spacing is ε, that is, the local carrier frequency is $f_c + \varepsilon/T$, then the complex envelope of the down converter output will be equal to $s(t)e^{j2\pi\varepsilon t/T}$, where $s(t)$ is the complex envelope of the transmitted signal. If the signal is sampled at N times the symbol rate, it is $s(k)e^{j2\pi\varepsilon k/N}$ in discrete-time form. Further assuming the timing offset is θ (in number of samples), then the received signal samples in an AWGN channel are

$$r(k) = s(k-\theta)e^{j2\pi\varepsilon(k-\theta)/N} + n(k)$$

where the signal power is $\sigma_s^2 \triangleq E\{|s(k)|^2\}$ and the noise power is $\sigma_n^2 \triangleq E\{|n(k)|^2\}$.

Assume that we observe $2N + L$ consecutive samples of the received signal (Figure 14.6). These samples contain a complete symbol which has $(N+L)$ samples, where L samples belong to the cyclic prefix. The position of this symbol in the observation interval is unknown, and we denote timing offset of the start of this symbol from the start of the observation interval as θ. Thus the samples in the cyclic prefix have indexes of

$$\mathcal{I} \triangleq \{\theta, ..., \theta + L - 1\}$$

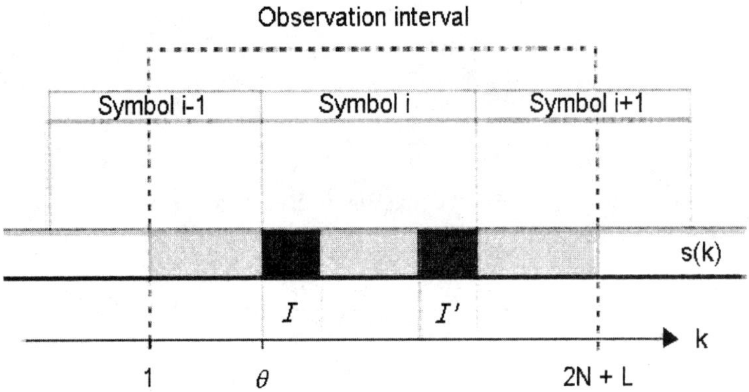

Figure 14.6 Set $\mathcal{I}$ contains the cyclic prefix, that is, the copies of the L data samples in $\mathcal{I}'$. From [14]. Copyright © 1997 IEEE.

and the samples with the indexes of

$$\mathcal{I}' \triangleq \{\theta + N, ..., \theta + N + L - 1\}$$

are the replicas of the samples of the cyclic prefix. All received samples in the observation interval can be organized in a vector

$$\mathbf{r} \triangleq [r(1), ..., r(2N + L)]^T$$

The samples in the cyclic prefix and their copies are pair-wise correlated. That is

$$\forall k \in \mathcal{I}: \quad E\{r(k)r^*(k+m)\} = \begin{cases} \sigma_s^2 + \sigma_n^2, & m = 0 \\ \sigma_s^2 e^{-j2\pi\varepsilon}, & m = N \\ 0 & \text{otherwise} \end{cases} \tag{14.13}$$

and the remaining samples $r(k), k \notin \mathcal{I} \cup \mathcal{I}'$ are mutually uncorrelated (see (13.13)).

The correlations given in (14.13) can be used to estimate the timing offset and the frequency offset. The peak of the correlation between samples in $\mathcal{I}$ and $\mathcal{I}'$ can be used to estimate timing offset. The fact that the correlation is $\sigma_s^2 e^{-j2\pi\varepsilon}$, for $m = N$ can be used to estimate the frequency offset. In fact, the precise estimates are found using the joint maximum likelihood estimations of delay θ and frequency offset ε as

[14]

$$\widehat{\theta}_{ML} = \arg\max_{\theta}\{\gamma(\theta) - \rho\Phi(\theta)\} \tag{14.14}$$

$$\widehat{\varepsilon}_{ML} = -\frac{1}{2\pi}\angle\gamma(\widehat{\theta}_{ML}) \tag{14.15}$$

where $\arg\max(x)$ is the argument of x that maximizes x; and $\angle$ denotes the argument of a complex number and

$$\rho = \frac{\sigma_s^2}{\sigma_s^2 + \sigma_n^2} = \frac{SNR}{SNR+1} \tag{14.16}$$

$$\gamma(m) \triangleq \sum_{k=m}^{m+L-1} r(k)r^*(k+N) \tag{14.17}$$

$$\Phi(m) \triangleq \frac{1}{2}\sum_{k=m}^{m+L-1} |r(k)|^2 + |r(k+N)|^2 \tag{14.18}$$

$\gamma(m)$ is a sum of L consecutive correlations between pairs of samples spaced N samples apart. $\Phi(m)$ is the power (or energy in a unit time) of the samples of the same L consecutive pairs of samples spaced N samples apart. If SNR is not too low, $\gamma(m)$ is likely to be maximum if $m = \theta$ and close to zero when m is far from θ. This intuitively justifies (14.14). Each term in $\gamma(m)$ has a phase that is proportional (approximately, due to noise) to the frequency offset ε (see (14.13)). Consequently, $\gamma(m)$ has a phase that is proportional (approximately, due to noise) to the frequency offset ε. Thus we have (14.15).

The synchronization system is shown in Figure 14.7 which implements (14.14) through (14.18). The inputs to this system are the complex baseband OFDM signal samples after the IF down convertor.

Figure 14.8 shows some simulation results given in [14]. Note that above a certain threshold (4 to 6), increasing L further will not improve performance. More simulation results show that there are error floors for time-dispersive channels (Figure 14.9).

A paper on cyclic-extension-based estimators for timing and frequency offset in 2000 [16] derived a new likelihood function for the joint timing and frequency-offset estimation and formulated three classes of unbiased estimators, namely, maximum-likelihood, minimum variance unbiased, and moment estimator. It is claimed that these estimators are more computationally efficient. Interested readers may refer to [16] for details.

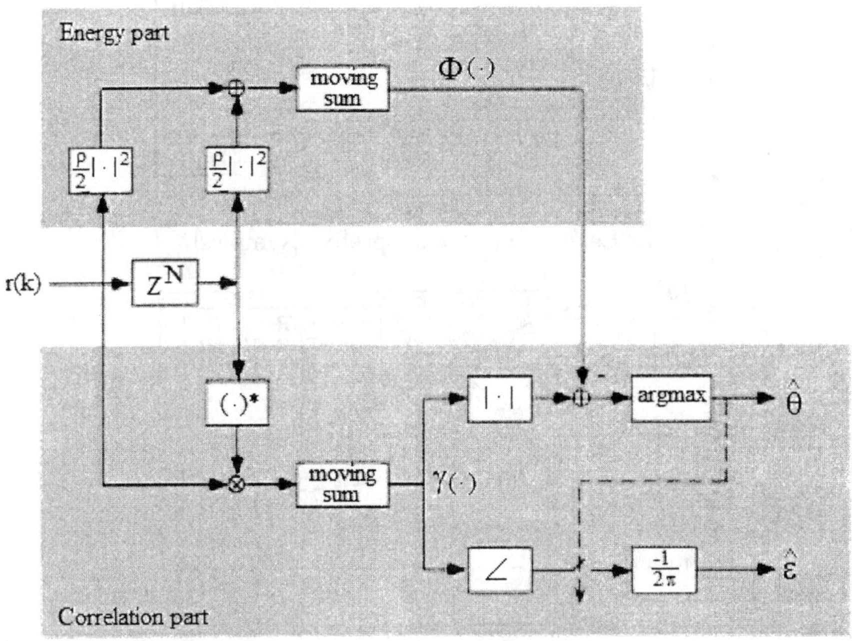

Figure 14.7 Structure of the estimator. From [14]. Copyright © 1997 IEEE.

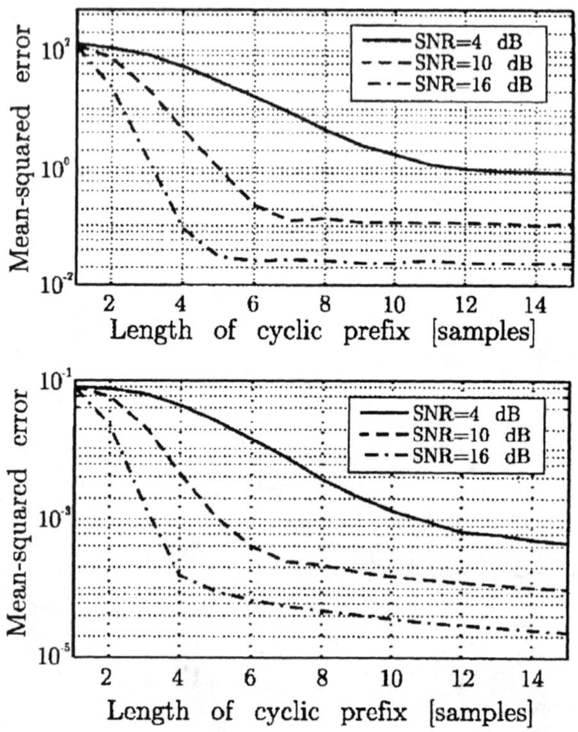

Figure 14.8 Performance of the time (top) and frequency (bottom) estimators for the AWGN channel (4, 10, and 16 dB). The dimensionless performance measure is expressed in squared units relative to the sample interval (top) and the intertone spacing (bottom). The number of subcarriers is $N = 256$. From [14]. Copyright © 1997 IEEE.

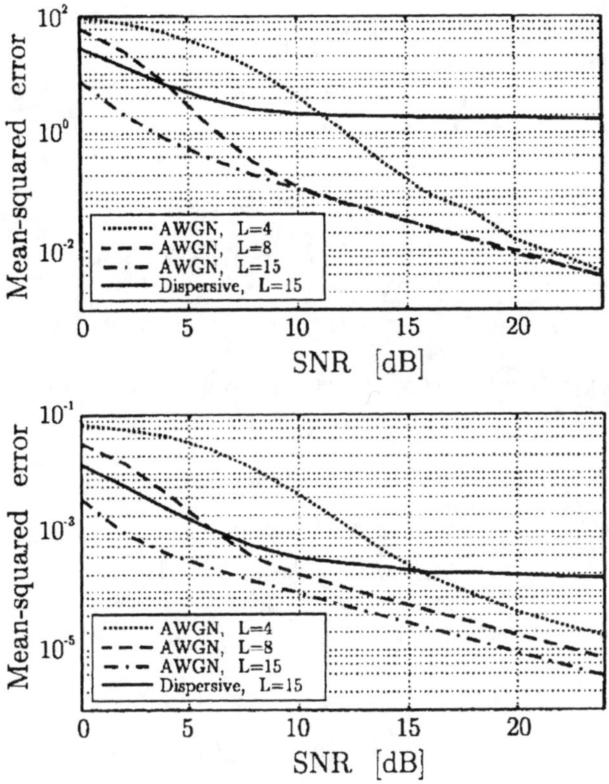

Figure 14.9 Performance of the time (top) and frequency (bottom) estimators for the AWGN channel ($L = 4$, $L = 8$, and $L = 15$) and the dispersive channel ($L = 15$): The dimensionless performance measure is expressed in squared units relative to the sample interval (top) and the intertone spacing (bottom). From [14]. Copyright © 1997 IEEE.

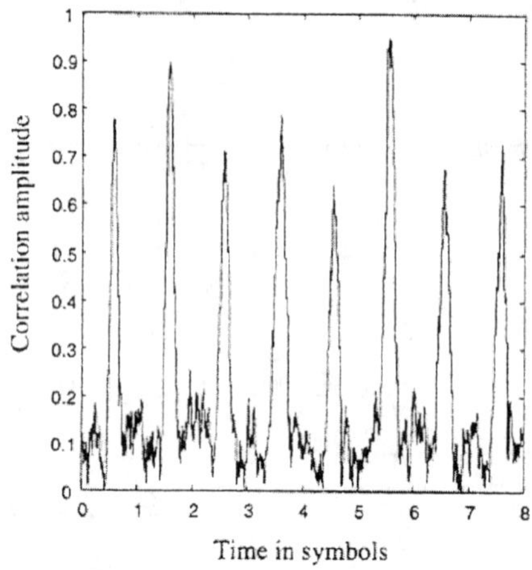

(a) Example of correlation output amplitude for eight OFDM symbols with 192 subcarriers and a 20% guard time.

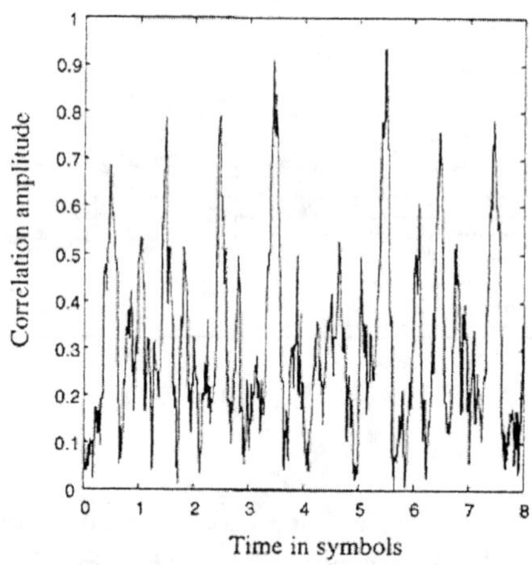

(b) Example of correlation output amplitude for eight OFDM symbols with 48 subcarriers and a 20% guard time.

Figure 14.10 Cyclic correlator output (a, b). From [3]. Copyright © 2000 Richard van Nee and Ramjee Prasad.

In comparison with pilot symbol methods, the advantage of the cyclic prefix method is that there are no extra symbols to send. However there are several disadvantages. First the peak value of the correlator output varies significantly from symbol to symbol (see Figure 14.10). The reason for this is that even though the average symbol power is constant, the power in the cyclic prefix varies from symbol to symbol due to the randomness of the data. Second, when the number of subcarriers N is small, the undesired sidelobes between the peaks are comparable to the peaks, and may even be higher than the peaks (see Figure 14.10). This is because the data are random, hence the correlation is a random variable. This randomness is more severe when N is small. Third, the cyclic prefix method can only be used to find symbol timing, it cannot detect the start of a frame. Thus it is only good for tracking in a continuous OFDM transmission and not good for detection of the start of a packet in packet transmission.

14.6 SYNCHRONIZATION USING PILOT SYMBOLS

Synchronization based on pilot (or training) symbols is a widely used and reliable method. It can detect the start of a frame, determine symbol timing, and estimate the carrier frequency offset at a small expense of bandwidth increase. It can be used for continuous transmission as well as for packet transmission.

Many schemes of pilot-symbol-assisted synchronization were proposed in the literature [17–31]. The earliest scheme appeared in a paper by Warner and Leung (1993) for an OFDM-FM system designed for the pure ALOHA mobile radio environment [17]. In the transmitter, a number of subcarriers are reserved for pilot tones with known phase and amplitude. In the receiver, frame synchronization is achieved by detecting the presence of signal power. Coarse synchronization alignment to within $\pm 1/2$ sample period is achieved with a correlator matched to the pilot tone, implemented in the frequency domain. The phase and magnitude distortions are estimated by transmitting a separate PN sequence and are used to compensate for the frequency-domain data after the FFT demodulator in an equalizer. Fine synchronization is performed on the equalizer's output, from which the time shift (within $\pm 1/2$ sample period) that maximizes the correlation with the reference signal is calculated and used to compensate for the phase shift for each sub-channel before extracting the data.

Classen and Meyr in 1994 proposed a two-stage frequency synchronizer in [18]. The first stage is the acquisition stage and the second stage is the tracking stage. Both are after the FFT demodulator. Coarse timing synchronization is assumed and not discussed in the paper. The frequency synchronizer is data-aided— that is, a pilot symbol set is transmitted and a correlation between received repeated pilot symbols

z_n and z_{n+D} for some integer D is used to extract the frequency offset.

Moose in 1994 proposed to use a two-symbol pilot sequence for frequency synchronization [1]. The second symbol is the replica of the first symbol. The frequency offset ε (normalized to subcarrier spacing) causes a phase difference between the FFT points of the first symbol and the second symbol, which can be extracted by correlation between the two FFT output sequences. A maximum likelihood estimator for ε is formed based on the ratio of the imaginary part and the real part of the correlation.

Schmidl and Cox in 1997 proposed a two-symbol training sequence for synchronization of frame, symbol timing, and estimation of carrier frequency offset [19]. The first training symbol consists of two specially constructed identical halves in the time domain. Correlation between these two halves is used to detect the start of the frame and the beginning of a symbol. The second training symbol contains a PN sequence on the odd frequencies to measure these subchannels and another PN sequence on the even frequencies to help determine frequency offset. Frequency offset is partially corrected using the first training symbol and is further corrected by a correlation with the second training symbol. Schmidl and Cox's scheme meets all synchronization needs effectively with a minimum amount of overhead. Later, improved versions of the Schmidl and Cox synchronizer were reported in [20–23].

Since the Schmidl and Cox scheme is an efficient scheme and a base of many later schemes in the literature, in the following, we present the Schmidl and Cox scheme in detail. After that we will present highlights of its newer versions proposed by other authors. At the end of this chapter, as a practical example, we will describe the pilot-symbol based synchronization for the IEEE 802.11a standard.

14.7 THE SCHMIDL-COX SYNCHRONIZER

The Schmidl-Cox synchronizer relies on two training symbols. The first one is used for frame and symbol timing, as well as coarse frequency offset estimation. The second one is used for fine frequency offset estimation.

The two halves of the first training symbol are made identical (in time order) by transmitting a pseudonoise (PN) sequence on the even frequencies, while zeros are used on the odd frequencies. The first training symbol can be used for two purposes. First the correlation between the two identical halves can be used for frame/symbol timing. Second, the two identical halves will remain identical after passing through the channel, except that there will be a phase difference between them caused by the carrier frequency offset. This fact can be used for a coarse estimate of the carrier frequency offset. Due to missing energy of odd frequency components, the even frequency components should be properly scaled up to maintain an approximately

freq. num. k	$c_{1,k}$	$c_{2,k}$	$v_k = \sqrt{2}c_{2,k}/c_{1,k}$
-4	$7+7j$	$5-5j$	$-j$
-3	0	$-5-5j$	
-2	$-7+7j$	$-5-5j$	j
-1	0	$-5+5j$	
0	$7+7j$	$-5-5j$	-1
1	0	$5+5j$	
2	$7-7j$	$-5+5j$	-1
3	0	$5-5j$	
4	$7+7j$	$5+5j$	1

Table 14.1 Illustration of use of PN sequences for training symbols. From [19]. Copyright © 1997 IEEE.

constant signal energy for each symbol. For example, if QPSK is the modulation, the even frequency components of this training symbol should be multiplied by $\sqrt{2}$ at the transmitter, or the four points of the QPSK constellation can be selected from a larger constellation, such as 64QAM, so that points with higher energy can be used. An equivalent method of generating the first training symbol is to generate a half-length symbol using a half-length IFFT and the resultant time-domain samples are repeated to form the second half of the symbol. This method does not require a PN sequence with zeros on the odd frequencies.

The second training symbol consists of two PN sequences; one is on the odd frequencies for measuring these subchannels, and another is on the even frequencies to help determine frequency offset. Table 14.1 is an example of using PN sequences in the training sequence for an OFDM signal with nine subcarriers with the points chosen from a subset of a 64QAM constellation. $\{c_{1,k}\}$ is for the first training symbol and $\{c_{2,k}\}$ is for the second training symbol. $\{v_k\}$ is a differentially encoded PN sequence which will be used in estimation of the carrier frequency offset. The selection of a particular PN sequence should not have much effect on the performance of the synchronization algorithms. Instead the PN sequence can be chosen on the basis of being easy to implement or having a low peak-to-average power ratio. Complex samples r_k are taken from the outputs of I-channel and Q-channel A/D converters. (Figure 14.11).

14.7.1 Frame/Symbol Timing Estimator

Now we will show how the first training symbol $u(t)$ can be used for frame/symbol timing. At the transmitter, the first half of the first training symbol is identical to the second half (in time order). At the receiver, the carrier frequency offset of Δf

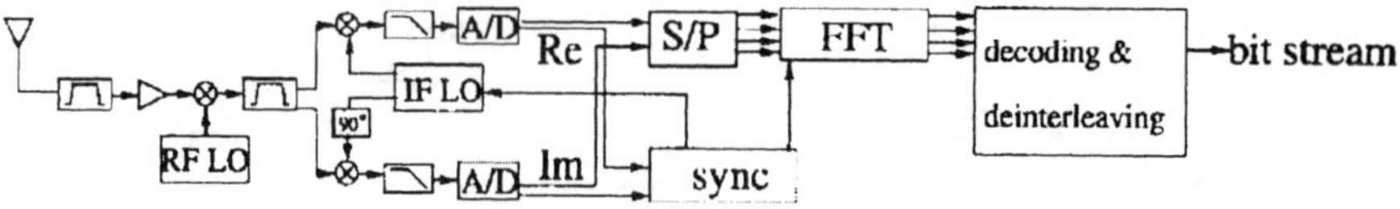

Figure 14.11 Block diagram of OFDM receiver. From [19]. Copyright © 1997 IEEE.

causes a phase shift of $2\pi t\Delta f$ on each sample. In addition, all samples have a phase shift $\theta(t)$ caused by the channel, and they are approximately the same assuming the channel is at worst slow-fading. If the received complex envelope at time t in the first half is $r(t) = u(t)\exp\{\theta(t) + 2\pi t\Delta f\}$, then $T/2$ seconds later the signal is $r(t + T/2) = u(t + T/2)\exp\{\theta(t + T/2) + 2\pi t\Delta f + \pi T\Delta f\}$. Since $u(t) = u(t + T/2)$ by design, we have

$$r^*(t)r(t+T/2) = |u(t)|^2\exp\{\pi T\Delta f + \theta(t+T/2) - \theta(t)\} \approx |u(t)|^2\exp\{\pi T\Delta f\} \tag{14.19}$$

It is seen that the effect of the channel ($\theta(t)$) is cancelled, and the result has a phase of $\phi = \pi T\Delta f$ that is proportional to the frequency offset.

Let there be L complex samples in one-half of the first training symbol (excluding the cyclic prefix); we can define a sum of the pairs of products as

$$P(d) = \sum_{m=0}^{L-1}(r^*_{d+m}r_{d+m+L}) \tag{14.20}$$

which can be implemented with the iterative formula

$$P(d+1) = P(d) + (r^*_{d+L}r_{d+2L}) - (r^*_d r_{d+L}) \tag{14.21}$$

$P(d)$ is the correlation function between the samples of the first half and the second half. At the start of the frame, the products of each of these pairs of samples will have approximately the same phase, so the magnitude of $P(d)$ will be a large value. Note that d is a time index of the first sample in a window of $2L$ samples. This window slides along in time as the receiver searches for the first training symbol. The received energy for the second half-symbol is defined by

$$R(d) = \sum_{m=0}^{L-1}|r_{d+m+L}|^2 \tag{14.22}$$

which can also be calculated iteratively. $R(d)$ may be used as part of an automatic gain control (AGC) loop. A *timing metric* can be defined as

$$M(d) = \frac{|P(d)|^2}{[R(d)]^2} \quad (14.23)$$

where the normalization by $[R(d)]^2$ is necessary as the received signal power is typically time variant.

Figure 14.12 shows a simulated example of the timing metric as a window slides past coincidence between the window and the first training symbol. The simulation is for the AWGN channel and for an OFDM signal with 1,000 subcarriers, a carrier frequency offset of 12.4 subcarrier spacings, and an signal-to-noise ratio (SNR) of 10 dB, where the SNR is the total signal (all the subcarriers) to noise power ratio. The timing metric reaches a plateau which has a length equal to the length of the guard interval minus the length of the channel impulse response since there is no ISI within this plateau to distort the signal. For the AWGN channel, the length of the plateau is equal to the length of the guard interval. For frequency selective channels, the length of the impulse response of the channel is shorter than the guard interval by design, so the length of the plateau is shorter than that for the AWGN channel. Figure 14.13 shows how the timing metric should evolve as the correlation window slides for an AWGN channel at high SNR. It starts the plateau when $d = -G$, where G is the length of the guard interval, and ends the plateau when $d = 0$. This is consistent with the simulation result in Figure 14.12. If the channel has an impulse response with length τ, the plateau should start when $d = -G + \tau$ and end when $d = 0$ for high SNR. Thus, we can determine the starting point of the frame as the rear edge of the plateau. However, when SNR is low, the boundary of the plateau will be blurred and some uncertainty of the starting point of the frame will exist. It was suggested that an average center of the plateau be used as the correct timing. The center is determined by finding the maximum of the metric, and the 90% points on the left and right sides of the maximum; the center is the average of these two points. Simulation results in [19] showed that this method results in no degradation in SNIR (signal-to-noise-plus-interference ratio) for an AWGN channel and 0.06 dB for an exponential channel at an $SNR = 40$ dB. See the end of Section 14.7.2 for a description of the exponential channel.

14.7.2 Performance of Timing Estimator

The performance of the Schmidl and Cox symbol timing estimator is evaluated by two parameters. The first one is the probability of missing the training sequence or false detection since the timing metric is not only used to determine symbol timing but also used to determine whether the training sequence has been received. The

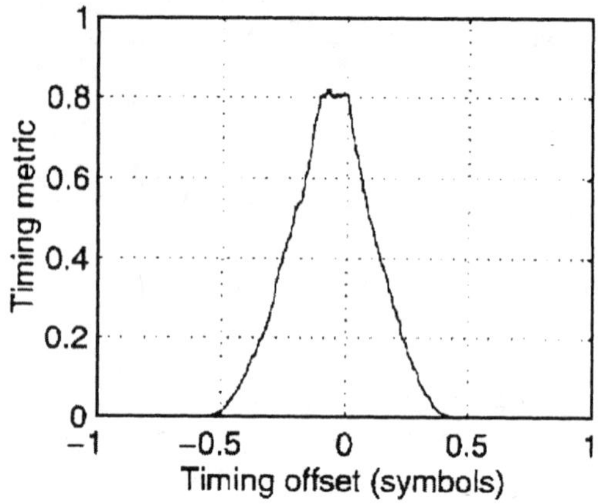

Figure 14.12 Example of the timing metric for the AWGN channel ($SNR = 10$ dB). From [19]. Copyright © 1997 IEEE.

second one is the degradation in SNIR due to the error in symbol timing.

Based on the fact that there are a large number of terms in the $P(d)$ and $R(d)$, it is found in [19] by the central limit theorem that the probability distributions of them are approximately Gaussian. At high SNRs, the distributions of the ratio of them ($q(d) = |P(d)/R(d)|$) and the square of $q(d)$, which is $M(d)$, are also approximately Gaussian.

Let each complex sample $r_m = s_m + n_m$ be made of a signal part and a noise part. Let the variance of the real and imaginary parts be

$$E[\text{Re}\{s_m\}^2] = E[\text{Im}\{s_m\}^2] \triangleq \sigma_s^2$$

$$E[\text{Re}\{n_m\}^2] = E[\text{Im}\{n_m\}^2] \triangleq \sigma_n^2$$

so that σ_s^2/σ_n^2 is the SNR. It is found [19] that at the optimum symbol timing point, d_{opt}, where SNR is high, the distribution of the timing metric $M(d_{opt})$ is approximately Gaussian with mean value

$$\mu_M = E[M(d_{opt})] = \frac{\sigma_s^4}{(\sigma_s^2 + \sigma_n^2)^2} = \frac{1}{(1 + SNR^{-1})^2} \qquad (14.24)$$

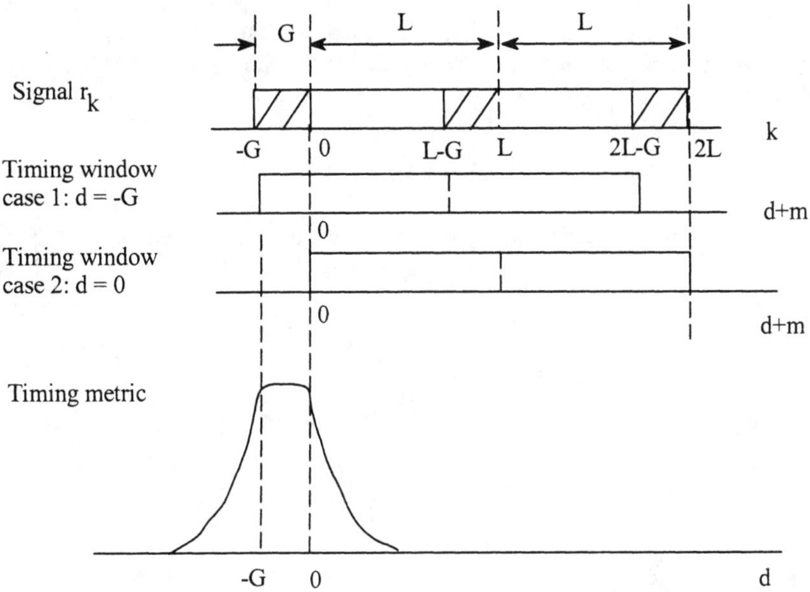

Figure 14.13 Timing metric of the Schmidl-Cox synchronizer.

and variance

$$\begin{aligned}\sigma_M^2 &= var[M(d_{opt})] = \frac{2\sigma_s^4[(1+\mu_M)\sigma_s^2\sigma_n^2 + (1+2\mu_M)\sigma_n^4]}{L(\sigma_s^2+\sigma_n^2)^4} \\ &= \frac{2SNR^2[(1+\mu_M)SNR + (1+2\mu_M)]}{L(SNR+1)^4}\end{aligned} \tag{14.25}$$

That is

$$M(d_{opt}) \sim N(\mu_M, \sigma_M^2) \tag{14.26}$$

where $\sim$ indicates "is distributed", and $N(\mu, \sigma^2)$ indicates a Gaussian random variable with mean μ and variance σ^2.

At a position outside the first training symbol, the terms in the sum $P(d_{outside})$ add with random phase since there is not a periodicity for samples spaced by L

samples. In other words, the correlation between samples spaced by L samples outside the first training symbol is essentially zero. Thus the samples can be considered to be composed of just noise terms and the statistics will be independent of the SNR. Based on this approximation, the distribution of the timing metric $M(d_{outside})$ is found [19] to be chi-square given by

$$M(d_{outside}) \sim \frac{1}{2L}\chi_2^2 \tag{14.27}$$

where χ_2^2 is a chi-square random variable with two degree of freedom. Its PDF is

$$f_{\chi_2^2}(r) = \begin{cases} \frac{1}{2}e^{-r/2}, & r \geq 0 \\ 0, & r < 0 \end{cases} \tag{14.28}$$

and its mean is 2 and its variance is 4.

Using the well-known PDF conversion formula for function $y = g(x)$ [27, p. 95]

$$f_y(y) = \sum_i \frac{f_x(x_i)}{|g'(x_i)|}$$

where $f_x(x)$ and $f_y(y)$ are PDFs of x and y, respectively, x_i are the roots of $y = g(x)$, and $g'(x)$ is the derivative of $g(x)$, we can obtain the PDF of $M(d_{outside})$ as

$$f_{M(d_{outside})}(r) = \begin{cases} L\exp\{-rL\}, & r \geq 0 \\ 0, & r < 0 \end{cases} \tag{14.29}$$

and mean and variance are

$$E[M(d_{outside})] = 1/L \tag{14.30}$$

$$var[M(d_{outside})] = 1/L^2 \tag{14.31}$$

Figure 14.14 shows the expected value of the timing metric with $L = 512$. At high SNR, it is approaching 1 (see also (14.24)) at the optimum timing point. It is almost zero (1/512 in this case) when at positions outside the first symbol. Simulation results in [19] also confirmed the theoretical results in Figure 14.14.

Probability of missing the training sequence or false detection can be calculated using equations (14.26) to (14.31). Assuming the detector threshold is γ, the probability of missing is given by

$$P_M = \int_{-\infty}^{\gamma} \frac{1}{\sqrt{2\pi}\sigma_M} \exp\left\{-\frac{(r-\mu_M)^2}{2\sigma_M^2}\right\} dr \tag{14.32}$$

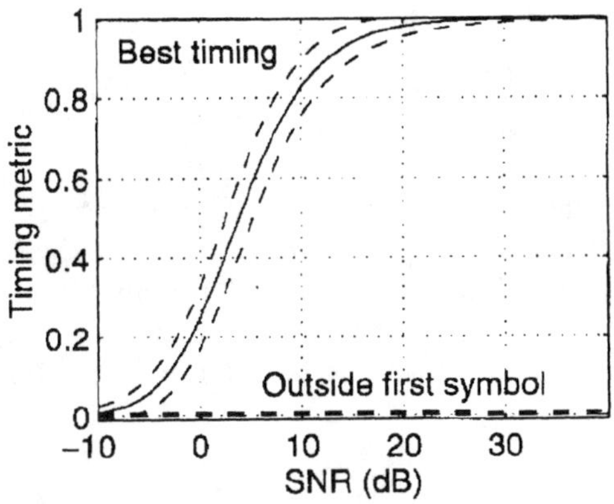

Figure 14.14 Expected values of timing metric with $L = 512$. Dashed lines indicate three standard deviations. From [19]. Copyright © 1997 IEEE.

For example, let there be 1,000 subchannels and let $L = 512$. If the system is designed to detect a signal if the SNR is at least 10 dB, then from (14.24) and (14.25), when the signal is present, we obtain the mean value as 0.826 and the variance as 5.581×10^{-4}. When the signal is not present, from (14.30) and (14.31), we have the mean value as 1.953×10^{-3} and the variance as 3.815×10^{-6}. If the desired probability of missing a training sequence is 1×10^{-3}, from (14.32), it is found that $\gamma = 0.753445$ (using a trial-and-error method with computer software like MathCAD).

When computing false detection probability, the number of symbols of a frame must be considered since each symbol could be taken as the training symbol, which contributes to the overall false detection probability. Assuming there are K symbols in one OFDM frame, since the sliding window length is half a symbol long, the false detection probability is

$$P_F = 2KP_{F1} \tag{14.33}$$

where the probability of false detection in one occasion is given by

$$P_{F1} = \int_{\gamma}^{\infty} f_{M(d_{outside})}(r)dr \tag{14.34}$$

Thus from (14.33), (14.34), and (14.29), we have

$$P_F = \begin{cases} 2Ke^{-\gamma L}, & \gamma \geq 0 \\ 2K, & \gamma \leq 0 \end{cases} \tag{14.35}$$

For the above example, $\gamma = 0.753445$, assuming $K = 100$, it is found that $P_F \cong 0$. For $P_F = 1 \times 10^{-3}$, the threshold is $\gamma_1 = 0.0238$. Comparing γ and γ_1, we see that there is a lot of space to move down the threshold so that P_M can be reduced for the same SNR or the synchronizer can operate at a lower SNR for the same P_M.

The second performance indicator of the Schmidl and Cox symbol timing estimator is the degradation in SNIR due to the error in symbol timing. SNIR can be defined as

$$\text{SNIR} = \frac{\sigma_s^2}{\sigma_n^2 + \sigma_i^2} \tag{14.36}$$

where σ_i^2 is the variance of the intersymbol interference (ISI) added by incorrect symbol timing. Two channels, the AWGN channel and a frequency selective channel, are simulated. The frequency selective channel has sixteen paths and an exponential power delay profile. Sixteen paths are chosen with path delays of 0, 4, 8, ..., 60 samples, so the impulse response of the channel is shorter than the guard interval. The amplitude of each path is given as

$$A_i = \exp(-\tau_i/60) \tag{14.37}$$

where A_i is the amplitude of the ith path and τ_i is the delay of the ith path in samples. Two methods to determine the symbol timing are compared on the basis of reduction in SNIR. As mentioned before, theoretically, we can determine the starting point of the frame as the rear edge of the timing metric plateau. However, when SNR is low, the edges of the plateau will be blurred. A simple practical method is to find the peak of the timing metric. The second method is to find the average center of the plateau as the correct timing. The center is determined by finding the peak of the metric, and the 90% points on the left and the right side of the peak, then the center is the average of these two points. Table 14.2 compares the two methods based on the SNIR reduction due to timing error. It is seen that the average center method is significantly better.

Note that Table 14.2 shows that SNIR reduction goes up when SNR increases for the peak-detection method. This seems to contradict to our common sense. However, this is possible. In fact, using (14.36), compared with SNR of the AWGN channel with perfect timing, the reduction in SNIR is

$$\text{SNIR reduction} = \frac{SNR}{SNIR} = \frac{\sigma_s^2/\sigma_n^2}{\sigma_s^2/(\sigma_n^2 + \sigma_i^2)} = 1 + \frac{\sigma_i^2}{\sigma_n^2}$$

	AWGN		Exponential	
SNR (dB)	peak	avg 90%	peak	avg 90%
0	0.0087	0.0007	0.0213	0.0034
10	0.0060	0.0000	0.0336	0.0003
20	0.0364	0.0000	0.1828	0.0014
30	0.3068	0.0000	1.2374	0.0119
40	0.9023	0.0000	4.1953	0.0599

Table 14.2 Reduction in SNIR (dB) due to errors in symbol timing. From [19]. Copyright © 1997 IEEE.

Thus, if the detection method is poor, σ_i^2 does not decrease as fast as σ_n^2 when SNR increases, the net effect is an increase in SNIR reduction. This is exactly what happened to the peak-detection method.

14.7.3 Frequency Offset Estimator

We have shown in (14.19) that the correlation between two samples $T/2$ apart in the first training symbol has an angle of

$$\phi = \pi T \Delta f \tag{14.38}$$

that is proportional to the carrier frequency offset Δf. The $P(d)$ is a sum of these correlations so that its angle is also ϕ. If $|\phi|$ can be guaranteed to be less than π, the carrier frequency offset can be estimated near the best timing point using

$$\widehat{\Delta f} = \frac{\widehat{\phi}}{\pi T} = \frac{\text{angle}(P(d))}{\pi T} \tag{14.39}$$

Otherwise, angle $(P(d))$ cannot reflect the total ϕ since the angle of a complex number is limited to the range of $[-\pi, \pi]$. In this case, the actual angle of $P(d)$ would be $\phi + 2\pi z$, where z is an integer, and the actual frequency offset would be

$$\frac{\phi}{\pi T} + \frac{2z}{T} \tag{14.40}$$

The fractional part can be estimated using (14.39). The integer part can be found using the second training symbol. First the two training symbols are partially frequency corrected by multiplying the samples by $\exp(-j2t\widehat{\phi}/\pi T)$. Then the FFT outputs of these two symbols will not have inter-subcarrier interference since the remaining frequency offset is an integer multiple of $1/T$, and the orthogonality between subcarriers is restored. However we still need to find out this integer part of the frequency offset and correct it since it degrades the SNR (see (14.1)).

After the two training symbols are partially frequency corrected, let their FFTs be $\{x_{1,k}\}$ and $\{x_{2,k}\}$, and let the differentially-encoded PN sequence on the even frequencies of the second training symbol be $\{v_k\}$ (as illustrated in Table 14.1). Note that $\{v_k\}$ is not transmitted but known to the receiver. At the output of the FFT, because of the uncompensated frequency shift of $2z/T$, ignoring noise and distortion, $\{x_{1,k}\}$ and $\{x_{2,k}\}$ will be the same as the transmitted data except that they will be shifted by $2z$ positions (frequency shifting property of FFT). That is, $x_{1,k} = c_{1,k-2z}$ and $x_{2,k} = c_{2,k-2z}$. Consequently, $\{v_k\}$ will be shifted by $2z$ positions. A correlation of $\{v_k\}$ with $\{x_{1,k}\}$ and $\{x_{2,k}\}$ for even k would make sense as a metric for finding z. The metric is given as [19]

$$B(g) = \frac{\left|\sum_{k\in X} x^*_{1,k+2g} v^*_k x_{2,k+2g}\right|^2}{2\left(\sum_{k\in X} |x_{2,k}|^2\right)^2} \tag{14.41}$$

where $X = \{-W, -W+2, ..., -4, -2, 2, 4, ..., W-2, W\}$ is the set of indices. W is the number of even frequencies (excluding 0) with the PN sequence. A typical term in $B(g)$ is

$$x^*_{1,k+2g} v^*_k x_{2,k+2g} = c^*_{1,k+2g-2z} \sqrt{2} \frac{c^*_{2,k}}{c^*_{1,k}} c_{2,k+2g-2z}$$

It is clear that when $g = z$,

$$x^*_{1,k+2g} v^*_k x_{2,k+2g} = c^*_{1,k} \sqrt{2} \frac{c^*_{2,k}}{c^*_{1,k}} c_{2,k} = \sqrt{2}|c_{2,k}|^2$$

and $B(g) = 1$. Otherwise, $B(g) < 1$. Thus, the number of even positions shifted (z) can be calculated by finding $\widehat{g}$ to maximize $B(g)$, with integer g spanning the range of possible frequency offsets. Then the complete frequency offset estimate is

$$\widehat{\Delta f} = \frac{\widehat{\phi}}{\pi T} + \frac{2\widehat{g}}{T} \tag{14.42}$$

14.7.4 Performance of Frequency Offset Estimator

The performance of the carrier frequency estimator can be evaluated in two parts. First, the variance of the estimate of the fractional part is found as

$$\text{var}[\widehat{\phi}/\pi] = \frac{1}{\pi^2 \cdot L \cdot \text{SNR}} \tag{14.43}$$

and the Cramér-Rao bound is also the same

$$\text{var}[\widehat{\phi}/\pi] \geq \frac{1}{\pi^2 \cdot L \cdot \text{SNR}} \tag{14.44}$$

Simulation results in [19] show the variance is very close to the bound. Since the carrier frequency offset estimate (in subcarrier spacing) is made up of the sum of the fractional part and an even integer, the variance of the fractional part is the variance of the entire estimate if the integer part estimate is correct. Second, for the integer part of the estimate, the distribution of the carrier frequency offset metric should be found out for both the correct and incorrect integer offsets to determine the probability of making an error in finding the integer part of the carrier frequency offset. The expected value and variance of $B(g)$ can be found in a similar manner as in the previous section. At the correct frequency offset estimate, from (14.41),

$$\mu_B = E[B(g_{correct})] = \frac{\sigma_s^4}{(\sigma_s^2 + \sigma_n^2)^2} \tag{14.45}$$

and

$$\text{var}[B(g_{correct})] = \frac{\sigma_s^4[(2 + 2\mu_B)\sigma_s^2\sigma_n^2 + (1 + 4\mu_B)\sigma_n^4]}{W(\sigma_s^2 + \sigma_n^2)^4} \tag{14.46}$$

At the incorrect frequency offset estimate,

$$E[B(g_{incorrect})] = \frac{1}{2W}\left(1 + \frac{\sigma_s^2}{\sigma_s^2 + \sigma_n^2}\right) < \frac{1}{W} \tag{14.47}$$

and

$$\text{var}[B(g_{incorrect})] = \frac{1}{4W^2}\left(1 + \frac{3\sigma_s^4 + 2\sigma_s^2\sigma_n^2}{(\sigma_s^2 + \sigma_n^2)^2}\right) < \frac{1}{W^2} \tag{14.48}$$

Figure 14.15 shows the expected value of carrier frequency offset metric $B(g)$ versus SNR for both a correct and incorrect frequency offset for the 1,000 subcarrier OFDM signal. The dashed lines indicate three standard deviations from each curve. It is seen that when the frequency offset estimate is correct, the metric is much larger than when the estimate is incorrect. This property of the metric assures a reliable estimation of the frequency offset.

14.8 THE COULSON SYNCHRONIZER

Coulson's modification is to replace the two pilot symbols modulated by two PN sequences in the Schmidl-Cox synchronizer with a single pilot symbol modulated

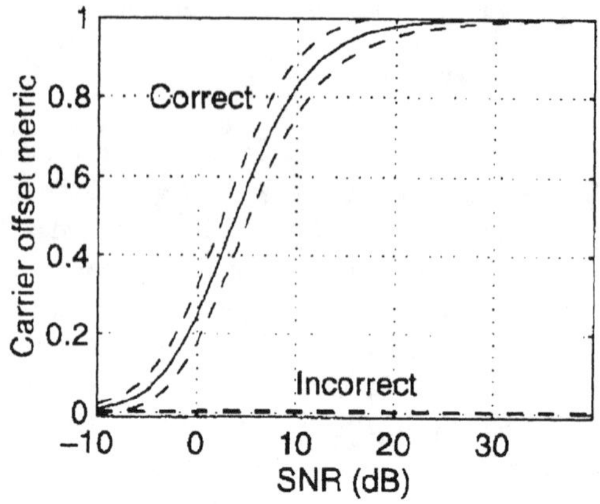

Figure 14.15 Expected value of carrier frequency offset metric with $W = 500$. Dashed lines indicate three standard deviations. From [19]. Copyright © 1997 IEEE.

by a maximum length sequence (MLS or m-sequence) [20,23]. Specifically, the pilot symbol comprises an L-length complex MLS of half the length of the OFDM symbol period (i.e., $L = N/2$), which is repeated. The MLS is synthesized in the frequency domain and occupies a similar bandwidth to OFDM data. As the MLS is half the length of an OFDM symbol, each MLS subcarrier has twice the bandwidth of an OFDM subcarrier. The advantages over the Schmidl-Cox synchronizer are first that the overhead is reduced, and second that all subcarriers are occupied by the constant-envelope MLS, thus maximizing the pilot symbol SNR at the receiver. Further, an important property of the MLS is that it has a constant envelope in both frequency domain and time domain. Thus the SNR of each sample is maximized and constant no matter in which domain it is processed. Another advantage of Coulson's synchronizer is that it can estimate sub-sample timing offset. However, it is likely to fail in the presence of large frequency offset [24] and needs higher implementation complexity due to its use of a matched filter. A training symbol with the same structure as the one proposed in [23] is exploited in [31], and the proposed synchronizer is also sensitive to large frequency offset [24]. In the following, highlights of Coulson's synchronizer are presented.

Assuming the channel is AWGN or at worst a flat fading channel and the channel parameters may be accurately approximated as constant over the duration of one OFDM symbol, the nth received sample can be written as

$$r_n = as(nT - \tau_s)\exp[2\pi\nu(nT - \tau_s) + \theta] + \eta(nT) \tag{14.49}$$

where a is the amplitude; $s(t)$ is the signal; T is the sampling period; τ_s is the sub-sample timing offset induced by the combination of channel and system distortions, $-1/2 \leq \tau_s/T \leq 1/2$; ν is the frequency offset consisting of the Doppler shift and RF local oscillator offset; θ is the phase offset; and $\eta(t)$ is complex AWGN having variance σ_w^2. All these parameters are time-variant from one symbol to the next. Each of them needs to be estimated, and the estimation needs to be updated through the reception process. The Coulson synchronizer estimates the frequency offset ν, the time offset τ_s, and the phase offset θ.

Just like the Schmidl-Cox synchronizer, the Coulson's synchronization algorithm starts with the cross correlation between OFDM time-domain samples separated by L positions in the sliding window of L length. We denote the received signal vector of L samples in the sliding window as

$$\mathbf{r}_l = [r_l, r_{l+1}, \ldots, r_{l+L-1}]^T \tag{14.50}$$

where $[\cdot]^T$ denotes transpose. The sample r_l is the first sample in the sliding window. We define $l = 0$ as the starting time of the OFDM packet (or frame), then l represents the packet timing offset in samples. The lth sample of the cross correlation is (see [23] for derivation)

$$P_l = \mathbf{r}_l^H \mathbf{r}_{l+L} \approx (L - |l|)a^2 S^2 \exp\{-j2\pi\nu LT\} + \text{noise term} \tag{14.51}$$

where $[\cdot]^H$ denotes Hermitian transpose and S is the magnitude of the MLS. It is seen that the signal part of P_l has the maximum magnitude when timing is correct ($l = 0$).

The timing metric is the normalized $|P_l|$, the same as in [19],

$$|\rho_l| = \frac{|\langle \mathbf{r}_l^H \mathbf{r}_{l+L} \rangle|}{\sqrt{\langle \mathbf{r}_l^H \mathbf{r}_l \rangle \langle \mathbf{r}_{l+L}^H \mathbf{r}_{l+L} \rangle}} = \frac{|P_l|}{\sqrt{R_l R_{l+L}}} \tag{14.52}$$

where

$$R_l = \mathbf{r}_l^H \mathbf{r}_l \tag{14.53}$$

is the autocorrelation of the received signal. So we need two correlators to calculate the timing metric.

The synchronization algorithm is as follows:

1. *Packet Detection*: When $|\rho_l| \geq T_{C1}$ where $0 < T_{C1} < 1$ is a predefined threshold, a packet is detected. Equivalently, this detection rule can be written as $X_l = |P_l| - T_{C1}\sqrt{R_l R_{l+L}} \geq 0$. Further, as R_l was found to be significantly correlated across l in real-world conditions, it was found empirically that it is

more efficient to use either

$$X_l = |P_l| - T_{C1} R_l \tag{14.54}$$

or

$$X_l^{'} = P_l^2 - T_{C1}^2 R_l^2 \tag{14.55}$$

2. *Frequency Offset Estimation*: When X_l becomes positive, the phase of the cross correlation P_l is used to estimate the frequency offset,

$$\widehat{\nu} = \frac{1}{2\pi LT} \arctan\left(\frac{\text{Im}\{P_l\}}{\text{Re}\{P_l\}}\right) \tag{14.56}$$

which is shown in [1] as the maximum likelihood estimate. Since X_l typically first becomes positive when $l < 0$, P_l typically will not be at its peak value, thus the initial estimation is regarded as a forward estimate, which will be subsequently improved upon as l approaches 0.
3. *Start of Packet Estimation*: The forward estimate $\widehat{\nu}$ is applied to a "clean" copy of the MLS to form a matched filter having the same frequency offset as the received signal. That is, the coefficients of the L-length matched filter are given by

$$n_l^* = s^*[lT] \exp(j2\pi\widehat{\nu}lT) \tag{14.57}$$

or in vector form as

$$\mathbf{n}^H = [s^*[0], s^*[1]e^{j2\pi\widehat{\nu}T}, ..., s^*[L-1]e^{j2\pi\widehat{\nu}(L-1)T}] \tag{14.58}$$

Then this matched filter is applied to the received signal to produce (see [23] for derivation)

$$\begin{aligned} Q_l &= \mathbf{n}^H \mathbf{r}_l \\ &\approx LaS^2 \text{sinc}\left(l - \frac{\tau_s}{T}\right) \exp\{-j2\pi\nu(lT - \tau_s) + \theta + \phi((L-1)/2L)\} \\ &\quad + \text{noise term} \end{aligned} \tag{14.59}$$

where $\phi = 2\pi(\widehat{\nu} - \nu)LT$ is the frequency offset estimation error. The second timing metric similar to that given by (14.52) is

$$|\rho_l^{'}| = \frac{|\langle \mathbf{n}^H \mathbf{r}_l \rangle|}{\sqrt{\langle \mathbf{n}^H \mathbf{n} \rangle \langle \mathbf{r}_l^H \mathbf{r}_l \rangle}} = \frac{|Q_l|}{\sqrt{LS}\sqrt{R_l}} \tag{14.60}$$

This metric is further modified to compensate for the effect of channel dispersion by replacing $|Q_l|$ with $|Q_l + Q_{l+1}|$

$$|\rho_l''| = \frac{|Q_l + Q_{l+1}|}{\sqrt{L}S\sqrt{R_l}} \tag{14.61}$$

which is most efficiently evaluated as the difference

$$Y_l = |Q_l + Q_{l+1}| - T_{C2}\sqrt{L}S\sqrt{R_l} \tag{14.62}$$

or a less computationally expensive DSP implementation version

$$Y_l' = |Q_l + Q_{l+1}|^2 - T_{C2}^2 L S^2 R_l \tag{14.63}$$

where T_{C2} is the threshold of correlation which, when exceeded, makes Y_l become positive. The most likely value for the start of packet, $l = 0$, is found by comparing the amplitudes of $|Q_l|$ and $|Q_{l+1}|$.

4. *Phase Offset Estimation*: From (14.59), it is seen that when the start of the packet ($l = 0$) is detected, Q_l is at its peak value Q_0 whose phase is predominantly determined by θ. Thus the phase of the peak Q_0 is used to estimate the phase offset as

$$\widehat{\theta} = \arctan\left(\frac{\text{Im}\{Q_0\}}{\text{Re}\{Q_0\}}\right) \tag{14.64}$$

which by analog with (14.56) is maximum likelihood.

5. *Timing Offset Estimation*: Having estimated v and θ, corrections can be applied to the received second MLS signal and the resulting signal can be used to estimate the subsample offset τ_s. First, the corrected signal samples are passed through an L-point FFT and the output $\{H_k\}$ is multiplied with MLS element by element as $H_k S_k^*$. The estimation of τ_s is given as [23]

$$\widehat{\tau}_s = \frac{LT}{2\pi}\frac{\mathbf{k}^T\boldsymbol{\beta} - \frac{1}{L}\mathbf{k}^T\mathbf{o}\mathbf{o}^T\boldsymbol{\beta}}{\mathbf{k}^T\mathbf{k} - \frac{1}{L}(\mathbf{k}^T\mathbf{o})^2} \tag{14.65}$$

where $\boldsymbol{\beta}$ is a column vector of elements $\beta_k = \arg\{H_k S_k^*\}$, $\mathbf{k} = [0, 1, ..., L-1]^T$, and $\mathbf{o} = [1, 1, ..., 1]^T$. This estimation is both minimum mean square error (MMSE) and ML [23].

The performance of the Coulson synchronizer was evaluated in [20] for a system with 100 subcarriers and 128-point FFT. No comparison with the Schmidl-Cox synchronizer is provided in [20]. However, we can compare the packet detection metric quite easily. The packet detection metric is depicted in Figure 14.16. Comparing Figure 14.16 with Figure 14.12, it is seen that the timing metrics both span a period of

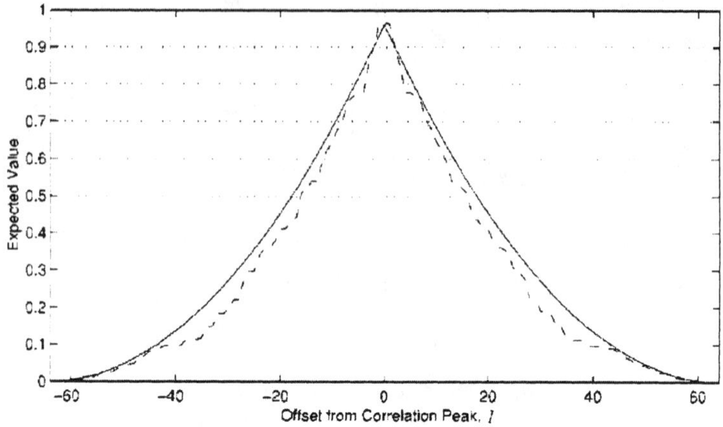

Figure 14.16 Packet detection metric $|\rho_l|^2$ (SNR/bit = 20 dB, 100 subcarriers, 128-point FFT). From [20]. Copyright © 2001 IEEE.

a symbol interval. In both cases, the peak's localizibility is poor, which incurs large peak detection variance under conditions with noise and interference. The Coulson's timing metric has no plateau. However, it is not because of its pilot symbol design; rather it is because no cyclic extension is assumed in the guard interval. Improvement of peak's localizibility can be achieved by using a different pattern of the pilot symbol, as will be described in the next three sections.

14.9 THE MINN-ZENG-BHARGAVA SYNCHRONIZER

In order to avoid the plateau of the timing metric in the Schmidl-Cox synchronizer, Minn et al. proposed two modified timing offset estimation methods [21]. The first method (method A) is called the sliding window method which uses two modifications: (1) all samples over one symbol period (excluding guard interval), instead of over one-half symbol period, are used in calculation of a half symbol energy required in the timing metric and (2) averaging of timing metrics over a window of guard interval length is used instead of 90% maximum points average. Specifically, the modified timing metric is given as

$$M_1(d) = \frac{1}{G+1} \sum_{k=-G}^{0} M_f(d+k) \tag{14.66}$$

where G is the number of guard samples, $M_f(d)$ is defined as

$$M_f(d) = \frac{|P(d)|^2}{R_f^2(d)} \tag{14.67}$$

where $P(d)$ is the same as defined in (14.20), and

$$R_f(d) = \frac{1}{2}\sum_{m=0}^{N-1} |r_{d+m}|^2 \tag{14.68}$$

The second method (method B) is called the training symbol method. Instead of having the pattern of time-domain samples of the training symbol as $[+A \quad +A]$, where A represents samples of length $L = N/2$, in the Schmidl-Cox synchronizer, the proposed method designs the samples of the training symbol (excluding cyclic prefix) in the form of

$$s = [+B \quad +B \quad -B \quad -B] \tag{14.69}$$

with B representing samples of length $L = N/4$ generated by $N/4$-point IFFT of a PN sequence. The timing metric is given as

$$M_2(d) = \frac{|P_2(d)|^2}{R_2^2(d)} \tag{14.70}$$

where

$$P_2(d) = \sum_{k=0}^{1}\sum_{m=0}^{L-1} r^*_{d+2Lk+m} r_{d+2Lk+m+L} \tag{14.71}$$

and

$$R_2(d) = \sum_{k=0}^{1}\sum_{m=0}^{L-1} |r_{d+2Lk+m+L}|^2 \tag{14.72}$$

In both methods, $P_2(d)$ and $R_2(d)$ or $P(d)$ and $R_f(d)$ can be calculated iteratively as demonstrated before (see (14.21)). The timing metrics of methods A and B as well as Schmidl and Cox's method are shown in Figure 14.17 under no noise and distortion conditions. It can be seen that the metrics of both methods A and B reach the peak at the correct timing point which is the rear edge of the plateau. However, method B has a much sharper metric curve; thus, the variance of detection of the correct timing point will be much smaller. Intuitively we can explain why these two metrics behave like this. In method A, the metric is an average of metric values over the guard interval. When $d = 0$, $M_1(d)$ is the average of $M_f(k)$, $k = -G, ..., 0$,

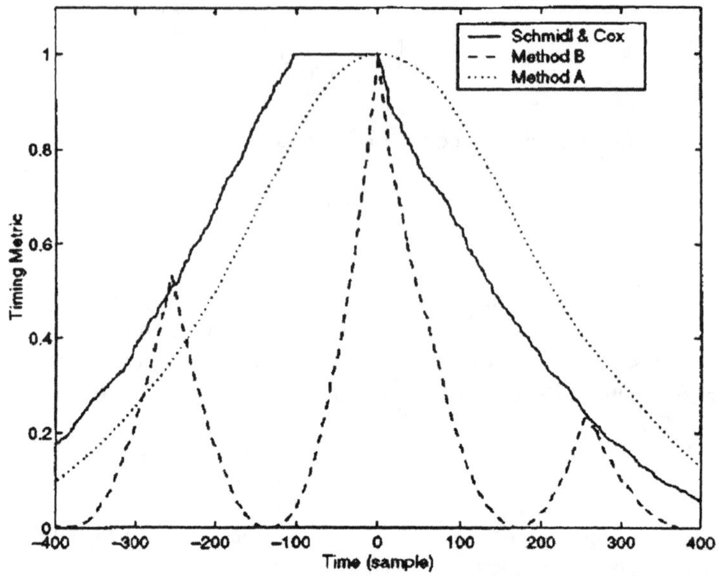

Figure 14.17 Timing metric under no noise and distortion condition. System parameters: 1,000 QPSK subcarriers, 1,024-point FFT, 102-sample guard interval, and frequency offset of 12.4 subcarier spacing. From [21]. Copyright © 2000 IEEE.

which are all equal and the results of correlating equal samples in the two halves of the correlation window, and thus $M_1(d)$ is the maximum. Deviated from $d = 0$, the correlations in $P(d)$ will have points before the guard interval (when $d < 0$) or after the training symbol (when $d > 0$). Thus $M_1(d)$ will gradually roll off. In method B, when $d = 0$, the correlations in $P_2(d)$ will have points of equal samples ($+$ with $+$, $-$ with $-$) and the metric is the maximum. Deviated from $d = 0$, the correlations in $P_2(d)$ will have points of equal samples ($+$ with $+$, $-$ with $-$) and some points of opposite-sign samples ($+$ with $-$ or $-$ with $+$). Due to the cancellation effect between positive correlations and negative correlations, the roll-off speed is much faster than that of method A. Simulations in [21] show that both proposed methods have a significantly smaller estimator variance than the Schmidl-Cox synchronizer under both AWGN and ISI channels. As an overall performance for both channels, method B with the modification of half-symbol energy calculation in (14.68) gives the best result.

14.10 THE SHI-SERPEDIN SYNCHRONIZER

Shi and Serpedin [22] proposed to use a pilot symbol pattern of

$$[+B \quad +B \quad -B \quad +B] \tag{14.73}$$

where B stands for a time-domain sequence of $N/4$ training samples with constant variance (power) (e.g., an m-sequence), and it can be generated with good approximation by using an $N/4$-point IFFT of an m-sequence. Shi and Serpedin tried all possible patterns in the set of $[\pm B \quad \pm B \quad \pm B \quad \pm B]$ and found that $[+B \quad +B \quad -B \quad +B]$ is the best in terms of lower false detection probability and higher correct acquisition probability. Furthermore, they found that the negative term $-B$ in (14.73) can be placed in any location without loss of performance.

In [22], the received signal samples are modeled as

$$r(n) = \alpha(n)x(n-\vartheta)\exp\{j(2\pi\varepsilon n/N + \theta)\} + w(n) \tag{14.74}$$

where ϑ stands for the timing offset, $\alpha(n)$ is the channel amplitude, ε is the carrier frequency offset (normalized to subcarrier spacing), θ denotes the phase offset, and $w(n)$ denotes the zero-mean white Gaussian noise with variance σ_w^2. The SNR is defined as $SNR = \alpha^2\sigma_s^2/\sigma_w^2$ where σ_s^2 is the variance (power) of the data of the training symbol (it is constant for every sample by design). Observation is assumed to include the whole training symbol as well as a prefix of length p and a suffix of length q. If we set the time of the start of the training symbol as $t = 0$, then we can write the observation sample vector as

$$R = [r(-p), r(-p+1), ..., r(0), ..., r(N-1), r(N), ..., r(N+q-1)]^T$$

Observation samples in a sliding window of N samples are taken to calculate the timing metric. If we denote the starting time of the window as l, then l is the timing offset of the window with respect to the training symbol. When $l < 0$, the window front edge is earlier than the first training sample. When $l > 0$, the window front edge is later than the first training sample. When $l = 0$, the window front edge is at the starting point of the training symbol, and correct timing is detected.

Defining subvectors

$$R_i = [r(l + (i-1)N/4), ..., r(l + iN/4 - 1)]^T, \quad i = 1, 2, 3, 4$$

using a method given in [16], the timing metric at any trial time instant l is shown as

$$Q(l) = \frac{P(l)}{V(l)} = \frac{|R_1^H R_2 - R_2^H R_3 - R_3^H R_4| + |R_2^H R_4 - R_1^H R_3| + |R_1^H R_4|}{\frac{3}{2}\sum_{i=1}^{4}|R_i|^2} \tag{14.75}$$

where

$$\begin{aligned} P(l) &= \underbrace{|R_1^H R_2 - R_2^H R_3 - R_3^H R_4|}_{P_1(l)} + \underbrace{|R_2^H R_4 - R_1^H R_3|}_{P_2(l)} + \underbrace{|R_1^H R_4|}_{P_3(l)} \\ &= P_1(l) + P_2(l) + P_3(l) \end{aligned} \quad (14.76)$$

and

$$V(l) = \frac{3}{2}\sum_{i=1}^{4} |R_i|^2 \quad (14.77)$$

The six terms in $P(l)$ are all possible correlations of the four parts of the training symbol. The negative terms are due to the negative sign of the third part. The terms in each group have a common phase due to the same index difference between the two R_i's in each correlation. The terms in $P_1(l)$ have a common phase of $\beta = \exp(-j\pi\varepsilon/2)$. The terms in $P_2(l)$ have a common phase of β^2. The terms in $P_3(l)$ have a common phase of β^3 [22].

The mean values and variances of $P(l)$ and $V(l)$ are shown approximately as [22]

$$\mu_P = \left(\frac{3N}{2} - 7|l|\right)\sigma_s^2\alpha^2, \quad \sigma_P^2 = (9N - 31|l|)\frac{\sigma_s^2\sigma_w^2\alpha^2}{2}$$

$$\mu_V = \frac{3N}{2}\left(\sigma_s^2\alpha^2 + \sigma_w^2\right), \quad \sigma_V^2 = 9N\sigma_w^2\left(\sigma_s^2\alpha^2 + \frac{\sigma_w^2}{2}\right)$$

for high SNR. The mean value of $Q(l)$ is then approximately as

$$\mu_Q = \left(1 - \frac{14|l|}{3N}\right)\cdot\rho \approx 1 - \frac{14|l|}{3N}, \text{ (high SNR)} \quad (14.78)$$

where $\rho = SNR/(SNR+1)$. For comparison, the timing metric of Coulson and Minn et al. are also presented in terms of the four subvectors. For training symbols in the form of $[+A \quad +A]$, a timing metric similar to (14.75) is proposed in [23][1]

$$Q'(l) = \frac{P'(l)}{V'(l)} = \frac{|R_1^H R_3 + R_2^H R_4|}{|R_1|^2 + |R_2|^2} \quad (14.79)$$

[1] $|R_1^H R_3 + R_2^H R_4|$ is $|R_1^H R_2 + R_3^H R_4|$ in [22] in this expression, which does not agree with (14.52).

and its mean is

$$\mu_{Q'} = \left(1 - \frac{2|l|}{N}\right) \cdot \rho$$

For a training symbol in the form of $[+B \quad +B \quad -B \quad -B]$ (e.g., Minn et al.), using the same method in [16], the timing metric is found as

$$Q_1(l) = \frac{|R_1^H R_2 - R_2^H R_3 + R_3^H R_4| + |R_2^H R_4 + R_1^H R_3| + |R_1^H R_4|}{3(|R_3|^2 + |R_4|^2)} \tag{14.80}$$

In [21], however, the timing metric for symbol $[+B \quad +B \quad -B \quad -B]$ is proposed as in (14.70) , which is equivalent to

$$Q_2(l) = \frac{|R_1^H R_2 + R_3^H R_4|}{0.5\left(\sum_{i=1}^{4} |R_i|^2\right)} \tag{14.81}$$

If only the last two terms in the denominator are considered, the above can be modified as

$$Q_3(l) = \frac{|R_1^H R_2 + R_3^H R_4|}{|R_3|^2 + |R_4|^2} \tag{14.82}$$

All these metrics are compared in Figures 14.18 and 14.19 for a 128-subcarrier OFDM signal. Figure 14.18 shows the mean values of $Q(l)$ and $Q'(l)$. It is seen that the timing metric of pattern $[+B \quad +B \quad -B \quad +B]$ is much better than that of pattern $[+A \quad +A]$ in terms of localizibility. The experimental curve of $Q(l)$ and $Q'(l)$ agree quite well for $N/6 \leq |l| < 0$, and the experimental curve shows bigger discrepancies beyond this range of l. In fact, from (14.78) we can calculate that when the mean value of $Q(l)$ reaches zero when $|l| = 3N/14$, the experimental results are nonzero. Figure 14.19 shows the mean values of $Q(l)$, $Q_1(l)$, $Q_2(l)$, and $Q_3(l)$. The timing metric $Q_1(l)$ of pattern $[+B \quad +B \quad -B \quad -B]$ has two undesirable lower peaks besides the main peak. $Q_2(l)$ of the same pattern has a large undesirable peak, while $Q_3(l)$ of the same pattern has four undesirable low peaks. From this comparison we can see that the proposed pattern $[+B \quad +B \quad -B \quad +B]$ is the best.

Only packet detection and estimation of the carrier frequency offset are considered in [22]. The packet detection method is similar to that in [23], which has two steps. In step one, upon computing $P(l)$ and $V(l)$ (see (14.76) and (14.77)), if $Q(l) > T_c$, where $0 < T_c < 1$ is a predefined threshold, or equivalently

$$Z(l) = P(l) - T_c V(l) > 0 \tag{14.83}$$

the receiver declares a coarse packet detection. The receiver then obtains the range

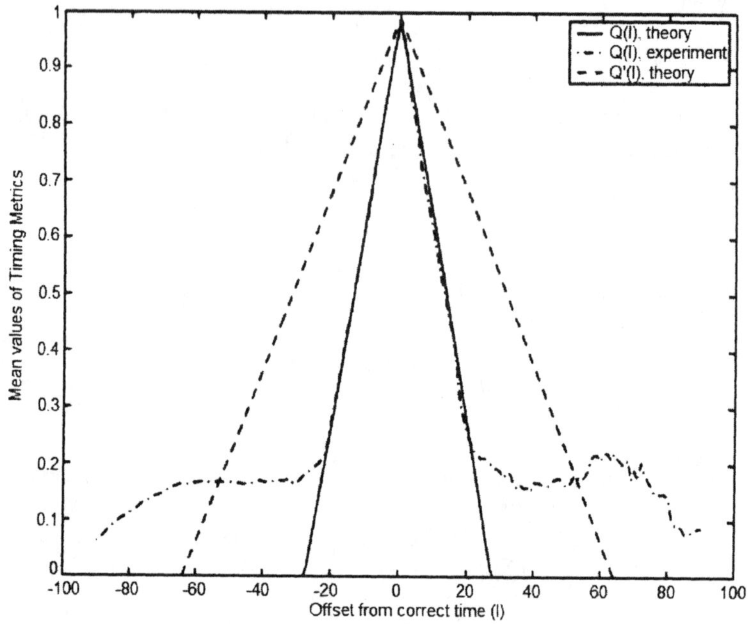

Figure 14.18 Mean values of timing metrics for different training symbols: $Q(l)$ and $Q'(l)$. SNR = 20 dB. From [22]. Copyright © 2004 IEEE.

Ω of the index l, in which $Q(l)$ is above the threshold. The second step is to obtain the fine estimate of the start of the packet and the estimate of the carrier frequency offset. The receiver searches the peak of $Q(l)$ (or simply $P(l)$ in a slow-varying channel) in range Ω. The start of the packet $l = d$ thus is estimated as

$$\widehat{d} = \arg\max_{l} Q(l) \tag{14.84}$$

and the carrier frequency offset is estimated as

$$\widehat{\varepsilon} = \frac{2}{\pi} \arg P_1(\widehat{d}) \tag{14.85}$$

The performance of a Shi-Serpedin synchronizer is characterized by coarse packet detection probability, fine packet detection probability, and mean square error of the frequency estimation.

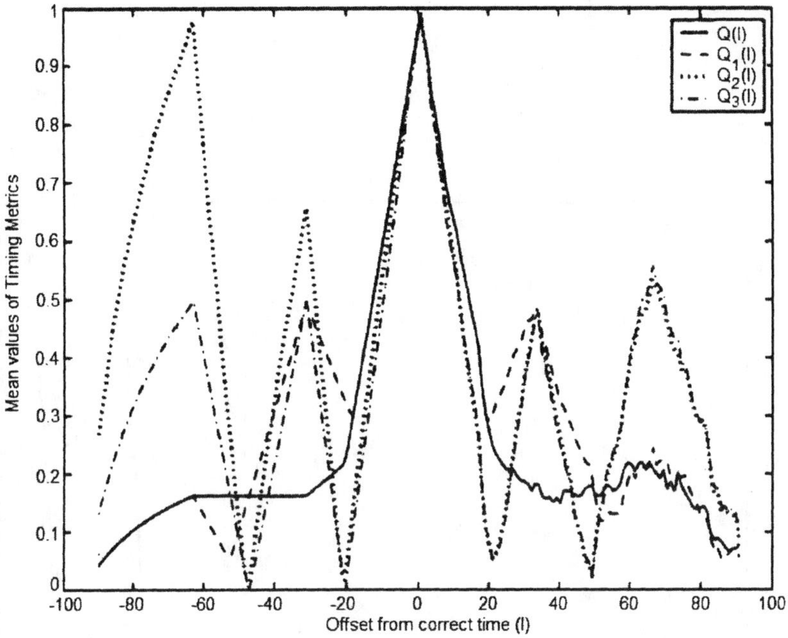

Figure 14.19 Mean values of timing metrics for different training symbols: $Q(l)$, $Q_1(l)$, $Q_2(l)$, and $Q_3(l)$. SNR = 20 dB. From [22]. Copyright © 2004 IEEE.

The coarse packet detection probability is given as

$$p_{1r} = \Pr(Z(l) > 0) \approx \frac{1}{2}\text{erfc}\left(-\frac{\mu_z}{\sqrt{2}\sigma_z}\right)$$

where $\text{erfc}(x)$ is the complementary error function and is defined in Section 4.3, and the mean value and variance of $Z(l)$ are

$$\mu_z = \mu_P - T_c\mu_V, \quad \sigma_z^2 = \sigma_P^2 + T_c^2\sigma_V^2 - 2T_c\text{cov}\{P, V\}$$

respectively, where $\text{cov}\{P, V\}$ is the covariance of $P(l)$ and $V(l)$. Figure 14.20 shows that for $SNR > 10$ dB, when $T_c = 0.6$, the coarse detection probability for $[+B \quad +B \quad -B \quad -B]$ is approaching 1 in the search range of the fine detection $[-L, 0]$, while for $[+A \quad +A]$, there is no threshold value to get robust (high probability) coarse packet detection in the range $[-L, 0]$.

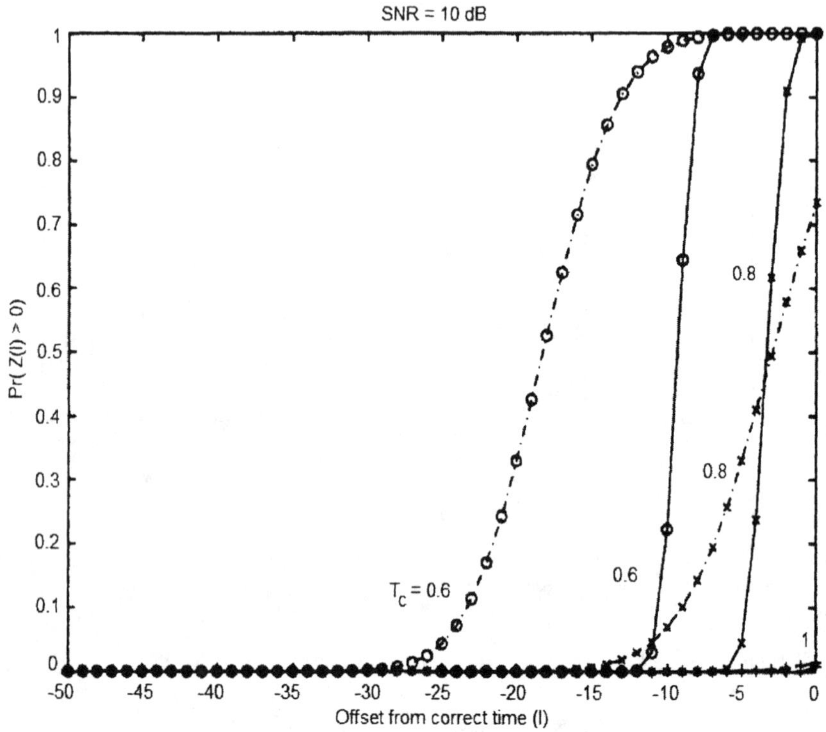

Figure 14.20 Probability of $Z(l) > 0$ for $E_s/N_o = 10$ dB. $N = 128$ and $L = 16$. L is the guard interval length. Solid line is for training symbol $[+B \quad +B \quad -B \quad +B]$. Dashed line is for training symbol $[+A \quad +A]$. From [22]. Copyright © 2004 IEEE.

In the absence of a signal, according to central limit theorem, $P_1(l), P_2(l)$, and $P_3(l)$ can be approximated as three Gaussian RV with variances $\sigma_i^2 = (1-i/4)N\sigma_w^4$, $i = 1, 2, 3$, respectively [22]. Thus $P(l)$ is the sum of three independent Rayleigh RV with the pdfs $f(x_i) = 2x_i/\sigma_i^2 \exp(-x_i^2/\sigma_i^2)$, $i = 1, 2, 3$. $V(l)$ can be viewed as a Gaussian RV with mean $\mu_V = 3N\sigma_w^2/2$ and variance $\sigma_V^2 = 9N\sigma_w^4/2$. A closed-form expression for the false alarm probability $P_F = \Pr(|P_1(l)|+|P_2(l)|+|P_3(l)| > T_c V(l))$ cannot be obtained. Instead, numerical results are shown in Figure 14.21. For $N = 128$, $T_c = 0.6$, $P_F = 1.38 \times 10^{-9}$, which is extremely small.

The fine packet detection probability is the probability that the detector finds that $P(0) > P(l), l \in [-L, -1]$. Since the most likely error is $P(-1) > P(0)$,we examine first $\Pr(P(0) < P(-1))$ or $\Pr(\Delta = P(0) - P(-1) < 0)$, where Δ can be approximated as a real Gaussian RV with $\mu_\Delta = 7\sigma_s^2\alpha^2$ and variance $\sigma_\Delta^2 = (41 + 12\cos(\beta))\sigma_s^2\sigma_w^2\alpha^2 < 53\sigma_s^2\sigma_w^2\alpha^2$ [22], where $\beta = \exp(-j\pi\varepsilon/2)$. It is found that

$$p_{2r} = \Pr(\Delta < 0) = \frac{1}{2}\text{erfc}\left(\frac{\mu_\Delta}{\sqrt{2}\sigma_\Delta}\right) < \frac{1}{2}\text{erfc}\left(7\sqrt{\frac{\text{SNR}}{106}}\right)$$

At high SNR, p_{2r} is very small (e.g., $p_{2r} < 0.0033$). Similarly, the error probability that the detector chooses other points in the range $[-L, -1]$ can be determined and is much smaller. Thus, the fine detection probability can be approximated as

$$P_{fd} \approx 1 - p_{2r}$$

For the $[+A \quad +A]$ training symbol and the metric $Q'(l)$ defined in (14.79), similar results can be obtained. The probability that the closest neighbor of $P'(0)$ is chosen is found as

$$p_{3r} = \Pr(\Delta' < 0) = \frac{1}{2}\text{erfc}\left(\sqrt{\frac{\text{SNR}}{6}}\right)$$

where $\Delta' = P'(0) - P'(-1)$. Since $7\sqrt{\text{SNR}/106} \approx 1.67\sqrt{\text{SNR}/6}$, $p_{3r} >> p_{2r}$. That is, the Shi-Serpedin synchronizer is more robust than the Schmidl-Cox synchronizer.

The above results and more comparisons in [22] show that the proposed training symbol and estimators perform better than the Schmidl-Cox and Minn-Zeng-Bhargava methods in terms of better detection probability and accuracy and extended identifiability range of the carrier frequency offset estimate. When compared with Coulson's method, the proposed scheme requires less implementation complexity, is more robust to large carrier frequency offsets, and exhibits nearly the same performance in terms of estimation accuracy. However, Coulson's method can estimate the subsample timing error and phase offset, while this method cannot.

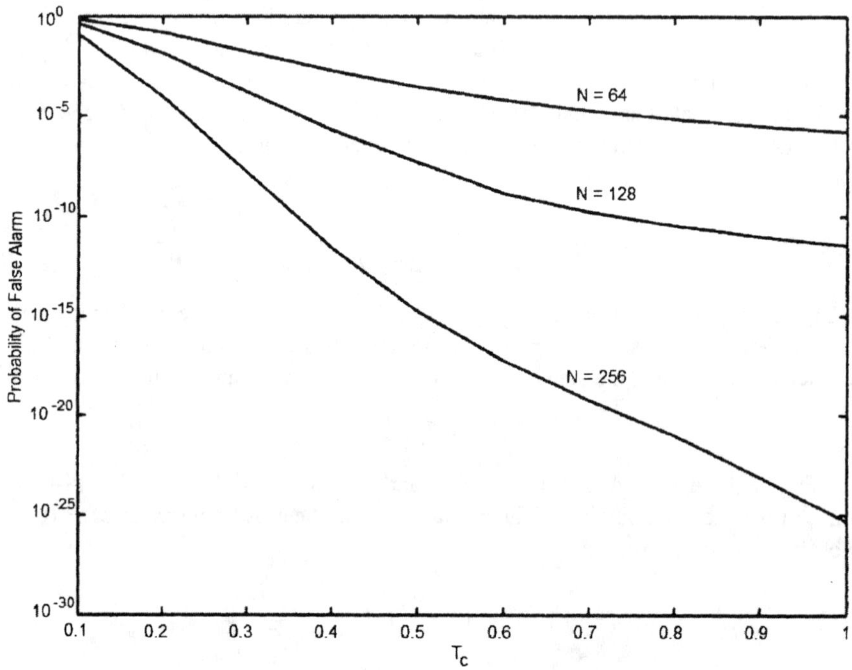

Figure 14.21 Probability of false alarm for various N and T_c. From [22]. Copyright © 2004 IEEE.

14.11 MORE TRAINING SYMBOL PATTERNS

Minn, Bhargava, and Letaief studied more training symbol patterns for OFDM timing by computer search in [5]. The proposed training symbols have the general form of $[\pm A, \pm A, ..., \pm A]$, where A represents the repeated part. Assuming the training symbol has L parts of M samples each, the timing metric at timing offset d is defined as

$$\Lambda_{\varepsilon}(d) = \left(\frac{L}{L-1}\frac{|P(d)|}{E(d)}\right)^2 \tag{14.86}$$

where $E(d)$ is the signal energy of the training symbol

$$E(d) = \sum_{i=0}^{M-1}\sum_{k=0}^{L-1}|r(d+i+kM)|^2$$

and $P(d)$ is the sum of the pair-wise correlation of the received samples $r(m)$, given by

$$P(d) = \sum_{k=0}^{L-2} b(k) \sum_{m=0}^{M-1} r^*(d+kM+m)r(d+(k+1)M+m)$$

where $b(k) = p(k)p(k+1), k = 0, 1, ..., L-2$, and $\{p(k), k = 0, 1, ..., L-1\}$ denote the signs of the repeated parts of the training symbol and will be called the training symbol pattern. The Schmidl and Cox metric is an approximation of the proposed metric for the case of $L = 2$ and pattern $p = [++]$.

Maintaining a low peak-to-average power ratio (PAPR) is also a consideration in choosing the training patterns in [5]. Golay complementary sequences (see Section 13.7.3) have very low PAPR (≤ 2). Thus the training patterns in [5] are formed as follows. The repeated part has $M = N/L$ samples (for N-point IFFT and L identical parts), and is generated by M-point IFFT of a length N_u/L Golay complementary sequence ($N_u \leq N$, N_u is the number of subcarriers, the rest are virtual subcarriers, (i.e., zeros, for inserting frequency guard bands)). Table 14.3 shows the computer search results from [5] for the best patterns under the Golay complementary constraints and 10% cyclic prefix constraint. These patterns are the best in the sense that $\sum_{d=-N}^{N}|P(d)|^2$ is the minimum, which leads to the steepest rolloff correlation metric trajectory. Figure 14.22 shows the corresponding metrics of the first pattern for each L value in Table 14.3. The Schmidl and Cox metric (S&C) is included for comparison.

L	Pattern
4	$(-+--)$, $(+++-)$ and their sign-inverted variants
8	$(++--+---)$, $(-++---+-)$ and their sign-inverted variants
16	$(+--+++--+-++-+--)$ $(--++-++----++++-)$ and their sign-inverted variants

Table 14.3 Best training symbol patterns. From [5]. Copyright © 2003 IEEE.

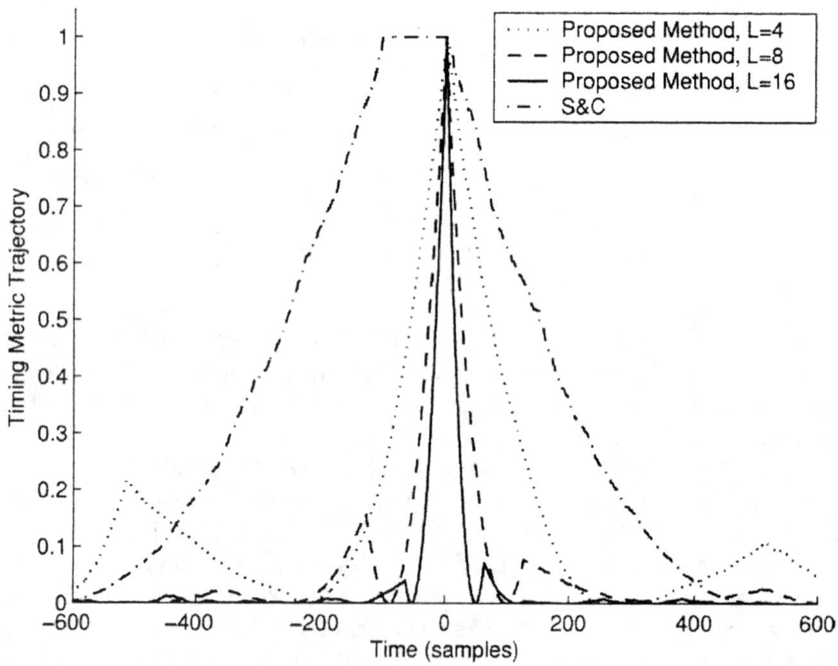

Figure 14.22 Timing metric trajectory under noiseless conditions assuming no channel distortion. (Time index 0 corresponds to the exact timing point.) From [5]. Copyright © 2003 IEEE.

14.12 SYNCHRONIZATION FOR THE IEEE 802.11A STANDARD

The OFDM of IEEE 802.11a is implemented by a 64-point IDFT, but only 52 subcarriers are used. The other 12 subcarriers are just null carriers. The nonzero subcarrier indexes range from -26 to 26, with the 0th subcarrier unused and 4 subcarriers for pilot symbols. Therefore the number of data subcarriers is actually 48. The channel separation is $1/3.2\ \mu s = 0.3125$ MHz. With a 0.8-μs guard interval, the OFDM symbol duration is 4 μs. So the OFDM symbol rate is $1/4\ \mu s = 0.25$ Mbps. The modulations used are BPSK, QPSK, 16QAM, and 64QAM. The coding rates are $1/2, 2/3$, and $3/4$. With various combinations of modulations and coding rates, the data rates supported are $6, 9, 12, 18, 24, 36, 48$, and 54 Mbps. For example, with the highest order modulation (64QAM) and code rate (3/4), the data rate is $\log_2(64)\times$ 0.25 Mbps$\times 3/4 \times 48 = 54$ Mbps. The carrier frequency of IEEE 802.11a is in the 5.2-GHz band.

As an example, we describe the OFDM synchronizer for the IEEE 802.11a WLAN standard [32]. It uses two 8 μs symbols for the detection of start of packet, timing, and frequency offset estimation. Figure 14.23 shows the OFDM training structure (preamble), where t_1 to t_{10} denote repetitions of a short training symbol and T_1 and T_2 are the identical two parts of a long training symbol with GI2 as its cyclic extension.

Figure 14.24 shows the training symbol distribution. The short training symbols are located on nonzero subcarriers whose indexes are a multiple of 4 (i.e., $\{-24, -20, -26, -12, -8, -4, 4, 8, 12, 16, 20, 24\}$). Thus, a short OFDM training symbol consists of 12 subcarriers, which are modulated by the elements of the sequence S, given by

$$\begin{aligned} S_{-26,26} = {} & \sqrt{13/6} \times \{0,0,1+j,0,0,0,-1-j,0,0,0,1+j, \\ & 0,0,0,-1-j,0,0,0,-1-j,0,0,0,1+j,0,0,0, \\ & 0,0,0,-1-j,0,0,0,-1-j,0,0,0,1+j, \\ & 0,0,0,1+j,0,0,0,1+j,0,0,0,1+j,0,0\} \end{aligned} \tag{14.87}$$

The multiplication by a factor of $\sqrt{13/6}$ is to normalize the average power of the resulting OFDM symbol, which utilizes 12 out of 52 subcarriers. The fact that only spectral lines of $S_{-26,26}$ with indices that are a multiple of 4 have nonzero amplitude results in a periodicity of $T_{FFT}/4 = 0.8\ \mu s$. That is, the output samples of the 64-point IFFT actually have a period of 16 samples. That is, we can obtain four short training symbols from the output of the 64-point IFFT. The remaining six short training symbols are obtained by periodic extensions. Note that the magnitude of time-domain samples of a short symbol is not constant [32].

The long training symbols are present on each nonzero subcarrier. A long OFDM

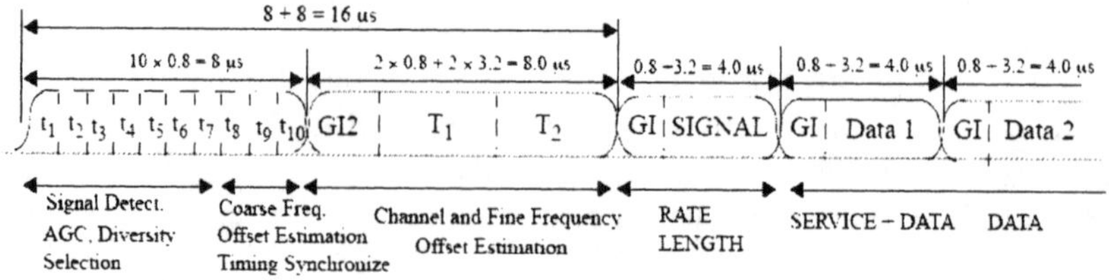

Figure 14.23 IEEE 802.11a frame structure. From [32]. Copyright © 1999 IEEE.

training symbol consists of 52 subcarriers (excluding a zero value at dc), which are modulated by the elements of the sequence L, given by

$$\begin{aligned} L_{-26,26} = \; & \{1, 1, -1, -1, 1, 1, -1, 1, -1, 1, 1, 1, 1, 1, 1, -1, -1, 1, \\ & 1, -1, 1, -1, 1, 1, 1, 1, 0, 1, -1, -1, 1, 1, -1, 1, -1, 1, \\ & -1, -1, -1, -1, -1, 1, 1, -1, -1, 1, -1, 1, -1, 1, 1, 1, 1\} \end{aligned} \tag{14.88}$$

The time-domain samples are obtained by 64-point IFFT and extending it to two periods plus a cyclic extension. Again, the time-domain samples of the long symbol do not have a constant magnitude [32]. However, both the short and long training symbols are designed such that the PAPR is approximately 3 dB, which is much lower that the PAPR of random data symbols. This guarantees the degradation of the training symbols due to nonlinearity in the amplifiers to be smaller than the data symbols.

After the preamble, in each OFDM symbol, four of the subcarriers are dedicated to pilot signals in order to make the coherent detection robust against frequency offsets and phase noise. These pilot signals are put in subcarriers −21, −7, 7, and 21. The pilots are BPSK modulated by a pseudo binary sequence to prevent the generation of spectral lines.

The first seven short symbols are intended for packet detection, AGC, and diversity selection. The last three short symbols are intended for coarse frequency offset estimation and timing synchronization. The reasons for using short training symbols are as follows [3]. The first reason to use the short training symbol is that it can detect the packet quickly and adjust the receiver gain quickly. Longer training symbols can also do the same. However, the short training symbols provide quicker measurements and adjustments. The second reason to use the short training symbols is that it makes it possible to do coarse carrier frequency offset estimation with a large unam-

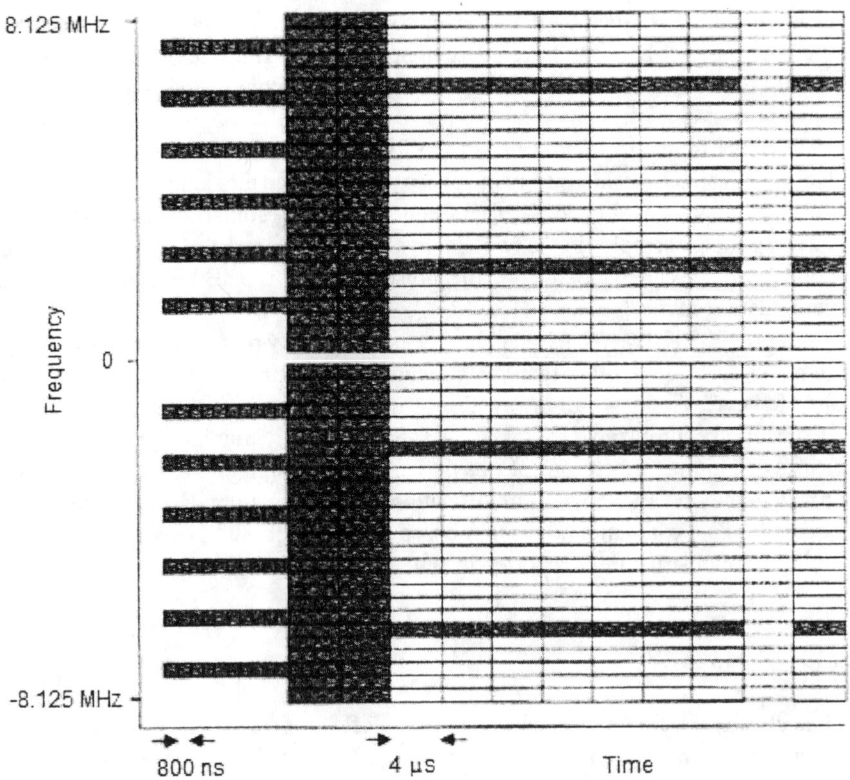

Figure 14.24 IEEE 802.11a pilot symbol distribution. From [3]. Copyright © 2000 Richard van Nee and Ramjee Prasad.

biguous range. For a repetitive signal with a duration of T, the maximum measurable unambiguous frequency offset range is $\Delta f = \pm 1/(2T)$, since a higher offset will result in a phase difference from one symbol to the next exceeding $\pm\pi$ which cannot be detected using a simple correlation method (see Section 14.7.3). For IEEE 802.11a, with a period of 800 ns, a frequency offset up to 625 kHz can be estimated. The IEEE 802.11a standard requires the maximum offset per user to be 20 ppm. This means that the offset seen by the receiver could be as high as 40 ppm, as the transmitter and receiver both can have up to 20 ppm offsets. Thus the maximum offset is 232 kHz for carrier frequency of 5.8 GHz, which is within the 625-kHz range that can be estimated.

The long training symbol can be used for estimations of timing, fine frequency offset, and channel. By measuring the phase drift between the samples that are a long time (3.2 μs) apart in T_1 and T_2, fine frequency offset estimation can be achieved. By averaging the samples in the two identical parts of T_1 and T_2, a coherent demodulation reference can by obtained with a noise level that is 3 dB lower than the noise level of data symbols [3].

The IEEE 802.11a WLAN standard does not specify particular synchronization techniques to be used, other than the structure and distribution of the training symbols. However, the long training symbol structure is similar to that of Schmidl-Cox or Coulson synchronizers ($[+A \quad + A]$), and the short training symbol structure consists of multiple repetitions of this pattern. Thus, Schmidl-Cox or Coulson synchronizers may be applied with modifications to this structure. A particular carrier frequency offset estimation algorithm based on the IEEE 802.11a WLAN standard training symbol structure is reported in [33].

14.13 SUMMARY

In this chapter we first presented the results of effect of carrier frequency offset and timing error on the degradations of the SNRs or BERs of the OFDM system. A feature of the effect of frequency offset on OFDM is that the SNR degradation is proportional to the E_s/N_o, which is not the case for single-carrier (SC) modulations. The most salient feature is that the degradation of OFDM due to frequency offset or phase noise is $N^2 E_s/N_o$ or N times greater than that of SC, respectively. That is, OFDM is more sensitive to frequency offset and phase noise than SC modulation. The effect of symbol timing errors was also discussed. It rotates the demodulated signal if the offset causes the start of the DFT frame to fall in the region between the end of the ISI of the previous symbol and the start of the current symbol. The phase rotation must be corrected and can be corrected by channel estimation. However, timing offsets beyond this range destroy orthogonality and cause severe symbol demodulation errors.

Sampling timing error causes phase rotation, too. However, sampling frequency error causes time-varying phase rotation and ICI, which has an irreversible effect on the demodulation errors. Next, synchronization methods, including coarse timing recovery, coarse frequency estimation, fine frequency correction, and fine timing correction were discussed. First we discussed synchronization methods that use cyclic extension for correlation, which in turn is used by the timing detector and frequency offset estimator. Then we presented a number of pilot-assisted correlation-based synchronization methods. We described in great detail the Schmidl-Cox synchronizer, which uses a $[+A \quad +A]$ pilot symbol pattern. The Schmidl-Cox synchronizer is the first such synchronizer proposed in the literature and thus is historically important, and the metric defined and the analytical method of the detection probability and false alarm probability are followed by many latter authors. However its timing metric has a plateau so that detection error is not satisfactory. Later authors tried to improve the Schmidl-Cox synchronizer and several representative algorithms were described in this chapter. One of them is the Shi-Serpedin synchronizer which uses a pilot pattern of $[+B \quad +B \quad -B \quad +B]$ and its timing metric rolloff is the steepest. In comparison with others, the Shi-Serpedin synchronizer has a better detection probability and accuracy, and extended identifiability range of the carrier frequency offset estimate. A detailed treatment of the Shi-Serpedin synchronizer was provided. After that, more good pilot patterns suggested by Minn et al. were briefly presented. Finally, to conclude the chapter, we presented the synchronization features of the IEEE 802.11a waveform as a practical example.

References

[1] Moose, P. H., "A technique for orthogonal frequency division multiplexing frequency offset correction," *IEEE Trans. Commun.*, vol. 42, no. 10, October 1994, pp. 2908–2914.

[2] Pollet, T., M. van Bladel, and M. Moeneclaey, "BER sensitivity of OFDM systems to carrier frequency offset and Wiener phase noise," *IEEE Trans. Commun.*, vol. 43, no. 2/3/4, February/March/April 1995, pp. 191–193.

[3] van Nee, R., and R. Prasad, *OFDM for Wireless Multimedia Communications*, Norwood, Massachusetts: Artech House, 2000.

[4] Hwang, W., H. Kang, and K. Kim, "Approximation of SNR degradation due to carrier frequency offset for OFDM in shadowed multipath channels," *IEEE Commun. Letters,* vol. 7, no. 12, December 2003, pp. 581–583.

[5] Minn, H., V. Bhargava, and K. Letaief, "A robust timing and frequency synchronization for OFDM systems," *IEEE Trans. Wireless Commun.*, vol. 2, no. 4, July 2003, pp. 822–839.

[6] Pollet, T., P. Spruyt, and M. Moeneclaey, "The BER performance of OFDM systems using non-synchronized sampling," *Proc. Globecom '94*, November 1994, pp. 253–257.

[7] Bahai, A. S. R., and B. R. Saltzberg, *Multi-Carrier Digital Communications: Theory and Applications of OFDM*, New York: Kluwer Academic/Plenum Publishers, 1999.

[8] Fazel, K., et al., "A concept of digital terrestrial television broadcasting," *Wireless Personal Communications*, vol. 2, 1995, pp. 9–27.

[9] de Bot, P., et al., "An overview of the modulation and channel coding schemes developed for digital terrestrial television broadcasting within the dTTb project," *Int. Broadcasting Conv.*, September 16–20, 1994, pp. 569–576.

[10] Nogami, H., and T. Nagashima, "A frequency and timing period acquisition technique for OFDM systems," *Proc. Personal, Indoor and Mobile Radio Commun. (PIRMC)*, September 27–29, 1995, pp. 1010–1015.

[11] Toutier, P., R. Monnier, and P. Lopez, "Multicarrier modem for HDTV terrestrial broadcasting," *Signal Processing: Image Communication*, vol. 5, no. 5–6, December 1993, pp. 379–403.

[12] Daffara, F., and O. Adami, "A new frequency detector for orthogonal multicarrier transmission techniques," *Proc. IEEE Vehicular Technology Conf. (VTC'95)*, Chicago, Illinois, July 15–28, 1995, pp. 804–809.

[13] van de Beek, et al., "Low complex frame synchronization in OFDM systems," *Proc. Int. Conf. Universal Personal Commun. (ICUPC)*, November 6–10, 1995, pp. 982–986.

[14] van de Beek, J.-J., M. Sandell, and P. Börjesson, "ML estimation of time and frequency offset in OFDM systems," *IEEE Trans. Signal Processing*, vol. 45, no. 7, July 1997, pp. 1800–1805.

[15] Sandell, M., J. J. van de Beek, and P. Börjesson, "Timing and frequency synchronization in OFDM systems using cyclic prefix," *Proc. Int. Symp. Synchronization,* Saalbau, Essen, Germany, December 14–15, 1995, pp. 16–19.

[16] Lashkarian, N., and S. Kiaei, "Class of cyclic-based estimators for frequency-offset estimation of OFDM systems," *IEEE Trans. Commun.,* vol. 48, December 2000, pp. 2139–2149.

[17] Warner, W., and C. Leung, "OFDM/FM frame synchronization for mobile radio data communications," *IEEE Trans. Vehicular Technology,* vol. 42, no. 3, August 1993, pp. 302–313.

[18] Classen, F., and H. Meyr, "Frequency synchronization algorithms for OFDM systems suitable for communication over frequency selective fading channels," *Proc. IEEE Vehicular Technology Conf. (VTC'94),* Stockholm, Sweden, June 8–10, 1994, pp. 1655–1659.

[19] Schmidl, T., and D. Cox, "Robust frequency and timing synchronization for OFDM," *IEEE Trans. Communications*, vol. 45, no. 12, December 1997, pp. 1613–1621.

[20] Coulson, A., "Maximum likelihood synchronization for OFDM using a pilot symbol: analysis," *IEEE J. Select. Areas Commun.*, vol. 19, December 2001, pp. 2495–2503.

[21] Minn, H., M. Zeng, and V. K. Bhargava, "On timing offset estimation for OFDM systems," *IEEE Commun. Letters*, vol. 4, July 2000, pp. 242–244.

[22] Shi, K., and E. Serpedin, "Coarse frame and carrier synchronization of OFDM systems: A new metric and comparison," *IEEE Trans. Wireless Commun.,* vol. 3, no. 4, July 2004, pp. 1271–1284.

[23] Coulson, A., "Maximum likelihood synchronization for OFDM using a pilot symbol: algorithm," *IEEE J. Select. Areas Commun.*, vol. 19, December 2001, pp. 2486–2494.

[24] Sourour, E., and G. E. Bottomley, "Effect of frequency offset on DS-SS acquisition in slowly fading channels," *Proc. IEEE Wireless Communications and Networking Conf.*, vol. 2, New Orleans, Louisiana, September 1999, pp. 569–573.

[25] Lambrette, U., M. Speth, and H. Meyr, "OFDM burst frequency synchronization by single carrier training data," *IEEE Commun. Lett.*, vol. 1, no. 2, 1997, pp. 46–48.

[26] Keller, T., et al., "Orthogonal frequency division multiplex synchronization techniques for frequency-selective fading channels," *IEEE J. Select. Areas Commun.*, vol. 19, June 2001, pp. 999–1008.

[27] Papoulis, A., *Probability, Random Variables, and Stochastic Processes,* 2nd ed., New York: McGraw-Hill, 1984.

[28] Muller-Weinfurtner, S., "On the optimality of metrics for coarse frame synchronization in OFDM: a comparison," *Proc. Int. Symp. Personal, Indoor and Mobile Radio Communications (PIMRC)*, vol. 2, Boston, Massachusetts, 1998, pp. 533–537.

[29] Morelli, M., and U. Mengali, "An improved frequency offset estimator for OFDM applications," *IEEE Commun. Lett.*, vol. 3, March 1999, pp. 75–77.

[30] Morelli, M., and U. Mengali, "Carrier-frequency estimation for transmmsion over selective channels," *IEEE Trans. Commun.*, vol. 48, September 2000, pp. 1580–1589.

[31] Tufvesson, F., O. Edfors, and M. Faulkner, "Time and frequency synchronization for OFDM using PN-sequence preambles," *Proc. 50th IEEE Vehicular Technology Conf.*, vol. 4, Amsterdam, the Netherlands, September 1999, pp. 2203–2207.

[32] IEEE Standard 802.11a, 1999.

[33] Li, J., G. Liu, and G. B. Giannakis, "Carrier frequency offset estimation for OFDM-based WLANs," *IEEE Signal Processing Lett.*, vol. 8, March 2001, pp. 80–82.

Selected Bibliography

- Cimini, L., and L. Greenstein, "Orthogonal frequency-division multiplexing," *Wiley Encyclopedia of Telecommunications*, J. G. Proakis, (ed.), New York: John Wiley & Sons, 2003.
- Classen, F., and H. Meyr, "Synchronization algorithms for an OFDM system for mobile communication," *ITG-Fachtagung*, October 26–28, 1994, pp. 105–113.
- Hanzo, L., W. Webb, and T. Keller, *Single- and Multi-Carrier Quadrature Amplitude Modulation: Principles and Applications for Personal Communications, WLANs and Broadcasting*, New York: IEEE Press, 2000.
- Litwin, L., "An introduction to multicarrier modulation," *IEEE Potentials*, April/May 2000, pp. 36–38.

Chapter 15

OFDM in Multipath Fading Channels

As previously discussed, under ideal conditions, an OFDM demodulator can separate all subcarriers completely without ICI and ISI. Thus, for OFDM transmission in an AWGN channel, the BER performance of each subcarrier is the same as that of a single-carrier modulation in an AWGN channel. All BER expressions of all types of modulations in the AWGN channel presented in previous chapters are applicable to OFDM in the AWGN channel.

When the channel is not AWGN (for example, if it is fading), the performance of OFDM will degrade. We have discussed in detail degradations of various modulations in fading channels in Chapter 11. A discussion of remedial measures (equalization and diversity techniques) are also presented there. However, those are for single-carrier modulations. Due to the multicarrier nature of OFDM, fading and multipath will incur not only intersymbol interference (ISI), which also happens for single-carrier modulations, but also interchannel (inter-subcarrier) interference (ICI) that is unique to multicarrier modulations. In this chapter, we will first examine the performance degradation of OFDM due to fading (Section 15.1). We will present analytical and/or simulation results for four types of channels: frequency-flat slow fading, frequency-selective slow fading, frequency-flat fast fading, and frequency-selective fast fading. Then we will discuss the remedial measures: channel estimation and equalization (Section 15.2). We will discuss in detail pilot-assisted linear minimum mean square estimator (LMMSE) or Wiener filter and its simplified variants. We also will discuss decision-directed estimators. Other remedial methods such as differential detection, diversity reception, coding, and multi-input-multi-output (MIMO) wireless links will be discussed in Section 15.3.

15.1 PERFORMANCE OF OFDM IN FADING CHANNELS

OFDM was initially conceived for frequency-selective fading channels. It is important to know how well OFDM can perform in various channel conditions. Based on that information, correctional measures can be devised.

Before we discuss performance of OFDM in fading channels, we need to establish the models for fading channels. In Chapter 11, we characterized the multipath fading channel by four different types: frequency selective fast fading, frequency nonselective (flat) fast fading, frequency selective slow fading, and frequency flat slow fading (see Table 11.1). The last two are more common than the first two in practice since fast fading only occurs for very low data rates at which the mobile can move a long distance and experience a wide range of signal strength in a symbol period. Slow fading is also called quasi-static or quasi-stationary fading, which means the complex multiplicative fading factor does not vary in a symbol period but it changes from symbol to symbol.

Mobile radio channels can be modeled differently, depending on the channel characteristics and system parameters: delay spread, data rate, modulation scheme, and vehicle speed. For example, in analysis and simulation of OFDM of 128 or 512 subcarriers in a mobile radio communication channel, Cimini modeled the channel as frequency flat fast fading with Rayleigh envelope distribution for a symbol rate of 7.5 kHz [1]. At a carrier frequency of 850 MHz, independent fades are about 7 inches apart, giving a fade every 6.6 ms at 60 mph. For $N = 128$, the OFDM symbol rate is 58.6 Hz. This makes symbol duration 17.1 ms, which means there are about three fades in one symbol, thus the channel is fast fading. If the data symbol rate is higher, say, 100 kHz, for the same other system parameters, the channel would become slow fading since there will be at most one fade in a symbol period. The channel can still be considered frequency flat in New York City since the rms delay spread is about 3 μs, which is much smaller than the OFDM symbol period (12.8 μs), but must be considered frequency selective in Newark where the rms delay spread can be as high as 25 μs, which is greater than the OFDM symbol period [1]. In fact, the frequency selective slow fading model is the most common model for mobile radio channels.

In the following, we will discuss demodulation of OFDM in these four types of channels. For all cases, we make the following assumptions. The carrier and timing synchronization are perfect. The transmitter filter (or windowing function) and the receiver filter frequency responses are flat within the signal bandwidth so that they do not distort signals but only filter the wideband noise. In other words, the effect of filtering on signals is assumed to be negligible so that it will not be included in the analysis. In the case of frequency selective fading, the OFDM signal has a guard interval that is longer than the maximum delay spread so that no ISI will occur. In addition, a cyclic extension is amended as a prefix to the OFDM symbol in the guard interval. As we have proved in Section 12.6, an empty guard interval can avoid ISI but not ICI. With the cyclic extension, ICI can also be avoided.

15.1.1 Frequency Flat Slow Fading Channel

As shown in Table 11.1, when rms excess delay σ_τ is much less than the symbol period, the multipath signals received by the receiver are not resolvable. They form a single-path signal with a fading envelope and random phase and the channel is frequency nonselective (flat). This frequency flat fading channel is characterized by a single signal with a complex multiplicative fading factor on the signal component. In general the fading factor is a random process. That is, it changes with time. However, if the change is slow in comparison with the symbol duration, we can approximately consider it as constant in a symbol period and it only changes from symbol to symbol. This is called quasi-stationary or quasi-static fading, or simply slow fading. For frequency flat slow fading channel, the fading factor on the received signal's complex envelope is constant over an OFDM symbol period, which can be written as

$$h = \alpha \exp(j\theta) \tag{15.1}$$

where α is the amplitude and θ is the phase. For narrow band radio transmission, h is usually modeled as a wide-sense stationary complex Gaussian random variable with zero mean or nonzero mean. As stated in Chapter 11 (Section 11.1.3), α is Rayleigh distributed if h has a zero mean or Rician distributed if h has a nonzero mean. In Rayleigh fading, θ is uniformly distributed in $[-\pi, \pi]$. In Rician fading, the PDF of θ is a very complex function [2, p. 50]:

$$f_\theta(x) = \frac{1}{2\pi}e^{-K} + \frac{1}{2}\sqrt{\frac{K}{\pi}}\cos(x)e^{-K\sin^2(x)}[1 + \text{erf}(\sqrt{K}\cos(x))]$$

for $x \in [-\pi, \pi]$, where K is the K factor defined in Section 11.1.3, and $\text{erf}(x)$ is the error function defined in Section B.1.2.3.

Recall that in Chapter 12 we stated that to satisfy the Nyquist sampling rate, the IDFT/DFT pair used to modulate and demodulate OFDM should be in the index range $[-N/2, N/2-1]$. In this chapter we use the IDFT/DFT pair defined in $[0, N-1]$, as most of the papers in the literature did. This will simplify the analysis and will not affect the analysis results. The reasons are stated in Note 2 of Section 12.2.2.

Thus the OFDM signal is

$$s_n = \frac{1}{N}\sum_{i=0}^{N-1} d_i \exp(j2\pi\frac{in}{N}), \quad n = 0, 1, ..., N-1 \tag{15.2}$$

The sampled received signal is

$$r_n = hs_n + n_n \tag{15.3}$$

where n_n is complex AWGN. The signal part of the DFT demodulator output is

$$X_k = \sum_{n=0}^{N-1} hs_n \exp(-j2\pi \frac{nk}{N}) = hd_k, \quad k = 0, 1, ..., N-1 \tag{15.4}$$

For simplicity, we will no longer indicate the index ranges of n and i in the discussion that follows.

The complete demodulator output is

$$Y_k = hd_k + W_k \tag{15.5}$$

where W_k is the DFT of the zero-mean complex Gaussian noise in the channel, which is still a zero-mean complex Gaussian random variable.

$$W_k = \sum_{n=0}^{N-1} n_n \exp(-j2\pi \frac{nk}{N}) \tag{15.6}$$

Note that fading factor h only affects the individual data symbol. Fading does not introduce any ISI and ICI. To recover the original data, h must be estimated and Y_k must be divided by h (equalization). Otherwise, the BER performance will degrade. We have analyzed this situation in Chapter 11 in the context of single-carrier modulation (see Section 11.2). The error probability (bit or symbol) is an average of the error probability for a particular signal to noise ratio, as given in (11.17). The results in Chapter 11 are all applicable for OFDM in this case (frequency flat slow fading).

15.1.2 Frequency Selective Slow Fading Channel

As mentioned in the beginning of this chapter, the frequency selective slow fading channel is the most common channel model for mobile wireless communications. Frequency selective fading channel is characterized by multiple resolvable signal paths in the channel. Assuming there are L signal paths, the channel impulse response in the ith symbol period is modeled as [1]

$$h(t) = \sum_{m=0}^{L-1} h_m(iT_s)\delta(t - \tau_m(iTs)) \tag{15.7}$$

where T_s is the symbol period

$$h_m(iT_s) = \alpha_m(iT_s)\exp(-j2\pi f_c\tau_m(iT_s))$$

is the random complex gain of the mth path signal in the ith symbol period, L is the number of signal paths, f_c is the carrier frequency, $\alpha_m(iT_s)$ is a real random attenuation factor, and $\tau_m(iT_s)$ is the mth path time delay which is also random. $\alpha_m(iT_s)$ and $\tau_m(iT_s)$ remain constant in a symbol, but change from symbol to symbol. This is, of course, an approximation for a slow changing fading channel. A more accurate model is the time-varying fading channel that will be discussed later. For notation simplicity, we will drop the time argument iT_s in the impulse response expression and rewrite it as

$$h(t) = \sum_{m=0}^{L-1} h_m\delta(t - \tau_m) \tag{15.8}$$

For mobile radio channels, $h_m = \alpha_m e^{-j2\pi f_c\tau_m}$ is usually modeled as a complex Gaussian random variable with zero mean. Then the amplitude gain α_m is Rayleigh distributed. In some cases h_m may have a nonzero mean, then α_m is Rician distributed. The paths may be uncorrelated or correlated [4,5]. The delays τ_m are uniformly and independently distributed over a length which is assumed to be equal to or less than the length of the cyclic prefix.

Excluding noise, the receive signal is

$$x(t) = s(t) * h(t) = \sum_{m=0}^{L-1} h_m s(t - \tau_m) \tag{15.9}$$

---

1 Regardless of actual signal paths, frequency selective channel can also be modeled as a tapped delay line with tap spacing $1/W$, i.e., the channel impulse response is

$$h(\tau, t) = \sum_{m=-\infty}^{\infty} h_m(t)\delta(\tau - \frac{m}{W})$$

where W is the signal bandwidth [3, p. 729]. This channel model achieves a resolution of $1/W$ in the multipath delay profile. It is based on the fact that if the bandpass signal's bandwidth is W, then its equivalent lowpass signal's bandwidth is $W/2$, and the signal can be represented by its samples at a sampling frequency of W.

In general, τ_m may not be an integer multiple of the sampling period. However, with little loss of accuracy, a common practice in literature is to approximate it as a multiple of the sampling period. When $x(t)$ is sampled, it becomes

$$x_n = \sum_{m=0}^{L-1} h_m s_{n-D_m} \tag{15.10}$$

where $D_m = \tau_m / T_{sample}$ is the delay in the number of sampling periods. Thus, the channel is modeled as a tapped delay line with the weights as $\{h_m\}$. Note that the delay spacing between two adjacent taps is the delay difference. For analysis purpose, the model only needs to show relative delays so that τ_0 is set to be zero and $D_0 = 0$. With this convention, a more convenient form of (15.10) is

$$x_n = \sum_{m=0}^{M-1} h_m s_{n-m} \tag{15.11}$$

where M is the maximum relative delay in number of sampling periods [6]. In this expression, $h_m \neq 0$ for taps corresponding to the L multipaths and $h_m = 0$ for taps between those nonzero taps. The vector $\mathbf{h} = [h_0, h_1, ..., h_{M-1}]^T$ is called the *discrete channel impulse response*. Note that we have modified the vector $\mathbf{h}$ defined earlier (it only has L nonzero components, no zero components); however, they are equivalent.

15.1.2.1 Time-Domain Fading Statistics

We will study the time-domain fading statistics using the discrete channel impulse response $\mathbf{h} = [h_0, h_1, ..., h_{M-1}]^T$.

To accommodate the Rician fading case, the most general channel impulse response model is assumed, which consists of a fixed complex component u_m and a random diffused component v_m with zero-mean complex Gaussian distribution, that is

$$h_m = u_m + v_m, \quad u_m = |u_m| e^{j\theta_m} \tag{15.12}$$

where θ_m is the phase of the fixed component and is uniformly distributed within $[0, 2\pi)$. The fading envelope $\alpha_m = |h_m|$ follows the Rician probability density, which is (see (11.14))

$$p(\alpha_m) = \frac{2(K_m^t + 1)\alpha_m}{\Omega_m^t} \exp\left(-K_m^t - \frac{(K_m^t + 1)\alpha_m^2}{\Omega_m^t}\right)$$

$$\times I_0\left(2\sqrt{\frac{K_m^t(K_m^t+1)}{\Omega_m^t}}\alpha_m\right),$$

$$\alpha_m \geq 0 \qquad (15.13)$$

where

$$K_m^t = \frac{|u_m|^2}{\text{var}\{v_m\}} \qquad (15.14)$$

is the Rician K factor and

$$\Omega_m^t = E\{\alpha_m^2\} = |u_m|^2 + \text{var}\{v_m\} \qquad (15.15)$$

is the fading power of the mth Rician path. The superscript t indicates time-domain parameters.

From (15.14) and (15.15) we can obtain

$$\text{var}\{v_m\} = \frac{\Omega_m^t}{1+K_m^t} \qquad (15.16)$$

When $K_m^t = 0$, the Rician pdf reduces to the Rayleigh pdf

$$p(\alpha_m) = \frac{2\alpha_m}{\Omega_m^t}\exp\left(-\frac{\alpha_m^2}{\Omega_m^t}\right), \quad \alpha_m \geq 0 \qquad (15.17)$$

The correlation between different paths is described by the path covariance matrix

$$\mathbf{C_h} = E\{\mathbf{v}\mathbf{v}^H\} = E\{(\mathbf{h}-E\{\mathbf{h}\})(\mathbf{h}-E\{\mathbf{h}\})^H\} = E\{\mathbf{h}\mathbf{h}^H\} - E\{\mathbf{h}\}E\{\mathbf{h}^H\}$$

where $\mathbf{v} = [v_0, v_1, ..., v_{M-1}]^T$ and $\mathbf{h} = [h_0, h_1, ..., h_{M-1}]^T$ and $(\cdot)^H$ denotes Hermitian transpose. Note that since only L of the M elements in $\mathbf{h}$ are nonzero, only L^2 elements of $\mathbf{C_h}$ are nonzero and the rest of the elements of $\mathbf{C_h}$ are zero. When the paths are uncorrelated, $\mathbf{C_h}$ reduces to a diagonal matrix whose mth element is the variance of the mth path. But there are only L nonzero elements, corresponding to the L paths.

15.1.2.2 Frequency-Domain Fading Statistics

We have seen (15.11) in Section 12.6 in the illustration of the usefulness of the cyclic extension. Using this discrete channel model, with the existence of the cyclic extension of at least $M-1$ samples, we have shown in Section 12.6 that the received samples are the results of the circular convolution of the transmitted signal and the channel impulse response

$$x_n = \sum_{m=0}^{N-1} h_m^c s_{n-m}^c \tag{15.18}$$

$\{s_m^c\}$ is the circularly extended $\{s_m\}$. $\{h_m^c\}$ is obtained by amending $(N-M)$ zeros to $\{h_m\}$ to form $\{h_m\} = \{h_0, ..., h_{M-1}, 0, ..., 0\}$ and then circularly extending it. By the convolution theorem of DFT,

$$\text{DFT}_k(\mathbf{x}) = \text{DFT}_k(\mathbf{h}) \times \text{DFT}_k(\mathbf{s}) = H_k d_k$$

where $\mathbf{x} = \{x_n\}_{n=0}^{N-1}$ and $\mathbf{s} = \{s_n\}_{n=0}^{N-1}$, and

$$H_k = \text{DFT}_k(\mathbf{h}) = \sum_{m=0}^{M-1} h_m \exp(-j2\pi \frac{km}{N}) \tag{15.19}$$

In the presence of AWGN, the demodulator output is

$$Y_k = H_k d_k + W_k \tag{15.20}$$

where W_k is the DFT of the zero-mean complex Gaussian noise in the channel, which is still a complex zero-mean Gaussian random variable. Equation (15.20) is very similar to (15.5) of the flat slow fading case. Again, fading does not introduce ISI and ICI. In the case of flat slow fading, the fading factor h in the time domain passes through the DFT demodulator and directly appears as a scaling factor of the data symbol d_k, and the scaling factor is the same for all subcarriers, while in the case of frequency selective slow fading, the scaling factors are different for different subcarriers and they are DFT of the channel impulse response. To recover the original data, $\{H_k\}$ must be estimated and X_k must be divided by H_k (equalization). Without equalization, there will be severe BER performance degradations, which will be different for different subcarriers, in general.

H_k is called a subcarrier fading factor or gain, for which Zhang et al. studied several cases: whether Rician paths exist and how many of them exist, and whether paths are correlated [4, 5]. The following are their results.

From (15.19) we have

$$H_k = \sum_{m=0}^{M-1} h_m \exp(-j2\pi \frac{km}{N})$$

$$= \underbrace{\sum_{m=0}^{M-1} u_m \exp(-j2\pi \frac{km}{N})}_{U_k} + \underbrace{\sum_{m=0}^{M-1} v_m \exp(-j2\pi \frac{km}{N})}_{V_k} \quad (15.21)$$

$$= U_k + V_k$$

where U_k and V_k are DFTs of u_m and v_m, respectively. U_k and V_k are a fixed complex component and a zero-mean complex Gaussian random variable, respectively, due to the statistical nature of their time-domain counterparts. Thus in general the subcarrier fading envelope $|H_k|$ can be modeled as a Rician random variable too.

The power of the fixed component U_k is

$$|U_k|^2 = |\mathrm{DFT}_k(\mathbf{u})|^2 = \left| \sum_{m=0}^{M-1} |u_m| \exp(-j(2\pi \frac{km}{N} - \theta_m)) \right|^2$$

From (15.14) and (15.15) we have

$$|u_m|^2 = \frac{K_m^t \Omega_m^t}{1 + K_m^t}$$

Thus we have

$$|U_k|^2 = |\mathrm{DFT}_k(\mathbf{u})|^2 = \left| \sum_{m=0}^{M-1} \sqrt{\frac{K_m^t \Omega_m^t}{1 + K_m^t}} \exp(-j(2\pi \frac{km}{N} - \theta_m)) \right|^2 \quad (15.22)$$

From (15.21), we can write $H_k = \mathbf{w}_k \mathbf{h}$, where $\mathbf{w}_k = [1, e^{-j2\pi k \cdot 1/N}, e^{-j2\pi k \cdot 2/N}$, ..., $e^{-j2\pi k \cdot (M-1)/N}]$. Then H_k is a linear transform of the multivariate Gaussian random vector $\mathbf{h}$ and it is still Gaussian with variance of [7]

$$\mathrm{var}(H_k) = \mathrm{var}(V_k) = \mathbf{w}_k \mathbf{C_h} \mathbf{w}_k^H \quad (15.23)$$

which is the diffused power of the mth component.

Using (15.22) and (15.23) we have the K-factor and fading power of the kth subcarrier as

$$K_k^f = \frac{|\mathrm{DFT}_k(\mathbf{u})|^2}{\mathbf{w}_k \mathbf{C_h} \mathbf{w}_k^H} \quad (15.24)$$

$$\Omega_k^f = |\mathrm{DFT}_k(\mathbf{u})|^2 + \mathbf{w}_k \mathbf{C_h} \mathbf{w}_k^H \quad (15.25)$$

In summary, for fading channels with Rician paths, the OFDM subcarriers at the DFT demodulator output also have Rician fading gains, and their fading parameters can be determined from the time-domain channel impulse response using (15.24) and (15.25).

15.1.2.3 Channels with Uncorrelated Paths

When the fading channel consists only of uncorrelated paths, the covariance matrix $\mathbf{C_h}$ is a diagonal matrix whose mth element is the variance of v_m, and the variance $\mathrm{var}(V_k)$ of the diffused component of the subcarrier fading coefficient can be reduced to

$$\mathrm{var}(V_k) = \mathbf{w}_k \mathbf{C_h} \mathbf{w}_k^H = \sum_{m=0}^{M-1} \mathrm{var}(v_m) = \sum_{m=0}^{M-1} \frac{\Omega_m^t}{1+K_m^t}$$

Note that there are only L nonzero elements in $\mathbf{C_h}$, corresponding to the L actual paths. As a result, the subcarrier K factor and power are given, respectively, as

$$K_k^f = \frac{\left|\sum_{m=0}^{M-1} \sqrt{\frac{K_m^t \Omega_m^t}{1+K_m^t}} \exp(-j(2\pi \frac{km}{N} - \theta_m))\right|^2}{\sum_{m=0}^{M-1} \frac{\Omega_m^t}{1+K_m^t}} \tag{15.26}$$

$$\Omega_k^f = \left|\sum_{m=0}^{M-1} \sqrt{\frac{K_m^t \Omega_m^t}{1+K_m^t}} \exp(-j(2\pi \frac{km}{N} - \theta_m))\right|^2 + \sum_{m=0}^{M-1} \frac{\Omega_m^t}{1+K_m^t} \tag{15.27}$$

Note that $\mathrm{var}(V_k)$ is the same for all subcarriers in this case. In the following, a few variations are studied.

- *Channels with All Rayleigh Paths.* This is the most common model for mobile channels. When all paths are Rayleigh, $K_k^f = 0$ and $\Omega_k^f = \mathrm{var}(V_k) = \sum_{m=0}^{M-1} \mathrm{var}(v_m)$ for all k. Thus all subcarriers will have the same power irrespective of the channel impulse response profile in the time domain. That is, all subcarriers will have Rayleigh pdfs with the same power.
- *Channels with One Rician and Multiple Rayleigh Paths.* This type of channel can be found in practice too [8,9]. In general, we can assume that the ith path is Rician with K factor K_i^t and power Ω_k^t, and the rest are Rayleigh. Usually the first arriving path is Rician. In this case, from (15.22), the fixed component power is

$$|U_k|^2 = \left|\sum_{m=0}^{M-1} \sqrt{\frac{K_m^t \Omega_m^t}{1+K_m^t}} \exp(-j(2\pi \frac{km}{N} - \theta_m))\right|^2 = \frac{K_i^t \Omega_i^t}{1+K_i^t}, \quad \forall k \tag{15.28}$$

This indicates that all subcarriers have the same fixed component. Consequently, K_k^f and Ω_i^f are all the same for all subcarriers, too. All subcarriers will have the same Rician pdf. Furthermore,

$$K_k^f = \frac{K_i^t \Omega_i^t / (1 + K_i^t)}{\frac{\Omega_i^t}{1+K_i^t} + \sum_{m=0, m \neq i}^{M-1} \frac{\Omega_m^t}{1+K_m^t}} < K_i^t \tag{15.29}$$

That is, the K factor of all subcarriers will be lower than the time-domain K factor of the Rician channel. This is due to the fact that the subcarrier variance consists of contributions from both the Rician and Rayleigh channels. In summary, if in the time domain, one path is Rician and the rest are Rayleigh, then in the frequency domain, all subcarriers are Rician with a K factor lower than the K factor in the time domain.

- *Channels with Multiple Rician and Rayleigh Paths.* Examples of this type of channel can be found in [10, 11]. Having more than one Rician path leads to a multiple-point U profile in (15.22). For example, if there are two Rician paths, then there are two nonzero terms in (15.22). The total composite signal is the vector sum of two signals. Due to the fact that the phase is a function of subcarrier index k, the composite signal will be different for different k. That is, the U_k varies with k, and K_k^f and Ω_k^f both are different for different subcarriers. In other words, signal strength is nonuniformly distributed across subcarriers. This is unlike the above two cases: all Rayleigh or one Rician plus Rayleigh, where signal strength is uniformly distributed across subcarriers. Results of an ATDMA picocell channel model with five Rician paths and one Rayleigh path in [4, 5] show good agreement between analytical results given by (15.26) and (15.27) and simulation results.

15.1.2.4 Channels with Correlated Paths

Practical examples of correlated paths can be found in [12–14]. In this case, the subcarrier K factors and power are still determined by (15.24) and (15.25), respectively. However, the channel covariance matrix is not diagonal, so that $\mathbf{w}_k \mathbf{C}_\mathbf{h} \mathbf{w}_k^H$ is a function of subcarrier index k. This will lead to different K factors and powers for different subcarriers. Hence we conclude that OFDM subcarriers in channels with correlated paths will always experience nonuniform fading in all cases, regardless of the distribution of Rician and Rayleigh paths.

15.1.2.5 Subcarrier BER Evaluation

Once the kth subcarrier's K factor and power are determined using the results

	path 1	path 2	path 3	path 4	path 5	path 6
Ω_m^t (dB)	0	-3.3	-9.3	-14.3	-20.3	-26.8
K_m^t (dB)	10	2.51	2.51	2.51	2.51	0
τ_m (ns)	0	77	186	299	404	513
m ($\approx \tau_m/50$ ns)	0	1	4	6	8	10
θ_m(deg)	0	10	30	20	40	—

Table 15.1 ATDMA picocell channel consisting of five Rician and one rayleigh paths with decreasing power profiles. ATDMA: advanced time division multiple access.

presented above for various channel conditions, the pdf of the subcarrier gain $|H_k|$, denoted as $p(|H_k|)$, can be determined using (15.13). Then the BER of the subcarrier at E_b/N_0 is given by

$$P_{b,k} = \int_0^\infty P_b(\gamma_k)p(|H_k|)d|H_k| \tag{15.30}$$

where $\gamma_k = |H_k|^2 E_b/N_o$, or if the pdf $p(\gamma_k)$ is known (for example, see (11.18) and (11.31)), we can have

$$P_{b,k} = \int_0^\infty P_b(\gamma_k)p(\gamma_k)d\gamma_k \tag{15.31}$$

where $P_b(\gamma_k)$ can be obtained using a proper BER expression developed in earlier chapters with E_b/N_o being replaced by γ_k. For example, if the modulation is QPSK, then

$$P_b(\gamma_k) = Q\left(\sqrt{2\gamma_k}\right) = Q\left(\sqrt{\frac{2|H_k|^2 E_b}{N_o}}\right) \tag{15.32}$$

As an illustration, consider the performance of an OFDM system with 64 subcarriers modulated with 16QAM and operating at $E_b/N_o = 15$ dB for all subcarriers [4,5]. The channel is the ATDMA picocell channel given by Table 15.1, where 50 ns is the sampling period, and the row labeled m is the sample indexes with nonzero channel gains. The path correlation matrix is given by

$$\mathbf{C_h} = \begin{bmatrix} 1 & 0.82 & 0.50 & 0.36 & 0.11 & 0 \\ 0.82 & 1 & 0.73 & 0.41 & 0.13 & 0.05 \\ 0.50 & 0.73 & 1 & 0.81 & 0.39 & 0.14 \\ 0.36 & 0.41 & 0.81 & 1 & 0.76 & 0.38 \\ 0.11 & 0.13 & 0.39 & 0.76 & 1 & 0.75 \\ 0 & 0.05 & 0.14 & 0.38 & 0.75 & 1 \end{bmatrix} \tag{15.33}$$

The exact BER expression for square M-ary QAM with Gray mapping is [15]

$$P_b = \frac{1}{\log_2 \sqrt{M}} \sum_{i=1}^{\log_2 \sqrt{M}} P_b(i) \tag{15.34}$$

where

$$\begin{aligned} P_b(i) &= \frac{1}{\sqrt{M}} \sum_{n=1}^{(1-2^{-i})\sqrt{M}-1} \left\{ (-1)^{\left\lfloor \frac{n2^{i-1}}{\sqrt{M}} \right\rfloor} \left(2^{i-1} - \left\lfloor \frac{n2^{i-1}}{\sqrt{M}} + \frac{1}{2} \right\rfloor \right) \right. \\ & \left. \cdot \operatorname{erfc}\left((2n+1)\sqrt{\frac{3\log_2 M \cdot \gamma}{2(M-1)}} \right) \right\} \end{aligned} \tag{15.35}$$

Using the information in Table 15.1 and (15.33), $p(|H_k|)$ can be calculated. Then replacing γ with γ_k in (15.35) and using (15.30) we can evaluate the BER for the kth subcarrier. The results are given in Figure 15.1. It is interesting to see that each subcarrier has a different BER and a few high BER subcarriers dominate the average BER. Thus an adaptive scheme should be developed so that the high-quality subcarriers are made to carry more bits or use less power than the low-quality subcarriers, which are called statistical bit loading and power control, respectively [4, 5].

15.1.3 Frequency Flat Fast Fading Channel

Frequency flat fast fading channel is characterized by a single signal with a complex multiplicative fading factor on the signal component, with the fading factor being time varying in a symbol period. In mobile communications, fast fading is caused by Doppler shift. Thus the fading factor can be modeled as

$$h(t) = \alpha(t) \exp(j\theta(t)) \tag{15.36}$$

where $\alpha(t)$ is the time-varying amplitude, and

$$\theta(t) = -2\pi[(f_c + f_D(t))\tau(t) - f_D(t)t] \tag{15.37}$$

is the time-varying phase [16]. $\alpha(t)$ and $\theta(t)$ are independent from each other. $f_D(t)$ is the Doppler frequency shift. $\tau(t)$ is the random delay. All of them are time varying due to the movement of the mobile unit. When the mobile unit is moving at a speed v, the Doppler shift f_D is modeled as a random process

$$f_D(t) = f_M \cos\varphi(t)$$

where $f_M = v/\lambda_c$ and λ_c is the carrier wavelength. $\varphi(t)$ is the angle of incidence

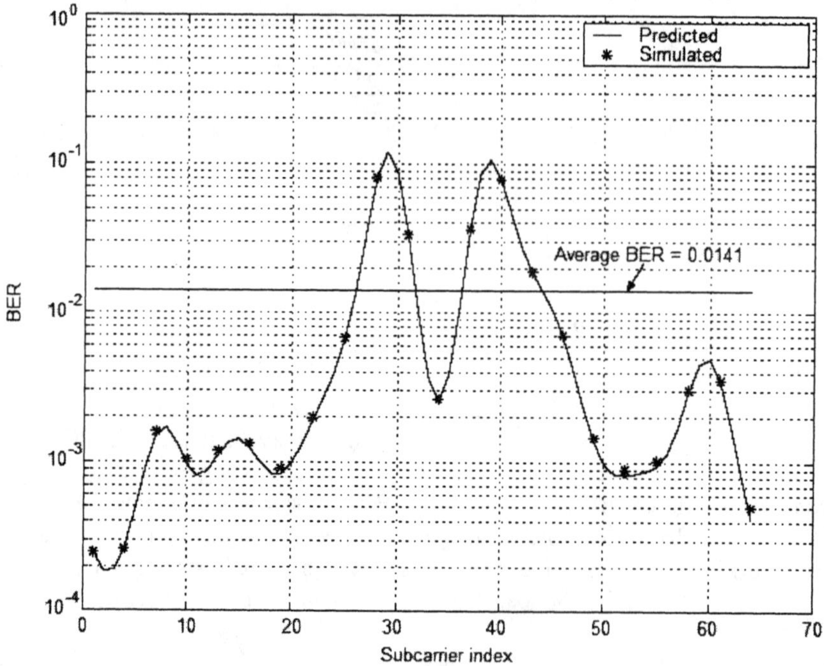

Figure 15.1 Simulated and predicted subcarrier BERs of a 64-subcarrier 16QAM OFDM system at $E_b/N_o = 15$ dB. Fading channel conditions are specified in Table 15.1 with path correlations given in (15.32). From [5].

of the plane wave. However, in a very short period such as a symbol interval, the amplitude, delay, and Doppler shift can be approximately considered constant. That is $\alpha(t) = \alpha$, $\tau(t) = \tau$, and

$$f_D = f_M \cos\varphi \tag{15.38}$$

where $\varphi(t) = \varphi$ is considered constant, and is uniformly distributed in $[-\pi, \pi]$ in an isotropic scattering channel. In this channel, it can be shown that the autocorrelation function of $h(t)$ is (Appendix 15A)

$$R(\xi) = E\{|h(t)|^2\} J_0(2\pi f_M \xi) \tag{15.39}$$

where $J_0(\cdot)$ is the zero-order Bessel function of the first kind (see Appendix 15A for the definition). The channel's discrete-time model is

$$h_n = \alpha_n \exp(j\theta_n) \tag{15.40}$$

where n indicates nth sample. Thus the discrete channel impulse response can be denoted as $\mathbf{h} = [h_0, h_1, ..., h_{N-1}]$.

Thus the signal part of the output of the DFT demodulator is [1]

$$\begin{aligned} X_k &= \frac{1}{N}\sum_{n=0}^{N-1} s_n h_n \exp(-j2\pi\frac{kn}{N}) \\ &= \frac{1}{N}\sum_{n=0}^{N-1}\left[\sum_{i=0}^{N-1} d_i \exp(j2\pi\frac{in}{N})\right] h_n \exp(-j2\pi\frac{kn}{N}) \\ &= \frac{1}{N}\sum_{i=0}^{N-1} d_i \left[\sum_{n=0}^{N-1} h_n \exp(j2\pi\frac{(i-k)n}{N})\right] \\ &= \sum_{i=0}^{N-1} H_{i-k} d_i \end{aligned} \tag{15.41}$$

where

$$H_{i-k} = \frac{1}{N}\sum_{n=0}^{N-1} h_n \exp(j2\pi\frac{(i-k)n}{N}) \tag{15.42}$$

If $h_n = 1$ for all n (the distortionless channel), H_{i-k} is simply the Kronecker delta function δ_{ik} and $X_k = d_k$. In the presence of fading, $H_{i-k} \neq \delta_{ik}$ and

$$X_k = H_0 d_k + \sum_{\substack{i=0 \\ i\neq k}}^{N-1} H_{i-k} d_i = H_0 d_k + c_k \tag{15.43}$$

where

$$H_0 = H_{k-k} = \frac{1}{N}\sum_{n=0}^{N-1} h_n \tag{15.44}$$

and

$$c_k = \sum_{\substack{i=0 \\ i\neq k}}^{N-1} H_{i-k} d_i \tag{15.45}$$

The second term represents ICI caused by loss of orthogonality. With noise term W_k, the demodulator output is

$$Y_k = H_0 d_k + c_k + W_k \tag{15.46}$$

Note that the fading factor H_0 is the same for all subcarriers, which is the average of all terms of the channel impulse response. Correction to the fading effect is needed to reduce the degradation caused by the fading. Without equalization, the performance is analyzed and simulated in terms of signal-to-interference ratio (SIR) in [16, p. 229] by Stüber, as described in the following.

When N is sufficiently large, the central limit theorem applies and the ICI can be modeled as a Gaussian random process. Since d_i and H_{i-k} are independent and $E\{d_i\} = 0$ (for QAM and PSK), then $E\{c_k\} = 0$. Assuming uncorrelated data, the autocorrelation of c_k is

$$\begin{aligned} E[c_k c_{k+r}^*] &= E\left[\sum_{\substack{i=0 \\ i\neq k}}^{N-1} \sum_{\substack{l=0 \\ l\neq k+r}}^{N-1} d_i d_l^* H_{i-k} H_{l-k-r}^*\right] \\ &= E_s \sum_{\substack{i=0 \\ i\neq k,k+r}}^{N-1} E[H_{i-k} H_{i-k-r}^*] \end{aligned}$$

where $E_s = E\{d_i d_i^*\} = E\{|d_i|^2\}$ is the average symbol energy. If we further assume $E[|h_k|^2] = 1$ and isotropic scattering, the autocorrelation becomes (see Appendix 15A)

$$\begin{aligned} E[c_k c_{k+r}^*] &= E_s\delta_r - \frac{E_s}{N^2}\sum_{n=0}^{N-1}\sum_{p=0}^{N-1} J_0(2\pi f_M T_s(n-p)) \\ &\quad \cdot \left[\exp(j2\pi\frac{pr}{N}) + (1-\delta_r)\exp(j2\pi\frac{nr}{N})\right] \end{aligned} \tag{15.47}$$

The variance of the ICI is

$$E[|c_k|^2] = E[c_k c_k^*] = E_s - \frac{E_s}{N^2}\sum_{n=0}^{N-1}\sum_{p=0}^{N-1} J_0(2\pi f_M T_s(n-p)) \tag{15.48}$$

Define $i = n - p$, then $-(N-1) \leq i \leq (N-1)$, and there are N terms with $i = 0$, and $(N-i)$ terms with index i and $(N-i)$ terms with index $-i$. Thus

$$E[|c_k|^2] = E_s - \frac{E_s}{N^2}\{N + 2\sum_{i=1}^{N-1}(N-i)J_0(2\pi f_M T_s i)\} \tag{15.49}$$

where the fact that $J_0(\cdot)$ is an even function is used.

From (15.49) we see that the ICI power is only a function of E_s, N, T_s, and f_M, and is independent of the signal constellation. The signal part in (15.46) is $d_k H_0$ and its average signal symbol energy is

$$\begin{aligned}
E' &= E\{|d_k H_0|^2\} = E_s E\{|H_0|^2\} \\
&= E_s E\{\frac{1}{N^2}\sum_{n=0}^{N-1}\sum_{p=0}^{N-1} h_n h_p^*\} \\
&= E_s \frac{1}{N^2}\sum_{n=0}^{N-1}\sum_{p=0}^{N-1} E\{|h(t)|^2\} J_0(2\pi f_M T_s(n-p)) \\
&= E_s \frac{1}{N^2}\sum_{n=0}^{N-1}\sum_{p=0}^{N-1} J_0(2\pi f_M T_s(n-p)) \\
&= \frac{E_s}{N^2}\{N + 2\sum_{i=1}^{N-1}(N-i)J_0(2\pi f_M T_s i)\}
\end{aligned} \tag{15.50}$$

which is the same as the second term in (15.49).

The (average) signal-to-ICI ratio (SIR) is defined as

$$SIR = \frac{E'}{E[|c_k|^2]} = \frac{\frac{1}{N^2}\{N + 2\sum_{i=1}^{N-1}(N-i)J_0(2\pi f_M T_s i)\}}{1 - \frac{1}{N^2}\{N + 2\sum_{i=1}^{N-1}(N-i)J_0(2\pi f_M T_s i)\}} \tag{15.51}$$

which is only a function of N and normalized Doppler shift $f_M T_s$. Note that if $E\{|h(t)|^2\} \neq 1$, it will appear in the numerator and the denominator, but they cancel each other. Therefore setting $E\{|h(t)|^2\} = 1$ does not affect the final SIR expression. When f_M is very small, $J_0(2\pi f_M T_s i) \cong 1$, and

$$N + 2\sum_{i=1}^{N-1}(N-i)J_0(2\pi f_M T_s i) \cong N + 2\sum_{i=1}^{N-1}(N-i) = N^2$$

then $E' \cong E_s$ and $E[|c_k|^2] \cong 0$.

Figure 15.2 shows the SIR as a function of the normalized maximum Doppler shift (f_M) for several values of N. SIR decreases with increased f_M and N.

The SIR will introduce an error floor in the BER curve. The complex amplitude of the useful signal, H_0, is the average of N zero-mean Gaussian (see (15.44)) random variables so that it is also Gaussian, and $|H_0|$ is Rayleigh. The signal-to-noise ratio for a particular fading amplitude $|H_0|$ is

$$\gamma_s = |H_0|^2 E_b/N_o$$

From (11.17) and (11.18), the average symbol error probability is

$$P_s = \int_0^\infty \frac{1}{\overline{\gamma}_s} \exp(-\gamma_s/\overline{\gamma}_s) P_s(\gamma_s) d\gamma_s \tag{15.52}$$

where $\overline{\gamma}_s$ is the average SNR. For 16QAM, from (9.34)

$$P_s(\gamma_s) = 3Q\left(\sqrt{\frac{\gamma_s}{5}}\right)\left[1 - \frac{3}{4}Q\left(\sqrt{\frac{\gamma_s}{5}}\right)\right] \tag{15.53}$$

Substituting the SIR for $\overline{\gamma}_s$ gives the error floor due to ICI. Figure 15.3 shows the calculated and simulated error floors due to ICI for 16QAM with $R_s = 5$ Msps, where the theoretical lines are calculated using (15.52).

15.1.4 Frequency Selective Fast Fading Channel

The channel impulse response of a frequency selective fast fading channel with L signal paths is given by

$$h(\tau, t) = \sum_{m=0}^{L-1} h_m(t)\delta(\tau - \tau_m(t)) \tag{15.54}$$

where $h_m(t)$ is the complex random channel gain and is defined by (15.36) to (15.38).

For most radio transmission, $h_m(t)$ is assumed as a wide-sense stationary (WSS) narrowband complex Gaussian process with zero mean, and is uncorrelated between different paths. The channel then is called wide-sense stationary uncorrelated scattering (WSSUS) channel. Further, $\tau_m(t)$ usually changes very little in a symbol period so that we can assume that it is constant in a symbol. Then, the channel impulse response can be simplified as

$$h(\tau, t) = \sum_{m=0}^{L-1} h_m(t)\delta(\tau - \tau_m) \tag{15.55}$$

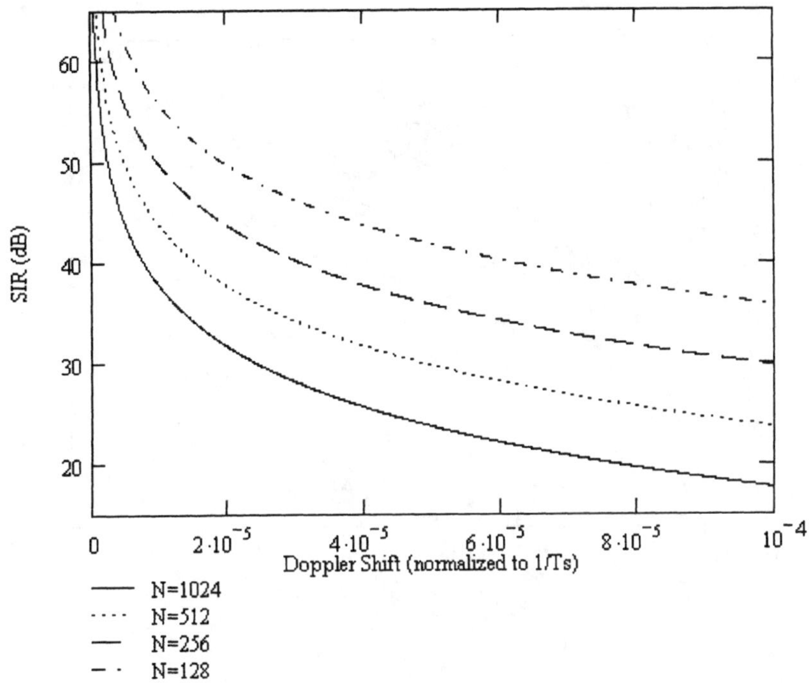

Figure 15.2 Signal-to-interference ratio due to ICI.

Excluding noise, the receiver input signal is

$$x(t) = \sum_{m=0}^{L-1} h_m(t)s(t - \tau_m)$$

Following an argument similar to the case of frequency selective slow fading, we obtain the sampled signal as [6]

$$x_n = \sum_{m=0}^{M-1} h_{m,n}s_{n-m} \tag{15.56}$$

where $h_{m,n}$ is the value of the channel impulse response of path m at sampling time instant n. MT_s is the maximum delay. The discrete channel impulse response now

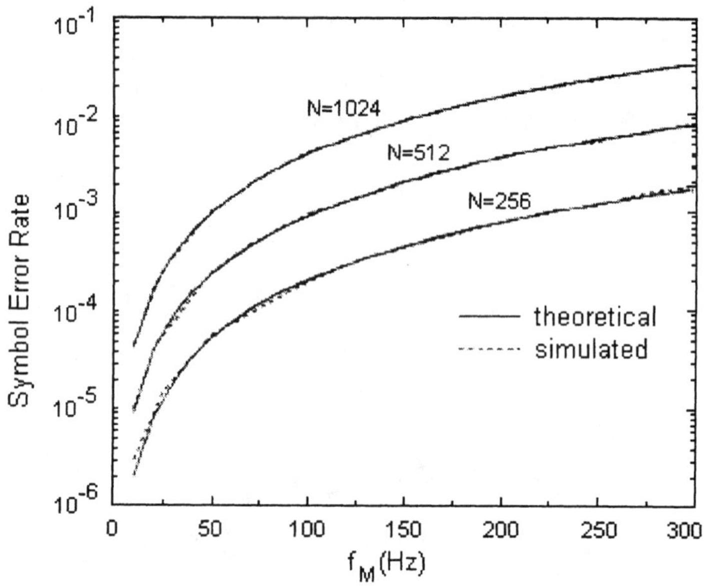

Figure 15.3 Error floor due to ICI for 16QAM. ($R_s = 5$ Msps). From [6]. Copyright © 1995 IEEE.

is a function of n (i.e., time-varying), and it can be described by a matrix as

$$\mathbf{H} = \begin{bmatrix} h_{0,0} & h_{0,1} & \dots & h_{0,N-1} \\ h_{1,0} & h_{1,1} & \dots & h_{1,N-1} \\ \vdots & \vdots & \vdots & \vdots \\ h_{M-1,0} & h_{M-1,1} & \dots & h_{M-1,N-1} \end{bmatrix} = \begin{bmatrix} \mathbf{h}_0 \\ \mathbf{h}_1 \\ \vdots \\ \mathbf{h}_{M-1} \end{bmatrix} \tag{15.57}$$

where $\mathbf{h}_i = [h_{i,0}, h_{i,1}, ..., h_{i,N-1}]$ is the channel impulse response for the ith path.

Excluding noise, the DFT demodulator output can be shown as (see Appendix 15B)

$$X_k = \sum_{i=0}^{N-1} \sum_{m=0}^{M-1} d_i H_{m,i-k} \exp(-j2\pi \frac{im}{N}) \tag{15.58}$$

where

$$H_{m,i-k} = \frac{1}{N}\sum_{n=0}^{N-1} h_{m,n}\exp(j2\pi\frac{n(i-k)}{N}) \tag{15.59}$$

We can further write (15.58) as

$$\begin{aligned} X_k &= \sum_{i=0}^{N-1} d_i \sum_{m=0}^{M-1} H_{m,i-k}\exp(-j2\pi\frac{im}{N}) \\ &= \sum_{i=0}^{N-1} \eta_{k,i} d_i \end{aligned} \tag{15.60}$$

where

$$\begin{aligned} \eta_{k,i} &= \sum_{m=0}^{M-1} H_{m,i-k}\exp(-j2\pi\frac{im}{N}) \\ &= \sum_{m=0}^{M-1}\left[\frac{1}{N}\sum_{n=0}^{N-1} h_{m,n}\exp(j2\pi\frac{n(i-k)}{N})\right]\exp(-j2\pi\frac{im}{N}) \\ &= \frac{1}{N}\sum_{m=0}^{M-1}\sum_{n=0}^{N-1} h_{m,n}\exp(-j2\pi\frac{i(m-n)+nk}{N}) \end{aligned} \tag{15.61}$$

Comparing (15.60) to (15.41), we see that they are the same except that the definition of H_{i-k} is replaced by $\eta_{k,i}$. For the flat fading case, H_{i-k} is the $(k-i)$th point of DFT of h_n. In the case of frequency selective fading, $\eta_{k,i}$ is the weighted sum of the DFT of $h_{m,n}$ due to the existence of the multiple paths. When there is only one path $(M = 1)$, (15.61) resets to (15.42).

Rewrite (15.60) as

$$X_k = \eta_{k,k}d_k + \sum_{\substack{i=0 \\ i\neq k}}^{N-1} \eta_{k,i}d_i = \eta_{k,k}d_k + c_k \tag{15.62}$$

where

$$\begin{aligned} \eta_{k,k} &= \sum_{m=0}^{M-1} H_{m,0}\exp(-j2\pi\frac{km}{N}) \\ &= \sum_{m=0}^{M-1}\left[\frac{1}{N}\sum_{n=0}^{N-1} h_{m,n}\right]\exp(-j2\pi\frac{km}{N}) \end{aligned} \tag{15.63}$$

is a multiplicative noise term which is the Fourier transform of the time averages of the channel impulse response, and

$$c_k = \sum_{\substack{i=0 \\ i \neq k}}^{N-1} \eta_{k,i} d_i = \sum_{\substack{i=0 \\ i \neq k}}^{N-1} \sum_{m=0}^{M-1} d_i H_{m,i-k} \exp(-j2\pi \frac{im}{N}) \tag{15.64}$$

is the ICI term. If $h_{m,n} = h_m$ (no time variation in the symbol period), then $H_{m,i-k} = h_m \delta_{ik}$ and $\eta_{k,k} = \sum_{m=0}^{M-1} h_m \exp(-j2\pi \frac{km}{N})$ and $c_k = 0$. There is no ICI.

Without equalization, the performance is given by Russell and Stüber in [6]. When N is sufficiently large, the central limit theorem applies and the ICI can be modeled as a Gaussian random process. Since d_i and $H_{m,i-k}$ are independent and $E\{d_i\} = 0$ (for QAM and PSK), then $E\{c_k\} = 0$. Assuming uncorrelated scattering, and uncorrelated data, the autocorrelation of c_k is shown as (see Appendix 15B)

$$E[c_k c_{k+r}^*] = E_s \sum_{\substack{i=0 \\ i \neq k,k+r}}^{N-1} \sum_{m=0}^{M-1} E[H_{m,i-k} H_{m,i-k-r}^*] \tag{15.65}$$

where $E_s = E\{d_i d_i^*\} = E\{|d_i|^2\}$ is the average symbol energy. If we further assume $\sum_{m=0}^{M-1} E[|h_{m,k}|^2] = 1$ and isotropic scattering, the autocorrelation becomes (see Appendix 15B)

$$\begin{aligned} E[c_k c_{k+r}^*] &= E_s \delta_r - \frac{E_s}{N^2} \sum_{n=0}^{N-1} \sum_{p=0}^{N-1} J_0(2\pi f_M T_s (n-p)) \\ &\quad \cdot \left[\exp(j2\pi \frac{pr}{N}) + (1-\delta_r) \exp(j2\pi \frac{nr}{N}) \right] \end{aligned} \tag{15.66}$$

which is the same as (15.47). Consequently, the variance of the ICI is also the same as in (15.48) and (15.49).

Unlike the frequency flat fast fading case, the multiplicative fading factor on subcarriers $\eta_{k,k}$ is a function of k. However, it can be shown (Appendix 15B) that all subcarriers have the same average signal symbol energy

$$\begin{aligned} E_k' &= E\{|d_k \eta_{k,k}|^2\} \\ &= \frac{E_s}{N^2} \{N + 2\sum_{i=1}^{N-1} (N-i) J_0(2\pi f_M T_s i)\} \end{aligned} \tag{15.67}$$

which is the same as (15.50). As a result, the SIR expression (15.51) and curves and error floors in Figures 15.2 and 15.3 also are applicable to this case [6]. Similar to the frequency-flat fast fading case, if $\sum_{m=0}^{M-1} E[|h_{m,k}|^2] \neq 1$, it will not affect the final form of SIR since it will appear in the numerator and the denominator and they will cancel each other.

15.1.5 Summary of Expressions of Fading Factors and ICI Terms

Table 15.2 summarizes the expressions of DFT demodulator output in all four cases. Table 15.3 summarizes the fading factor expressions and Table 15.4 summarizes the ICI expressions for all four types of fading channels. In all these tables FFSF is frequency flat slow fading, FSSF is frequency selective slow fading, FFFF is frequency flat fast fading, and FSFF is frequency selective fast fading. Note that when fading is slow there are no ICI terms. When fading is fast there are ICI terms. When $M = 1$, the expressions of both fading factors and ICI terms for a selective fading case degenerate back to a flat fading case.

15.2 CHANNEL ESTIMATION AND EQUALIZATION

To reduce the performance degradations due to fading, equalization (or correction) is necessary. This in turn requires estimation of the fading factors: h in (15.5), H_k in (15.20), H_0 in (15.46), and $\eta_{k,k}$ in (15.62). Recall that derivations of these fading factors are under the assumption that synchronizations are perfect, that is, the detection is coherent. Thus, channel estimation is a part of the coherent detection operation. To perform coherent detection, channel impairments other than fading factors, that is, carrier frequency offset and timing offset, must be estimated, which is addressed in Chapter 14. Often the estimation of fading factors is performed jointly with the estimation of frequency offset and timing offset. An alternative to coherent detection is differential detection which we discussed for single-carrier modulations in several occasions in previous chapters. For OFDM, differential detection can be done in time domain or frequency domain. In the first case, each symbol is compared with the previous symbol on the same subcarrier. This method is the same with the single-carrier case. In the case of frequency-domain differential detection, each symbol is compared with the symbol of the adjacent subcarrier within the same OFDM symbol.

Methods of channel estimation can be basically classified into two categories: pilot-assisted and decision-directed. Pilot-assisted channel estimation methods can be further divided into two subclasses: two dimensional or one dimensional. The two-dimensional method estimates the channel variations both in time and in fre-

Channel	DFT Demodulator Output (signal part)
FFSF	$X_k = hd_k$
FSSF	$X_k = H_k d_k$
FFFF	$X_k = H_0 d_k + c_k$
FSFF	$X_k = \eta_{k,k} d_k + c_k$
	$Y_k = X_k + W_k$ (total output with noise)
	$0 \leq k \leq N-1$

Table 15.2 Summary of expressions of DFT demodulator output.

Channel	Fading Factor
FFSF	h
FSSF	$H_k = \sum_{m=0}^{M-1} h_m e^{-j2\pi \frac{km}{N}}$
FFFF	$H_0 = \frac{1}{N}\sum_n h_n$
FSFF	$\eta_{k,k} = \frac{1}{N}\sum_{m=0}^{M-1}\left[\sum_n h_{m,n}\right] e^{-j2\pi\frac{km}{N}}$
	$0 \leq n \leq N-1$
M is the maximum delay spread in samples. $M \geq L$, the number of signal paths	

Table 15.3 Summary of expressions of fading factors.

Channel	ICI Term
FFSF	0
FSSF	0
FFFF	$c_k = \sum_{i \neq k} H_{i-k} d_i = \sum_{i \neq k} d_i \underbrace{\frac{1}{N}\left[\sum_n h_n e^{j2\pi\frac{(i-k)n}{N}}\right]}_{H_{i-k}}$
FSFF	$c_k = \sum_{i \neq k} \eta_{k,i} d_i = \sum_{i \neq k} d_i \underbrace{\sum_{m=0}^{M-1}\left[\frac{1}{N}\sum_n h_{m,n} e^{j2\pi\frac{(i-k)n}{N}}\right] e^{-j2\pi\frac{im}{N}}}_{\eta_{k,i}}$
	$0 \leq n \leq N, 0 \leq i \leq N-1$

Table 15.4 Summary of expressions of ICI terms.

quency. Pilot symbols are sparsely distributed across time slots and across subcarriers. The estimation of the fading factor for each subcarrier and at each symbol time is accomplished by interpolation in time and frequency simultaneously (two-dimensional interpolation). However the interpolations can be separated into two one-dimensional interpolations. The decision-directed channel estimation (DDCE) does not use any pilot symbol. Instead, modulation is removed from subcarriers using the previously demodulated symbol (the decision). Then all subcarriers can be used to estimate the channel. Of course this method is not as reliable as the pilot-assisted method which uses pilot symbols within, before, and after the current symbol. However the advantages are in the savings of the bandwidth occupied by the pilot symbols and the ability to tack the fast fading since DDCE can use every symbol as its input. Note that both pilot-assisted and decision-directed channel estimation methods may use the same estimation algorithm. The difference is the source of the input to the estimator.

15.2.1 Pilot-Assisted Channel Estimation

To estimate the channel in time domain and in frequency domain, pilot symbols are inserted in the time-frequency grid in certain patterns, depending on the estimation algorithms. The estimator uses the pilot symbols to estimate the channel for those particular frequencies and time periods. Then the channel responses on other frequencies and time periods are obtained by interpolations.

There are many proposed pilot patterns. For example, five different patterns examined by Tufvesson et al. in [17] are shown in Figure 15.4. The estimator is a Kalman filter and for this estimator, it is proved by analysis that pattern 3 has the lowest BER for QPSK for the same Doppler shift (Figure 15.5). Note that for other types of filters, such as the linear minimum mean squared error (LMMSE) filter or Wiener filter, the performance order might not be the same as in Figure 15.4. In [18], pattern 1 in Figure 15.4 is called block-type and compared to a pattern called comb-type pilot arrangement, which is the mirror image of pattern 1 (Figure 15.6). That is, it has pilots on a subcarrier for very L subcarriers and on each symbol in a block of OFDM symbols in a packet-switched transmission. Various combinations of the two pilot patterns with different estimators and interpolators are compared. It is shown by simulation in [18] that the comb-type pattern with LMMSE estimator and low-pass interpolation performs the best. This comb-type pattern is similar to pattern 3 in Figure 15.4, which is also proved to be the best for Kalman estimation.

Channel variation manifests itself as the fading factor given in (15.36), which shows that the channel parameters are the Doppler shift and delay spread. To estimate the Doppler shift and the delay spread, a sufficient density of pilot symbols

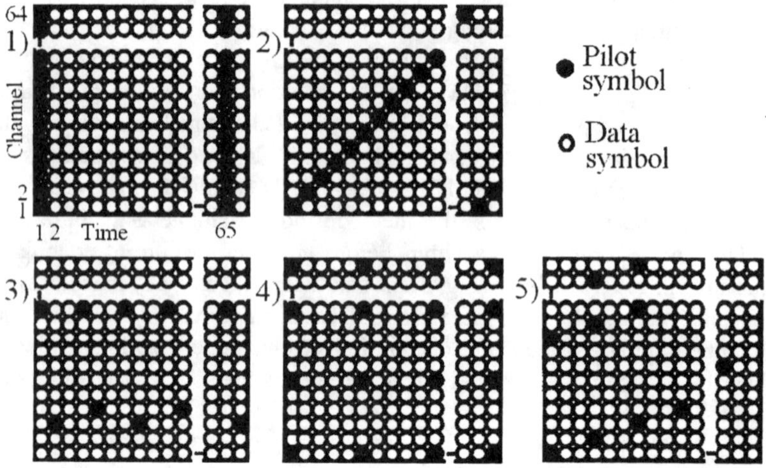

Figure 15.4 Pilot patterns analyzed in [17]. From [17]. Copyright © 1997 IEEE.

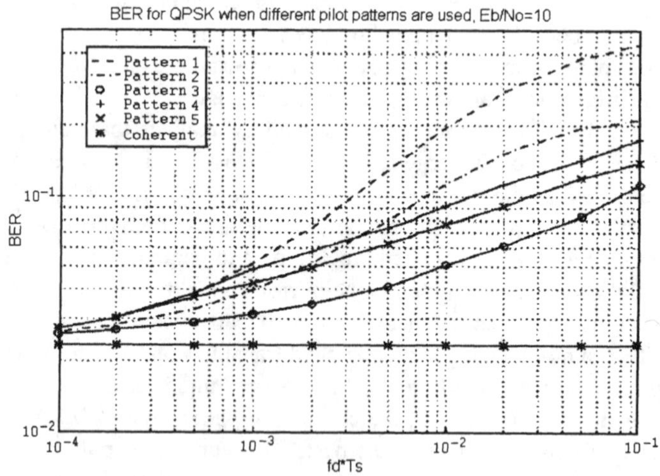

Figure 15.5 Resulting bit error rate as a function of Doppler value when the different pilot patterns are used. The same number of pilot symbols are used for all curves. From [17]. Copyright © 1995 IEEE.

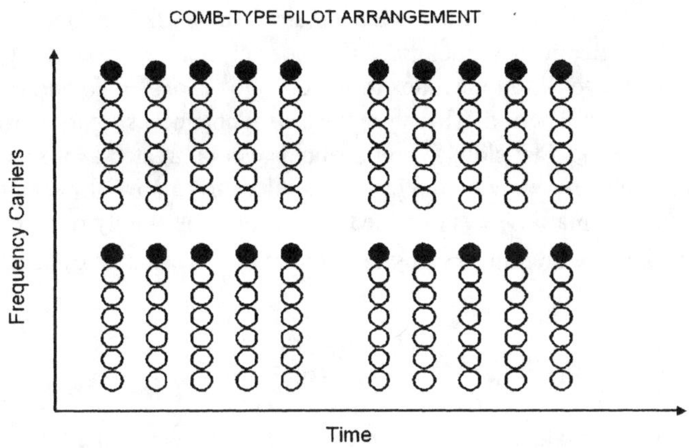

Figure 15.6 Comb-type pilot arrangement. From [18]. Copyright © 2002 IEEE.

is required. Assuming the maximum Doppler spread is f_M, then the channel variation bandwidth in the frequency domain is $2f_M$. Thus according to the sampling theorem, the sampling interval in the time domain should be $T_{sample} \leq 1/(2f_M)$, or $N_tT_s \leq 1/(2f_M)$, where N_t is the spacing between pilot symbols in the time domain. That is

$$N_t \leq \frac{1}{2f_MT_s} \tag{15.68}$$

In the time domain, the channel variation has a width of τ_M which is the maximum delay spread. By duality between Fourier transform and inverse Fourier transform, the sampling theorem requires that in frequency domain the sampling interval would be $F_{sample} \leq 1/\tau_M$, or $N_f\Delta F \leq 1/\tau_M$, where ΔF is the OFDM subcarrier spacing. That is

$$N_f \leq \frac{1}{\tau_M\Delta F} \tag{15.69}$$

In practice, oversampling by a factor of two is considered to be a rule of thumb [19].

15.2.2 Slow Fading Channel Estimation and Equalization

From (15.5) and (15.20) we know that the effect of flat slow fading is an identical complex fading factor h on all subcarriers' data after the demodulation and the effect of frequency-selective slow fading is a different complex fading factor H_k on each subcarrier's data after the demodulation. Considering channel variation from symbol to symbol, we need to add a time index to h and H_k to make them h_l and $H_{l,k}$, where l denotes the lth symbol period. Therefore the task of channel estimation is to estimate the factor h_l or $H_{l,k}$. Flat slow fading channels can be treated as a special case of the frequency-selective slow fading case. Therefore from now on we only consider $H_{l,k}$. Once the estimate $\widehat{H}_{l,k}$ is obtained, equalization is simply done by a one-tap equalizer. That is, by dividing $Y_{l,k}$ by $\widehat{H}_{l,k}$ to obtain the estimate of data:

$$\widehat{d}_{l,k} = \frac{Y_{l,k}}{\widehat{H}_{l,k}} \tag{15.70}$$

Even though equalization for slow fading channels is simple, the channel estimation is challenging. There are many channel estimation schemes proposed in the literature for slow fading channels. The basic form of these estimators is the linear minimum mean-squared error (LMMSE) estimator using the Wiener filtering. It is called an LMMSE or MMSE estimator by many authors [18, 20–24] and the Wiener estimator by others [19]. The Kalman filter is also a type of LMMSE estimator [25] and was also proposed for channel estimation, such as in [17]. The majority of the channel estimators for OFDM in the literature is the LMMSE estimator using Wiener filtering, due to its simplicity in comparison with the Kalman filter. We will use the commonly used name "LMMSE estimator" and discuss it in detail. The LMMSE estimator has high complexity when the number of subcarriers, N, is very large. Reduction of complexity can be achieved through various methods like lower-rank approximation of the estimator coefficient matrix and using DFT to perform the signal processing. We will discuss these complexity-reduced algorithms as well.

In the following we will present the estimators mainly in the context of pilot-assisted channel estimation. It is important to point out that these estimators can also be used for decision-directed channel estimation. However, due to the fact that all received symbols can be used as inputs of the decision-directed channel estimator, other simple but effective methods can also be used. These methods will be discussed in Section 15.2.5.

15.2.2.1 LMMSE (Wiener) Estimator

In the following we will present the formulation of the LMMSE estimator for slow fading channels. The formulation was originally presented by Edfors et al in [21] as a one dimensional (1-D) estimator. But this estimator can process two dimensional (2-D) signals as well.

The observations used as input to the estimator can be either obtained from pilot symbols (pilot-assisted estimation) or from receiver decisions (decision-directed estimation). The pattern of the observation points, from pilots or decisions, can be one dimensional (1-D) or two dimensional (2-D). The observation pattern used in [21] is the first pattern in Figure 15.4 (one dimensional). However, this need not be the only case. In fact, any pattern can be used provided that the observations are organized in a vector and channel fading factors are also organized in a vector accordingly [26]. That is, whether the LMMSE estimator is a 1-D or 2-D estimator depends on whether the observation pattern is 1-D or 2-D. The LMMSE estimates the channel fading factors on the observation symbols and the fading factors on other symbols will be obtained through interpolations in both frequency and time domain.

From (15.20), we can write the DFT demodulator output $\mathbf{y}$ in vector form for slow fading cases as

$$\mathbf{y} = \mathbf{D}\boldsymbol{\eta} + \mathbf{w} \tag{15.71}$$

where $\mathbf{y}$ is the observation vector

$$\mathbf{y} = [y_0, y_1, ..., y_{K-1}]^T$$

where K is the number of observations, and y_i is one of the observations, $\{Y_{l,k}\}$, of the lth symbol period and kth subcarrier frequency. The $\boldsymbol{\eta}$ is the channel fading vector,

$$\boldsymbol{\eta} = [\eta_0, \eta_1, ..., \eta_{K-1}]^T$$

where η_i is one of the fading factors, $\{H_{l,k}\}$, of the lth symbol period and kth subcarrier frequency. The sets of K observations $\{Y_{l,k}\}$ and fading factors $\{H_{l,k}\}$ can be organized in different ways for different observation patterns to form $\mathbf{y}$ and $\boldsymbol{\eta}$. For example, we can organize them into sub-vectors by frequency first and arrange these sub-vectors in time or vice versa. For instance, if pattern 4 in Figure 15.4 is used and we choose to use all observation points in the block of 64×64, we can have

$$\mathbf{y} = [(Y_{0,0}, Y_{0,5}, ..., Y_{0,60}), (Y_{5,0}, Y_{5,5}, ..., Y_{5,60}), ..., (Y_{60,0}, Y_{60,5}, ..., Y_{60,60})]^T$$

and

$$\boldsymbol{\eta} = [(H_{0,0}, H_{0,5}, ..., H_{0,60}), (H_{5,0}, H_{5,5}, ..., H_{5,60}), ..., (H_{60,0}, H_{60,5}, ..., H_{60,60})]^T$$

$\mathbf{D}$ is a diagonal matrix of the transmitted complex data on the observation points, which is known to the estimator either by training data or receiver decisions.

$$\mathbf{D} = \begin{bmatrix} d_0 & 0 & \cdots & 0 \\ 0 & d_1 & \cdots & 0 \\ 0 & 0 & \cdots & 0 \\ 0 & 0 & \cdots & d_{K-1} \end{bmatrix} \tag{15.72}$$

$\mathbf{w}$ is a vector of independent identically distributed (i.i.d.) complex zero-mean Gaussian noise with variance σ_n^2.

$$\mathbf{w} = [W_0, W_1, ..., W_{K-1}]^T$$

The noise $\mathbf{w}$ is assumed to be uncorrelated with the channel $\boldsymbol{\eta}$.

Before we derive the LMMSE estimator, using Figure 15.4 as an example, a discussion of the observation vector $\mathbf{y}$ is in order. If $\mathbf{y}$ is obtained from pilot symbols and the first pattern in Figure 15.4 is used, then $K = N = 64$, the number of subcarriers; and all pilots are in one symbol period. In other cases, K could be equal, smaller or greater than N, and Y_k could be from different rows and/or columns in the symbol grid. For example, if pattern 3 in Figure 15.4 is used, there are 6 pilot symbols per column for all 64 columns, so there are 384 pilot symbols in the 64×64 block, thus $K = 384$ or less.

The LMMSE is to estimate $\boldsymbol{\eta}$ by a linear function of $\mathbf{y}$ with the criterion of minimum mean squared error. That is, an estimate of $\boldsymbol{\eta}$ is given by $\widehat{\boldsymbol{\eta}} = \mathbf{Cy}$ where $\mathbf{C}$ is a $K \times K$ coefficient matrix, and the mean-squared error $E[(\boldsymbol{\eta} - \mathbf{Cy})^H(\boldsymbol{\eta} - \mathbf{Cy})]$ is minimized. According to [25, pp. 324–327], the LMMSE requires that $\mathbf{C}$ satisfy the orthogonal condition

$$E[(\boldsymbol{\eta} - \mathbf{Cy})\mathbf{y}^H] = \mathbf{0} \tag{15.73}$$

The above leads to the optimal coefficient matrix

$$\mathbf{C}_o = \mathbf{R}_{\boldsymbol{\eta}\mathbf{y}}\mathbf{R}_{\mathbf{yy}}^{-1} \tag{15.74}$$

and the LMMSE thus is

$$\widehat{\boldsymbol{\eta}}_{lmmse} = \mathbf{R}_{\boldsymbol{\eta}\mathbf{y}}\mathbf{R}_{\mathbf{yy}}^{-1}\mathbf{y} \tag{15.75}$$

where $\mathbf{R}_{\boldsymbol{\eta\eta}} = E\{\boldsymbol{\eta}\boldsymbol{\eta}^H\}$ is the channel autocorrelation matrix. And

$$\begin{aligned} \mathbf{R}_{\boldsymbol{\eta}\mathbf{y}} &= E\{\boldsymbol{\eta}\mathbf{y}^H\} = E\{\boldsymbol{\eta}(\mathbf{D}\boldsymbol{\eta} + \mathbf{w})^H\} \\ &= E\{\boldsymbol{\eta}\boldsymbol{\eta}^H\mathbf{D}^H\} + \mathbf{E}\{\boldsymbol{\eta}\mathbf{w}^H\} \\ &= E\{\boldsymbol{\eta}\boldsymbol{\eta}^H\mathbf{D}^H\} = \mathbf{R}_{\boldsymbol{\eta\eta}}\mathbf{D}^H \end{aligned} \tag{15.76}$$

Thus

$$\widehat{\eta}_{lmmse} = \mathbf{R}_{\eta\eta}\mathbf{D}^H\mathbf{R}_{\mathbf{yy}}^{-1}\mathbf{y} \tag{15.77}$$

Further,

$$\begin{aligned}\mathbf{R}_{\mathbf{yy}} &= E[(\mathbf{D}\boldsymbol{\eta} + \mathbf{w})(\mathbf{D}\boldsymbol{\eta} + \mathbf{w})^H] = E[\mathbf{D}\boldsymbol{\eta}\boldsymbol{\eta}^H\mathbf{D}^H] + E(\mathbf{w}\mathbf{w}^H) \\ &= \mathbf{D}\mathbf{R}_{\eta\eta}\mathbf{D}^H + \sigma_n^2\mathbf{I} \end{aligned}\tag{15.78}$$

where $\mathbf{I}$ is the $K \times K$ identity matrix and

$$\begin{aligned}\mathbf{R}_{\mathbf{yy}}^{-1} &= [\mathbf{D}\mathbf{R}_{\eta\eta}\mathbf{D}^H + \sigma_n^2\mathbf{I}]^{-1} \\ &= [\mathbf{D}(\mathbf{R}_{\eta\eta} + \mathbf{D}^{-1}\sigma_n^2\mathbf{I}(\mathbf{D}^H)^{-1})\mathbf{D}^H]^{-1} \\ &= (\mathbf{D}^H)^{-1}[\mathbf{R}_{\eta\eta} + \sigma_n^2\mathbf{I}(\mathbf{D}^H\mathbf{D})^{-1}]\mathbf{D}^{-1}\end{aligned}\tag{15.79}$$

Thus

$$\widehat{\eta}_{lmmse} = \mathbf{R}_{\eta\eta}[\mathbf{R}_{\eta\eta} + \sigma_n^2(\mathbf{D}\mathbf{D}^H)^{-1}]^{-1}\widehat{\eta}_{ls} \tag{15.80}$$

where

$$\widehat{\eta}_{ls} = \mathbf{D}^{-1}\mathbf{y} = \left[\frac{y_0}{d_0}, \frac{y_1}{d_1}, \ldots, \frac{y_{K-1}}{d_{K-1}}\right]^T \tag{15.81}$$

is the least square (LS) estimator which minimizes $(\mathbf{y} - \mathbf{D}\boldsymbol{\eta})^H(\mathbf{y} - \mathbf{D}\boldsymbol{\eta})$ [25]. $\widehat{\eta}_{ls}$ can be considered as a coarse estimate of $\boldsymbol{\eta}$, obtained by simply dividing the received signal y_k with the known data d_k. $\widehat{\eta}_{lmmse}$ thus is a more accurate estimate based on $\widehat{\eta}_{ls}$ and related to channel autocorrelation and noise variance. Both estimators (15.80) and (15.81) have their drawbacks. The LMMSE is more complex, whereas the LS estimator has a high mean-square error [20].

The LMMSE estimator requires inversion of the matrix $\mathbf{R}_{\eta\eta} + \sigma_n^2(\mathbf{D}\mathbf{D}^H)^{-1}$, which must be updated every time the data in $\mathbf{D}$ change. To simplify, we replace $(\mathbf{D}\mathbf{D}^H)^{-1}$ with the expectation $E\{(\mathbf{D}\mathbf{D}^H)^{-1}\}$. Simulation shows that the performance degradation is negligible [21]. Assuming the same signal constellation on all tones and equal probability on all constellation points, then

$$E\{(\mathbf{D}\mathbf{D}^H)^{-1}\} = E\left\{\begin{bmatrix} \frac{1}{|d_0|^2} & 0 & \cdots & 0 \\ 0 & \frac{1}{|d_1|^2} & \cdots & 0 \\ 0 & 0 & \cdots & 0 \\ 0 & 0 & \cdots & \frac{1}{|d_{K-1}|^2}\end{bmatrix}\right\} = E\{|1/d_k|^2\}\mathbf{I}$$

Defining the average SNR as $E\{|d_k|^2\}/\sigma_n^2$, then we obtain the simplified estimator

$$\widehat{\eta} = \mathbf{R}_{\eta\eta}\left(\mathbf{R}_{\eta\eta} + \frac{\beta}{SNR}\mathbf{I}\right)^{-1}\widehat{\eta}_{ls} \tag{15.82}$$

where

$$\beta = E\{|d_k|^2\}E\{|1/d_k|^2\}$$

is a constant depending on the signal constellation. For example, $\beta = 1$ for QPSK and $\beta = 17/9$ for 16QAM. Because $\mathbf{D}$ is no longer a factor in the matrix calculation, the matrix inversion $\left(\mathbf{R}_{\eta\eta} + \frac{\beta}{SNR}\mathbf{I}\right)^{-1}$ need not be calculated each time $\mathbf{D}$ changes.

To use the LMMSE in (15.82), we need to know the channel autocorrelation matrix $\mathbf{R}_{\eta\eta} = E\{\eta\eta^H\} = [R_{pq}]$, where $R_{pq} = E\{\eta_p\eta_q^*\}$ for $p = 0, 1, ..., K-1$ and $q = 0, 1, ..., K-1$. Assume that the pth element of the observation vector $\mathbf{y}$ is from the lth symbol period and kth subcarrier, and the qth element is from the l'th symbol period and k'th subcarrier, then

$$R_{pq} = E\{H_{l,k}H^*_{l',k'}\}$$

where (see (15.19))

$$H_{l,k} = \sum_{m=0}^{M-1} h_m(lT_s)e^{-j2\pi km/N} \tag{15.83}$$

where T_s is the symbol period. Recall that there are only L nonzero terms in all M channel fading coefficients (see (15.8)), and $m = \tau_m/T_{sample} = \tau_m N/T_s$ in (15.83); thus we can write it as

$$H_{l,k} = \sum_{m=0}^{L-1} h_m(lT_s)e^{-j2\pi k\tau_m/T_s}$$

Note that in general τ_m may not be an integer multiple of T_{sample}, but can be approximated as an integer multiple of T_{sample} with little loss of accuracy. In addition, a value of m may not be unique among all paths. If that happens, the sum $\sum_{m=0}^{L-1}$ will include those repeated terms.

Then

$$R_{pq} = E\{H_{l,k}H^*_{l',k'}\} = \sum_{m=0}^{L-1}\sum_{n=0}^{L-1} E\left\{h_m(lT_s)h_n^*(l'T_s)\right\} e^{-j2\pi(k\tau_m - k'\tau_n)/T_s} \tag{15.84}$$

For mobile radio channels, $h_m(t)$s are wide-sense stationary narrow-band complex Gaussian processes, which are uncorrelated for different paths. We further assume that $h_m(t)$ has the same normalized correlation function $r_t(\Delta t)$ for all m and, hence, the same normalized power spectrum. That is

$$r_m(\Delta t) \triangleq E\{h_m(t+\Delta t)h_m^*(t)\} = \sigma_m^2 r_t(\Delta t) \tag{15.85}$$

where σ_m^2 is the average power of the mth path. Thus

$$\begin{aligned} R_{pq} &= E\{H_{l,k}H_{l',k'}^*\} = \sum_{m=0}^{L-1} E\left\{h_m(lT_s)h_m^*(l'T_s)\right\} E\left\{e^{-j2\pi(k-k')\tau_m/T_s}\right\} \\ &= \sum_{m=0}^{L-1} \sigma_m^2 r_t[(l-l')T_s] E\left\{e^{-j2\pi(k-k')\tau_m/T_s}\right\} \\ &= r_t[(l-l')T_s] \sum_{m=0}^{L-1} \sigma_m^2 E\left\{e^{-j2\pi(k-k')\tau_m \Delta f}\right\} \\ &= \sigma_H^2 r_t[(l-l')T_s] r_f[(k-k')\Delta f] \triangleq r_H(l-l', k-k') \end{aligned} \tag{15.86}$$

where $\Delta f = 1/T_s$ is the subcarrier spacing, and

$$\sigma_H^2 \triangleq \sum_{m=0}^{L-1} \sigma_m^2 \tag{15.87}$$

is the total average power of the channel impulse response, and

$$r_f[(k-k')\Delta f] = \sum_{m=0}^{L-1} \frac{\sigma_m^2}{\sigma_H^2} E\left\{e^{-j2\pi(k-k')\tau_m \Delta f}\right\} \tag{15.88}$$

Equation (15.86) shows that the 2-D autocorrelation function of the fading factors $H_{l,k}$ can be separated into two 1-D autocorrelation functions, along the time axis and frequency axis, respectively. Note that $r_t(0) = r_f(0) = 1$.

If pattern 1 in Figure 15.4 is used, all observations are in the same symbol period, $l = l'$, and the correlation is along the frequency axis

$$R_{pq} = E\{H_{l,k}H_{l,k'}^*\} = \sigma_H^2 r_f[(k-k')\Delta f] \tag{15.89}$$

If the comb-type pattern in Figure 15.6 is used, all observations are from the same frequency, $k = k'$, and the correlation is along the time axis

$$R_{pq} = E\{H_{l,k}H_{l',k}^*\} = \sigma_H^2 r_t[(l-l')T_s] \tag{15.90}$$

For Clarke's (or Jakes') mobile channel model, from (15.39) we have[2]

$$r_t[(l-l')T_s] = J_0(2\pi f_M(l-l')T_s) \tag{15.91}$$

Also according to Clarke's (or Jakes') mobile channel model [27,28], the pdf of τ is approximately exponential: $p(\tau) = \frac{1}{\sigma_\tau}e^{-\tau/\sigma_\tau}$, where σ_τ is the rms value (also mean value) of τ, and

$$\begin{aligned} E\left\{e^{-j2\pi(k-k')\tau_m\Delta f}\right\} &= \int_0^\infty \frac{1}{\sigma_\tau}e^{-\tau/\sigma_\tau}e^{-j2\pi(k-k')\tau\Delta f}d\tau \\ &= \frac{1}{1+j2\pi\sigma_\tau(k-k')\Delta f} \end{aligned}$$

Subsequently, we have

$$\begin{aligned} r_f[(k-k')\Delta f] &= \frac{1}{1+j2\pi\sigma_\tau(k-k')\Delta f}\sum_{m=0}^{L-1}\frac{\sigma_m^2}{\sigma_H^2} \\ &= \frac{1}{1+j2\pi\sigma_\tau(k-k')\Delta f} \end{aligned} \tag{15.92}$$

where (15.87) is used.

15.2.2.2 2-D Asymmetric Wiener Filters

The estimator presented in the previous section is a *symmetric estimator* in the sense that the points whose fading factors are being estimated are the points of observations. The fading factors of other points must be obtained by interpolation. However, interpolation can be incorporated in channel estimation as proposed by Hoeher et al in [19,29]. The estimator proposed in [19,29] uses decimated observation points to estimate each and every point of the OFDM signal grid. It is called an *asymmetric Wiener filter* in the sense that the number of output points is not equal to the number of input points.

The derivation of the asymmetric Wiener filter is similar to that of the LMMSE estimator. Assume that the OFDM symbol grid has a size of $N_{symbol} = K_t \times N$, where K_t is the number of symbols along the time axis, and N is the number of subcarrier frequencies along the frequency axis. Let $\{l,k\}$ denote a symbol at symbol period l and subcarrier frequency k. Observations (pilots or receiver decisions) are

[2] We introduced Clarke's model for fast fading channels in Section 15.1.3 and used its time correlation for the ICI power calculation. In fact, Clarke's model applies to both fast and slow fading mobile channels. In slow fading channels, since the maximum Doppler shift f_M is very small, the time correlation is very strong within a symbol so that the fading factor is considered constant in a symbol.

denoted by set $\mathcal{P}$ and their locations are denoted by $\{l', k'\}$. Denote the number of observations as K.

From (15.20), the observation at location $\{l', k'\}$ can be written as

$$Y_{l',k'} = d_{l',k'} H_{l',k'} + W_{l',k'}$$

or in vector form we have

$$\mathbf{y} = \mathbf{D}\boldsymbol{\eta} + \mathbf{w}$$

where $\mathbf{y}$, $\boldsymbol{\eta}$, and $\mathbf{w}$ are all $K \times 1$ vector, and $\mathbf{D}$ is $K \times K$ as defined in (15.72). This is the same as in (15.71). What is different now is that we want to use the $K \times 1$ observation vector $\mathbf{y}$ to estimate not only the K observation points, but all other points in the grid. We organize all fading factors in a vector $\boldsymbol{\eta}_1$, which is $N_{symbol} \times 1$, with each element being a fading factor $H_{l,k}$ for some $\{l, k\}$. Then following the same steps in the previous section, we obtain the LMMSE estimator given by

$$\mathbf{C}_o = \mathbf{R}_{\boldsymbol{\eta}_1 \mathbf{y}} \mathbf{R}_{\mathbf{yy}}^{-1} \tag{15.93}$$

where $\mathbf{R}_{\boldsymbol{\eta}_1 \mathbf{y}}$ is the $N_{symbol} \times K$ cross-correlation matrix between $\boldsymbol{\eta}_1$ and $\mathbf{y}$. $\mathbf{R}_{\mathbf{yy}}$ is the $K \times K$ autocorrelation matrix of the observations. Subsequently, $\mathbf{C}_o$ is $N_{symbol} \times K$; each row of it is the Wiener filter coefficients for a particular point $\{l, k\}$.

Using (15.93) we can obtain further results similar to those in the previous section. Following (15.76) we obtain

$$\mathbf{R}_{\boldsymbol{\eta}_1 \mathbf{y}} = \mathbf{R}_{\boldsymbol{\eta}_1 \boldsymbol{\eta}} \mathbf{D}^H \tag{15.94}$$

where $\mathbf{R}_{\boldsymbol{\eta}_1 \boldsymbol{\eta}} = E[\boldsymbol{\eta}_1 \boldsymbol{\eta}^H]$ is an $N_{symbol} \times K$ cross-correlation matrix between $\boldsymbol{\eta}_1$ and $\boldsymbol{\eta}$. The autocorrelation matrix $\mathbf{R}_{\mathbf{yy}}$ and its inverse are the same as before (see (15.78) and (15.79)). Thus from (15.79) and (15.94) we have

$$\widehat{\boldsymbol{\eta}}_{1,lmmse} = \mathbf{R}_{\boldsymbol{\eta}_1 \boldsymbol{\eta}} [\mathbf{R}_{\boldsymbol{\eta\eta}} + \sigma_n^2 (\mathbf{D}\mathbf{D}^H)^{-1}]^{-1} \widehat{\boldsymbol{\eta}}_{ls} \tag{15.95}$$

where $\widehat{\boldsymbol{\eta}}_{ls}$ is given by (15.81).

Comparing (15.95) with (15.80) we see that the only difference is the first matrix $\mathbf{R}_{\boldsymbol{\eta}_1 \boldsymbol{\eta}}$ in (15.95) while it is $\mathbf{R}_{\boldsymbol{\eta\eta}}$ in (15.80). Consequently, by replacing $(\mathbf{D}\mathbf{D}^H)^{-1}$ with the expectation $E\{(\mathbf{D}\mathbf{D}^H)^{-1}\}$, we obtain a result similar to (15.82) as

$$\widehat{\boldsymbol{\eta}}_1 = \mathbf{R}_{\boldsymbol{\eta}_1 \boldsymbol{\eta}} \left(\mathbf{R}_{\boldsymbol{\eta\eta}} + \frac{\beta}{SNR} \mathbf{I} \right)^{-1} \widehat{\boldsymbol{\eta}}_{ls} \tag{15.96}$$

Again, the only difference from (15.82) is the first correlation matrix in the expression.

The result in (15.93) is given in [19] in the form of a row vector of $\mathbf{C}_o$:

$$\mathbf{c}_o^T(l,k) = \mathbf{r}^T(l,k)\mathbf{R}_{\mathbf{yy}}^{-1} \tag{15.97}$$

where $\mathbf{c}_o^T(l,k)$ is the lkth row of $\mathbf{C}_o$ and $\mathbf{r}^T(l,k)$ is the lkth row of $\mathbf{R}_{\eta_1\mathbf{y}}$. Note that $\mathbf{r}^T(l,k)$ is the cross-correlation vector between the lkth fading factor and all K observations. That is

$$\mathbf{r}^T(l,k) = [E\{H_{l,k}Y_{l',k'}^*\}], \quad \forall\{l',k'\} \in \mathcal{P}$$

From (15.97), it is easy to see that we can customize the filter design for each location $\{l,k\}$. That is, for each location, an optimal subset instead of the entire set of the observations can be used for estimation. In this case $\mathbf{R}_{\mathbf{yy}}$ and $\mathbf{r}^T(l,k)$ are different for each different location $\{l,k\}$. However, in this way, the complexity of the estimator can be significantly reduced. The subset can be chosen by a criterion of "nearest" distance in some sense, as will be described in the next section.

15.2.3 Reduction of Complexity of the LMMSE Estimator

The major drawback of the LMMSE estimator is its complexity. Particularly, the matrix inversion $\left(\mathbf{R}_{\eta\eta} + \frac{\beta}{SNR}\mathbf{I}\right)^{-1}$ incurs a large number of computations when the number of observations is large.

A first step to reduce the complexity is to choose a subset of K observations with $K < N_{grid}$, where N_{grid} is the number of total available observation points in an OFDM frame. The number of observations K in the subset must not be unnecessarily large for desired BER performance. Since not all N_{grid} points are used, we can choose an optimal subset for each location $\{l,k\}$ to be estimated. Hoeher et al. [19] proposed to choose K according to either of two criteria. (1) Optimize the subset for each location so that MMSE is minimized. This optimum rule can be rather complicated for practical block sizes: $\begin{pmatrix} N_{grid} \\ K \end{pmatrix}$ possibilities must be checked. (2) Search the K observation locations $\{l',k'\}$ that are nearest to the location $\{l,k\}$ which is to be estimated. "Nearest" could be in the sense of the sum of horizontal and vertical distances, $|l-l'|+|k-k'|$ (suitable if the normalized bandwidths are similar in both dimensions; using the Euclidean distance appeared to be worse), or some form of "weighted" distances, such as $|l-l'|f_M T_s + |k-k'|\tau_{\max}\Delta F$, which was actually used in the example in [19].

Other complexity reduction strategies include decomposing the 2-D filtering into two cascaded 1-D filterings in frequency direction and time direction, respectively [19,29], a DFT-based LMMS estimator which can be applied to a 1-D LMMSE estimator [20–22], a 2-D DFT-based LMMSE estimator which can be applied to a 2-D

LMMSE estimator [23], a singular value decomposition (SVD)-based lower rank approximation which can be applied to 1-D and 2-D filtering [21], parametric channel approximation [24], or a windowed DFT-based lower rank approximation [30]. We will describe some of the important ones next.

15.2.3.1 Cascaded 1-D Filtering

Replacing the 2-D Wiener filter with two cascaded 1-D Wiener filters in time domain and frequency domain can reduce the arithmetical complexity of the filter, the data and coefficient storage, as well as the delay [29,31]. The two filters can be in arbitrary order.

The following is an example presented in [29], which can illustrate the operation of the two cascaded 1-D Wiener filters. The estimator is pilot-assisted.. The OFDM frame has $K_t = 13$ time slots (symbols) and $N = 5$ frequencies. The spacing of pilot symbols is $N_t = 3$ and $N_f = 2$, respectively. The frame structure is shown in Figure 15.7. The order of filtering is frequency-domain first and time-domain second.

1. Step 1: the modulation is removed from the observations to obtain $Y'_{l,k} = Y_{l,k}/d_{l,k}$, which is the LM estimate of $H_{l,k}$. Assume the frequency-domain filter's order is chosen as $N_k - 1$, then the output of the filter is

$$\widetilde{H}_{l,k} = \sum_{i=0}^{N_k-1} c_k(i) Y'_{l,k-iN_f}; \quad \forall k; \quad l = \text{const}; \quad \forall l \in S_l$$

   where $\{c_k(i)\}$ is the set of N_k coefficients, and S_l is the set of time indices with training symbols. For example, $S_l = \{0, 3, 6, 9, 12\}$ in Figure 15.7. For each $l \in S_l$, this filtering operation estimates $H_{l,k}$ for all values of k along the frequency axis. That is, symbols whose fading factors are estimated include the training and data symbols. Thus, the filtering is asymmetrical.
2. Step 2: for a fixed k, using $\widetilde{H}_{l,k}; \forall l \in S_l$ as inputs to the second Wiener filter in the time domain of a chosen order of $N_l - 1$, the output is

$$\widehat{H}_{l,k} = \sum_{i=0}^{N_l-1} c_l(i) \widetilde{H}_{l-iN_t,k,}; \quad \forall l; \quad k = \text{const}; \quad \forall k \in S_k$$

   where $\{c_l(i)\}$ is the set of N_l coefficients. $S_k = \{0, 1, ..., N-1\}$ is the set of frequencies. In Figure 15.7, $S_k = \{0, 1, 2, 3, 4\}$. For each $k \in S_k$, this filtering operation estimates $H_{l,k}$ for all symbols along the time axis. Again, the filtering is asymmetrical since the inputs are only from those time slots whose

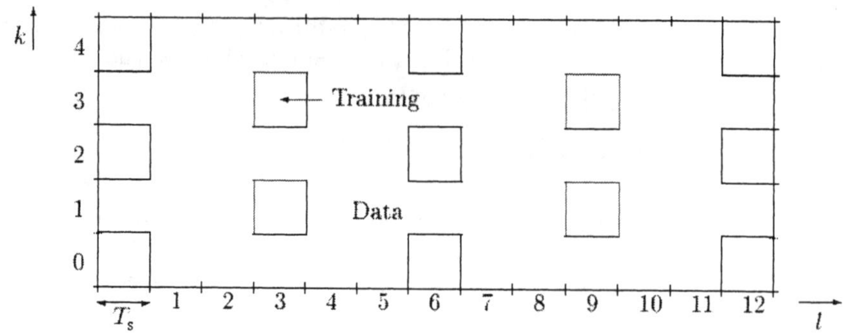

Figure 15.7 OFDM frame structure. From [29].

index $l \in S_l$, and the outputs are for all time slots.

According to [29], the coefficients $\{c_k(i)\}$ for the kth symbol are given by

$$\mathbf{c}_k^T = \mathbf{r}_k^T \mathbf{R}_k^{-1} \tag{15.98}$$

where $\mathbf{R}_k$ is the $N_k \times N_k$ autocorrelation matrix of the complex fading process of the channel, and $\mathbf{r}_k$ is the $N_k \times 1$ cross-correlation vector, which depends on the index k. No further detail was given in [29]. However we can see that $\mathbf{R}_k$ is independent of k because the channel is assumed WSSUS, and that $\mathbf{r}_k$ depends on the index k because it contains the cross-correlation values between the kth channel fading factor ($H_{l,k}$, l fixed) and N_k other surrounding $H_{l,k}$'s, l fixed. The coefficient given by (15.98) is not optimal since it is not the same as determined by (15.74) or (15.93), which eventually requires information of SNR. However, (15.98) is easier to use since it only involves a channel characteristic.

Comparison in [19,29] shows that for pilot-aided channel estimation, 2-D Wiener filtering outperforms 1-D Wiener filtering in terms of overhead and mean-square error performance. However, even though two cascaded 1-D filters are not equivalent to the 2-D filter, for a similar computation effort, the performance of $2 \times$ 1-D Wiener filtering is similar to 2-D Wiener filtering.

15.2.3.2 DFT-Based Lower Rank Approximation

For the observation symbol pattern 1 in Figure 15.4, $\boldsymbol{\eta} = [H_0, H_1, ..., H_{N-1}]^T$, is the sampled frequency response of a channel with short time duration compared to the OFDM symbol length and, hence, its associated cyclic impulse response $\mathbf{h}$ has only a few taps with significant power. If we perform the estimation in the time domain, we can reduce the complexity of the estimation by using this power concentration. This prompted the DFT-based scheme proposed by Van de Beek and Edfors et al. in [20, 32] which will be presented here. Note that this scheme is only applicable to the observation pattern 1 in Figure 15.4, which is called a block-type pattern in [18].

To derive this estimator, we rewrite (15.71) as

$$\mathbf{y} = \mathbf{DFh} + \mathbf{w} \tag{15.99}$$

where $\mathbf{Fh} = \boldsymbol{\eta} = [H_0, H_1, ..., H_{N-1}]^T$ and $\mathbf{F}$ is the DFT-matrix given by

$$\mathbf{F} = \begin{bmatrix} W_N^{00} & \cdots & W_N^{0(N-1)} \\ \vdots & \ddots & \vdots \\ W_N^{(N-1)0} & \cdots & W_N^{(N-1)(N-1)} \end{bmatrix} \tag{15.100}$$

with

$$W_N^{nk} = \exp(-j2\pi nk/N)$$

Using the same orthogonality principle given in (15.73), the LMMSE estimate of $\mathbf{h}$ is found to be

$$\widehat{\mathbf{h}}_{lmmse} = \mathbf{R_{hy}}\mathbf{R_{yy}^{-1}}\mathbf{y} \tag{15.101}$$

where

$$\begin{aligned} \mathbf{R_{hy}} &= E\{\mathbf{hy}^H\} = E\{\mathbf{h}(\mathbf{DFh}+\mathbf{w})^H\} \\ &= E\{\mathbf{h}(\mathbf{h}^H\mathbf{F}^H\mathbf{D}^H + \mathbf{w}^H)\} = E\{\mathbf{hh}^H\mathbf{F}^H\mathbf{D}^H\} + E\{\mathbf{hw}^H\} \\ &= \mathbf{R_{hh}}\mathbf{F}^H\mathbf{D}^H \end{aligned} \tag{15.102}$$

Thus

$$\widehat{\mathbf{h}}_{lmmse} = \mathbf{R_{hh}}\mathbf{F}^H\mathbf{D}^H\mathbf{R_{yy}^{-1}}\mathbf{y} \tag{15.103}$$

Further,

$$\begin{aligned} \mathbf{R_{yy}} &= E\{\mathbf{yy}^H\} = E[(\mathbf{DFh}+\mathbf{w})(\mathbf{DFh}+\mathbf{w})^H] \\ &= E[\mathbf{DFhh}^H\mathbf{F}^H\mathbf{D}^H] + E(\mathbf{ww}^H) \\ &= \mathbf{DFR_{hh}}\mathbf{F}^H\mathbf{D}^H + \sigma_n^2\mathbf{I} \end{aligned} \tag{15.104}$$

Then it can be shown (Appendix 15C)

$$\widehat{\mathbf{h}}_{lmmse} = \mathbf{R_{hh}}\mathbf{F}^H\mathbf{D}^H(\mathbf{DFR_{hh}}\mathbf{F^H}\mathbf{D^H} + \sigma_{\mathrm{n}}^2\mathbf{I})^{-1}\mathbf{y}$$

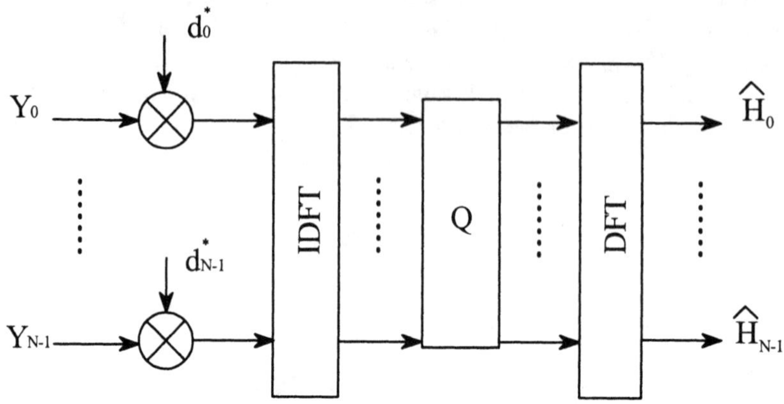

Figure 15.8 DFT-based channel estimator. From [20]. Copyright © 1995 IEEE.

$$= \mathbf{Q}\mathbf{F}^H\mathbf{D}^H\mathbf{y} \tag{15.105}$$

where

$$\mathbf{Q} = \mathbf{R_{hh}}(\mathbf{R_{hh}} + (\mathbf{F^H D^H D F})^{-1}\sigma_{\mathbf{n}}^2))^{-1}(\mathbf{F}^H\mathbf{D}^H\mathbf{D}\mathbf{F})^{-1} \tag{15.106}$$

and finally, the estimate of the frequency response is

$$\widehat{\eta}_{lmmse} = \mathbf{F}\widehat{\mathbf{h}}_{lmmse} = \mathbf{F}\mathbf{Q}\mathbf{F}^H\mathbf{D}^H\mathbf{y} \tag{15.107}$$

The estimator described by (15.107) has a structure as shown in Figure 15.8, where the processing before the IDFT block implements $\mathbf{D}^H\mathbf{y}$, and the IDFT block implements $\mathbf{F}^H$ and the DFT block realizes $\mathbf{F}$.

The above DFT-based LMMSE estimator requires the calculation of an $N \times N$ matrix $\mathbf{Q}$, which still incurs high complexity when N is large, similar to that required in the original LMMSE estimator given by (15.80). However, the DFT-based algorithm allows us to reduce the complexity by discarding the components with near zero values after the IDFT operation. Most of the energy of $\mathbf{h}$ is concentrated in, or near, the first L taps of the channel impulse response, where L is the number of samples in the guard interval (otherwise the guard interval should be modified). Therefore we can set the pqth element of $\mathbf{R_{hh}}$ $R_{pq} = 0$ for $p, q \notin [0, L-1]$, then $\mathbf{Q}$ is reduced to an $L \times L$ matrix $\mathbf{Q}'$. Denote the first L columns of the DFT-matrix $\mathbf{F}$

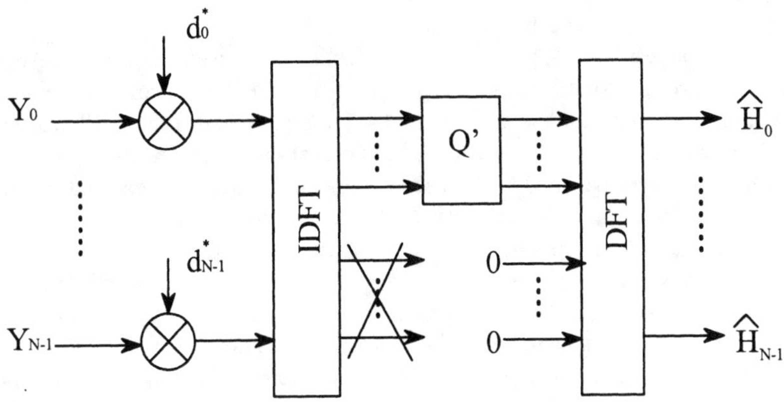

Figure 15.9 Reduced DFT-based channel estimator. From [20]. Copyright © 1995 IEEE.

as $\mathbf{T}$ and denote the upper left $L \times L$ corner of $\mathbf{R_{hh}}$ as $\mathbf{R'_{hh}}$, the modified LMMSE estimator becomes

$$\widehat{\eta}_{lmmse} = \mathbf{TQ'T}^H\mathbf{D}^H\mathbf{y} \tag{15.108}$$

where

$$\mathbf{Q}' = \mathbf{R'_{hh}}(\mathbf{R'_{hh}} + (\mathbf{T^H D^H DT})^{-1}\sigma_{\mathrm{n}}^2))^{-1}(\mathbf{T}^H\mathbf{D}^H\mathbf{DT})^{-1} \tag{15.109}$$

The modified estimator is shown in Figure 15.9. Since L is usually a small fraction of N, the complexity modified estimator will decrease considerably.

The most salient drawback of the DFT-based lower-rank approximation is the error floor caused by energy leakage. Recall that the channel fading factor vector η is the DFT of channel impulse response $\mathbf{h}$. When the channel delay τ_m is sample-spaced, according to DFT theory [33], no leakage of energy to taps outside the interval $[0, L]$ will occur and the estimator in (15.108) will not lose any information about the channel. Thus the modified estimator (15.108) is equivalent to the LMMSE estimator (15.107). However, more generally, the channel delay profile is nonsample-spaced or even continuous. Although L is chosen to be longer than the maximum delay, the channel fading factor vector η will have energy on all taps, not only the taps in the interval $[0, L]$. That is, there is energy leakage to taps outside the interval $[0, L]$. In this case, the estimator in (15.108) will lose information about the channel and its performance will be degraded compared to the LMMSE estimator (15.107). To reduce the loss, the number of taps maybe chosen to be longer than L, and a trade-off between complexity and performance is needed.

Simulation results in [20] show that LMMSE is always better than LE (least error estimator), and the more the taps are used the better the performance, as expected. If the number of taps is too small, an irreducible error floor of the BER will exist, which is due to the energy leakage to the taps that are not included in the filter.

Other more sophisticated DFT-based channel estimators are reported in the literature. In [22], the estimator uses both the frequency-domain and time-domain correlations, so that it performs better than the above estimator from [20], which uses only frequency-domain correlation. Yet its complexity can be reduced by the DFT-based algorithm with small eigenvalues of the frequency-domain correlation being discarded. In [23], an asymmetrical MMSE filter or interpolator is proposed and it uses the pilot symbols to estimate fading factors on all symbols. The interpolator is implemented through a two-dimensional FFT and IFFT pair. Robust designs which are insensitive to channel statistics are presented in both papers. In [30], a windowed DFT-based MMSE channel estimator is proposed. A Hanning window is first applied to the channel observation vector before it is processed by a DFT-based estimator. Windowing reduces the effect caused by the energy leakage in the channel observation vector.

15.2.3.3 SVD-Based Lower Rank Approximation

To avoid energy leakage in a DFT-based lower-rank approximation, Edfors et al. proposed an SVD-based optimal lower-rank approximation [21]. SVD is the singular value[3] decomposition [25] of the $\mathbf{R}_{\eta\eta}$ matrix

$$\mathbf{R}_{\eta\eta} = \mathbf{U}\mathbf{\Lambda}\mathbf{U}^H$$

where $\mathbf{U}$ is a unitary matrix (i.e., $\mathbf{U}\mathbf{U}^H = \mathbf{I}$) containing the singular vectors of $\mathbf{R}_{\eta\eta}$ and $\mathbf{\Lambda}$ is a diagonal matrix containing the singular values of $\mathbf{R}_{\eta\eta}$ $\lambda_1 \geq \lambda_2 \geq ... \geq \lambda_N$ on its diagonal. The pattern 1 in Figure 15.4 is assumed. That is

$$\mathbf{y} = [Y_0, Y_1, ..., Y_{N-1}]^T$$

and

$$\boldsymbol{\eta} = [H_0, H_1, ..., H_{N-1}]^T \tag{15.110}$$

[3] Consider an $N \times p$ matrix $\mathbf{H}$ and its Gram matrix $\mathbf{G} = \mathbf{H}^H\mathbf{H}$. The Gram matrix is Hermitian and nonnegative definite. If the eigenvalues of $\mathbf{G}$ are denoted as $\lambda_1^2, ..., \lambda_p^2$, then $\lambda_1, ..., \lambda_p$ are the singular values of $\mathbf{H}$ [25].

The optimal rank-p estimator is shown to be

$$\widehat{\eta}_p = \mathbf{U}\mathbf{\Delta}_p\mathbf{U}^H\widehat{\eta}_{ls} \tag{15.111}$$

where $\mathbf{\Delta}_p$ is a diagonal matrix with entries

$$\delta_k = \begin{cases} \frac{\lambda_k}{\lambda_k+\frac{\beta}{SNR}}, & k = 1, 2, .., p \\ 0, & k = p+1, .., N \end{cases} \tag{15.112}$$

Parameter p should be chosen to maintain acceptable estimation error. The smaller the p, the smaller the complexity, but the bigger the estimation error. It was shown that the magnitude of $\mathbf{R}_{\eta\eta}$ becomes small after about $L+1$ values, where L is the number of samples in the guard interval. This makes $\lambda_k \approx 0$ for $k > L$. Thus we can expect a good approximation when p is in the range of L, which is usually much smaller than the number of subcarriers N.

Simulation results in [21] show that the SVD-based lower-rank estimator's SER (symbol error rate) floor decreases rapidly when the rank p is greater than the guard interval length L. For an example in [21], by choosing $p = 8$ for $L = 4$, the SER curve exhibits an SER error floor of only 10^{-6}, which is better than the DFT-based lower-rank estimator.

Even though the above is presented in the context of a 1-D estimator (frequency domain), the method is obviously applicable to a 2-D LMMSE estimator, since the simplification process only involves the SVD of the channel correlation matrix. However, in [34], it is shown that such an estimator gives an inferior performance for a fixed complexity. Therefore it seems that separating the use of the frequency-domain correlation and time-domain correlation is the most efficient way of estimating the channel, as was pointed out in [29].

15.2.4 Estimator for Fast Fading Channels

For fast fading channels, the output of the DFT demodulator not only has a multiplicative fading factor $\eta_{k,k}$ (for frequency selective channel) or H_0 (for frequency flat channel) on the data, but also an ICI term c_k (see (15.44), (15.45), (15.63), and (15.64)). Both $\eta_{k,k}$ (or H_0) and c_k are determined by the time-domain channel impulse response matrix $\mathbf{H}$ (or vector $\mathbf{h}$ for frequency flat channel). The frequency flat case can be treated as a special case of the frequency selective case.

Some methods in the literature estimate $\eta_{k,k}$ and equalization is performed by a one-tap equalizer as for the case of slow fading channels. For example, gain $\eta_{k,k}$ is estimated using MMSE in [35]. In [36], $\eta_{k,k}$ is estimated using a 2D- or 1D-Wiener estimator which satisfies the MMSE criterion, too. Some other methods estimate the channel matrix (all $\eta_{k,i}$, for $i, k = 1, .., N$) and use it to estimate the data [37,38] or to reduce the ICI by substracting the estimated ICI from the received signal [39]. A method using linear approximation of channel variation for ICI reduction is proposed in [39], which outperforms the algorithms also based on the linear approximation described in [37,38]. The above are frequency-domain estimations.

Channel estimation can also be done in time domain. Time-domain equalization using linear phase interpolation (LPI) for **H** is described in [40]. **H** is estimated using MMSE in [41]. In [42], time-domain LE of **H** is obtained from pilots and LMMSE is used to reduce the noise of the LE and interpolation is used to estimate **H**.

In the following, we describe the method presented in [39] due to its low complexity and better performance than other similar algorithms in [37,38].

Adding a noise term W_k to (15.62) we can write the output of the OFDM demodulator as

$$Y_k = \eta_{k,k} d_k + \sum_{\substack{i=0 \\ i \neq k}}^{N-1} \eta_{k,i} d_i + W_k \tag{15.113}$$

where $\eta_{k,i}$ is given by (15.61) and can be written as

$$\begin{aligned} \eta_{k,i} &= \frac{1}{N} \sum_{m=0}^{M-1} \sum_{n=0}^{N-1} h_{m,n} \exp(-j2\pi \frac{i(m-n)+kn}{N}) \\ &= \frac{1}{N} \sum_{n=0}^{N-1} \left[\sum_{m=0}^{M-1} h_{m,n} \exp(-j2\pi \frac{im}{N}) \right] \exp(j2\pi \frac{-n(k-i)}{N}) \end{aligned}$$

Defining a *time variant transfer function* as

$$V_{i,n} = \sum_{m=0}^{M-1} h_{m,n} \exp(-j2\pi \frac{im}{N}) \tag{15.114}$$

Then we have the *channel matrix* in the frequency domain as

$$\eta_{k,i} = \frac{1}{N} \sum_{n=0}^{N-1} V_{i,n} \exp(j2\pi \frac{-n(k-i)}{N}) \tag{15.115}$$

Note that the average of $V_{i,n}$ during one OFDM symbol is

$$\overline{V}_k = \eta_{k,k} = \frac{1}{N} \sum_{n=0}^{N-1} V_{k,n} \tag{15.116}$$

If the time variation of $V_{i,n}$ during one OFDM symbol is small, a linear interpolation on each subcarrier i can be used as an approximation. For applying the approximation, at least one value of $V_{i,n}$ during each symbol period must be known. From a conventional one-tap channel estimation (that is, dividing Y_k of a pilot symbol by the known data d_k on the pilot symbol) we can obtain an estimate of $\eta_{k,k}$ which is an estimate of $\overline{V}_k$ due to the relation in (15.116). It is shown in [37] that it is a good choice to use the value of $\overline{V}_k$ as a representation of $V_{k,n}$ in the middle of the symbol. The proposed linear model of one OFDM symbol is then [39]

$$V_{k,n} \approx \overline{V}_k + \overline{V}'_k(n - \frac{N-1}{2}) \tag{15.117}$$

where $\overline{V}'_k$ is the derivative of $\overline{V}_k$. This derivative can be approximately calculated from $\overline{V}_k$ value of the previous symbol and the next symbol with

$$\overline{V}'_k = \frac{\overline{V}_k^{(next)} - \overline{V}_k^{(prev)}}{2N_s}$$

where N_s is the length of one OFDM symbol including the guard interval. Note that processing must be delayed by one symbol time since we need $\overline{V}_k^{(next)}$ to compute $\overline{V}'_k$.

Inserting (15.115) and (15.117) into (15.113) we have

$$Y_k = \overline{V}_k d_k + \sum_{\substack{i=0\\ i\neq k}}^{N-1} \left[\frac{1}{N}\sum_{n=0}^{N-1}\left(\overline{V}_i + \overline{V}'_i(n - \frac{N-1}{2})\right)\exp(j2\pi\frac{-n(k-i)}{N})\right] d_i + W_k \tag{15.118}$$

where the second term is

$$\begin{aligned}
&\sum_{\substack{i=0\\ i\neq k}}^{N-1} \frac{1}{N}\overline{V}_i \sum_{n=0}^{N-1} \exp(j2\pi\frac{-n(k-i)}{N})d_i \\
&+ \sum_{\substack{i=0\\ i\neq k}}^{N-1} \overline{V}'_i \frac{1}{N}\sum_{n=0}^{N-1} n\exp(j2\pi\frac{-n(k-i)}{N})d_i \\
&- \sum_{\substack{i=0\\ i\neq k}}^{N-1} \frac{1}{N}\overline{V}'_i \frac{N-1}{2}\sum_{n=0}^{N-1} \exp(j2\pi\frac{-n(k-i)}{N})d_i
\end{aligned} \tag{15.119}$$

Using the fact that

$$\sum_{n=0}^{N-1} \exp(j2\pi \frac{ni}{N}) = \begin{cases} N, & i = 0 \bmod N \\ 0, & i \neq 0 \bmod N \end{cases}$$

we can identify that the first and the third terms in (15.119) are both zero. Further, we define ζ_i as

$$\zeta_i = \frac{1}{N} \sum_{n=0}^{N-1} n \exp(j2\pi \frac{ni}{N}) = \begin{cases} (N-1)/2, & i = 0 \bmod N \\ -[1 - \exp(j\frac{2\pi}{N}i)]^{-1}, & i \neq 0 \bmod N \end{cases} \tag{15.120}$$

then the second term in (15.119) is

$$\sum_{\substack{i=0 \\ i \neq k}}^{N-1} \overline{V}'_i \frac{1}{N} \sum_{n=0}^{N-1} n \exp(j2\pi \frac{-n(k-i)}{N}) d_i = \sum_{i=0}^{N-1} \overline{V}'_i \zeta_{i-k} d_i - \overline{V}'_k \zeta_0 d_k \tag{15.121}$$

Thus we have

$$Y_k = \overline{V}_k d_k + \sum_{i=0}^{N-1} \overline{V}'_i \zeta_{i-k} d_i - \overline{V}'_k \zeta_0 d_k + W_k \tag{15.122}$$

Defining

$$\Xi = \begin{bmatrix} 0 & \zeta_1 & \cdots & \zeta_{N-1} \\ \zeta_{-1} & 0 & \cdots & \zeta_{N-2} \\ \vdots & \vdots & \ddots & \vdots \\ \zeta_{1-N} & \zeta_{2-N} & \cdots & 0 \end{bmatrix}$$

$$\overline{\mathbf{v}} = [\overline{V}_0, \overline{V}_1, ..., \overline{V}_{N-1}]^T$$

$$\overline{\mathbf{v}}' = [\overline{V}'_0, \overline{V}'_1, ..., \overline{V}'_{N-1}]^T$$

$$\mathbf{w} = [W_0, W_1, ..., W_{N-1}]^T$$

$$\mathbf{d} = [d_0, d_1, ..., d_{N-1}]^T$$

$$\mathbf{y} = [Y_0, Y_1, ..., Y_{N-1}]^T$$

Equation (15.122) can be written in a vector form as

$$\mathbf{y} = [\text{diag}(\overline{\mathbf{v}}) + \Xi\text{diag}(\overline{\mathbf{v}}')]\,\mathbf{d} + \mathbf{w} \tag{15.123}$$

where diag$(\mathbf{z})$ is a diagonal matrix composed of the elements of vector $\mathbf{z}$. Note that there are zeros instead of ζ_0 on the diagonal of Ξ, which corresponds to the fact that the term $\overline{V}'_k\zeta_0 d_k$ in (15.122) is subtracted from the sum $\sum_{i=0}^{N-1}\overline{V}'_i\zeta_{i-k}d_i$. That is, no ζ_0 is used in the calculation of $\mathbf{y}$.

The estimated channel matrix of size $N \times N$ is therefore

$$\widehat{\mathbf{\Lambda}} = \text{diag}(\overline{\mathbf{v}}) + \Xi\text{diag}(\overline{\mathbf{v}}') \tag{15.124}$$

which contains the elements $\eta_{k,i}$. Since Ξ is only related to OFDM parameters, it can be pre-calculated only once at initialization. The elements of $\overline{\mathbf{v}}$ are $\eta_{k,k}$ which can be estimated using a one-tap estimator from the received pilot signal. The elements of $\overline{\mathbf{v}}'$ are estimated using (15.117).

Once $\widehat{\mathbf{\Lambda}}$ is calculated, it can be inverted to equalize $\mathbf{y}$ to estimate the data

$$\widehat{\mathbf{d}} = \widehat{\mathbf{\Lambda}}^{-1}\mathbf{y} = \mathbf{d} + \widehat{\mathbf{\Lambda}}^{-1}\mathbf{w}$$

However, when N is large, inverting $\widehat{\mathbf{\Lambda}}$ incurs a big computational complexity. A more computationally efficient way of equalization is to cancel the ICI from $Y_k = \eta_{k,k}d_k + \sum_{\substack{i=0\\i\neq k}}^{N-1}\eta_{k,i}d_i + W_k$ and then use the one-tap equalizer to recover the data symbol. The ICI term $\sum_{\substack{i=0\\i\neq k}}^{N-1}\eta_{k,i}d_i$ involves other symbols on other subcarriers. Some of them may be pilot symbols and the rest may be data symbols. Pilot symbols are known to the receiver, but data symbols must be estimated. The temporary estimation of data symbols can be obtained by an MMSE equalization of the received symbols

$$\widetilde{d}_i = \frac{Y_i\widehat{\eta}^*_{i,i}}{|\widehat{\eta}_{i,i}|^2 + 1/SNR}$$

and the resulting ICI cancellation scheme is

$$\widehat{Y}_k = Y_k - \overbrace{\sum_{\substack{i=0\\i\neq k,p(i)=0}}^{N-1}\widehat{\eta}_{k,i}\widetilde{d}_i}^{\text{data symbols}} - \overbrace{\sum_{\substack{i=0\\i\neq k,p(i)=1}}^{N-1}\widehat{\eta}_{k,i}c_i}^{\text{pilot symbols}}$$

where c_i are known pilot symbols and $p(i)$ indicates pilot positions. The final data estimate is given by a one-tap equalizer as

$$\widehat{d}_k = \frac{\widehat{Y}_k}{\widehat{\eta}_{k,k}}$$

15.2.5 Decision-Directed Channel Estimation

Decision-directed channel estimation (DDCE) is an alternative to the pilot-assisted channel estimation. It uses previous demodulator decisions to help estimate channel fading factors of the current symbol period. DDCE can also be combined with pilot-assisted channel estimation to improve channel estimation. Contributions to DDCE or combined pilot-assisted and DDCE are reported in [43,44]. The methods of DDCE can be categorized into three groups: iterative filtering [43, 45], low-pass filtering [46–48], and LMMSE filtering [49, 50]. However, the first step of all of them is the same: dividing the received symbols with the corresponding demodulator decisions to obtain the raw estimates, which then are sent to the filter that follows to produce the final estimates. The low-pass filtering methods have no special features from ordinary low-pass filtering and we will not elaborate on them here. Interested readers may refer to corresponding articles for detail. We will present the iterative method next and briefly discuss the LMMSE method, which is largely the same as that for pilot-assisted channel estimation.

Wilson et al. [43] (1994) and Ran et al. [45] (2003) use the iterative method. The channel fading factor for the lth symbol period on the kth subcarrier is estimated by

$$\widehat{H}_{l,k} = \gamma \widetilde{H}_{l,k} + (1-\gamma)\widehat{H}_{l-1,k}$$

where

$$\widetilde{H}_{l,k} = \frac{Y_{l,k}}{\widehat{d}_{l,k}}$$

and $\widehat{d}_{l,k}$ is the data symbol estimate which is the decision device output whose input is the equalized received symbol

$$\widetilde{d}_{l,k} = \frac{Y_{l,k}}{\widetilde{H}_{l-1,k}},$$

$\gamma << 1$ is a positive updating factor which is given in the range $(0, 0.9)$ in [43]. However, in [45], for M-ary QAM, it is given by

$$\gamma = \gamma_{\max}\left[\frac{p(d/Y) - 1/M}{1 - 1/M}\right]^{\alpha}$$

where $\alpha = 1.5$ and $\gamma_{\max} = 0.2$. $p(d/Y)$ is the a posteriori probability of the detected symbol.

The advantage of this interactive estimator is its simplicity. Iteration is performed independently for each subcarrier. No channel statistics such as correlation function or signal-to-noise ratio are needed. Iteration is performed in time domain and time-domain correlation is implicitly utilized. However, on the other hand, this method has not reached potential performance since the frequency-domain correlation of the channel is not utilized.

Frenger et al describe a 2-D LMMSE decision-directed channel estimator which makes use of both time-domain and frequency-domain correlations [49]. To estimate the channel fading factors for the kth symbol period on all subcarriers, the estimator takes as inputs previously received symbols in K symbol periods on all subcarriers; the previous decisions on all these symbols are used as the divisors to obtain the LS estimate in (15.81). Then the LS estimate is used to obtain the LMMSE estimate. The rest of the estimator is the same as the pilot-assisted LMMSE estimator. In fact, when we presented the LMMSE estimators and their simplified approximations, we said that they can accept pilot symbols as well as data symbols. Thus, all estimators presented earlier can be used for DDCE.

In DDCE, since the divisors are not from pilot symbols, they are not as reliable, so the estimator's performance will suffer from error propagation. However, it does not require the overhead of the pilot-assisted estimators. Moreover, when there is a need to track fast channel variations, sparsely inserted pilots are insufficient. In this case, the decision-directed channel estimation will be the method of choice. To overcome the shortcomings of DDCE and pilot-assisted channel estimation, the two methods can be combined. Schemes in this regard are reported in [51–55].

15.3 OTHER REMEDIAL MEASURES AGAINST FADING

Equalization using the results of channel estimation is the most important measure against fading. There are other methods of combating fading effects. Some of them do not require channel estimation, so system complexity is reduced. Some others use more advanced schemes so that system performance is improved. In this section we will briefly introduce these methods.

15.3.1 Differential Detection

For single-carrier systems, as described in Chapter 4, differential modulations, such as DBPSK, DQPSK, and $\pi/4$-DQPSK, can be used to replace the coherent modulations in fading channels where carrier synchronization is difficult to achieve. Even though for a fixed BER performance differential detection requires 2 to 3 dB more SNR, the system complexity is reduced, and more importantly, the system becomes robust against channel impairments. Differential modulations can also be used for OFDM systems in fading channels. Due to the fact that OFDM is a two-dimensional modulation, differential encoding and detection can be done in the time domain and the frequency domain [26].

Just like single-carrier systems, time-domain differential encoding and detection are performed between consecutive symbols for each subcarrier separately. Differential encoding makes the lth symbol and the $(l+1)$th symbol have the phases of θ_l and $\theta_{l+1} = \theta_l + \phi_l$, respectively, where ϕ_l represents information bits. When they are received, their phase becomes $\theta_l + \beta_l$ and $\theta_l + \phi_l + \beta_{l+1}$, respectively, where β_l and β_{l+1} are the phase disturbances at the lth symbol and the $(l+1)$th symbol, respectively. The differential detector compares phases of two consecutive symbols and their phase difference is $\phi_l + (\beta_{l+1} - \beta_l)$, where $(\beta_{l+1} - \beta_l)$ is the phase noise which affects the detection error probability. When the channel is slow fading, $\beta_{l+1} \approx \beta_l$ and the phase noise is negligible. When the channel is fast fading, the phase noise will cause an increase in BER. Thus the OFDM signal symbol period should be small relative to the channel coherence time, or equivalently, the symbol rate should be higher than the maximum Doppler frequency shift (see (15.91) and Section 11.1).

Frequency-domain differential encoding and detection are performed between consecutive subcarriers for each symbol period. Similar to the time-domain case, there is phase noise $(\beta_{k+1} - \beta_k)$ in the differential detection result, where β_k and β_{k+1} are the phase disturbances on the kth subcarrier and $(k+1)$th subcarrier, respectively. To ensure small phase noise, subcarrier spacing should be small relative to the channel coherence bandwidth, or equivalently, the spacing should be much smaller than the $1/\sigma_\tau$, where σ_τ is the rms delay spread.

According to [26], both time-domain and frequency-domain differential detection have a maximum loss of 3 dB in SNR with respect to ideal coherent detection. However, there is also loss in SNR in practical coherent detection with respect to the ideal coherent detection because of imperfect channel estimation and because part of the signal power is spent on pilots. Thus the net loss of differential detection versus coherent detection is typically within the range of 1 to 2 dB.

15.3.2 Diversity Reception

In Chapter 11 we briefly discussed diversity reception techniques for single-carrier modulations in fading channels. They include space diversity, polarization diversity, frequency diversity, time diversity, and multipath diversity. All these diversity techniques can be used for OFDM. However, OFDM is a multi-carrier modulation;

due to its two-dimensional nature, diversity in time and frequency can be achieved separately, or simultaneously through error-control codes. The term coded OFDM (COFDM) is for OFDM schemes that are equipped with error-control codes in time, or frequency, or both. Some OFDM schemes exploit the spread signal energy in the multiple paths in the frequency-selective channel (multipath diversity). These schemes are called vector OFDM (VOFDM). If multiple antennas are equipped, then space-time, or space-frequency, and even space-time-frequency codes can be used to achieve various diversities, including multipath diversity. These schemes are called MIMO (multi-input-multi-output)-OFDM schemes. VOFDM can be considered as a special case of MIMO-OFDM. For VOFDM interested readers may refer to [56–59]. We will discuss COFDM and MIMO-OFDM next. A lot of research results of these schemes have been reported in the literature. However, a detailed description of them is beyond the scope of this book and only a brief introduction is given here.

15.3.3 Coded OFDM (COFDM)

Error-control coding is widely used in OFDM systems. Coded OFDM (COFDM) has been used in digital audio broadcasting [60], digital video broadcasting [61], and radio LANs like IEEE 802.11 and HIPERLAN [26].

Coding in single-carrier systems is a type of time diversity. However, in OFDM systems, coding is typically a type of frequency diversity. For frequency-selective channels, the channel frequency response may exhibit deep fades on some frequencies and strong amplitudes on other frequencies. When coded symbols are assigned to the subcarriers, in the receiver the information link provided by coding will make the correct symbols on the strong subcarriers help recover the erroneous symbols on the weaker subcarriers. Codes used for OFDM are mostly convolutional codes [62]. Some systems use concatenated Reed-Solomon and convolutional codes, such as the DVB-T (digital video broadcasting-terrestrial) system [61], and some systems use trellis coded modulation (TCM) [63].

For frequency-selective slow fading channels, the channel frequency response is reflected by the post-DFT fading factors which are the DFT of the channel impulse response (see Section 15.1.2.2). Since the frequency response of a typical channel is continuous, the adjacent subcarriers will have similar fading factors. A long sequence of small subcarrier fading factors will cause a decoding error since all codes have limited error-correction capability. To avoid this problem, consecutive code symbols should not be assigned to adjacent subcarriers. Thus, interleaving is often used in conjunction with COFDM [64,65].

Interleaving in OFDM can be done at bit level before modulation mapping [61, 64] or at symbol level after modulation mapping [65]. In the latter method, coded bits are mapped to modulation constellation points (coded symbols) first, then coded symbols are interleaved by any proper interleaving method (such as block (periodic) interleaving or convolutional interleaving [66]). The interleaved symbols are assigned to subcarriers. In the receiver, the symbols at the output of the DFT demodulator are first deinterleaved before they are sent to the decoder. Deinterleaving spreads consecutive weak symbols to multiple codewords containing isolated erroneous symbols which can be corrected by the error-control code.

From the above we can see that the role of interleaving for OFDM is different from the role of interleaving for single-carrier systems. For single-carrier transmission, interleaving is to combat bursty errors in the time domain. For OFDM, interleaving, whether at bit level or symbol level, has the effect of combating bursty errors in the frequency domain. Time-domain bursty interference has minimal effect on OFDM transmission since the bursty interference is a wide-band interference which has very little energy in each subband. In fact, this is one of the advantages of OFDM, as stated in the very beginning of Chapter 12. However, a time-domain bursty interference affects one or more OFDM symbols which contain many data symbols, deinterleaving in frequency domain spreads them to different codewords which can be corrected; thus interleaving in frequency domain also has the effect of combating time-domain bursty interference. To further reduce the effect of time-domain bursty interference, time-domain interleaving at sample level can be used [63].

15.3.4 MIMO-OFDM

Multiple-input-multiple-output (MIMO) wireless links have recently emerged as one of the most significant breakthroughs in modern digital communications. MIMO systems (single or multiple carriers) are described in two tutorial papers by Gesbert et al. [67] and Paulraji et al. [68]. MIMO-OFDM in particular is described in another tutorial paper by Stüber et al. [69].

In Chapter 11, we described the structure, the operation principle, and the capacity of a MIMO system. We also introduced the key techniques to exploit the high channel capacity of the MIMO wireless link: space-time diversity coding and spatial multiplexing. Similarly, space-time diversity coding and spatial multiplexing are also key techniques to achieve the capacity potential in a MIMO-OFDM system.

Figure 15.10 illustrates a typical $Q \times L$ MIMO-OFDM system [69]. The MIMO encoder performs space-time coding of one or more blocks of data symbols to produce Q blocks of coded symbols. Then they are sent to Q OFDM modulators, one for each block. Each OFDM modulator is connected to an antenna. In the receiver, each antenna is followed by an OFDM demodulator. Assuming the OFDM has N subcarriers, the OFDM transforms a frequency-selective channel into a collection of N separate flat-fading channels. With this $Q \times L$ MIMO-OFDM configuration, a

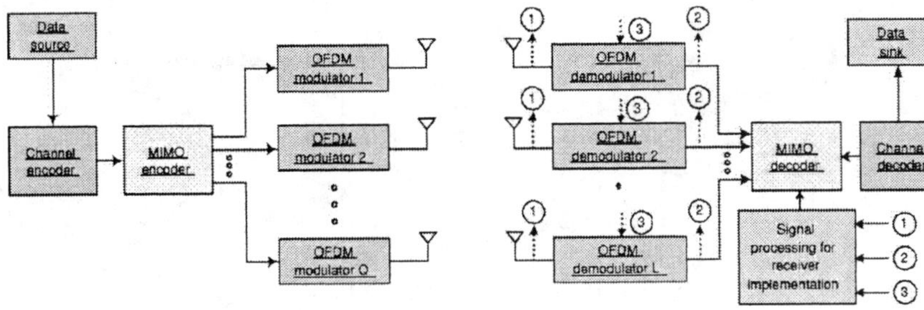

Figure 15.10 $Q \times L$ MIMO-OFDM system, where Q and L are the numbers of inputs and outputs, respectively. From [69]. Copyright © 2004 IEEE.

MIMO frequency-selective channel is transformed into a collection of N flat-fading MIMO channels, one for each subcarrier, with each having dimension $Q \times L$. As an example, Figure 15.11 shows a MIMO-OFDM system proposed in [70]. It has four transmitting antennas and p receiving antennas. In this case, at time n, each of two data blocks, $\{b_i[n, k] : k = 0, 1, ..., N-1, i = 1, 2\}$, is space-time coded into two different blocks $\{t_{2(i-1)+j}[n, k] : k = 0, 1, ..., N-1, j = 1, 2, i = 1, 2\}$. The OFDM signal for the ith transmitting antenna is modulated by $t_i[n, k]$ at the kth subcarrier of the nth OFDM block.

In the above configuration, space-time coding is performed on the data of subcarriers before modulation. Thus we say it employs a subcarrier-based space diversity. While it is effective since each subcarrier undergoes only flat-fading, the complexity of the system is great due to the multiple DFT blocks. To reduce the complexity a symbol-based space diversity is proposed for MIMO-OFDM [71]. It requires only one DFT block. Space-time coding is performed on the OFDM symbols after OFDM modulation. Figure 15.12 shows the system. Both systems in Figures 15.10 and 15.12 achieve the same order of diversity.

15.4 SUMMARY

In this chapter, we first studied the performance of OFDM in various fading channels. In a frequency flat slow fading channel, the channel effect is a constant gain on all subcarriers of the OFDM symbol at the output of the DFT demodulator. This constant gain is the same as the channel gain h. However this gain is random, varying from one symbol to next. For each subcarrier, this situation is the same as that for single-carrier modulations. Without correcting h, the BER expressions developed in Chapter 11

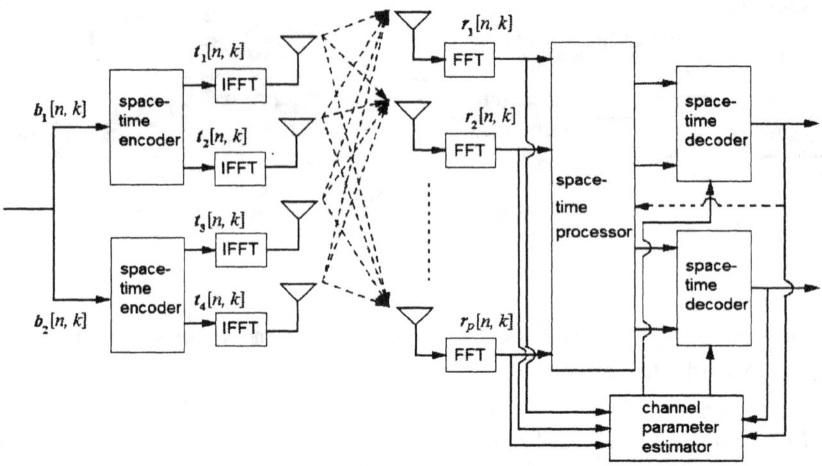

Figure 15.11 A MIMO-OFDM system. From [70]. Copyright © 2002 IEEE.

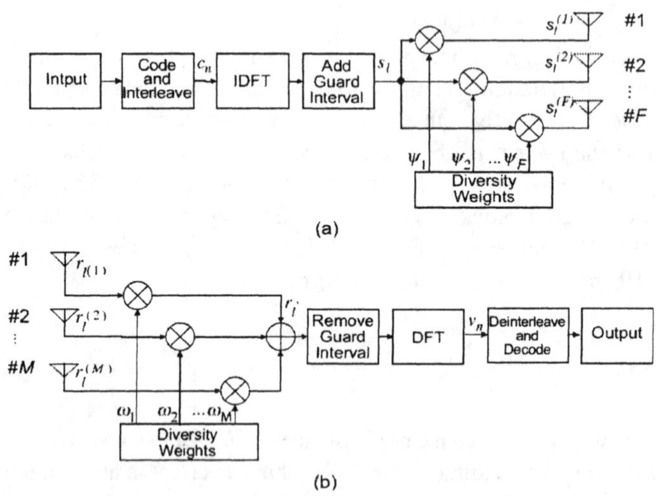

Figure 15.12 MIMO-OFDM using symbol-based space diversity. From [71]. Copyright © 2004 IEEE.

apply.

Frequency-selective slow fading is the most common model for mobile communications. Therefore we have studied this channel in a great detail. We studied its time domain (i.e., before DFT demodulation) statistics and frequency domain (i.e., after DFT demodulation) statistics. The frequency-domain subcarrier gain H_k is the Fourier transform of the channel impulse response. For uncorrelated multiple paths, we have the following three conclusions. (1) If the multiple paths are all Rayleigh in the time domain, then H_k is also Rayleigh for all subcarriers and all of them have the same power. (2) If in the time domain, one path is Rician and the rest are Rayleigh, then in the frequency domain, all subcarriers are Rician with the same K factor, but it is lower than the K factor in the time domain. (3) If there are multiple Rician and Rayleigh paths, frequency-domain signals are all Rician and their strength is nonuniformly distributed across subcarriers. For correlated multiple paths, frequency-domain OFDM subcarriers will always experience nonuniform fading in all cases, regardless of the distribution of Rician and Rayleigh paths. In all of the above cases in frequency-selective slow fading, once H_k is determined, without correcting H_k, the BER evaluation is similar to that for the single-carrier case for each subcarrier.

For the frequency-flat fast fading channel, all subcarriers have the same frequency-domain gain H_0 which is the time average of the channel impulse response. However, besides the multiplicative gain H_0 on the signal, there is an additive ICI term which acts as an additional noise term. An expression for signal to interference ratio (SIR) was derived.

For the frequency-selective fast fading channel, each subcarrier has a different frequency-domain gain $\eta_{k,k}$ which is the Fourier transform of the time averages of the channel impulse response. Again, as in the frequency-flat fast fading case, there is an additive ICI term. An expression of SIR was derived and it turns out to be the same as in the frequency-flat fast fading case.

After performance was discussed, we studied channel estimation. The emphasis was on the pilot-assisted channel estimation for frequency-selective slow fading channels. We discussed in great detail the LMMSE (Wiener) estimator, including the symmetrical Wiener estimator where the channel gains to be estimated are the gains on the pilot symbols, and the asymmetrical Wiener estimator where the channel gains to be estimated are not only the gains on the pilot symbols but also the gains on the data symbols. The asymmetrical Wiener estimator incorporates interpolation into the channel estimation.

The major problem of the Wiener estimator is its complexity. We discussed in detail the methods to reduce the complexity of the LMMSE estimator, including cascaded 1-D filtering, DFT-based lower rank approximation, and SVD-based lower rank approximation. Estimation for fast fading channels and decision-directed channel estimation were also discussed.

Finally, we briefly presented remedial measures other than channel estimation and equalization that use the results of the channel estimation. These methods include differential detection, diversity reception, channel coding, and MIMO link and space-time coding that works together with MIMO link.

APPENDIX 15A DERIVATION OF (15.39) AND (15.47)

Autocorrelation function of the complex channel gain:

The autocorrelation of $h(t)$ is derived here. The autocorrelation is

$$R(\xi) = E\{h(t)h^*(t+\xi)\} = E\{\alpha(t)\alpha(t+\xi)\exp(j\theta(t) - j\theta(t+\xi))\}$$

since $\alpha(t)$ and $\theta(t)$ are independent, we have

$$R(\xi) = E\{\alpha(t)\alpha(t+\xi)\}E\{\exp(j\theta(t) - j\theta(t+\xi))\}$$

Now we assume that in a short period (such as in a symbol interval), the Doppler shift and the path delay remain approximately constant, i.e., $f_D(t) = f_D$, and $\tau(t) = \tau$, then

$$\begin{aligned}\theta(t) - \theta(t+\xi) &= -2\pi[(f_c + f_D)\tau - f_D t] + 2\pi[(f_c + f_D)\tau - f_D(t+\xi)] \\ &= -2\pi f_D \xi\end{aligned}$$

Thus

$$\begin{aligned}R(\xi) &= E\{\alpha(t)\alpha(t+\xi)\}E\{\exp(-2\pi f_D \xi)\} \\ &= E\{\alpha(t)\alpha(t+\xi)\}\{E[\cos 2\pi f_D \xi] - jE[\sin 2\pi f_D \xi]\}\end{aligned}$$

When the mobile unit is moving at a speed v, the Doppler shift f_D is modeled as a random process and given by

$$f_D(t) = f_M \cos\varphi(t)$$

where $f_M = v/\lambda_c$ and λ_c is the carrier wavelength. $\varphi(t)$ is the angle of incidence of the plane wave. For a short period, $\varphi(t)$ can be considered constant, thus

$$f_D = f_M \cos\varphi$$

and φ is uniformly distributed in $[-\pi, \pi]$. That is, the plane wave arrives at the mobile unit antenna from all directions with equal probabilities. This is a reasonable assumption for macrocellular applications (Clarke's model [27] or Jakes' model [28]). Thus

$$\begin{aligned} E[\cos 2\pi f_D\xi] &= E[\cos(2\pi f_M \xi \cos\varphi)] \\ &= \frac{1}{2\pi}\int_{-\pi}^{\pi} \cos(2\pi f_M \xi \cos\varphi) d\varphi \\ &= \frac{1}{\pi}\int_{0}^{\pi} \cos(2\pi f_M \xi \cos\varphi) d\varphi \\ &\triangleq J_0(2\pi f_M \xi) \end{aligned}$$

where $J_0(\cdot)$ is called the zeroth-order Bessel function of the first kind. Similarly it is easy to see that $E[\sin 2\pi f_D \xi] = 0$. Thus

$$R(\xi) = E\{\alpha(t)\alpha(t+\xi)\} J_0(2\pi f_M \xi)$$

Further if we assume that $\alpha(t)$ is also approximately constant in a short period then we have

$$R(\xi) = E\{\alpha^2(t)\} J_0(2\pi f_M \xi) = E\{|h(t)|^2\} J_0(2\pi f_M \xi) \tag{15.125}$$

This is to say that the change of the autocorrelation is mainly determined by the Doppler shift.

Taking the Fourier transform of (15.125), we can obtain the PSD of the channel complex envelope as

$$\Phi(f) = E\{|h(t)|^2\} \frac{1}{2\pi f_M \sqrt{1 - \left(\frac{f}{f_M}\right)^2}} \tag{15.126}$$

This pdf can also be obtained directly from the Clarke model of the mobile radio channel [27, 28] and the autocorrelation (15.125) is obtained by the inverse Fourier transform of (15.126) [28].

Proof of (15.47):

Since

$$E[c_k c_{k+r}^*] = E_s \sum_{\substack{i=0 \\ i \neq k, k+r}}^{N-1} E[H_{i-k} H_{i-k-r}^*] \tag{15.127}$$

and

$$H_{i-k} = \frac{1}{N}\sum_{n=0}^{N-1} h_n \exp(j2\pi\frac{n(i-k)}{N})$$

we have

$$\begin{aligned}
&E[H_{i-k}H^*_{i-k-r}] \\
&= \frac{1}{N^2}\sum_{n=0}^{N-1}\sum_{p=0}^{N-1} E[h_n h_p^*]\exp(j2\pi\frac{n(i-k)}{N})\exp(j2\pi\frac{-p(i-k-r)}{N})
\end{aligned} \quad (15.128)$$

We need to find $E[h_n h_p^*]$, which is the autocorrelation between $h(nT_s)$ and $h(pT_s)$.

For the isotropic scattering mobile radio channel with a narrow-band signal and a maximum Doppler shift f_M, from (15.125) the autocorrelation function of the channel is

$$R(\tau) = E\{|h(t)|^2\}J_0(2\pi f_M \tau)$$

thus

$$E[h_n h_p^*] = R((n-p)T_s) = E\{|h(t)|^2\}J_0(2\pi f_M T_s(n-p)) \quad (15.129)$$

where T_s is the sampling period. If we further assume $E\{|h(t)|^2\} = E[|h_k|^2] = 1$, from (15.127), (15.128), and (15.129), we have

$$\begin{aligned}
&E[c_k c^*_{k+r}] \\
&= E_s \sum_{\substack{i=0 \\ i\neq k,k+r}}^{N-1}\left[\frac{1}{N^2}\sum_{n=0}^{N-1}\sum_{p=0}^{N-1} J_0(2\pi f_M T_s(n-p))e^{j2\pi\frac{n(i-k)}{N}}e^{j2\pi\frac{-p(i-k-r)}{N}}\right] \\
&= E_s \sum_{i=0}^{N-1}\left[\frac{1}{N^2}\sum_{n=0}^{N-1}\sum_{p=0}^{N-1} J_0(2\pi f_M T_s(n-p))e^{j2\pi\frac{n(i-k)}{N}}e^{j2\pi\frac{-p(i-k-r)}{N}}\right] \\
&\quad -E_s\frac{1}{N^2}\sum_{n=0}^{N-1}\sum_{p=0}^{N-1} J_0(2\pi f_M T_s(n-p))\exp(j2\pi\frac{pr}{N}) \quad (i \text{ equals } k) \\
&\quad -E_s\frac{1}{N^2}\sum_{n=0}^{N-1}\sum_{p=0}^{N-1} J_0(2\pi f_M T_s(n-p))\exp(j2\pi\frac{nr}{N}) \quad (i \text{ equals } k+r)
\end{aligned} \quad (15.130)$$

Since

$$\begin{aligned}
&\sum_{i=0}^{N-1} \exp(j2\pi\frac{n(i-k)}{N})\exp(j2\pi\frac{-p(i-k-r)}{N}) \\
&= \exp(j2\pi\frac{-nk+p(k+r)}{N})\sum_{i=0}^{N-1}\exp(j2\pi\frac{i(n-p)}{N}) \\
&= \exp(j2\pi\frac{-nk+p(k+r)}{N})N\delta_{np}
\end{aligned}$$

Thus

$$\begin{aligned}
E[c_k c_{k+r}^*] &= \frac{E_s}{N}\sum_{n=0}^{N-1}\sum_{p=0}^{N-1} J_0(2\pi f_M T_s(n-p))\exp(j2\pi\frac{-nk+p(k+r)}{N})\delta_{np} \\
&\quad -\frac{E_s}{N^2}\sum_{n=0}^{N-1}\sum_{p=0}^{N-1} J_0(2\pi f_M T_s(n-p))\exp(j2\pi\frac{pr}{N}) \\
&\quad -\frac{E_s}{N^2}\sum_{n=0}^{N-1}\sum_{p=0}^{N-1} J_0(2\pi f_M T_s(n-p))\exp(j2\pi\frac{nr}{N}) \\
&= \frac{E_s}{N}\sum_{p=0}^{N-1} J_0(0)\exp(j2\pi\frac{pr}{N}) \quad \text{(only n = p terms remain) } (J_0(0)=1) \\
&\quad -\frac{E_s}{N^2}\sum_{n=0}^{N-1}\sum_{p=0}^{N-1} J_0(2\pi f_M T_s(n-p))\exp(j2\pi\frac{pr}{N}) \\
&\quad -\frac{E_s}{N^2}\sum_{n=0}^{N-1}\sum_{p=0}^{N-1} J_0(2\pi f_M T_s(n-p))\exp(j2\pi\frac{nr}{N}) \\
&= E_s\delta_r - \frac{E_s}{N^2}\sum_{n=0}^{N-1}\sum_{p=0}^{N-1} J_0(2\pi f_M T_s(n-p))\left[e^{j2\pi\frac{pr}{N}} + e^{j2\pi\frac{nr}{N}}\right]
\end{aligned}$$

Note that the two terms in the brackets exist only when $r \neq 0$. When $r = 0$, the two terms should be replaced by 1 instead of 2 (see (15.130) where the last two terms are for $i = k$ and $i = k + r$. When $r = 0$, these two cases are the same case. Thus only one term should be kept). Thus we can write

$$E[c_k c_{k+r}^*] = E_s\delta_r - \frac{E_s}{N^2}\sum_{n=0}^{N-1}\sum_{p=0}^{N-1} J_0(2\pi f_D T_s(n-p))$$

$$\cdot \left[\exp(j2\pi\frac{pr}{N}) + (1-\delta_r)\exp(j2\pi\frac{nr}{N})\right]$$

which is (15.65).

APPENDIX 15B PROOF OF (15.58), (15.65), (15.66), AND (15.67)

Proof of (15.58):

Excluding noise, the DFT demodulator output is

$$\begin{aligned}
X_k &= \frac{1}{N}\sum_{n=0}^{N-1}\left\{x_n \exp(-j2\pi\frac{kn}{N})\right\} \\
&= \frac{1}{N}\sum_{n=0}^{N-1}\left\{\left[\sum_{m=0}^{M-1} h_{m,n}s_{n-m}\right]\exp(-j2\pi\frac{kn}{N})\right\} \\
&= \frac{1}{N}\sum_{n=0}^{N-1}\left\{\left[\sum_{m=0}^{M-1} h_{m,n}\left(\sum_{i=0}^{N-1} d_i \exp(j2\pi\frac{i(n-m)}{N})\right)\right]\exp(-j2\pi\frac{kn}{N})\right\} \\
&= \frac{1}{N}\sum_{n=0}^{N-1}\left\{\left[\sum_{m=0}^{M-1}\left(\sum_{i=0}^{N-1} h_{m,n}d_i \exp(j2\pi\frac{i(n-m)}{N})\exp(-j2\pi\frac{kn}{N})\right)\right]\right\} \\
&= \sum_{m=0}^{M-1}\left(\sum_{i=0}^{N-1} d_i\left[\frac{1}{N}\sum_{n=0}^{N-1} h_{m,n}\exp(j2\pi\frac{n(i-k)}{N})\right]\exp(-j2\pi\frac{im}{N})\right) \\
&= \sum_{i=0}^{N-1}\sum_{m=0}^{M-1} d_i H_{m,i-k}\exp(-j2\pi\frac{im}{N}) \qquad (15.131)
\end{aligned}$$

Proof of (15.65):

Assuming uncorrelated scattering, and uncorrelated data, the autocorrelation of c_k is

$$\begin{aligned}
&E[c_k c_{k+r}^*] \\
&= E\left[\sum_{\substack{i=0\\ i\neq k}}^{N-1}\sum_{\substack{l=0\\ l\neq k+r}}^{N-1}\sum_{m=0}^{M-1}\sum_{p=0}^{M-1} d_i d_l^* H_{m,i-k}H_{p,l-k-r}^*\exp(-j2\pi\frac{(i-l)m}{N})\right] \\
&= \sum_{\substack{i=0\\ i\neq k,k+r}}^{N-1}\sum_{m=0}^{M-1} E\left[d_i d_i^* H_{m,i-k}H_{m,i-k-r}^*\right] \text{ (expectations of the cross terms are 0)}
\end{aligned}$$

$$= E_s \sum_{\substack{i=0 \\ i\neq k,k+r}}^{N-1} \sum_{m=0}^{M-1} E[H_{m,i-k} H^*_{m,i-k-r}] \tag{15.132}$$

where $E_s = E\{d_i d_i^*\} = E\{|d_i|^2\}$ is the average symbol energy.

Proof of (15.66):

We have

$$\begin{aligned} &\sum_{m=0}^{M-1} E[H_{m,i-k} H^*_{m,i-k-r}\} \\ = &\sum_{m=0}^{M-1} \frac{1}{N^2} \sum_{n=0}^{N-1} \sum_{p=0}^{N-1} E[h_{m,n} h^*_{m,p}] \exp(j2\pi \frac{n(i-k)}{N}) \exp(j2\pi \frac{-p(i-k-r)}{N}) \end{aligned} \tag{15.133}$$

From (15.125) we have

$$E[h_{m,n} h^*_{m,p}] = E\{|h_m(t)|^2\} J_0(2\pi f_D T_s (n-p)) \tag{15.134}$$

If we assume $\sum_{m=0}^{M-1} E[|h_{m,k}|^2] = 1$, from (15.132) and (15.133) we have

$$\begin{aligned} &E[c_k c^*_{k+r}] \\ = &E_s \sum_{\substack{i=0 \\ i\neq k,k+r}}^{N-1} \left[\frac{1}{N^2} \sum_{n=0}^{N-1} \sum_{p=0}^{N-1} J_0(2\pi f_D T_s(n-p)) e^{j2\pi \frac{n(i-k)}{N}} e^{j2\pi \frac{-p(i-k-r)}{N}} \right] \end{aligned}$$

This is exactly the same as (15.130). The rest of the derivation is the same as in Appendix 15A and the final expression for the correlation is exactly the same.

Proof of (15.67):

$$\begin{aligned} E'_k &= E\{|d_k \eta_{k,k}|^2\} = E_s E\{|\eta_{k,k}|^2\} \\ &= E_s E \left\{ \sum_{m=0}^{M-1} \left[\frac{1}{N} \sum_{n=0}^{N-1} h_{m,n} \right] e^{-j2\pi \frac{km}{N}} \sum_{l=0}^{M-1} \left[\frac{1}{N} \sum_{p=0}^{N-1} h^*_{l,p} \right] e^{j2\pi \frac{kl}{N}} \right\} \\ &= E_s \frac{1}{N^2} \left\{ \sum_{m=0}^{M-1} \sum_{n=0}^{N-1} \sum_{l=0}^{M-1} \sum_{p=0}^{N-1} E\{h_{m,n} h^*_{l,p}\} \exp(-j2\pi \frac{k(m-l)}{N}) \right\} \end{aligned}$$

$$
\begin{aligned}
&= E_s \frac{1}{N^2}\left\{\sum_{m=0}^{M-1}\sum_{n=0}^{N-1}\sum_{p=0}^{N-1} E\{h_{m,n}h^*_{m,p}\}\right\} \text{ (expectations of cross terms are 0)} \\
&= E_s \frac{1}{N^2}\sum_{n=0}^{N-1}\sum_{p=0}^{N-1}\sum_{m=0}^{M-1} E\{|h_m(t)|^2\}J_0(2\pi f_M T_s(n-p)) \text{ (from (15.125))} \\
&= E_s \frac{1}{N^2}\sum_{n=0}^{N-1}\sum_{p=0}^{N-1} J_0(2\pi f_M T_s(n-p)) \text{ (since } \sum_{m=0}^{M-1} E[|h_{m,k}|^2] = 1) \\
&= \frac{E_s}{N^2}\{N + 2\sum_{i=1}^{N-1}(N-i)J_0(2\pi f_M T_s i)\} \text{ (from (15.50))}
\end{aligned}
$$

APPENDIX 15C PROOF OF (15.105)

$$
\begin{aligned}
\widehat{\mathbf{h}}_{lmmse} &= \mathbf{R_{hh}F}^H\mathbf{D}^H(\mathbf{DFR_{hh}F^HD^H} + \sigma_n^2\mathbf{I})^{-1}\mathbf{y} \\
&= \mathbf{R_{hh}F}^H\mathbf{D}^H(\mathbf{DF}(\mathbf{R_{hh}} + (\mathbf{DF})^{-1}\sigma_n^2(\mathbf{F^HD^H})^{-1})\mathbf{F^HD^H})^{-1}\mathbf{y} \\
&= \mathbf{R_{hh}F}^H\mathbf{D}^H(\mathbf{F^HD^H})^{-1}[\mathbf{R_{hh}} + (\mathbf{F^HD^HDF})^{-1}\sigma_n^2]^{-1}(\mathbf{DF})^{-1}\mathbf{y} \\
&= \mathbf{R_{hh}}[\mathbf{R_{hh}} + (\mathbf{F^HD^HDF})^{-1}\sigma_n^2]^{-1}(\mathbf{DF})^{-1}\mathbf{y} \\
&= \mathbf{R_{hh}}[\mathbf{R_{hh}} + (\mathbf{F^HD^HDF})^{-1}\sigma_n^2]^{-1}(\mathbf{DF})^{-1}(\mathbf{F}^H\mathbf{D}^H)^{-1}\mathbf{F}^H\mathbf{D}^H\mathbf{y} \\
&= \underbrace{\left[\mathbf{R_{hh}}[\mathbf{R_{hh}} + (\mathbf{F^HD^HDF})^{-1}\sigma_n^2]^{-1}(\mathbf{F}^H\mathbf{D}^H\mathbf{DF})^{-1}\right]}_{\mathbf{Q}_{lmmse}}\mathbf{F}^H\mathbf{D}^H\mathbf{y} \\
&= \mathbf{Q}_{lmmse}\mathbf{F}^H\mathbf{D}^H\mathbf{y}
\end{aligned}
$$

References

[1] Cimini, L. J., Jr. "Analysis and simulation of a digital mobile channel using orthogonal frequency division multiplexing," *IEEE Trans. Commun.*, vol. 33, no. 7, July 1985, pp. 665–675.

[2] Mark, J., and W. Zhuang, *Wireless Communications and Networking*, Upper Saddle River, New Jersey: Prentice Hall, 2003.

[3] Proakis, J., *Digital Communications, Second Edition*, New York: McGraw-Hill, 1989.

[4] Zhang, K., "Performance analysis and enhancement of multicarrier transmission systems," Ph.D. Thesis, School of Electrical and Electronic Engineering, Nanyang Technological University, Singapore, May 2004.

[5] Zhang, K., Z. Song, and Y. L. Guan, "Multicarrier channels with non-identical subcarrier fading distributions analysis and adaptation," *IEEE Communications Letters*, vol. 8, no. 6, June 2004, pp. 368–370.

[6] Russell, M., and G. L. Stüber, "Interchannel interference analysis of OFDM in a mobile environment," *Proc. IEEE Veh. Technol. Conf.*, 1995, pp. 820–824.

[7] Press, S. J., *Applied Multivariate Analysis*, New York: Holt, Rinehart and Winston, Inc., 1972.

[8] Kivinen, J., X. Zhao, and P. Vainikainen, "Empirical characterization of wide-band indoor radio channel at 5.3 GHz," *IEEE Trans. Antennas Propagat.*, vol. 49, no. 8, August 2001, pp. 1192–1203.

[9] Mohr, W., "Modeling of wideband mobile radio channels based on propagation measurements," *Proc. IEEE Int. Symposium on Personal, Indoor and Mobile Radio Communications (PIMRC95)*, 1995, pp. 397–401.

[10] Ojanpera, T., and R. Prasad, *Wideband CDMA for Third Generation Mobile Communications*, Norwood, Massachusetts: Artech House, 1998.

[11] Patzold, M., *Mobile Fading Channels*, New York: Wiley, 1981.

[12] Webb, W. T., and R. Steele, "Variable rate QAM for mobile radio," *IEEE Trans. Commun.*, vol. 43, no. 7, July 1995, pp. 2223–2230.

[13] Chung, S. T., and A. J. Goldsmith, "Adaptive multicarrier modulation for wireless systems," *Proc. IEEE Asilomar Conf. on Signals, Systems and Computers*, Pacific Grove, California, November 2000, pp. 1603–1607.

[14] Goldsmith, A. J., and S. G. Chua, "Variable-rate variable-power MQAM for fading channels," *IEEE Trans. Commun.*, vol. 45, no. 10, October 1997, pp. 1218–1230.

[15] Kondo, S., and L. B. Milstein, "Performance of multicarrier DS CDMA systems," *IEEE Trans. Commun.*, vol. 44, no. 2, February 1996, pp. 238–246.

[16] Stüber, G. L., *Principles of Mobile Communication*, Boston: Kluwer, 1996.

[17] Tufvesson, F., and T. Maseng, "Pilot assisted channel estimation for OFDM in mobile cellular systems," *Proc. 47th IEEE Veh. Technol. Conf.*, Phoenix, Arizona, May 1997, pp. 1639–1643.

[18] Coleri, S., et al., "Channel estimation techniques based on pilot arrangement in OFDM systems," *IEEE Trans. Broadcast.*, vol. 48, no. 3, 2002, pp. 223–229.

[19] Hoeher, P., S. Kaiser, and P. Robertson, "Two-dimensional pilot-symbol-aided channel estimation by Wiener filtering," *Proc. 1997 IEEE Int. Conf. Acoustics, Speech, and Signal Processing*, Munich, Germany, April 1997, pp. 1845–1848.

[20] van de Beek, et al., "On channel estimation in OFDM systems," *Proc. 45th IEEE Veh. Technol. Conf.*, Chicago, Illinois, July 1995, pp. 815–819.

[21] Edfors, O., et al., "OFDM channel estimation by singular value decomposition," *IEEE Trans. Commun.*, vol. 46, July 1998, pp. 931–939.

[22] Li, Y., L. J. Cimini, Jr., and N. R. Sollenberger, "Robust channel estimation for OFDM systems with rapid dispersive fading channels," *IEEE Trans. Commun.*, vol. 46, July 1998, pp. 902–915.

[23] Li, Y., "Pilot-Symbol-Aided Channel Estimation for OFDM in Wireless Systems," *IEEE Trans.*

Veh. Techno., vol. 49, no. 4, July 2000, pp. 1027–1215.

[24] Yang, B., et al., "Channel estimation for OFDM transmission in multipath fading channels based on parametric channel modeling," *IEEE Trans. Commun.*, vol. 49, no. 3, March 2001, pp. 467–479.

[25] Scharf, L. L. *Statistical Signal Processing: Detection, Estimation, and Time Series Analysis*, Reading Massachusetts: Addison Wesley, 1991.

[26] van Nee, R., and R. Prasad, *OFDM for Wireless Multimedia Communications*, Norwood, Massachusetts: Artech House, 2000.

[27] Clarke, R., "A statistical theory of mobile radio reception," *Bell System Tech. J.*, vol. 47, pp. 957–1000, 1968.

[28] Jakes, W. C., *Microwave Mobile Communications*, New York: IEEE Press, 1974.

[29] Hoeher, P., "TCM on frequency-selective land-mobile fading channels," *Proc. 5th Tirrenia Int. Workshop on Dig. Comm.*, Tirrenia, Italy, September 1991, pp. 317–328.

[30] Yang, B., Z. G. Cao, and K. B., Letaief, "Analysis of low-complexity windowed DFT-based MMSE channel estimator for OFDM systems," *IEEE Trans. Commun.*, vol. 49, no. 11, November 2001,
pp. 1977–1987.

[31] Hoeher, P., S. Kaiser, and P. Robertson, "Pilot-symbol-aided channel estimation in time and frequency," *Proc. 6th Communication Theory Mini-Conference in Conjunction with IEEE GLOBECOM '97*, Phoenix, Arizona, November 1997, pp. 90–96.

[32] Edfors, O., et al., "Analysis of DFT-based channel estimators for OFDM," *Technical Report*, Division of Signal Processing, Luleå University of Technology, S-971 b7 Luleå, Sweden, and School of Electrical and Computer Engineering, 1285 Electrical Engineering Building, Purdue University, West Lafayette, Indiana, 1996.

[33] Mitra, S., *Digital Signal Processing*, New York: McGraw-Hill, 1998.

[34] Sandell, M., and O. Edfors, "A comparative study of pilot-based channel estimators for wireless OFDM," Div. Signal Processing, Luleå Univ. Technology, Luleå, Sweden, Res. Rep. TULEA 1996, September 19, 1996.

[35] Park, S. Y., and C. G. Kang, "Performance of pilot-assisted channel estimation for OFDM system under time-varying multi-path Rayleigh fading with frequency offset compensation," *Proc. of IEEE Veh. Technol. Conf.* 2000, pp. 1245–1248.

[36] Shu, F., et al., "Time-frequency channel estimation for digital amplitude modulation broadcasting systems based on OFDM," *IEE Proc.-Commun.* vol. 150, no. 4, August 2003, pp. 259–264.

[37] Mostofi, Y., and D. Cox, "ICI mitigation for mobile OFDM receivers," *Proc. IEEE 38th ICC,* Anchorage, Alaska, May 2003, pp. 3351–3355.

[38] Linnartz, J. P. M. G., and A. Gorokhov, "New equalization approach for OFDM over dispersive and rapidly time varying channel," *CDROM Proc. PIMRC'00,* London, September 2000, pp. 1375–1379.

[39] Fischer, V., A. Kurpiers, and D. Karsunke, "ICI reduction method for OFDM systems," *Proceedings of the 8th International OFDM Workshop*, Hamburg, Germany, September 24–25, 2003.

[40] Han, K. Y., et al., "Time domain equalization using linear phase interpolation for OFDM in time variant multipath channels with frequency offset," *Proc. IEEE Veh. Technol. Conf.* 2000, pp. 1255–1258.

[41] Song, W. G., and J. T. Lim, "Pilot symbol aided channel estimation for OFDM with fast fading channels," *IEEE Trans. on Broadcasting,* vol. 49, no. 4, December 2003, pp. 398–402.

[42] Hutter, A. A., R. Hasholzner, and J. S. Hammerschmidt, "Channel estimation for mobile OFDM systems," *Proc. IEEE Veh. Technol. Conf.,* 1999. vol.1, pp. 305-309.

[43] Wilson, S. K., R. E. Khayata, and J. M. Cioffi, "16 QAM modulation with orthogonal frequency division multiplexing in a Rayleigh-fading environment," *Proc. IEEE Veh. Technol. Conf.,* June 1994, pp. 1660–1664.

[44] Liskov, N., and R. Curtis, "Performance of coherent phase and amplitude digital modulations with carrier recovery noise," *IEEE Trans. Commun.*, vol. 35, no. 9, September 1987, pp. 972–976.

[45] Ran, J., et al., "Decision-directed channel estimation method for OFDM systems with high velocities," *Proc. IEEE VTC2003-Spring,* April 2003, pp. 2358–2361.

[46] Al-Susa, E., and R. Ormondroyd, "A predictor-based decision feedback channel estimation method for COFDM with high resilience to rapid time-variations," *Proc. IEEE VCT1999-Fall,* September 1999, pp. 273–278.

[47] Chini, A., et al., "Filtered decision-feedback channel estimation for OFDM-based terrestrial broadcasting system," *IEEE Trans. Broadcasting*, vol. 44, no. 1, March 1998, pp. 2–11.

[48] Mignone, V., and A. Morello, "CD3-OFDM: A novel demodulation scheme for fixed and mobile receivers," *IEEE Trans. Commun.*, vol. 44, no. 9, September 1996, pp. 1144–1151.

[49] Frenger, P. K., and N. A. B. Svensson, "Decision-directed coherent detection in multicarrier systems on Rayleigh fading channels," *IEEE Trans. Veh. Tech.*, vol. 49, March 1999, pp. 490–498.

[50] Rim, M., J. Ahn, and Y.-S. Kim, "Decision-directed channel estimation for MQAM modulated OFDM systems," *Proc. IEEE Veh. Technol. Conf.,* 2002, pp. 1742–1746.

[51] Zhuang, A., and M. Renfors, "Combined pilot aided and decision directed channel estimation for the RAKE receiver," *Proc. IEEE VTC2000-Fall,* September 2000, pp. 710–713.

[52] Lee, W., and J. Zhu, "Channel estimation for high-speed packet-based OFDM communication systems," *Proc. The 5th IEEE International Symposium on Wireless Personal Multimedia Communications, 2002*, vol. 3, October 21–30, 2002, pp. 1293–1298.

[53] Rim, M., "Optimally combining decision-directed and pilot-symbol-aided channel estimation," *IEE Electronics Letters,* March 20, 2003, vol. 39, no. 6, pp. 558–560.

[54] Ito, M., et al., "An OFDM receiver with decision-directed channel estimation for the scattered pilot scheme in fast fading environments," *Proc. IEEE VTC2003-Spring,* April 2003, pp. 368–372.

[55] Kella, T., "Decision-directed channel estimation for supporting higher terminal velocities in OFDM based WLANs," *Proc. IEEE GLOBECOM'03*, vol. 3, December 1-5, 2003, pp. 1306–1310.

[56] Raleigh, G., and J. Cioffi, "Spatial-temporal coding for wireless communications," *Proc. IEEE 1996 Global Communications Conference*, November 1996, pp. 1809–1814.

[57] Raleigh, G., and V. Jones, "Multivariate modulation and coding for wireless communications,"

IEEE Journal on Selected Areas in Communications, vol. 17, May 1999, pp. 851–866.

[58] Ayanoglu, E., et al., "VOFDM broadband wireless transmission and its advantages over single carrier modulation," *Proc. IEEE International Conference on Communications, ICC 2001*, vol. 6, June 11-14, 2001, pp. 1660–1664.

[59] Xia, X., "Precoded and vector OFDM robust to channel spectral nulls and with reduced cyclic prefix length in single transmit antenna systems," *IEEE Trans. Commun.*, vol. 49, no. 8, August 2001, pp. 1363–1374.

[60] Shelswell, P., "The COFDM modulation system: the heart of digital audio broadcasting," *Electronics & Communications Engineering Journal*, June 1995, pp. 127–136.

[61] Reimers, U., "DVB-T: the COFDM-based system for terrestrial television," *Electronics & Communications Engineering Journal*, February 1997, pp. 28–32.

[62] IEEE Standard 802.11a, 1999.

[63] Rault, J., D. Castelain, and B. Le Floch, "The coded orthogonal frequency division multiplexing (COFDM) technique, and its application to digital radio broadcasting towards mobile receivers," *Proc. IEEE Global Telecommunications Conference, 1989,* GLOBECOM '89, November 27–30, 1989 pp. 428–432.

[64] Floch, B., and M. Alard, "Coded orthogonal frequency division multiplex," *Proceedings of IEEE*, vol. 83, no. 6, June 1995, pp. 982–996.

[65] Wesel, R., and J. Cioffi, "Fundamental of coding for broadcast OFDM," *Proc. IEEE 29th Asilomar Conference on Signals, Systems and Computers*, vol. 1, October 30–November 2, 1995, pp. 2–6.

[66] Sklar, B., *Digital Communications: Fundamentals and Applications,* Englewood Cliffs, New Jersey: Prentice-Hall, 1988.

[67] Gesbert, D., et al. "From theory to practice: an overview of MIMO space-time coded wireless systems," *IEEE Journal on Selected Areas in Communications*, vol. 21, no. 3, April 2003, pp. 281–301.

[68] Paulraji, A., et al., "An overview of MIMO communications—a key to gigabit wireless," *Proceedings of IEEE*, vol. 92, no. 2, February 2004, pp. 198–218.

[69] Stüber, G., et al., "Broadband MIMO-OFDM wireless communications," *Proceedings of IEEE*, vol. 92, no. 2, February 2004, pp. 271–294.

[70] Li, Y., J. Winters, and N. Sollenberger, "MIMO-OFDM for wireless communications: signal detection with enhanced channel estimation," *IEEE Trans. Commun.*, vol. 50, September 2002, pp. 1471–1477.

[71] Huang, D., and K. Letaief, "Symbol-based space diversity for coded OFDM systems," *IEEE Trans. Wireless Commun.*, vol. 3, no. 1, January 2004, pp. 117–127.

Selected Bibliography

- Bahai, A. S. R., and B. R. Saltzberg, *Multi-Carrier Digital Communications: Theory and Applications of OFDM,* New York: Kluwer Academic/Plenum Publishers, 1999.
- Cimini, L., and L. Greenstein, "Orthogonal frequency-division multiplexing," *Wiley Encyclopedia*

of Telecommunications, J. G. Proakis, (ed.), New York: John Wiley & Sons, 2003.

- Litwin, L., "An introduction to multicarrier modulation," *IEEE Potentials*, April/May, 2000, pp. 36–38.
- Meyr, H., M. Moeneclaey, and S. A. Fechtel, *Digital Communication Receivers*, New York: John Wiley & Sons, 1998.
- Rappaport, T. S., *Wireless Communications: Principles and Practice*, Upper Saddle River, New Jersey: Prentice Hall, 1996.
- Ruiz, A., J. M. Cioffi, and S. Kasturia, "Discrete multiple tone modulation with coset coding for the spectrally shaped channel," *IEEE Trans. Commun.*, vol. COM-40, no. 6, June 1992, pp. 1012–1029.
- Simon, M. K., and M. S. Alouini, *Digital Communication over Fading Channels*, New York: John Wiley & Sons, Inc, 2000.
- Steele, R., *Mobile Radio Communications*, Piscataway, New Jersey: IEEE, 1992.

Chapter 16

Wavelet OFDM and MASK-OFDM

OFDM schemes that are widely adopted in many industrial standards are all FFT-based. In an effort to search for more efficient modulation schemes, other multicarrier non-FFT-based schemes are emerging. Two salient ones are wavelet transform based OFDM (wavelet OFDM) and discrete cosine transform based OFDM which uses M-ary amplitude shift keying (MASK-OFDM). In this chapter, we will introduce these two groups of schemes. Both wavelet transform and cosine transform are well-known data compression methods for computer imaging and animation. Their uses in digital modulation only emerged recently. Compared to QAM-OFDM, WT-OFDM and MASK-OFDM have their advantages and disadvantages. Therefore, we cannot say one is absolutely superior than others. Whether a scheme is chosen or not should depend on the particular application in which it will be used.

In the following, we will describe various wavelet OFDM schemes first and MASK-OFDM at the end of this chapter. Section 16.1 presents the fundamentals of wavelets to prepare the reader for the sections of wavelet OFDM. In Section 16.2, we discuss multiscale modulation (MSM) in detail. Its special case, the fractal modulation, is presented in Section 16.3. The wavelet packet modulation (WPM), which can be considered as a generalization of MSM, is described in Section 16.4. Wavelet PAM, which is also a special case of MSM, is discussed in Section 16.5. The M-band wavelet modulation (MWM) is given in Section 16.6. Section 16.7 is devoted to the overlapped discrete wavelet multitone modulation (DWMT). Section 16.8 describes MASK-OFDM. Section 16.9 summarizes the chapter.

16.1 WAVELET FUNDAMENTALS

A *wavelet* $\psi(t)$ is a zero-mean continuous-time bandpass function. $\psi(t)$ is also a square-integrable function, or we say that $\psi(t) \in L^2(\mathbf{R})$ (a space) where the "L" signifies a Lebesque integral, the "2" denotes the integral of the square of the modulus of the function, and **R** indicates that the independent variable of integration t is a number over the whole real line.

By scaling and shifting the wavelet a family of *wavelet functions* can be generated.

$$\psi_{ab}(t) = \frac{1}{\sqrt{a}}\psi(\frac{t-b}{a})$$

where a is a scaling factor and b is a time-shifting parameter. Both a and b are real-valued. In general, the scaling and shifting parameters are discretized as

$$a = a_0^m, \quad b = nb_0a_0^m$$

where exponents m and n are integers. With the discretized parameters, the function family becomes

$$\psi_{mn}(t) = a_0^{-m/2}\psi(a_0^{-m}t - nb_0) \tag{16.1}$$

which forms the basis of the discrete wavelet transform (DWT) to be defined next. Note that although it is called DWT, the time variable is still continuous. DWT coefficients of a continuous-time function $f(t)$ are defined as

$$\begin{aligned} d_{mn} &= < f(t), \psi_{mn}(t) >= \int f(t)\psi_{mn}(t)dt \\ &= a_0^{-m/2}\int f(t)\psi(a_0^{-m}t - nb_0)dt \end{aligned} \tag{16.2}$$

When the DWT set is complete, the wavelet representation of $f(t)$ is

$$f(t) = \sum_m\sum_n d_{mn}\psi_{mn}(t) \tag{16.3}$$

From (16.2) and (16.3) we can see that the wavelet family $\{\psi_{mn}(t)\}$ must satisfy the orthonormal conditions. That is

$$< \psi_{mn}(t), \psi_{kl}(t) >= \delta_{mn}\delta_{kl}$$

where $\delta_{mn} = 1$ for $m = n$; 0 otherwise, is the Kronecker delta function.

In general, the complete representation of $f(t)$ requires a discrete wavelet family $\{\psi_{mn}(t)\}$ of an infinite number of functions. By introducing a low-pass-natured complementary *scaling function* $\phi(t)$, a function can be represented by using L finite resolutions of wavelet and scaling functions with the parameter values of $a_0 = 2$ and

$b_0 = 1$ as [1]

$$f(t) = \sum_{n=-\infty}^{\infty} c_{L,n} 2^{-L/2} \phi(2^{-L}t - n) + \sum_{m=1}^{L} \sum_{n=-\infty}^{\infty} d_{mn} 2^{-m/2} \psi(2^{-m}t - n) \quad (16.4)$$

where the scaling coefficients are similarly defined as

$$c_{L,n} =< f(t), \phi_{L,n}(t) >= 2^{-L/2} \int f(t)\phi(2^{-L}t - n)dt \quad (16.5)$$

where

$$\phi_{L,n}(t) = 2^{-L/2}\phi(2^{-L}t - n) \quad (16.6)$$

Equation (16.4) is called the multiresolution approximation (MRA) of a signal. Note that wavelet functions and scaling functions in (16.4) are orthogonal to each other. The scaling functions are all time-shifted versions of only one parent scaling function, that is, $\phi(2^{-L}t)$. In other words, the scaling functions are at only one scale. It is the coarsest approximation of $f(t)$. The wavelet functions have L scales (or resolutions), with $\psi(t/2)/\sqrt{2}$ being the finest resolution. The coefficients in (16.4) are called discrete wavelet transform (DWT) of the signal $f(t)$. Inversely, the $f(t)$ is the inverse discrete wavelet transform (IDWT) of the coefficients.

The scaling function has the property that the parent scaling function $\phi(t)$ can be expressed as a sum of shifted $\phi(2t)$ [2], that is

$$\phi(t) = \sqrt{2} \sum_{n} h_n \phi(2t - n) \quad (16.7)$$

where the coefficients h_n are a sequence of real or perhaps complex numbers called scaling function coefficients (or scaling filter or scaling vector). Equation (16.7) shows that $\phi(t)$ can be constructed by the weighted sum of its half-length translations.

The wavelet function can be expressed in terms of the scaling function as [2]

$$\psi(t) = \sqrt{2} \sum_{n} g_n \phi(2t - n) \quad (16.8)$$

where

$$g_n = (-1)^n h_{1-n} \quad (16.9)$$

Equation (16.8) shows that $\psi(t)$ can be constructed by the weighted sum of half-length translations of the scaling function.

Well-known examples of wavelets are Haar wavelets and sinc wavelets [2,3]. The Haar scaling function is a unit rectangular pulse of length 1 defined as

$$\phi(t) = \begin{cases} 1, & \text{for } 0 < t \le 1 \\ 0, & \text{otherwise} \end{cases} \tag{16.10}$$

and the Haar wavelet is a positive half-length pulse followed by a negative half-length pulse.

$$\psi(t) = \phi(2t) - \phi(2t-1) = \begin{cases} 1, & \text{for } 0 < t \le 1/2 \\ -1, & \text{for } 1/2 < t \le 1 \\ 0, & \text{otherwise} \end{cases} \tag{16.11}$$

Figure 16.1 shows the Haar scaling function and wavelet. Haar functions are a compact support (a finite duration) orthonormal wavelet system. However, their PSDs are squared sinc functions and have an infinite width. The spectrum of $\phi(t)$ is

$$\Phi(f) = \mathcal{F}\{\phi(t)\} = \text{sinc}(f)\exp(-j\pi f) = \frac{\sin(\pi f)}{\pi f}\exp(-j\pi f) \tag{16.12}$$

and the spectrum of $\psi(t)$ is

$$\begin{aligned} \Psi(f) &= \mathcal{F}\{\psi(t)\} = \mathcal{F}\{\phi(2t)\} - \mathcal{F}\{\phi(2(t-\frac{1}{2}))\} \\ &= \frac{1}{2}\Phi\left(\frac{f}{2}\right) - \frac{1}{2}\Phi\left(\frac{f}{2}\right)\exp(-j\pi f) \\ &= \frac{1}{2}\Phi\left(\frac{f}{2}\right)[1-\exp(-j\pi f)] \\ &= \frac{\sin(\pi f/2)}{\pi f}\exp(-j\pi f/2)[1-\exp(-j\pi f)] \end{aligned} \tag{16.13}$$

Figure 16.2 shows that the Haar scaling function is lowpass with energy concentrated in $[0,1]$. The PSD of $\psi(t)$ is bandpass with energy concentration moved higher in frequency. The PSD of $2\psi(2t)$ is $|\Psi(f/2)|^2$ and its energy concentration is shifted further to the higher frequency.

The sinc scaling function defined as

$$\phi(t) = \text{sinc}(\pi t) = \frac{\sin(\pi t)}{\pi t} \tag{16.14}$$

generates an orthogonal wavelet system. The corresponding wavelet function is

$$\psi(t) = 2\phi(2t) - \phi(t) \tag{16.15}$$

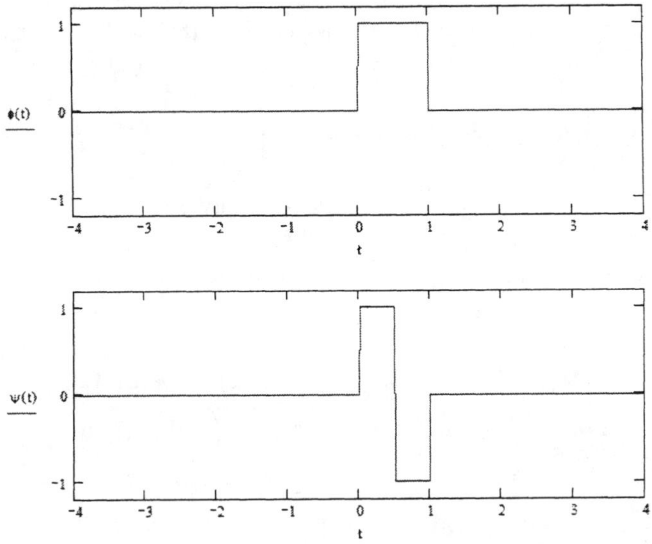

Figure 16.1 Haar scaling function and wavelet.

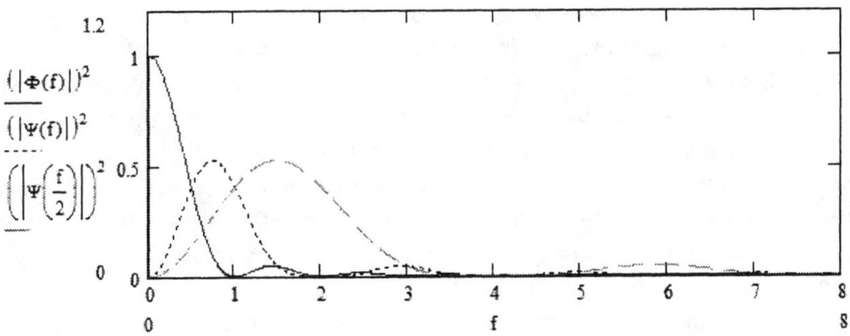

Figure 16.2 Haar PSDs of scaling function and wavelets.

Figure 16.4 shows the sinc scaling function and wavelet. Sinc functions have an infinite support and they form an orthonormal wavelet system. Their PSDs are rectangles with a finite support and a bandwidth of 1. In fact, the spectrum of $\phi(t)$ is

$$\begin{aligned}\Phi(f) &= \mathcal{F}\{\phi(t)\} = \mathcal{F}\{\text{sinc}(\pi t)\} \\ &= \Pi(f) = \begin{cases} 1, & -0.5 \leq f \leq 0.5 \\ 0, & \text{otherwise} \end{cases}\end{aligned} \tag{16.16}$$

which is low-pass with all energy in frequency interval $[0, 0.5]$ and the spectrum of $\psi(t)$ is

$$\begin{aligned}\Psi(f) &= \mathcal{F}\{\psi(t)\} = \mathcal{F}\{2\phi(2t)\} - \mathcal{F}\{\phi(t)\} \\ &= \Phi\left(\frac{f}{2}\right) - \Phi(f) = \Pi\left(\frac{f}{2}\right) - \Pi(f)\end{aligned} \tag{16.17}$$

which is bandpass with all energy in frequency interval $[0.5, 1]$. The PSD of $2\psi(2t)$ is $|\Psi(f/2)|^2$ and its energy concentration is shifted further to the higher frequency. Figure 16.3 shows the PSDs of sinc $\phi(t)$, $\psi(t)$, and $2\psi(2t)$. It can be seen that the PSDs are not overlapped. This is because of the infinite lengths of the sinc functions. For practical truncated sinc functions, overlapping in PSDs will happen.

The time-domain durations and frequency-domain bandwidth occupations of the scaling function and the wavelets can be visualized on a time-frequency (T-F) plane, called a T-F tiling diagram. The tiling diagram is based on the following facts. (1) The scaling function is a low-pass signal (represented by the bottom frequency band) and wavelets are bandpass signals (represented by upper bands). (2) The frequency band of the lowest frequency wavelet is located right next to the band of the low-pass scaling function and the bandwidth is the same as that of the scaling function. (3) The scaling function and wavelets have variable finite supports (durations). By the nature of Fourier transform, we know that a signal can not be limited simultaneously in time and frequency, so the lengths in time and frequency indicated by each block in the tiling diagram are approximations. However each block does indicate concentrations of energy in time and frequency.

For example, the tiling diagram would be exact in time domain for the Haar scaling function and wavelets (see Figure 16.1). However it would be an approximation (a poor one in this case) in frequency domain (see Figure 16.2). On the other hand, this tiling diagram would be exact in frequency domain for the sinc scaling function and wavelets (see Figure 16.3). However it would be an approximation (not too poor) in time domain (see Figure 16.4).

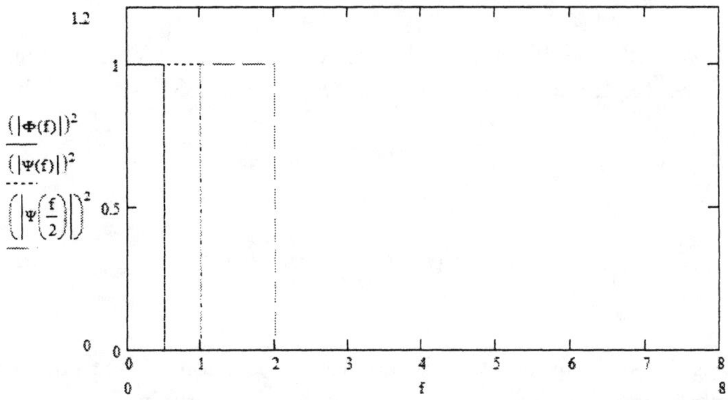

Figure 16.3 Sinc PSDs of scaling function and wavelets.

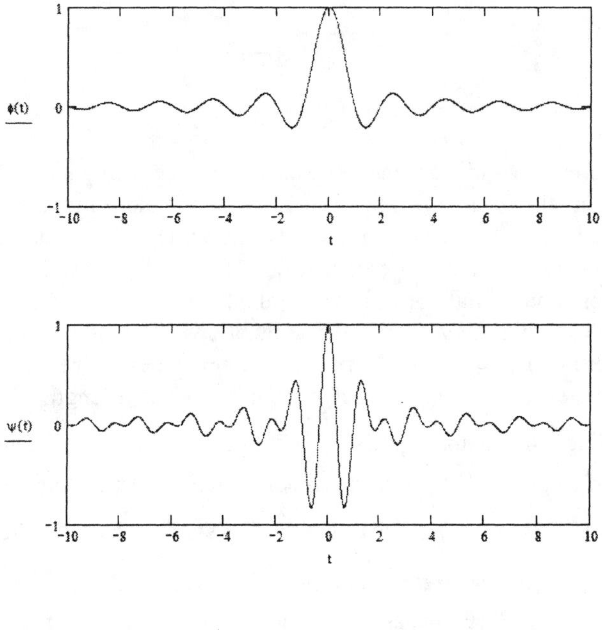

Figure 16.4 Sinc scaling function and wavelet.

16.2 MULTISCALE WAVELET MODULATION (MSM)

Multiscale wavelet modulation (MSM) was proposed in [4, 5] for distorting channels since it allows for frequency selective processing. The MSM signal is based on (16.4).

For comparison, a QAM signal is defined first as

$$m(t) = \sqrt{\frac{E_s}{T}} \sum_{k \in I} d_k \phi(\frac{t}{T} - k) \tag{16.18}$$

where E_s is the symbol energy, T is the symbol duration, d_k are complex-valued QAM-symbols, and $\phi(t)$ is the pulse shape with unit energy in T.

The MSM signal consists of one scaling function and $J - 1$ wavelet functions and is given by

$$\begin{aligned} m(t) &= \sqrt{2^{-(J-1)} \frac{E_s}{T}} \sum_j a_j^0 \phi(2^{-(J-1)} \frac{t}{T} - j) \\ &+ \sum_{n=1}^{J-1} \sqrt{2^{-(J-n)} \frac{E_s}{T}} \sum_j a_j^n \psi(2^{-(J-n)} \frac{t}{T} - j) \end{aligned} \tag{16.19}$$

where a_j^n are mutually independent complex-valued data symbols (or QAM symbols, as called in [4]) for $n = 0, 1, ..., J - 1$. Note that the superscript n is not an exponent. Instead it is to indicate that the data symbol is to be modulated on to the nth wavelet function. ϕ and ψ are defined in (16.7) and (16.8). Note that ϕ and ψ in (16.19) could have finite or infinite length. However for practical implementation we have to design them to have finite durations. The most compact pulse is $\psi(\frac{t}{2T})$ when letting $n = J - 1$ in (16.19). Thus we may design the length of $\psi(\frac{t}{2T})$ as $2T$. The longest symbol is for data symbol a_j^0, it has a length of $2^{J-1}T$ and the corresponding pulse shape is $\sqrt{2^{-(J-1)} \frac{E_s}{T}} \phi(2^{-(J-1)} \frac{t}{T} - j)$. There is also another symbol of length $2^{J-1}T$ for data symbol a_j^1, the corresponding pulse shape is $\sqrt{2^{-(J-1)} \frac{E_s}{T}} \psi(2^{-(J-1)} \frac{t}{T} - j)$. The rest of the data symbols have dyadic decreasing durations ($2^{J-n}T$) and the corresponding pulse shapes are $\sqrt{2^{-(J-n)} \frac{E_s}{T}} \psi(2^{-(J-n)} \frac{t}{T} - j)$, $n = 2, 3, ..., J - 2$. Note that as symbol durations increase, their amplitudes decrease. The result is that all symbols have the same average symbol energy if they all use the same signal constellation. In fact, the average symbol energy of the nth symbol is

$$
\begin{aligned}
& E\{|a_j^n|^2\} \int_0^{2^{(J-n)}T} \left[\sqrt{2^{-(J-n)}\frac{E_s}{T}}\psi(2^{-(J-n)}\frac{t}{T})\right]^2 dt \\
= \;& E\{|a_j^n|^2\}E_s \int_0^{2^{(J-n)}T} \frac{1}{2^{(J-n)}T}\psi^2(\frac{t}{2^{(J-n)}T})dt \\
= \;& E\{|a_j^n|^2\}E_s \int_0^1 \psi^2(x)dx
\end{aligned} \tag{16.20}
$$

where $x = \frac{t}{2^{(J-n)}T}$. This shows that all symbols, regardless of their lengths, have the same average symbol energy if they all use the same signal constellation. In particular, if we make

$$
E\{|a_j^n|^2\} \int_0^1 \psi^2(x)dx = 1 \tag{16.21}
$$

then the average symbol energy is E_s. For the scaling function symbol, (16.20) becomes $E\{|a_j^n|^2\}E_s \int_0^1 \phi^2(x)dx$. To have the same average symbol energy E_s, we require

$$
E\{|a_j^n|^2\} \int_0^1 \phi^2(x)dx = 1 \tag{16.22}
$$

Both (16.21) and (16.22) can be satisfied simultaneously (e.g., the Haar scaling function and wavelet).

By (16.7) and (16.8) we know that $m(t)$ can decomposed as weighted sum of the most compact pulse. Therefore, for independent data symbols, the power spectral density is determined by the most compact[1] pulse $\psi(\frac{t}{2T})$. From (16.8) we have $\psi(\frac{t}{2T}) = \sqrt{2}\sum_n g_n \phi(\frac{t}{T} - n)$. Thus, intuitively we can think that the power spectral density of $m(t)$ is determined by the Fourier transform of $\phi(\frac{t}{T})$. As a matter of fact, it was shown in [4, 5] (see Appendix 16A) that for independent i.i.d. data with zero-mean and unit variance, the power spectral density of the MSM signal $m(t)$ is

$$
S_{MSM}(f) = E_s|\Phi(fT2^{J-1})|^2 + E_s \sum_{n=1}^{J-1} |\Psi(fT2^{J-n})|^2 \tag{16.23}
$$

where

$$
\Phi(f) = \mathcal{F}\{\phi(t)\}, \; \Psi(f) = \mathcal{F}\{\Psi(t)\}
$$

---

[1] We do not use the term "the shortest pulse" since the wavelet (and scaling) pulse could have an infinite length, such as a sinc function, even though its energy is mostly confined in a short period.

Data	Pulse	T_s	PSD	Frequency band
a_j^3	$\sqrt{\frac{E_s}{2T}} a_j^3 \psi(\frac{t}{2T})$	$2T$	$E_s\lvert\Psi(2fT)\rvert^2$	upper 1/2
a_j^2	$\sqrt{\frac{E_s}{4T}} a_j^2 \psi(\frac{t}{4T})$	$4T$	$E_s\lvert\Psi(4fT)\rvert^2$	next 1/4 down from above
a_j^1	$\sqrt{\frac{E_s}{8T}} a_j^1 \psi(\frac{t}{8T})$	$8T$	$E_s\lvert\Psi(8fT)\rvert^2$	next 1/8 down from above
a_j^0	$\sqrt{\frac{E_s}{8T}} a_j^0 \phi(\frac{t}{8T})$	$8T$	$E_s\lvert\Phi(8fT)\rvert^2$	lower 1/8

Table 16.1 MSM parameters for $J = 4$.

are the Fourier transforms of $\phi(t)$ and $\psi(t)$, respectively. $\Phi(fT2^{J-1})$ and $\Psi(fT2^{J-n})$ are the Fourier transforms with scaling of frequencies. Equation (16.23) can be simplified to (see Appendix 16A)

$$S_{MSM}(f) = E_s|\Phi(fT)|^2 \tag{16.24}$$

This shows that the PSD of MSM is indeed determined by the Fourier transform of $\phi(\frac{t}{T})$.

From (16.23) we can see that $S_{MSM}(f)$ is a sum of PSDs of each individual pulse (a scaling function and $J-1$ wavelets).

The component symbols of an MSM signal are collectively called a supersymbol. The components of a supersymbol can be visualized as T-F tiling of the plane. Figure 16.5 is the tiling diagram for MSM, where each block represents the symbol duration and frequency bandwidth occupied by the symbol corresponding to the data as labeled.[2] As pointed out in the previous section, each block in the tiling diagram indicates energy concentrations in time and frequency of the corresponding scaling function or wavelets. In time domain, the length of each block might not be the same as the wavelet length, but it does represent the exact data symbol duration. In frequency, the length of each block represents the bandwidth of the majority of the symbol energy.

As an example, let $J = 4$. Then the blocks in Figure 16.5 have the characteristics shown in Table 16.1, where T_s is the symbol duration. Frequency bands are expressed as a fraction of the total bandwidth which is the bandwidth of $|\Phi(fT)|^2$. The table is arranged such that the positions of the rows of the table correspond to the positions of the frequency bands in Figure 16.5.

From this example we see that the shortest symbol duration is $2T$, instead of T. In this particular example, the total bandwidth occupied is the bandwidth of a symbol of duration T. 1/2 total bandwidth is occupied by a 2T-symbol, 1/4 by a

---

[2] Using the sinc wavelets as an example, the partition of the total band and position of each band are understood. (see Figure 16.3).

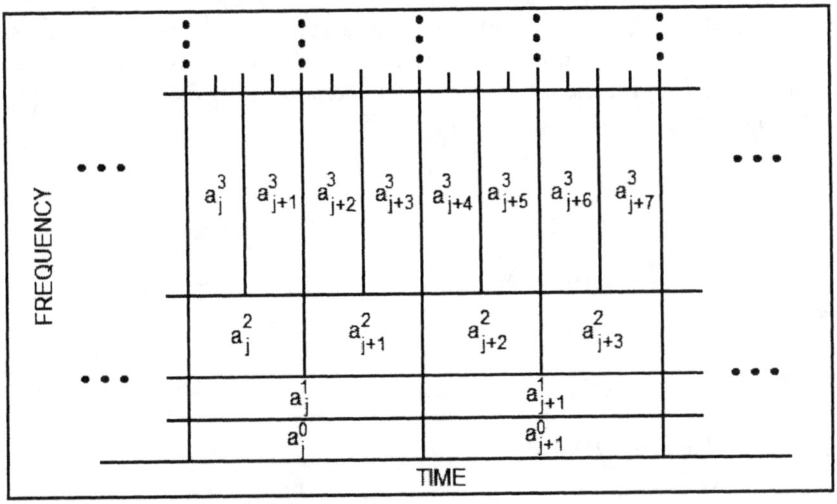

Figure 16.5 Tiling diagram for multiscale modulation. From [4]. Copyright © 1994 IEEE .

4T-symbol, and $2 \times 1/8$ by two 8T-symbols each.

The bandwidth efficiency of MSM is derived as follows. Let a_j^n be a member of a 2^{K_n} QAM constellation, where K_n is the number of bits per symbol. Then the bandwidth efficiency of MSM can be easily figured out from (16.19) as

$$\rho_{MSM} = \frac{\frac{1}{2^{(J-1)}T} K_0 + \sum_{n=1}^{J-1} \frac{1}{2^{(J-n)}T} K_n}{(1+\beta)/T}$$

The numerator is the bit rate in bits/second of the MSM and the denominator is the required bandwidth to transmit this signal, where β is the percent excess bandwidth from Nyquist signaling. The above can be simplified as

$$\rho_{MSM} = \frac{2^{-(J-1)} K_0 + \sum_{n=1}^{J-1} 2^{-(J-n)} K_n}{1+\beta} \tag{16.25}$$

When $K_n = K$ at each scale, the bandwidth efficiency of MSM is $K/(1+\beta)$ which is the same as the QAM signal with the same pulse shape $\phi(t/T)$. Thus theoretical bandwidth efficiency of MSM does not increase with respect to QAM. Practically, MSM is more robust against channel distortion since its low-pass subchannels will suffer from less ISI.

BER of MSM in AWGN channels can be evaluated as follows. From (16.19) we see that the MSM signal consists of orthogonal signals and thus demodulation is to project received signal on to basis functions (the scaling function and wavelets) and results will have no ISI and cross subchannel interference between symbols. Thus, the average probability of a symbol error in MSM is

$$\overline{P_s} = \sum_{n=0}^{J-1} \Pr(a^n) P_s^n \tag{16.26}$$

where $\Pr(a^n)$ is the relative frequency of a^n in a supersymbol, that is (see Figure 16.5)

$$\Pr(a^n) = \begin{cases} 2^{-(J-1)}, & n = 0 \\ 2^{-(J-n)}, & 1 \leq n \leq J-1 \end{cases} \tag{16.27}$$

Using (16.26) and (16.27) yields

$$\overline{P_s} = 2^{-(J-1)} P_s^0 + \sum_{n=1}^{J-1} 2^{-(J-n)} P_s^n \tag{16.28}$$

From (16.20) we know all symbols have the same average energy. If all symbols use the same modulation scheme, then $P_s^n = P_s$ for all n and $\overline{P_s} = P_s$. From (16.28) we see that the finer scale data have the most impact on error probability since they appear more often in the supersymbol.

To evaluate (16.28) we need to know P_s^n which will be known once the modulation on each symbol is determined. For example, if QPSK is used for all scales, then $P_s^n = Q(\sqrt{2E_b/N_o})$ where $E_b = E_s/2$ is the bit energy and $N_o/2$ is the two-sided PSD of the AWGN.

Finally, we consider implementation of MSM. Signal $m(t)$ can be decomposed into the weighted sum of $\phi(\frac{t}{T})$ (see (16.7) and (16.8)) with data symbols $\{a_j^n\}$ and a number of coefficients $\{h_n\}$, which can be realized by an IDWT (inverse discrete wavelet transform) with $\{a_j^n\}$ as inputs and $\phi(\frac{t}{T})$ as the pulse shaping function. Figure 16.6 shows the transceiver given in [5] (where $\phi(\frac{t}{T})$ is $\phi(t)$ in [4], which is an error).

16.3 FRACTAL MODULATION

Fractal modulation reported in [6](1992) probably is the earliest application of wavelets for modulations. The purpose is to use "self-similarity" between dyadic expansions of wavelets to create a type of frequency diversity so that a symbol is repeatedly

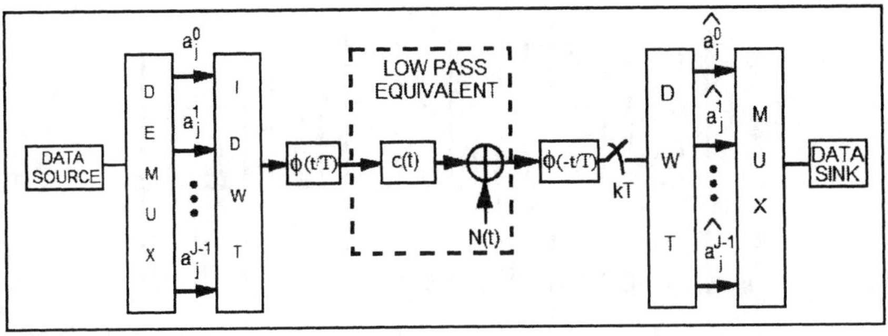

Figure 16.6 Transceiver processing for multiscale modulation. From [4]. Copyright © 1994 IEEE .

transmitted in multiple subbands using different symbol durations. If the channel is noisy and has unknown duration and bandwidth, fractal modulation is a candidate.

A fractal modulation signal is given by [6]

$$x(t) = \sum_m \sum_n \beta^{-m/2} q[n] \psi_n^m(t) \tag{16.29}$$

where

$$\psi_n^m(t) = 2^{m/2} \psi(2^m t - n)$$

where m and n are the dilation and translation indices, respectively. $q[n]$ is the data symbol modulated on wavelets $\psi_n^m(t)$ for $m = 0, 1, ..., M-1$. That is, the same data symbol is modulated onto M different dilations of the same wavelet, to within an amplitude factor. Figure 16.7 is a portrait of T-F tiling of a fractal signal.

By comparing (16.29) to (16.19) we see that the fractal modulation is in fact a special case of the MSM. By letting $a_j^n = a_j^m$ for $n \neq m$ in (16.19), MSM becomes a fractal modulation (although it is slightly different from (16.29) since it contains a scaling function as a basis function while (16.29) does not).

For details of fractal modulation, like wavelet design, bandwidth efficiency, and probability of error, the reader is referred to [6].

Fractal modulation is suitable to the case in which both the bandwidth and duration parameters are either unknown or not available to the transmitter. This case may arise in a range of both point-to-point and broadcast communication scenarios involving, for example, jammed and fading channels, multiple access channels, covert and low probability of intercept (LPI) communication, and broadcast communication to disparate receivers. Fractal modulation, due to its self-similarity between

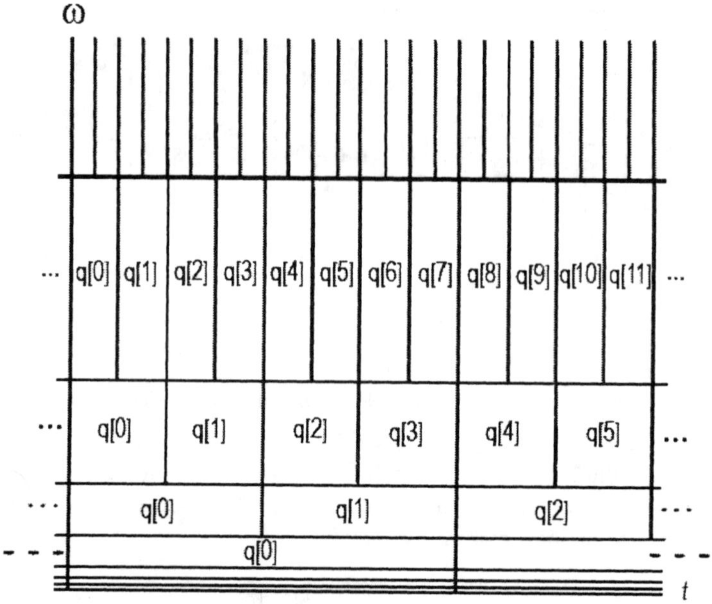

Figure 16.7 T-F tiling of fractal modulation. From [6]. Copyright © 1992 IEEE.

signal symbols at different symbol rates and different frequency bands, has the property that, at least in principle, an arbitrary short duration time-segment is sufficient to recover $q[n]$ given adequate receiver bandwidth or alternatively, using an arbitrarily narrow receiver bandwidth given sufficient duration. In practice, this means the receiver may have a very short time interval to receive the signal (say, due to fading), but if the receiver bandwidth is enough, it can receive the symbols at the high rates in the upper part of the T-F tiling diagram (see Figure 16.7) and recover $q[n]$. On the other hand, if the receiver has very narrow bandwidth (say, due to jamming) to receive the signal, but given long enough time duration, the receiver can receive the low rate symbols in the lower part of the T-F tiling diagram and recover $q[n]$.

16.4 WAVELET PACKET MODULATION (WPM)

Lindsey in 1997 proposed the *wavelet packet modulation* (WPM) in an attempt to reduce the effect of external interferences [7]. In this scheme, a packet of data symbols are modulated onto a set of orthonormal wavelets that are dyadic expansions of a parent wavelet. The wavelets are the pulse-shaping functions of the data symbols. They have different lengths and thus occupy different bandwidths. All these features are also possessed by the multiscale modulation (MSM) described before. However the wavelet packet modulation (WPM) has a new feature. That is, the T-F tiling of WPM is not arranged according to the order of the scales of the wavelets, instead, it is arranged according to the need of avoiding interferences in certain frequency bands or time durations. This gives the flexibility to the system designer to avoid various types of interferences. Thus WPM is in fact a generalization of MSM.

The wavelet packet functions are defined by

$$\begin{aligned} p_{2n}(t) &= \sqrt{2}\sum_{k\in Z} h_k p_n(2t-k) \\ p_{2n+1}(t) &= \sqrt{2}\sum_{k\in Z} g_k p_n(2t-k), \quad n \in Z_+ \end{aligned} \tag{16.30}$$

where Z is the integer set, Z_+ is the nonnegative integer set. h_k and g_k are scaling function coefficients and wavelet function coefficients defined in (16.7) and (16.9). In particular $p_0(t) = \phi(t)$ is the scaling function and $p_1(t) = \psi(t)$ is the wavelet function. Using these wavelet packet functions as pulse shapes, the WPM signal is given in [7] as

$$s_{WPM}(t) = \sum_{i=1}^{J}\sqrt{\frac{2^{l_i}E_s}{T}}\sum_{m=-\infty}^{\infty} a_m^i p_{n_i}\left(\frac{2^{l_i}t}{T} - m\right) \tag{16.31}$$

where a_m^i are the complex data symbol on the ith channel (so i is not an exponent) that has scale 2^{l_i} (l_i is an exponent, the channel symbol rate is 2^{-l_i}); $m \in Z$ is the time index; $l_i \in Z_-$ (nonpositive integer set) determines the scale 2^{l_i} of the wavelet; and $p_{n_i}(t)$ is the pulse shape wavelet whose spectral concentration band is determined by $n_i \in Z_+$.

The indexes l_i and n_i must satisfy the condition that the collection $\mathfrak{C} = \{(l_1, n_1), (l_2, n_2), ..., (l_J, n_J)\}$ defines a collection of dyadic intervals $I_i = [2^{l_i}n_i, 2^{l_i}(n_i+1))$ that form a disjoint covering of $[0, 1)$ (see the next example for illustration). When this condition is satisfied, the wavelets $2^{l_i/2}p_{n_i}\left(2^{l_i}t/T - m\right)$ form a complete orthonormal basis for a subspace of $L^2(\mathbf{R})$ denoted as W_0^0.

As an example, letting $J = 4$, we can have the set of (l_i, n_i) as shown in Table 16.2, which forms a disjoint covering of $[0, 1)$. Note that $\phi(t) = p_0(t)$ in Table 16.2. The WPM signal in Table 16.2 corresponds to the WPM tiling in Figure 16.15. The pulse of a_m^1 is $\phi(\frac{t}{2T})$ which is low-pass and its duration is $2T$. Dividing the total bandwidth (which is $1/T$) into two equal-width bands, $\phi(\frac{t}{2T})$ occupies the 0th band or lower $1/2$ of the total bandwidth. The pulse of a_m^2 is $p_4(\frac{t}{8T})$ which is bandpass and its duration is $8T$. Partitioning the total bandwidth into 8 bands, $p_4(\frac{t}{8T})$ occupies the fourth band which is right above the band of $\phi(\frac{t}{2T})$. The pulse of a_m^3 is $p_5(\frac{t}{8T})$ which is bandpass and its duration is $8T$. Partitioning the total bandwidth into eight bands, $p_5(\frac{t}{8T})$ occupies the fifth band which is right above the band of $p_4(\frac{t}{8T})$. The pulse of a_m^4 is $p_3(\frac{t}{4T})$ which is bandpass and its duration is $4T$. Partitioning the total bandwidth into four bands, $p_3(\frac{t}{4T})$ occupies the third band which is right above the band of $p_5(\frac{t}{8T})$.

Comparing Table 16.2 to Table 16.1, we see that both WPM and MSM use orthogonal wavelets for pulse shaping. The difference between them is that MSM uses only one scaling function $\phi(t)$ and one wavelet $\psi(t)$ and its dyadic expansions for pulse shaping, whereas WPM uses more than one scaling function and more than one of the wavelet functions defined in (16.30) for pulse shaping. This gives WPM the flexibility of arranging frequency bands in different orders, whereas MSM does not. In addition, it gives the WPM the flexibility of partitioning the frequency band differently, rather than the dyadic pattern. The next example shown in Table 16.3 and Figure 16.8 shows this feature of WPM. Figure 16.8(a) shows the example given by Table 16.2. The top line shows the dyadic intervals $I_i = [2^{l_i}n_i, 2^{l_i}(n_i+1))$ that form a disjoint covering of $[0,1)$. The bottom part is the F-tiling whose correspondence to the intervals I_i is apparent. For example, the symbol a_m^4 corresponds to interval $[\frac{3}{4},1)$ which is the upper $1/4$ in the $[0,1)$ interval and the corresponding frequency band is also the upper $1/4$ of the total frequency band. Figure 16.8(b) shows the example given by Table 16.3. In this example, the partition of the total bandwidth is different from that of Figure 16.8(a). This shows the flexibility of WPM, which gives the flexibility to the system designer to avoid various types of interferences, as mentioned in the beginning of this section.

WPM has the same power spectral density and bandwidth efficiency as those of MSM, which are proved in [7] in a similar way to that in [5].

The BER performance of WPM is not discussed in [7]. However, it is obvious that the same argument for MSM in Section 16.3 applies to WPM as well.

Implementation of WPM is through the IDWT/DWT pair as mentioned in [7]. Again, this is the same as that of MSM too.

16.5 WAVELET PAM

The purpose of the wavelet PAM proposed in [8] in 1996 is to improve the bandwidth efficiency of the system by adding dyadic expansions to the parent wavelet as pulse shapes. The baseband wavelet PAM signal is given as a sum of orthonormal wavelets:

Data	(l_i, n_i)	I_i	Pulse	T_s	Frequency band
a_m^4	$(-2, 3)$	$[\frac{3}{4}, 1)$	$\sqrt{\frac{E_s}{4T}}a_m^4 p_3(\frac{t}{4T})$	$4T$	upper $1/4$
a_m^3	$(-3, 5)$	$[\frac{5}{8}, \frac{3}{4})$	$\sqrt{\frac{E_s}{8T}}a_m^3 p_5(\frac{t}{8T})$	$8T$	next $1/8$ down
a_m^2	$(-3, 4)$	$[\frac{1}{2}, \frac{5}{8})$	$\sqrt{\frac{E_s}{8T}}a_m^2 p_4(\frac{t}{8T})$	$8T$	next $1/8$ down
a_m^1	$(-1, 0)$	$[0, \frac{1}{2})$	$\sqrt{\frac{E_s}{2T}}a_m^1 \phi(\frac{t}{2T})$	$2T$	lower $1/2$

Table 16.2 WPM parameters for $J = 4$.

Data	(l_i, n_i)	I_i	Pulse	T_s	Frequency band
a_m^6	$(-1, 1)$	$[\frac{1}{2}, 1)$	$\sqrt{\frac{E_s}{2T}}a_m^6 \psi(\frac{t}{2T})$	$2T$	upper $1/2$
a_m^5	$(-2, 1)$	$[\frac{1}{4}, \frac{1}{2})$	$\sqrt{\frac{E_s}{4T}}a_m^5 \psi(\frac{t}{4T})$	$4T$	next $1/4$ down
a_m^4	$(-4, 3)$	$[\frac{3}{16}, \frac{1}{4})$	$\sqrt{\frac{E_s}{16T}}a_m^4 p_3(\frac{t}{16T})$	$16T$	next $1/16$ down
a_m^3	$(-4, 2)$	$[\frac{1}{8}, \frac{3}{16})$	$\sqrt{\frac{E_s}{16T}}a_m^3 p_2(\frac{t}{16T})$	$16T$	next $1/16$ down
a_m^2	$(-4, 1)$	$[\frac{1}{16}, \frac{1}{8})$	$\sqrt{\frac{E_s}{16T}}a_m^2 \psi(\frac{t}{16T})$	$16T$	next $1/16$ down
a_m^1	$(-4, 0)$	$[0, \frac{1}{16})$	$\sqrt{\frac{E_s}{16T}}a_m^1 \phi(\frac{t}{16T})$	$16T$	lower $1/16$

Table 16.3 WPM parameters for $J = 6$.

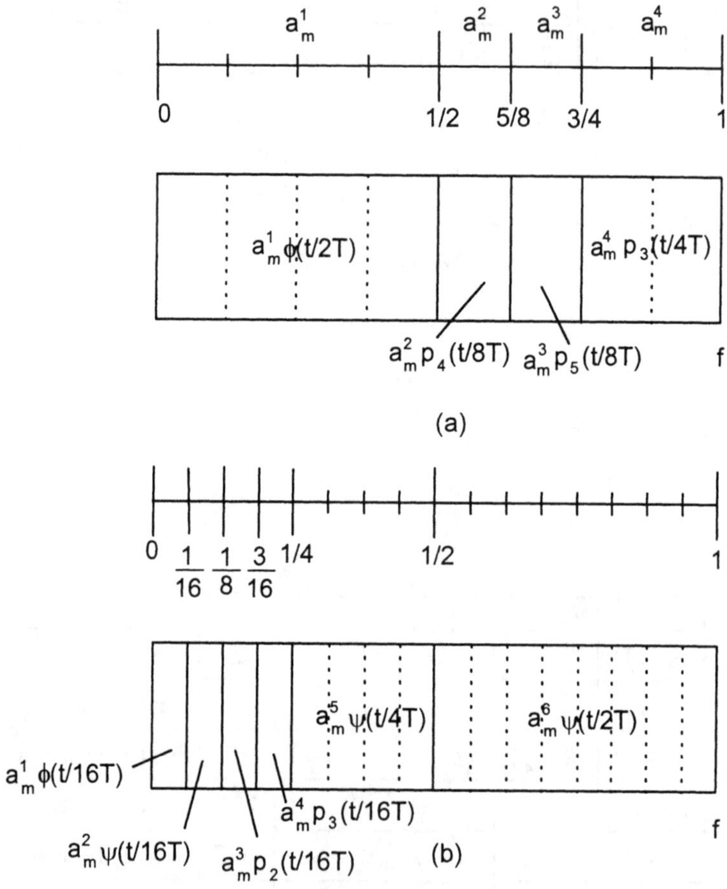

Figure 16.8 F-tiling of WPM: (a) for the example in Table 16.2 and (b) for the example in Table 16.3.

$$x(t) = \sum_{m=0}^{N} \sum_{n} a_{mn} \psi_{mn}(t) = \sum_{m=0}^{N} \sum_{n} a_{mn} 2^{-m/2} \psi(2^{-m}t - n) \tag{16.32}$$

where N is the number of data streams; $\psi(t)$ is the parent wavelet; and

$$\psi_m(t) = 2^{-m/2}\psi(2^{-m}t)$$

are its dyadic expansions. $\psi_{mn}(t)$ is shifted $\psi_m(t)$. This signal is very similar to the signal of MSM (see (16.19)) except that the scaling function is not used here. However, in the context of [8], the signal symbols $a_{mn} \in [-1, 1]$. That is, they are just binary instead of QAM symbols in MSM.

The bandwidth efficiency of (16.32) is compared to a raised cosine pulse with roll-off parameter $\beta = 0.2$. A 22-tap wavelet with compact spectrum chosen from [9] is used for $x(t)$. The pulse shapes and the PSDs of these two signals are shown in Figure 16.9 and Figure 16.10, respectively. The particular wavelet is chosen because of its narrow bandwidth compared to most other wavelets in the literature. The particular $\beta = 0.2$ of the raised cosine pulse is chosen because it will give a bandwidth compared favorably to the chosen wavelet. Therefore the effectiveness of wavelet PAM can be demonstrated after showing that wavelet PAM can achieve better bandwidth efficiency than the raised cosine.

The parent wavelet is $\psi_0(t) = \psi(t)$. Its waveform is shown in Figure 16.9 and its PSD is labeled $|\Psi_0(f)|^2$. The first dyadic expansion is $\psi_1(t) = \frac{1}{\sqrt{2}}\psi(\frac{t}{2})$ whose PSD ($|\Psi_1(f)|^2$) will be a shrunk version of that of $\psi(t)$ with half the bandwidth (by any practical definition, which is defined as 70 dB down from the maximum in [8]). Viewing $\psi(t)$ as a low-pass signal, then $\psi_1(t)$ is also a low-pass signal and its bandwidth is within the bandwidth of $\psi(t)$. The next dyadic expansion $\psi_2(t) = \frac{1}{2}\psi(\frac{t}{4})$ has a further shrunk PSD ($|\Psi_2(f)|^2$) and its bandwidth is also the bandwidth of $\psi(t)$, and so on. In summary, adding dyadic expansions of $\psi(t)$ to the signal $x(t)$ does not increase the bandwidth of the signal when it is viewed as a low-pass signal. When viewed as a bandpass signal the total bandwidth will increase slightly as more dyadic expansions are added, as shown in Figure 16.9(b).

The 70 dB (low-pass) bandwidth of $\psi(t)$ is 0.355, normalized to the sampling frequency, or $1.42/T$ Hz, referred to the symbol rate. The bandwidth of the raised cosine pulse with a $\beta = 0.2$ is 0.15, normalized to the sampling frequency, or $0.6/T$ Hz, referred to the symbol rate.

At first glance, the PSD of the wavelets compare unfavorably to that of the raised cosine. However, due to the fact that wavelets have zero mean values, single-side band (SSB) transmission can be used, so the required bandwidth of wavelet PAM can be reduced. Using SSB, the wavelet PAM, after modulation, has a bandwidth

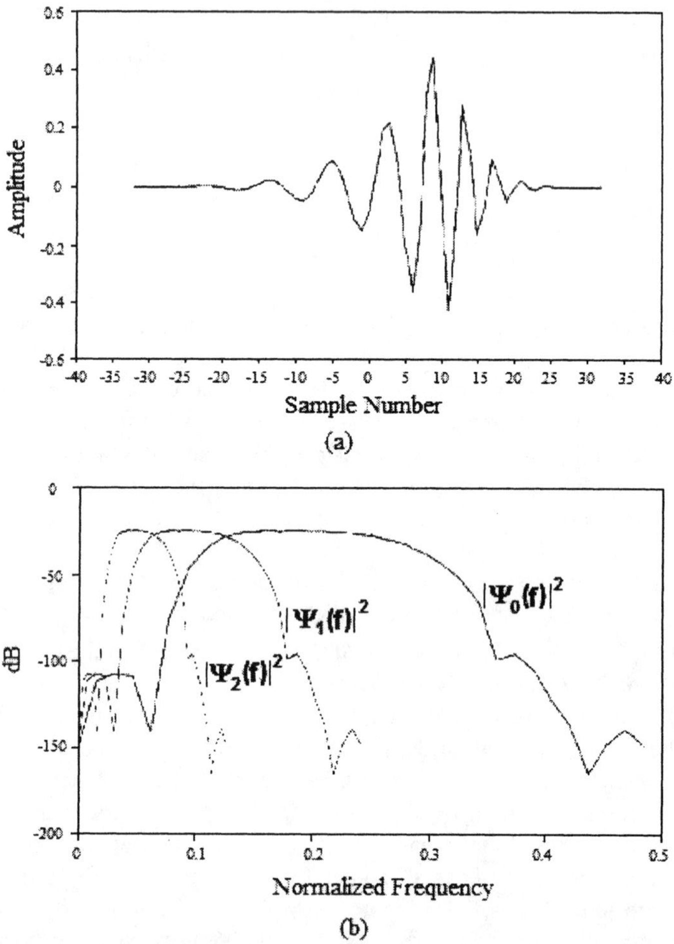

Figure 16.9 Parent wavelet and its power spectral density (PSD): (a) 22-tap wavelet, sampled four times/symbol period and (b) power spectrum of 22-tap wavelet. After [8]. Copyright © 1996 IEEE.

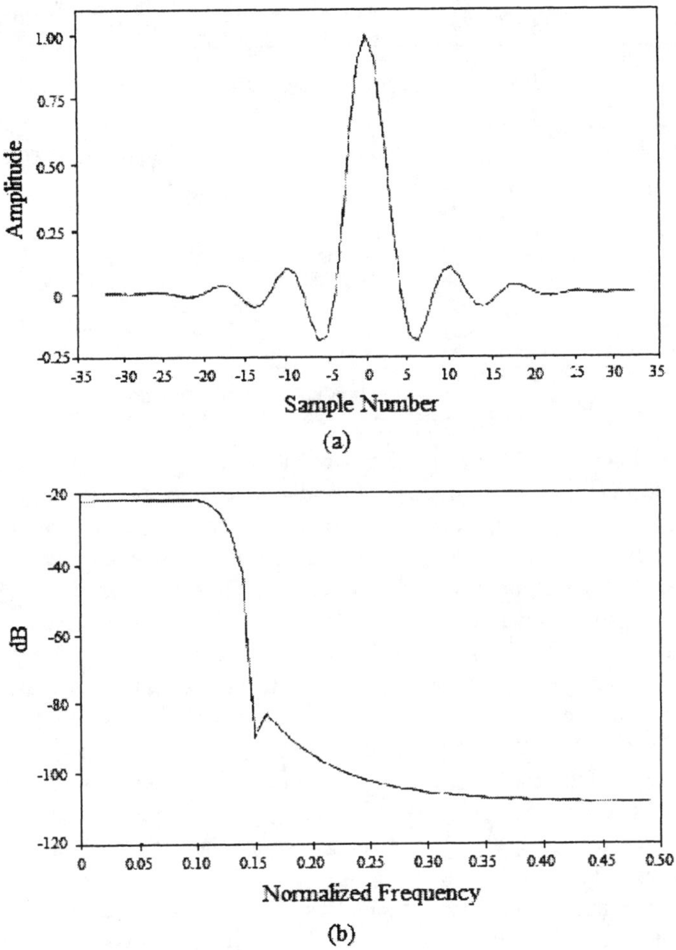

Figure 16.10 Raised cosine pulse and its PSD: (a) raised cosine pulse sampled four time/symbol period ($\beta = 0.2$) and (b) power spectral density for raised cosine pulse. From [8]. Copyright © 1996 IEEE.

System	Spectral Efficiency (low-pass)	Spectral Efficiency (bandpass)
Raised Cosine	0.83 bps/Hz	n.a.
Wavelet 1	0.70 bps/Hz	0.86 bps/Hz
Wavelet 1.5	1.06 bps/Hz	1.15 bps/Hz
Wavelet 1.75	1.23 bps/Hz	1.29 bps/Hz

Table 16.4 Spectral efficiency comparison of wavelet PAM

of $1.42/T$ Hz, while the raised cosine scheme has a bandwidth of $1.2/T$ Hz. The bandwidth efficiencies for BPSK are

$$\frac{1/T \text{ bps}}{1.42/T \text{ Hz}} = \frac{1}{1.42}\text{bps/Hz} = 0.70 \text{ bps/Hz}$$

for wavelet and

$$\frac{1/T \text{ bps}}{1.2/T \text{ Hz}} = \frac{1}{1.2}\text{bps/Hz} = 0.83 \text{ bps/Hz}$$

for the raised cosine, where bps is bits/second. This shows that the raised cosine still outperforms the wavelet in terms of bandwidth efficiency. However, by adding dyadic expansions to the signal, the wavelet scheme can further increase bandwidth efficiency. By adding the first dyadic expansion, the PAM can transmit extra 0.5 b/symbol period (the signal is called wavelet 1.5). The bandwidth efficiency would be $1.5/1.42$ bps/Hz, versus $1/1.2$ Hz for the raised cosine. By adding the first and the second dyadic expansions, the PAM can transmit extra 0.75 b/symbol period (the signal is called wavelet 1.75). The bandwidth efficiency would be $1.75/1.42$ bps/Hz, and so on. We can continue to add more dyadic expansions, reaching (in the limit) an efficiency of $2.0/1.42$ bps/Hz, while maintaining the same energy per bit.

Table 16.4 compares the bandwidth efficiencies of the raised cosine with $\beta = 0.2$ and various wavelet schemes using the wavelets given in Figure 16.9, using PAM modulation. From the table we can see that wavelet schemes are better when dyadic expansions are added.

Also listed are the bandwidth efficiencies of the wavelet schemes when wavelet signals are considered bandpass. From Figure 16.9 we can see that the bandwidth will be narrower when a wavelet is considered as bandpass thus the bandwidth efficiency will be better. In this case, even the parent wavelet has a better bandwidth efficiency than the raised cosine.

To use the wavelets as bandpass signals, the only change in the system is that we need a narrower bandpass filter after SSB modulation to just pass the spectrum within the wavelet signal's bandwidth.

As pointed out in [8], the extra capacity of the wavelet system can be used for channel coding instead of transmitting more bits per second. For example, if wavelet 1.5 is used, a rate 2/3 code can be used. A simple 4-state code can then deliver 3 dB coding gain over the raised cosine system with little bandwidth penalty ($1.42/T$ Hz versus $1.2/T$ Hz).

16.6 M-BAND WAVELET MODULATION (MWM)

In some applications, uniform partition of the channel bandwidth may be desired, for example the mitigation of a stationary narrowband interference. In [5], a uniform decomposition based on an M-band wavelet basis was proposed. M-band wavelets are a generalization of the dyadic (two-band) wavelets and they are given by [2]

$$\psi_n(t) = \sum_l \sqrt{M} h_n(l)\phi(Mt - l), \quad \text{for } n = 1, 2, ..., M-1$$

The M-band wavelet modulation (MWM) signal is given as [5]

$$\begin{aligned} m(t) &= \sqrt{\frac{E_s}{MT}} \sum_j a_j^0 \phi(\frac{t}{MT} - j) \\ &+ \sum_{n=1}^{M-1} \sqrt{\frac{E_s}{MT}} \sum_j a_j^n \psi_n(\frac{t}{MT} - j) \end{aligned} \tag{16.33}$$

This signal format is similar to that of MSM (see (16.19)) except for two differences: (1) all symbols have the same duration MT, and (2) there is not only one wavelet, but $M-1$ wavelets. Together with the scaling function, there are total M pulse shapes and the number of symbols in an MWM supersymbol is M.

The PSD of this signal can be found using a procedure similar to that of MSM [5] and it is

$$S_{MWM}(f) = E_s|\Phi(fTM)|^2 + E_s \sum_{n=1}^{M-1} |\Psi_n(fTM)|^2 \tag{16.34}$$

which can be simplified as

$$S_{MWM}(f) = E_s|\Phi(fT)|^2 \tag{16.35}$$

We can see from (16.34) that the scaling function and each wavelet occupy the same bandwidth and from (16.35) we see that the total bandwidth of the signal is the same as that of QAM and MSM. The total bandwidth is decomposed into M subbands.

It was also shown that MWM has the same bandwidth efficiency and the same probability of error as MSM [5].

However, implementation of MWM cannot be done by the IDWT/DWT pair as for MSM since it does not use dyadic wavelets. Instead, filter banks must be used in the modulator and demodulator [5].

16.7 OVERLAPPED DISCRETE WAVELET MULTITONE (DWMT) MODULATION

Sandberg et al. in 1995 proposed an overlapped discrete wavelet multitone modulation (DWMT) for high-speed copper wire communications [10]. Like conventional OFDM, the channel is divided into equal-bandwidth subchannels. However unlike conventional OFDM, the pulse waveforms for different symbol blocks are allowed to overlap in time. In addition, the *baseband* pulse is designed to minimize stop-band energy (thereby optimizing subchannel spectral containment), with the constraint that the set of *bandpass* pulses and their time-shifts by integer multiples of N provide a set of orthonormal waveforms (a Nyquist set) for transmission of the sequence of symbol blocks. The conventional OFDM pulse length is equal to the symbol block length and the baseband pulse is not optimized. Since the DWMT pulse is longer and its shape is optimized, DWMT has better spectral containment.

Specifically, the DWMT system is shown in Figure 16.11. Note that this figure also applies to conventional DMT (or OFDM). For DWMT, $N = M$, and for DMT, $N = M + \kappa$, where κ is the length of the cyclic extension. The data stream in each subchannel is up-sampled by a factor of N by inserting $N - 1$ zeros between two symbols of the symbol sequence. In each subchannel, modulation is implemented by passing zero-padded data symbols through a bandpass filter whose impulse response is f_l^m. Subchannel m is centered at the normalized radian frequency ω_m, and the lth sample of the impulse response is

$$f_l^m = p_l \cos(\omega_m l + \phi_m) \tag{16.36}$$

The baseband pulse p_l is common to all subchannels, and it has nonzero samples for the interval $0 \leq l \leq gN - 1$ only. The parameter $g > 1$ is the overlap factor. During each frame, each subchannel modulator accepts a new symbol, and a length-N segment of the subchannel waveform is produced as a convolution of the zero-padded symbol sequence with the filter impulse response, which is

$$x_l^m = \sum_{i=-\infty}^{\infty} s_i^m f_{l-iN}^m, \quad 0 \leq l \leq N - 1 \tag{16.37}$$

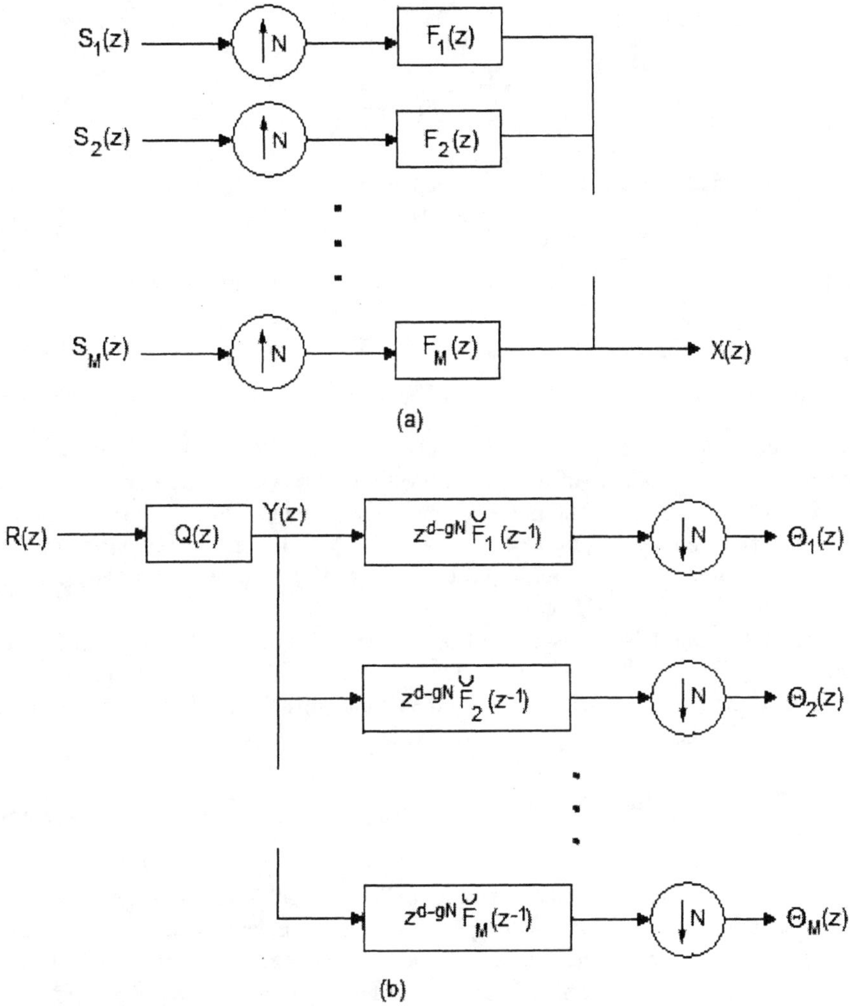

Figure 16.11 DWMT modem: (a) modulator and (b) demodulator. From [10]. Copyright © 1995 IEEE.

where s_i^m represents the symbol in symbol block i that is transmitted on subchannel m, with $i = 0$ being the current symbol. The frame duration is N/f_s, where f_s is the sampling frequency. The final DWMT signal is the sum of the M subchannel waveform segments:

$$x_l = \sum_{m=1}^{M} x_l^m = \sum_{i=-\infty}^{\infty} \sum_{m=1}^{M} s_i^m f_{l-iN}^m, \quad 0 \leq l \leq N-1 \tag{16.38}$$

In expanded form, with $f_l^m = 0$ outside the range $0 \leq l \leq gN-1$, (16.37) becomes

$$\begin{aligned} x_0^m &= s_0^m f_0^m + s_{-1}^m f_N^m + \ldots + s_{-(g-1)}^m f_{(g-1)N}^m \\ x_1^m &= s_0^m f_1^m + s_{-1}^m f_{1+N}^m + \ldots + s_{-(g-1)}^m f_{1+(g-1)N}^m \\ &\ldots \\ x_{N-1}^m &= s_0^m f_{N-1}^m + s_{-1}^m f_{2N-1}^m + \ldots + s_{-(g-1)}^m f_{gN-1}^m \end{aligned} \tag{16.39}$$

The above shows that the modulated signal of the mth subchannel has the following components: the products of the current symbol (s_0^m) with the subcarrier samples f_l^m for $l = 0, 1, \ldots N-1$, and the products of the previous symbols with the rest of the subcarrier samples f_l^m for $l = N, N+1, \ldots gN-1$. The second part is due to overlapping in time of the filter impulse response. If $g = 1$, the signal degenerates back to the conventional DMT.

The baseband pulse p_l used in a DWMT system is designed to minimize stopband energy (thereby optimizing subchannel spectral containment), with the constraint that the bandpass pulses f_l^m and their time-shifts by integer multiples of N form a set of orthonormal waveforms (a Nyquist set), that is,

$$\sum_l f_l^{m_1} f_{l-iN}^{m_2} = \beta \delta_i \delta_{m_1-m_2} \tag{16.40}$$

for some $\beta > 0$, where δ_l is the Kronecker delta function. Wavelets are natural candidates for this Nyquist set. This property ensures the independence of the signals between different subchannels ($\delta_{m_1-m_2}$) and different signal symbols (δ_i). Thus although signals for all subchannels are mixed and in addition there are overlappings in time among signal symbols, they still can be separated at the receiver by a correlation method, which is implemented by fast wavelet transform (FWT).

The pulses p_l loosely resemble in shape a DFT time-weighting function, such as a Hamming window. However, a time-weighting pulse, such as a Hamming window, which is designed to achieve a high level of spectral containment, without regard for (16.40), does not yield a Nyquist set of subchannel pulses, and significant irreducible

IS1 would result.

The frequencies (normalized to sampling rate $1/T_s$) for the tones in (16.36) are

$$\omega_m = \left(m - \frac{1}{2}\right)\frac{\pi}{M}, \text{ (DWMT)}$$

which are determined by the orthonormal requirement given by (16.40). The M frequencies are distinct and equally spaced. The normalized spacing is $\Delta\omega = \pi/M$, or $\Delta f = 1/(2M)$.

For comparison purposes, a DMT (OFDM) signal set is also presented, which is also defined by (16.38) and (16.36) with $g = 1$ and p_l being rectangular. The normalized frequencies of the DMT is

$$\omega_m = 2\left\lfloor\frac{m}{2}\right\rfloor\frac{\pi}{M}, \text{ (DMT)}$$

where $\lfloor x \rfloor$ denotes the integer floor of x. For $m = 1, 2, 3, 4, 5, ..., M$, $\lfloor\frac{m}{2}\rfloor = 0, 1, 1, 2, 2, ..., M/2$ (assuming M is even). Thus each of the frequencies repeats once except for the two at the two ends of the transmission band. Each pair of frequencies are used for a quadrature pair of tones with phases determined by [10]

$$\phi_m = \begin{cases} 0, & m = 1 \\ 0, & m \geq 2, m \text{ even} \\ -\frac{\pi}{2}, & m \geq 2, m \text{ odd} \end{cases}$$

There are $M/2 + 1$ distinct, equally spaced frequencies with a normalized spacing of $\Delta\omega = 2\pi/M$ or $\Delta f = 1/M$, which ensures orthogonality between tones.

For DMT systems, the multichannel modulation is performed in an efficient manner by IDFT. For DWMT systems, the multichannel modulation is performed in an efficient manner using an inverse M-band fast wavelet transform (IFWT).

The superior spectral containment of DWMT is illustrated in Figure 16.12(a, b) by showing the magnitudes $|F_m(e^{j\omega})|$ of frequency responses for six spectrally contiguous subchannels, for a DMT, and a DWMT system, respectively. The pulse shapes for each subchannel are the same so that the spectral shapes in all subchannels are the same except for centering at different subcarrier frequencies. Since tones are in quadrature pairs the DMT system, there are only three discernible magnitude-responses for the six bandpass pulses in Figure 16.12(a). The cyclic prefix length is $\kappa = 0$ for the DMT transmission. A particular cosine-modulated filter bank with $g = 8$ is considered for DWMT transmission. From the figures we can see that the first side lobe for a DMT subchannel is 13 dB down from the main lobe, whereas the first side lobe for an DWMT subchannel is 45 dB down. This 32 dB side lobe improvement is the effect of the longer pulse length in the DWMT system (8 times

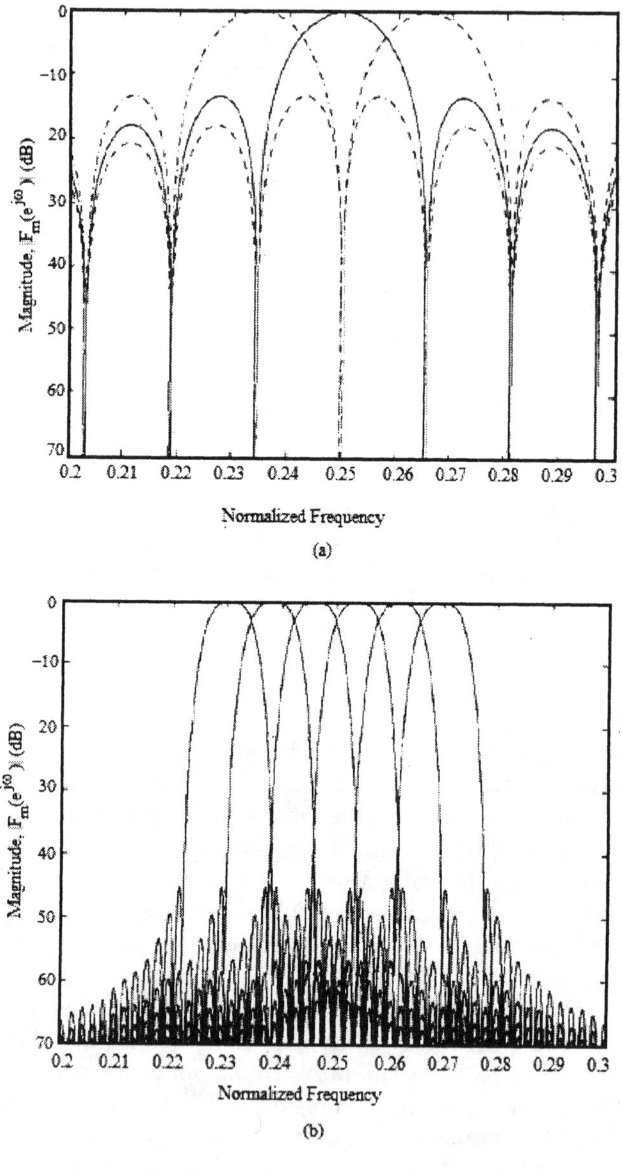

Figure 16.12 Frequency responses of six spectrally contiguous subchannel pulse sequences: (a) DMT transmission, and (b) DWMT transmission. From [10]. Copyright © 1995 IEEE.

longer than that of the DMT system here). In general, with DWMT, we can expect higher spectral concentration for larger g.

It was shown in [10] that DWMT systems are inherently more robust than other multicarrier implementations with regard to interchannel interference, and to narrow-band channel disturbances. The superior robustness follows as a consequence of the high degree of spectral containment that is present in the subchannels of a DWMT system. For details of the performance of DWMT, refer to [10].

16.8 DCT-BASED OFDM: MASK-OFDM

The conventional OFDM is implemented efficiently by the IDFT/DFT pair and the data symbols are complex, usually QAM symbols or MPSK symbols. The frequency separation between adjacent subcarrier frequencies is $1/T$, the symbol rate, for orthogonality between subcarriers. However, when data symbols are real valued, then the minimum frequency separation for orthogonality is only $1/(2T)$. This leads to the M-ary amplitude shift keying OFDM proposed in [11]. MASK-OFDM can be efficiently implemented by discrete cosine transform (DCT).

An MASK-OFDM signal is defined as

$$s(t) = \sum_{k=0}^{N-1} A_k \cos 2\pi \frac{k}{2T} t \tag{16.41}$$

A_k is one of the M-ary amplitudes. Each subcarrier frequency $f_k = k/2T$. The frequency separation is $1/(2T)$. The orthogonality of the signals can be easily verified.

Figure 16.13 shows the spectra of 4-channel OFDM systems with $1/T$ spacing (QAM or MPSK) and $1/(2T)$ spacing (MASK), respectively. From the figure we can see that the total null-to-null bandwidth is $(N+1)/T$ for QAM- or MPSK-OFDM and $(N+3)/(2T)$ for MASK-OFDM, respectively. The bandwidth savings is the ratio of the former over the latter: $2(N+1)/(N+3)$, which approaches 2 when N goes to infinity.

However, for the same modulation order M, coherent MASK has loss in power efficiency (i.e., increase in power for the same BER) compared with coherent MPSK or QAM. However coherent MASK still may be considered for OFDM since the bandwidth savings can be traded for power efficiency. For an approximately fixed bandwidth occupancy, when coherent MASK is used for OFDM, the number of bits per symbol ($k = \log_2 M$) can be reduced to half due to its half subcarrier frequency spacing in comparison with MPSK or QAM. That is, M can be reduced to $\sqrt{M}$. By doing so, the loss in MASK can be completely recovered against QAM and overly recovered against MPSK, as we will see shortly.

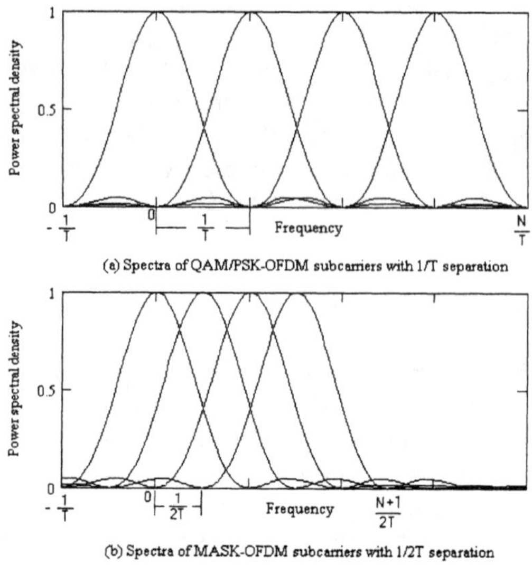

Figure 16.13 OFDM signal spectra. (a) Spectra of QAM/MPSK-OFDM implemented by IDFT. The subcarrier separation is 1/T. The spectra are complete. (b) Spectra of MASK-OFDM. The subcarrier separation is 1/(2T). Note that only the positive-frequency spectra are shown except for the first subcarrier. Mathematically, there are mirror images in negative frequency. However they can be eliminated by filtering at RF band. From [11]. Copyright © 2003 IEEE.

Assume that the QAM is the most popular square QAM with amplitudes of $\pm 1, \pm 3, ..., \pm(\sqrt{M}-1)$ on both I- and Q-channels, and the amplitudes of the MASK are $\pm 1, \pm 3, ..., \pm(M-1)$. The BER expressions of MASK and QAM with Gray bits-to-symbol mapping for the coherent (optimum) receiver in an AWGN channel are given as (see Chapters 8 and 9)

$$P_b \approx \frac{2(M-1)}{(\log_2 M)M} Q\left(\sqrt{\frac{6\log_2 M}{(M^2-1)}\frac{E_b}{N_o}}\right), \quad \text{(MASK)} \tag{16.42}$$

$$P_b \approx \frac{4(\sqrt{M}-1)}{(\log_2 M)\sqrt{M}} Q\left(\sqrt{\frac{3\log_2 M}{(M-1)}\frac{E_b}{N_o}}\right), \quad \text{(QAM)} \tag{16.43}$$

In (16.42), substituting M with $\sqrt{M}$ we obtain (16.43) exactly. That is, reducing

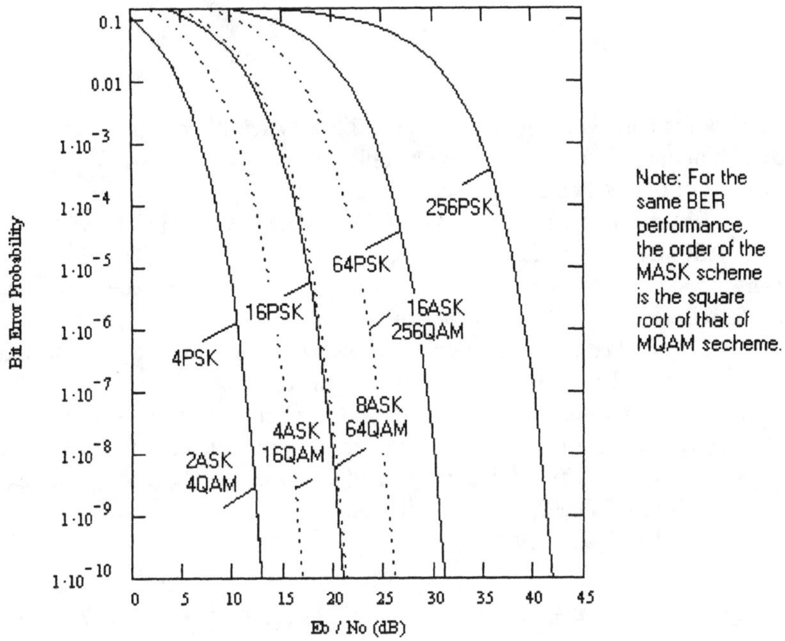

Figure 16.14 BER comparison between MASK, MQAM, and MPSK. From [11]. Copyright © 2003 IEEE.

MASK's order to $\sqrt{M}$ leads to the exact same power efficiency as that of QAM.

Reducing MASK's order to $\sqrt{M}$ leads to a better power efficiency than MPSK. The MPSK's BER for the coherent (optimum) receiver in an AWGN channel is given as (see Chapter 4)

$$P_b \approx \frac{2}{\log_2 M} Q\left(\sqrt{\frac{2kE_b}{N_o}} \sin\frac{\pi}{M}\right), \quad \text{(MPSK)}$$

Comparing MASK and MPSK using the BER expressions is not straightforward. They are compared in Figure 16.14. From the figure it is seen that reducing MASK's order to $\sqrt{M}$ leads to 0, 4, 10, and 16 dB improvement in power efficiency compared with 4, 16, 64, and 256PSK, respectively.

MASK's symbol rate is double that of QAM (since $\log_2 M = 2\log_2\sqrt{M}$). Thus in terms of QAM's symbol duration T, MASK-OFDM's null-to-null bandwidth becomes $(N+3)/T$, and the bandwidth ratio of MASK over QAM (or PSK) is

$$\text{BWR} = \frac{N+3}{N+1} = 1 + \frac{2}{N+1} \tag{16.44}$$

For $N = 8$, the bandwidth increase is about 22%. When N becomes very large, the bandwidth increase is negligible. For example, when $N = 256$, $BWR = 1.008$; the bandwidths of both schemes are essentially the same.

The peak-to-average-power ratio (PAPR) of MASK-OFDM is derived in [11]. A $\sqrt{M}$-ary ASK-OFDM has the exact PAPR of an M-ary QAM-OFDM. Thus by using $\sqrt{M}$-ary ASK-OFDM, not only can the power efficiency loss be recovered, but also the PAPR will remain the same in comparison with QAM-OFDM.

MASK-OFDM cannot be implemented by IDFT since its frequency separation is $1/(2T)$ instead of $1/T$. Fortunately, there is the discrete cosine transform (DCT) that comes to the rescue. The discrete cosine transform and inverse discrete cosine transform (IDCT) are a pair of orthogonal transforms which are popular in image compression coding. The DCT algorithm is the basis for the widespread coding standards such as JPEG, MPEG, and so forth.

The DCT/IDCT pair are [12]

$$X(n) = \frac{2}{N}\varepsilon(n)\sum_{k=0}^{N-1} x(k)\cos\frac{\pi n(2k+1)}{2N}, \quad n = 0, 1, ..., N-1, \text{ (DCT)} \tag{16.45}$$

$$x(k) = \sum_{n=0}^{N-1} \varepsilon(n)X(n)\cos\frac{\pi n(2k+1)}{2N}, \quad k = 0, 1, ..., N-1, \text{ (IDCT)} \tag{16.46}$$

where

$$\varepsilon(n) = \begin{cases} \frac{1}{\sqrt{2}}, & n = 0 \\ 1, & \text{otherwise} \end{cases} \tag{16.47}$$

Let $t = n \cdot \Delta t, T = N \cdot \Delta t$ in (16.41), where Δt is the sampling interval, the discrete-time MASK-OFDM signal is

$$s(n) = \sum_{k=0}^{N-1} A_k \cos\frac{\pi n(2k)}{2N} \tag{16.48}$$

Comparing (16.45) and (16.48) reveals that (16.45) is basically a frequency-shifted version of (16.48). The shift is $1/(4N)$ in discrete-time signal or $1/(4T)$ in continuous-time signal. Thus by allowing the first subcarrier frequency to be $1/(4T)$ instead of 0, the MASK-OFDM signal can be written in the form of the DCT (16.45).

In order to use the DCT, we redefine the MASK-OFDM signal as

$$s(t) = \varepsilon(t) \sum_{k=0}^{N-1} A_k \cos \frac{\pi(2k+1)t}{2T} \tag{16.49}$$

where

$$\varepsilon(t) = \begin{cases} \frac{1}{\sqrt{2}}, & 0 \le t \le \Delta t \\ 1, & \Delta t \le t \le T \end{cases} \tag{16.50}$$

The subcarrier frequencies now are $1/(4T), 3/(4T), 5/(4T), ..., (2N-1)/(4T)$. The separation is still $1/(2T)$. The total signal bandwidth is shifted up by $1/(4T)$, which is a small amount.

Note that $\varepsilon(t)$ is a symbol pulse shaping function which is almost rectangular, except that the amplitude is smaller for the first sampling interval. It can be easily shown that the spectrum of the pulse is given by

$$\varepsilon(f) = T\frac{\sin \pi fT}{\pi fT} e^{-j\pi fT} - (1 - \frac{1}{\sqrt{2}})\frac{T}{N}\frac{\sin \pi fT/N}{\pi fT/N} e^{-j\pi fT/N} \tag{16.51}$$

where the first term accounts for the original rectangular pulse and the second term is for the small notch in the first sampling interval. The second term is about $3.4N$ times smaller than the first one. Therefore for practical systems where N is usually big, the effect of $\varepsilon(t)$ on the signal spectrum is negligible. In addition, due to the fact that $\varepsilon(t)$ is almost rectangular, the effect of $\varepsilon(t)$ on the signal PAPR is also negligible for large N. Thus for practical systems where N is usually big, the conclusions on the spectrum and PAPR obtained above are still valid.

The discrete form of (16.49) is (16.45) with $x(k) = A_k$. That is

$$s(n) = \frac{2}{N}\varepsilon(n) \sum_{k=0}^{N-1} A_k \cos \frac{\pi n(2k+1)}{2N}, \quad n = 0, 1, ..., N-1 \tag{16.52}$$

(MASK-OFDM modulation)

where $2/N$ is just a constant mandated by the DCT expression, which has no effect on the signal shape and spectrum shape. The sampling frequency is still N/T. The DCT and IDCT are a pair of orthogonal transforms. That is, MASK-OFDM in the form of (16.49) can be generated by an N-point DCT using (16.52) and demodulated by the following N-point IDCT (16.53):

$$A_k = \sum_{n=0}^{N-1} \varepsilon(n)s(n)\cos\frac{\pi n(2k+1)}{2N}, \quad k = 0, 1, ..., N-1 \quad (16.53)$$

(MASK-OFDM demodulation)

Many fast algorithms exist for computing the DCT/IDCT efficiently [12, 13]. Lee's FCT algorithm [12] follows the concept of FFT by decomposing the N-point DCT or IDCT into two smaller N/2-point DCT or IDCT, and decomposing can be repeated as needed. Lee's algorithm requires $(N/2)\log_2 N$ *real* multiplications and $(3N/2)\log_2 N - N + 1$ *real* additions. Other FCT/IFCT algorithms either increase the speed further or have some new features.

An N-point FFT/IFFT needs $(N/2)\log_2 N$ *complex* multiplications and $N\log_2 N$ *complex* additions [14]. Recall that an N-subcarrier QAM/MPSK-OFDM requires a $2N$-point IFFT/FFT (see Section 12.2.2), which requires $N(\log_2 N + 1)$ complex multiplications and $2N(\log_2 N + 1)$ complex additions. If Lee's FCT algorithm is used for MASK-OFDM, not only is the number of multiplications and additions reduced to about half of that of QAM/MPSK-OFDM, but also the multiplications and additions are real instead of complex. This makes the modulation or demodulation of MASK-OFDM require about 1/4 of the computations of QAM/MPSK-OFDM. Since the symbol rate of MASK-OFDM is double that of QAM/PSK-OFDM, the computation speed reduction factor can only be about two. Note that the above comparison is made for the case where the channel is AWGN. If the new scheme is to be used in a time-dispersive channel, a guard time is to be inserted between symbols. In this case, due to the shorter symbol duration of MASK-OFDM, the guard time will occupy a higher percentage of the symbol duration, which makes the speed reduction factor lower than two. The degree of reduction depends on the specific system configuration, especially the percentage of the guard time length in the total symbol length. Nonetheless there is speed reduction, which could be substantial if the guard time is small in comparison with the total symbol length (guard time can be zero for some systems, for example, satellite communication systems). Speed reduction translates into either less hardware complexity and power consumption for the same data rate or higher data rates for the same hardware complexity and power consumption.

It should be emphasized that the analysis and results here are for AWGN channels. The performance of the new scheme is expected to degrade in time-dispersive channels and fading channels, as well as in cases of imperfect carrier and symbol synchronization. In these impaired channels or adverse conditions, it remains to be studied that to what extent the implementation advantages stated above can be preserved. Remedial measures, such as channel estimation and correction for time-dispersive channels and fading channels, that can be used to reduce the degradation are to be investigated.

16.9 SUMMARY

In this chapter we have studied several important wavelet based modulation schemes.

The earliest is fractal modulation. Fractal modulation is a type of diversity technique which transmits the same symbol using different symbol lengths in different subbands. Because of the "self-similarity" between dyadic expansions of wavelets, they are perfectly suitable as the pulse shapes for the fractal modulation. It arranges the symbol rates (hence the bandwidths of subchannels) according to the order of dyadic expansion of the wavelets. Because information is contained in multiple symbols with different lengths in different frequency bands, fractal modulation is suitable for transmission over noisy channels of simultaneously unknown duration and bandwidth.

The first scheme is the multiscale wavelet modulation (MSM). The purpose of MSM is just like conventional QAM-OFDM, that is, to overcome distortion caused by the frequency-selective fading channel by dividing total bandwidth into narrower subbands. However, unlike QAM-OFDM, MSM uses different time durations for symbols in different subchannels; in other words, MSM has nonuniform subbands. To ensure orthogonality, each symbol is shaped by a wavelet or scaling function. MSM includes the fractal modulation as a special case, although its goal is different from that of the fractal modulation. In this chapter we studied several important wavelet OFDM schemes.

The third scheme is wavelet pulse shaping of PAM signals which improves the bandwidth efficiency of the system by adding dyadic expansions to the parent wavelet as pulse shapes. Using a spectrally compact wavelet and its dyadic expansions, and coupled with the fact that wavelets have zero mean and they can be transmitted using single-side band (SSB) transmission, 50% more bandwidth efficiency can be obtained. The wavelet-PAM is a special case of MSM.

The fourth scheme is the wavelet packet modulation (WPM). Like MSM, WPM uses orthogonal wavelets for pulse shapes and thus has nonuniform subbands; unlike MSM, positions of subchannels of WPM are not arranged according to the order of dyadic expansions of the wavelet. This gives flexibility of assigning frequency bandwidth to different subchannels to satisfy different applications or to mitigate different interferences.

The above four are similar in the sense that they use dyadic wavelets of different lengths for pulse shaping of symbols in different subchannels. They all are multirate modulations with each subchannel carrying a different symbol rate and occupying a different bandwidth.

The fifth scheme is the overlapped discrete multitone modulation (DWMT). The purpose is to improve the spectral containment of the OFDM signal, which is the same as the purpose of MSM and wavelet PAM. However, unlike all above four schemes, and like conventional OFDM, DWMT divides the channel bandwidth into equal-bandwidth subbands. In addition, pulse lengths are longer than a symbol length and symbols are overlapped in time. However due to orthogonality between the wavelet pulses, symbols can still be separated at the receiver by using wavelet transform.

All above five schemes have multiple subchannels and the signals in the subchannels are orthogonal, so that they are OFDM schemes. And all of them use discrete wavelet transform (DWT). Thus we can call all of them DWT-OFDM or wavelet OFDM. In addition, all of them except for the DWMT use different symbol lengths. This is different from the conventional OFDM.

The sixth scheme is the uniform-subband wavelet scheme called M-band wavelet modulation (MWM). In this scheme, the scaling function and the wavelet are used as pulse shapes on symbols with the same duration. No dyadic expansions of the wavelet are used. Modulation and demodulation are implemented through filter banks instead of IDWT and DWT.

Figure 16.15 compares the T-F tiling of WPM and MSM together with two other schemes: QAM and MWM (M-band wavelet modulation). The fractal modulation and wavelet PAM have the same T-F tiling of MSM and DWMT is a type of MWM. Each rectangular area in the figure represents the product (BT) of the bandwidth (B) and the duration of a symbol (T). An interference environment consisting of time impulse (wideband) and tone jammer (narrow band but long duration) is assumed in the figure. The arrows indicate the positions of the tone jammer and the impulse. The shaded areas indicate symbols corrupted by the interferences. In QAM, the entire bandwidth is occupied by each symbol so the tone jammer will corrupt every symbol, although the impulse will only corrupt one symbol. In MWM, the situation is just the opposite: the impulse will corrupt every symbol and the tone jammer will corrupt one. In MSM, the tone jammer will corrupt all eight symbols in the upper part of the spectrum and the impulse will corrupt four symbols in the lower part of the spectrum as well as two symbols in the upper part. The total number of corrupted symbols is 12. The WPM, which is optimized for this particular interference, will suffer from the least amount of damage. Only five symbols will be corrupted.

In this chapter we also have introduced a DCT-based OFDM, the MASK-OFDM. Coherent MASK-OFDM uses a subcarrier frequency separation of $1/(2T)$ for orthogonality which reduces the bandwidth of the OFDM signal to about half for the same order of modulation. However, to achieve the same BER performance, the bandwidth savings must be traded for power efficiency. By reducing the modulation order from M to $\sqrt{M}$, with a slightly wider bandwidth, the MASK-OFDM can achieve a BER performance which is the same as or better than that of an Mth-order QAM- or MPSK-OFDM, respectively. The PAPR of the $\sqrt{M}$-ary ASK-OFDM is the same as that of the M-ary QAM-OFDM. MASK-OFDM can be digitally and

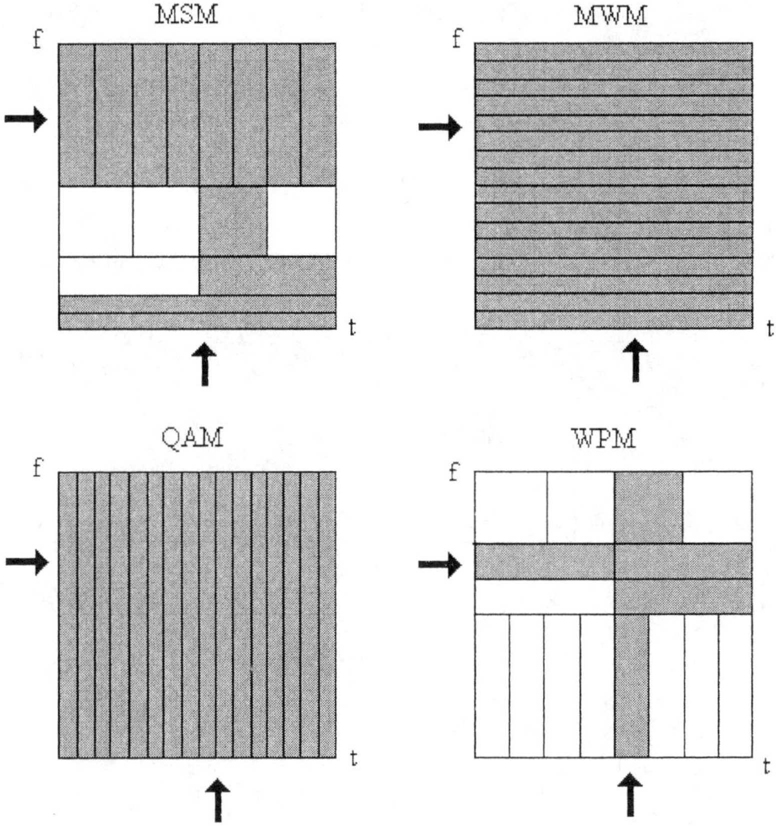

Figure 16.15 Comparison on modulation methods for a given interference environment consisting of time impulse and tone jammers. Gray-shaded areas indicate corrupted symbols. From [7]. Copyright © IEEE 1997.

efficiently implemented by an FCT/IFCT pair which requires less computational or circuit complexity in comparison with IFFT/FFT pair for QAM/MPSK-OFDM.

APPENDIX 16A POWER SPECTRUM OF MSM

The following derivation is from [5]. To derive the power spectral density of the multiscale modulation (MSM), we start with the scaling function

$$\phi(\frac{t}{T}) = \sqrt{2}\sum_n h_n \phi(2\frac{t}{T} - n) \tag{16.54}$$

where N is the number of coefficients. Taking the Fourier transform of (16.54) we have

$$\begin{aligned} T\Phi(fT) &= \sqrt{2}\sum_n h_n \mathcal{F}\{\phi(\frac{2}{T}(t - \frac{Tn}{2}))\} = \sqrt{2}\sum_n h_n \mathcal{F}\{\phi(\frac{2}{T}t)\}e^{-j2\pi f \frac{n}{2}T} \\ &= \sqrt{2}\sum_n h_n \left[\frac{T}{2}\Phi\left(\frac{fT}{2}\right)\right] e^{-j2\pi f \frac{n}{2}T} \\ &= \frac{T}{\sqrt{2}}\Phi\left(\frac{fT}{2}\right)\sum_n h_n e^{-j2\pi f \frac{n}{2}T} = \frac{T}{\sqrt{2}} H\left(\frac{fT}{2}\right)\Phi\left(\frac{fT}{2}\right) \end{aligned}$$

or

$$\Phi(fT) = \frac{1}{\sqrt{2}} H\left(\frac{fT}{2}\right)\Phi\left(\frac{fT}{2}\right) \tag{16.55}$$

where $\Phi\,(f)$ is the Fourier transform of $\phi(t)$ and

$$H\left(\frac{fT}{2}\right) = \sum_n h_n e^{-j2\pi f \frac{n}{2}T}$$

and H is the discrete-time Fourier transform of $\{h_n\}$.

Similarly the Fourier transform of the function $\psi(\frac{t}{T}) = \sqrt{2}\sum_n g_n \phi(2\frac{t}{T} - n)$ is

$$\Psi(fT) = \frac{1}{\sqrt{2}} G\left(\frac{fT}{2}\right)\Phi\left(\frac{fT}{2}\right)$$

A relation between $H(f)$ and $G(f)$ is [5]

$$|H(f)|^2 + |G(f)|^2 = 2 \tag{16.56}$$

which is commonly referred to as the power complementary property. Using this property we have

$$\begin{aligned}|\Phi(f)|^2+|\Psi(f)|^2 &= \frac{1}{2}\left|H\left(\frac{f}{2}\right)\right|^2\left|\Phi\left(\frac{f}{2}\right)\right|^2+\frac{1}{2}\left|G\left(\frac{f}{2}\right)\right|^2\left|\Phi\left(\frac{f}{2}\right)\right|^2\\ &= \frac{1}{2}\left|\Phi\left(\frac{f}{2}\right)\right|^2\left(\left|H\left(\frac{f}{2}\right)\right|^2+\left|G\left(\frac{f}{2}\right)\right|^2\right)\\ &= \left|\Phi\left(\frac{f}{2}\right)\right|^2\end{aligned}$$

Relation (16.56) is used in the last step above. Thus we have

$$\left|\Phi\left(\frac{f}{2}\right)\right|^2=|\Phi(f)|^2+|\Psi(f)|^2 \tag{16.57}$$

The MSM signal $m(t)$ in (16.19) can be written as

$$m(t)=x_0(t)+x_1(t)+\ldots+x_{J-1}(t)$$

where

$$x_0(t)=\sqrt{2^{-(J-1)}\frac{E_s}{T}}\sum_j a_j^0\phi(2^{-(J-1)}\frac{t}{T}-j)$$

$$x_n(t)=\sqrt{2^{-(J-n)}\frac{E_s}{T}}\sum_j a_j^n\psi(2^{-(J-n)}\frac{t}{T}-j),\quad n=1,2,\ldots,N-1$$

Assuming i.i.d. zero-mean wide-sense stationary complex data, $\{x_n(t)\}_{n=0}^{J-1}$ are independent and WSS. Then the autocorrelation is

$$R_m(\tau)=E\{m(t)m^*(t-\tau)\}=\sum_{n=0}^{J-1}R_n(\tau)$$

where all cross terms vanished since they are zero and

$$R_n(\tau)=E\{x_n(t)x_n^*(t-\tau)\}$$

The PSD of $m(t)$ is the Fourier transform of $R_m(\tau)$ and it is (see (A.18))[3]

$$\begin{aligned} S_{MSM} &= \sum_{n=0}^{J-1} \mathcal{F}\{R_n(\tau)\} \\ &= \frac{1}{T2^{(J-1)}} \sigma_0^2 |T2^{J-1} \Phi(fT2^{J-1})|^2 \\ &+ \sum_{n=1}^{J-1} \frac{1}{T2^{J-n}} \sigma_n^2 |T2^{J-n} \Psi(fT2^{J-n})|^2 \end{aligned}$$

where

$$\sigma_0^2 = 2^{-(J-1)} \frac{E_s}{T} \sigma_a^2$$

$$\sigma_n^2 = 2^{-(J-n)} \frac{E_s}{T} \sigma_a^2$$

Thus

$$\begin{aligned} S_{MSM} &= \sum_{n=0}^{J-1} \mathcal{F}\{R_n(\tau)\} \\ &= \frac{1}{T2^{(J-1)}} 2^{-(J-1)} \frac{E_s}{T} \sigma_a^2 |T2^{J-1} \Phi(fT2^{J-1})|^2 \\ &+ \sum_{n=1}^{J-1} \frac{1}{T2^{J-n}} 2^{-(J-n)} \frac{E_s}{T} \sigma_a^2 |T2^{J-n} \Psi(fT2^{J-n})|^2 \\ &= E_s \sigma_a^2 |\Phi(fT2^{J-1})|^2 + E_s \sum_{n=1}^{J-1} \sigma_a^2 |\Psi(fT2^{J-n})|^2 \end{aligned}$$

Now assuming unit variance of data, then using (16.57) for the sum of the term involving the scaling function and the $n = 1$ term of series involving the wavelets we have

$$\begin{aligned} \frac{S_{MSM}}{E_s} &= |\Phi(fT2^{J-1})|^2 + \sum_{n=1}^{J-1} |\Psi(fT2^{J-n})|^2 \\ &= |\Phi(fT2^{J-1})|^2 + |\Psi(fT2^{J-1})|^2 + \sum_{n=2}^{J-1} |\Psi(fT2^{J-n})|^2 \end{aligned}$$

3 The period for $\phi(2^{-(J-1)} \frac{t}{T} - j)$ is $2^{(J-1)}T$ and the period for $\psi(2^{-(J-n)} \frac{t}{T} - j)$ is $2^{(J-n)}T$.

$$= |\Phi(fT2^{J-2})|^2 + \sum_{n=2}^{J-1} |\Psi(fT2^{J-n})|^2$$

By repeatedly using (16.57), we finally reach

$$S_{MSM} = E_s|\Phi(fT)|^2$$

which is the same as the QAM.

References

[1] Akansu, A., et al., "Wavelet and subband transforms: fundamentals and communication applications frequency," *IEEE Communications Magazine*, vol. 35, December 1997, pp. 104–115.

[2] Burrus, C. S., R. A. Gopinath, and H. Guo, *Introduction to Wavelets and Wavelet Transforms*, Upper Saddle River, New Jersey: Prentice Hall, 1998.

[3] Boggess, A., and F. J. Narcowich, *A First Course in Wavelets with Fourier Analysis*, Upper Saddle River, New Jersey: Prentice Hall, 2001.

[4] Jones, W., "Multi-scale wavelet modulation," *Proceedings of the 26th Southeastern Symposium on System Theory*, March 20–22, 1994, pp. 576–580.

[5] Jones, W. W., "A unified approach to orthogonally multiplexed communication using wavelet bases and digital filter banks," Ph.D. dissertation, Ohio Univ., Athens, August 1994.

[6] Wornell, G. W., and A. V. Oppenheim, "Wavelet based representations for a class of self-similar signals with application to fractal modulation," *IEEE Trans. Info. Theory*, vol. 38, no. 2, March 1992, pp. 785–800.

[7] Lindsey, A. R., "Wavelet packet modulation for orthogonally transmultiplexed communication," *IEEE Trans. on Signal Processing*, vol. 45, no. 5, May 1997, pp. 1336–1339.

[8] Livingston, J. N., and C. Tung, "Bandwidth efficient PAM signaling using wavelets," *IEEE Trans. Communications*, vol. 44, December 1996, pp. 1629–1631.

[9] Vetterli, M., and C. Herley, "Wavelets and filter banks: theory and design," *IEEE Trans. Signal Processing*, vol. 40, no. 9, September 1992.

[10] Sandberg, S., and M. Tzannes, "Overlapped discrete multitone modulation for high speed copper wire communications," *IEEE JSAC*, vol. 13, no. 9, Dec. 1995, pp. 1571–1585.

[11] Xiong, F. "M-ary amplitude shift keying OFDM system," *IEEE Trans. on Communications*, vol. 51, no. 10, October 2003, pp. 1638–1642.

[12] Lee, B. G., "A new algorithm to compute the discrete cosine transform," *IEEE Trans. on Acoustics, Speech, and Signal Processing*, vol. 32, no. 6, Dec. 1984, pp. 1243–1245,.

[13] Hou, H. S., "A fast recursive algorithm for computing the discrete cosine transform," *IEEE Transactions on Acoustics, Speech, and Signal Processing*, vol. 35, no. 10, October 1987, pp. 1455–1461.

[14] Mitra, S., *Digital Signal Processing*, New York: McGraw-Hill, 1998.

Selected Bibliography

- Daubechies, I., *Ten Lectures on Wavelets*, Philadelphia Pennsylvania: SIAM, 1992.
- Mallat, S. G., "A theory for multiresolution signal decomposition: the wavelet representation," *IEEE Trans. on Pattern Analysis and Machine Intelligence*, vol. 11, no. 7, July 1989, pp. 647–692.
- Newlin, H., "Developments in the use of wavelets in communications systems," *IEEE Conf. Proc. MILCOM 98*, vol. 1, 1998, pp. 343–349.

Appendix A

Power Spectral Densities of Signals

In this appendix we deal with the problem of finding spectra of signals and power spectral densities (PSDs) of noise and modulated signals. The reader is assumed to have an understanding of the characteristics of deterministic signals and their Fourier transform or series, characteristics of random processes, their correlation functions, and their power spectral densities. Based on this knowledge, we concentrate on finding spectra of bandpass signals and PSDs of bandpass random processes, and PSDs of digitally modulated signals in baseband or passband. Specifically, formulas of PSDs of baseband digital signals (line codes), quadrature modulations (MPSK, MSK, and so forth), and continuous phase modulations (CPM) are derived.

A.1 BANDPASS SIGNALS AND SPECTRA

Consider a signal in the form

$$s(t) = a(t)\cos[2\pi f_c t + \theta(t)] \tag{A.1}$$

where $a(t)$ is the amplitude (envelope), $\theta(t)$ is the phase, and f_c is the carrier frequency of $s(t)$. For signals of practical interest, $a(t)$ and $\theta(t)$ are slow-changing functions of t, compared with the carrier frequency. Thus the bandwidth occupied by the signal is small relative to the carrier frequency. Such a signal $s(t)$ is called a narrowband bandpass signal, or simply, a bandpass signal. The signal can be rewritten as

$$\begin{aligned} s(t) &= a(t)\cos\theta(t)\cos 2\pi f_c t - a(t)\sin\theta(t)\sin 2\pi f_c t \\ &= x(t)\cos(2\pi f_c t) - y(t)\sin(2\pi f_c t) \end{aligned}$$

where

$$x(t) = a(t)\cos\theta(t)$$

$$y(t) = a(t)\sin\theta(t)$$

are called inphase and quadrature components, respectively. We will show that the frequency components of $x(t)$ and $y(t)$ are concentrated near $f = 0$. Thus these two signal components are called low-pass signals.

By defining the *complex envelope* of $s(t)$ as

$$\widetilde{s}(t) = x(t) + jy(t)$$

we can write

$$s(t) = \text{Re}[\widetilde{s}(t)e^{j2\pi f_c t}]$$

where Re[ ] denotes the real part of the content of the brackets.

The Fourier transform of the signal $s(t)$ is

$$\begin{aligned} S(f) &= \int_{-\infty}^{\infty} s(t)e^{-j2\pi ft}dt \\ &= \int_{-\infty}^{\infty} \text{Re}[\widetilde{s}(t)e^{j2\pi f_c t}]e^{-j2\pi ft}dt \\ &= \frac{1}{2}\int_{-\infty}^{\infty} [\widetilde{s}(t)e^{j2\pi f_c t} + \widetilde{s}^*(t)e^{-j2\pi f_c t}]e^{-j2\pi ft}dt \\ &= \frac{1}{2}[\widetilde{S}(f - f_c) + \widetilde{S}^*(-f - f_c)] \end{aligned} \tag{A.2}$$

where * denotes conjugate, and $\widetilde{S}(f)$ is the Fourier transform of the complex envelope $\widetilde{s}(t)$. Note that $S(f)$ is in general a complex function. Its magnitude $|S(f)|$ is called the amplitude spectrum and its argument is called the phase spectrum.

Since the frequency components in the spectrum of $s(t)$ are concentrated around f_c, the above expression indicates that the frequency components of $\widetilde{S}(f)$ must be in the neighborhood of zero frequency. In other words, the complex envelope $\widetilde{s}(t)$ is a low-pass (or baseband) signal. Complex envelope is also called equivalent baseband or low-pass signal. Equation (A.2) is a very important expression which shows that the spectrum of a bandpass signal is the frequency-shifted version of the spectrum of its complex envelope. Therefore in many cases it suffices to determine the spectrum of the complex envelope.

The PSD of a signal is defined as the signal power in a unit frequency bandwidth (i.e., watts per Hz). The PSD definition is

$$\Psi_s(f) \triangleq |S(f)|^2$$

Note that PSD is always a real function. Since $|S(f)|$ is the amplitude spectrum and is in the unit of volts per Hz, $|S(f)|^2$ is in the unit of watts per Hz (on a resistance of 1 Ω, so called normalized power).

For the bandpass signal, (A.2) can be converted into power spectral density form. Since $\widetilde{S}(f-f_c)$ and $\widetilde{S}^*(-f-f_c)$ do not overlap, the PSD of $s(t)$ is

$$\begin{aligned}
\Psi_s(f) &= |S(f)|^2 = S(f)S^*(f) \\
&= \frac{1}{2}[\widetilde{S}(f-f_c) + \widetilde{S}^*(-f-f_c)] \cdot \frac{1}{2}[\widetilde{S}^*(f-f_c) + \widetilde{S}(-f-f_c)] \\
&= \frac{1}{4}[\widetilde{S}(f-f_c)\widetilde{S}^*(f-f_c) + \widetilde{S}^*(-f-f_c)\widetilde{S}(-f-f_c)] \\
&= \frac{1}{4}[|\widetilde{S}(f-f_c)|^2 + |\widetilde{S}(-f-f_c)|^2] \\
&= \frac{1}{4}[\Psi_{\widetilde{s}}(f-f_c) + \Psi_{\widetilde{s}}(-f-f_c)] \qquad \text{(A.3)}
\end{aligned}$$

where

$$\Psi_{\widetilde{s}}(f) = |\widetilde{S}(f)|^2$$

is the PSD of the equivalent baseband signal.

We will see in the next section that for a random process we cannot define a spectrum, but the power spectral density can be defined for characterizing its frequency domain characteristics.

A.2 BANDPASS STATIONARY RANDOM PROCESS AND PSD

The major noises, thermal noise and shot noise, in an electronic system are white noise. That is, their power spectral densities are flat over a wide range of frequencies. In bandpass systems, due to filters in the system, the noise entering the system becomes bandlimited.

Many bandpass signals of practical interest, such as FSK, PSK signals, binary, or M-ary, may well be represented by (A.1). However, either $a(t)$ or $\theta(t)$ or both of them may be modulated by information-carrying symbols. Due to the random nature of the information symbols, signal $s(t)$ is a random process instead of a deterministic signal.

Thus the description and properties of a bandpass random process, be it noise or signal, and its spectral density are of interest. In this section we will establish the relation between the bandpass stationary random process and its equivalent baseband stationary random process, and the relation between their power spectral densities.

Consider a wide-sense stationary random process $n(t)$ with zero mean and a PSD of $\Psi_n(f)$. The PSD is concentrated about $\pm f_c$, and is zero outside a certain interval, where f_c is called the carrier frequency. The process is called a narrow-band bandpass process if the width of the spectral density is much smaller than f_c. Similar to narrow-band bandpass signals, a narrow-band bandpass process can be represented by any one of the three different forms [1,2]

$$\begin{aligned} n(t) &= a(t)\cos[2\pi f_c t + \theta(t)] \\ &= x(t)\cos(2\pi f_c t) - y(t)\sin(2\pi f_c t) \\ &= \mathrm{Re}[\widetilde{n}(t)e^{j2\pi f_c t}] \end{aligned} \tag{A.4}$$

where $a(t)$ and $\theta(t)$ are the random amplitude and phase, $x(t)$ and $y(t)$ are the in-phase and quadrature components which are also random, and $\widetilde{n}(t)$ is the complex envelope of $n(t)$,

$$\widetilde{n}(t) = x(t) + jy(t)$$

which is random too.

The autocorrelation function of $n(t)$ is

$$\begin{aligned} R_{nn}(\tau) &= E[n(t)n(t-\tau)] \\ &= E\{[x(t)\cos(2\pi f_c t) - y(t)\sin(2\pi f_c t)] \\ &\quad \cdot[x(t-\tau)\cos(2\pi f_c(t-\tau)) - y(t-\tau)\sin(2\pi f_c(t-\tau))]\} \\ &= R_{xx}(\tau)\cos(2\pi f_c t)\cos(2\pi f_c(t-\tau)) \\ &\quad + R_{yy}(\tau)\sin(2\pi f_c t)\sin(2\pi f_c(t-\tau)) \\ &\quad - R_{xy}(\tau)\cos(2\pi f_c t)\sin(2\pi f_c(t-\tau)) \\ &\quad - R_{yx}(\tau)\sin(2\pi f_c t)\cos(2\pi f_c(t-\tau)) \\ &= \frac{1}{2}[R_{xx}(\tau) + R_{yy}(\tau)]\cos 2\pi f_c \tau \\ &\quad + \frac{1}{2}[R_{xx}(\tau) - R_{yy}(\tau)]\cos 2\pi f_c(2t-\tau) \\ &\quad - \frac{1}{2}[R_{yx}(\tau) - R_{xy}(\tau)]\sin 2\pi f_c \tau \\ &\quad - \frac{1}{2}[R_{yx}(\tau) + R_{xy}(\tau)]\sin 2\pi f_c(2t-\tau) \end{aligned}$$

where

$$R_{xx}(\tau) = E\{x(t)x(t-\tau)\}$$

$$R_{yy}(\tau) = E\{y(t)y(t-\tau)\}$$

$$R_{xy}(\tau) = E\{x(t)y(t-\tau)\}$$

$$R_{yx}(\tau) = E\{y(t)x(t-\tau)\}$$

Since $n(t)$ is stationary, its autocorrelation function must not be a function of time t. Therefore the second and fourth term in the above expression must be zero. This leads to

$$R_{xx}(\tau) = R_{yy}(\tau) \tag{A.5}$$

and

$$R_{yx}(\tau) = -R_{xy}(\tau) \tag{A.6}$$

and

$$R_{nn}(\tau) = R_{xx}(\tau)\cos 2\pi f_c\tau + R_{xy}(\tau)\sin 2\pi f_c\tau \tag{A.7}$$

Using the definitions one can check that the autocorrelation functions $R_{xx}(\tau)$ and $R_{yy}(\tau)$ are even functions. From (A.6) we can derive an important property of the cross-correlation function of $x(t)$ and $y(t)$

$$R_{yx}(-\tau) = E[y(t)x(t+\tau)] = E[x(\xi)y(\xi-\tau)] = R_{xy}(\tau) = -R_{yx}(\tau) \tag{A.8}$$

Thus $R_{yx}(\tau)$ is an odd function of τ. Due to the relation of (A.6), $R_{xy}(\tau)$ is also an odd function of τ. This in turn verifies that $R_{nn}(\tau)$ is an even function of τ, as it should be from its definition.

The autocorrelation function of the complex envelope $\widetilde{n}(t)$ is

$$\begin{aligned} R_{\widetilde{n}\widetilde{n}}(\tau) &= \frac{1}{2}E[\widetilde{n}(t)\widetilde{n}^*(t-\tau)] \\ &= \frac{1}{2}E[[x(t)+jy(t)][x(t-\tau)-jy(t-\tau)]] \\ &= \frac{1}{2}[R_{xx}(\tau)+R_{yy}(\tau)-jR_{xy}(\tau)+jR_{yx}(\tau)] \end{aligned}$$

Using (A.5) and (A.6) we have

$$R_{\widetilde{n}\widetilde{n}}(\tau) = R_{xx}(\tau) + jR_{yx}(\tau) \tag{A.9}$$

This relates the autocorrelation function of the complex envelope to the autocorrelation functions of the quadrature components. Further, from (A.9) and (A.8) we can see that

$$R_{\widetilde{n}\widetilde{n}}(\tau) = R^*_{\widetilde{n}\widetilde{n}}(-\tau) \tag{A.10}$$

The power spectral density of it is the Fourier transform of the autocorrelation function (Wiener-Khinchine relation):

$$\begin{aligned}\Psi_{\tilde{n}}(f) &= \int_{-\infty}^{\infty} R_{\tilde{n}\tilde{n}}(\tau)e^{-j2\pi f\tau}d\tau = \int_{-\infty}^{\infty} R^*_{\tilde{n}\tilde{n}}(-\tau)e^{-j2\pi f\tau}d\tau \\ &= \left[\int_{-\infty}^{\infty} R_{\tilde{n}\tilde{n}}(-\tau)e^{j2\pi f\tau}d\tau\right]^* = \Psi^*_{\tilde{n}}(f)\end{aligned} \quad \text{(A.11)}$$

This means that $\Psi_{\tilde{n}}(f)$ is a real-valued function of f.

Using (A.7) and (A.9) we have

$$R_{nn}(\tau) = \text{Re}[R_{\tilde{n}\tilde{n}}(\tau)e^{j2\pi f_c\tau}] \quad \text{(A.12)}$$

This indicates that the autocorrelation function of the bandpass stationary random process is completely determined by the autocorrelation function of its complex envelope. Consequently the power spectral density of $n(t)$ is also completely determined by the power spectral density of $\tilde{n}(t)$ as follows.

$$\begin{aligned}\Psi_n(f) &= \mathcal{F}\{\text{Re}[R_{\tilde{n}\tilde{n}}(\tau)e^{j2\pi f_c\tau}]\} \\ &= \mathcal{F}\{\frac{1}{2}[R_{\tilde{n}\tilde{n}}(\tau)e^{j2\pi f_c\tau} + R^*_{\tilde{n}\tilde{n}}(\tau)e^{-j2\pi f_c\tau}]\} \\ &= \frac{1}{2}[\Psi_{\tilde{n}}(f - f_c) + \Psi^*_{\tilde{n}}(-f - f_c)]\end{aligned}$$

where we have used the frequency shifting and conjugate function properties of the Fourier transform ($\mathcal{F}$) (see [1]). Further, since $\Psi_{\tilde{n}}(f)$ is real-valued (see (A.11)), the above equation can be finalized as

$$\Psi_n(f) = \frac{1}{2}[\Psi_{\tilde{n}}(f - f_c) + \Psi_{\tilde{n}}(-f - f_c)] \quad \text{(A.13)}$$

This equation indicates that the PSD of a bandpass stationary random process, be it noise or signal, is completely determined by the PSD of its complex envelope or equivalent baseband signal. It consists of frequency shifted (also scaled by $1/2$) $\Psi_{\tilde{n}}(f)$ centered at f_c and $-f_c$ respectively. Thus when we examine the PSD of a bandpass stationary random process, it is sufficient to examine the PSD of its complex envelope or equivalent baseband signal.

A.3 POWER SPECTRAL DENSITIES OF DIGITAL SIGNALS

Digital signals are essentially random in nature. For example, in digital telephony the digital signals are digitized voice, which are random. In digital television the digital signals are digitized image and voice, which are also random. The bandwidth occupied by a digital signal is of the most concern to system design engineers. In this section we derive a general formula for the PSD of digital signals. This general formula includes cases of correlated data and uncorrelated data. Therefore it can be used for a wide range of applications involving baseband signals and bandpass signals.

Let the baseband digital signal be represented by

$$s(t) = \sum_{n=-\infty}^{\infty} a_n g(t - nT) \tag{A.14}$$

where a_n are discrete random data symbols, $g(t)$ is a signal of duration T (i.e., nonzero only in $[0, T]$). Let us name $g(t)$ as the *symbol function*. It could be any signal with a Fourier transform. For example it could be a baseband symbol-shaping pulse or a burst of carrier at passband. The random sequence $\{a_n\}$ could be binary or nonbinary.

Now to find out the power spectral density of the signal in (A.14), we first truncate the signal to get

$$s_N(t) = \sum_{n=-N}^{N} a_n g(t - nT) \tag{A.15}$$

Next we assume it is not random and take the Fourier transform of both sides of (A.15), the spectrum of this truncated signal is found as

$$A_N(f) = G(f) \sum_{n=-N}^{N} a_n e^{-j\omega nT}$$

where $\omega = 2\pi f$. The power spectral density of the original signal in (A.14) is obtained by taking the statistical average and time limit of $|A_N(f)|^2$ as follows.

$$\begin{aligned}
\Psi_s(f) &= \lim_{N\to\infty} \frac{1}{(2N+1)T} E\{|A_N(f)|^2\} \\
&= \lim_{N\to\infty} \frac{|G(f)|^2}{(2N+1)T} E\left\{\left|\sum_{n=-N}^{N} a_n e^{-j\omega nT}\right|^2\right\} \\
&= \lim_{N\to\infty} \frac{|G(f)|^2}{(2N+1)T} E\left\{\left(\sum_{n=-N}^{N} a_n e^{-j\omega nT}\right)\left(\sum_{n=-N}^{N} a_n e^{-j\omega nT}\right)^*\right\} \\
&= |G(f)|^2 \lim_{N\to\infty} \frac{1}{(2N+1)T} \sum_{n=-N}^{N} \sum_{m=-N}^{N} E\{a_n a_m\} e^{j(m-n)\omega T}
\end{aligned}$$

$$= |G(f)|^2 \lim_{N\to\infty} \frac{1}{(2N+1)T} \sum_{n=-N}^{N} \sum_{k=n+N}^{n-N} R(k)e^{-jk\omega T}$$

where $k = n-m$, $E\{a_n a_m\} = E\{a_n a_{n-k}\} = R(k)$ is the autocorrelation function of the data bits. Note that $R(-k) = R(k)$. Realizing the inner summations become the same regardless of the value of index n when $N \to \infty$, we obtain

$$\begin{aligned}\Psi_s(f) &= \frac{|G(f)|^2}{T} \lim_{N\to\infty} \left(\frac{2N+1}{(2N+1)} \sum_{k=n+N}^{n-N} R(k)e^{-jk\omega T} \right) \\ &= \frac{|G(f)|^2}{T} \sum_{k=\infty}^{-\infty} R(k)e^{-jk\omega T}\end{aligned}$$

Equivalently this is

$$\Psi_s(f) = \frac{|G(f)|^2}{T} \sum_{k=-\infty}^{\infty} R(k)e^{-jk\omega T} \tag{A.16}$$

Now we discuss two possible cases of $R(k)$.

A.3.1 Case 1: Data Symbols Are Uncorrelated

Assume a_n has a mean of $E\{a_n\} = m_a$ and a variance of σ_a^2, then

$$\begin{aligned}R(k) &= \begin{cases} E\{a_n^2\}, & k=0 \\ E\{a_n\}E\{a_{n-k}\}, & k \neq 0 \end{cases} \\ &= \begin{cases} \sigma_a^2 + m_a^2, & k=0 \\ m_a^2, & k \neq 0 \end{cases}\end{aligned}$$

Substitute this for $R(k)$ in (A.16) we have

$$\Psi_s(f) = \frac{|G(f)|^2}{T} \left(\sigma_a^2 + m_a^2 \sum_{k=-\infty}^{\infty} e^{-jk\omega T} \right)$$

By revoking the Poisson sum formula [1, p. 62]

$$\sum_{k=-\infty}^{\infty} e^{-jk\omega T} = \frac{1}{T} \sum_{k=-\infty}^{\infty} \delta(f - \frac{k}{T})$$

where $\delta(f)$ is the impulse function, we have

$$\Psi_s(f) = \frac{|G(f)|^2}{T}\left(\sigma_a^2 + \frac{m_a^2}{T}\sum_{k=-\infty}^{\infty}\delta(f-\frac{k}{T})\right) \tag{A.17}$$

That is, for uncorrelated data

$$\Psi_s(f) = \underbrace{\frac{\sigma_a^2|G(f)|^2}{T}}_{\text{continuous spectrum}} + \underbrace{(\frac{m_a}{T})^2\sum_{k=-\infty}^{\infty}|G(\frac{k}{T})|^2\delta(f-\frac{k}{T})}_{\text{discrete spectrum}} \tag{A.18}$$

The first term is the continuous part of the spectrum which is a scaled version of the PSD of the symbol-shaping pulse. The second term is the discrete part of the spectrum which has spectral lines at frequencies k/T (i.e., multiples of the data rate). The spectral lines have an envelope of the shape of the PSD of the symbol-shaping pulse. Each spectral line has a strength of $(m_a/T)^2|G(k/T)|^2$.

In digital communications, the most common case is that the data bits are binary (±1), equiprobable, stationary, and uncorrelated. That is

$$s(t) = \sum_{n=-\infty}^{\infty} a_n g(t-nT)$$

where $a_n = +1$ or -1 with equal probabilities, and $E\{a_n a_m\} = E\{a_n\}E\{a_m\}$ for $n \neq m$. It is easy to see that its mean is zero:

$$m_a = E\{a_n\} = 0.5(1) + 0.5(-1) = 0$$

and its variance is

$$\sigma_a^2 = E\{a_n^2\} + m_a^2 = E\{a_n^2\} = 0.5(+1)^2 + 0.5(-1)^2 = 1$$

Further, since a_n are uncorrelated and stationary, then

$$\begin{aligned} R(k) &= E\{a_n a_{n-k}\} \\ &= \begin{cases} E\{a_n^2\} = \sigma_a^2 = 1, & k=0 \\ E\{a_n a_{n-k}\} = E\{a_n\}E\{a_{n-k}\} = 0\cdot 0 = 0, & k \neq 0 \end{cases} \end{aligned}$$

Now refer to (A.18), note that $m_a = 0$ and $\sigma_a^2 = 1$, the second term becomes 0 and the PSD of the signal is

$$\Psi_s(f) = \frac{|G(f)|^2}{T} \tag{A.19}$$

This is a very important expression which will be used often in evaluating the PSD of digital modulated signals. It shows that the PSD of a binary (± 1), equiprobable, stationary, and uncorrelated data sequence is just equal to the energy spectral density $|G(f)|^2$ of the symbol-shaping pulse $g(t)$ divided by the symbol duration.

Example 1 Data bits are binary (± 1), equiprobable, stationary, and uncorrelated. The symbol-shaping pulse is the rectangular pulse: $g(t) = 1$ for $(-T/2, T/2)$, and 0 elsewhere. Then from a Fourier transform table,

$$G(f) = T\left(\frac{\sin \pi f T}{\pi f T}\right)$$

From (A.19) its PSD is

$$\Psi_s(f) = \frac{|G(f)|^2}{T} = T\left(\frac{\sin \pi f T}{\pi f T}\right)^2$$

A.3.2 Case 2: Data Symbols Are Correlated

For the general case where there is correlation between that data, let $\widetilde{a}_n$ be the corresponding data that have been normalized to have unity variance and zero mean, that is $\widetilde{a}_n = (a_n - m_a)/\sigma_a$, thus $a_n = \sigma_a \widetilde{a}_n + m_a$, then

$$\begin{aligned} R(k) &= E\{a_n a_{n-k}\} = E\{(\sigma_a \widetilde{a}_n + m_a)(\sigma_a \widetilde{a}_{n-k} + m_a)\} \\ &= \sigma_a^2 E\{\widetilde{a}_n \widetilde{a}_{n-k}\} + \underbrace{E\{\sigma_a m_a \widetilde{a}_n\}}_{0} + \underbrace{E\{\sigma_a m_a \widetilde{a}_{n-k}\}}_{0} + E\{m_a^2\} \\ &= \sigma_a^2 \rho(k) + m_a^2 \end{aligned}$$

where

$$\rho(k) = E\{\widetilde{a}_n \widetilde{a}_{n-k}\}$$

is the autocorrelation coefficient. Thus for correlated data

$$\begin{aligned} \Psi_s(f) &= \frac{|G(f)|^2}{T}\left(\sigma_a^2 \sum_{k=-\infty}^{\infty} \rho(k) e^{-jk\omega T} + m_a^2 \sum_{k=-\infty}^{\infty} e^{-jk\omega T}\right) \\ &= \underbrace{\sigma_a^2 R_d |G(f)|^2 w_p(f)}_{\text{continuous spectrum}} + \underbrace{(m_a R_d)^2 \sum_{k=-\infty}^{\infty} |G(kR_d)|^2 \delta(f - kR_d)}_{\text{discrete spectrum}} \end{aligned} \quad \text{(A.20)}$$

where $R_d = 1/T$ is the data rate, and the Poisson sum formula is used for the second term conversion and

$$w_p(f) = \sum_{k=-\infty}^{\infty} \rho(k)e^{-jk\omega T} = \mathcal{F}\left\{\sum_{k=-\infty}^{\infty} \rho(k)\delta(\tau - kT)\right\}$$

is a spectral weight function which can be obtained from the Fourier transform ($\mathcal{F}$) of the autocorrelation coefficient impulse train. Comparing (A.20) with (A.18) shows that the PSD of the correlated digital signal has the same discrete spectral part of the uncorrelated one. However, its continuous spectral part has an additional weight function which is determined by the autocorrelation coefficient impulse train of the data sequence.

In this section we have derived the PSD of a baseband digital signal by using time and the statistical average of the Fourier transform of the truncated signal. Another method is to use the Wiener-Khinchine relation (Section A.2). That is, we first find the autocorrelation function of the digital signal $g(t)$. Then we take the Fourier transform to find the PSD expression. See [2, p. 191] for details.

A.4 POWER SPECTRAL DENSITIES OF DIGITAL BANDPASS SIGNALS

We have mentioned that symbol function $g(t)$ could be any function with a Fourier transform defined on $[0, T]$. Even though the spectral density expressions derived in the previous section are used mainly for a variety of baseband digital signals, they can also be used for some bandpass signals. For example, $g(t)$ could be a burst of carrier, sinusoidal, or square waveform. If $g(t)$ is a burst of sinusoidal signal, the signal $s(t)$ would be an ASK signal. The spectrum of $g(t)$ can be easily found as

$$\begin{aligned} G(f) &= \mathcal{F}\{\cos(2\pi f_c t)\}, \quad 0 \le t \le T \\ &= \frac{1}{2}T[\text{sinc}(f - f_c)T + \text{sinc}(f + f_c)T]e^{-j\pi f T} \end{aligned}$$

and the PSD of the ASK signal is

$$\begin{aligned} \Psi_s(f) &= \frac{|G(f)|^2}{T} = \frac{1}{T}\left(\frac{1}{2}T[\text{sinc}(f - f_c)T + \text{sinc}(f + f_c)T]\right)^2 \\ &\approx \frac{1}{4}T\{\text{sinc}^2(f - f_c)T + \text{sinc}^2(f + f_c)T\} \end{aligned}$$

which appears as two squared sinc functions centered at f_c and $-f_c$, respectively.

For bandpass signals with data symbols embedded in the phase or frequency of the carrier, the signal cannot be written as the form of (A.14). We must find another way to compute the spectra.

Many bandpass signals of practical interest, such as FSK, PSK signals, binary, or M-ary, can be considered as a sum of an inphase component and a quadrature component. There is a rather easy way to compute the PSD of such a signal. We will discuss the method here. For other bandpass signals, such as a continuous phase modulation (CPM) signal, finding the PSD is much more complicated. We will discuss the method in the next section.

Now consider a carrier modulated bandpass signal in the form of

$$s(t) = x(t)\cos(2\pi f_c t) - y(t)\sin(2\pi f_c t)$$

where $x(t)$ and $y(t)$ are the inphase and quadrature components, determined by data symbol sequences $\{x_k\}$ and $\{y_k\}$, respectively.

$$x(t) = \sum_{k=-\infty}^{\infty} x_k p(t - kT)$$

$$y(t) = \sum_{k=-\infty}^{\infty} y_k q(t - kT)$$

The signals $p(t)$ and $q(t)$ are baseband pulse-shaping functions defined on $[0, T]$, whose Fourier transforms,

$$P(f) = \mathcal{F}\{p(t)\}$$

and

$$Q(f) = \mathcal{F}\{q(t)\}$$

exist.

We assume that $\{x_k\}$ and $\{y_k\}$ are independent, identically distributed random sequences with zero means

$$m_x = m_y = 0$$

and each sequence member has mean square value

$$E\{x_k^2\} = \sigma_x^2$$

and

$$E\{y_k^2\} = \sigma_y^2$$

respectively. Thus $\{x_k\}$ and $\{y_k\}$ are discrete stationary random processes. According to (A.13), it suffices to find the PSD of the complex envelope. The complex envelope of the signal $s(t)$ is

$$\widetilde{s}(t) = x(t) + jy(t)$$

To calculate the PSD of $\widetilde{s}(t)$, we calculate its autocorrelation function first.

$$\begin{aligned} R_{\widetilde{s}}(\tau) &= E\{\widetilde{s}(t)\widetilde{s}^*(t-\tau)\} \\ &= E\{[x(t)+jy(t)][x(t-\tau)-jy(t-\tau)]\} \\ &= R_x(\tau) + R_y(\tau) \end{aligned}$$

where

$$R_x(\tau) = E\{x(t)x(t-\tau)\}$$

and

$$R_y(\tau) = E\{y(t)y(t-\tau)\}$$

and the cross-correlation terms vanished since they are zero. From the Wiener-Khinchine theorem, we have

$$\begin{aligned} \Psi_{\widetilde{s}}(f) &= \mathcal{F}\{R_{\widetilde{s}}(\tau)\} = \mathcal{F}\{R_x(\tau)\} + \mathcal{F}\{R_y(\tau)\} \\ &= \Psi_x(f) + \Psi_y(f) \end{aligned}$$

where $\Psi_x(f)$ and $\Psi_y(f)$ are the PSD of the inphase and quadrature component, respectively. Comparing the expressions of $x(t)$ and $y(t)$ with (A.14) we realize that we can use (A.18) to find $\Psi_x(f)$ and $\Psi_y(f)$. That is

$$\Psi_x(f) = \frac{\sigma_x^2 |P(f)|^2}{T}$$

and

$$\Psi_y(f) = \frac{\sigma_y^2 |Q(f)|^2}{T}$$

thus

$$\Psi_{\widetilde{s}}(f) = \frac{\sigma_x^2 |P(f)|^2}{T} + \frac{\sigma_y^2 |Q(f)|^2}{T} \tag{A.21}$$

This expression can be used for all quadrature modulated signals, including M-ary PSK, MSK, QAM, and so forth.

A.5 POWER SPECTRAL DENSITIES OF CPM SIGNALS

CPM signal is defined by

$$s(t) = A\cos(2\pi f_c t + \Phi(t, \mathbf{a})), \quad -\infty \leq t \leq \infty \tag{A.22}$$

The transmitted M-ary symbols a_i are embedded in the phase

$$\Phi(t, \mathbf{a}) = 2\pi h \sum_{k=-\infty}^{\infty} a_k q(t - kT) \tag{A.23}$$

with

$$q(t) = \int_{-\infty}^{t} g(\tau) d\tau. \tag{A.24}$$

The M-ary data a_k may take any of the M values: $\pm1, \pm3, ..., \pm(M-1)$,where M usually is a power of 2. The phase is proportional to the parameter h which is called the modulation index. Function $g(t)$ is the *frequency shape pulse*. The function $g(t)$ usually has a smooth pulse shape over a finite time interval $0 \leq t \leq LT$, and is zero outside. Function $q(t)$ is the *phase function*. For $g(t)$ defined on $[0, LT]$, $q(t) = 0$ for $t < 0$, and $q(t)$ reaches the maximum at $t = LT$ and remains at the maximum thereafter. Usually $g(t)$ is normalized to have $q(t \geq LT) = 1/2$.

We assume that symbols $\{a_k\}$ are statistically independent and identically distributed with prior probabilities $\Pr(a_n) = P_n$, $n = \pm1, \pm3, ..., \pm(M-1)$, where $\sum P_n = 1$.

The CPM signal can be written as

$$\begin{aligned} s(t) &= A\,\mathrm{Re}\{\exp[j2\pi f_c t + j\Phi(t, \mathbf{a})]\} \\ &= A\,\mathrm{Re}\{\exp[j\Phi(t, \mathbf{a})]\exp[j2\pi f_c t]\} \end{aligned}$$

The complex envelope of the CPM signal is (omitting the constant amplitude A, which will not affect the shape of the autocorrelation function and the power spectral density)

$$\tilde{s}(t) = \exp[j\Phi(t, \mathbf{a})]$$

We now present a simple, fast, and reliable general numerical method of computing the PSD of the general CPM signal. This method was first given in [3] and appeared in [2,4] as well.

First we find the autocorrelation of $\tilde{s}(t)$.

$$R_{\tilde{s}}(t+\tau; t) = E\{\exp[j\Phi(t+\tau, \mathbf{a})][\exp[j\Phi(t, \mathbf{a})]^*\}$$

$$\begin{aligned} &= E\{\exp[j\Phi(t+\tau,\mathbf{a}) - j\Phi(t,\mathbf{a})]\} \\ &= E\left\{\exp\left[j2\pi h \sum_{k=-\infty}^{\infty} a_k[q(t+\tau-kT) - q(t-kT)]\right]\right\} \end{aligned}$$

The sum in the exponent can be expressed as a product of exponents

$$R_{\tilde{s}}(t+\tau;t) = E\left\{\prod_{k=-\infty}^{\infty} \exp\left[j2\pi h a_k[q(t+\tau-kT) - q(t-kT)]\right]\right\}$$

Now we carry out the expectation over the data symbols. Since they are statistically independent, we have

$$R_{\tilde{s}}(t+\tau;t) = \prod_{k=-\infty}^{\infty}\left\{\sum_{\substack{n=-(M-1)\\ n \text{ odd}}}^{M-1} P_n \exp\left[j2\pi h n[q(t+\tau-kT) - q(t-kT)]\right]\right\} \tag{A.25}$$

where P_n is the a priori probability of the symbol n. Then we take the time-average of the autocorrelation function in $[0, T]$. (In fact averaging over any time interval of duration T would give the same result.)

$$R_{\tilde{s}}(\tau) = \frac{1}{T}\int_0^T R_{\tilde{s}}(t+\tau;t)dt \tag{A.26}$$

Seemingly, there are an infinite number of factors in the product of (A.25), but in fact there are only a finite number of terms that have nonzero exponents. Note that the integration interval is only in $[0, T]$. Since $q(t)$ is 0 for $t < 0$ and equal to $1/2$ for $t \geq LT$, for most k, the two terms in the exponent in (A.25) will both be zero or both be $1/2$. That is, they cancel to zero for all but a finite number of factors.

Suppose $\tau \geq 0$ and let $\tau = \xi + mT$ where $0 \leq \xi < T$ and $m = 0, 1, 2, ...,$ then for the exponent to be unconditionally zero, we have to have

$$\begin{cases} t+\xi+mT-kT < 0 \\ t-kT < 0 \end{cases} \tag{A.27}$$

or

$$\begin{cases} t+\xi+mT-kT > LT \\ t-kT > LT \end{cases} \tag{A.28}$$

Notice that $0 \leq t < T$ and $0 \leq \xi < T$, then $0 \leq t+\xi < 2T$. From (A.27) we have $k > m+2$. From (A.28) we obtain $k < -L$. Thus for the exponent to be nonzero we need

$$1 - L \leq k \leq m + 1 \tag{A.29}$$

For other values of the index k, the exponent is zero, the sum of P_n becomes 1. Thus (A.26) reduces to

$$\begin{aligned} R_{\tilde{s}}(\tau) &= R_{\tilde{s}}(\xi + mT) \\ &= \frac{1}{T}\int_0^T \prod_{k=1-L}^{m+1} \left\{ \sum_{\substack{n=-(M-1) \\ n \text{ odd}}}^{M-1} P_n \exp[j2\pi h n q_d(t, m, k, \xi)] \right\} dt \end{aligned} \tag{A.30}$$

where

$$q_d(t, m, k, \xi) = q(t + \xi - (k - m)T) - q(t - kT)$$

Note that we have assumed that $\tau \geq 0$ in deriving the above expression. However, we can use a property of autocorrelation functions to find $R_{\tilde{s}}(\tau)$ for $\tau \leq 0$. It is well known that [5]

$$R_{\tilde{s}}(-\tau) = R_{\tilde{s}}^*(\tau) \tag{A.31}$$

This property can be used to find $R_{\tilde{s}}(\tau)$ for $\tau \leq 0$ from values of $R_{\tilde{s}}(\tau)$ for $\tau \geq 0$.

Before we proceed to find the PSD of the CPM signal by taking the Fourier transform of $R_{\tilde{s}}(\tau)$, we need to study (A.30) for $\tau = \xi + mT \geq LT$. The results will be used in finding the PSD.

We divide the range of k, in which the exponent is nonzero, given in (A.29), into two segments: $1 - L \leq k \leq 0$ and $1 \leq k \leq m + 1$. For $1 - L \leq k \leq 0$, we have $t + \tau - kT > LT$ for $t \in [0, T]$. Then $q(t + \tau - kT) = q(LT)$ which is a constant. For $1 \leq k \leq m+1$ we have $t - kT \leq 0$, then $q(t - kT) = 0$ always. Thus we have

$$\begin{aligned} R_{\tilde{s}}(\tau) &= R_{\tilde{s}}(\xi + mT) \\ &= \frac{1}{T}\int_0^T \prod_{k=1-L}^{0} \left\{ \sum_{\substack{n=-(M-1) \\ n \text{ odd}}}^{M-1} P_n \exp\left[j2\pi h n[q(LT) - q(t - kT)]\right] \right\} \\ &\quad \times \prod_{k=1}^{m+1} \left\{ \sum_{\substack{n=-(M-1) \\ n \text{ odd}}}^{M-1} P_n \exp\left[j2\pi h n q(t + \tau - kT)\right] \right\} dt \end{aligned} \tag{A.32}$$

Now notice that

$$\begin{aligned} & \exp\left[j2\pi hnq(t+\tau-kT)\right] \\ = & \exp\left[j2\pi hnq(t+\xi+mT-kT)\right] \\ = & \exp\left[j2\pi hnq(t+\xi-(k-m)T)\right] \\ = & \exp\left[j2\pi hnq(t+\xi-iT)\right] \end{aligned}$$

where $i = k - m$. Using index i, $\prod_{k=1}^{m+1}$ becomes $\prod_{i=1-m}^{1} = \prod_{i=1-m}^{-L} \prod_{i=-L+1}^{1}$. For indexes $1-m \le i \le -L, q(t+\xi-iT) = q(LT)$ always. Thus (A.32) becomes

$$\begin{aligned} R_{\tilde{s}}(\tau) = & \frac{1}{T}\int_0^T \prod_{k=1-L}^{0} \left\{ \sum_{\substack{n=-(M-1) \\ n \text{ odd}}}^{M-1} P_n \exp\left[j2\pi hn[q(LT)-q(t-kT)]\right] \right\} \\ & \times C_a^{m-L} \prod_{k=-L+1}^{1} \left\{ \sum_{\substack{n=-(M-1) \\ n \text{ odd}}}^{M-1} P_n \exp\left[j2\pi hnq(t+\xi-kT)\right] \right\} dt, \end{aligned}$$

$$(\text{for } m \ge L) \tag{A.33}$$

where

$$C_a = \sum_{\substack{n=-(M-1) \\ n \text{ odd}}}^{M-1} P_n \exp\left[j2\pi hnq(LT)\right] \tag{A.34}$$

is a constant, independent of τ and C_a^{m-L} represents the product $\prod_{i=1-m}^{-L}$. The integrand in (A.33) is independent of m, and only depends on ξ. We can write the autocorrelation function as

$$R_{\tilde{s}}(\tau) = R_{\tilde{s}}(\xi+mT) = C_a^{m-L}\psi(\xi), \qquad m \ge L, \qquad 0 \le \xi < T \tag{A.35}$$

where $\psi(\xi)$ denotes the integral in (A.33) which only need be computed in $[0,T]$. Thus $R_{\tilde{s}}(\tau)$ is separable in the arguments ξ and m. To compute $R_{\tilde{s}}(\tau)$ one only needs to compute $\psi(\xi)$ in $[0,T]$ and multiply it with the constant C_a^{m-L} which is the geometrical decaying factor from symbol interval to symbol interval.

Now we proceed to find the power spectral density of the CPM signal by taking the Fourier transform of $R_{\tilde{s}}(\tau)$

$$\Psi_{\tilde{s}}(f) = \mathcal{F}\{R_{\tilde{s}}(\tau)\} = \int_{-\infty}^{\infty} R_{\tilde{s}}(\tau)\, e^{-j2\pi f\tau} d\tau$$

Using (A.31), this can be simplified as

$$\Psi_{\tilde{s}}(f) = 2\,\mathrm{Re}\left[\int_0^\infty R_{\tilde{s}}(\tau)\, e^{-j2\pi f\tau} d\tau\right] \tag{A.36}$$

in which $\tau \geq 0$. Therefore $R_{\tilde{s}}(\tau)$ need be computed only for positive τ values. The integral in the above equation can be divided into two parts

$$\int_0^\infty R_{\tilde{s}}(\tau)\, e^{-j2\pi f\tau} d\tau = \int_0^{LT} R_{\tilde{s}}(\tau)\, e^{-j2\pi f\tau} d\tau + \int_{LT}^\infty R_{\tilde{s}}(\tau)\, e^{-j2\pi f\tau} d\tau \tag{A.37}$$

where the second integral is

$$\begin{aligned}
\int_{LT}^\infty R_{\tilde{s}}(\tau)\, e^{-j2\pi f\tau} d\tau &= \sum_{m=L}^\infty \int_0^T C_a^{m-L}\psi(\xi)e^{-j2\pi f(\xi+mT)}d\xi \\
&= \sum_{m=L}^\infty e^{-j2\pi fmT} \int_0^T C_a^{m-L}\psi(\xi)e^{-j2\pi f\xi}d\xi \\
&= \left(\sum_{m=L}^\infty e^{-j2\pi fmT} C_a^{m-L}\right)\int_0^T \psi(\xi)e^{-j2\pi f\xi}d\xi \\
&= e^{-j2\pi fLT}\left(\sum_{n=0}^\infty e^{-j2\pi fnT} C_a^{n}\right)\int_0^T \psi(\xi)e^{-j2\pi f\xi}d\xi
\end{aligned} \tag{A.38}$$

From (A.34) we know $|C_a| \leq 1$. Consider $|C_a| < 1$ first. The sum in (A.38) is geometric and converges (the well-known sum formula for a geometric series $\sum_{n=0}^\infty x^n = 1/(1-x)$, $|x| < 1$, is used)

$$\sum_{n=0}^\infty e^{-j2\pi fnT} C_a^n = \frac{1}{1 - C_a e^{-j2\pi fT}} \tag{A.39}$$

Further, let $m = L$ in (A.35) we have

$$\psi(\xi) = R_{\tilde{s}}(\xi + LT)$$

Then (A.38) can be expressed in terms of $R_{\tilde{s}}(\tau)$ as

$$\int_{LT}^\infty R_{\tilde{s}}(\tau)\, e^{-j2\pi f\tau} d\tau = \frac{e^{-j2\pi fLT}}{1 - C_a e^{-j2\pi fT}} \int_0^T R_{\tilde{s}}(\tau + LT)\, e^{-j2\pi f\tau} d\tau \tag{A.40}$$

Then (A.36) can be expressed in the following final form

$$\begin{aligned}\Psi_{\tilde{s}}(f) &= 2\,\mathrm{Re}\left\{\int_0^{LT} R_{\tilde{s}}(\tau)\,e^{-j2\pi f\tau}d\tau \right.\\ &\left. +\frac{e^{-j2\pi fLT}}{1-C_a e^{-j2\pi fT}}\int_0^{T} R_{\tilde{s}}(\tau+LT)\,e^{-j2\pi f\tau}d\tau\right\} \qquad \text{(A.41)}\end{aligned}$$

From this expression we see that $R_{\tilde{s}}(\tau)$ only has to be computed over the interval $[0,(L+1)T]$. The $R_{\tilde{s}}(\tau)$ is most easily computed numerically by using (A.30). Then using (A.34) and (A.41) the PSD of a general CPM signal can be found for the case $|C_a|<1$.

A special case is that $P_n = 1/M, n = \pm1, \pm3, ..., \pm(M-1)$. For this case, all quantities are real-valued, we have

$$R_{\tilde{s}}(\tau) = \frac{1}{T}\int_0^T \prod_{k=1-L}^{[\tau/T]} \frac{1}{M}\frac{\sin 2\pi hM[q(t+\tau-kT)-q(t-kT)]}{\sin 2\pi h[q(t+\tau-kT)-q(t-kT)]} \qquad \text{(A.42)}$$

and

$$\begin{aligned}\Psi_{\tilde{s}}(f) &= 2\int_0^{LT} R_{\tilde{s}}(\tau)\cos 2\pi f\tau d\tau + \frac{2}{1+C_a^2-2C_a\cos 2\pi fT}\cdot\\ &\left\{(1-C_a\cos 2\pi fT)\int_{LT}^{(L+1)T} R_{\tilde{s}}(\tau)\cos 2\pi f\tau d\tau\right.\\ &\left. -C_a\sin 2\pi fT\int_{LT}^{(L+1)T} R_{\tilde{s}}(\tau)\sin 2\pi f\tau d\tau\right\} \qquad \text{(A.43)}\end{aligned}$$

Now consider the $|C_a|=1$ case. The analysis here basically follows the one given in [3]. First we note that when $q(LT)=0$, then $C_a=1$. But this is not our case since we have normalized $q(t)$ so that $q(LT)=1/2$ in the above analysis. For $q(LT)=1/2$,

$$C_a = \sum_{\substack{n=-(M-1)\\ n\text{ odd}}}^{M-1} P_n \exp[j\pi hn]$$

When h is not an integer, there are M cases that make $|C_a|=1$ for M-ary symbols. That is, if the a priori probability of one symbol is 1 and the a priori probability of other symbols is 0, then

$$C_a = e^{jhn\pi}, \qquad \Pr(n) = 1 \tag{A.44}$$

These are extreme cases that have very little practical value. When h is an integer, for M-ary symbols,

$$\begin{aligned} C_a &= \sum_{\substack{n=-(M-1) \\ n \text{ odd}}}^{M-1} P_n \exp[j\pi h n] \\ &= \begin{cases} -1, & h \text{ odd integer} \\ +1, & h \text{ even integer} \end{cases} \end{aligned} \tag{A.45}$$

These two cases make $R_{\tilde{s}}(\tau)$ periodical outside $|\tau| = LT$, see (A.35). It is clear from (A.35) and (A.45) that when h is odd, the period is $2T$, when h is even, the period is T.

Extend the periodic part of $R_{\tilde{s}}(\tau)$ to also cover the interval $[-LT, LT]$, and call this part $R_{dis}(\tau)$, then the rest of $R_{\tilde{s}}(\tau)$ can be called $R_{con}(\tau)$, thus

$$R_{\tilde{s}}(\tau) = R_{con}(\tau) + R_{dis}(\tau)$$

where $R_{dis}(\tau)$ is strictly periodical and $R_{con}(\tau)$ is not. It is obvious that $R_{con}(\tau) \equiv 0$, for $|\tau| \geq LT$ from the above definition. Thus $R_{con}(\tau)$ yields the continuous part of the PSD and $R_{dis}(\tau)$ yields the discrete part of the PSD.

$$\Psi_{\tilde{s}}(f) = 2\,\mathrm{Re}\left[\int_0^{LT} R_{con}(\tau)\, e^{-j2\pi f\tau} d\tau\right] + F_{dis}(f) \tag{A.46}$$

where $F_{dis}(f)$ is the discrete PSD of $R_{dis}(\tau)$ that should be computed using formulas of Fourier coefficients instead of (A.41). When h is even since the period of $R_{dis}(\tau)$ is T, the discrete frequency components appear at $f = \pm k/T$, $k = 0, 1, 2, ...$ When h is odd, $R_{dis}(\tau) = \pm\psi(\xi)$ alternatively for every T seconds, that is, it is a periodic function with *odd half-wave symmetry* [6, p. 104]. Its spectrum would only have odd harmonics at $f = \pm(2k+1)/2T$, $k = 0, 1, 2, ...$

The property that a CPM signal with integer index has discrete frequency components can be used to recover the carrier and symbol timing in CPM receivers. However, some of the components may have zero amplitude depending on the specific shape of the frequency pulse.

References

[1] Haykin, S., *Communication Systems*, 3rd ed., New York: John Wiley, 1994.

[2] Proakis, J. G., *Digital Communications,* 2nd ed., New York: McGraw-Hill, 1989.

[3] Aulin, T., and C-E. Sundberg, "An easy way to calculate the power spectrum for digital FM," *Proceedings of the IEE, Part F: Communication, Radar and Signal Processing*, vol. 130(6), 1983, pp. 519–526.

[4] Anderson, J., T. Aulin, and C.-E. Sundberg, *Digital Phase Modulation,* New York: Plenum Publishing Company, 1986.

[5] Whalen, A. D., *Detection of Signals in Noise*, New York and London: Academic Press, 1971.

[6] McGillem, C., and G. Cooper, *Continuous and Discrete Signal and System Analysis*, 3rd ed., Philadelphia, Pennsylvania: Saunders College Publishing, 1991.

Appendix B

Detection of Signals

B.1 DETECTION OF DISCRETE SIGNALS

Detection of discrete signals can be modeled by the classical hypothesis test problem. In this section we first examine the binary hypothesis test problem. We will address the M-ary hypothesis problem in later sections.

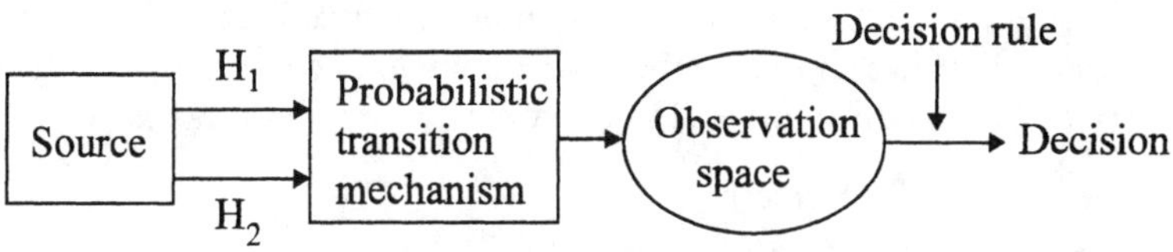

Figure B.1 Binary hypothesis test model.

B.1.1 Binary Hypothesis Test

The binary hypothesis model is shown in Figure B.1. The source generates an output which is one of two possible choices. We refer to them as *hypotheses*, H_1 and H_2. In binary digital communications, these two outputs of the source are data 1 and 0. Therefore H_1 and H_2 can be associated with 1 and 0 being transmitted, respectively. The signal passes through a *probabilistic transition mechanism* which generates a point in the *observation space* based on the hypothesis and according to some probability law. In digital communications, it is the noisy channel which transports the signal but also adds noise to the signal. The observation space in general is N-dimensional. Each hypothesis maps into a point in the observation space. Each point is an N-dimensional vector:

$$\mathbf{r} \triangleq \begin{bmatrix} r_1 \\ r_2 \\ \vdots \\ r_N \end{bmatrix}$$

The probabilistic transition mechanism generates points in accord with the two known conditional probability densities $p(\mathbf{r}/H_1)$ and $p(\mathbf{r}/H_2)$.[1] The object is to use this information to develop a suitable decision rule, based on which the two observations are compared and one of the hypotheses is chosen. To do this we must examine various criteria for making decisions.

B.1.2 Decision Criteria

With the binary hypothesis problem we know that either H_1 or H_2 is true. When we make decisions we could choose H_1 or H_2. But a third choice is possible, that is, don't know. The third choice is actually used in digital communications. That is called *erasure*. However, we confine our discussion here to decision rules that require a choice of two hypotheses. In this case, there are four actions:

1. H_1 true; choose H_1 (correct);
2. H_1 true; choose H_2 (erroneous);
3. H_2 true; choose H_2 (correct);
4. H_2 true; choose H_1 (erroneous).

The first and third are correct choices. The second and the fourth are erroneous choices. The decision criterion should attach some relative importance to the four possible actions. The method of processing the received data ($\mathbf{r}$) will depend on the criterion we select.

B.1.2.1 Bayes Criterion

Bayes criterion is based on the average cost of four actions. We first define four costs as

1. $C_{11} \triangleq$ cost of action 1;

[1] $p(\mathbf{r}/H_i), i = 1, 2$ are the probability densities of $\mathbf{r}$ under the hypotheses $H_i, i = 1, 2$. Note that these two densities both are represented by the same function name p, but they are actually two different functions. The difference is indicated by the arguments. We will use this type of notation in the rest of this appendix.

2. $C_{21} \triangleq$ cost of action 2;
3. $C_{22} \triangleq$ cost of action 3;
4. $C_{12} \triangleq$ cost of action 4.

Note that C_{ij} denotes the cost of choosing hypothesis i while hypothesis j is true. Then we define the total cost as

$$\begin{aligned} \mathcal{R} \triangleq\ & C_{11}P_1 \Pr(\text{choose } H_1/H_1 \text{ is true}) \\ & +C_{21}P_1 \Pr(\text{choose } H_2/H_1 \text{ is true}) \\ & +C_{22}P_2 \Pr(\text{choose } H_1/H_2 \text{ is true}) \\ & +C_{12}P_2 \Pr(\text{choose } H_2/H_2 \text{ is true}) \end{aligned} \tag{B.1}$$

where $P_1 = \Pr(H_1)$ and $P_2 = \Pr(H_2)$ are *a priori probabilities* of the hypotheses.

For each observation $\mathbf{r}$ we must choose H_1 or H_2. This is equivalent to divide the entire observation space Z into two subspaces Z_1 and Z_2. If $\mathbf{r}$ falls into Z_1 we say H_1 is true. If $\mathbf{r}$ falls into Z_2 we say H_2 is true. Thus the cost can be written as

$$\begin{aligned} \mathcal{R} =\ & C_{11}P_1 \int_{Z_1} p(\mathbf{r}/H_1)d\mathbf{r} + C_{21}P_1 \int_{Z_2} p(\mathbf{r}/H_1)d\mathbf{r} \\ & +C_{22}P_2 \int_{Z_2} p(\mathbf{r}/H_2)d\mathbf{r} + C_{12}P_2 \int_{Z_1} p(\mathbf{r}/H_2)d\mathbf{r} \\ =\ & P_1C_{11} \int_{Z_1} p(\mathbf{r}/H_1)d\mathbf{r} + P_1C_{21} \int_{Z-Z_1} p(\mathbf{r}/H_1)d\mathbf{r} \\ & +P_2C_{22} \int_{Z-Z_1} p(\mathbf{r}/H_2)d\mathbf{r} + P_2C_{12} \int_{Z_1} p(\mathbf{r}/H_2)d\mathbf{r} \end{aligned} \tag{B.2}$$

Observing that

$$\int_Z p(\mathbf{r}/H_1)d\mathbf{r} = \int_Z p(\mathbf{r}/H_2)d\mathbf{r} = 1$$

therefore

$$\begin{aligned} \mathcal{R} =\ & P_1C_{21} + P_2C_{22} \\ & + \int_{Z_1} \{P_2(C_{12} - C_{22})p(\mathbf{r}/H_2) - P_1(C_{21} - C_{11})p(\mathbf{r}/H_1)\}d\mathbf{r} \end{aligned} \tag{B.3}$$

We assume that the cost of a wrong decision is higher than the cost of a correct decision, that is,

$$C_{21} > C_{11}$$

$$C_{12} > C_{22}$$

In (B.3), to minimize $\mathcal{R}$ we should assign all the values of $\mathbf{r}$ which make the integral $\int_{Z_1}(\cdot)d\mathbf{r} < 0$ to Z_1 and assign all the values of $\mathbf{r}$ which make the integral $\int_{Z_1}(\cdot)d\mathbf{r} > 0$ to Z_2. Values of $\mathbf{r}$ which make $\int_{Z_1}(\cdot)d\mathbf{r} = 0$ have no effect on the cost and may be assigned arbitrarily (they are assigned to H_1 here). Thus the decision regions are defined by: If

$$P_2(C_{12} - C_{22})p(\mathbf{r}/H_2) \geq P_1(C_{21} - C_{11})p(\mathbf{r}/H_1) \tag{B.4}$$

assign $\mathbf{r}$ to Z_2 (i.e., H_2 is true), otherwise assign $\mathbf{r}$ to Z_1 (i.e., H_1 is true). Expression (B.4) can be written as

$$\frac{p(\mathbf{r}/H_2)}{p(\mathbf{r}/H_1)} \underset{H_1}{\overset{H_2}{\gtrless}} \frac{P_1(C_{21} - C_{11})}{P_2(C_{12} - C_{22})}$$

The left-hand-side quantity is called *likelihood ratio:*

$$\Lambda(\mathbf{r}) \triangleq \frac{p(\mathbf{r}/H_2)}{p(\mathbf{r}/H_1)} \tag{B.5}$$

and the right-hand-side quantity is called *threshold*:

$$\eta \triangleq \frac{P_1(C_{21} - C_{11})}{P_2(C_{12} - C_{22})} \tag{B.6}$$

Then the Bayes test becomes

$$\Lambda(\mathbf{r}) \underset{H_1}{\overset{H_2}{\gtrless}} \eta \tag{B.7}$$

or

$$\ln \Lambda(\mathbf{r}) \underset{H_1}{\overset{H_2}{\gtrless}} \ln \eta \tag{B.8}$$

since $ln(\cdot)$ is a monotonic function. In the following we study several cases which are important in digital communication.

B.1.2.2 Minimum Probability of Error Criterion

A reasonable assignment of costs would be like this: let the costs for errors be equal and the costs for correct decisions be zero. Then we have $C_{12} = C_{21} = 1$ and

$C_{11} = C_{22} = 0$. The total cost from (B.2) is

$$\mathcal{R} = P_1 \int_{Z_2} p(\mathbf{r}/H_1)d\mathbf{r} + P_2 \int_{Z_1} p(\mathbf{r}/H_2)d\mathbf{r} \tag{B.9}$$

which is the total error probability. Since the Bayes test minimizes $\mathcal{R}$, this set of cost assignments is called *minimum error probability criterion*. Under this criterion, the Bayes test becomes

$$\Lambda(\mathbf{r}) \underset{H_1}{\overset{H_2}{\gtrless}} \frac{P_1}{P_2} \tag{B.10}$$

or

$$\ln \Lambda(\mathbf{r}) \underset{H_1}{\overset{H_2}{\gtrless}} \ln \frac{P_1}{P_2} \tag{B.11}$$

B.1.2.3 Maximum A Posteriori Probability Criterion (MAP)

The minimum error probability criterion is also the *maximum a posteriori probability* (MAP) criterion.

Proof: The a posteriori probabilities are $\Pr(H_1/\mathbf{r})$ and $\Pr(H_2/\mathbf{r})$. Denote the probability density of r as $p(\mathbf{r})$. The MAP test is

$$\begin{aligned}
\frac{\Pr(H_2/\mathbf{r})}{\Pr(H_1/\mathbf{r})} &= \frac{\Pr(H_2/\mathbf{r})p(\mathbf{r})}{\Pr(H_1/\mathbf{r})p(\mathbf{r})} \\
&= \frac{p(H_2,\mathbf{r})}{p(H_1,\mathbf{r})} \\
&= \frac{p(\mathbf{r}/H_2)\Pr(H_2)}{p(\mathbf{r}/H_1)\Pr(H_1)} \\
&= \frac{p(\mathbf{r}/H_2)P_2}{p(\mathbf{r}/H_1)P_1} \underset{H_1}{\overset{H_2}{\gtrless}} 1
\end{aligned}$$

or

$$\Lambda(\mathbf{r}) \underset{H_1}{\overset{H_2}{\gtrless}} \frac{P_1}{P_2}$$

which is the minimum error probability test.

B.1.2.4 Maximum Likelihood Criterion

Now we further assume that $P_1 = P_2$, that is, the two hypotheses are equally likely. This case is the most frequent case in digital communications. The minimum error probability test becomes

$$\Lambda(\mathbf{r}) \underset{H_1}{\overset{H_2}{\gtrless}} 1$$

which is

$$p(\mathbf{r}/H_2) \underset{H_1}{\overset{H_2}{\gtrless}} p(\mathbf{r}/H_1)$$

This means the decision rule is to compare likelihood functions and choose the largest: *maximum likelihood criterion.*

B.1.3 *M* Hypotheses

Now we consider M hypotheses. For M hypotheses, the risk is

$$\mathcal{R} = \sum_{i=1}^{M} \sum_{j=1}^{M} P_j C_{ij} \int_{z_i} p(\mathbf{r}/H_j) d\mathbf{r} \tag{B.12}$$

where the first summation is over choices and the second summation is over hypotheses. The optimum Bayes detector is the one that partitions the observation space Z into regions Z_i such that the risk is minimized. The risk expression can be written as

$$\begin{aligned}
\mathcal{R} &= \sum_{i=1}^{M} \sum_{j=1}^{M} C_{ij} \int_{Z_i} p(\mathbf{r}/H_j) P_j d\mathbf{r} \\
&= \sum_{i=1}^{M} \sum_{j=1}^{M} C_{ij} \int_{Z_i} \Pr(H_j/\mathbf{r}) p(\mathbf{r}) d\mathbf{r} \\
&= \sum_{i=1}^{M} \int_{Z_i} \left(\sum_{j=1}^{M} C_{ij} \Pr(H_j/\mathbf{r}) \right) p(\mathbf{r}) d\mathbf{r}
\end{aligned}$$

Let

$$\beta_i(\mathbf{r}) = \sum_{j=1}^{M} C_{ij} \Pr(H_j/\mathbf{r})$$

Then

$$\begin{aligned}\mathcal{R} &= \sum_{i=1}^{M} \int_{Z_i} \beta_i(\mathbf{r}) p(\mathbf{r}) d\mathbf{r} \\ &= \int_{Z_1} \beta_1(\mathbf{r}) p(\mathbf{r}) d\mathbf{r} + \int_{Z_2} \beta_2(\mathbf{r}) p(\mathbf{r}) d\mathbf{r} + \cdots + \int_{Z_M} \beta_M(\mathbf{r}) p(\mathbf{r}) d\mathbf{r}\end{aligned}$$

Each particular **r** will be included in only one integral. The Bayes detector is supposed to assign it to the region Z_i where it will make the smallest contribution to the risk. Clearly this is done by choosing the smallest $\beta_i(\mathbf{r})$ and assigning **r** to that region. Thus the detection rule becomes: compute $\beta_i(\mathbf{r})$ and choose the smallest.

Now consider the costs, $C_{ii} = 0$ and $C_{ij} = C, i \neq j$. Then

$$\beta_i(\mathbf{r}) = C \sum_{\substack{j=1 \\ j \neq i}}^{M} \Pr(H_j/\mathbf{r})$$

or

$$\beta_i(\mathbf{r}) = C[1 - \Pr(H_i/\mathbf{r})]$$

It is clear that choosing the largest $\Pr(H_i/\mathbf{r})$ is equivalent to choosing the smallest $\beta_i(\mathbf{r})$. So now the test becomes a maximum a posteriori probability test. If we further let $C = 1$, looking at the risk expression (B.12) we see that $\mathcal{R}$ now is equal to the average error probability. So this is also a minimum error probability test.

Since

$$\Pr(H_i/\mathbf{r}) = \frac{p(\mathbf{r}/H_i)P_i}{p(\mathbf{r})}$$

and $p(\mathbf{r})$ is the same to all hypotheses, the alternative of this test is to compute $p(\mathbf{r}/H_i)P_i$ and choose the largest. In the following we apply this result to the general Gaussian problem [1, pp. 96–97] [2, pp. 91–92].

For a general Gaussian problem, there are M hypotheses and the observations are N-dimensional vectors which are Gaussian random variables. The density of the vector under the hypothesis H_i is

$$p(\mathbf{r}/H_i) = \frac{1}{(2\pi)^{N/2}|\mathbf{K}_i|^{1/2}} \exp[-\frac{1}{2}(\mathbf{r}^T - \boldsymbol{\mu}_i^T)\mathbf{K}_i^{-1}(\mathbf{r} - \boldsymbol{\mu}_i)] \qquad \text{(B.13)}$$

where $\boldsymbol{\mu}_i$ is the mean vector of $\mathbf{r}$ under hypothesis H_i.

$$\boldsymbol{\mu}_i \triangleq E\{\mathbf{r}/H_i\} = \begin{bmatrix} \mu_{1i} \\ \mu_{2i} \\ \vdots \\ \mu_{Ni} \end{bmatrix}, \quad i = 1, 2, ..., M$$

and $\mathbf{K}_i$ is the covariance matrix under hypothesis H_i, defined as

$$\mathbf{K}_i = \begin{bmatrix} E\{(r_1 - \mu_{1i})^2\} & \cdots & E\{(r_1 - \mu_{1i})(r_N - \mu_{Ni})\} \\ \vdots & & \vdots \\ E\{(r_N - \mu_{Ni})(r_1 - \mu_{1i})\} & \cdots & E\{(r_N - \mu_{Ni})^2\} \end{bmatrix} \tag{B.14}$$

which is an $N \times N$ matrix.

From above the optimum detector computes

$$P_i p(\mathbf{r}/H_i) = \frac{P_i}{(2\pi)^{N/2}|\mathbf{K}_i|^{1/2}} \exp[-\frac{1}{2}(\mathbf{r}^T - \boldsymbol{\mu}_i^T)\mathbf{K}_i^{-1}(\mathbf{r} - \boldsymbol{\mu}_i)]$$

and choose the largest. Taking the logarithm and neglecting common terms we have this decision rule: Compute

$$l_i = \ln P_i - \frac{1}{2}\ln|\mathbf{K}_i| - \frac{1}{2}(\mathbf{r}^T - \boldsymbol{\mu}_i^T)\mathbf{K}_i^{-1}(\mathbf{r} - \boldsymbol{\mu}_i) \tag{B.15}$$

and choose the largest.

B.2 DETECTION OF CONTINUOUS SIGNALS WITH KNOWN PHASES

Binary signals are the most common type of signals, both in baseband modulation and carrier modulation. M-ary signals are also used in modulation, especially in carrier modulation. In this section we derive the optimum receiver and its error performance for binary signals first. Then we progress to address the issues for M-ary signals.

B.2.1 Detection of Binary Signals

B.2.1.1 Receiver Structure

A digital signal waveform with binary signaling consists of two kinds of signals $s_1(t)$

and $s_2(t)$ for $nT \leq t \leq (n+1)T, n = 0, 1, 2....$ To simplify analysis without loss of generality, we set time duration as $0 \leq t \leq T$. From the point of view of detection theory, we say we have two hypotheses:

$$\begin{aligned} H_1 &: s_1(t), \quad 0 \leq t \leq T \text{ , is sent} \\ H_2 &: s_2(t), \quad 0 \leq t \leq T \text{ , is sent} \end{aligned}$$

The energy of them is

$$E_1 = \int_0^T s_1^2(t)dt$$

and

$$E_2 = \int_0^T s_2^2(t)dt$$

In general these two signals may be correlated. We define

$$\rho_{12} \triangleq \frac{1}{\sqrt{E_1 E_2}} \int_{-\infty}^{\infty} s_1(t)s_2(t)dt = \rho_{21}$$

as the *correlation coefficient* of $s_1(t)$ and $s_2(t)$. $|\rho_{12}| \leq 1$.

The received signal is

$$r(t) = s_i(t) + n(t), \quad i = 1, 2, \quad 0 \leq t \leq T$$

where $s_i(t)$ is one of the two possible signals, and $n(t)$ is the additive white Gaussian noise with zero mean and a two-sided spectral density of $N_o/2$. This implies that the autocorrelation function of $n(t)$ is

$$E\{n(t)n(\tau)\} = \frac{N_o}{2}\delta(t - \tau)$$

(see [2, p. 156]), where $\delta(t)$ is the Dirac delta function.

Since $n(t)$ is a Gaussian process, the received signal $r(t)$ is also a Gaussian process with $s_i(t)$ as its mean value. In order to use the hypothesis test results described above, we need to represent $r(t)$ by a set of discrete quantities. Similar to the concept of the Fourier series, we expand $r(t)$ as a weighted sum of a set of orthonormal basis functions [1, p. 178]

$$r(t) = \sum_{i=1}^{N} r_i \phi_i(t) \tag{B.16}$$

where $\phi_i(t)$ are orthonormal basis functions and

$$r_i = \int_0^T r(t)\phi_i(t)dt \tag{B.17}$$

This can be verified by simply plugging in the sum expression of $r(t)$ into the integral and using the orthonormal property of the basis functions. The vector $[r_1, r_2, \cdots, r_N]$ is the desired set of discrete quantities, called projections of $r(t)$ onto $\phi_i(t)$. It can be shown that since $\phi_i(t)$ are orthogonal, r_i are Gaussian and independent of each other. The orthogonal basis functions $\phi_i(t)$ can be found by the *Gram-Schmidt* procedure.

In this procedure, we choose the normalized first signal as the first coordinate or basis function

$$\phi_1(t) = \frac{s_1(t)}{\sqrt{E_1}}$$

Clearly $\phi_1(t)$ has unit energy, that is, it is *normalized.*

$$s_1(t) = \sqrt{E_1}\phi_1(t) = s_{11}\phi_1(t)$$

where coefficient $s_{11} = \sqrt{E_1}$ is the *projection* of $s_1(t)$ on the coordinate $\phi_1(t)$. Next we define the second basis function as

$$\phi_2(t) = c_2[s_2(t) - c_1\phi_1(t)]$$

where c_1 must be chosen to satisfy orthogonality of $\phi_2(t)$ with $\phi_1(t)$ and c_2 must be chosen to satisfy normality of $\phi_2(t)$. It is easy to find that

$$c_2 = \frac{1}{\sqrt{E_2 - s_{21}^2}} \tag{B.18}$$

$$c_1 = s_{21} = \int_0^T s_2(t)\phi_1(t)dt \tag{B.19}$$

The remaining $\phi_i(t)$ consist of an arbitrary orthonormal set whose members are orthogonal to $\phi_1(t)$ and $\phi_2(t)$. However, only the first two coefficients $[r_1, r_2]$ depend on transmitted signals. Therefore we only need to check these two *sufficient statistics*

$$\begin{aligned} r_i &= \int_0^T r(t)\phi_i(t)dt \\ &= \begin{cases} \int_0^T [s_1(t) + n(t)]\phi_i(t)dt, & H_1 \\ \int_0^T [s_2(t) + n(t)]\phi_i(t)dt, & H_2 \end{cases} \end{aligned}$$

$$= \begin{cases} s_{i1} + n_i, & H_1 \\ s_{i2} + n_i, & H_2 \end{cases}, \quad i = 1, 2$$

where

$$n_i = \int_0^T n(t)\phi_i(t)dt, \quad i = 1, 2$$

They are Gaussian random variables with zero-means and are independent of each other. Their variances are identical, that is,

$$\begin{aligned} E\{n_i^2\} &= E\left\{\left(\int_0^T n(t)\phi_i(t)dt\right)^2\right\} \\ &= E\left\{\int_0^T n(t)\phi_i(t)dt \int_0^T n(\tau)\phi_i(\tau)d\tau\right\} \\ &= \int_0^T \int_0^T E\{n(t)n(\tau)\}\phi_i(t)\phi_i(\tau)dtd\tau \\ &= \int_0^T \int_0^T \frac{N_o}{2}\delta(t-\tau)\phi_i(t)\phi_i(\tau)dtd\tau \\ &= \frac{N_o}{2}, \quad i = 1, 2 \end{aligned}$$

The mean values of $[r_1, r_2]$ are[2]

$$H_1 \quad : \quad E\{r_i/H_1\} = \int_0^T s_1(t)\phi_i(t)d = s_{i1}, \quad i = 1, 2$$

$$H_2 \quad : \quad E\{r_i/H_2\} = \int_0^T s_2(t)\phi_i(t)d = s_{i2}, \quad i = 1, 2$$

and their covariance matrixes are (from (B.14))

$$\begin{aligned} \mathbf{K}_i &= \left[\begin{array}{cc} E\{n_1^2\} & E\{n_1 n_2\} \\ E\{n_2 n_1\} & E\{n_2^2\} \end{array}, \quad i = 1, 2 \right] \\ &= \begin{bmatrix} \frac{N_o}{2} & 0 \\ 0 & \frac{N_o}{2} \end{bmatrix}, \quad i = 1, 2 \end{aligned}$$

which are identical regardless of H_i. Substituting this into (B.13), we can form the

---

2 Note that we use μ_{ij} for mean values in general cases. Here symbol s_{ij} instead of μ_{ij} is used to indicate that they are projections of signals onto the basis functions.

likelihood ratio

$$\Lambda[r(t)] = \frac{p(\mathbf{r}/H_1)}{p(\mathbf{r}/H_2)} = \frac{\frac{1}{\pi N_o}\exp\{-\frac{1}{N_o}[(r_1 - s_{11})^2 + (r_2 - s_{21})^2]\}}{\frac{1}{\pi N_o}\exp\{-\frac{1}{N_o}[(r_1 - s_{12})^2 + (r_2 - s_{22})^2]\}}$$

Taking the logarithm and canceling common terms, we have the likelihood ratio test

$$\ln \Lambda[r(t)] = -\frac{1}{N_o}\sum_{i=1}^{2}(r_i - s_{i1})^2 + \frac{1}{N_o}\sum_{i=1}^{2}(r_i - s_{i2})^2 \underset{H_2}{\overset{H_1}{\gtrless}} \ln \eta$$

where the initial threshold $\eta = P_2/P_1$ for minimum error probability criterion. Realizing that each sum is just the squared magnitude of the difference vector, the above expression can be written as

$$\ln \Lambda[r(t)] = -\frac{1}{N_o}|\mathbf{r} - \mathbf{s}_1|^2 + \frac{1}{N_o}|\mathbf{r} - \mathbf{s}_2|^2 \underset{H_2}{\overset{H_1}{\gtrless}} \ln \eta$$

or

$$\mathbf{r}^T(\mathbf{s}_1 - \mathbf{s}_2) \underset{H_2}{\overset{H_1}{\gtrless}} \frac{N_o}{2}\ln \eta + \frac{1}{2}(|\mathbf{s}_1|^2 - |\mathbf{s}_2|^2) \triangleq \gamma \tag{B.20}$$

The left-hand side is the sufficient statistic and the right-hand side is the final threshold.

From Parseval's theorem [3, p. 32],

$$\begin{aligned} l &\triangleq \mathbf{r}^T(\mathbf{s}_1 - \mathbf{s}_2) = \mathbf{r}^T\mathbf{s}_1 - \mathbf{r}^T\mathbf{s}_2 \\ &= \int_0^T r(t)s_1(t)dt - \int_0^T r(t)s_2(t)dt \end{aligned} \tag{B.21}$$

and

$$\gamma = \frac{N_o}{2}\ln \eta + \frac{1}{2}(E_1 - E_2) \tag{B.22}$$

Thus the receiver can be implemented as in Figure B.2 using two correlators. A single correlator implementation is shown in Figure B.3. The threshold γ can be reduced to simpler form for special cases. When the two signals have the same energy,

$$\gamma = \frac{N_o}{2}\ln \frac{P_2}{P_1}$$

which is simply determined by the a priori probabilities and the noise spectral density. If further the two a priori probabilities are equal, then $\gamma = 0$. This is the familiar threshold used in binary communication systems.

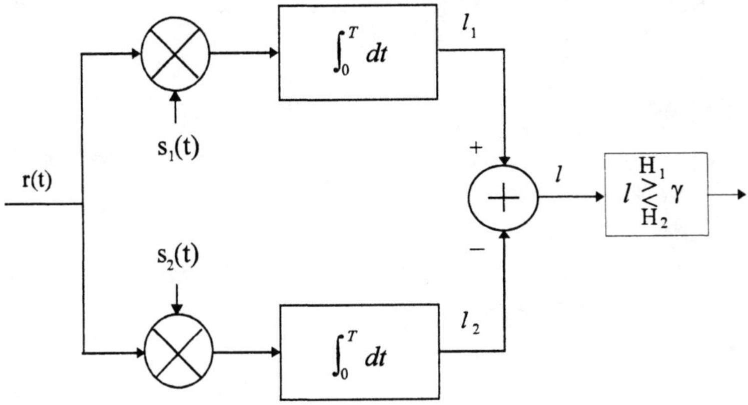

Figure B.2 Optimum binary receiver: two correlator implementation.

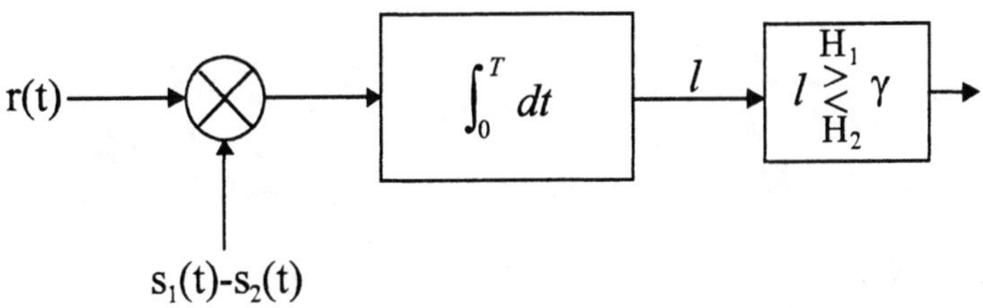

Figure B.3 Optimum binary receiver: one correlator implementation.

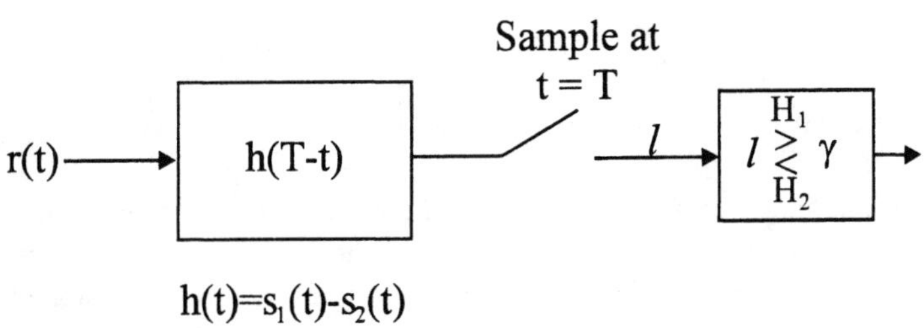

Figure B.4 Optimum binary receiver: matched filter implementation.

A correlator with reference signal $s(t)$ can be implemented as a matched filter with an impulse response

$$h(t) = s(T - t)$$

and sampled at time T. This can be verified by simply checking the output of the matched filter at $t = T$. The output of the matched filter is

$$\begin{aligned} z(t) &= r(t) * h(t) = \int_0^t r(\tau)h(t-\tau)d\tau \\ &= \int_0^t r(\tau)s[T-(t-\tau)]d\tau \\ &= \int_0^t r(\tau)s(T-t+\tau)d\tau \end{aligned}$$

When $t = T$,

$$\begin{aligned} z(T) &= \int_0^T r(\tau)s(T-T+\tau)d\tau \\ &= \int_0^T r(\tau)s(\tau)d\tau \end{aligned}$$

This is exactly the correlation between $r(t)$ and $s(t)$. From the derivation we can see that the correlator output is equivalent to the matched filter output *only* at $t = T$, the sampling moment. Figure B.4 shows the matched filter implementation of the optimum binary receiver. Note that the matched filter impulse response is

$$h(t) = s_1(T - t) - s_2(T - t) \tag{B.23}$$

B.2.1.2 Matched Filter Maximizes Signal-to-Noise Ratio

We derived the matched filter receiver from the correlator receiver which, in turn, is derived from the minimum error probability criterion. One would naturally relate minimum error probability with maximum signal-to-noise ratio. It is indeed true that the optimum receiver's output has the maximum signal-to-noise ratio. This can be proved by checking the matched-filter receiver.

For an arbitrary signal $s(t)$, we want to choose a filter which can maximize the output signal-to-noise ratio at sampling time $t = T$. Assume the filter transfer function is $H(f)$, and the signal spectrum is $S(f)$, then the output signal is

$$a(t) = \int_{-\infty}^{\infty} H(f)S(f)e^{j2\pi ft}df$$

The output noise spectral density is $|H(f)|^2\frac{N_o}{2}$ (see [2, p. 47]) and the output noise autocorrelation function is

$$R(\tau) = \int_{-\infty}^{\infty} |H(f)|^2\frac{N_o}{2}e^{j2\pi f\tau}df$$

The noise variance is therefore

$$\sigma^2 = R(0) = \frac{N_o}{2}\int_{-\infty}^{\infty} |H(f)|^2df$$

The output signal power to average noise power ratio at $t = T$ is

$$\left(\frac{S}{N}\right)_T = \frac{\left|\int_{-\infty}^{\infty} H(f)S(f)e^{j2\pi fT}df\right|^2}{\frac{N_o}{2}\int_{-\infty}^{\infty} |H(f)|^2df}$$

By using *Schwartzś inequality* [3, p. 618] it is easy to see

$$\begin{aligned}\left(\frac{S}{N}\right)_T &\leq \frac{\int_{-\infty}^{\infty} |H(f)|^2df \int_{-\infty}^{\infty} |S(f)|^2df}{\frac{N_o}{2}\int_{-\infty}^{\infty} |H(f)|^2df} \\ &= \frac{2}{N_o}\int_{-\infty}^{\infty} |S(f)|^2df\end{aligned}$$

Equality of Schwartz's inequality holds only when

$$H(f) = kS^*(f)e^{-j2\pi fT}$$

or

$$h(t) = \begin{cases} ks(T-t), & 0 \leq t \leq T \\ 0, & \text{elsewhere} \end{cases}$$

This is the matched filter.

B.2.1.3 Error Probability

Next we proceed to derive the error probability of the optimum receiver. It is conveniently done using the one correlator receiver. We need to check the probability density of output l first. Denote

$$s_d(t) \triangleq s_1(t) - s_2(t)$$

which is called the difference signal. Its energy is

$$\begin{aligned} E_d &= \int_0^T s_d^2(t)dt \\ &= \int_0^T s_1^2(t)dt - 2\int_0^T s_1(t)s_2(t)dt + \int_0^T s_2^2(t)dt \\ &= E_1 - 2\rho_{12}\sqrt{E_2 E_1} + E_2 \end{aligned} \tag{B.24}$$

Under hypothesis H_1

$$\begin{aligned} l &= \int_0^T r(t)s_d(t)dt \\ &= \int_0^T [s_1(t) + n(t)]s_d(t)dt \\ &= \int_0^T s_1(t)s_d(t)dt + \int_0^T n(t)s_d(t)dt \\ &= \int_0^T s_1^2(t)dt - \int_0^T s_1(t)s_2(t)dt + \int_0^T n(t)s_d(t)dt \\ &= E_1 - \rho_{12}\sqrt{E_1 E_2} + n = \mu_1 + n \end{aligned}$$

where

$$\mu_1 \triangleq E_1 - \rho_{12}\sqrt{E_1 E_2} \tag{B.25}$$

and $n = \int_0^T n(t)s_d(t)dt$ is a zero-mean Gaussian random variable. Its variance is

$$\begin{aligned} \sigma^2 &= E\{n^2\} = E\left\{\left[\int_0^T n(t)s_d(t)dt\right]^2\right\} \\ &= \int_0^T \int_0^T E\{n(t)n(\tau)\}s_d(t)s_d(\tau)dtd\tau \\ &= \frac{N_o}{2}\int_0^T \int_0^T \delta(t-\tau)s_d(t)s_d(\tau)dtd\tau \\ &= \frac{N_o}{2}\int_0^T s_d^2(t)dt = \frac{N_o}{2}E_d \end{aligned} \tag{B.26}$$

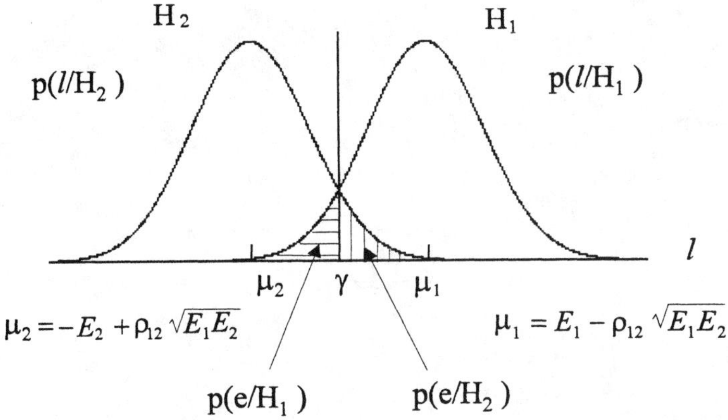

Figure B.5 Decision space.

Similarly under hypothesis H_2

$$l = -E_2 + \rho_{12}\sqrt{E_1E_2} + n = \mu_2 + n$$

where

$$\mu_2 \triangleq -E_2 + \rho_{12}\sqrt{E_1E_2} \tag{B.27}$$

and n has the same density as before.

Having seen $l = \mu_i + n$, $i = 1, 2$, and n is a Gaussian random variable with zero mean and a variance of $E_dN_o/2$, we can see that l is a Gaussian random variable with mean μ_i and a variance of $E_dN_o/2$. Thus the conditional densities of l are

$$p(l/H_1) = \frac{1}{\sqrt{\pi N_oE_d}}\exp\{-\frac{(l-\mu_1)^2}{N_oE_d}\}$$

$$p(l/H_2) = \frac{1}{\sqrt{\pi N_oE_d}}\exp\{-\frac{(l-\mu_2)^2}{N_oE_d}\}$$

where l is a value of the random variable l. Figure B.5 shows these two densities (assuming $\rho_{12} \leq 0$). When H_2 is true the receiver will make an error if l is on the right side of the threshold and vice versa. The average bit error probability is

$$P_b = P_1P(e/H_1) + P_2P(e/H_2)$$

where

$$\begin{aligned} P(e/H_2) &= \Pr(l > \gamma/H_2) = \int_{\gamma}^{\infty} p(l/H_2)dl \\ &= \int_{\gamma}^{\infty} \frac{1}{\sqrt{\pi N_o E_d}} \exp\{-\frac{(l-\mu_2)^2}{N_o E_d}\}dl \\ &= \int_{(\gamma-\mu_2)/\sqrt{N_o E_d/2}}^{\infty} \frac{1}{\sqrt{2\pi}} \exp\{-\frac{x^2}{2}\}dx \\ &= Q(\frac{\gamma-\mu_2}{\sqrt{N_o E_d/2}}) \end{aligned} \quad \text{(B.28)}$$

where

$$Q(x) \triangleq \int_x^{\infty} \frac{1}{\sqrt{2\pi}} \exp\{-\frac{u^2}{2}\}du$$

is called *Q-function.*[3]

Similarly

$$\begin{aligned} P(e/H_1) &= \Pr(l < \gamma/H_1) = 1 - \Pr(l > \gamma/H_1) \\ &= 1 - Q(\frac{\gamma-\mu_1}{\sqrt{N_o E_d/2}}) \end{aligned} \quad \text{(B.29)}$$

Now we discuss an important case where two signals are equally likely. Then $\eta = 1$

---

3 Van Trees [1, p. 37] denoted this as $\text{erfc}_*(x)$. But it is not the same as $\text{erfc}(x)$ which is called *complementary error function.*

$$\text{erfc}(x) \triangleq 1 - \text{erf}(x)$$

where $\text{erf}(x)$ is the *error function* defined as

$$\text{erf}(x) \triangleq \frac{2}{\sqrt{\pi}} \int_0^x \exp(-u^2)du$$

which is usually tabulated in mathematics handbooks. Thus $Q(x)$ is related to $\text{erfc}(x)$ by

$$Q(x) = \frac{1}{2}\text{erfc}(\frac{x}{\sqrt{2}})$$

or

$$\text{erfc}(x) = 2Q(\sqrt{2}x)$$

See Appendix C for values of the Q-function and erf(x).

and

$$\gamma = \frac{N_o}{2}\ln\eta + \frac{1}{2}(E_1 - E_2) = \frac{1}{2}(E_1 - E_2) = \frac{1}{2}(\mu_1 + \mu_2) \tag{B.30}$$

The threshold is in the midway between the two mean values in Figure B.5. $P(e/H_1)$ and $P(e/H_2)$ are the small shaded areas. From the symmetry of the figure we can easily see that

$$P(e/H_1) = P(e/H_2)$$

It can also be verified from the error probability expressions using a property of the Q function: $Q(x) = 1 - Q(-x)$. Thus

$$P_b = \frac{1}{2}(P(e/H_1) + P(e/H_2)) = P(e/H_2) = P(e/H_1)$$

Thus substituting (B.30) into (B.28) we have

$$P_b = Q\left(\frac{\mu_1 - \mu_2}{\sqrt{2N_oE_d}}\right) \tag{B.31}$$

Now substituting (B.25), (B.27), and (B.24) into (B.31) we obtain the final result

$$P_b = Q\left(\sqrt{\frac{E_d}{2N_o}}\right) \tag{B.32}$$

This expression shows that the larger the distance (E_d) between the two signals $s_1(t)$ and $s_2(t)$, the smaller the P_b. This is intuitively convincing since the larger the distance, the easier for the detector to distinguish them. It is also important to note that the error probability *only* depends on the difference signal's energy, not its shape.

It is also revealing to see the relation between P_b with the individual signal energies. Substituting (B.24) into (B.32),

$$P_b = Q\left(\sqrt{\frac{E_1 + E_2 - 2\rho_{12}\sqrt{E_2E_1}}{2N_o}}\right) \tag{B.33}$$

This expression indicates that P_b depends not only on the individual signal energies, but also on the correlation between them. It is interesting to discover that when $\rho_{12} = -1$ (i.e., when two signals are *antipodal*), P_b is the minimum. If $\rho_{12} = 0$, the signals are *orthogonal*. P_b is not minimum for orthogonal signals.

Now we consider a common case in digital communications. Two signals are equally likely, have equal energies (E_b) and are antipodal. Then (B.33) becomes

$$P_b = Q\left(\sqrt{\frac{2E_b}{N_o}}\right) \tag{B.34}$$

If these two signals are equiprobable and equal energy, but orthogonal, then

$$P_b = Q\left(\sqrt{\frac{E_b}{N_o}}\right) \tag{B.35}$$

B.2.2 Detection of M-ary Signals

B.2.2.1 Receiver Structure

Assuming M hypotheses, the received signal under hypothesis H_i is

$$H_i : r(t) = s_i(t) + n(t), \qquad 0 \leq t \leq T, \qquad i = 1, 2, ..., M$$

where $s_i(t)$ is the signal and $n(t)$ is the white Gaussian noise with zero mean and a PSD of $N_o/2$. The signal energy is

$$\int_0^T s_i^2(t)dt = E_i, \qquad i = 1, 2, ..., M$$

and M signals are correlated

$$\frac{1}{\sqrt{E_i E_j}} \int_0^T s_i(t)s_j(t)dt = \rho_{ij}, \qquad i, j = 1, 2, ..., M \tag{B.36}$$

We have shown in Section B.2.1.1 that $r(t)$ can be represented by a vector $[r_1, r_2, \cdots, r_N]$ whose elements are projections of $r(t)$ onto a set of orthonormal functions $\phi_i(t), i = 1, 2, \cdots, N$. r_i will be Gaussian and independent of each other. Like in the binary case, the orthogonal basis functions $\phi_i(t)$ can be found by the *Gram-Schmidt procedure*. They are

$$\phi_1(t) = s_1(t)/\sqrt{E_1}$$

$$\phi_1(t) = c_2[s_2(t) - c_1\phi_1(t)]$$

where c_2, c_1 are given in (B.18) and (B.19), respectively. We can find $\phi_3(t)$ similarly by defining

$$\phi_3(t) = c_3[s_3(t) - c_1\phi_1(t) - c_2\phi_2(t)]$$

where c_1 and c_2 are determined by requiring orthogonality and c_3 is determined by requiring $\phi_3(t)$ to be normalized. The resulting $\phi_i(t)$ are orthogonal and normalized. They are *orthonormal* functions. We proceed this way until one of two things happens:

1. M orthonormal functions are obtained.
2. $N(< M)$ orthonormal functions are obtained and the remaining signals can be represented by linear combinations of these orthonormal functions.[4]

Then we use this set of orthonormal functions to generate N coefficients

$$r_i = \int_0^T r(t)\phi_i(t)dt, \qquad i = 1, 2, ..., N$$

r_i are statistically independent Gaussian random variables with variance of $N_o/2$. Their mean values depend on hypotheses,

$$\begin{aligned} s_{ij} &= E\{r_i/H_j\} = \int_0^T s_j(t)\phi_i(t)dt \\ i &= 1, 2, ..., N\ ; j = 1, 2, ..., M \end{aligned}$$

This is the general Gaussian problem with equal $|\mathbf{K}_i|$ for all i and also $\mathbf{K}_i^{-1} = \frac{2}{N_o}\mathbf{I}$, where $\mathbf{K}_i$ is defined in (B.14). When the criterion is minimum error probability, using (B.15) and dropping the term $\frac{1}{2}\ln|\mathbf{K}_i|$ which is independent of any hypothesis, we have this sufficient statistic

$$l'_j = \ln P_j - \frac{1}{N_o}\sum_{i=1}^{N}(r_i - s_{ij})^2, \qquad j = 1, 2, ..., M \tag{B.37}$$

and determine H_j is true when l'_j is the largest. Expanding the above we have

$$l'_j = \ln P_j - \frac{1}{N_o}\left[\sum_{i=1}^{N} r_i^2 + \sum_{i=1}^{N} s_{ij}^2 - \sum_{i=1}^{N} 2r_i s_{ij}\right]$$

The term $\sum_{i=1}^{N} r_i^2$ is independent of any hypothesis, which can be dropped. From

[4] In many cases the signal set is in the form of linear combinations of orthonormal functions already, such as MPSK, QAM, and M-ary PAM, then the Gram-Schmidt procedure is unnecessary.

Parseval's theorem

$$\sum_{i=1}^{N} s_{ij}^2 = \int_0^T s_j^2(t)dt = E_j$$

$$\sum_{i=1}^{N} r_i s_{ij} = \int_0^T r(t)s_j(t)dt$$

Thus the new sufficient statistic is

$$l_j = \int_0^T r(t)s_j(t)dt + B_j \tag{B.38}$$

where the bias term

$$B_j = \frac{N_o}{2}\ln P_j - \frac{1}{2}E_j \tag{B.39}$$

The detection rule is to choose the largest of l_j. The receiver is shown in Figure B.6. Its matched filter equivalence is shown in Figure B.7. It is worthwhile to point out that the decision rule of (B.38) does not require that the M signals are orthogonal. When signals are equally likely and have the same energy, then the bias term in l_j can be dropped.

If $N < M$, the optimum receiver can be implemented using N correlators [4, pp. 185–186]. This is due to the fact that [1, p. 169]

$$s_j(t) = \sum_{i=1}^{N} s_{ij}\phi_i(t)$$

Then

$$\begin{aligned}\int_0^T r(t)s_j(t)dt &= \int_0^T r(t)\sum_{i=1}^{N} s_{ij}\phi_i(t)dt \\ &= \sum_{i=1}^{N} s_{ij}\int_0^T r(t)\phi_i(t)dt = \sum_{i=1}^{N} s_{ij}r_i\end{aligned}$$

where s_{ij} are known. Thus the receiver only needs to compute r_i using N correlators and weight the output with the signal coefficients s_{ij} to form the correlation (Figure B.8). The matched filter equivalence is shown in Figure B.9.

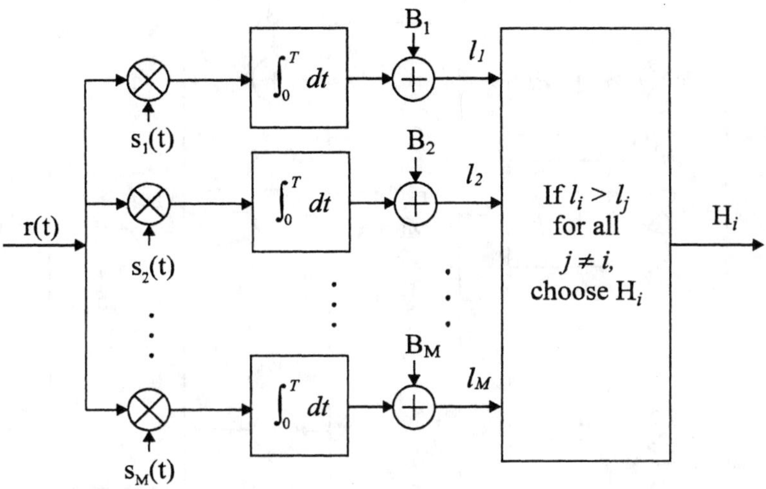

Figure B.6 Coherent M-ary detector using M correlators.

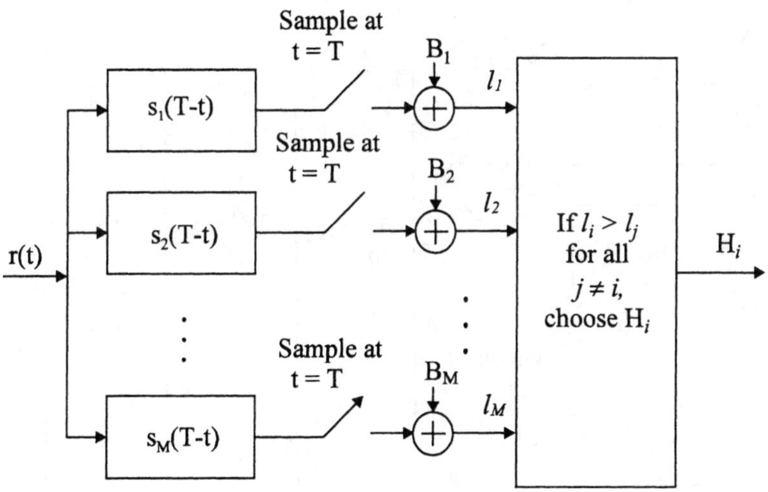

Figure B.7 Coherent M-ary detector using M matched filters.

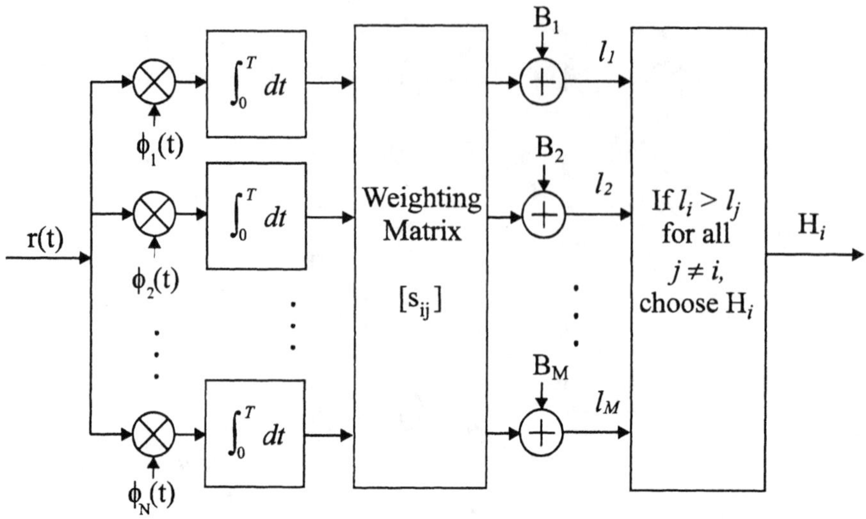

Figure B.8 Coherent M-ary detector using N correlators.

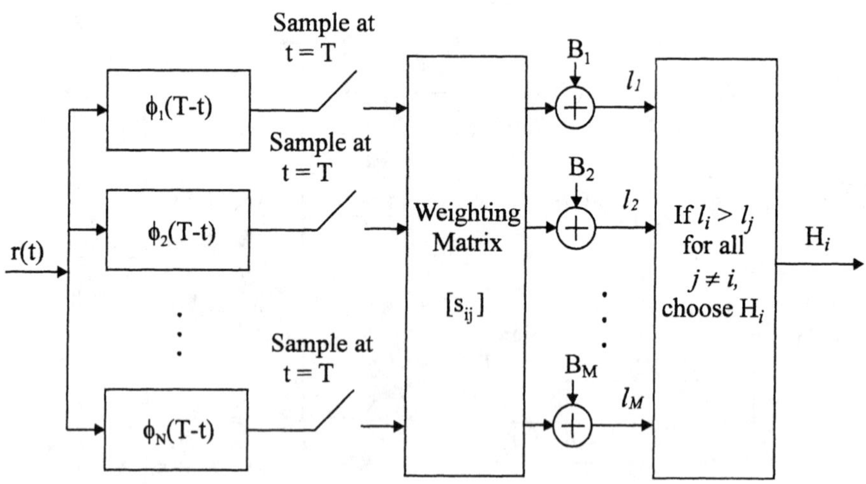

Figure B.9 Coherent M-ary detector using N matched filters.

B.2.2.2 Error Probability

Now we will derive the error probability of this important receiver for the symmetrical case: M signals are orthogonal to each other, each signal has the same symbol energy and a priori probability. That is, $P_j = P$ and $E_j = E_s$ for all j. In this case the bias term in (B.38) can be dropped and

$$l_j = \int_0^T r(t)s_j(t)dt$$

Since the problem is symmetrical we can assume that H_1 is true. An error occurs if $l_j > l_1, j \neq 1$.

$$P(e/H_1) = 1 - \Pr(\text{all } l_j < l_1 : j \neq 1/H_1) \tag{B.40}$$

Because of symmetry, the average symbol error probability is equal to the above probability. Note that the $l_j (j \neq 1)$ have the same density conditioned on H_1, and are also independent of each other. Thus

$$\begin{aligned} P_s &= P(e/H_1) \\ &= 1 - \int_{-\infty}^{\infty} p(l_1/H_1) \left[\int_{-\infty}^{l_1} p(l_j/H_1) dl_2 \right]^{M-1} dl_1 \end{aligned} \tag{B.41}$$

To find the densities of l_1 and l_j under hypothesis H_1, we note that

$$\begin{aligned} l_j &= \int_0^T r(t)s_j(t)dt \\ &= \int_0^T [s_1(t) + n(t)]s_j(t)dt \\ &= \int_0^T s_1(t)s_j(t)dt + \int_0^T n(t)s_j(t)dt \\ &= \begin{cases} E_s + n, & j = 1 \\ n, & j \neq 1, \end{cases} \end{aligned}$$

where n is a Gaussian random variable with zero mean and a variance $\frac{N_o}{2}E_s$ (see (B.26)).

Thus the densities are

$$p(l_1/H_1) = \frac{1}{\sqrt{\pi N_o E_s}} \exp\{-\frac{(l_1 - E_s)^2}{N_o E_s}\}$$

and

$$p(l_j/H_1) = \frac{1}{\sqrt{\pi N_o E_s}} \exp\{-\frac{l_j{}^2}{N_o E_s}\}$$

Substituting these densities into (B.41) and normalizing the variables we obtain

$$\begin{aligned} P_s &= 1 - \int_{-\infty}^{\infty} \frac{1}{\sqrt{2\pi}} \exp\{-\frac{1}{2}\left(x - \sqrt{\frac{2E_s}{N_o}}\right)^2\} \\ &\cdot \left[\int_{-\infty}^{x} \frac{1}{\sqrt{2\pi}} \exp\{-\frac{y^2}{2}\}dy\right]^{M-1} dx \end{aligned}$$

(for orthogonal equal energy signal set) (B.42)

This expression cannot be analytically integrated. However, we can derive a very simple bound. In order to obtain a bound for more general cases, we assume the signal set is not symmetrical. Then we replace l_1 in B.40 with l_i and rewrite the P_s as

$$P_s = \Pr(\text{any } l_j > l_i : j \neq i/H_i)$$

Now several l_j can be greater than l_i. The events are not mutually exclusive. Thus

$$P_s \leq \sum_{j \neq i} \Pr(l_j > l_i)$$

where each term in the sum is just the probability of error for two signals, which can be written in terms of their Euclidean distance as in (B.32). Thus

$$P_s \leq \sum_{j \neq i} Q\left(\frac{d_{ij}}{\sqrt{2N_o}}\right) \tag{B.43}$$

where

$$d_{ij}^2 = \int_{-\infty}^{\infty} [s_i(t) - s_j(t)]^2 dt$$

For an *orthogonal and equal energy* signal set, all distances between any two signals are equal. The distance $d = \sqrt{2E_s}$, thus (B.43) becomes

$$P_s \leq (M-1)Q\left(\sqrt{\frac{E_s}{N_o}}\right)$$

(for equal energy and orthogonal signal set) (B.44)

For fixed M this bound becomes increasingly tight as E_s/N_o is increased. In fact, it becomes a good approximation for $P_s \leq 10^{-3}$. For $M = 2$, it becomes the exact expression.

B.3 DETECTION OF CONTINUOUS SIGNALS WITH UNKNOWN PHASES

In the above discussion of detection of M-ary signals (binary signal is just a special case), we have made a fundamental assumption: the signal phases are known to the receiver. This requires the receiver to have the mechanism to track the carrier phase so that the local reference signal phase is synchronized with that of the received signal. This is called *coherent* detection which is optimum in an AWGN channel. In practical channels, such as fading channels, multipath channels, the received signal phase is very difficult or even impossible to track. Thus the detection process may have to disregard the phase information to avoid complex circuits, at some expense of performance degradation. This is called *noncoherent* detection. When the received signal phase is not tracked by the receiver, the signal phase is unknown to the receiver. The results of this section can by applied to binary and M-ary noncoherent FSK and DPSK.

B.3.1 Receiver Structure

Assuming M hypotheses with a priori probabilities $P_i = \Pr(H_i)$, the received signal under hypothesis H_i is

$$H_i : r(t) = s_i(t, \theta) + n(t), \qquad 0 \leq t \leq T, \qquad i = 1, 2, ..., M$$

where $s_i(t, \theta)$ is the signal with an unknown phase θ and $n(t)$ is the white Gaussian noise with zero mean and a PSD of $N_o/2$. We assume the unknown phase is random with a PDF $p_\theta(\theta)$.

The procedure of deriving the optimum receiver for M-ary signals with unknown phases is similar to that of the known signals. First we use the Gram-Schmidt procedure to find a set of orthonormal basis functions and expand $r(t)$ onto these basis functions. The resultant vector $\mathbf{r}$ will be used to form the likelihood function $p(\mathbf{r}/H_i)$. The decision rule for minimum error probability criterion is to compute P_i $p(\mathbf{r}/H_i)$ and choose the largest. The only new feature here is that the signal phase is random. Therefore the likelihood function is conditioned on the random phase θ, and the unconditional likelihood is

$$p(\mathbf{r}/H_i) = \int_{-\infty}^{\infty} p_\theta(\theta) p(\mathbf{r}/H_i, \theta) d\theta \tag{B.45}$$

Because components of $\mathbf{r}$ are independent Gaussian variables with equal variance, therefore $|\mathbf{K}_i|$ are all equal for all i, and in addition $\mathbf{K}_i^{-1} = \frac{2}{N_o}\mathbf{I}$, $\mathbf{I}$ is the identity matrix. Thus the likelihood of (B.13) can be denoted simply as

$$\begin{aligned} p(\mathbf{r}/H_i, \theta) &= K \exp[-\frac{1}{N_o}(\mathbf{r}^T - \mathbf{s}_i^T(\theta))(\mathbf{r} - \mathbf{s}_i(\theta))] \\ &= K \exp[-\frac{1}{N_o}\sum_{j=1}^{N}(r_j - s_{ji}(\theta))^2] \\ &= K \exp\left\{-\frac{1}{N_o}\sum_{j=1}^{N}[r_j^2 + s_{ji}^2(\theta) - 2r_j s_{ji}(\theta)]\right\} \\ &= K' \exp\left\{-\frac{1}{N_o}\sum_{j=1}^{N}[s_{ji}^2(\theta) - 2r_j s_{ji}(\theta)]\right\} \end{aligned} \tag{B.46}$$

where K and K' are constants. The term $\exp[-\frac{1}{N_o}\sum_{j=1}^{N} r_j^2]$ is common to all hypotheses, thus is included in the constant K'. The two sum terms in the exponent can be written as two integrals by using Parseval's theorem as we did from (B.37) through (B.38). Thus

$$p(\mathbf{r}/H_i, \theta) = K' \exp\left\{-\frac{1}{N_o}[\int_0^T s_i^2(t,\theta)dt - 2\int_0^T r(t)s_i(t,\theta)dt]\right\} \tag{B.47}$$

The unconditional likelihood is

$$p(\mathbf{r}/H_i) = K' \int_{-\infty}^{\infty} p_\theta(\theta)d\theta \exp\left\{-\frac{1}{N_o}[\int_0^T s_i^2(t,\theta)dt - 2\int_0^T r(t)s_i(t,\theta)dt]\right\} \tag{B.48}$$

Now we assume that θ is uniformly distributed on $[0, 2\pi]$, that is

$$p_\theta(\theta) = \frac{1}{2\pi}, \quad 0 \le \theta \le 2\pi$$

The decision rule is to maximize $P_i p(\mathbf{r}/H_i)$ over all i. From above it is seen that this is to maximize

$$\begin{aligned} & P_i \int_0^{2\pi} \frac{1}{2\pi} d\theta \exp\left\{ -\frac{1}{N_o} [\int_0^T s_i^2(t,\theta)dt - 2\int_0^T r(t)s_i(t,\theta)dt] \right\} \\ = \; & P_i e^{-\frac{E_i}{N_o}} \int_0^{2\pi} \frac{1}{2\pi} d\theta \exp\left\{ \frac{2}{N_o} \int_0^T r(t)s_i(t,\theta)dt] \right\} \qquad \text{(B.49)} \end{aligned}$$

where $E_i = \int_0^T s_i^2(t,\theta)dt$ is the energy of the ith signal. Now we assume that $s_i(t)$ is the *carrier-modulated* signal, that is,

$$s_i(t) = A_i(t)\cos[2\pi f_c t + \varphi_i(t)]$$

where $A_i(t)$ is the amplitude modulation and $\varphi_i(t)$ is the phase modulation. If the signal is frequency modulated, we can embed it into the phase modulation term. The received signal with unknown phase will be

$$s_i(t,\theta) = A_i(t)\cos[2\pi f_c t + \varphi_i(t) + \theta]$$

Thus (B.49) can be written as

$$\begin{aligned} & P_i e^{-\frac{E_i}{N_o}} \int_0^{2\pi} \frac{1}{2\pi} d\theta \exp\left\{ \frac{2}{N_o} \int_0^T r(t)s_i(t,\theta)dt] \right\} \\ = \; & P_i e^{-\frac{E_i}{N_o}} \int_0^{2\pi} \frac{1}{2\pi} d\theta \exp\left\{ \frac{2}{N_o} \int_0^T r(t)A_i(t)\cos[2\pi f_c t + \varphi_i(t) + \theta]dt \right\} \\ & \qquad \text{(B.50)} \end{aligned}$$

Now we define

$$\begin{aligned} z_{c_i} &= \int_0^T r(t)A_i(t)\cos[2\pi f_c t + \varphi_i(t)]dt \\ &= \int_0^T r(t)s_i(t)dt \qquad \text{(B.51)} \end{aligned}$$

and

$$\begin{aligned} z_{s_i} &= \int_0^T r(t)A_i(t)\sin[2\pi f_c t + \varphi_i(t)]dt \\ &= \int_0^T r(t)s_i(t,\frac{\pi}{2})dt \end{aligned}$$

where

$$
\begin{aligned}
s_i(t, \frac{\pi}{2}) &= A_i(t)\sin[2\pi f_c t + \varphi_i(t)] \\
&= A_i(t)\cos[2\pi f_c t + \varphi_i(t) - \frac{\pi}{2}]
\end{aligned}
$$

Then (B.50) becomes

$$
\begin{aligned}
& P_i e^{-\frac{E_i}{N_o}} \int_0^{2\pi} \frac{1}{2\pi} \exp\left(\frac{2}{N_o}[z_{c_i}\cos\theta - z_{s_i}\sin\theta]\right) d\theta \\
= \ & P_i e^{-\frac{E_i}{N_o}} \int_0^{2\pi} \frac{1}{2\pi} \exp[\frac{2}{N_o} z_i \cos(\theta + \delta)] d\theta \qquad \text{(B.52)}
\end{aligned}
$$

where

$$z_i = \sqrt{z_{c_i}^2 + z_{s_i}^2}$$

and

$$\delta = \tan^{-1}\left(\frac{z_{s_i}}{z_{c_i}}\right)$$

The integral in (B.52) is related to the zeroth-order modified Bessel function defined as

$$I_0(x) = \int_0^{2\pi} \frac{1}{2\pi} \exp[x\cos(\theta + \delta)] d\theta, \quad \delta \text{ is any constant}$$

By using this function (B.52) can be written as

$$P_i e^{-\frac{E_i}{N_o}} I_0(\frac{2z_i}{N_o})$$

Taking the logarithm we can equivalently state the rule as maximizing

$$l_i \triangleq \ln I_0(\frac{2z_i}{N_o}) - \frac{E_i}{N_o} + \ln P_i \qquad \text{(B.53)}$$

Because for fixed E_i/N_o, $\ln(\cdot)$ and $I_0(\cdot)$ both are monotonic functions, this is to maximize

$$l_i \triangleq \frac{2z_i}{N_o} - \frac{E_i}{N_o} + \ln P_i \qquad \text{(B.54)}$$

Figure B.10 is the receiver realizing the above decision rule. Figure B.11 is the matched filter-envelope detector version.

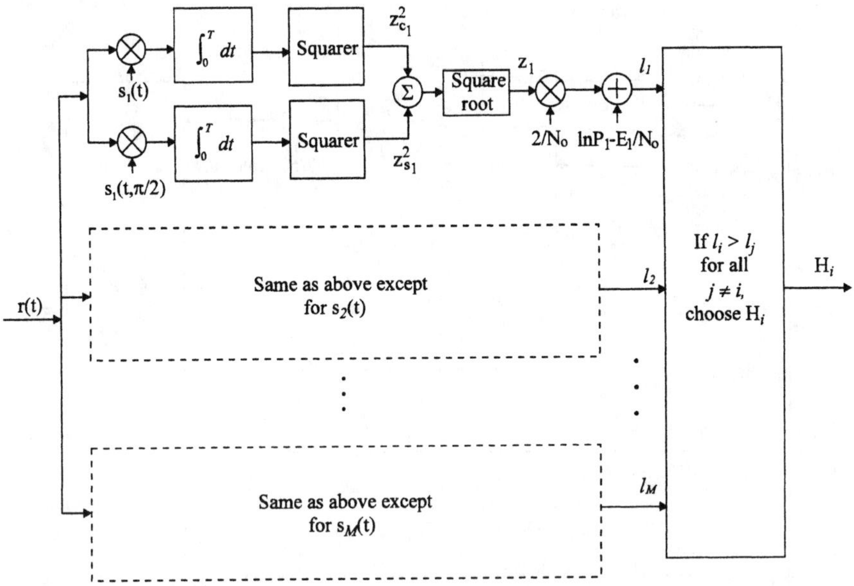

Figure B.10 Optimum detector for M-ary signals with unknown phases: correlator form.

If we further assume that the signals are equally likely and have the same energy, which is common in noncoherent communications, then the decision rule is to maximize z_i or z_i^2:

$$\begin{aligned} l_i^2 &\triangleq z_i^2 = z_{c_i}^2 + z_{s_i}^2 \\ &= \left(\int_0^T r(t)s_i(t)dt\right)^2 + \left(\int_0^T r(t)s_i(t,\frac{\pi}{2})dt\right)^2 \end{aligned} \tag{B.55}$$

The block diagram is shown in Figure B.12. Note that the two reference signals $s_i(t)$ and $s_i(t, \frac{\pi}{2})$ are orthogonal. Figure B.13 is the matched filter-envelope detector version, where

$$h_i(t) = s_i(T-t), \qquad 0 \le t \le T$$

Proof: The output of the matched filter is

$$z_i(t) = \int_0^t r(\tau)s_i(T-t+\tau)d\tau$$

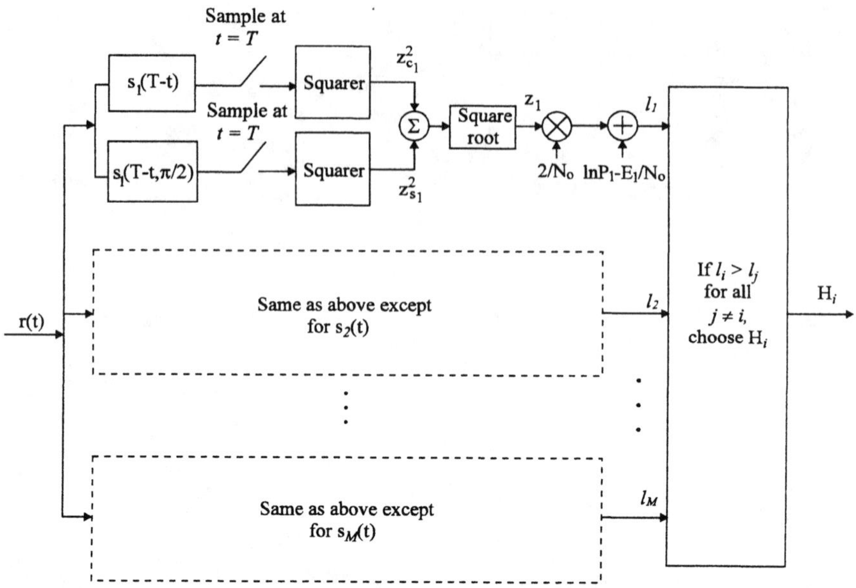

Figure B.11 Optimum detector for M-ary signals with unknown phases: matched filter form.

Substituting the expression of $s_i(\tau)$ into the above equation we have

$$
\begin{aligned}
& z_i(t) \\
= & \int_0^t r(\tau)A_i(T-t+\tau)\cos[2\pi f_c(T-t+\tau)+\varphi_i(T-t+\tau)]d\tau \\
= & \int_0^t r(\tau)A_i(T-t+\tau)\cos 2\pi f_c(T-t)\cos[2\pi f_c\tau+\varphi_i(T-t+\tau)]d\tau \\
& -\int_0^t r(\tau)A_i(T-t+\tau)\sin 2\pi f_c(T-t)\sin[2\pi f_c\tau+\varphi_i(T-t+\tau)]d\tau \\
= & \int_0^t r(\tau)A_i(T-t+\tau)\cos[2\pi f_c\tau+\varphi_i(T-t+\tau)]d\tau\cos 2\pi f_c(T-t) \\
& -\int_0^t r(\tau)A_i(T-t+\tau)\sin[2\pi f_c\tau+\varphi_i(T-t+\tau)]d\tau\sin 2\pi f_c(T-t)
\end{aligned}
$$

The envelope at $t = T$ is

$$\left\{ \left[\int_0^t r(\tau) A_i(T-t+\tau) \cos[2\pi f_c \tau + \varphi_i(T-t+\tau)] d\tau \right]^2 + \left[\int_0^t r(\tau) A_i(T-t+\tau) \sin[2\pi f_c \tau + \varphi_i(T-t+\tau)] d\tau \right]^2 \right\}^{1/2}_{t=T}$$

$$= \left\{ \left[\int_0^T r(\tau) A_i(\tau) \cos[2\pi f_c \tau + \varphi_i(\tau)] d\tau \right]^2 + \left[\int_0^T r(\tau) A_i(\tau) \sin[2\pi f_c \tau + \varphi_i(\tau)] d\tau \right]^2 \right\}^{1/2}$$

$$= \left\{ \left[\int_0^T r(\tau) s_i(\tau) d\tau \right]^2 + \left[\int_0^T r(\tau) s_i(\tau, \frac{\pi}{2}) d\tau \right]^2 \right\}^{1/2}$$

which is exactly the output of the correlator version.

B.3.2 Receiver Error Performance

B.3.2.1 Binary Orthogonal Signals

We first derive the symbol error probability $P_s = P_b$ for the binary case. Then the results can be easily extended to the M-ary case.

The noncoherent receiver for binary signals is symmetrical in structure. The error probability of choosing $s_1(t)$ when $s_2(t)$ is actually transmitted is the same as that of choosing $s_2(t)$ when $s_1(t)$ is actually transmitted. For equally likely signals the average P_b is equal to either one of these two probabilities. Therefore in the following analysis we assume that $s_1(t)$ is transmitted.

Note that the receiver in Figure B.12 or Figure B.13 is for any signal sets with equal energy and equal a priori probability; orthogonality is not required. Here we further assume that the two signals are *orthogonal*. This is due to the fact that binary orthogonal signals have the smallest error probability among the class of binary signals with constrained peak energy and unknown phase [5].

Since $s_1(t)$ is orthogonal to $s_2(t)$ (also to $s_2(t, \frac{\pi}{2})$) the integrators outputs z_{c_2} and z_{s_2} are Gaussian random variables with zero mean and a variance of $\sigma^2 = \frac{N_o}{2} E_s$

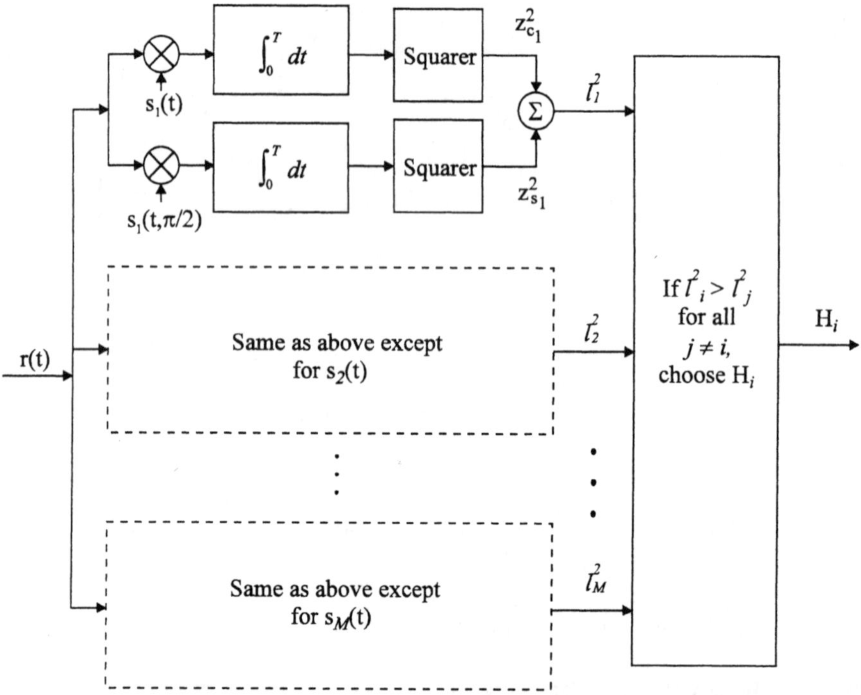

Figure B.12 Correlator receiver for M-ary signals with unknown phases. Signals are equally likely and with equal energy.

(refer to (B.26)). z_{c_2} and z_{s_2} are independent of each other since $s_i(t)$ and $s_i(t, \frac{\pi}{2})$ are orthogonal [1]. $l_2 = \sqrt{z_{c_2}^2 + z_{s_2}^2}$. It is well known that the PDF of the root-sum square of two independent Gaussian random variables is Rayleigh. That is

$$p(l_2/H_1) = \frac{l_2}{\sigma^2} e^{-l_2^2/2\sigma^2}, \quad l_2 \geq 0$$

z_{c_1} and z_{s_1} are also independent Gaussian random variables with the same variance, but their mean values are $E_s \cos\theta$ and $E_s \sin\theta$, respectively. The PDF of $l_1 = \sqrt{z_{c_1}^2 + z_{s_1}^2}$ is the well-known Rician distribution [4, pp. 322–324]

$$p(l_1/H_1) = \frac{l_1}{\sigma^2} I_0(\frac{\mu l_1}{\sigma^2}) e^{-(l_1^2+\mu^2)/2\sigma^2}, \quad l_1 \geq 0$$

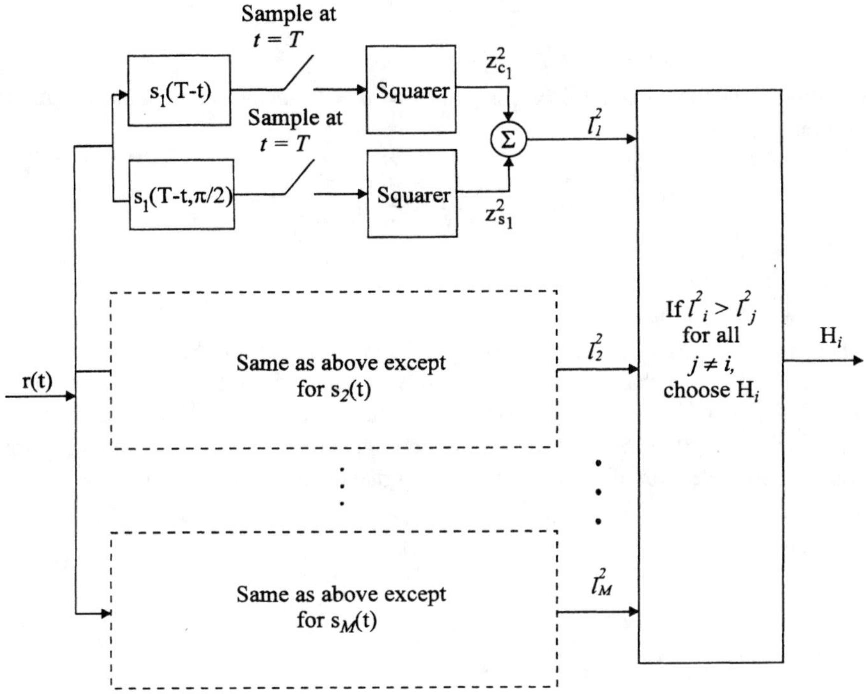

Figure B.13 Matched-filter receiver for M-ary signals with unknown phases. Signals are equally likely and with equal energy.

where $\mu = E_s = E_b$. The probability of error, given that $s_1(t)$ is sent, is

$$\begin{aligned} P(e/H_1) &= \Pr(l_2 \geq l_1/H_1) \\ &= \int_0^\infty \int_{l_1}^\infty p(l_1, l_2/H_1) d_{l_2} d_{l_1} \end{aligned}$$

The random variables l_1, l_2 are independent because their noise parts are independent, which in turn follows the orthogonality of the two signals. Therefore

$$\begin{aligned} P(e/H_1) &= \int_0^\infty p(l_1/H_1)[\int_{l_1}^\infty p(l_2/H_1) d_{l_2}] d_{l_1} \\ &= \int_0^\infty p(l_1/H_1)[\int_{l_1}^\infty \frac{l_2}{\sigma^2} e^{-l_2^2/2\sigma^2} d_{l_2}] d_{l_1} \end{aligned}$$

$$= \int_0^\infty \frac{l_1}{\sigma^2} I_0(\frac{\mu l_1}{\sigma^2}) e^{-(l_1^2+\mu^2)/2\sigma^2} [e^{-l_1^2/2\sigma^2}] d_{l_1}$$

Combine the exponents, let $y = \sqrt{2}l_1$ and $\xi = \mu/\sqrt{2}$, the above integration can be written as

$$\begin{aligned} P(e/H_1) &= \frac{1}{2} e^{-\mu^2/4\sigma^2} \int_0^\infty \frac{y}{\sigma^2} I_0(\frac{\xi y}{\sigma^2}) e^{-(y^2+\xi^2)/2\sigma^2} dy \\ &= \frac{1}{2} e^{-\mu^2/4\sigma^2} \end{aligned}$$

where the integration is one since the integrand is exactly the Rician density. We have already stated that this error probability is equal to the average symbol error probability which in the binary case is also the bit error probability. Substituting values of μ and σ^2 into the above expression, we have the bit error probability for noncoherent detection of binary orthogonal signals in an AWGN channel as

$$P_b = \frac{1}{2} e^{-E_b/2N_o} \tag{B.56}$$

B.3.2.2 M-ary Orthogonal Signals

For M-ary signals, the noncoherent receiver will produce M sufficient statistics: $l_i, i = 1, 2, \cdots, M$. The decision device will compare them and choose the largest. We again assume that $s_1(t)$ is transmitted. With M-ary orthogonal signals, l_1 will be Rician and the rest will be Rayleigh. All l_i are independent. Again due to symmetry of the problem the symbol error probability will be the same as the error probability of losing detection of $s_1(t)$ when it is sent.

$$P_s = P(e/H_1) = 1 - P(c/H_1)$$

where $P(c/H_1)$ is the probability of correct detection. If we fix l_1, the conditional probability is

$$\begin{aligned} P(c/H_1, l_1) &= \Pr(\text{all } l_i \leq l_1, i = 2, 3, \cdots M/H_1) \\ &= \left(\int_0^{l_1} \frac{l}{\sigma^2} e^{-l^2/2\sigma^2} dl \right)^{M-1} \\ &= [1 - e^{-l_1^2/2\sigma^2}]^{M-1} \end{aligned}$$

Now averaging this over the Rician distribution of l_1, we have

$$P(c/H_1) = \int_0^\infty \frac{l_1}{\sigma^2} I_0(\frac{\mu l_1}{\sigma^2}) \exp[\frac{l_1^2 + \mu^2}{2\sigma^2}][1 - e^{-l_1^2/2\sigma^2}]^{M-1} dl_1$$

Expanding the term raised to power $M-1$ by binomial theorem and integrating term by term, we obtain the symbol error probability of noncoherent detection for M-ary orthogonal signals as

$$P_s = 1 - P(c/H_1) = \sum_{k=1}^{M-1} \frac{(-1)^{k+1}}{k+1} \binom{M-1}{k} \exp[-\frac{kE_s}{(k+1)N_o}] \qquad \text{(B.57)}$$

where $\binom{M-1}{k}$ is the binomial coefficient, defined by

$$\binom{M-1}{k} = \frac{(M-1)!}{(M-1-k)!k!}$$

The leading term of (B.57) provides an upper bound as

$$P_s \leq \frac{M-1}{2} \exp[-\frac{E_s}{2N_o}] \qquad \text{(B.58)}$$

For fixed M this bound becomes increasingly close to the actual value of P_s as E_s/N_o is increased. In fact when $M = 2$, the upper bound becomes the exact expression.

References

[1] Van Trees, H. L., *Detection, Estimation, and Modulation Theory, Part I*, New York, John Wiley & Sons, Inc., 1968.

[2] Whalen, A. D., *Detection of Signals in Noise*, New York and London: Academic Press, 1971.

[3] Ziemer, R. E., and W. H. Tranter, *Principles of Communications, Systems, Modulations, and Noise*, 4th ed., Boston and Toronto: Houghton Mifflin Company, 1995.

[4] Simon, K. M., S. M. Hinedi, and W. C. Lindsey, *Digital Communication Techniques, Signal Design and Detection*, Englewood Cliffs, New Jersey: Prentice Hall, 1995.

[5] Viterbi, A. J., *Principles of Coherent Communications*, New York: McGraw-Hill Book Company, 1966.

Appendix C

Mathematical Tables

C.1 TRIGONOMETRY IDENTITIES

$\sin(-\theta) = -\sin\theta$	$\cos(-\theta) = \cos\theta$
$\sin 2\theta = 2\sin\theta\cos\theta$	$\cos 2\theta = \cos^2\theta - \sin^2\theta$
$\sin^2\theta = \frac{1}{2}(1 - \cos 2\theta)$	$\cos^2\theta = \frac{1}{2}(1 + \cos 2\theta)$
$\sin(A + \frac{\pi}{2}) = \cos A$	$\sin(A - \frac{\pi}{2}) = -\cos A$
$\cos(A + \frac{\pi}{2}) = -\sin A$	$\cos(A - \frac{\pi}{2}) = \sin A$
$\sin(A + B) = \sin A\cos B + \cos A\sin B$	$\sin^2\theta + \cos^2\theta = 1$
$\sin(A - B) = \sin A\cos B - \cos A\sin B$	
$\cos(A + B) = \cos A\cos B - \sin A\sin B$	
$\cos(A - B) = \cos A\cos B + \sin A\sin B$	
$\sin A\sin B = \frac{1}{2}\cos(A - B) - \frac{1}{2}\cos(A + B)$	
$\cos A\cos B = \frac{1}{2}\cos(A - B) + \frac{1}{2}\cos(A + B)$	
$\sin A\cos B = \frac{1}{2}\sin(A - B) + \frac{1}{2}\sin(A + B)$	

C.2 FOURIER TRANSFORM PAIRS

Fourier Transform (FT): $X(f) = \int_{-\infty}^{\infty} x(t)e^{-j2\pi ft}dt$

Inverse Fourier Transform (IFT): $x(t) = \int_{-\infty}^{\infty} X(f)e^{j2\pi ft}df$

Function	Time Domain Signal $x(t)$	Spectrum $X(f)$
Rectangular	$\Pi(\frac{t}{T}) \triangleq \begin{cases} 1, & \lvert t\rvert \leq T/2 \\ 0, & \lvert t\rvert > T/2 \end{cases}$	$T\text{sinc}(Tf) = T\frac{\sin(\pi Tf)}{\pi Tf}$
Triangular	$\Lambda(\frac{t}{T}) \triangleq \begin{cases} 1 - \frac{\lvert t\rvert}{T}, & \lvert t\rvert \leq T \\ 0, & \lvert t\rvert > T \end{cases}$	$T\text{sinc}^2(Tf) = T\left(\frac{\sin(\pi fT)}{\pi fT}\right)^2$
Unit step	$u(t) \triangleq \begin{cases} 1, & t > 0 \\ 0, & t < 0 \end{cases}$	$\frac{1}{2}\delta(f) + \frac{1}{j2\pi f}$
Signum	$\text{sgn}(t) \triangleq \begin{cases} +1, & t > 0 \\ -1, & t < 0 \end{cases}$	$\frac{1}{j2\pi f}$
Constant	1	$\delta(f)$
Impulse	$\delta(t)$	1
Sinc	$\text{sinc}(at)$	$\frac{1}{\lvert a\rvert}\Pi\left(\frac{f}{a}\right)$
Sinusoid	$\cos(2\pi f_0 t + \varphi)$	$\frac{1}{2}e^{j\varphi}\delta(f - f_0) + \frac{1}{2}e^{-j\varphi}\delta(f + f_0)$
Exponential, one-sided	$e^{-at}u(t), \quad a > 0$	$\frac{1}{a+j2\pi f}$
Exponential, two-sided	$e^{-a\lvert t\rvert}, \quad a > 0$	$\frac{2a}{a^2+(2\pi f)^2}$
Gaussian	$e^{-a^2t^2}$	$\frac{\sqrt{\pi}}{a}e^{-(\pi f/a)^2}$
Impulse Train	$\sum_{n=-\infty}^{\infty} \delta(t - nT_s)$	$f_s \sum_{m=-\infty}^{\infty} \delta(f - mf_s)$ $f_s = 1/T_s$

C.3 FOURIER TRANSFORM THEOREMS

Operation	Function	Fourier Transform
Linearity	$ax(t) + by(t)$	$aX(f) + bY(f)$
Scale Change	$x(at)$	$\frac{1}{\lvert a \rvert}X(f/a)$
Time Reversal	$x(-t)$	$X(-f)$
Complex Conjugate	$x^*(t)$	$X^*(-f)$
Duality	$X(t)$	$x(-f)$
Time Shift	$x(t-\tau)$	$X(f)e^{-j2\pi f\tau}$
Frequency Translation	$x(t)e^{j2\pi f_0 t}$	$X(f - f_0)$
Modulation	$x(t)\cos(2\pi f_0 t)$	$\frac{1}{2}X(f - f_0) + \frac{1}{2}X(f + f_0)$
Convolution	$x(t) * y(t)$	$X(f)Y(f)$
Multiplication	$x(t)y(t)$	$X(f) * Y(f)$
Time Differentiation	$\frac{d^n x(t)}{dt^n}$	$(j2\pi f)^n X(f)$
Time Integration	$\int_{-\infty}^{t} x(\tau)d\tau$	$(j2\pi f)^{-1}X(f) + \frac{1}{2}X(0)\delta(f)$

C.4 DISCRETE FOURIER TRANSFORM THEOREMS

Discrete Fourier Transform (DFT):

$$X(k) = \sum_{n=0}^{N-1} x(n) W_N^{kn}, \quad k = 0, 1, ..., N-1$$

$$W_N = e^{-j2\pi/N}$$

Inverse Discrete Fourier Transform (IDFT):

$$x(n) = \frac{1}{N} \sum_{k=0}^{N-1} X(k) W_N^{-kn}, \quad n = 0, 1, ..., N-1$$

Circular Convolution:

$$x(n) \otimes y(n) = \sum_{m=0}^{N-1} x(m) y(n-m)$$

Operation	Function	Discrete Fourier Transform
Linearity	$ax(n) + by(n)$	$aX(k) + bY(k)$
Time Shift	$x(n-m)$	$W_N^{mk} X(k)$
Frequency Shift	$x(n)e^{j2\pi nm/N}$	$X(k-m)$
Duality	$N^{-1}X(n)$	$x(-k)$
Circular Convolution	$x(n) \otimes y(n)$	$X(k)Y(k)$
Multiplication	$x(n)y(n)$	$N^{-1}X(k) \otimes Y(k)$

C.5 Q-FUNCTION AND ERROR FUNCTIONS

Q-function is defined as

$$Q(x) = \frac{1}{\sqrt{2\pi}} \int_{x}^{\infty} e^{-u^2/2} du$$

Van Trees [1, p. 37] denoted this as $\text{erfc}_*(x)$. But it is not the same as $\text{erfc}(x)$ which is called *complementary error function*.

$$\text{erfc}(x) \triangleq 1 - \text{erf}(x)$$

where $\text{erf}(x)$ is the *error function* defined as

$$\text{erf}(x) \triangleq \frac{2}{\sqrt{\pi}} \int_{0}^{x} \exp(-u^2) du$$

Another definition of error function is given in [2], which we denote as $\text{erf}_*(x)$,

$$\text{erf}_*(x) = \frac{1}{\sqrt{2\pi}} \int_{0}^{x} e^{-u^2/2} du$$

It is easily shown

$$\text{erf}_*(x) = \frac{1}{2} \text{erf}(\frac{x}{\sqrt{2}})$$

Table of $Q(x) = \frac{1}{\sqrt{2\pi}} \int_x^\infty e^{-u^2/2} du, \qquad x = 0$ to 3

x	$Q(x)$	x	$Q(x)$	x	$Q(x)$	x	$Q(x)$
0.05	0.48006	0.80	0.21186	1.55	0.06057	2.30	0.01072
0.10	0.46017	0.85	0.19766	1.60	0.05480	2.35	9.38671×10^{-3}
0.15	0.44038	0.90	0.18406	1.65	0.04947	2.40	8.19754×10^{-3}
0.25	0.42074	0.95	0.17106	1.70	0.04457	2.45	7.14281×10^{-3}
0.20	0.40129	1.00	0.15866	1.75	0.04006	2.50	6.20967×10^{-3}
0.30	0.38209	1.05	0.14686	1.80	0.03593	2.55	5.38615×10^{-3}
0.35	0.36317	1.10	0.13567	1.85	0.03216	2.60	4.66119×10^{-3}
0.40	0.34458	1.15	0.12507	1.90	0.02872	2.65	4.02459×10^{-3}
0.45	0.32636	1.20	0.11507	1.95	0.02559	2.70	3.46697×10^{-3}
0.50	0.30854	1.25	0.10565	2.00	0.02275	2.75	2.97976×10^{-3}
0.55	0.29116	1.30	0.09680	2.05	0.02018	2.80	2.55513×10^{-3}
0.60	0.27425	1.35	0.08851	2.10	0.01786	2.85	2.18596×10^{-3}
0.65	0.25785	1.40	0.08076	2.15	0.01578	2.90	1.86581×10^{-3}
0.70	0.24196	1.45	0.07353	2.20	0.01390	2.95	1.58887×10^{-3}
0.75	0.22663	1.50	0.06681	2.25	0.01222	3.00	1.34990×10^{-3}

Table of $Q(x) = \frac{1}{\sqrt{2\pi}} \int_x^\infty e^{-u^2/2} du, \qquad x = 3$ to 6

x	$Q(x)$	x	$Q(x)$	x	$Q(x)$	x	$Q(x)$
3.05	1.144×10^{-3}	3.80	7.235×10^{-5}	4.55	2.682×10^{-6}	5.30	5.790×10^{-8}
3.10	9.676×10^{-4}	3.85	5.906×10^{-5}	4.60	2.112×10^{-6}	5.35	4.398×10^{-8}
3.15	8.164×10^{-4}	3.90	4.810×10^{-5}	4.65	1.660×10^{-6}	5.40	3.332×10^{-8}
3.25	6.871×10^{-4}	3.95	3.908×10^{-5}	4.70	1.301×10^{-6}	5.45	2.518×10^{-8}
3.20	5.770×10^{-4}	4.00	3.167×10^{-5}	4.75	1.017×10^{-6}	5.50	1.899×10^{-8}
3.30	4.834×10^{-4}	4.05	2.561×10^{-5}	4.80	7.933×10^{-7}	5.55	1.428×10^{-8}
3.35	4.041×10^{-4}	4.10	2.066×10^{-5}	4.85	6.173×10^{-7}	5.60	1.072×10^{-8}
3.40	3.369×10^{-4}	4.15	1.662×10^{-5}	4.90	4.792×10^{-7}	5.65	8.022×10^{-9}
3.45	2.803×10^{-4}	4.20	1.335×10^{-5}	4.95	3.711×10^{-7}	5.70	5.990×10^{-9}
3.50	2.326×10^{-4}	4.25	1.069×10^{-5}	5.00	2.867×10^{-7}	5.75	4.462×10^{-9}
3.55	1.926×10^{-4}	4.30	8.540×10^{-6}	5.05	2.209×10^{-7}	5.80	3.316×10^{-9}
3.60	1.591×10^{-4}	4.35	6.807×10^{-6}	5.10	1.698×10^{-7}	5.85	2.458×10^{-9}
3.65	1.311×10^{-4}	4.40	5.413×10^{-6}	5.15	1.302×10^{-7}	5.90	1.818×10^{-9}
3.70	1.078×10^{-4}	4.45	4.294×10^{-6}	5.20	9.964×10^{-8}	5.95	1.341×10^{-9}
3.75	1.078×10^{-4}	4.50	3.398×10^{-6}	5.25	7.605×10^{-8}	6.00	9.866×10^{-10}

Table of $\mathrm{erf}(x) \triangleq \frac{2}{\sqrt{\pi}} \int_0^x \exp(-u^2)du$

x	erf(x)	x	erf(x)	x	erf(x)	x	erf(x)
0.05	0.05637	0.80	0.74210	1.55	0.97162	2.30	0.99886
0.10	0.11246	0.85	0.77067	1.60	0.97635	2.35	0.99911
0.15	0.16800	0.90	0.79691	1.65	0.98038	2.40	0.99931
0.25	0.22270	0.95	0.82089	1.70	0.98379	2.45	0.99947
0.20	0.27633	1.00	0.84270	1.75	0.98667	2.50	0.99959
0.30	0.32863	1.05	0.86244	1.80	0.98909	2.55	0.99969
0.35	0.37938	1.10	0.88021	1.85	0.99111	2.60	0.99976
0.40	0.42839	1.15	0.89612	1.90	0.99279	2.65	0.99982
0.45	0.47548	1.20	0.91031	1.95	0.99418	2.70	0.99987
0.50	0.52050	1.25	0.92290	2.00	0.99532	2.75	0.99990
0.55	0.56332	1.30	0.93401	2.05	0.99626	2.80	0.99992
0.60	0.60386	1.35	0.94376	2.10	0.99702	2.85	0.99994
0.65	0.64203	1.40	0.95229	2.15	0.99764	2.90	0.99996
0.70	0.67780	1.45	0.95970	2.20	0.99814	2.95	0.99997
0.75	0.71116	1.50	0.96611	2.25	0.99854	3.00	0.99998

C.6 RELATIONS BETWEEN Q-FUNCTION AND ERROR FUNCTIONS

$Q(x)$ and $\mathrm{erfc}(x)$ are related by

$$Q(x) = \frac{1}{2}\mathrm{erfc}(\frac{x}{\sqrt{2}})$$

or

$$\mathrm{erfc}(x) = 2Q(\sqrt{2}x)$$

$Q(x)$ and $\mathrm{erf}(x)$ are related by

$$Q(x) = \frac{1}{2}\left[1 - \mathrm{erf}(\frac{x}{\sqrt{2}})\right]$$

or

$$\mathrm{erf}(x) = 1 - 2Q(\sqrt{2}x)$$

C.7 APPROXIMATION OF Q-FUNCTION

An upper bound of Q-function is

$$Q(x) < \frac{1}{x\sqrt{2\pi}} e^{-x^2/2} \triangleq Q_u(x)$$

Figure C.1 shows the $Q(x)$ and its upper bound.

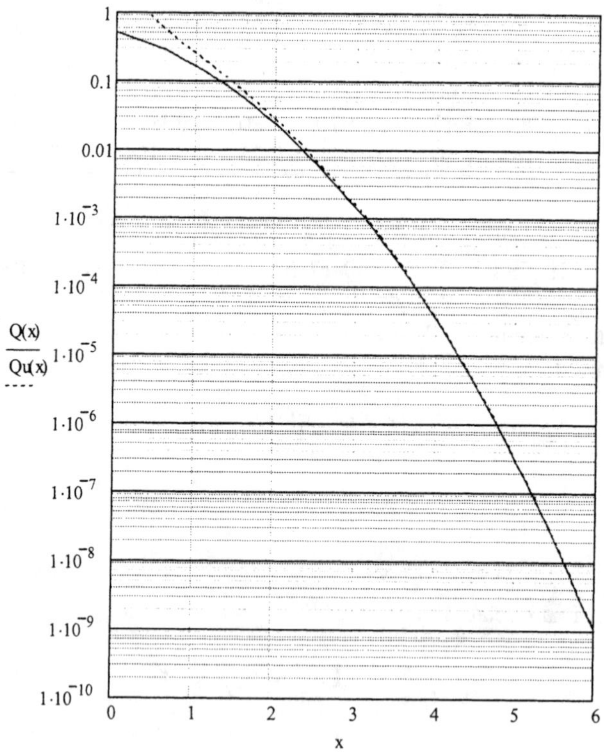

Figure C.1 $Q(x)$ and its upper bound $Q_u(x)$.

References

[1] Van Trees, H. L., *Detection, Estimation, and Modulation Theory, Part I*, New York: John Wiley & Sons, Inc., 1968.

[2] Papoulis, A., *Probability, Random Variables, and Stochastic Processes*, 2nd ed., New York: McGraw-Hill, 1984.

Acronyms

ac	alternate current
ACI	adjacent channel interference
ACTS	Advanced Communications Technology Satellite (of NASA)
A/D	analog-to-digital (converter)
ADC	analog-to-digital converter
ADSL	asymmetrical digital subscriber line
AGC	automatic gain control
AM	amplitude modulation
AMF	average matched filter
AMI	alternative mark inversion
AMI-NRZ	alternative mark inversion-nonreturn-to-zero
AMI-RZ	alternative mark inversion-return-to-zero
AMPS	advance mobile phone service
ASK	amplitude shift keying
AT&T	American Telephone and Telegraph
ATDMA	advanced time division multiple access
AWGN	additive white Gaussian noise
BER	bit error rate
Bi-Φ-L	biphase level
Bi-Φ-M	biphase mark
Bi-Φ-S	biphase space
BFSK	binary frequency shift keying
BNZS	binary *n* zero substitution
BPF	bandpass filter
BPSK	binary phase shift keying
CCI	cochannel interference
CCITT	Consultative Committee for International Telephone and Telegraph
C/D	power ratio between the main-path signal and the delayed signal

CDMA	code division multiple access
CDPD	cellular digital packet data (system)
CHDB	compatible high density bipolar
C/I	carrier-to-interference ratio
CMI	coded mark inversion
C/N	carrier-to-noise ratio
COFDM	coded orthogonal frequency division multiplexing
CPFSK	continuous phase frequency shift keying
CPM	continuous phase modulation
CR	carrier recovery
CSMA/CD	carrier sense multiple access/collision detection
D/A	digital-to-analog (converter)
DAC	digital-to-analog converter
DAR	decision-aided reconstruction
dB	decibel
DBPSK	differential binary phase shift keying (differentially encoded and differentially demodulated binary phase shift keying)
dc	direct current
DCT	discrete cosine transform
DDCE	decision-directed channel estimation
DDCR	decision-directed carrier recovery
DE	diagonal encoding
DEBPSK	differentially encoded binary phase shift keying (demodulation may not be differential)
DEMPSK	differentially encoded *M*-ary phase shift keying
DEQPSK	differentially encoded quadrature phase shift keying
DFCE	decision feedback channel estimation
DFE	decision feedback equalizer
DFT	discrete Fourier transform
DM	delay modulation
DmB1M	differential *m* binary with 1 mark inversion
DMI	differential mode inversion
DMPSK	differential *M*-ary phase shift keying
DMSK	duobinary minimum shift keying
DMT	discrete multitone (modulation)
DPPM	differential pulse position modulation
DPSK	differential phase shift keying (usually it refers to DBPSK, but can be used as a generic name for all differential phase shift keying schemes)
DQPSK	differential quadrature phase shift keying
DS	digital sum

DSFSK	double sinusoidal frequency shift keying
DSI	dummy sequence insertion
DSV	digital sum variation
DVB	digital video broadcasting
DVB-T	digital video broadcasting-terrestrial
DWMT	discrete wavelet multitone modulation
DWT	discrete wavelet transform
ETACS	European total access communication system
FCT	fast cosine transform
FDM	frequency division multiplexing
FFFF	frequency flat fast fading
FFSF	frequency flat slow fading
FFSK	fast frequency shift keying
FFT	fast Fourier transform
FIR	finite impulse response (filter)
FM	frequency modulation
FSFF	frequency selective fast fading
FSSF	frequency selective slow fading
FSK	frequency shift keying
FWT	fast wavelet transform
GMSK	Gaussian minimum shift keying
GSM	global system for mobile communication
HDB	high density bipolar
HDTV	high definition television
HE	horizontal encoding
HF	high frequency
HPA	high power amplifier
Hz	hertz
IBM	International Business Machines (Corporation)
IBO	input back off
ICI	interchannel interference
IDCT	inverse discrete cosine transform
IDFT	inverse discrete Fourier transform
IF	intermediate frequency
IFCT	inverse fast cosine transform
IFFT	inverse fast Fourier transform
IJF-OQPSK	intersymbol-interference/jitter-free OQPSK
IPC	inner product calculator
ISD	integrate-sample-dump (circuit)
ISI	intersymbol interference
LAN	local area network

LCE-MH	multi-*h* signal with a depth of correlation of *L*
LE	linear equalizer
LMMSE	linear minimum mean square estimation
LO	local oscillator
LOS	line of sight
LPF	low-pass filter
LRC	raised cosine pulse of length *L*
LREC	rectangular pulse of length *L*
LSRC	spectrally raised cosine pulse of length *L*
MAM	*M*-ary amplitude modulation
MAP	maximum a posteriori probability
MASK	*M*-ary amplitude shift keying
MCM	multi-carrier modulation
MFSK	*M*-ary frequency shift keying
MHPM	multi-*h* phase modulation
MIMO	multi-input-multi-output
MISO	multi-input-single-output
ML	maximum likelihood
MLS	maximum length sequence
MLSD	maximum likelihood sequence detection
MLSE	maximum likelihood sequence estimation
MMSE-DFE	minimum mean square error decision feedback equalizer
MMSE-LE	minimum mean square error linear equalizer
MPPM	multipulse pulse position modulation
MPSK	*M*-ary phase shift keying
MQORC	modifed quadrature overlapped raised-cosine (modulation)
MRA	multiresolution approximation
MSK	minimum shift keying
MSM	multiscale modulation
MWM	M-band wavelet modulation
NASA	National Aeronautics and Space Administration
NCO	numerically-controlled oscillator
NRZ	nonreturn-to-zero
NRZ-L	nonreturn-to-zero-level
NRZ-M	nonreturn-to-zero-mark
NRZ-S	nonreturn-to-zero-space
OFDM	orthogonal frequency division multiplexing
OOK	binary on-off keying (binary ASK)
OQPSK	offset quadrature phase shift keying
PAM	pulse amplitude modulation
PAPR	peak-to-average power ratio

PCM	pulse code modulation
PDF	probability density function
π/4-QPSK	π/4 quadrature phase shift keying
PIM	pulse interval modulation
PIWM	pulse interval and width modulation
PEP	peak envelope power
PFDmB(m+1)B	partially-flipped differential *m* bits with (m+1)th check bit
PFM	pulse frequency modulation
PLL	phase lock loop
PN	pseudorandom noise
PPM	pulse position modulation
PSAM	pilot symbol assisted modulation
PSD	power spectral density
PSK	phase shift keying
PT	pseudoternary
PTM	pulse time modulation
PTS	partial transmit sequence
PWM	pulse width modulation
QAM	quadrature amplitude modulation
QBL	quasi-bandlimited (modulation)
QORC	quadrature overlapped raised cosine (modulation)
QOSRC	quadrature overlapped squared raised cosine (modulation)
QPSK	quadrature phase shift keying
Q^2PSK	quadrature quadrature phase shift keying
RCA	Radio Corporation of America
RC-QPSK	raised-cosine quadrature phase shift keying
ROM	read only memory
RX	receiver
RZ	return to zero
SC	single channel
SDR	signal-to-distortion ratio
SER	symbol error rate
SFSK	sinusoidal frequency shift keying
SHPM	single-*h* phase modulation
SIMO	single-input-multi-output
SIR	signal-to-interference ratio
SISO	single-input-single-output
SLM	selective mapping
SMSK	serial minimum shift keying
SNR	signal-to-noise ratio
SQAM	superposed quadrature amplitude modulation

SQORC	staggered quadrature overlapped raised cosine (modulation)
SSB	single side band
STBC	space-time block code
STC	space-time code
STTC	space-time trellis code
SVD	singular value decomposition
SWFM	square wave frequency modulation
TCM	trellis coded modulation
TD	transform decomposition
TD-FFT	transform decomposition fast Fourier transform
TFM	tamed frequency shift keying
TSI-OQPSK	two-symbol-interval offset quadrature phase shift keying
TTIB	transparent tone in band
TWTA	traveling wave tube amplifier
TX	transmitter
USDC	United States digital cellular (system)
VA	Viterbi algorithm
VCO	voltage-controlled oscillator
VE	vertical encoding
VOFDM	vector orthogonal frequency division multiplexing
W-CDMA	wide-band code division multiple access
WLAN	wireless local area network
WPM	wavelet packet modulation
WSS	wide sense stationary
WSSUS	wide sense stationary uncorrelated scattering
XPSK	cross-correlated quadrature phase shift keying
ZF-DFE	zero-forcing decision feedback equalizer
ZF-LE	zero-forcing linear equalizer

About the Author

Fuqin Xiong received a B.Sc. in communication engineering in 1970 and an M.Sc. in communication and electronic system engineering in 1982 from Tsinghua University, Beijing, China, and a Ph.D. degree in 1989 from the Department of Electrical Engineering, University of Manitoba, Winnipeg, Canada.

From 1970 to 1978 Dr. Xiong was an assistant lecturer, and from 1983 to 1984 a lecturer in the Department of Radio-Electronics Engineering, Tsinghua University, Beijing, China. From 1984 to 1985 he was a visiting scholar at the Department of Electrical Engineering, University of Manitoba. Dr. Xiong joined the Department of Electrical and Computer Engineering, Cleveland State University, in 1990 as an assistant professor. He was promoted to associate professor with tenure in 1995. In 1997, while on sabbatical leave, he spent the spring semester at City University of Hong Kong and the summer in Tsinghua University, Beijing, China. In 2002, Dr. Xiong was promoted to full professor. In the fall semester of 2003, he spent his sabbatical leave at Nanyang Technological University of Singapore.

Dr. Xiong's research interests are in communication engineering, particularly in modulation and coding techniques for digital communications, including satellite and mobile communications. He has published numerous articles in technical journals such as *IEEE Transactions* and *IEE Proceedings*. He also served as a reviewer for *IEEE Transactions on Communications* and *IEE Proceedings of Communications*.

Dr. Xiong has been directing several research projects on digital modulation techniques sponsored by NASA Glenn Research Center, Cleveland, Ohio. In the four consecutive summers of 2000 to 2003, he was awarded the NASA faculty fellowship and conducted research at NASA Glenn Research Center. Dr. Xiong is a senior member of the IEEE.

Index

Recent Titles in the Artech House Telecommunications Library

Vinton G. Cerf, Senior Series Editor

Access Networks: Technology and V5 Interfacing, Alex Gillespie

Achieving Global Information Networking, Eve L. Varma et al.

Advanced High-Frequency Radio Communications, Eric E. Johnson et al.

ATM Interworking in Broadband Wireless Applications, M. Sreetharan and S. Subramaniam

ATM Switches, Edwin R. Coover

ATM Switching Systems, Thomas M. Chen and Stephen S. Liu

Broadband Access Technology, Interfaces, and Management, Alex Gillespie

Broadband Local Loops for High-Speed Internet Access, Maurice Gagnaire

Broadband Networking: ATM, SDH, and SONET, Mike Sexton and Andy Reid

Broadband Telecommunications Technology, Second Edition, Byeong Lee, Minho Kang, and Jonghee Lee

The Business Case for Web-Based Training, Tammy Whalen and David Wright

Centrex or PBX: The Impact of IP, John R. Abrahams and Mauro Lollo

Chinese Telecommunications Policy, Xu Yan and Douglas Pitt

Communication and Computing for Distributed Multimedia Systems, Guojun Lu

Communications Technology Guide for Business, Richard Downey, Seán Boland, and Phillip Walsh

Community Networks: Lessons from Blacksburg, Virginia, Second Edition, Andrew M. Cohill and Andrea Kavanaugh, editors

Component-Based Network System Engineering, Mark Norris, Rob Davis, and Alan Pengelly

Computer Telephony Integration, Second Edition, Rob Walters

Customer-Centered Telecommunications Services Marketing, Karen G. Strouse

Deploying and Managing IP over WDM Networks, Joan Serrat and Alex Galis, editors

Desktop Encyclopedia of the Internet, Nathan J. Muller

Digital Clocks for Synchronization and Communications, Masami Kihara, Sadayasu Ono, and Pekka Eskelinen

Digital Modulation Techniques, Second Edition, Fuqin Xiong

E-Commerce Systems Architecture and Applications, Wasim E. Rajput

Engineering Internet QoS, Sanjay Jha and Mahbub Hassan

Error-Control Block Codes for Communications Engineers, L. H. Charles Lee

Essentials of Modern Telecommunications Systems, Nihal Kularatna and Dileeka Dias

FAX: Facsimile Technology and Systems, Third Edition, Kenneth R. McConnell, Dennis Bodson, and Stephen Urban

Fundamentals of Network Security, John E. Canavan

Gigabit Ethernet Technology and Applications, Mark Norris

The Great Telecom Meltdown, Fred R. Goldstein

Guide to ATM Systems and Technology, Mohammad A. Rahman

A Guide to the TCP/IP Protocol Suite, Floyd Wilder

Home Networking Technologies and Standards, Theodore B. Zahariadis

Implementing Value-Added Telecom Services, Johan Zuidweg

Information Superhighways Revisited: The Economics of Multimedia, Bruce Egan

Installation and Maintenance of SDH/SONET, ATM, xDSL, and Synchronization Networks, José M. Caballero et al.

Integrated Broadband Networks: TCP/IP, ATM, SDH/SONET, and WDM/Optics, Byeong Gi Lee and Woojune Kim

Internet E-mail: Protocols, Standards, and Implementation, Lawrence Hughes

Introduction to Telecommunications Network Engineering, Second Edition, Tarmo Anttalainen

Introduction to Telephones and Telephone Systems, Third Edition, A. Michael Noll

An Introduction to U.S. Telecommunications Law, Second Edition, Charles H. Kennedy

IP Convergence: The Next Revolution in Telecommunications, Nathan J. Muller

LANs to WANs: The Complete Management Guide, Nathan J. Muller

The Law and Regulation of Telecommunications Carriers, Henk Brands and Evan T. Leo

Managing Internet-Driven Change in International Telecommunications, Rob Frieden

Marketing Telecommunications Services: New Approaches for a Changing Environment, Karen G. Strouse

Mission-Critical Network Planning, Matthew Liotine

Multimedia Communications Networks: Technologies and Services, Mallikarjun Tatipamula and Bhumip Khashnabish, editors

Next Generation Intelligent Networks, Johan Zuidweg

Open Source Software Law, Rod Dixon

Performance Evaluation of Communication Networks, Gary N. Higginbottom

Performance of TCP/IP over ATM Networks, Mahbub Hassan and Mohammed Atiquzzaman

The Physical Layer of Communications Systems, Richard A. Thompson, David Tipper, Prashant Krishnamurthy, and Joseph Kabara

Practical Guide for Implementing Secure Intranets and Extranets, Kaustubh M. Phaltankar

Practical Internet Law for Business, Kurt M. Saunders

Practical Multiservice LANs: ATM and RF Broadband, Ernest O. Tunmann

Principles of Modern Communications Technology, A. Michael Noll

A Professional's Guide to Data Communication in a TCP/IP World, E. Bryan Carne

Programmable Networks for IP Service Deployment, Alex Galis et al., editors

Protocol Management in Computer Networking, Philippe Byrnes

Pulse Code Modulation Systems Design, William N. Waggener

Reorganizing Data and Voice Networks: Communications Resourcing for Corporate Networks, Thomas R. Koehler

Security, Rights, and Liabilities in E-Commerce, Jeffrey H. Matsuura

Service Assurance for Voice over WiFi and 3G Networks, Richard Lau, Ram Khare, and William Y. Chang

Service Level Management for Enterprise Networks, Lundy Lewis

SIP: Understanding the Session Initiation Protocol, Second Edition, Alan B. Johnston

Smart Card Security and Applications, Second Edition, Mike Hendry

SNMP-Based ATM Network Management, Heng Pan

Spectrum Wars: The Policy and Technology Debate, Jennifer A. Manner

Strategic Management in Telecommunications, James K. Shaw

Strategies for Success in the New Telecommunications Marketplace, Karen G. Strouse

Successful Business Strategies Using Telecommunications Services, Martin F. Bartholomew

Telecommunications Cost Management, S. C. Strother

Telecommunications Department Management, Robert A. Gable

Telecommunications Deregulation and the Information Economy, Second Edition, James K. Shaw

Telecommunications Technology Handbook, Second Edition, Daniel Minoli

Telemetry Systems Engineering, Frank Carden, Russell Jedlicka, and Robert Henry

Telephone Switching Systems, Richard A. Thompson

Understanding Modern Telecommunications and the Information Superhighway, John G. Nellist and Elliott M. Gilbert

Understanding Networking Technology: Concepts, Terms, and Trends, Second Edition, Mark Norris

Understanding Voice over IP Security, Alan B. Johnston and David M. Piscitello

Videoconferencing and Videotelephony: Technology and Standards, Second Edition, Richard Schaphorst

Visual Telephony, Edward A. Daly and Kathleen J. Hansell

Wide-Area Data Network Performance Engineering, Robert G. Cole and Ravi Ramaswamy

Winning Telco Customers Using Marketing Databases, Rob Mattison

WLANs and WPANs towards 4G Wireless, Ramjee Prasad and Luis Muñoz

World-Class Telecommunications Service Development, Ellen P. Ward

For further information on these and other Artech House titles, including previously considered out-of-print books now available through our In-Print-Forever® (IPF®) program, contact:

Artech House
685 Canton Street
Norwood, MA 02062
Phone: 781-769-9750
Fax: 781-769-6334
e-mail: artech@artechhouse.com

Artech House
46 Gillingham Street
London SW1V 1AH UK
Phone: +44 (0)20 7596-8750
Fax: +44 (0)20 7630-0166
e-mail: artech-uk@artechhouse.com

Find us on the World Wide Web at: www.artechhouse.com